SYSTEMIC DISEASES

PART III

HANDBOOK OF
CLINICAL NEUROLOGY

Editors

PIERRE J. VINKEN   GEORGE W. BRUYN

Executive Editor

KENNETH ELLISON DAVIS

Editorial Advisory Board

R.D. ADAMS, S.H. APPEL, E.P. BHARUCHA,
H.L. KLAWANS†, C.D. MARSDEN†, H. NARABAYASHI,
A. RASCOL, L.P. ROWLAND, F. SEITELBERGER

VOLUME 71

ELSEVIER

AMSTERDAM • LAUSANNE • NEW YORK • OXFORD • SHANNON •
SINGAPORE • TOKYO

SYSTEMIC DISEASES

PART III

Editors

PIERRE J. VINKEN GEORGE W. BRUYN

Volume Editors

M.J. AMINOFF C.G. GOETZ

REVISED SERIES 27

ELSEVIER

AMSTERDAM • LAUSANNE • NEW YORK • OXFORD • SHANNON •
SINGAPORE • TOKYO

ELSEVIER SCIENCE B.V.
Sara Burgerhartstraat 25
P.O. Box 211, 1000 AE Amsterdam, The Netherlands

First edition 1998

Library of Congress Cataloging in Publication Data
This volume forms Part III in a set comprising 3 volumes entitled "Systemic Diseases Part I, II and III", respectively; the original CIP-data appear in Volume 63 (Rev.Ser. 19), "Systemic Diseases Part I", which was published in 1993.

ISBN: 0-444-81290-3

Foreword to Systemic Diseases – Part III

This is the last of the three volumes in this revised series of the Handbook directed at the neurological problems associated with systemic illness and complementing volumes 38 and 39 ("Neurological Manifestations of Systemic Diseases" Parts I and II) published in the first series in 1979 and 1980 respectively. They provide comprehensive updates of the accumulated literature, taking into account the many developments that have occurred in clinical medicine over the last quarter-century. In particular, the advent of new imaging techniques and therapeutic approaches have created another dimension, provided new solutions for old problems, but which, at the same time, have introduced new complications and dilemmas that are discussed here. Over this time, advances in the neurosciences have added to the general understanding of disease processes and have suggested more rational approaches to clinical management.

The "Systemic Diseases" volumes are especially important at a time when the practice of medicine is changing. Increasing subspecialization has made it difficult for neurologists to keep abreast of general developments in general medicine. At the same time, neurologists are being asked more and more to participate in the care of patients with general medical problems, serving as consultants to advise about the appropriate management of neurological complications occurring in general medical disorders. It is our hope that this volume and its two companion volumes will be of help to readers by providing practical guidance, a digest of the pertinent literature, and a summary of the various problems for which consultation may be requested.

We thank the various contributors to this volume, who have taken time from their busy schedules to participate in this endeavour. We are also grateful to the editorial team for their care and attention in the production of this volume. It is our hope that this work will serve its intended purpose to advance the care of patients with neurological manifestation of systemic disease.

October 1998

M.J.A.
C.G.G.
P.J.V.
G.W.B.

List of contributors

S. Al Deeb
*Department of Clinical Neurosciences, Division of Neurology, Riyadh
Armed Forces Hospital, P.O. Box 7897, Riyadh 11159, Saudi Arabia*   585

E.L. Alexander
206 Longwood Avenue, Baltimore, MD 21210, U.S.A.   59

C. Balmaceda
*The Neurological Institute, Columbia Presbyterian Medical Center,
710 West 168th Street, Room 425, New York, NY 10032, U.S.A.*   611

J.R. Berger
*Department of Neurology, University of Kentucky, College of Medicine,
Chandler Medical Center, KY Clinic L445, Lexington,
KY 40536-0284, U.S.A.*   261

T.P. Bleck
*Neuroscience Intensive Care Unit, University of Virginia, Charlottesville,
VA 22908, U.S.A.*   569

T.H. Brannagan
*Department of Neurology, Neurological Institute, Columbia-Presbyterian
Medical Center, New York, NY 10032, U.S.A.*   431

G.A.W. Bruyn
*Department of Rheumatology, Medical Center Leeuwarden, Henri
Dunantweg 2, 8934 AD Leeuwarden, The Netherlands*   15, 35

S.M. Chang
*Neuro-Oncology Service, University of California, San Francisco,
350 Parnassus Avenue, San Francisco, CA 94117, U.S.A.*   641

L.M. Cher
*Department of Neurology and Neuro-Oncology, Austin and Repatriation
Medical Centre, Heidelberg, Victoria 3161, Australia*   673

B.A. Cohen
*Department of Neurology, Northwestern University School of Medicine,
Chicago, IL, U.S.A.*   261

A. Das
*Department of Neuro-Oncology, Massachusetts General Hospital,
1 Hawthorne Place, Boston, MA 02114, U.S.A.*   673

D.M. Escolar
*Department of Neurology, The George Washington University Medical
Center, 2150 Pennsylvania Avenue N.W., Suite 7-406, Washington,
DC 20037, U.S.A.*   501

M.R. Fetell
*The Neurological Institute, Columbia Presbyterian Medical Center,
710 West 168th Street, Room 425, New York, NY 10032, U.S.A.*   611

D.S. Goodin
*Department of Neurology, Room M-794, University of California,
San Francisco, San Francisco, CA 94143-0114, U.S.A.*   209

P.H. Gordon
*Department of Neurology, Neurological Institute, Columbia-Presbyterian
Medical Center, 710 West 168th Street, New York, NY 10032, U.S.A.*   431

J.E. Greenlee
*Department of Neurology, University of Utah School of Medicine,
50 North Medical Drive, Salt Lake City, UT 84132, U.S.A.*   399

D.R. Gress
*NeuroCritical Care and NeuroVascular Service, Department of Neurology,
University of California, San Francisco, San Francisco, CA 94143-0124,
U.S.A.*   525

T. Heiman Patterson
*Department of Neurology, Allegheny University of The Health Sciences,
Philadelphia, PA 19102, U.S.A.*   547

J.C. Hemphill III
*NeuroCritical Care and NeuroVascular Service, Department of Neurology,
University of California, San Francisco, San Francisco, CA 94143-0124,
U.S.A.*   525

J.W. Henson
Neurology Service, Massachusetts General Hospital, 1 Hawthorne Place, Boston, MA 02114, U.S.A. 673

F.H. Hochberg
Neurology Service, Massachusetts General Hospital, 1 Hawthorne Place, Boston, MA 02114, U.S.A. 673

B.A. Janssen
Department of Rheumatology, Queen Elizabeth Hospital, Edgbaston, University of Birmingham, Birmingham B15 2TH, U.K. 35

J.J. Kelly, Jr.
Department of Neurology, The George Washington University Medical Center, 2150 Pennsylvania Avenue N.W., Suite 7-406, Washington, DC 20037, U.S.A. 501

A. Krumholz
Department of Neurology, N4W46, University of Maryland Medical Center, 22 South Greene Street, Baltimore, MD 21201-1595, U.S.A. 463

N. Latov
Department of Neurology, The Neurological Institute, Columbia-Presbyterian Medical Center, 710 West 168th Street, Room 425, New York, NY 10032, U.S.A. 431

R.P. Lisak
Departments of Neurology and Immunology and Microbiology, Wayne State University School of Medicine, Detroit Medical Center, 6E University Health Center, 4201 St. Anne, Detroit, MI 48201, U.S.A. 3

W.J. Litchy
Department of Neurology, Health Partners, 2220 Riverside Avenue South, Minneapolis, MN 55454, U.S.A. 129

F.D. Lublin
Department of Neurology, Allegheny University Hospital – MCP Division, Philadelphia, PA 19129, U.S.A. 547

H.M. Markusse
Department of Rheumatology, Dr. Daniel den Hoed Clinic, Groene Hilledÿk 301, 3075 EA Rotterdam, The Netherlands 15

P.M. Moore
Department of Neurology, Wayne State University School of Medicine, Detroit Medical Center, 6E University Health Center, 4201 St. Antoine, Detroit, MI 48291, U.S.A. 149

M.K. Nicholas
Neuro-Oncology Service, Brain Tumor Research Center, University of California, San Francisco, 350 Parnassus Avenue, San Francisco, CA 94117, U.S.A. 641

R.K. Olney
Department of Neurology, Box 0114, Room M-794, University of California, San Francisco, 505 Parnassus, San Francisco, CA 94143-0114, U.S.A. 99

G.J. Parry
Department of Neurology, University of Minnesota Medical School, 420 Delaware Street S.E., Minneapolis, MN 55455-0323, U.S.A. 353

K. Peterson
Department of Neurology and Neurological Sciences, Stanford University Medical Center, Stanford, CA 94305-5235, U.S.A. 335

R.W. Price
Department of Neurology, Room 4M62, San Francisco General Hospital, 1001 Potrero Avenue, San Francisco, CA 94110-3518, U.S.A. 235

K.M. Shannon
Department of Neurological Sciences, Rush Medical College, Rush–Presbyterian–St. Luke's Medical Center, 1725 W. Harrison Street, Suite 1106, Chicago, IL 60643, U.S.A. 121

D.M. Simpson
Clinical Neurophysiology Laboratories and Neuro-AIDS Research Program, The Mount Sinai Medical Center, Box 1052, 1 Gustave L. Levy Place, New York, NY 10029, U.S.A. 367

E.J. Skalabrin
Stanford Stroke Center, Stanford University Medical Center, 701 Welch Road, Building B, Suite 325, Palo Alto, CA 94304-1705, U.S.A. 569

Y.T. So
Department of Neurology, L226, Oregon Health Sciences University, 3181 S.W. Sam Jackson Park Road, Portland, OR 97201, U.S.A. 191

B.J. Stern
Department of Neurology, Emory University School of Medicine, Atlanta, GA 30322, U.S.A. 463

J.W. Swanson
Department of Neurology, Mayo Clinic and Mayo Medical School, 200 First Street S.W., Rochester, MN 55905, U.S.A. 173

Volume Editors

M.J. Aminoff
   *Clinical Neurophysiology Laboratories, Department of Neurology,
   University of California, San Francisco, 505 Parnassus Street, Room 348M,
   San Francisco, CA 94143-0216, U.S.A.*

and

C.G. Goetz
   *Section of Movement Disorders, Department of Neurological Sciences,
   Rush-Presbyterian-St. Luke's Medical Center, 1725 West Harrison Street,
   Suite 1106, Chicago, IL 60612, U.S.A.*

Contents

I. Collagen Vascular Diseases and Vasculitis

Handbook of Clinical Neurology, Vol. 27 (71): Systemic Diseases, Part III
M.J. Aminoff and C.G. Goetz, editors
© 1998 Elsevier Science B.V. All rights reserved

CHAPTER 1

Etiology, pathophysiology and classification of the connective tissue disorders

ROBERT P. LISAK

Departments of Neurology and Immunology and Microbiology, Wayne State University School of Medicine, Detroit Medical Center, Detroit, MI, U.S.A.

The collagen-vascular or connective tissue diseases encompass a large of number of disorders that are quite heterogeneous in their manifestations including the pattern of involvement of the nervous system. Indeed two such disorders, dermatomyositis and polymyositis, are by definition neurologic disorders since voluntary muscle is the principal organ system affected. The thread of commonalty in these diseases is the prominent role of immunopathologic, and often true autoimmune, mechanisms in their pathogenesis. By tradition autoimmune diseases that are highly organ-specific such as myasthenia gravis, Lambert–Eaton myasthenic syndrome (LEMS), multiple sclerosis (MS), thyroiditis, and some forms of diabetes mellitus are appropriately not classified as collagen-vascular diseases. The traditional collagen-vascular diseases, including the primary and secondary vasculitides, can and frequently do involve multiple organ systems, although not in every patient. It is of some interest that patients with organ-specific autoimmune diseases not infrequently have clinical or serological manifestations of one or more of the collagen-vascular diseases or one or more of the other organ-specific autoimmune diseases. There are also patients with more than one of the classic collagen-vascular diseases as well as patients who have overlapping features of several of these disorders without fulfilling the criteria for any one of them.

There are many classifications of the collagen-vascular diseases/connective tissue diseases. Table 1 offers one such classification. Since the etiology of most of these disorders is basically unknown and the pathogenic mechanisms are incompletely understood (Lisak et al. 1988), it is not surprising that alternate classifications abound in the literature.

IMMUNOPATHOGENIC MECHANISMS

Immunopathologic mechanisms can be thought of as immunologic reactions that are abnormal because they are either directed at self-antigens ('true' autoimmunity) or, although directed at appropriate antigens, are abnormal because they persist or take place in a host tissue that becomes damaged by the release of immune molecules. Coombs and Gell (1968) originally described four basic immunopathologic mechanisms. As more has been learned about the immune system their scheme has been revised to list other mechanisms or to incorporate these newer mechanisms as subcategories of the original four that were delineated.

One modified classification is presented in Table 2. If one views antibody-mediated reactions that modify surface membranes or receptors as a totally separate category, there is no definite normal equivalent function for such reactions. However, if one considers these as variety of a type II reaction (i.e., antibody 'mistakenly' directed against a self-antigen) in which a simple complement-dependent cytotoxic reaction does not occur but there is instead a

TABLE 1

Collagen-vascular/connective tissue diseases.

1. Systemic lupus erythematosus (SLE)

2. Rheumatoid arthritis (RA)

3. Scleroderma (systemic sclerosis)

4. Dermatomyositis

5. Polymyositis

6. Sjögren's syndrome
    a) Primary
    b) Secondary (i.e., with SLE or RA)

7. Primary vasculitides
    a) Polyarteritis nodosa
    b) Hypersensitivity angiitis
    c) Churg–Strauss vasculitis
    d) Leukocytoclastic vasculitis
    e) Giant cell (temporal) arteritis-polymyalgia
        rheumatica
    f) Wegener's granulomatosus
    g) Cryoglobulinemias
        i)   primary (essential)
        ii) secondary
    h) Takayasu arteritis (aortic arch or pulseless
        disease)
    i) Isolated angiitis of the CNS (granulo-
        matous vasculitis of the CNS)
    j) Behçet's disease
    k) Lymphoid granulomatosis
    l) Other 'primary' CNS vasculitis

8. Secondary vasculitides
    a) Other collagen-vascular diseases
    b) Known infections
    c) Neoplasia
    d) Substance (drug) abuse

9. Mixed connective tissue disease

10. Juvenile rheumatoid arthritis

11. Relapsing polychondritis

12. Ankylosing spondylitis

13. Reiter's disease

14. Cogan's syndrome

15. Overlap syndromes

Modified from Lisak (1986). See Table 3 for additional vasculitides.

modulation of the antigen (receptor or other membrane component), the current schema maintains the logic of the original classification of immunopathologic mechanisms.

Type I or reaginic reactions are normally important in parasite killing and when pathologic are clearly responsible for a great many of the classic 'allergic' respiratory diseases and inhalation allergies and are involved in anaphylaxis. The symptoms of reaginic reactions are caused by the local or systemic release of histamine, prostaglandins and leukotrienes, and by substances released via degranulation of leukocytes, etc.

Many autoimmune diseases are caused by the various subtypes of type II reactions, antibodies directed against self-antigens. As noted earlier, the physiologic function of these antibody-mediated reactions is to help protect the host from infectious agents. The abnormality is the direction of antibodies at self-antigens. Among the type II-mediated disorders are some of the autoimmune hemolytic anemias and certain phases of autoimmune myasthenia gravis (Lopate and Pestronk 1994). The activation of the complement cascade by the interaction of antigen with its cognate antigen leads to membrane damage. In addition, this type of reaction can lead to an accumulation of inflammatory cells that, in turn, can contribute to the tissue damage by phagocytosis and by release of toxic mediators. Antibodies directed against normal cell surface components leading to activation and subsequent cell sequestration are very important in the pathogenesis of some hemolytic anemias and in idiopathic thrombocytopenic purpura. These last two subtypes of pathologic mechanisms are very important in the evolution of many of the manifestations of the collagen-vascular diseases covered in this volume, especially systemic lupus erythematosus (SLE) (Moore and Lisak 1995). Blockade of ligand binding to its receptor by antibodies to a site on or near the actual ligand binding site is an important pathogenic mechanism in some patients with myasthenia gravis (Howard et al. 1987) as well as in patients with diabetes mellitus and acanthocytosis, secondary to antibodies to the insulin receptor. It is also likely, but not proven, that antibodies to receptors for neurotransmitters are responsible for some autoimmune diseases of the nervous system or even certain neurological manifestations of SLE. Specific down-regulation of a surface receptor by an antibody binding to two adjacent receptor molecules is clearly an important mechanism in myasthenia gravis and explains some of the reduction of acetylcholine receptor at the myasthenic neuromuscular junction

TABLE 2

Immunopathologic mechanisms.

Immune mechanism	Reactants	Mediators	Normal function
Reaginic (type I)	IgE, mast cell or basophil and antigen	Vasoactive amines, arachidonic acid metabolites, inflammatory cells	Parasite killing
Antibody to self-antigens (type II)	IgG or IgM and antigen	Complement	Infections
1. Direct cytotoxic		Inflammatory cells	
2. Alteration of membranes and surface receptors a) sequestration via cell surface activation b) block of ligand binding c) receptor down-regulation d) receptor stimulation		Immune complex between antigen on circulating cell and antibody, reacts with Fc receptor on RE cell	Infections
Immune complex deposition (type III)	IgG or IgM + antigen (complement)	Complement; inflammatory cells	Antigen clearance
Cell-mediated reactions (type IV)			
1. T-cell mediated a) Delayed hypersensitivity	T-cells (helper-inducer and antigen, type II MHC dependent	Lymphocytes and lymphokines monocyte-macrophages and monokines	Infections with obligate intracellular organisms
b) T-cell cytotoxic	T-cells (cytotoxic) and antigen, type I MHC dependent	Lymphocytes and products, secondary phagocytosis	? Tumor necrosis, infection (obligate intracellular organisms)
2. ADCC	K-cells (lymphocytes) and macrophages; Fc portion of Ig	K-cells and macrophages	? Infections; tumor destruction
3. NK-cell reactions	NK lymphocytes	NK lymphocytes	Tumor destruction and infections

IgG = immunoglobulin G; IgM = immunoglobulin M; MHC = major histocompatibility complex; ADCC = antibody-dependent cell-mediated cytotoxic; K-cells = killer cells; NK lymphocytes = natural killer cells. Modified from Lisak (1994a).

where IgG deposition is found but there is not membrane destruction or evidence of complement activation (Drachman et al. 1978).

Antibody-induced down-regulation of one or more of the voltage-gated calcium channels seems to be the important pathogenic mechanism in the production of LEMS, both in association with small cell tumors and in the idiopathic variety (Erlington and Newsom-Davis 1994; Lennon 1994a; Lennon et al. 1995). Antibody inhibition of cell function without cell destruction may be important in the pathogenesis of certain CNS complications of SLE (Moore and Lisak 1995). Antibodies to components of neuronal cells have been reported in several of the paraneoplastic neurologic disorders (Dalmau and Posner 1994; Lennon 1994b). While these associations are often quite strong, the role of the reported antibodies, which are often to cytoplasmic or nuclear components in the production of the clinical and pathologic changes, is not known. Damage to myelin or other nerve components seems to be important in several of the demyelinating and axonal neuropathies. Rather than down-regulating a receptor, antibodies may actually lead to receptor stimulation such as has been described for autoimmune hyperthyroidism and some instances of anti-insulin receptor antibodies.

Deposition of circulating immune complexes, type III reaction, is a very important immune mechanism in the production of immunopathologic/autoimmune diseases and is very important in the collagen-vascular diseases, including several of the primary and secondary vasculitides (Moore 1989, 1994; Jennette et al. 1994a). Ordinarily the formation of immune complexes is an important physiologic process which assists in the clearance of foreign antigens from the organism. The abnormality of type III reactions lies in the deposition of complexes in host tissue with subsequent tissue damage. Among these are serum sickness, SLE nephritis, polyarteritis nodosa and some uncommon drug-related hemolytic anemias. It is now clear that immune complex deposition is probably a very important mechanism in the production of both juvenile and adult onset dermatomyositis (Whitaker and Engel 1972; Kissel et al. 1986; Emslie-Smith and Engel 1990). In some instances the nature of the antigen complexed with the antibody that forms the pathogenic complexes is known, but in many it is not. Tissue damage is mediated by the activation of the complement cascade as well as by the effect of the infiltrating inflammatory cells and their released mediators. Immune complex deposition, without accompanying inflammation, can lead to damage to vessels (vasculopathy) with subsequent microhemorrhages and ischemia in the target organ (Berden et al. 1982; Hanly et al. 1992; Moore and Lisak 1995).

Type IV mechanisms, the last general category of immunopathologic mechanisms, are the cell-mediated reactions. The classical normal T-cell-mediated reactions (initiated and controlled by T-cells) are of major importance as a defense mechanism against obligate intracellular microbial organisms such as viruses, mycobacteria and fungi and may also be important in tumor surveillance. The occurrence of opportunistic infections in patients with iatrogenic or disease-induced suppression of the cell-mediated T-

cell-directed system, such as is seen in HIV-infected patients, offers dramatic proof of the importance of the normal functioning of this system. While there are many experimental diseases that can be induced in animals that are clearly initiated by cell-mediated mechanisms or entirely caused by such mechanisms, the evidence for such mechanisms in human autoimmune diseases or in the collagen-vascular diseases is indirect or based on animal models of the human diseases. A T-helper cell-mediated immunopathologic reaction to myelin basic protein and perhaps to other myelin antigens seems to be the primary immune mechanism responsible for acute disseminated encephalomyelitis (ADEM; also called post-infectious, or parainfectious, or post-vaccination or post-immunization encephalomyelitis) (Cohen and Lisak 1987; Lisak 1994b). T-helper cell reactions may also be important in allograft rejection. A cytotoxic T-cell reaction seems to be involved in certain forms of hepatitis, allograft rejection and in graft versus host reactions. While indirect, there is also strong evidence that a cytotoxic T-cell reaction is important in polymyositis and in certain forms of inclusion body myositis (Engel and Arahata 1984). Although there is increasing indirect evidence that a T-cell reaction to one or more myelin components is important in MS, certain forms of uveitis, retinal diseases and some varieties of primary vasculitis, this has not been proven. In part, the restriction of T-cell reactions by the MHC (Zinkernagel and Doherty 1974; Unanue 1991) makes it very difficult to prove directly that a human disease is T-cell mediated. It is much easier to satisfy Witebsky's postulates for antibody-mediated diseases.

Antibody-dependent cell-mediated cytotoxic (ADCC) reactions have generated much interest as putative pathologic reactions in autoimmune diseases, including MS, but its role in the body's normal defense mechanisms is not clear. As a physiologic mechanism it has been implicated in resistance to infection and tumor destruction. Pathologically it is probably also involved in allograft rejection. The specificity of these reactions lies in the antibody binding to its cognate antigen but the tissue destruction is caused by killer cells (K-cells) which bind to the Fc portion of the immunoglobulin molecule (the antibody) rather than the antigen–antibody reaction calling forth the complement cascade and other mediators of type II immunopathologic reactions. Even less is known about natural killer (NK) cell-mediated reactions, which have been suggested as being physiologically important in tumor destruction and in early phases of the response to infection. It is not known if any immunopathologic diseases are mediated by such NK-cell reactions.

It is important to understand that more than one immunopathologic reaction may be important in an individual disease. Several different immunopathologic reactions are triggered by antibodies to the acetylcholine receptor in myasthenia gravis (Lopate and Pestronk 1994). In SLE some of the disease is caused by the deposition of circulating immune complexes consisting of autoantibodies and autoantigens, double-stranded DNA and anti-DNA being a prime example. Other parts of SLE are caused by antibodies binding to a surface antigen on a red cell, platelet or white cell. Indeed in SLE there is evidence that different neurological complications/syndromes are caused by different mechanisms and probably by mechanisms directed at different antigens (Lisak et al. 1988; Moore and Lisak 1995). In the rheumatoid arthritis (RA) joint there is evidence of both cell-mediated and immune complex-mediated reactions (Zvaifler 1983; Firestein and Zvaifler 1990; Panayi et al. 1992). In experimental autoimmune encephalomyelitis, a model for ADEM and MS, and experimental autoimmune neuritis, a model for Guillain–Barré syndrome, there is increasing evidence that the disease is initiated by T-cells but that the subsequent demyelination is due to antibodies to myelin components, often to a different myelin antigen than the initiating T-cell response.

THE PROBLEM OF VASCULITIS

The differing concepts of what constitutes vasculitis and how to classify the vasculitides result in considerable confusion. The term vasculitis should mean that a blood vessel has histologic evidence both of damage to elements of the vessel and of evident inflammation. Perivascular inflammatory cells with a normal blood vessel, even if activated when studied for evidence of up-regulation of adhesion molecules, do not constitute a vasculitic lesion, although in some diseases both type of lesions may be found, especially in different organs. Perivascular inflammatory lesions are classic for cell-mediated immune reactions. A vessel that shows damage but no evidence of inflammation is best described as a vasculopathy even if the vascu-

lopathy is possibly the product of a healed or treated vasculitis. In such an instance other evidence, such as historical from the literature, clinical, or pathologic (from adjacent tissue areas), is required to imply that this was a vasculitis. The lack of an inflammatory reaction at any time in a vasculopathy does not preclude an immunopathogenesis because direct damage from immune complex deposition can occur without calling forth a vigorous inflammatory reaction.

The most logical way to classify the vasculitides is by etiology or pathogenesis; this is not possible since knowledge of the etiology and pathogenesis of many of these disorders is imperfect. One simple division of the vasculitides is into those that are generalized and those that are limited to one organ. The best example of a vasculitis limited to one organ is isolated central nervous system (CNS) vasculitis (angiitis). Over the 20 or more years that this disease has been recognized as an entity, the hypothesis that this is truly limited to the CNS, and usually just the brain, has continued to hold (Moore 1989, 1994). However, while isolated vasculitis of the peripheral nervous system (PNS) may truly exist (Kissel et al. 1985; Dyck et al. 1987; Hawke et al. 1991), most patients followed over a long period of time eventually manifest involvement of other organs (Said et al. 1988). Thus whether isolated vasculitis of the PNS exists as a clear-cut and separate entity is still unclear. There are some patients who seem to have a vasculitis limited to the skin, but it is not always possible to identify such patients when they first present, and many, especially in those with a chronic or recurring vasculitis, eventually develop a systemic or multiorgan syndrome. Nevertheless certain well described systemic vasculitides do show a proclivity to involve certain organs more frequently than others, such as is seen in Wegener's granulomatosis.

A second classification is primary or secondary vasculitis (see Tables 1 and 3). In some patients with collagen-vascular diseases such as SLE or RA, a systemic vasculitis may develop although immune-mediated damage occurs in other organs in the same patient via other immune mechanisms. As an example, a patient with SLE with hemolytic anemia and thrombocytopenia caused by antibodies to cognate antigens on red cells and platelets could develop a systemic vasculitis which becomes manifest as a mononeuropathy multiplex. In addition, a true vasculitis can be caused by infectious agents such as bacteria and fungi (tuberculosis, syphilis, aspergillosis being but a few examples). In addition patients with certain malignancies can develop a vasculitis; a CNS vasculitis, seen rarely in patients with Hodgkin's disease, is one such example (Giang 1994) (see Table 3). It might be argued reasonably that many of the idiopathic vasculitides, be they generalized/systemic, organ specific, or localized, may eventually be shown to result from direct infection of a vessel (Younger et al. 1988), such as is seen with the vasculitis of the carotid artery which may follow herpes zoster ophthalmicus (Hilt et al. 1983) or to develop indirectly through an autoimmune response (types II, III, or IV) perhaps involving molecular mimicry between infectious and host antigens.

Another approach to classification is to consider

TABLE 3

Classification of the vasculitides.

Direct infection of vessels
    Bacterial
    Mycobacterial
    Spirochetal
    Rickettsial
    Fungal
    Viral

Immunologically mediated
    Immune complex-mediated
        Cryoglobulinemia
        Henoch–Schönlein purpura
        Lupus vasculitis
        Rheumatoid vasculitis
        Serum sickness
        Infection-induced immune complex vasculitis
            Bacterial
            Viral
        Paraneoplastic vasculitis (some)
        Drug-induced vasculitis (some)
        Behçet's disease
    Direct antibody attack-mediated
        Goodpasture's syndrome
        Kawasaki's disease
    Antineutrophilic cytoplasmic autoantibody-
        mediated
        Wegener's granulomatosis
        Microscopic polyangiitis
        Churg–Strauss syndrome
        Drug-induced (some)
    Cell-mediated vasculitis
        Allograft cellular vascular rejection
        Giant cell arteritis
        Takayasu arteritis
        Isolated CNS vasculitis

Modified from Jennette et al. (1995).

the type of vessel that is involved, but it is possible that both arteries and veins can be involved in the same disorder. Yet another approach is by the size of the vessels involved (Jennette et al. 1994a, b). While there are some problems of overlap, there is considerable merit to such approaches. Many patients have relatively circumscribed clinical syndromes in which the involved type or portion of a vessel, such as the branching points (bifurcations and trifurcations), is relatively limited, and the pattern helps identify a particular disorder. In some instances a local vessel becomes part of the name of the disease, e.g., temporal arteritis rather than giant cell arteritis. This can lead to errors in diagnosis and management. While some of the headache of temporal arteritis is due to inflammation of the temporal arteries, the visual problems, neuropathy, jaw claudication, and polymyalgia rheumatica are clearly caused by ischemia and inflammation in other arteries.

Another approach is to classify on the basis of the predominant types of inflammatory cells or the histologic pattern of the inflammation (Table 3). There was a historical view that vasculitides in which granulocytes predominate evolve into diseases in which more chronic inflammatory cells, lymphocytes and monocytes, predominate. It is now clear that in many instances this is not the case and that in many of the vasculitides the mononuclear cells are the critical inflammatory cells from the onset. It has been argued that dividing the vasculitides into granulocytic and mononuclear predominate forms serves a useful purpose since the etiologies and pathogenesis are very likely to be different (Moore 1989, 1994; Jennette et al. 1994b). There is very good evidence for immune complex deposition in the highly necrotic and granulocytic vasculitides such as polyarteritis nodosa and leukoclastic vasculitis, although there are exceptions since there is good evidence for immune complex deposition in dermatomyositis, in which disorder the predominant inflammatory cells are mononuclear. In addition there are other forms of vasculitis that share a mononuclear or neutrophilic predominance and yet are mediated by different immune mechanisms (Table 3).

A final approach was mentioned earlier, that is, to classify by etiology when the etiology seems reasonably secure or there is a clear-cut association. There is strong evidence that the lesions in Goodpasture's syndrome are mediated by antibodies directed against basement membrane and some evidence that Kawasaki's disease may be mediated by antibodies to endothelial cells (Leung et al. 1986) (Table 3). The need for the target antigen to be induced by cytokines (see below) suggests that there is an additional inflammatory stimulus involved in the disease pathogenesis. Finally several of the primary vasculitides are associated with antibodies to cytoplasmic components of neutrophils although the exact manner in which these antibodies contribute to the pathogenesis of these disorders is not clear (Gross et al. 1995). Two major patterns of anticytoplasmic antibodies (ACPA) (Van der Woude et al. 1985) have been observed with different disease associations: Wegener's granulomatosis with classic granular cytoplasmic pattern with accentuation of fluorescence within the nuclear lobes (cANCA) and microscopic angiitis with the APCA in a perinuclear pattern (pANCA). The specificities are such that the validity of using the presence of such antibodies to diagnose unequivocally the presence and exact type or variety of vasculitis is still an open question (Gross et al. 1995).

CYTOKINES AND CHEMOKINES

Over the last 10 years it has become increasingly apparent that cytokines are critically important in the pathogenesis of inflammatory, autoimmune and infectious disorders (Durum and Oppenheim 1991) and may even play a role in secondary damage in disorders traditionally thought to be degenerative in nature. Cytokines have been defined as products of cells that are secreted or released and act locally to influence other cells. While originally thought of as products of immune/inflammatory cells and acting on cells of the immune system or on cells that were targets of an injurious immune reaction (target cells for a cytotoxic reaction), it is now clear that cytokines can be produced and released or secreted by many cell types and that most cytokines can act on many different cell types. They not only act locally on cell types nearby, including cells of the type that secreted the cytokine (paracrine), but may act on the secreting cell itself (autocrine). In addition, while the physiologic role of cytokines involves local secretion, it is clear that when administered systemically or secreted in very large amounts in some pathologic states, cytokines can act on cells at a distance. They differ from hormones, which generally act at a distance in their physiologic role, and from neurotransmitters in that cytokine synthesis and release generally requires induction or at least up-regulation.

Cytokines are probably very important in the pathogenesis of the collagen-vascular diseases as well as in other autoimmune disorders. Part of their importance relates to their regulatory role in up- and down-regulation of the cells of the immune system. Thus cytokines such as interleukin-1 (IL-1) and interleukin-2 (IL-2) are important in the afferent limb of the immune response with their activity resulting in expansion of effector cell populations, many of which have specificity to a particular antigen. Others such as IL-6 and IL-10 have important roles in proliferation and maturation of B-cells, the cells that become the immunoglobulin-secreting cells. IL-12 is thought to be important in pushing pluripotential T-cells towards development into Th1 (helper–inducer) status, while IL-4 influences such cells along the Th2 (helper–suppressor) line of development.

The role of cytokines in autoimmune and other inflammatory responses is not limited to interactions among immune/inflammatory cells. Among the effects of IL-1 and tumor necrosis factor-α (TNF-α) are changes in sleep cycling, production of pyrexia and loss of appetite, and they are also involved in stimulating production of acute-phase reactants and participation in septic shock. Thus some of the fatigue and weight loss occurring in patients with inflammatory disorders may be mediated, in part, by the action of cytokines (Dinarello and Wolff 1993).

Local pathologic damage is also influenced and directly mediated by cytokines. Activation of vascular endothelial cells with up-regulation of adhesion molecules such as the selectins, integrins, and cellular adhesion molecules (CAMS) is largely caused by cytokines such as IL-1, TNF-α, and interferon-γ (IFN-γ) (Stoolman 1989; Pober and Cotran 1990; Shimizu et al. 1992). Induction of major histocompatibility complex (MHC) class II molecules on vascular endothelium, which generally follows the induction of the adhesion molecules, is mediated by IFN-γ, and its action is enhanced by TNF-α. While IFN-α and IFN-β are able to increase levels of MHC class I antigens on some cells they may inhibit IFN-γ induction of MHC class II, as does transforming growth factor-β.

The activation of vascular endothelium plays a

primary role in the trafficking of inflammatory cells to different organs as well as in increasing their adherence to the endothelium and their ability to migrate across the vessel into the parenchyma of the different organs. The presence of MHC class II molecules on a cell is necessary for it to serve as an antigen-presenting cell for CD4[+] (helper) T-cells, although there are some CD4[+] T-cell-mediated reactions that are cytotoxic. Thus induction of MHC class II on both professional antigen-presenting cells such as macrophages, but also on other cells such as vascular endothelium, B-cells, microglia, and Langerhans' cells of the skin could contribute to propagation of an immune reaction, including an autoimmune/immunopathologic reaction, in the target organ. It could also contribute to a CD4[+]-mediated cytotoxic reaction. CD8[+] cytotoxic reactions are generally restricted by MHC class I, and therefore up-regulation of MHC class I, as well as of accessory/adhesion molecules, on a cell membrane could lead to necrosis (Shevach 1991). There is also evidence that TNF-α may be an important mediator of demyelination within the CNS (Selmaj and Raine 1988). IL-1 can alter retinal electrophysiology prior to overt pathologic changes in the retina (Brosnan et al. 1989). Thus cytokines are not only important in autoimmune/immunopathologic reactions by modulating (amplifying and down-regulating) the immune response but by causing tissue damage and by producing local and systemic symptoms.

Chemokines are another interesting series of modulators of the immune system and have a common structural motif of four cysteine residues that form two intramolecule disulfide bridges (Miller and Krangel 1992). These substances are a subset of cytokines and are able to cause chemical attraction of certain mobile inflammatory cells (principally PMNs, monocyte/macrophages and eosinophils) to the source of the chemical (chemokinesis). Chemokines help modulate leukocyte adhesion, direct the leukocytes through the vessels into the parenchyma and within the parenchyma and then activate the inflammatory cells. The latter process results in the release of toxic substances that contribute to the tissue damage seen in the inflammatory disorders (Furie and Randolph 1995). It is of interest that IL-8 is a chemokine. Most chemokines are produced in response to stimulation including stimulation by other proinflammatory substances such as IL-1.

TOLERANCE AND AUTOIMMUNITY

The development of autoantibodies or autoreactive cells that are pathogenic represents a failure of normal immune mechanisms. Several mechanisms have been proposed to explain the avoidance of autoimmunity and conversely how failure of one or more of these mechanisms can lead to an autoimmune disease. The original proposal by Burnet (1959) was that exposure to self during development made the immune system incapable of reacting to such antigens because of the deletion (abortion) of the autoreactive cells (clonal deletion theory). This is unlikely to be the usual explanation given the presence of autoreactive T-cells and autoantibodies found in normal individuals (Burns et al. 1983).

A second mechanism invokes suppressor cells, some of which are antigen specific, which suppress the autoimmune cells (suppressor cell theory). A degree of failure of suppressor cell mechanisms would allow for the emergence of autoimmune T- and B-cells leading to an autoimmune reaction (Miller and Schwartz 1982; Schwartz 1991). There is much experimental evidence to recommend this mechanism but it is hard to demonstrate unequivocally, especially in humans.

A third postulated mechanism involves networks of idiotypes which are antibodies formed to other antibodies (Jerne 1974). The first antibody (immunoglobulin molecule) acts as an antigen and the second antibody is directed to the antigen binding site (idiotype) of the first antibody. The second antibody is called an anti-idiotypic antibody. This anti-idiotypic antibody can itself act as an antigen eliciting an anti-anti-idiotypic antibody. It has also been suggested that similar networks exist for antigen-specific cells. Failure of this network would allow an autoimmune antibody or cell to emerge. The system probably becomes operative when the immune system is activated but whether abnormalities of such networks lead to autoimmunity is not certain.

Sequestration of an antigen during development has been postulated as another mechanism leading to tolerance. Nervous system, ocular and other antigens develop in organs with restricted access to circulating cells. In addition, since T-cells recognize antigen in the context of MHC, the relative paucity of MHC in certain organ systems, such as the nervous system, may further protect the organism from the emergence of autoreactive immune cells (Cowing 1985). If an appropriate cell is later exposed to the antigen in the context of MHC during life, the autoimmune cell could emerge and trigger an autoimmune disease. The exposure could be to the self-antigen itself or to cross-reacting epitopes present in exogenous agents including viruses and bacteria (Lisak et al. 1968; Fujinami and Oldstone 1985; Stefansson et al. 1985) This latter mechanism is termed molecular mimicry (Fujinami 1994). However, the presence of apparently non-pathogenic T-cells reactive with autoantigens in the circulation of normal individuals (Burns et al. 1983) suggests that other peripheral mechanisms must also be involved in preventing autoimmunity. One such possible mechanism is that of antigen-induced tolerance (Nossal 1983). The exposure of antigen-specific cells to antigen probably in the absence of MHC or in the absence of accessory molecules suppresses the antigen-specific cells (Feldman et al. 1985).

AUTOANTIBODIES

As noted, autoantibodies can be demonstrated in the serum of normal individuals, usually in low titer and more frequently with advancing age (Schwartz 1991; Zweiman and Lisak 1991). The emergence of high titers of autoantibodies, especially those that are of the IgG type and of high avidity, may arise through direct stimulation of B-cells or via T-cell activation (by exposure to specific antigen or cross-reacting epitopes or by polyclonal T-cell stimulation) or through lack of sufficient antigen-specific or antigen-non-specific suppression. B-cell tolerance is experimentally harder to achieve and harder to maintain than T-cell tolerance. Therefore with loss or reduction of suppressor mechanisms, perhaps including failure of the idiotype network, B-cells that recognize autoantigens may go on to produce higher titers of antibodies, some of which become pathogenic. In this scenario, the potential to produce autoantibodies is present within the germ-line. An alternate mechanism is somatic cell mutation promoted by gene rearrangements with failure to suppress such mutant autoantibody-producing cells when they emerge. It is known that autoantibodies can be produced by malignant B-cell clones and that certain 'benign' paraproteins, including IgM which are often not T-cell dependent, are indeed autoantibodies.

IMMUNOGENETICS

The immune system, like most other systems, is strongly influenced by genes. The immune response to any antigen, including presumably autoantigens, is the product of the effect of environmental stimuli on a series of genes that controls specificity of the immune response. These genes include: (1) the familiar MHC genes, which control specificity of antigen recognition and presentation, leading to activation and perhaps suppression of specific immune responses; (2) non-MHC genes, including those that control production and levels of immune mediators such as complement components; and (3) genes that control the specific recognition of antigenic epitopes by T-cells, called T-cell receptors (TCR), and B-cells (specific immunoglobulin molecules). It is of interest that the TCR and MHC are members of the immunoglobulin superfamily. Because the MHC has been shown to be critical in controlling the specificity of the immune response and to influence the susceptibility or relative resistance of various inbred strains of experimental animals to develop naturally occurring and induced autoimmune diseases, there has been much interest in looking for associations between autoimmune/immunopathologic diseases and particular MHC alleles. In some instances, such as HLA-B27 in ankylosing spondylitis, the association is extremely strong; in others, associations are much weaker or have not been demonstrated. Associations of an autoimmune disease and an MHC allele may be different in different populations. This has been described for several disorders. Indeed in myasthenia gravis the association varies between ethnic/racial groups and with thymic histology (Lisak 1994a). Newer molecular biological methods have replaced the earlier serologic (for MHC I) or mixed lymphocyte reaction (for MHC II), allowing the detection of an increased number of loci and alleles. Whether these techniques will lead to the identification of MHC alleles specific for the development of, or at least an increased susceptibility to develop, specific autoimmune/immunopathologic diseases is not clear. What is known is that it is possible to induce autoimmune disease in inbred strains of animals thought to be resistant to that disease by virtue of their MHC type (Shaw et al. 1992). The demonstration that there is an association between particular TCR families in different species and the response to autoantigens such as myelin basic protein has led to a search for associations between increased frequency of particular TCR families and autoimmune diseases. Given the highly 'outbred' nature of humans it is not surprising that to date the demonstration of such a disease, autoantigen, or even single epitope usage/association is still lacking. It is important to note that TCR (both the αβ and δγ) is the product of gene rearrangements, as is immunoglobulin synthesis. Thus this is one set of genes that could be different in monozygotic twins and perhaps explain the lack of complete concordance of autoimmune diseases in such twinships.

REFERENCES

BERDEN, J.H.M., L. HANG, P.J. MCCONAHEY and F.J. DIXON: Analysis of vascular lesions in murine SLE. J. Immunol. 130 (1982) 1699–1705.

BROSNAN, C.F., M.S. LITWAK, C.E. SCHROEDER, K. SELMAJ, C.S. RAINE and J.C. AREZZO: Preliminary studies of cytokine-induced functional effects on the visual pathways in the rabbit. J. Neuroimmunol. 25 (1989) 227–239.

BURNET, F.M.: The clonal selection theory of acquired autoimmunity. New York, Cambridge University Press (1959).

BURNS, J.B., A. ROSENZWEIG, B. ZWEIMAN and R.P. LISAK: Isolation of myelin basic protein-reactive T-cell lines from normal human blood. Cell. Immunol. 81 (1983) 435–440.

COHEN, J.A. and R.P. LISAK: Acute disseminated encephalomyelitis. In: J.A. Aarli, W.H.M. Behan and P.O. Behan (Eds.), Clinical Neuroimmunology. Oxford, Blackwell Scientific Publications (1987) 192–213.

COOMBS, R.R.A. and P.G.H. GELL: Classification of allergic reactions responsible for clinical hypersensitivity and disease. In: P.G.H. Gell and R.R.A. Coombs (Eds.), Clinical Aspects of Immunity. Philadelphia, F.A. Davis (1968) 575.

COWING, C.: Does T-cell restriction to Ia limit the need for self-tolerance? Immunol. Today 6 (1985) 72–74.

DALMAU, J. and J.B. POSNER: Neurologic paraneoplastic antibodies (anti-Yo; anti-Hu; anti-Ri). The case for a nomenclature based on antibody and antigen specificity. Neurology 44 (1994) 2241–2246.

DINARELLO, C.A. and S.M. WOLFF: The role of interleukin-1 in disease. New Engl. J. Med. 328 (1993) 106–113.

DRACHMAN, D.B., C.W. ANGUS, R.N. ADAMS, J.D. MICHELSON and G.J. HOFFMAN: Myasthenic antibodies cross-link acetylcholine receptors to accelerate degradation. New Engl. J. Med. 298 (1978) 1116–1122.

DURUM, S.K. and J.J. OPPENHEIM: Proinflammatory cytokines and immunity. In: W.E. Paul (Ed.), Fundamental Immunology, 3rd Edit. New York, Raven Press (1991) 801–836.

DYCK, P.J., T.J. BENSTEAD, D.L. CONN, C.S. STEVENS, A.J. WINDEBANK and P.A. LOW: Nonsystemic vasculitic neuropathy. Brain 110 (1987) 843–854.

EMSLIE-SMITH, A. and A.G. ENGEL: Microvascular changes in early and advanced dermatomyositis: a quantitative study. Ann. Neurol. 27 (1990) 343–356.

ENGEL, A.G. and K. ARAHATA: Monoclonal antibody analysis of mononuclear cells in myopathies. II. Phenotypes of autoinvasive cells in polymyositis and inclusion body myositis. Ann. Neurol. 16 (1984) 209–215.

ERLINGTON, G. and J. NEWSOM-DAVIS: Clinical presentation and current immunology of the Lambert–Eaton myasthenic syndrome. In: R.P. Lisak (Ed.), Handbook of Myasthenia Gravis and Myasthenic Syndromes. New York, Marcel Dekker (1994) 81–102.

FELDMANN, M., E.D. ZANDERS and J.R. LAMB: Tolerance in T-cell clones. Immunol. Today 6 (1985) 58–62.

FIRESTEIN, G.S. and N.J. ZVAIFLER: How important are T cells in rheumatoid synovitis? Arthr. Rheum. 33 (1990) 768–773.

FUJINAMI, R.S.: Molecular mimicry: virus modulation of the immune response. In: R.R. McKendall and W.G. Stroop (Eds.), Handbook of Neurovirology. New York, Marcel Dekker (1994) 103–114.

FUJINAMI, R.S. and M.B.A. OLDSTONE: Amino acid homology between the encephalitogenic site of myelin basic protein and virus: mechanism for autoimmunity. Science 230 (1985) 1043–1045.

FURIE, M.B. and G.J. RANDOLPH: Chemokines and tissue injury. Am. J. Pathol. 146 (1995) 1287–1301.

GIANG, D.W.: Central nervous system vasculitis secondary to infections, toxins, and neoplasms. Semin. Neurol. 14 (1994) 313–319.

GROSS, W.L., E. CSERNOK and U. HELMCHEN: Antineutrophil cytoplasmic autoantibodies, autoantigens, and systemic vasculitis. A.P.M.I.S. 103 (1995) 81–97.

HANLY, J.G., N.M.G. WALSH and V. SANGALANG: Brain pathology in systemic lupus erythematosus. J. Rheumatol. 19 (1992) 732–741.

HAWKE, S.H.B., L. DAVIES, R. PAMPHLETT, Y.-P. GUO, J.D. POLLARD and J.G. MCLEOD: Vasculitic neuropathy: a clinical and pathological study. Brain 114 (1991) 2175–2190.

HILT, D.C., D. BUCHHOLZ, A. KRUMHOLZ, H. WEISS and J.S. WOLINSKY: Herpes zoster ophthalmicus and delayed contralateral hemiparesis caused by cerebral angiitis: diagnosis and management approaches. Ann. Neurol. 14 (1983) 543–553.

HOWARD, F.M., V.A. LENNON, J. FINLEY, J. MATSUMOTO and L.R. EVELBACK: Clinical correlations of antibodies that bind, block, or modulate human acetylcholine receptors in myasthenia gravis. Ann. N.Y. Acad. Sci. 505 (1987) 526–538.

JENNETTE, J.C. AND R.J. FALK: Update on the pathobiology of vasculitis. Monogr. Pathol. 37 (1995) 156–172.

JENNETTE, J.C., R.J. FALK, K. ANDRASSY, P.A. BACON, J. CHURG, W.L. GROSS, E.C. HAGEN, G.S. HOFFMAN, C.G. HUNDER ET AL.: Nomenclature of systemic vasculitides: the proposal of an international consensus conference. Arthr. Rheum. 37 (1994a) 187–192.

JENNETTE, J.C., R.J. FALK and D.M. MILLING: Pathogenesis of vasculitis. Semin. Neurol. 14 (1994b) 291–299.

JERNE, N.K.: Towards a network theory of the immune system. Ann. Immonol. Inst. Pasteur (Paris) 125 (1974) 373.

KISSEL, J.T., A.P. SLIVKA, J.R. WARMOLTS and J.R. MENDELL: The clinical spectrum of necrotizing angiopathy of the peripheral nervous system. Ann. Neurol. 18 (1985) 251–257.

KISSEL, J.T., J.R. MENDELL and K.W. RAMMOHAN: Microvascular deposition of complement membrane attack complex in dermatomyositis. New Engl. J. Med. 314 (1986) 329–334.

LENNON, V.A.: Serological diagnosis of myasthenia gravis and the Lambert–Eaton myasthenic syndrome. In: R.P. Lisak (Ed.), Handbook of Myasthenia Gravis and Myasthenic Syndromes. New York, Marcel Dekker (1994a) 149–164.

LENNON, V.A.: Paraneoplastic autoantibodies: the case for a descriptive generic nomenclature. Neurology 44 (1994b) 2236–2240.

LENNON, V.A., T.J. KRYZER, G.E. GRIESMANN, P.E. O'SUILLEBHAIN, A.J. WINDEBANK, A. WOPPMANN, G.P. MILKANICH and E.H. LAMBERT: Calcium-channel antibodies in the Lambert–Eaton syndrome and other paraneoplastic syndromes. New Engl. J. Med. 332 (1995) 1467–1474.

LEUNG, D.Y.M., T. COLLINS, L.A. LAPIERRE, R.S. GEHA and J.S. POBER: Immunoglobulin M antibodies present in the acute phase of Kawasaki syndrome lyse cultured vascular endothelial cells stimulated with gamma interferon. J. Clin. Invest. 77 (1986) 1428–1435.

LISAK, R.P.: Neurological manifestations of collagen vascular disease. In: A.K. Asbury, G.M. McKhann and W.I. McDonald (Eds.), Diseases of the Nervous System. Clinical Neurobiology, 1st Edit. Philadelphia, W.B. Saunders (1986) 1499–1509.

LISAK, R.P.: The immunology of neuromuscular disease. In: J. Walton, G. Karpati and D. Hilton-Jones (Eds.), Disorders of Voluntary Muscle, 6th Edit. Edinburgh, Churchill Livingstone (1994a) 381–414.

LISAK, R.P.: Immune-mediated parainfectious encephalomyelitis. In: R.R. McKendall and W.G. Stroop (Eds.), Handbook of Neurovirology. New York, Marcel Dekker (1994b) 173–186.

LISAK, R.P., R.G. HEINZE, G.A. FALK and M.W. KIES: Search for anti-encephalitogenic antibody in human demyelinative diseases. Neurology 18 (1968) 122–128.

LISAK, R.P., P.M. MOORE, A.I. LEVINSON and B. ZWEIMAN: Neurological complications of collagen vascular diseases. Ann. N.Y. Acad. Sci. 540 (1988) 115–121.

LOPATE, G. and A. PESTRONK: The myasthenic neuromuscular junction. In: R.P. Lisak (Ed.), Handbook of Myasthenia Gravis and Myasthenic Syndromes. New York, Marcel Dekker (1994) 225–237.

MILLER, M.D. and M.S. KRANGEL: Biology and biochemistry of the chemokines; a family of chemotactic and inflammatory cytokines. Crit. Rev. Immunol. 12 (1992) 17–46.

MILLER, K.B. and R.S. SCHWARTZ: Autoimmunity and suppressor T lymphocytes. Adv. Intern. Med. 27 (1982) 281–313.

MOORE, P.M.: Immune mechanisms in the primary and secondary vasculitides. J. Neurol. Sci. 93 (1989) 129–145.

MOORE, P.M.: Vasculitis of the central nervous system. Semin. Neurol. 14 (1994) 307–312.

MOORE, P.M. and R.P. LISAK: Systemic lupus erythematosus-immunopathogenesis of neurologic dysfunction. Semin. Immunopathol. 17 (1995) 43–60.

NOSSAL, G.J.V.: Cellular mechanisms of immunologic tolerance. In: W.E. Paul, C.F. Fathman and H. Metzger (Eds.), Annu. Rev. Immunol. (1983) 33–62.

PANAYI, G.S., J.S. LANCHBURY and G.H. KINGSLEY: The importance of the T-cell in initiating and maintaining the chronic synovitis of rheumatoid arthritis. Arthr. Rheum. 35 (1992) 729–235.

POBER, J.S. and R.S. COTRAN: The role of endothelial cells in inflammation. Transplantation 50 (1990) 537–544.

SAID, G., C. LACROIX-CIAUDO, H. FUJIMURA, C. BLAS and N. FAUX: The peripheral neuropathy of necrotizing arteritis: a clinicopathological study. Ann. Neurol. 23 (1988) 461–465.

SCHWARTZ, R.S.: Autoimmunity and autoimmune diseases. In: W.E. Paul (Ed.), Fundamental Immunology, 3rd Edit. New York, Raven Press (1991) 1033–1097.

SELMAJ, K. and C.S. RAINE: TNF mediates myelin and oligodendrocyte damage in vitro. Ann. Neurol. 23 (1988) 339–346.

SHAW, M.K., C. KIM, H. KHANG-LOON, R.P. LISAK and H.Y. TSE: A combination of adoptive transfer and antigenic challenge induces consistent murine experimental autoimmune encephalomyelitis in C57/BL6 mice and other reputed resistant strains. J. Neuroimmunol. 39 (1992) 139–150.

SHEVACH, E.M.: Accessory molecules. In: W.E. Paul (Ed.), Fundamental Immunology, 3rd Edit. New York, Raven Press (1991) 531–576.

SHIMIZU, Y., W. NEWMAN, Y. TANAKA and S. SHAW: Lymphocyte interaction with endothelial cells. Immunol. Today 13 (1992) 106–112.

STEFANSSON, K., M.E. DIEPERINK, D.P. RECHMAN, C.M. GOMEZ and L.S. MARTON: Sharing of antigenic determinants between the nicotinic acetylcholine receptor and protein in *Escherichia coli, Proteus vulgaris,* and

Klebsiella pneumoniae. Possible role in the pathogenesis of myasthenia gravis. New Engl. J. Med. 312 (1985) 221–225.

STOOLMAN, L.M.: Adhesion molecules controlling lymphocyte migration. Cell 190 (1989) 907–914.

UNANUE, E.R.: Macrophages, antigen-presenting cells, and the phenomena of antigen handling and presentation. In: W.E. Paul (Ed.), Fundamental Immunology, 3rd Edit. New York, Raven Press (1991) 111–144.

VAN DER WOUDE, F.J., N. RASMUSSEN, S. LOBATTO, A. WIIK, L.A. PERMIN, L.A. VAN ES, M. VAN DER GIESSEN, G.K. VAN DER HEM and T.H. THE: Autoantibodies against neutrophils and monocytes: tool for diagnosis and marker of disease activity in Wegener's granulomatosis. Lancet i (1985) 425–429.

WHITAKER, J.N. and W.K. ENGEL: Vascular deposits immu-noglobulin and complement in idiopathic inflammatory myopathy. New Engl. J. Med. 286 (1972) 333–338

YOUNGER, D.S., A.P. HAYS, J.C. BRUST and L.P. ROWLAND: Granulomatous angiitis of the brain. An inflammatory reaction of diverse etiology. Arch. Neurol. 45 (1988) 514–518.

ZINKERNAGEL, R.M. and P.C. DOHERTY: Restriction of in vitro T-cell mediated cytotoxicity in lymphocytic choriomeningitis within a syngeneic or semi-syngeneic system. Nature 248 (1974) 701–702.

ZVAIFLER, N.J.: Pathogenesis of the joint disease of rheumatoid arthritis. Am. J. Med. 75 (1983) 3–12.

ZWEIMAN, B. and R.P. LISAK: Autoantibodies: autoimmunity and immune complexes. In: J.B. Henry (Ed.), Clinical Diagnosis and Management by Laboratory Methods, 18th Edit. Philadelphia, W.B. Saunders (1991) 885–911.

Handbook of Clinical Neurology, Vol. 27 (71): Systemic Diseases, Part III
M.J. Aminoff and C.G. Goetz, editors

CHAPTER 2

Nervous system involvement in rheumatoid arthritis

G.A.W. BRUYN[1] and H.M. MARKUSSE[2]

[1]*Department of Rheumatology, Medical Center Leeuwarden, 8934 AD Leeuwarden, The Netherlands, and*
[2]*Department of Rheumatology, Dr. Daniel Den Hoed Clinic, 3075 EA Rotterdam, The Netherlands*

Rheumatoid arthritis is a systemic autoimmune disease of unknown aetiology characterised by inflammation of synovial tissue leading to chronic destruction of joints. Both cellular and humoral mechanisms are involved in the initial inflammatory response, as evidenced by the finding of activated B and T cells, rheumatoid factors, immune complexes, and cytokines in serum and synovial fluid of patients with rheumatoid arthritis. In addition to these immunologically related phenomena, several lines of evidence suggest that non-immune pathways play an important role later in the disease (Zvaifler and Firestein 1994).

The annual incidence of rheumatoid arthritis is between 2 and 4 per 10,000 adults. Women account for 75% of cases and disease onset is between 35 and 50 years of age in 80% of cases. The American Rheumatism Association has proposed criteria for the classification of rheumatoid arthritis which are listed in Table 1 (Arnett et al. 1988).

Rheumatoid arthritis is not confined to the joints. Extra-articular manifestations are evident in 10–20% of patients. Pleuropericarditis, nodule formation, vasculitis and nervous system involvement are among the most frequently observed extra-articular manifestations.

Nervous system involvement may occur at several, not mutually exclusive, levels: (a) the peripheral nervous system; (b) the autonomic nervous system; (c) the central nervous system (CNS); (d) muscle/ neuromuscular junction.

TABLE 1

The 1987 revised criteria for the classification of rheumatoid arthritis.

Morning stiffness
Arthritis of 3 or more joint areas
Arthritis of hand joints
Symmetric arthritis
Rheumatoid nodules
Serum rheumatoid factor
Radiographic changes

For classification purposes, a patient is said to have rheumatoid arthritis if at least 4 of these 7 criteria are fulfilled.

THE PERIPHERAL NERVOUS SYSTEM

In general, peripheral nerves in patients with rheumatoid arthritis may be affected by the disease in two different patterns: an ischaemic neuropathy due to thrombosis or vasculitis, and an entrapment neuropathy due to compression by swollen synovium, aponeurosis, or bony exostoses.

The neuropathy observed in rheumatoid arthritis may take any of three forms: a mild sensory polyneuropathy, a more severe combined sensorimotor neuropathy, and a mononeuritis multiplex (Beckett and Dinn 1972; Peyronnard et al. 1982; Conn and Dyck 1984). The variable clinical expression can be explained by the acuteness of the ischaemia, the presence of collateral circulation, and the degree of nervous tissue sensitivity to ischaemia (Moore and Cupps 1983).

Polyneuropathy

The neuropathy most commonly encountered in rheumatoid arthritis is a distal sensory neuropathy of the lower limbs. The neuropathy occurs in about 10% of patients with rheumatoid arthritis and the gender distribution is similar to that of the underlying disease (Conn and Dyck 1984). It presents with numbness in toes or feet, develops insidiously and shows a symmetrical and stock-like distribution. Patients complain of paraesthesias, dysaesthesias, or burning feet of variable severity. Pain or dysaesthesia may also be present in the calf muscles. Although this type of neuropathy is not characterised by muscle weakness, both sensory and motor conduction are frequently disturbed and many patients exhibit abnormalities in nerve conduction studies without any clinical manifestation of peripheral nerve involvement (Dudley Hart and Golding 1960). There are indications that large myelinated nerve fibres are the first to be affected (Bekkelund et al. 1996). The pathogenesis of this type of polyneuropathy is not clear. Both thrombosis and endarteritis of vasa nervorum resulting in ischaemia of nerves have been found in sensory neuropathy of rheumatoid arthritis. Prognosis for recovery is good and only on rare occasions does progression occur to more severe polyneuropathy.

The more severe types of polyneuropathy, i.e., the mixed sensorimotor neuropathy and the mononeuritis multiplex, are present in patients with rheumatoid arthritis and vasculitis, i.e., an inflammation with polymorphonuclear cells of small and medium-sized blood vessels (Fig. 1). It occurs in about 50% of patients with rheumatoid vasculitis (Schmid et al. 1961). The average age of affected patients is about 56 years, with an equal distribution among both sexes, in contrast with the female predominance in rheumatoid arthritis (Dudley Hart and Golding 1960; Ferguson and Slocumb 1961; Scott et al. 1981; Wees et al. 1981; McCombe et al. 1991). The clinical features of patients with rheumatoid vasculitis generally involve long-standing erosive disease (over 10 years), weight loss, anorexia, malaise, fever, hepatosplenomegaly, rashes, cutaneous ulcers, digital infarctions on nail-bed, nail-fold, or finger-pulp, rheumatoid nodules, and laboratory abnormalities such as an elevated erythrocyte sedimentation rate, anaemia, eosinophilia, severe thrombocytosis, hypocomplementaemia, and high

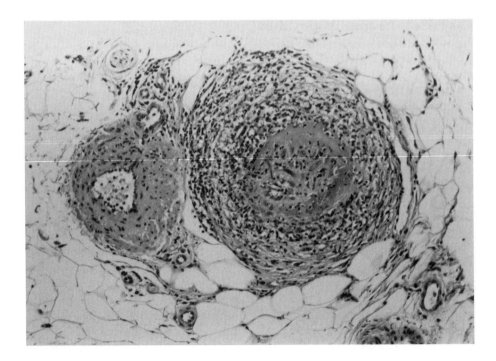

Fig. 1. Histology of vasculitis with fibrinoid necrosis in muscle biopsy of a patient with sensorimotor neuropathy (haematoxylin–eosin, × 400). Courtesy of Dr. R. Kibbelaar.

titres of circulating IgM and IgG rheumatoid factors (Pallis and Scott 1965; Conn and Dyck 1984). Mononeuritis multiplex in patients with disease of short duration, and even as the presenting symptom of rheumatoid vasculitis, has been described occasionally (Chang et al. 1984; McCombe et al. 1991).

Severe sensorimotor neuropathy may start acutely and asymmetrically, i.e., as a mononeuritis, and then progress to a diffuse polyneuropathy (Conn and Dyck 1984; Nuver-Zwart et al. 1986). The nerves most commonly affected are the peroneal, posterior tibial, and ulnar nerves (Chang et al. 1984). Mononeuritis multiplex generally develops later in the course of the illness than symmetric sensory polyneuropathy (Fleming et al. 1976).

Neurological examination reveals a loss of vibration sense together with an altered appreciation of light touch and pinprick. Position sense is preserved. Foot drop or wrist drop may occur. The ankle jerk may be lost but is usually retained, even in patients with wasted calf muscles. Lesions in the hands are less common, but do occur in more severely affected cases.

The diagnosis is confirmed by sensory nerve conduction abnormalities in conventional nerve conduction and somatosensory evoked potential (SSEP) studies. Electrophysiological studies show evidence of axonal degeneration or advanced demyelination as documented by marked sensory and motor nerve conduction abnormalities and widespread denervation of involved muscles (Wees et al. 1981; Peyronnard et al. 1982). On sural nerve biopsy, histopathological examination shows an absence of, or only mild proliferative endarterial lesions in conjunction with, segmental demyelination and some areas of axonal degeneration in cases of distal sensory neuropathy. In patients with the combined sensorimotor or mononeuritis multiplex type of neuropathy, vasculitis of the epineurial arteries may be found (Peyronnard et al. 1982), with evidence of an immune complex mediated mechanism: deposition of IgG, IgM, complement factor C3, and fibrin may be demonstrated along the walls of medium-sized arteries (Conn et al. 1972; Van Lis and Jennekens 1977). Massive wallerian degeneration of nerve fibres is usually present in sensorimotor neuropathy and mononeuritis multiplex (Puéchal et al. 1995). However, other authors have emphasised that the histology of both the neural and the vascular abnormalities of various types of neuropathy rather reflects a continuum than a clearcut demarcation. Even in some severe cases of mono-

neuritis multiplex, a bland intimal thickening and thrombosis of arteries may be present (Pallis and Scott 1965; Weller et al. 1970; Conn et al. 1972).

The peripheral neuropathy of rheumatoid arthritis has to be distinguished from that occurring in other connective tissue diseases, but particularly from polyarteritis nodosa. Polyarteritis nodosa may also present with various arthritic and neurologic symptoms including mononeuritis multiplex. No distinction can be made histologically between the necrotising arteritis found in rheumatoid disease and that of polyarteritis nodosa. However, small digital infarctions are usually not found in polyarteritis nodosa nor are the other stigmata of long-standing rheumatoid disease, such as joint deformations and nodules.

The natural course of the mixed sensorimotor neuropathy is variable: few cases remain constant without treatment. Some improve or resolve spontaneously. Most cases, however, worsen without treatment (Dudley Hart and Golding 1960). Patients die of infection, systemic vasculitis, or both (Dudley Hart and Golding 1960; Pallis and Scott 1965; Scott et al. 1981). In a recent retrospective study, the 5-year survival rate for patients with mononeuritis multiplex was 57% and did not differ from that of patients with distal symmetric sensory or sensorimotor neuropathy (Puéchal et al. 1995). There is a poor prognosis for those who present with cutaneous vasculitis, extensive neuropathy, a depressed complement C4 level, and advanced age (Puéchal et al. 1995).

Treatment is of the underlying disease. Rheumatoid vasculitis is commonly treated with a combination of corticosteroids and cytostatic agents. The combination of prednisone (1 mg/kg/day) and azathioprine (2 mg/kg/day) or cyclophosphamide (1–2 mg/kg/day) is usually effective. One study reported a reduced frequency of polyneuropathy in patients with rheumatoid vasculitis who were treated with low-dose methotrexate (Kaye et al. 1996). The value of plasmapheresis is unproven, but in anecdotal cases with high levels of circulating immune complexes plasmapheresis has been found worthwhile (Nuver-Zwart et al. 1986).

Entrapment neuropathy

Entrapment neuropathy in rheumatoid arthritis is common. Impingement of the median nerve in the carpal tunnel is the most frequently encountered

TABLE 2

Common nerve entrapment syndromes in rheumatoid arthritis.

Carpal tunnel syndrome
Ulnar neuropathy at the elbow or wrist
Posterior interosseous nerve syndrome
Femoral neuropathy
Peroneal neuropathy
Tarsal tunnel syndrome

entrapment neuropathy in these patients, occurring in 23–69% of patients (Chamberlain and Corbett 1970; Fleming et al. 1976). Other peripheral nerves that may also become entrapped are listed in Table 2.

Carpal tunnel syndrome
Carpal tunnel syndrome occurs in the general population with an estimated prevalence of 22 per 10,000 adults (De Krom et al. 1992). In patients with rheumatoid arthritis, its prevalence is about twice as high as in the general population. Carpal tunnel syndrome may be the presenting symptom of rheumatoid arthritis due to median nerve compression in the wrist by tenosynovitis of the flexor tendons of the fingers (Chamberlain and Corbett 1970; Fleming et al. 1976). In one prospective study of extra-articular features in 102 patients with rheumatoid arthritis, it was both one of the most frequent – occurring in 20% of cases – and one of the earliest extra-articular manifestations (Fleming et al. 1976). In the carpal tunnel, the median nerve is volarly confined by the strong carpal ligament, and adjacent to two flexor tendon sheaths. The tendon sheaths swell as a result of the rheumatoid inflammatory process, thereby compressing the median nerve.

As the median nerve is a mixed nerve, symptoms are both sensory and motor. The patient complains of pain and occasionally paraesthesias which worsen at night. Numbness or paraesthesia is felt in the thumb, index finger, middle finger and ring finger. Weakness of the thenar muscles, especially the abductor pollicis brevis, occurs in advanced cases. On physical examination, a volar mass bulging at the wrist may be present and relates to synovitis of the tendon sheaths. Tinel's sign (gentle tapping of the median nerve at the wrist produces a sensation of pain or paraesthesias in the territory of the median nerve) may be present. The Phalen manoeuvre may be positive: forced flexion of the wrist reproduces symptoms. There may be atrophy of the thenar eminence and weakness of abductor pollicis brevis. Pain may extend up to the arm and shoulder, so that the differential diagnosis may include root compression and the thoracic outlet syndrome.

The diagnosis is established by electromyography (EMG) and nerve conduction studies, which are both highly specific and sensitive in making the correct diagnosis. Radiculopathy due to cervical spine disease, diffuse peripheral neuropathy or proximal median nerve involvement can be excluded by electrodiagnostic studies. About half the patients with carpal tunnel syndrome also have conduction abnormalities of the contralateral median nerve. The compression of the median nerve can be visualised by ultrasound, computed tomography (CT) and magnetic resonance imaging (MRI). However, these investigations have no role in management and are unnecessary in clinical practice.

Conservative therapeutic measures including reduced use of the wrist, placement of a wrist splint and administration of non-steroidal inflammatory drugs should be adopted first. Corticosteroid injections in the carpal tunnel are sometimes beneficial. If these measures fail, surgical division of the volar carpal ligament is the treatment of choice.

Ulnar neuropathy at the elbow
As the ulnar nerve passes through the ulnar groove at the elbow, it is subject to several types of mechanical injury relating to compression, stretch or angulation. Such injuries are especially likely to occur in the presence of joint deformity. A cubital tunnel syndrome occurs when the ulnar nerve becomes entrapped just below the elbow by a bulging synovium and the taut aponeurosis of origin of the flexor carpi ulnaris. The point of constriction is generally 1.5–3.5 cm distal to the medial epicondyle. With ulnar neuropathies arising at or about the elbow, patients complain of numbness that typically involves the little finger and medial side of the hand. Pain and tenderness may occur at the elbow and radiate toward the hand. An increase in paraesthesias often occurs with elbow flexion. Weakness of the flexor carpi ulnaris, flexor digitorum profundus of the fourth and fifth digits and intrinsic hand muscles occurs. Clawing of the fingers is less visible in compression at the elbow than if the compression is at the wrist. Electrodiagnostic studies are important in

confirming ulnar nerve involvement and localising the lesion. Both motor and sensory testing are required to make the diagnosis.

Ulnar nerve syndromes at the wrist
The ulnar nerve may become compressed in Guyon's canal at the wrist. Within the canal, the ulnar nerve divides into superficial sensory and deep motor branches. The clinical deficit depends on the level of the lesion. When the lesion is at or proximal to Guyon's canal, it causes a motor and sensory deficit. There is weakness of all intrinsic hand muscles and the hypothenar muscles, and sensory abnormalities involve the palmar ulnar aspect of the hand, both sides of the little finger, and the ulnar border of the ring finger. By contrast, a distal lesion affecting the deep branch causes only a motor deficit that affects all the intrinsic muscles of the hand but spares the hypothenar muscles. There is no sensory deficit. The lesion is at the distal end of the canal. With more proximal involvement of the deep branch, all ulnar-innervated hand muscles are weak but there is no sensory loss. When the lesion involves only the superficial sensory branch, there is only a sensory deficit limited to the palmar ulnar aspect of the hand, both sides of the little finger, and the ulnar border of the ring finger. Typically, the ulnar aspect of the hand is spared on the dorsal surface, since that is supplied by the dorsocutaneous branch above the canal.

Posterior interosseous nerve syndrome
The posterior interosseous nerve is a pure motor branch of the radial nerve. It runs anterior to the elbow, and antecubital cysts due to effusion in the joint may entrap it (McDonald and Smith 1996). The syndrome is characterised by a partial inability of the fingers to extend at the metacarpophalangeal joints. On examination, pressure over the anterior aspect of the elbow may produce a shooting pain, and an antecubital swelling may sometimes be visible. Dorsiflexion of the hand is weak and is associated with a radial deviation of the hand because of paralysis of the extensor carpi ulnaris. Distinction from rupture of the extensor tendons of the fingers may be difficult. An EMG will confirm the diagnosis, whilst echography, arthrography, CT scan or MRI may permit visualisation of the synovial cyst. Treatment with corticosteroid injections in the elbow joint or a surgical decompression synovectomy are both effective (Ishikawa and Hirohata 1990).

Femoral, tibial or peroneal nerve entrapment
The femoral nerve is infrequently entrapped by synovial cyst from the hip joint; this can be visualised by echography, angiography, CT scan or MRI (Létourneau et al. 1991). The tibial or peroneal nerve may become entrapped by a Baker's cyst extending from the knee joint.

Tarsal tunnel syndrome
The posterior tibial nerve runs beneath the flexor retinaculum on the medial aspect of the ankle. At this level, it may become compressed by three tendon sheaths in the tarsal tunnel, which is formed by the retinaculum and the medial malleolus. The tarsal tunnel syndrome is characterised by a burning pain over the medial side of the heel, the sole of the foot and the plantar surface of the toes. Wasting and weakness of the intrinsic muscles of the foot may ensue as a late sign. Asymptomatic patients with rheumatoid arthritis show abnormalities on neurophysiological testing indicative of this syndrome in 5–25% of cases. Conservative treatment with non-steroidal anti-inflammatory medication or, occasionally, an injection of a steroid preparation into the tarsal tunnel may be beneficial. Surgical decompression of the posterior tibial nerve is rarely necessary.

Brown's syndrome
Tenosynovitis of the superior oblique tendon or sheath may occur (Knopf 1989), giving rise to a tendon entrapment syndrome. The syndrome mimicks a palsy of the inferior oblique muscle, demonstrating diplopia on upward and inward gaze. Symptoms may start off with a clicking sound in the orbit near the nose, but later pain and diplopia in the vertical position are present as well (Killian et al. 1977). Brown's syndrome usually responds well to oral corticosteroid administration.

THE AUTONOMIC NERVOUS SYSTEM

Clinical signs of involvement of the autonomic nervous system in patients with rheumatoid arthritis are well known. Cold, clammy hands are a sign of active rheumatoid arthritis; in addition, erythema palmare may frequently be found in rheumatoid patients. The first scientific indication of a dysfunctional autonomic nervous system came from a study of the sweating reflex after injection with nicotine in 100 patients with rheumatoid arthritis (Kalliomäki et al. 1963). Subsequently, Bennett and Scott (1965) found an association between polyneuropathy and autonomic neuropathy. Studies using various non-invasive cardiovascular tests to investigate autonomic dysfunction in rheumatoid arthritis are relatively rare. Recently Toussirot et al. (1993) studied 50 patients with rheumatoid arthritis. All patients underwent cardiovascular tests of heart variation in deep breathing, Valsalva manoeuvre, and orthostatic change in posture. Autonomic dysfunction (defined as two out of three tests abnormal) was found in 60% of the patients, although clinical manifestations of autonomic dysfunction were absent. No correlations were found between autonomic dysfunction and disease duration, the presence of rheumatoid factors in serum, disease activity and the degree of joint destruction. On examination none of the patients showed neurological abnormalities, and symptoms of orthostatic hypotension, or gastrointestinal, sweating, genital or urogenital disorders were not noted. These results are in concordance with previous studies (Edmonds et al. 1979; Leden et al. 1983). A recent study, however, reported a diminished autonomic nervous system responsiveness in recent-onset rheumatoid arthritis (Geenen et al. 1996), exclusively affecting the sympathetic activity, while parasympathetic activity was normal.

THE CENTRAL NERVOUS SYSTEM (CNS)

The CNS may also be affected in rheumatoid arthritis. Cervical myelopathy due to compression of the spinal cord by osseous structures or inflammatory tissue is the most frequently observed manifestation. Inflammation of CNS structures has also been described and may be related to cerebral vasculitis or direct involvement of CNS tissues by inflammatory cells.

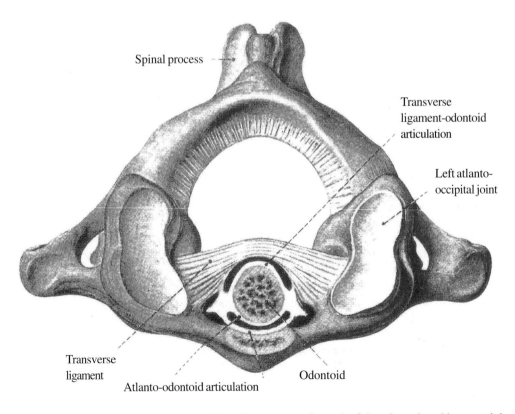

Spinal process

Transverse ligament-odontoid articulation

Left atlanto-occipital joint

Transverse ligament

Atlanto-odontoid articulation

Odontoid

Fig. 2. Anatomical specimen outlining the relations between anterior arch of the atlas, odontoid peg, and the transverse ligament.

Cervical myelopathy

Synovitis of the axial joints, especially those in the cervical spine, may occur in rheumatoid arthritis. Proliferative synovitis or pannus involving the atlanto-axial, atlanto-odontoid, and atlanto-occipital joints and the synovial-lined bursa lying between the odontoid peg and the transverse ligament occurs in about half of cases, and involvement of the cervical joints below C2 accounts for the remainder (Boyle 1971). Thoracic or lumbar intervertebral joints only rarely become involved and will not be discussed further. Cervical deformities are usually associated with severe destructive peripheral disease, the presence of rheumatoid nodules, and a high titre of rheumatoid factor. Associations between cervical subluxations and other clinical features of rheumatoid arthritis, e.g., disease duration, gender, and steroid therapy, are less clear.

Anatomy

In the normal situation, the separation between the anterior arch of the atlas and the odontoid peg measures less than 3 mm. This distance is maintained when the neck is flexed, as the transverse ligament tightly confines the odontoid peg within the articular notch on the anterior arch of the atlas (Fig. 2). Loosening of the transverse ligament results in horizontal atlanto-axial subluxation, i.e., the atlas slips forward relative to the odontoid process of the axis when the neck is flexed. Thus, complete severance of the transverse ligament results in a subluxation of no more than 4–5 mm, whilst destruction of additional ligaments, e.g., ligaments of the atlanto-axial joints and between the odontoid process and the occiput (i.e., the apical and alar ligaments), is required for occurrence of a subluxation of over 5 mm (Fielding et al. 1974). When this occurs and the neck is flexed, the spinal cord may be compressed against the posterior arch of the atlas (Fig. 3). This is called anterior atlanto-axial subluxation.

In addition, the formation of rheumatoid pannus around the odontoid process may occur, which further compromises the available space in the spinal canal. On MRI, so-called pseudotumours may be visible (Kenez et al. 1993), which are formed of pannus and granulation tissue (Fig. 4). The periodontoid pannus tissue may also erode the odontoid peg, making it vulnerable to minor trauma. A severely eroded or fractured dens may result in movement of the atlas posterior to the axis. This so-called posterior atlanto-axial subluxation is rare because severe destruction is necessary to allow such instability.

Unilateral synovitis of one of the occipito-atlanto-axial joints, i.e., the lateral masses, may produce the syndrome of a non-reducible tilting of the head towards the affected side (Halla et al. 1982; Halla and Hardin 1990). If both lateral masses of the axis become eroded, the skull is allowed to descend on to the cervical spine and the odontoid process to enter the foramen magnum. Compression due to the dens, therefore, is not targeted at the cervical spinal cord, but at the brain stem. This deformity has been termed vertical subluxation, cranial settling, atlanto-axial impaction, or basilar invagination (Figs. 5 and 6).

Subaxial subluxation occurs about as often as atlanto-axial subluxation (Meikle and Wilkinson 1971; Winfield et al. 1981). Although the pathogenesis is not entirely clear, the primary foci are probably the small neurocentral joints of Luschka, from which pannus spreads forward to destroy the adjacent intervertebral disc. This inflammatory process may lead to an unstable spine, radiologically shown as a staircase or stepladder deformity (Figs. 7, 8 and 9). Inflammation of the intervertebral disc may also lead to a stable fibrotic ankylosis of the two adjacent vertebrae. However, as the spinal canal is narrower in the middle and lower portions of the cervical spine, subluxation is much more critical at the lower levels. Compression of the spinal cord may occur at any level or even at multiple levels. Combinations of the above-described deformities, i.e., atlanto-axial subluxation, basilar invagination, and subaxial subluxation, may also occur.

Epidemiology

Dependent upon the rigidity of definition and patient selection, atlanto-axial subluxations can be shown radiologically in between 17 and 86% of patients with rheumatoid arthritis (Meikle and Wilkinson 1971; Stevens et al. 1971; Smith et al. 1972) and has been noted in 30–46% of necropsy studies (Eulderink and Meijers 1976). Deformities of the cervical spine may be seen as early as 1 year after onset of disease, and the majority of patients (83%) with subluxation show the first evidence of subluxation within 2 years of disease presentation (Winfield et al. 1981). About 25–80% of patients showing radiological signs of atlanto-axial subluxation develop radiological progression (Smith et al. 1972; Pellici et al. 1981; Winfield et al.

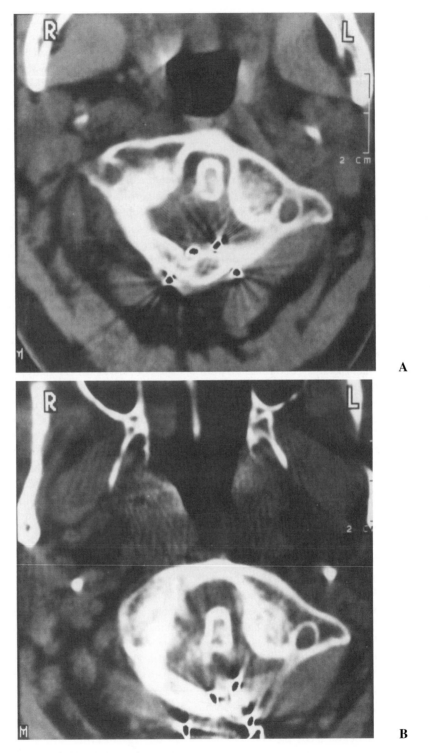

Fig. 3. (A) Transverse CT scan at the level of the atlas. The neck is in neutral position. The odontoid peg is positioned near the anterior arch of the atlas. Posteriorly, several surgical wires witness a prior fusion procedure. (B) The neck is in flexed position. Anterior displacement of the atlas in relation to the axis is obvious, with impingement of the spinal cord between odontoid peg and the posterior arch of the atlas.

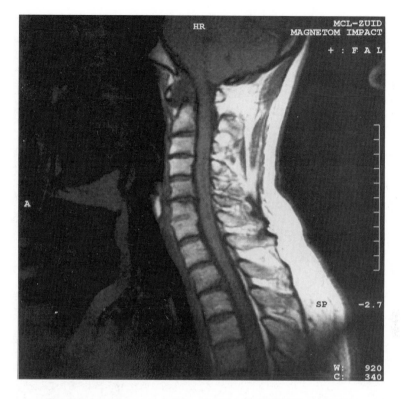

Fig. 4. Pannus tissue around the odontoid peg (MRI, T1 weighted).

1983; Rana 1989). Basilar invagination of the dens occurs in about 3–13% of patients with rheumatoid arthritis (Winfield et al. 1981). About 10% of patients with a subluxation will eventually develop signs of cervical myelopathy (Pellici et al. 1981; Winfield et al. 1981, 1983; Watt and Cummins 1990). This complication never occurs early in the course of disease (Winfield et al. 1981). Of 100 patients with rheumatoid arthritis followed prospectively, 34 developed cervical subluxations after a mean of 9.5 years of follow-up. Among these 34 patients, nine subluxations deteriorated relentlessly; only one developed cervical myelopathy (Winfield et al. 1983).

Pathophysiology
The pathophysiology of chronic compression myelopathy is incompletely understood. Both a mechanical compression (Al-Mefty et al. 1993) and ischaemia of the spinal cord may be involved.

Histology
In rheumatoid arthritis, most histopathological changes of the spinal cord are found in the dorsal white matter and include oedema, axonal swellings or balloons, and necrosis (Henderson et al. 1993). The cuneate fascicle is generally more affected than the gracile (Nakano et al. 1978; Henderson et al. 1993). Ventral changes include central chromatolysis of the ventral horn cells, usually at multiple levels. Vasculitis is rare in the spinal cord (Henderson et al. 1993).

Signs and symptoms
Patients present with neck pain in 40–88% of cases. Pain frequently results from irritation of the second cervical root and involves the occiput and the upper cervical spine. Other presenting symptoms of patients with cervical myelopathy are paraesthesias and numbness of arms and legs, sensations of hot and cold, jumping legs, and generalised muscle weakness most commonly starting in the hands. Several of these symptoms may easily be mistaken for a peripheral neuropathy. Neurological signs and symptoms include spastic quadriparesis, episodic loss of consciousness, loss of dexterity, gait disturbance, dizziness with flexion of the neck, and Lhermitte's sign (electric shocks produced by

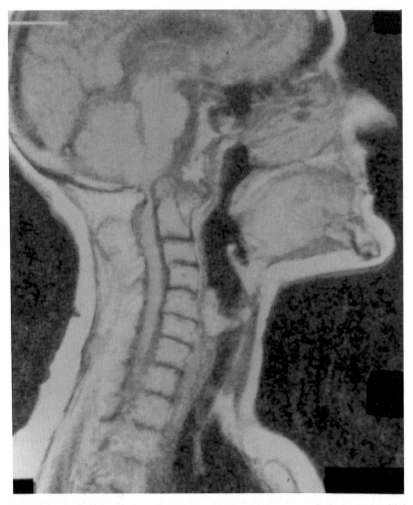

Fig. 5. MRI of the cervical spine (T1 weighted) showing anterior displacement of atlas on axis with impingement of the spinal cord, and a beginning cranial settling. Courtesy of Professor F.C. Breedveld.

movements of the neck). Abnormal plantar responses are an important indication of cervical myelopathy in rheumatoid arthritis. Clinically unsuspected respiratory insufficiency may be common in patients with severe medullary compression (Howard et al. 1994). Symptomatic depression of the respiratory centre may be characterised by Ondine's curse: this may be manifest by florid nightmares and a terror of going to sleep. Other brain stem pathology, including cranial neuropathy, is rarely encountered (Menezes et al. 1985). Weakness in both hands, occipital headache, dysphagia, nausea and incontinence are signs of impending compression, which may result in sudden death (Mikulowski et al. 1975; Parish et al. 1990).

Diagnosis

The degree of atlanto-axial subluxation is routinely evaluated by lateral plain radiographs in flexion and extension. The atlanto-axial distance is measured in the flexed position, with distances greater than 3 mm defined as a subluxation. A distance of 10 mm or more suggests that there is compression of the spinal cord. However, patients with a very large cervical canal may have no neurological manifestations with a C1–C2 subluxation of ≥ 10 mm. Other patients with a narrower canal may have symptoms with a smaller separation. Another drawback of plain radiographs is the low sensitivity for patients with cranial settling. SSEP studies are specific but not sensitive and may fail to diagnose significant atlanto-axial subluxations

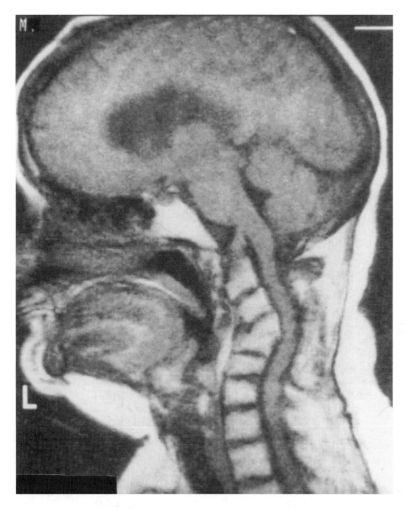

Fig. 6. MRI of the cervical spine (T1 weighted) showing progressive cranial settling: odontoid peg has protrused through the foramen magnum. Courtesy of Professor F.C. Breedveld.

(Rosa et al. 1993). CT scan may delineate the bony structures, but administration of contrast is required to visualise the spinal cord. MRI is currently the technique of choice for imaging the spinal cord. Good resolution of the posterior fossa, foramen magnum, spinal cord, cerebrospinal fluid, bones and pannus is obtained by using MRI (Breedveld et al. 1987; Dvorak et al. 1989). Dynamic MRI with the patient flexing and extending the head may be more sensitive, but this manoeuvre is not yet standard procedure in most centres (Dvorak et al. 1989; Roca et al. 1993).

Surgery

The indications for surgical stabilisation are intractable pain and clinical evidence of myelopathy. In general, surgery produces symptomatic relief in patients with neck or radicular pain. The outcome of surgery for patients with a neurological deficit is less certain (McRorie et al. 1996). However, once cervical myelopathy has developed, natural progression may be rapid with a poor prognosis. Of 43 patients studied by Meijers et al. (1984), 34 were judged fit enough to sustain a surgical fusion. After a follow-up of 5 years, 10 were alive and 24 had died. All nine patients who were managed non-operatively died within 1 year; four deaths were directly attributable to spinal cord compression. Thus, surgery can significantly improve chances for survival in these patients.

Since there is a poor correlation between the degree of subluxation and neurological complications,

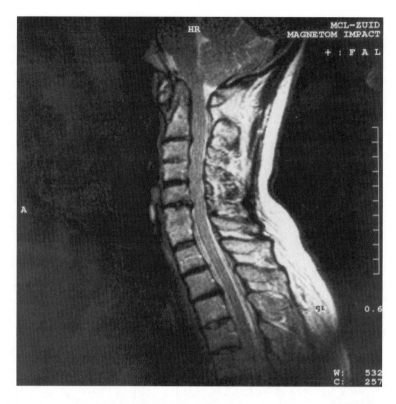

Fig. 7. MRI of the cervical spine (T2 weighted) showing impingement of spinal cord by inflammatory tissue at the level of vertebral disc C5–C6.

operative fusion in neurologically asymptomatic patients with atlanto-axial subluxation is contentious. Some authors advocate early 'prophylactic' fusion for patients in relatively good health if mobile atlanto-axial subluxation is greater than 6 mm (Papadopoulos et al. 1991). These advocates hope to stabilise the subluxation and prevent progression. One also points out that recurrence of cervical instability after fusion occurs in only 5.5% of patients who only show atlanto-axial subluxation, whereas the recurrence rate increases to 36% in patients with more widespread disease, e.g., patients with both horizontal and vertical atlanto-axial subluxation who required fusion from the occiput to C3 (Agarwal et al. 1992). Finally, for those who require surgery in end-stage disease, a gloomy prognosis exists (Casey and Crockard 1995a,b). The largest series to date (Casey et al. 1996) prospectively reports the outcomes of two groups who underwent surgery for the cervical spine – patients who were still ambulant at the time of presentation of myelopathy and patients who had lost the ability to walk. There was a significantly higher

proportion of surgical complications, poorer survival, and limited prospects for functional recovery for the latter group, making a strong case for early surgical intervention. Adversaries of early surgery point to the perioperative mortality of about 10% (Zoma et al. 1987; McRorie et al. 1996). In addition, the natural course of C1–C2 slip is not known. It is also not known whether progression may be retarded by surgery. Given these pros and cons, the sole presence of an atlanto-axial slip is at present no indication for surgery, but patients threatened by cranial settling, i.e., those who show the radiological illusion of improvement of their anterior atlanto-axial subluxation, may be good candidates for early fusion (Agarwal et al. 1992). However, the poor prognosis with late treatment clearly warrants a controlled trial of prophylactic surgery.

The key targets of operation are the relief of spinal cord compression and associated neurological dysfunction, the relief of pain, and the stabilisation of the cervical spine in as normal a position as possible (Zoma et al. 1987). Simple posterior fusion is an

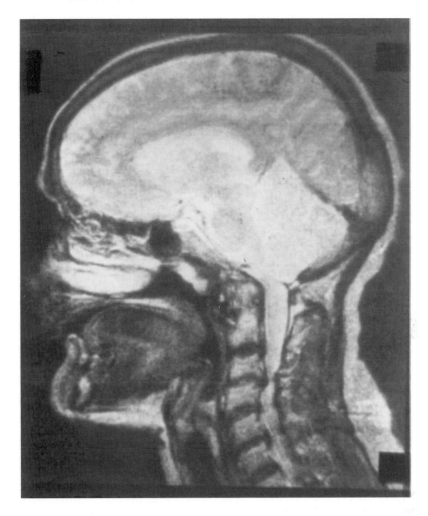

Fig. 8. MRI of the cervical spine (T1 weighted) showing a typical stepladder deformity of the cervical spine with impingement of the spinal cord at multiple levels. Courtesy of Professor F.C. Breedveld.

appropriate approach in the absence of pannus behind the dens and a satisfactory reposition of the C1–C2 subluxation. In situations in which pannus behind the dens causes persistent compression, transoral removal of the odontoid peg and associated inflammatory tissue has been advocated (Crockard et al. 1985). This procedure includes removal of the dens, C1 arch, anterior longitudinal ligament, apical ligament, alar ligaments, transverse ligament and tectorial membrane. Since 90% of patients with rheumatoid arthritis and a history of transoral decompression developed an unstable craniovertebral junction – even those who had stable spines preoperatively – this operation should only be performed along with posterior occipitocervical fusion (Dickman et al.

1992). When subatlanto-axial subluxations occur, anterior subaxial decompression with or without posterior fusion is the method of preference (Crockard et al. 1985). Other investigators, however, have concluded that anterior decompression is rarely necessary (Heywood et al. 1988). Since no controlled trial has been performed in which results of decompression and fusion are compared with those of fusion alone, preferences for one or the other operative technique are not based on scientific evidence.

Overall, there is currently no alternative to surgery. Collars do not prevent progression to subluxation. Pannus tissue may, however, partially respond to administration of high-dose intravenous corticosteroids (Louthrenoo et al. 1992) and this therapy

Fig. 9. Post-mortem specimen of the cervical spine showing severe impingement of the spinal cord by an anterior slip of C4 on C5. Courtesy of Professor F. Eulderink.

may therefore be worthwhile before a major surgical procedure is undertaken.

With early surgical intervention, early postoperative immobilisation and improved anaesthetic and perioperative technique, postoperative mortality is currently less than 10% and improvement of neurological function and loss of pain are to be expected (Crockard et al. 1985; Zoma et al. 1987; Papadopoulos et al. 1991).

The best surgical approach has yet to be defined, including the level of fusion (C1–C2 or lower), whether the occiput should be included, and the necessity of internal decompression with removal of the pannus. Different types of fusion with wires, bone graft, atlanto-axial fixator, and screws have all been reported, but no comparative studies have been done. The question whether C1–C2 surgery

facilitates lower cervical spine subluxations has yet to be resolved.

Brain involvement

Inflammation of CNS structures has been described infrequently and may relate to cerebral vasculitis and direct involvement of the CNS tissues by inflammatory cells and rheumatoid nodules. It reportedly occurs in patients with active, rheumatoid-factor-positive erosive articular disease of long duration.

Any CNS involvement may be symptomatic or asymptomatic. The significance of a stroke – and its possible relationship to rheumatoid arthritis – may not always be appreciated, leading to an underestimation of the frequency of CNS manifestations. Almost 50% of patients with CNS involvement have an altered mental status; motor deficits, seizures, and cranial nerve palsies are each found in about 25% of cases (Ramos and Mandybur 1975; Weisman and Zwaifler 1975; Gupta and Ehrlich 1976; Kim and Collins 1981; Beck and Corbett 1983; Bathon et al. 1989).

The diagnosis of inflammatory CNS involvement is usually made at autopsy. Pathological findings in brain, spinal cord or meninges include rheumatoid nodules, non-specific meningeal inflammation, and vasculitis. A combination of these abnormalities may be present. Rheumatoid nodules are the most frequent pathological finding in the brain; they are located in the meninges and, interestingly, do not extend into the brain parenchyma. CNS nodules are largely asymptomatic (Kim and Collins 1981; Kim et al. 1982; Jackson et al. 1984; Bathon et al. 1989). Spinal nodulosis in rheumatoid arthritis can lead to nerve root compression, spinal cord stenosis and paraparesis (Friedman 1970; Markenson et al. 1979). The next most commonly observed abnormality is a non-specific inflammatory infiltrate of the leptomeninges or pachymeninges. The infiltrate is composed chiefly of plasma cells (73% in one series), and multinucleated giant cells (36% of cases). Vasculitis was found in 37% of cases in an autopsy series and involved vessels of brain and spinal cord parenchyma, as well as meninges. The infiltrate was usually a lymphocytic vasculitis (Bathon et al. 1989).

Normal pressure hydrocephalus (NPH) may be considered another rare extra-articular complication of rheumatoid arthritis. The condition is clinically characterised by gait disturbance, dementia and urinary incontinence. Examination of the CSF usually reveals slightly elevated protein levels without increase in cells. It may be hypothesised that in rheumatoid arthritis NPH is due to a resorption disturbance of the cerebrospinal fluid as a result of rheumatoid inflammation of the meninges. Patients with NPH associated with rheumatoid arthritis may improve after treatment with corticosteroids as well as after spinal fluid drainage by a ventricular shunt, even when the hydrocephalus has been long standing (Rasker et al. 1985; Markusse et al. 1995).

For diagnosing pachymeningitis, contrast-enhanced MRI seems to be useful (Yuh et al. 1990). The sensitivity of a combination of CSF studies and CT scanning or MRI varies between 92 and 100% in cases of vasculitis (Stone et al. 1994).

MYOPATHY AND MYASTHENIA IN RHEUMATOID ARTHRITIS

Muscle weakness and wasting frequently occur in rheumatoid arthritis and may be caused by immobility, disuse, joint pain and synovitis. It has been suggested, however, that the amount of muscle atrophy is discordant with the degree of disuse in patients with this disease. Intramuscular infiltrates mostly consisting of lymphocytes and plasma cells have been described in rheumatoid arthritis. These infiltrates ('nodular myositis') have also been described in other diseases and are therefore not specific for rheumatoid arthritis (Sokoloff et al. 1951). In addition, arteritis of the small arteries has been found in 8% of random muscle biopsies of patients with rheumatoid arthritis without extra-articular symptoms (Sokoloff et al. 1951), whereas interstitial inflammatory infiltrates are found in approximately 25% of muscle biopsies of patients with rheumatoid vasculitis (Puéchal et al. 1995). More recently, Voskuyl et al. (1998) found perivascular infiltrates in randomly obtained muscle biopsies in 75% of patients with rheumatoid vasculitis, versus 14% found in patients with rheumatoid arthritis.

Elevations of muscle enzymes are rare and generally suggest the presence of another abnormality such as polymyositis, hypothyroidism, muscle trauma or drug toxicity. However, in rheumatoid arthritis, myositis indistinguishable from idiopathic polymyositis may occur, possibly representing an overlap between the two diseases (Halla et al. 1984).

Drug agents used in the treatment of rheumatoid arthritis may also cause myopathy. Corticosteroids may lead to a gradually progressive proximal muscular weakness, in which the serum muscle enzymes are normal. EMG shows typical abnormalities and treatment consists of lowering the dosage of corticosteroids. D-Penicillamine is a second-line anti-rheumatic drug that may lead to a gradually progressive myasthenia gravis (Schrader et al. 1972; Carter et al. 1986; Héraud et al. 1994). Features of D-penicillamine-induced myasthenia gravis are a drug dose of over 500 mg daily, predominance of oculomotor symptoms in 90% of cases, the presence of antibodies to acetylcholine receptors in 75% of cases and to striated muscle in 53%, an increased prevalence of the DR1 DW35 HLA haplotype, and resolution of the myasthenia after cessation of D-penicillamine

(Carter et al. 1986). D-Penicillamine as well as another second-line antirheumatic drug, chloroquine, may rarely give rise to an inflammatory myositis (Whisnant et al. 1963; Schrader et al. 1972).

REFERENCES

AGARWAL, A.K., W.C. PEPPELMAN, D.R. KRAUS, B. POLLOCK, B.L. STOLZER, C.H. EISENBEIS and W.F. DONALDSON: Recurrence of cervical spine instability in rheumatoid arthritis following previous fusion: can disease progression be prevented by early surgery? J. Rheumatol. 19 (1992) 1364–1370.

AL-MEFTY, O., H.L. HARKEY, I. MARAWI, D.E. HAINES, D.F. PELER, H.I. WILNER, R.R. SMITH, H.R. HOLADAY, J.L. HAINING, W.F. RUSSELL, B. HARRISON and T.H. MIDDLETON: Experimental chronic compressive cervical myelopathy. J. Neurosurg. 79 (1993) 550–561.

ARNETT, F.C., S.M. EDWORTHY, D.A. BLOCH, D.J. MCSHANE, J.F. FRIES, N.S. COOPER, L.A. HEALEY, S.R. KAPLAM, M.H. LIANG, H.S. LUTHRA, T.A. MEDSGER, D.M. MITCHELL, D.H. NEUSTADT, R.S. PINALS, J.G. SCHALLER, J.T. SHARP, R.L. WILDER and G.G. HUNDER: The American Rheumatism Association 1987 revised criteria for the classification of rheumatoid arthritis. Arthr. Rheum. 31 (1988) 315–324.

BATHON, J.M., L.W. MORELAND and A.G. DIBARTOLOMEO: Inflammatory central nervous system involvement in rheumatoid arthritis. Sem. Arthr. Rheum. 18 (1989) 258–266.

BECK, D.O. and J.J. CORBETT: Seizures due to central nervous system rheumatoid meningovasculitis. Neurology 33 (1983) 1058–1061.

BECKETT, V.L. and J.J. DINN: Segmental demyelination in rheumatoid arthritis. Quart. J. Med. 41 (1972) 71–80.

BEKKELUND, S.I., S.I. MELLGREN, A. PROVEN and G. HUSBY: Quantified neurological examination with emphasis on motor and sensory functions in patients with rheumatoid arthritis and controls. Br. J. Rheumatol. 35 (1996) 1116–1121.

BENNETT, P.H. and J.T. SCOTT: Autonomic neuropathy in rheumatoid arthritis. Ann. Rheum. Dis. 24 (1965) 161–168.

BOYLE, A.C.: The rheumatoid neck. Proc. Roy. Soc. Med. 64 (1971) 1161–1165.

BREEDVELD, F.C., P.R. ALGRA, C.J. VIELVOYE and A. CATS: Magnetic resonance imaging in the evaluation of patients with rheumatoid arthritis and subluxation of the cervical spine. Arthr. Rheum. 30 (1987) 624–630.

CARTER, H., C. JOB-DESLANDRE, F. DELRIEU, A. KAHAN and C.J. MENKÈS: Myasthénie induite par la D-penicillamine au cours du traitement de la polyarthrite rhumatoïde. Rev. Rhum. Mal. Ostéoartic. 53 (1986) 241–244.

CASEY, A.T.H. and H.A. CROCKARD: Preoperative spinal cord area predicts surgical outcome for atlantoaxial subluxation in rheumatoid arthritis – an analysis of 134 myelopathic patients. J. Neurosurg. 82 (1995a) 368A.

CASEY, A.T.H. and H.A. CROCKARD: In the rheumatoid patient: surgery to the cervical spine. Br. J. Rheumatol. 34 (1995b) 1078–1086.

CASEY, A.T.H., H.A. CROCKARD, J.M. BLAND, J. STEVENS, R. MOSKOVICH and A.O. RANSFORD: Surgery of the rheumatoid cervical spine for the non-ambulant myelopathic patient – too much, too late? Lancet 347 (1996) 1004–1007.

CHAMBERLAIN, M.A. and M. CORBETT: Carpal tunnel syndrome in early rheumatoid arthritis. Ann. Rheum. Dis. 29 (1970) 149–152.

CHANG, R.W., C.L. BELL and M. HALLETT: Clinical characteristics and prognosis of vasculitic mononeuropathy multiplex. Arch. Neurol. 41 (1984) 618–622.

CONN, D.L. and P.J. DYCK: Angiopathic neuropathy in connective tissue diseases. In: P.J. Dyck, P.K. Thomas, E.H. Lambert and R. Bunge (Eds.), Peripheral Neuropathy. Philadelphia, W.B. Saunders (1984) 2027–2043.

CONN, D.L., F.C. MCDUFFIE and P.J. DYCK: Immunopathologic study of sural nerves in rheumatoid arthritis. Arthr. Rheum. 15 (1972) 135–143.

CROCKARD, H.A., W.K. ESSIGMAN, J.M. STEVENS, J.L. POZO, A.O. RANSFORD and B.E. KENDALL: Surgical treatment of cervical cord compression in rheumatoid arthritis. Ann. Rheum. Dis. 44 (1985) 809–816.

DE KROM, M.C.T.F.M., P.G. KNIPSCHILD, A.D.M. KESTER, C.T. THIJS, P.F. BOEKKOOI and F. SPAANS: Carpal tunnel syndrome: prevalence in the general population. J. Clin. Epidemiol. 45 (1992) 373–376.

DICKMAN, C.A., J. LOCANTRO and R.G. FESSLER: The influence of transoral odontoid resection on stability of the craniovertebral junction. J. Neurosurg. 77 (1992) 525–530.

DUDLEY HART, F. and J.R. GOLDING: Rheumatoid neuropathy. Br. Med. J. (1960) 1594–1600.

DVORAK, J., D. GROB, H. BAUMGARTNER, N. GESCHWEND, W. GRAUER and S. LARSSON: Functional evaluation of the spinal cord by magnetic resonance imaging in patients with rheumatoid arthritis and instability of upper cervical spine. Spine 14 (1989) 1057–1064.

EDMONDS, M.E., T.C. JONES, W.A. SAUNDERS and R.D. STURROCK: Autonomic neuropathy in rheumatoid arthritis. Br. Med. J. 2 (1979) 173–175.

EULDERINK, F. and K.A.E. MEIJERS: Pathology of the cervical spine in rheumatoid arthritis: a controlled study in 44 spines. J. Pathol. 120 (1976) 91–96.

FERGUSON, R.H. and C.H. SLOCUMB: Peripheral neuropathy in rheumatoid arthritis. Bull. Rheum. Dis. 9 (1961) 251–254.

FIELDING, J.W., G.V.B. COCHRAN, J.F. LAWSING and M. HOHL:

Tears of the transverse ligament of the atlas – a clinical and biomechanical study. J. Bone Joint Surg. 56A (1974) 1683–1691.

FLEMING, A., S. DODMAN, J.M. CROWN and M. CORBETT: Extra-articular features in early rheumatoid disease. Br. Med. J. 1 (1976) 1241–1243.

FRIEDMAN, H.: Intraspinal rheumatoid nodule causing nerve root compression: case report. J. Neurosurg. 32 (1970) 689–691.

GEENEN, R., G.L.R. GODAERT, J.W.G. JACOBS, M.L. PETERS and J.W.J. BIJLSMA: Dimished autonomic nervous system responsiveness in RA of recent onset. J. Rheumatol. 23 (1996) 258–264.

GUPTA, V.P. and G.E. EHRLICH: Organic brain syndrome in rheumatoid arthritis following corticosteroid withdrawal. Arthr. Rheum. 19 (1976) 1333–1338.

HALLA, J.T. and J.G. HARDIN: The spectrum of atlantoaxial facet joint involvement in rheumatoid arthritis. Arthr. Rheum. 3 (1990) 325–329.

HALLA, J.T., S. FALLAHI and J.G. HARDIN: Nonreducible rotational head tilt and lateral mass collapse: a prospective study of frequency, radiographic findings, and clinical features in patients with rheumatoid arthritis. Arthr. Rheum. 25 (1982) 1316–1324.

HALLA, J.T., W.J. KOOPMAN, S. FALLAHI, S.J. OH, R.E. GAY and R.E. SCHROHENLOHER: Rheumatoid myositis: clinical and histological features and possible pathogenesis. Arthr. Rheum. 27 (1984) 737–743.

HENDERSON, F.C., J.F. GEDDES and H.A. CROCKARD: Neuropathology of the brainstem and spinal cord in end stage rheumatoid arthritis: implications for treatment. Ann. Rheum. Dis. 52 (1993) 629–637.

HÉRAUD, A., C. DROMER, E. FATOUT, B. FOURNIÉ and A. FOURNIÉ: Penicillamine-induced myasthenia gravis with urinary incontinence in a patient with rheumatoid arthritis. Rev. Rhumatol. Mal. Ostéoartic. 61 (1994) 54–55.

HEYWOOD, A.W.B., I.D. LEARMONTH and M. THOMAS: Cervical spine instability in rheumatoid arthritis. J. Bone Joint Surg. 70B (1988) 702–707.

HOWARD, R.S., F. HENDERSON, N.P. HIRSCH, J.M. STEVENS, B.E. KENDALL and H.A. CROCKARD: Respiratory abnormalities due to craniovertebral junction compression in rheumatoid disease. Ann. Rheum. Dis. 53 (1994) 134–136.

ISHIKAWA, H. and K. HIROHATA: Posterior interossus nerve syndrome associated with rheumatoid synovial cysts of the elbow joint. Clin. Orthop. 254 (1990) 134–139.

JACKSON, C.G., R.L. CHESS and J.R. WARD: A case of rheumatoid nodule formation within the central nervous system and review of the literature. J. Rheumatol. 11 (1984) 237–240.

KALLIOMÄKI, J.L., H.A. SAARRIMAA and P. TOIVANEN: Axon reflex sweating in rheumatoid arthritis. Ann. Rheum. Dis. 22 (1963) 46–49.

KAYE, O., C.C. BECKERS, P. PAQUET, J.E. ARRESE, G.E. PIÉRARD and M.G. MALAISE: The frequency of cutaneous vasculitis is not increased in patients with rheumatoid arthritis treated with methotrexate. J. Rheumatol. 23 (1996) 253–257.

KENEZ, J., L. TUROCZY, P. BARSI and R. VERES: Retro-odontoid 'ghost' pseudotumours in atlanto-axial instability caused by rheumatoid arthritis. Neuroradiology 35 (1993) 367–369.

KILLIAN, P.J., B. MCLAIN and O.J. LAWLESS: Brown's syndrome – an unusual manifestation of rheumatoid arthritis. Arthr. Rheum. 20 (1977) 1080–1084.

KIM, R.C. and G.H. COLLINS: The neuropathology of rheumatoid disease. Prog. Hum. Pathol. 12 (1981) 5–15.

KIM, R.C., G.H. COLLINS and J.E. PARISI: Rheumatoid nodule formation within the choroid plexus: report of a second case. Arch. Pathol. Lab. Med. 106 (1982) 83–84.

KNOPF, H.L.S.: An unusual case of painful ophthalmoplegia in a patient with rheumatoid arthritis. Ann. Ophthalmol. 21 (1989) 412–413.

LEDEN, I., A. ERIKSSON, B. LILJA, G. STURFELT and G. SUNDKVIST: Autonomic nerve function in rheumatoid arthritis. Scand. J. Rheumatol. 12 (1983) 166–170.

LÉTOURNEAU, L., M. DESSUREAULT and S. CARETTE: Rheumatoid iliopsoas bursitis presenting as unilateral femoral nerve palsy. J. Rheumatol. 18 (1991) 462–462.

LOUTHRENOO, W., E.L. ZAGER, B. FREUNDLICH and R.B. ZURIER: Intravenous corticosteroid therapy of cervical cord compression in rheumatoid arthritis. Clin. Exp. Rheumatol. 10 (1992) 173–175.

MARKUSSE, H.M., P.H.E. HILKENS, M.J. VAN DEN BENT and CH.J. VECHT: Normal pressure hydrocephalus associated with rheumatoid arthritis responding to prednisone. J. Rheumatol. 22 (1995) 342–343.

MARKENSON, J.A., J.S. MCDOUGAL, P. TSAIRIS, M.D. LOCKSHIN and C.L. CHRISTIAN: Rheumatoid meningitis: a localized immune process. Ann. Intern. Med. 90 (1979) 786–789.

MCCOMBE, P.A., A.C. KLESTOV, A.E. TANNENBERG, J.B. CHALK and M.P. PENDER: Sensorimotor peripheral neuropathy in rheumatoid arthritis. Clin. Exp. Neurol. 28 (1991) 146–153.

MCDONALD, S.P. and M.D. SMITH: An uncommon cause of finger drop in a patient with rheumatoid arthritis. Ann. Rheum. Dis. 55 (1996) 728–730.

MCRORIE, E.R., P. MCLOUGHLIN, T. RUSSELL, I. BEGGS, G. NUKI and N.K. HURST: Cervical spine surgery in patients with rheumatoid arthritis: an appraisal. Ann. Rheum. Dis. 55 (1996) 99–104.

MEIJERS, K.A.E., A. CATS, H.P.H. KREMER, W. LUYENDIJK, G.J. ONVLEE and R.T.W.M. THOMEER: Cervical myelopathy in rheumatoid arhritis. Clin. Exp. Rheumatol. 2 (1984) 239–245.

MEIKLE, J.A. and M. WILKINSON: Rheumatoid arthritis of the cervical spine: radiological assessment. Ann. Rheum. Dis. 30 (1971) 154–161.

MENEZES, A.H., J.C. VAN GILDER, C.R. CLARK and G. EL-KHOURY: Odontoid upward migration in rheumatoid arhritis. An analysis of 45 patients in cranial settling. J. Neurosurg. 63 (1985) 500–509.

MIKULOWSKI, P., F.A. WOLHEIM, P. ROTMIL and I. OLSEN: Sudden death in rheumatoid arthritis with atlanto-axial dislocation. Acta Med. Scand. 198 (1975) 445–451.

MONEIM, M.S.: Ulnar nerve compression at the wrist. Hand Clin. 8 (1992) 337–344.

MOORE, P.M. and T.R. CUPPS: Neurological complications of vasculitis. Ann. Neurol. 14 (1983) 155–167.

NAKANO, K.K., W.C. SCHOENE, R.A. BAKER and D.M. DAWSON: The cervical myelopathy associated with rheumatoid arthritis: analysis of 32 patients with two postmortem cases. Ann. Neurol. 3 (1978) 319–341.

NUVER-ZWART, H.H., A.M.T. BOERBOOMS and T.J.M. DE WITTE: Plasma exchange in a patient with severe rheumatoid arthritis complicated by mononeuritis multiplex and cutaneous ulcers. Neth. J. Med. 29 (1986) 79–84.

PALLIS, C.A. and J.T. SCOTT: Peripheral neuropathy in rheumatoid arthritis. Br. Med. J. 1 (1965) 1141–1147.

PAPADOPOULOS, S.M., C.A. DICKMAN and V.K.H. SONNTAG: Atlantoaxial stabilization in rheumatoid arthritis. J. Neurosurg. 74 (1991) 1–7.

PARISH, D.C., J.A. CLARK, S.M. LIEBOWITZ and W.C. HICKS: Sudden death in rheumatoid arthritis from vertical subluxation of the odontoid process. J. Am. Med. Assoc. 82 (1990) 297–304.

PELLICI, P.M., C.S. RANAWAT, P. TSAIRIS and W.J. BRYAN: A prospective study of the progression of rheumatoid arthritis in the cervical spine. J. Bone Joint Surg. 65A (1981) 342–346.

PEYRONNARD, J.-M., L. CHARRON, F. BEAUDET and F. COUTURE: Vasculitic neuropathy in rheumatoid disease and Sjögren syndrome. Neurology 32 (1982) 839–845.

PUÉCHAL, X., G. SAID, P. HILLIQUIN, J. COSTE, C. JOB-DES-LANDRE, C. LACROIX and C.J. MENKES: Peripheral neuropathy with necrotizing vasculitis in rheumatoid arthritis: a clinicopathologic and prognostic study in thirty-two patients. Arthr. Rheum. 38 (1995) 1618–1629.

RAMOS, M. and T.I. MANDYBUR: Cerebral vasculitis in rheumatoid arthritis. Arch. Neurol. 32 (1975) 271–275.

RANA, N.A.: Natural history of atlanto-axial subluxation in rheumatoid arthritis. Spine 14 (1989) 1054–1056.

RASKER, J.J., E.N.H. JANSEN, J. HAAN and J. OOSTROM: Normal-pressure hydrocephalus in rheumatic patients: a diagnostic pitfall. New Engl. J. Med. 312 (1985) 1239–1241.

ROCA, A., W.K. BERNREUTER and G.S. ALARÇON: Functional magnetic resonance imaging should be included in the evaluation of the cervical spine in patients with rheumatoid arthritis. J. Rheumatol. 20 (1993) 1485–1488.

ROSA, C., M. ALVES, M.V. QUEIROS, F. MORGADO and A. DE MENDONÇA: Neurologic involvement in patients with rheumatoid arthritis with atlantoaxial subluxation – a clinical and neurophysiological study. J. Rheumatol. 20 (1993) 248–252.

SCHMID, F.R., N.S. COOPER, M. ZIFF and C. MCEWEN: Arteritis in rheumatoid arthritis. Am. J. Med. 30 (1961) 56–83.

SCHRADER, P.L., H.A. PETERS and D.S. DAHL: Polymyositis and penicillamine. Arch. Neurol. 27 (1972) 456–457.

SCOTT, D.G.I., P.A. BACON and C.R. TRIB: Systemic rheumatoid vasculitis: a clinical and laboratory study of 50 cases. Medicine 60 (1981) 288–297.

SHEA, J.D. and E.J. MCCLAIN: Ulnar nerve compression syndromes at and below the wrist. J. Bone Joint Surg. 51A (1969) 1095–1103.

SMITH, P.H., R.T. BENN and J. SHARP: Natural history of rheumatoid cervical luxations. Ann. Rheum. Dis. 31 (1972) 431–439.

SOKOLOFF, L., S.L. WILENS and J.J. BUNIM: Arteritis of striated muscle in rheumatoid arthritis. Am. J. Pathol. 27 (1951) 157–161.

STEINER, J.W. and A.J. GELBLOOM: Intracranial manifestations in two cases of systemic rheumatic disease. Arthr. Rheum. 2 (1959) 537–545.

STEVENS, J.C., N.E.F. CARTLIDGE, M. SUANDRES, A. APPLEBY, M. HALL and D.A. SHAW: Atlanto-axial subluxation and cervical myelopathy in rheumatoid arthritis. Quart. J. Med. 159 (1971) 394–408.

STONE, J.H., M.G. POMPER, R. ROUBENOFF, T.J. MILLER and D.H. HELLMAN: Sensitivities of non-invasive tests for CNS vasculitis: a comparison of lumbar puncture, CT and MRI. J. Rheumatol. 21 (1994) 1277–1282.

TOUSSIROT, E., G. SERRATRICE and P. VALENTIN: Autonomic nervous system involvement in rheumatoid arthritis. J. Rheumatol. 20 (1993) 1508–1514.

VAN LIS, J.M. and F.G. JENNEKENS: Immunofluorescence in a case of rheumatoid neuropathy. J. Neurol. Sci. 33 (1977) 313–321.

VOSKUYL, A.E., S.G. VAN DUINEN, A.H. ZWINDERMAN, F.C. BREEDVELD and J.M.W. HAZES: The diagnostic value of perivascular infiltrates in muscle biopsy specimens for the assessment of rheumatoid vasculitis. Ann. Rheum. Dis. 57 (1998) 114–117.

WATT, I. and B. CUMMINS: Management of rheumatoid neck. Ann. Rheum. Dis. 49 (1990) 805–807.

WEES, S.J., I.N. SUNWOO and S.J. OH: Sural nerve biopsy in systemic necrotizing vascultis. Am. J. Med. 71 (1981) 525–532.

WEISMAN, M. and N. ZWAIFLER: Cryoglobulinemia in rheumatoid arthritis: significance in serum of patients with rheumatoid vasculitis. J. Clin. Invest. 56 (1975) 725–739.

WELLER, R.O., F.E. BRUCKNER and M.A. CHAMBERLAIN: Rheumatoid neuropathy: a histological and electrophysiological study. J. Neurol. Neurosurg. Psychiatry 33 (1970) 592–604.

WHISNANT, J.P., R.E. ESPINOSA, R.E. KIERLAND and E.G. LAMBERT: Chloroquine neuromyopathy. Proc. Mayo Clin. 38 (1963) 501–513.

WINFIELD, J., D. COOKE, A.S. BROOK and M. CORBETT: A prospective study of the radiological changes in the cervical spine in rheumatoid disease. Ann. Rheum. Dis. 40 (1981) 109–114.

WINFIELD, J., A. YOUNG, P. WILLIAMS and M. CORBETT: Prospective study of the radiological changes in hands, feet and cervical spine in adult rheumatoid disease. Ann. Rheum. Dis. 42 (1983) 613–618.

YUH, W.T., J.M. DREW, M. RIZZO, T.J. RYALS, Y. SATO and W.E. BELL: Evaluation of pachymeningitis by contrast enhanced MR imaging in a patient with rheumatoid arthritis. Am. J. Neuroradiol. 1990 (11) 1247–1248.

ZOMA, A., R.D. STURROCK, W.D. FISHER, P.A. FREEMAN and D.L. HAMBLEN: Surgical stabilisation of the rheumatoid cervical spine. J. Bone Joint Surg. 69B (1987) 8–12.

ZVAIFLER, N.J. and G.S. FIRESTEIN: Pannus and pannocytes. Alternative models of joint destruction in rheumatoid arthritis. Arthr. Rheum. 37 (1994) 783–789.

Handbook of Clinical Neurology, Vol. 27 (71): Systemic Diseases, Part III
M.J. Aminoff and C.G. Goetz, editors

Nervous system involvement in systemic lupus erythematosus, including the antiphospholipid antibody syndrome

B.A. JANSSEN[1] and G.A.W. BRUYN[2]

[1]*Department of Rheumatology, University of Birmingham, Birmingham, U.K., and*
[2]*Department of Rheumatology, Medisch Centrum Leeuwarden, Leeuwarden, The Netherlands*

Systemic lupus erythematosus (SLE) is an auto-immune disorder that affects multiple organ systems, including the joints, skin, kidneys, lungs, heart and nervous system. Involvement of the nervous system was recognised even in the earliest accounts of SLE. Moriz Kaposi (1872) described two patients who suffered from recurrent stupor and coma, in addition to arthritis and lymphadenopathy. William Osler (1903) found both stupor and recurrent hemiplegia in his patients with SLE. The latter also emphasised the systemic nature of this 'erythematous' disease. These observations were expanded by other investigators and yet SLE was considered for many years to be a rare disease. It was particularly Dubois (Dubois 1956; Dubois and Tuffanelli 1964) who emphasised that SLE was not an infrequent disorder and that neuropsychiatric (NP) manifestations occur commonly.

Although the cause of the disorder is not known, it is generally accepted that much of its pathology is mediated by autoantibodies directed against macro-molecular components of cell nuclei. Antinuclear antibodies are present in about 95% of patients. Other antibodies, such as those against native or double-stranded DNA and to Sm, a ribonuclear protein antigen, are present in a smaller percentage of patients, but have a higher specificity for SLE.

Another enigmatic syndrome, first recognised in patients with SLE, has been unravelled only during the past decade. This syndrome affects patients with and without SLE and is centered around the presence of so-called antiphospholipid antibodies (aPL). Their association with arterial and venous thrombosis, recurrent miscarriages and thrombocytopenia, has resulted in the description of a new syndrome termed the antiphospholipid antibody syndrome (Boey et al. 1983; Hughes 1993). Nervous system involvement occurs not infrequently in patients with aPL; thus, neurologists will be involved in the care of many of these patients.

EPIDEMIOLOGY

SLE has an incidence rate of 1.5/100,000 per year for males and 6.5/100,000 per year for females in the Western world (Hopkinson 1992). Formerly it was thought that most cases occurred in women in their childbearing years, but a recent review noted that the highest incidence occurs in women of middle age (Hopkinson 1992). Prevalence rates are 3.7/100,000 for males and 45.4/100,000 for females, with an overall prevalence of 24.6 cases per 100,000 population (Hopkinson et al. 1993). The diagnosis of SLE is made on clinical and serological grounds. For classification purposes, the patient should fulfill at least four of the revised criteria published by the American College of Rheumatology (Tan et al. 1982). These criteria, which were primarily intended as guidelines for clinical studies and not rigid diagnostic criteria, are listed in Table 1. The proposed definition of NP SLE in these criteria is, however, inadequate, given that only two clinical manifestations, psychosis and seizures, have been included. This omission has been recognised, but so far has not been corrected (Singer and Denburg 1990).

Most authors report a prevalence rate of central nervous system (CNS) involvement in patients with SLE of approximately 50%, although the range is large, varying from 18 to 75% (Dubois and Tuffanelli 1964; Estes and Christian 1971; Feinglass et al. 1976; Gibson and Myers 1976; Abel et al. 1980; Hochberg et al. 1985; Futrell et al. 1992; Sibley et al. 1992; Miguel et al. 1994). These series are summarised in Table 2. The variation in prevalence amongst studies stems from differences in study populations and design, in addition to variations in definitions and diagnostic methods, which result from the lack of a gold standard for what constitutes CNS lupus.

Symptoms and signs of cerebral SLE usually develop within the first 2 years of the disease, although they may occasionally precede or coincide with disease onset (Feinglass et al. 1976). As yet, there are no good clinical predictors of who will develop CNS involvement, although certain features including cutaneous vasculitis, thrombocytopenia, low complement, renal disease, lymphopenia, generally active disease, and number of disease manifestations have been found to be associated with active CNS disease (Feinglass et al. 1976; Gibson and Myers 1976; Abel et al. 1980; Janssen 1992).

CNS features may occur in isolation or as part of a generalised disease flare; they may manifest once or may be recurrent. Relapse may involve the same symptoms or may have completely different manifes-

TABLE 1

The 1982 revised criteria of the American College of Rheumatology for the classification of systemic lupus erythematosus.

Malar rash
Discoid rash
Photosensitivity
Oral ulcers
Non-erosive arthritis
Serositis (pleuritis, pericarditis or peritonitis)
Renal disorder (proteinuria greater than 0.5 g/day or cellular casts)
Neurologic disorder (seizures or psychosis)
Haematologic disorder (haemolytic anaemia with reticulocytosis; or leukopenia with less than 4000 cells/µl on two or more occasions; or, lymphopenia with less than 1500 cells/µl on two or more occasions; or thrombocytopenia (less than 100,000 cells/µl)
Immunologic disorder (positive LE cell preparation; or presence of antibodies to double-stranded DNA; or anti-Sm antibody)
Antinuclear antibody

For classification purposes, a patient is considered to have SLE if at least 4 of these 11 criteria are fulfilled.

tations. Generally, CNS events occur independently of systemic activity; in a minority of patients, they accompany active extra-CNS disease (Kaell et al. 1986; Janssen 1992).

NEUROPATHOLOGY

The most common histopathological finding in the CNS is not a true arteritis, but a so-called 'bland' vasculopathy with vessel occlusion and associated microinfarction (Fig. 1). The involved vessels –

TABLE 2

Incidence of nervous system disease in patients with SLE.

Author(s) and year	Patients (n)	Type of study (prospective/ retrospective)	Nervous system manifestations (%)	Psychiatric manifestations (%)	Mortality (%)
Dubois and Tuffanelli (1964)	520	retrospective	26	12	19
Estes and Christian (1971)	150	prospective	48	42	19
Feinglass et al. (1976)	140	retrospective	51	17	10
Gibson and Myers (1976)	80	prospective	51	33	15
Abel et al. (1980)	180	prospective	77	52	23
Hochberg et al. (1985)	150	prospective	39	16	NA
Futrell et al. (1992)	91	retrospective/ prospective	72	38	12
Miguel et al. (1994)	43	prospective	85	63	NA

n denotes number; NA denotes data not available.

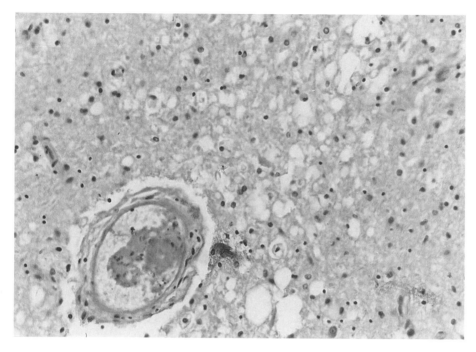

Fig. 1. Vasculopathy in the brain: a partially thrombosed cerebral vessel with incipient cavitation of the surrounding tissue. An occasional lymphocyte is seen lying in the vessel wall. (Haematoxylin and eosin, × 250.) Courtesy of Prof. F. Eulderink.

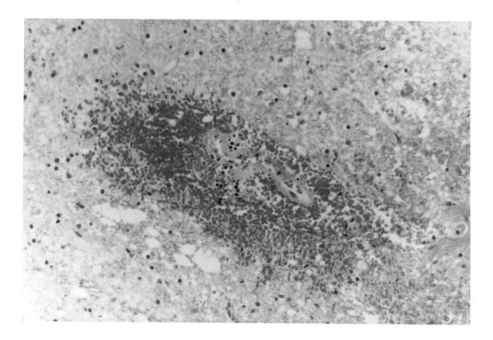

Fig. 2. Perivascular haemorrhage in the midbrain. (Haematoxylin and eosin, × 200.) Courtesy of Surgeon Commander R.W. Smith (with permission of the BMJ Publishing Group, Ann. Rheum. Dis. 53 (1994) 327–333).

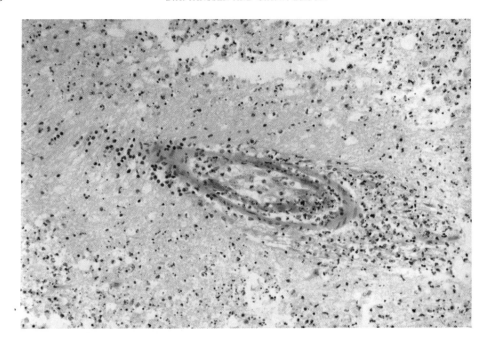

Fig. 3. Vasculitis and fibrinoid necrosis of a cerebellar vessel with an inflammatory infiltrate in the surrounding brain. (Haematoxylin and eosin, × 200.) Courtesy of Surgeon Commander R.W. Smith (with permission of the BMJ Publishing Group, Ann. Rheum. Dis. 53 (1994) 327–333).

usually small arterioles or capillaries with a diameter of <100 μm – typically demonstrate thickening of the wall with endothelial proliferation, perivascular gliosis, and hyalinisation, in the absence of a significant cellular infiltrate (Johnson and Richardson 1968). Other less common pathological features include haemorrhage (Fig. 2), which may be intracerebral, subarachnoid or subdural, and large vessel infarction or thrombosis. In about 10% of autopsies a genuine vasculitis is observed (Fig. 3). Similar findings have been described in two subsequent clinicopathological studies (Ellis and Verity 1979; Hanly et al. 1992).

More recently, the histopathological features of patients with the antiphospholipid antibody syndrome have been described in detail (Hughson et al. 1993). The consistent lesion is a non-inflammatory thrombotic occlusion of vessels of all sizes, either arterial, venous, or occasionally both. Distinct lesions, including endothelial swelling, hyperplasia and mitosis of endothelial cells, and a novel angiomatoid thrombovascular complex, were found in both cortical and subcortical arteries. In addition, fragments of platelet membrane were found in the walls of small cortical and meningeal vessels. These findings suggest that the

characteristic lesions found in the cerebral vessels of patients with SLE, notably thickening and irregularity, are the result of repeated episodes of thrombosis and are associated with circulating aPL (Ellison et al. 1993; Belmont et al. 1996).

Histological abnormalities in the peripheral nervous systems of lupus patients have not been studied extensively, but include inflammatory obliteration of blood vessels in the epineurium and perineurium with intimal fibrosis of small arteries, and perivascular lymphocytic infiltration without vessel wall abnormalities (Johnson and Richardson 1968). Demyelination, axonal degeneration and amorphous material within peripheral nerves have also been described (Johnson and Richardson 1968; Rechthand et al. 1984).

AETIOLOGY AND PATHOGENESIS

The current view of autoimmunity is that a susceptible host can generate pathogenic B and T cell populations in response to self-antigen of a given cell type. In turn, these B and T cells and their products, i.e., various autoantibodies, superantigens, immune complexes, and cytokines, may cause tissue damage. In organ-specific autoimmune diseases, e.g., myasthenia gravis or chronic thyroiditis, one can identify the pathogenic autoantibody and T cell. In lupus, however, one finds a diversity of autoantibodies. Although correlations between autoantibodies and tissue damage have been demonstrated in lupus nephritis (Pisetsky 1993) with anti-DNA autoantibodies appearing to cause renal injury directly, their role in other clinical manifestations, e.g., CNS injury, remains controversial at least. Also, this view does not account for the particular tropism of lupus autoantibodies for nuclear antigens such as DNA, histones, Sm, SS-A, SS-B, RNP, and some enzymes related to nucleic-acid function.

In order to understand lupus autoimmunity, it has been proposed that a genetically determined defect in the tolerance of autoimmune B and T cells may underlie the immunological disturbances in SLE. Various genetic factors play a role in the susceptibility of individuals to develop autoimmune disease, e.g., the genes of the major histocompatibility complex (MHC) (Batchelor 1993). Increased frequencies of the HLA alleles A1, B8, DR2, and DR3 have been described in Caucasian subjects with SLE (Reveille et al. 1983). Other studies have demonstrated an association between complement deficiencies and increased disease susceptibility (De Juan et al. 1993). The strongest genes linked to SLE susceptibility are those encoding deficiencies of proteins of the classic pathway of complement, especially C1q, C2 and C4 (Loos and Heinz 1986). Over 75% of patients with a deficiency of C1q, C1r, C1s, or C4 develop SLE. Recent interest has focused on the genetically determined role of the *Fas* gene, which is important in the regulation of apoptosis, or programmed cell death (Watanabe-Fukunaga et al. 1992). Apoptosis is crucial in the regulatory turnover of cells, including lymphocytes. In an autoimmune mouse model, mutations of the *Fas* gene resulted in defective apoptosis (Mountz et al. 1994). Cheng et al. (1994) were the first to identify the link between *Fas*-mediated apoptosis and autoimmunity in humans. They showed that an increased concentration of soluble *Fas* molecule is present in the serum of patients with SLE. However, other investigators have questioned the specificity of this finding as they also found increased serum levels in other diseases (Elon et al. 1996; Okubo et al. 1996). New light on this possible mechanism was recently shed by observations that most of the lupus autoantigens are packed as oligonucleosomes and clustered in surface blebs of apoptotic cells, suggesting that these surface structures are important immunogenic particles. These blebs are shed from the surface of the apoptotic cell, are taken up by antigen-presenting cells through a nucleosome receptor, processed and presented to T cells that help nucleosome-specific B cells to produce pathogenic autoantibodies (Koutouzov et al. 1996).

In addition to genetic factors, numerous other factors have been shown to influence both susceptibility to and severity of SLE, including sex hormones (Roubinian et al. 1977; Ansar Ahmed et al. 1985; Da Silva 1995) and environmental factors, such as ultraviolet radiation, drugs, dietary factors, stress, or infections (Hess and Farhey 1995). Although the mechanism is still unclear, it is feasible that these factors may adversely affect apoptosis and hence influence an individual's susceptibility.

The pathogenesis of cerebral lupus is even more difficult to explain than that of SLE. Given the autoimmune nature of SLE and the inability of the histopathological findings to account for the clinical manifestations, several hypotheses have been advanced to explain the diverse CNS manifestations. These can be divided into those based on: (1) immune-complex vasculitis; (2) neurone-reactive autoantibodies; (3) thrombosis associated with aPL; and (4) cytokine-enhanced immunological disturbances.

Immune complex vasculitis

Vasculitis in SLE is most commonly due to the local deposition of DNA:anti-DNA immune complexes in blood vessel walls (Belmont et al. 1996). These DNA:anti-DNA complexes have also been identified in the choroid plexus of patients with NP lupus, both by immunofluorescence and by electron microscopy (Atkins et al. 1972), although the authors failed to examine the choroid tissue from patients without CNS involvement. Similar findings were reported in a controlled study, but choroid tissue from rheumatoid arthritis patients as well as SLE patients without CNS involvement also showed immunoglobulin deposition (Boyer et al. 1980). In addition, serological evidence of a true vasculitis in the brains of lupus patients is not a consistent finding. Investigations for complement-dependent inflammatory injury have yielded conflicting data (Hadler et al. 1973; Jongen et al. 1990). Furthermore, high titres of DNA autoantibodies have not been found in the cerebrospinal fluid (CSF) (Winfield et al. 1978). Moreover, true vasculitis is found in only about 10% of brain specimens of patients with cerebral lupus (Johnson and Richardson 1968). The conclusion drawn from these studies is that the highly vascular choroid plexus may act as a non-specific trap for circulating immune complexes. Since animal studies have shown that the permeability of the blood–brain barrier may be altered by high levels of circulating immune complexes (Harbeck et al. 1979), it is possible that immune complexes may play a permissive rather than a direct pathogenic role, by allowing ingress of systemically produced autoantibodies that would otherwise not gain access to the CNS.

Neurone-reactive antibodies

Neuronal antibodies are an attractive alternative explanation for the often diffuse and reversible manifestations of cerebral lupus, which occasionally completely lack histopathological abnormalities (Johnson and Richardson 1968). Animal studies have supported this concept, as administration of antibodies to brain constituents can induce seizures or behavioural changes (Simon and Simon 1975; Karpiak et al. 1976). Furthermore, inbred mouse strains that develop an autoimmune disease reminiscent of SLE have been shown to develop cognitive impairment over time (Vogelweid et al. 1991, 1994). In lupus patients, numerous investigators have reported a heterogeneous population of neurone-reactive autoantibodies in both serum and CSF. Their mechanism of action is unclear, but is postulated to be via antibody interference of neuronal signalling, as cell death in the absence of vascular abnormalities has not been demonstrated histologically (Johnson and Richardson 1968). Neurone-reactive autoantibodies can be subdivided into lymphocytotoxic antibodies (LCA) cross-reactive with brain tissue (Bluestein and Zvaifler 1976; Bresnihan et al. 1977; Bluestein 1979; Denburg et al. 1994a), and neuronal antibodies (NA), which are targeted directly against membrane neuronal antigens (Bluestein et al. 1981). However, the relationship between these antibodies and activity of CNS disease (Bluestein et al. 1981; Long et al. 1990; Denburg et al. 1994a), the specific neuronal antigens involved, and the mechanism(s) by which this interaction is transmitted to the cell nucleus (Hanson et al. 1992; Hanly and Hong 1993; Denburg and

Behmann 1994) remain controversial. Recently, a significant association was found between reactivities to several lymphocyte membrane antigens and cognitive impairment (Long et al. 1990; Denburg et al. 1994a).

A separate category of neurone-reactive autoantibodies is that composed of antibodies directed against intracytoplasmic constituents, e.g., autoantibodies to the ribosomal P protein, antineurofilament antibodies, and anti-SS-A/Ro or anti-SS-B/La antibodies (Mevorach et al. 1993). The high specificity of antiribosomal P protein antibodies for lupus psychosis was first reported by Bonfa et al. (1987) and later confirmed by Schneebaum et al. (1991). Since then, other investigators have questioned the specificity of these antibodies, as they found anti-P antibodies in 50% of SLE patients without psychosis (Van Dam et al. 1991; reviewed by Teh and Isenberg 1994). In a later study of Isshi and Hirohata (1996), it was shown that the major reason for the conflicting results in the literature regarding the association of anti-P with lupus psychosis is the purity of the ribosomal P peptides used in the assays. A good explanation why these anti-P antibodies are only found in serum and not in CSF is not yet provided. Antibodies to neurofilament proteins, which form part of the cytoskeleton of neuronal cells, have been demonstrated to occur in NP SLE, but again are not specific to this entity (Robbins et al. 1988). They may also occur in other disorders such as Alzheimer's disease and Creutzfeldt–Jakob disease.

Antiphospholipid antibodies (aPL)

Antibodies to phospholipids are a diverse group that include lupus anticoagulant and anticardiolipin antibodies. Recent data indicate that the antigenic targets of these antibodies are circulating phospholipid-binding plasma proteins, like β_2-glycoprotein I, prothrombin or complexes of these proteins with phospholipids (reviewed by Roubey 1996) instead of cardiolipin or phospholipid-containing platelet membranes. The prevalence of aPL is around 1–2% in the normal population, and approximately 25–60% in patients with SLE (Love and Santoro 1990). There is controversy as to whether aPL is an independent risk factor for stroke in otherwise healthy young people (Antiphospholipid Antibodies in Stroke Study Group 1993; Muir et al. 1994). In lupus patients, however, these antibodies are clearly associated with venous and arterial thromboses including stroke, recurrent fetal loss, and thrombocytopenia (Hughes 1993). This complex of clinical features has been termed 'the antiphospholipid antibody syndrome (APS)' (Table 3). The disorder may be classified as primary, i.e., if there is no accompanying autoimmune disease, or as secondary, when patients have SLE or a lupus-like disease.

While all four IgG subclasses are found among aPL, only IgG2 is significantly associated with thrombotic complications (Sammaritano et al. 1997). In addition, the receptor Fc γ-RIIA-H131, capable of recognizing IgG2 and present on platelets, monocytes and endothelial cells, is associated with thrombosis in these patients (Sammaritano et al. 1997).

Cerebral infarction, particularly recurrent stroke, is associated with aPL (Boey et al. 1983; Harris et al. 1984; Futrell and Millikan 1989; Love and Santoro 1990; Montalban et al. 1991). Thrombosis may occur in both large and small vessels in the brain. In an unselected group of 59 patients with lupus, occlusive cerebrovascular lesions developed exclusively in 4 of the 34 patients with aPL (Sturfelt et al. 1987). When cerebral angiographic studies were performed in these patients, they revealed large-artery stenosis or occlusion in about 50% of cases and were negative in the remainder (Levine and Welch 1987a). Although initial studies emphasised the association with focal events such as stroke (Harris et al. 1984; Levine and Welch 1987a), chorea (Bruyn and Padberg 1984; Asherson et al. 1987), optic ischaemia (Levine and Welch 1987b; Digre et al. 1989), or transverse myelitis (Alarçon-Segovia et al. 1989; Lavalle et al. 1990), subsequent studies have also found an association with more diffuse manifestations, such as headache (Levine and Welch 1987b; Briley et al. 1989), seizures (Herranz et al. 1994), and multi-infarct dementia (Asherson et al. 1989) – although notably not with psychosis (Blaser et al. 1993). Since aPL are also frequently associated with left-sided cardiac valve lesions, an increased risk of cardiogenic

emboli is conceivable in these patients (Chartash et al. 1989). Although several autopsy series did not reveal any relationship between microinfarction and endocarditis (Johnson and Richardson 1968; Hanly et al. 1992), clinical studies showed a high association between cardiac valve lesions and stroke (Futrell and Millikan 1989; Futrell et al. 1992).

Endothelial cell swelling is one of the most characteristic cerebral and renal histopathological features in both primary and secondary APS (Levine et al. 1990; Hughson et al. 1993). It has been proposed that the endothelial swelling is caused by the incorporation of fragments of platelet thrombi into the vascular endothelium (Ellison et al. 1993). Thus, repeated episodes of thrombosis in these small vessels may lead to incorporation of platelet fragments and to thickening and irregularity of small vessels. In long-standing NP SLE, small vessel abnormalities may lead to the microinfarction found at autopsy (Ellison et al. 1993). It is not clear whether these antibodies primarily target the brain and not other organs, and how these antibodies interact with the endothelial cells of the cerebral vasculature (Hess et al. 1993; Hess 1994).

Cytokines

Cytokines may play a role both in the development and persistence of CNS symptoms (Bruyn 1995). Interferons have been shown to increase the severity of autoimmune disease in animals (Heremans et al. 1978). Furthermore, interferons may themselves directly precipitate psychosis, as has been demonstrated in patients with hepatitis B (McDonald et al. 1987) who were treated with interferon-α. Intrathecal synthesis of interferon-α (Lebon et al. 1983) and interleukin-6 (Hirohata and Miyamoto 1990) has also been reported in SLE patients with psychosis (Shiozawa et al. 1990). Cytokines, especially IL-1 and TNF-α, may have a key role in induction of complement-dependent leuko-occlusive vasculopathy, although its clinical relevance in CNS disease has yet to be determined (Belmont et al. 1996).

CLINICAL FEATURES

NP features of SLE are protean. Table 4 summarises the frequency of the various NP manifestations found in the larger studies.

TABLE 3

Clinical features of the antiphospholipid antibody syndrome.

Recurrent arterial or venous thrombosis
Recurrent miscarriages
Thrombocytopaenia

TABLE 4

Frequency of neuropsychiatric manifestations in patients with SLE.

	Feinglass et al. ($n = 140$)	Gibson and Myers ($n = 80$)	Bennahum and Messner ($n = 54$)	Futrell et al. ($n = 91$)	Abel et al. ($n = 180$)	Janssen ($n = 213$)
Psychiatric manifestations	24 (17%)	27 (33%)	38 (59%)	25 (28%)	42 (39%)	70 (33%)
Seizure	17 (12%)	20 (25%)	12 (20%)	22 (24%)	8 (4.4%)	15 (7%)
Stroke	16 (12%)	10 (12.5%)	7 (13%)	14 (15%)	3 (1.6%)	38 (18%)
Cranial nerve palsy	16 (12%)	4 (5%)	1 (2%)	4 (4%)	11 (6%)	11 (5%)
Chorea	2 (1%)	2 (2.5%)	1 (2%)	1 (1%)	0 (0%)	3 (1%)
Transverse myelitis	1 (1%)	3 (3%)	1 (2%)	3 (3%)	0 (0%)	2 (1%)
Cerebellar ataxia	5 (4%)	2 (2.5%)	3 (6%)	1 (1%)	0 (0%)	1 (1%)
Peripheral neuropathy	15 (11%)	2 (2.5%)	N A	3 (3%)	14 (7.7%)	5 (2%)
Headache	N A	17 (17%)	10 (18.5%)	N A	38 (21%)	60 (28%)

n denotes number; NA denotes not available.

Disorders of mental function

Analysis of the literature is complicated by a lack of uniformity of psychiatric definitions in addition to changing nomenclature, variable methodology, a systematic bias relating psychiatric symptoms to disease, and a confusing tendency to lump neurological and psychiatric symptoms together (reviewed by Iverson 1993). Disorders of mental function are probably the commonest cerebral manifestation of SLE and have been reported in up to 75% of lupus patients (Dubois and Tuffanelli 1964; Estes and Christian 1971; Feinglass et al. 1976; Gibson and Myers 1976; Miguel et al. 1994). In two recent studies, the prevalence of psychiatric disorders amongst ambulatory patients and hospitalised patients with SLE was 20 and 63%, respectively (Hay et al. 1992). Psychiatric abnormalities may range from mild subjective cognitive dysfunction to full-blown delirium, dementia, organic hallucinosis, delusional syndromes, or major depressive syndromes. Episodes of psychiatric illness are often accompanied by neurological symptoms or signs, and especially by migraine or seizures (Feinglass et al. 1976; Miguel et al. 1994). Psychiatric abnormalities in lupus patients may be due to (1) activity of cerebral lupus; (2) a primary psychiatric disorder; or (3) complications of SLE, such as uraemia, hypertension, electrolyte disturbances, infection, or iatrogenic causes. Psychiatric symptoms, e.g., anxiety, may also be due to psychological distress and coping reactions to a chronic disease.

Mild cognitive dysfunction is far more common than the more severe psychiatric manifestations

(Estes and Christian 1971; Feinglass et al. 1976; Abel et al. 1980; Kaell et al. 1986; Hay et al. 1994; Carbotte et al. 1995). Symptoms are largely subjective and consist of impairment of verbal and visual memory, concentration, conceptual reasoning, motor speed, or mood changes. Even though symptoms are usually mild, impaired cognitive function may be a major source of emotional distress to affected patients. The extent of cognitive impairment has only been appreciated in the past decade, when neuropsychological testing became routine. Initial psychometric studies by Carbotte et al. (1986) demonstrated cognitive impairment in 87% of female SLE patients with active cerebral lupus, and in 81% of those with inactive cerebral lupus. Of greater significance was the finding of cognitive impairment in 42% of SLE patients with no history of NP SLE. Subsequent investigations have reported much lower prevalences of around 20% (Hanly et al. 1992; Hay et al. 1992, 1994), reflecting selection bias in the former studies. All these studies, however, indicated that cognitive impairment is present subclinically in a significant percentage of patients. When symptoms such as delirium and dementia manifest clinically, the designation 'organic brain syndrome' or 'organic mental syndrome' is used (DSM-III-R). Miguel et al. (1994) reported such symptoms in 16% of their patients.

More severe disorders of mental function, e.g., psychosis, occur with a prevalence ranging from 5 to 63% (Feinglas et al. 1976; Gibson and Myers 1976; Abel et al. 1980; Hochberg et al. 1985; Sibley et al. 1992; Miguel et al. 1994). Psychosis may manifest as schizophrenia or paranoia, severe depression, mania, or catatonia. The duration of psychosis can range from a few hours to several months (O'Connor 1959). Histological examination of the brains of these patients is either strikingly normal or shows disseminated encephalomalacia compatible with microinfarction of the cerebral cortex (O'Connor 1959; Johnson and Richardson 1968). In the study of O'Connor (1959), 10 patients with psychosis were autopsied. Six of these patients had mainly psychiatric symptoms and four had both psychiatric and neurological syndromes. On microscopic examination the following findings were noted, mostly in the cerebral cortex: haemorrhage; foci of necrosis; thickened and hyalinised arterial walls; evidence of cellular degeneration; areas of infiltration with large mononuclear and/or polymorphonuclear cells; corpora amylacea; and dark-staining granular material throughout the brain.

Guze (1967) studied 75 unselected SLE patients, of whom 12 had psychiatric complications related to their SLE. These 12 patients were admitted to hospital 24 times over a 6-year period for psychiatric reasons: affective illness was the most frequently encountered disorder (present in 10 admissions), followed by organic brain syndrome (nine) and, less frequently, by a schizophreniform disorder (five). In a recent survey using current psychiatric nomenclature, Miguel et al. (1994) found 27 patients (63%) with psychiatric manifestations amongst a group of 46 lupus patients admitted to hospital with a disease flare. Among the various psychiatric diagnoses listed, an organic mood disorder with depressive symptoms was the most frequent diagnosis (in 19 patients), followed by delirium (three patients), dementia and organic mental disorder not otherwise specified (two patients each), and organic hallucinosis (one patient).

Corticosteroids may induce psychosis, although this effect has probably been overestimated in the past. In diseases other than SLE, corticosteroids may induce mild alterations in mood, but uncommonly produce major psychiatric disorders. Corticosteroids played a major role in causing psychosis in only 2 of 140 SLE patients reported by Sergent et al. (1975). The majority of NP features in that series occurred while patients were on no steroids or on low-dose therapy. In a recent prospective study, low- to moderate-dose steroid therapy has not been shown to induce psychosis in lupus patients (Denburg et al. 1994b).

A chronic disease such as SLE may have a nega-

tive impact on patients' psychological status, as may a host of environmental factors relating to chronic illness such as financial problems, marital and family distress and reduced vocational opportunities. The incidence of divorce amongst women with SLE is well over 50%. Suicide is common, but it is controversial whether suicide attempts occur more often in SLE patients than in the general population (Matsukawa et al. 1994). The loss of valued activities further increases feelings of helplessness, anxiety, or depression. Kremer et al. (1981) found no association between disease activity and non-organic psychopathology. Similarly, Hay et al. (1992) found no correlation between disease activity and psychiatric abnormalities; however, the latter authors did note a significant correlation between psychiatric abnormalities and social stress. Such correlations obviously have important therapeutic implications.

Seizures

Both generalised and partial seizures may be observed in SLE, with the generalised form being more common (Dubois and Tuffanelli 1964; Feinglass et al. 1976; Sibley et al. 1992). Seizures may infrequently be the initial presentation of SLE or may precede the diagnosis by several years. Seizures mostly occur during an exacerbation and do not subsequently recur, although a recurrent convulsive disorder may develop in a minority of cases (Dubois and Tuffanelli 1964; Johnson and Richardson 1968). The reported prevalence of seizures in SLE patients ranges from 7 to 74%, but a fair estimation is about 15% (Dubois and Tuffanelli 1964; Feinglass et al. 1976; Gibson and Myers 1976; Hochberg et al. 1985; Sibley et al. 1992). A few patients with SLE may have idiopathic (primary generalised) epilepsy, as Futrell et al. (1992) noted a family history of seizures in 2 of their 91 patients. Differentiation of seizures caused by disease from those secondary to other causes, e.g., hypertension, uraemia, infection, or corticosteroid treatment, may be difficult. Anticonvulsants such as hydantoins and primidone are claimed to unmask SLE or to induce a lupus-like syndrome (reviewed by Lee and Chase 1975), although these clinical disorders usually regress following withdrawal of the drug. Neuropathological examination of SLE patients with seizures frequently shows cortical abnormalities, including large haemorrhages and infarcts, as well as multiple small ischaemic and haemorrhagic lesions (Johnson and Richardson 1968).

Headache syndromes

According to several authors, chronic or recurrent headache is one of the most frequent symptoms of SLE, reported in up to 50% of patients (Abel et al. 1980; Worrall et al. 1990; Janssen 1992). A prospective study of chronic or recurrent headache found a prevalence of 68%, about equally divided between vascular (migraine) headache and tension headache (Vazquez-Cruz et al. 1990). Headache episodes most commonly occur in isolation, although they can occur in combination with neurological or psychiatric features in about 20% of cases (Vazquez-Cruz et al. 1990). They usually occur simultaneously with exacerbations of extra-CNS SLE and abate as other evidence of disease activity subsides (Brandt and Lessell 1978).

Strokes

Stroke is one of the best defined manifestations of cerebral lupus. It is one of the few NP features of SLE that may be fatal. Strokes occur in approximately 6–15% of cases (Feinglass et al. 1976; Futrell and Millikan 1989; Sibley et al. 1992). Patients who have had a stroke have a risk of recurrence of over 60% (Futrell and Millikan 1989). The major period of risk is during the first 20 years of the disease, with a peak incidence during the first year (Futrell and Milikan 1989). In one series of 30 patients with large-vessel occlusive disease, the average age at stroke onset was 35 years, and the diagnosis of SLE was made on average 4.4 years earlier (Mitsias and Levine 1994). Furthermore, 86% of patients had active SLE at the time of their stroke. In addition, the short-term mortality rate was 40%. The mechanism of stroke may be heterogeneous, including cardiogenic embolism originating from lupus valve disease, an increased tendency for thrombosis related to aPL (Futrell and Millikan 1989), cerebral vasculitis, cervical arterial dissection, and premature atherosclerosis (Mitsias and Levine 1994). Petri et al. (1996) showed that homocysteine may be an independent risk factor for stroke and thrombotic events in SLE. However, despite the high prevalence of premature atherosclerosis in SLE patients, autopsy data have

shown that atherosclerotic lesions as the cause of stroke are rare (Johnson and Richardson 1968; Gibson and Myers 1976; Hanly et al. 1992).

Brainstem disorders and cranial neuropathies

Brainstem lesions occur not infrequently and commonly manifest as ophthalmoplegia, although other cranial neuropathies also occur (Bennahum and Messner 1975). The histological basis of these lesions is usually vaso-occlusive disease of small vessels that results in ischaemia. Symptoms produced by lesions in the brainstem reflect the variety of anatomical structures involved (Bennahum and Messner 1975). The clinical presentation of optic nerve involvement is variable. The patient may present with acute painless visual loss, with either optic disc swelling (optic neuropathy) or with a normal appearing disc (retrobulbar neuropathy). In some cases, there is associated orbital pain and a central scotoma on visual field examination (Jabs et al. 1986). Optic nerve involvement is not infrequently associated with transverse myelitis. Bilateral forms of internuclear ophthalmoplegia have been reported (Jackson et al. 1986; Cogen et al. 1987). Rosenstein et al. (1989) reported an isolated, pupil-sparing third nerve palsy as the initial manifestation of SLE. The various neuro-ophthalmological manifestations of SLE are listed in Table 5. Other cranial nerve palsies, including isolated

TABLE 5

Neuro-ophthalmological manifestations of SLE.

Orbital lesions
Myopathy of orbital muscles
Orbital pseudotumour
Tolosa–Hunt syndrome
Anterior visual pathway papilloedema
   pseudotumour cerebri
   optic neuropathy
   retinal microangiopathy
   retinal artery occlusion
   retinal vein occlusion
Retrochiasmal pathway
   visual field defects
   cerebral blindness
   visual hallucinations and transient blindness
   with or without migraine
Brainstem lesions
   unilateral internuclear ophthalmoplegia
   bilateral internuclear ophthalmoplegia
   isolated third nerve palsy
   isolated sixth nerve palsy

---

sixth nerve palsy (Sedwick and Burde 1983) or la-
ryngeal nerve palsy (Espana et al. 1990; Gordon and
Dunn 1990), as well as trigeminal sensory neuropathy
(Lundberg and Werner 1972), have been reported.

Movement disorders

Movement disorders including chorea, hemiballismus, athetosis, ataxia and parkinsonism have all been
documented in SLE, but occur infrequently (Johnson
and Richardson 1968). Chorea is the most common
movement disorder and occurs in less than 2% of patients with cerebral lupus (Bruyn and Padberg 1984).
Clinically, it does not differ from Sydenham's chorea.
The chorea may be unilateral, although it is generalised in most instances.

The choreiform movements usually last for 8–16
weeks but occasionally persist for years (Bouchez et
al. 1985); they usually respond well to corticosteroids. Haloperidol is usually added in patients refractory to corticosteroids (Cervera et al. 1997). The
majority of lupus patients and chorea have aPL
(Asherson et al. 1987; Cervera et al. 1993). In this
respect it is somewhat surprising that Bruyn and
Padberg (1984) failed to find any vascular pathological lesion of the basal ganglia in 8 of 10 patients they
studied with chorea. These pathological data correspond to findings of CT/MRI scanning, which are either normal or show abnormalities in different
cerebral areas (Cervera et al. 1997). Therefore, the
pathological significance of aPL remains unclear in
this condition.

Peripheral and autonomic neuropathies

Peripheral nerve involvement in SLE may manifest as
mononeuritis multiplex, a sensorimotor distal symmetrical polyneuropathy, or a radiculoneuropathy with
elevated CSF protein and predominantly proximal
motor loss resembling Guillain–Barré syndrome
(Millette et al. 1986). Variants of these forms may
occur (Rechthand et al. 1984). The evolution of lupus
polyneuropathy is subacute to chronic, developing
over weeks to months, rather than days. The course
may be remitting and relapsing. Mild to moderate
transient sensory or sensorimotor neuropathies occur
most commonly, usually in the setting of active lupus
(Estes and Christian 1971; Feinglass et al. 1976;
Abel et al. 1980). Mononeuritis multiplex involving

the sciatic, peroneal, tibial, median, ulnar and radial
nerves have all been reported (Johnson and
Richardson 1968; Feinglass et al. 1976; Abel et al.
1980). Autonomic neuropathy occurs infrequently
(Omdal et al. 1994), but may be dramatic (Hoyle et
al. 1985).

Myelopathies

Transverse myelitis is an uncommon but potentially
life-threatening complication of SLE. The occurrence
of transverse myelitis is characterised by a rapidly
evolving paraplegia and frequently, by fever, sensory
loss and sphincter impairment (Andrianakos et al.
1975). The interval between the presenting neurological symptom and the maximum deficit ranges from
hours to months, but is usually less than 24 hours.
Spinal cord involvement occurs most commonly at
the mid- to lower thoracic level, although higher levels
are occasionally affected. Major arteries of the spinal
cord usually show no evidence of occlusion
(Andrianakos et al. 1975); vascular changes including
thrombotic occlusion of the small leptomeningeal vessels, perivascular round-cell infiltration, and vasculitis, cause ischaemic necrosis in most cases.
Associations with optic neuritis (Jabs et al. 1986) and
aPL (Lavalle et al. 1990) have been reported. The
prognosis of this disorder is poor, with a fatality rate
of over 50%, the survivors being severely disabled.
Early diagnosis and aggressive treatment with high-
dose intravenous administration of prednisone alone
(Harisdangkul et al. 1995) or in combination with cyclophosphamide appear to improve the prognosis
(Barile and Lavalle 1992). Data on anticoagulation of
these patients are still lacking.

Other syndromes

Several neurological syndromes, although rare, are
thought to be associated with SLE. These include
lupoid sclerosis, myasthenia gravis, Brown's syndrome, cerebral venous thrombosis, pseudotumor
cerebri and benign intracranial hypertension. Lupoid
sclerosis is a term that has been applied to patients
with SLE who coincidentally manifest symptoms and
signs typical of multiple sclerosis. Autopsy studies
have revealed areas of demyelination and lesions
characteristic of SLE (Johnson and Richardson
1968). Myasthenia gravis may occur either before or

after the onset of SLE and may be associated with thymoma. Brown's syndrome is characterised by limitation of elevation of the adducted eye. It is due to a mechanical limitation of movement of the superior oblique tendon. Clinically, it simulates a palsy of the inferior oblique muscle with vertical diplopia on upward and inward gaze (Alonso-Valdivielso et al. 1993). Cerebral venous thrombosis is an infrequently reported manifestation of NP lupus. Headache is the main symptom whereas focal symptoms and papilloedema may be absent. MRI is the investigation of choice to make the diagnosis (Laversuch et al. 1995). Drug-induced sterile meningitis, due to some non-steroidal anti-inflammatory drugs, azathioprine or trimethoprim-sulfamethoxasole (Escalante and Stimmler 1992), has been described. Aseptic meningitis is characterised by headache, fever, nuchal rigidity and non-specific CSF abnormalities, and usually responds to corticosteroid therapy. Even rarer syndromes, including one case exhibiting Klüver–Bucy syndrome, have been reported (Oliveira et al. 1989).

DIAGNOSTIC TESTS IN NEUROPSYCHIATRIC LUPUS

Before diagnosing CNS SLE, other causes for the neurological signs and symptoms must be considered. In particular, infection, hypertension, intracranial bleeding, renal failure, drug effects and cardiogenic emboli should be ruled out. Unfortunately, there is no specific diagnostic test for NP SLE, and the diagnosis remains a clinical one.

CSF examination

CSF abnormalities are commonly found in SLE but are non-specific. A modest, usually lymphocytic, pleocytosis and raised protein levels (up to 1200 mg/l) are present in roughly one-third of patients with NP SLE (Feinglass et al. 1976). Futrell et al. (1992) performed 47 lumbar punctures in 30 patients with cerebral lupus. Pleocytosis was found in three and an isolated elevation of protein concentration was found in another three patients. Elevated IgG/albumin was present in four patients. No correlation could be made between CSF abnormalities and any specific manifestation of cerebral lupus. In lupus polyneuropathy, however, albumino-cytologic dissociation is usually present (Feinglass et al. 1976; Rechthand et

al. 1984; Millette et al. 1986). The clinical relevance of examining the CSF is therefore not to diagnose cerebral lupus, but to rule out CNS infection. It should be borne in mind that CSF cultures occasionally may be sterile despite CNS infection (Futrell et al. 1992).

Neuropsychometric testing (NPT)

Cognitive impairment can be identified and quantified in a standardised and repeatable manner by NPT. It is the only method available to the clinician for quantifying one of the most frequently observed disorders of mental function in SLE, that is, cognitive dysfunction. The extent of cognitive dysfunction has been extensively evaluated over the past years using this technique, particularly by the groups of Denburg (Carbotte et al. 1986, 1995), Ginsburg et al. (1992) and Hanly et al. (1992). Limitations of NPT include their inability to test severely ill patients, the time-consuming nature of the tests, practice and learning effects that will bias follow-up studies, and inherent ethnic and cultural biases in these tests.

Electroencephalography (EEG)

EEG is of little diagnostic value in NP lupus. EEG findings are frequently abnormal, showing diffuse slowing without any distinctive pattern, unless focal features are clinically apparent. Diagnostic sensitivity is in the order of 17–71% (Feinglass et al. 1976; Ritchlin et al. 1992), but specificity is invariably low. Recently, the technique of quantified EEG (QEEG) has been applied in a prospective study of 52 patients with SLE (Ritchlin et al. 1992). QEEG sensitivity appeared to be high (87%), whereas the specificity was only moderate (75%). The technique proved to be particularly useful in the subgroup with various forms of affective disorders and depression in the absence of neurological signs. Since this technique is not available in most hospitals, experience is limited.

Angiography

Cerebral angiography is an insensitive diagnostic technique in CNS lupus, except where symptoms or signs of major vessel haemorrhage, vasculitis or thrombosis develop (Sanders and Hogenhuis 1986). The procedure may be helpful pre-operatively in the setting of cerebral haemorrhage. As angiography is an invasive technique with low sensitivity, it is not used routinely in the assessment of NP SLE.

Computerised tomography (CT)

The first studies on CT scanning in NP lupus were conducted in the late 1970s (Bilaniuk et al. 1977; Gonzalez-Scarano et al. 1979). CT is most useful in detecting gross morphological abnormalities such as large intracranial infarcts, intracranial haemorrhages, and cerebral atrophy. Cerebral atrophy is a non-specific abnormality frequently found on CT scanning, even of asymptomatic SLE patients. As yet, it is unclear whether the atrophy is due to the disease itself or to long-term corticosteroid therapy (Ostrov et al. 1982). As the predominant neuropathology in NP lupus is diffuse microinfarction, CT is frequently normal, even in the presence of focal neurological signs or symptoms (Carette et al. 1982).

Magnetic resonance imaging (MRI)

MRI is more sensitive than CT in demonstrating particular morphologic abnormalities such as oedema and small infarcts (Vermess et al. 1983; McCune et al. 1988; Sibbitt et al. 1989). Three predominant patterns of MRI abnormalities are recognised in SLE patients with NP symptoms: large areas of increased intensity involving the white matter, multiple small hyperintense areas of the white matter, and focal hyperintensity of the grey matter. Cerebral atrophy is the foremost lesion found in SLE patients without NP symptoms (Chinn et al. 1997). Large hyperintense areas in the white matter are compatible with cerebral infarction and are also readily diagnosed by CT. The hyperintense white matter lesions are frequently clinically asymptomatic, are associated with aPL, and probably represent microinfarction (Molad et al. 1992; Ishikawa et al. 1994). The nature of the grey matter abnormalities is unclear. They could either represent areas of oedema or reversible ischaemia, as resolution after corticosteroid treatment has been reported (Sibbitt et al. 1989). In one study of 28 SLE patients with acute CNS events (Sibbitt et al. 1989), MRI was able to detect cerebral lesions in all patients with focal symptoms, whereas CT abnormalities were found in only 38%. MRI also showed lesions in all patients with seizures, whereas in patients with

cognitive dysfunction or psychosis, it had a sensitivity of only 20% (McCune et al. 1988). Whether the prevalence of silent brain MRI abnormalities is increased in SLE patients without CNS lupus is no longer a matter of debate. An early MRI study (Jarek et al. 1994) found that few asymptomatic patients without a history of cerebral lupus showed focal white matter lesions. Subsequently, a prospective study by Gonzalez-Crespo et al. (1995) found that 10 out of 20 SLE patients without a history of CNS events showed multiple punctate areas of increased signal in the white matter. These results were confirmed in another study of Kozora et al. (1998), who demonstrated MRI abnormalities of the brain in a large majority of lupus patients who had no evidence of CNS involvement, even when tested neuropsychologically.

Few asymptomatic patients without a history of cerebral lupus will show focal white matter lesions (Jarek et al. 1994). However, a prospective study by Gonzalez-Crespo et al. (1995) found that 50% of 20 SLE patients without a history of CNS events showed multiple punctate areas of increased signal in the white matter. The latter results were confirmed by Chinn et al. (1997) who found the small white matter lesions in both SLE patients with and without prior CNS event, as well as in healthy control subjects.

MRI may also be useful in diagnosing the level of spinal cord involvement in patients with transverse myelitis (Kenik et al. 1987). Limitations of MRI include its poor specificity: the MRI findings alone do not distinguish SLE from, for example, multiple sclerosis, or from ischaemic or hypertensive cerebral abnormalities.

Radionuclear scanning

Initial reports of a very high sensitivity of technetium brain scans for detecting active CNS disease in SLE patients (Tan et al. 1978) were not confirmed by other studies. Brain scans appeared abnormal only in the setting of obvious neurological deficits in one study (Small et al. 1977), whereas another study showed infrequent scan abnormalities in active CNS lupus (Gibson and Myers 1976). In addition, Pinching et al. (1978) showed pathological scan results in all patients, even in those without clinical signs of CNS involvement. More recently, however, single photon emission computerised tomography (SPECT) has been used to demonstrate multiple small focal perfusion defects of cerebral blood flow in patients with NP lupus (Kushner et al. 1990; Nossent et al. 1991). In a study of 20 patients with SLE who were clinically suspected of having 'active' NP involvement, Nossent et al. (1991) found a sensitivity of almost 90% but a low specificity. Rubbert et al. (1993) showed abnormal cortical perfusion in 90% of patients with mild symptoms of CNS disease, i.e., cognitive deficits, compared with 10% in SLE patients without CNS symptoms. Although these studies indicate that SPECT probably does not sufficiently discriminate between active and inactive cerebral lupus, the technique may become an important tool in differentiating mild CNS disease from functional conditions such as depression.

Positron emission tomography (PET)

PET detects reduced cerebral metabolism utilising radio-labelled deoxyglucose. PET scanning has been able to detect abnormalities of glucose uptake and a localised disturbance of cerebral blood flow in active NP lupus (Hiraiwa et al. 1983; Stoppe et al. 1990). Two patterns are recognised: (1) focal glucose hypometabolism; and (2) a diffuse non-homogeneous distribution pattern of high and low glucose metabolism. In a study of 10 SLE patients with neurological symptoms and three patients with no NP involvement, PET scanning showed abnormalities in all 10 with active NP involvement, but was normal in the latter three. The pathological changes on PET corresponded to the clinical features (Stoppe et al. 1990). Since PET scanning requires proximity to a cyclotron, its utility will remain limited.

TREATMENT

No single effective treatment exists for NP lupus. Patients may have different neurological symptoms from the same underlying pathogenetic mechanism or, conversely, similar symptoms despite different pathogenetic mechanisms. For example, the inflammatory vascular occlusion caused by vasculitis and the non-inflammatory thrombotic vasculopathy occurring in patients with aPL may both result in an identical clinical picture. However, whilst corticosteroids may be effective in treating the vasculitis, they do not successfully treat stroke related to thrombosis (Fessell 1980; Futrell and Millikan 1989). Thus, corticosteroid ad-

ministration in the setting of a stroke appears only justified if there is sufficient associated systemic activity (Futrell and Millikan 1989).

No controlled drug trials have been conducted in NP SLE, although the findings from recent trials in lupus nephritis are being applied to treat the more severe CNS manifestations (Austin et al. 1986). The combination of heterogeneity of pathogenetic mechanisms and clinical features, the lack of sensitive diagnostic tests and measures of disease activity, and the spontaneous reversibility of the NP features make interpretation of any therapeutic trial very difficult. Therapy thus remains largely empirical. The three mainstays of treatment of cerebral lupus are corticosteroids, cytotoxic agents, and anticoagulants.

Corticosteroids

Corticosteroids, which have both potent anti-inflammatory and immunosuppressant effects, are the first line of treatment for the moderate to severe (non-thrombotic) NP manifestations of SLE. The best theoretical indications for corticosteroid therapy are those syndromes where vasculitis or antineuronal antibodies can be demonstrated in the CNS. In addition, high-dose corticosteroids are probably indicated in cases of coma (Fessell 1980), seizures, psychosis, chorea, and transverse myelitis, although this again has not been established in controlled studies. Corticosteroids do not predictably lower serum levels of aPL and do not appear to protect against strokes (Fessell 1980; Futrell and Millikan 1989). The optimal corticosteroid dosage is unknown, although as a general rule, oral prednisone is started at a dose of 1 mg/kg/day for 4–6 weeks, before being tapered. Although intravenous (i.v.) pulse therapy has been used extensively in the treatment of lupus, particularly in renal lupus (Cathcart et al. 1976), there are only a few published reports on its use in non-renal SLE (Eyanson et al. 1980; Dutton et al. 1982). However, patient numbers were small, the studies were uncontrolled, and in one study, the diagnosis of SLE was not clearly established (Dutton et al. 1982). Nevertheless, the trend appeared to be for a better response to i.v. than oral corticosteroids, where the latter had failed. A recent report also claimed success in the treatment of transverse myelitis, where pulse corticosteroids were followed by cyclophosphamide for an average of 6 months (Barile and Lavalle 1992).

Although concern about steroid-induced psychosis has probably been overstated in the past, the association with infection remains significant (O'Connor 1959; Sergent et al. 1975).

Azathioprine and cyclophosphamide

Cytotoxic agents, e.g., cyclophosphamide and azathioprine, have numerous effects on leucocytes including their ability to suppress circulatory B and T lymphocytes as well as autoantibody formation by B-lineage cells in lymphoid tissues. Corticosteroids, azathioprine, and cyclophosphamide have in common their potent ability to induce apoptosis (Carson et al. 1992; Mountz et al. 1994). The use of cytotoxic agents in renal lupus has been extensively studied in the past two decades. Treatment of lupus nephritis with oral azathioprine in combination with prednisone was superior to prednisone alone in preserving renal function. There are, however, no controlled data on use of this drug in the management of CNS lupus. Cyclophosphamide has also proved to be effective in lupus nephritis in combination with prednisone (Austin et al. 1986; Steinberg and Steinberg 1991). In one open prospective trial, the effect of monthly intravenous infusions of cyclophosphamide to nine patients with either lupus nephritis, cerebral lupus, or both, was studied (McCune et al. 1988). Although CNS disease improved in a small number of patients, there were insufficient data to permit conclusions about treatment of the neurological disease. In another study on refractory CNS lupus in nine patients, six of whom had been unresponsive to high-dose corticosteroid therapy (Boumpas et al. 1991), it was found that six had a complete and two a partial recovery when treated with intravenous cyclophosphamide in combination with prednisone; side effects, however, were common. Neuwelt et al. (1995) reported on the outcome of 31 patients with severe NP lupus treated with intravenous cyclophosphamide. Patients with fewer cerebral manifestations showed a better clinical improvement, whereas patients with the organic brain syndrome had a poor prognosis. From these studies and personal observations, it appears that i.v. pulse cyclophosphamide therapy (dose 0.5–1 g/m^2) may be the best therapy for lupus patients with progressive deterioration in level of consciousness, acute psychosis, coma, or transverse myelitis.

Anticoagulation and antiplatelet agents

Lupus patients with a history of thrombotic stroke have a very high risk of recurrence, and the current trend is thus to anticoagulate these patients for life (Asherson et al. 1989; Futrell and Millikan 1989). Similarly, patients with a first episode of cerebral ischaemia in the presence of aPL should be anticoagulated, as they too have a high risk of further thrombotic events (Nencini et al. 1992; Khamashta et al. 1995). To date, there are no convincing data to suggest that lupus patients with asymptomatic aPL should be prophylactically anticoagulated, as only a minority will develop complications (Asherson and Cervera 1993). Whether anticoagulation has a role in the treatment of other neurological manifestations which have been reported in association with aPL, such as seizures (Herranz et al. 1994), chorea (Asherson et al. 1987), migraine (Briley et al. 1989), amaurosis fugax (Digre et al. 1989), cerebral vein thrombosis (Uziel et al. 1994), and transverse myelitis (Lavalle et al. 1990), remains unclear. The question regarding the type and duration of treatment is also remains a vexed one. Although controlled prospective studies have not been performed, retrospective studies have shown superior efficacy of high-dose warfarin over aspirin and low-dose warfarin (Rosove and Brewer 1992; Derksen et al. 1993; Khamashta et al. 1995). Another report has suggested that low molecular weight heparin may be effective, although follow-up was too short to permit any definite conclusions to be drawn (Rosove and Brewer 1992). The optimal duration of treatment is unknown, but is probably life-long (Rosove and Brewer 1992; Derksen et al. 1993; Khamashta et al. 1995). Derksen et al. (1993) showed that patients with a thromboembolic event in the presence of PL had a 55% risk of recurrent venous thrombosis within 1 year after withdrawal of oral anticoagulant drugs. In conclusion, in patients with major thrombosis such as stroke, high-dose warfarin is the treatment, aiming for an international normalised ratio level between 3.5 and 4.0. Warfarin resistance may be encountered in patients who are receiving concurrent azathioprine (Singleton and Conyers 1992).

Additional therapies include intravenous administration of immune globulin, anti-idiotypic antibodies, low-dose methotrexate, cyclosporine, total lymphoid irradiation, and plasmapheresis. Experience with these therapies is limited. Severe refractory CNS disease, however, may warrant the addition of these treatment modalities. Anecdotal successes of plasmapheresis for cerebral lupus have been reported (Evans et al. 1981), but the experience for other indications, e.g., nephritis, has been disappointing (Jones et al. 1981; Lewis et al. 1992).

PROGNOSIS

The natural course of SLE is itself highly variable and unpredictable. The 10-year survival rate for patients with SLE has improved greatly and now reaches 75–90% (Wallace et al. 1981; Swaak et al. 1989; Pistiner et al. 1991; Janssen 1992; Ward et al. 1995). This is a significant improvement over the 50% mortality rates reported in the 1950s (Merrill and Shulman 1955).

CNS involvement is the third most common cause of death in lupus patients, after infection and renal failure (Rosner et al. 1982; Janssen 1992). A recent prospective study of an inception cohort showed SLE as the most common cause of death, being responsible in 34%, before infection in 22% and cardiovascular disease in 16% (Ward et al. 1995). Among those who died of SLE, one-third had severe multisystemic illness, one-third died of nephritis, and one-fourth died of CNS disease (Ward et al. 1995). However, there are few studies to support the concept that CNS lupus is an independent risk factor for a poor prognosis. In a recent series, the mortality rate amongst patients with and without CNS disease was 21 and 18%, respectively (Sibley et al. 1992). Even when one analysis showed a reduced survival rate amongst lupus patients with seizures and psychosis, stepwise linear regression demonstrated urine protein and high serum creatinine as independent predictors for poor survival but not seizures or psychosis (Ginzler et al. 1982).

The prognosis of separate neurological or psychiatric manifestations has not so far been systematically studied. Although not fatal, mild symptoms of psychiatric SLE, such as cognitive impairment or mood disorders, may respond to low-dose corticosteroids or even abate spontaneously (Sibley et al. 1992; Denburg et al. 1994b; Hay et al. 1994). Coma occurs rarely and has an ominous prognosis, although there are anecdotal reports of good responses to i.v. pulse methylprednisolone in combination with cyclophosphamide (Sibley et al. 1992). Lupus-related seizures usually respond to anticonvulsant treatment and may never recur once the acute episode has subsided (Dubois and Tuffanelli 1964; Sergent et al. 1975; Sibley et al. 1992). In 11 of the 12 patients with seizures described by Sibley et al. (1992), anticonvulsant therapy could be stopped after a mean of 6 years (range 1–13 years). In the series of Johnson and Richardson (1968), the three patients who had died from status epilepticus all had structural lesions, e.g., intracerebral haemorrhage.

Chorea is usually a self-limiting disorder, although in some cases it persists for years. The disorder may respond to treatment with anticoagulation or corticosteroids. Resistant cases may respond to haloperidol. Cranial nerve palsies respond to corticosteroids in about 50% of cases (Sergent et al. 1975; Sibley et al. 1992); in the remainder a residual deficit persists.

The course of ischaemic stroke in patients with lupus does not appear to differ from that of patients with strokes due to atherosclerosis. Particularly in patients with aPL, the risk of recurrent stroke is high. Although no controlled trials have been performed, the current view is that these patients should be anticoagulated for life, as discussed earlier.

In summary, whether cerebral lupus is the result of one unifying pathogenetic mechanism or of multiple diverse processes is as yet unclear. In our view, cerebral lupus should be regarded as an assembly of various disorders. Thus, autoantibodies including those directed against neuronal surface antigens and intracytoplasmic antigens play an important role in the diffuse manifestations of cerebral lupus, such as psychosis. Other antibodies such as aPL appear to play a pathogenetic role in focal neurological disease, such as stroke and transverse myelitis. The demarcation between diffuse and focal disease, however, is often not clear cut. Newer techniques, including MR spectroscopy, PET and SPECT scans may shed light on the disturbed perfusion of cerebral tissue and cellular metabolism, even in the context of normal anatomy (Chinn et al. 1997; Colamuissi et al. 1997).

The prognosis of SLE itself, as well as its cerebral manifestations, has improved greatly over recent decades. The course of most cerebral manifestations is frequently self-limiting, and rarely fatal. Clinical features with significant morbidity and mortality such as stroke, status epilepticus, coma and transverse myelitis warrant an aggressive therapeutic approach and, as with all the NP lupus syndromes, require controlled studies to determine optimal therapy.

REFERENCES

ABEL, T., D.D. GLADMAN and M.B. UROWITZ: Neuropsychiatric lupus. J. Rheumatol. 7 (1980) 325–333.

ALARÇON-SEGOVIA, D., M. DELZE, C.V. ORIA, J. SANCHEZ-GUERRERO and L. GOMEZ-PACHECO: Antiphospholipid antibodies and the antiphospholipid syndrome in systemic lupus erythematosus: a prospective analysis of 500 consecutive patients. Medicine (Baltimore) 68 (1989) 353–365.

ALONSO-VALDIVIELSO, J.L., B. ALVAREZ LARIO, J. ALEGRE LOPEZ, M.J. SEDANO TOUS and A. BUITRAGO GOMEZ: Acquired Brown's syndrome in a patient with systemic lupus erythematosus. Ann. Rheum. Dis. 52 (1993) 63–64.

ANDRIANAKOS, A.A., J. DUFFY, M. SUZUKI and J.T. SHARP: Transverse myelopathy in systemic lupus erythematosus: report of three cases and review of the literature. Ann. Intern. Med. 83 (1975) 616–624.

ANSAR AHMED, S., W.J. PENHALE and N. TALAL: Sex hormones, immune responses, and autoimmune disease. Mechanisms of sex hormones action. Am. J. Pathol. 121 (1985) 531–551.

ANTIPHOSPHOLIPID ANTIBODIES IN STROKE STUDY GROUP: Anticardiolipin antibodies are an independent risk factor for first ischemic stroke. Neurology 43 (1993) 2069–2073.

ASHERSON, R.A. and R. CERVERA: Antiphospholipid syndrome. J. Invest. Dermatol. 100 (1993) 21S–27S.

ASHERSON, R.A., R.H.W.M. DERKSEN, E.N. HARRIS, B.N. BOUMA, A.E. GHARAVI, L. KATER and G.R.V. HUGHES: Chorea in systemic lupus erythematosus and 'lupus-like' disease: association with antiphospholipid antibodies. Semin. Arthritis Rheum. 4 (1987) 253–259.

ASHERSON, R.A., M.A. KHAMASHTA, A. GIL, J. VAZQUEZ, O. CHAN, E. BAGULEY and G.R.V. HUGHES: Cerebrovascular disease and antiphospholipid antibodies in systemic lupus erythematosus, lupus-like disease, and the primary antiphospholipid syndrome. Am. J. Med. 86 (1989) 391–399.

ATKINS, C.J., J.J. KONDON, F.P. QUISMORIO and G.J. FRIOU: The choroid plexus in systemic lupus erythematosus. Ann. Intern. Med. 76 (1972) 65–72.

AUSTIN, H.A., J.H. KLIPPEL, J.E. BALOW, N.G. LE RICHE, A.D. STEINBERG, P.H. PLOTZ and J.L. DECKER: Therapy of lupus nephritis. Controlled trial of prednisone and cytotoxic drugs. N. Engl. J. Med. 314 (1986) 614–619.

BARILE, L. and C. LAVALLE: Transverse myelitis in systemic lupus erythematosus – the effect of IV pulse methylprednisolone and cyclophosphamide. J. Rheumatol. 19 (1992) 370–372.

BATCHELOR, J.R.: Systemic lupus erythematosus and

genes within the HLA region. Br. J. Rheumatol. 32 (1993) 13–15.

BELMONT, H.M., S.B. ABRAMSON and J.T. LIE: Pathology and pathogenesis of vascular injury in systemic lupus erythematosus: interactions of inflammatory cells and activated endothelium. Arthr. Rheum. 39 (1996) 9–22.

BENNAHUM, D.A. and R.P. MESSNER: Recent observations on central nervous systemic lupus erythematosus. Semin. Arthritis Rheum. 4 (1975) 253–266.

BILANIUK, L.T., S. PATEL and R.A. ZIMMERMAN: Computed tomography of systemic lupus erythematosus. Radiology 124 (1977) 119–121.

BLASER, K.U., M.A. KHAMASHTA, M.T. HERRANZ and G.R.V. HUGHES: Psychiatric disorders in patients with systemic lupus erythematosus: lack of association with antiphospholipid antibodies (letter). Br. J. Rheumatol. 32 (1993) 646–647.

BLOCK, S.R., M.D. LOCKSHIN, J.B. WINFIELD, M.E. WEKSLER, M. IMAMURA, R.J. WINCHESTER, R.C. MELLORS and C.L. CHRISTIAN: Immunologic observations on 9 sets of twins either concordant or discordant for SLE. Arthr. Rheum. 19 (1976) 545–554.

BLUESTEIN, H.G.: Heterogeneous neurocytotoxic antibodies in systemic lupus erythematosus. Clin. Exp. Immunol. 35 (1979) 210–217.

BLUESTEIN, H.G. and N.J. ZVAIFLER: Brain-reactive lymphocytotoxic antibodies in the serum of patients with systemic lupus erythematosus. J. Clin. Invest. 57 (1976) 509–516.

BLUESTEIN, H.G., G.W. WILLIAMS and A.D. STEINBERG: Cerebrospinal fluid antibodies to neuronal cells: association with neuropsychiatric manifestations of systemic lupus erythematosus. Am. J. Med. 70 (1981) 240–246.

BOEY, M.L., C.B. COLACO, A.E. GHARAVI, K.B. ELKON, S. LOIZOU and G.R.V. HUGHES: Thrombosis in systemic lupus erythematosus: striking association with the presence of circulating lupus anticoagulant. Br. Med. J. 287 (1983) 1021–1023.

BONFA, E., S.J. GOLOMBEK, L.D. KAUFMAN, S. SKELLY, H. WEISSBACH, N. BROT and K.B. ELKON: Association between lupus psychosis and anti-ribosomal P protein antibodies. N. Engl. J. Med. 317 (1987) 265–271.

BOUCHEZ, B., G. ARNOTT, P.Y. HATRON, A. WATTEL and B. DEVULDER: Chorée et lupus érythémateux disséminé avec anticoagulant circulant. Trois cas. Rev. Neurol. 141 (1985) 571–577.

BOUMPAS, D.T., H. YAMADA, N.J. PATRONAS, D. SCOTT, J.H. KLIPPEL and J.E. BALOW: Pulse cyclophosphamide for severe neuropsychiatric lupus. Quart. J. Med. 81 (1991) 975–984.

BOYER, R.S., N.C.J. SUN, A. VERITY, K.M. NIES and J.S. LOUIE: Immunoperoxidase staining of the choroid plexus in systemic lupus erythematosus. J. Rheumatol. 7 (1980) 645–650.

BRANDT, K.D. and S. LESSELL: Migrainous phenomena in systemic lupus erythematosus. Arthr. Rheum. 21 (1978) 7–16.

BRESNIHAN, B., M. OLIVER, G. GRIGOR and G.R.V. HUGHES: Brain reactivity of lymphocytotoxic antibodies in systemic lupus erythematosus with and without cerebral involvement. Clin. Exp. Immunol. 30 (1977) 333–337.

BRILEY, D.P., B.M. COULL and S.H. GOODNIGHT: Neurological disease associated with antiphospholipid antibodies. Ann. Neurol. 25 (1989) 221–227.

BRUYN, G.A.W.: Controversies in lupus: nervous system involvement. Ann. Rheum. Dis. 54 (1995) 159–167.

BRUYN, G.W. and G. PADBERG: Chorea in systemic lupus erythematosus. A critical review. Eur. Neurol. 23 (1984) 435–448.

CARBOTTE, R.M., S.D. DENBURG and J.A. DENBURG: Prevalence of cognitive impairment in systemic lupus erythematosus. J. Nerv. Ment. Dis. 174 (1986) 357–364.

CARBOTTE, R.M., S.D. DENBURG and J.A. DENBURG: Cognitive deficit associated with rheumatic diseases: neuropsychological perspectives. Arthr. Rheum. 38 (1995) 1363–1374.

CARETTE, S., M.B. UROWITZ, H. GROSMAN and E. ST. LOUIS: Cranial computerized tomography in systemic lupus erythematosus. J. Rheumatol. 9 (1982) 855–859.

CARSON, D.A., D.B. WASSON, L.M. ESPARZA, C.J. CARRERA, T.J. KIPPS and H.B. COTTAM: Oral antilymphocyte activity and induction of apoptosis by 2-chloro-2′-arabino-fluoro-2′-deoxyadenosine. Proc. Natl. Acad. Sci. U.S.A. 89 (1992) 2970–2974.

CATHCART, E.S., B.A. IDELSON, M.A. SCHEINBERG and W.G. COUSER: Beneficial effects of methylprednisolone pulse therapy in diffuse proliferative nephritis. Lancet i (1976) 163–166.

CERVERA, R., M.A. KHAMASHTA, J. FONT, G.D. SEBASTIANI, A. GIL, P. LAVILLA, I. DOMENECH, A.O. AYDINTUG, A. JEDRYKA-GORAL, E. DE RAMON, M. GALEAZZI, H.J. HAGA, A. MATHIEU, F. HOUSSIAU, M. INGELMO and G.R.V. HUGHES: Clinical and immunologiocal patterns of disease expression in a cohort of 1000 patients. Medicine (Baltimore) 72 (1993) 113–124.

CERVERA, R., R.A. ASHERSON, J. FONT, M. TIKLY, L. PALLARES, A. CHAMORRO and M. INGELMO: Chorea in the antiphospholipid syndrome: clinical, radiologic, and immunologic characteristics of 50 patients from our clinics and the recent literature. Medicine (Baltimore) 76 (1997) 203–212.

CHARTASH, E.K., D.M. LANS, S.A. PAGET, T. QAMAR and M.D. LOCKSHIN: Aortic insufficiency and mitral regurgitation in patients with systemic lupus erythematosus and the antiphospholipid antibody syndrome. Am. J. Med. 86 (1989) 407–412.

CHENG, J., T. ZHOU, C. LIU, J.P.D. SHAPIRO, M.J. BRAUER, M.C. KIEFER, P.J. BARR and J.D. MOUNTZ: Identification of a soluble form of the fas molecule that protects cells from fas-mediated apoptosis. Science 263 (1994) 1759–1761.

CHINN, R.J.S., I.D. WILKINSON, M.A. HALL-CRAGGS, M.N.J. PALEY, E. SHORTALL, S. CARTER, B.E. KENDALL, D.A. ISENBERG, S.P. NEWMAN and M.J.G. HARRISON: Magnetic resonance imaging of the brain and cerebral proton spectroscopy in patients with systemic lupus erythematosus. Arthr. Rheum. 40 (1997) 36–46.

COGEN, M.S., L.B. KLINE and E.R. DUVALL: Bilateral internuclear ophthalmoplegia in systemic lupus erythematosus. J. Clin. Neuro-Ophthalmol. 7 (1987) 69–73.

COLAMUSSI, P., F. TROTTA, R. RICCI, C. CITTANTI, M. GOVONI, G. BARBARELLA, M. GIGANTI, L. UCCELLI, C. TREVISAN and A. PIFFANELLI: Brain perfusion SPECT and proton magnetic resonance spectroscopy in the evaluation of two systemic lupus erythematosus patients with mild neuropsychiatric manifestations. Nucl. Med. Commun. 18 (1997) 269–273.

DA SILVA, J.A.P.: Sex hormones, glucocorticoids and autoimmunity: facts and hypotheses. Ann. Rheum. Dis. 54 (1995) 6–16.

DE JUAN, D., J.M. MARTIN-VILLE, J.J. GOMEZ-REINO, J.L. VICARIO, A. CORELL, J. MARTINEZ-LASO, D. BENMAMMER and A. ARNAIZ-VILLENA: Differential contribution of C4 and HLA-DQ genes to systemic lupus erythematosus susceptibility. Hum. Genet. 91 (1993) 579–584.

DENBURG, J.A. and S.A. BEHMANN: Lymphocyte and neuronal antigens in neuropsychiatric lupus: presence of an elutable, immunoprecipitable lymphocyte/neuronal 52 kD reactivity. Ann. Rheum. Dis. 53 (1994) 304–308.

DENBURG, S.D., S.A. BEHMANN, R.M. CARBOTTE and J.A. DENBURG: Lymphocyte antigens in neuropsychiatric systemic lupus erythematosus. Relationship of lymphocyte antibody specificities to clinical disease. Arthr. Rheum. 37 (1994a) 369–375.

DENBURG, S.D., R.M. CARBOTTE and J.A. DENBURG: Corticosteroids and neuropsychological functioning in patients with systemic lupus erythematosus. Arthr. Rheum. 37 (1994b) 1311–1320.

DERKSEN, R.H.W.M., P.G. DE GROOT, L. KATER and H.K. NIEUWENHUIS: Patients with antiphospholipid antibodies and venous thrombosis should receive long term anticoagulant treatment. Ann. Rheum. Dis. 52 (1993) 689–692.

DIGRE, K.B., F.J. DURCAN, D.W. BRANCH, D.M. JACOBSON, M.W. VARNER and J.R. BARINGER. Amaurosis fugax associated with antiphospholipid antibodies. Ann. Neurol. 25 (1989) 228–232.

DUBOIS, E.L.: The effect of the LE cell test on the clinical picture of systemic lupus erythematosus. Ann. Intern. Med. 38 (1953) 1265–1294.

DUBOIS, E.L. and D.L. TUFFANELLI: Clinical manifestations of systemic lupus erythematosus. Computer analysis of 520 cases. J. Am. Med. Assoc. 190 (1964) 104–111.

DUTTON, J.J., R.M. BURDE and T.G. KLINGELE: Autoimmune retrobulbic optic neuritis. Am. J. Ophthalmol. 94 (1982) 11–14.

ELON, K.B., P.H. KRAMMER, D.H. LYNCH, N. GOEL and M.F. SELDIN: Elevated levels of soluble Fas in systemic lupus erythematosus. Arthr. Rheum. 9 (1996) 1612–1613.

ELLIS, S.G. and M.A. VERITY: Central nervous system involvement in systemic lupus erythematosus: a review of neuropathologic findings: 57 cases 1955–1977. Semin. Arthritis Rheum. 8 (1979) 212–221.

ELLISON, D., K. GATTER, A. HERYET and M. ESIRI: Intramural platelet deposition in cerebral vasculopathy of systemic lupus erythematosus. J. Clin. Pathol. 46 (1993) 37–40.

ESCALANTE, A. and M.M. STIMMLER: Trimethoprim-sulfamethoxasole induced meningitis in systemic lupus erythematosus. J. Rheumatol. 19 (1992) 800–802.

ESPANA, A., J.M. GUTIERREZ, C. SORIA, L. GILA and A. LEDO: Recurrent laryngeal palsy in systemic lupus erythematosus. Neurology 40 (1990) 1143–1144.

ESTES, D. and C.L. CHRISTIAN: The natural history of systemic lupus erythematosus by prospective analysis. Medicine (Baltimore) 50 (1971) 85–95.

EVANS, D.T.P, M. GILES, D.J. HORNE, A.J.F. APICE, A. RIGLAR and B.H. TOH: Cerebral lupus erythematosus responding to plasmapheresis. Postgrad. Med. J. 57 (1981) 247–251.

EYANSON, S., M.H. PASSO, M.A. ALDO-BENSON and M.D. BENSON: Methylprednisolone pulse therapy for non-renal lupus erythematosus. Ann. Rheum. Dis. 39 (1980) 377–380.

FEINGLASS, E.J., F.C. ARNETT, C.A. DORSCH, T.M. ZIZIC and M.B. STEVENS: Neuropsychiatric manifestations of systemic lupus erythematosus: diagnosis, clinical spectrum, and relationship to other features of the disease. Medicine (Baltimore) 55 (1976) 323–339.

FESSELL, W.J.: Megadose corticosteroid therapy in systemic lupus erythematosus. J. Rheumatol. 7 (1980) 486–500.

FIELDER, A.H.L., M.J. WALPORT, J.R. BATCHELOR, R.I. RYNES, C.M. BLACK, I.R. DODI and G.R.V. HUGHES: Family study of the major histocompatibility complex in patients with systemic lupus erythematosus: importance of null alleles of C4A and C4B in determining disease susceptibility. Br. Med. J. 286 (1983) 425–428.

FUTRELL, N. and C. MILLIKAN: Frequency, etiology, and prevention of stroke in patients with systemic lupus erythematosus. Stroke 20 (1989) 583–591.

FUTRELL, N., N.R. SCHULTZ and C. MILLIKAN: Central nervous system disease in patients with systemic lupus erythematosus. Neurology 42 (1992) 1649–1657.

GIBSON, T. and A.R. MYERS: Nervous system involvement in systemic lupus erythematosus. Ann. Rheum. Dis. 35 (1976) 398–406.

GINSBURG, K.S., E.A. WRIGHT, M.G. LARSON, A.H. FOSSEL, M. ALBERT, P.H. SCHUR and M.H. LIANG: A controlled study of the prevalence of cognitive dysfunction in randomly selected patients with systemic lupus erythematosus. Arthr. Rheum. 35 (1992) 776–782.

GINZLER, E.M., H.S. DIAMOND, W. WEINER, M. SCHLESINGER, J.F. FRIES, C. WASNER, T.A. MEDSGER, G. ZIEGLER, J.H. KLIPPEL, N.M. HADLER, D.A. ALBERT, E.V. HESS, G. SPENCER-GREEN, A. GRAYZEL, D. WORTH, B.H. HAHN and E.V.A. BARNETT: A multicenter study of outcome in systemic lupus erythematosus. I. Entry variables as predictors of prognosis. Arthr. Rheum. 25 (1982) 601–611.

GONZALEZ-CRESPO, M.R., F.J. FLANCO, A. RAMES, I. MATEO, M.A. LOPEZ-PINO and J.J. GOMEZ-REINA: Magnetic resonance of the brain in systemic lupus erythematosus. Br. J. Rheumatol. 34 (1995) 1055–1060.

GONZALEZ-SCARANO, F., R.P. LISAK, L.T. BILANIUK, R.A. ZIMMERMAN, P.C. ARKINS and B. ZWEIMAN: Cranial computed tomography in the diagnosis of systemic lupus erythematosus. Ann. Neurol. 5 (1979) 158–165.

GORDON, T. and E.C. DUNN: Systemic lupus erythematosus and right recurrent laryngeal nerve palsy. Br. J. Rheumatol. 29 (1990) 308–309.

GUZE, S.B.: The occurrence of psychiatric illness in systemic lupus erythematosus. Am. J. Psychiatry 123 (1967) 1562–1570.

HADLER, N.M., R.D. GERWIN, M.M. FRANK, J.N. WHITAKER, M. BAKER and J.L. DECKER: The fourth component of complement in the cerebrospinal fluid in SLE. Arthr. Rheum. 16 (1973) 507–521.

HANLY, J.G., S. RAJARAMAN, S. BEHMANN and J.A. DENBURG: A novel neuronal antigen identified by sera from patients with systemic lupus erythematosus. Arthr. Rheum. 31 (1988) 492–499.

HANLY, J.G. and C. HONG: Antibodies to brain integral membrane proteins in systemic lupus erythematosus. J. Immunol. Meth. 161 (1993) 107–118.

HANLY, J.G., J.D. FISK, G. SHERWOOD, E. JONES, J. VERRIER JONES and B. EASTWOOD: Cognitive impairment in patients with systemic lupus erythematosus. J. Rheumatol. 19 (1992) 562–567.

HANLY, J.G., N.M.G. WALSH and V. SANGALANG: Brain pathology in systemic lupus erythematosus. J. Rheumatol. 19 (1992) 732–741.

HANSON, V.G., M. HOROWITZ, D. ROSENBLUTH, H. SPIERA and S. PUSZKIN: Systemic lupus erythematosus patients with central nervous system involvement show autoantibodies to a 50-kD neuronal membrane protein. J. Exp. Med. 176 (1992) 565–573.

HARBECK, R.J., A.A. HOFFMAN, S.A. HOFFMAN and D.W. SHUCARD: Cerebrospinal fluid and the choroid plexus during acute immune complex disease. Clin. Immunol. Immunopathol. 13 (1979) 413–425.

HARISDANGKUL, V., D. DOORENBOS and S.H. SUBRAMONY: Lupus transverse myelopathy: better outcome with early recognition and aggressive high-dose intravenous corticosteroid pulse treatment. J. Neurol. 242 (1995) 326–331.

HARRIS, E.N., A.E. GHARAVI, R.A. ASHERSON, M.L. BOEY and G.R.V. HUGHES: Cerebral infarction in systemic lupus: association with anticardiolipin antibodies. Clin. Exp. Rheumatol. 2 (1984) 47–51.

HAY, E.M., D. BLACK, A. HUDDY, F. CREED, B. TOMENSON, R.M. BERNSTEIN and P.J. LENNOX HOLT: Psychiatric disorder and cognitive impairment in systemic lupus erythematosus. Arthr. Rheum. 35 (1992) 411–416.

HAY, E.M., A. HUDDY, D. BLACK, P. MBAYA, B. TOMENSON, R.M. BERNSTEIN, P.J. LENNOX HOLT and F. CREED: A prospective study of psychiatric disorder and cognitive function in systemic lupus erythematosus. Ann. Rheum. Dis. 53 (1994) 298–303.

HEREMANS, H., A. BILLIAU, A. COLOMBATTI, J. HILGERS and P. DE SOMER: Interferon treatment of NZB mice: accelerated progression of autoimmune disease. Infect. Immunol. 21 (1978) 925.

HERRANZ, M.T., G. RIVIER, M.A. KHAMASHTA, K.U. BLASER and G.R.V. HUGHES: Association between antiphospholipid antibodies and epilepsy in patients with systemic lupus erythematosus. Arthr. Rheum. 4 (1994) 568–571.

HESS, D.C.: Models for central nervous system complications of antiphospholipid syndrome. Lupus 3 (1994) 253–257.

HESS, E.V. and Y. FARHEY: Etiology, environmental relationships, epidemiology, and genetics of systemic lupus erythematosus. Curr. Opin. Rheumatol. 7 (1995) 371–375.

HESS, D.C., J.C. SHEPPARD and R.J. ADAMS: Increased immunoglobulin binding to cerebral endothelium in patients with antiphospholipid antibodies. Stroke 24 (1993) 994–999.

HIRAIWA, M., C. NONAKA, A. TOSHIAKI and I. MASAAKI: PET in systemic lupus erythematosus: relation of cerebral vasculitis to PET findings. Am. J. Nucl. Radiol. 4 (1983) 541–543.

HIROHATA, S. and T. MIYAMOTO: Elevated levels of interleukin-6 in cerebrospinal fluid from patients with systemic lupus erythematosus and central nervous system involvement. Arthr. Rheum. 33 (1990) 644–649.

HOCHBERG, M.C.: Epidemiology of systemic lupus erythematosus. In: R.G. Lahita (Ed.), Systemic Lupus Erythematosus. New York, Churchill-Livingstone (1992) 103–117.

HOCHBERG, M.C., R.E. BOYD, J.M. AHEARN, F.C. ARNETT, W.B. BIAS, T.T. PROVOST and M.B. STEVENS: Systemic lupus erythematosus: a review of clinico-laboratory features and immunogenetic markers in 150 patients

with emphasis on demographic subsets. Medicine (Baltimore) 64 (1985) 285–295.

HOPKINSON, N.D.: Epidemiology of systemic lupus erythematosus. Ann. Rheum. Dis. 12 (1992) 1292–1294.

HOPKINSON, N.D., M. DOHERTY and R.J. POWELL: The prevalence and incidence of systemic lupus erythematosus in Nottingham, UK, 1989–1990. Br. J. Rheumatol. 32 (1993) 110–115.

HOYLE, C., D.J. EWING and A.C. PARKER: Acute autonomic neuropathy in association with systemic lupus erythematosus. Ann. Rheum. Dis. 44 (1985) 420–424.

HUGHES, G.R.V.: The antiphospholipid syndrome: ten years on. Lancet 342 (1993) 341–344.

HUGHSON, M.D., G.A. MCCARTY, C.M. SHOLER and R.A. BRUMBACK: Thrombotic cerebral arteriopathy in patients with the antiphospholipid syndrome. Mod. Pathol. 6 (1993) 644–653.

ISHIKAWA, O., K. OHNISHI, Y. MIYACHI and H. ISHIZAKA: Cerebral lesions in systemic lupus erythematosus detected by magnetic resonance imaging. Relationship to anticardiolipin antibody. J. Rheumatol. 21 (1994) 87–90.

ISSHI, K. and S. HIROHATA: Association of anti-ribosomal P protein antibodies with neuropsychiatric lupus erythematosus. Arthr. Rheum. 39 (1996) 1483–1490.

IVERSON, G.L.: Psychopathology associated with systemic lupus erythematosus: a methodological review. Semin. Arthritis Rheum. 22 (1993) 242–251.

JABS, D.A., N.R. MILLER, S.A. NEWMAN, M.A. JOHNSON, G.G. HEINER and M.B. STEVENS: Optic neuropathy in systemic lupus erythematosus. Arch. Ophthalmol. 104 (1986) 564–568.

JACKSON, G., M. MILLER, G. LITTLEJOHN, R. HELME and R. KING: Bilateral internuclear ophthalmoplegia in systemic lupus erythematosus. J. Rheumatol. 13 (1986) 1161–1162.

JANSSEN, B.A.: Systemic lupus erythematosus – clinical and serological features in 213 patients. Master of Medicine Thesis, University of Sydney, 1992.

JAREK, M.J., S.G. WEST, M.R. BAKER and K.M. RAK: Magnetic resonance imaging in systemic lupus erythematosus patients without a history of neuropsychiatric lupus erythematosus. Arthr. Rheum. 37 (1994) 1609–1613.

JOHNSON, R.T. and E.P. RICHARDSON: The neurological manifestations of systemic lupus erythematosus. Medicine (Baltimore) 47 (1968) 337–369.

JONES, J.V., M.F. ROBINSON, R.K. PARCIANY, L.F. LAYTER and B. MACLEOD: Therapeutic plasmapheresis in systemic lupus erythematosus: effect on immune complexes and antibodies to DNA. Arthr. Rheum. 24 (1981) 1113–1120.

JONGEN, P.J.H., A.M.TH. BOERBOOMS, K.J.B. LAMERS, B.C.

RAES and G. VIERWINDEN: Diffuse CNS involvement in systemic lupus erythematosus: intrathecal synthesis of the 4th component of complement. Neurology 40 (1990) 1593–1596.

KAELL, D.T., M. SHETTY, B.C.P. LEE and M.D. LOCKSHIN: The diversity of neurologic events in systemic lupus erythematosus. Arch. Neurol. 43 (1986) 273–276.

KAPOSI, M.K.: Neue Beiträge zur Kenntnis des Lupus erythematosus. Arch. Dermatol. Syph. 4 (1872) 37–78.

KARPIAK, S.E., L. GRAF and M.M. RAPPAPORT: Antiserum to brain gangliosides produces recurrent epileptiform activity. Science 194 (1976) 735–737.

KENIK, J.G., K. HROHN, R.B. KELLY, M. BIERMAN, M.D. HAMMEKE and J.A. HURLEY: Transverse myelitis and optic neuritis in systemic lupus erythematosus: a case report with magnetic resonance imaging findings. Arthr. Rheum. 30 (1987) 947–950.

KHAMASHTA, M.A., M.J. CUADRADO, F. MUJIC, N.A. TAUB, B.J. HUNT and G.R.V. HUGHES: The management of thrombosis in the antiphospholipid-antibody syndrome. N. Engl. J. Med. 332 (1995) 993–997.

KOUTOUZOV, S., A. CABRESPINES, Z. AMOURA, H. CHABRE, C. LOTTON and J.F. BACH: Binding of nucleosomes to a cell surface receptor: redistribution and endocytosis in the presence of lupus antibodies. Eur. J. Immunol. 26 (1996) 472–486.

KOZORA, E., S.G. WEST, B.L. KOTZIN, L. JULIAN, S. PORTER and E. BIGLER: Magnetic resonance imaging abnormalities and cognitive deficits in systemic lupus erythematosus patients without overt central nervous system disease. Arthr. Rheum. 41 (1998) 41–47.

KREMER, J.M., R.I. RYNES, L.E. BARTHOLOMEW, L.D. RODICHOK, E.W. PELTON, E.A. BLOCK and R.J. SILVER: Non-organic non-psychotic psychopathology (NONPP) in patients with systemic lupus erythematosus. Semin. Arthr. Rheum. 11 (1981) 182–189.

KUSHNER, M.J., M. TOBIN, F. FAZEKAS, J. CHAWLUK, D. JAMIESON, B. FREUNDLICH, S. GRENELL, L. FREEMEN and M. REIVICH: Cerebral blood flow variations in CNS lupus. Neurology 40 (1990) 99–102.

LAVALLE, C., S. PIZARRO, C. DRENKARD, J. SANCHEZ- GUERRERO and D. ALARÇON-SEGOVIA: Transverse myelitis: a manifestation of SLE strongly associated with antiphospholipid antibodies. J. Rheumatol. 17 (1990) 34–37.

LAVERSUCH, C.J., M.M. BROWN, A. CLIFTON and B.E. BOURKE: Cerebral venous thrombosis and acquired protein S deficiency: an uncommon cause of headache in systemic lupus erythematosus. Br. J. Rheumatol. 34 (1995) 572–575.

LEBON, P., G.R. LENOIR, A. FISCHER and A. LAGRUE: Synthesis of intrathecal interferon in systemic lupus erythematosus with neurological complications. Br. Med. J. 287 (1983) 1165–1167.

LEE, S.L. and P.H. CHASE: Drug-induced systemic lupus erythematosus: a critical review. Semin. Arthritis Rheum. 5 (1975) 83–103.

LEVINE, S.R. and K.M.A. WELCH: Cerebrovascular ischemia associated with lupus anticoagulant. Stroke 18 (1987a) 257–263.

LEVINE, S.R. and K.M.A. WELCH: The spectrum of neurologic disease associated with antiphospholipid antibodies. Arch. Neurol. 44 (1987b) 876–883.

LEVINE, S.R., M.J. DEEGAN, N. FUTRELL and K.M.A. WELCH: Cerebrovascular and neurologic disease associated with antiphospholipid antibodies: 48 cases. Neurology 40 (1990) 1181–1189.

LEWIS, E.J., L.G. HUNSICKER, S. LAN, R.D. ROHDE and J. LACHIN: A controlled trial of plasmapheresis therapy in severe lupus nephritis. N. Engl. J. Med. 326 (1992) 1373–1379.

LONG, A.A., S.D. DENBURG, R.M. CARBOTTE, D.P. SINGAL and J.A. DENBURG: Serum lymphocytotoxic antibodies and neurocognitive function in systemic lupus erythematosus. Ann. Rheum. Dis. 49 (1990) 249–253.

LOOS, M. and H.P. HEINZ: Component deficiencies. 1. The first component: C1q, C1r, C1s. Prog. Allergy 39 (1986) 212–231.

LOVE, P.E. and S.A. SANTORO: Antiphospholipid antibodies: anticardiolipin and the lupus anticoagulant in systemic lupus erythematosus (SLE) and in non-SLE disorders. Prevalence and clinical significance. Ann. Intern. Med. 112 (1990) 682–698.

LUNDBERG, P.O. and I. WERNER: Trigeminal sensory neuropathy in systemic lupus erythematosus. Acta Neurol. Scand. 48 (1972) 330–340.

MACFADYEN, D.J., R.J. SCHNEIDER and I.A. CHISHOLM: A syndrome of brain, inner ear and retinal microangiopathy. Can. J. Neurol. Sci. 14 (1987) 315–318.

MATSUKAWA, Y., S. SAWADA, T. HAYAMA, H. USUI and T. HORIE: Suicide in patients with systemic lupus erythematosus. Lupus 3 (1994) 31–37.

MCCUNE, W.J., A. MACGUIRE, A. AISEN and S. GEBARSKI: Identification of brain lesions in neuropsychiatric systemic lupus erythematosus by magnetic resonance imaging. Arthr. Rheum. 31 (1988) 159–166.

MCCUNE, W.J., J. GOLBUS, W. ZELDES, P. BOHLKE, R. DUNNE and D.A. FOX: Clinical and immunological effects of monthly administration of intravenous cyclophosphamide in severe systemic lupus erythematosus. N. Engl. J. Med. 318 (1988) 1423–1431.

MCDONALD, E.M., A.H. MANN and H.C. THOMAS: Interferon as mediators of psychiatric morbidity: an investigation of recombinant alpha-interferon in hepatitis-B carriers. Lancet ii (1987) 1175–1177.

MERRILL, M. and L.E. SHULMAN: Determination of prognosis in chronic disease, illustrated by systemic lupus erythematosus. J. Chron. Dis. 1 (1955) 12–32.

MEVORACH, D., E. RAZ, O. SHALEV, I. STEINER and E. BEN-CHETRIT: Complete heart block and seizures in an adult with systemic lupus erythematosus. Arthr. Rheum. 36 (1993) 259–262.

MIGUEL, E.C., R.M.R. PEREIRA, C.A. DE BRAGANÇA PEREIRA, L. BAER, R.E. GOMES, L.C.F. DE SA, R. HIRSCH, N.G. DE BARROS, J.M. DE NAVARRO and V. GENTIL: Psychiatric manifestations of systemic lupus erythematosus: clinical features, symptoms, and signs of central nervous system activity in 43 patients. Medicine (Baltimore) 73 (1994) 224–232.

MILLETTE, T.J., S.H. SUBRAMONY, A.S. WEE and V. HARIS-DANGHOL: Systemic lupus erythematosus presenting with recurrent acute demyelinating polyneuropathy. Eur. Neurol. 25 (1986) 397–402.

MITSIAS, P. and S.R. LEVINE: Large cerebral vessel occlusive disease in systemic lupus erythematosus. Neurology 44 (1994) 385–393.

MOLAD, Y., Y. SIDI, M. GORNISH, M. LERNER, J. PINKHAS and A. WEINBERGER: Lupus anticoagulant: correlation with magnetic resonance imaging of brain lesions. J. Rheumatol. 19 (1992) 556–561.

MONTALBAN, J., A. CODINA, J. ORDI, M. VILARDELL, M. KHAMASHTA and G.R.V. HUGHES: Antiphospholipid antibodies in cerebral ischemia. Stroke 22 (1991) 750–753.

MOUNTZ, J.D., W. JIANGUO, J. CHENG and T. ZHOU: Autoimmune disease. A problem of apoptosis. Arthr. Rheum. 37 (1994) 1415–1420.

MUIR, K.W., I.B. SQUIRE, W. ALWAN and K.R. LEES: Anticardiolipin antibodies in an unselected stroke population. Lancet 344 (1994) 452–456.

NENCINI, P., M.C. BARUFFI, R. ABBATE, G. MASSAI, L. AMADUCCI and D. INZITARI: Lupus anticoagulant and anticardiolipin antibodies in young adults with cerebral ischemia. Stroke 23 (1992) 189–193.

NEUWELT, C.M., S. LACKS, B.R. KAYE, J.B. ELLMAN and D.G. BORENSTEIN: Role of intravenous cyclophosphamide in the treatment of severe neuropsychiatric systemic lupus erythematosus. Am. J. Med. 98 (1995) 32–41.

NOSSENT, J.C., A. HOVESTADT, D.H.W. SCHÖNFELD and A.J.G. SWAAK: Single-photon-emission computed tomography of the brain in the evaluation of cerebral lupus. Arthr. Rheum. 34 (1991) 1397–1403.

O'CONNOR, J.F.: Psychoses associated with systemic lupus erythematosus. Ann. Intern. Med. 51 (1959) 526–536.

OKUBO, M., H. ISHIDA and R. KASUKAWA: Elevated levels of soluble Fas in systemic lupus erythematosus. Arthr. Rheum. 9 (1996) 1612–1613.

OLVEIRA, V., J.M. FERRO, J.P. FOREID, T. COSTA and A. LEVY: Klüver–Bucy syndrome in systemic lupus erythematosus. J. Neurol. 236 (1989) 55–56.

OMDAL, R., R. JORDE, S.I. MELLGREN and G. HUSBY: Autonomic function in systemic lupus erythematosus. Lupus 3 (1994) 413–417.

OSLER, W.: On the visceral manifestations of the erythema group of skin diseases. Trans. Assoc. Am. Phys. 18 (1903) 599–624.

OSTROV, S.G., R.M. QUENCER, N.B. GAYLIS and R.D. ALTMAN: Cerebral atrophy in systemic lupus erythematosus: steroid or disease-induced phenomenon. Am. J. Neuroradiol. 3 (1982) 21–23.

PETRI, M., R, ROUBENOFF, G.E. DALLAL, M.R. NADEAU, J. SELHUB and I.H. ROSENBERG: Plasma homocysteine as a risk factor for atherothrombotic events in systemic lupus erythematosus. Lancet 348 (1996) 1120–1124.

PINCHING, A.J., R.L. TRAVERS and G.R.V. HUGHES: Oxygen-15 brain scanning for detection of cerebral involvement in systemic lupus erythematosus. Lancet i (1978) 898–901.

PISETSKY, D.S.: Autoantibodies and their significance. Curr. Opin. Rheumatol. 5 (1993) 549–556.

PISTINER, M., D.J. WALLACE, S. NESSIM, A.L. METZGER and J.R. KLINENBERG: Lupus erythematosus in the 1980s: a survey of 570 patients. Semin. Arthritis Rheum. 21 (1991) 55–64.

RECHTHAND, E., D.R. CORNBLAT, B.J. STERN and J.O. MEYERHOFF: Chronic demyelinating polyneuropathy in systemic lupus erythematosus. Neurology 34 (1984) 1375–1377.

REVEILLE, J.D., W.B. BIAS, J.A. WINKELSTEIN, T.T. PROVOST, C.A. DORSCH and F.C. ARNETT: Familial systemic lupus erythematosus: immunogenetic studies in eight families. Medicine (Baltimore) 62 (1983) 21–35.

RITCHLIN, C.T., R.J. CHABOT, K. ALPER, J. BUYON, H.M. BELMONT, R. ROUBEY and S.B. ABRAMSON: Quantified electroencephalography. A new approach to the diagnosis of cerebral dysfunction in systemic lupus erythematosus. Arthr. Rheum. 35 (1992) 1330–1342.

ROBBINS, M.L., S.E. KORNGUTH, C.L. BELL, K. KALINKE, D. ENGLAND, P. TURSKI and F.M. GRAZIANO: Antineurofilament antibody evaluation in neuropsychiatric systemic lupus erythematosus. Arthr. Rheum. 31 (1988) 623–628.

ROSENSTEIN, E.D., J. SOBELMAN and N. KRAMER: Isolated, pupil-sparing third nerve palsy as initial manifestation of systemic lupus erythematosus. J. Clin. Neuro-Ophthalmol. 9 (1989) 285–288.

ROSNER, S., E.M. GINZLER, H.S. DIAMOND, W. WEINER, M. SCHLESINGER, J.F. FRIES, C. WASNER, T.A. MEDSGER, G. ZIEGLER, J.H. KLIPPEL, N.M. HADLER, D.A. ALBERT, E.V. HESS, G. SPENCER-GREEN, A. GRAYZEL, D. WORTH, B.H. HAHN and E.V.A. BARNETT: A multicenter study of outcome in systemic lupus erythematosus. II. Causes of death. Arthr. Rheum. 25 (1982) 612–617.

ROSOVE, M.H. and P.M.C. BREWER: Antiphospholipid thrombosis: clinical course after the first thrombotic events in 70 patients. Ann. Intern. Med. 117 (1992) 303–308.

ROUBEY, R.A.S.: Immunology of the antibody antiphospholipid antibody syndrome. Arthr. Rheum. 39 (1996) 1444–1454.

ROUBINIAN, J.R., R. PAPOIAN and N. TALAL: Androgenic hormones modulate autoantibody responses and improve survival in murine lupus. J. Clin. Invest. 69 (1977) 1066–1070.

RUBBERT, A., J. MARIENHAGEN, K. PIRNER, B. MANGER, J. GEBMEIER, A. ENGELHARDT, F. WOLF and J.R. KALDEN: Single-photon-emission computed tomography analysis of cerebral blood flow in the evaluation of central nervous system involvement in patients with systemic lupus erythematosus. Arthr. Rheum. 36 (1993) 1253–1262.

SAMMARITANO, L.R., N. SONIA, R. SOBEL, S.K. LO, R. SIMANTOV, R. FURIE, A. KAELL, R. SILVERSTEIN and J.E. SALMON: Anticardiolipin IgG subclasses: association of IgG2 with arterial and/or venous thrombosis. Arthr. Rheum. 40 (1997) 1998–2006.

SANDERS, E.A.J.M. and L.A.H. HOGENHUIS: Cerebral vasculitis as presenting symptom of systemic lupus erythematosus. Acta Neurol. Scand. 74 (1986) 75–77.

SCHNEEBAUM, A.B., J.D. SINGLETON, S.G. WEST, J.K. BLODGETT, L.G. ALLEN, J.C. CHERONIS and B.L. KOTZIN: Association of psychiatric manifestations with antibodies to ribosomal P proteins in systemic lupus erythematosus. Am. J. Med. 90 (1991) 54–62.

SEDWICK, L. and R. BURDE: Isolated sixth nerve palsy as initial manifestation of systemic lupus erythematosus. J. Clin. Neuro-Ophthalmol. 3 (1983) 109–110.

SERGENT, J.S., M.D. LOCKSHIN, M.S. KLEMPNER and B.A. LIPSKY: Central nervous system disease in systemic lupus erythematosus. Am. J. Med. 58 (1975) 644–655.

SHIOZAWA, S., Y. KUROKI, M. KIM, S. HIROHATA and T. OGINO: Interferon-α in lupus psychosis. Arthr. Rheum. 35 (1990) 417–422.

SIBBITT, W.L., R.R. SIBBITT, R.H. GRIFFEY, C. ECKEL and A.D. BANKHURST: Magnetic resonance and computed tomographic imaging in the evaluation of acute neuropsychiatric disease in systemic lupus erythematosus. Ann. Rheum. Dis. 48 (1989) 1014–1022.

SIBLEY, J.T., W.P. OLSZYNSKI, W.E. DECOTEAU and M.B. SUNDARAM: The incidence and prognosis of central nervous system disease in systemic lupus erythematosus. J. Rheumatol. 19 (1992) 47–52.

SIMON, J. and O. SIMON: Effect of passive transfer of anti-brain antibodies to a normal recipient. Exp. Neurol. 47 (1975) 523–534.

SINGER, J. and J.A. DENBURG: Diagnostic criteria for neuropsychiatric systemic lupus erythematosus: the results of a consensus meeting. J. Rheumatol. 17 (1990) 1397–1402.

SINGLETON, J.D. and L. CONYERS: Warfarin and azathioprine: an important drug interaction. Am. J. Med. 92 (1992) 217.

SMALL, P., M.F. MASS, P.F. KOHLER and R.J. HARBECK: Central nervous system involvement in SLE. Diagnostic profile and clinical features. Arthr. Rheum. 20 (1977) 869–878.

STEINBERG, A.D. and S.C. STEINBERG: Long-term preservation of renal function in patients with lupus nephritis receiving treatment that included cyclophosphamide versus those treated with prednisone only. Arthr. Rheum. 34 (1991) 945–950.

STOPPE, G., K. WILDHAGEN, J.W. SEIDEL, G.-J. MEYER, O. SCHOBER, P. HEINTZ, H. KUNKEL, H. DEICHER and H. HUNDESHAGEN: Positron emission tomography in neuropsychiatric lupus erythematosus. Neurology 40 (1990) 304–306.

STURFELT, G., O. NIVED, R. NORBERG, R. THORSTENSSON and K. KROOK: Anticardiolipin antibodies in patients with systemic lupus erythematosus. Arthr. Rheum. 30 (1987) 382–388.

SWAAK, A.J.G., J.C. NOSSENT, W. BRONSVELD, A. VAN ROOYEN, E.J. NIEUWENHUYS, L. THEUNS and R.J.Y. SMEENK: Systemic lupus erythematosus. I. Outcome and survival: Dutch experience with 110 patients studied prospectively. Ann. Rheum. Dis. 48 (1989) 447–454.

TAN, E.M., A.S. COHEN, J.F. FRIES, A.T. MASI, D.J. MCSHANE, N.F. ROTHFIELD, J.G. SCHALLER, N. TALAL and R.J. WINCHESTER: The 1982 revised criteria for the classification of systemic lupus erythematosus. Arthr. Rheum. 25 (1982) 1271–1277.

TAN, R.F., D.D. GLASMAN, M.B. UROWITZ and N. MILNE: Brain scan findings in central nervous system involvement in systemic lupus erythematosus. Ann. Rheum. Dis. 37 (1978) 357–362.

TEH, L. and D.A. ISENBERG: Antiribosomal P protein antibodies in systemic lupus erythematosus. Arthr. Rheum. 37 (1994) 307–315.

UZIEL, Y., R.M. LAXER, R. SCHNEIDER, S. BLASER, M. ANDREW and E.D. SILVERMAN: Cerebral vein thrombosis in childhood systemic lupus erythematosus. Arthr. Rheum. 37, Suppl. 6 (1994) R40.

VAN DAM, A., H. NOSSENT, J. DE JONG, J. MEILOF, E. TER BORG, T. SWAAK and R.M.T. SMEENK: Diagnostic value of antibodies against ribosomal phosphoproteins. A cross sectional and longitudinal study. J. Rheumatol. 18 (1991) 1026–1034.

VAZQUEZ-CRUZ, J., H. TRABOULSSI, H A. RODRIQUEZ-DE LA SERA, C. GELI, C. ROIG and C. DIAZ: A prospective study of chronic or recurrent headache in systemic lupus erythematosus. Headache 30 (1990) 232–235.

VERMESS, M., R.M. BERNSTEIN, G.M. BYDDER, R.E. STEINER, L.R. YOUNG and G.R.V. HUGHES: Nuclear magnetic resonance (NMR) imaging of the brain in systemic lupus erythematosus. J. Comput. Assist. Tomogr. 7 (1983) 461–467.

VIA, C.S. and B.S. HANDWERGER: B-cell and T-cell function in systemic lupus erythematosus. Curr. Opin. Rheumatol. 5 (1993) 570–574.

VOGELWEID, C.M., J.C. JOHNSON, C.L BESCH-WILLIFORD, J. BASLER AND S.E. WALKER: Inflammatory central nervous system disease in lupus-prone MRL/1pr mice: comparative histologic and immunohistochemical findings. J. Neuroimmunol. 35 (1991) 89–99.

VOGELWEID, C.M., D.C. WRIGHT, J.C. JOHNSON, J.E. HEWETT and S.E. WALKER: Evaluation of memory, learning ability, and clinical neurologic function in pathogen-free mice with systemic lupus erythematosus. Arthr. Rheum. 37 (1994) 889–897.

WALLACE, D.J., T. PODELL, J. WEINER, J.R. KLINENBERG, S. FOROUZESH and E.L. DUBOIS: Systemic lupus erythematosus – survival patterns: experience with 699 patients. J. Am. Med. Assoc. 245 (1981) 943–948.

WARD, M.M., E. PYUN and S. STUDENSKI: Causes of death in systemic lupus erythematosus. Arthr. Rheum. 38 (1995) 1492–1499.

WATANABE-FUKUNAGA, R., C.I. BRANNAN, N.G. COPELAND, N.A. JENKINS and S. NAGATA: Lymphoproliferation disorder in mice explained by defects in Fas antigen that mediates apoptosis. Nature 356 (1992) 314–317.

WINFIELD, J.B., C.M. BRUNNER and D. KOFFLER: Serologic studies in patients with systemic lupus erythematosus and central nervous system dysfunction. Arthr. Rheum. 21 (1978) 289–294.

WORRALL, J.E., M.L. SNAITH, J.R. BATCHELOR and D.A. ISENBERG: SLE: a rheumatological review of the clinical features, serology and immunogenetics of 100 SLE patients during long-term follow-up. Quart. J. Med. 275 (1990) 319–330.

Handbook of Clinical Neurology, Vol. 27 (71): Systemic Diseases, Part III
M.J. Aminoff and C.G. Goetz, editors

CHAPTER 4

Neurologic disease in Sjögren's syndrome

ELAINE L. ALEXANDER

206 Longwood Avenue, Baltimore, MD 21210, U.S.A.

Neurologic disease in Sjögren's syndrome (SS) can affect any part of the central nervous system (CNS) or peripheral nervous system (PNS), as well as skeletal muscle (Alexander 1986a, 1987b, 1992, 1993a; Debacker and Dehaene 1995) resulting in a wide range of neurologic damage and dysfunction. Although PNS disease as a potential neurologic manifestation of SS has been recognized for many years (Alexander 1986a, 1987b), CNS disease was not identified as a neurologic complication of SS until the clinical description from John's Hopkins Medical Institutions by Alexander et al. (1981). Subsequently, a multidisciplinary team of clinicians and investigators spanning many subspecialities at a highly specialized tertiary referral center sequentially described the clinical spectrum of CNS and PNS manifestations of SS, autoantibody profiles and their clinical relevance, the peripheral (cutaneous) vasculopathy associated with neurologic disease, neurodiagnostic evaluation (including electrophysiology, neuroimaging, and angiographic studies, as well as cerebrospinal fluid (CSF) analysis), immunogenetics, pathology, immunopathogenesis and, finally, treatment and management.

As these multifaceted data were analyzed, an impressive body of evidence has developed which indicates that neurologic disease in SS is but another manifestation of an underlying immunologically mediated disorder of the cellular and humoral immune system. What has emerged, and continues to emerge, is a cohesive picture of a multi-pronged immunologic

attack focused on small blood vessels of the CNS, PNS and muscle (i.e., a small vessel mononuclear inflammatory vasculopathy). Evidence exists for the participation of multiple, and cascading, inflammatory or immunologic insults to the nervous system including: (i) abnormal trafficking of mononuclear cells (predominantly lymphocytes, and less commonly, plasma cells and macrophages/monocytes) across the endothelium with breach of the blood–brain and blood–nerve barriers; (ii) direct infiltration of mononuclear cells into perivascular spaces and surrounding nervous system tissue with attendant damage to surrounding cells and tissue; (iii) synthesis of molecules (e.g., cytokines and excitatory neurotoxins) by infiltrating inflammatory cells that mediate tissue damage; (iv) vascular endothelial cell and vessel wall damage; (v) complement pathway activation. The association of certain autoantibodies (i.e., anti-Ro(SS-A); anti-endothelial cell) with vascular pathology implicates these autoantibodies in disease pathogenesis in some SS patients.

Initially, the description of neurologic disease in SS was met with healthy skepticism, since it could not be comprehended how such a broad spectrum of neurologic disease in a disorder as common as SS (which affects approximately 3% of the adult population) could have been overlooked for so many years. The potential reasons or explanations for this dilemma have been clarified. Subsequently, investigators worldwide, in small series and in numerous case reports, have been confirming essentially all reported aspects of the clinical disease expression, neurodiagnostic evaluation, immunologic associations, and pathology which have been described by the Hopkins cohort. This chapter will cover the journey from the initial clinical observations through the wide spectrum of neurologic presentations of SS, with practical consideration of the diagnosis and evaluation of the SS patient with neurologic disease. The chapter will end with a synthesis of our current state of understanding of the pathology and immunopathogenesis of neurologic disease in SS.

SUMMARY OF THE CLINICAL ENTITY SJÖGREN'S SYNDROME

Clinical definition

SS syndrome is a chronic autoimmune disorder characterized by multiple abnormalities of the cellular (i.e., lymphoproliferative) and humoral immune response and immune dysregulation. SS is defined clinically by the presence of the sicca complex (i.e., dryness syndrome). The term 'autoimmune exocrinopathy' (Talal 1966) has been used to describe the sicca manifestations associated with progressive infiltration, impairment of function, and destruction of salivary, lacrimal, and other exocrine glands by mononuclear inflammatory cells (predominantly lymphocytes and plasma cells). The syndrome has 'glandular' (i.e., non-systemic), as well as 'extraglandular' (i.e., systemic) manifestations.

SS occurs most commonly in women (female:male ratio = 9:1), but also has been described in men (Molina et al. 1986). Men as well as women can develop extraglandular complications but, interestingly, men tend to be seronegative for one or more autoantibodies (present in approximately 40–50% of women with primary SS). Although traditionally considered a disease of middle-aged-to-elderly women, there is a growing awareness that SS also can affect young adults, adolescents and even children (Berman et al. 1990; Anaya et al. 1995; Ohtsuka et al. 1995; Nagahiro et al. 1996).

SS can occur alone (primary SS) or may be associated with another connective tissue disorder (secondary SS). Secondary SS is most commonly associated with rheumatoid arthritis (RA) and, less commonly, with systemic lupus erythematosus (SLE),

progressive systemic sclerosis, or overlap syndromes, particularly SS/lupus overlap syndrome. Alternatively, SS may be associated with benign (pseudolymphoma or angioblastic lymphadenopathy with dysproteinemia) or malignant (lymphoma or lymphosarcoma) lymphoproliferative disorders. SS patients have a 44-fold increased risk of developing lymphoma, usually non-Hodgkin's B cell, as compared to normal controls (Kassan et al. 1978).

The major clinical manifestations of the sicca complex are related to involvement of the exocrine glands and include keratoconjunctivitis sicca (dry eyes), xerostomia (dry mouth) and, less commonly, recurrent or chronic episodes of major salivary gland enlargement. SS patients with well documented SS may not spontaneously report sicca symptoms. Therefore, a careful and thorough review of systems targeted in SS is necessary for adequate clinical evaluation of SS. Other exocrine glands also may be affected, giving rise to dryness of the mucous membranes of the nasopharynx, oropharynx, and upper respiratory tract, resulting in laryngitis sicca, tracheobronchitis, recurrent otitis media, chronic sinusitis, accelerated dental caries, and gingivitis with receding gums. Dry skin (xerosis or cutaneous sicca) and vaginal dryness (vaginitis sicca) are common features of SS also related to exocrine gland dysfunction.

Common systemic features include fatigue, malaise, low grade fever, lymphadenopathy, myalgias, and arthralgias. 'Extraglandular complications' can involve any organ in the body with associated mononuclear inflammatory infiltrates (perivascular and parenchymal) and organ dysfunction. One or more organ-specific autoimmune disorders, indistinguishable from those occurring as isolated entities (e.g., Grave's disease, Hashimoto's thyroiditis, pancreatitis, pernicious anemia, chronic active hepatitis, primary biliary sclerosis, inflammatory bowel disease, celiac disease, interstitial pulmonary fibrosis, and interstitial nephritis), may occur. Cytopenias, such as autoimmune thrombocytopenia, lymphopenia, neutropenia, and Coomb's positive autoimmune hemolytic anemia, have been observed.

SS is a common rheumatic disorder affecting, conservatively, 2–3% of the adult population (an estimated 4 million patients within the United States). In a study of 103 elderly women in a residential community (Strickland et al. 1987), 2% were documented as having definite primary SS, 12% possible SS, and

7% had anti-Ro(SS-A) antibodies. Only one patient had RA and none had SLE. In a study of 705 Swedes between the ages of 52 and 72 years, 3% had definite SS (Jacobsson et al. 1989), and approximately one-third had one or more objective features of subclinical SS (Jacobsson et al. 1992). A similar frequency of SS has been documented in Greece (Drosos et al. 1988). Neurologic disease can occur in primary or secondary SS (G.E. Alexander et al. 1981; Alexander et al. 1982; Hashii et al. 1989; Ishii et al. 1993). Likewise, neurologic disease can occur in seronegative as well as seropositive SS patients (Hietaharju et al. 1990; Escudero et al. 1992, 1995; Alexander 1993).

Prevalence of neurologic disease in SS

The actual prevalence of neurologic disease in the general SS population is unknown (Snaith 1990; Alexander 1991, 1992; Moutsopoulos et al. 1993). Frequency estimates vary depending on several variables, including the SS population studied, the location of the study (i.e., highly specialized tertiary referral centers, less specialized academic centers, or clinical practice settings), the interest and expertise of the evaluating investigators or clinicians, and the extent and sophistication of the neurologic evaluation.

PNS disease in SS is generally considered to be more common than CNS disease. CNS disease is often accompanied by PNS disease (Molina et al. 1985a). PNS disease is reported to occur between 10 and 20% of SS patients (Bunim 1961; Kaltreider and Talal 1969; Shearn 1971; Whaley et al. 1973; G.E. Alexander et al. 1981; Alexander et al. 1982; Alexander 1987b). In prospective studies, however, the frequency has been considerably higher (Hietaharju et al. 1990, 1992; Mauch et al. 1994).

CNS disease was initially estimated to occur in approximately 20–25% of SS patients referred to a highly specialized tertiary referral center (G.E. Alexander et al. 1981; Alexander et al. 1982, 1983). Subsequently, in prospective studies, CNS disease has been reported in SS populations in frequencies ranging from a very low percentage up to 70% (Alexander 1986a; Ya-Xin and Chang-Hua 1986; Binder et al. 1988; Drosos et al. 1989; Andonopoulos et al. 1990; Hietaharju et al. 1990, 1993a, b; Escudero et al. 1992, 1995; Volk et al. 1994). Neuropsychiatric disturbances are reported to occur in 28–42% of SS patients (Stoltze et al. 1969; Drosos et al. 1989; Mu-

kai et al. 1990).

Several authors have emphasized that sicca manifestations may be mild or absent at the time SS patients present with PNS on CNS manifestations (Olsen et al. 1989, 1991; Hietaharju et al. 1990; Alexander 1992, 1993; Escudero et al. 1992, 1995; Provost et al. 1993). In other words, subclinical SS or a forme fruste of SS may underlie neurologic disease in SS (Olsen et al. 1989, 1991). A prospective study of 100 consecutive admissions to an university neurology inpatient service revealed a 3% prevalence of definite SS. Awareness of these observations should prompt the clinician to appropriately evaluate patients with neurologic disease of obscure etiology for SS.

Diagnosis

Although a number of criteria have been proposed (Fox et al. 1984, 1986b; Homma et al. 1986; Skopouli et al. 1986; Manthorpe et al. 1992; Vitali et al. 1993), currently there are no generally accepted criteria for the diagnosis of SS. A European consortium has proposed diagnostic criteria for definite and probable SS (Vitali et al. 1993).

In the patient who is symptomatic for the sicca complex, the diagnosis is routinely established by a combination of ocular and oral diagnostic studies. Ophthalmologic examination includes a Schirmer's test and a Rose Bengal dye test (Kincaid 1987). The Schirmer's test is considered positive if less than 5 mm of filter paper is moistened within 5 min (Kincaid 1987). Characteristic Rose Bengal staining as defined by Holms (1949) or Van Bijsterveld (1969) is required for a positive examination.

The oral component of SS involvement is most commonly documented with a minor salivary gland biopsy graded by the scale proposed by Greenspan et al. (1974). Usually, the presence of two or more foci of 50 lymphocytes per aggregate is required for a positive biopsy (Greenspan et al. 1974), although some criteria require only one (Vitali et al. 1993). Lesser degrees of inflammation and gland destruction are suggestive, but not diagnostic of SS. The significance of salivary gland atrophy, fibrosis and fatty infiltration is controversial (Greenspan et al. 1974).

Baseline and stimulated salivary flow rates, scintigraphy, sialography, and lacrimal osmolality determinations also are used in the evaluation of SS, but are less sensitive. In most instances, a positive ocular examination (either positive Schirmer's test or a Rose Bengal dye test) and a positive minor salivary gland biopsy are required for the diagnosis of definite SS (Manthorpe et al. 1986). In the patient with abnormal (but not diagnostic) studies the diagnosis of probable, rather than definite, SS may be considered (Manthorpe et al. 1986; Vitali et al. 1993). Neurologic disease can occur in patients with both definite and probable SS (Hietaharju et al. 1990).

The question of the significance of the presence, method of detection, and levels of autoantibodies (particularly, antibodies to Ro(SS-A)/La(SS-A) peptides) in SS is another important and, often, poorly understood issue in the diagnosis of SS (see Antibody seroreactivity in CNS-SS). Most proposed criteria have not included the presence of autoantibodies as a diagnostic criteria for SS, whereas other criteria have (Fox et al. 1986a; Vitali et al. 1993, 1996). If autoantibodies (anti-Ro(SS-A)/La(SS-B), antinuclear antibodies (ANA), or rheumatoid factor (RF)) are present in a patient suspected of SS, they are useful as confirmatory documentation of SS or a related chronic inflammatory disorder. Importantly, however, the absence of autoantibodies does not exclude the diagnosis of SS.

Recently, new preliminary criteria for SS have been proposed by the European consortium (Vitali et al. 1993, 1996). Four of six criteria are required for the diagnosis of definite and three for probable SS. These criteria have several features that distinguish them from previous criteria. Firstly, they are less stringent; secondly, they do not require objective documentation of both ocular and oral involvement; thirdly, a positive minor salivary gland biopsy is not required for diagnosis of definite SS; fourthly, only one lymphocytic focus is required for a diagnostic minor salivary gland biopsy. Finally, the presence of autoantibodies (particularly anti Ro(SS-A/La (SS-B) antibodies) is a diagnostic criterion.

CLINICAL DESCRIPTION OF THE NEUROLOGIC SYNDROMES

Neurologic spectrum of CNS disease

The clinical manifestations of CNS disease in SS are multiple and diverse (Alexander 1986a, 1992; Alexander and Provost 1987) and span the entire neuroaxis, including the brain, spinal cord, and optic nerves

TABLE 1

Spectrum of central nervous system disease in Sjögren's syndrome.

Neurologic abnormality
  Brain
    Focal
        Motor deficits
        Sensory deficits
        Aphasia/dysarthria
        Seizure disorders
        Brainstem syndromes
        Cerebellar syndromes
        Gait disturbances
        Extrapyramidal motor syndrome
        Headache
        Neurovascular syndromes
        Intracerebral hemorrhage
        Subarachnoid hemorrhage
        Cortical blindness
        Gaze disturbances

    Non-focal (diffuse)
        Subacute or acute encephalopathy
        Aseptic meningitis, often recurrent
        Cognitive dysfunction
        Dementia
        Affective psychiatric abnormalities

  Spinal cord
    Transverse myelitis
    Chronic progressive myelitis
    Brown-Séquard syndrome
    Neurogenic bladder
    Lower motor neuron disease

(Table 1). Neurologic disease can be categorized as focal (i.e., discrete neurologic deficits) and/or non-focal (i.e., psychiatric and/or cognitive dysfunction) neurologic disease. The neurologic abnormalities may be subtle, usually of insidious onset or, less often, protean and of acute or subacute presentation. Early in the course of neurologic disease, neurologic symptoms or deficits are characteristically transient and usually resolve with return to baseline function. With time, however, CNS disease manifestations may become multifocal, recurrent, additive, fixed, and, eventually, in some cases, chronic and progressive. There may be long disease-free intervals between successive neurologic events.

Focal CNS-SS disease

The spectrum of focal neurologic involvement is well described in the Hopkins publications and recently confirmed by others throughout the world (Ya-Xin and Chang-Hua 1986; Konttinen et al. 1987; Hashii et al. 1989; Berman et al. 1990; Hietaharju et al. 1990, 1993a,b; Kohriyama et al. 1990; Lahoz Rallo et al. 1990; Bakchine et al. 1991; Nagao et al. 1991; Escudero et al. 1992; Tesar et al. 1992; Yoshiiwa et al. 1992; Ishii et al. 1993; Lafforgue et al. 1993; Ménage et al. 1993; Escudero et al. 1995; Giordano et al. 1995; Lyu et al. 1995). Focal neurologic deficits include motor and/or sensory loss (i.e., monoparesis, hemiparesis, paraparesis, quadriparesis, hemisensory, monosensory or truncal sensory deficits), aphasia/dysarthria, and visual loss. CNS-SS patients can have neurologic presentations and disease manifestations indistinguishable from multiple sclerosis (MS) (Alexander et al. 1986b; Rutan et al. 1986; Konttinen et al. 1987; Lyu et al. 1995). Major stroke syndromes of acute onset are distinctly unusual, but do occur in SS (Alexander 1993; Brasoni et al. 1994). Seizure disorders (accompanied by abnormal electroencephalogram (EEG), see below) are usually petit mal (absence) or temporal lobe (psychomotor) in type. Grand mal seizures, focal motor seizures, status epilepticus, and epilepsy partialis continua (Bansal et al. 1987) have been observed, but less commonly.

Although cerebral inflammatory vascular disease (IVD) in SS usually affects small blood vessels (see Pathology/immunopathogenesis), SS patients can develop cerebral vasculitis of medium-to-larger extracranial and intracranial cerebral blood vessels (Case Records 1985; Ferreiro and Robalino 1987; Sato et al. 1987; Berman et al. 1990; Kohriyama et al. 1990; Lahoz Rallo et al. 1990; Provost et al. 1991; Alexander 1993; Giordano et al. 1995). The clinical presentations and cerebral angiography may be indistinguishable from polyarteritis nodosa (PAN), temporal arteritis, or Takayasu's arteritis. There may be extensive collateralization (Provost et al. 1993; Nagahiro et al. 1996) with Moa Moa phenomenon.

Movement disorders include tremors (involuntary and intentional), dystonias, pseudoathetosis, choreoathetosis, dyskinesias/akinesias, gait disturbances, ataxia (truncal and appendicular), and Parkinson-like syndrome (Nagao et al. 1991; Fabre et al. 1996). A case of astasia–abasia (Lafforgue et al. 1993) has been reported. Brainstem deficits include eye movement disorders and involvement of one or more cranial nerve nuclei (multiple cranial palsies). Cerebellar deficits occur in association with other fo-

cal neurologic deficits, but also may occur alone, resulting in isolated spinocerebellar degeneration syndrome (Alexander et al. 1982; Bakchine et al. 1991; Terao et al. 1994). Significant impairment in cognitive function and psychiatric abnormalities in SS may have an organic basis (see below) (Malinow et al. 1985; Caselli et al. 1991; Ranzenbach et al. 1991).

The spinal cord also may be involved in CNS-SS disease (G.E. Alexander et al. 1981; Alexander et al. 1982, 1986; Konttinen et al. 1987; Bakchine et al. 1991; Ménage et al. 1993; Giordano et al. 1995; Harada et al. 1995; Lyu et al. 1995). Acute transverse myelopathy, often recurrent, and chronic progressive myelopathy can occur. Neurogenic bladder may accompany these clinical presentations or occur alone. Less commonly, Brown-Séquard syndrome and lower motor disease, resembling ALS, have been observed. Optic nerve involvement, consistent with optic neuritis, can occur alone or accompany spinal cord involvement.

Non-focal CNS disease

Non-focal (diffuse) manifestations of brain involvement include subacute or acute encephalopathy and aseptic meningoencephalitis, often recurrent (Alexander 1982; Drosos et al. 1989; Vrethem et al. 1990; Caselli et al. 1991, 1993; Garraty et al. 1993; Kawashima et al. 1993; Mauch et al. 1994; Tsuji et al. 1994; Utset et al. 1994; Giordano et al. 1995). Psychiatric and cognitive dysfunction are part of the neurologic spectrum of CNS-SS (Malinow et al. 1985; Selnes et al. 1985; Alexander 1986a, 1987, 1992, 1993; Robson et al. 1986; Tarter 1989; Vitali et al. 1989; Hietaharju et al. 1990; Mukai et al. 1990; Caselli et al. 1991, 1993; Creange et al. 1992; Garraty et al. 1993; Giordano et al. 1995). SS patients often spontaneously complain of changes in cognitive function characterized by recent memory impairment, problems with attention and concentration, and a decline in the performance of work or daily living tasks (which may be more severe than the degree of impairment documented on formal cognitive function testing).

Formal cognitive function testing has confirmed abnormalities in attention and concentration, a decrement in intelligence quotient in verbal exceeding performance, and, less frequently, dysnomia (Malinow et al. 1985; Selnes et al. 1985; Caselli et al. 1991, 1993; Belin et al. 1994; Mauch et al. 1994). The former features are characteristic of the category of dementia classified as subcortical, which is seen in such disorders as Parkinson's or Huntington's disease. SS patients may also develop cognitive function/dysfunction indistinguishable from multi-infarct dementia or Alzheimer's disease (Malinow et al. 1985; Alexander 1986, 1992; Caselli et al. 1991, 1993). Affective psychiatric disturbances also have been observed (Malinow et al. 1985; Drosos et al. 1989; Vitali et al. 1989).

The presence of abnormal neuroimaging studies has suggested an organic etiology for cognitive and psychiatric dysfunction in CNS-SS (see Neurodiagnostic studies) (Alexander 1986; Alexander et al. 1986, 1994). Multiple, predominantly white matter, subcortical and periventricular regions of increased signal intensity have been observed on brain magnetic resonance imaging (MRI) scans in CNS-SS patients with non-focal (i.e., psychiatric and/or dysfunction in the absence of focal CNS disease), as well as focal CNS disease (Alexander et al. 1986, 1988, 1994).

Furthermore, in preliminary studies, Technetium 99m-HMPAO brain scans, which measure cerebral blood flow, demonstrate multiple areas of cortical hypoperfusion in patients with cognitive and/or psychiatric dysfunction, some of whom have normal brain MRI scans (Alexander, unpublished observations). These studies suggest that there are both cortical and subcortical abnormalities of cerebral perfusion in CNS-SS that could contribute to cognitive impairment and psychiatric disturbances. Subtle cognitive dysfunction may be an early manifestation of CNS disease in SS and often accompanies focal CNS disease.

These observations have important implications for the potential diagnosis and treatment of dementia. Dementia associated with SS may be indistinguishable from dementias of other etiologies (see above). Approximately 80% of persons aged 65 and over are estimated to have cognitive dysfunction and 4–5% have severe dementia. Even if only a relatively small proportion of SS patients (approximately 10%) eventually develop dementia or cognitive dysfunction, alone or in association with focal CNS disease, potentially, the number of U.S. patients with SS-related cognitive dysfunction could be over 400,000. It is likely that the cognitive dysfunction or dementia of SS is misdiagnosed as Alzheimer's disease or multi-infarct dementia. Unlike these disorders, however, CNS-SS is a potentially treatable and reversible form of dementia (Alexander 1991; Caselli et al. 1991, 1993). Since SS represents a treatable cause of dementia, patients presenting with dementia should be systematically evaluated for SS.

Paraneoplastic-like syndromes

Two of the more common paraneoplastic neurologic syndromes associated with carcinoma (i.e., subacute sensory neuronopathy and cerebellar degeneration) have been observed in SS in the apparent absence of malignancy (Alexander et al. 1982; Malinow et al. 1986; Mellgren et al. 1989; Griffin et al. 1990; Bakchine et al. 1991; Terao et al. 1994). Antineuronal antibodies against the 38 kDa Hu antigen, Purkinje cells, and 34 and 63 kDa Yo antigens have been observed in these paraneoplastic syndromes associated with carcinoma (Graus et al. 1986).

Whether SS patients with these paraneoplastic-like neurologic syndromes also have antibodies with similar specificities is not clear. Anti-Hu antibodies have not been detected in SS patients with sensory neuronopathy (Griffin et al. 1990; Alexander and Posner, unpublished observations). On the other hand, other investigators (Moll et al. 1993) have detected antineuronal antibodies in SS patients by indirect immunofluorescence against rat cerebral and cerebellar cortices and dorsal root ganglia, as well as by Western blot analysis of cerebellar tissue. Of 45 consecutive SS patients, 25 (60%) had neurologic disease (11 major; 14 minor). Antineuronal antibodies were present in 6/11 (55%) of SS patients with major, and in 4/34 (11%) with minor neurologic disease. Four basic staining patterns were observed. Reactivity was present against the following antigens: (1) 38 kDa, (2) 38 and 70 kDa nuclear S, (3) neuronal specific cytoplasmic, (4) 52 kDa mitochondria, (5) 52 kDa axon or neurofilament antigens.

It was initially proposed that antibodies against the 38 kDa protein, seen in some SS patients with sensory neuropathy, were directed against the Hu antigen (Moll et al. 1993), classically associated with small cell lung carcinoma and paraneoplastic neurologic disease (Graus 1986). These SS antisera, however, did not react with a HuD fusion protein (Vecht et al. 1995), suggesting that the antibody does not have Hu specificity.

Reactivity against 38 kDa and 45 kDa proteins

(mainly cytoplasmic) also was observed on Southern blot analyses of serum from an SS patient with cerebellar degeneration (Terao et al. 1994). These antibodies are similar in immunoreactivity to those detected in a patient with paraneoplastic cerebellar degeneration and limbic encephalitis (Tsukamoto et al. 1993).

CNS disease in children and adolescents

CNS disease can occur in children and adolescents (Berman et al. 1990; Garraty et al. 1993; Ohtsuka et al. 1995; Nagahiro et al. 1996). A 10-year-old girl with primary SS and anti-Ro(SS-A) antibodies developed optic neuropathy and CNS disease characterized by multifocal neurologic deficits. Brain MRI showed multiple diffuse focal regions of increased signal intensity in both white and gray matter. Cerebral angiography documented multifocal abnormalities of the anterior and posterior inferior cerebellum anterior bilaterally consistent with small vessel angiitis.

A 9-year-old girl with primary SS had multifocal CNS disease resembling MS (Ohtsuka et al. 1995). An adolescent young adult (18 years old) with primary SS, anti-Ro(SS-A) antibodies and fatal recurrent aseptic meningoencephalitis has been reported (Garraty et al. 1993). A 17-year-old girl with multiple occlusions of major cerebral arteries has been reported (Nagahiro et al. 1996).

Diagnostic studies in CNS-SS

The three most useful non-invasive neurodiagnostic studies in evaluating SS patients suspected of having active CNS disease involvement are brain MRI scans, electrophysiologic studies, and CSF analyses (Alexander 1987; for review see Alexander 1992, 1993). These diagnostic modalities, on the one hand, provide anatomic and functional evidence for multifocal neurologic involvement and, on the other hand, for active cerebral inflammation. At the present time, the utilization of all three of these modalities is recommended, as each provides different and complementary types of information about the neuroanatomic basis of brain lesions, extent of functional impairment, and severity of disease activity.

Neuroimaging studies
Neuroradiologic studies (i.e., brain/spinal cord computed tomography (CT) scans, brain MRI scans,

TABLE 2

Clinical neurodiagnostic tests in central and peripheral nervous system disease in Sjögren's syndrome.

Neuroradiology
Computed axial tomography (CAT)
Cerebral angiography
Magnetic resonance imaging (MRI)
Spectroscopy

Electrophysiology
Electroencephalography (EEG)
Evoked response testing
PSVERs
BAERs
SSERs
Nerve conduction studies
Electromyography (EMG)
Quantitative sensory testing

Cerebrospinal fluid (CSF) analysis
Cell count and differential
Total protein
IgG index
Oligoclonal bands
Cytology

Spinal cord
MRI with gadolinium enhancement

spectroscopy, and cerebral angiography) may provide definitive information about the presence, nature, and extent of neurologic involvement in CNS-SS (Alexander 1987, 1992, 1993; Alexander et al. 1988, 1994).

Brain CT scans are relatively insensitive in documenting neuroanatomic abnormalities in CNS-SS. Brain CT scans (double-dose enhanced studies with delayed imaging) detect neuroanatomic abnormalities in fewer than 20% of CNS-SS cases (Alexander 1987, 1992, 1993; Alexander et al. 1988, 1989). The abnormalities observed on brain CT include large infarcts, intracerebral hemorrhages and, rarely, subarachnoid hemorrhage. Spinal cord CT scanning has shown spinal subarachnoid hemorrhage secondary to the rupture of the anterior spinal artery (Alexander et al. 1982). The large infarcts on brain CT scans correspond to large regions of increased signal intensity on brain MRI scans (see below).

Brain MRI scans are more sensitive than CT scans in evaluating CNS disease in SS (Alexander et al. 1988). Approximately 80% of SS patients with progressive focal neurologic dysfunction have brain MRI scans with multiple small regions of increased

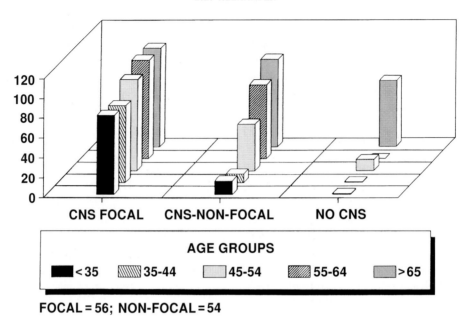

FOCAL = 56; NON-FOCAL = 54

Fig. 1. Age distribution of abnormal brain MRIs in CNS-SS. CNS focal/non-focal: multiple regions of increased signal intensity on T2- and proton-weighted images. No CNS disease: one or two regions of increased signal intensity.

signal intensity on T2- and proton density-weighted images, predominantly in subcortical and periventricular white matter (Figs. 1–3). A subset of CNS-SS patients have one or more larger (>10 mm) regions of increased signal intensity (Fig. 1) and these patients almost invariably have antibodies to Ro(SS-A) protein (Alexander et al. 1994). A smaller proportion of CNS-SS patients have cortical or cerebellar atrophy or ventricular dilatation in the absence of regions of increased signal intensity.

A significant proportion (approximately 50%) of SS patients with non-focal CNS disease also demonstrate an age-related increment in multiple small regions of increased signal intensity.

In our experience, SS patients without clinical evidence of CNS disease have a very low frequency of abnormal MRI scans with more than one region of increased signal intensity (Alexander et al. 1988). Others (Manthorpe et al. 1992; Pierot et al. 1993), however, described small regions of increased signal intensity on brain MRI scans within the basal ganglia and white matter of cerebral hemispheres in up to 60% of SS patients without clinical evidence of CNS disease. These abnormalities may be secondary to subclinical disease.

Evidence, to date, suggests that regions of increased signal intensity on brain MRI/CT scans in CNS-SS almost invariably are fixed and usually do not resolve with therapy. Unlike the situation in MS (McFarland et al. 1992; Miller 1994), regions of increased signal intensity in CNS-SS usually are not enhanced by contrast agents (i.e., gadolinium) (Alexander et al., unpublished observations). This observation would suggest that the blood–brain barrier (BBB) is intact in CNS-SS, This, however, may not be the case. The albumin quotient is usually normal in active CNS-SS, indicating that the BBB is grossly intact. Brain histopathology from CNS-SS patients, however, commonly shows perivascular inflammation and disruption of the endothelial cell barrier of small cerebral blood vessels with perivascular edema, myelin pallor, and extravasation of erythrocytes. These observations suggest that in CNS disease in SS, there may be disruption of the BBB of small cerebral vessels, which is not reflected in the abnormalities detected by gadolinium-enhanced brain scans.

The exact pathologic correlate(s) of regions of increased signal intensity on brain MRI scans, in general, cannot be established. The multiple regions of increased signal intensity on brain MRI scans in SS

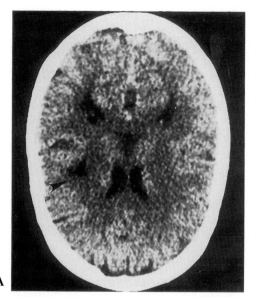 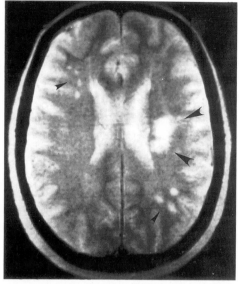

A   B

Fig. 2. 50-year-old white woman with anti-Ro(SS-A) antibody positive SS first presented with mild cognitive dysfunction and subsequently developed severe dementia. Within 6 months she had multiple strokes, including a dense right hemiparesis with aphasia. Somatosensory evoked responses were abnormal (left tibial nerve). CSF analysis showed an elevated protein, elevated IgG index, two oligoclonal bands, and reactive lymphoid cells (including cells resembling monocytoid B cells). Left internal carotid cerebral angiography showed subtle small vessel changes consistent with angiitis. CT of the brain (A) shows a left parietal lucency (infarct: arrow), but no other lesions. MRI (B) shows multiple regions of increased signal intensity (i.e., more than 10), and two larger (i.e., >10 mm) confluent left parietal regions of increased signal intensity in the centrum semiovale, very close to but not touching the ventricular wall (arrowheads). The larger regions correspond to the lucency seen on CT. At least seven additional small lesions are present in the left occipital and right frontoparietal regions (arrows). The patient responded to corticosteroid and monthly pulse intravenous cyclophosphamide therapy (1-year course) and has remained in remission.
    Repetition time 2409; echo time 80; axial. (Modified from Alexander et al. 1988 with permission.)

are consistent with infarction, ischemia, edema, or demyelination. Correlative neuropathologic studies in CNS-SS, however, suggest that the MRI abnormalities are commonly associated with dilated perivascular (Virchow–Robin) spaces, perivascular myelin pallor and disorganization, microinfarcts and, less commonly, macroinfarcts (Alexander 1987, 1992; Ranzenbach et al. 1991, 1992). Similar multiple regions of increased signal intensity (i.e., unidentified bright objects (UBOs)) on brain MRI scans are often dismissed as a normal consequence of aging. Previous studies, however, have shown that such white matter hyperintensities (other than 'normal' periventricular capping and grade I lateral ventricle rims) are rare in healthy individuals less than 45 years of age (Kozachuk et al. 1990). Thus, in CNS-SS patients, multiple regions of increased signal intensity on brain MRI are probably related to the underlying patho-

physiologic process affecting predominantly the small blood vessels of the brain.

Finally, preliminary studies of Tech 99m-HMPAO brain SPECT scans, which measure cerebral blood flow/metabolism, show a very high frequency of regional cortical hypoperfusion abnormalities in SS patients with both focal and non-focal CNS disease (Alexander et al., unpublished observations; Belin 1994). Some CNS-SS patients with normal brain MRI scans may have abnormal SPECT brain scans. These studies suggest that alterations in cerebral blood flow may be an early marker of CNS involvement in CNS-SS.

Cerebral angiography

Cerebral angiography is performed for two main reasons in CNS-SS (i): to exclude other etiologies of CNS disease (i.e., arteriosclerosis, and cerebrovas-

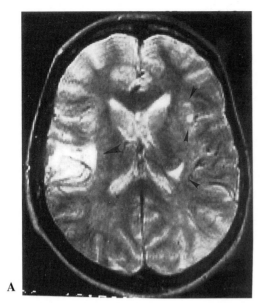 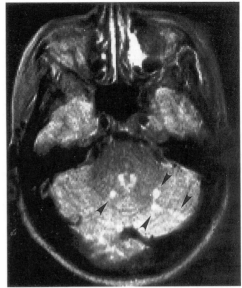

A   B

Fig. 3. 60-year-old woman had long-standing anti-Ro(SS-A) antibody negative primary SS, an affective psychiatric disorder and more than 10 distinct clinical events involving brain and spinal cord that had occurred sequentially over 20 years. The patient had recurrent episodes of urticarial cutaneous lesions that often coincided with an acute neurologic event. Histopathologic examination of the skin lesions showed mononuclear (i.e., lymphocytic) small vessel vasculopathy and frank necrotizing arteritis. Cerebellar ataxia (truncal and appendicular) was profound. An EEG showed slowing in the right posterior hemisphere and left temporal lobe with sharp wave activity. Somatosensory evoked response testing (right median nerve) was abnormal. CSF analysis showed a mildly elevated IgG index and one (oligo) clonal band. Cerebral angiography was negative. CT showed lucencies in the right parietal lobe, bilateral basal ganglia, and posterior limb of the internal capsule (not shown). MRI shows more than 10 regions of increased signal intensity. (A) Atrophy and encephalomalacia of the right temporoparietal region in the area of the sylvian fissure. Smaller lesions are also present in the left thalamus and frontal regions. Increased signal intensity surrounds the anterior poles of the lateral ventricles. (B) A minimum of four white matter lesions involving the cerebellar hemispheres posterolateral to the fourth ventricle on either side (arrowheads). Repetition time 3000; echo time 64; axial. Modified from Alexander et al. 1988 with permission.)

cular disease, AV malformations, other vascular abnormalities, and congenital aneurysms), and (ii) to establish a potential diagnosis of CNS vasculopathy/vasculitis. Cerebral angiography in CNS-SS detects abnormalities of small cerebral arteries consistent with, but not diagnostic of, small vessel angiitis in approximately 20% of highly selected cases of SS patients with CNS disease (G.E. Alexander et al. 1981; Alexander et al. 1982, 1988; Alexander 1986b, 1992, 1993; Berman et al. 1990).

Cerebral angiograms in CNS-SS show one or more focal regions of narrowing (stenosis): post-stenotic dilatation, vessel occlusion, with or without collateralization, delayed emptying of vessels, or anastomotic channels. In a small number of cases, medium or large sized vessels are affected (Ferreiro and Robalino 1987; Provost et al. 1991; Alexander

1993; Giordano et al. 1995; Nagahiro et al. 1996). CNS-SS patients with small to medium sized artery abnormalities on cerebral angiography have a significantly increased frequency of severe focal CNS disease, multiple and large regions of increased signal intensity in brain MRI imaging, and anti-Ro(SS-A) antibodies (Alexander et al. 1994). Carotid artery occlusion with extensive collateralization resembling Moa Moa phenomenon has been observed in anti-Ro(SS-A) positive individuals (Provost et al. 1991). Multiple saccular aneurysms with a predilection for branching points, consistent with PAN, however, have not been described in CNS-SS.

It is important to emphasize that a normal brain MRI scan and/or cerebral angiogram does not exclude the presence of active CNS-SS. The presence of an inflammatory meningeal/cerebral vasculopathy in several CNS-SS patients with normal brain MRIs and cerebral angiograms has been documented at autopsy or brain biopsy (Malinow et al. 1985; Alexander et al. 1986a, 1988, 1992, 1993; Caselli et al. 1991). This observation raises the issue of indications for meningeal/brain biopsy in the evaluation of CNS-SS. There are several situations in which a meningeal/brain biopsy is recommended to: (1) establish a definitive diagnosis in cases in which the diagnosis is unclear; (2) distinguish CNS-SS from other CNS disease, such as MS or Alzheimer's disease; (3) exclude other etiologies including infection, malignancy and lymphoma; (4) evaluate individuals who have an immunologically reactive CSF analysis (protein elevation, moderate to marked pleocytosis, elevated IgG index, and oligoclonal bands); and (5) evaluate patients who have a rapid deterioration in their clinical status and who will need aggressive therapeutic intervention

Electrophysiologic studies

Electrophysiologic studies including EEG, multimodality evoked response (MMER) testing and blink reflex, are clinically useful in evaluating and following patients with CNS-SS (G.E. Alexander et al. 1981;

TABLE 3

Features associated with anti-Ro antibodies in Sjögren's syndrome patients with central nervous system disease.

Clinical
   Major focal CNS events
   Intracerebral hemorrhage
   Subarachnoid hemorrhage
   CNS disease-related death

Neuroimaging studies – MRI/CT
   Multiple (> 10) MRI regions of increased signal
      intensity
   Large MRI/CT infarcts or hemorrhage

Cerebral angiography
   Small vessel angiitis
   Larger vessel arteritis
   Aneurysms
   Vascular occlusion
   Collateralization

Histopathology
   Small vessel vasculopathy*
   Cerebral vasculitis

*Also usually present in anti-Ro antibody negative SS patients.

Alexander 1987b, 1992, 1993; Alexander et al. 1988). EEG is abnormal in approximately one-third of CNS-SS patients with severe progressive disease (G.E. Alexander et al. 1981; Alexander et al. 1982; Malinow et al. 1985; Alexander 1986b, 1987b, 1992). Patients with focal neurologic deficits may show focal slow wave activity, decreased amplitude, or spikes (sharp waves). In patients with petit mal or temporal lobe epilepsy, EEGs, including sleep studies, may show seizure discharges. In patients with encephalopathy or progressive dementia, there may be diffuse slowing.

MMER testing may provide information about brain function in CNS-SS (Malinow et al. 1985; Alexander 1986a, 1987b, 1992). MMERs measure the integrity of the neuronal circuitry from the periphery to the cerebral cortex via the visual (pattern stimulated visual evoked responses (PSVERs)), auditory (brainstem auditory evoked responses (BAERs)), and peripheral nerve (somatosensory evoked responses (SERs)) pathways. The blink reflex detects abnormalities of the fifth nerve (peripheral or central), reflected clinically in trigeminal neuropathy. Abnormalities of any one of these MMERs indicate a functional interruption of the involved neuroanatomic pathway. One or more MMER studies are abnormal in approximately one-half to two-thirds of SS patients with progressive, focal CNS disease (G.E. Alexander et al. 1981; Alexander et al. 1982; Alexander 1992, 1993). EEGs and MMERs may detect subclinical abnormalities in CNS-SS which antedate the development of clinical manifestations or regions of increased signal intensity on brain MRI scans. Therefore, these studies may be useful in diagnosing early CNS involvement and in objectively following patients and monitoring response to therapy.

Cerebrospinal fluid analysis

CSF analysis reflects the presence of an immune response and disease activity in CNS-SS, excludes other etiologies of CNS disease, and provides information about the immunopathogenesis of CNS disease (G.E. Alexander et al. 1981; Alexander et al. 1982; Alexander 1986a,b, 1987b, 1992, 1993; Vrethem et al. 1990). The CSF total protein is usually normal, but may be mildly elevated. On occasion it is significantly elevated. The CSF IgG/total protein ratio is often mildly elevated. More importantly, the IgG index, indicative of the intrathecal synthesis of IgG, is elevated in approximately 50% of patients with active focal disease. One or more oligoclonal bands, detected by agarose gel electrophoresis, is observed in a similar proportion of patients. Most commonly, there are one or two bands, but multiple bands (up to seven) have been observed (Alexander et al. 1986a). It is not yet known against which antigen(s) these antibodies are directed.

Mononuclear polymorphic CSF pleocytosis, usually mild, occurs in a subset of SS patients with CNS disease (G.E. Alexander et al. 1981; Alexander et al. 1982, 1983, 1986b; De la Monte et al. 1983; Alexander 1987b). A substantial number of SS patients with active CNS disease and very abnormal brain MRI scans, however, do not have CSF pleocytosis. In fact, the CSF analysis may be entirely normal in clinically active CNS disease confirmed by brain MRI and/or brain autopsy/biopsy. In the case of aseptic meningitis, a larger number of mononuclear cells (i.e., similar in type to those observed in the CNS-SS patients with a mild mononuclear pleocytosis) are present within the CSF (Alexander et al. 1982).

Cytologic examination of the inflammatory cells demonstrates small round lymphocytes, reactive lymphoid cells, plasma cells, and atypical mononuclear cells with distinctive cleft/reniform nuclei and scant pale cytoplasm. These latter mononuclear cells appear identical morphologically (by light and electron microscopy) to a type of lymphocyte classified as a monocytoid B cell. Monocyte B lymphocytes normally reside in the region of the gastrointestinal tract termed mucosa associated lymphoid tissue (MALT). Monocytoid B lymphocytes have the capacity for transformation into non-Hodgkin's B cell lymphoma, both in SS (Shin et al. 1991), and in other disorders, including AIDS (Sohn et al. 1985; Sheibani et al. 1990). The spectrum of mononuclear inflammatory cells described in the CSF in CNS-SS, including the atypical mononuclear cells, is also observed on histopathologic examination of brain tissue (see below) within inflammatory infiltrates in the meninges, in and around blood vessels and, less commonly, within the brain parenchyma of CNS-SS patients (De la Monte et al. 1983).

Anti-Ro(SS-A) antibody seroactivity in CNS-SS

Prior to considering the relationship between the presence of autoantibodies in SS and the develop-

ment of CNS disease in SS, the general significance of autoantibodies in SS will be reviewed. A number of different autoantibodies can occur in patients with SS (Harley et al. 1986) including, most commonly, ANA, RF, and antibodies to small molecular weight ribonucleoproteins called Ro(SS-A) and La(SS-B). A subset of seropositive SS patients also may have hyperglobulinemia and, less commonly, cryoglobulinemia.

Antibodies to nRNP occur uncommonly; and to native DNA on Sm, very rarely in SS. In contrast, these autoantibodies occur in a substantial proportion of patients with classical SLE. ANA and RF may occur in a number of rheumatic or autoimmune disorders. Antibodies to Ro(SS-A)/La(SS-B), however, are more restricted in expression and occur primarily in patients with primary SS, the family of closely related disorders described as the presumed lupus variants (PLVS) (i.e., subacute cutaneous lupus erythematosus (SCLE), Sjögren's/lupus erythematosus (SS/LE) overlap syndrome, and neonatal lupus syndrome (NLE)), and in classical SLE.

SS patients who are seroreactive by standard serologic techniques (see below) for anti-Ro(SS-A)/La(SS-B) antibodies usually have a positive ANA (with a speckled staining pattern) and often have RF. Patients who are seronegative for anti-Ro(SS-A)/La(SS-B) autoantibodies are usually seronegative for other autoantibodies or have a low titer ANA (Molina et al. 1985b). It must be emphasized, however, that seronegativity for autoantibodies, including anti-Ro(SSA)/La(SSB) antibodies, does not exclude the diagnosis of SS. Autoantibody profiles are adjuvant diagnostic markers and, in some cases, are predictive of certain disease manifestations in SS.

Seroreactivity for antibodies to Ro(SS-A)/La(SS-B) is partially linked to the sensitivity and specificity of the assay employed for detection of antibodies. Historically, gel double immunodiffusion (Ouchterlony) has been used to detect precipitating antibodies to Ro(SS-A) (employing human spleen extract)/La(SS-B) (using calf thymus extract) peptides (Clark et al. 1969). Between 40 and 60% of SS patients have precipitating antibodies to Ro(SS-A) protein, and approximately one-third of these individuals also have anti-La(SS-B) antibodies. Because only a small number of normal individuals (Maddison et al. 1981) have anti-Ro(SS-A)/La(SS-B) antibodies by this technique, this assay is specific and detects significant

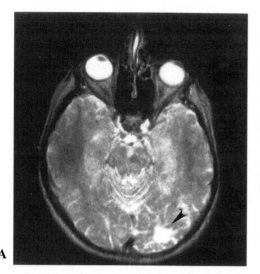

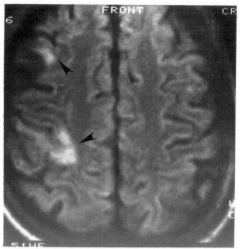

A

B

Fig. 4. 52-year-old white woman had anti-Ro(SS-A) antibody positive primary SS with progressive dementia and recurrent mild left hemiparesis and hemisensory deficits. Visual and right tibial nerve somatosensory evoked response studies were abnormal. MRI shows (A) left occipital lesion and (B) right anterior and posterior parietal subcortical regions of increased signal intensity (arrowheads). Repetition time 2.0; echo time 120; axial. (Modified from Alexander et al. 1988 with permission.)

titers of precipitating antibodies. Gel immunodiffusion, however, has been considered by some investigators (Clark et al. 1969; Harley et al. 1992) to be relatively insensitive.

Solid-phase assays (i.e., enzyme linked immunoabsorbent assay (ELISA)), employing immunoabsorbent purified bovine Ro(SS-A) and rabbit thymus La(SS-B) antigens or recombinant Ro(SS-A)/La(SS-B) antigens, have been developed (Harley et al. 1992). These assays purportedly are 100-fold more sensitive than gel double diffusion (Harley et al. 1992). In relationship to previous studies, however, these assays appear to detect high levels of anti-Ro(SS-A)/La(SS-B) antibodies in approximately 40–60% of SS patients (comparable to the proportion of individuals with precipitating antibodies detected by gel immunodiffusion) and low titers in the remainder. Approximately 15–17% of healthy normal individuals, however, have comparable low levels of these antibodies detected by ELISA (Harley et al. 1986; Gaither et al. 1987). Thus, the ELISA is more sensitive, but less specific, than gel double immunodiffusion for the detection of anti-Ro(SS-A)/La(SS-B) antibodies. The potential clinical significance, if any, of low titer anti-Ro(SS-A)/La(SS-B) antibodies is not known. One recent study suggests that low titer antibodies may be associated with sicca symptoms in healthy mature adults (Jacobsson et al. 1992). In the case of ANA and RF, however, low autoantibody titers usually are clinically insignificant.

The important points about seroreactivity in SS patients can be summarized as follows: First, approximately 40–60% of primary SS patients have significant titers of precipitating autoantibodies to Ro(SS-A) peptide and a smaller proportion (approximately one-third) of these patients also have antibodies to La(SS-B) peptide. Second, anti-La(SS-B) antibodies almost never occur in the absence of anti-Ro(SS-A) antibodies. Third, if patients have anti-Ro(SS-A)/La(SS-B) antibodies, they are more likely to have significant titers of other autoantibodies (i.e., ANA and RF). Fourth, approximately half of SS patients are seronegative for significant titers of antibodies to either Ro(SS-A) or La(SS-B). Fifth, patients who are seronegative for anti-Ro(SS-A)/La(SS-B) antibodies usually do not have significant titers of other autoantibodies. Finally, SS patients tend to remain seropositive or seronegative over time (i.e., they do not 'seroconvert').

These general observations about anti-Ro(SS-A)/La(SS-B) antibodies are relevant to the occurrence and manifestations of CNS disease in SS. The absence of anti-Ro(SS-A)/La(SS-B) autoantibodies

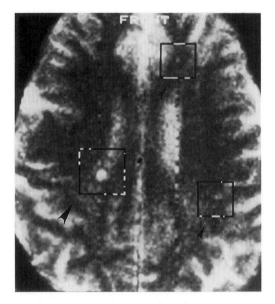

Fig. 5. 52-year-old white woman had anti-Ro(SS-A) antibody positive primary SS and a 2-year history of severe affective psychiatric disorder and progressive cognitive dysfunction without focal CNS disease. Visual and somatosensory (right tibial) evoked response testing and blink reflex were abnormal. Results of CSF analysis, EEG and CT of the brain were within normal limits. MRI, however, shows regions of increased signal intensity in the right parietal periventricular region (larger box) and left frontal and parietal areas (smaller boxes). A follow-up study 3 months later showed new small lesions in the brainstem (midbrain and pons). Repetition time 2.0; echo time 120; axial. (Modified from Alexander et al. 1988 with permission.)

does not exclude the diagnosis of SS, nor exclude the potential development of CNS disease in SS. Both seropositive and seronegative SS patients can develop significant CNS disease, unexplained by other etiologies (Alexander et al. 1983, 1986b, 1994; Molina et al. 1985a; Alexander 1987b, 1992; Hietaharju et al. 1991; Escudero et al. 1992, 1995). Both seropositive and seronegative CNS-SS patients may have abnormal brain MRI scans, which show multiple regions of increased signal intensity consistent with infarction/ischemia/edema. Preliminary brain histopathologic studies indicate that both anti-Ro(SS-A) positive and negative CNS-SS patients may have a small vessel ischemic/hemorrhagic cerebral vasculopathy (see Pathology). There are potential differences, however, between anti-Ro(SS-A) antibody positive and negative CNS-SS patients (i.e., the characteristics and severity of neurologic disease, neurologic manifestations, clinical course, neuroimaging studies, and histopathology) (see below).

CNS disease in anti-Ro(SS-A) antibody positive individuals

Although CNS disease in SS may occur in patients with or without antibodies to anti-Ro(SS-A)/La(SS-B) peptides (Alexander et al. 1983, 1986a, b, 1994; Alexander 1987b, 1992) CNS-SS patients with anti-Ro(SS-A) antibodies, as a group, have unique neurologic features. SS patients with CNS disease who have anti-Ro(SS-A) antibodies may have very serious progressive CNS disease (Alexander 1982a, 1987b, 1992, 1993; Sato et al. 1987; Ranzenbach et al. 1991, 1992; Garraty et al. 1993; Alexander et al. 1994; Giordano et al. 1995).

Conversely, those patients with catastrophic CNS disease, as a group, are highly enriched for the presence of anti-Ro(SS-A) antibodies. In a highly selected group of CNS-SS patients, anti-Ro(SS-A) antibodies are increased in frequency in SS patients with, in contrast to those without: (1) serious focal progressive CNS disease; (2) serious complications such as major intracerebral or subarachnoid hemorrhage; (3) abnormal brain MRI or CT scan showing multiple and large infarcts/ischemia; and (4) abnormal cerebral angiography consistent with vasculitis/vasculopathy.

The SS patient group that directly expires secondary to CNS disease also is highly enriched for the presence of anti-Ro(SS-A) antibodies. The small group of CNS-SS patients that has histopathologic documentation of frank angiitis (i.e., inflammation of small, medium, or large caliber cerebral blood vessels), in addition to a small vessel mononuclear inflammatory vasculopathy, essentially is all anti-Ro(SS-A) antibody positive (see Immunopathogenesis).

Another case has been published (Giordano et al. 1995) which illustrates the association of anti-Ro(SS-A) antibodies, severe multifocal CNS disease, fatal subarachnoid hemorrhage, and necrotizing arteritis of small and medium sized cerebral blood vessels. An anti-Ro(SS-A) antibody positive woman with multifocal CNS disease had a brain MRI scan showing multiple regions of increased signal intensity, consistent with infarction, and diffuse cerebral and cerebellar atrophy. She developed an acute, diffuse subarachnoid hemorrhage. Cerebral angiography showed bilateral occlusion of the superior cerebellar arteries with fusiform basilar artery dilatation. Histopathology showed multiple old infarcts as well as active necrotizing arteritis (see Pathology).

Therefore, we conclude that anti-Ro(SS-A) antibody positive, in contrast to negative, CNS-SS patients may beat risk for more serious and extensive CNS disease. We also postulate that anti-Ro(SS-A) antibodies may play a role in the immunopathogenesis of CNS-SS (see Immunopathogenesis).

DIFFERENTIAL DIAGNOSIS

The differential diagnosis of neurologic disease in SS (see Table 4) is extensive and involves disorders affecting the central and PNS, as well as, muscle. Etiologies include other chronic inflammatory disorders, autoimmune diseases, vasculitides, neurodegenerative diseases, and disorders associated with vascular abnormalities, infection, and neoplasia.

CNS disease in SS commonly affects the white matter of brain and spinal cord and, less commonly, the optic nerve. Thus, it is not unexpected that MS-like syndromes occur in some SS patients. The clinical and neurodiagnostic features of such patients are indistinguishable from classical MS (Alexander et al. 1986b, 1988; Hashii et al. 1989; Alexander 1992, 1993; Harada et al. 1995).

Neurologic manifestations of SS patients with disease mimicking MS include hypesthesia, spasticity, hyperreflexia, or both, ataxia, truncal or appendicular, hemiparesis or hemiplegia, dysarthria, cerebellar speech, nystagmus, and internuclear ophthalmoplegia. SS patients with involvement of the spinal cord may have paraparesis, quadriparesis, acute or subacute transverse myelopathy, chronic progressive myelopathy, neurogenic bladder, urinary or rectal incontinence, or Brown-Séquard syndrome. If the optic nerve is involved in SS, patients may develop a monocular or bilateral visual loss or optic neuritis/neuropathy.

A subset of patients with primary SS has been described with clinical neurologic features (Alexander et al. 1986, 1988; Alexander 1992, 1993; Ménage et al. 1993; Harada et al. 1995; Ohtsuka et al. 1995), brain MRI scans, electrophysiologic studies, and CSF immunologic parameters which may be indistinguishable from those observed in MS

(Poser et al. 1983; Rose 1976) (Tables 5 and 6). Clinically, SS patients with neurologic disease resembling MS sustain multiple neurologic events over time involving the brain, spinal cord, and optic tract that are often transient or recurrent (i.e., relapsing, remitting). In some SS patients, neurologic disease may be progressive and cumulative (i.e., chronic, progressive). CNS-SS patients with MS-like disease also share characteristics with individuals described as having lupoid sclerosis (Fulford et al. 1972), who have neurologic manifestations resembling MS and autoimmune serologic features, but do not meet diagnostic criteria for SLE (Tan et al. 1982).

Brain MRI scans in CNS-SS patients with MS-like neurologic disease may show striking similarities to MRI studies in MS patients (McFarland et al. 1992). Approximately two-thirds of SS patients with active CNS disease have multiple focal areas of increased signal intensity on T2- and proton density-weighted brain MRI images (Alexander et al. 1986b, 1988).

These abnormalities are located predominantly in periventricular and subcortical white matter. These brain MRI lesions may be neuroradiologically indistinguishable in appearance from those produced by the presence of BBB disruption or plaques in MS. CNS-SS patients, however, do not have plaques on brain histopathology. Brain or spinal cord MRIs in MS often demonstrate enhancing lesions with gadolinium contrast (McFarland et al. 1992). In CNS-SS, however, enhancing brain MRI lesions are uncommon (Alexander et al., unpublished observations).

Electrophysiologic studies such as MMER testing in CNS-SS often are abnormal (see below) and indistinguishable from those observed in MS (Alexander et al. 1982; Alexander 1993). Approximately 50% of CNS-SS patients have one or more abnormal MMERs (i.e., PSVERs, BAERs, or SSERs). As in MS, functional abnormalities delineated by MMERs often establish the presence of subclinical or multifocal disease in CNS-SS.

CSF analyses from both CNS-SS and MS patients often reveal evidence of an immune mediated process (Alexander et al. 1986a, b). In both CNS-S and MS there is breach of the BBB by inflammatory mononuclear cells (predominantly lymphocytes) and the intrathecal synthesis of IgG. More than 50% of SS patients with well-established, progressive, active CNS disease that resembles MS have abnormal

TABLE 4

Differential diagnosis of neurologic disease in
Sjögren's syndrome.

CNS disease – brain, spinal cord or both
   Multiple sclerosis (MS) and other demyelinating
   diseases
      MS
      Optic neuritis/atrophy
      Transverse myelitis/myelopathy
      Chronic progressive myelopathy
   Chronic inflammatory or autoimmune disorders
      Systemic lupus erythematosus (SLE)
      Sjögren's/lupus overlap syndrome (SS/LE)
      Rheumatoid arthritis (RA)
      Progressive systemic sclerosis (PSS)
      Mixed connective tissue disease (MCTD)
      Overlap syndromes
      MS
   Vasculitides – inflammatory diseases of brain
   vessels
      Isolated CNS vasculitis
      Polyarteritis nodosa (PAN)
      Retinal vasculitis
      Wegener's granulomatosis
      Temporal arteritis/polymyalgia rheumatica
      Aorta branch arteritis (Takayasu's)
      SLE
      Antiphospholipid antibody arteriopathy
   Chronic meningoencephalitis
      Infectious meningitis
         Viral
         Sarcoid
         Syphilis
         Lyme disease
         HIV
      Connective tissue disease
      Recurrent aseptic meningitis (Mollaret's)
   Infectious
      Neurosyphilis
      Lyme disease
      Sarcoid
      HIV
   Malignancy
      Primary cerebral lymphoma
      Carcinomatous meningitis
   Neurodegenerative diseases
      Parkinson's disease
      Alzheimer's disease
      Amyotrophic lateral sclerosis
   Paraneoplastic syndromes
      Cerebellar degeneration syndrome
      Meningoencephalitis
      Encephalopathy
   Dementia (cognitive dysfunction)
      Attention/concentration deficits
      Alzheimer's disease
      Subcortical dementia
      Multi-infarct dementia
      Binswanger's disease
      AIDS
      Acute/subacute encephalopathy
      Parkinson's disease with dementia
   Cerebrovascular diseases
      Cerebral or spinal subarachnoid hemorrhage

   Intracerebral hemorrhage
   Moa Moa/collaterization
   Binswanger's disease
   Stroke
   Cerebral aneurysm
   Vascular malformations
Seizure disorders
   Generalized
      Petit mal (absence)
      Grand mal
   Partial or focal
      Simple
      Complex

Peripheral nervous system
   Cranial nerve/cranial nerve roots
      I   Anosmia, hyposmia, dysosmia
      II   Optic neuropathy
      III, IV, VI   Extraocular muscle and pupillary
                     abnormalities
      V   Trigeminal neuropathy
      VII   Bell's palsy
      VIII   Neurosensory learning loss
                     Vestibular dysfunction
                     Menière's disease
      IX   Glossopharyngeal neuralgia
                     Hypogeusia, dysgeusia (also VII)
      Bulbar palsy
      Multiple cranial palsies (III, V, X, XII)
   Peripheral nerve
      Sensory
      Motor
      Mixed sensory/motor
      Mononeuritis multiplex
      Brachial plexopathy
      Lumbar plexopathy
      Occipital neuropathy
   Autonomic
      Shy–Drager syndrome
   Inflammatory neuropathy
      Acute demyelinating neuropathy
      Chronic demyelinating neuropathy
      Chronic inflammatory demyelinating
         polyradiculoneuropathy (CIDP)
      Paraneoplastic syndromes
   Spinal and cranial ganglia
      Cranial
         Trigeminal
         Gasserian
      Dorsal root ganglia
         Sensory neuronopathy
         Paraneoplastic syndromes

Muscle
   Inflammatory muscle disease
      Myositis
      Polymyositis
      Dermatomyositis
      Myositis in rheumatic disorders
   Polymyalgia rheumatica
   Inclusion body myositis
   Mitochondrial myopathies
   Myasthenia gravis
   Chronic fatigue syndrome
   Fibromyalgia/fibrositis
   Metabolic myopathies
      Hypokalemic periodic paralysis

CSF analyses (mild mononuclear pleocytosis, an elevated IgG index, and one or more clonal bands). In both MS and CNS-SS, there is evidence for activation of the complement pathway in the CSF (Sanders et al. 1987). Soluble C5b-9 (SC5b-9), the neoantigen indicative of activation of the terminal complement pathway, is found in the serum and CSF of patients with active CNS-SS and MS (Sanders et al. 1986).

The available evidence suggests that SS patients with neurologic disease resembling MS have neurologic manifestations secondary to their underlying chronic inflammatory or autoimmune disorder (i.e., SS). Both CNS-SS and MS have perivascular/vascular mononuclear inflammatory infiltrates, around and within small blood vessels which result in abnormalities of subcortical and periventricular white matter. Thus, it is not surprising that the clinical presentations and neurologic manifestations of CNS-SS

TABLE 5

Neurologic manifestations in Sjögren's syndrome with central nervous system disease mimicking multiple sclerosis.

Clinical
  Brain or spinal cord
    Weakness in one or more limbs
      Hemiparesis/hemiplegia
    Sensory abnormalities
      Paresthesia
      Dysesthesia
      Hypesthesia
    Spasticity, hyperreflexia, or both
    Incoordination
    Ataxia, sensory or cerebellar
    Cranial neuronopathy
    Dysarthria
    Cerebellar speech
    Nystagmus
    Internuclear ophthalmoplegia, other gaze palsies
    Cognitive dysfunction
    Affective psychiatric disturbance

  Spinal cord
    Spastic/ataxic paraparesis/quadriparesis
    Acute or subacute transverse myelopathy
    Chronic progressive myelopathy
    Bowel and bladder dysfunction
    Brown-Séquard syndrome
    Truncal band-like sensation
    Lhermitte sign

  Optic tract
    Monocular or bilateral visual loss
    Optic or retrobulbar neuritis

disease in some SS patients may be indistinguishable from those observed in MS. There is, however, no histopathologic evidence that SS patients with MS-like neurologic syndromes have both SS and MS. Conversely, MS patients with SS features are most likely to have SS with MS-like CNS disease, rather than two autoimmune disorders (i.e., SS and MS) (Noseworthy et al. 1989; Sandberg-Wollheim et al. 1992).

IMMUNOGENETICS OF SS

Primary SS is an autoimmune disorder which has well defined associations with certain HLA class II alloantigens coded by immune response genes (Goldstein and Arnett 1987). Almost all patients with primary SS have the supertypic specificity, HLA-DRw52 (Wilson et al. 1984). HLA-DR3, an 'autoimmune alloantigen' in linkage disequilibrium with HLA-DRw52, also is increased in frequency in primary SS (approximately 50%) compared to normal control populations with a modest relative risk of 3.0 (Wilson et al. 1984). In all studies, however, there has been a difference of approximately 40–50% between the frequency of HLA-DRw52 and HLA-DR3. Thus, the supertypic specificity HLA-DRw52 has been more closely linked to primary SS than any specific HLA-DR alloantigens. In other words, HLA-DR3, while increased in frequency in SS compared to normal controls, does not constitute all of the HLA-DR alloantigen associations in SS.

SS patients share common HLA class II DRBI alloantigens. The combined frequency of HLA-DR3, DR5, or DRw6 alloantigens, each identical for five amino acids ($_9$EYSTS$_{13}$) in the first hypervariable region (HVRI) of the DRβI chain and each, in turn, in linkage disequilibrium with the supertypic specificity HLA-DRw52, is increased in both primary SS and each of the presumed lupus variants (PLVs), compared to normal controls (Alexander et al. 1992b). Individuals with either HLA-DR5 or DRw6 alone, in the absence of HLA-DR3, as well as individuals with HLA-DR3 may have either primary SS or one of the PLVs and may synthesize anti-Ro(SS-A) antibodies. In contrast, HLA-DR2 or DQw1, both associated with classical SLE and lupus nephritis, are not associated significantly with primary SS or any of the PLVs. Thus, it appears that primary SS and each of the

TABLE 6

Central nervous system disease in Sjögren's syndrome mimicking multiple sclerosis – comparative features.

Clinical feature	Sjögren's syndrome	Multiple sclerosis
Neurologic presentation		
Age at presentation (years)	9–80	<50
Clinical course		
Relapsing, remitting	+	+
Chronic progressive	+	+
Clinical signs		
Optic nerve	+	+
Optic neuritis	+	+
Brain		
Multifocal sensory or motor deficits	+	+
Truncal or appendicular ataxia	+	+
Cognitive dysfunction	+	+
Psychiatric dysfunction	+	+
Spinal cord		
Transverse myelitis	+	+
Bowel and bladder dysfunction	+	+
Peripheral nervous system		
Peripheral nerves	may accompany CNS	rare
Dorsal root ganglia	may accompany CNS	–
Muscle (myositis, myopathy)	may accompany CNS	–
Neurodiagnostic		
Neuroradiologic – scans and angiography		
MRI brain scan		
Gray matter	uncommon	uncommon
White matter	+	+
Periventricular	+	+
Subcortical	+	+
MRI with contrast (enhanced lesions)	usually –	+
SPECT Technetium ^{99}M HMPAO	+	?
Computed axial tomography	<20%	<20%
Cerebral angiography	<20%	–
Electrophysiologic		
Nerve conduction studies	+	usually –
Electromyography	+	usually –
EEG	+	–
MMER	+	+
PSVERs	+	+
BAERs	+	+
SERs	+	+
Median	+	+
Tibial	+	+
Cerebrospinal fluid analysis		
Mild protein elevation	+	+
CSF IgG/albumin index	+	+
Oligoclonal bands		
1–2	common	uncommon
>3	less common (up to 7)	common
Mild mononuclear pleocytosis	+	+
Monocytoid B cells	+	–
Albumin quotient	–	–
SC5b-9	+	+
Serologic studies		
Erythrocyte sedimentation rate	20%	–
Antinuclear antibodies	+50%	+30%
Rheumatoid factor	+40%	–
Anti-Ro(SS-A)/La(SS-B) antibodies	+40%	10–15%
Immune complexes	+	+
Cryoglobulins	+	–
Immunogenetic studies		
HLA-A16; B8; DR3, DR5(11), or DRw6; DQw2; DRw52	+	–
HLA-A3; DR2, DR4, or DRw6, DQw1	–	+

+ = elevated or abnormal.

PLVs, in contrast to classical SLE, are closely associated immunogenetically with respect to the expression class II DRβI alloantigens.

Molecular genetic typing of HLA-DRB1 and DRB3 genes shows that the predominant disease association in both primary SS and PLVs is with DRB1 genes encoding the DR3, DR5, or DRw6 $_9$EYSTS$_{13}$-containing peptides, but not with DRB3 genes determining the DRw52 specificities (Alexander et al. 1992b). A combination of DQA1/DQB1 genes in linkage disequilibrium with DRB1 genes encoding DRβI $_9$EYSTS$_{13}$ peptides, likewise, are highly associated with disease expression in both primary SS and PLVs (Alexander et al. 1992b). Multiple copies of these DQA1/DQB1 genes are even more highly associated with both primary SS and PLVs (Alexander et al. 1992a). These observations strongly suggest that primary SS and the PLVs are nosologically more closely related to each other than to classical SLE.

These immunogenetic observations have relevance to CNS disease in SS. CNS disease occurs not only in primary SS, but also in SS/LE overlap syndrome patients, and eventually may develop in the mothers of infants with NLE (usually many years after the birth of the NLE infant). The clinical features and expression of CNS disease are similar between these clinical subgroups.

There are no specific HLA alloantigens (Alexander et al., unpublished observation; Hietaharju et al. 1990) or DRB1, DRB3, or DQA1/DQB1 genes associated with CNS disease in SS (Alexander et al., unpublished observations). There does, however, appear to be an HLA class II association with in vitro inducible tumor necrosis factor-α (TNF-α) synthesis (Jacob et al. 1992). CNS-SS patients, who are DR3 or DR4, have high inducible TNF-α levels and more serious CNS disease. Conversely, CNS-SS patients who are DR1, DR2, or DRw6 have low inducible TNF-α levels and a relatively low frequency of CNS disease. High levels of CSF TNF-α have been associated with CNS disease in MS and other neurologic disorders (Sharief et al. 1991). These observations suggest that, in addition to disease susceptibility genes (i.e., DR3, DR5 or DRw6), other class II-associated (i.e., TNFα) genes may influence disease expression in SS (i.e., CNS disease) via selective inducibility of cytokines (i.e., TNF-α synthesis) (see Immunopathogenesis).

NEUROPATHOLOGY

Pathology and immunopathogenesis of CNS disease in SS

During the past several years, the understanding of the immunopathogenesis of neurologic disease occurring in SS has been greatly advanced by several new observations (Table 7). The analysis of brain MRI scans and cerebral angiography demonstrates abnormalities consistent with small cerebral vessel disease. Histopathology confirms the presence of small cerebral blood vessel disease. CNS-SS can be characterized as a mononuclear inflammatory ischemic/hemorrhagic small vessel cerebral vasculopathy, almost always accompanied by pleomorphic mononuclear inflammatory infiltrates within the meninges and choroid plexus (i.e., chronic meningoencephalitis) (G.E. Alexander et al. 1981; Alexander et al. 1983; De la Monte et al. 1983, 1985; Alexander 1986a, 1987b, 1992, 1993a, b; Kohriyama et al. 1990; Bakchine et al. 1991, 1993; Caselli et al. 1991, 1993; Garraty et al. 1993; Giordano et al. 1995; Nagahiro et al. 1996). In some CNS-SS patients, usually but not always, anti-Ro(SS-A) autoantibody positive individuals, there also is frank vasculitis (i.e., angiitis) (G.E. Alexander et al. 1981; Alexander et al. 1982a, d; Alexander 1986a, b, 1992, 1993; Sato et al. 1987; Lahoz Rallo et al. 1990; Ranzenbach et al. 1991, 1992; Giordano et al. 1995; Nagahiro et al. 1996). Although, there is perivascular myelin pallor and disorganization, classical demyelinating lesions (i.e., plaques) characteristic of MS are not observed.

SS patients with isolated chronic encephalopathy, resembling Alzheimer's dementia but resolving following administration of corticosteroids, have the same mononuclear parenchymal vasculopathy with involvement of the leptomeninges (Caselli et al. 1993). Histopathologic features of Alzheimer's disease or other neurologic diseases, however, have not been present on brain autopsy or biopsy specimens from SS patients.

The cerebral inflammatory vasculopathy has several distinctive features. Small cerebral blood vessels are more commonly involved than medium or large caliber cerebral blood vessels. The venous system (postcapillary venules, venules and veins) is involved invariably. Less commonly, the arterial system (usually small arteries or arterioles) also may be involved. There is a striking predilection for inflammatory cells to infiltrate blood vessels within the white matter, particularly in the subcortical/periventricular regions. Gray (cortical and basal ganglia) matter involvement also can occur, but is always accompanied by vessel abnormalities within the white matter. Mononuclear inflammatory infiltrates surround and in some cases directly invade the vessel wall. Inflammatory infiltrates can extend into the surrounding brain parenchyma. These histopathologic features are characteristic of cerebral vasculopathy, rather than frank vasculitis.

The inflammatory infiltrates are composed predominantly of lymphocytes and plasma cells (Table 7). The lymphocytes are pleomorphic in appearance and include small round lymphocytes, larger lymphocytes (i.e., lymphoblastoid cells), and atypical mononuclear cells (with distinctive reniform (clefted) nuclei and sparse pale cytoplasm) (Alexander et al. 1982d, 1986a, b; De la Monte et al. 1983, 1985; Alexander 1987). The morphology of the latter mononuclear cells is identical to that of monocytoid B cells (derived from MALT). Monocytoid B cells infiltrate salivary glands in SS (Schmid et al. 1989; Sheibani et al. 1990; Shin et al. 1991) and occur in non-Hodgkin's B cell lymphoma in SS (Shin et al. 1991). The morphologic spectrum of cerebral inflammatory cells is indistinguishable from the pleomorphism of cells observed in the CSF of CNS-SS patients (Alexander and Alexander 1983; De la Monte et al. 1985; Alexander 1986b).

Preliminary lymphocyte phenotyping studies indicate that the infiltrating lymphocytes are predominantly T cells with some B cells and plasma cells. Monocytes/macrophages, often containing hemosiderin, are commonly present at the site of subarachnoid or intracerebral micro-hemorrhages (see below). These observations suggest that, in CNS-SS, chronic inflammation is associated with microhemorrhages in the meninges and brain parenchyma. In CNS-SS cases, with macro-infarcts on macrohemorrhages associated with major subarachnoid on intracerebral events, hemosiderin-laden macrophages/monocytes are numerous.

Granulocytes or neutrophilic leukocytes are not a characteristic feature of the meningeal/choroidal or parenchymal infiltrates of CNS-SS. In anti-Ro(SS-A) positive individuals with frank necrotizing lymphocytic angiitis, rarely, a classical picture of neutrophilic necrotizing arteritis also may be observed in one or more vessels (Alexander et al. 1982a). Eosinophils, multinucleate giant cells, or granulomata have not been observed within the brains of CNS-SS patients.

The cerebral vascular endothelial cell is clearly a target of pathologic assault in CNS-SS. On the one hand, endothelial cells can be hypertrophied with the morphology of high capillary venules. Hypertrophied endothelial cells can occlude the blood vessel lumen and result in small vessel occlusive disease. On the other hand, endothelial cells may appear damaged, with disorganization of the cytoskeletal structure observed by electron microscopy (Alexander et al., unpublished observations). Blood vessels actually may appear 'denuded' of endothelial cells. Endothelial cells have been observed with only tenuous foot process attachments to the basement membrane and also free in the blood vessel lumen. Abnormal cerebral blood vessels with damaged endothelial cells, blood vessel wall thickening and fibrosis, however, may also be associated with perivascular edema, hemorrhage, or infarction in the absence of striking inflammatory infiltrates (i.e., paucicellular inflammatory infiltrates) (Bakchine et al. 1991). Not uncommonly, the architecture of small cerebral blood vessels is destroyed in the absence of apparent inflammatory cells (bland or non-inflammatory lesions).

The perivascular/vascular inflammatory infiltrates and endothelial cell damage may result in two main pathologic processes in CNS-SS – leaky or occluded blood vessels. In the former case, blood vessels may be leaky to intravascular fluid or erythrocytes. Extravasated fluid results in perivascular edema with myelin pallor and disorganization. Extravasation of erythrocytes results in hemorrhage. In the second case, occlusion of blood vessels results in ischemia or infarcts. In CNS-SS, the meningeal and perivascular/vascular mononuclear inflammatory cerebral infiltrates are often associated with micro- and, less commonly, with macro-ischemic infarcts. Microhemorrhages within the meninges (i.e., acute and chronic subarachnoid hemorrhages) are very common (Alexander and Alexander 1983; De la Monte et al. 1985; Alexander 1987a, 1992, 1993; Giordano et al. 1995), and are less common within the brain parenchyma. Rarely, major subarachnoid or intracerebral hemorrhage occurs, almost always in anti-Ro(SS-A) antibody positive individuals (Alexander and Alexander 1983; Ranzenbach et al. 1991, 1992; Alexander 1992, 1993) (see CNS disease in anti-Ro(SS-A) antibody positive individuals).

Other investigators have confirmed the presence of necrotizing arteritis in anti-Ro(SSA) antibody positive SS patients with catastrophic CNS disease (Giordano et al. 1995). An anti-Ro(SS-A) antibody positive woman who succumbed to a fatal cerebral subarachnoid hemorrhage demonstrated severe necrotizing neutrophilic vasculitis of the basilar and superior cerebellar arteries with thrombosis. Mononuclear (lymphocytic) infiltrates were present within and around blood vessels in the meninges and parenchyma.

These observations strongly suggest that mononuclear inflammatory cells and anti-Ro(SS-A) autoantibodies may participate in the immunopathogenesis of inflammatory vascular damage in CNS-SS (see below). Although, we do not know the immunologic event(s) or mechanisms which initiate and perpetuate cerebral vascular inflammation and damage in CNS-SS, the complement pathway is activated in both seropositive and seronegative CNS-SS (Sanders et al. 1987). Furthermore, class II restricted inducible tumor necrosis factor α (TNF-α) synthesis is associated with CNS-SS (Jacob et al. 1992, see Immunogenetics). Local TNF-α (and other cytokines) synthesis may play a role in mediating vascular permeability, as well as endothelial cell and blood vessel damage in CNS-SS.

Other investigators also have observed histopathological abnormalities consistent with a small vessel cerebral vasculopathy in CNS-SS (Shearn 1971; Pittsley and Talal 1980; Rutan et al. 1986; Ferreiro and Robalino 1987; Ingram et al. 1987; Sato et al. 1987; Caselli et al. 1991). Ingram et al. (1987) have documented a bland diffuse vasculopathy in two patients with SS and devastating CNS disease, each of whom had both anti-Ro(SS-A) and anti-phospholipid antibodies. Thrombotic thrombocytopenic purpura (Steinberg et al. 1971; Noda et al. 1990) has been reported in anti-Ro(SS-A) antibody positive CNS-SS.

Association of CNS-SS with peripheral inflammatory vascular disease

An association of CNS disease with coexistent peripheral (i.e., skin, nerve, muscle) IVD (i.e.,

vasculitis) has been documented (Alexander and Provost 1981, 1991; G.E. Alexander et al. 1981; Alexander et al. 1982a, 1983; Molina et al. 1985a, b, 1986; Alexander 1986b, 1987a, 1992, 1993) (see Alexander 1992 for review). Approximately one-half to two-thirds of SS patients with biopsy documented peripheral IVD develop focal or non-focal CNS disease (Molina et al. 1985a). Peripheral IVD may be either of two histopathologic types: mononuclear (MIVD) (i.e., lymphocytic) or neutrophilic (NIVD) (i.e., leukocytoclastic) (Molina et al. 1985b; Tsokos et al. 1987). Patients with recurrent IVD tend to have the same histopathologic type of IVD recurring over time (Molina et al. 1985b; Alexander 1987).

The two histopathologic types of peripheral IVD have consistent differential serologic associations (Molina et al. 1985b). Patients with NIVD are usually seropositive for multiple autoantibodies (see above), including anti-Ro(SS-A) and, less commonly, anti-La(SS-B) antibodies. Patients with MIVD are seronegative for these autoantibodies. Patients with NIVD also have higher levels of circulating immune complexes and decreased serum complement levels compared to patients with MIVD (Molina et al. 1985b). Both histopathologic types of peripheral IVD have elevated levels of SC5b-9, suggesting that the complement pathway is activated.

Anti-Ro(SS-A) antibodies are also present in increased frequency in SS patients with systemic vasculitis (Alexander et al., unpublished observations; Lahoz Rallo et al. 1990). Both histopathologic types (i.e., NIVD and MIVD) of peripheral or systemic IVD are associated with PNS and CNS disease in SS (Molina et al. 1985a, b; Alexander 1986b, 1992). Of interest, anti-Ro(SS-A) antibody positive patients with peripheral NIVD and CNS disease usually have necrotizing lymphocytic, not neutrophilic, vasculopathy/vasculitis of the CNS. A case has been reported (Giordano et al. 1995), however, in which an anti-Ro(SS-A) antibody positive woman also had necrotizing neutrophilic cerebral arteritis. This observation suggests a selective trafficking of lymphocytes to the CNS in SS.

These observations have led us to recommend that SS patients with peripheral or systemic IVD should be carefully assessed for clinical evidence of concomitant CNS and, if clinically indicated, evaluated by appropriate neurodiagnostic studies. The co-existence of CNS with peripheral and systemic

IVD suggests that concomitant cerebral vascular inflammation might be at least a partial etiology for the CNS complications of SS (see Immunopathogenesis).

Autoantibodies in the immunopathogenesis of CNS-SS

Several observations strongly implicate anti-Ro(SS-A) antibodies in the immunopathogenesis of vascular injury observed in CNS-SS. The presence of anti-Ro(SS-A) antibodies (Alexander et al. 1982a, 1994; Sato et al. 1987; Ranzenbach et al. 1991, 1992; Garraty et al. 1993; Giordano et al. 1995) is highly associated with: (1) serious focal progressive CNS complications; (2) large 'lesions' on brain MRI/CT scans consistent with infarcts or ischemia; (3) abnormal cerebral angiography consistent with small vessel angiitis; (4) histopathology indicative of frank angiitis (in addition to diffuse small vessel mononuclear inflammatory vasculopathy); and (5) the coexistence of necrotizing peripheral and/or systemic IVD in SS patients with active CNS disease.

Several pieces of experimental evidence support the hypothesis that anti-Ro(SS-A) antibodies may be involved in mediating vascular damage in CNS-SS. Vascular endothelial cells appear to express Ro(SS-A)/La(SS-B) antigens. In in vitro studies, we have demonstrated that anti-Ro(SS-A) antibodies preferentially stain the plasma membranes and cytoplasm of proliferating cultured human umbilical vein and bovine retinal endothelial cells (Alexander et al., unpublished observations). Furthermore, anti-Ro(SS-A) antibody positive sera from CNS-SS patients detect 60 kDa and 50 kDa proteins in Western blots of cultured endothelial cells. The molecular weight of these proteins is similar to the 60 kDa (Ben-Chetrit et al. 1988; Deutscher et al. 1988) and 52 kDa Ro(SS-A) (Ben-Chetrit et al. 1988) proteins that have been recently cloned and sequenced. The CNS-SS sera also detect the 60 kDa (Ro(SS-A)) and 52 kDa (La(SS-B)) cloned peptides. These observations strongly suggest that endothelial cells may express Ro(SS-A) or closely related antigens. It is attractive to speculate that anti-Ro(SS-A) antibodies may mediate or potentiate endothelial cell injury and impact on endothelial cell proliferation and vessel regeneration.

Other autoantibodies, which have been implicated in the pathogenesis of CNS disease or vasculitis occurring in other rheumatic disorders, are either absent from the serum or CSF or present in very low frequency within the serum of patients with CNS-SS. Antiribosomal P antibodies, which have been associ-

TABLE 7

Histopathologic observations in central and peripheral nervous system disease in Sjögren's syndrome – a mononuclear inflammatory vascular disease (IVD).

Pathologic feature	Central nervous system	Peripheral nervous system		
	Brain and/or spinal cord	Dorsal root ganglia	Nerve	Muscle
Ischemic/hemorrhagic lesions				
Micro/macro infarcts	+	n.o.	+	n.o.
Micro/macro hemorrhages	+	no	+	+
Mononuclear inflammatory infiltrates		+	+	+
Leptomeninges	+	n.a.	n.a.	n.a.
Choroid plexus	+	n.a.	n.a.	n.a.
Vascular/perivascular	+	n.a.	n.a.	n.a.
Parenchyma	+	n.a.	n.a.	n.a.
Inflammatory vascular disease (IVD)				
Vasculopathy	common	+	+	+
Vasculitis	rare	+	+	+
IVD – histopathologic type				
Mononuclear	+	+	+	+
Neutrophilic	rare	n.o.	uncommon	uncommon

n.a. = not applicable; n.o. = not observed to date.

ated with psychosis (Bonfa et al. 1987) and depression (Schneebaum et al. 1991) in SLE, are absent from primary SS patients and present in only approximately 10% of patients with SS/LE overlap syndrome (Spezialetti et al. 1993). There is no correlation, however, between the presence of anti-ribosomal P antibodies and non-focal CNS disease, including psychosis or depression, occurring in SS (Spezialetti et al. 1993). Likewise, antineuronal antibodies, against cultured neuroblastoma cell lines, associated with diffuse lupus neuropsychiatric manifestations (Bluestein et al. 1981), are absent from sera and CSF of CNS-SS patients (Spezialetti et al. 1993). SS patients with sensory neuronopathy, unlike patients with paraneoplastic syndrome (Horwich et al. 1977), do not have antineuronal antibodies (Alexander and Posner, unpublished observations).

The antiphospholipid antibody syndrome, which occurs in SLE and in isolated cases, may be complicated by major occlusive cerebral vascular disease (Briley et al. 1989), which can be a rare neurologic complication of CNS-SS. Antiphospholipid antibodies are distinctly uncommon in SS, occurring in only 4.7% of primary SS patients and in 13% of patients with secondary SS (Alexander et al., unpublished observations). Antiphospholipid antibodies are not associated with CNS disease in SS. Two SS patients have been reported (Ingram et al. 1987), however, with both antiphospholipid and anti-Ro(SS-A) autoantibodies, each of whom developed multi-organ systemic disease including devastating neurologic complications, presumed to be secondary to multiple arterial thromboses. Both patients had a non-inflammatory vasculopathy of large and small vessels associated with high titers of antiphospholipid antibodies. The concurrence of both high titer anti-Ro(SS-A) and antiphospholipid antibodies in these unique patients may have contributed to the pathophysiology of the diffuse vasculopathy.

IgG antineutrophilic cytoplasmic antibodies have been associated with active Wegener's granulomatosis, systemic vasculitis, and necrotizing crescentic glomerulonephritis (Falk and Jennette 1988). These antibodies have been detected in less than 2% of SS patients (Alexander et al., unpublished observations). Antibodies to GMP-140, a novel endothelial cell granule protein antigen, have been reported in a spectrum of patients with small vessel vasculitis (McCarty et al. 1990). Antibodies to GMP-140 were present in approximately 23% of SS patients, but are not associated significantly with CNS disease or IVD of either histopathologic type (Alexander and McCarty, unpublished observations).

These observations have several important potential ramifications for the immunopathogenesis of CNS-SS. First, CNS-SS is not a potpourri of other rheumatic disorders which potentially can develop CNS disease (i.e., SLE, antiphospholipid syndrome, Wegener's granulomatosis, PAN, etc.). Second, a number of autoantibody systems which have been implicated in the pathogenesis of CNS disease in other disorders are not implicated in CNS-SS (i.e., antiribosomal P, -neuronal, -phospholipid or -cardiolipin (ACA), -neutrophilic cytoplasmic (ANCA) and -GMP140 antibodies).

If autoantibodies do play a role in vascular injury in CNS disease in SS, the available clinical and experimental evidence would suggest that anti-Ro(SS-A) autoantibodies would be the most likely candidates to participate in the immunopathogenesis. Furthermore, these observations suggest that, with respect to the potential role of autoantibodies in the immunopathogenesis of CNS disease, in SS, and other disorders, mechanisms of vascular damage may differ between SS and other types of vasculitis and SLE.

THERAPY IN CNS-SS

Although the most effective therapy for CNS-SS has not been established in controlled trials, there are guidelines for indications and an approach to therapy based on current treatment modalities used in other purported autoimmune or chronic inflammatory neurologic disorders.

The first issue is the appropriate selection of patients for therapy. Other etiologies of CNS disease in SS patients need to be evaluated and rigorously excluded. The following types of SS patients should be considered for therapy: acute or subacute life threatening neurologic disease with serious neurologic deficits, chronic progressive neurologic disease with multiple events over time resulting in cumulative neurologic impairment; subarachnoid or intracerebral hemorrhage, aseptic meningitis, angiographically or biopsy confirmed vasculitis, and acute or subacute encephalopathy or dementia. The indication for treatment of chronic progressive dementia or cognitive impairment is less straightforward.

The second point is the appropriate evaluation and monitoring of patients selected for therapy. CNS-SS should not be treated empirically. SS patients, with either focal or non-focal CNS disease, should be thoroughly evaluated with the following non-invasive neurodiagnostic tests (see section): brain CT or MRI scan, complete CSF analysis, EEG, and MMER. The objective non-invasive neurodiagnostic test(s) which are abnormal should be followed serially to monitor disease activity and response to therapy. Angiography and/or brain/meningeal biopsy may be indicated in certain clinical situations. Treatment of patients on purely clinical grounds, without objective confirmation of neurologic disease, should be undertaken with great caution.

The following observations on the corticosteroid treatment of CNS-SS are based upon the experience of the author who has developed a conservative therapeutic approach. Oral corticosteroids alone (in moderate to high doses) are effective in some, but not all, CNS-SS patients. The complications of corticosteroid therapy in CNS-SS are standard and significant. Pulse intravenous (i.v.) corticosteroids have been employed with variable success on a short-term basis. Based on the immunopathology of CNS-SS, it is highly unlikely that active CNS-SS will be treated effectively and definitively with low-dose or brief courses of oral or pulse corticosteroids. In our experience, azathioprine has not been effective in the treatment of CNS-SS. Other investigators have used methotrexate, cyclosporine, and i.v. γ-globulin with anecdotal success. Plasmapheresis and immunosuppressive therapy have been used with modest success for those rare patients with severe systemic disease (including glomerulonephritis, systemic vasculitis, or CNS disease), most of whom are anti-Ro(SS-A) antibody positive with hyperglobulinemia or cryoglobulinemia.

Patients with progressive focal neurologic dysfunction in whom there are objective parameters of CNS disease activity and progression should receive monthly i.v. pulse cyclophosphamide therapy for at least 12 months until disease stabilization or improvement. This therapy may be repeated every 3 months during the second year. The initial dose is $0.75 \, g/m^2$ with modification to maintain the white blood cell count at approximately 3000 cells/mm^2 at the nadir (7–10 days following therapy). The drug is best used in conjunction with corticosteroids (initially in divided doses) which are then consolidated and tapered to a maintenance schedule over several months. This protocol is patterned after the NIH protocol for the treatment of nephritis in SLE (Felson and Anderson 1984). In our experience the regimen has been well tolerated with no significant side effects. This protocol appears to be effective in stabilizing or improving neurologic disease for prolonged periods of time (up to 10 years). Obviously, a randomized prospective study would need to be performed to critically assess the efficacy of monthly i.v. pulse cyclophosphamide therapy in CNS-SS.

Intravenous pulse cyclophosphamide therapy of CNS-SS was selected for several empirical reasons. Oral cyclophosphamide therapy is effective in the treatment of CNS-SS, but has been associated with the development of bladder cancer. The potential for development of other malignancies or lymphoma in immunocompromised SS patients is significant. Furthermore, sustained daily therapy is associated with an increased risk of bone marrow suppression (often necessitating discontinuation of drug), infections, hemorrhagic cystitis, alopecia and infertility. The NIH protocol for the treatment of nephritis in SLE (Felson and Anderson 1984), using monthly i.v. pulse cyclophosphamide, provides extensive experience with the drug and a low frequency of serious complications.

There are indications, however, for short-term oral or i.v. cyclophosphamide therapy. In CNS-SS patients who are seriously ill, whose neurologic status is deteriorating rapidly, and in whom there is objective documentation of active progressive neurologic disease, daily oral or i.v. cyclophosphamide therapy may be indicated to achieve rapid immunosuppression and to induce remission. After this goal has been obtained and the patient stabilized, monthly i.v. pulse cyclophosphamide therapy can be substituted.

Potentially, CNS-SS may be a treatable and reversible cause of dementia (Alexander 1987b, 1992, 1993; Caselli et al. 1991, 1993; Creange et al. 1992). To date, we have treated a small number of elderly patients with progressive, advanced dementia (and strikingly abnormal brain MRI scans), and the results have been variable. Less severe dementia of shorter duration, occurring in patients with concomitant progressive focal CNS disease, appears to be more responsive to immunosuppressive therapy.

Other investigators have reported the successful treatment of dementia in CNS-SS (Caselli et al. 1991). The preliminary observations indicate that progressive dementia of relatively recent onset in SS may be treated successfully with immunosuppressive therapy and thus may be reversible. Empirically, it would be reasonable to assume that therapeutic intervention would be most likely to be effective relatively early in the course of the disease.

PERIPHERAL NERVOUS SYSTEM DISEASE

Although PNS disease has been recognized for many years as a systemic complication of SS (Kaltreider and Talal 1969; Shearn 1971; Pittsley and Talal 1980; G.E. Alexander et al. 1981; Alexander 1987b), recently there has been a growing awareness of the multiplicity of the clinical presentations of peripheral nerve dysfunction in SS and a growing understanding of the pathophysiology of PNS. Henrik Sjögren, in his original monograph (Sjögren 1935), first described an SS patient with peripheral neuropathy characterized by bilateral facial nerve palsy and transient sensory changes. Subsequently, other investigators have reported a range of PNS manifestations. Peripheral nervous disease in SS has been reviewed in depth previously (Alexander 1987b; Kaplan and Schaumburg 1991).

TABLE 8

Peripheral nervous system disease in Sjögren's syndrome.

Peripheral neuropathies
   Sensory
   Motor
   Mixed sensory/motor
   Motor (mononeuropathy or mononeuritis multiplex)
   Autonomic
   Entrapment syndromes
   Cranial
   Brachial plexopathy
   Lumbar plexopathy
   Occipital

Ganglia
   Spinal: pure sensory neuronopathy
   Cranial
      Gasserian – trigeminal neuropathy
      Hypoglossal

Clinical manifestations

The range of clinical manifestations of PNS disease in SS is secondary to the involvement of various components of the PNS including cranial and spinal nerves and their nerve roots and ganglia. The sensory, motor, and autonomic components of the PNS, alone or in combination, may be affected in SS.

Sensory neuropathy

Among the most common PNS manifestations of primary SS is a distal symmetrical pansensory polyneuropathy which tends to affect the lower, more than upper, extremities (G.E. Alexander et al. 1981; Alexander et al. 1982; Alexander 1987b, 1993; Mellgren et al. 1989; Kaplan et al. 1990; Inoue et al. 1991; Gemignani et al. 1994). The symptoms are predominantly anesthesia, paresthesia, and pain in a stocking-glove distribution. Usually symptoms are relatively mild and non-disabling.

Pain, however, may be more severe, resulting in a distal small fiber painful neuropathy, with features indistinguishable from idiopathic distal small fiber painful sensory neuropathy (ISFPN) (Windebank et al. 1990). Since a subset of idiopathic painful neuropathies have mononuclear inflammatory infiltrates on nerve biopsy, it is possible that a subset of idiopathic painful neuropathies actually have SS, either missed on diagnosis or subclinical. The pain in SS neuropathy may be neuropathic and also mediated by the sympathetic nervous system (Galer et al. 1992). In the less common mixed polyneuropathy, there is usually a sensory predominance characterized by painful paresthesias and dysesthesias.

Pure sensory neuropathy (neuronopathy)

Relatively recently, a pure sensory neuropathy (neuronopathy), due to neuronal damage in the gasserian and spinal ganglia, has been recognized as a PNS complication of SS (Malinow et al. 1986; Alexander 1987b, 1992, 1993; Mellgren et al. 1989; Griffin et al. 1990; Gemignani et al. 1994). There has been a growing recognition of SS cases occurring in populations of patients previously considered to have idiopathic sensory ataxic neuronopathy (ISAN) (Sobue et al. 1993, 1995). Sicca symptoms may be mild or absent at the time of presentation with sensory neuropathy (Font et al. 1990; Griffin et al. 1990). SS is probably the underlying etiology of neuropathy in a substantial number of ISAN patients and should be included in the differential diagnosis.

Sensory neuronopathy in SS is characterized by dysesthesias and paresthesias involving the trunk and extremities, severe global sensory loss, sensory ataxia, and pseudoathetosis. There is loss of position and vibratory sensation, with preservation of pain and temperature. Sensory neuronopathy may be asymmetrical or patchy in distribution. Motor strength is preserved, but reflexes are profoundly depressed.

There is wide variability in the initial clinical presentation, severity, and rapidity of progression of pure sensory neuronopathy in SS. The clinical course ranges from acute onset to subtle presentation with insidious progression. Approximately 40% of patients improve spontaneously. Response to immunosuppressive therapy is variable and unpredictable.

Damage to ganglia neurons has been demonstrated by electrophysiologic and histopathologic studies. In neuronopathy in SS, abnormal blink reflex and cutaneous-induced masseter silent period are associated with normal jaw jerks (Valls-Solé et al. 1990). These findings suggest that the lesion in neuronopathy in SS involves damage to the gasserian ganglia, not to the trigeminal axons, since an axonal lesion would be expected to involve the large axons from muscle spindle receptors.

Histopathologic studies confirm the loss of ganglia neurons. Spinal ganglionitis with dorsal root inflammation and fibrosis, as well as small vessel vasculopathy, have been implicated in immunopathogenesis (Malinow et al. 1986; Mellgren et al. 1989; Griffin et al. 1990).

SS patients with sensory neuronopathy can be either seroreactive for antibodies to Ro(SS-A)/La(SS-R) or seronegative (Font et al. 1990; Griffin et al. 1990; Sobue et al. 1993, 1995; Oobayashi and Miyawaki 1995). In our experience, anti-Hu antibodies are negative (Alexander, unpublished observations; Posner, unpublished observations; Griffin et al. 1990) (see Paraneoplastic syndrome for further discussion). Antibodies to dorsal root ganglia neurons have been described in sensory neuropathy in SS (Satake et al. 1995).

Prior to its recognition as a PNS complication of SS, sensory neuronopathy was most often associated with paraneoplastic syndromes (most commonly, small cell lung and ovarian carcinoma) (Horwich et al. 1977). SS patients with sensory neuronopathy attributed to their underlying chronic inflammatory disorder have not had evidence of carcinoma upon extensive evaluation and extended observation. Carcinomatous sensory neuronopathy differs from pure sensory neuronopathy of SS in its aggressive progressive course, prominence of small fiber loss, association with oculomotor and cerebellar manifestations, and by the presence of specific antinuclear antibodies against neurons (i.e., anti-Hu) (Horwich et al. 1977). Both are characterized by destructive lymphocytic infiltrates of dorsal root ganglia (i.e., dorsal root ganglionitis) with a loss of ganglia neuron cell bodies (Horwich et al. 1977; Malinow et al. 1986).

Autonomic neuropathy

There is growing recognition of autonomic neuropathy as a PNS complication of SS. Autonomic neuropathy can occur alone or in association with sensory neuropathy in SS (Gudesblatt et al. 1985; Low et al. 1988; Mellgren et al. 1989; Font et al. 1990; Griffin et al. 1990; Kaplan and Schaumburg 1991; Galer et al. 1992; Alexander 1993; Kumazawa et al. 1993; Sobue et al. 1993; Denislic and Meh 1994; Gemignani et al. 1994; Oobayashi and Miyawaki 1995). Autonomic neuropathy is characterized by a mixture of sympathetic and parasympathetic dysfunction. The most common problems include pupillary and lid abnormalities (tonic pupils, light – near dissociation (Adie's pupil), and Horner's syndrome), circulatory reflex abnormalities (orthostatic hypotension without compensatory reflex tachycardia, supine hypertension, and resting tachycardia), motility disorders of bowel and bladder, genital organ dysfunction (erectile failure and retrograde ejaculation), and autonomic insufficiency. Esophageal and gastric dysmotility are observed commonly in SS and also may be secondary to autonomic dysfunction. Autonomic insufficiency could contribute to decreased lacrimation, salivation, and sweating abnormalities (i.e., segmental or complete anhidrosis and hyperhidrosis) – clinical features of the sicca syndrome. What relative role autonomic neuropathy plays in the pathogenesis of the sicca complex in SS or modulation of an existing destructive inflammatory response in exocrine glands is unknown.

Autonomic dysfunction is a common accompaniment of ataxic sensory neuronopathy in SS (Mellgren et al. 1989; Font et al. 1990; Griffin et al. 1990; Kumazawa et al. 1993; Sobue et al. 1993; Oobayashi and Miyawaki 1995), suggesting that sympathetic ganglia, as well as cranial and dorsal root ganglia, are involved in the pathologic process. In fact, two anti-Ro(SS-A) antibody positive SS patients with sensory neuronopathy demonstrated anhidrosis segmentally distributed along the dermatomes of the spinal segment and had an absent cholinergic sweat test. These observations imply involvement of post-ganglionic sympathetic ganglion cells in sudomotor and, presumably, vasomotor dysfunction in SS.

Motor neuropathy

Motor involvement of the PNS in SS is less common than sensory involvement. If motor involvement predominates, patients may demonstrate motor dysfunction (e.g., mononeuropathy or mononeuritis multiplex (MM)). MM is an unusual complication of SS even in the presence of necrotizing systemic vasculitis (Massey 1980; Alexander et al. 1982d; Kaplan and Schaumburg 1991; Serradell and Gaya 1993; Gemignani et al. 1994). In several cases, MM has been documented in association with necrotizing angiitis of the sural nerve and the presence of anti-Ro(SS-A) antibodies (Kaltreider and Tatal 1969; Alexander et al. 1983; Kaplan et al. 1990).

Ascending motor polyneuropathies (i.e., acute or chronic relapsing inflammatory polyneuropathy or Guillain–Barré syndrome) have been observed in SS (Alexander et al., unpublished observations; Gross 1987; Barnes et al. 1988). Severe motor impairment accompanied by variable sensory loss, has been documented by nerve conduction studies. Sural nerve biopsy shows dropout of large myelinated fibers (Gross 1987) or prominent remyelination on fiber studies (Marbini et al. 1982).

Entrapment neuropathy

Other common PNS manifestations of SS are 'entrapment syndromes', often in the carpal, ulnar and tarsal distributions. These abnormalities usually occur in the absence of demonstrable arthritis, synovitis, or obvious inflammation. While these neuropathies are presumed to be secondary to nerve entrapment, definitive proof for this etiology is often lacking.

While the presumptive entrapment neuropathies can usually be managed by conservative measures (i.e., splints, physical therapy and salicylates or nonsteroidal anti-inflammatory drugs), on occasion corticosteroids or surgical intervention may be necessary. Brachial plexopathies also have been observed (Alexander et al. 1982d; Alexander 1987b).

CRANIAL NEUROPATHY

Optic neuropathy

Recently, optic nerve (CN I) involvement in SS has been increasingly recognized (Alexander 1986a, b, 1987b, 1992, 1993; Wise and Agudelo 1988; Tesar et al. 1992; Harada et al. 1995). Optic nerve disease in SS may occur alone, but more commonly is associated with multifocal CNS disease involving brain or spinal cord in a neurologic syndrome resembling MS (Alexander et al. 1986b). Clinical presentations include acute retrobulbar optic neuritis, ischemic optic neuropathy, and insidious visual loss with optic atrophy. Subclinical optic neuropathy can occur with characteristic funduscopic findings. Visual evoked response testing often demonstrates decreased amplitude (consistent with ischemia) and, less commonly, prolonged latency (consistent with demyelination), or both.

Trigeminal neuropathy

Cranial nerve involvement is observed in patients with SS and may be peripheral or central (see below). The most well recognized cranial nerve syndrome is trigeminal sensory neuropathy, which may be the sole neurologic abnormality in SS (Kaltreider and Talal 1969; Alexander 1993). The symptoms consist of unilateral or, less commonly, bilateral numbness or paresthesia in the distribution of the maxillary and/or mandibular division of the trigeminal (CN V) nerve. The ophthalmic division is involved less commonly. Pain may be present, but usually is not severe. Most often the corneal reflex is spared. Motor function is normal. Trigeminal neuropathy may be secondary to a localized ganglionitis affecting the gasserian ganglion. Altered taste perception of the tongue may be detected ipsilaterally. This divisional impairment suggests a peripheral, rather than central, process. There are practical clinical implications of the presence of trigeminal neuropathy, which can exacerbate sicca symptoms by worsening lacrimal and taste deficits. The risk of corneal or lingual ulceration is increased. Taste perception is altered due to loss of pain, touch, and temperature sensation in the mucous membranes.

Other cranial nerves

Other cranial nerve deficits may also occur, but are less common. Facial nerve involvement (Bell's palsy) may compromise autonomic secretory function and further impair the existing decrement in tear and saliva production. Neurosensory hearing loss and peripheral vestibular dysfunction have been observed in SS (Alexander and Maddox, unpublished observations; Rutan et al. 1986; McCombe et al. 1992). Other less common cranial neuropathies include the cranial nerves I, III, IV, VI, and IX. Bulbar palsy and recurrent cranial polyneuropathy has been reported (Serradell and Gaya 1993).

Other ocular manifestations

Retinal vasculitis
Another potential manifestation of IVD in SS is retinal vasculitis. Retinal vasculitis, associated with anti-Ro(SS-A) antibodies, has been observed in SS (Farmer et al. 1985; Alexander et al., unpublished observations). Symptomatically mild SS may be associated with severe retinal vasculitis characterized by progressive, irreversible retinal ischemia, optic disk and retinal neovascularization, vitreous hemorrhage, traction retinal detachment, and anterior segment neovascularization (Farmer et al. 1985). Thus, inflammatory occlusion of the retinal arterioles may be a rare cause of visual loss in SS.

Anterior acute and chronic uveitis (iritis)
The most common ophthalmologic feature of SS is keratoconjunctivitis sicca. Chronic uveitis, an inflammatory disorder of the uveal tract of the eye, also has been described as a feature of primary SS (Rosenbaum and Bennett 1987). In addition, severe acute anterior uveitis (iritis) has been described in an anti-Ro(SS-A) antibody positive patient (Bridges and Burns 1992).

Electrophysiology

Nerve conduction velocities and electromyography show abnormalities consistent with peripheral neuropathy in a high proportion of SS cases clinically suspected of having peripheral neuropathy (Mellgren et al. 1989; Inoue et al. 1991; Kaplan and Schaumburg 1991; Serradell and Gaya 1993; Gemignani et al. 1994). Neurogenic changes on EMG are present in approximately 70% of cases (Mellgren et al. 1989). Sensory amplitudes are most commonly abnormal (ulnar > sural > median) followed by motor amplitudes (peroneal > tibial). Conduction velocities, distal latencies, and F-wave latencies are prolonged in a smaller proportion of cases (Mellgren et al. 1989).

In SS patients with predominant sensory neuronopathies, sensory nerve action potential (SNAP) amplitudes may be reduced or absent (Griffin et al. 1990). Motor abnormalities are restricted to evidence of minimal denervation on EMG or mild alteration in amplitude or conduction in individual nerves. Sensory amplitudes may progressively decline. Conduction velocities in affected nerves remain normal, until the amplitude is reduced severely (Griffin et al. 1990).

Pathology

Histopathology of PNS tissue from SS patients with peripheral neuropathy has been examined by a number of investigators (Kaltreider and Talal 1969; Pittsley and Talal 1980; Peyronnard et al. 1982; Malinow et al. 1986; Alexander 1987b, 1993; Mellgren et al. 1989; Griffin et al. 1990; Heylen et al. 1990; Vrethem et al. 1990; Inoue et al. 1991; Alexander et al. 1992; Serradell and Gaya 1993; Gemignani et al. 1994).

Sural nerve biopsies in SS peripheral neuropathy are characterized by axonal degeneration which can be focal or multifocal (Mellgren et al. 1989; Griffin et al. 1990; Heylen et al. 1990; Gemignani et al. 1994). Evidence for axonal degeneration is: relative preservation of conduction velocities, unequivocal severe decrease in density of myelinated fibers, and the marked increase in the proportion of fibers undergoing axonal degeneration. Necrotizing mononuclear (i.e., lymphocytic) vasculopathy/vasculitis of small vessels (i.e., epineural arterioles and venules) of nerves may be involved in fiber degeneration (Mellgren et al. 1989; Griffin et al. 1990; Heylen et al.

1990; Gemignani et al. 1994). There is a decrease in the mean density of myelinated fibers, compared to controls (Mellgren et al. 1989; Griffin et al. 1990; Gemignani et al. 1994). The larger myelinated fibers are preferentially reduced (Mellgren et al. 1989; Griffin et al. 1990; Inoue et al. 1991; Gemignani et al. 1994). Fiber loss may be diffuse. The mean percentage of teased fibers undergoing axonal degeneration is much higher in SS patients than in controls (Mellgren et al. 1989). Some investigators have seen evidence of remyelination and regeneration (Gemignani et al. 1994).

At early stages of disease, there is abundant Wallerian-like degeneration. In latter stages, denervated Schwann cell bands and increased endoneurial collagen are prominent. Another striking feature of peripheral nerve biopsies in SS is the presence of mononuclear (predominantly lymphocytic) inflammatory cells in and around (vascular/perivascular) small epineural blood vessels (i.e., arterioles and venules) (Mellgren et al. 1989; Griffin et al. 1990; Inoue et al. 1991; Serradell and Gaya 1993). In some cases, frank necrotizing vasculitis has been documented (Mellgren et al. 1989). Prominent alterations of the endoneurial microvessels with thickening and reduplication of the basal lamina are reported (Inoue et al. 1991; Gemignani et al. 1994). Endothelial cell and intimal proliferation, recanalization and perivascular hemosiderin in macrophages (Mellgren et al. 1989) are identical to those observed in the nervous system of SS patients with CNS disease and in the skin of SS patients with cutaneous vasculitis/vasculopathy (G.E. Alexander et al. 1981; Alexander et al. 1982; Alexander 1986b, 1987a, b, 1992, 1993; Molina et al. 1985a, b). Mellgren et al. (1989) have postulated that the mononuclear inflammatory vasculopathy/vasculitis of small vessels of nerves may be involved in the pathogenesis of fiber degeneration in SS.

In SS patients with a predominant sensory neuronopathy, there is histopathologic evidence of destruction of the spinal ganglia and posterior nerve roots (Malinow et al. 1986; Griffin et al. 1990). Malinow et al. (1986) reported the first evidence of spinal or dorsal root ganglionitis as the immunopathogenic mechanism for sensory neuropathy in SS. Mononuclear (i.e., predominantly lymphocytic) inflammatory infiltrates are around individual neurons and throughout the ganglia, as well as in and around blood vessels (Malinow et al. 1986; Griffin et al. 1990). There is a varying degree of

neuronal loss and degenerating neurons with vacuolization (Malinow et al. 1986; Griffin et al. 1990). Electron microscopy shows individual neurons whose proximal axons are larger and contain whorls of disorganized, closely spaced neurofilaments (Griffin et al. 1990). Small outer bulbs surrounding such fibers suggest previous local demyelination (Griffin et al. 1990). There is also loss of fibers in the dorsal roots associated with inflammatory infiltrates (Malinow et al. 1986; Griffin et al. 1990).

Most of the inflammatory cells are T lymphocytes with smaller numbers of macrophages. The majority of T lymphocytes appear to be CD8 T positive cytotoxic/suppressor cells (Leu-2a) (Griffin et al. 1990). These cells may be involved in injury and death of ganglion cells in sensory neuronopathies.

Thus, in parallel with the histopathologic observations in the CNS, mononuclear perivascular/vascular inflammatory infiltrates are prominent features of the histopathology of the neuropathy and neuronopathy of SS.

Diagnosis of peripheral nervous system disease

Quantitative sensory testing
Sensitive techniques which evaluate sensory nerve dysfunction may be clinically useful in evaluating PNS disease in SS. Although sensory neuropathy is a common neuropathy in SS patients with PNS, quantitative sensory testing has not yet been utilized routinely in diagnosis and evaluation. Quantitative vibratory, thermal and pain perception thresholds were performed in six women with SS who complained of pain and sensory symptoms and demonstrated, in four, abnormal vibratory thresholds indicative of small fiber dysfunction (Denislic and Meh 1994).

Neuroimaging studies
Spinal cord MRI studies may identify abnormalities in patients with clinical features consistent with spinal cord involvement such as transverse myelopathy or chronic progressive myelopathy. One or more regions of increased signal intensity, consistent with ischemia, infarction or demyelination have been observed.

Spinal cord MRI studies of SS patients with sensory neuronopathy have demonstrated regions of high intensity on T2-weighted images in the posterior column of the spinal cord in both the fasciculi cuneatus and gracilis (Sobue et al. 1995). This neuroanatomic localization in the posterior columns is consistent with the distribution of sensory loss extending to the arms, legs and trunk observed in SS sensory neuronopathy. Thus, spinal cord MRI may provide important information regarding central axon involvement of the sensory ganglion observed in SS-associated neuronopathy.

Nerve biopsies

Sural nerve biopsies can be performed for several clinical indications. Sural nerve biopsies may exclude other potential etiologies of PNS dysfunction in patients with SS. In addition, sural nerve biopsies may provide information about the underlying immunopathologic process in PNS disease in SS. In the case of mild distal sensory peripheral neuropathies, nerve biopsies are not necessary or recommended. Biopsies may be helpful, however, in several clinical settings. In a patient with a neuropathy who has systemic complications suspected to be secondary to vascular inflammation or in patients with CNS involvement, a peripheral nerve biopsy may document the presence of peripheral IVD. A positive biopsy provides indirect, although presumptive, evidence that the systemic complications are occurring in the setting of established vascular inflammation and suggests that it is highly likely that there is systemic vasculitis. This is helpful when it is not possible or safe to obtain tissue from an involved organ. It should be emphasized that because of the potential focal nature of vascular inflammation, a negative nerve biopsy does not exclude the presence of a vasculitic neuropathy or coexistent systemic vascular complications.

MUSCLE DISEASE

Musculoskeletal symptom complex

Non-specific musculoskeletal complaints are very common and may be prominent and incapacitating symptoms in SS. There is often a transient, self-limited symptom complex resembling a 'flu-like' syndrome which can be very disturbing and disabling to patients. This symptom complex is characterized by recurrent episodes of low-grade fever, fatigue, malaise, arthralgias, myalgias, lymphadenopathy and, in some cases, salivary gland en-

largement. The pathogenesis of this musculoskeletal symptom complex in SS is unknown. The high frequency of mononuclear inflammatory cells infiltrating muscle fibers and vessels in muscle tissue obtained from SS patients without clinical evidence of myositis (Bunim 1961; Whaley et al. 1973; Bloch et al. 1992), however, suggests that there may be an inflammatory etiology (see below) for the musculoskeletal symptom complex. This symptom complex resembles the musculoskeletal symptoms characteristic of chronic fatigue syndrome or fibromyalgia/fibrositis (Alexander et al. 1992c).

Inflammatory myopathy

Frank inflammatory myopathy is a feature of SS and there is a wide spectrum of clinical presentations. Proximal myopathies are most common. Distal involvement, however, may be prominent (Bunim 1961) or, rarely almost exclusive. Most commonly, the myopathy of SS is mild and insidious (i.e., 'low grade'). The muscle enzymes may be normal or only mildly elevated. The electromyogram may be only mildly abnormal.

At the other end of the spectrum is a clinical presentation indistinguishable from polymyositis with characteristic serum enzyme elevations and/or electromyographic features (Bunim et al. 1961; Bohan et al. 1977; Alexander et al. 1982b). Dermatomyositis has been observed in SS. We have seen patients who have had childhood dermatomyositis and later developed SS (Alexander et al., unpublished observations). Inclusion body myositis has been reported in SS (Chad et al. 1982).

Inflammatory myopathies are probably under-recognized and therefore under-diagnosed in SS. Limited information is available on the prevalence of inflammatory muscle disease (i.e., myositis) in primary SS. In several series, the prevalence of clinical myositis ranges from 2.5 to 10% (Bunim 1961; Talal 1966; G.E. Alexander et al. 1981; Alexander et al. 1982b, d, 1986; Molina et al. 1986; Alexander 1987b, 1993; Bloch et al. 1992).

The frequency of subclinical histopathologic evidence of muscle inflammation on muscle biopsy, however, is much higher. Bunim (1961) reported focal lymphocytic myositis (and IVD, see below) in 14/19 (74%) SS patients lacking clinical evidence of myopathy (4 moderate to severe and 11 mild inflammatory myositis). In the subsequent studies of Bloch and Bunim (1965), 9 of 23 (39%) patients with primary SS, without symptoms of muscle disease showed focal 'chronic' (mononuclear) myositis. A recent study of muscle biopsies in 15 asymptomatic SS patients showed inflammatory infiltrates in 11 (73%) (Vrethem et al. 1990). We have reported biopsy documented myositis in 5 of 30 (17%) patients with primary SS, only two of whom had clinical features characteristic of myositis (Alexander et al. 1982b). As mentioned above, this high frequency of lymphocytic inflammatory infiltrates within muscle of SS patients, without clinical or laboratory evidence of a frank myopathy, may be related to their prominent musculoskeletal symptoms (see above).

Hypokalemic periodic paralysis

A myopathy associated with hyperchloremic, hypokalemic, metabolic acidosis secondary to distal-type II renal tubular dysfunction has been reported in SS. This disorder may present as hypokalemic periodic paralysis (Shioji et al. 1970; Raskin et al. 1981; Christensen 1985; Pun and Wang 1989; Poux et al. 1992; Yoshiiwa et al. 1992; Nakhoul et al. 1993; Chang et al. 1995).

Myasthenia gravis

Myasthenia gravis, an autoimmune muscle disease, has been reported in a woman with seropositive RA and SS who received penicillamine (Van Offel et al. 1987). Mild or subclinical evidence of SS may occur in patients with classical myasthenia gravis (Lindahl et al. 1986).

Immunopathogenesis of muscle disease in SS

Muscle biopsy in SS myositis shows diffuse or focal mononuclear inflammatory infiltrates which are predominantly lymphocytic, with a variable number of macrophages and plasma cells (Bunim 1961; Alexander et al. 1982b; Ringel et al. 1982; Finol et al. 1989; Vrethem et al. 1990; Bloch et al. 1992; Kraus et al. 1994). Muscle fiber necrosis and degeneration, if present, are observed most commonly in and adjacent to areas of inflammation. There is variation in fiber size with internal nuclei. Immunopathologic mechanisms of muscle damage appear to be related

to mononuclear (i.e., lymphocytic) infiltrates within muscle, but precise information about how the inflammatory infiltrates cause muscle damage and dysfunction is not known. The phenotype of lymphocytes infiltrating muscle has not been established. It is possible that CD8+ cytotoxic T lymphocytes or natural killer cells injure muscle cells by direct cellular cytotoxicity or antibody-dependent cellular cytotoxicity (ADCC).

A striking feature of myositis in SS is the accompanying inflammatory perivascular/vascular infiltrates (Alexander et al. 1982b; Ringel et al. 1982; Alexander 1987a, b, 1992; Vrethem et al. 1990; Finol et al. 1994; Kraus et al. 1994). Lymphocytes and plasma cells surround and invade small vessels and capillaries. The presence of IVD within and around blood vessels in muscle, either alone or in association with myositis, is well established in SS (Table 1). This topic is discussed in reviews on IVD in SS (Alexander 1986b; 1987a) and by others (Finol et al. 1989; Kraus et al. 1994). In some cases, frank destructive (i.e., necrotizing) infiltration of vessel walls by mononuclear cells has been observed.

Ultrastructural studies of muscle biopsies (Finol et al. 1989) show varying degrees of muscle atrophy, loss of sarcolemmal ultrastructure, and electron-dense deposits. Capillaries are proliferative with distortion or occlusion of lumens and thickened and convoluted basement membranes (Alexander et al. 1982b; Alexander 1987a; Finol et al. 1989). Immunocytochemistry of affected tissue shows deposition of immunoglobulin (IgG and IgM) and complement (C3) within vessel walls (Alexander et al. 1982b; Ringel et al. 1982), suggesting the deposition of circulating immune complexes and activation of the complement pathway. Small vessel injury could also be perpetrated by deposition of autoantibodies or circulating immune complexes and activation of the complement pathway. Myositis in SS has been accompanied by peripheral IVD of skin (Kraus et al. 1994) or nerve (Vrethem et al. 1990) or systemic vasculitis suggesting a more generalized inflammation of blood vessels (Alexander 1987a). Thus, inflammatory muscle disease, as well as central and PNS disease in SS, is accompanied by prominent mononuclear (i.e., lymphocytic) perivascular/vascular inflammatory infiltrates (i.e., mononuclear inflammatory vasculopathy).

REFERENCES

ALEXANDER, E.L.: Central nervous system (CNS) manifestations of primary Sjögren's syndrome: an overview. Scand. J. Rheumatol. 61 (1986a) 161–165.

ALEXANDER, E.L: Immunopathologic mechanisms of inflammatory vascular disease in primary Sjögren syndrome. Scand. J. Rheumatol. 61 (1986b) 280–285.

ALEXANDER, E.L: Inflammatory vascular disease in Sjögren syndrome. In: N. Talal, H.M. Moutsopoulos and S.S. Kassan (Eds.), Sjögren's Syndrome: Clinical and Immunological Aspects. Heidelberg, Springer-Verlag (1987a) 102–.

ALEXANDER, E.L.: Neuromuscular complications of primary Sjögren's syndrome. In: N. Talal, H.M. Moutsopoulos and S.S. Kassan (Eds.), Sjögren's Syndrome: Clinical and Immunological Aspects. Heidelberg, Springer-Verlag (1987b) 61–.

ALEXANDER, E.L.: CNS disease in Sjögren's syndrome (CNS-SS) – fact or fantasy? Rheumatol. Now 7 (1991) 7–11.

ALEXANDER, E.L.: Central nervous system disease in Sjögren's syndrome. New insights into immunopathogenesis. Rheum. Dis. Clin. N. Am. 18 (1992) 637–672.

ALEXANDER, E.L.: Neurologic disease in Sjögren's syndrome: mononuclear inflammatory vasculopathy affecting central/peripheral nervous system and muscle. Rheum. Dis. Clin. N. Am. 19 (1993a) 869–908.

ALEXANDER, E.L.: Skin manifestations of Sjögren's syndrome. In: T.B. Fitzpatrick, A.Z. Eisen, K. Wolf et al. (Eds.), Dermatology in General Medicine, 4th Edit. New York, McGraw-Hill (1993b) 2211–2221.

ALEXANDER, E.L. and G.E. ALEXANDER: Aseptic meningoencephalitis in primary Sjögren's syndrome. Neurology 33 (1983) 593–598.

ALEXANDER, E.L. and T.T. PROVOST: Ro(SS-A) and La(SS-B) antibodies. Semin. Immunopathol. 4 (1981) 253–273.

ALEXANDER, E.L. and T.T. PROVOST: Sjögren's syndrome: association of cutaneous vasculitis with nervous system disease. Arch. Dermatol. 123 (1987) 801–810.

ALEXANDER, E.L. and T.T. PROVOST: Cutaneous manifestations of Sjögren's syndrome, In: R.E. Jordon (Ed.), Immunologic Diseases of the Skin. East Norwalk, CT, Appleton (1991) 401–.

ALEXANDER, E.L., C. CRAFT, C. DORSCH, R.L. MOSER, T.T. PROVOST and G.E. ALEXANDER: Necrotizing arteritis and spinal subarachnoid hemorrhage in Sjögren syndrome. Ann. Neurol. 11 (1982a) 632–635.

ALEXANDER, E.L., L. JOSIFEK, T.T. PROVOST and G.E. ALEXANDER: Myositis/vasculitis in primary Sjögren's syndrome. Arthr. Rheum. 25, Suppl. 4 (1982b) S15.

ALEXANDER, E.L., T.J. HIRSCH, F.C. ARNETT, T.T. PROVOST and M.B. STEVENS: Ro(SS-A) and La (SS-B) antibodies in the clinical spectrum of Sjögren's syndrome. J. Rheumatol. 9 (1982c) 239–246.

ALEXANDER, E.L., T.T. PROVOST, M.B. STEVENS and G.E. ALEXANDER: Neurologic complications of primary Sjögren's syndrome. Medicine (Baltimore) 61 (1982d) 247–257.

ALEXANDER, E.L., F.C. ARNETT, T.T. PROVOST and M.B. STEVENS: Sjögren's syndrome: association of anti-Ro(SS-A) antibodies with vasculitis hematologic abnormalities, and serologic hyperreactivity. Ann. Intern. Med. 98 (1983) 155–159.

ALEXANDER, E.L., J.E. LIJEWSKI, M.S. JERDAN and G.E. ALEXANDER: Evidence for an immunopathogenetic basis for central nervous system disease in primary Sjögren's syndrome. Arthr. Rheum. 29 (1986a) 1223–1231.

ALEXANDER, E.L., K. MALINOW and J.E. LIJEWSKI: Primary Sjögren's syndrome with central nervous system dysfunction mimicking multiple sclerosis. Ann. Intern. Med. 104 (1986b) 323–330.

ALEXANDER, E.L., T.T. PROVOST and W.B. BIAS: Unique immunogenetic associations distinguish Sjögren's syndrome central nervous system (CNS-SS) from multiple sclerosis (MS). Arthr. Rheum. 29 (1986c) S63.

ALEXANDER, E.L., S.S. BEALL, B. GORDON, O.A SELNES, G.D. YANNAKAKIS, N. PATRONAS, T.T. PROVOST and H.F. MCFARLAND: Magnetic resonance imaging of cerebral lesions in patients with the Sjögren's syndrome. Ann. Intern. Med. 108 (1988) 815–823.

ALEXANDER, E.L., J. MCNICHOLL, R.M. WATSON, W. BIAS, M. REICHLIN and T.T. PROVOST: The immunogenetic relationship between Ro(SS-A)/La(SS-B) positive Sjögren's/lupus erythematosus overlap syndrome and the neonatal lupus syndrome. J. Invest Dermatol. 93 (1989) 751–756.

ALEXANDER, E.L., A. ANSARI and J.R. PLITT: DQAI/DQBI trans allele heterozygosity and multiple gene effect promote anti-Ro(SS-A)/La(SS-B) antibody response (ARo/LaAb-R) in primary Sjögren's syndrome (SS) and the presumed lupus variants (PLV). Arthr. Rheum. 35 (1992a) S14.

ALEXANDER, E.L., Z. FRONEK, J.R. PLITT, S. HSU, H. ERLICH, T. BUGAWAN, L. STEINMAN and A. ANSARI: HLA-DR/DR5/DRw6, not DDRB3/DQA1/DQB1, are disease susceptibility genes in both primary Sjögren's syndrome (SS) and the 'presumed lupus variants' (PLV). FASEB J. 6 (1992b) A1446.

ALEXANDER, E.L., A.J. KUMAR and W.E. KOZACHUK: The chronic fatigue syndrome controversy. Ann. Intern. Med. 117 (1992c) 343–344.

ALEXANDER, E.L., M.R. RANZENBACH, A.J. KUMAR, W.E.

KOZACHUK, A.E. ROSENBAUM, N. PATRONAS, J.B. HARLEY and M. REICHLIN: Anti-Ro(SS-A) antibodies in central nervous system disease associated with Sjögren's syndrome (CNS-SS): clinical, neuroimaging, and angiographic correlates. Neurology 44 (1994) 899–908.

ALEXANDER, G.E., T.T. PROVOST. M.B. STEVENS and E.L. ALEXANDER: Sjögren's syndrome: central nervous system manifestations. Neurology 31 (1981) 1391–1396.

ANAYA, J.M., N. OGAWA and N. TALAL: Sjögren's syndrome in childhood. J. Rheumatol. 22 (1995) 1152–1158.

ANDONOPOULOS, A.P., G. LAGOS, A.A DROSOS and H.M. MOUTSOPOULOS: The spectrum of neurological involvement in Sjögren's syndrome. Br. J. Rheumatol. 29 (1990) 21–23.

BAKCHINE, S., C. DUYCKAERTS, L. HASSINE, M.P. CHAUNU, E. TURELL, B. WECHSLER and F. CHAIN: Central and peripheral nervous system lesions in primary Sjögren's syndrome. Clinico-pathological study of one case. Rev. Neurol. 147 (1991) 368–375.

BANSAL, S.K., I.M. SAWHNEY and J.S. CHOPRA: Epilepsia partialis continua in Sjögren's syndrome. Epilepsia 28 (1987) 362–363.

BARNES, D., S.R. HAMMANS and N.J. LEGG: Chronic relapsing inflammatory polyneuropathy complicating sicca syndrome. J. Neurol. Neurosurg. Psychiatry 51 (1988) 159–160.

BELIN, C., C. MORONI, N. CAILLAT-VIGNERON, J.L. DUMAS, M. BAUDIN and L. GUILLEVIN: Neuropsychological abnormalities in Sjögren syndrome. Neurology 44 (1994) A276.

BEN-CHETRIT, E., E.K. CHAN, K.F. SULLIVAN and E.M. TAN: A 52 kDa protein is a novel component of the SS-A/Ro antigenic particle. J. Exp. Med. 167 (1988) 1560–1571.

BEN-CHETRIT, E., B.J. GANDY, E.M. TAN and K.F. SULLIVAN: Isolation and characterization of a cDNA clone encoding the 60 kDa component of the human SS-A/Ro ribonucleoprotein autoantigen. J. Clin. Invest. 83 (1989) 1284–1292.

BERMAN, J.L., S. KASHII, M.S. TRACHTMAN and R.M. BURDE: Optic neuropathy and central nervous system disease secondary to Sjögren's syndrome in a child. Ophthalmology 97 (1990) 1606–1609.

BINDER, A., M.L. SNAITH and D. ISENBERG: Sjögren's syndrome: a study of its neurological complications. Br. J. Rheumatol. 27 (1988) 275–280.

BLOCH, K.J., W.W. BUCHANAN, M.J. WOHL and J.J. BUNIM: Sjögren's syndrome: a clinical, pathological, and serological study of 62 cases. Medicine (Baltimore) 71 (1992) 386–401.

BLUESTEIN, H.G., G.W. WILLIAMS and A.D. STEINBERG: Cerebrospinal fluid antibodies to neuronal cells: association with neuropsychiatric manifestations of systemic lupus erythematosus. Am. J. Med. 70 (1981) 240–246.

BOHAN, A., J.B. PETER, R.L. BOWMAN and C.M. PEARSON: A computer assisted analysis of 153 patients with polymyositis and dermatomyositis. Medicine (Baltimore) 56 (1977) 255–286.

BONFA, E., S.J. GOLOMBEK, L.D. KAUFMAN, S. SKELLY, H. WEISSBACH, N. BROT and K.B. ELKON: Association between lupus psychosis and anti-ribosomal P protein antibodies. N. Engl. J. Med. 317 (1987) 265–271.

BRANDT, K.D., S. LESSELL and A.S. COHEN: Cerebral disorders of vision in systemic lupus erythematosus. Ann. Intern. Med. 83 (1975) 163–169.

BRAGONI, M., V. DI PIERO, R. PRIORI, G. VALESINI and G.L. LENZI: Sjögren's syndrome presenting as ischemic stroke. Stroke 25 (1994) 2276–2279.

BRIDGES, A.J. and R.P. BURNS: Acute iritis associated with primary Sjögren's syndrome and high-titer anti-SS-A/Ro and anti-SS-B/La antibodies; treatment with combination immunosuppressive therapy. Arthr. Rheum. 35 (1992) 560–563.

BRILEY, D.P., B.M. COULL and S.H. GOODNIGHT: Neurological disease associated with antiphospholipid antibodies. Ann. Neurol. 25 (1989) 221–227.

BUNIM, J.J.: A broader spectrum of Sjögren's syndrome and its pathogenetic implications. Ann. Rheum. Dis. 20 (1961) 1–.

CASELLI, R.J., B.W. SCHEITHAUER, C.A. BOWLES, M.R. TRENERRY, F.B. MEYER, J.S. SMIGIELSKI and M. RODRIGUEZ: The treatable dementia of Sjögren's syndrome. Ann. Neurol. 30 (1991) 98–101.

CASELLI, R.J., B.W. SCHEITHAUER, J.D. O'DUFFY, G.C. PETERSON, B.F. WESTMORELAND and P.A. DAVENPORT: Chronic inflammatory meningoencephalitis should not be mistaken for Alzheimer's disease. Mayo Clin. Proc. 68 (1993) 846–853.

CASE RECORDS OF THE MASSACHUSETTS GENERAL HOSPITAL: Weekly clinicopathological exercises. Case 2-1985. A 57-year-old woman with fever, pain in the legs, anemia, and hypergammaglobulinemia. N. Engl. J. Med. 312 (1985) 103–112.

CHAD, D., P. GOOD, L. ADELMAN, W.G. BRADLEY and J. MILLS: Inclusion body myositis associated with Sjögren's syndrome. Arch. Neurol. 39 (1982) 186–188.

CHANG, Y.C., C.C. HUANG, Y.Y. CHIOU and C.Y. YU: Renal tubular acidosis complicated with hypokalemic periodic paralysis. Pediatr. Neurol. 13 (1995) 52–54.

CHRISTENSEN, K.S.: Hypokalemic paralysis in Sjögren's syndrome secondary to renal tubular acidosis. Scand. J. Rheumatol. 14 (1985) 58–60.

CREANGE, A., D. LAPLANE, K. HABIB, N. ATTAL and V. ASSUERUS: Dementia disclosing primary Gougerot–Sjögren's syndrome. Rev. Neurol. 148 (1992) 376–380.

DE BACKER, H. and I. DEHAENE: Central nervous system disease in primary Sjögren's syndrome. Acta Neurol. Belg. 95 (1995) 142–146.

DE LA MONTE, S.M., G.M. HUTCHINS and P.K. GUPTA: Polymorphous meningitis with atypical mononuclear cells in Sjögren's syndrome. Ann. Neurol. 14 (1983) 455–461.

DE LA MONTE, S.M., P.K. GUPTA and G.M. HUTCHINS: Polymorphous exudates and atypical mononuclear cells in the cerebrospinal fluid of patients with Sjögren's syndrome. Acta Cytol. 29 (1985) 634–637.

DENISLIC, M. and D. MEH: Neurophysiological assessment of peripheral neuropathy in primary Sjögren's syndrome. Clin. Invest. 72 (1994) 822–829.

DEUTSCHER, S.L., J.B. HARLEY and J.D. KEENE: Molecular analysis of the 60 kDa human Ro ribonucleoprotein. Proc. Natl. Acad. Sci. USA 85 (1988) 9479–9483.

DROSOS, A.A., A.P. ANDONOPOULOS, J.S. COSTOPOULOS, C.S. PAPADIMITRIOU and H.M. MOUTSOPOULOS: Prevalence of primary Sjögren's syndrome in an elderly population. Br. J. Rheumatol. 27 (1988) 123–127.

DROSOS, A.A., A.P. ANDONOPOULOS, G. LAGOS, N.V. ANGELOPOULOS and H.M. MOUTSOPOULOS: Neuropsychiatric abnormalities in primary Sjögren's syndrome. Clin. Exp. Rheumatol. 7 (1989a) 207–209.

DROSOS, A.A., N.V. ANGELOPOULOS, A. LIAKOS and H.M. MOUTSOPOULOS: Personality structure disturbances and psychiatric manifestations in primary Sjögren's syndrome. J. Autoimmun. 2 (1989b) 489–493.

ESCUDERO, D., A. OLIVE and P. LATORRE: Prevalence of Sjögren's syndrome in a neurology impatient population. Am. J. Med. 92 (1992) 341–.

ESCUDERO, D., P. LATORRE, M. CODINA, J. COLL-CANTI and J. COLL: Central nervous system disease in Sjögren's syndrome. Ann. Méd. Interne (Paris) 146 (1995) 239–242.

FABRE, N., J.M. FAUCHEUX, Y. ROLAND, A. CANTAGREL and G. GERAUD: Corticosteroid-responsive Parkinsonism associated with primary Sjögren's syndrome. Report of two cases. Neurology 46 (1996) A299–300.

FALK, R.J. and J.C. JENNETTE: Anti-neutrophil cytoplasmic autoantibodies with specificity for myeloperoxidase in patients with systemic vasculitis and idiopathic necrotizing and crescentric glomerulonephritis. N. Engl. J. Med. 318 (1988) 1651–1657.

FARMER, S.G., J.L. KINYOUN, J.L. NELSON and M.H. WENER: Retinal vasculitis associated with autoantibodies to Sjögren's syndrome A antigen. Am. J. Ophthalmol. 100 (1985) 814–821.

FELSON, D.T. and J. ANDERSON: Evidence for the superiority of immunosuppressive drugs and prednisone alone in lupus nephritis. N. Engl. J. Med. 311 (1984) 1528–1533.

FERREIRO, J.E. and B.D. ROBALINO: Primary Sjögren's syndrome with diffuse cerebral vasculitis and lymphocytic interstitial pneumonitis. Am. J. Med. 82 (1987) 1227–1232.

FONT, J., J. VALLS, R. CERVERA, A. POU, M. INGELMO and F. GRAUS: Pure sensory neuropathy in patients with primary Sjögren's syndrome: clinical, immunological,

and electromyographic findings. Ann. Rheum. Dis. 49 (1990) 775–778.

FOX, R.I., F.V. HOWELL, R.C. BONE and P. MICHELSON: Primary Sjögren's syndrome: clinical and immunopathologic features. Semin. Arthr. Rheum. 14 (1984) 77–105.

FOX, R.I., C. ROBINSON, J. CURD, P. MICHELSON R.C. BONE and F.V. HOWELL: First international symposium on Sjögren's syndrome: suggested criteria for classification. Scand. J. Rheumatol. 61, Suppl. (1986a) 28–30.

FOX, R.I., C.A. ROBINSON, J.G. CURD, F. KOZIN and F.V. HOWELL: Sjögren's syndrome: proposed criteria for classification. Arthr. Rheum. 29 (1986b) 577–585.

FULFORD, K.W., R.D. CATTERAL, J.J.DELHANTY, D. DONIACH and M. KREMER: A collagen disorder of the nervous system presenting as multiple sclerosis. Brain 95 (1972) 373–386.

GAITHER, K.K., O.F. FOX, H. YAMAGATA, M.J. MAMULA, M. REICHLIN and J.B. HARLEY: Implications of anti-Ro/Sjögren syndrome A antigen autoantibody in normal sera for autoimmunity. J. Clin. Invest. 79 (1987) 841–846.

GALER, B.S., M.C. ROWBOTHAM, K.V. MILLER, A. WALTON and H.L. FIELDS: Treatment of inflammatory, neuropathic and sympathetically maintained pain in a patient with Sjögren's syndrome. Pain 50 (1992) 205–208.

GARRATY, R.P., P.A. MCKELVIE and E. BYRNE: Aseptic meningoencephalitis in primary Sjögren's syndrome. Acta Neurol. Scand. 88 (1993) 309–311.

GEMIGNANI, F., A. MARBINI, G. PAVESI, S. DIVITTORIO, P. MANGANELLI, G. CENACCHI and D. MANCIA: Peripheral neuropathy associated with primary Sjögren's syndrome. J. Neurol. Neurosurg. Psychiatry 57 (1994) 983–986.

GIORDANO, M.J., D. COMMINS and D.L. SILBERGELD: Sjögren's cerebritis complicated by subarachnoid hemorrhage and bilateral superior cerebellar artery occlusion: case report. Surg. Neurol. 43 (1995) 48–51.

GOLDSTEIN, R. and F.C. ARNETT: The genetics of rheumatic disease in man. Immunology of the rheumatic diseases. Rheum. Dis. Clin. N. Am. 13 (1987) 487–510.

GRAUS, F., K.B. ELKON, C. CORDON CARDO and J.B. POSNER: Sensory neuronopathy and small-cell lung cancer: antineuronal antibody that also reacts with the tumor. Am. J. Med. 80 (1986) 45–52.

GRAUS, F., A. POU, KANTERIWICZ and N.E. ANDERSON: Sensory neuropathy and Sjögren's syndrome: clinical and immunologic study of two patients. Neurology 38 (1988) 1637–1639.

GREENSPAN, J.S., T.E. DANIELS, N. TALAL and R.A. SYLVESTER: The histopathology of Sjögren's syndrome in labial salivary gland biopsies. Oral Surg. Oral Med. Oral Pathol. 37 (1974) 217–229.

GRIFFIN, J.W., D.R. CORNBLATH, E. ALEXANDER, J. CAMPBELL,

P.A. LOW and S. BIRD: Ataxic sensory neuropathy and dorsal root ganglionitis associated with Sjögren's syndrome. Ann. Neurol. 27 (1990) 304–315.

GROSS, M.: Chronic relapsing inflammatory polyneuropathy complicating sicca syndrome. J. Neurol. Neurosurg. Psychiatry 50 (1987) 939–940.

GUDESBLATT, M., A.D. GOODMAN, A.E. RUBENSTEIN, A.N. BENDER and H.S. CHOI: Autonomic neuropathy associated with autoimmune disease. Neurology 35 (1985) 261–264.

HARADA, T., T. OHASHI, R. MIYAGISHI, H. FUKUDA, K. YOSHIDA, Y. TAGAWA and H. MATSUDA: Optic neuropathy and acute transverse myelopathy in primary Sjögren's syndrome. Jpn. J. Ophthalmol. 39 (1995) 162–165.

HARLEY, J.B., E.L. ALEXANDER, W.B. BIAS, O.F. FOX, T.T. PROVOST, M. REICHLIN, H. YAMAGATA and F.C. ARNETT: Anti-Ro(SS-A) and anti-La(SS-B) in patients with Sjögren's syndrome. Arthr. Rheum. 29 (1986) 196–206.

HARLEY, J.B., R.H. SCOFIELD and M. REICHLIN: Anti-Ro in Sjögren's syndrome and systemic lupus erythematosus. In: R. Fox (Ed.), Rheumatic Disease Clinics of North America. Philadelphia, PA, W.B. Saunders (1992) 337–358.

HASHII, M., K. KOMAI, S. MATSUBARA, Y. IDE and M. TAKAMORI: Spastic paraplegia in Sjögren's syndrome associated with mixed connective tissue disease (MCTD) – a case report. Rinsho Shinkeigaku 29 (1989) 1052–1054.

HEYLEN, A., J.P. DEVOGELAER, H. NOEL and C. NAGANT DE DEUXCHAISNES: Axonal polyneuropathy without vasculitis, follicular lymphoma and primary sicca syndrome. A rare association. Clin. Rheumatol. 9 (1990) 84–87.

HIETAHARJU, A., U. YLI-KERTTULA, V. HAKKINEN and H. FREY: Nervous system manifestations in Sjögren's syndrome. Acta Neurol. Scand. 81 (1990) 144–152.

HIETAHARJU, A., M. KORPELA, J. ILONEN and H. FREY: Nervous system disease, immunological features, and HLA phenotype in Sjögren's syndrome. Ann. Rheum. Dis. 51 (1992) 506–509.

HIETAHARJU, A., S. JAASKELAINEN, M. HIETARINTA and H. FREY: Central nervous system involvement and psychiatric manifestations in systemic sclerosis (scleroderma): clinical and neurophysiological evaluation. Acta Neurol. Scand. 87 (1993a) 382–387.

HIETAHARJU, A., V. JÄNTTI, M. KORPELA and H. FREY: Nervous system involvement in systemic lupus erythematosus, Sjögren's syndrome and scleroderma. Acta Neurol. Scand. 88 (1993b) 299–308.

HOLMS, S.: Keratoconjunctivitis sicca and the sicca syndrome. Acta Ophthalmol. 33, Suppl. (1949) 13–.

HOMMA, M., T. TOJO and M. AKIZUKI: Criteria for Sjögren's syndrome in Japan. Scand. J. Rheumatol. 61, Suppl. (1986) 26–27.

HORWICH, M.S., L. CHO, R.S. PORRO ET AL.: Subacute sensory neuropathy: a remote effect of carcinoma. Ann. Neurol. 2 (1977) 7–19.

INGRAM, S.B., S.H. GOODNIGHT, JR. and R.M. BENNETT: An unusual syndrome of a devastating noninflammatory vasculopathy associated with anticardiolipin antibodies: report of two cases. Arthr. Rheum. 30 (1987) 1167–1172.

INOUE, A., C. KOH, N. TSUKADA and N. YANAGISAWA: Peripheral neuropathy associated with Sjögren's syndrome: pathologic and immunologic study of two patients. Jpn. J. Med. 30 (1991) 452–457.

ISHII, A., E. OGUNI, T. YOSHIZAWA, H. MIZUSAWA and R. MURAKI: A case of progressive systemic sclerosis with Sjögren's syndrome presenting with coma, convulsion and bilateral thalamic hypodensity on computed tomography. Clin. Neurol. 33 (1993) 966–970.

JACOB, C.O., J.R. PLITT and Z. FRONEK: HLA-class II associated tumor necrosis factor α (TNF-α) synthesis in central nervous system disease in Sjögren's syndrome (CNS-SS). Arthr. Rheum. 35 (1992) S121.

JACOBSSON, L.T., T.E. AXELL, B.U. HANSEN, V.J. HENRICSSON, A. LARSSON and K. LIEBERKIND: Dry eyes or mouth – an epidemiological study in Swedish adults, with special reference to primary Sjögren's syndrome. J. Autoimmun. 2 (1989) 521–527.

JACOBSSON, L., B.U. HANSEN, R. MANTHORPE, K. HARDGRAVE, B. NEAS and J.B. HARLEY: Association of dry eyes and dry mouth with anti-Ro/SS-A and anti-La/SS-B autoantibodies in normal adults. Arthr. Rheum. 35 (1992) 1492–1501.

KALTREIDER, H.B. and N. TALAL: The neuropathy of Sjögren's syndrome. Trigeminal nerve involvement. Ann. Intern. Med. 70 (1969) 751–762.

KAPLAN, J.G. and H.H. SCHAUMBURG: Predominantly unilateral sensory neuronopathy in Sjögren's syndrome. Neurology 41 (1991) 948–949.

KAPLAN, J.G., R. ROSENBERG, E. REINITZ ET AL.: Invited review: peripheral neuropathy in Sjögren's syndrome. Muscle Nerve 13 (1990) 570–579.

KASSAN, S.S., T.L. THOMAS and H.M. MOUTSOPOULOS: Increased risk of lymphoma in sicca syndrome. Ann. Intern. Med. 89 (1978) 888–892.

KATTAH, J., T. CUPPS, G. DI CHIRO and H.J. MANZ: An unusual case of central nervous system vasculitis. J. Neurol. 234 (1987) 344–347.

KAWASHIMA, N., R. SHINDO and M. KOHNO: Primary Sjögren's syndrome with subcortical dementia. Intern. Med. (Japan) 32 (1993) 561–564.

KAZUHIKO, K., G. SOBUE, K. YAMAMOTO ET AL.: Segmental anhidrosis in the spinal dermatomes in Sjögren's syndrome-associated neuropathy. Neurology 43 (1993) 1820–1823.

KINCAID, M.C.: The eye in Sjögren's syndrome. In: N.

Talal, H.M. Moutsopoulos and S.S. Kassan (Eds.), Sjögren's Syndrome. Heidelberg, Springer-Verlag (1987) 25–.

KOHRIYAMA, K., A. KOHNO and S. ARIMORI: A case of temporal arteritis associated with polymyagia rheumatica and subclinical Sjögren's syndrome. Clin. Neurol. 30 (1990) 272–280.

KOMAROFF, A.L. and D. BUCHWALD: Symptoms and signs of chronic fatigue syndrome. Rev. Infect. Dis. 13, Suppl. (1991) S8–S11.

KONTTINEN, Y.T., E. KINNUNEN, M. VON BONDSDORFF, P. LILLQVIST, I. IMMNEN, V. BERGROTH, M. SEGERBERG-KONTTINEN and C. FRIMAN: Acute transverse myelopathy successfully treated with plasmapheresis and prednisone in a patient with primary Sjögren's syndrome. Arthr. Rheum. 30 (1987) 339–344.

KOZACHUK, W.E., C. DECARLI, M.B. SCHAPIRO, E.E. WAGNER, S.I. RAPOPORT and B. HORWITZ: White matter hyperintensities in dementia of the Alzheimer's type and in healthy subjects without cerebrovascular risk factors: a magnetic resonance imaging study. Arch. Neurol. 47 (1990) 1306–1310.

KRAUS, A., G. CERVANTES, E. BAROJAS and D. ALARCON SEGOVIA: Retinal vasculitis in mixed connective tissue disease. A fluoroangiographic study. J. Rheumatol. 12 (1985) 1122–1124.

KRAUS, A., M. CIFUENTES, A.R. VILLA, J. REYES and D. ALARCON-SEGOVIA: Myositis in primary Sjögren's syndrome. Report of 3 cases. J. Rheumatol. 21 (1994) 649–653.

KUMAZAWA, K., G. SOBUE, K. YAMAMOTO and T. MITSUMA: Segmental anhidrosis in the spinal dermatomes in Sjögren syndrome-associated neuropathy. Neurology 43 (1993) 1820–1823.

LAFFORGUE, P., E. TOUSSIROT, F. BILLÉ and P.C. ACQUAVIVA: Astasia – abasia revealing a primary Sjögren's syndrome. Clin. Rheumatol. 12 (1993) 261–264.

LAHOZ RALLO, C., J.R. ARRIBAS LOPEZ, F. ARNALICH FERNANDEZ, A. MONEREO ALONSO, M.C. LLANOS CHAVARRI and J. CAMACHO SILES: Vasculitis necrotizante tipo panarteritis nodosa en un sindrome de Sjögren primario de larga evolución. Ann. Intern. Med. 7 (1990) 528–530.

LINDAHL, G., A. LEFVERT and E. HEDFORS: Periduct lymphocytic infiltrates in salivary glands in myasthenia gravis patients lacking Sjögren's syndrome. Clin. Exp. Immunol. 66 (1986) 95–102.

LOW, P.A., L.E. MERTZ, R.G. AUGER, A.M. DILLION, R.D. FEALEY, S.S. JARADEH, B.R. YOUNGE, W.J. LITCHY and P.J. DYCK: The autonomic neuropathies of Sjögren's syndrome. Neurology 38, Suppl. 1 (1988) 104.

LYU, R.K., S.T. CHEN, L.M. TANG and T.C. CHEN: Acute transverse myelopathy and cutaneous vasculopathy in primary Sjögren's syndrome. Eur. Neurol. 60 (1995) 359–362.

MADDISON, P.J., T.T. PROVOST and M. REICHLIN: Serological findings in patients with 'ANA negative' systemic lupus erythematosus. Medicine (Baltimore) 60 (1981) 87–94.

MALINOW, K.L., R. MOLINA, B. GORDON, O.A. SELNES, T.T. PROVOST and E.L. ALEXANDER: Neuropsychiatric dysfunction in primary Sjögren's syndrome. Ann. Intern. Med. 103 (1985) 344–350.

MALINOW, K.L., G.D. YANNAKAKIS, S.M. GLUSMAN, D.W. EDLOW, J. GRIFFIN, A. PESTRONK, D.L. POWELL, R. RAMSEY-GOLDMAN, B.H. EIDELMAN and T.A. MEDSGER, JR.: Subacute sensory neuropathy secondary to dorsal root ganglionitis in primary Sjögren's syndrome. Ann. Neurol. 20 (1986) 535–537.

MANTHORPE, R., P. OXHOLM, J.U. PRAUSE and M. SCHIODT: The Copenhagen criteria for Sjögren's syndrome. Scand. J. Rheumatol. 61, Suppl. (1986) 19–21.

MANTHORPE, R., T. MANTHORPE and S. SJÖBERG: Magnetic resonance imaging of the brain in patients with primary Sjögren's syndrome. Scand. J. Rheumatol. 21 (1992) 148–149.

MARBINI, A., F. GEMIGNANI, P. MANGANELLI, E. GOVONI, M.M. BRAGAGLIA and U. AMBANELLI: Hypertrophic neuropathy in Sjögren's syndrome. Acta Neuropathol. (Berlin) 57 (1982) 309–312.

MASSEY, E.W.: Sjögren's syndrome and mononeuritis multiplex (Letter). Ann. Intern. Med. 63 (1980) 87.

MAUCH, E., C. VÖLK, G. KRATZSCH, H. KRAPF, H.H. KORNHUBER, H. LAUFEN and K.J. HUMMEL: Neurological and neuropsychiatric dysfunction in primary Sjögren's syndrome. Acta Neurol. Scand. 89 (1994) 31–35.

MCCARTY, G.A., K.A. LISTER, M. REICHLIN and R. MCEVER: Autoantibodies to a novel endothelial cell granule membrane protein antigen in vasculitis. Clin. Res. 38 (1990) 316A.

MCCOMBE, P.A., G.L. SHEEAN, D.B. MCLAUGHLIN and M.P. PENDER: Vestibular and ventilatory dysfunction in sensory and autonomic neuropathy associated with primary Sjögren syndrome. J. Neurol. Neurosurg. Psychiatry 55 (1992) 1211–1212.

MCFARLAND, H., J.A. FRANK, P.S. ALBERT, M.E. SMITH, R. MARTIN, J.O HARRIS, N. PATRONAS, H. MALONI and D.E MCFARLIN: Using gadolinium-enhanced magnetic resonance imaging lesions to monitor disease activity in multiple sclerosis. Ann. Neurol. 32 (1992) 758–766.

MELLGREN, S.I., D.L. CONN, J.C. STEVENS and P.J. DYCK: Peripheral neuropathy in primary Sjögren's syndrome. Neurology 39 (1989) 390–394.

MÉNAGE, P., B. DE TOFFOL, D. DEGENNE, D. SAUDEAU, P. BARDOS and A. AUTRET: Syndrome de Gougerot-Sjögren primitif atteinte neurologique centrale évoluant par poussées. Rev. Neurol. (Paris) 149 (1993) 554–556.

MILLER, D.H.: Magnetic resonance in monitoring the treatment of multiple sclerosis. Ann. Neurol. 36 (1994) S91–S94.

MIRO, J., J.L. PENA-SAGREDO, J. BERCIANO, S. INSUA, C. LENO and R. VELARDE: Prevalence of primary Sjögren syndrome in patients with multiple sclerosis. Ann. Neurol. 27 (1990) 582–584.

MOLINA, R., T.T. PROVOST and E.L. ALEXANDER: Peripheral inflammatory vascular disease in Sjögren's syndrome: association with nervous system complications. Arthr. Rheum. 28 (1985a) 1341–1347.

MOLINA, R., T.T. PROVOST and E.L. ALEXANDER: Two types of inflammatory vascular disease in Sjögren's syndrome: differential association with seroreactivity to rheumatoid factor and antibodies to Ro(SS-A) and with hypocomplementemia. Arthr. Rheum. 28 (1985b) 1251–1258.

MOLINA, R., T.T. PROVOST, F.C. ARNETT, W.B. BIAS, M.C. HOCHBERG, R.W. WILSON and E.L. ALEXANDER: Primary Sjögren's syndrome (SS) in men. Clinical, serologic, and immunogenetic features. Am. J. Med. 80 (1986) 23–31.

MOLL, J.W.B., H.M. MARKUSSE, J.J. PIJNENBURG, C.J. VECHT and S.C HENZEN-LOGMANS: Antineuronal antibodies in patients with neurologic complications of primary Sjögren's syndrome. Neurology 43 (1993) 2574–2581.

MONTECUCCO, C., D.M. FRANCIOTTA, R.CAPORALI, F. DEGENNARO, A. CITTERIO and G.V. MELZI D'ERIL: Sicca syndrome and anti-SSA(Ro) antibodies in patients with suspected or definite multiple sclerosis. Scand. J. Rheumatol. 18 (1989) 407–412.

MOUTSOPOULOS, H.M., J.H. SARMAS and N. TALAL: Is central nervous system involvement a systemic manifestation of primary Sjögren's syndrome? Rheum. Dis. Clin. N. Am. 19 (1993) 909–912.

MUKAI, M., A. SAGAWA, Y. BABA, Y. AMASAKI, K. KATSUMATA, M. YOSHIKAWA, T. NAKABAYASKI, I. WATANABE, I. YASUDA and A. FUJISAKU: Neuro-psychiatric symptom associated with primary Sjögren's syndrome. Ryumachi 30 (1990) 109–118.

NAGAO, T., K. TAKAGI, H. HASHIDA, T. MASAKI and M. SAKUTO: A case of progressive systemic sclerosis and Sjögren's syndrome complicated by parkinsonism with special reference to the beneficial effect of corticosteroid. Rinsho Shinkeigaku 31 (1991) 1238–1240.

NAGAHIRO, S., A. MANTANI, K. YAMADA, Y. USHIO, C.M. LOFTUS and R.G. DACY, JR.: Multiple cerebral arterial occlusions in a young patient with Sjögren's syndrome: case report. Neurosurgery 38 (1996) 592–595.

NAKHOUL, F., Y. PLAVNIC, H. LICHTIG and O.S. BETTER: Hypokalemic flaccid paralysis as the presenting symptom of autoimmune interstitial nephropathy. Isr. J. Med. Sci. 29 (1993) 300–303.

NODA, M., M. KITAGAWA, F. TOMODO and H. IIDA: Thrombotic thrombocytopenic purpura as a complicating factor in a case of polymyositis and Sjögren's syndrome. Am. J. Clin. Pathol. 94 (1990) 217–221.

NOSEWORTHY, J.H., B.H. BASS, M.K. VANDERVOORT, G.C.

EBERS, G.P. RICE, B.G. WEINSHENKER, C.J. MCLAY and D.A. BELL: The prevalence of primary Sjögren's syndrome in a multiple sclerosis population. Ann. Neurol. 25 (1989) 95–98.

OHTSUKA, T., Y. SAITO, M. HASEGAWA, M. TATSUNO, S.TAKITA, M. ARITA and K. OKUYAMA: Central nervous system disease in a child with primary Sjögren's syndrome. J. Pediatr. 127 (1995) 961–963.

OLSEN, M.L., F.C. ARNETT and D. ROSENBAUM: Sjögren's syndrome and other rheumatic disorders presenting to a neurology service. J. Autoimmun. 2 (1989) 477–483.

OLSEN, M.L., S. O'CONNOR, F.C. ARNETT, D. ROSENBAUM, J.C. GROTTA and N.B. WARNER: Autoantibodies and rheumatic disorders in a neurology inpatient population: a prospective study. Am. J. Med. 90 (1991) 479–488.

OOBAYASHI, Y. and S. MIYAWAKI: Ataxic sensory and autonomic neuropathies associated with primary Sjögren's syndrome: a case report. Ryumachi 35 (1995) 107–111.

PEYRONNARD, J.M., L. CHARRON, F. BEAUDET and F. COUTURE: Vasculitic neuropathy in rheumatoid disease and Sjögren's syndrome. Neurology 32 (1982) 839–845.

PIEROT, L., C. SAUVE, J.M. LEGER, N. MARTIN, A.C. KOEGER, B. WECHSLER and J. CHIRAS: Asymptomatic cerebral involvement in Sjögren's syndrome: MRI findings of 15 cases. Neuroradiology 35 (1993) 378–380.

PITTSLEY, R.A. and N. TALAL: Neuromuscular complications of Sjögren's syndrome. In: P.J. Vinken and G.W. Bruyn (Eds.), Handbook of Clinical Neurology, Vol 3. Amsterdam, Elsevier Science (1980) 419–433.

POSER, C.M., D.W. PATY, L. SCHEINBERG, W.I. MCDONALD, F.A. DAVIS, G.C. EBERS, K.P. JOHNSON, W.A. SIBLEY, D.H. SILBERBERG and W.W. TOURTELLOTTE: New diagnostic criteria for multiple sclerosis: guidelines for research protocols. Ann. Neurol. 13 (1983) 227–231.

POUX, J.M., P. PEYRONNET, Y. LE MEUR, J.P. FAVEREAU, J.P. CHARMES and C. LEROUX-ROBERT: Hypokalemic quadriplegia and respiratory arrest revealing primary Sjögren's syndrome. Clin. Neuropathol. 37 (1992) 189–191.

PROVOST, T.T., N. TALAL, W. BIAS, J.B. HARLEY, M. REICHLIN and E. ALEXANDER: Ro(SS-A) positive Sjögren's/lupus erythematosus (SS/LE) overlap patients are associated with the HLA-DR3 and/or DRw6 phenotypes. J. Invest. Dermatol. 91 (1988a) 369–371.

PROVOST, T.T., N. TALAL, J.B. HARLEY, M. REICHLIN and E. ALEXANDER: The relationship between anti-Ro(SS-A) antibody positive Sjögren's syndrome and anti-Ro(SS-A) antibody positive lupus erythematosus. Arch. Dermatol. 124 (1988b) 63–71.

PROVOST, T.T., L.S. LEVIN, R.M. WATSON, M. MAYO and H. RATRIE, III: Detection of anti-Ro(SS-A) antibodies by gel double diffusion and a 'sandwich' ELISA in systemic and subacute cutaneous lupus erythematosus and Sjögren's syndrome. J. Autoimmun. 4 (1991a) 87–96.

PROVOST, T.T., H. MOSES, E.L. MORRIS, J. ALTMAN, J.B. HARLEY, E. ALEXANDER and M. REICHLIN: Cerebral vasculopathy associated with collateralization resembling Moya Moya phenomenon and with anti-Ro/SS-A and anti-La/SS-B antibodies. Arthr. Rheum. 34 (1991b) 1052–1055.

PUN, K.K. and C.L. WANG: Hypokalemic periodic paralysis due to the Sjögren's syndrome in Chinese patients. Ann. Intern. Med. 110 (1989) 405–406.

RANZENBACH, M.R., J.R. PLITT, A.J. KUMAR, N. PATRONAS, E. KOO, G.E. ALEXANDER and E.L. ALEXANDER: CNS disease in Sjögren's syndrome (CNS-SS): role of anti-Ro(SS-A) antibodies. FASEB J. 5 (1991) A1386.

RANZENBACH, M.R., A. KUMAR, A. ROSENBAUM, N. PATRONAS, J.B. HARLEY, M. REICHLIN and E. ALEXANDER: Anti-Ro(SS-A) autoantibodies (A-RoAb) in the immunopathogenesis of serious focal CNS disease in Sjögren's syndrome (CNS-SS). Arthr. Rheum. 35 (1992) S168.

RASKIN, R.J., J.T. TESAR and O.F. LAWLESS: Hypokalemic periodic paralysis in Sjögren's syndrome. Arch. Intern. Med. 141 (1981) 1671–1673.

RINGEL, S.P., J.Z. FORSTOT, E.M. TAN, C. WEHLING, R.C. GRIGGS and D. BUTCHER: Sjögren's syndrome and polymyositis or dermatomyositis. Arch. Neurol. 39 (1982) 157–163.

ROBSON, S.C., P. KLEMP and O.L. MEYERS: Central nervous system manifestations of Sjögren's syndrome. A case report. S. Afr. Med. J. 69 (1986) 196–197.

ROSENBAUM, J.T. and R.M. BENNETT: Chronic anterior and posterior uveitis and primary Sjögren's syndrome. Am. J. Ophthalmol. 104 (1987) 346–352.

ROSSI, P., V. FOSSALUZZA and F. TOSATO: Coexistence of Sjögren's syndrome associated systemic sclerosis with Adie's syndrome. J. Rheumatol. 13 (1986) 823–825.

RUTAN, G.A., A.J. MARTINEZ, J.T. FIESHKO and D.H. VAN THIEL: Primary biliary cirrhosis, Sjögren's syndrome, and transverse myelitis. Gastroenterology 90 (1986) 206–210.

SANDBERG-WOLLHEIM, M., T. AXELL, B.U. HANSEN, V. HENRICSSON, E. INGESSON, L. JACOBSSON, LARSSON, K. LIEBERKIND and R. MANTHORPE: Primary Sjögren's syndrome in patients with multiple sclerosis. Neurology 42 (1992) 845–847.

SANDERS, M.E., C.L. KOSKI, D. ROBBINS ET AL.: Activated terminal complement in cerebrospinal fluid in Guillain–Barré syndrome and multiple sclerosis. J. Immunol. 135 (1986) 1–4.

SANDERS, M.E., E.L. ALEXANDER and C.L. KOSKI: Detection of activated terminal complement (C5b-9) in cerebral spinal fluid from patients with central nervous system

involvement of primary Sjögren's syndrome or systemic lupus erythematosus. J. Immunol. 138 (1987) 2095–2099.

SATAKE, M., Y. TAKEO, T. TWAKI, T. YAMADA and T. KOBAYASHI: Anti-dorsal root ganglion neuron antibody in a case of dorsal root ganglionitis associated with Sjögren's syndrome. J. Neurol. Sci. 132 (1995) 122–125.

SATO, K., N. MIYASAKA, K. NISHIOKA ET AL.: Primary Sjögren's syndrome associated with systemic necrotizing vasculitis: a fatal case. Arthr. Rheum. 30 (1987) 717–718.

SCHMID, U., K. LENNERT and F. GLOOR: Immunosialadenitis (Sjögren's syndrome) and lymphoproliferation. Clin. Exp. Rheumatol. 7 (1989) 175–180.

SCHNEEBAUM, A.B., J.D. SINGLETON, S.G. WEST ET AL.: Association or psychiatric manifestations with antibodies to proteins in systemic lupus erythematosus. Am. J. Med. 90 (1991) 54–62.

SELNES, O., B. GORDON, K. MALINOW ET AL.: Cognitive dysfunction in primary Sjögren's syndrome. Neurology 35 (1985) S179.

SERRADELL, A. and J. GAYA: Trois cas de neuropathies périphériques rares associées au syndrome de Gougerot–Sjögren primitif. Rev. Neurol. 149 (1993) 481–484.

SHARIEF, M.K., M. PHIL and HENTGES: Association between tumor necrosis factor-α and disease progression in patients with multiple sclerosis (Letter). N. Engl. J. Med. 325 (1991) 467–472.

SHEARN, M.: Sjögren's Syndrome. Philadelphia, PA, W.B. Saunders (1971).

SHEIBANI, K., J. BEN-EZRA, W.G. SWARTZ ET AL.: Monocytoid B cell lymphoma in a patient with human immunodeficiency virus infection. Arch. Pathol. Lab. Med. 114 (1990) 1264–1267.

SHELDON, J.J., R. SIDDHARTHAN and J. TOBIAS: Magnetic resonance imaging of multiple sclerosis: comparison with clinical and CT examinations in 74 patients. Am. J. Neuroradiol. 6 (1985) 683–.

SHIN, S.S., K. SHEIBANI, A. FISHLEDER ET AL.: Monocytoid B cell lymphoma in patients with Sjögren's syndrome: a clinicopathologic study of 13 patients. Hum. Pathol. 22 (1991) 422–430.

SHIOJI, R., T. FURUYAMA, S. ONODERA, H. SAITO and Y. SASAKI: Hypokalemic periodic paralysis in Sjögren's syndrome. Am. J. Med. 48 (1970) 456–.

SJÖGREN, H.: Zur Kenntnis der Keratoconjunctivitis sicca. II. Allgemeine Symptomatologie und Ätiologie. Acta Ophthalmol. 13, Suppl. (1935) 1–.

SKOPOULI, F.N., A.A. DROSOS, T. PAPAIOANNOU ET AL.: Preliminary diagnostic criteria for Sjögren's syndrome. Scand. J. Rheumatol. 61, Suppl. (1986) 22–25.

SNAITH, M.L.: The neurological complications and associations of Sjögren's syndrome. B.S.S.A. April–June (1990) 2–4.

SOBUE, G., R. YASUDA, T. KACHI ET AL.: Chronic progressive sensory ataxic neuropathy: clinicopathological features of idiopathic and Sjögren's syndrome-associated cases. Neurology 240 (1993) 1–7.

SOBUE, G., R. YASUDA, K. KUMAZAWA ET AL.: MRI demonstrates dorsal column involvement of the spinal cord in Sjögren's syndrome-associated neuropathy. Neurology 45 (1995) 592–593.

SOHN, C.C., K. SHEIBANI, C.D. WINBERG ET AL.: Monocytoid B lymphocytes: their relation to the patterns of the acquired immunodeficiency syndrome (AIDS) and AIDS related lymphadenopathy. Hum. Pathol. 16 (1985) 979–985.

SPEZIALETTI, R., J. PETER and H.G. BLUESTEIN: Neuropsychiatric disease in Sjögren's syndrome – anti-ribosomal P and anti-neuronal antibodies. Am. J. Med. 95 (1993) 153–160.

STEINBERG, A.D., W.T. GREEN, JR. and N. TALAL: Thrombotic thrombocytopenic purpura complicating Sjögren's syndrome. J. Am. Med. Assoc. 215 (1971) 757–761.

STOLTZE, C.A., D.G. HANLON, G.L. PEASE ET AL.: Keratoconjunctivitis sicca and Sjögren's syndrome. Arch. Intern. Med. 106 (1969) 513–521.

STRICKLAND, R.W., J.T. TESAR, B.H. BERNE ET AL.: The frequency of sicca syndrome in an elderly female population. J. Rheumatol. 14 (1987) 766–771.

TALAL, N.: Sjögren's syndrome. Bull. Rheum. Dis. 16 (1966) 404–407.

TAN, E.M., A.S. COHEN, J.F. FREIS ET AL.: The 1982 revised criteria for the classification of systemic lupus erythematosus. Arthr. Rheum. 25 (1982) 1271–1277.

TERAO, Y., K. SAKAI, S. KATO ET AL.: Antineuronal antibody in Sjögren's syndrome masquerading as paraneoplastic cerebellar degeneration. Lancet 343 (1994) 790–.

TESAR, J.T., V. MCMILLAN, R. MOLINA ET AL.: Optic neuropathy and central nervous system disease associated with primary Sjögren's syndrome. Am. J. Med. 92 (1992) 686–692.

TSOKOS, M., S. LAZAROU and H. MOUTSOPOULOS: Vasculitis in primary Sjögren's syndrome. Am. J. Clin. Pathol. 88 (1987) 26–31.

TSUJI, T., S. OHTA, K. MATSUNAGA ET AL.: A case of aseptic meningoencephalitis, in a patient with secondary Sjögren syndrome with systemic lupus erythematosus. Clin. Neurol. 34 (1994) 646–650.

TSUKAMOTO, T., R. MOCHIZUKI, H. MOCHIZUKI ET AL.: Paraneoplastic cerebellar degeneration and limbic encephalitis in a patient with adenocarcinoma of the colon. J. Neurol. Neurosurg. Psychiatry 56 (1993) 713–716.

TWIJNSTRA, A., J. VERSCHUUREN, T.N. BRYNE ET AL.: Prolonged or remitting anti-Hu-associated paraneoplastic sensory neuronopathy (PSN) and encephalomyelitis (PEM). Neurology 45, Suppl. (1995) A321.

UTSET, T.O., M. GOLDER, G. SIBERRY, N. KIRI, R.M. CRUM and M. PETRI: Depressive symptoms in patients with systemic lupus erythematosus: association with central nervous system lupus and Sjögren's syndrome. J. Rheumatol. 21 (1994) 2039–2045.

VALLS-SOLÉ, J., F. GRAUS, J. FONT ET AL.: Normal proprioceptive trigeminal afferents in patients with Sjögren's syndrome and sensory neuronopathy. Ann. Neurol. 28 (1990) 786–790.

VAN BIJSTERVELD, O.P.: Diagnostic tests in the sicca syndrome. Arch. Ophthalmol. 82 (1969) 10–.

VAN OFFEL, J.F., L.M. FRANCKX, L.S. DECLERCK, P. EVENS, R. MERCELIS, A NEETENS and W.J. STEVENS: Differential diagnosis in a patient with secondary Sjögren's syndrome and neuromuscular complications. Br. J. Rheumatol. 26 (1987) 470–471.

VANDERZANT, C., M. BRUMBERG, A. MACGUIRE ET AL.: Isolated small-vessel angiitis of the central nervous system. Arthr. Neurol. 45 (1988) 683–687.

VARTDAL, F., L.M. SOLLID, B. VANDVIK ET AL.: Patients with multiple sclerosis carry DQB1 genes which encode shared polymorphic amino acid sequences. Hum. Immunol. 25 (1989) 103–110.

VECHT, C.J, H. HOOIJKAAS, P. SPENKELINK ET AL.: Anti-Hu anti-Ro/SS-A and neurological complications in primary Sjögren's syndrome. Neurology 45, Suppl. (1995) A256.

VITALI, C., A. TAVONI, R. NERI, P. CASTROGIOVANNI, G. PASERO and S. BOMBARDIERI: Fibromyalgia features in patients with primary Sjögren's syndrome. Evidence of a relationship with psychological depression. Scand. J. Rheumatol. 18 (1989) 21–27.

VITALI, C., S. BOMBARDIERI, H.M. MOUTSOPOULOS ET AL.: Preliminary criteria for the classification of Sjögren's syndrome: results of a prospective concerted action supported by the European community. Arthr. Rheum. 36 (1993) 340–347.

VITALI, C., S. BONBARDIERI, H.M. MOUTSOPOULOS, J. COLL, R. GERLI, P.Y. HATRON, L. KATER, Y.T. KONTTINEN, R. MANTHORPE, O. MEYER, M. MOSCA, P. OSTUNI, R.A. PELLERITO, Y. PENNEC, S.R. PORTER, A. RICHARDS, B. SAUVEZIE, M. SCHIODT, M. SCIUTO, Y. SHOENFELD, F.N. SKOPOULI, J.S. SMOLEN, F. SOROMENHO, M. TISHER, M. TOMSIC, J.P. VAN DE MERWE, M.J. WATTIAUX and C.M. YEOMAN: Assessment of the European classification criteria for Sjögren's syndrome in a series of clinically defined cases: results of a prospective multicentre study. Ann. Rheum. Dis. 55 (1996) 116–121.

VRETHEM, M., J. ERNEURUDH, F. LINDSTRÖM ET AL.: Immunoglobulins within the central nervous system in primary Sjögren's syndrome. J. Neurosci. 100 (1990a) 186–192.

VRETHEM, M., B. LINDVAL and H. HOLMGREN: Neuropathy and myopathy in primary Sjögren's syndrome: neurophysiological, immunological and muscle biopsy results. Acta Neurol. Scand. 82 (1990b) 126–131.

WHALEY, K., J. WEBB, B.A. MCAVOY ET AL.: Sjögren's syndrome. 2. Clinical associations and immunological phenomena. Q. J. Med. 42 (1973) 513–548.

WILSON, R.W., T.T. PROVOST, W.B. BIAS ET AL.: Sjögren's syndrome: influence of multiple HLA-D region alloantigens on clinical serologic expression. Arthr. Rheum. 27 (1984) 1245–1253.

WINDEBANK, A.J., M.D. BLEXRUD, P.J. DYCK ET AL.: The syndrome of acute sensory neuropathy: clinical features and electrophysiologic and pathologic changes. Neurology 40 (1990) 584–591.

WISE, C.M. and C.A. AGUDELO: Optic neuropathy as an initial manifestation of Sjögren's syndrome. J. Rheumatol. 15 (1988) 799–802.

YA-XIN, F. and W. CHANG-HUA: Neurological complications of Sjögren's syndrome. Chin. Med. J. 99 (1986) 751–754.

YOSHIIWA, A., T. NABATA, S. MORIMOTO ET AL.: A case of Hashimoto's thyroiditis associated with renal tubular acidosis, Sjögren's syndrome and empty sella syndrome. Folia Endocrinol. Jpn. 68 (1992) 1215–1223.

Handbook of Clinical Neurology, Vol. 27 (71): Systemic Diseases, Part III
M.J. Aminoff and C.G. Goetz, editors

CHAPTER 5

Neurological abnormalities in scleroderma

RICHARD K. OLNEY

Department of Neurology, School of Medicine, University of California, San Francisco, CA, U.S.A.

Scleroderma refers to a group of disorders that share the common feature of sclerosis or thickening of the skin. In its broadest usage, scleroderma includes both disorders that are thought to be predominantly intrinsic medical diseases and disorders that are thought to be produced predominantly as reactions to environmental or extrinsic factors. Primary scleroderma refers to the first group of disorders, which contains cases without known cause that are believed to represent an autoimmune disease. The terms secondary scleroderma and scleroderma-like disorders refer to the latter group of disorders. Considerable controversy surrounds the concept of secondary scleroderma with regard both to the degree of similarity secondary cases share with primary cases and to which environmental or extrinsic factors are relevant to the etiopathogenesis of scleroderma-like syndromes. Because of this diversity of opinion, in contrast to the many points of consensus about primary scleroderma, the issues of secondary scleroderma will be discussed separately in the last section of this chapter. Most of this chapter will focus on primary scleroderma and the neurological abnormalities with which it is associated.

PRIMARY SCLERODERMA

Primary scleroderma is divided into systemic sclerosis and localized scleroderma, with several subtypes

TABLE 1

The various types of primary scleroderma.

---

1. Systemic sclerosis
    (a) Pre-scleroderma
    (b) Systemic sclerosis with limited cutaneous involvement (limited cutaneous systemic sclerosis)
    (c) Systemic sclerosis with diffuse cutaneous involvement (diffuse cutaneous systemic sclerosis)
    (d) Systemic sclerosis sine scleroderma
    (e) Overlap syndromes

2. Localized scleroderma
    (a) Morphea
    (b) Generalized morphea
    (c) Linear scleroderma

---

of each form (Table 1). The basic similarities between these two major categories of scleroderma are similarity in induration, sclerosis or atrophy of skin lesions and partially shared pathological features of biopsied skin lesions. However, the clinical distribution of skin lesions is quite distinct and the clinical course of these two categories of scleroderma is marked by striking differences and minimal similarities. Cases with systemic sclerosis do not later develop features of localized scleroderma, but cases of localized scleroderma may uncommonly make a transition into systemic sclerosis (Birdi et al. 1993; Dehen et al. 1994; Mayorquin et al. 1994).

Systemic sclerosis

Systemic sclerosis includes an incompletely differentiated form, two common differentiated forms, one rare differentiated form, and overlap syndromes. The most widely accepted diagnostic criteria for systemic sclerosis apply to the diagnosis of the two common differentiated forms, which are diffuse cutaneous systemic sclerosis and limited cutaneous systemic sclerosis. According to these criteria, the diagnosis of systemic sclerosis can be made by the presence of proximal scleroderma (tightness, thickening and non-pitting induration proximal to the metacarpophalangeal or metatarsophalangeal joints) or the presence of two or more of the following three criteria: (1) sclerodactyly, (2) digital pitting scars of fingertips or loss of substance of the distal finger pad, and (3) bilateral basilar pulmonary fibrosis (Subcommittee for Scleroderma Criteria 1980). In a comparison of 264 patients with definite systemic sclerosis and 413 patients with systemic lupus erythematosus, polymyositis/dermatomyositis or isolated Raynaud's phenomenon, these criteria were found to have a 97% sensitivity and 98% specificity (Subcommittee for Scleroderma Criteria 1980). Although widely accepted conceptually, consensus criteria have not been agreed upon for the diagnosis of the incompletely differentiated form (pre-scleroderma), the rare differentiated form (systemic sclerosis sine scleroderma), and overlap syndromes. The prevalence of systemic sclerosis may be as high as 19–75/100,000 based on a recent population-based survey, but 0.1–14/100,000 is the usual prevalence rate that has been reported (Maricq et al. 1989; LeRoy and Silver 1993). The prevalence is much lower in children (Denardo et al. 1994; Pelkonen et al. 1994). Women are affected far more commonly than men, with the highest ratio being 15:1 during young adult years (LeRoy and Silver 1993).

Pre-scleroderma

The incompletely differentiated form is often referred to as pre-scleroderma (LeRoy et al. 1988). The presence of Raynaud's phenomenon is an integral part of the concept of pre-scleroderma, but is insufficient by itself to suggest the probable subsequent evolution into systemic sclerosis. Among patients with systemic sclerosis, 80–95% have Raynaud's phenomenon, usually as the initial symptom (Subcommittee for Scleroderma Criteria 1980; Isenberg and

Black 1995). However, approximately 5% of the general population has Raynaud's phenomenon and only 5–10% eventually develop rheumatologic diseases including systemic sclerosis, so that the presence of Raynaud's phenomenon is a sensitive but non-specific early clinical feature of systemic sclerosis (Subcommittee for Scleroderma Criteria 1980; Gerbracht et al. 1985; Isenberg and Black 1995).

The likelihood of a patient with Raynaud's phenomenon later developing systemic sclerosis is markedly increased if certain abnormalities are noted by nailfold capillaroscopy and if certain, specific antinuclear antibodies are present (Kallenberg et al. 1988; LeRoy et al. 1988; Zufferey et al. 1992; Black and Denton 1995). In a 6-year follow-up study of 46 patients who had Raynaud's phenomenon but did not fulfill the criteria for systemic sclerosis, the presence of the anticentromere type of antinuclear antibody identified patients who developed limited cutaneous systemic sclerosis with 60% sensitivity and 98% specificity, and the presence of the antitopoisomerase I (also referred to as Scl-70) type of antinuclear antibody identified patients who developed diffuse cutaneous systemic sclerosis with 38% sensitivity and 100% specificity (Kallenberg et al. 1988). In a 6-year follow-up study of 30 patients who had Raynaud's phenomenon but did not fulfill the criteria for systemic sclerosis, the capillaroscopic findings of avascularity or more than one megacapillary per digit identified patients who developed systemic sclerosis with 100% sensitivity and 70% specificity for avascularity or 88% specificity for megacapillaries (Zufferey et al. 1992). The combined presence of one of these two specific types of antinuclear antibodies and the presence of avascularity or more than one megacapillary per digit by capillaroscopy is highly predictive of the subsequent development of systemic sclerosis (LeRoy et al. 1988; Black and Denton 1995).

Limited cutaneous systemic sclerosis

The term limited cutaneous systemic sclerosis is largely synonymous with the CREST (calcinosis, Raynaud's phenomenon, esophageal dysmotility, sclerodactyly and telangiectasia) syndrome, because the latter designation describes the common features that distinguish limited from diffuse cutaneous systemic sclerosis. These patients have a prodromal period that usually lasts several years but rarely may continue for several decades, during which time they have Raynaud's phenomenon but do not fulfill the criteria for systemic sclerosis (Subcommittee for Scleroderma Criteria 1980; Gerbracht et al. 1985; Isenberg and Black 1995). Although abnormalities of nailfold capillaroscopy and anticentromere antibody activity are predictive of its future development, the diagnosis of limited cutaneous systemic sclerosis is not made unless the patient fulfills the consensus diagnostic criteria, usually with development of sclerodactyly that is associated with digital pitting scars of fingertips or loss of substance of the distal finger pad. The diagnosis of systemic sclerosis can also be made with development of sclerodactyly that is associated with bilateral basilar pulmonary fibrosis; some of these patients will remain in the limited cutaneous systemic sclerosis category, but many will also rapidly develop proximal scleroderma and have diffuse cutaneous systemic sclerosis. In patients with limited cutaneous systemic sclerosis, telangiectasias will soon appear in 80%, usually on the limbs distally and on the face after the development of Raynaud's phenomenon and sclerodactyly (Medsger and Steen 1993). Subcutaneous calcifications, referred to as calcinosis, occur on fingers and other sites of minor trauma in about one-half of patients with limited cutaneous systemic sclerosis, in contrast to their rare occurrence in patients with diffuse disease (Medsger and Steen 1993). Lower esophageal hypomotility is present in about three-quarters of patients, usually soon after the appearance of sclerodactyly if cine-esophagrams are performed (Medsger and Steen 1993). However, abnormal esophageal motility is also common in other connective tissue diseases (Lapadula et al. 1994). Arthralgias are common, but tendon friction rubs are rare (Medsger and Steen 1993; Black and Denton 1995). An antinuclear antibody of any type is present in 95%, and the anticentromere antibody is present in 50% (Medsger and Steen 1993). Patients with limited cutaneous systemic sclerosis and HLA haplotype DQB1-0501 are more likely than those with DQB1-0201 to develop anticentromere antibodies (Morel et al. 1995).

In contrast to diffuse cutaneous systemic sclerosis, patients with limited cutaneous systemic sclerosis are usually considered to have a favorable prognosis (Barnett et al. 1988; Clements et al. 1990; Czirjak et al. 1993; White et al. 1995). Over 70% of patients survive beyond 6 or 10 years (Barnett et al. 1988;

Clements et al. 1990). Those who challenge the more favorable prognosis of limited disease still find 6-year survival rates over 70% (Lee et al. 1992). A broad consensus exists for the understanding that prognosis most directly depends on the rate at which pulmonary, renal or cardiac involvement develops (Altman et al. 1991; Lee et al. 1992; Bulpitt et al. 1993; White et al. 1995). Among patients with any form of systemic sclerosis who had renal involvement, survival has been improved from 15 to 76% at 1 year by using angiotensin-converting enzyme inhibitors to treat the associated hypertension (Steen et al. 1990). Patients with limited cutaneous systemic sclerosis seem to be more likely to develop pulmonary arterial hypertension than those with diffuse disease and less likely to have pulmonary interstitial fibrosis, myocarditis or myocardial fibrosis (Medsger and Steen 1993). Treatments for cardiac and pulmonary complications are not yet very effective (Medsger and Steen 1993; Black and Denton 1995).

Diffuse cutaneous systemic sclerosis

Patients with diffuse cutaneous systemic sclerosis usually have a prodromal period of less than 1 year, during which time they have Raynaud's phenomenon, but do not fulfill the criteria for systemic sclerosis (Subcommittee for Scleroderma Criteria 1980; Gerbracht et al. 1985; Isenberg and Black 1995). Although abnormalities of nailfold capillaroscopy and antitopoisomerase I antibody activity have predictive value, the diagnosis of diffuse cutaneous systemic sclerosis is not made unless the patient develops proximal scleroderma (Subcommittee for Scleroderma Criteria 1980). The diagnosis of systemic sclerosis can also be made with development of sclerodactyly that is associated with bilateral basilar pulmonary fibrosis; many of these patients will also develop proximal scleroderma over subsequent months, diagnostic of diffuse cutaneous systemic sclerosis. Telangiectasias are not unusual late in the course of diffuse disease, but early telangiectasias or calcinosis at any stage are quite uncommon (Medsger and Steen 1993). Lower esophageal hypomotility is present in about three-quarters of patients early in the course of disease, especially if cine-esophagrams are performed (Medsger and Steen 1993). Esophageal hypomotility usually becomes more severe over time, but intestinal hypomotility develops in only a minority of patients (Medsger and Steen 1993). Arthralgias or arthritis is more common in diffuse than limited disease. Tendon friction rubs occur in about two-thirds of patients with diffuse cutaneous systemic sclerosis, but is very uncommon in limited disease (Medsger and Steen 1993; Black and Denton 1995). An antinuclear antibody of any type is present in 95%, and the antitopoisomerase I antibody is present in 40% (Medsger and Steen 1993).

In patients who have the most prominent proximal scleroderma, survival is as low as 40% at 6 years (Clements et al. 1990) and 21% at 10 years (Barnett et al. 1988), without consideration of internal involvement. The proximal spread of scleroderma may be inhibited by treatment with D-penicillamine; this treatment may also reduce internal involvement and improve long-term prognosis (Jimenez and Sigal 1991). As with limited disease, prognosis is most directly dependent on the rate at which pulmonary, renal or cardiac involvement develops (Altman et al. 1991; Lee et al. 1992; Bulpitt et al. 1993; Czirjak et al. 1993; White et al. 1995). Among patients with any form of systemic sclerosis who had renal involvement, survival has been improved from 15 to 76% at 1 year by using angiotensin-converting enzyme inhibitors to treat the associated hypertension (Steen et al. 1990). Pulmonary interstitial fibrosis and pulmonary arterial hypertension are the most common causes of death (Medsger and Steen 1993; Black and Denton 1995). The risk of death from pulmonary hypertension is markedly increased when it develops in a patient with systemic sclerosis who has HLA haplotype DRw52 (Langevitz et al. 1992). Effective treatments have not been established for these pulmonary complications (Medsger and Steen 1993; Black and Denton 1995), but the use of immunosuppressive therapy is often tried and occasionally seems beneficial (McCune et al. 1994). Approximately 50% of patients with diffuse cutaneous systemic sclerosis have electrocardiographic abnormalities, but only 10–20% develop myocarditis or myocardial fibrosis (Follansbee et al. 1993; Medsger and Steen 1993). In these patients, arrhythmia or congestive heart failure are common causes for death (Follansbee et al. 1993). Treatment of myocarditis with corticosteroids or intravenous pulse methylprednisolone is not of proven benefit; it is occasionally beneficial in anecdotal cases (Clemson et al. 1992; Follansbee et al. 1993; Kerr and Spiera 1993; Black and Denton 1995). Other treatments, such as the usage of digi-

talis, antiarrhythmics and diuretics, are non-specific (Clemson et al. 1992; Follansbee et al. 1993; Black and Denton 1995).

Systemic sclerosis sine scleroderma
Systemic sclerosis sine scleroderma represents less than 2% of patients with systemic sclerosis (Medsger and Steen 1993; Black and Denton 1995). These are patients who have the typical complications of internal involvement from systemic sclerosis, but do not have sclerosis of the skin.

Overlap syndromes
The concept of an overlap syndrome is widely accepted when it is defined as the occurrence of features that fulfill the diagnostic criteria for two or more discrete types of connective tissue disease in the same patient. The most common overlap syndrome that includes systemic sclerosis is the concurrence of systemic sclerosis and polymyositis/dermatomyositis (Tuffanelli and Winkelmann 1961; Clements et al. 1978; Mimori 1987; Lotz et al. 1989). With this overlap syndrome, systemic sclerosis typically precedes polymyositis by several years (Ringel et al. 1990; Averbuch-Heller et al. 1992). This syndrome is discussed later in this chapter under the heading of Polymyositis/dermatomyositis. A less common overlap syndrome that includes systemic sclerosis is the concurrence of systemic sclerosis and systemic lupus erythematosus. With this overlap, either disorder may occur first (Asherson et al. 1991; Gendi et al. 1992).

Controversy surrounds the concept of an 'overlap' syndrome when it is defined as a distinct syndrome that shares some, but not all features of two or more discrete types of connective tissue disease. The most notable and controversial 'overlap' syndrome that includes systemic sclerosis is mixed connective tissue disease (Mukerji and Hardin 1993). Mixed connective tissue disease was first described in 1972 and was believed to represent a distinct overlap syndrome with partial features of systemic sclerosis, systemic lupus erythematosus and polymyositis/dermatomyositis, in which there was a high titer of extractable nuclear antigen and its ribonucleoprotein component (Sharp et al. 1972; Bennett et al. 1978). Although some authors continue to find that it often maintains distinct features over time (Lundberg and Hedfors 1991), other studies with long-term follow-up have not substantiated the existence of mixed connective tissue disease as a distinct disease with overlapping features, because the disorder in most patients evolves into definite systemic sclerosis or definite systemic lupus erythematosus without persistence of the features that suggested the other of these two diseases (Kallenberg 1994; Van den Hoogen et al. 1994; Gendi et al. 1995).

Pathogenesis

Systemic sclerosis is a disease of unknown cause in which mononuclear leukocyte adhesion and endothelial abnormalities that cause increased vascular permeability seem to be the earliest detectable pathological events, with subsequent interstitial infiltration by the mononuclear leukocytes (Black and Denton 1995). Serum levels of soluble adhesion molecules, such as endothelial leukocyte adhesion molecule-1 (ELAM-1), are often increased in patients with the recent onset of systemic sclerosis (Carson et al. 1993; Gruschwitz et al. 1995). However, this is not specific for systemic sclerosis, because ELAM-1 levels are also increased with recent onset systemic lupus erythematosus and various vasculitides (Carson et al. 1993). The mononuclear leukocytes are consistently interleukin-2 producing, CD4-positive T-lymphocytes, but it is uncertain whether these T-lymphocytes cause the initial endothelial abnormalities or are an immunologic response to it (Prescott et al. 1992; Black and Denton 1995). From a clinical perspective, the onset of the disease is usually associated with intense vasospasm of the extremities and internal organs. The possibility exists that this vasospasm causes or contributes to the endothelial abnormalities through the local effects of endothelial cellular products, such as cytokines, cell adhesion molecules or nitric oxide (Swerlick and Lawley 1993; Black and Denton 1995). The increased deposition of extracellular matrix, or fibrosis, follows the interstitial infiltration by mononuclear leukocytes. Many different cytokines have been implicated as mediators of fibrosis (Black and Denton 1995). Among the many cytokines that are expressed at high levels during fibrosis, transforming growth factor-b has attracted much interest, because it can stimulate transcription of collagen genes (Varga and Jimenez 1995). Furthermore, peripheral mononuclear cells from patients with systemic sclerosis produce greater amounts of transforming growth factor-b under controlled in vitro conditions than do mononuclear cells from normal subjects (Ota et al. 1995). However, in the tight-skin mouse model for scleroderma, persistent expression of collagen I and III genes has been demonstrated during the fibrotic process, without expression of transforming growth factor-b mRNA (Pablos et al. 1995).

Systemic sclerosis has been reported rarely to develop in two or more members of a family. The predisposition to immune-mediated disease caused by certain shared HLA haplotypes may partially explain familial occurrence in affected identical twins (Cook et al. 1993) and in some affected first-degree relatives in other families (Sasaki et al. 1991; Hietarinta et al. 1993; McColl and Buchanan 1994; Manolios et al. 1995). However, the familial occurrence of systemic sclerosis is not explained sufficiently or necessarily by shared HLA haplotypes. In some families in which shared HLA haplotypes seem relevant, other family members have the shared haplotype without systemic sclerosis, to suggest that the HLA haplotype may predispose to, but is not sufficient by itself to cause familial systemic sclerosis (Hietarinta et al. 1993; Manolios et al. 1995). Such shared HLA haplotypes are different in different families (Manolios et al. 1995). Furthermore, in other families in which more than one member has systemic sclerosis, HLA haplotypes are not shared by affected members (De Juan et al. 1994; Manolios et al. 1995).

Localized scleroderma

Localized scleroderma includes morphea, generalized morphea, and linear scleroderma. Localized forms of scleroderma usually affect children and young adults, who are more often females than males. Although present on the trunk, proximal limb or face, sclerodermatous lesions that may be classified as a localized form of scleroderma are specifically excluded as being a form of 'proximal scleroderma' that fulfills the major criteria for the diagnosis of diffuse cutaneous systemic sclerosis (Subcommittee for Scleroderma Criteria 1980). Furthermore, patients with localized forms of scleroderma rarely have even one of the two or more minor features that support the diagnosis of systemic sclerosis (Subcommittee for Scleroderma Criteria 1980) and rarely have Raynaud's phenomenon or other common symptoms of systemic sclerosis, except for a localized skin lesion.

Biopsies of skin lesions in patients with localized forms of scleroderma will reveal a similar thinning of the epidermis with loss of rete pegs and often increased dermal thickness as are seen in skin biopsies of patients with systemic sclerosis; however, the diffuse vascular changes and symptomatic visceral involvement are not seen (Falanga et al. 1986; Dehen et al. 1994; Uziel et al. 1994).

Morphea

Morphea is defined clinically as one to several circumscribed areas of thickened or atrophic skin that are associated with pigmentary change (Dehen et al. 1994; Uziel et al. 1994). The skin lesion typically begins on the trunk as a small plaque with an ivory-colored, warm, edematous center and a violaceous border. The border reflects the area of active inflammation. Over months or a few years, the size of the sclerotic plaque stabilizes, and it loses the violaceous border. It then becomes either soft and atrophic or firm and sclerotic and is either hypopigmented or hyperpigmented. During the initial months of development, the skin lesion often becomes bound to the underlying subcutaneous tissue.

Morphea is approximately 2–3 times more common in females than males. It may begin at any age, with the age of onset in one series having a mean of 32 years and a range from 1 month to 74 years (Dehen et al. 1994). A polyclonal hypergammaglobulinemia and the presence of an antinuclear antibody each occurs in nearly half of patients. However, anti-double-stranded DNA antibodies, anti-Scl-70 and anticentromere antibodies are rarely, if ever, detected in morphea (Dehen et al. 1994). Asymptomatic abnormalities of esophageal motility and pulmonary gas exchange each have been found in 17% of patients with morphea, but are symptomatic only in those uncommon cases that progress to systemic sclerosis (Dehen et al. 1994). Transition from morphea to systemic sclerosis has been reported in individual case reports and in 1–6% of patients in large series (Birdi et al. 1993; Dehen et al. 1994).

Generalized morphea

Generalized morphea is separated traditionally from morphea by the presence of multiple morphea lesions that cover a larger proportion of the body surface area (Dehen et al. 1994; Uziel et al. 1994). However, the number of lesions and the percentage of involved to total body surface area that are required for this diagnosis are not well defined. Generalized morphea rarely covers more than 20% of the total body surface area (Dehen et al. 1994). Generalized morphea is the least common of the three major forms of localized scleroderma, encompassing 3–19% of cases (Uziel et al. 1994). Patients who more rapidly develop generalization of morphea appear to be at increased risk of developing asymptomatic esophageal or pulmonary involvement, but are not clearly at increased risk of developing systemic sclerosis (Dehen et al. 1994).

Linear scleroderma

Linear scleroderma is defined clinically as an area of thickened or atrophic skin that is associated with pigmentary change and that has a shape with one axis much longer than the other. Such a linear lesion typically begins on a limb or the face. A lesion that begins on the face is called an 'en coup de sabre' lesion. In one large series, 44 of 53 patients (83%) experienced the onset of linear scleroderma before the age of 25 years, and females were affected 4 times more often than males (Falanga et al. 1986). Morphea often coexists with linear scleroderma (Dehen et al. 1994). Antinuclear antibodies, anti-single-stranded DNA antibodies and eosinophilia were present in approximately half of patients, usually those with active disease for over 2 years and more extensive involvement (Falanga et al. 1986). In spite of the common occurrence of the immune abnormalities, none of the 53 patients in this series later developed systemic sclerosis, with a mean follow-up time of 10 years. Anecdotal case reports have documented the rare transition of linear scleroderma into systemic sclerosis (Mayorquin et al. 1994). In about half of patients with linear scleroderma in a limb, disability results from a joint contracture beneath a longitudinal skin lesion that crosses a joint; these patients often had hypergammaglobulinemia (Falanga et al. 1986).

NEUROLOGICAL COMPLICATIONS OF PRIMARY SCLERODERMA

In the past, primary scleroderma has been considered to be the diffuse connective tissue disease that has the least frequent neurological abnormalities. The incidence of neurological complications had been reported to be as low as 0.8% in a study that did not classify myopathy as a neurological manifestation (Tuffanelli and Winkelmann 1961). Furthermore, in early series with more common neurological complications, these features were usually thought to be coincidental, iatrogenic or secondary to other organ involvement, rather than directly caused by scleroderma (Gordon and Silverstein 1970). Over the past decade, this idea has changed. In one retrospective study of 50 patients with systemic sclerosis who were seen over a 10-year period, 20 patients (40%) developed one or more neurological manifestations, with no identifiable cause other than scleroderma in half of these patients (Averbuch-Heller et al. 1992). Myopathy was the most common form of neurological involvement and occurred in 11 of the 50 patients (22%). Peripheral neuropathy was nearly as common and detected in 9 patients (18%). Other neurological manifestations included myelopathy in 4 patients (8%), cerebrovascular disease in 3 (6%) and myasthenia gravis in 1 (2%). A different group has also reported a high incidence of neurological complications in systemic sclerosis (Hietaharju et al. 1993a, b). These authors have suggested that patients with systemic sclerosis who have anti-U1RNP antibodies, and perhaps those with anti-Scl-70 antibodies, are at higher risk of developing neurological complications (Hietarinta et al. 1994).

Myopathy

Myopathy is the most common neurological manifestation of primary scleroderma. It has been described in up to 96% of patients (Clements et al. 1978), but the prevalence of myopathy depends on the design of the study, the features required to suggest the diagnosis, and the proportion of patients with systemic sclerosis rather than systemic or localized scleroderma. In the series with the highest prevalence of myopathy, 24 patients with systemic sclerosis followed at one institution were examined prospectively with manual strength testing, electromyography, muscle enzyme levels and, if the preceding raised the possibility but not the certainty of myopathy, a muscle biopsy; 23 of the 24 patients (96%) were found to have myopathy (Clements et al. 1978). At the other end of the prevalence spectrum, myopathy was identified in only 9 of 130 patients (7%) in one series based on a retrospective review of inpatient records for documentation of weakness or abnormal muscle biopsy

results (Gordon and Silverstein 1970). Most groups report an intermediate prevalence. Symptoms of weakness, fatigability, or myalgia have been reported in 50–83% of patients with scleroderma (Medsger et al. 1968; Hietaharju et al. 1993b). Objective clinical and laboratory signs of myopathy have been reported in 19–53% of patients (Medsger et al. 1968; Averbuch-Heller et al. 1992; Hietaharju et al. 1993b). Electromyographic (EMG) studies on groups of patients with systemic sclerosis and localized scleroderma demonstrate that myopathy is very common in systemic sclerosis and that muscle involvement is more common and diffuse in systemic sclerosis than in localized scleroderma (Hausmanowa-Petrusewicz and Kozminska 1961; Hausmanowa-Petrusewicz et al. 1982).

Myopathies that have been described in primary scleroderma can be separated into seven categories for the purpose of discussion: (1) myositis or polymyositis, (2) myopathy with vasculopathy, (3) a 'simple' or non-progressive myopathy, (4) inclusion body myositis, (5) corticosteroid myopathy, (6) disuse myopathy, and (7) focal myopathy with localized scleroderma (Table 2). The differences in prevalence rates for all types of myopathy taken together are explained largely by the differences in prevalence of the 'simple', non-progressive myopathy. The vast majority of described myopathies fit into one of the first three categories, which have overlapping features. For example, based on one's bias to 'lump or split', the cases of myopathy with vasculopathy described by Ringel et al. (1990) could be included in the myositis category as dermatomyositis or maintained in a separate category. Furthermore, because the first two categories are based on pathological features and the third on the clinical course, some cases may fit into more than one category, especially if additional data were available. For example, the cases described as

TABLE 2

Myopathies that are associated with primary scleroderma.

1. Myositis or polymyositis
2. Myopathy with vasculopathy
3. 'Simple' or non-progressive myopathy
4. Inclusion body myositis
5. Corticosteroid myopathy
6. Disuse myopathy
7. Focal myopathy with localized scleroderma

'simple' or non-progressive myopathy by Clements et al. (1978), Averbuch-Heller et al. (1992) and Hietaharju et al. (1993b), which do not include analysis of the microvasculature, have benign clinical features that suggest overlap with most of the patients described by Russell and Hanna (1983), who had microvascular abnormalities on muscle biopsies. Myopathy due to fibrosis is not included among these categories (Medsger et al. 1968), because many of these cases may have represented myositis or myopathy with vasculopathy and because more recent series have not found that fibrosis is a likely explanation for myopathy in scleroderma (Clements et al. 1978; Ringel et al. 1990).

Polymyositis/dermatomyositis

The most acutely disabling form of myopathy that develops in patients with primary scleroderma is polymyositis or dermatomyositis. Most of these patients have a true overlap syndrome, fulfilling criteria for both systemic sclerosis and polymyositis/dermatomyositis. Among patients with polymyositis/dermatomyositis, 4–26% also have primary scleroderma (Mimori 1987; Lotz et al. 1989). Among patients with primary scleroderma, 5–12% develop polymyositis/dermatomyositis (Tuffanelli and Winkelmann 1961; Clements et al. 1978; Mimori 1987). These myopathies occur commonly in patients with systemic sclerosis but rarely and focally in patients with localized scleroderma (Schwartz et al. 1981). Systemic sclerosis typically precedes muscle involvement by several years (Ringel et al. 1990; Averbuch-Heller et al. 1992). The myopathic features of polymyositis, dermatomyositis and myositis with systemic sclerosis are virtually identical (Clements et al. 1978; Mimori 1987; Ringel et al. 1990). These patients develop progressive weakness acutely or subacutely that usually is more prominent in the proximal than the distal limb muscles and affects neck flexion (Mimori 1987). Myalgia is common, especially with activity. In patients with the concurrence of systemic sclerosis and myositis, violaceous scaling patches over the knuckles (Gottron's sign) and a heliotropic rash of the upper eyelids that are characteristic of dermatomyositis are seen in about one-fourth of patients (Mimori 1987).

Elevation of serum muscle enzyme levels is usually present and helps to distinguish this type of myopathy from the others that occur in association with systemic sclerosis. Serum levels of creatine kinase and aldolase are typically elevated mildly or moderately (Clements et al. 1978; Mimori 1987; Ringel et al. 1990; Averbuch-Heller et al. 1992; Hietaharju et al. 1993b). Markedly elevated levels that are more than 10-fold above the upper limit of normal are not rare (Clements et al. 1978; Averbuch-Heller et al. 1992; Hietaharju et al. 1993b).

Non-specific immunologic abnormalities include the usual elevation of the erythrocyte sedimentation rate and the common presence of antinuclear antibodies (Mimori 1987; Ringel et al. 1990). Less sensitive but more specific abnormalities include the presence of anti-PM-Scl (polymyositis-scleroderma) antibodies in North American and British patients and anti-Ku antibodies in Japanese patients. In one series of British patients, the anti-PM-Scl antibody was found in about 10% of 256 patients with polymyositis/dermatomyositis and in about 3% of 879 patients with systemic sclerosis; however, among the 32 patients who had the antibody, all had systemic sclerosis and 28 patients (88%) had polymyositis (Marguerie et al. 1992). In other series, about 50% of patients with this overlap syndrome have been found to have the anti-PM-Scl antibody, which is directed primarily against a nucleolar 100 kDa protein (Bluthner and Bautz 1992; Ge et al. 1994). The development of this antibody is associated with the presence of HLA DR3 and DQw2 alleles (Oddis et al. 1992). In a similar manner for a series of Japanese patients that included 240 with diffuse cutaneous systemic sclerosis and 105 with polymyositis/dermatomyositis, 10 of 26 patients (38%) with both diseases had anti-Ku antibodies and none had anti-PM-Scl antibodies (Mimori 1987).

EMG studies are useful to confirm the presence of myopathy as the cause for fatigability and weakness, particularly in patients with pain from myalgia or arthralgia that may be limiting effort and obscuring the presence of weakness. The results of EMG studies are abnormal in 92–100% of patients with myositis and systemic sclerosis (Clements et al. 1978; Mimori 1987; Ringel et al. 1990; Averbuch-Heller et al. 1992). Two types of abnormalities are usually recorded. One typical abnormality is the finding of brief, small, polyphasic motor unit action potentials with rapid recruitment. This finding reflects a loss in the number of normally functioning muscle fibers per motor unit and hence a myopathic pathophysiology for clinical weakness. The other frequent abnormality is recording of fibrillation potentials, which are produced by the spontaneous discharge of single muscle fibers that are not connected to their nerve supply. The presence of fibrillation potentials in myopathy correlates with elevation of serum muscle enzyme levels and with pathological evidence for degeneration/regeneration of muscle fibers. The combination of these two EMG findings in a patient with systemic sclerosis is strongly supportive of myositis.

Pathologically, most patients with myositis and systemic sclerosis have endomysial mononuclear inflammatory infiltration and degeneration/regeneration of muscle fibers (Clements et al. 1978; Ringel et al. 1990; Averbuch-Heller et al. 1992; Hietaharju et al. 1993b). These pathological features are similar to those seen in the muscles of patients who have polymyositis without systemic sclerosis. However, differences between polymyositis, dermatomyositis and myositis with scleroderma have been seen in infiltrating cell type with a quantitative analysis using monoclonal antibody markers and histochemical stains (Arahata and Engel 1984). The predominant

perimysial exudate consisted of T-cells and macrophages in myositis with scleroderma (Fig. 1); T-cells and macrophages invaded non-necrotic muscle fibers less frequently in myositis with scleroderma than in polymyositis. The percentage of B-cells at all sites was significantly higher in dermatomyositis than in myositis with systemic sclerosis.

Patients with myositis and systemic sclerosis respond well to immunosuppressive treatment in a manner similar to adults with polymyositis or dermatomyositis alone (Clements et al. 1978; Mimori 1987; Averbuch-Heller et al. 1992; Marguerie et al. 1992). In most patients, treatment with high-dose daily corticosteroids (such as oral prednisone 1 mg/kg/day) is effective in inducing a remission. Occasionally, treatment in patients with dysphagia or with severe weakness and markedly elevated creatine kinase levels is initiated with i.v. methylprednisolone 0.5–1.0 g/day for 3 consecutive days. Azathioprine or methotrexate are commonly begun 1 or 2 weeks after the initiation of corticosteroid therapy, so that the corticosteroids may be tapered more rapidly several months later. No long-term, randomized treatment trials are avail-

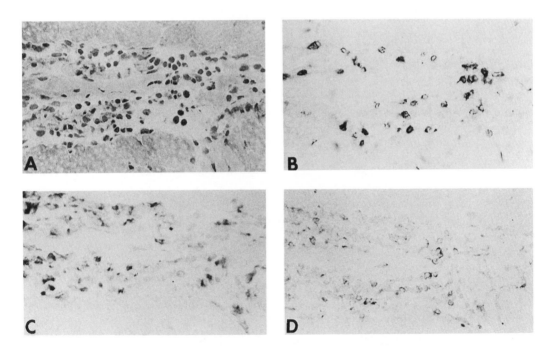

Fig. 1. Perimysial inflammatory cells in systemic sclerosis: 2 μm thick serial sections stained trichromatically (A), and reacted for the T8 (B), acid phosphatase (C), and T4 (D) markers (all × 260). Nearly all the inflammatory cells are accounted for by T-cells and macrophages. More T8⁺-cells than T4⁺-cells, and more T-cells than macrophages are present. (From Arahata and Engel 1984.)

able, but one series of 25 patients with a median follow-up of 8 years has reported that two-thirds of patients were receiving less than 7.5 mg/day of prednisolone and were doing well with regard to the myositis (Marguerie et al. 1992).

Myopathy with vasculopathy

Abnormalities of intramuscular capillaries have been recognized as a common feature of myopathy in scleroderma since 1968 (Norton et al. 1968; Jerusalem et al. 1974; Russell and Hanna 1983). Ultrastructural features of these capillary lesions include reduplication of capillary basement membrane (BM), an increase in mean capillary area, and an increase in the mean number of endothelial cells per capillary cross-section (Jerusalem et al. 1974).

In one detailed study of muscle histology from patients with diffuse cutaneous systemic sclerosis who were not selected for the presence of muscular symptoms or signs, strength of the 28 patients was measured quantitatively and the patients were divided into 2 groups, 19 with normal strength and 9 with weakness (Russell and Hanna 1983). Only one-third of patients with weakness and one-fourth of patients with normal strength had elevation of serum muscle enzyme levels. EMG studies identified evidence of myopathy with small, brief polyphasic motor unit action potentials more often in weak (86%) than strong (46%) patients; fibrillation potentials were recorded from only one patient with weakness. On histologic examination, none of the patients had endomysial inflammatory infiltration or muscle fiber

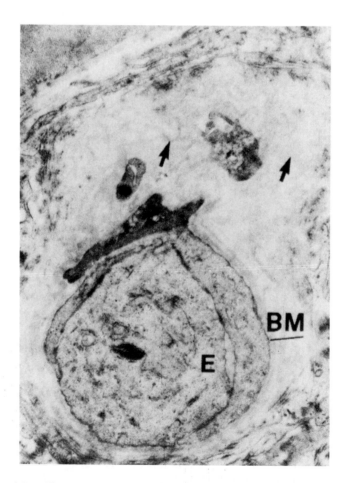

Fig. 2. An endomysial capillary with a swollen endothelial cell (E) occluding the lumen from a patient with systemic sclerosis. The basement membrane (BM) is laminated and concentrated on the side of the capillary with the arrows. (From Russell and Hanna 1983.)

degeneration/regeneration to suggest myositis. The percentage of capillaries with swollen endothelial cells, thickened BMs, or surrounding 'felt-like' material was significantly greater in strong patients with systemic sclerosis than in controls and significantly greater in weak than strong patients (Fig. 2). Thus, microvascular abnormalities within the muscle seemed relevant to the pathogenesis of myopathy in systemic sclerosis.

In a different series, the primary criteria for inclusion were: (1) the diagnosis of scleroderma or an overlap connective tissue disease with scleroderma, (2) clinical weakness, and (3) the performance of a muscle biopsy (Ringel et al. 1990). Symmetric proximal weakness developed after the onset of scleroderma in all 14 patients. Creatine kinase and erythrocyte sedimentation rate were each elevated in two-thirds. EMG studies revealed small, short, polyphasic motor unit action potentials with fibrillation potentials in all 14 patients. Muscle biopsy was abnormal in all 14 patients. In 7 of 14 (50%), the fundamental pathology related to vasculopathy. These seven patients had intimal proliferation in endomysial and perimysial vessels. Five also had perivascular inflammatory infiltrates, one had necrotizing vasculitis, and two had perifascicular atrophy. Endomysial and perimysial connective tissue was only mildly increased, arguing against fibrosis as an etiopathogenetic mechanism. These seven patients with myopathy and vasculopathy share many features in common with dermatomyositis, especially the five with perivascular inflammatory infiltrate.

In one quantitative study of quadriceps muscle biopsies in 12 patients with myopathy in systemic sclerosis and in 15 with polymyositis or other myopathies, type 2 fiber atrophy, reduction in capillary density and decreased capillary : fiber ratio were reduced more in myopathy with systemic sclerosis than in other categories (Scarpelli et al. 1992).

Thus, abnormalities of the microvascular supply seem to be frequently related to the development of myopathy in patients with scleroderma.

'Simple' or non-progressive myopathy
The term 'simple' myopathy was coined by Clements et al. (1978) to describe the most common form of myopathy that their patients developed. In their prospective study, 19 of 24 patients (79%) with systemic sclerosis had this myopathy. At initial evaluation, proximal muscle strength was estimated by manual testing to be mildly reduced (between 62 and 89% of normal for age and sex) in 16 and within normal limits (between 90 and 99% of normal) in the other 3 patients. Creatine kinase levels were normal in 12, minimally elevated in 6 (between 1 and 2 times the upper limit of normal) and moderately elevated in 1 (approaching 10 times the upper limit). With EMG studies, 18 of 19 had an increased incidence of polyphasic motor unit action potentials, but amplitude and duration were normal in all. Furthermore, no patient had spontaneous fibrillation potentials. Weakness, elevated creatine kinase level or polyphasic motor unit action potentials were seen in all 19 patients. Only 3 underwent a muscle biopsy, which revealed fibrosis and variation in muscle fiber size but not inflammatory infiltration. Over the course of long-term follow-up (mean duration of 28 months), strength remained stable with 18 patients receiving no treatment and 1 patient receiving a 4-month course of corticosteroids for uncertain indication.

Descriptions of myopathy in systemic sclerosis since 1978 either have not mentioned 'simple' or non-progressive myopathy or have recognized it in a smaller proportion of patients. In one series, 4 of 11 patients (36%) were diagnosed to have a simple myopathy based on minimal weakness, mild elevation of creatine kinase levels, EMG abnormalities limited to polyphasic motor unit action potentials of normal size, and a stable course for strength without immunosuppressive treatment (Averbuch-Heller et al. 1992). In another series, only 2 of 32 patients (6%) were said to have a simple or non-progressive myopathy (Hietaharju et al. 1993b). Histopathological features are not described for the 6 patients in these last two series.

Inclusion body myositis
In contrast to patients with polymyositis or dermatomyositis, patients with inclusion body myositis tend to be older, to develop weakness more insidiously and to have more common involvement of distal muscles even though proximal weakness predominates (Lotz et al. 1989). Muscle enzyme levels are normal or only mildly elevated, and EMG abnormalities are similar to polymyositis. The diagnosis is suggested by the light microscopic findings of muscle fibers with rimmed vacuoles, mononuclear inflammatory infiltrate that is predominantly endomysial, and invasion of non-

necrotic muscle fibers with mononuclear cells. The diagnosis is confirmed by the electron microscopic finding of characteristic filamentous inclusions in the nucleus or sarcoplasm. Treatment with prednisone, azathioprine or methotrexate rarely produces significant long-term benefit.

The uncommon association of inclusion body myositis with any connective tissue disease has been described only recently, and scleroderma was not one of the associated diseases in this large series (Lotz et al. 1989). Inclusion body myositis has been described anecdotally in three cases associated with systemic sclerosis (Helm et al. 1990; Averbuch-Heller et al. 1992). In one of these cases in which the diagnosis was confirmed by both light and electron microscopy, high-dose corticosteroids produced rapid clinical improvement in a 22-year-old man (Averbuch-Heller et al. 1992). This case suggests that a trial of corticosteroids may be warranted in patients with systemic sclerosis who develop inclusion body myositis.

Corticosteroid myopathy

Because the development of myopathy is a well known complication of corticosteroid therapy with little regard to the indications for its usage, this type of myopathy has been reported in few cases of systemic sclerosis (Gordon and Silverstein 1970) and is undoubtedly far more common than the limited number of reports reflect. Corticosteroid myopathy presents with mild proximal weakness. Creatine kinase levels are in the normal range, unless a second disease is present. EMG examination produces normal results, unless the myopathy is severe or of long standing. When abnormal, small short motor unit action potentials with rapid recruitment are seen with a reduction in the peak-to-peak amplitude of the interference pattern. Fibrillation potentials are not produced by corticosteroid myopathy. If the creatine kinase level is elevated or fibrillation potentials are recorded, alternative diagnoses such as myositis should be considered. Muscle biopsy is indicated to evaluate corticosteroid myopathy only when alternative diagnostic possibilities are under strong consideration. The pathological feature of corticosteroid myopathy is atrophy of type 2 more than type 1 fibers.

Disuse myopathy

Disuse myopathy has features similar to those of corticosteroid myopathy, so it is not diagnosed unless corticosteroids have not been used or weakness has developed after discontinuation of corticosteroids. Weakness is proximal or about a joint with pain, contracture or fibrosis. Creatine kinase levels are in the normal range. EMG examination produces normal results, unless the myopathy is severe or of long standing. When abnormal, small short motor unit action potentials with rapid recruitment are seen with a reduction in the peak-to-peak amplitude of the interference pattern. Fibrillation potentials are not produced by disuse myopathy. If the creatine kinase level is elevated or fibrillation potentials are recorded, alternative diagnoses such as myositis should be considered. Muscle biopsy is indicated to evaluate disuse myopathy only when alternative diagnostic possibilities are under strong consideration. The pathological feature of disuse myopathy is atrophy of type 2 more than type 1 fibers. In recent series that focus on myopathies, this category is diagnosed uncommonly (Hietaharju et al. 1993b). Some patients with 'simple' or non-progressive myopathy who have normal creatine kinase levels may also fit under this designation.

Focal myopathy with localized scleroderma

Myopathy has also been reported with the 'en coup de sabre' lesion (Schwartz et al. 1981). The involved muscle was subjacent to the cutaneous lesion and revealed both type 1 and type 2 fiber atrophy with mild mononuclear infiltration. Although it is not clear that the patients had focal weakness clinically, EMG evidence for focal myopathy without fibrillation potentials has been identified as a common finding subjacent to localized skin lesions in morphea (Hausmanowa-Petrusewicz and Kozminska 1961).

Peripheral neuropathy

Although once thought to be a rare and coincidental complication of scleroderma, peripheral neuropathy, especially distal axonal polyneuropathy and trigeminal neuropathy, is now recognized in 10–20% of patients with scleroderma, especially systemic sclerotic forms (Averbuch-Heller et al. 1992; Hietaharju et al. 1993b). Peripheral neuropathies associated with pri-

TABLE 3

Neuropathies that are associated with primary
scleroderma.

1. Distal axonal polyneuropathy
2. Trigeminal sensory neuropathy
3. Vasculitic neuropathy
4. Other neuropathies reported in primary
   scleroderma
   (a) compression neuropathies
   (b) brachial neuritis
   (c) optic neuropathy

mary scleroderma include distal axonal polyneuropathy and trigeminal sensory neuropathy as common
types and vasculitic neuropathy as a rare type; other
neuropathies have been reported in cases of primary
scleroderma, but do not have a well described incidence (Table 3).

Distal axonal polyneuropathy

Distal axonal polyneuropathy is the most common
type of generalized neuropathy that develops in scleroderma and has been observed in the range of 12–
14% with systemic sclerosis (Averbuch-Heller et al.
1992; Hietaharju et al. 1993b). In these patients, the
presenting complaints usually include paresthesias
and blunted sensation in the toes or feet (Olney
1992). Loss of position sense also may be noticed as
imbalance. Patients who develop sharp stabbing,
lancinating or burning pain in the feet are usually
found to have a peripheral neuropathy. In contrast to
vasculitic neuropathy, sensory symptoms in distal axonal polyneuropathy begin in the toes or feet symmetrically and gradually spread proximally over time
in a length-dependent manner. By the time symptoms
begin in the finger tips symmetrically, distal lower limb
symptoms have spread up to the mid-calf level. Distal
weakness is not usually a presenting complaint, but
may be noted on examination.

The neurologic signs of distal axonal polyneuropathy usually include depressed or absent ankle tendon
reflexes and decreased threshold for perception of
vibration in the toes and at the ankle (Olney 1992). A
decreased threshold for perception of pain, warm
and cold is common as well. Weakness of toe movements and reduced bulk of intrinsic foot muscles is
occasionally seen early in the course of distal axonal
polyneuropathy, but is uncommon until the upper
edge of the stocking decrease in sensation has spread
proximally to the mid-calf level.

Clinical evidence for altered sensory perception
has been seen in up to half of 29 patients for tactile
thresholds in the feet, with reduced thresholds for
temperature in 5 and vibration in 1 (Schady et al.
1991). The disproportionately greater involvement of
tactile sensation may be explained, at least in part, by
increased firmness of the skin from scleroderma. In a
different series, quantitative vibratory detection
thresholds were as sensitive as any clinical sign or
laboratory test at documenting distal axonal polyneuropathy (Hietaharju et al. 1993b).

Nerve conduction studies reveal symmetrically
reduced or absent sensory nerve action potentials in
the feet, normal or reduced compound muscle action
potentials in the feet, and normal or mildly reduced
nerve conduction velocities when present (Hietaharju
et al. 1993b). EMG studies reveal large, long motor
unit action potentials in distal lower limb muscles with
reduced interference patterns (Hietaharju et al.
1993b). Sural nerve biopsies reveal axonal degeneration with increased endoneurial connective tissue and mild microvascular changes (Corbo et al.
1993).

The pathogenesis for distal axonal polyneuropathy in scleroderma is uncertain, but does not usually
involve vasculitis. Visceral involvement was common
among 7 of 50 patients (14%) with this neuropathy
and systemic sclerosis in one series, with renal failure
in one and with gastrointestinal involvement in five
others; however, except for uremia in the one, results
of metabolic evaluation did not reveal a causal
relationship between the visceral involvement and the
polyneuropathy (Averbuch-Heller et al. 1992).

The differential diagnosis for distal axonal polyneuropathy in a patient with systemic sclerosis includes other causes for the neuropathy, including
diabetes mellitus, alcoholism, renal failure, hypothyroidism, vitamin B_{12} deficiency, toxic medications,
environmental toxins, monoclonal protein disorders
and inherited neuropathies. Thus, the history needs to
include detailed questions about alcohol consumption, prescription medications, non-prescription medications, potential toxic exposures at work and home
and family history for neuropathic symptoms and foot
deformities. The laboratory evaluation includes fast-
ing blood sugar, liver enzyme tests, creatinine, blood
urea nitrogen, thyroid stimulating hormone level,
serum vitamin B_{12} level, and serum protein electrophoresis at a minimum. If the preceding are normal or

do not suggest another etiology, the distal axonal polyneuropathy is presumed to be related to systemic sclerosis. In routine practice, nerve biopsy is not indicated, unless asymmetric onset of symptoms, asymmetric clinical or electrophysiological signs or rapid progression raise a suspicion for vasculitis.

Treatment for distal axonal polyneuropathy is symptomatic for pain and rehabilitative for functional difficulties. Corticosteroids or other immunosuppressive treatments are not known to have benefit.

Trigeminal sensory neuropathy

Trigeminal sensory neuropathy is characterized by slowly progressive facial numbness that may be either unilateral or bilateral and is often associated with paresthesias or pain (Hagen et al. 1990). In one large series of patients with trigeminal sensory neuropathy, 15 of 81 patients (19%) had systemic sclerosis (Hagen et al. 1990). Among patients with systemic sclerosis, trigeminal neuropathy has been described in the range of 2 in 50 (4%) to 2 in 32 (6%) (Averbuch-Heller et al. 1992; Hietaharju et al. 1993b). Furthermore, this neuropathy is often the presenting symptom of systemic sclerosis (Farrel and Medsger 1982; Lecky et al. 1987; Hietaharju et al. 1993b). The sensory symptoms often begin in a small unilateral patch around the mouth or in the distribution of the maxillary division (Lecky et al. 1987; Hagen et al. 1990). The involved territory usually expands gradually and unilaterally. The maximal deficit is usually reached over several months to 2 years (Hagen et al. 1990). Contralateral sensory symptoms may develop several years later or never occur (Lecky et al. 1987; Hagen et al. 1990). Pain and paresthesias each develop in 50–60% of patients (Hagen et al. 1990).

With blink reflex studies, an afferent delay (i.e., delayed ipsilateral R1 and bilateral R2) or an absent response is recorded in about one-half of patients (Lecky et al. 1987; Hagen et al. 1990). Pathological data are quite limited, but support degeneration of peripheral myelinated axons from a lesion of, or distal to, the gasserian ganglion (Lecky et al. 1987). The cause of trigeminal sensory neuropathy is uncertain, but may involve either localized vasculitis or fibrosis (Teasdall et al. 1980; Lecky et al. 1987). The development of trigeminal sensory neuropathy does not herald the onset of systemic vasculitis or generally indicate the need to initiate immunosup-

pressive therapy (Hagen et al. 1990). In one series, prednisone treatment of 38 patients resulted in improvement of pain in only two, and facial numbness did not change in any patient (Hagen et al. 1990). However, the diagnosis of trigeminal sensory neuropathy may lead to the diagnosis and treatment of systemic sclerosis.

Vasculitic neuropathy

Vasculitic neuropathy is defined by the pathogenetic mechanism that produces injury to the nerve fibers: inflammatory occlusion of blood vessels produces ischemic infarction of one or more nerves (Olney 1992). Vasculitic neuropathy is usually suspected clinically in a patient with a mononeuropathy multiplex and known connective tissue disease. However, it may be the initial manifestation of connective tissue disease and present as a generalized polyneuropathy with little or no asymmetry rather than a mononeuropathy multiplex. Thus, the possibility of vasculitic neuropathy needs to be considered in many patients with neuropathy of undefined cause, especially in those patients in whom symptoms and signs have developed with asymmetry or without following a length-dependent distribution or in whom functionally significant deficits have developed rapidly.

Pain is a more common feature of vasculitic neuropathy than of distal axonal polyneuropathy (Olney 1992). Pain may be an acute deep ache that is poorly localized but proximal within an affected limb or may be a distal burning. Paresthesias, decreased sensory thresholds, and weakness develop acutely or subacutely in nearly all patients, often asymmetrically. On neurological examination, most patients have weakness and abnormal sensation for pain and temperature, whereas impairment of vibration and position sense is uncommon unless the neuropathy is advanced or confluent. The peroneal nerve is the one that is most frequently affected by vasculitis, and the ulnar nerve is the most commonly involved one in the upper limb.

Nerve conduction studies document a process characterized by axonal loss with decreased amplitude of sensory nerve and compound muscle action potentials and with normal or mildly reduced conduction velocities. Detection of asymmetric or non-length-dependent axonal loss particularly supports vasculitic neuropathy. These signs include: (1) 2-fold or greater difference in amplitude between right- and left-sided responses of the same nerve, (2) a low amplitude response for one but not another nerve within a limb, or (3) a low amplitude response for an upper limb nerve, if amplitude is normal for at least one lower limb nerve. With needle EMG studies, one sign of a non-length-dependent axonal loss is the presence of acute partial denervation (reduced recruitment of motor unit action potentials with or without fibrillation potentials) in some but not other proximal muscles.

In patients with known or suspected systemic sclerosis who develop an asymmetric peripheral neuropathy, aggressive evaluation is usually undertaken. Although this is an uncommon neuropathy in the setting of systemic sclerosis, vasculitic neuropathy is the most serious and life-threatening one. In large series of vasculitic neuropathy, the incidence of systemic sclerosis as an associated connective tissue disease ranged from 1 to 3% (Harati and Niakan 1986; Said et al. 1988). Mononeuritis multiplex apparently due to vasculitis has been described in one case among 125 patients with systemic sclerosis in one prospective series (Lee et al. 1984), in case reports of systemic sclerosis (Herrick et al. 1994a; Leichenko et al. 1994), and in a single case associated with morphea (Goldsmith et al. 1991).

Nerve biopsy is usually performed to confirm this pathogenetic mechanism and thereby initiates long-term immunosuppressive treatment. Additionally, a broad group of laboratory tests is performed to assess the possibility of an alternative associated systemic disease. These tests often include complete blood count with differential and platelet count, erythrocyte sedimentation rate, antineutrophil cytoplasmic antibody (ANCA), antinuclear antibody (ANA), rheumatoid titer, complement levels, hepatitis B surface antigen, chemistry tests of renal and liver function, and urinalysis.

Treatment of vasculitic neuropathy is usually initiated with corticosteroids, either oral prednisone 1 mg/kg/day or i.v. methylprednisolone 1 g/day for 3 days followed by oral prednisone (Olney 1992). If the vasculitis is systemic and necrotizing, oral cyclophosphamide 2 mg/kg/day is often added and continued for 1 or more years. Recovery from the sensory and motor deficits is likely in survivors, with meaningful improvement in 28% at 3 months, 60% at 6 months, and 86% at 1 year (Chang et al. 1984).

Other neuropathies

All patients who present with focal neuropathies do not have vasculitic neuropathy. Entrapment or compression neuropathies offer an alternative explanation in the limbs. Cases of carpal tunnel syndrome and ulnar neuropathy have been described (Lee et al. 1984; Berth-Jones et al. 1990; Averbuch-Heller et al. 1992). Fibrosis associated with scleroderma offers an intuitively obvious potential explanation for entrapment neuropathies, but an increased incidence of carpal tunnel syndrome or other entrapments relative to a control population has not been well documented in prospective series, probably because of the relatively low incidence of scleroderma. A case of brachial neuritis has been reported (Hietaharju et al. 1993b). Unilateral or bilateral optic neuropathy may develop in association with limited cutaneous systemic sclerosis (Boschi et al. 1993; Hietaharju et al. 1993b). With these various types of neuropathy, the association with scleroderma may be the chance occurrence of the two diseases, rather than a causal relationship.

Autonomic dysfunction

Symptoms or signs of autonomic dysfunction are common in studies in which they have been carefully sought. The autonomic abnormalities sometimes suggest an autonomic neuropathy. For example, a mean reduction in normal heart rate variability was seen in 8 patients with systemic sclerosis with limited skin thickening, but not in 9 patients with systemic sclerosis with diffuse skin thickening, compared with 17 age- and sex-matched control subjects (Hermosillo et al. 1994). In studies that have utilized the sympathetic skin response to assess autonomic function in the limb, abnormalities have been found in two-thirds of 32 patients with systemic sclerosis or localized scleroderma in one study (Raszewa et al. 1991) and in one-quarter of 16 patients in another (Schady et al. 1991).

Other series have reported symptoms related to autonomic dysfunction that are due to sclerosis of the affected organs rather than the sympathetic or parasympathetic nerves that innervate them. In one study of urinary bladder function in nine women with scleroderma, three had abnormal residual urine volumes and four had cystometrographic evidence for detrussor areflexia (Lazzeri et al. 1995). In all four of the latter patients, pathological examination of the detrussor muscle revealed arterial and capillary abnormalities as the probable explanation for the muscular dysfunction. Sexual dysfunction due to vaginal dryness is common among women with systemic sclerosis, but may be related more to cutaneous involvement with loss of secretory glands than autonomic neuropathy reducing secretory activity (Bhadauria et al. 1995). Erectile dysfunction has been reported in men with systemic sclerosis (Lotfi et al. 1995; Nehra et al. 1995). In one particularly well-studied case, the cause for erectile dysfunction was corporeal veno-occlusive disease that was produced by excessive accumulation of extracellular matrix (Nehra et al. 1995).

Myasthenia gravis

Several anecdotal reports have been published, in which patients have developed myasthenia gravis either before (Gordon and Silverstein 1970; Bhalla et al. 1993) or after (Averbuch-Heller et al. 1992) systemic sclerosis. These may represent chance associations of two different autoimmune diseases, but the possibility of HLA B8/DR3 predisposing to both has been proposed (Bhalla et al. 1993).

Myelopathy

Myelopathy with spastic quadriparesis and sensory impairment has been reported in 4 of 50 patients (8%) with systemic sclerosis in one series (Averbuch-Heller et al. 1992). The ages of these four patients ranged from 22 to 43 years, and duration of systemic sclerosis varied from 1 to 20 years. The gastrointestinal tract had become involved with systemic sclerosis in three of four, but vitamin B_{12} and vitamin E levels were normal in all four. None had neuroradiologic evidence for compression or other structural abnormality. Cerebrospinal fluid studies revealed a minimal elevation of protein in one of four (to 0.686 g/l with the upper limit normal being 0.650 g/l), but no pleocytosis in any of the four. Thus, a reasonable explanation for myelopathy other than scleroderma was not found in any of the four patients.

Cerebrovascular disease

In one series, 3 of 32 patients (6%) had hemiparesis with or without additional signs from cerebrovascular disease, resulting in completed strokes in two and a transient ischemic attack in one (Averbuch-Heller et al. 1992). These three women ranged in age from 50 to 56 years and did not have risk factors for cerebrovascular disease, including normal carotid Doppler studies and no detectable antiphospholipid antibodies. In another series, 2 of 32 patients had transient ischemic attacks (Hietaharju et al. 1993a). These two women were 49 and 58 years old and had symptoms from involvement of the left carotid or vertebrobasilar arteries.

Other central nervous system abnormalities

Among 32 patients with systemic sclerosis, 5 (16%) were found to have prominent central nervous system (CNS) or psychiatric symptoms (Hietaharju et al. 1993a). In addition to the two described in the preceding paragraph with transient ischemic attacks, clinical evidence included features of encephalopathy, psychosis, anxiety disorder, and generalized major motor seizure disorder. None of these five patients had abnormalities on EEG or visual evoked potential studies, which were performed as part of the prospective study design.

Systemic necrotizing vasculitis is a rare but reported complication of systemic sclerosis (Kamouchi et al. 1991; Ishikawa et al. 1993). Vasculitis in systemic sclerosis is commonly associated with anticardiolipin antibodies (Herrick et al. 1994a, b), including one case with peripheral nerve and CNS involvement (Herrick et al. 1994a). In some cases, CNS abnormalities including headache, encephalopathy, seizures, or subarachnoid hemorrhage are produced by vasculitic involvement of cerebral vessels (Estey et al. 1979; Pathak and Gabor 1991; Ishida et al. 1993). Treatment with corticosteroids or cyclophosphamide has been beneficial to some of these cases (Estey et al. 1979; Pathak and Gabor 1991).

Epilepsy has been described as a complication of the 'en coup de sabre' form of linear scleroderma. In one recent case report with neuropathological examination of the intracranial lesion subjacent to the area of forehead and scalp involvement, band-like sclerosis of the leptomeninges and associated vessels was found, with calcified and anomalous vessels in the underlying parenchyma (Chung et al. 1995). This led to the suggestion that the 'en coup de sabre' lesion may be more appropriately classified as a neuro-cutaneous syndrome rather than a localized form of scleroderma.

SECONDARY SCLERODERMA AND SCLERODERMA-LIKE DISORDERS

Eosinophilic fasciitis

Eosinophilic fasciitis is a well-accepted, scleroderma-like disorder that is characterized by eosinophilia and diffuse fasciitis (Medsger and Steen 1993). The fasciitis involves arms and legs, but spares the hands and feet, which helps to separate it from systemic sclerosis. Furthermore, Raynaud's phenomenon is not present, and antinuclear antibody is absent. Full-thickness biopsy of skin, fascia and muscle typically reveals the diagnostic finding of inflammation and sclerosis from the subdermis through deep fascia, and occasionally superficial muscle (Medsger and Steen 1993). Patients often respond well to corticosteroid treatment.

Toxic oil syndrome

The toxic oil syndrome occurred as an epidemic related to the ingestion of adulterated rapeseed cooking oil in Spain during 1981 (Alonso-Ruiz et al. 1993). Approximately 20,000 people acutely developed combinations of myalgia, rash, eosinophilia or pulmonary edema. During the intermediate phase (2–4 months after exposure), subcutaneous edema and skin tenderness were common, pulmonary hypertension and abnormal liver function often developed, and myalgia persisted. During the chronic phase, Raynaud's phenomenon and scleroderma-like skin lesions were common and often associated with myalgia, arthralgia, contractures, and polyneuropathy. With follow-up for over 8 years, nearly half of patients continued to have at least mild complaints (Alonso-Ruiz et al. 1993). The toxic oil syndrome represents one of the best studied examples of a secondary scleroderma-like disorder.

Eosinophilia–myalgia syndrome

The eosinophilia–myalgia syndrome was recognized in 1990 among patients who ingested adulterated tryptophan (Clauw et al. 1990; Connolly et al. 1990).

Severe myalgias and marked peripheral eosinophilia were the most consistent features (Medsger and Steen 1993). This syndrome more closely resembles eosinophilic fasciitis than scleroderma (Freundlich et al. 1990).

Neurological complications have been described in some patients with this syndrome, including myopathies, axonal polyneuropathies and, least commonly but most severely, demyelinating polyneuropathies (Verity et al. 1991; Donofrio et al. 1992). The demyelinating polyneuropathy has been described in single cases and small series. In one of the small series, two of three patients died from progressive polyneuropathy unresponsive to treatment with plasmapheresis, corticosteroids and, in one, cyclophosphamide (Donofrio et al. 1992). Autopsy revealed patchy perivascular infiltrates and fibrosis in nerve and muscle. A milder, less progressive polyneuropathy and a myopathy have been described more commonly (Verity et al. 1991). Muscle involvement is manifest with myalgia more than weakness. The creatine kinase level is normal or slightly increased. Muscle biopsy reveals non-eosinophilic epimysial and perimysial infiltration and type 2 fiber atrophy (Verity et al. 1991).

Silicone breast implants.

A number of cases of scleroderma have been reported in women with silicone breast implants (Teuber et al. 1995). This is the most recently described secondary scleroderma syndrome and one of the most controversial at present. The presence of antinuclear and anticentromere antibodies has been reported to be more common in women with silicone breast implants than in the general population, suggesting that silicone has induced an autoimmune disease (Cuellar et al. 1995). However, other authors who have compared patients with silicone breast implants to others without implants have not found an increased incidence for rheumatologic diseases including systemic sclerosis (Kallenberg 1994; Goldman et al. 1995; Hochberg et al. 1995; Sanchez-Guerrero et al. 1995).

Organic solvent and occupational chemical exposure

Several organic solvents have been reported to produce diffuse or localized scleroderma-like disorders

(LeRoy and Silver 1993; Czirjak et al. 1994; Dunnill and Black 1994). Potentially relevant chemicals include vinyl chloride, trichloroethylene and tetrachloroethylene. In a long-term follow-up study of gold miners, exposure to silica and other minerals has been shown to produce an excess incidence of secondary scleroderma (Steenland and Brown 1995). Patients with scleroderma-like disorders from exposure to chemicals generally have a better prognosis than those with primary scleroderma (Czirjak et al. 1993). Association of these cases with polyneuropathy or other neurological complications is uncommon (Bottomley et al. 1993). The similarity of these cases to primary scleroderma is controversial.

REFERENCES

ALONSO-RUIZ, A., M. CALABOZO, F. PEREZ-RUIZ and L. MANCEBO: Toxic oil syndrome. A long-term follow-up of a cohort of 332 patients. Medicine (Baltimore) 72 (1993) 285–295.

ALTMAN, R.D., T.A. MEDSGER, JR., D.A. BLOCH and B.A. MICHEL: Predictors of survival in systemic sclerosis (scleroderma). Arthr. Rheum. 34 (1991) 403–413.

ARAHATA, K. and A.G. ENGEL: Monoclonal antibody analysis of mononuclear cells in myopathies. I: Quantitation of subsets according to diagnosis and sites of accumulation and demonstration and counts of muscle fibers invaded by T cells. Ann. Neurol. 16 (1984) 193–208.

ASHERSON, R.A., H. ANGUS, J.A. MATHEWS, O. MEYERS and G.R. HUGHES: The progressive systemic sclerosis/systemic lupus overlap: an unusual clinical progression. Ann. Rheum. Dis. 50 (1991) 323–327.

AVERBUCH-HELLER, L., I. STEINER and O. ABRAMSKY: Neurologic manifestations of progressive systemic sclerosis. Arch. Neurol. 49 (1992) 1292–1295.

BARNETT, A.J., M.H. MILLER and G.O. LITTLEJOHN: A survival study of patients with scleroderma diagnosed over 30 years (1953–1983): the value of a simple cutaneous classification in the early stages of the disease. J. Rheumatol. 15 (1988) 276–283.

BENNETT, R.M., D.M. BONG and B.H. SPARGO: Neuropsychiatric problems in mixed connective tissue disease. Am. J. Med. 65 (1978) 955–962.

BERTH-JONES, J., P.A. COATES, R.A. GRAHAM-BROWN and D.A. BURNS: Neurological complications of systemic sclerosis – a report of three cases and review of the literature. Clin. Exp. Dermatol. 15 (1990) 91–94.

BHADAURIA, S., D.K. MOSER, P.J. CLEMENTS, R.R. SINGH, P.A. LACHENBRUCH, R.M. PITKIN and S.R. WEINER: Genital tract abnormalities and female sexual function impairment in systemic sclerosis. Am. J. Obstet. Gynecol. 172 (1995) 580–587.

BHALLA, R., W.I. SWEDLER, M.B. LAZAREVIC, H.S. AJMANI and J.L. SKOSEY: Myasthenia gravis and scleroderma. J. Rheumatol. 20 (1993) 1409–1410.

BIRDI, N., R.M. LAXER, P. THORNER, M.J. FRITZLER and E.D. SILVERMAN: Localized scleroderma progressing to systemic disease. Case report and review of the literature. Arthr. Rheum. 36 (1993) 410–415. (Published erratum appeared in Arthr. Rheum. 36 (1993) 1182.)

BLACK, C.M. and C.P. DENTON: The management of systemic sclerosis. Br. J. Rheumatol. 34 (1995) 3–7.

BLUTHNER, M. and F.A. BAUTZ: Cloning and characterization of the cDNA coding for a polymyositis-scleroderma overlap syndrome-related nucleolar 100-kD protein. J. Exp. Med. 176 (1992) 973–980.

BOSCHI, A., B. SNYERS and M. LAMBERT: Bilateral optic neuropathy associated with the crest variant of scleroderma. Eur. J. Ophthalmol. 3 (1993) 219–222.

BOTTOMLEY, W.W., R.A. SHEEHAN-DARE, P. HUGHES and W.J. CUNLIFFE: A sclerodermatous syndrome with unusual features following prolonged occupational exposure to organic solvents. Br. J. Dermatol. 128 (1993) 203–206.

BULPITT, K.J., P.J. CLEMENTS, P.A. LACHENBRUCH, H.E. PAULUS, J.B. PETER, M.S. AGOPIAN, J.Z. SINGER, V.D. STEEN, D.O. CLEGG, C.M. ZIMINSKI ET AL.: Early undifferentiated connective tissue disease: III. Outcome and prognostic indicators in early scleroderma (systemic sclerosis). Ann. Intern. Med. 118 (1993) 602–609.

CARSON, C.W., L.D. BEALL, G.G. HUNDER, C.M. JOHNSON and W. NEWMAN: Serum ELAM-1 is increased in vasculitis, scleroderma, and systemic lupus erythematosus. J. Rheumatol. 20 (1993) 809–814.

CHANG, R.W., C.L. BELL and M. HALLETT: Clinical characteristics and prognosis of vasculitic mononeuropathy multiplex. Arch. Neurol. 41 (1984) 618–621.

CHUNG, M.H., J. SUM, M.J. MORRELL and D.S. HOROUPIAN: Intracerebral involvement in scleroderma en coup de sabre: report of a case with neuropathologic findings. Ann. Neurol. 37 (1995) 679–681.

CLAUW, D.J., D.J. NASHEL, A. UMHAU and P. KATZ: Tryptophan-associated eosinophilic connective-tissue disease. A new clinical entity? J. Am. Med. Assoc. 263 (1990) 1502–1506.

CLEMENTS, P.J., D.E. FURST, D.S. CAMPION, A. BOHAN, R. HARRIS, J. LEVY and H.E. PAULUS: Muscle disease in progressive systemic sclerosis: diagnostic and therapeutic considerations. Arthr. Rheum. 21 (1978) 62–71.

CLEMENTS, P.J., P.A. LACHENBRUCH, S.C. NG, M. SIMMONS, M. STERZ and D.E. FURST: Skin score. A semiquantitative measure of cutaneous involvement that improves prediction of prognosis in systemic sclerosis. Arthr. Rheum. 33 (1990) 1256–1263.

CLEMSON, B.S., W.R. MILLER, J.C. LUCK and J.A. FERISS: Acute myocarditis in fulminant systemic sclerosis. Chest 101 (1992) 872–874.

CONNOLLY, S.M., S.R. QUIMBY, W.L. GRIFFING and R.K. WINKELMANN: Scleroderma and L-tryptophan: a possible explanation of the eosinophilia-myalgia syndrome. J. Am. Acad. Dermatol. 23 (1990) 451–457.

COOK, N.J., A.J. SILMAN, J. PROPERT and M.I. CAWLEY: Features of systemic sclerosis (scleroderma) in an identical twin pair. Br. J. Rheumatol. 32 (1993) 926–928.

CORBO, M., R. NEMNI, S. IANNACCONE, A. QUATTRINI, M. LODI, L. PRADERIO, M. COMOLA, I. LORENZETTI, G. COMI and N. CANAL: Peripheral neuropathy in scleroderma. Clin. Neuropathol. 12 (1993) 63–67.

CUELLAR, M.L., E. SCOPELITIS, S.A. TENENBAUM, R.F. GARRY, L.H. SILVEIRA, G. CABRERA and L.R. ESPINOZA: Serum antinuclear antibodies in women with silicone breast implants. J. Rheumatol. 22 (1995) 236–240.

CZIRJAK, L., Z. NAGY and G. SZEGEDI: Survival analysis of 118 patients with systemic sclerosis. J. Intern. Med. 234 (1993) 335–337.

CZIRJAK, L., E. POCS and G. SZEGEDI: Localized scleroderma after exposure to organic solvents. Dermatology 189 (1994) 399–401.

DEHEN, L., J.C. ROUJEAU, A. COSNES and J. REVUZ: Internal involvement in localized scleroderma. Medicine (Baltimore) 73 (1994) 241–245.

DE JUAN, M.D., J. BELZUNEGUI, I. BELMONTE, J. BARADO, M. FIGUEROA, J. CANCIO, S. VIDAL and E. CUADRADO: An immunogenetic study of familial scleroderma. Ann. Rheum. Dis. 53 (1994) 614–617.

DENARDO, B.A., L.B. TUCKER, L.C. MILLER, I.S. SZER and J.G. SCHALLER: Demography of a regional pediatric rheumatology patient population. Affiliated Children's Arthritis Centers of New England. J. Rheumatol. 21 (1994) 1553–1561.

DONOFRIO, P.D., C. STANTON, V.S. MILLER, L. OESTREICH, D.S. LEFKOWITZ, F.O. WALKER and E.W. ELY: Demyelinating polyneuropathy in eosinophilia-myalgia syndrome. Muscle Nerve 15 (1992) 796–805.

DUNNILL, M.G. and M.M. BLACK: Sclerodermatous syndrome after occupational exposure to herbicides – response to systemic steroids. Clin. Exp. Dermatol. 19 (1994) 518–520.

ESTEY, E., A. LIEBERMAN, R. PINTO, M. MELTZER and J. RANSOHOFF: Cerebral arteritis in scleroderma. Stroke 10 (1979) 595–597.

FALANGA, V., T.A. MEDSGER, M. REICHLIN and P. RODNAN: Linear scleroderma. Ann. Intern. Med. 104 (1986) 849–857.

FARREL, D.A. and T.A. MEDSGER: Trigeminal neuropathy in progressive systemic sclerosis. Am. J. Med. 73 (1982) 57–62.

FOLLANSBEE, W.P., T.R. ZERBE and T.A. MEDSGER: Cardiac and skeletal muscle disease in systemic sclerosis

(scleroderma): a high risk association. Am. Heart J. 125 (1993) 194–203.

FREUNDLICH, B., V.P. WERTH, A.H. ROOK, C.R. O'CONNOR, H.R. SCHUMACHER, J.J. LEYDEN and P.D. STOLLEY: L-Tryptophan ingestion associated with eosinophilic fasciitis but not progressive systemic sclerosis. Ann. Intern. Med. 112 (1990) 758–762.

GE, Q., Y. WU, E.P. TRIEU and I.N. TARGOFF: Analysis of the specificity of anti-PM-Scl autoantibodies. Arthr. Rheum. 37 (1994) 1445–1452.

GENDI, N., T. GORDON, S.B. TANNER and C.M. BLACK: The evolution of a case of overlap syndrome with systemic sclerosis, rheumatoid arthritis and systemic lupus erythematosus. Br. J. Rheumatol. 31 (1992) 783–786.

GENDI, N.S., K.I. WELSH, V.W.J. VAN, R. VANCHEESWARAN, J. GILROY and C.M. BLACK: HLA type as a predictor of mixed connective tissue disease differentiation. Ten-year clinical and immunogenetic followup of 46 patients. Arthr. Rheum. 38 (1995) 259–266.

GERBRACHT, D.D., V.D. STEEN, G.L. ZIEGLER, T.A. MEDSGER, JR. and G.P. RODNAN: Evolution of primary Raynaud's phenomenon (Raynaud's disease) to connective tissue disease. Arthr. Rheum. 28 (1985) 87–92.

GOLDMAN, J.A., J. GREENBLATT, R. JOINES, L. WHITE, B. AYLWARD and S.H. LAMM: Breast implants, rheumatoid arthritis, and connective tissue diseases in a clinical practice. J. Clin. Epidemiol. 48 (1995) 571–582.

GOLDSMITH, P.C., C.J. FOWLER, M.L. SNAITH and P.M. DOWD: Morphoea and mononeuritis multiplex. J. Roy. Soc. Med. 84 (1991) 233–234.

GORDON, R.M. and A. SILVERSTEIN: Neurological manifestations in progressive systemic sclerosis. Arch. Neurol. 22 (1970) 126–134.

GRUSCHWITZ, M.S., O.P. HORNSTEIN and P. VON DEN DRIESCH: Correlation of soluble adhesion molecules in the peripheral blood of scleroderma patients with their in situ expression and with disease activity. Arthr. Rheum. 38 (1995) 184–189.

HAGEN, N.A., J.C. STEVENS and C.J. MICHET, JR.: Trigeminal sensory neuropathy associated with connective tissue diseases. Neurology 40 (1990) 891–896.

HARATI, Y. and E. NIAKAN: The clinical spectrum of inflammatory-angiopathic neuropathy. J. Neurol. Neurosurg. Psychiatry 49 (1986) 1313–1316.

HAUSMANOWA-PETRUSEWICZ, I. and A. KOZMINSKA: Electromyographic findings in scleroderma. Arch. Neurol. 4 (1961) 281–287.

HAUSMANOWA-PETRUSEWICZ, I., S. JABLONSKA, M. BLASZCZYK and B. MATZ: Electromyographic findings in various forms of progressive systemic sclerosis. Arthr. Rheum. 25 (1982) 61–65.

HELM, T.N., R. VALENZUELA, W.F. BERGFELD, J. GUITART and C.J. POOLOS: In vivo complement C3 binding to the intercellular substance and cytoplasm of epidermal basal cells in a patient with scleroderma and inclusion body myositis. J. Am. Acad. Dermatol. 23 (1990) 753–754.

HERMOSILLO, A.G., R. ORTIZ, J. DABAGUE, J.M. CASANOVA and M. MARTINEZ-LAVIN: Autonomic dysfunction in diffuse scleroderma vs CREST: an assessment by computerized heart rate variability. J. Rheumatol. 21 (1994) 1849–1854.

HERRICK, A.L., P. OOGARAH, T.B. BRAMMAH, A.J. FREEMONT and M.I. JAYSON: Nervous system involvement in association with vasculitis and anticardiolipin antibodies in a patient with systemic sclerosis. Ann. Rheum. Dis. 53 (1994a) 349–350.

HERRICK, A.L., P.K. OOGARAH, A.J. FREEMONT, R. MARCUSON, M. HAENEY and M.I. JAYSON: Vasculitis in patients with systemic sclerosis and severe digital ischaemia requiring amputation. Ann. Rheum. Dis. 53 (1994b) 323–326.

HIETAHARJU, A., S. JAASKELAINEN, M. HIETARINTA and H. FREY: Central nervous system involvement and psychiatric manifestations in systemic sclerosis (scleroderma): clinical and neurophysiological evaluation. Acta Neurol. Scand. 87 (1993a) 382–387.

HIETAHARJU, A., S. JAASKELAINEN, H. KALIMO and M. HIETARINTA: Peripheral neuromuscular manifestations in systemic sclerosis (scleroderma). Muscle Nerve 16 (1993b) 1204–1212.

HIETARINTA, M., S. KOSKIMIES, O. LASSILA, E. SOPPI and A. TOIVANEN: Familial scleroderma: HLA antigens and autoantibodies. Br. J. Rheumatol. 32 (1993) 336–338.

HIETARINTA, M., O. LASSILA and A. HIETAHARJU: Association of anti-U1RNP- and anti-Scl-70-antibodies with neurological manifestations in systemic sclerosis (scleroderma). Scand. J. Rheumatol. 23 (1994) 64–67.

HOCHBERG, M.C., R. MILLER and F.M. WIGLEY: Frequency of augmentation mammoplasty in patients with systemic sclerosis: data from the Johns Hopkins–University of Maryland Scleroderma Center. J. Clin. Epidemiol. 48 (1995) 565–569.

ISENBERG, D.A. and C. BLACK: ABC of rheumatology. Raynaud's phenomenon, scleroderma, and overlap syndromes. Br. Med. J. 310 (1995) 795–798.

ISHIDA, K., T. KAMATA, H. TSUKAGOSHI and Y. TANIZAKI: Progressive systemic sclerosis with CNS vasculitis and cyclosporin A therapy. J. Neurol. Neurosurg. Psychiatry 56 (1993) 720.

ISHIKAWA, O., T. TAMURA, K. OHNISHI, Y. MIYACHI and K. ISHII: Systemic sclerosis terminating as systemic necrotizing angiitis. Br. J. Dermatol. 129 (1993) 736–738.

JERUSALEM, F., M. RAKUSA, A.G. ENGEL and R.D. MACDONALD: Morphometric analysis of skeletal muscle capillary ultrastructure in inflammatory myopathies. J. Neurol. Sci. 23 (1974) 391–402.

JIMENEZ, S.A. and S.H. SIGAL: A 15-year prospective study

of treatment of rapidly progressive systemic sclerosis with D-penicillamine. J. Rheumatol. 18 (1991) 1496–1503.

KALLENBERG, C.G.: Overlapping syndromes, undifferentiated connective tissue disease, and other fibrosing conditions. Curr. Opin. Rheumatol. 6 (1994) 650–654.

KALLENBERG, C.G., A.A. WOUDA, M.H. HOET and W.J. VAN VENROOIJ: Development of connective tissue disease in patients presenting with Raynaud's phenomenon: a six year follow up with emphasis on the predictive value of antinuclear antibodies as detected by immunoblotting. Ann. Rheum. Dis. 47 (1988) 634–641.

KAMOUCHI, M., M. YOSHINARI, H. GOTO, T. ISHITSUKA, K. MURAI, K. TASHIRO and M. FUJISHIMA: Disseminated intravascular coagulation in a patient with progressive systemic sclerosis associated with necrotizing angiitis and generalized lymphadenopathy. Acta Haematol. (Basel) 86 (1991) 203–205.

KERR, L.D. and H. SPIERA: Myocarditis as a complication in scleroderma patients with myositis. Clin. Cardiol. 16 (1993) 895–899.

LANGEVITZ, P., D. BUSKILA, D.D. GLADMAN, G.A. DARLINGTON, V.T. FAREWELL and P. LEE: HLA alleles in systemic sclerosis: association with pulmonary hypertension and outcome. Br. J. Rheumatol. 31 (1992) 609–613.

LAPADULA, G., P. MUOLO, F. SEMERARO, M. COVELLI, D. BRINDICCI, G. CUCCORESE, A. FRANCAVILLA and V. PIPITONE: Esophageal motility disorders in the rheumatic diseases: a review of 150 patients. Clin. Exp. Rheumatol. 12 (1994) 515–521.

LAZZERI, M., P. BENEFORTI, G. BENAIM, C. CORSI, V. CIAMBRONE, E. MARRAPODI, G. MINCIONE and D. TURINI: Vesical dysfunction in systemic sclerosis (scleroderma). J. Urol. 153 (1995) 1184–1187.

LECKY, B.R.F., R.A.C. HUGHES and N.M.F. MURRAY: Trigeminal sensory neuropathy. Brain 110 (1987) 1463–1485.

LEE, P., J. BRUNI and S. SUKENIK: Neurological manifestations in systemic sclerosis (scleroderma). J. Rheumatol. 11 (1984) 480–483.

LEE, P., P. LANGEVITZ, C.A. ALDERDICE, M. AUBREY, P.A. BAER, M. BARON, D. BUSKILA, J.P. DUTZ, I. KHOSTANTEEN, S. PIPER ET AL.: Mortality in systemic sclerosis (scleroderma). Quart. J. Med. 82 (1992) 139–148.

LEICHENKO, T., A.L. HERRICK, S.M. ALANI, R.C. HILTON and M.I. JAYSON: Mononeuritis in two patients with limited cutaneous systemic sclerosis. Br. J. Rheumatol. 33 (1994) 594–595.

LEROY, E.C. and R.M. SILVER: Systemic sclerosis and related syndromes: A. Epidemiology, pathology, and pathogenesis. In: H.R. Schumacher (Ed.), Primer on the Rheumatic Diseases. Atlanta, GA, Arthritis Foundation (1993) 118–120.

LEROY, E.C., C. BLACK, R. FLEISCHMAJER, S. JABLONSKA, T. KRIEG, T.A. MEDSGER, JR., N. ROWELL and F. WOLLHEIM:

Scleroderma (systemic sclerosis): classification, subsets and pathogenesis. J. Rheumatol. 15 (1988) 202–205.

LOTFI, M.A., J. VARGA and I.H. HIRSCH: Erectile dysfunction in systemic sclerosis. Urology 45 (1995) 879–881.

LOTZ, B.P., A.G. ENGEL, H. NISHINO, J.C. STEVENS and W.J. LITCHY: Inclusion body myositis: observations in 40 patients. Brain 112 (1989) 727–747.

LUNDBERG, I. and E. HEDFORS: Clinical course of patients with anti-RNP antibodies. A prospective study of 32 patients. J. Rheumatol. 18 (1991) 1511–1519.

MANOLIOS, N., H. DUNCKLEY, T. CHIVERS, P. BROOKS and H. ENGLERT: Immunogenetic analysis of 5 families with multicase occurrence of scleroderma and/or related variants. J. Rheumatol. 22 (1995) 85–92.

MARGUERIE, C., C.C. BUNN, J. COPIER, R.M. BERNSTEIN, J.M. GILROY, C.M. BLACK, A.K. SO and M.J. WALPORT: The clinical and immunogenetic features of patients with autoantibodies to the nucleolar antigen PM-Scl. Medicine (Baltimore) 71 (1992) 327–336.

MARICQ, H.R., M.C. WEINRICH, J.E. KEIL, E.A. SMITH, F.E. HARPER, A.I. NUSSBAUM, E.C. LEROY, A.R. MCGREGOR, F. DIAT and E.J. ROSAL: Prevalence of scleroderma spectrum disorders in the general population of South Carolina. Arthr. Rheum. 32 (1989) 998–1006.

MAYORQUIN, F.J., T.L. MCCURLEY, J.E. LEVERNIER, L.K. MYERS, J.A. BECKER, T.P. GRAHAM and T. PINCUS: Progression of childhood linear scleroderma to fatal systemic sclerosis. J. Rheumatol. 21 (1994) 1955–1957.

MCCOLL, G.J. and R.R. BUCHANAN: Familial CREST syndrome. J. Rheumatol. 21 (1994) 754–756.

MCCUNE, W.J., D.K. VALLANCE and J.P.R. LYNCH: Immunosuppressive drug therapy. Curr. Opin. Rheumatol. 6 (1994) 262–272.

MEDSGER, T.A. and V. STEEN: Systemic sclerosis and related syndromes: B. Clinical features and treatment. In: H.R. Schumacher (Ed.), Primer on the Rheumatic Diseases. Atlanta, GA, Arthritis Foundation (1993) 120–127.

MEDSGER, T.A., G.P. RODNAN, J. MOOSSY and J.W. VESTER: Skeletal muscle involvement in progressive systemic sclerosis (scleroderma). Arthr. Rheum. 11 (1968) 554–568.

MIMORI, T.: Scleroderma-polymyositis overlap syndrome: Clinical and serological aspects. Int. J. Dermatol. 26 (1987) 419–425.

MOREL, P.A., H.J. CHANG, J.W. WILSON, C. CONTE, D. FALKNER, D.J. TWEARDY and T.A. MEDSGER, JR.: HLA and ethnic associations among systemic sclerosis patients with anticentromere antibodies. Hum. Immunol. 42 (1995) 35–42.

MUKERJI, B. and J.G. HARDIN: Undifferentiated, overlapping, and mixed connective tissue diseases. Am. J. Med. Sci. 305 (1993) 114–119.

NEHRA, A., S.J. HALL, G. BASILE, E.B. BERTERO, R. MORELAND,

P. TOSELLI, L.M.A. DE and I. GOLDSTEIN: Systemic sclerosis and impotence: a clinicopathological correlation. J. Urol. 153 (1995) 1140–1146.

NORTON, W.L., E.R. HURN, D.C. LEWIS and M. ZIFF: Evidence of vascular injury in scleroderma and systemic lupus erythematosus: quantitative study of the microvascular bed. J. Lab. Clin. Med. 71 (1968) 919–933.

ODDIS, C.V., Y. OKANO, W.A. RUDERT, M. TRUCCO, R.J. DUQUESNOY and T.A. MEDSGER, JR.: Serum autoantibody to the nucleolar antigen PM-Scl. Clinical and immunogenetic associations. Arthr. Rheum. 35 (1992) 1211–1217.

OLNEY, R.K.: AAEM minimonograph no. 38: neuropathies in connective tissue disease. Muscle Nerve 15 (1992) 531–542.

OTA, H., S. KUMAGAI, A. MORINOBU, H. YANAGIDA and K. NAKAO: Enhanced production of transforming growth factor-beta (TGF-beta) during autologous mixed lymphocyte reaction of systemic sclerosis patients. Clin. Exp. Immunol. 100 (1995) 99–103.

PABLOS, J.L., E.T. EVERETT, R. HARLEY, E.C. LEROY and J.S. NORRIS: Transforming growth factor-beta1 and collagen gene expression during postnatal skin development and fibrosis in the tight-skin mouse. Lab. Invest. 72 (1995) 670–678.

PATHAK, R. and A.J. GABOR: Scleroderma and central nervous system vasculitis. Stroke 22 (1991) 410–413.

PELKONEN, P.M., H.J. JALANKO, R.K. LANTTO, A.L. MÄKELÄ, M.A. PIETIKAINEN, H.A. SAVOLAINEN and P.M. VERRONEN: Incidence of systemic connective tissue diseases in children: a nationwide prospective study in Finland. J. Rheumatol. 21 (1994) 2143–2146.

PRESCOTT, R.J., A.J. FREEMONT, C.J. JONES, J. HOYLAND and P. FIELDING: Sequential dermal microvascular and perivascular changes in the development of scleroderma. J. Pathol. 166 (1992) 255–263.

RASZEWA, M., I. HAUSMANOWA-PETRUSEWICZ, M. BLASZCZYK and S. JABLONSKA: Sympathetic skin response in scleroderma. Electromyogr. Clin. Neurophysiol. 31 (1991) 467–472.

RINGEL, R.A., J.E. BRICK, J.F. BRICK, L. GUTMANN and J.E. RIGGS: Muscle involvement in the scleroderma syndromes. Arch. Intern. Med. 150 (1990) 2550–2552.

RUSSELL, M.L. and W.M. HANNA: Ultrastructure of muscle microvasculature in progressive systemic sclerosis: relation to clinical weakness. J. Rheumatol. 10 (1983) 741–747.

SAID, G., C. LACROIX-CIAUDO, H. FUJIMURA, C. BLAS and N. FAUX: The peripheral neuropathy of necrotizing arteritis: a clinicopathological study. Ann. Neurol. 23 (1988) 461–465.

SANCHEZ-GUERRERO, J., G.A. COLDITZ, E.W. KARLSON, D.J. HUNTER, F.E. SPEIZER and M.H. LIANG: Silicone breast implants and the risk of connective-tissue diseases and symptoms. N. Engl. J. Med. 332 (1995) 1666–1670.

SASAKI, T., K. DENPO, H. ONO and H. NAKAJIMA: HLA in systemic scleroderma (PSS) and familial scleroderma. J. Dermatol. 18 (1991) 18–24.

SCARPELLI, M., R. MONTIRONI, D. TULLI, S. SISTI, G. MATERA, G.C. MAGI and Y. COLLAN: Quantitative analysis of quadriceps muscle biopsy in systemic sclerosis. Pathol. Res. Pract. 188 (1992) 603–606.

SCHADY, W., A. SHEARD, A. HASSELL, L. HOLT, M.I. JAYSON and P. KLIMIUK: Peripheral nerve dysfunction in scleroderma. Quart. J. Med. 80 (1991) 661–675.

SCHWARTZ, R.A., A.S. TEDESCO, L.Z. STERN, A.M. KAMINSKA, J.M. HARALDSEN and D.A. GREKIN: Myopathy associated with sclerodermal facial hemiatrophy. Arch. Neurol. 38 (1981) 592–594.

SHARP, G.C., W.S. IRVIN, E.M. TAN, R.G. GOULD and H.R. HOLMAN: Mixed connective tissue disease – an apparently distinct rheumatic disease syndrome associated with a specific antibody to an extractable nuclear antigen (ENA). Am. J. Med. 52 (1972) 148–159.

STEEN, V.D., J.P. COSTANTINO, A.P. SHAPIRO and T.A. MEDSGER, JR.: Outcome of renal crisis in systemic sclerosis: relation to availability of angiotensin converting enzyme (ACE) inhibitors. Ann. Intern. Med. 113 (1990) 352–357.

STEENLAND, K. and D. BROWN: Mortality study of gold miners exposed to silica and nonasbestiform amphibole minerals: an update with 14 more years of follow-up. Am. J. Indust. Med. 27 (1995) 217–229.

SUBCOMMITTEE FOR SCLERODERMA CRITERIA OF THE AMERICAN RHEUMATISM ASSOCIATION DIAGNOSTIC and THERAPEUTIC CRITERIA COMMITTEE: Preliminary criteria for the classification of systemic sclerosis (scleroderma). Arthr. Rheum. 23 (1980) 581–590.

SWERLICK, R.A. and T.J. LAWLEY: The role of microvascular endothelial cells in inflammation. J. Invest. Dermatol. 100 (1993) 111S–115S.

TEASDALL, R.D., R.A. FRAYHA and L.E. SHULMAN: Cranial nerve involvement in systemic sclerosis (scleroderma): a report of 10 cases. Medicine (Baltimore) 59 (1980) 149–159.

TEUBER, S.S., S.H. YOSHIDA and M.E. GERSHWIN: Immunopathologic effects of silicone breast implants. West. J. Med. 162 (1995) 418–425.

TUFFANELLI, D. and R.K. WINKELMANN: Systemic scleroderma: a clinical study of 727 patients. Arch. Dermatol. 84 (1961) 359–371.

UZIEL, Y., B.R. KRAFCHIK, E.D. SILVERMAN, P.S. THORNER and R.M. LAXER: Localized scleroderma in childhood: a report of 30 cases. Semin. Arthr. Rheum. 23 (1994) 328–340.

VAN DEN HOOGEN, F.H., P.E. SPRONK, A.M. BOERBOOMS, H. BOOTSMA, D.J. DE ROOIJ, C.G. KALLENBERG and L.B. VAN DE PUTTE: Treatment of systemic sclerosis. Curr. Opin. Rheumatol. 6 (1994) 637–641.

VARGA, J. and S.A. JIMENEZ: Modulation of collagen gene expression: its relation to fibrosis in systemic sclerosis and other disorders. Ann. Intern. Med. 122 (1995) 60–62.

VERITY, M.A., K.J. BULPITT and H.E. PAULUS: Neuromuscular manifestations of L-tryptophan-associated eosinophilia-myalgia syndrome: a histomorphologic analysis of 14 patients. Hum. Pathol. 22 (1991) 3–11.

WHITE, B., E.A. BAUER, L.A. GOLDSMITH, M.C. HOCHBERG, L.M. KATZ, J.H. KORN, P.A. LACHENBRUCH, E.C. LEROY, M.P. MITRANE, H.E. PAULUS ET AL.: Guidelines for clinical trials in systemic sclerosis (scleroderma). I. Disease-modifying interventions. The American College of Rheumatology Committee on Design and Outcomes in Clinical Trials in Systemic Sclerosis. Arthr. Rheum. 38 (1995) 351–360.

ZUFFEREY, P., M. DEPAIRON, A.M. CHAMOT and M. MONTI: Prognostic significance of nailfold capillary microscopy in patients with Raynaud's phenomenon and scleroderma-pattern abnormalities. A six-year follow-up study. Clin. Rheumatol. 11 (1992) 536–541.

Handbook of Clinical Neurology, Vol. 27 (71): Systemic Diseases, Part III
M.J. Aminoff and C.G. Goetz, editors

CHAPTER 6

Neurologic problems in mixed connective tissue disease

KATHLEEN M. SHANNON

*Department of Neurological Sciences, Rush Medical College, Rush–Presbyterian–St. Luke's Medical Center,
Chicago, IL, USA*

Unlike other connective tissue diseases, the diagnostic entity mixed connective tissue disease was born in the laboratory. Sharp described patients in whom a new type of antibody to ribonuclease-sensitive extractable nuclear antigen (ENA) accompanied clinical features of more than one connective tissue disease (Sharp et al. 1971, 1972). He called the syndrome mixed connective tissue disease (MCTD). Reporting a series of 25 patients in 1972, he defined four serological characteristics: (1) high titers of hemagglutinating antibody to saline ENA; (2) marked sensitivity of this antibody to ribonuclease with moderate sensitivity to trypsin and resistance to deoxyribonuclease; (3) high titers of speckled pattern on fluorescent antinuclear antibody testing; and (4) absence of Smith (Sm) antibody. Each patient had clinical features of at least two connective tissue disorders, among them systemic lupus erythematosus (SLE), scleroderma and polymyositis. The disorder was thought to have a benign prognosis, to be exquisitely steroid responsive, to spare the central nervous and renal systems and to have a low incidence of systemic vasculitis (Leibfarth and Persellin 1976).

The autoantigen Sharp described has been better characterized over time. Saline-extractable ENA appears to represent two moieties: ribonucleoprotein (RNP) and the Sm antigen, a glycoprotein. The ribonuclease-sensitive ENA autoantigen in MCTD is the small nuclear RNP made up of the uridine-rich U1-RNA and its associated 68 kDa peptide. This U1-

68 kDa RNP has been shown to be involved in splicing of premessenger RNA (Padgett et al. 1983). While high titers of anti-RNP antibodies appear to be associated with MCTD, the presence of anti-Sm correlates highly with the diagnosis of SLE.

Sharp's original observations and the concept that MCTD represents a distinct entity have been subjected to criticism. More widespread organ involvement, including renal and central nervous system (CNS) complications, has been well described (Black and Isenberg 1992). Reported mortality rates between 13 and 28.2% belie the suggestion that the disorder is benign (Black and Isenberg 1992; Gendi et al. 1995). Moreover, most patients with the clinical diagnosis of MCTD will evolve a syndrome compatible with SLE or scleroderma over a period of 5–10 years (Nimelstein et al. 1980; Black and Isenberg 1992; Van den Hoogen et al. 1994). The U1-68 kDa RNP antibody lacks specificity for MCTD, being well reported in SLE, scleroderma, myositis, and rheumatoid arthritis (Black and Isenberg 1992). When present in these conditions, it is predictive of less widespread pathology and more benign prognosis than the patient group at large (Ginsburg et al. 1983). It has thus been proposed that MCTD is merely an evolutionary phase in the development of other connective tissue diseases or that it is a mild form of SLE. According to this theory, it is the extent and severity of organ system involvement rather than the diagnostic entity itself that relates to the species of autoantibodies. The U1-68 kDa RNP antibody may merely be a predictor of milder and the Sm and anti-double-stranded DNA of more malignant course in various connective tissue disorders (Ginsburg et al. 1983; Mukerji and Hardin 1993). Nevertheless, most clinicians still use the diagnosis of MCTD for patients with the typical clinical syndrome who have antibodies to ENA without anti-double-stranded DNA or Sm antibodies (Kallenberg 1994).

MEDICAL CONDITION

Epidemiology

The prevalence of MCTD is unknown. It is intermediate in frequency between SLE and scleroderma (Black and Isenberg 1992). Women are more frequently affected than men by a factor of 4–79:1 (Sharp et al. 1972; Nakae et al. 1987). The disorder most commonly begins in the second and third decades of life, but has been reported in children and the elderly as well (Alarçon-Segovia 1994). The antibody to RNP is seen in about 5% of the normal population. The propensity to produce the antibody is probably genetically determined. Anti-RNP antibodies are associated with the HLA-DR2 and HLA-DR4 haplotypes in MCTD as well as normal controls (Hietarinta et al. 1994).

Clinical features of MCTD

Patients with MCTD have features of more than one connective tissue disease. Table 1 summarizes common clinical findings in three series of MCTD patients.

Nearly all patients complain of arthralgia and myalgia. Arthralgia is usually accompanied by non-deforming arthritis. A few patients have a rheumatoid-type arthritis with ulnar deviation and subcutaneous nodules. Seventy-two percent of patients have clinical and laboratory evidence of inflammatory myositis. Nearly all have Raynaud's phenomenon. Hand and finger swelling lends a tapered or sausage-like appearance to the digits. Skin thickening and tightening similar to scleroderma are generally restricted to the distal extremities. Digital ischemia and ulceration may be seen. Pitting edema of the forehead and erythematous malar or violaceous eyelid rash may be present. Disorders of esophageal motility like those seen in scleroderma affect more than 75% of patients. Lymphadenopathy is common sometimes to a degree suggestive of lymphoma. Fever, hepatosplenomegaly, pleuritis, pericarditis, and hair changes are seen in 20–40% of cases. Not well recognized in the early literature, pulmonary disease is now known to be common in MCTD. In a prospective study, 80% of MCTD patients had pulmonary disease, but 69% of those affected were asymptomatic (Sullivan et al. 1986). Pulmonary hypertension is the most common cause of death from MCTD (Mukerji and Hardin 1993). Renal disease is evident in some patients with MCTD. Proteinuria, nephrotic syndrome, hematuria, glomerulonephritis or proliferative vascular lesions may be seen (Koboyashi et al. 1985). Frequently associated laboratory abnormalities include leukopenia, anemia and hypergammaglobulinemia (Sharp 1987; Mukerji and Hardin 1993). Table 2 summarizes the frequency of neurologic complications in MCTD.

TABLE 1

Clinical features of MCTD.

Clinical feature	Sharp et al. (1972) (n = 25) (%)	Sharp et al. (1972) (n = 100) (%)	Lazaro et al. (1989) (n = 27) (%)
Arthritis/arthralgia	96	95	89
Swollen hands	88	66	93
Sclerodermatous changes	–	33	44
Raynaud's phenomenon	84	85	89
Myositis	72	63	70
Esophageal dysmotility	77	75	30
Pulmonary disease	–	43	15
Renal disease	–	5	11
Lymphadenopathy	68	39	–
Hepatomegaly	28	15	–

TABLE 2

Frequency of neurologic complications in MCTD.

	Bennett et al. (1978) ($n = 20$)	Lazaro et al. (1989) ($n = 27$)
Central nervous system		
Headache	6	1
Aseptic meningitis	(4)	(1)
'Febrile'	(2)	
Seizures	2	
Psychosis	2	
Cerebellar ataxia	1	1
Altered consciousness	1	
Transverse myelitis		1
Optic neuritis		1
Peripheral nervous system		
Trigeminal neuropathy	2	3
Peripheral neuropathy	2	1

Serology

Patients with MCTD typically have a positive anti-nuclear antibody in a speckled pattern. There is usually a high titer (>1:1600) of anti-RNP antibody (Sharp et al. 1972, 1976; Alarçon-Segovia 1994). Rheumatoid factor is frequently present and may fluctuate over the disease course. Cold-reactive lymphocytotoxic and anticardiolipin antibodies have also been described (Alarçon-Segovia 1994). When anti-double-stranded DNA or Sm antibodies accompany anti-RNP antibodies, the diagnosis is most likely SLE. When anti-Scl-70 antibodies are present, scleroderma is the probable diagnosis (Leibfarth and Persellin 1976; Bennett 1990; Alarçon-Segovia 1994). Table 3 summarizes published clinical and serologic criteria for the diagnosis of MCTD.

Course and prognosis

MCTD tends to be a variable illness. Symptoms may emerge and recede throughout the disease course. However, the majority of patients with MCTD will evolve typical clinical features of another CTD within 5 years of diagnosis (Nimelstein et al. 1980; Black and Isenberg 1992; Kallenberg 1994). In most patients, this disease evolution is in the direction of typical scleroderma, though some develop the picture of SLE. Although initially believed to be a benign condition, long-term follow-up of more than 200 patients in several series revealed that 13–28.2% had

died by 13 years' follow-up (Black and Isenberg 1992; Gendi et al. 1995). A more benign prognosis appears to be associated with the presence of Raynaud's phenomenon, more limited and milder organ involvement and the HLA-DR2 or -DR4 phenotype (Gendi et al. 1995).

PATHOGENESIS OF ANTIBODIES IN MCTD

Studies in MCTD and healthy control patients with the HLA-DR2 and HLA-DR4 genotypes demonstrate human T-cell clones reactive with small nuclear ribonucleoprotein antigens (Hoffman et al. 1993). This genotype thus appears to confer the propensity to lose self-tolerance to these antigens. The event or series of events which then triggers the autoimmune process remains unknown. It has been hypothesized that in this histocompatibility setting, autoreactive T-cells might be generated by antigenic similarity between microbial antigens and an autoantigen. These T-cells might activate B cells which can bind the intact RNP particle, processing and presenting it, resulting in the formation of autoreactive T-cell clones (Kallenberg 1994). Recently, it has been suggested that heat-shock proteins are highly conserved in the evolutionary process. Antigenic similarity between microbial and human heat-shock proteins might be a link between microbial infection and the subsequent development of autoimmunity in MCTD. In this regard, Mairesse et al. (1993) have recently demonstrated autoantibodies to the 73 kDa heat-shock protein in MCTD.

TABLE 3

Clinical and serological criteria for the diagnosis of MCTD.

Sharp et al. (1972)	Kasukawa et al. (1987)	Alarçon-Segovia and Cardiel (1989)
Criteria		
Major criteria	*Common symptoms*	Anti-RNP ⩾ 1:1600
Myositis	Raynaud's	Clinical
Pulmonary involvement	swollen hands	hand edema
Raynaud's or esophageal	Anti-RNP antibody	synovitis
hypomotility	Mixed findings	myositis
Swollen hands or	SLE-like	Raynaud's
Sclerodactyly	polyarthritis	acrosclerosis
Anti-ENA > 1:10,000,	lymphadenopathy	
Anti-U1-RNP +	facial erythema	
Anti-Sm	pericarditis/pleuritis	
Minor criteria	leukopenia	
Alopecia	thrombocytopenia	
Leukopenia	PSS-like findings	
Anemia	sclerodactyly	
Pleuritis	pulmonary findings	
Pericarditis	esophageal	
Arthritis	Hypomotility	
Trigeminal neuropathy	PM-like findings	
Malar rash	muscle weakness	
Thrombocytopenia	increased CPK	
Mild myositis	myogenic EMG	
Swollen hands		
Requirements for diagnosis		
Definite	1 of 2 common symptoms	Anti-RNP ⩾ 1:1600
4 major	anti-RNP antibody	⩾ clinical signs
Anti U1-RNP ⩾ 1:4000	1 or more findings in at	synovitis/myositis
Anti-Sm negative	least 2 disease categories	
Probable		
3 major (2 of first 3)		
2 minor		
Anti-U1-RNP ⩾ 1:1000		
Possible		
3 major *or*		
2 major or 1 major,		
3 minor		
Anti-U1-RNP ⩾ 1:100		

NEUROLOGICAL FEATURES

Frequency of nervous system involvement

Although peripheral nervous system symptoms have been described in virtually every series of MCTD patients, the CNS was thought to be relatively spared by the disorder. In fact, this was thought to be a distinguishing clinical characteristic of the syndrome (Sharp et al. 1972, 1976). More recent work has served to characterize better the peripheral nervous disorders and to emphasize that CNS sequelae, though rare, occur. Reports are largely anecdotal and data on the frequency of such sequelae are sparse.

In his follow-up of the 14 surviving patients in Sharp's original series of 25, Nimelstein et al. (1980) found nervous system involvement in three (21%). In Sharp et al.'s (1976) series of 100 patients, 10% had nervous system complaints. In Lazaro et al.'s (1989) series of 27 patients, neurologic signs or symptoms occurred in four (15%). In contrast, in a series of 20 patients reported by Bennett et al. (1978) neuropsychiatric disorders occurred in 11 (55%).

The types of disorders of the CNS in MCTD are

similar to those seen in SLE. In contrast, the peripheral nervous system events mirror those seen in scleroderma. This observation is not surprising when one considers the overlap of clinical symptoms among these autoimmune disorders.

Central nervous system manifestations

These usually occur during phases of the illness in which there is evidence of high systemic disease activity. The most common disorders are headache, aseptic meningitis, psychosis, and convulsions. Ischemic and hemorrhagic cerebrovascular disorders and transverse myelitis have been reported less commonly. Many of the CNS complications of MCTD are steroid responsive.

Headache is the most common neurologic symptom in MCTD. The headache may have a vascular quality (Bronshvag et al. 1978) or may accompany fever and nuchal rigidity. In Bennett et al.'s (1978) series of 20 patients, six had the latter picture. Aseptic meningitis was documented in four by lymphocytic pleocytosis or elevated cerebrospinal fluid (CSF) protein. There is a single case report of headache, bilateral papilledema and mild CSF pleocytosis in a patient who was found to have hypertrophic cranial pachymeningitis (Fujimoto et al. 1993).

Psychosis was seen in 15% of MCTD patients in Bennett et al.'s (1978) series. The psychosis was characterized by paranoid ideation and occurred in the context of other CNS dysfunction. Two of the three affected patients had associated delirium, one with generalized seizures and the other with aseptic meningitis. In all cases, the symptoms occurred during an exacerbation of systemic symptoms. Successful treatment of the systemic flare with corticosteroids improved the symptoms in two patients.

Two of Bennett's 20 patients had one or more generalized seizures. In one case, seizures occurred in the setting of a transient focal deficit thought to be ischemic. In the second, seizures complicated delirium with paranoid psychosis. The EEG revealed a focal slow wave abnormality in the former and generalized slowing and sharp activity in the latter patient (Bennett et al. 1978).

Cerebellar ataxia was seen in one case in each of Bennett and Lazaro's series (Bennett et al. 1978; Lazaro et al. 1989). In the case reported by Bennett, psychosis and convulsions occurred concurrently. In Lazaro's case ataxia was seen in conjunction with dysarthria, facial weakness and transient loss of vision in one eye.

There are two reported cases of hemorrhagic cerebrovascular disease in association with MCTD. The first, a fatal parieto-occipital hemorrhage, occurred in a 13-year-old girl. Generalized seizures and aseptic meningitis were also present (Graf et al. 1993). Toyoda et al. (1994) reported a fatal putaminal and intraventricular hemorrhage in a 51-year-old woman with MCTD. She had no other risk factors for intracerebral hemorrhage.

Transient, presumably ischemic, symptoms have been reported sporadically in MCTD. Bennett et al. (1978) reported transient ischemia of the internal capsule in a patient who also had trigeminal neuropathy, aseptic meningitis and seizure activity. A 9-year-old girl developed right hemiparesis related to ischemic infarction of the left hemisphere. Total occlusion of the left suprasellar internal carotid artery was demonstrated by angiogram (Graf et al. 1993).

Optic neuritis suggestive of demyelinating disease was reported in a single case (Flechtner and Baum 1994). The optic neuritis was associated with transverse myelitis and followed a relapsing and remitting course. Another woman with documented transverse myelitis had a remote history suggestive of bilateral optic neuritis (Weiss et al. 1978).

Weiss et al. (1978) reported a single patient in whom transverse myelitis developed. The patient had laboratory although not clinically typical signs of MCTD. In the case reported by Flechtner and Baum (1994), relapsing and remitting transverse myelitis was suggestive of multiple sclerosis.

McKenna et al. (1986) reported the occurrence of generalized chorea in a 15-year-old girl with MCTD. The presence of anti-double-stranded antibodies in this girl suggests, however, a diagnosis of SLE, in which chorea is well recognized.

Peripheral nervous system manifestations

The peripheral nervous system is a more frequent target of MCTD than the CNS. Peripheral nervous system events are not obviously linked to signs of systemic disease activity and are not often responsive to adequate treatment of the systemic illness.

Connective tissue disease is a well-recognized cause of trigeminal neuropathy. Prior to Sharp's original description of MCTD, Ashworth and Tait (1971) reported six women with connective tissue disease and trigeminal neuropathy. In retrospect, several of the patients would probably meet clinical criteria for MCTD (Dent and Johnson 1992). In Lecky's series of 22 patients with trigeminal neuropathy, among nine with connective tissue disease were three with MCTD (Lecky et al. 1987). Trigeminal neuropathy occurs in about 10% of MCTD patients (Sharp et al. 1976; Bennett et al. 1978; Nimelstein et al. 1980; Farrell and Medsger 1982).

Trigeminal neuropathy can antedate other clinical signs of MCTD, but more commonly occurs after other symptoms are established. The onset of trigeminal neuropathy does not correlate with signs of systemic disease activity (Hagen et al. 1990). It may be unilateral or bilateral (Searles et al. 1978; Farrell and Medsger 1982; Hagen et al. 1990; Alfaro-Giner et al. 1992). In 83% of cases, the maxillary and mandibular branches are involved together. Ophthalmic branch symptoms are rare. Early symptoms are usually numbness and paresthesias in the distribution of the maxillary or mandibular branches of the nerve. Burning and lancinating pain are common. Although a few patients have trigger points, the syndrome does not generally resemble trigeminal neuralgia (Hagen et al. 1990). Patients may complain of changes in oral sensation, involuntary tongue biting or weakness of chewing. The symptoms gradually progress, usually reaching peak severity in 1–24 months (Hagen et al. 1990). Nearly all patients have demonstrable hypesthesia to light touch, pain and temperature in the symptomatic nerve distribution (Vincent and Van Houzen 1980). About half have ageusia or dysgeusia. Some patients may have involvement of the motor root with masseter atrophy (Bennett et al. 1978; Hagen et al. 1990). Although dysfunction of regional cranial nerves is rare, associations with facial neuropathy have been reported (Hagen et al. 1990; Alfaro-Giner et al. 1992). CSF may be normal or show a lymphocytic pleocytosis or increased protein (Lecky et al. 1987). Blink reflex may show abnormalities in R1 on the affected side or R1 on the affected side and R2 bilaterally (Lecky et al. 1987; Hagen et al. 1990). No data are available on the histopathology of this condition.

Once established, symptoms tend to persist. Some patients have improvement in pain and numbness, but complete recovery is very rare (Hagen et al.

1990). Symptoms usually do not respond to analgesics, carbamazepine or corticosteroid therapy (Bennett et al. 1978; Lecky et al. 1987).

Polyneuropathy has been reported in up to 10% of patients with MCTD. Sensory loss in a stocking–glove distribution is typical. Like trigeminal neuropathy, peripheral neuropathy in MCTD is thought to be relatively resistant to corticosteroid treatment (Bennett et al. 1978).

Vincent and Van Houzen (1980) reported a 52-year-old man whose MCTD presented with clinical and electrodiagnostic signs of bilateral carpal tunnel syndrome coincident with trigeminal neuropathy.

There are three reported cases of MCTD presenting as myasthenia gravis with anti-acetylcholine antibodies (Yasuda et al. 1993). This might be explained by an increased prevalence of additional autoimmune disorders in patients with established connective tissue disease or may be a chance association.

The importance of myositis in MCTD is underscored by its inclusion as a diagnostic criterion in virtually all series of patients (Kasukawa et al. 1987; Sharp 1987; Alarçon-Segovia and Cardiel 1994). Proximal muscle weakness is seen in more than 50–100% of patients (Sharp et al. 1972, 1976; Bennett 1990). Concurrent changes of overlying skin mimic dermatomyositis in some patients (Lazaro et al. 1989). Elevations of serum creatine phosphokinase occur in about half of those with symptoms. Electromyographic changes of myopathy and biopsy-proven inflammatory myositis are common (Lazaro et al. 1989).

NEUROPATHOLOGIC FEATURES OF MCTD

There is a paucity of neuropathologic data in MCTD. Neuropathologic correlates of headache, aseptic meningitis, convulsion and psychosis are essentially unknown. Examination of necropsy material in hemorrhagic cerebrovascular disease showed cytoplasmic vacuolation and reactive atypical nuclei with fibrinoid necrosis but no vasculitis in one case (Graf et al. 1993), and no obvious explanation in the other case (Toyoda et al. 1994). In a single case of transverse myelitis, the thoracic cord was thinned with widespread loss of axons and myelin sheaths, astrocytosis and macrophage formation. The blood vessel walls were thin, with inflammatory cell infiltrates. Nerve biopsy in a patient with MCTD and peripheral neuropathy revealed vasculitis, neuronal infarcts and connective tissue fibrosis with damage to axons and myelin sheaths (Currie and Bradshaw 1979).

PATHOPHYSIOLOGY

The pathophysiology of nervous system involvement in MCTD remains unknown. Clinical evidence suggests there may be more than one mechanism of disease. The CNS complications of MCTD resemble those seen in SLE, occur in the context of active systemic disease and are largely steroid responsive. As in SLE, possible mechanisms of dysfunction include anti-neuronal antibodies, immune complex deposition and non-inflammatory vasculopathy. On the other hand, the peripheral nervous system complications resemble those seen in scleroderma, occur irrespective of systemic disease activity, and are largely steroid resistant. The pathophysiologic mechanisms of trigeminal neuralgia are unknown but may relate to extension of cutaneous or subcutaneous facial fibrosis to the nerve or vasculopathy. Carpal tunnel syndrome may result from local irritation or compression due to synovitis in the wrist. Sensory neuropathy may be mediated by vasculopathy (Bennett et al. 1978).

TREATMENT

The treatment of nervous system involvement in MCTD is based on anecdotal data rather than controlled clinical trials. CNS disorders, such as headache, aseptic meningitis, psychosis, convulsion, optic neuritis and transverse myelitis, have been reported by various authors to be responsive to corticosteroids (Bennett et al. 1978; Lazaro et al. 1989), or plasmapheresis or immunosuppressive drugs (Flechtner and Baum 1994). Disorders of the peripheral nervous system including trigeminal and peripheral neuropathy are believed relatively steroid resistant (Bennett et al. 1978; Vincent and Van Houzen 1980; Lecky et al. 1987; Hagen et al. 1990). Myositis is believed to mirror the systemic signs of MCTD and be quite sensitive to corticosteroid therapy (Sharp et al. 1976).

REFERENCES

ALARÇON-SEGOVIA, D.: Mixed connective tissue disease and overlap syndromes. Clin. Dermatol. 12 (1994) 309–316.

ALARÇON-SEGOVIA, D. and M.H. CARDIEL: Comparison between 3 diagnostic criteria for mixed connective tissue disease. Study of 593 patients. J. Rheumatol. 16 (1989) 328–334.

ALFARO-GINER, A., M. PENARROCHA-DIAGO and J.V. BAGAN-SEBASTIAN: Orofacial manifestations of mixed connective tissue disease with an uncommon serologic evolution. Oral Surg. Oral Med. Oral Pathol. 73 (1992) 441–444.

ASHWORTH, B. and G.B. TAIT: Trigeminal neuropathy in connective tissue disease. Neurology 21 (1971) 609–614.

BENNETT, R.M.: Scleroderma overlap syndromes. Rheum. Dis. Clin. N. Am. 16 (1990) 185–198.

BENNETT, R.M., D.M. BONG and B.H. SPARGO: Neuropsychiatric problems in mixed connective tissue disease. Am. J. Med. 65 (1978) 955.

BLACK, C. and D.A. ISENBERG: Mixed connective tissue disease – goodbye to all that. Br. J. Rheumatol. 31 (1992) 695–700.

BRONSHVAG, M.M., S.D. PRYSTOWSKY and D.C. TRAVIESA: Vascular headaches in mixed connective tissue disease. Headache 18 (1978) 154–160.

CURRIE, D.M. and D.C. BRADSHAW: Polyneuropathy in mixed connective tissue disease presenting as progressive systemic sclerosis (scleroderma): case report and literature review (Abstr.). Arch. Phys. Med. Rehab. 60 (1979) 594.

DENT, T.H.S. and V.W. JOHNSON: Aspects of mixed connective tissue disease: a review. J. Roy. Soc. Med. 85 (1992) 744–746.

FARRELL, D.A. and T.A. MEDSGER: Trigeminal neuropathy in progressive scleroderma. Am. J. Med. 73 (1982) 57–62.

FLECHTNER, K.M. and K. BAUM: Mixed connective tissue disease: recurrent episodes of optic neuropathy and transverse myelopathy. Successful treatment with plasmapheresis. J. Neurol. Sci. 126 (1994) 146–148.

FUJIMOTO, M., J. KIRA, H. MURAI, T. YOSHIMURA, K. TAKIZAWA and I. GOTO: Hypertrophic cranial pachymeningitis associated with mixed connective tissue disease; a comparison with idiopathic and infectious pachymeningitis. Intern. Med. 32 (1993) 510–512.

GENDI, N.S., K.I. WELSH, W.J. VAN VENROOIJ, R. VAN-CHEESWARAN, J. GILROY and C.M. BLACK: HLA type as a predictor of mixed connective tissue disease differentiation. Ten-year clinical and immunogenetic followup of 46 patients. Arthr. Rheum. 38 (1995) 259–266.

GINSBURG, W.W., D.L. CONN, T.W. BUNCH and F.C. MCDUFFIE: Comparison of clinical and serologic markers in systemic lupus erythematosus and overlap syndrome: a review of 247 patients. J. Rheumatol. 10 (1983) 235–241.

GRAF, W.D., J.M. MILSTEIN and D.D. SHERRY: Stroke and mixed connective tissue disease. J. Child Neurol. 8 (1993) 256–259.

HAGEN, N.A., J.C. STEVENS and C.J. MICHET: Trigeminal sensory neuropathy associated with connective tissue diseases. Neurology 40 (1990) 891–896.

HIETARINTA, M., J. ILONEN, O. LASSILA and A. HIETAHARJU: Association of HLA antigens with anti-SCL-70-antibodies and clinical manifestations of systemic sclerosis (scleroderma). Br. J. Rheumatol. 33 (1994) 323–326.

HOFFMAN, R.W., Y. TAKEDA, G.C. SHARP, D.R. LEE, D.L. HILL, H. KANEOKA and C.W. CALDWELL: Human T cell clones reactive against U-small nuclear ribonucleoprotein autoantigens from connective tissue disease patients and healthy individuals. J. Immunol. 151 (1993) 6460–6469.

KALLENBERG, C.G.: Overlapping syndromes, undifferentiated connective tissue disease, and other fibrosing conditions. Curr. Opin. Rheumatol. 6 (1994) 650–654.

KASUKAWA, R., T. TOJO and S. MIYAWAKI: Preliminary diagnostic criteria for classification of mixed connective tissue diseases and anti-nuclear antibodies. Amsterdam, Elsevier (1987) 41–47.

KOBOYASHI, S., M. NAGASE, M. KIMURA, K. OHYAMA, M. IKEYA and N. HONDA: Renal involvement in mixed connective tissue disease. Am. J. Nephrol. 5 (1985) 282–289.

LAZARO, M.A., J.A. MALDONADO COCCO, L.J. CATOGGIO, S.M. BABINI, O.D. MESSINA and O. GARCIA MORTEO: Clinical and serologic characteristics of patients with overlap syndrome: is mixed connective tissue disease a distinct clinical entity. Medicine (Baltimore) 68 (1989) 58–65.

LECKY, B.R.F., R.A.C. HUGHES and N.M.F. MURRAY: Trigeminal sensory neuropathy. Brain 110 (1987) 1463–1485.

LEIBFARTH, J.H. and R.H. PERSELLIN: Characteristics of patients with serum antibodies to extractable nuclear antigens. Arthr. Rheum. 19 (1976) 851–856.

MAIRESSE, N., M.F. KAHN and T. APPELBOOM: Antibodies to the constitutive 73-kd heat shock protein: a new marker of mixed connective tissue disease? Am. J. Med. 95 (1993) 595–600.

MCKENNA, F., H. ECCLES and V.C. NEUMANN: Neuropsychiatric disorders in mixed connective tissue disease. Br. J. Rheumatol. 25 (1986) 225–226.

MUKERJI, B. and J.G. HARDIN: Undifferentiated, overlapping, and mixed connective tissue diseases. Am. J. Med. Sci. 305 (1993) 114–119.

NAKAE, K., F. FURUSAWA, R. KASUKAWA ET AL.: A nationwide epidemiological survey on diffuse collagen diseases: estimation of prevalence rate in Japan. In: R. Kasukawa and G.C. Sharp (Eds.), Mixed Connective Tissue Disease and Anti-nuclear Antibodies. Amsterdam, Excerpta Medica (1987) 9–13.

NIMELSTEIN, S.J., S. BRODY, D. MCSHANE and H.R. HOLMAN: Mixed connective tissue disease: a subsequent evaluation of the original 25 patients. Medicine (Baltimore) 59 (1980) 239.

PADGETT, R.A., S.M. MOUNT, J.A. STEITZ and G.C. SHARP: Splicing of messenger RNA precursors is inhibited by antisera to small nuclear ribonucleoprotein. Cell 35 (1983) 101.

SEARLES, R.P., E.K. MLADINICH and R.P. MESSNER: Isolated trigeminal sensory neuropathy: early manifestation of mixed connective tissue disease. Neurology 28 (1978) 1286.

SHARP, G.C.: Mixed connective tissue disease. In: E. Braunwald, K.J. Isselbacher, R.G. Petersdorf, J.B. Martin, A.S. Fauci and R.K. Root (Eds.), Harrison's Principles of Internal Medicine. New York, McGraw-Hill (1987) 265.

SHARP, G.C., W.S. IRVIN, R.L. LAROQUE, C. VELEZ, V. DALY, A.D. KAISER and H.R. HOLMAN: Association of autoantibodies to different nuclear antigens with clinical patterns of rheumatic disease and responsiveness to therapy. J. Clin. Invest. 50 (1971) 350.

SHARP, G.C., W.S. IRVIN, E.M. TAN, R.G. GOULD and H.R. HOLMAN: Mixed connective tissue disease – an apparently distinct rheumatic disease syndrome associated with a specific antibody to an extractable nuclear antigen (ENA). Am. J. Med. 52 (1972) 148–159.

SHARP, G.C., W.S. IRVIN, C.M. MAY, H.R. HOLMAN, F.C. MCDUFFIE, E.V. HESS and F.R. SCHMID: Association of antibodies to ribonucleoprotein and Sm antigens with mixed connective-tissue disease, systemic lupus erythematosus and other rheumatic diseases. New Engl. J. Med. 295 (1976) 1149–1154.

SULLIVAN, W.E., D.J. HURST, C.E. HARMON, J.H. ESTHER, G.A. AGIA, J.D. MALTBY, S.B. LILLARD, C.N. HELD, J.F. WOLFE, E.V. SUNDERAHAN, H.R. MARIG and G.S. SHARP: A prospective evaluation emphasizing pulmonary involvement in patients with mixed connective tissue disease. Medicine (Baltimore) 63 (1986) 92–107.

TOYODA, K., H. TSUJI, S. SADOSHIMA, C. HORIMOTO and M. FUJISHIMA: Brain hemorrhage in mixed connective tissue disease. A case report. Angiology 45 (1994) 967–971.

VAN DEN HOOGEN, F.H., P.E. SPRONK, A.M. BOERBOOMS, H. BOOTSMA, D.J. DE ROOIJ, C.G. KALLENBERG and L.B. VAN DE PUTTE: Long-term follow-up of 46 patients with anti-(U1)snRNP antibodies. Br. J. Rheumatol. 33 (1994) 1117–1120.

VINCENT, F.M. and R.N. VAN HOUZEN: Trigeminal sensory neuropathy and bilateral carpal tunnel syndrome: the initial manifestation of mixed connective tissue disease. J. Neurol. Neurosurg. Psychiatry 43 (1980) 458–460.

WEISS, T.D., J.S. NELSON, R.M. WOOLSEY, H. ZUCKNER and A.R. BALDASSARE: Transverse myelitis in mixed connective tissue disease. Arthr. Rheum. 21 (1978) 982–986.

YASUDA, M., M. LOO, S. SHIOKAWA, T. WADA, Y. SUENAGA and M. NOBUNAGA: Mixed connective tissue disease presenting myasthenia gravis. Int. J. Med. 32 (1993) 633–637.

Handbook of Clinical Neurology, Vol. 27 (71): Systemic Diseases, Part III
M.J. Aminoff and C.G. Goetz, editors

CHAPTER 7

Polymyositis and dermatomyositis

WILLIAM J. LITCHY

*Department of Neurology, Health Partners, Minneapolis, MN, U.S.A.; and
Department of Neurology, University of Minnesota, Minneapolis, MN, U.S.A.*

Polymyositis (PM) and dermatomyositis (DM) are the two major subgroups of a heterogeneous group of disorders called inflammatory myopathies. Although there are other myopathies that have characteristics of the inflammatory myopathies, including inclusion body myositis (IBM), bacterial and viral myopathies and some toxic muscle diseases (Wright et al. 1994), most patients have either PM or DM (Dalakas 1991, 1992; Carpenter and Karpati 1992; Garlepp and Mastaglia 1996). For further coverage of these disorders, the reader is referred to *Vol. 66: Myopathies*, published in 1992 as Vol. 16 in the revised series of the *Handbook of Clinical Neurology*.

PM and DM are different diseases, although they have similar clinical features and laboratory abnormalities, and often respond to similar treatments. The primary clinical feature that distinguishes the two disorders is the involvement of the skin in DM. There are pathological and immunological features that make DM and PM different diseases. PM involves primary inflammation of the skeletal muscle with fiber necrosis and loss, with frequent involvement of other organs. DM involves inflammation of the blood vessels with secondary skeletal muscle fiber necrosis (Emslie-Smith and Engel 1990). Although other organs can be involved, skin is the primary organ affected.

In addition to clinical features, the diagnosis of PM and DM is based on blood studies, electromyography (EMG) and muscle biopsy (Griggs et al. 1995). Blood studies provide information about the severity of disease, the activity of the myopathy, and informa-

tion about genetic associations (Duncan et al. 1990). EMG provides information about the activity, severity and distribution of the myopathy as well as the efficacy of treatment. The muscle biopsy provides more specific diagnostic information about the inflammatory myopathy, and differentiation between DM and PM and other myopathies can usually be made with the results of the biopsy.

CLINICAL

Epidemiology

The annual incidence of inflammatory myopathy varies from 0.1 to 0.93 per 100,000 population (Pearson and Kurland 1969). Unfortunately, there are no reports separating the incidence of PM and DM in adults. In population-based studies in Olmsted County, Minnesota, the incidence is 0.6/100,000 (Kurland et al. 1969). In spite of PM and DM occurring in all age groups, the frequency of the disease predominates in the fifth and sixth decades of life (Medsger et al. 1970; Pachman and Cooke 1980; Thompson 1982; Pachman 1986; Nagai et al. 1992). Females in all age groups have this disorder more frequently (Medsger et al. 1970). People of African-American descent have an incidence more than twice that of the Caucasian population (Medsger et al. 1970).

Examples of familial DM and PM are uncommon. However, there have been reports of identical twins with DM. Other family studies have linked patients with DM and PM with relatives with other autoimmune disorders (Lewkonia and Buxton 1973). An association with HLA antigens has been reported. HLA-B8 is associated with childhood DM, as well as adult PM and HLA-B7, and HLAw6 has been associated with PM in African Americans.

Adult polymyositis and dermatomyositis

Skeletal muscle
The most frequent initial complaint is the insidious onset of weakness involving proximal muscles. Although the weakness is most frequently symmetrical, asymmetric weakness and isolated focal weakness of muscles may be the presenting symptom. Pain and/or tenderness of skeletal muscle is a frequent complaint and can be an important factor in the debilitating features of this disorder (Bohan and Peter 1975a, b; Bohan et al. 1977). Weakness of neck flexor muscles, often out of proportion to the weakness of other more proximal muscles, may be a clinically distinguishing feature (Pearson and Bohan 1977). In more severe forms of the disease, respiratory muscles are affected and the patient may complain of dyspnea with minimal exertion. During the treatment phase the complaints of respiratory distress may continue while the strength in the appendicular muscles has improved.

Dysphagia is a frequent complaint in patients who have the disease for some time, although rarely it may be one of the initial complaints. The symptoms associated with dysphagia depend on the muscles involved. Most frequently, the striated muscles of the pharynx, hypopharynx, and upper third of the esophagus are involved. In other cases, the lower esophagus may be more involved, resulting in reflux esophagitis. Cricopharyngeal muscles may also be involved, producing coughing episodes during swallowing (Rosenbaum et al. 1996).

Rheumatological symptoms
Arthralgias are often reported in patients who have muscle pain and/or muscle weakness (Citera et al. 1994). Arthritis has been described in 25–65% of patients with PM. It is often a non-specific inflammatory process affecting hands, wrists, and knees. Arthritis can lead to joint contractures in some patients. Raynaud's phenomenon, without scleroderma, is occasionally found in patients with either DM or PM.

Cardiac
There is an increased incidence of cardiac involvement in patients with PM and DM not correlated with the severity of disease (Askari 1988; Leib et al. 1994). The increased incidence of electrocardiographic abnormalities include disorders of bundle branch block, atrioventricular conduction defects and atrial dysrhythmias (Taylor et al. 1993).

Pulmonary involvement
Dyspnea is often associated with involvement of the diaphragm muscle (Sano et al. 1994). Unlike the findings in patients with amyotrophic lateral sclerosis, the contractile properties of the diaphragm muscle in PM are often affected out of proportion to the effect on the compound muscle action potential generated by phrenic nerve stimulation (personal observation). Primary disorders of the lungs, including interstitial lung disease, as well as secondary diseases related to aspiration, are also observed. The frequency of interstitial lung disease varies from 5 to 47% in different reports. Of note, the Jo-1 antibody has been used as a marker for interstitial lung disease associated with PM (Targoff et al. 1989).

Skin involvement
The involvement of the skin is associated with adult and childhood DM. The skin lesions include the classical heliotropic rash, a reddish-purple edematous erythema in the periorbital region. Other skin lesions include reddish-purple keratotic abnormalities on the extensor surface of the finger joints, as well as erythema over the elbows and knees. All of these lesions can result in scaling, pigmentation and depigmentation of the skin.

Prognosis
The 5-year survival rate has been reported in 70–93% in patients with adult DM or PM (Basset-Seguin et al. 1990). In one study, appropriate therapy resulted in a complete cure in 33% of the patients. Poor prognostic signs include inadequate therapy, older age, cardiac disease, interstitial lung disease, respiratory weakness and high sedimentation rate (DeVere 1975; Hendriksson and Sandstedt 1982; Benbassat et al. 1985; Lilley et al. 1988; Hendriksson and Lindvall 1990; Fafalak et al. 1994).

Childhood dermatomyositis

Clinically, childhood DM is characterized by generalized muscle weakness that is often more severe in the pelvic and pectoral muscles, erythematous skin lesions over the extensor surfaces of joints, muscle pain and/or tenderness, and a malar rash (Malleson 1982). The erythematous rash, temporally related to and often preceding the muscle weakness, appears in the periorbital region and extensor surfaces, similar to adult DM. The skin abnormalities and even the muscle weakness can be intensified by sun exposure. Calcifications in the subcutaneous tissue and muscle occur in up to 50% of patients with childhood DM.

Other organs can also be involved in childhood DM (Tymms and Webb 1985). Primary interstitial lung disease as well as vasculitis of the gastrointestinal tract can occur in this disease. In more severe forms of the disease, dyspnea, dysphagia and severe contractures can add to the disability associated with the disease in children.

Prognosis

The disease activity varies in duration, often lasting several months, and may even progress for many years before becoming inactive. Signs of childhood DM may go away completely, although in rare circumstances the disease progresses rapidly to death as a result of respiratory failure. The average duration of disease activity was 6 months for children who recovered and 40 months for those who had continued symptoms (Bowyer et al. 1983; Fafalak et al. 1994; Rose et al. 1996).

Polymyositis and dermatomyositis associated with malignancy

The association of PM and DM with malignancy remains open for discussion (Stone 1993). The incidence of an associated malignancy ranges from 6 to 45% for DM, and between 0 and 28% for PM, with males and females equally affected (Bohan and Peter 1975a,b; Callen 1988). The observation of a malignancy can occur before or after the onset of muscle disease (Barnes 1976; Zantos et al. 1994). The removal of the tumor sometimes results in improvement of muscle weakness, and the frequency of the type of neoplasm differs from the population in that ovary and stomach tumors are more common

than colorectal tumors (Manchul et al. 1985). Although there is no solid statistical evidence to make the association with PM, evidence is strong that the association exists (Bohan and Peter 1975a,b; Lakhanpal et al. 1986; Masi and Hochberg 1988; Richardson and Callen 1989; Leon-Monzon et al. 1994).

Overlap syndromes

There is a subpopulation of patients in whom PM and DM is associated with other connective tissue disorders (Reichlin et al. 1984). These overlap syndromes include: scleroderma myositis (Mimori 1987; Cook et al.), Sjögren's syndrome (Ringel et al. 1982), systemic lupus erythematosus, rheumatoid arthritis, antisynthetase syndrome and mixed connective tissue disease (Vilppula 1972).

PATHOLOGY

Although there are pathological features of skeletal muscle common to almost all inflammatory myopathies, there are characteristic abnormalities specific to PM and DM. Common features include: gross appearance, fiber necrosis, fiber regeneration, fiber diameter variation, increased connective tissue with fibrosis, and inflammation (Hohlfeld et al. 1993).

The inflammation consists of invasion of mononuclear cells (lymphocytes, plasma cells, histiocytes) of perimysial, perivascular and end-endomysial sites. The sites of cell accumulation and the immunophenotypic profile of cells is dependent on the specific inflammatory muscle disease, DM or PM. Although inflammatory cells are the hallmark of inflammatory myopathy, there are some large series reporting a small percentage of muscle biopsies containing necrotic and regenerating fibers without inflammatory cells.

Dermatomyositis

The muscle biopsy abnormalities in childhood and adult DM are similar, except that the pathological features are more pronounced in the childhood form. The main abnormalities are found in relation to blood vessels of the connective tissues, muscle, skin, gastrointestinal tract, fat, and small nerves (Banker). The presence of fibrin thrombi with subsequent infarction or ischemic injury is most prominent in a muscle. Perivascular collections of inflammatory cells with intimal hyperplasia of arteries and veins are the predominant changes that occur in the acute stage of the disease (Banker). Endomysial edema, scattered atrophic muscle fibers, and necrotic and degenerating fibers are also found in the acute stages of DM (Fig. 1).

The amount of inflammation varies with each person. Inflammatory cells are concentrated in the perimysium, where they are either perivascular or found diffusely (Fig. 1). There is a high percentage of B cells and CD4+ T cells in the infiltrate, and the pro-

portion of B cells and CD4 cells is highest at the perivascular sites (Engel and Arahata 1984; Tokano et al. 1993).

Intramuscular blood vessels often show endothelial hyperplasia. Immune complexes containing IgG, IgM, and complement (C3M) have been detected within the walls of intermuscular arteries and veins (Whitaker and Engel 1972; Engel et al. 1990).

Perifascicular atrophy, i.e., atrophy of the peripheral component of muscle fascicles, is a feature of DM (Banker). It does occur in some other connective tissue diseases, but never in PM. Perifascicular

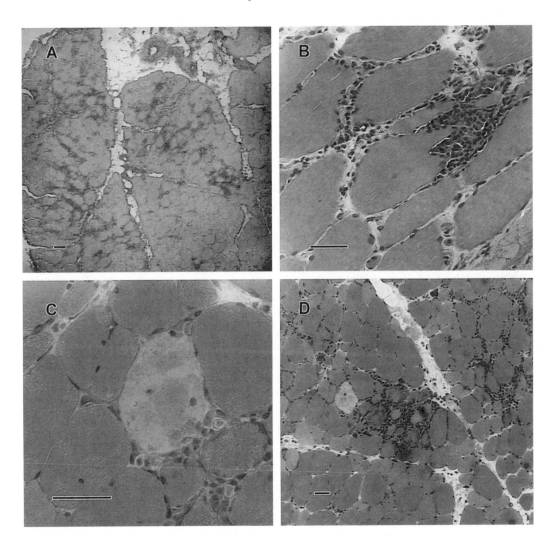

Fig. 1. Acute DM. Serial sections of muscle from a patient with active untreated DM. H&E-stained (A and C) and Gomori trichrome-stained (B) sections show open spaces due to tissue edema, perimysial and perivascular mononuclear inflammation, and atrophic, necrotic and degenerating fibers. NADH staining (D) is irregular with some moth-eaten fibers. Scale bars are all 50 μm.

atrophy involves both type I and type II muscle fibers. In the chronic stages of DM, there is often marked perifascicular atrophy as well as perimysial and endomysial fibrosis, vacuolar degeneration, and regenerating muscle fibers (Fig. 2).

The abnormalities in DM noted with the electron microscope, particularly the abnormalities of intravascular blood vessels, are characteristic of DM. The angiopathy demonstrates alterations in the endothelial cells of capillaries, arterioles, veins and small arteries (Banker and Victor 1966). There are microtubular inclusions frequently circumscribed by endoplasmic reticulum (Fig. 3). There are numerous signs of capillary degeneration, necrosis and regeneration. Multivesicular bodies and autophagic vacuoles are frequently observed, as are lipid bodies (Emslie-Smith and Engel 1990). It is not uncommon to see capillaries lined by degenerating and necrotic endothelial cells and occluded by thrombi.

Ultrastructural changes can also occur in nonatrophic and atrophic perivascular fibers. These changes include mitochondrial loss, focal myofibrillar degeneration, and areas absent of myofibrils. These areas may be filled with debris, clusters of mitochondria, and sarcotubular components. Other regions can show signs of local regeneration.

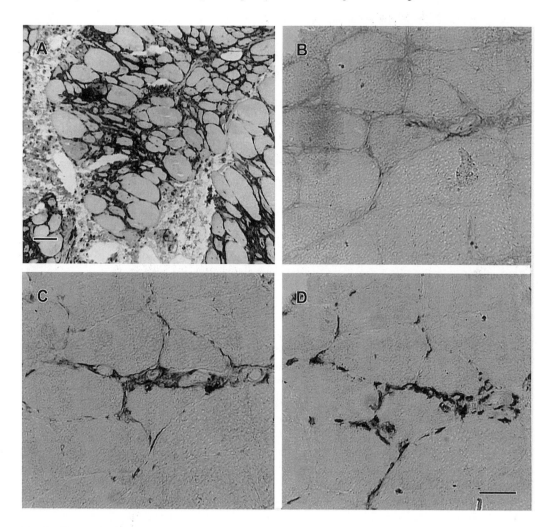

Fig. 2. Chronic DM. Muscle from a patient with chronic DM. Gomori trichrome-stained (A and D) and H&E-stained (B and C) sections show perimysial and endomysial fibrosis indicative of disease chronicity. There is marked perifascicular atrophy, with vacuolar degeneration and regenerating features in the affected fibers. Scale bars are 100 μm (A and B) and 50 μm (C and D).

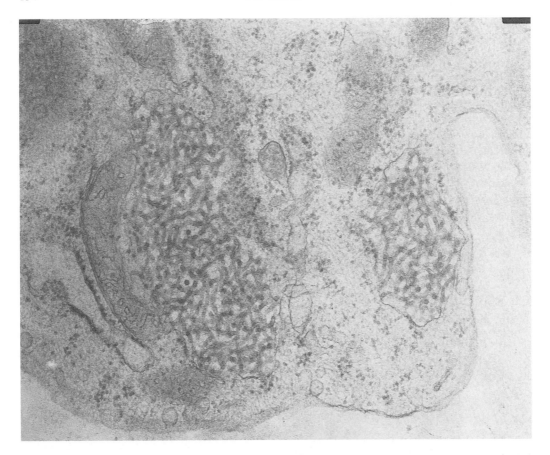

Fig. 3. Vascular endothelium in DM. An electron micrograph of muscle from a patient with DM shows the characteristic microtubular inclusions and swelling within the endothelium.

Polymyositis

The general abnormalities, including necrotic and regenerating fibers, are present in PM as they are in DM. More specific features of PM include the distribution and immunophenotypic profile of the mononuclear cells.

Inflammatory cells occur at perivascular, perimysial and endomysial sites (Fig. 4). The endomysial inflammatory cells contain a high percentage of CD8+ T cells and very few B cells. The endomysial inflammatory cells will attack the non-necrotic fibers and then surround necrotic muscle fibers; CD8+ T cells, accompanied by a lesser number of macrophages, invade and destroy necrotic fibers (Fig. 5).

Muscle biopsy findings in PM differ from DM in that there are no perivesicular atrophy, microvascular injury, or endothelial microtubule inclusions in PM.

The characteristics of the endomysial inflammatory cells are also different in the two disorders.

LABORATORY STUDIES

The accurate diagnosis of inflammatory myopathy is essential for developing a treatment plan and in determining the likelihood that there will be response to treatment and prognosis. Patients with PM or DM often respond very well to appropriate doses of corticosteroids. On the other hand, patients with IBM respond poorly to treatments frequently used for PM and DM. Other prognostic indicators for recovery for patients with DM and PM include the age of the patient, a long delay in making the diagnosis and beginning the best treatment, the presence of specific myositis-specific autoantibodies, and associated malignancy, vasculitis or other systemic disease.

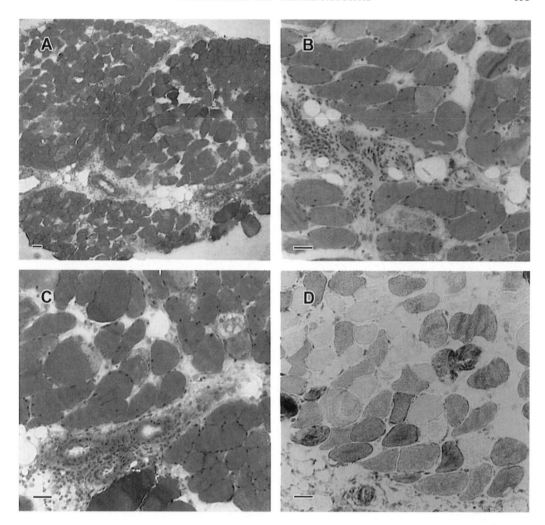

Fig. 4. Polymyositis. Muscle from a patient with untreated PM. H&E-stained (A, B and C) and Gomori trichrome-stained (D) sections show patchy endomysial inflammation with perimysial and endomysial fibrosis. Inflammation attacking non-necrotic fibers (A and D), as well as occasional necrotic fibers with scant adjacent inflammation (C) and (D), are characteristic of the disease. Scale bars are 100 μm (A) and 50 μm (B, C and D).

Although numerous laboratory studies are available for the evaluation of the patient suspected of having an inflammatory myopathy, a muscle biopsy, with appropriate histological studies, electron microscopy, as well as histochemical and cell-marker analysis, remains the best approach to the specific diagnosis of an inflammatory myopathy. Other laboratory studies, including serum enzyme studies, antibody-specific tests, EMG and imaging, are also valuable adjuncts to the diagnosis of patients. Each, in their own right, may provide different and useful information for the diagnosis and management of the patient and be useful in predicting the prognosis and efficacy of treatment.

Serum muscle enzymes

Serum enzymes are not specific diagnostic studies for DM or PM, but an increase in the serum concentration of certain muscle enzymes may help establish the diagnosis as well as indicate the degree to which muscle fibers are affected. The elevation of creatine kinase (CK), aldolase and glutamic oxaloacetic transaminase (SGOT), are often most affected and best reflect the activity of disease. Other muscle enzymes, however, may also be elevated. The muscle enzymes are released from muscle fibers during degeneration. Increases in the serum CK level are more specific

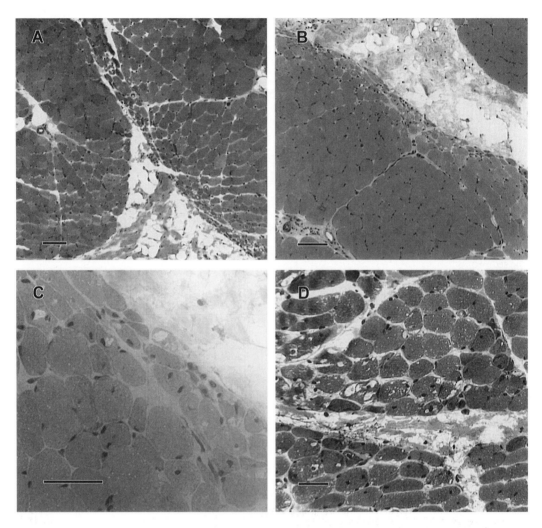

Fig. 5. Inflammatory cell markers in PM. Muscle from a patient with untreated PM. Immunoperoxide staining identifies T lymphocytes using anti-UCHL1 antibodies (A and B), B lymphocytes using anti-L26 antibodies (B) and CD8⁺ T lymphocytes using anti-CD8 antibodies (D). On low-power magnification (A), many endomysial T cell lymphocytes are evident. Serial sections (B, C and D) show no B cells (B) but several T cells (C) attacking an identified non-necrotic fiber, with invading CD8⁺ T cells (D). Scale bars are 100 μm (A) and 50 μm (B, C and D).

indicators of muscle injury than the increase of the levels of other enzymes, as other enzymes are present in liver and/or erythrocytes.

The serum CK concentration has proven to be a very sensitive indicator of muscle parenchymal involvement. However, in some patients it may not be elevated. This is particularly true in children with DM or in patients who have end-stage muscle disease where the muscle mass has been depleted by disease. In the active stages of PM and DM, muscle enzymes are often elevated. At times of remission or inactivity, the enzyme levels often return to normal. When the serum

CK is elevated, serial evaluation represents the most effective laboratory guide for monitoring the status of the disease as well as the effectiveness of treatment. However, it should be kept in mind that there are other causes of elevated serum CK that can confound disease monitoring. For example, needle EMG will often elevate the serum CK level for 2–3 days and, if identified, will mislead the clinician caring for the patient. Serum for enzyme studies should also be obtained before any EMG studies or other invasive studies are performed.

There are three isozymic forms of serum CK. MM

(skeletal muscle), BB (central nervous system), and MB (cardiac muscle) can be measured, although there is only an increase in MM and BB isozymes in muscle degeneration and regeneration.

Serum myoglobin is a sensitive index of the integrity of muscle fibers and has been used as a guide for treatment. The serum myoglobin level will frequently increase more rapidly than serum CK (Ringel et al. 1982; Caccamo et al. 1993). Markedly elevated serum myoglobin levels can produce secondary renal failure, a rare complication of active inflammatory myopathy (Rose et al. 1996).

Erythrocyte sedimentation rate (ESR)

The sedimentation rate has often been used as a laboratory study for the evaluation of patients with inflammatory myopathy. However, it is normal in more than half of the patients with PM and DM, and there is no correlation between the degree of elevation and the degree of weakness (Gran et al. 1993; Stonecipher et al. 1993). ESR is not a useful measure for evaluating the efficacy of treatment, prognosis, or the degree of improvement in a patient with this disease.

Antibody tests

Specific serum antibody studies can provide additional evidence for the clinical diagnosis of some inflammatory muscle diseases, particularly indicating subgroups of PM (Love et al. 1991).

The myositis-specific antibodies occur in a minority of patients with PM and DM but, because only one of each antibody appears in a given patient and because they are associated with distinct HLA haplotypes in patterned diseases, they are important (Hohlfeld et al. 1991; Kanai et al. 1993; Miller 1993). The myositis-specific antibodies fall into two major groups and their detection depends on specific immunodetection techniques. The categories are anticytoplasmic antibodies against translational components, namely antisynthetase, and anti-signal recognition particle (SRP) antibodies, usually associated with relatively severe muscle and systemic diseases. The second are antibodies against MI-2 and Mas antigens, usually associated with relatively mild muscle disease.

The anti-SRP antibodies are detected in less than 5% of PM patients. The patients with this antibody usually have severe myositis that is often acute in onset and associated with myalgias and cardiac involvement. Five-year survival rates in individuals with this antibody are 25%.

The MI-2 antibody is directed against a nuclear protein with an unknown function. Less than 10% of the patients with DM were found to have this antibody in one study. Positive patients usually have a florid rash but respond well to corticosteroids (Marguerie et al. 1992).

The Mas antibody precipitates an RNA molecule of unknown function.

ELECTROMYOGRAPHY AND NERVE CONDUCTION STUDIES

EMG observations in inflammatory myopathies are generally similar, although subtle abnormalities may help to distinguish between some of these disorders (Wilbourn 1993). In general, however, DM and PM cannot be distinguished by electrophysiological features and require other laboratory studies and clinical features to make the distinction. Besides providing diagnostic information about DM and PM, the EMG can extend knowledge about the inflammatory myopathy. Measuring the electrophysiological characteristics of the involved muscles can provide information about the distribution of abnormalities, the severity of abnormalities, the activity of disease, prognostic information, as well as the effectiveness of treatment. The EMG may be particularly useful in patients not responding to corticosteroid treatment, or when they are relapsing and getting progressively weaker in spite of receiving a treatment that should be effective.

Nerve conduction studies

Motor nerve conduction studies are performed by stimulating a nerve and recording the evoked compound muscle action potential from a muscle innervated by the stimulated nerve. The attributes measured include the compound muscle action potential amplitude and area, the distal latency, and the conduction velocity. The compound muscle action potential amplitude and area are the summated electrical activity generated by the activated muscle fibers. In muscle disease, a reduction of the amplitude and area compared to a reference population indicates a loss of muscle fibers as a result of the inflammatory process. In early and/or mild disease, the response may be normal. In more severe disease, the amplitude is often reduced proportionally to the severity of the disease. The compound muscle action potential, a reflection of the numbers of muscle fibers, may be reduced in severe DM and PM, and this is often an indicator of a poor prognosis (Wilbourn 1993).

The nerve conduction velocity is a measure of the rate of the translational movement of the compound action potential along the nerve. Changes in the nerve conduction velocity are most often due to abnormalities of the nerve or its myelin covering. In PM and DM, the nerve conduction velocity is normal. A slow conduction velocity in a patient with DM or PM is usually an indication of a disorder of the nerve superimposed on the inflammatory myopathy. The conduction velocity is often normal in patients with DM and PM.

Distal latency, the time between stimulation of the distal nerve and the response of muscle, is normal in inflammatory myopathy (Wilbourn 1993). An abnormal response is an indication that the patient has another disorder.

Repetitive stimulation of the motor nerve, a clinical electrophysiological technique which assesses the integrity of the neuromuscular junction, is almost always normal in patients with inflammatory myopathies. However, in the evaluation of a progressively weak patient with suspected inflammatory myopathy, repetitive stimulation is useful to distinguish between primary muscle and the neuromuscular junction disorders that also produce the weakness.

Long-latency reflexes including F wave and H reflex studies should be normal in patients with inflammatory myopathy.

Sensory nerve conduction studies are normal in DM and PM unless the patient has another disorder that affects the sensory axons.

Needle electromyography

Needle EMG is useful in the evaluation of patients with DM and PM. The needle recording electrode, usually a monopolar or concentric needle, is inserted into a muscle and the electrical activity generated by the muscle fibers is recorded. The electrical activity is categorized as insertional activity, spontaneous activity and voluntary activity.

Insertional activity

Insertional activity is the electrical activity generated by the movement of the needle through muscle. It represents the mechanical effect of the needle on the muscle fibers during movement. In normal muscle, the insertional activity is limited to <500 ms. In PM and DM, when the disease is active, insertional activity is frequently increased, representing an instability of the muscle membrane during this stage of the disease. An increase in the insertional activity is the first abnormality of the needle examination that will be observed in a patient with an inflammatory myopathy.

Spontaneous activity

There are several types of spontaneous activity generated by muscle, depending on the disease process. The spontaneous discharges are recognized by their pattern of firing, as well as by their configuration. Three forms of spontaneous activity are most commonly observed in DM and PM: fibrillation potentials, myotonic discharges, and complex repetitive discharges.

Fibrillation potentials. Fibrillation potentials are regular, spontaneous discharges of single muscle fibers, or portions of muscle fibers, that are not innervated. They occur in muscle fibers that have never been innervated, or have lost their innervation, and may occur as early as 3–4 days after denervation. Denervated fibers and muscle fibers that are regenerating will often have fibrillations until the nerve terminals connect with them.

Fibrillation potentials are triphasic potentials, 50–300 μV in amplitude and 1–5 ms in duration. They are usually regularly firing, although irregularly firing fibrillation may occur (Buchthal 1982). Fibrillation potentials are observed in a variety of disease processes that disrupt the connection between nerve terminal and muscle fiber and are frequently found in DM and PM. The rate of firing and density of fibrillation potentials are affected by temperature. A decrease in the temperature of the muscle will reduce the frequency of firing and the density of the potentials to the degree that they may not even be observed. Caffeine and some other medications may also affect the presence of the fibrillation potentials. Therefore, these attributes are not a good measure of the amount of muscle fiber separating from the nerve terminals.

The presence of fibrillation potentials, on the other hand, is a sensitive indicator of an active myopathic process in inflammatory myopathies. However, they may not be seen in all muscles in patients with inflammatory myopathy. It has been reported that in inflammatory myopathies, the presence of fibrillation potentials is more likely to be found in proximal muscles, particularly paraspinal and shoulder girdle muscles. This correlates with the distribution of weakness observed clinically. In patients with inflammatory myopathy resistant to treatment, or who are getting progressively worse on corticosteroids, an EMG may be useful in assessing the mechanism of the progression of disease. In patients with progressive muscle disease and no fibrillation potentials, the likelihood is that the progression is due to a steroid-induced myopathy rather than active inflammatory myopathy. The presence of fibrillation potentials, however, does not exclude that some component of the weakness is not due to the use of steroids.

Myotonic discharges. Myotonic discharges are also generated by single muscle fibers. The morphology of the waveforms can be either triphasic or biphasic. They differ from fibrillation potentials in their pattern and rate of firing. The rate is more rapid, ranging from 2 to 100 Hz, and it is more variable than fibrillation potentials with a waxing and waning pattern. The amplitude of the waveforms also waxes and wanes. The changes in firing rate and morphology result in a characteristic sound to the discharges.

Myotonic discharges result from alterations of the ion channels in the muscle fiber membrane. Like fibrillation potentials, they are not specific for a single disease and may even occur in neurogenic disorders. They are seen, occasionally, in inflammatory myopathies, but are more frequently seen in the myotonic syndromes, particularly myotonic dystrophy.

Complex repetitive discharges (CRDs). CRDs are another form of spontaneous discharge that occur in muscle disease. CRDs are discharges of groups of single muscle fibers activated ephaptically and firing in a regular pattern. The firing rate is 2–60 Hz, with both an abrupt start and stop. The configuration depends upon the firing synchrony of the individual muscle fibers contributing to the discharge. CRDs are observed in inflammatory myopathies, but usually in more chronic forms of the disease.

Voluntary potentials

Voluntary muscle activity represents the electrical activity of motor units. Individual muscle fibers contain 50–800 motor units, with each unit having 50–2000 muscle fibers. The morphology of the voluntarily activated motor unit potential (MUP), as well as the firing pattern, can provide information about the disease process.

In healthy individuals, single muscle fiber potentials are recorded as triphasic waves. The size and shapes of MUPs in normal muscle vary significantly, depending upon the areas of the muscle examined, the individual muscle, age, muscle temperature, and the recording needle used. Normal MUPs usually range from 100 µV to 5 mV in amplitude, and from 3 to 15 ms in duration (Buchthal 1953).

Changes occur in the voluntary MUP with disease. Attributes of the MUP that reflect disease include the amplitude, duration, number of phases/turns, and the firing pattern. The firing pattern, including the firing frequency, recruitment pattern and the activation pattern, is also changed in an affected muscle. The MUP amplitude reflects primarily the electrical potentials of muscle fibers near the recording electrode, while the MUP duration depends on electrical activity from muscle fibers that are both close and at a distance from the recording needle electrode. The number of phases, or turns, in a motor unit reflects the synchrony of firing of the individual muscle fibers.

The characteristic changes in MUPs in myopathies are the reduction in amplitude and a decrease in duration of the MUPs. Because of the variation in the electrical characteristics of individual muscle fibers, the desynchronization of firing and MUPs will often be polyphasic and will have many turns (Lambert et al. 1954).

The presence of short-duration, low-amplitude, polyphasic MUPs often represents a primary disease of muscle (Sandstedt et al. 1982). However, diseases of nerves, under some circumstances, can produce similar findings, and caution must be used when interpreting the electrophysiological results. On the other hand, long-duration, high-amplitude MUPs, often referred to as 'neurogenic', can also be seen in primary muscle diseases. IBM is often recognized as a disease with long-duration complex MUPs. Initially, this was interpreted as having simultaneous nerve injury. However, more recently, studies have demonstrated that chronic inflammatory muscle disease, including PM, IBM, or DM, will often have large MUPs (Lotz et al. 1989). Because of this, the use of the words myogenic and neurogenic to describe MUPs is discouraged.

Value of electromyography

EMG studies can be used to confirm a clinical diagnosis, help in determining the severity of disease, and to help direct the muscle biopsy study (Streib et al. 1979). Care should be taken when doing an EMG study to avoid muscles that will potentially be used for a muscle biopsy. Damage produced by needle insertion can produce artifacts that may obscure the manifestations of the disease or produce changes that are misleading. In general, EMG studies should be confined to one side of the body, allowing the other side to be used for muscle biopsy.

Serum muscle enzyme studies should not be performed immediately after an EMG study. Elevation in serum CK can be present for 72 h or more after an extensive needle EMG examination. If serum CK levels are needed, they should be obtained prior to the electrophysiological studies.

Needle EMG can also be used for the assessment of the efficacy in treatment. The absence of fibrillation potentials in a patient being treated suggests that the treatment is effective, as there is electrophysiologically no evidence of denervated muscle fibers. However, it should be noted that in some muscles reinnervation may never be complete, and fibrillation potentials may be present in spite of the fact that treatment has been effective in the disease.

Imaging studies

Magnetic resonance imaging studies in DM and PM show a change in the intensity of the signal in affected muscles as a result of perimuscular edema, and inflammatory changes in subcutaneous fat (Hernandez et al. 1993; Pitt et al. 1993). These abnormalities do provide non-invasive assessment of muscle injury and could be used to monitor the effects of therapy. However, to date, these studies have been primarily reserved for research, as they are expensive and have not proven better than a careful clinical evaluation in serum enzyme studies.

Muscle biopsy

Muscle biopsy remains the most specific study for the diagnosis and characterization of PM and DM.

Skin biopsies have also been used in the diagnosis of these disorders. Muscle biopsies are also useful for understanding the character and distribution of the inflammatory change, characterizing the cellular involvement and assessing the degree of parenchymal involvement. A proper selection of the muscle to biopsy is critical to obtain useful information. A muscle that is severely affected, or a muscle that is not affected, may not provide the information needed for diagnosis or for characterizing the disease.

There are many pathological features that are common to all inflammatory myopathies. At the same time, there are features specific for PM or DM. The features of the muscle biopsy are reviewed in the Pathology section.

THERAPY

In spite of the fact that the primary pathological mechanisms of muscle injury are different in PM and DM, and that numerous natural history studies on these disorders have not separated the two disorders, the treatment modalities are similar for both (Boyd and Neldner 1994; Dalakas 1994a; Griggs et al. 1995). Attempts to find specific mechanism-directed treatments for these disorders are still forthcoming. To date, the mainstays of treatment are directed at the suppression of the immunological mechanisms.

Determining the effectiveness of treatments has been limited both by assessment measures and by the fact that diseases with underlying different pathological mechanisms have frequently been combined in studies of therapeutic efficacy. Methods of assessing the treatment have included measures of muscle performance, biochemical studies, EMG, and studies of function, including respiratory abilities. Each of these measures can be helpful in assessing the effectiveness of treatment but, from a practical standpoint, serum CK remains the most frequently used measure in practice. The measurement of muscle strength is limited because it does not assess the activity of disease. EMG, although useful for determining the effectiveness of treatment, is not as sensitive, and the values measured often are delayed compared to the clinical response, and muscle biopsy is often a big procedure which provides only a small sample of muscle that is potentially affected.

Goals of therapeutic care plan

The approach to the diagnosis and treatment of the patient suspected of having DM or PM is important for the eventual outcome of care. When a patient has symptoms suspicious for an inflammatory disease, the clinician should make the diagnosis as quickly as possible. Physicians treating patients with muscle disease should have access to individuals experienced in performing and reading muscle biopsies and in performing EMG studies on these patients, and a laboratory capable of performing the necessary tests on blood.

Goals of treatment must include an accurate diagnosis. The diagnosis must be based on clinical suspicion, with confirmatory muscle biopsy studies, and supported by EMG and ancillary blood tests. Measures of treatment efficacy must be established before treatment is initiated. Efficacy must be monitored at regular intervals through the treatment phase. Monitoring for side effects of treatment, and developing alternate treatment plans when intolerable side effects occur, is essential. Treatment should be reduced, and then discontinued, when the patient responds to the treatment.

The following are treatment options available.

Corticosteroids

Although a number of treatment modalities are available, corticosteroids, specifically prednisone, remain the choice for initial therapy in patients with PM and DM (Winkelmann et al. 1968; Engel et al. 1972; Laxer et al. 1987; Matsubara et al. 1994; Mastaglia et al. 1997). Corticosteroids have both anti-inflammatory and immunosuppressive actions and reduce the numbers of circulating monocytes, lymphocytes and T cells, with T cells more affected than B cells. The recommended dose of prednisone, the most frequently used corticosteroid, varies, but generally is 50–75 mg/day (1 mg/kg body weight) in divided doses. Depending upon the author, higher or lower doses have been recommended. This treatment should continue until the CK returns to normal, and this may take 1–2 months or longer. At the end of this period of time, the CK may normalize, but the improvement in muscle strength will frequently lag behind the improvement in CK. Nonetheless, monitoring muscle strength should be done to follow the course of the disease over a longer period of time.

If there are no untoward side effects with prednisone, the medication can be continued. After the CK returns to normal, a tapering schedule should begin. Initially, tapering at increments of 10–15 mg/day every 3–4 weeks is reasonable providing that serum CK remains normal. If the serum CK rises, it may be necessary to increase the prednisone dose. If the response to prednisone is favorable, the goal of reaching a maintenance dose by 6 months is reasonable.

Using a treatment protocol of alternate-day prednisone for maintenance, 10–20 mg every other day is reasonable and preferred and reduces the side effects of the medication. Studies in PM and DM and in childhood DM patients have demonstrated that side effects are reduced and the outcome similar to every-day treatment when alternate-day treatment is used.

The morbidity with the use of prednisone can be serious and may result in the use of alternate treatments. Side effects include hypertension, glucose intolerance, cataracts, osteoporosis, aseptic necrosis of the hip, growth arrest in children, weight gain, and cushingoid appearance. Steroid-induced myopathy, a frequent effect of the use of prednisone, may obscure the clinical evaluation of patients with PM or DM. In some cases, although the CK may be normalizing, the patient may be getting weaker. This could be the result of a steroid-induced muscle disorder. It is often clinically difficult to determine if a failure to improve or a progression of weakness is the result of the disease or the treatment.

Patients who have been improving with treatment or are in remission may have a relapse of the disease. This can occur particularly when the reduction in the dose of corticosteroids is too rapid. A rise in the serum CK level often occurs before clinical deterioration, a reason for closely following the serum CK. When a relapse occurs, it may be necessary to increase the dose of prednisone. However, the relapse may not be a result of too rapid a taper, but may be due to the ineffectiveness of the medication, and alternate forms of treatment may have to be considered at that time.

Corticosteroids alone may not be an effective treatment for patients with DM or PM. When a patient does not respond to corticosteroids, initially it is important to reassess the diagnosis. It may also be necessary to make sure that the patient does not have a steroid-induced myopathy. The choice of a second drug is empirical and often is based on personal experiences.

Azathioprine

6-Mercaptopurine, a metabolite of azathioprine, inhibits DNA and RNA synthesis in lymphocytes that have entered a phase of proliferation and differentiation. Its major effect is on T cells, but may also be effective in antibody-mediated disorders which are T cell dependent.

Some clinicians believe that azathioprine is the immunosuppressive agent of choice in patients with DM or PM. However, it becomes effective only after several months of therapy and is not useful for the rapid control of severe, acute disease. In those patients that require long-term treatment for myopathy, azathioprine may be useful because of its steroid-sparing side effects (Bunch et al. 1980).

The recommended oral dose is 2–3 mg/day in three divided doses. A small number of patients have an entity of idiosyncratic reaction to the drug manifested as one or more of the following: fever, anorexia, vomiting, abdominal pain, myalgia, skin rash, or headache. Because about 20% of the patients may develop a transient leukopenia, blood counts and liver function tests should be performed every 1–2 months. There has been concern about the oncogenetic effects of azathioprine, but the risk is low.

Methotrexate

Methotrexate is a folic acid analog inhibiting the enzyme dihydrofolate reductase and preventing the conversion of folic acid to tetrahydrofolate, resulting in reduced DNA synthesis and proliferating lymphocytes. A proportion of patients refractory to prednisone do respond to methotrexate (Joffe et al. 1993). The initial dose is oral methotrexate 5–10 mg/week, which can be increased to 20 mg/week if necessary.

The toxic effects of methotrexate include acute interstitial pneumonitis, hepatotoxicity, fever, stomatitis, skin rash, bone marrow suppression and gastrointestinal symptoms. Methotrexate is contraindicated in patients with severe obesity, alcoholism, diabetes mellitus, renal failure, peptic ulcer disease and liver disease. Because it is a folate acid analog, it may be useful to supplement patients with 5 mg of folic acid once weekly after the methotrexate dose.

Cyclosporin

Cyclosporin inhibits the activation of T cells and the secretion of cytokines including interleukins 2, 3, 4 and 5, and gamma interferon by CD4[+] T cells. The overall effect is to prevent the proliferation and differentiation of cytotoxic T cells. The drug is administered in divided daily doses starting with 3 mg/kg/day, and the dose is adjusted according to clinical response and occurrence of side effects. Although available for a decade, there have been no controlled studies to indicate that it is better than prednisone or superior to azathioprine as a supplement in the treatment of PM or DM (Heckmatt et al. 1989; Emslie-Smith and Engel 1990; Pistoia et al. 1993).

The major side effect of cyclosporin is nephrotoxicity that may result in hypertension and azotemia. Other side effects include headache, gingival hypertrophy, hirsutism, thrombocytopenia and hemolytic anemia. Patients are also at increased risk for developing lymphomas and other malignancies, particularly those of the skin.

Cyclophosphamide

The value of cyclophosphamide in refractory DM and PM has not been evaluated in clinical trials. The reports of the use of intravenous cyclophosphamide have indicated mixed results. The drug can be administered intravenously as a pulse dose (500–1000 mg every 1–4 weeks) or as an oral daily dose (2–2.5 mg/kg).

Cyclophosphamide commonly induces a leukopenia requiring interruption of the drug treatment. Thrombocytopenia, anemia, eosinophilia and pancytopenia are other complications that one has to be concerned about. In general, the side effects of this drug are more severe than other immunosuppressive agents, and the use of this drug should be reserved for the most recalcitrant patients.

Intravenous immunoglobulin (i.v. Ig)

Intravenous immunoglobulin has become a frequently employed therapy for a variety of syndromes treated with immunosuppression. These include myasthenia gravis, Guillain–Barré syndrome and chronic inflammatory demyelinating polyradiculoneuropathy (CIDP) (Bodemer et al. 1990; Soueidan and Dalakas 1993; Amato et al. 1994). One clinical trial of i.v. Ig by Dalakas et al. (1993) has demonstrated a beneficial effect in patients with drug-resistant DM. Eleven of 12 patients who received i.v. Ig improved, and muscle function returned to normal in 75%. Improvement was found to correlate with the dissolution of deposits of C5b-9 reduction in ICAM-1 expression in capillaries, and a return of endomysial capillary numbers to normal in five patients. It was suggested that i.v. Ig interrupts the formation of membrane-attack complex and its deposition of capillaries by inhibiting the deposition of activated C4b and C3b fragments (Basta and Dalakas 1994; Dalakas 1994b). Similar positive results were not found in a study where it was the initial treatment in patients with PM or DM (Lang et al. 1991). The role of i.v. Ig has still not been established in the treatment of patients with inflammatory myopathies, but at this time appears to be a second-line treatment (Cherin et al. 1991, 1994; Sussman et al. 1994; Sussman and Pruzanski 1995; Sansome and Dubowitz 1995).

The dose of i.v. Ig is usually 400 mg/kg/day given daily for 5 days. This has been followed by 3-day treatments for 3–6 months, depending upon the response. When improvement does occur, it will usually be apparent after the first or second course of i.v. Ig (Jann et al. 1992).

I.v. Ig therapy is generally considered to be a safe form of treatment, although expensive. The most frequent side effects reported are headache, fever, nausea, skin rash, pruritus, and myalgias (Dalakas 1994a; Duhem et al. 1994). Some of these side effects may be associated with too rapid infusion of the drug. More serious complications include acute or chronic renal failure (Pasatiempo et al. 1994), anaphylactic reactions, transmission of hepatitis C, and aseptic meningitis (Reimold and Weinblatt 1994; Bertorini et al. 1996).

Plasmapheresis and leukapheresis

Plasmapheresis has been used to decrease circulating levels of pathogenic autoantibodies or immune complexes when leukapheresis has been tried to mitigate cell-mediated immune responses (Dau 1981).

There is evidence of efficacy of plasmapheresis in childhood DM and in older patients refractory to other immunosuppressants. However, a recent control trial with corticosteroid-resistant PM or DM compared plasma exchange with leukapheresis and with shampheresis, finding that plasmapheresis and leukapheresis were no more effective than sham treatment.

Total-body irradiation

Low-dose irradiation with 150 rads has been reported to induce worthwhile sustained remissions in a number of patients with refractory PM who are inca-

pacitated (Kelly et al. 1988; Dalakas 1994b). The rationale is that irradiation depresses peripheral lymphocytes and affects lymphocyte function for a long period of time. Adverse side effects are common with lymphoid irradiation, but are mild and transient with low-dose whole-body irradiation, although there has been a report of fatal bone marrow suppression following this regimen. Total-body irradiation is not curative because all patients require other therapeutic agents (Hubbard et al. 1982). At this time, whole-body irradiation does not appear to be a treatment of choice in this disorder.

Miscellaneous treatments

In addition to therapeutic agents, a number of measures can be used to help patients with PM or DM, particularly those with severe disease. Limiting activity, particularly strenuous activity, bed rest during active disease, and even getting extra rest, may help and are measures that may prevent further injury of muscle. On the other hand, as the disease is under control, physical therapy, moderate exercise, and a defined exercise program may benefit the patient. Some patients have dysphagia that results in weight loss. At a time when cell destruction is occurring, a negative caloric balance is even worse for the patient. High-protein diets may be helpful in making sure patients maintain appropriate weights. If necessary, a percutaneous gastrostomy may be used to preclude weight loss.

Acknowledgements

John Day, M.D., Ph.D., Assistant Professor of Neurology, University of Minnesota, Minneapolis, MN, U.S.A. provided the muscle biopsy illustrations. Trudy M. Aldrich provided secretarial support for preparation of the manuscript.

REFERENCES

AMATO, A.A., R.J. BAROHN, C.E. JACKSON, E.J. PAPPERT, Z. SAHENK and J.T. KISSEL: Inclusion body myositis: treatment with intravenous immunoglobulin. Neurology 444 (1994) 1516–1518.

ASKARI, A.D.: The heart in polymyositis and dermatomyositis. Mt. Sinai J. Med. 55 (1988) 479–482.

BANKER, B.Q.: Dermatomyositis of Childhood Ultrastructural Alterations of Muscle and Intramuscular Blood Vessels 46–75.

BANKER, B.Q. and M. VICTOR: Dermatomyositis (systemic

angiopathy) of childhood. Medicine (Baltimore) 45 (1966) 261–289.

BARNES, B.E.: Dermatomyositis and malignancy. A review of the literature. Ann. Intern. Med. 84 (1976) 68–76.

BASSET-SEGUIN, N., J.C. ROUJEAU, R. GHERARDI, J.C. GUILLAUME, J. REVUZ and R. TOURAINE: Prognostic factors and predictive signs of malignancy in adult dermatomyositis. A study of 32 cases. Arch. Dermatol. 126 (1990) 633–637.

BASTA, M. and M.C. DALAKAS: High-dose intravenous immunoglobulin exerts its beneficial effect in patients with dermatomyositis by blocking endomysial deposition of activated complement fragments. J. Clin. Invest. 94 (1994) 1729–1735.

BENBASSAT, J., D. GEFEL, K. LARHOLT, S. SUKENIK, V. MORGENSTERN and A. ZLOTNICK: Prognostic factors in polymyositis/dermatomyositis. A computer-assisted analysis of ninety-two cases. Arthr. Rheum. 28 (1985) 249–255.

BERTORINI, T.E., A.M. NANCE, L.H. HORNER, W. GREENE, M.S. GELFAND and J.H. JASTER: Complications of intravenous gammaglobulin in neuromuscular and other diseases. Muscle Nerve 19 (1996) 388–391.

BODEMER, C., D. TEILLAC, M. LE BOURGEOIS, B. WECHSLER and Y. DE PROST: Efficacy of intravenous immunoglobulins in sclerodermatomyositis. Br. J. Dermatol. 123 (1990) 545–546.

BOHAN, A. and J.B. PETER: Medical progress – polymyositis and dermatomyositis (1st of 2 parts). N. Engl. J. Med. 292 (1975a) 344–347.

BOHAN, A. and J.B. PETER: Polymyositis and dermatomyositis. N. Engl. J. Med. 292 (1975b) 403–407.

BOHAN, A., J.B. PETER, R.L. BOWMAN and C.M. PEARSON: A computer-assisted analysis of 153 patients with polymyositis and dermatomyositis. Medicine (Baltimore) 56 (1977) 255–286.

BOWYER, S.L., C.E. BLANE, D.B. SULLIVAN and J.T. CASSIDY: Childhood dermatomyositis: factors predicting functional outcome and development of dystrophic calcification. J. Pediatr. 103 (1983) 882–888.

BOYD, A.S. and K.H. NELDNER: Therapeutic options in dermatomyositis/polymyositis. Int. J. Dermatol. 33 (1994) 240–250.

BUCHTHAL, F.: Muscle action potentials in polymyositis. Neurology 3 (1953) 424–436.

BUCHTHAL, F.: Fibrillations: clinical electrophysiology. In: W.J. Culp and J. Ochoa (Eds.), Abnormal Nerves and Muscles as Impulse Generators. New York, Oxford University Press (1982) 632–662.

BUNCH, T.W., J.W WORTHINGTON, J.J. COMBS, D.M. ILSTRUP and A.G. ENGEL: Azathioprine with prednisone for polymyositis. A controlled, clinical trial. Ann. Intern. Med. 92 (1980) 365–369.

CACCAMO, D.V., C.Y. KEENE, J. DURHAM and D. PEVEN: Fulminant rhabdomyolysis in a patient with dermatomyositis. Neurology 43 (1993) 844–845.

CALLEN, J.P.: Malignancy in polymyositis/dermatomyositis. Clin. Dermatol. 6 (1988) 55–63.

CARPENTER, S. and G. KARPATI: The pathological diagnosis of specific inflammatory myopathies. Brain Pathol. 2 (1992) 13–19.

CHERIN, P., S. HERSON, B. WECHSLER, J.C. PIETTE, O. BLETRY, A. COUTELLIER, J.M. ZIZA and P. GODEAU: Efficacy of intravenous gammaglobulin therapy in chronic refractory polymyositis and dermatomyositis: an open study with 20 adult patients. Am. J. Med. 91 (1991) 162–168.

CHERIN, P., J.C. PIETTE, B. WECHSLER, O. BLETRY, J.M. ZIZA, R. LARAKI, P. GODEAU and S. HERSON: Intravenous gamma globulin as first line therapy in polymyositis and dermatomyositis: an open study in 11 adult patients. J. Rheumatol. 21 (1994) 1092–1097.

CITERA, G., M.A. GONI, J.A. MALDONADA COCCO and E.J. SCHEINES: Joint involvement in polymyositis/dermatomyositis. Clin. Rheumatol. 13 (1994) 70–74.

COOK, C.D., F.S. ROSEN and B.Q. BANKER: Dermatomyositis and Focal Scleroderma 979–1016.

DALAKAS, M.C.: Polymyositis, dermatomyositis and inclusion-body myositis. N. Engl. J. Med. 325 (1991) 1487–1498.

DALAKAS, M.C.: Inflammatory and toxic myopathies. Curr. Opin. Neurol. Neurosurg. 5 (1992) 645–654.

DALAKAS, M.C.: High-dose intravenous immunoglobulin and serum viscosity: risk of precipitating thromboembolic events. Neurology 44 (1994a) 223–226.

DALAKAS, M.C.: Current treatment of the inflammatory myopathies. Curr. Opin. Rheumatol. 6 (1994b) 594–601.

DALAKAS, M.C., I. ILLA, J.M. DAMBROSIA, S.A. SOUEIDAN, D.P. STEIN, C. OTERO, S.T. DINSMORE and S. MCCROSKY: A controlled trial of high-dose intravenous immune globulin infusions as treatment for dermatomyositis. N. Engl. J. Med. 329 (1993) 1993–2000.

DAU, P.C.: Plasmapheresis in idiopathic inflammatory myopathy. Arch. Neurol. 38 (1981) 544–552.

DEVERE, R.A.W.G.B.: Polymyositis: its presentation, morbidity and mortality. Brain 98 (1975) 637–666.

DUHEM, C., M.A. DICATO and F. RIES: Side-effects of intravenous immune globulins. Clin. Exp. Immunol. 97 (1994) 79–83.

DUNCAN, A.G., J.B. RICHARDSON, J.B. KLEIN, I.N. TARGOFF, T.M. WOODCOCK and J.P. CALLEN: Clinical, serologic, and immunogenetic studies in patients with dermatomyositis. Acta Derm. Venereol. (Stockh.) 71 (1990) 312–316.

EMSLIE-SMITH, A.M. and A.G. ENGEL: Microvascular changes in early and advanced dermatomyositis: a quantitative study. Ann. Neurol. 27 (1990) 343–356.

ENGEL, A.G. and K. ARAHATA: Monoclonal antibody analysis of mononuclear cells in myopathies. II. Phenotypes of autoinvasive cells in polymyositis and inclusion body myositis. Ann. Neurol. 16 (1984) 209–215.

ENGEL, A.G., K. ARAHATA and A. EMSLIE-SMITH: Immune

Effector Mechanisms in Inflammatory Myopathies. (1990) 141–157.

ENGEL, W.K., A. BORENSTEIN, D.C. DEVIVO, R.J. SCHWARTZMAN and J.R. WARMOLTS: High single dose alternate-day prednisone in treatment of dermatomyositis/polymyositis complex. Trans. Am. Neurol. Assoc. 97 (1972) 272–275.

ENGEL, W.K., A.S. LIGHTER and A.P. GALDI: Polymyositis: remarkable response to total body irradiation. Lancet i (1981) 658.

FAFALAK, R.G., M.G.E. PETERSON and L.J. KAGEN: Strength in polymyositis and dermatomyositis: best outcome in patients treated early. J. Rheumatol. 21 (1994) 643–648.

GARLEPP, M.J. and F.L. MASTAGLIA: Inclusion body myositis. J. Neurol. Neurosurg. Psychiatry 60 (1996) 251–255.

GRAN, J.T., G. MYKLEBUST and S. JOHANSEN: Adult idiopathic polymyositis without elevation of creatine kinase. Case report and review of the literature. Scand. J. Rheumatol. (1993) 94–96.

GRIGGS, R.C., J.R. MENDELL and R.G. MILLER: Evaluation and Treatment of Myopathies. Philadelphia, FA Davis (1995) 277.

HECKMATT, J., C. SAUNDERS, A.M. PETERS, M. ROSE, N. HASSON, N. THOMPSON, G. CAMBRIDGE, S.A. HYDE and V. DUBOWITZ: Cyclosporin in juvenile dermatomyositis. Lancet i (1989) 1063–1066.

HENDRIKSSON, K.G. and B. LINDVALL: Polymyositis and dermatomyositis 1990 – diagnosis, treatment and prognosis. Prog. Neurobiol. 35 (1990) 181–193.

HENDRIKSSON, K.G. and P. SANDSTEDT: Polymyositis – treatment and prognosis. A study of 107 patients. Acta Neurol. Scand. 65 (1982) 280–300.

HERNANDEZ, R.J., D.B. SULLIVAN, T.L. CHENEVERT and D.R. KEIM: MR imaging in children with dermatomyositis: musculoskeletal findings and correlation with clinical and laboratory findings. Am. J. Roentgenol. 161 (1993) 359–366.

HOHLFELD, R., A.G. ENGEL, K., II and M.C. HARPER: Polymyositis mediated by T lymphocytes that express the g/d receptor. N. Engl. J. Med. 324 (1991) 877–881.

HOHLFELD, R., N. GOEBELS and A.G. ENGEL: Cellular mechanisms in inflammatory myopathies. In: F.L. Mastaglia (Ed.), Inflammatory Myopathies. Baillière's Clinical Neurology. London, Baillière Tindell (1993) 617–635.

HUBBARD, W.N., M.J. WALPORT, K.E. HALNAN, R.P. BEANEY and G.R.V. HUGHES: Remission from polymyositis after total body irradiation. Br. Med. J. 284 (1982) 1915–1916.

JANN, S., S. BERETTA, M. MOGGIO, L. ADOBBATI and G. PELLEGRINI: High-dose intravenous human immunoglobulin in polymyositis resistant to treatment. J. Neurol. Neurosurg. Psychiatry 55 (1992) 60–62.

JIMENEZ, C., P.C. ROWE and D. KEENE: Cardiac and central nervous system vasculitis in a child with dermatomyositis. J. Child Neurol. 9 (1994) 297–300.

JOFFE, M.M., L.A. LOVE, R.L. LEFF, D.D. FRASER, I.N. TARGOFF, J.E. HICKS, P.H. PLOTZ and F.W. MILLER: Drug therapy of the idiopathic inflammatory myopathies: predictors of response to prednisone, azathioprine and methotrexate and a comparison of their efficacy. Am. J. Med. 94 (1993) 379–387.

KANAI, Y.T., H. TSUDA, H. HASHIMOTO, K. OKAMURA and S. HIROSE: HLA-DP positive T cells in patients with polymyositis/dermatomyositis. J. Rheumatol. 20 (1993) 77–79.

KELLY, J.J., H. MADOC-JONES, L.S. ADELMAN, P.L. ANDRES and T.L. MUNSAT: Response to total body irradiation in dermatomyositis. Muscle Nerve 11 (1988) 120–123.

KURLAND, L.T., W.A. HAUSER, R.H. FERGUSON ET AL: Epidemiologic features of diffuse connective tissue disorders in Rochester, Minnesota, 1951–1967 with special reference to systemic lupus erythematosus. Mayo Clin. Proc. 44 (1969) 649–657.

LAKHANPAL, S., T.W. BUNCH, D.M. ILSTRUP and L.J. MELTON, III: Polymyositis-dermatomyositis and malignant lesions: does an association exist? Mayo Clin. Proc. 61 (1986) 645–653.

LAMBERT, E.H., G.P. SAYRE and L.M. EATON: Electrical activity of muscle in polymyositis. Transact. Am. Neurol. Assoc. 79 (1954) 64–69.

LANG, B.A., R.M. LAXER, G. MURPHY, E.D. SILVERMAN and C.M. ROIFMAN: Treatment of dermatomyositis with intravenous gammaglobulin. Am. J. Med. 91 (1991) 169–172.

LAXER, R.M., L.D. STEIN and R.E. PETTY: Intravenous pulse methylprednisolone treatment of juvenile dermatomyositis. Arthr. Rheum. 30 (1987) 328–334.

LEIB, M.L., J.G. ODEL and M.J. COONEY: Orbital polymyositis and giant cell myocarditis. Ophthalmology 101 (1994) 950–954.

LEON-MONZON, M., I. ILLA and M.C. DALAKAS: Polymyositis in patients infected with human T-cell leukemia virus type I: the role of the virus in the cause of the disease. Ann. Neurol. 36 (1994) 643–649.

LEWKONIA, R.M. and P.H. BUXTON: Myositis in father and daughter. J. Neurol. Neurosurg. Psychiatry 36 (1973) 820–825.

LILLEY, H., X. DENNETT and E. BYRNE: Biopsy proven polymyositis in Victoria 1982–1987: analysis of prognostic factors.

LOTZ, B.P., A.G. ENGEL, H. NISHINO, J.C. STEVENS and W.J. LITCHY: Inclusion body myositis, observations in 40 patients. Brain 112 (1989) 727–747.

LOVE, L.A., R.L. LEFF, D.D. FRASER, I.N. TARGOFF, M. DALAKAS, P.H. PLOTZ and F.W. MILLER: A new approach to the classification of idiopathic inflammatory myopathy: myositis-specific autoantibodies define useful homogenous patient groups. Medicine (Baltimore) 70 (1991) 360–374.

MALLESON, P.: Juvenile dermatomyositis: a review. J. Roy. Soc. Med. 75 (1982) 33–37.

MANCHUL, L.A., A. JIN, K.I. PRITCHARD, J. TENENBAUM, N.F. BOYD, P. LEE, T. GERMANSON and D.A. GORDON: The frequency of malignant neoplasms in patients with polymyositis-dermatomyositis – a controlled study. Arch. Intern. Med. 145 (1985) 1835–1839.

MARGUERIE, C., C.C. BUNN, J. COPIER, R.M. BERNSTEIN, J.M. GILROY, C.M. BLACK, A.K. SO and M.J. WALPORT: The clinical and immunogenetic features of patients with autoantibodies to the nucleolar antigen PM-Scl. Medicine (Baltimore) 71 (1992) 327–336.

MASI, A.T. and M.C. HOCHBERG: Temporal association of polymyositis-dermatomyositis with malignancy: methodologic and clinical considerations. Mt. Sinai J. Med. 55 (1988) 471–478.

MASTAGLIA, F.L., B.A. PHILLIPS and P. ZILKO: Treatment of inflammatory myopathies. Muscle Nerve (1997) 651.

MATSUBARA, S., Y. SAWA and M. TAKAMORI: Letters to the Editor. Pulsed intravenous methylprednisolone combined with oral steroids as the initial treatment of inflammatory myopathies. J. Neurol. Neurosurg. Psychiatry 57 (1994) 1008.

MEDSGER, T.A., JR., W.N. DAWSON, JR. and A.T. MASI: The epidemiology of polymyositis. Am. J. Med. 48 (1970) 715–723.

MEHREGAN, D.R. and W.P.D. SU: Cyclosporine treatment for dermatomyositis/polymyositis. Mayo Clin. Dept. Dermatol. 51 (1993) 59–61.

MILLER, F.W.: Myositis-specific autoantibodies. Touchstones for understanding the inflammatory myopathies. J. Am. Med. Assoc. 270 (1993) 1846–1849.

MILLER, F.W., S.F. LEITMAN, M.E. CRONIN, J.E. HICKS, R.L. LEFF, R. WESLEY, D.D. FRASER, M. DALAKAS and P.H. PLOTZ: Controlled trial of plasma exchange and leukapheresis in polymyositis and dermatomyositis. N. Engl. J. Med. 326 (1992) 1380–1384.

MILLER, G., J.Z. HECKMATT and V. DUBOWITZ: Drug treatment of juvenile dermatomyositis. Arch. Dis. Child. 58 (1983) 445–450.

MIMORI, T.: Scleroderma – polymyositis overlap syndrome. Clinical and serologic aspects. Int. J. Dermatol. 26 (1987) 419–425.

NAGAI, T., T. HASEGAWA, M. SAITO, S. HAYASHI and I. NONAKA: Infantile polymyositis: a case report. Brain Dev. 14 (1992) 167–169.

PACHMAN, L.M.: Juvenile dermatomyositis. Pediatr. Clin. N. Am. 33 (1986) 1097–1117.

PACHMAN, L.M. and N. COOKE: Juvenile dermatomyositis: a clinical and immunologic study. J. Pediatr. 96 (1980) 226–234.

PASATIEMPO, A.M.G., J.A. KROSER, M. RUDNICK and B.I. HOFFMAN: Acute renal failure after intravenous immunoglobulin therapy. J. Rheumatol. 21 (1994) 347–349.

PEARSON, C.M.: Polymyositis 63–82.

PEARSON, C.M. and A. BOHAN: The spectrum of polymyositis and dermatomyositis. Med. Clin. N. Am. 61 (1977) 439–457.

PISTOIA, V., A. BUONCOMPAGNI, R. SCRIBANIS, L. FASCE, G. ALPIGIANI, G. CORDONE, M. FERRARINI, C. BORRONE and F. COTTAFAVA: Cyclosporin A in the treatment of juvenile chronic arthritis and childhood polymyositis-dermatomyositis. Results of a preliminary study. Clin. Exp. Rheumatol. 11 (1993) 203–208.

PITT, A.M., J.L. FLECKENSTEIN, G.G. GREENLEE JR, D.K. BURNS, W.W. BRYAN and R. HALLER: MRI-guided biopsy in inflammatory myopathy: initial results. Magn. Reson. Imaging 11 (1993) 1093–1099.

PLOTZ, P.H., M. DALAKAS, R.L. LEFF, L.A. LOVE, F.W. MILLER and M.E. CRONIN: Current concepts in the idiopathic inflammatory myopathies: polymyositis, dermatomyositis and related disorders. Ann. Intern. Med. 111 (1989) 143–157.

REICHLIN, M., P.J. MADDISON, I. TARGOFF, T. BUNCH, F. ARNETT, G. SHARP, E. TREADWELL and E.M. TAN: Antibodies to a nuclear/nucleolar antigen in patients with polymyositis overlap syndromes. J. Clin. Immunol. 4 (1984) 40–44.

REIMOLD, A.M. and M.E. WEINBLATT: Tachyphylaxis of intravenous immunoglobulin in refractory inflammatory myopathy. J. Rheumatol. 21 (1994) 1144–1146.

RICHARDSON, J.B. and J.P. CALLEN: Dermatomyositis and malignancy. Med. Clin. N. Am. 73 (1989) 1211–1220.

RINGEL, S.P., J.Z. FORSTOT, E.M. TAN, C. WEHLING, R.C. GRIGGS and D. BUTCHER: Sjögren's syndrome and polymyositis or dermatomyositis. Arch. Neurol. 39 (1982) 157–163.

ROSE, A.L. and J.N. WALTON: Polymyositis: a survey of 89 cases with particular reference to treatment and prognosis. Brain 89 747–768.

ROSE, M.R., J.T. KISSEL, L.S. BICKLEY and R.C. GRIGGS: Sustained myoglobinuria: the presenting manifestation of dermatomyositis. Am. Acad. Neurol. 47 (1996) 119–123.

ROSENBAUM, R.B., S.M. CAMPBELL and J.T. ROSENBAUM: Clinical Neurology of Rheumatic Diseases (1996).

SANDSTEDT, P.E.R., K.G. HENDRIKSSON and L.E. LARSSON: Quantitative electromyography in polymyositis and dermatomyositis, a long-term study. Acta Neurol. Scand. 65 (1982) 110–121.

SANO, M., M. SUZUKI, M. SATO, T. SAKAMOTO and M. UCHIGATA: Case report. Fatal respiratory failure due to polymyositis. Intern. Med. 33 (1994) 185–187.

SANSOME, A. and V. DUBOWITZ: Intravenous immunoglobulin in juvenile dermatomyositis – four year review of nine cases. Arch. Dis. Child. 72 (1995) 24–28.

SILBERT, P.L., W.V. KNEZEVIC and D.T. BRIDGE: Cerebral infarction complicating intravenous immunoglobulin therapy for polyneuritis cranialis. Neurology 42 (1992) 257–258.

SOUEIDAN, S.A. and M.C. DALAKAS: Treatment of inclusion body myositis with high-dose intravenous immunoglobulin. Neurology 43 (1993) 876–879.

STIGLBAUER, R., W. GRANINGER, L. PRAYER, J. KRAMER, H. SCHURAWITZKI, K. MACHOLD and H. IMHOF: Polymyositis: MRI-appearance at 1.5 T and correlation to clinical findings. Clin. Radiol. 48 (1993) 244–248.

STONE, O.J.: Dermatomyositis/polymyositis associated with internal malignancy: a consequence of how neoplasms alter generalized extracellular matrix in the host. Med. Hypoth. 41 (1993) 48–51.

STONECIPHER, M.R., J.L. JORIZZO, W.L. WHITE, F.O. WALKER and E. PRICHARD: Cutaneous changes of dermatomyositis in patients with normal muscle enzymes: dermatomyositis sinusitis. J. Am. Acad. Dermatol. 28 (1993) 951–956.

STREIB, E.W., A.J. WILBOURN and H. MITSUMOTO: Spontaneous electrical muscle fiber activity in polymyositis and dermatomyositis. Muscle Nerve 2 (1979) 14–18.

SUSSMAN, G.L. and W. PRUZANSKI: Editorial – the role of intravenous infusions of gamma globulin in the therapy of polymyositis and dermatomyositis. J. Rheumatol. 21 (1994) 990–992.

SUSSMAN, G.L. and W. PRUZANSKI: Treatment of inflammatory myopathy with intravenous gamma globulin. Curr. Opin. Rheumatol. 7 (1995) 510–515.

TARGOFF, I.N., F.C. ARNETT, L. BERMAN, C. O'BRIEN and M. REICHLIN: Anti-KJ: a new antibody associated with the syndrome of polymyositis and interstitial lung disease. J. Clin. Invest. 84 (1989) 162–172.

TAYLOR, A.J., D.C. WORTHAM, J.R. BURGE and K.M. ROGAN: The heart in polymyositis: a prospective evaluation of 26 patients. Clin. Cardiol. 16 (1993) 802–808.

THOMPSON, C.E.: Infantile myositis. Dev. Med. Child Neurol. 24 (1982) 307–313.

TOKANO, Y., T. OBARA, H. HASHIMOTO, K. OKUMURA and S. HIROSE: Soluble CD4, CD8 in patients with polymyositis/dermatomyositis. Clin. Rheumatol. 12 (1993) 368–374.

TYMMS, K.E. and J. WEBB: Dermatopolymyositis and other connective tissue diseases: a review of 105 cases. J. Rheumatol. 12 (1985) 1140–1148.

VILPPULA, A.: Muscular disorders in some collagen diseases. Acta Med. Scand. 540 (1972) 1–47.

WHITAKER, J.N.: Inflammatory myopathy: a review of etiologic and pathogenetic factors (invited review). Muscle Nerve 5 (1982) 573–592.

WHITAKER, J.N. and W.K. ENGEL: Vascular deposits of immunoglobulin and complement in idiopathic inflammatory myopathy. N. Engl. J. Med. 286 (1972) 333–338.

WILBOURN, A.J.: The electrodiagnostic examination with myopathies. J. Clin. Neurophysiol. 10 (1993) 132–148.

WINKELMANN, R.K., D.W. MULDER, E.H. LAMBERT, F.M. HOWARD and G.R.E. GIESSNER: Dermatomyositis-polymyositis: comparison of untreated and cortisone-treated patients. Proc. Mayo Clin. 43 (1968) 545–556.

WRIGHT, G.D., C. WILSON and A.L. BELL: Case report. D-Penicillamine induced polymyositis causing complete heart block. Clin. Rheumatol. 13 (1994) 80–82.

ZANTOS, D., Y. ZHANG and D. FELSON: The overall and temporal association of cancer with polymyositis and dermatomyositis. J. Rheumatol. 21 (1994) 1855–1859.

Handbook of Clinical Neurology, Vol. 27 (71): Systemic Diseases, Part III
M.J. Aminoff and C.G. Goetz, editors

CHAPTER 8

Neurologic manifestations of the systemic vasculitides

PATRICIA M. MOORE

Department of Neurology, Wayne State University School of Medicine, Detroit, MI, U.S.A.

The vasculitides are a group of diseases and disorders sharing the central feature of inflammation of the blood vessel wall with attendant tissue ischemia. Because involvement of the blood vessel is intrinsic to inflammation of any type, vasculitis may be a manifestation of diverse diseases. When inflammation targets the vasculature and tissue injury results from ischemia, the disease itself is called a vasculitis. Many varieties of vasculitis exist. Some are named on the basis of distinctive clinical features; others are recognized on the basis of a known etiology. Classification of the primary and secondary vasculitides still depends on clinical and histologic characteristics, although recent advances in understanding immunopathogenic mechanisms offer additional diagnostic tools.

Clinically, preferential involvement of certain organs renders many of the diseases characteristic. Histologically, the type and size of vessel, the character of the inflammatory infiltrate, and the presence of necrosis, aneurysm formation, and cicatrization in the vessel wall contribute distinct information. Recent studies of adhesion molecules, cytokines and their receptors, and neuro-peptides add to the histopathologic repertoire. Factors which contribute to the tissue ischemia include physical disruption of the vessel wall from the cellular infiltrate, hemorrhage from the altered wall competence, increased coagulation from changes in the normally anticoagulant endothelial cell surface, and increased vasomotor reactivity from released neuro-peptides.

Historically, the modern era of study of systemic vasculitis began over 100 years ago when two physicians, Kussmaul and Maier (1866), expressed keen interest in the post-mortem tissue of a patient who had greatly taxed their diagnostic acumen. Papers over the next century detailed clinical features of patients with sundry disorders, but the identification of a prominent cellular infiltration in the vessel wall enabled numerous authors to distinguish among diseases and name them (Kernohan and Woltman 1938; Zeek et al. 1948; Zeek 1952, 1953). Corticosteroids and later immunosuppressant medications ushered a third era of interest in vasculitis (Baggenstoss et al. 1951; Fauci et al. 1978a, b; Leib et al. 1979). More recent studies have reexamined the initial interpretations and natural history of the disease, and have inaugurated debate on the risks and benefits of aggressive therapy in individual diseases. Recent symposia on the criteria for diagnosis of vasculitis both clarified certain issues and stimulated debate over others (Bloch et al. 1990; Fries et al. 1990; Hunder et al. 1990; Smoller et al. 1990; Jennette et al. 1994). The clinician and scientist are aided by acquaintance with current understanding of the immunopathogenic mechanisms present in some of the vasculitic syndromes.

In the idiopathic vasculitides, we can often identify immediate processes that result in vascular damage but remain unaware of the inciting events (Hasler 1984; Savage and Ng 1986). Recent understanding of the cell-mediated interactions between T-cells and endothelial cells exemplifies this. Both cells can 'activate' each other and participate in the normal physiologic processes of immune surveillance and clearing the body of microbial pathogens. Pertinent yet unanswered questions include: which cell is the initial stimulus in vasculitis? Why do processes which are normally subject to tight regulatory control persist? Why do leukocytes which normally traverse the vessel wall remain in the vessel wall? Why does the endothelium, which normally facilitates development of inflammation without self-injury, become a target for injury in vasculitis? Future therapies depend on our answers to these questions.

Leukocyte–endothelial interactions

Inflammation begins with leukocyte adhesion to endothelial cells and subsequent recruitment of additional leukocytes to the region. The location of the vasculitis, type of inflammatory infiltrate and persistence of vascular inflammation are determined by adhesion molecules, cytokines, leukotriens, activated complement components and microbial products. Subsequent steps of inflammation, penetration in the vessel wall, and release of injurious products, vary with individual immunopathogenic mechanisms.

The vascular endothelium, a highly specialized, metabolically active monolayer of cells, contributes to functional specialization of different organs, maintains thromboresistance and vascular tone, directs lymphocyte circulation, and regulates inflammation and immune interactions. The dynamic interactions between endothelial cells, leukocytes and platelets contribute to numerous physiologically important mechanisms (Savage and Cooke 1993). In the development of inflammation, a pivotal step involves leukocyte recruitment and attachment in the presence of blood flow. Leukocyte attachment to the endothelium and infiltration of tissue are mediated by a multiple receptor-ligand system belonging to three families of related proteins: the selectins, the integrins, and the

immunoglobulin superfamily. The spatial and temporal development of selectins, chemoattractants, adhesion molecules and integrins results in recruitment of leukocytes to a specific tissue site (Braquet et al. 1989; Luscinskas et al. 1989; Moore 1989a; Osborn 1990; Argenbright and Barton 1992; Springer 1994).

The initial attachment of leukocytes to endothelial cells is through the binding of selectins to carbohydrate moieties on the leukocyte cell surface. This loose binding enables rolling of the leukocyte along the endothelial cell surface. Endothelial P selectin expressed at high densities or leukocyte L selectin can mediate the initial capture of leukocytes from the flowing blood (Lawrence and Springer 1991; Picker et al. 1991; Okada et al. 1994). Subsequently, E-selectin mediates rolling. At this stage, another set of molecules, the chemokines, recruit cells along a concentration gradient. Individual chemokines are fairly specific for leukocyte types; they amplify and increase the diversity of cells in the infiltrate. MIP-1a appears to be a chemoattractant for B-cells, cytotoxic T-cells and CD4 positive T-cells. MIP-1b is a chemoattractant for eosinophils, monocytes, and T-cells. Numerous other cloned chemokines, including IL-8, also contribute to endothelial/leukocyte adhesion (Matsushima and Oppenheim 1989; Rollins et al. 1990).

Integrins are the next step in the development of inflammation and mediate the adhesion (arrest) stage. Platelet activating factor (PAF), a biologically active phospholipid, is coexpressed with P-selectin at the time of the endothelial cell's activation and appears to signal activation of integrins on the leukocyte surface. Leucocyte β1 and β2 integrins binding to their ligands such as VCAM and ICAM (members of the immunoglobulin superfamily of adhesion molecules) mediate firmer adhesion of the leukocyte to the endothelium. Intercellular adhesion molecule (ICAM-1) is at least one ligand for the CD18 family of leukocyte integrins. There is some constitutive expression on several non-hematopoietic cells. Pro-inflammatory cytokines such as interleukin-1 (IL-1), IL-8, tumor necrosis factor (TNF), MCP-1, PAF, LTB4, and C5a upregulate expression of these receptor ligand molecules on endothelial cells and leukocytes. After firm adhesion, leukocytes then traverse the vessel wall. This last step, diapedesis of the leukocytes between endothelial cells involves homotypic adhesion of PECAM-1 (CD31) expressed on both leukocytes and endothelium.

Tissue injury depends on the ultimate location of fully activated leukocytes. In neutrophil-mediated tissue injury, neutrophils, following a chemoattractant gradient, completely traverse the wall and enter the tissue parenchyma. In this scenario, the final stages of activation, degranulation, and generation of toxic oxygen metabolites occur in the tissue with minimal changes in the vessel wall. However, if neutrophils are fully activated within the vessel wall (as they are in many cases of vasculitis) then the release of lytic granules (including collagenases, proteases, and elastases) and toxic radicals (including hydroxyl radicals and hydrogen peroxide) injures the vessel wall itself (Sacks et al. 1978; Blann and Scott 1991; Das Neves et al. 1991; Rothlein et al. 1994). The most severe injuries cause mural necrosis leading to hemorrhage and thrombosis.

T-lymphocyte-mediated interactions in the vessel wall also occur, but are less well defined. T-cell-mediated vascular inflammation may be antigen-specific or antigen-non-specific (Stevens et al. 1982; Pober et al. 1984; Dustin and Springer 1988; Springer 1994). Antigen-specific adhesion of T-lymphocytes to endothelial cells (such as seen in transplantation rejection and graft versus host disease) requires binding of T-cell antigen receptors (TCR) and major histocompatability complex (MHC) molecule receptors (CD4 and CD8) on T-lymphocytes to antigen in the antigen presenting groove of MHC molecules on endothelial cells. The presence of lymphocytes in an inflammatory lesion, however, does not identify an antigen-specific process. Cytokine initiated and amplified activation of either lymphocytes or endothelial cells provides a mechanism for cellular attachment in the absence of antigen. As described above, families of adhesion molecules (selectin, integrins, and IgCAMs) promote lymphocyte endothelial binding in the absence of a specific antigen. Another characteristic of lymphocytes, distinguishing them from neutrophils, is their egress from the tissue and reentry into the circulation (Cox and Ford 1982). These memory lymphocytes respond more quickly to stimuli and are often more refractory to deletion.

Other cells participate in the cell-mediated vasculitides. Important cells, particularly in the progression of inflammation from acute to chronic, are the mononuclear cells which when activated become macro-

phages (Snyder et al. 1982; Merrill and Chen 1991). These cells also release cytokines which recruit more monocytes, macrophages, and lymphocytes to the site of injury. Notably, they also possess regulatory functions and may crucially downregulate the inflammatory responses.

Platelets contribute to vascular damage by mechanisms distinct from their role in coagulation. Their cell surface receptors include class I MHC, P selectin, IgG receptors, low affinity IgE receptors, and receptors for von Willebrand factor and fibrinogen (Johnson 1991). A wide variety of substances activate platelets including epinephrine, adenosine diphosphate, collagen, serotonin, membrane attack complex of complement, vasopressin, PAF, and immune complexes. Platelets then release a variety of pro-inflammatory mediators that generate complement activation and augment neutrophil-mediated injury.

The eosinophil, characteristically present in lesions of patients with Churg–Strauss angiitis, may also participate in the pathogenesis of vascular injury (Tai et al. 1984).

Immune complex-mediated mechanisms

Immune (antigen–antibody) complexes are a normal part of the immune response and are usually cleared from the body without causing vascular inflammation. Immune complexes localized in vessel walls, either by deposition from the circulation or by in situ formation, may be pathogenic (Cochrane and Weigle 1958; Cochrane 1963a, b). Conditions conducive to pathogenicity are certain biophysical properties of the complexes or particularly receptive features of a vessel wall (increased vascular permeability secondary to platelets, complement activation, or presence of mast cells). Complexes at almost equivalence of antibodies and antigens are more likely to precipitate; those with a negative charge are more likely to interact with vessel walls.

Immune complexes can then initiate a series of events recruiting an inflammatory response. The Fc portion of the IgG and IgM antibody molecules in the complexes engages Fc receptors on neutrophils and monocytes, both attaching these cells to the site of immune complex localization and inducing degranulation and release of pro-inflammatory molecules. Immune complexes also activate complement components which induce a variety of inflammatory events. C2a and C3a increase vascular permeability and neutrophil degranulation. C5a attract neutrophils and monocytes to the region (Kniker and Cochrane 1968; Cochrane 1971; Mannik 1982, 1987; Fligiel et al. 1984; Joselov and Mannik 1984; Henson and Johnston 1987). The membrane attack complex, C5b-9, injures matrix materials and cells in the vessel wall. The results of these events include necrosis of the vessel wall, an exudative inflammatory response, and usually healing with prominent scarring.

Types of antigens identified in immune complexes include both heterologous (sulfonamides, mouse monoclonal antibodies, microbial antigens) and auto-antigens (nuclear antigens and rheumatoid factor). Antigens in in situ complex formation are less completely studied, but DNA complexes in the kidney are described.

Immune complex-mediated vasculitis is characterized histologically by mixed neutrophil and mononuclear cell infiltrates with prominent necrosis and evidence of immunoglobulin, complement, and fibrin deposition.

Autoantibody-mediated mechanisms

Until recently, investigators thought autoantibodies played a minor role in the pathogenesis of vasculitis (Kallenberg 1993). Several specific situations now illustrate real or potential immunopathogenic mechanisms. In Goodpasture's syndrome (vasculitis of the pulmonary alveolar capillaries and renal glomeruli) autoantibodies to type IV collagen in the capillary basement membrane bind to their target activate complement and leukocytes resulting in vascular injury.

In vitro studies of Kawasaki's disease, an acute viral vasculitis of children, reveal that antibodies to endothelial cells bind to neoantigens induced by IL-1 and TNF on cultured endothelial cells and lyse their targets (Leung et al. 1986). The in vivo role is not as clear.

Anti-neutrophil cytoplasmic antibodies (ANCA) are a group of antibodies reactive with the neutrophils (Van der Woude et al. 1985; Falk and Jennette 1988; Lockwood et al. 1988; O'Donoghue et al. 1989; Gaskin et al. 1993; Gross et al. 1993). ANCA have two histologic patterns, cANCA and pANCA, which correlate with two different autoantigens – myelo-

peroxidase and PC3 respectively. cANCA are strongly associated with Wegener's granulomatosis and microscopic polyarteritis (Tervaert et al. 1991). Their role in the pathogenesis of disease is not certain, but proposed mechanisms do account for many features of disease and are being actively investigated. Binding of ANCA to neutrophils or monocytes in vitro stimulates the cells to undergo a respiratory burst. This generates toxic oxygen metabolites and secretes pro-inflammatory mediators such as LTB4, IL-8, and MCP-1 that recruit more neutrophils and monocytes. The neutrophils also degranulate, releasing lytic enzymes which may injure the vascular endothelium. ANCA-associated vasculitides are characterized histologically by a neutrophil-rich inflammatory infiltrate.

Features of the CNS vasculature

Biochemical and immunologic features of endothelial cells and smooth muscle cells vary regionally. Small and large muscled arteries possess distinctive characteristics as do the microvascular and venous system. Endothelial cells of the central nervous system (CNS) are physically and biochemically conspicuously distinctive. Their tight junctions between cells, paucity of micropinocytic vessels, and high levels of γ-glutamyltranspeptidase are three examples. Several other properties of cerebral endothelial cells result in differences in inflammation of the cerebral blood vessels. Although cerebral endothelia are capable of expressing MHC class I molecules (restriction elements for antigen presentation to the T-cell receptor of $CD8^+$ T-cells) and MHC class II (restriction elements for $CD4^+$ helper T-cells), they appear to do this less often than the endothelium of the systemic vasculature (Pober et al. 1984, 1986; Wong et al. 1984; Fabry et al. 1992, 1993). A low constitutive expression of adhesion molecules also contributes to the lower proclivity of cerebral vessels for vasculitis. For example, ICAM1 appears to be expressed constituently only at low levels on brain endothelium in vivo in contrast to other tissue endothelium. Molecules regulating inflammation such as TGF-β, which downregulate adhesion of leukocytes, may play a more prominent role in the CNS than the systemic vaculature.

Lymphocyte traffic through the CNS is normally limited. The level of lymphocyte adhesion to brain endothelium was less than 5% compared with 15–20%

in other organs (Hart et al. 1990; Hickey 1991; Hickey et al. 1991). Activated lymphocytes do traverse the cerebral endothelium and enter the CNS and, at least in vitro, the cerebral endothelium responds to a spectrum of cytokines by expression adhesion molecules and MHC as do other tissues. Another notable feature of the cerebrovasculature is that leukocytes entering into the parenchyma cross through rather than between endothelial cells.

It thus appears that the CNS endothelium readily presents adhesion molecules, but may leave antigen presentation to parenchymal brain cells (Hickey and Kimura 1988). This may have survival advantages because inflammation within the cerebral vessel wall disrupts the tight endothelial junctions and integrity of the blood–brain barrier. Supporting evidence for this may be seen in the vasculitides. Inflammation of the CNS vessels does not occur often in systemic vasculitis. Many of the central neurologic complications in PAN, for example, appear later in the course of disease and might result from hypertensive or chronic vaso-occlusive changes rather than segmental inflammation of the vessel wall. Another disease that supports this assertion that inflammation of the CNS vasculature is tightly regulated is systemic lupus erythematosus (SLE). Degenerative or vaso-occlusive vasculopathy is present in the CNS, but inflammatory vascular disease rarely occurs despite the presence and deposition of circulating immunoglobulin which cause inflammation elsewhere.

Mechanisms which exacerbate or modify tissue injury associated with inflammation

Coagulation
Several events, intimately connected with but temporally dispersed from the initial events, contribute to the clinical features of vasculitis. Of these, coagulation is best studied. A number of pro- and anticoagulant properties associated with the endothelium normally exert a net anticoagulant effect. Additional endothelial anti-platelet and fibrinolytic properties contribute to the maintenance of thromboresistance. During inflammation, the balance changes and the endothelial surface exerts a net procoagulant effect (Bevilacqua et al. 1984, 1986; Rossi et al. 1985; Stern et al. 1985). The cytokines IL-1 and TNF have prominent procoagulant effects on the endothelium. Further, tissue factor, the principal procoagulant of human brain,

resides in specific regions of the nervous system and is increased during inflammation. Acute phase reactants, including IL-6, also induce fibrinogen. This confluence of procoagulant effects serves a physiologic function as reduction of blood flow through the inflamed vasculature appears to reduce the cascade which recruits additional cells to the area. Nonetheless, excessive coagulation would perpetuate tissue damage from ischemia. We do not yet know the optimal balance or when therapeutic intervention with anticoagulants is wise.

Vascular tone
Maintenance of vascular tone is, similarly, carefully regulated under normal circumstances. Intrinsic modulation of vascular tone depends, in part, on elaboration of both vasorelactants and vasoconstrictors (Goligorsky et al. 1994). Endothelins, which are powerful vasoconstrictors, and nitric oxide, which is a potent vasodilator, are part of a balanced system that regulates blood flow in the brain and other organs (Anonymous 1986; Hallenbeck and Dutka 1990; Goligorsky et al. 1994; Yamasaki et al. 1995). Endothelins can provoke a long lasting vasoconstriction in cerebral vessels of all sizes, including the microcirculation (Zhang et al. 1993). Under physiologic conditions, endothelins do not penetrate the blood–brain barrier or influence permeability. However, endothelins are produced by brain endothelial cells, smooth muscle cells, astrocytes and neurons. Potential sources within the brain for this vasoconstrictor are numerous. Nitric oxide, a free radical with high lipid solubility and an extremely short half-life of only a few seconds, is also produced widely within the brain by endothelial cells, astrocytes, and neurons. In inflammation, the cytokine- (particularly IL-1) associated release of endothelin induces an overriding vasoconstriction (Luscher et al. 1993). The functional effect, to reduce flow through the injured vessels, is likely an appropriate physiologic response. Persistent endothelin release also stimulates vascular smooth muscle proliferation. Thus, excessive or persistent vasoconstriction may add to ischemic tissue injury (Brenner et al. 1989; Gibbons and Dzau 1994; Hamann and Del Zoppo 1994).

Regulation of vascular inflammation
Vascular inflammation is normally transient. Mechanisms of persistent vasculitis include the obvious

cause, persistence of antigenic stimulation, but also other abnormalities that are less well defined. Better understanding of regulatory mechanisms in vasculitis could provide more effective, less toxic therapies. Current clinical studies of IL-8 and GM-CSF, which induce rapid shedding of L selectin from leukocytes and TGF-β which, in many tissues, downregulates the expression of adhesion molecules, may prove useful.

CLINICAL FEATURES

Idiopathic vasculitides

Polyarteritis nodosa

Polyarteritis nodosa, a systemic necrotizing vasculitis, affects medium-sized muscular arteries throughout the body with the notable exceptions of the spleen and lungs. The varying severity and often restricted disease are notable (Griffith and Vural 1951; Rose and Spencer 1957; Scott et al. 1982; Ronco et al. 1983; Savage et al. 1985; Guillevin et al. 1988; Lightfoot et al. 1990). Males develop the disease somewhat more frequently than females. The age of affected individuals is wide, 14–80 years, with a mean in the range of 40–50 years. Systemic symptoms of fever, malaise, and weight loss often herald the disease. Over half the patients have either arthralgias or a rash. The skin lesions may be either erythematous, purpuric, or vasculitic. Punched-out ulcers around the ankles are well described. Renal involvement is high, occurring in over 70% of patients, although an abnormal urinary sediment is more frequent than uremia. Hypertension develops in at least half the patients. Gastrointestinal changes occur in about 45% of patients and are a prominent cause of morbidity and mortality. Abdominal pain, hemorrhage, pancreatitis, and gut infarction may be prominent.

Histologic characteristics are segmental, transmural vascular inflammation with a mixture of lymphomononuclear cells and variable numbers of neutrophils and eosinophils (Lie 1990). Fibrinoid necrosis is typical but not diagnostic. Infiltration is followed by intimal proliferation and thrombosis. A temporal spectrum of vascular changes can be seen in vessels throughout the body. Strikingly, active necrotizing lesions and proliferative fibrotic healing lesions coexist in close proximity. The lesions have a predilection for vessel bifurcations.

Neurologically, both the central and peripheral nervous system abnormalities occur but the frequency, tempo, and histology vary (Moore and Fauci 1981). Peripheral neuropathies occur in 50–60% of patients and are the presenting manifestation of disease. Dysesthesias or frank pain is mentioned by some authors. Most frequent are the classic mononeuritis multiplex and polyneuropathies, either pattern occurring in over half the patients with neuropathy. The relative frequencies of mononeuropathy multiplex and polyneuropathy vary in different series (Lovshin and Kernohan 1948; Moore and Fauci 1981; Chang et al. 1984; Bouche et al. 1986). An ascending sensorimotor quadriparesis occurs resulting from infarction of watershed blood vessels supplying nerves in the mid-arm and mid-thigh (Sunderland 1945). The clinical presentation is that of an extensive mononeuropathy multiplex. Cutaneous neuropathies, frequently evident upon careful sensory examination, may not be symptomatic. More troublesome diagnostically are the plexopathies and radiculopathies. These are relatively infrequent and usually attributed to vasculitis only in patients with other known manifestations of disease. In sural nerve biopsies of carefully selected patients, histologic evidence of vascular inflammation in the vasonervorum and active axonal degeneration with asymmetrical involvement appears between or within fascicles. The cellular infiltrate consists mainly of macrophages and T-lymphocytes, particularly the CD4+ subset. Infiltrating cells exhibit immunologic activation markers such as IL-2R, transferrin receptor, and MHC class II antigen expression (Said et al. 1988; Cid et al. 1994).

CNS abnormalities develop in 40% of patients. Frequent presentations of CNS disease in PAN include encephalopathy, focal and multifocal lesions of the brain and spinal cord, subarachnoid hemorrhage, seizures, strokes, and cranial neuropathies (Parker and Kernohan 1949; Ford and Siekert 1965; Tervaert and Kallenberg 1993). These usually occur later in the course of disease than the peripheral neuropathies. Hypertension sometimes accompanies or follows the encephalopathy and the additional marginal ischemia may further compromise neurologic function. Visual symptoms are numerous; blurred vision and visual loss may result from inflammation of the choroidal, retinal, or brain parenchymal arteries (Goldsmith 1946; Kinyoun et al. 1987; Akova et al. 1993). Ocular manifestations which occur in 10–

20% of patients with PAN, include scleritis, peripheral ulcerative keratitis, non-granulomatous uveitis, retinal vasculitis, central retinal artery occlusion, and pseudotumor of the orbit (Herson and Sampson 1949). Involvement of the choroidal vessels is more frequent than the retinal vessels. Vasculitis of the optic nerve, chiasm, tract, and occipital cortex is well described. Diplopia may result from inflammation of the arteries supplying cranial nerves III, IV, or VI (Goldstein and Wexler 1937; Kirkali et al. 1991; Topaloglu et al. 1992).

Laboratory evidence of systemic inflammation is reflected non-specifically as anemia, leukocytosis, thrombocytosis, C-reactive proteins and elevated sedimentation rate. Antinuclear antibodies may be present in 20% of patients but are characteristically of low titer and non-specific pattern. Although its role in the pathogenesis of PAN is not clear, about 20–30% of patients have a hepatitis B antigenemia. In England, the incidence of hepatitis infection is lower than 10% for reasons that are not clear. ANCA, particularly to myeloperoxidase, are associated with microscopic polyarteritis possibly aiding diagnosis. Abnormalities in the urine sediment are common, even if the BUN and creatinine are normal. Creatinine clearance is the best measure of renal function and may indicate otherwise unapparent renal disease. Angiography is a useful diagnostic test and often suggests the diagnosis in patients when biopsy has been unrewarding. Notable and prominent radiographic changes are aneurysms particularly in the hepatic and renal vasculature. Segmental narrowings of vessels, variations in caliber, and pruning of the vascular tree all occur.

Although criteria are evolving, the diagnosis of PAN remains largely dependent on the classic methods of angiography and biopsy. Visceral angiography reveals evidence of aneurysms (typically small, multiple and diffuse) in about 65% of patients and is an important diagnostic study. The triad that alerts a physician to a diagnosis of PAN is systemic inflammation, angiographic evidence of enteric vascular diseases and histologic evidence of vasculitis, often in a peripheral nerve. The diagnosis of PAN is, thus, often substantiated by the neurologic disease.

Churg–Strauss angiitis
Churg and Strauss, in 1951, described distinctive features in the autopsy of 13 patients who died after an illness characterized by fever, asthma, eosinophilia, and a systemic illness (Churg and Strauss 1951). Churg–Strauss angiitis appears as a vasculitis of small- to medium-sized vessels with clinically distinctive features. The disease is often heralded by rhinitis and then increasingly severe asthma (Chumbley et al. 1977; Aupy et al. 1983; Masi et al. 1990). This prodrome may precede the development of eosinophilia and systemic vasculitis by 2–20 years. Clinical and hematologic features distinguish it from PAN. Early features may include anemia, weight loss, heart failure, recurrent pneumonia and bloody diarrhea. Pulmonary involvement is typical in CSS and rare in PAN. Similarly, the eosinophilia which is characteristic in CSS is not a feature of PAN (Lanham et al. 1984; Guillevin et al. 1988). Cutaneous manifestations include palpable purpura, erythema, and subcutaneous nodules. Peripheral nerves and kidneys are typically involved in 65 and 60%.

Histologically, medium and small vessels are affected. Debate continues over the necessity for strict histologic criteria (necrotizing vasculitis, tissue infiltration by eosinophils, and extravascular granuloma) to establish a diagnosis. Study of the potential mechanisms for vascular and tissue injury center on the eosinophil (Tai et al. 1984). The two diagnostically essential lesions are angiitis and extravascular necrotizing granulomas usually with eosinophilic infiltrates (Masi et al. 1990). In any single biopsy specimen, however, the changes may appear very similar to PAN.

Neurologic abnormalities are similar to those in PAN, but encephalopathies occurring early in the course of the disease are more frequent, probably reflecting the small size of vessels involved (Lichtig et al. 1989). CNS abnormalities include memory loss, confusion, seizures, subarachnoid hemorrhage, and chorea (Chang et al. 1993; Kok et al. 1993; Sehgal et al. 1995). Visual abnormalities are a prominent part of disease (Weinstein et al. 1983; Acheson et al. 1993). In the absence of histologic evidence of vasculitis in the brain, however, the frequency of cerebrovascular inflammatory disease remains conjectural. Peripheral neuropathies predominate (50–75% of patients) over CNS changes (25%). Peripheral neuropathies classically present as mononeuropathies multiplex but polyneuropathies also occur (O'Donovan et al. 1992; Liou et al. 1994). The histologic features of vasculitis in the peripheral nerve blood vessel may contain the typical features of eosinophils and granulomas but as often appear histologically similar to PAN.

Laboratory features reflect general systemic inflammation. Although the sedimentation rate is elevated, and ANA may be present in low titer no autoantibodies are diagnostic of the disease. ANCA are infrequently present. Thus, the clinical features again provide important information for diagnosis. Characteristically, the triad of asthma, eosinophilia and vasculitis in two extrapulmonary organs defines the disease.

Hypersensitivity vasculitis (HSV), the most frequently encountered of all the vasculitides, is a heterogeneous group of clinical syndromes characterized by inflammation of small vessels, typically venules (Winkelmann and Ditto 1964; Sams et al. 1976; Calabrese and Clough 1982; Mackel and Jordan 1982; Hodge et al. 1987; Zax et al. 1990). The predominant target organ is the skin; this feature unifies these diseases. In many instances, the vessel inflammation can be identified as a response to a precipitating antigen such as a drug, foreign protein, or microbe (Parish and Rhodes 1967; Parish 1971; Mullick et al. 1979; Mackel and Jordan 1982; Marks et al. 1984; Ong et al. 1988). Endogenous antigens, such as tumor antigens or serum proteins, can also serve as the sensitizing antigen.

Clinically, lesions appear as purpura or urticaria. Histologically, the presence of fragmentation and phagocytosis of nuclear debris (leukocytoclasia) is a typical pathologic feature. Damage to the vessel wall appears to result from deposition of immune complexes with activation of the complement cascade. Although the ensuing infiltrate is usually neutrophilic, a distinct lymphocytic subset may exist (Sams et al. 1976; Hodge et al. 1987; Calabrese et al. 1990). Henoch–Schoenlein purpura (HSP), palpable purpura and colicky abdominal in children, often reveals evidence of IgA in the walls of the arterioles and glomeruli (Mills et al. 1990). Isolated cutaneous lesions may occur in cryoglobulinemia, connective tissue diseases, and malignancies. Histologic and clinical features are monophasic. Myalgias and arthralgias occur in half the patients. ANCA autoantibodies are not present and ANAs present only in low titers. One-third of the patients do have elevated IgA on immunoelectropheresis. Involvement of the nervous system varies. In serum sickness encephalopathies,

seizures, and brachial plexopathies occur, although the incidence is unknown (Park and Richardson 1953; Lawley et al. 1984). In other HSV, subarachnoid hemorrhage or seizures are occasionally reported (Lewis and Philpott 1956). In most cases of cutaneous venulitis, the nervous system is not affected.

To the neurologist and internist, the prominent clinical dilemma in the diagnosis and therapy of HSV is whether the vasculitis will remain restricted to the skin or is the presenting manifestation of systemic vasculitis such as PAN, Churg–Strauss, or Wegener's. For this reason, evaluation of patients with HSV includes analysis of renal function, immunoglobulins and autoantibodies which may suggest alternate diseases. In all of these, the main goal is to identify any inciting causes, remove the cause, and observe the patients. Evidence of systemic disease (renal, cardiac, gastrointestinal, neurologic) warrants glucocorticoid or immunosuppressive therapy.

Isolated angiitis of the CNS (IAC), an idiopathic, recurrent vasculitis, is an uncommon disease of the CNS characterized by inflammation of small- and medium-sized vessels (Hughes and Brownell 1966; Kolodny et al. 1968; Jellinger 1977; Cupps et al. 1983; Kristoferitsch et al. 1984; Moore 1989b; Crane et al. 1991). This disease is usually diagnosed by a combination of clinical, angiographic, and histologic features. Rigorous evaluations are necessary to exclude the numerous secondary cause of CNS vascular inflammation and alternate causes of vasculopathy. Historically, the disease was called granulomatous angiitis on the basis of granulomata present on post-mortem examination. Histologic features ante mortem reveal that granulomata are a variable and often absent feature. The current term is isolated angiitis of the CNS (IAC). Some authors use the term 'primary angiitis of the CNS', but this often includes a variety of disorders and histologic confirmation is often lacking. The disease has been recognized with increasing frequency in the past decade, possibly because current treatment methods are so effective.

Symptoms and signs are restricted to the nervous system and typically include headaches, encephalopathies, strokes, cranial neuropathies, and myelopathies. Headache is prominent in at least half the patients. The clinical situation that should suggest the diagnosis are new onset of headaches and encephalopathy, particularly in association with multifocal signs. Notably, symptoms and laboratory evidence of systemic inflammation are absent.

Neurodiagnostic studies, including computed tomographic (CT) scan and magnetic resonance imaging (MRI), are often non-specifically abnormal (Miller et al. 1987). Cerebrospinal fluid (CSF) analysis is abnormal in only half the patients and even then the abnormalities may be a mild pleocytosis or protein elevation. Angiography is the most sensitive diagnostic study, although an occasional patient with only small-vessel disease may have a normal angiogram. Of greater concern to accurate diagnosis is the fact that the angiographic features are not specific for vasculitis; similar abnormalities may occur in non-inflammatory vasculopathies as well as vasculitis secondary to infections, drugs, and neoplasia. Angiography often shows single or multiple areas of beading along the course of a vessel, abrupt vessel terminations, hazy vessel margins, and neovascularization (Stein et al. 1987; Hellman et al. 1992; Alhalabi and Moore 1994).

The pathogenesis of the CNS vascular inflammation is not known. A cell-mediated process appears most likely. Classically, appropriate clinical features, absence of evidence of systemic inflammation, and angiographic and histologic data are important. In addition, because an occasional self-limited vasculopathy occurs, evidence of recurrent or persistent disease should be confirmed.

Isolated PNS vasculitis

Vasculitis restricted to the peripheral nervous system (PNS) is reported and in some series comprises one-third of all patients with a vasculitic neuropathy (Kissel et al. 1985, 1989; Harati and Niakan 1986; Hawke et al. 1991; Engelhardt et al. 1993). A major difficulty in establishing this as a distinct clinical diagnosis is that vasculitis of the peripheral nerve is often the presenting feature of systemic vasculitis. Since the distribution and histology of isolated vasculitis of the PNS is identical to that of PAN, it is difficult to determine whether patients, if untreated, would go on to develop systemic disease. Until more information on etiology or pathogenic mechanisms in the PNS vasculature are available, careful and repeated studies of these patients including creatinine clearances and fluorescein angiography are prudent.

Other vasculitides

There are several reports and series of patients with neurologic abnormalities not readily classifiable in the disease groups described above. A small-vessel vasculitis occurs and is restricted to the CNS, skin, and muscle (Miller et al. 1984). Retinocochlear encephalopathy, another disorder of small vessels, is more likely a microvasculopathy than a vasculitis (Bogousslavsky et al. 1989). The syndrome consists of a subacute encephalopathy (often with early psychiatric features), sensorineural hearing loss, and retinal arteriolar occlusions. This hearing loss in association with vascular disease must be distinguished from Cogan's syndrome (non-syphilitic interstitial keratitis and vestibuloauditory symptoms). Some patients with Cogan's syndrome have a vasculitic component that is predominantly an aortitis (Cheson et al. 1976; Haynes et al. 1980).

Secondary vasculitides

Vasculitis of the nervous systems secondary to a known cause or underlying process is both frequent and clinically important. Patients with CNS vasculitis from secondary causes far exceed patients with a primary, idiopathic vasculitis. A high index of suspicion enables a clinician to promptly institute therapy for a vasculitis that may be secondary to infection, neoplasia, or a toxin.

Infections

The range of infections that induce vascular inflammation is broad including bacteria, fungi, viruses and protozoa. Mechanisms of immune complex-mediated inflammation range from toxic disruption of endothelium, infection of the endothelial cells, and immune complex-mediated inflammation. The prolific inflammation induced by bacteria includes prominent accumulation of cells in the vessel wall with associated thrombosis and hemorrhage (Dodge and Swartz 1965; Lyons and Leeds 1967; Igarashi et al. 1984). This accounts for the frequent strokes that are part of acute bacterial meningitis and are responsible for much of the neurologic sequelae. Other infectious agents frequently causing a vasculitis but more difficult to detect are fungi. Aspergillosis, cryptococcus, *Coccidioides immitis*, *Histoplasma capsulatum* and mucormycoses, in particular, all infiltrate cerebral vessels (Martin et al. 1954; Schigenaga et al. 1975;

Kobayashi et al. 1977; Koeppen et al. 1981; Tija et al. 1985; Walsh et al. 1985; De la Torre and Gorraez 1989; Wheat et al. 1990; Williams et al. 1992). Neuro-borelliosis also results in vasculitis (Meurers et al. 1990; Miklossy et al. 1990). The clinical features range from subtle changes in cognition to fatal hemorrhagic infarctions. Other common infectious agents such as tuberculosis (Lehrer 1966; Teoh et al. 1989) and syphillis (Rabinov 1968) can cause a vasculitis. Strokes associated with viral infection demonstrate more pleomorphic features. Histologically, inflammation and necrosis of the cerebral blood vessel wall certainly do occur with viruses such as herpes simplex, herpes zoster, cytomegalovirus, toxoplasmosis, human immunodeficiency virus and varicella (Walker et al. 1973; Hirose and Hamashima 1978; Linnemann and Alvira 1980; Hilt et al. 1983; Powers 1986; Huang and Chou 1988). However, viruses such as herpes zoster may also cause vaso-occlusive disease without inflammatory changes.

Toxins

Another conspicuous cause of vasculitis is toxins (Citron et al. 1970; Rumbaugh et al. 1971; Reichlin 1982; Krendel et al. 1990; Williams et al. 1992). The cutaneous disorder, HSV, is associated with a wide variety of inciting agents but CNS complications are few. CNS vasculitis is well reported after a variety of illicit drugs, notably those with a prominent sympathomimetic effect, such as amphetamines. However, it remains difficult to determine the precise causative agent in a package which contains numerous additive and diluents. Cocaine and crack cocaine do cause stroke and have occasionally been associated with a vasculitis.

Neoplasia

The association of vasculitis of the nervous system and neoplasia is intriguing but most notable clinically for the importance of distinguishing between a vascular occlusion associated with encasement of the vessel by tumor and inflammation of a vessel remote from the neoplasm (Rubenstein 1966; Petito et al. 1978; Johnson et al. 1979; Greer et al. 1988; Roux et al. 1995). Hodgkin's disease is associated with a vasculitis which resolves with the treatment of the underlying disease. Other lymphomas, including lymphomatoid granulomatosis and angioendothelosis, are vasocentric and may present with CNS or PNS ab-

normalities illustrating the importance of histology in accurate diagnosis and therapy.

Connective tissue diseases

The connective tissue diseases are systemic inflammatory diseases in which a component of disease is often a vasculitis. However, with the possible exception of Sjögren's disease, vasculitis of the CNS rarely occurs (Skowronski and Gatter 1974; Ramos and Mandybur 1975; Estey et al. 1979; Watson 1979). In SLE, neurologic abnormalities are frequent but histologic evidence of vasculitis is rare. The pathogeneses of the neurologic abnormalities are undefined but multiple contributions from autoantibodies reactive with neuronal tissue, ischemia from coagulopathies, and behavioral changes from activation of the HPA axis all appear likely.

In the PNS, lupus vasculitis accounts for only 1% of patients presenting to a neurologist with a vasculitic neuropathy. Interestingly, among these patients, there is a higher percentage of sensory neuropathies than sensorimotor polyneuropathies. A vasculitic neuropathy may occur in association with rheumatoid arthritis, although a compressive neuropathy is more frequently encountered (Pallis and Scott 1965; Scott et al. 1981; Peyronnard et al. 1982; Koo et al. 1984). Similarly in the non-specific systemic inflammatory disease such as associated with cryoglobulins, a mononeuropathy multiplex, polyneuropathy or autonomic neuropathy may occur associated with vascular inflammation.

In Sjögren's disease, both CNS and PNS abnormalities occur, although the histology is not convincingly vasculitis. In one series of nerve biopsies, however, 8 out of 11 patients had findings consistent or highly suggestive of vasculitis; other patients had a perivascular inflammatory response. An alternative, distinctive neuropathy in Sjögren's is not vasculitis but a dorsal root ganglionitis. These patients present with a sensory neuropathy and ataxia usually associated with autonomic insufficiency (Malinow et al. 1986).

DIAGNOSTIC CONSIDERATIONS AND NEURORADIO-GRAPHIC STUDIES

Clinical neurologic abnormalities merit neurodiagnostic studies, but these are of variable utility in detecting underlying vascular inflammation. CSF analysis is neither sensitive nor specific for vasculitis but may identify underlying causes of vascular inflammation such as infections or tumors. Neither qualitative nor quantitative abnormalities of CSF synthesis are demonstrably present in patients with systemic or isolated CNS angiitis (Hirohata et al. 1993). Electroencephalography (EEG) is also neither specific nor sensitive for vasculitis but occasionally provides a supportive clue in the small-vessel vasculitides affecting the brain. Abnormal EEGs in patients with mild encephalopathies may also provide a monitor for therapy.

MRI often detects vascular disease but the signals result from fluid changes in ischemic tissue (Sole-Llenas and Pons-Tortella 1978; Miller et al. 1987; Harris et al. 1994). Thus, the result of vascular inflammation rather than the inflammation itself is identified. Nonetheless, the presence of multifocal vascular disease often suggests further studies. Magnetic resonance angiography (MRA) holds the promise of providing more information about vascular abnormalities prior to infarction but current studies do not have sufficient resolution to exclude many cases of CNS vasculitis.

Cerebral angiography remains the gold standard for detecting intracranial vasculopathies. There are, however, limitations (Ferris and Levine 1973; Travers et al. 1979; Garner et al. 1990; Hellman et al. 1992; Alhalabi and Moore 1994). It is an invasive, expensive procedure. Its utility also varies with the underlying vasculitis. Although angiography is abnormal in 75–90% of patients with primary CNS vasculitis, vascular inflammation may occur in vessels beyond the resolution of angiography as demonstrated in several biopsy proven cases of IAC. Its utility in systemic vasculitis remains undetermined, because it is seldom abnormal in polyarteritis nodosa or Wegener's, although it has been useful in lymphomatoid granulomatosis. Angiography of other regions of the body is often useful and is occasionally the critical diagnostic study in systemic vasculitis. Aortic arch arteriography is central to a diagnosis of Takayasu's arteritis. Polyarteritis nodosa may be revealed by renal or mesenteric angiography performed for the evaluation of persistent visceral pain or hemorrhage.

Cortical/leptomeningeal biopsy is an important part of diagnosis in the intracranial vasculitides because no other study distinguishes between inflammatory and non-inflammatory causes of angiographically demonstrable vascular irregularities. Further, in those cases with vascular inflammation, biopsy may provide the only information on an underlying cause requiring alternate therapy such as infection. A major limitation of biopsy remains the false negative rate which remains at 20–40%. Because the vessel samples by biopsy are of different size than those recognized by angiography or MRI, these latter studies do not yet provide a useful guide for biopsy site.

Diagnosis of vasculitis in individual patients often evolves over time as the pattern and extent of clinical facets appear. For this reason, a clinician must periodically reevaluate the diagnosis. A repeated, careful history with direct questions about renal, gastrointestinal, cardiac, pulmonary, and CNS functions is essential. Further studies, including urinalysis, creatinine clearance and BUN, are useful.

THERAPY AND PROGNOSIS

In current nosology, the vasculitides are divided into groups on the basis of clinical and pathologic features. Perhaps the single most important division to the clinician is that of primary (or idiopathic) and secondary vasculitis. Prompt identification of the secondary vasculitides results in better therapy. Clinical and pathologic features distinguish among the primary vasculitides. Early in any of the diseases the diagnosis may be enigmatic. As the diseases progress, the diagnosis is clearer. Several reviews on therapy in vasculitis are available (Kaeser 1984; Clements and Davis 1986; Carcassi 1989; Conn 1989; De Jesus and Talal 1990; DeVita et al. 1991; Kissel and Rammohan 1991; Omdal et al. 1993).

Treatment of vasculitis ranges from the simple measure of removing the cause of the chronic inflammation, to the corticosteroid/cyclophosphamide immunosuppressive mainstays of some diseases, to experimental measures in refractory vasculitis. As information on the mechanisms of vascular inflammation increases, more specific, less toxic therapies can be developed.

Removal of the underlying cause of vasculitis is practical and effective in many instances. Specifically, vasculitis secondary to infections responds to the appropriate antimicrobial agent; vasculitis secondary to drugs/medications usually responds to removal of the inciting agent; and vasculitis secondary to neoplasia often responds to treatment of the underlying tumor. For these reasons, a careful search for a cause of the chronic infection is the first step in evaluation of vasculitis.

Many of the vasculitides are identifiable by clinical and histologic features, but a treatable cause cannot be identified. Classic therapy involves a shotgun approach to the immune system decreasing large populations of immune effector cells. The advantage is efficacy; a major disadvantage is reduction in immune cells necessary for fighting infection and, in some cases, for tumor surveillance.

Corticosteroids remain a mainstay in the therapy of vasculitis and, in some types of vasculitis, corticosteroids alone are effective. Corticosteroids affect the immune system at numerous sites (Morand and Goulding 1993). Intracellularly, steroids bind to a cytoplasmic receptor protein and enter the cell where they alter the rate of ribosomal and mRNA synthesis. Among the effects are inhibition of synthesis of IL-1, IL-2, IL-6 and GM-CSF, decreased production of proinflammatory mediators such as eicosamide, and diminished lipocortin action (Goulding and Guyre 1992). The diminished secretion of IL-1 results in the reduced expression of adhesion molecules such as ICAM-1 on the endothelial cell surface. Inhibition of IL-2 diminishes the expansion of effector T-cell clones. Intercellularly, corticosteroids effect inhibition of recruitment and migration of neutrophils, lymphocytes and macrophages to the sites of inflammation and suppression of macrophage chemotaxis, cytotoxicity, and soluble mediator function. Thus, in vivo, corticosteroids affect both primary and secondary humoral antibody formation and cell-mediated immunity. The specific effect whereby they ameliorate vasculitic lesions, if indeed there is only one, is uncertain.

The efficacy of glucocorticoids, usually prednisone or methylprednisolone, varies among the vasculitic syndromes. Undoubtedly, it is the single most effective treatment in temporal arteritis. In some hypersensitivity vasculitides and in microangiopathic renal vasculitis, prednisone therapy alone may suffice. In other disorders, either the high dosage is unsustainable because of the considerable side effects or it is effective only as adjunct therapy. It is ineffective in well established Wegener's granulomatosis. Patients with polyarteritis nodosa may develop new neurologic side effects, while their disease is apparently quiescent on prednisone. In these disorders, experience with the underlying disease will enable the physician to determine when to use another agent such as cyclophosphamide.

Side effects of corticosteroids are numerous and frequent. In addition to the well described side effects of acute and chronic corticosteroid therapy, there are specific effects which impact on vascular diseases, particularly a potential to augment vasoconstriction and platelet aggregation. This may complicate treatment of vasculitic syndromes (Conn et al. 1988; Conn 1989). More recent studies of the effects of chronically elevated corticosteroids on glucocorticoid receptors in the hippocampus and associated changes in cognition underlie the need for careful use of these medications (Newcomer et al. 1994). There is currently increased use of immunosuppressive agents for their steroid sparing effects.

Cyclophosphamide, an alkylating nitrogen mustard, crosslinks DNA, thus interfering with cell division and diminishing clonal expansion of B- and T-lymphocytes. Immunologic effects include suppression of immunoglobulin production, diminished antigen-induced proliferation of helper and effector T-cells, and reduced cytotoxicity of macrophages. Early evaluations of cyclophosphamide suggested that acute high doses principally suppress humoral immunity while chronic doses affect cell-mediated immune mechanisms. More recently, influences of cyclophosphamide on endothelial cells are also being explored.

Cyclophosphamide was effectively added to therapeutic immunosuppressive regimens of systemic necrotizing vasculitis in 1979, dramatically reducing the mortality of Wegener's granulomatosis. Its efficacy in treatment of IAC and polyarteritis nodosa is well established. It is also a useful adjunct either to reduce the dosage and side effects of prednisone or for corticosteroid failures in other diseases. It is most often prescribed either in a daily oral dose or as an intermittent intravenous bolus. Oral cyclophosphamide is usually prescribed at a dose of 1–2 mg/kg/day. Hydration is important to prevent hemorrhagic cystitis and potentially reduce incidence of bladder malignancies. Monitoring of white blood cell count for a limiting factor of neutropenia >1500 minimizes the likelihood of infection.

More recently, bolus intravenous cyclophosphamide has been tried with the aim of reducing total dosage and side effects. Different protocols of pulse cyclophosphamide administration have been employed. Pulse cyclophosphamide has been effective in lupus nephritis (McCune et al. 1988) and systemic vasculitis (Scott and Bacon 1984) but, despite initial improvement, there may be a higher relapse rate in Wegener's than with standard cyclophosphamide therapy (Hoffman et al. 1990).

Cyclosporine is a cyclic endecapeptide extracted from the fungus *Tolypocladium inflatum*. Cyclosporine's immune effects are selective; it inhibits production of IL-2 by T-lymphocytes (Kahan 1989). It has dramatically impacted on the treatment of organ transplantation; the efficacy of cyclosporin in various autoimmune diseases is under investigation. The side effects of nephrotoxicity and hypertension limit its widespread use in human vasculitis, although further studies are needed to investigate a possible therapeutic benefit.

Azathioprine, a purine analog, inhibits protein and antibody synthesis in vitro. Both immunoglobulin production and cell-mediated effector function are diminished by azathioprine. Its role in the therapy of systemic vasculitides is restricted to those patients who cannot tolerate cyclophosphamide. It appears to be less effective than cyclophosphamide in inducing remission, but may be used later in the course in place of cyclophosphamide to sustain a remission.

Methotrexate is a folic acid analog that acts by inhibiting dihydrofolate reductase. Cells are depleted of folate with the resulting inhibition of the purine synthesis and blockage of DNA formation. In animals and in vivo human studies, methotrexate suppresses both cell-mediated and humoral immunity. Despite demonstrated efficacy in the therapy of rheumatoid arthritis, methotrexate is not established in the treatment of vasculitis. In several patients with IAC, methotrexate did not control a relapse. Recent studies in several centers suggest that methotrexate in higher dosages than previously used may be effective therapy in systemic vasculitis.

Pheresis (plasma exchange and leukapheresis), the selective removal of plasma or cellular components from the blood, has a simple rationale. Physically removing from the blood substances that cause a disease should improve a patient's condition.

Pheresis does remove immune complexes and facilitates the body's ability to further remove complexes. Pheresis must, however, be combined with an immunosuppressive regimen to prevent a rebound occurrence of immune complexes. Although it is demonstrably effective in controlled studies of Goodpasture's syndrome, certain hyperviscosity states, thrombotic thrombocytopenic purpura, Guillain–Barré, and myasthenia gravis, its efficacy in connective tissue and vasculitic diseases is unsubstantiated (Campion 1992). Pheresis has been used in polyarteritis nodosa, particularly that associated with hepatitis B viremia (Guillevin et al. 1991) and, in a series of patients with rapidly progressive renal disease, pheresis with immunosuppression appeared to be more effective than immunosuppression alone (Lockwood et al. 1979; Sessa et al. 1993). In the absence of controlled studies, it is difficult to determine if pheresis improves the morbidity and mortality over the standard treatment of prednisone and cyclophosphamide.

Intravenous immunoglobulin (i.v. Ig), initially used about 12 years ago for the treatment of primary immunodeficiency disease, effectively supplies antiinfectious agent antibodies by passive immunization. It was incidentally noted at the time to improve thrombocytopenia. Although the exact mechanism of this effect remains unknown, several possible explanations include: (1) an effect on the Fc receptor of phagocytic cells and B-lymphocytes reducing their effector functions; (2) production of anti-idiotypic antibodies; or (3) prevention of activated complement components from reaching their targets (Schwartz 1990). I.v. Ig is a demonstrably effective therapy for Kawasaki's syndrome (a childhood vasculitis associated with high cytokine levels and coronary arteritis). Some recent studies show that i.v. Ig may be effective in some patients with systemic vasculitis, particularly those with refractory vasculitis (Tuso et al. 1992; Jayne and Lockwood 1993). The occasional reversible impairment of renal function requires further study (Jayne and Lockwood 1993).

Non-steroidal anti-inflammatory medications (NSAIDs) have anti-inflammatory, analgesic, antipyretic, and platelet inhibitory actions. Those currently available come from a variety of chemical classes which affect their distribution in the body and, to some extent, their therapeutic performance. The similarities among them are many and, as a group, they are potentially useful adjuncts in autoimmune disease. A major mechanism of action of NSAIDs is the inhibition of cyclooxygenase activity and therefore the synthesis of prostaglandins (Brooks and Day 1991). Although not yet studied systematically, this effect may reduce some of the chronic changes in vasculitis.

Anti-prostaglandin. In the diseased blood vessel, endothelium-dependent smooth muscle relaxation is impaired. Both vasoconstriction and secondary platelet aggregation accelerate. An endothelial peptide, endothelin, is a smooth muscle contractor; increased levels of endothelin are present in some diseases. Anti-inflammatory and anti-vasospastic effects of PGE1 infusion have been reported in several cases (Hauptman et al. 1991). The size of the vessel involved appears to be a restricting factor in the efficacy of anti-PGE1 therapy. There would be limited effects in peripheral neuropathy due to vasonervorum infarction and other small vessels too small to contain muscle cells. Side effects are unusual but noteworthy. The vasodilatory effect could result in hypotension (potentially serious in patients with ischemic bowel) and steal syndromes.

Anti-platelet agents and anticoagulation. In certain of the vasculitides, there is evidence of a coagulopathy. Active Takayasu's disease is associated with hyperfibrinogenemia and hypofibrinolytic activity. Studies in patients with Kawasaki's disease reveal thrombocytosis, diminished fibrinolytic activity, and increased platelet derived β-thromboglobulin. Less data exist for other vasculitides, but thrombosis is a common clinical and histologic feature. Practically, glucocorticoid therapy inhibits endothelial production of prostacyclin, but has little effect on platelet thromboxane. Thus, although corticosteroids reduce the inflammation in the infiltrate in the vessel wall, platelet activation, thrombus deposition, and vasoconstriction continue. A potentially simple solution with minimal side effects would be to continue low-dose aspirin during immunosuppressive therapy. This, however, has not been rigorously studied. Anticoagulation with heparin or coumadin therapy has both theoretical and practical limitations. First, parts of the coagulation in the vessels may be a physiologic effect

to protect the vessel and tissue. In addition, many of the vasculitides demonstrate histologic evidence of perivascular hemorrhage. The risk of intracranial hemorrhage with these medications is potentially high.

Dapsone appears to suppress myeloperoxidase H_2O_2 halide-mediated cytotoxicity in polymorphonuclear leukocytes. Thus, dapsone would be and is most effective in leukocytoclastic vasculitis which contains a prominent polymorphonuclear cell infiltrate (Holtman et al. 1990). Its efficacy in dermatitis herpetiformis and erythema elevatum diutinum was established in the 1950s. It has some use in vasculitis complicating rheumatoid arthritis, probably because of the role of polymorphonuclear cells in immune complex-mediated vasculitis, but its variable efficacy in the visceral manifestations relegates this therapy to situations where other therapy fails. There is no anticipated benefit in the cell-mediated vasculitides. Small studies did not show a benefit of dapsone therapy in temporal arteritis.

Immunotherapy. Future strategies are directed at interrupting the inflammatory or proliferative cascade at key points. The goal is to down-regulate the process to the point of disease remission. There is clear scientific rationale but prominent clinical limitations in the application of these focused therapies.

Monoclonal antibody therapy. Monoclonal antibodies directed against surface markers on lymphocytes are potentially useful as selective therapeutic agents in a variety of immunologically mediated diseases (Isaacs et al. 1992). The most widely used monoclonal antibody is OKT3, which is directed at the CD3 antigen of the T-cell receptor complex on virtually all T-cells. Although an effective therapy in some malignancies, the acute toxicity and the long-term broad immunosuppression it induces markedly limit its utility. Antibodies directed at crucial components of the early phase of the immune response, including anti-class II MHC antibodies, anti-interleukin receptor antibodies, and anti-CD4 antibodies, have a more restricted effect, and therefore fewer side effects and are potentially useful. Such monoclonal antibodies have successfully induced a short-term remission in one patient with systemic vasculitis refractory to conventional immunosuppression (Mathieson et al. 1990). Monoclonal antibodies to the

IL-2 receptor might also modify recurrent inflammation. Generation of γ-interferon and NK cell function are all highly dependent on IL-2 synthesis and IL-2 receptor-driven endocytosis. Early studies of anti-IL-2Ra (expressed only on the surface of activated T-cells) has shown a benefit in therapy of human T-cell leukemia with minimal immunosuppressive side effects. Studies in inflammatory diseases have not yet been reported. Antibodies to cell adhesion molecules such as ICAM reduce histologic evidence of inflammation in experimental studies; results in human studies are pending.

Cytokines, or their inhibitors, in pharmacologic dosages, could downregulate inflammatory and proliferative pathways. Thus far, inhibitors of IL-1 and interferon-γ have had minimal success as therapeutic agents. Interferon-β has limited immunosuppressive effects and is used as a therapy in multiple sclerosis. Studies in systemic inflammatory diseases are pending. Another cytokine, interferon-α, is therapeutically effective in hemangiomatosis and, at least anecdotally, in Behçet's disease, refractory to conventional immunosuppression (Durand et al. 1993). The effects of interferon-α include inhibition of locomotion of capillary endothelium in vitro and angiogenesis in mice and may prove useful in minimizing chronic vessel changes in patients with vasculitis. Clinical effects of TGF-β are untested, but a prominent and severe immunosuppression is anticipated.

Treatment of patients with vasculitis requires expertise with a spectrum of vasculitides and autoimmune diseases, so that there will be an appreciation for the potential of overlap syndromes or evolution of a specific diagnosis over time. Experience in the use of immunosuppressant medications is also important because of the wide range of potential and actual side effects. Side effects occur with any of these agents, and the physician should be thoroughly acquainted with all potential side effects before initiating treatment. Most patients do well on the standard regimens. More difficult are those patients who only partially respond to treatment. The physician must determine whether the clinical effects are from persistent inflammation or other causes. Excluding infection should remain a high priority. Ischemia may result not only from the inflammation, but also from chronic changes in the vessel wall accompanied by thrombosis and/or hemorrhage.

PROGNOSIS

The prognosis of vasculitis depends on the underlying disease. Some disorders such as HSV retain a favorable prognosis with a mortality rate of less than 1%. The prognoses in other disorders such as Wegener's granulomatosis and IAC have improved with combination cyclophosphamide/corticosteroid therapy from almost uniformly fatal to 5-year survivals of 70–90%. In the systemic necrotizing vasculitides, polyarteritis nodosa and Churg–Strauss angiitis, the 5-year survival improved from 13% untreated to 48% with corticosteroid therapy to 80% with cyclophosphamide. Nonetheless, recent studies reveal that the 5-year survival rate in large clinics hovers around 58% (Abu-Shakra et al. 1994). The relatively poor prognosis stems from several features: accurate diagnosis and aggressive therapy may be initiated too late, delayed occlusive (and untreatable) vascular disease may supplant the initial inflammatory component, and death may occur from other disorders such as infections. A recent study evaluating the prognostic factors associated with morbidity revealed that cardiac and renal involvement were associated with an increased relative risk of dying from the disease within 5 years, while cutaneous lesions and peripheral neuropathies had a better outcome (Fortin et al. 1995).

In summary, study of the vasculitides illustrates recent advances and convergence of information in the biology of inflammation, clinical diagnostic studies, and newer therapies. Over the next decade, we anticipate that more focused, less toxic treatments will result from our studies of immunopathogenesis today.

REFERENCES

ABU-SHAKRA, M., H. SMYTHE, J. LEWTAS, E. BADLEY, D. WEBER and E. KEYSTONE: Outcome of polyarteritis nodosa and Churg–Strauss syndrome – an analysis of twenty-five patients. Arthr. Rheum. 37 (1994) 1798–1803.

ACHESON, J.F., O.C. COCKERELL, C.R. BENTLEY and M.D. SANDERS: Churg–Strauss vasculitis presenting with severe visual loss due to bilateral sequential optic neuropathy. Br. J. Ophthalmol. 77 (1993) 118–119.

AKOVA, Y.A., N.S. JABBUR and C.S. FOSTER: Ocular presentation of polyarteritis nodosa. Ophthalmology 100 (1993) 1775–1781.

ALHALABI, M. and P.M. MOORE: Serial angiography in isolated angiitis of the central nervous system. Neurology 44 (1994) 1221–1226.

ARGENBRIGHT, L.W. and R.W. BARTON: Interactions of leukocyte integrins with intercellular adhesion molecule 1 in the production of inflammatory vascular injury in vivo. J. Clin. Invest. 89 (1992) 259–272.

AUPY, M., C. VITAL, C. DEMINIERE and P. HENRY: Angéite granulomateuse allergique (syndrome de Churg et Strauss) revelée par une multinevrite. Rev. Neurol. 139 (1983) 651–656.

BAGGENSTOSS, A.H., R.M. SHICK and H.F. POLLEY: The effect of cortisone on the lesions of periarteritis nodosa. Am. J. Pathol. 27 (1951) 537–551.

BEVILACQUA, M.P., J.S. POBER, G.R. MAJEAU, R.S. COTRAN and M.A. GIMBRONE, JR.: Interleukin 1 (IL-1) induces biosynthesis and cell surface expression of procoagulant activity in human vascular endothelial cells. J. Exp. Med. 160 (1984) 618–623.

BEVILACQUA, M.P., J.S. POBER, G.R. MAJEAU, W. FIERS, R.S. COTRAN and M.A. GIMBRONE, JR.: Recombinant tumor necrosis factor induces procoagulant activity in cultured human vascular endothelium: characterization and comparison with the actions of interleukin 1. Proc. Natl. Acad. Sci. U.S.A. 83 (1986) 4533–4537.

BLANN, A.D. and D.G.I. SCOTT: Activated, cytotoxic lymphocytes in systemic vasculitis. Rheumatol. Int. 11 (1991) 69–72.

BLOCH, D.A., B.A. MICHEL, G.G. HUNDER ET AL.: The American College of Rheumatology 1990 criteria for the classification of vasculitis. Patients and methods. Arthr. Rheum. 33 (1990) 1068–1073.

BOGOUSSLAVSKY, J., J.M. GAIO, L.R. CAPLAN, F. REGLI, M. HOMMEL, T.R. HEDGES, M. FERRAZZINI and P. POLLACK: Encephalopathy, deafness and blindness in young women: a distinct retinocochleocerebral arteriolopathy? J. Neurol. Neurosurg. Psychiatry 52 (1989) 43–46.

BOUCHE, P., J.M. LEGER, M.A. TRAVERS, H.P. CATHALA and P. CASTAIGNE: Peripheral neuropathy in systemic vasculitis: clinical and electrophysiologic study of 22 patients. Neurology 36 (1986) 1598–1602.

BRAQUET, P., D. HOSFORD, M. BRAQUET, R. BOURGAIN and F. BUSSOLINO: Role of cytokines and platelet-activating factor in microvascular immune injury. Int. Arch. Allergy Appl. Immunol. 88 (1989) 88–100.

BRENNER, B.M., J.L. TROY and B.J. BALLERMANN: Endothelium-dependent vascular responses. J. Clin. Invest. 84 (1989) 1373–1378.

BROOKS, P.M. and R.O. DAY: Nonsteroidal antiinflammatory drugs – differences and similarities. N. Engl. J. Med. 324 (1991) 1716–1725.

CALABRESE, L.H. and J.D. CLOUGH: Hypersensitivity vasculitis group (HVG): a case oriented review of a continuing clinical spectrum. Cleve. Clin. Quart. 49 (1982) 17–42.

CALABRESE, L.H., B.A. MICHEL, D.A. BLOCH, W.P. AREND, S.M. EDWORTHY, A.S. FAUCI, J.F. FRIES, G.G. HUNDER, R.Y. LEAVITT and J.T. LIE: The American College of Rheumatology 1990 criteria for the classification of hypersensitivity vasculitis. Arthr. Rheum. 33 (1990) 1108–1113.

CAMPION, E.W.: Desperate diseases and plasmapheresis. N. Engl. J. Med. 326 (1992) 1425–1427.

CARCASSI, M.U.: Cytotoxic drugs in systemic autoimmune diseases. Clin. Exp. Rheumatol. 7 (1989) 181–186.

CHANG, R.W., C.L. BELL and M. HALLETT: Clinical characteristics and prognosis of vasculitic mononeuropathy multiplex. Arch. Neurol. 41 (1984) 618–621.

CHANG, Y., S. KARGA, J. GOATES and D. HOROUPIAN: Intraventricular and subarachnoid hemorrhage resulting from necrotizing vasculitis of the choroid plexus in a patient with Churg–Strauss syndrome. Clin. Neuropathol. 12 (1993) 84–87.

CHESON, B.D., A.Z. BLUMING and J. ALROY: Cogan's syndrome: a systemic vasculitis. Am. J. Pathol. 60 (1976) 549–555.

CHUMBLEY, L.C., E.G. HARRIS and R.A. DEREMEE: Allergic granulomatosis and angiitis (Churg–Strauss syndrome). Report and analysis of 30 cases. Mayo Clin. Proc. 52 (1977) 477–484.

CHURG, J. and L. STRAUSS: Allergic granulomatosis, allergic angiitis, and periarteritis nodosa. Am. J. Pathol. 27 (1951) 277–301.

CID, M., J.M. GRAU, J. CASADEMONT, E. CAMPO COLL-VINENT, A. LOPEZ-SOTO, M. INGELMO and A. URBANO-MARGUEZ: Immunochemical characterization of inflammatory cells and immunologic markers in muscle and nerve biopsy specimens from patients with systemic polyarteritis nodosa. Arthr. Rheum. 37 (1994) 1055–1061.

CITRON, B.P., M. HALPERN, M. MCCARRON, G D. LUNDBERG, R. MCCORMICK, I.J. PINCUS, D. TATTER and B.J. HAVERBACK: Necrotizing angiitis with drug abuse. N. Engl. J. Med. 283 (1970) 1003–1011.

CLEMENTS, P.J. and J. DAVIS: Cytotoxic drugs: their clinical applications to the rheumatic diseases. Semin. Arthr. Rheum. 15 (1986) 231–254.

COCHRANE, C.G.: Studies on the localization of circulating antigen-antibody complexes and other macromolecules in vessels. I. Structural studies. J. Exp. Med. 118 (1963a) 489–502.

COCHRANE, C.G.: Studies on the localization of circulating antigen-antibody complexes and other macromolecules in vessels. II. Pathogenic and pharmacodynamic studies. J. Exp. Med. 118 (1963b) 503–513.

COCHRANE, C.G.: Mechanisms involved in the deposition of immune complexes in tissues. J. Exp. Med. 134 (1971) 75s–89s.

COCHRANE, C G. and W.O. WEIGLE: The cutaneous reaction to soluble antigen-antibody complexes. A comparison with the Arthus phenomenon. J. Exp. Med. 108 (1958) 591–604.

CONN, D.L.: Update on systemic necrotizing vasculitis. Mayo Clin. Proc. 64 (1989) 535–543.

CONN, D.L., R.B. TOMPKINS and W.L. NICHOLS: Glucocorticoids in the management of vasculitis – a double edge sword? J. Rheumatol. 15 (1988) 1181–1183.

COX, J.H. and W.L. FORD: The migration of lymphocytes across specialized vascular endothelium. IV. Prednisolone acts at several points on the recirculation pathways of lymphocytes. Cell. Immunol. 66 (1982) 407–422.

CRANE, R., L.D. KERR and H. SPIERE: Clinical analysis of isolated angiitis of the central nervous system. A report of 11 cases. Arch. Intern. Med. 151 (1991) 2290–2294.

CUPPS, T.R., P.M. MOORE and A.S. FAUCI: Isolated angiitis of the central nervous system. Prospective diagnostic and therapeutic experience. Am. J. Med. 74 (1983) 97–105.

DAS NEVES, F.C., J. SUASSUNA and M. LEONELLI: Cell activation and the role of cell-mediated immunity in vasculitis. Contrib. Nephrol. 94 (1991) 13–21.

DE JESUS and N. TALAL: Practical use of immunosuppressive drugs in autoimmune rheumatic diseases. Crit. Care Med. 18 (1990) 132–137.

DE LA TORRE, F.E. and M.T. GORRAEZ: Toxoplasma-induced occlusive hypertrophic arteritis as the cause of discrete coagulative necrosis in the CNS. Hum. Pathol. 20 (1989) 604.

DEVITA, S., R. NER and S. BOMBARDIERI: Cyclophosphamide pulses in the treatment of rheumatic diseases: an update. Clin. Exp. Rheumatol. 9 (1991) 179–193.

DODGE, P.R. and M.N. SWARTZ: Bacterial meningitis – a review of selected aspects. II. Special neurologic problems, postmeningitic complications and clinicopathological correlations (concluded). N. Engl. J. Med. 272 (1965) 1003–1010.

DURAND, J.M., G. KAPLANSKI, H. TELLE, J. SOUBEYRAND and F. PAULO: Beneficial effects of interferon-α2b in Behçet's disease. Arthr. Rheum. 36 (1993) 1025.

DUSTIN, M.L. and T.A. SPRINGER: Lymphocyte function-associated antigen-1 (LFA-1) interaction with intercellular adhesion molecule-1 (1CAM-1) is one of at least three mechanisms for lymphocyte adhesion to cultured endothelial cells. J. Cell Biol. 107 (1988) 321–331.

ENGELHARDT, A., H. LORLER and B. NEUNDORFER: Immunohistochemical findings in vasculitic neuropathies. Acta Neurol. Scand. 87 (1993) 318–321.

ESTEY, E., A. LIEBERMAN, R. PINTO, M. MELTZER and J. RANSOHOFF: Cerebral arteritis in scleroderma. Stroke 10 (1979) 595–599.

FABRY, Z., M.M. WALDSCHMIDT, D. HENDRICKSON, J. KEINER, L. LOVE-HOMAN, F. TAKEI and M.N. HART: Adhesion molecules on murine brain microvascular endothelial cells; expression and regulation of ICAM-1 and Lgp 55. J. Neuroimmunol. 36 (1992) 1–11.

FABRY, Z., K.M. FITZSIMMONS, J.A. HERLEIN, T.O. MONINGER, M.B. DOBBS and M.N. HART: Production of the cytokines interleukin 1 and 6 by murine brain microvessel endothelium and smooth muscle pericytes. J. Neuroimmunol. 47 (1993) 23–34.

FALK, R.J. and J.C. JENNETTE: Anti-neutrophil cytoplasmic autoantibodies with specificity for myeloperoxidase in patients with systemic vasculitis and idiopathic necrotizing and crescentic glomerulonephritis. N. Engl. J. Med. 318 (1988) 1651–1657.

FAUCI, A.S., J.L. DOPPMANN and S.M. WOLFF: Cyclophosphamide-induced remissions in advanced polyarteritis nodosa. Am. J. Med. 64 (1978a) 890–894.

FAUCI, A.S., B.F. HAYNES and P. KATZ: The spectrum of vasculitis. Clinical, pathologic, immunologic, and therapeutic considerations. Ann. Intern. Med. 89 (1978b) 660–676.

FERRIS, E.J. and H.L. LEVINE: Cerebral arteritis: classification. Radiology 109 (1973) 327–441.

FLIGIEL, S.E.G., P.A. WARD, K.J. JOHNSON and G.O. TILL: Evidence for a role of hydroxyl radical in immune-complex-induced vasculitis. Am. J. Pathol. 115 (1984) 375–382.

FORD, R.G. and R.G. SIEKERT: Central nervous system manifestation of periarteritis nodosa. Neurology 15 (1965) 114–122.

FORTIN, P.R., M.G. LARSON, A.K. WATTERS, C.A. YEADON, D. CHOQUETTE and J.M. ESDAILE: Prognostic factors in systemic necrotizing vasculitis of the polyarteritis nodosa group – a review of 45 cases. J. Rheumatol. 22 (1995) 78–84.

FRIES, J.F., G.G. HUNDER, D.A. BLOCH, B.A. MICHEL, W.P. AREND, L.H. CALABRESE, A.S. FAUCI, R.Y. LEAVITT, J.T. LIE and R.W. LIGHTFOOT, JR.: The American College of Rheumatology 1990 criteria for the classification of vasculitis. Summary. Arthr. Rheum. 33 (1990) 1135–1136.

GARNER, B.F., P. BURNS, R.D. BUNNING and R. LAURENO: Acute blood pressure elevation can mimic arteriographic appearance of cerebral vasculitis (a postpartum case with relative hypertension). J. Rheumatol. 17 (1990) 93–97.

GASKIN, G., A.N. TURNER, J.J. RYAN, A.J. REES and C.D. PUSEY: Significance of autantibodies to the purified proteinase 3 in systemic vasculitis. ANCA-associated vasculitides. Immunol. Clin. Aspects 336 (1993) 287–289.

GIBBONS, G.H. and V.J. DZAU: The emerging concept of vascular remodeling. N. Engl. J. Med. 330 (1994) 1431–1438.

GOLDSMITH, J.: Periarteritis nodosa with involvement of the choroidal and retinal arteries. Am. J. Pathol. 29 (1946) 435–446.

GOLDSTEIN, I. and D. WEXLER: Bilateral atrophy of the optic nerve in periarteritis nodosa. Arch. Ophthalmol. 18 (1937) 767–773.

GOLIGORSKY, M.S., H. TSUKAHARA, H. MAZAZINE, T.T. ANDERSEN, A.B. MALIK and W.F. BAHOU: Termination of endothelin signaling: role of nitric oxide. J. Cell. Physiol. 158 (1994) 485–494.

GOULDING, N.J. and P.M. GUYRE: Regulation of inflammation by lipocortin 1. Immunol. Today 13 (1992) 295–297.

GREER, J.M., S. LONGLEY, N.L. EDWARDS, G.J. ELFENBEIN and R.S. PANUSH: Vasculitis associated with malignancy. Experience with 13 patients and literature review. Medicine (Baltimore) 67 (1988) 220–230.

GRIFFITH, G.C. and I.L. VURAL: Polyarteritis nodosa. A correlation of clinical and postmortem findings in seventeen cases. Circulation 3 (1951) 481–491.

GROSS, W.L., W.H. SCHMITT and E. CSERNOK: ANCA and associated diseases: immunodiagnostic and pathogenetic aspects. Clin. Exp. Immunol. 91 (1993) 1–12.

GUILLEVIN, L., D. LE THI HONG, P. GODEAU, P. JAIS and B. WECHSLER: Clinical findings and prognosis of polyarteritis nodosa and Churg–Strauss angiitis: a study in 165 patients. Br. J. Rheumatol. 27 (1988) 258–264.

GUILLEVIN, L., B. JARROUSSE, C. LOK, F. LHOTE, J.P. JAIS, D. LE THI HONG DU and A. BUSSEL: Longterm followup after treatment of polyarteritis nodosa and Churg–Strauss angiitis with comparison of steroids, plasma exchange and cyclophosphamide to steroids and plasma exchange. A prospective randomized trial of 71 patients. J. Rheumatol. 18 (1991) 567–574.

HALLENBECK, J.M. and A.J. DUTKA: Background review and current concepts of reperfusion injury. Arch. Neurol. 47 (1990) 1245–1254.

HAMANN, G.F. and G.J. DEL ZOPPO: Leukocyte involvement in vasomotor reactivity of the cerebral vasculature. Stroke 25 (1994) 2117–2119.

HARATI, Y. and A. NIAKAN: The clinical spectrum of inflammatory angiopathic neuropathy. J. Neurol. Neurosurg. Psychiatry 49 (1986) 1313–1316.

HARRIS, K.G., D.D. TRAN, W.J. SICKELS, S.H. CORNELL and W.T.C. YUH: Diagnosing intracranial vasculitis: the roles of MR and angiography. Am. J. Neuroradiol. 15 (1994) 317–330.

HART, M.N., F. ZSUZSANNA, M. WALDSCHMIDT and M. SANDOR. Lymphocyte interacting adhesion molecules on brain microvascular cells. Mol. Immunol. 27 (1990) 1355–1359.

HASLER, F.: Vasculitis: immunologic aspects. Eur. Neurol. 23 (1984) 389–393.

HAUPTMAN, H.W., S. RUDDY and W.N. ROBERT: Reversal of the vasospastic component of lupus vasculopathy by infusion of prostaglandin. Eur. J. Rheumatol. 18 (1991) 1747–1752.

HAWKE, S.H., L. DAVIES, R. PAMPHLETT, Y.P. GUO, J.D. POLLARD and J.G. MCLEOD: Vasculitis neuropathy. A clinical and pathologic study. Brain 114 (1991) 2175–2190.

HAYNES, B.F., M.I. KAISER-KUPFER, P. MASON and A.S. FAUCI: Cogan syndrome: studies in thirteen patients, long-term follow-up, and review of the literature. Medicine (Baltimore) 59 (1980) 426–441.

HELLMAN, D.B., R. ROUBENOFF, R.A. HEALY and H. WANG: Central nervous system angiography: safety and predictors of a positive result in 125 consecutive patients evaluated for possible vasculitis. J. Rheumatol. 19 (1992) 568–572.

HENSON, P.M. and R.B. JOHNSTON, JR.: Tissue injury in inflammation. Oxidants, proteinases, and cationic proteins. J. Clin. Invest. 79 (1987) 669–674.

HERSON, R.N. and R. SAMPSON: The ocular manifestations of polyarteritis nodosa. Quart. J. Med. 18 (1949) 123–132.

HICKEY, W.F.: Migration of hematogenous cells through the blood–brain barrier and the initiation of CNS inflammation. Brain Pathol. 1 (1991) 97–105.

HICKEY, W.F. and H. KIMURA: Perivascular microglial cells are bone marrow-derived and present antigens in vivo. Science 234 (1988) 290–292.

HICKEY, W.F., B.L. HSU and H. KIMURA: T-lymphocyte entry into the central nervous system. J. Neurosci. Res. 28 (1991) 254–260.

HILT, D.C., D. BUCHHOLZ, A. KRUMHOLZ, H. WEISS and J.S. WOLINSKY: Herpes zoster ophthalmicus and delayed contralateral hemiparesis caused by cerebral angiitis: diagnosis and management approaches. Ann. Neurol. 14 (1983) 543–553.

HIROHATA, S., K. TANIMOTO and K. ITO: Elevation of cerebrospinal fluid interleukin-6 activity in patients with vasculitides and central nervous system involvement. Clin. Immunol. Immunopathol. 66 (1993) 225–229.

HIROSE, S. and Y. HAMASHIMA: Morphological observations on the vasculitis in the mucocutaneous lymph syndrome. Eur. J. Pediatr. 129 (1978) 17–27.

HODGE, S.J., J.P. CALLEN and E. EKENSTAM: Cutaneous leukocytoclastic vasculitis: correlation of histopathological changes with clinical severity and course. J. Cutan. Pathol. 14 (1987) 279–284.

HOFFMAN, G.S., R.Y. LEAVITT, T.A. FLEISHER, J.R. MINOR and A.S. FAUCI: Treatment of Wegener's granulomatosis with intermittent high-dose intravenous cyclophosphamide. Am. J. Med. 89 (1990) 403–410.

HOLTMAN, J.H., D.H. NEUSTADT, J. KLEIN and J.P. CALLEN: Dapsone is an effective therapy for the skin lesions of subacute cutaneous lupus erythematosus and urticarial vasculitis in a patient with CZ deficiency. J. Rheumatol. 17 (1990) 1222–1225.

HUANG, T.E. and S.M. CHOU: Occlusive hypertrophic arteritis as the cause of discrete necrosis in CNS toxoplasmosis in the acquired immunodeficiency syndrome. Hum. Pathol. 19 (1988) 1210–1214.

HUGHES, J.T. and B. BROWNELL: Granulomatous giant-celled angiitis of the central nervous system. Neurology 16 (1966) 293–298.

HUNDER, G.G., W.P. AREND, D.A. BLOCH, L.H. CALABRESE, A.S. FAUCI, J.F. FRIES, R.Y. LEAVITT, J.T. LIE, R W. LIGHTFOOT, JR. and A.T. MASI: The American College of Rheumatology 1990 criteria for the classification of vasculitis. Introduction. Arthr. Rheum. 33 (1990) 1065–1067.

IGARASHI, M., R.C. GILMARTIN, B. GERALD, F. WILBURN and J.T. JABBOUR: Cerebral arteritis and bacterial meningitis. Arch. Neurol. 41 (1984) 531–535.

ISAACS, J.D., M.R. CLARK, J. GREENWOOD and H. WALDMAN: Therapy with monoclonal antibodies. J. Immunol. 148 (1992) 3062–3071.

JAYNE, D.R.W. and C.M. LOCKWOOD: High-dose pooled immunoglobulin in the therapy of systemic vasculitis. J. Autoimmun. 6 (1993) 207–219.

JELLINGER, K.: Giant cell granulomatous angiitis of the central nervous system. J. Neurol. 215 (1977) 175–90.

JENNETTE, J.C., R.J. FALK, K. ANDRASSY, P.A. BACON, J. CHURG, W.L. GROSS, C. HAGEN, G.S. HOFFMAN, G.G. HUNDER, C.G.M. KALLENBERG, R.T. MCCLUSKEY, R.A. SINICO, A.J. REES, L.A. VAN ES, R. WALDHERR and A. WIIK: Nomenclature of systemic vasculitides. Proposal of an international consensus conference. Arthr. Rheum. 37 (1994) 187–192.

JOHNSON, P.C., L.A. ROLAK, R.H. HAMILTON and J.F. LAGUNA: Paraneoplastic vasculitis of nerve: a remote effect of cancer. Ann. Neurol. 5 (1979) 437–444.

JOHNSON, R.J.: Platelets in inflammatory glomerular injury. Semin. Nephrol. 11 (1991) 276–284.

JOSELOV, S.A. and M. MANNIK: Localization of preformed circulating immune complexes in murine skin. J. Invest. Dermatol. 82 (1984) 335–340.

KAESER, H.E.: Cytostatic drugs in the treatment of severe vasculitides. Eur. Neurol. 23 (1984) 407–409.

KAHAN, B.D.: Drug therapy. N. Engl. J. Med. 321 (1989) 1725–1738.

KALLENBERG, C.G.M.: Autoantibodies in vasculitis: current perspectives. Clin. Exp. Rheumatol. 11 (1993) 355–360.

KERNOHAN, J.W. and H.W. WOLTMAN. Periarteritis nodosa: a clinicopathologic study with special reference to the nervous system. Arch. Neurol. Psychiatry 39 (1938) 655–686.

KINYOUN, J.L., R.E. KALINA and M.L. KLEIN: Choroidal involvement in systemic necrotizing vasculitis. Arch. Ophthalmol. 105 (1987) 939–942.

KIRKALI, P., R. TOPALOGLU, T. KANSU and A. BADDALOGLU: Third nerve palsy and internuclear ophthalmoplegia in periarteritis nodosa. J. Pediatr. Ophthalmol. Strabismus 1 (1991) 45–46.

KISSEL, J.T. and K.W. RAMMOHAN: Pathogenesis and therapy of nervous system vasculitis. Clin. Neuropharmacol. 14 (1991) 28–48.

KISSEL, J.T., A.P. SLIVKA, J.R. WARMOLTS and J.R. MENDELL: The clinical spectrum of necrotizing angiopathy of the peripheral nervous system. Ann. Neurol. 18 (1985) 251–257.

KISSEL, J.T., J.L. RIETHMAN, J. OMERZA, K.W. RAMMOHAN and J.R. MENDELL: Peripheral nerve vasculitis: immune characterization of the vascular lesions. Ann. Neurol. 25 (1989) 291–297.

KNIKER, W.T. and C.G. COCHRANE: The localization of circulating immune complexes in experimental serum sickness. The role of vasoactive amines and hydrodynamic forces. J. Exp. Med. 127 (1968) 119–136.

KOBAYASHI, R.M., M. COEL, G. NIWAYAMA and D. TRAUNER: Cerebral vasculitis in coccidioidal meningitis. Ann. Neurol. 1 (1977) 281–284.

KOEPPEN, A.H., L.S. LANSING, S. PENG and R.S. SMITH: Central nervous system vasculitis in cytomegalovirus infection. J. Neurosci. 51 (1981) 395–410.

KOK, J., A. BOSSERAY, J. BRION and M. MICOUD: Chorea in a child with Churg–Strauss syndrome. Stroke 24 (1993) 1263–1264.

KOLODNY, E.H., J.J. REBEIZ, V.S. CAVINESS, JR. and E.P. RICHARDSON, JR.: Granulomatous angiitis of the central nervous system. Arch. Neurol. 19 (1968) 510–524.

KOO, E.H., N. SOLBRIG and E.W. MASSEY: Granulomatous angiitis: its protean manifestations and response to treatment. Neurology 34, Suppl. 1 (1984) 202.

KRENDEL, D.A., S.M. DITTER, M.R. FRANKEL and W.K. ROSS: Biopsy-proven cerebral vasculitis associated with cocaine abuse. Neurology 40 (1990) 1092–1094.

KRISTOFERITSCH, W., K. JELLINGER and F. BOCK: Cerebral granulomatous angiitis with atypical features. Neurology 231 (1984) 38–40.

KUSSMAUL, A. and R. MAIER: Über eine bisher nicht beschriebene eigenthumliche Arterienerkrankung (Periarteritis nodosa), die mit Morbus Brightii und rapid fortschreitender allgemeiner Muskellähmung einhergeht. Dtsch. Arch. Klin. Med. 1 (1866) 484–498.

LANHAM, J.G., K.B. ELKON, C.D. PUSEY and G.R. HUGHES: Systemic vasculitis with asthma and eosinophilia: a clinical approach to the Churg–Strauss syndrome. Medicine (Baltimore) 63 (1984) 65–81.

LAWLEY, T.J., L. BIELORY, P. GASCON, K.B. YANCEY, N. YOUNG and M.M. FRANK: A prospective clinical and immunologic analysis of patients with serum sickness. N. Engl. J. Med. 311 (1984) 1407–1413.

LAWRENCE, M.B. and T.A. SPRINGER: Leukocytes roll on a selectin at physiological flow rates: distinction from and prerequisite for adhesion through integrins. Cell 65 (1991) 859–873.

LEHRER, H.: The angiographic triad in tuberculous meningitis. A radiographic and clinicopathologic correlation. Radiology 87 (1966) 829–835.

LEIB, E.S., C. RESTIVO and H.E. PAULUS: Immunosuppressive and corticosteroid therapy of polyarteritis nodosa. Am. J. Med. 67 (1979) 941–947.

LEUNG, D.Y.M., R.S. GEHA, J.W. NEWBURGER, J.C. BURNS, W. FIERS, L.A. LAPIERRE and J.S. POBER: Two monokines, interleukin 1 and tumor necrosis factor, render cultured vascular endothelial cells susceptible to lysis by antibodies circulating during Kawasaki syndrome. J. Exp. Med. 164 (1986) 1958–1972.

LEWIS, I.C. and M.G. PHILPOTT: Neurological complications in the Schönlein–Henoch syndrome. Arch. Dis. Child. 31 (1956) 369–371.

LICHTIG, C., R. LUDATSCHER, E. EISENBERG and E. BENTAL: Small blood vessel disease in allergic granulomatous angiitis (Churg–Strauss syndrome). J. Clin. Pathol. 42 (1989) 1001–1002.

LIE, J.T.: Illustrated histopathologic classification criteria for selected vasculitis syndrome. Arthr. Rheum. 33 (1990) 1074–1087.

LIGHTFOOT, R.W., JR., B.A. MICHEL, D.A. BLOCH ET AL.: The American College of Rheumatology 1990: criteria for the classification of polyarteritis nodosa. Arthr. Rheum. 33 (1990) 1088–1093.

LINNEMANN, C.C., JR. and M.M. ALVIRA: Pathogenesis of varicella-zoster angiitis in the CNS. Arch. Neurol. 37 (1980) 239–340.

LIOU, H., P. YIP, Y. CHANG and H. LIU: Allergic granulomatosis and angiitis (Churg–Strauss syndrome) presenting as prominent neurologic lesions and optic neuritis. J. Rheumatol. 21 (1994) 2380–2384.

LOCKWOOD, C.M., S. WORLLEDGE, A. NICHOLAS, C. COTTON and D.K. PETERS: Reversal of impaired splenic function in patients with nephritis or vasculitis (or both) by plasma exchange. N. Engl. J. Med. 300 (1979) 524.

LOCKWOOD, C.M., D. BAKES, S. JONES and C.O.S. SAVAGE: Auto-immunity and systemic vasculitis. Contrib. Nephrol. 61 (1988) 141–148.

LOVSHIN, L.L. and J.W. KERNOHAN: Peripheral neuritis in periarteritis nodosa: a clinicopathologic study. Arch. Intern. Med. 82 (1948) 321–338.

LUSCHER, T.F., C.M. BOULANGER, Z. YANG, G. NOLL and Y. DOHI: Interactions between endothelium-derived relaxing and contracting factors in health and cardiovascular disease. Circulation 87 (1993) 36–44.

LUSCINSKAS, F.W., A.F. BROCK, M.A. ARNAOUT and M.A. GIMBRONE, JR.: Endothelial-leukocyte adhesion molecule-1 dependent and leukocyte (CD11/CD18)-dependent mechanisms contribute to polymorphonuclear leukocyte adhesion to cytokine-activated human vascular endothelium. J. Immunol. 142 (1989) 2257–2263.

LYONS, E.L. and N.E. LEEDS: The angiographic demonstration of arterial vascular disease in purulent meningitis. Report of a case. Radiology 88 (1967) 935–938.

MACKEL, S.E. and R.E. JORDAN: Leukocytoclastic vasculitis. Arch. Dermatol. 118 (1982) 296–301.

MALINOW, K., G.D. YANNAKAKIS and S.M. GLUSMAN: Subacute sensory neuronopathy secondary to dorsal root ganglionitis in primary Sjögren's syndrome. Ann. Neurol. 20 (1986) 535–537.

MANNIK, M.: Pathophysiology of circulating immune complexes. Arthr. Rheum. 25 (1982) 783–787.

MANNIK, M.: Experimental models for immune complex-mediated vascular inflammation. Acta Med. Scand. Suppl. 715 (1987) 145–155.

MARKS, C.R., R.F. WILLKENS, K.R. WILSKE and P.B. BROWN: Small-vessel vasculitis and methotrexate. Ann. Intern. Med. 100 (1984) 916–917.

MARTIN, F.P., J.M. LUKEMAN, R.F. RANSON and L.J. GEPPERT: Mucormycosis of the central nervous system associated with thrombosis of the internal carotid artery. J. Pediatr. 44 (1954) 437–444.

MASI, A.T., G.G. HUNDER, J.T. LIE, B.A. MICHEL, D.A. BLOCH, W.P. AREND, L.H. CALABRESE, S.M. EDWORTHY, A.S. FAUCI and R.Y. LEAVITT: The American College of Rheumatology 1990 criteria for the classification of Churg–Strauss syndrome (allergic granulomatosis and angiitis). Arthr. Rheum. 33 (1990) 1094–1100.

MATHIESON, P.W., S.P. COBBOLD, G. HALE, M.R. CLARK, D.B.G. OLIVEIRA, C.M. LOCKWOOD and H. WALDMANN: Monoclonal-antibody therapy in systemic vasculitis. N. Engl. J. Med. 323 (1990) 250–254.

MATSUSHIMA, K.V. and J.J. OPPENHEIM: Interleukin-8 and MCAF. Novel inflammatory cytokines induced by TNF and IL-1. Cytokine 1 (1989) 2–10.

MCCUNE, W.J., J. GOLBUS, W. ZELDES, P. BOHLKE, R. DUNNE and D.A. FOX: Clinical and immunologic effects of monthly administration of intravenous cyclophosphamide in severe systemic lupus erythematosus. N. Engl. J. Med. 318 (1988) 1423–1431.

MERRILL, J.E. and I.S.Y. CHEN: HIV-1, macrophages, glial cells, and cytokines in AIDS nervous system disease. FASEB J. 5 (1991) 2391–2397.

MEURERS, B., W. KOHLEPP, R. GOLD, E. ROHRBACH and H.G. MERTENS: Histopathologic findings of the central and peripheral nervous system in neuroborreliosis: a report of three cases. J. Neurol. 237 (1990) 113–116.

MIKLOSSY, J., T. KUNTZE, J. BOGOUSSLAVSKY, F. REGLI and R.C. JANZER: Meningovascular form of neuroborreliosis: similarities between neuropathological findings in a case of Lyme disease and those occurring in tertiary neurosyphilis. Acta Neuropathol. (Berl.) 80 (1990) 568–572.

MILLER, D.H., L.F. HAAS, C. TEAGUE and T.J. NEALE: Small vessel vasculitis presenting as neurological disorder. J. Neurol. Neurosurg. Psychiatry 47 (1984) 791–794.

MILLER, D.H., I.E. ORMEROD, A. GIBSON, E.P. DU BOULAY, P. RUDGE and W.I. MCDONALD: MR brain scanning in patients with vasculitis: differentiation from multiple sclerosis. Neuroradiology 29 (1987) 226–231.

MILLS, J.A., B.A. MICHEL, D.A. BLOCH, L.H. CALABRESE, G.G. HUNDER, W.P. AREND, S.M. EDWORTHY, A.S. FAUCI, R.Y. LEAVITT and J.T. LIE: The American College of

Rheumatology 1990 criteria for the classification of Henoch–Schonlein purpura. Arthr. Rheum. 33 (1990) 1114–1121.

MOORE, P.M.: Immune mechanisms in the primary and secondary vasculitides. J. Neurol. Sci. 93 (1989a) 129–145.

MOORE, P.M.: Diagnosis and management of isolated angiitis of the central nervous system. Neurology 39 (1989b) 167–173.

MOORE, P.M. and A.S. FAUCI: Neurologic manifestations of systemic vasculitis. A retrospective and prospective study of the clinicopathologic features and responses to therapy in 25 patients. Am. J. Med. Sci. 71 (1981) 517–524.

MORAND, E.F. and N.J. GOULDING: Glucocorticoids in rheumatoid arthritis-mediators and mechanisms. Br. J. Rheumatol. 32 (1993) 816–819.

MULLICK, F.G., H.A. MCALLISTER, B.M. WAGNER and J.J. FENOGLIO, JR.: Drug related vasculitis. Clinicopathologic correlations in 30 patients. Hum. Pathol. 10 (1979) 313–325.

NEWCOMER, J.W., S. CRAFT, T. HERSHEY, K. ASKINS and M.E. BARDGETT: Glucocorticoid-induced impairment in declarative memory performance in adult humans. J. Neurosci. 14 (1994) 2047–2053.

O'DONOGHUE, D.J., C.D. SHORT, P.E.C. BRENCHLEY, W. LAWLEY and F.W. BALLARDIE: Sequential development of systemic vasculitis with anti-neutrophil cytoplasmic antibodies complicating anto-glomerular basement membrane disease. Clin. Nephrol. 32 (1989) 251–255.

O'DONOVAN, C.A., M. KEOGAN, H. STAUNTON, O. BROWNE and M.A. FARREL: Peripheral neuropathy in Churg–Strauss syndrome associated with IgA-C3 deposits. Ann. Neurol. 32 (1992) 411.

OKADA, Y., B.R. COPELAND, E. MORI, M.-M. TUNG, B.W. THOMAS and G.J. DEL ZOPPO: P-selectin and intercellular adhesion molecular-1 expression after focal brain ischemia and reperfusion. Stroke 25 (1994) 202–211.

OMDAL, R., G. HUSBY and W. KOLDINGSNES: Intravenous and oral cyclophosphamide pulse therapy in rheumatic diseases: side effects and complications. Clin. Exp. Rheumatol. 11 (1993) 283–288.

ONG, A.C.M., C.E. HANDLER and J.M. WALKER: Hypersensitivity vasculitis complicating intravenous streptokinase therapy in acute myocardial infarction. Int. J. Cardiol. 21 (1988) 71–73.

OSBORN, L.: Leukocyte adhesion to endothelium in inflammation. Cell 62 (1990) 3–6.

PALLIS, C.P. and J.T. SCOTT: Peripheral neuropathy in rheumatoid arthritis. Br. Med. J. 1 (1965) 1141–1147.

PARISH, W.E.: Studies on vasculitis. I. Immunoglobulins BIC, C reactive protein, and bacterial antigens in cutaneous vasculitic lesions. Clin. Allergy 1 (1971) 97–109.

PARISH, W.E. and E.L. RHODES: Bacterial antigens and aggregated gamma globulin in the lesions of nodular vasculitis. Br. J. Dermatol. 79 (1967) 131–147.

PARK, A.M. and J.C. RICHARDSON: Cerebral complications of serum sickness. Neurology 3 (1953) 227–283.

PARKER, H.L. and J.W. KERNOHAN: The central nervous system in periarteritis nodosa. Mayo Clin. Proc. 24 (1949) 43–48.

PETITO, C.K., G.J. GOTTLIEB, J.H. DOUGHERTY and F.A. PETITO: Neoplastic angioendotheliosis: ultrastructural study and review of the literature. Ann. Neurol. 3 (1978) 393–399.

PEYRONNARD, J.M., L. CHARRON, F. BEAUDET and F. COUTURE: Vasculitic neuropathy in rheumatoid disease and Sjögren syndrome. Neurology 32 (1982) 839–845.

PICKER, L.J., R.A. WARNOCK, A.R. BURNS, C.M. DOERSHUCK, E.L. BERG and E.C. BUTCHER: The neurophil selectin LECAM-1 presents carbohydrate ligands to the vascular selectins ELAM-1 and GMP-140. Cell 66 (1991) 921–933.

POBER, J.S., M.A. GIMBRONE, JR., R.S. COTRAN, C.S. REISS, S.J. BURAKOFF, W. FIERS and K.A. AULT: Ia expression by vascular endothelium is inducible by activated T cells and by human γ-interferon. J. Exp. Med. 157 (1983) 1339–1353.

POBER, J.S., M.A. GIMBRONE, JR., T. COLLINS, R.S. COTRAN, K.A. AULT, W. FIERS, A.M. KRENSKY, C. CLAYBERGER, C.S. REISS and S.J. BURAKOFF: Interactions of T lymphocytes with human vascular endothelial cells: Role of endothelial cell surface antigens. Immunobiology 168 (1984) 483–494.

POBER, J.S., T. COLLINS, M.A. GIMBRONE, JR., P. LIBBY and C.S. REISS: Inducible expression of class II major histocompatibility complex antigens and the immunogenicity of vascular endothelium. Transplantation 41 (1986) 141–146.

POWERS, J.M.: Herpes zoster maxillaris with delayed occipital infarction. J. Clin. Neuroophthalmol. 6 (1986) 113–115.

RABINOV, K.R.: Angiographic findings in a case of brain syphillis. Radiology 80 (1968) 622–624.

RAMOS, M. and T.I. MANDYBUR: Cerebral vasculitis in rheumatoid arthritis. Arch. Neurol. 32 (1975) 271–275.

REICHLIN, M.: Clinical and immunological significance of antibodies to Ro and La in systemic lupus erythematosus. Arthr. Rheum. 25 (1982) 767–772.

ROLLINS, B.J., T. YOSHIMERA, E.J. LEONARD and J.S. POBER: Cytokine-activated human endothelial cells synthesis and secrete a monocyte chemoattractant, MCP-1/JE. Am. J. Pathol. 136 (1990) 1229–1241.

RONCO, P., P. VERROUST, M. MIGNON, O. KOURILSKY, P. VANHILLE, A. MEYRIER, J.P. MERY and L. MORELMAROGER: Immunopathological studies of polyarteritis nodosa and Wegener's granulomatosis: a report of 43 patients and 51 renal biopsies. Quart. J. Med. 52 (1983) 212.

ROSE, G.A. and H. SPENCER: Polyarteritis nodosa. Quart. J. Med. 26 (1957) 43–81.

ROSSI, V., F. BREVIARIO, P. GHEZZI, E. DEJANA and A. MANTOVANI: Prostacyclin synthesis induced in vascular cells by interleukin-1. Science 229 (1985) 174–176.

ROTHLEIN, R., T.K. KISHIMOTO and E. MAINOLFI: Cross-linking of ICAM-1 induces co-signaling of an oxidative burst from mononuclear leukocytes. J. Immunol. 152 (1994) 2488–2495.

ROUX, S., M. GROSSIN, M. DE BRANDT, E. PALAZZO, F. VACHON and M.F. KAHN: Angiotropic large cell lymphoma with mononeuritis multiplex mimicking systemic vasculitis. J. Neurol. Neurosurg. Psychiatry 58 (1995) 363–366.

RUBENSTEIN, M.K.: Mononeuritiis in association with malignancy. Bull. Los Angeles Neurol. Soc. 31 (1966) 157–163.

RUMBAUGH, C.L., R.T. BERGERON, H.C. FANG and R. MCCORMICK: Cerebral angiographic changes in the drug abuse patient. Radiology 101 (1971) 335–344.

SACKS, T., C. MOLDOW, P. CRADDOCK, T. BOWERS and H. JACOB: Oxygen radicals mediate endothelial cell damage by complement-stimulated granulocytes. An in vitro model of immune vascular damage. J. Clin. Invest. 61 (1978) 1161–1167.

SAID, G., C. LACROIS-CIAUDO and H. FUJIMURA: The peripheral neuropathy of necrotizing arteritis: a clinico-pathologic study. Ann. Neurol. 23 (1988) 461–465.

SAMS, W.M., JR., E.G. THORNE, P. SMALL, M.F. MASS, R.M. MCINTOSH and R.E. STANFORD: Leukocytoclastic vasculitis. Arch. Dermatol. 112 (1976) 219–226.

SAVAGE, C.O.S. and S.P. COOKE: The role of the endothelium in systemic vasculitis. J. Autoimmun. 6 (1993) 237–249.

SAVAGE, C.O.S. and Y.C. NG: The aetiology and pathogenesis of major systemic vasculitides. Postgrad. Med. J. 62 (1986) 627–636.

SAVAGE, C.O.S., C.G. WINEARLS, D.J. REES EVANS and C.M. LOCKWOOD: Microscopic polyarteritis: presentation, pathology and prognosis. Quart. J. Med. 220 (1985) 467–483.

SCHIGENAGA, K., M. OKABE and K. ETHON: An autopsy case of aspergillus infection of the brain. Kumamoto Med. J. 28 (1975) 135–144.

SCHWARTZ, S.A.: Intravenous immunoglobulin (IVIG) for the therapy of autoimmune disorders. J. Clin. Immunol. 10 (1990) 81–89.

SCOTT, D.G.I. and P.A. BACON: Intravenous cyclophosphamide plus methylprednisolone in treatment of systemic rheumatoid vasculitis. Am. J. Med. 76 (1984) 377–384.

SCOTT, D.G.I., P.A. BACON and C.R. TRIBE: Systemic rheumatoid vasculitis: a clinical and laboratory study of 50 cases. Medicine (Baltimore) 60 (1981) 288–297.

SCOTT, D.G.I., P.A. BACON, P.J. ELLIOT, C.R. TRIBE and T.B. WALLINGTON: Systemic vasculitis in a district general hospital 1972–1980; clinical and laboratory features, classification, and prognosis in 80 cases. Quart. J. Med. 51 (1982) 292–311.

SEHGAL, M., J. SWANSON, R. DEREMEE and T. COLBY: Neurologic manifestations of Churg–Strauss syndrome. Mayo Clin. Proc. 70 (1995) 337–341.

SESSA, A., M. MERONI and G. BATTINI: Treatment and prognosis of renal and systemic vasculitis. Contrib. Nephrol. 94 (1993) 72–80.

SKOWRONSKI, T. and R.A. GATTER: Cerebral vasculitis associated with rheumatoid disease: a case report. J. Rheumatol. 1 (1974) 473.

SMOLLER, B.R., N.S. MCNUTT and F. CONTRERAS: The natural history of vasculitis. Arch. Dermatol. 126 (1990) 84–89.

SNYDER, D.S., D.I. BELLLER and E.R. UNANUE: Prostaglandins modulate macrophage Ia expression. Nature 299 (1982) 163–165.

SOLE-LLENAS, J. and E. PONS-TORTELLA: Cerebral angiitis. Neuroradiology 15 (1978) 1–11.

SPRINGER, T.A.: Traffic signals for lymphocyte re-circulation and leukocyte emigration: the multistep paradigm. Cell 76 (1994) 301–314.

STEIN, R.L., C.R. MARTINO, D.M. WEINERT, M. HUEFTLE and G.M. KAMMER: Cerebral angiography as a guide for therapy in isolated central nervous system vasculitis. J. Am. Med. Assoc. 257 (1987) 2193–2195.

STERN, D.M., I. BANK, P.P. NAWROTH, J. CASSIMERIS, W. KISIEL, J.W. FENTON II, C. DINARELLO, L. CHESS and E.A. JAFFE: Self-regulation of procoagulant events on the endothelial cell surface. J. Exp. Med. 162 (1985) 1223–1235.

STEVENS, S.K., I.L. WEISSMAN and E.C. BUTCHER: Differences in the migration of B and T lymphocytes: organ-selective localization in vivo and the role of lymphocyte-endothelial cell recognition. J. Immunol. 128 (1982) 844–851.

SUNDERLAND, S.: Blood supply of the nerves of the upper limb in man. Arch. Neurol. Psychiatry 53 (1945) 91–115.

TAI, P.C., M.E. HOLT, P. DENNY, A.R. GIBBS, B.D. WILLIAMS and C.J.F. SPRY: Deposition of eosinophil cationic protein in granulomas in allergic granulomatosis and vasculitis: the Churg–Strauss syndrome. Br. Med. J. 289 (1984) 400–402.

TEOH, R., M.J. HUMPHRIES, J.C. CHAN, H.K. NG and G. O'MAHONY: Intranuclear ophthalmoplegia in tuberculous meningitis. Tubercle 70 (1989) 61–64.

TERVAERT, J.W.C. and C. KALLENBERG: Neurologic manifestations of systemic vasculitis. Rheum. Dis. Clin. North Am. 19 (1993) 913–940.

TERVAERT, J.W.C., P.C. LIMBURG, J.D. ELEMA, M.G. HUITEMA, G. HORST, T.H. THE and C.G.M. KALLENBERG: Detection of autoantibodies against myeloid lysosomal enzymes: a useful adjunct to classification of patients with biopsy-proven necrotizing arteritis. Am. J. Med. 91 (1991) 59–65.

TIJA, D., U.K. YEOW and C.B. TAN: Cryptococcal meningitis. J. Neurol. Neurosurg. Psychiatry 48 (1985) 853–858.

TOPALOGLU, R., N. BESBAS, U. SAARCI, A. BAKKALOGLU and A. ONER: Cranial nerve involvement in childhood polyarteritis nodosa. Clin. Neurol. Neurosurg. 94 (1992) 11–13.

TRAVERS, R.L., D.J. ALLISON, R.P. BRETTLE and G.R.V. HUGHES: Polyarteritis nodosa. A clinical and angiographic analysis of 17 cases. Semin. Arthr. Rheum. 8 (1979) 184–189.

TUSO, P., A. MOUDGIL, J. HAY, D. GOODMAN, E. KANI, R. KOGYANA and S.C. JORDAN: Treatment of antineutrophil cytoplasmic autoantibody-positive systemic vasculitis and glomerulonephritis with pooled intravenous gammaglobulin. Am. J. Kidney Dis. 20 (1992) 504–508.

VAN DER WOUDE, F.J., N. RASMUSSEN, S. LOBATTO, A. WIIK, H. PERMIN, L. A. VAN ES, M. VAN DER GIESSEN, G. K. VAN DER HEM and T.H. THE: Autoantibodies against neutrophils and monocytes: tool for diagnosis and marker of disease activity in Wegener's granulomatosis. Lancet (1985) 425–429.

WALKER, R.J., III, T.E. GAMMEL and M.B. ALLEN: Cranial arteritis associated with herpes zoster. Case report with angiographic findings. Radiology 107 (1973) 109–110.

WALSH, T.J., D.B. HIER and L.R. CAPLAN: Aspergillosis of the central nervous system: clinicopathological analysis of 17 patients. Ann. Neurol. 18 (1985) 574–582.

WATSON, P.: Intracranial hemorrhage with vasculitis in rheumatoid arthritis. Arch. Neurol. 36 (1979) 58.

WEINSTEIN, J.M., H. CHUI, S. LANE, J. CORBETT and J. TOWFIGHI: Churg–Strauss syndrome (allergic granulomatous angiitis). Neuro-ophthalmologic manifestations. Arch. Ophthalmol. 101 (1983) 1217–1220.

WHEAT, L.J., B.E. BATTEIGER and B. SATHAPATAYAVONGS: Histoplasma capsulatum infections of the central nervous systema; a clinical review. Medicine (Baltimore) 69 (1990) 244–260.

WILLIAMS, P.L., R. JOHNSON and D. PAPPAGIANIS: Vasculitic and encephalitic complications associated with *Coccidioides immitus* infection of the central nervous system in humans: report of 10 cases and review. Clin. Infect. Dis. 14 (1992) 673–682.

WINKELMANN, R.K. and W.B. DITTO: Cutaneous and visceral syndromes of necrotizing or 'allergic' angiitis: study of 38 cases. Medicine (Baltimore) 43 (1964) 59–89.

WONG, G.H.W., P.F. BARTLESS, I. CLARK-LEWIS, F. BATTYE and J.W. SCHRADER: Inducible expression of H-2 and Ia antigens on brain cells. Nature 310 (1984) 688–691.

YAMASAKI, Y., N. MATSUURA, H. SHOZUHARA, H. OONDERA, Y. ITOYAMA and K. KOGURE: Interleukin-1 as a pathogenetic mediator of ischemic brain damage in rats. Stroke 26 (1995) 676–681.

ZAX, R.H., S.J. HODGE and J.P. CALLEN: Cutaneous leukocytoclastic vasculitis. Arch. Dermatol. 126 (1990) 69–72.

ZEEK, P.M.: Periarteritis nodosa: a critical review. Am. J. Clin. Pathol. 22 (1952) 777–790.

ZEEK, P.M.: Periarteritis nodosa and other forms of necrotizing angiitis. N. Engl. J. Med. 248 (1953) 764–772.

ZEEK, P.M., C.C. SMITH and J.C. WEETER: Studies on periarteritis nodosa. III. The differentiation between the vascular lesions of periarteritis nodosa and of hypersensitivity. Am. J. Pathol. 24 (1948) 889–917.

ZHANG, Z.G., M. CHOPP, C. ZALOGA, J.S. POLLOCK and U. FORSTERMANN: Cerebral endothelial nitric oxide synthase expression after focal cerebral ischemia in rats. Stroke 24 (1993) 2016–2021.

Handbook of Clinical Neurology, Vol. 27 (71): Systemic Diseases, Part III
M.J. Aminoff and C.G. Goetz, editors

CHAPTER 9

Neurological disorders in Wegener's granulomatosis

JERRY W. SWANSON

Department of Neurology, Mayo Clinic and Mayo Medical School, Rochester, MN, U.S.A.

Wegener's granulomatosis is a systemic disorder that is characterized by necrotizing granulomas and vasculitis which involves both arteries and veins. It characteristically affects the upper and lower respiratory tract, often in association with focal necrotizing glomerulitis. Untreated the disease can be rapidly fatal, especially once functional renal impairment is evident. The mean survival of untreated Wegener's granulomatosis has been reported as 5 months with an 82% 1 year mortality and a 90% 2 year mortality (Walton 1958). It is generally acknowledged that the disease was first described by Klinger (1931). It was Wegener (1936, 1939), though, who described three patients with a nasal granuloma, renal disease, and periarteritis nodosa, and recognized the disorder as a distinct clinical pathologic entity. Decades later, the concept of a limited form of the disorder characterized by absence of renal involvement was advanced by Carrington and Liebow (1966). The precise incidence of Wegener's granulomatosis is unknown but has been estimated at 0.4 cases per 100,000 (Kurland et al. 1984). The male–female ratio is approximately 1:1 (Hoffman et al. 1992a; Nishino et al. 1993a). The mean age of Wegener's granulomatosis has been reported to be in the fifth or sixth decade of life, although ages have ranged from the first decade to the ninth decade (Hoffman et al. 1992a; Nishino et al. 1993a).

The manifestations of Wegener's granulomatosis are protean. Initial clinical presentations and subsequent system involvement span a large range as illustrated by the NIH series (Table 1).

Most patients seek medical care because of upper or lower airway symptoms or a combination of these symptoms (90%). Nasal, sinus, tracheal, or ear problems are presenting symptoms in nearly three-fourths of patients and occur in over 90% of patients at some point during the course of the illness.

Pulmonary manifestations including infiltrates, nodules, or both, are noted to be evident initially in almost 50% of patients. Common symptoms at presentation include cough, hemoptysis, and pleuritis, and during the course of the illness, nearly two-thirds of all exacerbations are associated with these symptoms. In more than a third, chest X-rays show asymptomatic infiltrates or nodules. More than 80% eventually develop lung disease (Hoffman et al. 1992a).

Renal disease is a hallmark of generalized Wegener's granulomatosis. Almost 20% of patients have evidence of glomerulonephritis at the time of presentation; this is almost always asymptomatic. The urinary findings are characteristic of acute glomerulonephritis with proteinuria, hematuria, and erythrocyte casts. More than three-fourths of patients ultimately develop renal involvement, most commonly within the first 2 years of the disorder (Hoffman et al. 1992a).

Ocular involvement has been documented to be an early manifestation in 15% of patients. More than 50% develop ocular involvement at some point

TABLE 1

Frequency of findings in 158 patients with Wegener's granulomatosis*,**.

Condition	Onset (%)	Total (%)
Kidney		
Glomerulonephritis	18	77
Ear, nose, and throat	73	92
Sinusitis	51	85
Nasal disease	36	68
Otitis media	25	44
Hearing loss	14	42
Subglottic stenosis	1	16
Ear pain	9	14
Oral lesions	3	10
Lung	45	85
Pulmonary infiltrates	25	66
Pulmonary nodules	24	58
Hemoptysis	12	30
Pleuritis	10	28
Eye	15	52
Conjunctivitis	5	18
Dacryocystitis	1	18
Scleritis	6	16
Proptosis	2	15
Eye pain	3	11
Vision loss	0	8
Retinal lesions	0	4
Corneal ulcers	0	1
Iritis	0	2
Other		
Arthralgias/arthritis	32	67
Fever	23	50
Cough	19	46
Skin abnormalities	13	46
Weight loss		
(>10% body weight)	15	35
Peripheral neuropathy	1	15
CNS disease	1	8
Pericarditis	2	6

*Data from Hoffman et al. 1992a, with permission.
**Less than 1% had parotid, pulmonary artery, breast, or lower genitourinary (e.g., urethral, cervical, vaginal, and testicular) involvement or concurrent lymphadenopathy, sarcoid, or hypertension.

during the course of the illness (Hoffman et al. 1992a). Ocular manifestations include orbital inflammation (often with proptosis), scleritis, corneal ulceration, keratitis, conjunctival injection, periorbital lid edema, nasal lacrimal duct obstruction, retinal involvement, uveitis, and involvement of the optic, oculosympathetic, oculomotor, trochlear, and abducens nerves (Haynes et al. 1977; Bullen et al. 1983; Hoffman et al. 1992a; Nishino and Rubino 1993).

Symptoms suggestive of musculoskeletal involvement are seen in approximately one-third of patients initially and approximately two-thirds of patients at some point during the course of the illness. The most common symptoms are arthralgias or myalgias. Deformities of the joints are remarkable because of their absence. Dermatologic manifestations are seen in approximately 15% of patients initially and in over 45% at some time during the illness. A variety of lesions have been noted including ulcers, vesicles, papules, palpable purpura, and subcutaneous nodules. Fever and weight loss occur commonly during the illness as well.

Miscellaneous organ involvement includes heart (pericarditis, cardiac muscle, coronary vessels), parotid gland, breast, urethra, bone, prostate, and female reproductive tract (Fauci et al. 1983; Hoffman et al. 1992a).

The characteristic histopathology of Wegener's granulomatosis is necrotizing granulomatous vasculitis (Godman and Churg 1954). The vasculitis is usually characterized by fibrinoid necrosis of small arteries and veins, with initial infiltration of polymorphonuclear leukocytes followed by mononuclear cells. Healing is accompanied by fibrosis. All stages of abnormalities may be observed at a given time. Granulomata are usually well developed, and giant cells are evident. Granulomas may directly involve, lie adjacent to, or be spatially separated from a vascular lesion. Prominent areas of necrosis may be evident, especially in lung. Involvement of the upper respiratory tract, lower respiratory tract, and kidneys is characteristic, but virtually any organ system can be involved (Lie 1989, 1990) (Fig. 1).

Renal lesions are diverse. The earliest renal lesion is usually a focal and segmental glomerulitis. This may progress to diffuse necrotizing glomerulonephritis. Segmental necrotizing glomerulonephritis is suggestive of vasculitis but is not specific for Wegener's granulomatosis. Necrotizing or granulomatous vasculitis, while suggestive of Wegener's granulomatosis, is inconclusive in the absence of respiratory tract involvement.

DIAGNOSTIC CRITERIA

Traditional diagnosis has been based upon involvement of at least two organ systems plus histopathologic evidence of necrotizing granulomatous

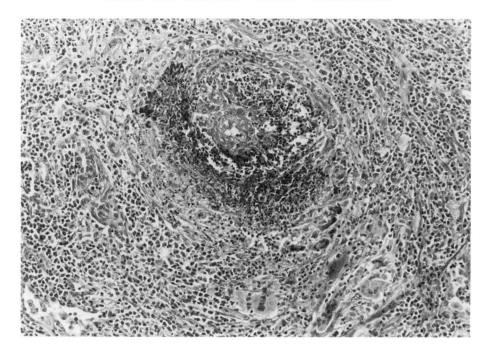

Fig. 1. Pulmonary necrotizing granulomatous vasculitis in Wegener's granulomatosis. The vessel is extensively damaged, and numerous giant cells are present. Adjacent alveolar spaces are filled with a mixture of acute and chronic inflammatory cells (H. & E., × 200). (Courtesy of Dr. J.E. Parisi, Mayo Clinic.)

vasculitis (Fauci et al. 1983). In 1990, the American College of Rheumatology proposed new diagnostic criteria (Leavitt et al. 1990). According to these criteria, a diagnosis of Wegener's granulomatosis is made if at least two of the following four criteria are present: nasal or oral inflammation, abnormal findings on chest radiography, abnormal urinary sediment, and granulomatous inflammation in biopsy specimens. The presence of two or more of these criteria yielded a sensitivity of 88.2% and a specificity of 92.0%. Another set of criteria has also been extensively employed. In this approach, a diagnosis is made if at least one of three systems is involved: (1) ear, nose and throat (E); (2) lung (L); (3) kidney (K); and if biopsy examination shows characteristic features of granulomatous inflammation, typically with vasculitis. The ELK system has been found to be reliable (DeRemee et al. 1976). Antineutrophil cytoplasmic antibodies (c-ANCA) are highly specific for Wegener's granulomatosis and have been used to augment the ELK system when specific pathologic findings are not available (DeRemee 1993).

NEUROLOGIC MANIFESTATIONS OF WEGENER'S GRANULOMATOSIS

Neurologic involvement has recently been reported to be 33.6% (109 of 324 patients) by Nishino et al. (1993a). This contrasts with another recently reported series in which 23% of 158 patients had neurologic involvement (Hoffman 1992a) and two older reports of 25.7% (Anderson et al. 1975) and 54% (Drachman 1963). Drachman's report (1963) was a review of the literature, much of which antedated the development of modern treatment with cytotoxic agents, and survival was typically only a few months. The spectrum of neurologic involvement in this disorder nevertheless remains broad (Table 2).

Drachman (1963) categorized neurologic involvement as follows: (1) granulomatous lesions encroaching on the nervous system by contiguous invasion from nasal or paranasal granulomas; (2) remote granulomatous lesions; (3) vasculitis involving neural structures. Nervous system complications can also occur as a result of treatment (e.g., corticosteroid myopathy).

TABLE 2

Neurological involvement in 324 patients with Wegener's granulomatosis.

	No. of patients
Peripheral neuropathy	53
Cranial neuropathy	21
Ophthalmoplegia	16
Cerebrovascular events	13
Seizures	10
Cerebritis	5
Miscellaneous	33
Total no. of patients*	109

*Thirty-five patients had more than one neurological manifestation. (From Nishino et al. 1993a, with permission.)

Peripheral neuropathy

The most frequent neurologic involvement is peripheral neuropathy. The frequency of peripheral nervous system involvement, excluding cranial neuropathies, has been variably reported as 10.8% (Anderson et al. 1975), 15% (Hoffman et al. 1992a), 16% (Nishino et al. 1993a), 21% (Drachman 1963), and 28.6% (Walton 1958).

The mean age of patients with peripheral neuropathy has been reported as 55 years, with a male–female ratio of 1.4:1 (Nishino et al. 1993a). Although peripheral neuropathy is occasionally a major presenting symptom of Wegener's granulomatosis, it more commonly occurs after non-neurologic symptoms of the disorder are present (Hoffman et al. 1992a; Nishino et al. 1993a). Renal involvement occurs in more than 80% of patients who have peripheral neuropathy, a significantly higher percentage than in those patients without peripheral neuropathy. This may be due to a similar underlying mechanism for each of these, i.e., small vessel vasculitis (Nishino et al. 1993a).

Mononeuropathy multiplex (multiple mononeuropathy) is the most common neuropathic pattern which is typical of neuropathy due to vasculitis of any type including Wegener's granulomatosis. A distal symmetrical sensorimotor peripheral neuropathy may also be seen (Table 3). Although peripheral neuropathy most commonly occurs after non-neurologic features of Wegener's granulomatosis are established, it may be a major presenting symptom of the disorder (Dickey and Andrews 1990; Hoffman et al. 1992a; Nishino et al. 1993a).

TABLE 3

Pattern of involvement in patients with peripheral neuropathy.

	No. of patients
Multiple mononeuropathy	42
Distal symmetrical polyneuropathy	6
Unclassified peripheral neuropathy	5
Total no. of patients	53

(From Nishino et al. 1993a, with permission.)

In the majority of patients, peripheral neuropathy improves or stabilizes with corticosteroid and cyclophosphamide therapy, but on occasion the neuropathy worsens despite treatment (Nishino et al. 1993a).

The histopathology of the peripheral neuropathy is typical of that seen in necrotizing vasculitis which accompanies a variety of distinctive vasculitic syndromes. Necrotizing vasculitis is accompanied by inflammatory cell infiltration (primarily mononuclear) and necrosis of the walls of the nutrient blood vessels. In peripheral nerve, the process involves epineurial arterioles and venules; this is presumably related to the paucity of vessels of this size in perineurium and endoneurium. Necrotizing vasculitis is segmental or 'patchy', involving only a portion of a vessel's longitudinal dimension and sometimes only a portion of the transverse dimension. Usually vascular changes are present at several locations before fiber degeneration occurs. Fibrinoid material is often present in necrotic areas of vessel walls. Other histopathologic findings in nerve that suggest necrotizing vasculitis are: (1) hemorrhage in epineurium, perineurium or endoneurium; (2) hemosiderin-filled macrophages near inflamed vessels; (3) mural or perivascular inflammation, sometimes accompanied by intimal proliferation; (4) thrombotic occlusion and recanalization of epineurial vessels. Perivascular inflammatory cells without mural infiltration or necrosis are a frequent finding, but this is relatively non-specific. Myelinated fiber density is typically reduced, sometimes in a sector or central fascicular pattern. Teased fiber preparations usually show active axonal degeneration (Chalk et al. 1993a). Granulomas are rarely seen in peripheral nerve biopsy specimens (Fig. 2).

It is rarely possible to make a diagnosis of a specific vasculitic syndrome on the basis of a nerve biopsy specimen. Nevertheless, the caliber of vessel involvement may allow assignment to one of two broad groups: (1) involvement of larger epineurial arterioles (100–250 μm) is typical in Wegener's granulomatosis, Churg–Strauss syndrome (allergic granulomatosis and angiitis), polyarteritis nodosa, and rheumatoid vasculitis; and (2) involvement of smaller epineurial arterioles (less than 100 μm) is more commonly seen in Sjögren's syndrome, systemic lupus erythematosus, and non-systemic vasculitis of nerve (Chalk et al. 1993a).

Cranial neuropathy and external ophthalmoplegia

Cranial nerves II–XII have been reported to be affected in Wegener's granulomatosis. Cranial nerve involvement was 9.4% in one large series (Fauci et al. 1983) and 6.5% in the Mayo Clinic series (Nishino et al. 1993a); the latter series, though, categorized external ophthalmoparesis separately since it was difficult to determine whether this resulted from physical block of movement of the extraocular muscles related to orbital pseudotumor, from vasculitis of extraocular muscles, or from ocular motor nerve damage due to nerve impingement by the pseudotumor. As noted in Table 4, the optic, abducens and facial nerves are the most commonly affected cranial nerves. Multiple cranial nerves can be affected, most often unilaterally (Nishino et al. 1993a). Cranial nerves are usually involved by contiguous spread of granulomatous lesions from sinuses in the middle ear

TABLE 4

Cranial nerve involvement.

Nerve	No. of patients
2nd	10
3rd	2
4th	2
5th	5
6th	8
7th	8
8th	2
9th	1
10th	1
12th	2
Total no. of patients*	21

*More than one cranial nerve was affected in eight patients. (From Nishino et al. 1993a, with permission.)

to the retropharyngeal area and base of the skull (Murty 1990). Vasculitis may also involve cranial nerves. Multiple cranial neuropathy may arise on the basis of granulomatous involvement of the pachymeninges (Cogan 1955; Drachman 1963). In one reported case, a patient who presented with multiple unilateral cranial neuropathies had evidence of gadolinium enhancement on MRI along the ipsilateral tentorium. A meningeal biopsy showed evidence of granulomatous inflammation consistent with Wegener's granulomatosis (Fig. 3). Confirmatory histopathology from a nasal biopsy was also attained (Nishino et al. 1993b).

Hearing loss has been reported to occur in 35% of patients and is usually conductive in origin on the basis of otitis media (Hoffman et al. 1992a). Sensorineural hearing loss can also occur, usually due to cochlear involvement and typically in association with a conductive component (McCaffrey et al. 1980). Vertigo occurs rarely (Murty 1990; Nishino et al. 1993a). The reason that vestibular involvement occurs infrequently compared to cochlear involvement is uncertain.

Optic nerve involvement occurs most often on the basis of orbital involvement by contiguous granulomatous extension with acute optic nerve compression. Less commonly, vasculitis can affect the optic nerve (Haynes et al. 1977; Bullen et al. 1983). Vasculitic involvement of the optic nerve is occasionally accompanied or preceded by attacks of transient monocular blindness (Bullen et al. 1983). Clinically, edema of the optic nerve is typically present at the time of visual loss if vasculitis is the underlying mechanism. Vasculitis of the posterior ciliary arteries can produce ischemic optic neuropathy. Central retinal artery occlusion due to vasculitis can also cause visual loss. Optic neuropathy inferior visual field loss, orbital pain and optic nerve edema have been observed without orbital involvement and attributed to granulomatous involvement of the optic nerve (Haynes et al. 1977; Bullen et al. 1983).

External ophthalmoparesis usually occurs as a result of contiguous orbital granulomatous involvement and is characteristically associated with pain, proptosis, and sometimes erythema of the eyelid. It may occur bilaterally. Vasculitic involvement of cranial nerves III, IV, and VI may also produce ophthalmoparesis. Imaging with computed tomography (CT) or magnetic resonance imaging (MRI) can usually

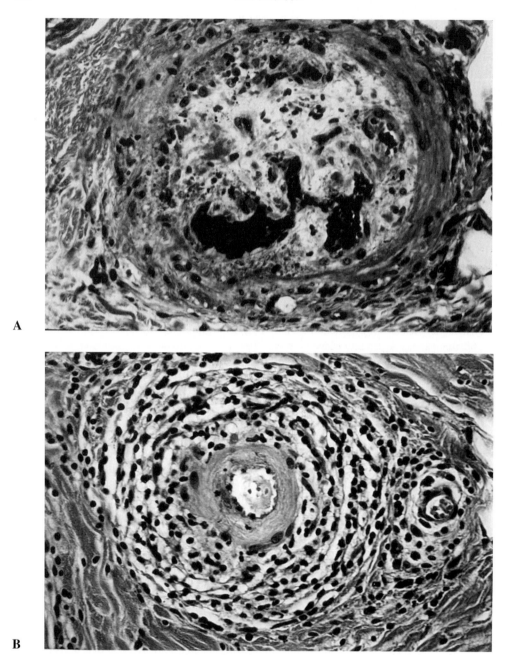

Fig. 2. Peripheral nerves involved by necrotizing vasculitis in Wegener's granulomatosis. (A) Large epineurial arteriole obliterated by granulation tissue demonstrating early recanalization. Note the prominent necrosis of the tunica media (left half of the arteriole) (H. & E., × 720). (From Dyck et al. 1972.) (B) Media of this arteriole is absent. There is perivascular inflammation consisting primarily of mononuclear inflammatory cells (H. & E., × 160).

demonstrate orbital granulomatous involvement when this is the etiology for ophthalmoparesis (Asmus et al. 1993; Duncker et al. 1993).

Pathologically, non-infectious granulomatous vasculitis is the most common cause of proptosis and ophthalmoparesis. Granulomatous vasculitis can

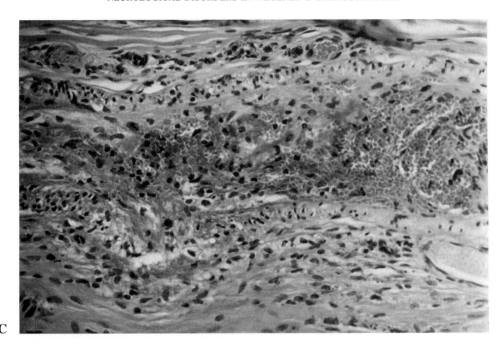

Fig. 2. (C) Epineurial hemorrhage in sural nerve which is an occasional phenomenon seen in necrotizing vasculitis. The darkened area horizontally placed at the center of the picture consists of the products of blood breakdown and indicates that the hemorrhage did not occur at the time of surgery (H. & E., × 160). (Courtesy of Dr. P.J. Dyck, Mayo Clinic.)

involve any orbital structure, though, including optic nerve, optic vessels, and extraocular muscles. Purulent sinusitis may also produce proptosis and lead to cavernous sinus thrombosis.

Horner's syndrome (oculosympathetic paresis) is a rare manifestation of Wegener's granulomatosis. Both small vessel vasculitis and direct granulomatous invasion of the oculosympathetic pathway have been postulated to be potential underlying mechanisms. Horner's syndrome may occur as an isolated neurologic manifestation of Wegener's granulomatosis. It can also occur in association with mononeuropathy multiplex and multiple cranial neuropathies (Nishino and Rubino 1993).

Myopathy

Muscle weakness is most commonly the result of motor nerve involvement and hence the weakness usually has a peripheral nerve pattern. This is often accompanied by sensory disturbances related to sensory nerve fiber involvement. Although arthralgias and myalgias are frequent manifestations, myopathy occurs rarely. It was identified in less than 1% of the

Mayo Clinic series (Nishino et al. 1993a) whereas Drachman's (1963) review reported myopathy in 4%. Improvement in muscle strength tends to occur with immunosuppressive treatment. Corticosteroid myopathy should be considered when myopathy arises in patients undergoing treatment with these agents.

Pathologic findings of muscle disease are quite variable. Various arteritic lesions including active and healed necrotizing arteritis may be present (Fig. 4). Focal infarction of muscle may be evident. A diffuse necrotizing myopathy has also been described (Stern et al. 1965). Granulomatous changes may sometimes be evident as well (Finkelman et al. 1993; Banker 1994). In addition, neurogenic atrophy may occur as a result of motor nerve involvement (Stern et al. 1965).

Myelopathy

Myelopathy is a rare manifestation. There is one reported case of a patient who presented with myelopathy which pathologically appeared to be secondary to inflammatory changes in the dura with

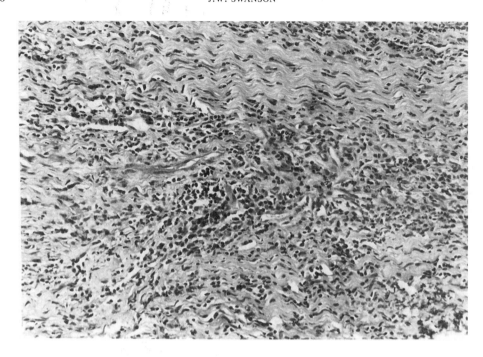

Fig. 3. Involvement of dura by Wegener's granulomatosis. The dura is expanded by dense, predominately perivascular collections of chronic inflammatory cells including scattered giant cells (H. & E., × 250). (Courtesy of Dr. J.E. Parisi, Mayo Clinic.)

secondary spinal cord compression (Nishino et al. 1993b). A case of cauda equina syndrome has been reported, but pathologic and imaging verification was not obtained (Martens 1982).

Cerebrovascular disease

Thirteen of 109 patients with neurologic involvement (6.2% of those who developed neurologic involvement) had cerebrovascular events thought to be related to Wegener's granulomatosis. Of these, 12 patients had cerebral infarction and one had a subdural hematoma without a history of trauma. In five of these patients, other features of cerebral vasculitis (e.g., encephalopathy, seizures) were present (Nishino et al. 1993a). Intracerebral hemorrhage and subarachnoid hemorrhage have rarely been reported (Drachman 1963). Cortical vein thrombosis is another rare manifestation (Drachman 1963; Mickle et al. 1977).

There are multiple potential mechanisms for cerebrovascular events in patients with Wegener's granulomatosis. These include vasculitis, necrotizing granuloma, defects of coagulation, and hypertension secondary to renal involvement by the disease

(Nishino et al. 1993a). Cerebral angiography is typically negative in central nervous system (CNS) vasculitis, probably because the small vessels (50–300 μm) are usually affected and these are below the sensitivity of the technique (Nishino et al. 1993a).

Pathologically proven cerebrovascular manifestations have included infarction, hemorrhage (intraparenchymal and subarachnoid), and cerebral venous thrombosis. In one autopsy case, interhemispheric subarachnoid hemorrhage and cerebral infarction resulted from vasculitis of the anterior cerebral artery (Drachman 1963). In another case, subarachnoid hemorrhage apparently arose in or near the choroid plexus; there was acute granulocytic exudate with fibrin beneath the ventricular wall and in the choroid plexus (Tuhy et al. 1958). Another patient who suffered a fatal caudate hemorrhage with extension into the ventricular system was found at postmortem examination to have intracranial necrotizing arteritis with aneurysmal dilatation at the site of the hemorrhage (Lucas et al. 1976). In a case of bilateral anterior cerebral hemisphere infarction, the pathologic findings included large arterial mural thrombosis and fibrinoid necrosis as well as necrotizing vasculitis

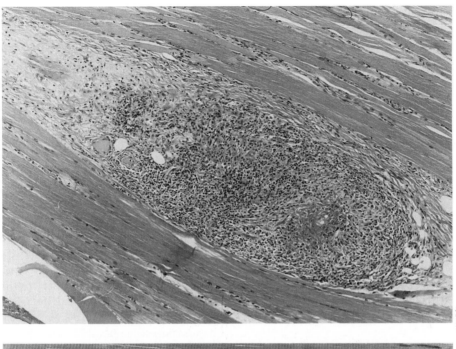

A

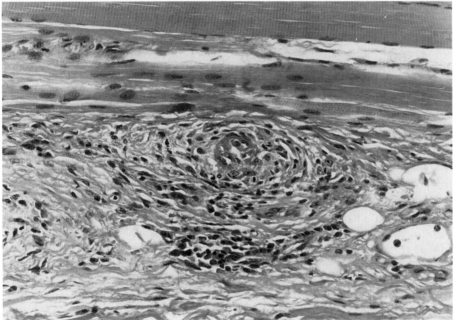

B

Fig. 4. Necrotizing arteritis in skeletal muscle associated with Wegener's granulomatosis. (A) Skeletal muscle with large interstitial inflammatory focus, comprised of necrotic vessels with extensive collections of chronic inflammatory cells (H. & E., × 125). (B) The wall and perivascular spaces surrounding this perimysial vessel are infiltrated and expanded by chronic inflammatory cells (H. & E., × 400). (Courtesy of Dr. J.E. Parisi, Mayo Clinic.)

and granulomatous lesions in small arteries and veins diffusely in the frontal area in association with a suppurative meningitis. This suggested invasion of a granulomatous lesion from the nasal cavity by contiguity (Satoh et al. 1988). In another patient with cortical vein thrombosis, biopsy revealed necrotizing cerebral thrombophlebitis (Mickle et al. 1977). Cavernous sinus invasion from the infratemporal fossa presumably via the basal foramina has also been documented to result in internal carotid artery occlusion (Goldberg et al. 1983). One patient who died from acute hemorrhage of the thalami had pathologic evidence of small blood vessel fibrinoid necrosis with associated inflammatory reaction (MacFayden 1960). Fred et al. (1964) reported a patient with cerebral infarctions who at autopsy had diffuse cerebral arteritis characterized by focal necrotizing granulomas in many vessel walls, an aneurysm of the left vertebral artery resulting from arteritic involvement, and fibrinoid necrosis of several small vessels.

Seizures

Seizures occur as a result of vasculitis, cerebral granulomatous involvement, or progressive renal or pulmonary failure or other systemic processes related to the underlying disease (Nishino et al. 1993a). Wegener's granulomatosis has rarely presented with a seizure prior to the diagnosis of the disorder. In one reported case, this was convincingly attributed to remote granulomatous involvement based upon MRI (Miller and Miller 1993). The MRI findings improved with corticosteroid and cyclophosphamide treatment.

Diabetes insipidus

Diabetes insipidus has been rarely reported; it may be the presenting manifestation. Only one case was found in 324 patients reported from the Mayo Clinic (Nishino et al. 1993a). A suprasellar mass may be demonstrated on CT X-ray scanning or MRI (Rosete et al. 1991). Diabetes insipidus is postulated to be secondary to granulomatous inflammation of the posterior pituitary or vasculitis involving the arterial supply to the posterior pituitary (Haynes and Fauci 1978). This complication has been documented to respond to treatment with cyclophosphamide (Rosete et al. 1991), but amelioration with treatment does not always occur (Hurst et al. 1983).

Headache

Headache is a frequent complaint. This is most commonly related to sinus involvement by the disorder or superimposed bacterial sinusitis. Headache may also accompany orbital pseudotumor. Headache is less commonly related to intracranial involvement and when it is, other neurologic manifestations are typically present (Drachman 1963; Nishino et al. 1993a).

Temporal arteritis

Of 345 consecutive patients seen at the Mayo Clinic with Wegener's granulomatosis, five had histologic confirmation of vasculitis of the temporal artery (Nishino et al. 1993c). All five patients were older than 60 years of age and had jaw claudication, visual loss, headache (sometimes accompanied by diplopia), or polymyalgia rheumatica at the time of initial evaluation. Of the four patients who had an erythrocyte sedimentation rate (ESR) available for review, each of the results was elevated. Temporal artery biopsy showed giant cell arteritis in one patient and non-giant cell arteritis in the other four patients. Four patients subsequently developed renal failure, and three had pulmonary lesions. One case did not fulfill American College of Rheumatology criteria, but c-ANCA were present (Nishino et al. 1993c). Another well-substantiated case of giant cell arteritis in a patient with Wegener's granulomatosis was reported by Small and Brisson (1991).

These cases likely represent cases of Wegener's granulomatosis presenting in a manner that mimicked giant cell arteritis. Alternatively, these cases may represent independent expression of the two disorders, or expression of a disease process caused by the other disorder, or a separate predisposing factor might underlie development of each of these clinical processes (Cupps 1993). Until the pathogenesis of each of these disorders is known, their relationship will remain uncertain.

DIFFERENTIAL DIAGNOSIS

The diagnosis should be suspected in the setting of an inflammatory, non-infectious systemic disorder that involves the upper and lower respiratory tract and is associated with renal involvement. This picture should be distinguished from Churg–Strauss syndrome (allergic granulomatosis and angiitis) which can produce similar organ involvement but is associated with prominent eosinophilia and asthma. It must also be separated from pulmonary hemorrhagic syndromes such as Goodpasture's syndrome and angiocentric immunoproliferative syndromes such as lymphomatoid granulomatosis.

Diagnosis of limited Wegener's granulomatosis can be a difficult process since destructive upper airway disease can be secondary to infection (e.g., fungi, mycobacteria, syphilis, actinomycosis), malignancy (e.g., extranodal lymphoma, squamous cell carcinoma), or nasal inhalation of noxious substances (e.g., cocaine). Pulmonary granulomatous infections (e.g., fungi, mycobacteria) may be associated with vasculitis and necrosis. Accordingly, lung biopsy specimens should be submitted for special stains and cultures to exclude these infections. A c-ANCA determination may aid in the differential diagnosis (see below).

A specific histopathologic diagnosis is usually most reliably obtained from the tissue obtained at open lung biopsy, although tissue from upper airway sites (paranasal sinuses, nose, and subglottic region) can often provide confirmation. As noted, renal biopsy specimens can also provide supportive evidence but are usually non-specific.

The diagnosis is usually evident or should at least be suspected because of accompanying symptoms or signs by the time neurologic manifestations arise. The most common neurologic manifestation is peripheral neuropathy, most often as mononeuropathy multiplex. Occasionally this may be a heralding symptom of the disorder. Vasculitic syndromes in addition to Wegener's granulomatosis that need to be considered with this clinical picture include polyarteritis nodosa, Churg–Strauss syndrome (allergic granulomatosis and angiitis), rheumatoid arthritis, Sjögren's syndrome, systemic lupus erythematosus, hypersensitivity vasculitis, giant cell arteritis, Behçet's syndrome, and non-systemic vasculitis of the peripheral nerve. Other conditions that can also produce the picture of mononeuropathy multiplex are diabetes mellitus, sarcoidosis, leprosy, inherited susceptibility to pressure palsies, lymphomatoid granulomatosis, human immunodeficiency virus infection, malignant infiltration of nerve trunks, Lyme disease, multifocal motor neuropathy with conduction block, neurofibromatosis, and subacute bacterial endocarditis (Chalk et al. 1993a).

Cranial neuropathies are common, but rarely are an isolated presenting feature of the disorder. If a systemic disorder is suspected, a wide array of vasculitic disorders such as those which produce multiple mononeuropathies should be considered. Sarcoidosis frequently affects the cranial nerves. Additionally, tubercular and fungal meningitis should be considered.

Wegener's granulomatosis is unique among the systemic vasculitic syndromes in its ability to produce ophthalmoparesis with proptosis. When these clinical features are present but there is no evidence of systemic illness, processes such as Graves' ophthalmopathy and retro-orbital mass lesions (e.g., abscess, tumor, vascular process) must be seriously considered.

Encephalopathy almost always occurs in the context of previously identified Wegener's granulomatosis. In a patient treated with immunosuppressive therapy such as cyclophosphamide and corticosteroids, the evolution of focal or diffuse CNS dysfunction should be followed by a careful search for CNS infection, metabolic derangements, and hypertensive encephalopathy before a diagnosis of cerebritis is established.

Non-specific laboratory findings

The ESR is nearly always elevated in active generalized Wegener's granulomatosis; elevations in excess of 100 mm/h are often seen. The ESR may be normal in limited forms of the disorder. Leukocytosis and thrombocytosis are often present. Conversely, leukopenia or thrombocytopenia is rarely seen in untreated patients. Normochromic, normocytic anemia may also be present (Fauci and Wolff 1973); hypochromic anemia can be seen with alveolar hemorrhage. Elevation of serum immunoglobulins may be present and usually consists of IgG and IgM (Fauci et al. 1983). Rheumatoid factor is present in over 50%. Abnormal urinary sediment and significant proteinuria are frequently noted (Hoffman et al. 1992a).

Antineutrophil cytoplasmic antibodies (ANCA)

Until antineutrophil cytoplasmic antibodies (ANCA) were first reported to be associated with Wegener's granulomatosis by Van der Woude et al. (1985), there were no laboratory findings that were highly specific for the disorder. This finding is based upon the following facts. Polymorphonuclear granulocytes contain primary and secondary granules. The primary granules contain myeloperoxidase, several serine proteases which include neutrophil elastase, cathepsin G, and proteinase 3 and lysozymes and other microbicidal enzymes. Monocytes contain similar granules. ANCA reacts to proteins with neutrophil granules as well as monocyte lysosomes. In sera from patients with Wegener's granulomatosis, indirect immunofluorescent techniques usually show a coarse granular staining pattern of cytoplasmic neutrophil granules (cytoplasmic, or c-ANCA). The antigen responsible for this pattern has been found to be 29 kDa serine proteinase, proteinase 3 (Goldschmeding et al. 1989; Niles et al. 1989). Microscopic polyarteritis has also been associated with this pattern.

A second immunofluorescent staining pattern in ethanol-fixed neutrophils stains in a perinuclear location (perinuclear or p-ANCA) (Falk and Jennette 1988). The antigen most frequently recognized by the sera from these patients is myeloperoxidase. p-ANCA and antimyeloperoxidase rarely occur in Wegener's granulomatosis. It has frequently been identified in patients with microscopic polyarteritis. p-ANCA has also been observed in a large number of other inflammatory disorders including polyarteritis nodosa, rheumatoid arthritis, systemic erythematosus, antiglomerular basement membrane disease, Crohn's disease, and chronic liver disorders.

With the preceding kept in mind, the sensitivity of c-ANCA for detecting generalized, active, histologically confirmed Wegener's granulomatosis is 80% or greater, and the specificity is approximately 97%. In patients without renal involvement, the sensitivity is only 60–70%. Patients with limited disease and initially negative tests for ANCA have subsequently developed c-ANCA (Nolle et al. 1989; Specks et al. 1989; Galperin and Hoffman 1994). c-ANCA has also occasionally been seen with tuberculosis and undifferentiated non-Hodgkin's lymphoma (Davenport 1992). Despite this, indirect immunofluorescence is useful in the evaluation of disorders consistent with Wegener's granulomatosis. However, it follows from the above findings that a negative test in the appropriate clinical setting (when histopathologic studies of involved tissues reveal usual features and infection has been excluded), this finding does not exclude the diagnosis of Wegener's granulomatosis.

Although some reports have suggested that a rise in c-ANCA titer predictably antedates exacerbations of Wegener's granulomatosis, a study by Kerr et al. (1993) did not confirm this. In this report, serial ANCA titers were prospectively obtained in 72 patients and compared with variation in disease activity. Changes in serial titers temporally correlated with a change in disease activity in only 64%. Twenty-four percent of patients had a rise in c-ANCA titer preceding clinical exacerbations. Forty-four percent of patients who had a rise in c-ANCA titers either had continued remission or were clinically improving while c-ANCA titers increased. Accordingly, a rise in c-ANCA titer alone should not be considered adequate evidence of a clinical exacerbation, and treatment decisions should not be predicated on this finding. Careful monitoring of such patients for clinical activity of the disease is nonetheless prudent.

In two cases of Tolosa–Hunt syndrome (idiopathic granulomatous involvement of the cavernous sinus), c-ANCA have been identified. Neither patient had evidence of systemic inflammatory involvement despite clinical histories spanning 9 years and 4 years, respectively. It has been speculated that Tolosa–Hunt syndrome may represent a form of localized Wegener's granulomatosis. Whether these disorders are pathogenetically related or not will require further understanding of each of these disorders (Montecucco et al. 1993).

In one study, assay for ANCA was performed in 166 consecutive patients with peripheral neuropathy to see if the presence might predict a vasculitic neuropathy. None of these patients (including those with the diagnosis of vasculitic neuropathy and one with Wegener's granulomatosis) had a positive c-ANCA. Some patients with positive p-ANCA or antimyeloperoxidase titers had non-vasculitic neuropathies (Chalk et al. 1993b). The authors concluded that the utility of ANCA as a serologic test for vasculitic neuropathy was limited by non-specificity.

Cerebrospinal fluid studies

Cerebrospinal fluid study results are often normal even in the presence of CNS involvement by Wegener's granulomatosis (Fred et al. 1964). Abnormalities that have been reported in some patients with intracranial involvement have included elevation of opening pressure, pleocytosis (primarily mononuclear cells), and elevated total protein (Fred et al. 1964; Atcheson and Van Horn 1977; Nishino et al. 1993b; Weinberger et al. 1993). Subarachnoid hemorrhage when present is accompanied by characteristic cerebrospinal fluid changes.

Imaging studies

Both CT and MRI can be helpful in the assessment of patients with Wegener's granulomatosis. Inflammation of the nasal cavity and paranasal sinuses is the most common finding. Granulomatous involvement of the paranasal sinuses and orbits can usually be most sensitively demonstrated on MRI; characteristically, there is loss of signal intensity on T1- and T2-weighted images, and there may be enhancement of these lesions following administration of gadolinium. MRI rather frequently shows hyperintense abnormalities in cerebral white matter on T2-weighted images. These findings are non-specific and can be seen in a variety of disorders. Bony erosion or destruction of the skull is best demonstrated with CT scanning (Asmus et al. 1993; Duncker et al. 1993). When either cerebral infarction or intracranial hemorrhage is present, changes characteristic of these processes can be observed.

Meningeal enhancement on MRI with intravenous contrast has been described in a patient with multiple cranial neuropathies. Granulomatous angiitis was documented on meningeal biopsy (Nishino et al. 1993b). In another patient, MRI of the head showed meningeal and gyral enhancement as well as increased frontal lobe white matter signal; meningeal biopsy revealed arachnoid thickening, lymphocytic infiltration and necrotizing granulomas; cortical biopsy showed necrotizing vasculitis of the small vessels (Weinberger et al. 1993). The radiographic findings improved with immunosuppressive treatment.

Another patient who presented with seizures and was found to have contrast-enhancing intraparenchymal masses on CT and MRI also had radiographic improvement with treatment (Miller and Miller 1993).

As noted previously, cerebral angiography is typically normal even in the presence of CNS vasculitic involvement.

PATHOGENESIS

Since the seminal reports of Klinger (1931) and Wegener (1936), it has been theorized that stimulation of airways leads to disease expression. Bronchoalveolar lavage studies show that neutrophilic alveolitis is a consequence of airway stimulation in Wegener's granulomatosis (Hoffman et al. 1991). It has been postulated that an inflammatory response in patients with the capacity to produce antibodies to proteinase 3 may result in pulmonary and systemic inflammatory responses (Hoffman 1993). Although the stimuli might be infectious processes, non-infectious inhaled substances that produce neutrophilic alveolitis might trigger the disorder in patients who produce ANCA. For example, pulmonary exposure to silica can produce neutrophilic infiltrates (Bolton et al. 1981). Attempts have been made to show an association of infection with Wegener's granulomatosis (Falk et al. 1990), but these studies have not been confirmed (Hoffman et al. 1992a). Furthermore, analysis of bronchoalveolar lavage fluid in patients with recent onset Wegener's granulomatosis have not revealed infectious agents (Hoffman et al. 1991).

Studies examining the association between Wegener's granulomatosis and HLA antigens have not shown consistent results. A study of HLA alleles in 83 patients did not identify unique markers as having an increased frequency in Wegener's granulomatosis (Galperin and Hoffman 1994). If there is a genetic predisposition to development of Wegener's granulomatosis, this is not apparent from HLA antigen studies thus far.

A role for ANCA in the etiology of Wegener's granulomatosis has been postulated. One such theory suggests that circulating ANCAs can enhance activity of neutrophils and monocytes, leading to their adherence to vessel walls with ensuing injury (Jennette et al. 1993). Further investigations are needed to prove or refute the role of ANCAs in the pathogenesis of the disorder.

TREATMENT

The prognosis of Wegener's granulomatosis was generally poor in untreated or ineffectively treated patients. This was especially true after functional renal impairment developed. A small group of patients, though, probably less than 10% of cases, has been recognized to have a more indolent course; these patients usually remain free of renal involvement for a prolonged period of time.

The development of the standard treatment regimen comprised initially of daily oral therapy with cyclophosphamide, 2 mg/kg, and prednisone, 1 mg/kg, for 1 month. In desperate situations, treatment may begin with up to 1 g of methylprednisolone daily for 3 days and up to 4 mg/kg of cyclophosphamide for a few days. The doses are subsequently reduced to the more conventional range. Once the patient experiences marked improvement consisting of suppression of the disease with stabilization of renal function and at least some resolution of pulmonary infiltrates, prednisone is tapered to alternate days over the following 2 months. If remission continues, the prednisone is tapered over several months. Cyclophosphamide is continued for at least 1 year after the patient achieves complete remission and is typically tapered by 25 mg increments every 2–3 months until discontinued or until disease recurrence requires an increase in the dose. Treatment with the standard protocol is restarted if a relapse occurs. This protocol has led to marked improvement or partial remission in 91% of patients and complete remission in 75% of patients. The median time to complete remission is 12 months, although fewer than 10% of patients do not achieve remission for much longer periods of time, sometimes as long as 5 years. Fifty percent of complete remissions are followed by relapse. Although the approach is designed to rapidly control the disease and to minimize side effects, morbidity occurs in almost all patients. Disease-related morbidity has been noted in 86% of patients (Table 5) and includes chronic renal insufficiency in 42%, hearing loss in 35%, cosmetic and functional nasal deformities in 28%, tracheal stenosis in 13% and visual loss in 8%. Many patients suffer more than one permanent type of morbidity. Visual loss has been noted to occur primarily as a result of a retro-orbital Wegener's granulomatosis usually as a result of a pseudotumor or mass lesion. Permanent morbidity attributed to disease plus medical or surgical treatment or both includes chronic sinus dysfunction in 47% and pulmonary insufficiency in 17%. Finally, some forms of morbidity appear to be solely related to treatment toxicity. This includes cyclophosphamide-induced cystitis in 43%, bladder cancer in 3–4%, and

TABLE 5

Permanent morbidity in 158 patients with Wegener's granulomatosis*.

Condition	Patients, %
Infertility	>57 **
Chronic sinus disease	47
Cyclophosphamide cystitis	43 ***
Chronic renal insufficiency	42
Dialysis	11
Renal transplant	5
Hearing loss	35
Partial unilateral or bilateral	33
Complete unilateral	1
Complete bilateral	1
Nasal deformity	28
Mild	10
Severe	18
Pulmonary insufficiency	17
Tracheal stenosis	13
Osteoporosis-related fractures	11
Total malignancies	10 §
Visual loss	8
Partial unilateral or bilateral	5
Complete unilateral	2
Complete bilateral	1
Aseptic necrosis	3
Bladder cancer	4.2
Myelodysplasia	2

*Data from Hoffman et al. 1992a, with permission.
**Survey of women with reproductive potential prior to cyclophosphamide therapy.
***43% of patients treated with cyclophosphamide (141), 38% of 158 patients.
§10% of patients treated with a cytotoxic agent (145), 8% of 158 patients.

infections requiring hospitalization and intravenous antibiotics in 46%. Half of the infections occur while patients are on daily prednisone, while 21% occur while on alternate-day prednisone compared with only 12% while on no immunosuppressive therapy (Hoffman et al. 1992a; Hoffman 1994). Of particular note is that the combination of corticosteroids and cyclophosphamide increases the risk of herpes zoster infection 20 fold (Hoffman et al. 1992a).

Thirteen percent of patients in the NIH cohort died as a result completely or partially related to active Wegener's granulomatosis, chronic morbidity from previously active disease, complications of treatment, or a combination of these factors. The significant disease- and treatment-related morbidity led to the search for other treatments.

Intermittent high-dose intravenous cyclophosphamide often leads to remission which is followed by relapse despite continued therapy (Hoffman et al. 1992a).

Methotrexate administered once weekly in doses of 0.15–0.30 mg/kg with concurrent corticosteroid administration has led to remission in two-thirds of patients within 12 months. Almost two-thirds of these patients have been able to discontinue corticosteroids. This regimen had reduced toxicity compared to cyclophosphamide and may provide a reasonable alternative to treatment if the remissions are sustained (Hoffman et al. 1992b).

Trimethoprim-sulfamethoxazole in a dose of one double-strength tablet twice daily has been reported to be an effective treatment for disease confined to the upper or lower respiratory tract (without evidence of disseminated vasculitis or renal involvement). This was initially reported by DeRemee et al. (1985) but has subsequently been observed by others (Georgi et al. 1991). A prospective controlled trial has shown that this combination of agents when used in patients who are in remission yielded fewer upper airway and nasal relapses; it did not reduce renal, pulmonary, or nervous system relapses (Stegeman et al. 1996). The mechanism by which this treatment has its therapeutic effect is as yet uncertain.

Control of systemic disease often seems to prevent development and progression of the neurologic manifestations (Moore and Fauci 1981) and may allow improvement of neurologic deficits related to peripheral neuropathy and intracranial involvement (Atcheson and Van Horn 1977; Haynes and Fauci 1978; Moore and Cupps 1983; Finkelman et al. 1993; Miller and Miller 1993; Nishino et al. 1993a; Weinberger et al. 1993).

Development of vasculitis has been reported to occur rarely in patients while undergoing treatment with cyclophosphamide. This may respond to treatment with high-dose intravenous corticosteroids with or without high-dose intravenous cyclophosphamide (Kroneman and Pevzner 1986).

REFERENCES

ANDERSON, J.M., D.G. JAMIESON and J.M. JEFFERSON: Nonhealing granuloma and the nervous system. Quart. J. Med. 44 (1975) 309–323.

ASMUS, R., H. KOLTZE, C. MUHLE, R.P. SPIELMANN, G. DUNCKER, B. NOLLE, A. BEIGEL and W.L. GROSS: MRI of the head in Wegener's granulomatosis. In: L.G. Gross (Ed.), ANCA-Associated Vasculitides: Immunological and Clinical Aspects. New York, Plenum Press (1993) 319–321.

ATCHESON, S.G. and G. VAN HORN: Subacute meningitis heralding a diffuse granulomatous angiitis (Wegener's granulomatosis?). Neurology 27 (1977) 262–264.

BANKER, B.Q.: Other inflammatory myopathies. In: A. G. Engel and C. Franzini-Armstrong (Eds.), Myology: Basic and Clinical. New York, McGraw-Hill (1994) 1461–1484.

BOLTON, W.K., P.M. SURATT and B.C. STURGILL: Rapidly progressive silicon nephropathy. Am. J. Med. 71 (1981) 823–828.

BULLEN, C.L., T.J. LIESEGANG, T.J. MCDONALD and R.A. DEREMEE: Ocular complications of Wegener's granulomatosis. Ophthalmology 90 (1983) 279–290.

CARRINGTON, C.B. and A.A. LIEBOW: Limited forms of angiitis and granulomatosis of Wegener's type. Am. J. Med. 41 (1966) 497–527.

CHALK, C.H., P.J. DYCK and D.L. CONN: Vasculitic neuropathy. In: P.J. Dyck and P.K. Thomas (Eds.), Peripheral Neuropathy. Philadelphia, PA, W.B. Saunders (1993a) 1424–1436.

CHALK, C.H., H.A. HOMBURGER and P.J. DYCK: Anti-neutrophil cytoplasmic antibodies in vasculitic peripheral neuropathy. Neurology 43 (1993b) 1826–1827.

COGAN, D.G.: Corneoscleral lesions in periarteritis nodosa and Wegener's granulomatosis. Trans. Am. Ophthalmol. Soc. 53 (1955) 321–342.

CUPPS, T.R.: Vasculitis masquerading as vasculitis. Mayo Clin. Proc. 68 (1993) 194–196.

DAVENPORT, A.: 'False positive' perinuclear and cytoplasmic antineutrophil cytoplasmic antibody results leading to misdiagnosis of Wegener's granulomatosis and/or microscopic polyarteritis. Clin. Nephrol. 37 (1992) 124–130.

DEREMEE, R.A.: The nosology of Wegener's granulomatosis utilizing the ELK format augmented by c-ANCA. In: W.L. Gross (Ed.), Associated Vasculitides: Immunological Clinical Aspects. New York, Plenum Press (1993) 209–215.

DEREMEE, R.A., T.J. MCDONALD, E.G. HARRISON, JR. and D.T. COLES: Wegener's granulomatosis: anatomic correlates, a proposed classification. Mayo Clin. Proc. 51 (1976) 777–781.

DEREMEE, R.A., T.J. MCDONALD and L.H. WEILAND: Wegener's granulomatosis: observations on treatment with antimicrobial agents. Mayo Clin. Proc. 60 (1985) 27–32.

DICKEY, W. and W.J. ANDREWS: Wegener's granulomatosis presenting as peripheral neuropathy: diagnosis confirmed by serum anti-neutrophil antibodies. J. Neurol. Neurosurg. Psychiatry 53 (1990) 269–270.

DRACHMAN, D.D.: Neurological complications of Wegener's granulomatosis. Arch. Neurol. 8 (1963) 145–155.

DUNCKER, G., B. NOLLE, R. ASMUS, H. KOLTZE and R. ROCHELS: Orbital involvement' in Wegener's granulomatosis. In: W.L. Gross (Ed.), ANCA-Associated Vasculitides: Immunological and Clinical Aspects. New York, Plenum Press (1993) 315–317.

DYCK, P.J., D.L. CONN and H. OKAZAKI: Necrotizing angiopathic neuropathy: 3-dimensional morphology of fiber degeneration related to sites of occluded vessels. Mayo Clin. Proc. 47 (1972) 461–475.

FALK, R.J. and J.C. JENNETTE: Antineutrophil cytoplasmic antibodies with specificity for myeloperoxidase in patients with systemic vasculitis and idiopathic necrotizing and crescentic glomerulonephritis. N. Engl. J. Med. 318 (1988) 1651–1657.

FALK, R.J., S. HOGAN, T.S. CAREY and J.C. JENNETTE: The clinical course of patients with antineutrophil cytoplasmic autoantibody associated glomerulonephritis cytoplasmic autoantibody associated glomerulonephritis and systemic vasculitis. Ann. Intern. Med. 113 (1990) 656–663.

FAUCI, A.S. and S.M. WOLFF: Wegener's granulomatosis: pathology and a review of the literature. Medicine (Baltimore) 52 (1973) 535–561.

FAUCI, A.S., B.F. HAYNES, P. KATZ and S.M. WOLFF: Wegener's granulomatosis: prospective clinical and therapeutic experience with 85 patients for 21 years. Ann. Intern. Med. 98 (1983) 76–85.

FINKELMAN, R., T. MUNSAT, H. MANDELL, L. ADELMAN and E. LOGIGIAN: Neuromuscular manifestations of Wegener's granulomatosis: a case report. Neurology 43 (1993) 617–618.

FRED, H.L., E.C. LYNCH, S.D. GREENBERG and A. GONZALEZ-AGULO: A patient with Wegener's granulomatosis exhibiting unusual clinical and morphologic features. Am. J. Med. 37 (1964) 311–319.

GALPERIN, C. and G.S. HOFFMAN: Antineutrophil cytoplasmic antibodies in Wegener's granulomatosis and other diseases: clinical issues. Cleve. Clin. J. Med. 61 (1994) 416–427.

GEORGI, J., M. ULMER and W.L. GROSS: Co-trimoxazol bei Wegenerscher Granulomatose – eine prospektive Studie. Immunität Infektion 19 (1991) 97–98.

GODMAN, G.C. and J. CHURG: Wegener's granulomatosis: pathology and review of the literature. Arch. Pathol. 58 (1954) 533–553.

GOLDBERG, A.L., A.L. TIEVSKY and S. JAMSHIDI: Wegener granulomatosis invading the cavernous sinus: a CT demonstration. J. Comput. Assist. Tomogr. 7 (1983) 701–703.

GOLDSCHMEDING, R., C.E. VAN DER SCHOOT, D. TEN BOKKEL HUININK, C.E. HACK, M.E. VAN DEN ENDE, C.G. KALLENBERG and A.E. VON DEM BORNE: Wegener's granulomatosis antibodies identify a novel diisopropylphluorophosphate-binding protein in the lysosomes of normal human neutrophils. J. Clin. Invest. 84 (1989) 1577–1587.

HAYNES, B.F. and A.S. FAUCI: Diabetes insipidus associated with Wegener's granulomatosis successfully treated with cyclophosphamide. N. Engl. J. Med. 299 (1978) 764.

HAYNES, B.F., M.L. FISHMAN, A.S. FAUCI and S.M. WOLFF: The ocular manifestations of Wegener's granulomatosis. Fifteen years experience and review of the literature. Am. J. Med. 63 (1977) 131–141.

HOFFMAN, G.S.: Treating Wegener's granulomatosis: how far have we come? Cleve. Clin. J. Med. 61 (1994) 414–415.

HOFFMAN, G.S., J.M.G. SECHLER, J.I. GALLIN, J.H. SHELHAMER, A. SUFFREDINI, F.P. OGNIBENE, R.J. BALTARO, T.A. FLEISHER, R.Y. LEAVITT, W.D. TRAVIS, M.F. BARILE, M. TSOKOS, R.P. HOLMAN, S.E. STRAUS and A.S. FAUCI: Bronchoalveolar lavage analysis in Wegener's granulomatosis. Am. Rev. Resp. Dis. 143 (1991) 401–407.

HOFFMAN, G.S., G.S. KERR, R.Y. LEAVITT, C.W. HALLAHAN, R.S. LEBOVICS, W.D. TRAVIS, M. ROTTEM and A.S. FAUCI: Wegener's granulomatosis: an analysis of 158 patients. Ann. Intern. Med. 116 (1992a) 488–498.

HOFFMAN, G.S., R.Y. LEAVITT, G.S. KERR and A.S. FAUCI: Treatment of Wegener's granulomatosis with methotrexate. Arthr. Rheum. 35 (1992b) 1322–1329.

HURST, N.P., N.A. DUNN and T.M. CHALMERS: Wegener's granulomatosis complicated by diabetes. Ann. Rheum. Dis. 42 (1983) 600–601.

JENNETTE, J.C., B.H. EWERT and R.J. FALK: Do antineutrophil cytoplasmic autoantibodies cause Wegener's granulomatosis and other forms of necrotizing vasculitis? Rheum. Dis. Clin. North Am. 19 (1993) 1–14.

KERR, G.S., T.A. FLEISHER, C.W. HALLAHAN, R.Y. LEAVITT, A.S. FAUCI and G.S. HOFFMAN: Limited prognostic value of anti-neutrophil cytoplasmic antibody in patients with Wegener's granulomatosis. Arthr. Rheum. 36 (1993) 365–371.

KLINGER, H.: Granzformen der Periarteritis nodosa. Frankf. Z. Pathol. 42 (1931) 455–480.

KRONEMAN, O.C. and M. PEVZNER: Failure of cyclophosphamide to prevent cerebritis in Wegener's granulomatosis. Am. J. Med. 80 (1986) 526–527.

KURLAND, L.T., T.Y. CHUANG and G.H. HUNDER: The epidemiology of systemic arteritis. In: R.C. Lawrence and L.E. Shulman (Eds.), Epidemiology of the Rheumatic Diseases. New York, Gower (1984) 196.

LEAVITT, R.Y., A.S. FAUCI, D.A. BLOCH ET AL.: The American College of Rheumatology 1990 criteria for the classification of Wegener's granulomatosis. Arthr. Rheum. 33 (1990) 1101–1107.

LIE, J.T.: Systemic and isolated vasculitis: a rational approach to classification and pathologic diagnosis. Pathol. Ann. 24 (1989) 25–114.

LIE, J.T.: Illustrated histopathologic classification criteria for selected vasculitis syndromes. Arthr. Rheum. 33 (1990) 1074–1087.

LUCAS, F.V., S.P. BENJAMIN and M.C. STEINBERG: Cerebral vasculitis in Wegener's granulomatosis. Cleve. Clin. Quart. 43 (1976) 275–281.

MACFAYDEN, D.J.: Wegener's granulomatosis with discrete lung lesions and peripheral neuritis. Can. Med. Assoc. J. 83 (1960) 760–764.

MARTENS, J.: Spinal cord involvement in Wegener's granulomatosis. Clin. Rheumatol. 1 (1982) 221.

MCCAFFREY, T.V., T.J. MCDONALD, G.W. FACER and R.A. DEREMEE: Otologic manifestations of Wegener's granulomatosis. Otolaryngol. Head Neck Surg. 88 (1980) 586–593.

MICKLE, J.P., J.E. MCLENNAN, J.G. CHI and C.W. LIDDEN: Cortical vein thrombosis in Wegener's granulomatosis (Case report). J. Neurosurg. 46 (1977) 248–251.

MILLER, K.S. and J.M. MILLER: Wegener's granulomatosis presenting as a primary seizure disorder with brain lesions demonstrated by magnetic resonance imaging. Chest 103 (1993) 316–318.

MONTECUCCO, C., R. CAPORALI, C. PACCHETTI and M. TURLA: Is Tolosa–Hunt syndrome a limited form of Wegener's granulomatosis? Report of two cases with antineutrophil cytoplasmic antibodies. Br. Soc. Rheumatol. 32 (1993) 640–641.

MOORE, P.M. and T.R. CUPPS: Neurological complications of vasculitis. Ann. Neurol. 14 (1983) 155–167.

MOORE, P.M. and A.S. FAUCI: Neurologic manifestations of systemic vasculitis. A retrospective and prospective study of the clinicopathologic features and responses to therapy in 25 patients. Am. J. Med. 71 (1981) 517–524.

MURTY, G.E.: Wegener's granulomatosis: otorhinolaryngological manifestations. Clin. Otolaryngol. 15 (1990) 385–393.

NILES, J.L., R.T. MCCLUSKEY, M.F. AHMAD and M.A. ARNAOUT: Wegener's granulomatosis autoantigen is a novel serine proteinase. Blood 74 (1989) 1888–1893.

NISHINO, H. and F.A. RUBINO: Horner's syndrome in Wegener's granulomatosis: report of four cases. J. Neurol. Neurosurg. Psychiatry 56 (1993) 897–899.

NISHINO, H., F.A. RUBINO, R.A. DEREMEE, J.W. SWANSON and J.E. PARISI: Neurological involvement in Wegener's granulomatosis: an analysis of 324 consecutive patients at the Mayo Clinic. Ann. Neurol. 33 (1993a) 4–9.

NISHINO, H., F.A. RUBINO and J.E. PARISI: The spectrum of neurologic involvement in Wegener's granulomatosis. Neurology 43 (1993b) 1334–1337.

NISHINO, H., R.A. DEREMEE, F.A. RUBINO and J.E. PARISI: Wegener's granulomatosis associated with vasculitis of the temporal artery: report of five cases. Mayo Clin. Proc. 68 (1993c) 115–121.

NOLLE, B., U. SPECKS, J. LUDEMAN, M.S. ROHRBACH, R.A. DEREMEE and W.L. GROSS: Anticytoplasmic autoantibodies: their immunodiagnostic value in Wegener's granulomatosis. Ann. Intern. Med. 111 (1989) 28–40.

ROSETE, A., A.R. CABRAL, A. KRAUS and D. ALARÇON-SEGOVIA: Diabetes insipidus secondary to Wegener's granulomatosis: report and review of the literature. J. Rheumatol. 18 (1991) 761–765.

SATOH, J., N. MIYASAKA, T. YAMADA, T. NISHIDO, M. OKUDA, T. KUROIWA and R. SHIMOKAWA: Extensive cerebral infarction due to involvement of both anterior cerebral arteries by Wegener's granulomatosis. Ann. Rheum. Dis. 47 (1988) 606–611.

SMALL, P. and M.L. BRISSON: Wegener's granulomatosis presenting as temporal arteritis. Arthr. Rheum. 34 (1991) 220–223.

SPECKS, U., C.L. WHEATLEY, T.J. MCDONALD, M.S. ROHRBACH and R.A. DEREMEE: Anticytoplasmic autoantibodies in the diagnosis and follow-up of Wegener's granulomatosis. Mayo Clin. Proc. 64 (1989) 28–36.

STEGEMAN, C.A., J.W. COHEN TERVAERT, P.E. DEJONG and C.G.M. KALLENBERG: Trimethoprim-sulfamethoxazole (cotrimoxazole) for the prevention of relapses of Wegener's granulomatosis. N. Engl. J. Med 335 (1996) 16–20.

STERN, G.M., A.V. HOFFBRAND and H. URICH: The peripheral nerves and skeletal muscles in Wegener's granulomatosis: a clinico-pathological study of four cases. Brain 88 (1965) 151–164.

TUHY, J.E., G.L. MAURICE and N.R. NILES: Wegener's granulomatosis. Am. J. Med. 25 (1958) 638–646.

VAN DER WOUDE, F.J., S. LOBATTO, H. PERMIN, M. VAN DER GIESSEN, N. RASMUSSEN and A. WIIK: Autoantibodies against neutrophils and monocytes: tool for diagnosis and marker of disease activity in Wegener's granulomatosis. Lancet i (1985) 425–429.

WALTON, E.W.: Giant cell granuloma of the respiratory tract (Wegener's granulomatosis). Br. Med. J. 2 (1958) 265–270.

WEGENER, F.: Über generalisierte, septische Efaberkrankungen. Verh. Dtsch. Ges. Pathol. 29 (1936) 202–210.

WEGENER, F.: Über eine eigenartige rhinogene Granulomatose mit besonderer Beteiligung des Arterien Systems und der Nieren. Beitr. Pathol. Anat. Allg. Pathol. (1939) 36–68.

WEINBERGER, L.M., M.L. COHEN, B.F. REMLER, M.H. NAHEEDY and R.J. LEIGH: Intracranial Wegener's granulomatosis. Neurology 43 (1993) 1831–1834.

Handbook of Clinical Neurology, Vol. 27 (71): Systemic Diseases, Part III
M.J. Aminoff and C.G. Goetz, editors

CHAPTER 10

Giant cell arteritis, polymyalgia rheumatica and Takayasu arteritis

YUEN T. SO

Department of Neurology, Oregon Health Sciences University, Portland, OR, U.S.A.

Vasculitis encompasses a complex spectrum of disorders with diverse clinical manifestations. For the most part, the etiology is unknown, and its pathogenesis is incompletely understood. The classification of vasculitis is largely based on the clinical and histopathologic features. The primary or idiopathic vasculitides are first separated from those associated with systemic diseases such as rheumatoid arthritis, systemic lupus erythematosus and the infectious illnesses. The primary vasculitides are further differentiated on the basis of the caliber of the involved blood vessels, the histopathologic features and the associated clinical features. Despite considerable overlap in clinical manifestations and few pathognomonic clinical or histopathologic findings, the current system permits one to make a satisfactory classification of individual vasculitic syndromes (Arend et al. 1990; Bloch et al. 1990; Hunder et al. 1990).

This chapter discusses two vasculitic disorders, giant cell arteritis and Takayasu arteritis. Both affect predominantly large and medium-sized arteries, although arterioles, veins and venules are involved to a smaller extent. Granulomatous angiitis of the central nervous system affects blood vessels of similar caliber. The three syndromes are differentiated on the basis of the age predilection and the anatomic distribution of vasculitic involvement.

GIANT CELL ARTERITIS

References to this syndrome may be found as early as Ali Ibn Isa's treatise in tenth century Bagdad. Hutchinson provided the first modern clinical report in 1889, describing an elderly man with bilaterally inflamed and swollen temporal arteries that progressed to pulseless and presumably thrombosed cords. Almost half a century passed before Horton et al. established the disease as a distinct nosologic entity in 1932. They described the salient clinical features as well as the first biopsy findings in two patients. The disease was named temporal arteritis on the basis of the focal nature of the symptomatology. Subsequent investigators recognized quickly that blood vessels other than the temporal arteries may be involved. The name giant cell arteritis was coined to emphasize the distinctive histopathology. Other terms, such as cranial arteritis, arteritis of the aged and Horton's disease, had also been used but did not gain widespread acceptance. Despite the fact that giant cells are not invariably present on biopsy in a given patient, giant cell arteritis is the name most commonly used in recent literature and will be used throughout this chapter.

Epidemiology

Even though it is rare before the age of 50, giant cell arteritis is probably the commonest systemic arteritis in individuals of European descent. Only a few case reports describe its occurrence in patients of African heritage. About two-thirds of the cases are seen in patients 70 years or older. Women have an approximately three-fold increase in risk compared to men. A population-based estimate from Olmsted County, MN suggests an incidence of 17 per 100,000 population of age 50 or older (Machado et al. 1988). The Olmsted County incidence is among the highest reported. It is similar to those reported from northern Europe and Iceland (Baldursson et al. 1994), but is considerably higher than those of other locales, such as Shelby County, TN (1.58/100,000) (Smith et al. 1983) and Israel (0.49/100,000) (Friedman et al. 1982). Likely explanations for the discrepancies include differences in methodology of case ascertainment, differences in health care, as well as unknown environmental or genetic factors that may influence the incidence.

A number of familial occurrences of giant cell arteritis have been reported suggesting a possible genetic influence (Mathewson and Hunder 1986). Like other autoimmune diseases, it has been associated with several HLA haplotypes such as HLA-DR4 (Hunder et al. 1977). Some authors have described nearly simultaneous occurrence of the disorder in husband and wife (Galetta et al. 1990) and in family members living as neighbors. Although these anecdotal reports may reflect an environmental factor, they may simply be a result of chance occurrence.

Pattern of vasculitic involvement

An appreciation of the pattern of vascular involvement is essential to understand the clinical manifestations of giant cell arteritis. For all practical purposes, the disease is a generalized vasculitis and not just temporal arteritis. Postmortem studies showed a characteristic pattern of medium and large arteries involvement.

There is a high frequency of arteritis in the aortic arch and its branches, including the carotid and vertebral arteries. The superficial temporal, ophthalmic, vertebral and posterior ciliary arteries are frequently affected (Wilkinson and Russell 1972). Vertebral artery involvement is almost exclusively extracranial with a sharply defined border about 5 mm beyond the point of dural perforation. Carotid artery involvement

is also usually extracranial with the siphon being the commonest site of disease (Gilmour 1941; Wilkinson and Russell 1972; Howard et al. 1984; Vincent and Vincent 1986). Other vessels, such as the aorta, coronary, mesenteric, renal, iliac and femoral arteries, may also be affected by arteritis. The vasculitic topography in giant cell arteritis correlates with the presence of elastin in blood vessels. It is possible that elastin or modification of elastin in the course of aging somehow plays an antigenic role in the pathogenesis of this disease.

Clinical manifestations

Virtually any portion of the neuraxis may be affected in giant cell arteritis. Depending on the site of inflammatory or vasculitic involvement, a remarkably diverse spectrum of systemic and neurologic manifestations are possible in patients. However, in most patients, the characteristic and often unique pattern of vascular lesions gives rise to a remarkably stereotypic syndrome. Some symptoms and signs are so prevalent in giant cell arteritis that the complete absence of them would cast serious doubt on the diagnosis (Table 1).

Constitutional symptoms
Fever, weight loss, fatigue, malaise and anorexia are present in the majority of patients with giant cell arteritis. Weight loss is present in 55–71% of patients

TABLE 1

Common clinical and neurologic manifestations of giant cell arteritis.

Very common but non-specific
Headache
Weight loss, malaise or fever
Common, specific for giant cell arteritis when the symptom is typical
Polymyalgia rheumatica
Jaw claudication
Temporal artery tenderness
Non-specific and less common, indistinguishable from similar symptoms seen in other neurologic diseases
Visual loss
Diplopia
Stroke
Neuropsychiatric impairment
Neuropathy

(Russell 1959; Huston et al. 1978). Intermittent fever is seen in about 20–83% (Russell 1959; Paulley and Hughes 1960; Huston et al. 1978; Goodman 1979). In a literature review of 111 patients aged 65 or older with fever of unknown origin, giant cell arteritis accounted for 18 cases (Esposito and Gleckman 1978). This disorder is therefore an important cause of unexplained fever in the elderly (Ghose 1976). The constitutional symptoms underscore the systemic nature of the disease. In rare occasions, these symptoms may precede the more specific manifestations of giant cell arteritis. Since malignancy, infection, thyroid disease and depression all mimic some aspects of the disorder, diagnostic confusion is common especially early in the disease course.

Polymyalgia rheumatica
The term refers to a sensation of stiffness and achiness experienced primarily over the axial skeleton and the proximal aspects of the limbs. It may occur as part of the syndrome of giant cell arteritis or it may present in isolation without clinical evidence of arteritis (see below). It is present in about half of all patients in most large series and is one of the most prevalent symptoms in giant cell arteritis (Russell 1959; Paulley and Hughes 1960; Huston et al. 1978; Goodman 1979; Hunder et al. 1990).

The onset of polymyalgia rheumatica is typically insidious although rarely it may be acute. The pain and stiffness affect muscles of the neck and torso, shoulder and hip girdles, and less commonly proximal portion of the arms and thighs (Chuang et al. 1982). The clinical involvement is invariably bilateral, although symptoms may be asymmetric especially in the early stage of the disease (Hamrin 1972). The pain is often severe and functionally limiting, and commonly causes sleep disturbances. Stiffness is typically worse on arising in the morning and after periods of inactivity and improves slowly with activities. Motion of the axial skeleton or the affected limbs is often limited. This limitation seems to be primarily a consequence of pain, as a full range of motion is often possible with slow passive movements.

Careful examination may reveal mild tenderness and swelling of joints as well as evidence of synovitis (Chuang et al. 1982; Chou and Schumacher 1984). The knees and sternoclavicular joints are frequently affected. Some patients may also experience symptoms in the distal joints such as wrists, hands, elbows and ankles. An inflammatory joint effusion may be present (Chou and Schumacher 1984) but, as a rule, there is no clinical or radiological evidence of joint deformity. Polymyalgia rheumatica is sometimes difficult to distinguish from rheumatoid arthritis. The infrequent metacarpal-phalangeal involvement, the predilection for the proximal joints and the non-deforming nature of the synovitis and arthritis favor a diagnosis of polymyalgia rheumatica.

Headache
About 59–90% of patients complain of headache (Huston et al. 1978; Goodman 1979; Hunder et al. 1990; Berlit 1992). Headache is frequently the presenting complaint. The pain may be particularly severe along the superficial temporal arteries. The involved arteries may be indurated and tender. The scalp away from the inflamed vessels is often tender as well, perhaps as a consequence of ischemia to the skin and subcutaneous tissues. Pressure from a pillow, combing hair or other stimulation of the scalp or face is often uncomfortable (Russell 1959).

The classic pattern of unilateral or bilateral temporal pain is only present in about half of all patients with headache. In others, the pain may be generalized, or it may be localized over the frontal, occipital or other cranial regions (Paulley and Hughes 1960; Seymour 1987). More unusual locations of pain such as lingual or otalgic discomfort may also be seen. The diffuse pattern reflects involvement of other vessels, such as the occipital, facial and lingual arteries. The character of headache may be boring, burning, lacinating, throbbing, sharp or dull (Russell 1959; Seymour 1987). The headache history may be further complicated by pre-existing migraine or tension headache. Thus, both the location and the character of head pain are of limited diagnostic utility. In a patient older than 50 years, any new headache or recent changes in a previously stable headache should arouse the clinician's suspicion.

Jaw claudication
About one-third to two-thirds of patients complain of pain, weakness or tiredness of jaw muscles with repeated chewing or prolonged talking (Russell 1959; Huston et al. 1978; Hunder et al. 1990). These are likely to be ischemic symptoms secondary to vasculitis and thrombosis of the facial arteries. Jaw

claudication may be mistaken for diseases of the temporomandibular joint. True claudication appears during use of the muscles and disappears promptly with rest (Hamilton et al. 1971). When the characteristic history is obtained, jaw claudication is highly specific for the diagnosis of giant cell arteritis. Similar claudication occurs less commonly in the tongue and pharyngeal muscles with prolonged deglutination and chewing. If left untreated, loss of taste ensues, followed by beefy red enlargement or even necrotic ulceration of the tongue.

Visual loss

Blindness is the most dreaded complication of giant cell arteritis. In Goodman's review of 819 patients from 14 series published between 1946 and 1978, the incidence varied widely from 7 to 60% (Goodman 1979). Overall, 36% suffered from blindness with the highest incidence reported by the early series. In another series of 166 patients diagnosed between 1981 and 1983 with biopsy-proven temporal arteritis, 10% had amaurosis fugax and only 8% had permanent blindness (Caselli et al. 1988b). The trend of declining incidence of blindness is likely due to earlier recognition and treatment of this disorder.

Anterior ischemic optic neuropathy (AION) accounts for 80–90% of patients with ocular complications (Table 2) (Turner et al. 1974; Liu et al. 1994). In most cases, vision deteriorates suddenly

and painlessly (Liu et al. 1994). The blindness may not be noticed by patients until the contralateral eye is covered (Russell 1959). In about 10–20% of cases, transient reversible blindness or amaurosis fugax precedes permanent visual loss (Russell 1959; Hollenhorst et al. 1960; Huston et al. 1978; Liu et al. 1994). In others, visual symptoms may begin as premonitory floaters or scotoma and then progress to blindness over a few days. Bilateral visual loss occurs in about one-third to one-half of patients and may be either sequential or simultaneous (Beri et al. 1987; Liu et al. 1994). Blindness of the contralateral eye typically appears within 1–2 weeks of visual loss of the first eye. Prompt recognition and initiation of treatment are therefore critical.

The impairment of vision is typically severe. About 20% of affected eyes have no light perception, and another 50% have visual acuity of 20/200 or worse (Liu et al. 1994). Color perception may be reduced. Afferent pupillary defect is common. In eyes testable with perimetry, central, altitudinal and arcuate scotomas are the predominant patterns of field defects. Fundoscopic examination in these cases shows pallid optic disk swelling and flame-shaped hemorrhages (Keltner 1982; Liu et al. 1994). Less commonly, the disk may be normal suggesting a retrobulbar location of ischemia (Cohen and Damaske 1975; Hayreh 1981), or the retina may be the primary site of ischemia (Cohen and Damaske

TABLE 2

Causes of blindness in giant cell arteritis.

Types of visual loss	Vascular anatomy	Funduscopic findings	Other clinical findings
Anterior ischemic optic neuropathy (AION)	Posterior ciliary arteries	Acute, swollen disks; chronic, optic atrophy	Central, altitudinal, or arcuate scotomas
Prechiasmal or chiasmal optic neuropathy	Anterior cerebral, anterior or posterior communicating arteries	Normal disk	Same as AION
Branch retinal artery occlusion	Central retinal artery	Retinal edema, cherry red spot	Visual loss does not respect vertical or horizontal meridians
Anterior segment ischemia	Long posterior and anterior ciliary arteries	Normal fundus	Corneal edema Keratopathy Chemosis Non-reactive pupil Chronic: cataracts, rubeosis iridis
Cortical ischemia	Intracranial arteries or emboli from vertebral arteries	Normal fundus	Hemianopic field defects Spares pupillary reactivity

1975). In the latter, a visible opacity of the retina due to the presence of edema is discernible 24–48 h after occlusion of the central retinal artery. The fovea retains its normal color since it is devoid of overlying neurons, and the color contrast gives rise to the characteristic appearance of 'cherry red spot'.

Giant cell arteritis is an uncommon cause of AION, accounting for 5–20% of all cases (Guyer et al. 1985; Bastiaensen et al. 1986; Beri et al. 1987). When compared to non-arteritic AION, patients with arteritis as a group are older, are more often female, and have more severe and more frequently bilateral visual loss (Beri et al. 1987). There is no reliable fundoscopic means to differentiate arteritic from non-arteritic AION. The fundoscopic appearance is also similar to that of optic neuritis. The diagnosis has to be made on the basis of other clinical accompaniments.

Ischemia to central visual pathways may also give rise to visual deficits under rare circumstances. Cortical blindness from infarction of the occipital cortex is the best recognized syndrome (Chisholm 1975). This may be a result of either intracranial arteritis or artery-to-artery embolism from damaged extracranial vessels (Butt et al. 1991). Another cause of acute visual loss is ischemia to the iris, ciliary body and other adjacent structures (Kelly et al. 1987). This condition is rare since the anterior segment of the optic globe is supplied dually by the posterior and anterior ciliary arteries. In well established cases, anterior segment ischemia leads to corneal clouding, striate keratopathy, flare in the anterior chambers and an irregular non-reactive pupil.

Other ophthalmic manifestations
Diplopia or ophthalmoparesis occurs in 10–15% of patients (Russell 1959; Huston et al. 1978). Careful examination of patients with diplopia usually reveals ocular motility abnormalities. Diplopia may be the sole complaint at presentation (Dimant et al. 1980). Ptosis sometimes accompanies ophthalmoparesis suggesting oculomotor nerve involvement. Like diabetic third-nerve palsy, the pupil is generally spared. On the other hand, unlike diabetic third-nerve palsy, the ophthalmoparesis in giant cell arteritis may fluctuate daily (Dimant et al. 1980). Recovery is usually complete and may occur even without treatment.

Diplopia is potentially due to involvement of either the cranial nerves or the extraocular muscles. The former is probably more common. Either oculomotor or abducens nerve palsy may be evident on neurologic examination (Russell 1959; Sibony and Lessell 1984). A neurogenic cause is further suggested by several reported cases of oculomotor synkinesis (Barricks et al. 1977; Sibony and Lessell 1984). An example of synkinesis is lid retraction with attempted downgaze or adduction and is probably caused by misdirected regeneration of oculomotor nerve fibers after previous denervation of lid muscles. Ischemia of eye movement muscles is uncommon and, when present, is often accompanied by chemosis, proptosis and lid swelling (Barricks et al. 1977).

Central nervous system complications
Cerebrovascular disease, including strokes, transient ischemic attacks and rarely cerebral hemorrhage, have been reported in approximately 7% of patients (Hollenhorst et al. 1960; Caselli and Hunder 1993). It is a major cause of death in patients with giant cell arteritis. The disease may affect the internal carotid artery (Russell 1959; Cull 1979; Howard et al. 1984; Vincent and Vincent 1986), the vertebrobasilar circulation (Crompton 1959; Wilkinson and Russell 1972; Monteiro et al. 1984; Säve-Söderbergh et al. 1986; McLean et al. 1993; Sheehan et al. 1993), or the circle of Willis (Wilkinson and Russell 1972). Depending on the localization of the ischemia, a combination of aphasia, hemiparesis, hemisensory loss, ataxia, hemianopia or cortical blindness may be present. 'Top of the basilar' and lateral medullary syndromes are also well described. The ratio of carotid to vertebrobasilar ischemic events is about 2:1 (Caselli et al. 1988b). Since the proportion of vertebrobasilar ischemia in the general population is much lower, giant cell arteritis may have a predilection for the posterior circulation.

Acute or subacute cerebral symptoms such as depression, agitation, confusion or cognitive impairments are also recognized complications (Goodman 1979; Caselli 1990). Mental symptoms are frequently present without signs of focal ischemia. Cognitive decline sometimes manifests during the taper of corticosteroid treatment of giant cell arteritis and may be the only symptom of relapse of arteritis (Caselli 1990). The exact prevalence is difficult to ascertain as cognitive decline may be overlooked in the elderly population, and the frequent coexistence of fever may complicate mental status evaluation. Depression and mental confusion were present in 20 and 11% of patients respectively in Russell's series of 35 patients (Russell 1959). In the Mayo Clinic series, 3% of patients had neuropsychiatric symptoms (Caselli et al. 1988b). Prompt institution of treatment may lead to at least partial recovery of mental function. Giant cell arteritis is therefore an important though uncommon cause of treatable dementia.

A diverse spectrum of other neurologic symptoms and signs has been described in association with giant cell arteritis. For the most part, they are a consequence of focal ischemic or inflammatory lesions in various parts of the central nervous system (Table 3). These manifestations are rare and are not discussed further.

Peripheral neuropathies

In Caselli's series, peripheral neuropathy was diagnosed in 23 of 166 (14%) patients (Caselli et al. 1988a,b). Mononeuropathies, either simplex or multiplex, are the best recognized neuropathic disorders. The disease may affect any of the major peripheral nerves. Reports have documented lesions of the median (Russell 1959; Warrell et al. 1968; Caselli et al. 1988a), radial (Warrell et al. 1968; Mulcahy et al. 1984), peroneal (Russell 1959; Meadows 1966; Caselli et al. 1988a) and sciatic nerves (Massey and Weed 1978), as well as the brachial plexus and the spinal roots (Meadows 1966; Sanchez et al. 1983; Caselli et al. 1988a). Symptoms and signs of neuropathy typically develop in the presence of other systemic and neurologic symptoms. The neuropathic deficits sometimes appear suddenly, suggesting an ischemic or vascular etiology (Russell

TABLE 3

Unusual neurologic manifestations of giant cell arteritis.

Anosmia (Schon 1988; Berlit 1992)
Parkinsonism (Santambrogio et al. 1989)
Myelopathy (Caselli et al. 1988b)
Nuchal rigidity (Paulley and Hughes 1960)
SIADH (Gentric et al. 1988)
Tongue numbness or gangrene (Arnung and Nielsen 1979; Caselli et al. 1988b)
Tremor (Caselli et al. 1988b)
Trismus (Desser 1969; Manganelli et al. 1992)
Vertigo, tinnitus and deafness (Russell 1959; Paulley and Hughes 1960)
Vertebral artery dissection (Sheehan et al. 1993)

1959; Mulcahy et al. 1984). Other patients have a subacute or chronic evolution of symptoms. A commonly associated mononeuropathy is a median neuropathy localized to the wrist. A causal relationship cannot be inferred as some of these cases may represent carpal tunnel entrapment unrelated to giant cell arteritis.

Half of the patients with neuropathy in the Mayo Clinic series (Caselli et al. 1988a) had a generalized polyneuropathy. This presumably represents a confluent mononeuropathy multiplex, as is seen in other vasculitides or rheumatologic disorders. The available reports provide only a sketchy description of its clinical features. Evolution of neuropathic symptoms may be chronic or subacute. Nerve conduction studies show both axonal loss and conduction slowing. Electromyographic examination may reveal fibrillation potentials most prominently in the distal muscles (Caselli et al. 1988a).

Other systemic manifestations

Aside from jaw and tongue claudication, patients may experience claudication or even gangrene of the limbs (Hamilton et al. 1971; Klein et al. 1975; Huston et al. 1978; Caselli et al. 1988a). Bruits and diminished peripheral pulses are encountered occasionally on examination. Symptoms and signs sometimes improve with corticosteroid therapy. Arteritis may also affect the aorta and may present dramatically as an unsuspected cause of aortic dissection or aneurysmal rupture (Paulley and Hughes 1960; Harris 1968; Klein et al. 1975; Säve-Söderbergh et al. 1986). Myocardial infarction from arteritis of the coronary arteries is uncommon but well recognized. Renal artery involvement is usually limited to the finding of erythrocyte casts in the urine, but may rarely result in renal insufficiency (Klein et al. 1975). Respiratory symptoms such as cough, sore throat and hoarseness may occur in as many as 9% of patients with giant cell arteritis (Larson et al. 1984). Little is known about the origin of these symptoms although they are likely due to ischemia or inflammation of the laryngeal and pharyngeal tissues. Respiratory symptoms usually resolve promptly with corticosteroid treatment.

Histopathology

Pathologic confirmation of giant cell arteritis is obtained most frequently from temporal artery biopsy. The histologic appearance of giant cell arteritis is highly variable. About half of the abnormal biopsies show the classic picture of granulomatous inflammation with scattered giant cells. Inflammatory infiltrates are typically seen at the intima-media junction of the vessel wall and extend through the media into the adventitia. Giant cells are not found in the other 50% of abnormal samples (Lie 1990). Instead, there is a predominantly lymphocytic panarteritis infiltrated with some neutrophils and eosinophils. The type of histopathologic appearance does not depend on the duration of disease or the erythrocyte sedimentation rate. In all biopsy samples, partial occlusion of the arterial lumen is common. The internal elastic lamella of the arterial wall is frequently fragmentated, a finding that may persist long after the active phase of the disease.

A short course of corticosteroid treatment probably reduces the diagnostic yield of temporal artery biopsy only slightly if at all. In one study, the incidence of abnormal biopsy fell from 80% in patients who did not receive steroid to about 60% in those who were treated for less than 1 week prior to biopsy (Lie 1987). In other studies, the rate of positive findings did not change when compared to biopsies taken before treatment (Chmelewski et al. 1992; Achkar et al. 1994). Biopsy may show vasculitis even after 2 weeks of corticosteroid treatment. Another potential cause of false negative biopsy is inadequate sampling of the temporal artery. The inflammatory lesions may be focal and segmental even in a clinically symptomatic artery (Albert et al. 1976; Klein et al. 1976). Isolated foci of arteritis interspersed with normal arteries ('skip lesions') are present in about one-third of patients. Foci of arteritis as short as 330 μm in length have been described (Klein et al. 1976).

Laboratory findings

Normal values for erythrocyte sedimentation rate (ESR) increase with age (Hayes and Stinson 1976). The upper normal limit in subjects aged 50 or older is about 30–40 mm/h (Kulvin 1972; Hayes and Stinson 1976). In patients with active arteritis, the ESR is almost always elevated, and the degree of elevation is usually marked. In Russell's (1959) and Hollenhorst et al.'s (1960) series, the ESR was 50 mm/h or more (Westergren method) in 89 and 97% of patients, and

it exceeded 80 mm/h in 50 and 78% of patients respectively. All 42 patients in a Mayo Clinic series had elevated ESR, with a median of 96 mm/h (Huston et al. 1978). The mean ESR was 95 mm/h in a more recent series of 19 patients (Berlit 1992). Even when extremely elevated, the ESR is entirely non-specific. In large series of patients with ESR of 100 mm/h or more, only 2–3 % had evidence of arteritis (Zarcharski and Kyle 1967; Payne 1968). The majority of patients had neoplastic, infectious or other inflammatory diseases.

Although the ESR is without question a useful diagnostic test, caution is needed in the interpretation of normal or mildly elevated values. Of 28 patients with ESR reported by Russell (1959), one had a rate of 27 and two had rates between 30 and 49 mm/h. Five of 31 patients with biopsy-proven giant cell arteritis had ESR of 40 mm/h or less (Hedges et al. 1983). In another review of 80 patients with giant cell arteritis or polymyalgia rheumatica, 18 (22.5%) initially had an ESR of less than 30 mm/h (Ellis and Ralston 1983). Four of these 80 patients experienced visual loss and had ESR only in the range of 11–40 mm/h. Furthermore, it is likely that published studies overestimate the diagnostic sensitivity of ESR. Since an elevated ESR has long been recognized as an essential feature of giant cell arteritis, a clinician may not consider the diagnosis or pursue a temporal artery biopsy in a patient with normal ESR. In the absence of elevated ESR, even disabling complications like blindness may be attributed to other diseases. Thus, patients with normal ESR are likely to be under-represented in retrospective reviews. There are many reports of normal ESR in patients with otherwise typical giant cell arteritis (Kansu et al. 1977; Biller et al. 1982; Papadakis and Schwartz 1986; Wong and Korn 1986; Jundt and Mock 1991; Wise et al. 1991). A particularly common setting is a patient treated with low-dose corticosteroids for a previous history of polymyalgia rheumatica (Papadakis and Schwartz 1986; Wise et al. 1991). Low-dose corticosteroid probably suppresses ESR even when it is insufficient to control the other manifestations of arteritis.

A mild normochromic or hypochromic anemia is present in the majority of patients (Russell 1959; Hollenhorst et al. 1960; Esposito and Gleckman 1978; Huston et al. 1978; Berlit 1992). The hemoglobin concentration is usually in the range of 9–12 g/dl. Red blood cell volume may be mildly decreased. The anemia is resistant to iron therapy and improves only with corticosteroid treatment. Platelet count may be increased. About one-third of patients have a mild leukocytosis (Huston et al. 1978). Elevated transaminase and alkaline phosphatase levels are common. Serum albumin may be decreased. Serum protein electrophoresis may reveal an increased level of α_2-globulin. The serum levels of other acute phase reactants such as C-reactive protein, fibrinogen and haptoglobin rise during active disease and return to normal after treatment (Andersson et al. 1986a).

Diagnostic considerations

Immunosuppressive treatment in an elderly population carries a significant risk. For this reason, histologic confirmation is often needed to guide decision making regarding long-term therapy. Biopsy is usually obtained from the temporal artery. Rarely, the occipital or facial artery may be more prominently affected and may be biopsied. Due to the possibility of skip lesions, the biopsy should remove an adequate length of the artery, 2–3 cm at a minimum and 4–6 cm if the artery is clinically normal. Moreover, the specimens need to be examined at 5-mm intervals. The morbidity of the surgery is low, although scalp ischemia may rarely occur. It is common practice in some centers to obtain a biopsy on the contralateral side if the frozen-section examination does not reveal vasculitis (Hall et al. 1983). Bilateral biopsy probably increases the diagnostic yield only slightly. Out of 87 patients with bilateral biopsies, seven showed vasculitis only on the second biopsy and 80 had normal biopsy on both sides (Klein et al. 1976).

Only about one-third of all patients who undergo temporal artery biopsy have positive findings (Klein et al. 1976; Hall et al. 1983; Hedges et al. 1983; Roth et al. 1984; Chmelewski et al. 1992). If only patients with typical clinical features of giant cell arteritis are considered, the yield of positive biopsy improves to over 80% (Hall et al. 1983; Hedges et al. 1983; Berlit 1992). The concurrent presence of new-onset headache, abnormal temporal artery on examination and jaw claudication in particular is highly predictive of a positive biopsy (Vilaseca et al. 1987). A negative temporal artery biopsy reduces the likelihood of giant cell arteritis considerably. In one study, only 8 of 88

patients with negative biopsy were eventually diagnosed to have the disease over a median follow-up period of 70 months (Hall et al. 1983). Hedges et al. (1983) reported a similarly low false-negative rate; 3 of their 63 patients with negative biopsy eventually developed giant cell arteritis after a mean follow-up of 4.8 years. The most common alternative diagnosis was polymyalgia rheumatica without arteritis (Hall et al. 1983), a related condition that does not have the risk of blindness and other serious neurologic complications (Table 4). Important alternative diagnoses to consider are non-inflammatory neurologic diseases such as migraine headaches, cerebrovascular diseases, connective tissue disorders especially rheumatoid arthritis, and various occult malignancies.

In 1990, the American College of Rheumatology (ACR) compared the clinical findings in 214 patients with giant cell arteritis to 593 patients with other forms of vasculitis (Hunder et al. 1990) and developed empirical criteria for diagnosis of the disease. One scheme used a set of five diagnostic criteria (Table 5). The presence of three or more criteria provided a sensitivity of 93.5% and a specificity of 91.2% (Hunder et al. 1990). Another scheme employed a classification tree using six criteria (Table 5), again achieving sensitivity and specificity of over 90%. Abnormal tenderness over the temporal artery or the scalp is particularly useful to separate giant cell arteritis from other vasculitides. It is important to recognize the values and limitations of the ACR study. In the absence of a perfect criterion for the diagnosis of giant cell arteritis, the selection of cases and controls in the ACR study was based on the expert opinion of subcommittee members (Bloch et al. 1990). Therefore, sensitivity and specificity should not be interpreted in the traditional manner when an independent gold standard is available to evaluate a test or a set of diagnostic criteria. The criteria used in the ACR schemes (Table 5) in essence reflect consensus opinion. Furthermore, the study was designed to test only the distinction of giant cell arteritis from other vasculitides. Its general applicability needs independent confirmation. Nevertheless, the study highlights the distinctive clinical features of giant cell arteritis. It is distinguishable from other vasculitic disorders with reasonable diagnostic certainty.

Present data suggest that temporal artery biopsy has a high diagnostic sensitivity of 80–90%. Either a

TABLE 4

Ultimate diagnosis in 263 patients with negative temporal artery biopsy (Hall et al. 1983; Hedges et al. 1983; Roth et al. 1984; Chmelewski et al. 1992).

	No. (%)
Polymyalgia rheumatica or giant cell arteritis*	67 (25)
Atherosclerotic or embolic diseases	45 (17)
Connective tissue or inflammatory diseases	29 (11)–36 (14)**
Non-arteritic optic neuropathy	30 (11)
Malignancy	25 (9)
Headache	20 (8)
Diabetes mellitus	18 (7)
Infection	13 (5)

*The majority of patients probably had polymyalgia rheumatica. The exact number of patients with giant cell arteritis could not be determined from two references (Roth et al. 1984; Chmelewski et al. 1992).
**The uncertainty is due to different methods of categorization used by investigators.

TABLE 5

American College of Rheumatology criteria for classification of giant cell arteritis (Hunder et al. 1990).

Traditional format – a patient is classifed as having giant cell arteritis if at least three of the following five criteria are met:
  1. Development of symptoms or findings at age 50 or older;
  2. New headache;
  3. Temporal artery tenderness to palpation or decreased pulsation unrelated to arteriosclerosis;
  4. Artery biopsy specimen showing vasculitis with either mononuclear cell predominance or granulomatous inflammation usually with multinucleated giant cells;
  5. ESR $\geqslant$ 50 mm/h (Westergren).

Classification tree – criteria 1–4 above, plus the following (see Hunder et al. 1990):
  5. Tender areas or nodules over the scalp away from arteries;
  6. Claudication of jaw, tongue or swallowing.

positive or negative biopsy therefore contributes significantly to the clinical management in patients (Hall et al. 1983; Vilaseca et al. 1987). As mentioned earlier, histological confirmation of the diagnosis of vasculitis is important in patients susceptible to side effects of corticosteroids. These include obese or elderly patients and those with diabetes or osteoporosis. Temporal artery biopsy is also useful when the pre-biopsy probability of giant cell arteritis is low (Nadeau 1988). If a biopsy is negative in these patients without the typical clinical presentation, it is reasonable to search for alternative diagnosis and withhold treatment during a period of close monitoring. On the other hand, in patients with the typical clinical features, the false negative rate of biopsy is unacceptably high (Nadeau 1988). These patients need to be treated even when histological confirmation is not possible.

Treatment and prognosis

Corticosteroids have been in use since the early 1950s and have remained the drug of choice. Most of the symptoms of giant cell arteritis respond to 10 or 20 mg of prednisone, but larger doses are probably necessary to eliminate the risk of blindness. Most authors suggest the equivalent of 40–80 mg of prednisone as an initial treatment. As soon as a presumptive diagnosis is made, treatment should be instituted without waiting for confirmation from a temporal artery biopsy. With treatment, dramatic diminution of headache, stiffness, myalgia and jaw claudication may occur within 24–72 h. ESR begins to fall within days of treatment and normalizes within 2 weeks in about 50% and within 4 weeks in over 70% of patients (Andersson et al. 1986a). Despite the rapid resolution of symptoms and signs, it is important to realize that active arteritis persists through the first few weeks of steroid treatment. For instance, the percentage of positive temporal artery biopsy was unaffected by 2 or more weeks of steroid treatment (Chmelewski et al. 1992). This is also illustrated by a case report of a patient who died of an unrelated cause after 6 weeks of successful steroid treatment. Despite prompt and complete normalization of symptoms and ESR, persistent arteritis was demonstrable at autopsy (Evans et al. 1994). A treating physician should therefore resist the temptation to lower the steroid dosage too quickly.

Despite the lack of a controlled clinical trial, there is wide acceptance of the efficacy of steroid. Birkhead et al. (1957) provided the earliest comprehensive review of treatment experience. When compared with historical controls from the pre-steroid era, the patients treated in the early 1950s had a lower incidence of blindness. Reported cases of visual loss occurred typically before initiation of treatment; blindness is uncommon once a patient is under treatment (Cohen 1973; Chmelewski et al. 1992; Myles et al. 1992; Aiello et al. 1993; Liu et al. 1994). If visual loss is to appear after starting steroids, it usually occurs under one of the following circumstances: within a few days of starting treatment, non-compliance with medication, or daily dosages of 20 mg or less of prednisone (Hollenhorst et al. 1960; Meadows 1966; Cohen 1973; Huston et al. 1978; Liu et al. 1994).

There is no consensus on the starting dose of corticosteroid. Daily prednisone doses of 20 mg may be sufficient in many patients (Delecoeuillerie et al. 1988; Lundberg and Hedfors 1990; Myles et al. 1992), but there are rare reports of blindness in patients receiving 40–60 mg of prednisone per day (Rosenfeld et al. 1986; Liu et al. 1994). Certainly in patients who present with threatening visual or neurologic symptoms, the use of high-dose corticosteroids (the equivalence of 40 mg or more of prednisone) is warranted. Some authors even advocate intravenous methylprednisolone at about 1000 mg/day, as it offers greater bioavailability over oral prednisone (Model 1978; Rosenfeld et al. 1986; Matzkin et al. 1992; Liu et al. 1994).

Although some authors suggested that visual loss is largely irreversible (Birkhead et al. 1957; Meadows 1966), the prognosis is not as grim as was once believed. On the basis of a literature review, Schneider et al. (1971) observed that about 15% of treated patients had improvement in vision. Subsequent series of treated patients reported improvement in 15–34% (Turner et al. 1974; Aiello et al. 1993; Liu et al. 1994). Prognosis is dependent on the severity of visual deficits. Eyes with no light perception rarely recover (Meadows 1966; Wilkinson and Russell 1972; Liu et al. 1994). In a series of 41 treated patients, improvement was seen in none of 13 eyes with no light perception, 2 of 15 with perception of light or hand motion, and 17 of 35 with vision of counting fingers or better (Liu et al. 1994).

There is no consensus on the optimal long-term management of giant cell arteritis. Despite prompt response to steroids, mild symptoms persist in most patients, confounding the withdrawal of treatment. The incidence of steroid-related side effects is high, largely due to the elderly age of this population. There are no sound data to allow risk-to-benefit consideration in treatment design. Some authors suggested switching to alternate-day prednisone to minimize side effects (Andersson et al. 1986b). Others found alternate-day prednisone inadequate to suppress symptoms (Hunder et al. 1975). The course and duration of the disease vary considerably among different centers. Most patients seen at the Mayo Clinic were not taking steroids after 1 year (Hunder et al. 1975). In contrast, 76% of 49 patients reported by Berlit (1992) were still taking steroids at an average follow-up of 3.4 years. In another study, 43% of patients were on steroids after 5 years (Andersson et al. 1986b). The relapse rate also varies from 26% (Hunder et al. 1975) to 50% or over (Andersson et al. 1986b; Berlit 1992). Most relapses occur in the first year of disease. About half of the relapses appear within the first month and almost all occur within 1 year of cessation of treatment. There has been limited experience with the use of steroid-sparing agents such as azathioprine, methotrexate, cyclophosphamide and dapsone. Until further data are available, steroids remain the only proven treatment.

We recommend a minimum of 40–60 mg of oral prednisone for at least the first month in patients with neurologic or ocular symptoms. Low-dose prednisone such as 20 mg/day may be used in patients with primarily polymyalgia rheumatica or constitutional symptoms. In severe cases such as impending visual loss or central nervous system ischemia, 250–500 mg of methylprednisolone may be given intravenously 2–

4 times/day. This should be continued for 3–5 days before switching to conventional oral steroid therapy. Once symptoms and ESR are satisfactorily controlled for 1 month or more, steroid dosage may be tapered slowly. It is generally prudent to bring the daily prednisone dosage below 20 mg by the third or fourth month. Patients who relapse frequently develop polymyalgia rheumatica. Tell-tale symptoms of polymyalgia rheumatica should be monitored closely along with complaints such as headache and jaw claudication. An elevated ESR signals a relapse but it is not always present. Relapses can be treated with a short course of high-dose oral prednisone or intravenous methylprednisolone, followed by rapid taper to a dose slightly higher than that used prior to relapse.

POLYMYALGIA RHEUMATICA

Polymyalgia rheumatica is a clinical syndrome characterized by a systemic inflammatory state along with proximal aching and stiffness. It is a frequent manifestation of giant cell arteritis, but it also occurs frequently without arteritis. In the absence of arteritis, the clinical syndrome is non-specific and objective physical findings are seldom impressive. Because of the ambiguity in case definition, the epidemiologic data available are not as accurate as those on giant cell arteritis. Nevertheless, it is clear that polymyalgia rheumatica is very common. Its incidence increases with age. It rarely occurs before the age of 50. The estimated annual incidence is 13–20/100,000 in persons 50–59 years old, and 78–135/100,000 in persons 70–79 years of age (Chuang et al. 1982; Salvarani et al. 1995; Schaufelberger et al. 1995).

The symptoms and signs of polymyalgia rheumatica have been discussed in connection with giant cell arteritis. Aside from the stiffness and aching in the torso and proximal limbs, patients may experience malaise, weight loss or low-grade fever. ESR is elevated in most but not all patients. Electromyographic examination of proximal muscles may be normal, or may reveal either acute denervation (Bromberg et al. 1990) or changes suggestive of a myopathy (Buchthal 1970). It is evident that these clinical features and laboratory findings are not specific. As discussed earlier, the clinical symptoms may be difficult to distinguish from early rheumatoid arthritis. The clinician needs to exclude other mimicking conditions such as inflammatory myopathies, fibromyalgia, depression, as well as those listed in Table 4.

The pathogenesis of polymyalgia rheumatica is unknown. Although some symptoms may be attributed to a synovitis, not all patients have evidence of synovitis (Chou and Schumacher 1984; Al-Hussaini and Swannell 1985; Kyle et al. 1990). There should not be any symptoms or physical findings of vasculitis and end-organ ischemia. Patients with polymyalgia rheumatica and clinical evidence of vasculitis have giant cell arteritis by definition. This distinction between polymyalgia rheumatica and giant cell arteritis can usually be made on the basis of clinical presentation, occasionally with validation from temporal artery biopsy.

There is no uniform agreement on whether to perform temporal artery biopsy on patients with no symptoms attributable to arteritis. Several factors suggest that the majority of patients with polymyalgia rheumatica can be managed without temporal artery biopsy. Firstly, the diagnostic yield of biopsy is low compared to patients with symptomatic giant cell arteritis. Most studies suggest that about 15–20% of patients with polymyalgia rheumatica will develop biopsy-proven arteritis (Kogstad 1965; Wilske and Healey 1967; Miller and Stevens 1978; Chuang et al. 1982), although higher estimates have been reported (Fauchald et al. 1972). Secondly, the clinical management of polymyalgia rheumatica and giant cell arteritis is similar. Finally, the overall risk of serious vascular events is low regardless of the result of temporal artery biopsy, and this risk is further lowered with prompt initiation of steroid treatment.

Most patients with polymyalgia rheumatica respond quickly, usually within 1 week, to 10–20 mg of oral prednisone (Ayoub et al. 1985; Delecoeuillerie et al. 1988; Kyle and Hazleman 1989; Lundberg and Hedfors 1990). The duration of treatment is unpredictable. Relapse is common with attempts to taper the dose. Many patients require at least low-dose corticosteroids for over 1 year. Prolonged therapy longer than 4 years is not uncommon (Ayoub et al. 1985). Although almost all patients do well, they must be monitored carefully for development of symptoms of arteritis. In rare instances, vascular complications of giant cell arteritis develop in patients on low-dose corticosteroids despite apparent control of polymyalgia rheumatica and ESR (Papadakis and Schwartz 1986; Wise et al. 1991).

Takayasu described the syndrome that bears his name in 1908, although there were earlier descriptions in the 19th century. The disease is a chronic arteritis of unknown cause that affects large vessels, particularly the aorta, its main branches, and the pulmonary artery. It has also been known by a variety of names including pulseless disease, reverse coarctation, aortic arch syndrome and occlusive thromboaortopathy, all in reference to the severe stenotic lesions in the carotid and subclavian circulation.

The earliest and the largest series of patients came from Japan, China and other Asian countries (Nakao et al. 1967; Ishikawa 1978; Zheng et al. 1990). Despite increasing recognition in Africa, Europe and North America, the disorder is rare outside Asia. Statistics on its prevalence are scarce. The only epidemiologic data in North America provided an incidence rate of 2.6/million, but the figure was based on only three cases occurring over 13 years in Olmsted County, MN (Hall et al. 1985). In two series totaling 92 patients seen either at the Mayo Clinic or the National Institutes of Health, Caucasians comprised 74% and Asians 10% (Hall et al. 1985; Kerr et al. 1994). The disorder is characteristically a disease of young women. The female:male ratio is about 9:1. The majority of patients present in their second to fourth decades (Arend et al. 1990). About one-third of the patients are under the age of 20 (Arend et al. 1990; Kerr et al. 1994).

Clinical manifestations

Conceptually, it is useful to emphasize two stages of the disease process: an inflammatory stage followed by a pulseless or sclerotic stage (Nasu 1975; Hall et al. 1985). In any given patient, however, the clinical course is highly variable, and the two stages frequently overlap. The initial inflammatory stage is characterized by non-specific symptoms of fever, myalgias, arthralgias and weight loss. Many patients are anemic, and most have accelerated ESR. The pathology in the involved vessels consists of lymphocytic and granulomatous inflammation. The disease slowly progresses to the pulseless stage during which multiple stenotic lesions and saccular aneurysms develop. Audible bruits are present over the abdominal aorta or the carotid arteries in 80–94%

of patients (Hall et al. 1985; Kerr et al. 1994). Pulses may be lost over any of the major arteries in the neck and the limbs. Absence of at least one arterial pulse is demonstrable in the vast majority of patients.

End-organ involvement results from the extensive and severe vascular stenosis. Over half of the patients have hypertension. The disorder is responsible for about two-thirds of all renovascular hypertension in some Asian countries (Chugh and Sakhuja 1992). Retinal vessel anastomoses resulting from chronic arterial stenosis may be evident on fundoscopic examination. Palpitation, dyspnea and angina pectoris may result from systemic and pulmonary hypertension, aortic incompetence, and secondary left- or right-heart failure. Claudication of the limbs is common and affects the upper limbs more frequently than the lower (Hall et al. 1985; Arend et al. 1990; Kerr et al. 1994).

The most common neurologic symptom is headache which occurs in about half of all patients (Cupps and Fauci 1981; Kerr et al. 1994). Most other neurologic complications are attributable to cerebrovascular diseases. Postural lightheadedness, dizziness or even syncope occur in about one-third of patients (Hall et al. 1985; Kerr et al. 1994). These symptoms reflect the severity of the occlusive disease, especially that of the vertebral arteries (Kerr et al. 1994). Transient ischemic attack, stroke, cerebral hemorrhage, subclavian steal syndrome and hypertensive encephalopathy occur infrequently in less than 10% of patients (Cupps and Fauci 1981; Kerr et al. 1994). Despite its relative infrequency, cerebrovascular disease is the most important cause of serious morbidity or death in Takayasu arteritis (Ishikawa 1978; Rose and Sinclair-Smith 1980). Another well-recognized though uncommon neurologic complication is the subclavian steal syndrome. When the subclavian artery proximal to the origin of the vertebral artery is severely stenotic, exercise of a limb may induce diversion of perfusion from the posterior circulation (Yoneda et al. 1977; Hennerici et al. 1988).

Angiography is the most useful diagnostic procedure. It is often necessary to fully evaluate the aortic arch, the brachiocephalic branches, the superior mesenteric and renal arteries. Almost all patients have arterial abnormalities at multiple sites. The angiographic findings include irregularities of the vessel walls, focal stenosis and aneurysms. Similar abnormalities may be visualized with magnetic resonance imaging, particularly in the aortic long-axis images (Yamada et al. 1993). In patients with symptomatic subclavian steal syndrome, Doppler studies may show reversal of blood flow in the vertebral arteries.

Laboratory findings

ESR is elevated in 45–83% of patients (Hall et al. 1985; Ishikawa 1988; Arend et al. 1990; Chugh and Sakhuja 1992). Mild anemia is seen in about half of the patients. Chest radiographs may reveal widening of the mediasternum, calcification or other irregularities of the aorta, or pulmonary arterial changes. Serum renin level may be elevated in patients with renovascular hypertension.

Treatment and prognosis

In many patients, prednisone in the range of 30–100 mg/day results in prompt improvement in systemic inflammatory symptoms and normalization of the ESR (Fraga et al. 1972; Hall et al. 1985; Ishikawa 1991; Kerr et al. 1994). Arterial pulses may return after several months in about half of the patients. Some patients need an additional cytotoxic agent, either because of unresponsiveness to corticosteroid or inability to taper the steroid once the disease is controlled. Adjunctive therapies using low-dose cyclophosphamide of 0.5–2 mg/kg/day or methotrexate of 0.3 mg/kg/week have been used with apparent success in small open-label studies (Shelhamer et al. 1985; Hoffman et al. 1994). In patients with fixed stenotic lesions, angioplasty or surgical intervention are necessary to restore blood flow. The prognosis for long-term survival is excellent, with reported 5-year survival ranging from 91 to 97% (Hall et al. 1985; Subramanyan et al. 1989; Hoffman et al. 1994).

REFERENCES

ACHKAR, A.A., J.T. LIE, G.G. HUNDER, W.M. O'FALLON and S.E. GABRIEL: How does previous corticosteroid treatment affect the biopsy findings in giant cell (temporal) arteritis? Ann. Intern. Med. 120 (1994) 987–992.

AIELLO, P.D., J.C. TRAUTMANN, T.J. MCPHEE, A.R. KUNSELMAN and G.G. HUNDER: Visual prognosis in giant cell arteritis. Ophthalmology 100 (1993) 550–555.

AL-HUSSAINI, A.S. and A.J. SWANNELL: Peripheral joint involvement in polymyalgia rheumatica: a clinical study of 56 cases. Br. J. Rheumatol. 24 (1985) 27–30.

ALBERT, D.M., M.C. RUCHMAN and J.L. KELTNER: Skip areas in temporal arteritis. Arch. Ophthalmol. 94 (1976) 2072–2077.

ANDERSSON, R., B.E. MALMVALL and B.A. BENGTSSON: Acute phase reactants in the initial phase of giant cell arteritis. Acta Med. Scand. 220 (1986a) 365–367.

ANDERSSON, R., B.E. MALMVALL and B.A. BENGTSSON: Long-term corticosteroid treatment in giant cell arteritis. Acta Med. Scand. 220 (1986b) 465–469.

AREND, W.P., B.A. MICHEL, D.A. BLOCH, G.G. HUNDER, L.H. CALABRESE, S.M. EDWORTHY, A.S. FAUCI, R.Y. LEAVITT, J.T. LIE, R.W. LIGHTFOOT, JR. ET AL.: The American College of Rheumatology 1990 criteria for the classification of Takayasu arteritis. Arthr. Rheum. 33 (1990) 1129–1134.

ARNUNG, K. and I.L. NIELSEN: Temporal arteritis and gangrene of the tongue. Acta Med. Scand. 206 (1979) 239–240.

AYOUB, W.T., C.M. FRANKLIN and D. TORRETTI: Polymyalgia rheumatica. Duration of therapy and long-term outcome. Am. J. Med. 79 (1985) 309–315.

BALDURSSON, O., K. STEINSSON, J. BJORNSSON and J.T. LIE: Giant cell arteritis in Iceland. Arthr. Rheum. 37 (1994) 1007–1012.

BARRICKS, M.E., D.B. TRAVIESA, J.S. GLASER and I.S. LEVY: Ophthalmoplegia in cranial arteritis. Brain 100 (1977) 209–221.

BASTIAENSEN, L.A.K., R.W.M. KEUNEN, C.C. TIJSSEN and J.J. VANDONINCK: Anterior ischemic optic neuropathy: sense and nonsense in diagnosis and treatment. Doc. Ophthalmol. 61 (1986) 205–210.

BERI, M., M.R. KLUGMAN, J.A. KOHLER and S.S. HAYREH: Anterior ischemic optic neuropathy. VII. Incidence of bilaterality and various influencing factors. Ophthalmology 94 (1987) 1020–1028.

BERLIT, P.: Clinical and laboratory findings with giant cell arteritis. J. Neurol. Sci. 111 (1992) 1–12.

BILLER, J., J. ASCONAPE, M.E. WEINBLATT and J. TOOLE: Temporal arteritis associated with normal sedimentation rate. J. Am. Med. Assoc. 247 (1982) 486–487.

BIRKHEAD, N.C., H.P. WAGENER and R.M. SCHICK: Treatment of temporal arteritis with adrenal corticosteroids. J. Am. Med. Assoc. 163 (1957) 821–827.

BLOCH, D.A., B.A. MICHEL, G.G. HUNDER, D.J. MCSHANE, W.P. AREND, L.H. CALABRESE, S.M. EDWORTHY, A.S. FAUCI, J.F. FRIES, R.Y. LEAVITT, J.T. LIE, R.W. LIGHTFOOT, A.T. MASI, J.A. MILLS, M.B. STEVENS, S.L. WALLACE and N.J. ZVAIFLER: The American College of Rheumatology 1990 criteria for the classification of vasculitis. Patients and methods. Arthr. Rheum. 33 (1990) 1068–1073.

BROMBERG, M.B., P.D. DONOFRIO and B.M. SEGAL: Steroid-responsive electromyographic abnormalities in polymyalgia rheumatica. Muscle Nerve 13 (1990) 138–141.

BUCHTHAL, F.: Electrophysiologic abnormalities in metabolic myopathies and neuropathies. Acta Neurol. Scand. Suppl. 43 (1970) 129–176.

BUTT, Z., J.F. CULLEN and E. MUTLUKAN: Pattern of arterial involvement of the head, neck, and eyes in giant cell arteritis: three case reports. Br. J. Ophthalmol. 75 (1991) 368–371.

CASELLI, R.J.: Giant cell (temporal) arteritis: a treatable cause of multi-infarct dementia. Neurology 40 (1990) 753–755.

CASELLI, R.J. and G.G. HUNDER: Neurologic aspects of giant cell (temporal) arteritis. Rheum. Dis. Clin. N. Am. 19 (1993) 941–953.

CASELLI, R.J., J.R. DAUBE, G.G. HUNDER and J.P. WHISNANT: Peripheral neuropathic syndromes in giant cell (temporal) arteritis. Neurology 38 (1988a) 685–689.

CASELLI, R.J., G.G. HUNDER and J.P. WHISNANT: Neurologic disease in biopsy-proven giant cell (temporal) arteritis. Neurology 38 (1988b) 352–359.

CHISHOLM, H.: Cortical blindness in cranial arteritis. Br. J. Ophthalmol. 59 (1975) 332–333.

CHMELEWSKI, W.L., K.M. MCKNIGHT, C.A. AGUDELO and C.M. WISE: Presenting features and outcomes in patients undergoing temporal artery biopsy. A review of 98 patients. Arch. Intern. Med. 152 (1992) 1690–1695.

CHOU, C.T. and H.R. SCHUMACHER, JR.: Clinical and pathologic studies of synovitis in polymyalgia rheumatica. Arthr. Rheum. 27 (1984) 1107–1117.

CHUANG, T.Y., G.G. HUNDER, D.M. ILSTRUP and L.T. KURLAND: Polymyalgia rheumatica: a 10-year epidemiologic and clinical study. Ann. Intern. Med. 97 (1982) 672–680.

CHUGH, K.S. and V. SAKHUJA: Takayasu's arteritis as a cause of renovascular hypertension in Asian countries. Am. J. Nephrol. 12 (1992) 1–8.

COHEN, D.N.: Temporal arteritis. Improvement in visual prognosis and management with repeat biopsies. Trans. Am. Acad. Ophthalmol. Otolaryngol. 77 (1973) 74–85.

COHEN, D.N. and M.M. DAMASKE: Temporal arteritis: a spectrum of ophthalmic complications. Ann. Ophthalmol. 7 (1975) 1045–1054.

CROMPTON, M.R.: The visual changes in temporal (giant-cell) arteritis: report of a case with autopsy findings. Brain 82 (1959) 377–390.

CULL, R.E.: Internal carotid artery occlusion caused by giant cell arteritis. J. Neurol. Neurosurg. Psychiatry 42 (1979) 1066–1067.

CUPPS, T.R. and A.S. FAUCI. The Vasculitides. Philadelphia, PA, W.B. Saunders (1981).

DELECOEUILLERIE, G., P. JOLY, A. COHEN DE LARA and J.B. PAOLAGGI: Polymyalgia rheumatica and temporal arteritis: a retrospective analysis of prognostic features and different corticosteroid regimens (11 year survey of 210 patients). Ann. Rheum. Dis. 47 (1988) 733–739.

DESSER, E.J.: Miosis, trismus, and dysphagia. An unusual presentation of temporal arteritis. Ann. Intern. Med. 71 (1969) 961–962.

DIMANT, J., D. GROB and N.G. BRUNNER: Ophthalmoplegia, ptosis, and miosis in temporal arteritis. Neurology 30 (1980) 1054–1058.

ELLIS, M.E. and S. RALSTON: The ESR in the diagnosis and management of the polymyalgia rheumatica/giant cell arteritis syndrome. Ann. Rheum. Dis. 42 (1983) 168–170.

ESPOSITO, A.L. and R.A. GLECKMAN: Fever of unknown origin in the elderly. J. Am. Geriatr. Soc. 26 (1978) 498–505.

EVANS, J.M., K.P. BATTS and G.G. HUNDER: Persistent giant cell arteritis despite corticosteroid treatment. Mayo Clin. Proc. 69 (1994) 1060–1061.

FAUCHALD, P., O. RYGVOLD and B. OYSTESE: Temporal arteritis and polymyalgia rheumatica: clinical and biopsy findings. Ann. Intern. Med. 77 (1972) 845–852.

FRAGA, A., G. MINTZ, L. VALLE and G. FLORES-IZQUIERGO: Takayasu's arteritis: frequency of systemic manifestations and favorable response to maintenance steroid therapy with adrenocorticosteroids. Arthr. Rheum. 15 (1972) 617–624.

FRIEDMAN, G., B. FRIEDMAN and J. BENBASSAT: Epidemiology of temporal arteritis in Israel. Isr. J. Med. Sci. 18 (1982) 241–244.

GALETTA, S.L., E.C. RAPS, A.E. WULC, M.G. FARBER, G.L. PLOCK, C.W. NICHOLS and H.M. FRIEDMAN: Conjugal temporal arteritis. Neurology 40 (1990) 1839–1842.

GENTRIC, A., E. BACCINO, D. MOTTIER, S. ISLAM and J. CLEDES: Temporal arteritis revealed by a syndrome of inappropriate secretion of antidiuretic hormone. Am. J. Med. 85 (1988) 559–560.

GHOSE, M.K.: Arteritis of the aged (giant cell arteritis) and fever of unexplained origin. Am. J. Med. 60 (1976) 429–436.

GILMOUR, J.R.: Giant-cell chronic arteritis. J. Pathol. 53 (1941) 263–277.

GOODMAN, B.W.: Temporal arteritis. Am. J. Med. 67 (1979) 839–852.

GUYER, D.R., N.R. MILLER, C.L. AUER and S.L. FINE: The risk of cerebrovascular and cardiovascular disease in patients with anterior ischemic optic neuropathy. Arch. Ophthalmol. 103 (1985) 1136–1142.

HALL, S., S. PERSELLIN, J.T. LIE, P.C. O'BRIEN, L.T. KURLAND and G.G. HUNDER: The therapeutic impact of temporal artery biopsy. Lancet ii (1983) 1217–1220.

HALL, S., W. BARR, J.T. LIE, A.W. STANSON, F.J. KAZMIER and G.G. HUNDER: Takayasu arteritis. A study of 32 North American patients. Medicine (Baltimore) 64 (1985) 89–99.

HAMILTON, C.R.S., W.M. SHELLEY and P.A. TUMULTY: Giant

cell arteritis. Including temporal arteritis and polymyalgia rheumatica. Medicine (Baltimore) 50 (1971) 1–27.

HAMRIN, B.: Polymyalgia arteritica with morphological changes in the large arteries. Acta Med. Scand. Suppl. 533 (1972) 4–164.

HARRIS, M.: Dissecting aneurysm of the aorta due to giant cell arteritis. Br. Heart J. 30 (1968) 840–844.

HAYES, G.S. and I.N. STINSON: Erythrocyte sedimentation rate and age. Arch. Ophthalmol. 94 (1976) 939–940.

HAYREH, S.S.: Posterior ischemic optic neuropathy. Ophthalmologica 182 (1981) 29–41.

HEDGES, T.R.D., G.L. GIEGER and D.M. ALBERT: The clinical value of negative temporal artery biopsy specimens. Arch. Ophthalmol. 101 (1983) 1251–1254.

HENNERICI, M., C. KLEMM and W. RAUTENBERG: The subclavian steal phenomenon: a common vascular disorder with rare neurologic deficits. Neurology 38 (1988) 669–673.

HOFFMAN, G.S., R.Y. LEAVITT, G.S. KERR, M. ROTTEM, M.C. SNELLER and A.S. FAUCI: Treatment of glucocorticoid-resistant or relapsing Takayasu arteritis with methotrexate. Arthr. Rheum. 37 (1994) 578–582.

HOLLENHORST, R.W., J.R. BROWN, H.P. WAGNER and R.M. SCHICK: Neurologic aspects of temporal arteritis. Neurology 10 (1960) 490–498.

HOWARD, G.F., S.U. HO, K.S. KIM and J. WALLACH: Bilateral carotid artery occlusion resulting from giant cell arteritis. Ann. Neurol. 15 (1984) 204–207.

HUNDER, G.G., S.G. SHEPS, G.L. ALLEN and J.W. JOYCE: Daily and alternate-day corticosteroid regimens in treatment of giant cell arteritis. Comparison in a prospective study. Ann. Intern. Med. 82 (1975) 613–618.

HUNDER, G.G., H.I. TASWELL, A.A. PINEDA and L.R. ELVEBACK: HLA antigens in patients with giant cell arteritis and polymyalgia rheumatica. J. Rheumatol. 4 (1977) 321–323.

HUNDER, G.G., D.A. BLOCH, B.A. MICHEL, M.B. STEVENS, W.P. AREND, L.H. CALABRESE, S.M. EDWORTHY, A.S. FAUCI, R.Y. LEAVITT, J.T. LIE ET AL.: The American College of Rheumatology 1990 criteria for the classification of giant cell arteritis. Arthr. Rheum. 33 (1990) 1122–1128.

HUSTON, K.A., G.G. HUNDER, J.T. LIE, R.H. KENNEDY and L.R. ELVEBACK: Temporal arteritis: a 25 year epidemiologic, clinical, and pathologic study. Ann. Intern. Med. 88 (1978) 162–167.

ISHIKAWA, K.: Natural history and classiciation of occlusive thromboaortopathy (Takayasu's disease). Circulation 57 (1978) 27–35.

ISHIKAWA, K.: Diagnostic approach and proposed criteria for the clinical diagnosis of Takayasu's arteriopathy. J. Am. Coll. Cardiol. 12 (1988) 964–972.

ISHIKAWA, K.: Effects of prednisolone therapy on arterial angiographic features in Takayasu's disease. Am. J. Cardiol. 68 (1991) 410–413.

JUNDT, J.W. and D. MOCK: Temporal arteritis with normal erythrocyte sedimentation rates presenting as occipital neuralgia. Arthr. Rheum. 34 (1991) 217–219.

KANSU, T., J.J. CORBETT, P. SAVINO and N.J. SCHATZ: Giant cell arteritis with normal sedimentation rate. Arch. Neurol. 34 (1977) 624–625.

KELLY, S.P., D.A. ROBERTSON and C.K. ROSTRON: Preventable blindness in giant cell arteritis. Br. Med. J. 294 (1987) 431–432.

KELTNER, J.L.: Giant-cell arteritis: signs and symptoms. Ophthalmology 89 (1982) 1101–1110.

KERR, G.S., C.W. HALLAHAN, J. GIORDANO, R.Y. LEAVITT, A.S. FAUCI, M. ROTTEM and G.S. HOFFMANN: Takayasu arteritis. Ann. Intern. Med. 120 (1994) 919–929.

KLEIN, R.G., G.G. HUNDER, A.W. STANSON and S.G. SHEPS: Large artery involvement in giant cell (temporal) arteritis. Ann. Intern. Med. 83 (1975) 806–812.

KLEIN, R.G., R.J. CAMPBELL, G.G. HUNDER and J.A. CARNEY: Skip lesions in temporal arteritis. Mayo Clin. Proc. 51 (1976) 504–510.

KOGSTAD, O.A.: Polymyalgia rheumatica. Acta Med. Scand. 187 (1965) 591–598.

KULVIN, S.M.: Erythrocyte sedimentation rates in the elderly. Arch. Ophthalmol. 88 (1972) 617–618.

KYLE, V. and B.L. HAZLEMAN: Treatment of polymyalgia rheumatica and giant cell arteritis. I. Steroid regimens in the first two months. Ann. Rheum. Dis. 48 (1989) 658–661.

KYLE, V., J. TUDOR, E.P. WRAIGHT, G.A. GRESHAM and B.L. HAZLEMAN: Rarity of synovitis in polymyalgia rheumatica. Ann. Rheum. Dis. 49 (1990) 155–157.

LARSON, T.S., S. HALL, N.G.G. HEPPER and G.G. HUNDER: Respiratory tract symptoms as a clue to giant cell arteritis. Ann. Intern. Med. 101 (1984) 594–597.

LIE, J.T.: The classification and diagnosis of vasculitis in large and medium-sized blood vessels. Pathol. Ann. 22 (1987) 125–162.

LIE, J.T.: Illustrated histopathologic classification criteria for selected vasculitis syndromes. American College of Rheumatology Subcommittee on Classification of Vasculitis. Arthr. Rheum. 33 (1990) 1074–1087.

LIE, J.T., D.D. FAILONI and D.C. DAVIS: Temporal arteritis with giant cell aortitis, coronary arteritis, and myocardial infarction. Arch. Pathol. Lab. Med. 110 (1986) 857–860.

LIU, G.T., J.S. GLASER, N.J. SCHATZ and J.L. SMITH: Visual morbidity in giant cell arteritis. Clinical characteristics and prognosis for vision. Ophthalmology 101 (1994) 1779–1785.

LUNDBERG, I. and E. HEDFORS: Restricted dose and duration of corticosteroid treatment in patients with polymyalgia rheumatica and temporal arteritis. J. Rheumatol. 17 (1990) 1340–1345.

MACHADO, E.B., C.J. MICHET, D.J. BALLARD, G.G. HUNDER, C.M. BEARD, C.P. CHU and W.M. O'FALLON: Trends in incidence and clinical presentation of temporal arteritis in Olmsted County, Minnesota, 1950–1985. Arthr. Rheum. 31 (1988) 745–749.

MANGANELLI, P., L. MALVEZZI and A. SAGINARIO: Trismus and facial swelling in a case of temporal arteritis. Clin. Exp. Rheumatol. 10 (1992) 102–103.

MASSEY, E.W. and T. WEED: Sciatic neuropathy with giant-cell arteritis. N. Engl. J. Med. 298 (1978) 917.

MATHEWSON, J.A. and G.G. HUNDER: Giant cell arteritis in two brothers. J. Rheumatol. 13 (1986) 190–192.

MATZKIN, D.C., T.L. SLAMOVITS, R. SACHS and R.M. BURDE: Visual recovery in two patients after intravenous methylprednisolone treatment of central retinal artery occlusion secondary to giant-cell arteritis. Ophthalmology 99 (1992) 68–71.

MCLEAN, C.A., M.F. GONZALES and J.P. DOWLING: Systemic giant cell arteritis and cerebellar infarction. Stroke 24 (1993) 899–902.

MEADOWS, S.P.: Temporal or giant cell arteritis. Proc. R. Soc. Med. 59 (1966) 329–333.

MILLER, L.D. and M.B. STEVENS: Skeletal manifestations of polymyalgia rheumatica. J. Am. Med. Assoc. 240 (1978) 27–29.

MODEL, D.G.: Reversal of blindness in temporal arteritis with methylprednisolone. Lancet i (1978) 340.

MONTEIRO, M.L., J.R. COPPETO and P. GRECO: Giant cell arteritis of the posterior cerebral circulation presenting with ataxia and ophthalmoplegia. Arch. Ophthalmol. 102 (1984) 407–409.

MULCAHY, F., L.D. JUBY and G.N. CHANDLER: Giant cell arteritis presenting with peripheral neuropathy. Postgrad. Med. J. 60 (1984) 670–671.

MYLES, A.B., T. PERERA and M.G. RIDLEY: Prevention of blindness in giant cell arteritis by corticosteroid treatment. Br. J. Rheumatol. 31 (1992) 103–105.

NADEAU, S.E.: Temporal arteritis: a decision-analytic approach to temporal artery biopsy. Acta Neurol. Scand. 78 (1988) 90–100.

NAKAO, K., M. IKEDA, S.I. KIMATA, H. NIITANI, M. MIYAHARA, Z.I. ISHIMI, K. HASHIBA, Y. TAKEDA, T. OZAWA, S. MATSU-SHITA and M. KURAMOCHI: Takayasu's arteritis: clinical report of eighty-four cases and immunological studies of seven cases. Circulation 35 (1967) 1141–1155.

NASU, Y.: Takayasu's trunkoarteritis in Japan. A statistical observation in 76 autopsy cases. Pathol. Microbiol. 43 (1975) 140–146.

PAPADAKIS, M.A. and N.D. SCHWARTZ: Temporal arteritis after normalization of erythrocyte sedimentation rate in polymyalgia rheumatica. Arch. Intern. Med. 146 (1986) 2283–2284.

PAULLEY, J.W. and J.P. HUGHES: Giant-cell arteritis, or arteritis of the aged. Br. Med. J. 2 (1960) 1562–1567.

PAYNE, R.W.: Causes of the grossly elevated sedimentation rate. Practitioner 200 (1968) 415–417.

ROSE, A.G. and C.C. SINCLAIR-SMITH: Takayasu's arteritis:

a study of 16 autopsy cases. Arch. Pathol. Lab. Med. 104 (1980) 231–237.

ROSENFELD, S.I., G.S. KOSMORSKY, T.G. KLINGELE and R.M. BURDE: Treatment of temporal arteritis with ocular involvement. Am. J. Med. 80 (1986) 143–145.

ROTH, A.M., L. MILSOW and J.L. KELTNER: The ultimate diagnoses of patients undergoing temporal artery biopsies. Arch. Ophthalmol. 102 (1984) 901–903.

RUSSELL, R.W.R.: Giant-cell arteritis. A review of 35 cases. Quart. J. Med. 28 (1959) 471–488.

SALVARANI, C., S.E. GABRIEL, W.M. O'FALLON and G.G. HUNDER: Epidemiology of polymyalgia rheumatica in Olmsted County, Minnesota, 1970–1991. Arthr. Rheum. 38 (1995) 369–373.

SANCHEZ, M.C., J. ARENILLAS, A. GUTIERREZ, J. ALONSO and J. ALVAREZ: Cervical radiculopathy: a rare symptom of giant cell arteritis. Arthr. Rheum. 26 (1983) 207–209.

SANTAMBROGIO, L., G. BELLOMO, M. MERCURI, G. ALAGIA and C. CIUFFETTI: Temporal arteritis presenting as an extrapyramidal disorder. Acta Neurol. Scand. 81 (1989) 361–362.

SÄVE-SÖDERBERGH, J., B. MALMVALL, R. ANDERSSON and B. BENGTSSON: Giant cell arteritis as a cause of death. J. Am. Med. Assoc. 255 (1986) 493–496.

SCHAUFELBERGER, C., B.A. BENGTSSON and R. ANDERSSON: Epidemiology and mortality in 220 patients with polymyalgia rheumatica. Br. J. Rheumatol. 34 (1995) 261–264.

SCHNEIDER, H.A., A.A. WEBER and P.H. BALLEN: The visual prognosis in temporal arteritis. Ann. Ophthalmol. 3 (1971) 1215–1230.

SCHON, R.: Involvement of smell and taste in giant cell arteritis. J. Neurol. Neurosurg. Psychiatry 51 (1988) 1594.

SEYMOUR, S.: The headache of temporal arteritis. J. Am. Geriatr. Soc. 35 (1987) 163–165.

SHEEHAN, M.M., C. KEOHANE and C. TWOMEY: Fatal vertebral giant cell arteritis. J. Clin. Pathol. 46 (1993) 1129–1131.

SHELHAMER, J.H., D.J. VOLKMAN, J.E. PARRILLO, T.J. LAWLEY, M.R. JOHNSTON and A.S. FAUCI: Takayasu's arteritis and its therapy. Ann. Intern. Med. 103 (1985) 121–126.

SIBONY, P.A. and S. LESSELL: Transient oculomotor synkinesis in temporal arteritis. Arch. Neurol. 41 (1984) 87–88.

SMITH C.A., W.J. FIDLER and R.S. PINALS: The epidemiology of giant cell arteritis: report of a ten-year study in Shelby County, Tennessee. Arthr. Rheum. 26 (1983) 1214–1219.

SUBRAMANYAN, R., J. JOY and K.G. BALAKRISHNAN: Natural history of aortoarteritis (Takayasu's disease). Circulation 80 (1989) 429–437.

TURNER, R.G., J. HENRY, A.I. FRIEDMANN and D. GERAINT JAMES: Giant cell arteritis. Postgrad. Med. J. 50 (1974) 265–269.

VILASECA, J., A. GONZALEZ, M.C. CID, J. LOPEZ-VIVANCOS and A. ORTEGA: Clinical usefulness of temporal artery biopsy. Ann. Rheum. Dis. 46 (1987) 282–285.

VINCENT, F.M. and T. VINCENT: Bilateral carotid siphon involvement in giant cell arteritis. Neurosurgery 18 (1986) 773–776.

WARRELL, D.A., S. GODFREY and E.G. OLSEN: Giant cell arteritis with peripheral neuropathy. Lancet i (1968) 1010–1013.

WILKINSON, I.M.S. and R.W.R. RUSSELL: Arteries of the head and neck in giant cell arteritis. Arch. Neurol. 27 (1972) 378–391.

WILSKE, K.R. and L.A. HEALEY: Polymyalgia rheumatica: a manifestation of systemic giant-cell arteritis. Ann. Intern. Med. 66 (1967) 77–86.

WISE, C.M., C.A. AGUDELO, W.L. CHMELEWSKI and K.M. MCKNIGHT: Temporal arteritis with low erythrocyte sedimentation rate: a review of five cases. Arthr.

Rheum. 34 (1991) 1571–1574.

WONG, R.L. and J.H. KORN: Temporal arteritis without an elevated erythrocyte sedimentation rate. Case report and review of the literature. Am. J. Med. 80 (1986) 959–964.

YAMADA, I., F. NUMANO and S. SUZUKI: Takayasu arteritis: evaluation with MR imaging. Radiology 188 (1993) 89–94.

YONEDA, S., T. NUKADA, K. TADA, M. IMAIZUMI, T. TAKANO and H. ABE: Subclavian steal in Takayasu's arteritis. A hemodynamic study by means of ultrasonic Doppler flowmetry. Stroke 8 (1977) 264–268.

ZARCHARSKI, L.R. and R.A. KYLE: Significance of extreme elevation of erythrocyte sedimentation rate. J. Am. Med. Assoc. 202 (1967) 264–266.

ZHENG, D.Y., L.S. LIU and D.J. FAN: Clinical studies in 500 patients with aortoarteritis. Chin. Med. J. 103 (1990) 536–540.

Handbook of Clinical Neurology, Vol. 27 (71): Systemic Diseases, Part III
M.J. Aminoff and C.G. Goetz, editors

CHAPTER 11

Behçet's disease

DOUGLAS S. GOODIN

Department of Neurology, University of California, San Francisco, CA, U.S.A.

In a series of publications beginning in 1936, Professor Hulûsi Behçet (1889–1948), a Turkish dermatologist at the University of Istanbul, described an illness that he termed the triple symptom complex and that now bears his name (Behçet 1937, 1938, 1939, 1940; Mutlu and Scully 1994). His initial paper in the western literature described the syndrome in two patients (a man and a woman) whom he had seen in his dermatology clinic over the preceding two decades (Behçet 1937). In subsequent publications (Behçet 1938, 1939, 1940) he added further cases to his series and described the essential features of what he felt was a new disease: 'It is an illness which lasts many years and in which the three following types of symptoms are prominent: firstly transient aphthous changes in the mouth, secondly ulcerations on the genitalia, and thirdly attacks of iritis, although the latter symptom is not always present' (Behçet 1940). Prior to Behçet's seminal work in defining the triple symptom complex as a distinct disease entity, there had been scattered case reports of the illness (e.g., Shigeta 1924; Adamantiades 1931; Whitwell 1934) as well as descriptions dating back to antiquity (Feigenbaum 1956). Thus, in the third book of endemic diseases attributed to Hippocrates, an illness quite reminiscent of Behçet's triple symptom complex is described: 'There were other fevers also, which I shall describe in due course. Many had aphthae and sores in the mouth. Fluxes about the genitals were

copious; sores, tumors internal and external; the swellings which appear in the groin. Watery inflammation of the eyes, chronic and painful. Growths on the eyelids, external and internal, in many cases destroying the sight' (cited by Feigenbaum 1956).

Although Behçet was aware of the association of the triple symptom complex with the occurrence of erythema nodosum (Behçet 1940), it was not until the work of others in the 1940s and 1950s that the multisystem nature of the disease process began to be appreciated. For example, involvement of the nervous system was first described by Knapp (1941) in a 29-year-old woman with recurrent hypopyon iritis, vitreous and retinal hemorrhages and a fluctuating ataxic syndrome over 7 years (cited by Pallis and Fudge 1956). The pathergy test (see below) was first reported by Jensen (1941). The articular manifestations, although not fully appreciated, were first noted in a review by France et al. (1951). Such multisystem involvement is now known to include mucocutaneous, urogenital, cardiovascular, articular, pleuropulmonary, gastrointestinal and neurological disturbances. The involvement of the nervous system in this disease is of particular importance because it accounts for a significant percentage of the morbidity and mortality from Behçet's disease (Pallis and Fudge 1956; Schotland et al. 1963; Wolf et al. 1965; Chajek and Fainaru 1975; O'Duffy 1979; Shakir et al. 1990; Bohlega et al. 1993).

EPIDEMIOLOGY

Behçet's disease has a world-wide distribution although it has a higher prevalence in some geographic areas, such as the Middle East, the Mediterranean Basin, and Japan. For example, in a nationwide survey from Japan the prevalence of Behçet's disease was estimated at 13.5 cases/ 100,000 population (Nakae et al. 1993). Similarly the prevalence in Iran has been estimated to be between 16 and 100 cases/100,000 population (Gharibdoost et al. 1993). By contrast, a survey from Yorkshire, U.K., estimated the prevalence as 0.64 cases/100,000 population (Chamberlain 1977). Similar to the low prevalence in the U.K., O'Duffy (1978) reports 'a point prevalence (sic) in the U.S. of 1/300,000/year'. In Portugal, the prevalence has been estimated at 1.5 cases/100,000 population (Crespo et al. 1993). Although a detailed analysis of the prevalence and incidence rates are not available from many other parts of the world, the maldistribution of the number of patients reported in the literature from different geographic regions is striking. Thus, the number of patients reported in studies from the Middle East and Japan is often an order of magnitude (sometimes two) larger than the number of patients in reports from the U.S., the U.K. or other parts of the world (Chamberlain 1977; O'Duffy 1978; Yazici et al. 1984; Serdaroglu et al. 1989; Assaad-Khalil et al. 1993; Crespo et al. 1993; Gharibdoost et al. 1993; Kötter et al. 1993; Nakae et al. 1993; Scherrer et al. 1993; Al Dalaan et al. 1994). Indeed, in a recent report from the International Study Group for Behçet's Disease (ISGBD), the number of patients recruited from the Middle East, the Mediterranean Basin and Japan outnumbered those from the U.S., the U.K. and France by 870 to 44 (ISGBD 1990).

Even within the same ethnic population of the same country, the prevalence of the disease may not be evenly distributed. Thus, the surveys of Japan (Yamamoto et al. 1974; Nakae et al. 1993) have consistently reported that Behçet's disease is most prevalent in the northern latitudes of Japan with 30.5 cases/100,000 population in Hokkaido (Nakae et al. 1993), and least prevalent in the south with 0.99 cases/100,000 population in Kyushu (Nakae et al. 1993). Such observations are reminiscent of the geographic distribution of multiple sclerosis (MS),

another putative autoimmune disease, in which increasing latitude (either north or south of the equator) has consistently been found to be positively correlated with disease prevalence (Mathews et al. 1984). However, the evidence from Japan regarding the importance of increased latitude on disease prevalence has not been confirmed in other countries. Moreover, a consideration of the world-wide distribution of Behçet's disease does not particularly suggest a general latitude effect similar to that seen in MS. Thus, the high prevalence of Behçet's disease in the Middle East (and other regions close to the equator) is distinctly dissimilar to the world-wide distribution of MS. This suggests that increasing latitude is not a consistently important factor in determining disease prevalence in Behçet's disease and provides evidence that other environmental factors are responsible.

In most large series the prevalence of Behçet's disease in men exceeds that in women. Certainly this is true of the reports from the Middle East and in early reports from Japan, where the percentage of males in reports of over 100 cases ranges from 54.5 to 91% (Yamamoto et al. 1974; Chajek and Fainaru 1975; Yazici et al. 1984; Serdaroglu et al. 1989; Assaad-Khalil et al. 1993; Gharibdoost et al. 1993). Reports from the U.K., the U.S., and other parts of the world, by contrast, have found a higher prevalence in women (Chamberlain 1977; O'Duffy 1978; Crespo et al. 1993; Kötter et al. 1993; Scherrer et al. 1993). These latter studies generally have involved small sample sizes and, consequently, it is possible that the true sex ratio may have been somewhat distorted in these reports. Nevertheless, even if the lowest estimate of the male preponderance (54.5%) from the Middle East and Japan were used for comparison, it would be statistically unlikely to observe O'Duffy's (1978) male:female ratio of 12:37 (32.4%; $P = 0.01$) or Chamberlain's (1977) ratio of 12:32 (37.5%; $P = 0.05$). Moreover, in a study comparing German patients with Behçet's disease with patients from the Mediterranean Basin, Kötter et al. (1993) found a marked difference between the percentage of males in the German patients with Behçet's disease (33%) compared to that from the Mediterranean area (75%). This difference is also significant ($P = 0.02$). Thus, it seems reasonable to conclude that the sex ratio is truly different in different parts of the world. It is not clear what might account for these sex ratio differences. It seems unlikely that this is due to a difference in the criteria used to make the diagnosis of Behçet's disease because some of the disparate reports have used the same set of diagnostic standards (e.g., Chamberlain 1977; O'Duffy 1978; Yazici et al. 1984; Kötter et al. 1993; Al Dalaan et al. 1994). Also the epidemiology of the disease may be changing in Japan (Yamamoto et al. 1974; Nakae et al. 1993). The most recent Japanese survey from 1991 estimates that the prevalence of Behçet's disease is increasing compared to earlier surveys done in 1972 and 1984. Moreover, in the 1991 survey women with the complete syndrome were affected almost as often as men (male cases:female cases = 1.07) whereas in the earlier 1972 survey the male:female ratio in patients with the complete syndrome was 1.78. Such a change, if true, suggests that there has been a change in some environmental factor or factors in Japan that are related pathogenetically to Behçet's disease.

There are also other apparent differences in Behçet's disease reported from different parts of the world. For example, the pathergy test (see below), with a few exceptions (e.g., Nakae et al. 1993; Al Dalaan et al. 1994), is positive in the majority of patients reported from the Middle East and Japan (Yazici et al. 1980, 1984; ISGBD 1990; Gharibdoost et al. 1993; O'Neill et al. 1993). By contrast, this test is generally normal in patients reported from the U.S.,

the U.K., or other parts of the world (Chamberlain 1977; O'Duffy 1990; Crespo et al. 1993; Sherrer et al. 1993). Also there is a significant linkage of Behçet's disease to the HLA-B5 phenotype in patients from the Middle East and Japan (Yazici et al. 1980, 1984; Gharibdoost et al. 1993; Mizuki et al. 1993, 1994; Nakae et al. 1993; Al Dalaan et al. 1994) whereas this linkage has not been found in the U.S. and U.K. (e.g., O'Duffy et al. 1976; Chamberlain 1977).

Behçet's disease has its peak incidence in the age range of 20–30 years for both men and women (Fig. 1), although rarely it may begin before the age of 15 or after the age of 45 (Yamamoto et al. 1974; Chajek and Fainaru 1975; Gharibdoost et al. 1993; Al Dalaan et al. 1994). When the disease occurs in childhood there is a suggestion that the clinical manifestations of the disease may differ from those observed in adults (e.g., Lang et al. 1990; Shafaie et al. 1993). Thus, in a report from Iran, Shafaie et al. (1993) found a lower prevalence of oral and genital aphthous ulcers and a higher prevalence of both anterior and posterior uveitis in children compared to adults. Interestingly, in an earlier review of published case reports, Lang et al. (1990) concluded that ocular lesions may be less frequent in children whereas the mucocutaneous lesion were equally prevalent. What accounts for such discrepant observations in children is not clear but it may be due, in part, to

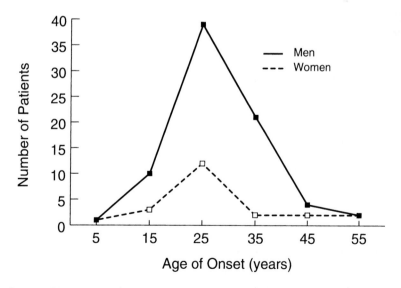

Fig. 1. Ages of onset of Behçet's disease in men and women. Data compiled from the studies of Chajek and Fainaru (1975), Banna and El Ramahi (1991), and Jorizzo et al. (1995). Points are plotted at the midpoint of each decade of life and represent the aggregate number of patients with an age of onset in the decade.

regional differences in the clinical expression of Behçet's disease which, as discussed earlier, has also been noted in adults.

PATHOGENESIS

The etiology of Behçet's disease is not known. Behçet believed that it was caused by a virus and reported the presence of elementary bodies in the oral and genital ulcers in each of the cases he studied (Behçet 1940). Subsequent authors, however, have failed to demonstrate inclusion bodies (e.g., Jensen 1941; Schotland et al. 1963). Moreover, early claims of the transmissibility of the illness to rabbits (Alm and Öberg 1945) and reports of viral isolation (e.g., Sezer 1953; Evans et al. 1957) have not been confirmed. Reports continue about the association of Behçet's disease with a variety of organisms such as tuberculosis (e.g., Tang et al. 1991), *Borrelia burgdorferi* (e.g., Isogai et al. 1991), human immunodeficiency virus (e.g., Buskila et al. 1991), *Streptococcus sanguis* (e.g., Yokota et al. 1992; Isogai et al. 1993), herpes simplex (e.g., Lee et al. 1993), varicella zoster (e.g., Pedersen et al. 1993), or hepatitis C (e.g., Munke et al. 1995).

Early suggestions were also made that the fundamental problem in Behçet's disease may be a vascular disorder producing a hypercoagulable state that can be reflected by such changes as elevated fibrinogen and factor VIII levels and decreased fibrinolytic activity in the serum of patients during relapses (Pallis and Fudge 1956; Schotland et al. 1963; Wolf et al. 1965; O'Duffy 1990). Certainly, vascular occlusive disease is a well described feature of Behçet's disease. Nevertheless, this complication is not a universal finding in Behçet's disease and most recent authors believe that these vascular lesions and serum alterations are the result, not the cause, of the illness (e.g., O'Duffy 1990).

Most authors agree that the primary pathologic change in Behçet's disease is a vasculitis, possibly on an autoimmune basis, as demonstrated by the frequently observed perivascular infiltration by lymphocytes and mononuclear cells in the affected tissues (O'Duffy 1990; Jorizzo et al. 1995). In support of the autoimmune hypothesis are several recent studies reporting autoantibodies directed at endogenous proteins in the serum of patients with Behçet's disease. These have included such proteins as keratin (e.g., Balcon et al. 1993), retinal antigens such as the S-antigen (e.g., Yamamoto et al. 1993, 1994), nuclear, smooth muscle, and cardiolipin antigens (e.g., Taylor et al. 1993; Al Dalaan et al. 1994) and corneal epithelium antigens (e.g., Soylu et al. 1993).

There have also been reports linking Behçet's disease and the exposure to various environmental pollutants such as organophosphates, inorganic copper, and DDT (e.g., Ishikawa et al. and Nishiyama et al., cited by O'Duffy 1978). These reports, however, have not been confirmed.

Genetic factors appear to play at least some role in the pathogenesis of Behçet's disease. In those parts of the world where Behçet's disease is most prevalent there is a significant association with the HLA-B5 phenotype, especially the HLA-B51 subtype (Yazici et al. 1980; Crespo et al. 1993; Freire et al. 1993; Gharibdoost et al. 1993; Nakae et al. 1993; Al Dalaan et al. 1994; Mizuki et al. 1994). Also, the occasional familial occurrence (2–20%) of Behçet's disease reported in large series would suggest a genetic mechanism (Yamamoto et al. 1974; Chajek and Fainaru 1975; ISGBD 1990; Hamza and Bardi 1993). In addition, in the absence of any clear difference in environmental exposure between men and women, the apparent male preponderance of Behçet's disease would seem to favor genetic mechanisms. It is unlikely, however, that only genetic mechanisms are responsible for Behçet's disease. Thus, the fact that not all affected individuals are the same phenotype or sex, the fact that the sex ratio is different in different geographic regions or may change over time in one region, and the fact that only a minority of patients have a family history of Behçet's disease all suggest that an environmental factor or factors are at least partially, if not largely, responsible. Perhaps the most reasonable hypothesis is that the disease is caused by an environmental exposure in a genetically susceptible individual.

DIAGNOSIS

Several different sets of diagnostic criteria have been proposed for establishing the diagnosis of Behçet's disease. The first was that of Mason and Barnes (1969); subsequently four other sets of criteria, each with its own adherents, were proposed (Behçet's Disease Research Committee of Japan 1974;

TABLE 1

Diagnostic criteria for Behçet's disease developed by the International Study Group*.

1. Recurrent oral ulceration – minor aphthous, major aphthous or herpetiform ulceration observed by physician or patient which recurred at least three times in one 12-month period.
2. In addition to any two of the following:
   A. Recurrent genital ulceration – aphthous ulceration or scarring observed by physician or patient.
   B. Eye lesions – anterior uveitis, posterior uveitis, or cells in the vitreous on slit lamp examination; or retinal vasculitis observed by an ophthalmologist.
   C. Skin lesions – erythema nodosum observed by physician or patient, pseudofolliculitis or papulopustular lesions; or acneiform nodules observed by a physician in post-adolescent patients not on corticosteroid treatment.
   D. Positive pathergy test – read by a physician at 24–48 h.

Findings applicable only in absence of other clinical explanations.
*International Study Group for Behçet's Disease 1990.

O'Duffy 1974a,b; Zhang 1980; Dilsen et al. 1986). Following the 1985 International Conference on Behçet's disease in London, an International Study Group (including representatives from all but one of the groups that developed the competing diagnostic criteria) was established to evaluate the sensitivity and specificity of the different criteria and to develop an international standard. The results of this effort were published in the Lancet (ISGBD 1990). The study group recruited 914 patients from 12 centers in seven countries and the diagnosis of Behçet's disease was accepted if it had been made by an experienced clinician based on any one of the pre-existing sets of diagnostic standards. A new set of diagnostic standards was then developed based on a random 60% sample of the 914 patients. The different sets of diagnostic criteria were then compared with each other for sensitivity on the remaining 40% sample and for specificity against a control group consisting of patients with recurrent oral aphthous ulcerations due to other connective tissue diseases. The diagnostic criteria developed by the ISGBD (shown in Table 1) were found to be superior to any of the previously existing sets of diagnostic criteria, having a sensitivity of 95% and a specificity of 98% against the control group. In their report the ISGBD acknowledged that

Behçet's disease may rarely occur (< 3% of cases) in the absence of recurrent oral aphthous ulcerations, but nevertheless felt that this clinical feature was so characteristic of the illness that a confident diagnosis of Behçet's disease (at least for research purposes) could not be made in its absence.

NON-NEUROLOGICAL MANIFESTATIONS OF BEHÇET'S DISEASE

Behçet's disease is a multi-system illness that has a particular predilection for involvement of several different organ systems outside the nervous system.

Oral and genital mucosal manifestations

Recurrent aphthous ulcers occur on the mucosal membranes of the lips, tongue, cheeks and gingiva and less commonly on the palate, pharynx and tonsils. These lesions can be shallow or deep, have a smooth border and are generally 2–10 mm in diameter (Fig. 2). They can occur singly or in crops and can be quite painful. Typically these lesions heal spontaneously in 7–14 days and recur at irregular intervals between several days and many months (see Shimizu et al. 1979 for a review). These recurrent oral aphthous ulcers are seen in over 95% of cases of Behçet's disease and are the earliest clinical manifestation of the disease in the large majority (Chajek and Fainaru 1975; ISGBD 1990; Gharibdoost et al. 1993).

Recurrent genital ulcers similar to the oral lesions may appear on the scrotum or vulva and, less commonly, on the penis, vaginal mucosa, and perianally. These lesions can also be painful although they are often painless in women and may therefore go undetected by the patient. Unlike the oral lesions which generally heal without scarring, the genital lesions often leave scars that may be detected on physical examination even when the disease is inactive. Genital aphthous ulcers occur in 60–90% of cases but these are only infrequently the earliest manifestation of the illness (Chajek and Fainaru 1975; Shimizu et al. 1979; ISGBD 1990; Gharibdoost et al. 1993).

Cutaneous manifestations

Several types of cutaneous lesion can occur during the course of Behçet's disease. These include papulopustular lesions (either spontaneous or induced),

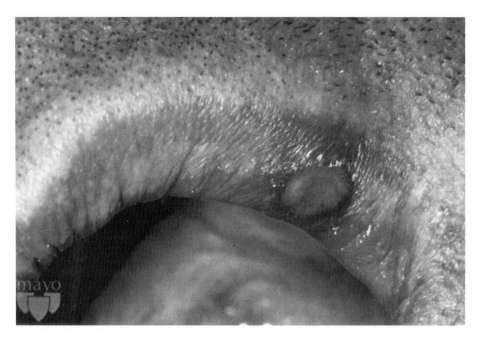

Fig. 2. Oral aphthous ulcer in a patient with Behçet's disease. (Figures courtesy of Dr. J. Desmond O'Duffy, Department of Internal Medicine, Mayo Clinic, Rochester, MN, U.S.A.)

pseudofolliculitis, acneiform nodules, and erythema nodosum (Chajek and Fainaru 1975; Shimizu et al 1979; O'Duffy 1990; Gharibdoost et al. 1993; Jorizzo et al. 1995). These lesions are included as part of the international diagnostic criteria (see Table 1). Some authors, however, have argued that occurrence of acneiform nodules should be excluded from the diagnostic criteria because, on histologic examination, these lesions generally contain only non-specific follicular-based pathologic changes which they claim can be found in normal post-adolescent individuals (Jorizzo 1986; Jorizzo et al. 1995). Accordingly, these authors recommend that the diagnostic criteria for Behçet's disease require histologic examination of skin lesions in order to demonstrate the vessel-based pathology (i.e., either a neutrophilic vascular reaction or leukocytoclastic vasculitis) that is characteristic of Behçet's disease (Jorizzo 1986; Jorizzo et al. 1995). Skin lesions (especially the papulopustular lesions and pseudo-folliculitis) occur in 70–90% of patients with Behçet's disease, but only rarely are these lesions the initial manifestation of the illness (Chajek and Fainaru 1975; Shimizu et al. 1979). Many patients have several different skin lesions at any one time beginning either synchronously or asynchronously. The lesions tend to heal spontaneously within 1–2 weeks and recur at irregular intervals similar to the mucosal lesions.

Erythema nodosum consists of slightly raised, indurated, erythematous lesions, most often on the legs. They occur in 20–45% of patients with Behçet's disease (Chajek and Fainaru 1975; Shimizu et al. 1979; ISGBD 1990; Gharibdoost et al. 1993). Its association with this illness was noted initially by Behçet (1940). The lesions of erythema nodosum are typically ovoid in shape and range in size between 1 and several cm^2. Often several lesions are present simultaneously. The lesions can be somewhat tender. They resolve spontaneously after 10–14 days, although they often leave some residual hyper-pigmentation at the lesion site which persists much longer. As with the other skin lesions in Behçet's disease, erythema nodosum tends to recur at irregular intervals.

Cutaneous hypersensitivity to minor external trauma (the pathergy test) is a characteristic feature of Behçet's disease, particularly in the Middle East and Japan (Yazici et al. 1980, 1984; ISGBD 1990; Gharibdoost et al. 1993; O'Neill et al. 1993). The test is performed by making a needle prick aseptically in an avascular area on the forearm. If the test is posi-tive, this is followed in 24–48 h by erythema, swelling and the formation of a papule at the site of the needle prick (Shimizu et al. 1979; Yazici et al. 1980). As discussed earlier, the sensitivity of the test varies between 8 and 84%, depending in part upon geographic considerations (Chamberlain 1977; O'Duffy 1979; Yazici et al. 1980, 1984; Crespo et al. 1993; Gharibdoost et al. 1993; Scherrer et al. 1993). In the group of patients studied by the ISGBD (1990), the sensitivity of the pathergy test was 58%. The specificity of the test, by contrast, is considerably higher, estimated by the ISGBD (1990) at 90% and by Yazici et al. (1980) at 97%.

Ocular manifestations

Ocular lesions are the most serious of the common problems to affect patients with Behçet's disease because they lead, not infrequently, to visual loss. Ocular involvement occurs in 60–80% of patients with Behçet's disease, with visual loss occurring in 20–30% (Chajek and Fainaru 1975; Shimizu et al. 1979; ISGBD 1990; Gharibdoost et al. 1993; Sakamoto et al. 1995). The involvement can occur separately in the anterior or the posterior segments of the uveal tract or it can occur simultaneously in both segments (Fig. 3). In the anterior segment the most common manifestation is an iridocyclitis, often with hypopyon (the condition of having leukocytes in the anterior chamber of the eye). It is generally a transient condition, lasting only a few days and usually does not have any permanent effect on vision (Chajek and Fainaru 1975; Shimizu et al. 1979).

Other anterior segment involvement consists of conjunctivitis or corneal ulceration, often in combination with the iridocyclitis (Chajek and Fainaru 1975). Involvement of the posterior segment includes choroiditis, inflammation of or hemorrhage into the vitreous, retinal vessel thrombosis or vasculitis (Fig. 4), and optic papillitis (Chajek and Fainaru 1975). These posterior conditions, especially retinal vasculitis, generally last longer and are much more likely to result in permanent impairment of vision than involvement of the anterior segment (Chajek and Fainaru 1975; Shimizu et al. 1979; O'Duffy 1990; Sakamoto et al. 1995). Like the mucocutaneous manifestations of Behçet's disease, the uveitis tends to recur at irregular intervals of between several weeks and several months.

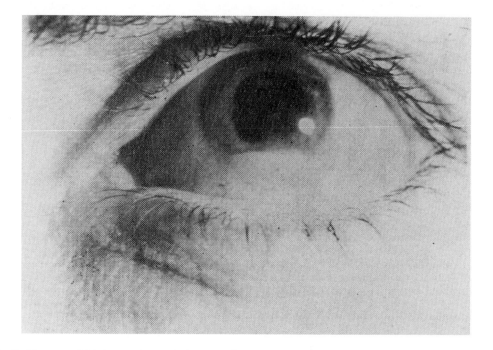

Fig. 3. Hypopyon iritis in a patient with Behçet's disease. (Reproduced with permission from Alemá 1978.)

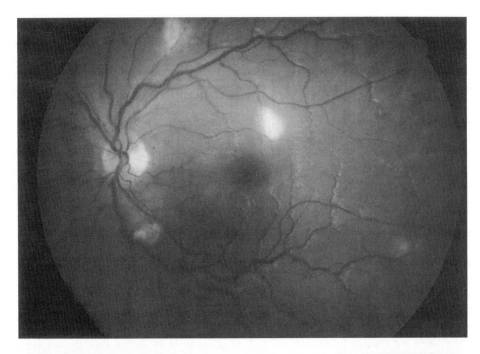

A

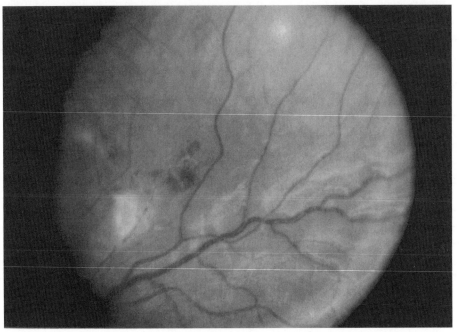

B

Fig. 4. (A) Fundus photograph of left eye of a 21-year-old female with Behçet's disease. White area near fovea represents retinal edema secondary to small vessel occlusion. Areas superior and inferior to optic nerve also represent retinal ischemia associated with small vessel occlusion. (B) Enlargement of area superior to the optic nerve in (A). Occlusive perivasculitis can be seen surrounding arteriole with retinal edema and hemorrhage. Figures courtesy of Dr. David Herman, Department of Ophthalmology, Mayo Clinic, Rochester MN, U.S.A.

Vascular manifestations

Thrombophlebitis in Behçet's disease can affect either the superficial or deep veins and occurs in approximately 10–40% of patients (Chajek and Fainaru 1975; Shimizu et al. 1979; ISGBD 1990; O'Duffy 1990; Gharibdoost et al. 1993; Al Dalaan et al. 1994). In some patients, the superficial thrombophlebitis can be induced by any venepuncture, similar to the hypersensitivity of the skin induced by the pathergy test discussed above (Chajek and Fainaru 1975). In some series, the deep veins are affected more commonly than the superficial veins (e.g., Gharibdoost et al. 1993; Al Dalaan et al. 1994) and, when larger veins such as the superior or inferior vena cava are obstructed, major clinical sequelae may result (Shimizu et al. 1979; O'Duffy 1990; Gharibdoost et al. 1993; Al Dalaan et al. 1994). Large artery aneurysms or occlusions are also described in Behçet's disease (Shimizu et al. 1979; Numan et al. 1994). These have been described as occurring in up to 34% of cases (e.g., Lakhanpal et al. 1985; Al Dalaan et al. 1994) although others found them distinctly uncommon (e.g., ISGBD 1990; Gharibdoost et al. 1993).

Articular manifestations

Involvement of the joints in Behçet's disease, particularly the larger joints, occurs in approximately 30–50% of patients and can present as either arthralgia or arthritis (Chajek and Fainaru 1975; Shimizu et al. 1979; Gharibdoost et al. 1993; Al Dalaan et al. 1994). The joint involvement is usually asymmetric and recurrent. Destruction of the joints and atrophy of the bone can occur but is quite rare (Shimizu et al. 1979).

Other non-neurologic manifestations

Several other systemic manifestations have also been reported in Behçet's but will not be considered separately because these manifestations are either infrequent or inconsistent between reports. Gastrointestinal involvement similar to that found in inflammatory bowel disease has been described and has led to the characterization of the entero-Behçet's subtype of the disease (e.g., Shimizu et al. 1975). This manifestation is generally reported to be rare

(e.g., Gharibdoost et al. 1993; Al Dalaan et al. 1994). Genito-urinary involvement with transient hematuria or proteinuria may occur (e.g., Shimizu et al. 1979; Gharibdoost et al. 1993). Recurrent epididymitis with pain and swelling lasting 1–2 weeks occurs in 4–8% of patients in most series (e.g., Chajek and Fainaru 1975; Shimizu et al. 1979; ISGBD 1990; Gharibdoost et al. 1993; Al Dalaan et al. 1994). Hepatic involvement is rare although a Budd–Chiari syndrome is occasionally reported, resulting from thrombotic occlusion of the hepatic vein (e.g., Shimizu et al. 1979; O'Duffy 1990; Al Dalaan et al. 1994). Cardiopulmonary involvement may also occur with pleural effusions, hemoptysis, or pericarditis (e.g., Shimizu et al. 1979; Gharibdoost et al 1993; Al Dalaan et al. 1994).

NERVOUS SYSTEM INVOLVEMENT IN BEHÇET'S DISEASE

Involvement of the central nervous system (CNS) is well known in Behçet's disease. Most reports indicate that CNS disease is only occasionally the presenting symptom of the illness (Schotland et al. 1963; Chajek and Fainaru 1975; O'Duffy and Goldstein 1976; Serdaroglu et al. 1989; Al Dalaan et al. 1994). Two large reports, one from Saudi Arabia (Bohlega et al. 1993) and another from Tunisia (Hentati et al. 1993), however, put the frequency of neurologic presentation considerably higher (15 and 48% of cases with CNS involvement respectively). The reason for this discrepancy is not known. In addition, the reported prevalence of this so-called 'neuro-Behçet' subtype of the disease varies considerably (3–49%) (Chajek and Fainaru 1975; ISGBD 1990; Bohlega et al. 1993; Gharibdoost et al. 1993; Hentati et al. 1993). The reason for such a wide discrepancy in the estimated prevalence of neuro-Behçet's disease is not clear. Part of the difference may be due to methodological discrepancies such as the inclusion (by some authors) of patients with uncomplicated headache but without other evidence of nervous system involvement as part of neuro-Behçet's disease (e.g., Al Dalaan et al. 1994). However, methodological differences seem inadequate to explain such a wide range of estimates. It is more likely that the prevalence of CNS involvement, as with other clinical aspects of Behçet's disease, varies between different geographic regions. Thus, in a recent report by O'Neill et al. (1993), the prevalence of CNS involvement was found to be approximately 3- to 5-fold lower ($P < 0.05$) in most Middle Eastern countries and Japan compared to countries in Western Europe and North America.

In other respects the epidemiology of neuro-Behçet's disease seems consistent with the disease as a whole. Thus, in areas such as the U.S. where female cases outnumber the male cases, the women with neuro-Behçet's disease also outnumber the men with this form of the disease (e.g., O'Duffy and Goldstein 1976). Conversely, in reports from the Middle East, the men with neuro-Behçet's disease outnumber the women (e.g., O'Duffy et al. 1976; Chamberlain 1977; Assaad-Khalil et al. 1993; Al Dalaan et al. 1994). Similarly, the HLA type associated with neuro-Behçet's disease is similar to that for the total Behçet's disease population in the country from which the reports originate (e.g., O'Duffy et al. 1976; Chamberlain 1977; Assaad-Khalil et al. 1993; Al Dalaan et al. 1994).

The onset of CNS involvement can occur at almost any time during the course of Behçet's disease. In the review of Wolf et al. (1965), CNS involvement was found to occur anywhere from 2 weeks to 27 years following the first appearance of the clinical manifestations of Behçet's disease and, as mentioned earlier, it may rarely antedate the other symptoms (e.g., O'Duffy and Goldstein 1976; Al Dalaan et al. 1994). In the majority of the cases, however, CNS involvement occurs within the first 8 years of the illness (Pallis and Fudge 1956; Schotland et al. 1963; O'Duffy and Goldstein 1976; Assaad-Khalil et al. 1993), with the average time from the onset of the systemic features of Behçet's disease to the appearance of the neurologic symptoms being about 4–6 years (Schotland et al. 1963; Chajek and Fainaru 1975; Assaad-Khalil et al. 1993).

Neuropathologic findings

The pathologic findings that have been observed in patients with neuro-Behçet's disease can be divided into changes that occur diffusely and those that occur more focally (Rubenstein and Urich 1963). The diffuse changes that occur are primarily related to a chronic low-grade meningoencephalitis (Fig. 5) with perivascular and meningeal infiltration by lympho-

cytes, plasma cells and macrophages (Pallis and Fudge 1956; McMenemey and Lawrence 1957; Wadia and Williams 1957; Rubenstein and Urich 1963; Schotland et al. 1963; Alemá 1978; Lakhanpal et al. 1985). It is the focal changes, however, that are the most characteristic neuropathologic features of Behçet's disease. These consist of small necrotic foci (Fig. 6) demonstrating loss of all tissue elements and often centered around blood vessels (Pallis and Fudge 1956; McMenemey and Lawrence 1957; Wadia and Williams 1957; Rubenstein and Urich 1963; Schotland et al. 1963; Alemá and Bignami 1966; Alemá 1978). These lesions often have accumulated microglial phagocytes and, in the end stages, only glial scarring, with or without cavitation, is evident (Rubenstein and Urich 1963). Demyelination is generally minimal and cytoplasmic or nuclear inclusion bodies are absent (McMenemey and Lawrence 1957; Schotland et al. 1963). These necrotic lesions are scattered throughout the gray and white matter of the CNS although they have a predilection for involving certain areas of the nervous system in contrast to others. Thus, these necrotic foci are most frequently found in the brainstem, internal capsule, and cerebral peduncles; they tend to spare the cerebral cortex, the cerebral white matter, the cerebellum and the sensory tracts (Rubenstein and Urich 1963; Alemá and Bignami 1966; Shimizu et al. 1979). In addition to the necrotic lesions, focal cuffing of blood vessels by lymphocytes is seen in scattered locations but with a predilection for the subependymal vessels, and in the brainstem and basal ganglia (Rubenstein and Urich 1963). Wallerian degeneration, especially in the corticospinal tracts, can often be seen below the necrotic foci in the brainstem.

Pathologic examination of the blood vessels is generally normal. However, thrombosis of the dural sinuses, especially of the superior sagittal sinus, does occur in neuro-Behçet's disease (Serdaroglu et al. 1989; Shakir et al. 1990; Kansu et al. 1991; Hentati et al. 1993; Al Dalaan et al. 1994; Daif et al. 1995). Despite the occurrence of arterial occlusions in the peripheral vasculature in Behçet's disease, arterial occlusion within the CNS is extremely rare (Pallis and Fudge 1956; O'Duffy and Goldstein 1976; Alemá 1978; Serdaroglu et al. 1989; Shakir et al. 1990; Banna and El Ramahi 1991; Kansu et al. 1991; Al Dalaan et al. 1994).

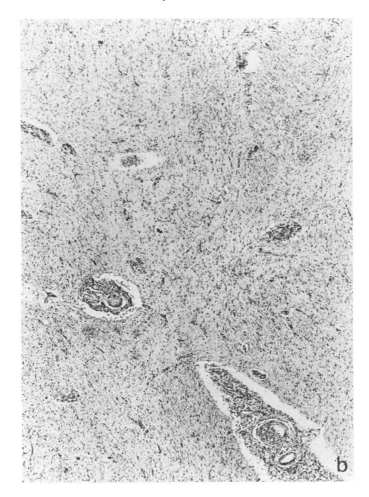

Fig. 5. Infiltration of the hemispheric white matter demonstrating perivascular cuffing by inflammatory cells. (Reproduced with permission from Alemá 1978.)

Neuroimaging studies

Computed tomography (CT) has been found by several authors to be unrevealing in most patients with neuro-Behçet's disease (Serdaroglu et al. 1989; Banna and El Ramahi 1991; Al Dalaan et al. 1994). It may occasionally show low-density lesions or provide evidence of dural sinus thrombosis but in most cases it is normal (Serdaroglu et al. 1989; Banna and El Ramahi 1991; Al Dalaan et al. 1994). Some authors, by contrast, have claimed that CT is abnormal in a much higher percentage of cases: Assaad-Khalil et al. (1993) found that CT was abnormal in 62% of patients with neuro-Behçet's disease, with low-density cerebral lesions in 34%, atrophy in 37%, and sinus thrombosis in 5.7% of the cases. The reason for such a discrepancy between reports is not clear. Magnetic resonance imaging (MRI) is more often abnormal (Fig. 7) and shows abnormal foci of high signal intensity on T2-weighted images (Besana et al. 1989; Banna and El-Ramahi 1991; Akman-Demir et al. 1995). Some of these lesions may also appear as low signal intensity on T1-weighted images (Banna and El Ramahi 1991), thereby resembling the radiographic appearance of MS plaques (e.g., Uhlenbrock and Sehlen 1989). Superior sagittal sinus thrombosis also can be documented on MRI in most patients (Daif et al. 1995). Angiography is abnormal in patients with dural sinus thrombosis but otherwise is usually normal and does not demonstrate large vessel involvement in patients with

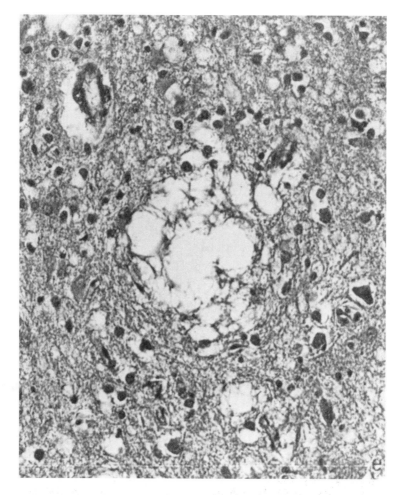

Fig. 6. Neurotic focus in the cerebrum of a patient with Behçet's disease. (Reproduced with permission from Alemá 1978.)

neuro-Behçet's disease (Banna and El Ramahi 1991; Al Dalaan et al. 1994). On rare occasions, the angiogram may demonstrate arterial occlusion secondary to a large vessel vasculitis (e.g., Nishimura et al. 1991). Also on rare occasions angiography may demonstrate an avascular mass secondary to a cerebral infarction that can mimic a brain abscess (e.g., Kozin et al. 1977).

Evoked potentials

The technique of recording evoked cerebral potentials to different sensory stimuli has also been used to study function in patients with neuro-Behçet's disease. Abnormalities have been detected in 33–94% of such patients who have been studied with

auditory, somatosensory and visual evoked potentials (Besana et al. 1989; Benli et al. 1993; Spadoro et al. 1993). The most common abnormality reported has been in the visual evoked potential (VEP) which shows either an absent response or a prolongation of the P100 latency especially with smaller (15 min) check sizes (Besana et al. 1989). In most cases, however, the patients demonstrating VEP abnormalities have clinically apparent ocular involvement such as retinal vasculitis or optic papillitis, so that the evoked potential findings add little to their management. Also the majority of auditory evoked potential abnormalities have been found in patients with clinical signs or symptoms of brainstem dysfunction (Spadaro et al. 1993). Somatosensory evoked potentials, especially those elicited by stimulation of lower

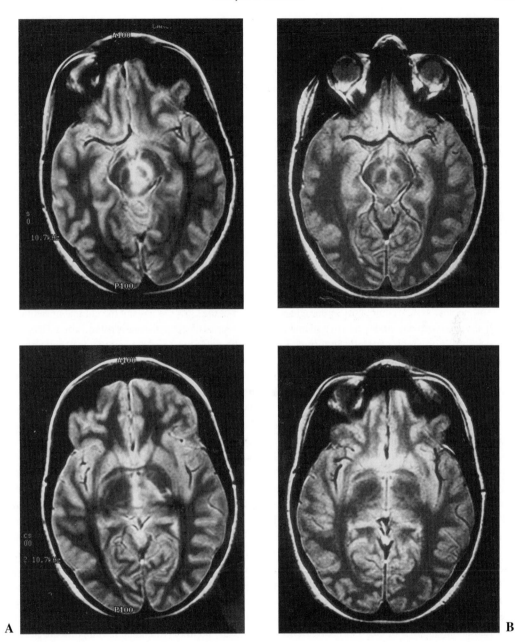

A   B

Fig. 7. MRI from a 30-year-old woman with a 5-year history of Behçet's disease manifested by recurrent oral and genital aphthous ulcerations and episodic lower extremity synovitis. Her neurological history began in March 1995 when she developed diplopia, a left-sided ptosis, and blurring of vision. CSF examination at that time showed nine lymphocytes and her erythrocyte sedimentation rate was 49 mm/h. An MRI scan (T2-weighted) was done on 4/7/95 which demonstrated abnormally high signal in the medial dorsal midbrain at the level of the red nucleus extending into the ventral medial thalamic nuclei (Panel A). The patient recovered following treatment with intravenous steroids and was neurologically normal. A follow-up MRI scan was done on 5/2/95 which demonstrated resolution of the abnormal T2-weighted signal in her brainstem and diencephalon (Panel B). (Figures courtesy of J. Desmond O'Duffy, Department of Internal Medicine, Mayo Clinic, Rochester, MN, U.S.A.)

extremity nerves such as the posterior tibial, peroneal, or sural, may be abnormal in up to 50% of patients with neuro-Behçet's disease who lack signs or symptoms of sensory dysfunction in the limbs (Spadaro et al. 1993). Such a result is interesting given the relative paucity of clinical sensory abnormalities in patients with neuro-Behçet's disease (Wadia and Williams 1957; Rubenstein and Urich 1963; O'Duffy and Goldstein 1976; Assaad-Khalil et al. 1993; Al Dalaan et al. 1994). Nevertheless, evoked potential studies are non-specific and it is unclear how the detection of subclinical lesions in these afferent pathways aids in the diagnosis or management of individual patients.

Cerebrospinal fluid

Except in patients with uncomplicated headache as their sole nervous system complaint (Serdaroglu et al. 1989), the cerebrospinal fluid (CSF) in patients with neuro-Behçet's disease is generally abnormal. Estimates for the prevalence of CSF abnormality in patients with neuro-Behçet's disease vary between 53 and 86% of cases (Wolf et al. 1965; O'Duffy and Goldstein 1976; Serdaroglu et al. 1989; Bohlega et al. 1993; Hentati et al. 1993). The abnormalities consist of an elevated protein content and a pleocytosis. The pleocytosis is typically lymphocytic although polymorphonuclear cells are often intermixed and may sometimes predominate (Wolf et al. 1965; O'Duffy and Goldstein 1976; Serdaroglu et al. 1989; Bohlega et al. 1993). Cell counts are generally below 100 cells/mm³ but counts up to 5100 have been recorded (Wolf et al. 1965). The protein level is generally less than 100 mg/dl (Wolf et al. 1965; O'Duffy and Goldstein 1976; Hentati et al. 1993) but in a few instances it has been considerably higher. Oligoclonal IgG bands are infrequently present, being reported in only 8–42% of cases of neuro-Behçet's disease (Sharief et al. 1991; Bohlega et al. 1993; McLean et al. 1995). The IgG index, however, may be elevated in over 50% of patients (Sharief et al. 1991). One study reported that the presence of oligoclonal IgG bands did not correlate with the activity of the disease whereas IgA and IgM bands were most prevalent during active disease and least prevalent during remissions (Sharief et al. 1991). This finding requires confirmation and its implication on the pathogenesis of neuro-Behçet's disease is unclear.

The CSF glucose is generally normal. CSF pressure is usually normal except in patients with dural sinus thrombosis or pseudotumor cerebri.

Other laboratory abnormalities

There are few consistent laboratory abnormalities in patients with either Behçet's disease or the neuro-Behçet's subtype. The erythrocyte sedimentation rate is often elevated but this is commonly found in patients with Behçet's disease both with and without nervous system involvement. (e.g., Gharibdoost et al. 1993).

CLINICAL MANIFESTATIONS OF NEURO-BEHÇET'S DISEASE

As mentioned earlier, Behçet's disease can affect any part of the CNS. As a result the clinical presentation of neuro-Behçet's disease is often varied. Headache is the single most common symptom in patients with neuro-Behçet's disease (Pallis and Fudge 1956; Wadia and Williams 1957; Schotland et al. 1963; O'Duffy and Goldstein 1976; Serdaroglu et al. 1989; Assaad-Khalil et al. 1993), but this symptom is non-specific and can be due to multiple causes. Indeed, in one study (Serdaroglu et al. 1989) uncomplicated headache without other neurologic signs or symptoms was not accompanied by either the CSF or CT changes that are often associated with neuro-Behçet's disease. In this study these uncomplicated headaches were found generally to antedate or occur simultaneously with the onset of Behçet's disease and they were often migrainous in character. These observations suggest that uncomplicated headache may not portend other nervous system involvement in patients with Behçet's disease. In neuro-Behçet's disease, other symptoms or signs such as hemiparesis, papilledema, ataxia, cranial nerve palsies, signs of meningeal irritation, depression, personality and cognitive changes, memory disturbances and seizures occur in different combinations during the course of illness (Table 2).

Pallis and Fudge (1956) proposed a division of patients with neuro-Behçet's disease into three distinct clinical syndromes – a brainstem syndrome, a meningomyelitis syndrome, and a confusional syndrome. The validity of this classification scheme has been questioned (e.g., Assaad-Khalil et al. 1993) and

TABLE 2

Frequency of occurrence of the neurologic
manifestations of neuro-Behçet's disease.*

Neurologic manifestation	% involved
Headache	70. 1
Reactive depression	65.5
Hemiparesis/hemiplegia	25.5
Meningitis	19.5
Pseudobulbar palsy	19.5
Cerebellar dysfunction	18.3
Paresthesias	16.1
Personality change	10.3
Pseudotumor cerebri	9.1
Paraplegia	9.1
Cranial nerve palsy	9.1
Epileptic seizures	6.9
Memory disturbance	5.7
Dural sinus thrombosis	5.7
Alteration of consciousness	4.6
Paranoid schizophrenia	4.6
Hypomania	2.3

* Based on a survey of 87 patients with neuro-Behçet's disease (Assaad-Khalil et al. 1993).

a review of Pallis and Fudge's cases does not generally support their proposed classification scheme. Indeed, most of their cases (across clinical categories) had a mixture of meningeal signs and symptoms, varied neurologic manifestations and a remitting and relapsing or a relapsing progressive course.

Other authors have also attempted to classify the neurologic syndromes that occur in Behçet's disease. Thus, Shakir et al. (1990) divided cases into those patients with elevated intracranial pressure, those with cerebrovascular accidents, and those with spinal cord syndromes. This classification scheme, however, was based on a small sample (10 patients) and does not account for the large number of patients in the literature who have evidence of parenchymal brain and brainstem disease but without cerebrovascular accidents. Similarly, Bohlega et al. (1993) grouped their patients into those with dural sinus thrombosis, those with strokes in a major arterial distribution, and those with parenchymal brainstem or spinal cord syndromes. Again, such a scheme does not account for the patients with cerebral disease reported in the literature and, in any event, large vessel occlusion with subsequent stroke is very rare in Behçet's disease.

Rather than dividing patients from small samples into categories, however, it seems more appropriate to classify patients according to the known pathology of the illness, recognizing that, on occasion, different pathologies may co-exist in the same patient. The three most important pathologic changes in the nervous system in Behçet's disease are: (a) diffuse perivascular infiltration of the leptomeninges; (b) small multifocal necrotic lesions in gray and white matter of the brain, brainstem, and spinal cord; and (c) thrombosis of the dural sinuses. The clinical syndromes of neuro-Behçet's disease, therefore, might be anticipated to reflect these pathological changes as well as the tendency of the illness to be a recurrent, remitting-relapsing disease in its other clinical manifestations.

MS-like syndrome

Review of the published cases on neuro-Behçet's disease indicates that the large majority of patients with neuro-Behçet's disease have a clinical course that bears a striking resemblance to the course of MS (e.g., Pallis and Fudge 1956; Schotland et al. 1963; O'Duffy and Goldstein 1976; Besana et al. 1989; Serdaroglu et al. 1989; Hentati et al. 1993). Thus, patients with neuro-Behçet's disease often present with a multifocal remitting-relapsing, relapsing-progressive, or chronic progressive syndrome involving the brain, brainstem, and spinal cord. For example, among the 38 patients reported by Schotland et al. (1963), 24/38 (63%) had a fluctuating course, 10/38 (26%) had a chronic progressive course, and 4/38 (11 %) had an apparently benign monophasic course. The experience of Hentati et al. (1993) was similar with 55/141 (39%) having a relapsing course, 79/141 (56%) a progressive course, and 7/141 (5%) a benign course. Despite the similar clinical presentation to MS, however, there are important differences as well. Patients with neuro-Behçet's disease are more likely to have signs and symptoms of meningeal irritation than patients with MS and, consequently, the CSF pleocytosis that occurs in neuro-Behçet's is generally more conspicuous than that seen during an acute attack of MS. Visual impairment is common in both diseases but in neuro-Behçet's disease this symptom is generally due to posterior uveitis and not to retrobulbar neuritis (Motomura et al. 1980; Kansu et al. 1989, 1991). Also mental and speech disturbances are common during attacks of neuro-Behçet's disease whereas these symptoms are rare in MS (Motomura et al. 1980). Conversely, sensory signs and symptoms are common in MS (e.g., Mathews et al. 1984) but are unusual in neuro-Behçet's disease (Wadia and Williams 1957; Rubenstein and Urich 1963; O'Duffy and Goldstein 1976; Assaad-Khalil et al. 1993; Al Dalaan et al. 1994). Paroxysmal symptoms such as L'Hermitte's sign, tonic seizures, and paroxysmal dysarthria-ataxia, although uncommon in MS, are thought to be highly characteristic of that disorder (see Mathews et al. (1984) for a review). Such symptoms, by contrast, are much less common in neuro-Behçet's disease (Motomura et al. 1980) although they are occasionally reported (e.g., Akman-Demir et al. 1995; Falga-Tirado et al. 1995). Neuro-Behçet's disease also tends to be more progressive than MS with shorter intervals between relapses. Both MS and neuro-Behçet's disease demonstrate similar-appearing lesions on MRI although the distribution of lesions is different in the two disorders. Neuro-Behçet's disease tends to involve the brainstem and spare the periventricular white matter whereas MS does the converse (Banna and El-Ramahi 1991). Also neuro-Behçet's disease involves the gray matter more often than MS – a finding that may be better demonstrated by SPECT imaging than by MRI (e.g., Watanabe et al. 1995). Both conditions can lead to multimodality evoked potential abnormalities but these are more common in MS (Besana et al. 1989; Benli et al. 1993; Spadaro et al. 1993).

There are also important epidemiological differences between the two disorders. Thus, the male preponderance of Behçet's disease, the high prevalence of Behçet's disease in areas with a low prevalence of MS and vice versa, the different HLA linkage (e.g., Mathews et al. 1984; Moore and O'Duffy 1986), and the occurrence of antecedent recurrent oral and genital ulcers in Behçet's disease generally make distinction between the two disorders easy. However, in some individual cases the distinction may be quite difficult.

Dural sinus thrombosis and intracranial hypertension

Dural sinus thrombosis is well described in Behçet's disease (e.g., Serdaroglu et al. 1989; Shakir et al. 1990; El Ramahi and Al Kawi 1991; Wechsler et al. 1992; Assaad-Khalil et al. 1993; Bohlega et al. 1993; Hentati et al. 1993; Al Dalaan et al. 1994; Daif

et al. 1995). The frequency of this complication has varied between reports of a low of less than 6% (Assaad-Khalil et al. 1993; Hentati et al. 1993) to a high of 30–40% (Shakir et al. 1990; Wechsler et al. 1992). The reason for this variation is unclear although it may relate, in part, to whether investigators have routinely used angiography to establish the diagnosis. For example, in the study of Assaad-Khalil et al. (1993), where CT was used, pseudotumor cerebri was reported in 9.1% of patients whereas dural sinus thrombosis was found in only 5.7%. However, because CT is normal in approximately half of patients with proven dural sinus thrombosis (e.g., Wechsler et al. 1992; Daif et al. 1995), the estimate of 5.7% is almost certainly low. Nevertheless, this cannot be the only reason for the discrepancy because, in a series of 152 patients with neuro-Behçet's disease described by Hentati et al. (1993), the prevalence of intracranial hypertension was only 3.5%. Thus, even though these authors did not perform angiograms, the prevalence of symptomatic dural sinus thrombosis in this series must have been even lower than 3.5%.

The signs and symptoms of dural sinus thrombosis are not specific to Behçet's disease. The large majority of patients present with signs and symptoms of increased intracranial pressure such as headache, papilledema, and vomiting. These are the only presenting features in over half of the patients (e.g., Wechsler et al. 1992; Daif et al. 1995). Other signs and symptoms that occur less commonly are focal neurologic deficits due to venous infarction (either hemorrhagic or ischemic), focal seizures, cranial nerve palsies, stupor and coma. Hypercoagulable states and infections are common causes of dural sinus thrombosis in many parts of the world but, in at least some countries where Behçet's disease has a high prevalence, this disease has been reported to be the single most common condition associated with dural sinus thrombosis (Daif et al. 1995). The superior sagittal sinus is the most commonly affected (Wechsler et al. 1992; Daif et al. 1995). Other sinuses such as the lateral sinus, the straight sinus, the transverse sinus, or the cavernous sinus may also be affected, either together with the superior sagittal sinus or, less commonly, alone. In either case the presenting symptoms and signs are similar (e.g., Wechsler et al. 1992).

The CSF pressure is generally elevated but may be normal in a small percentage of patients (e.g.,

Wechsler et al. 1992; Daif et al. 1995). Occasional patients may have pseudotumor cerebri and present with the signs and symptoms of intracranial hypertension such as headache and papilledema and yet have no angiographic or other abnormalities to explain their elevated CSF pressure (e.g., Wilkins et al. 1986; Shakir et al. 1990; Banna and El Ramahi 1991; El Ramahi and Al Kawi 1991; Bosch et al. 1995). CT scans, as mentioned earlier, are often normal in patients with proven dural sinus thrombosis and are therefore of little use in excluding the diagnosis. MRI is much more likely to demonstrate the thrombosis but even this method can sometimes give false-negative results (Daif et al. 1995). It is therefore important to use angiography (either primarily or secondarily) to evaluate patients with signs and symptoms of increased intracranial pressure.

Treatment of dural sinus thrombosis in Behçet's disease has generally consisted of heparin in conjunction with oral corticosteroid medication and/ or immunosuppressive drugs such as cyclophosphamide, azathioprine, or chlorambucil (El Ramahi and Al Kawi 1991; Wechsler et al. 1992; Daif et al. 1995). Some patients have been treated with corticosteroids or heparin alone and many have been maintained on long-term anticoagulants or antiplatelet agents (Wechsler et al. 1992). Most patients do well regardless of the treatment regimen selected (Wechsler et al. 1992; Daif et al. 1995)

Aseptic meningitis

As mentioned earlier, signs and symptoms of meningeal irritation are frequent accompaniments of an acute attack of neuro-Behçet's disease. During such attacks there are generally also signs and symptoms of parenchymal disease of the brain, brainstem or spinal cord. On occasion, however, these other manifestations may be absent and the patient may present with the syndrome of either monophasic or recurrent aseptic meningitis (e.g., O'Duffy and Goldstein 1976; Serdaroglu et al. 1989; Assaad-Khalil et al. 1993). Some authors have estimated that this is the initial neurologic syndrome in 40–45% of patients with CNS involvement (e.g., O'Duffy and Goldstein 1976), although in a prospective series of patients followed by Serdaroglu et al. (1989) this presentation was found in only 2 of their 17 patients (12%) with neuro-Behçet's disease. Also, on

occasion, recurrent episodes of aseptic meningitis may precede other evidence of nervous system involvement by several years (e.g., O'Duffy and Goldstein 1976).

Neuropsychiatric syndromes

Neuropsychiatric syndromes have long been known to occur in patients with neuro-Behçet's disease and, in fact, were the basis for one of the three clinical categories of the disease suggested by Pallis and Fudge (1956). The individual manifestations are quite varied clinically but include such disturbances as personality changes, dementia, psychosis, memory disturbances and alteration of consciousness (Pallis and Fudge 1956; Schotland et al. 1963; Serdaroglu et al. 1989; Assaad-Khalil et al. 1993; Bohlega et al. 1993). Excluding a reactive depression from chronic disease, however, these syndromes are relatively infrequent manifestations of neuro-Behçet's disease (Pallis and Fudge 1956; Schotland et al. 1963; Serdaroglu et al. 1989; Assaad-Khalil et al. 1993; Bohlega et al. 1993).

Seizures

Epileptic seizures are generally reported to occur in 4–8% of cases in most large series of Behçet's disease (e.g., Assaad-Khalil et al. 1993; Hentati et al. 1993; Al Dalaan et al. 1994). These seizures can be either focal or generalized (e.g., Schotland et al. 1963) and generally respond well to anticonvulsant medications.

Peripheral nervous system and muscle involvement

Involvement of the peripheral nervous system (PNS) or muscle is very rare in Behçet's disease. There are, however, occasional reports of PNS dysfunction that can take several different clinical forms such as mononeuropathy multiplex, polyneuropathy, or radiculopathy (see Takeuchi et al. 1989, for a review). When nerve biopsies have been done on patients with these clinical manifestations (mostly on patients with mononeuropathy multiplex), the pathology has generally demonstrated a vasculitis (Takeuchi et al. 1989). PNS involvement may also be a complication of some of the medications used to treat Behçet's disease such as thalidomide (e.g., Hamza et al. 1993). There are also occasional reports of autonomic nervous system dysfunction in patients with Behçet's disease. Thus, Dilsen et al. (1993) reported an increase in sympathetic nervous system function in patients with Behçet's disease as measured by a change in the galvanic skin response (GSR) in different experimental conditions. Surprisingly, however, the patients with the neuro-Behçet's form of the disease did not show this effect. The implication of this observation is unclear.

There are also occasional reports of myositis occurring in Behçet's disease and this complication may be more prevalent in children than adults with this illness (e.g., Lang et al. 1990).

TREATMENT

The earliest agents used in the treatment of Behçet's disease were therapies directed against infections such as syphilis and tuberculosis (e.g., Whitwell 1934; Behçet 1937). These therapies, however, were generally unhelpful in ameliorating the signs and symptoms of the illness. Other early treatments including vitamins, bismuth, gold, arsenic, iodides, smallpox vaccine, estrogens, antibiotics, antihistamines and transfusions were similarly unsuccessful (France et al. 1951; Wadia and Williams 1957). By the 1950s corticosteroids or ACTH were found by some authors to be useful in reducing the recurrences of oral and genital aphthous ulcerations and in hastening the recovery from uveitis although the benefit of this treatment in patients with meningoencephalitis was less clear (e.g., Wadia and Williams 1957). The value of oral or topical corticosteroids, however, remained somewhat controversial (e.g., Wolf et al. 1965; O'Duffy et al. 1971). By the 1970s the use of immunosuppressive therapy was introduced and drugs such as chlorambucil and cyclophosphamide were reported to be helpful (e.g., O'Duffy 1978; Shimizu et al. 1979; Shahram et al. 1993). Indeed, these agents are still the principal form of treatment of Behçet's disease, often in combination with oral corticosteroids. A variety of other immunosuppressive agents (e.g., azathioprine, cyclosporine, and, more recently, FK506) have also been reported to be useful (e.g., Jorizzo 1986; O'Duffy et al. 1990; Mochizuki et al. 1993). It is unclear which treatment

regimen is the best. For example, O'Duffy (1990) reports that chlorambucil (0.1 mg/kg/day) will control most patients with posterior uveitis or meningoencephalitis (i.e., those complications causing the majority of the morbidity in Behçet's disease). The dose of chlorambucil can generally be reduced over time but prolonged treatment is often necessary (O'Duffy 1990). Others have argued, by contrast, that, because of the risks of infertility and chromosomal damage from chlorambucil, this therapy should be reserved for refractory cases and that initial treatment should consist of prednisone (1 mg/kg/day) in conjunction with azathioprine (Jorizzo 1986).

Other drugs such as colchicine and thalidomide have also been used successfully to suppress the aphthous ulcerations (e.g., Matsumura and Mizushima 1975; Mizushima et al. 1977; O'Duffy 1977; Jorizzo 1986; Denman et al. 1993) but the effect of these agents on the CNS manifestations is not known. Most recently, favorable experience has been reported using interferons (α-2a, α-2b and γ) in patients with Behçet's disease (e.g., Mahrle and Schulze 1990; Alpsoy et al. 1994; Hamuryudan et al. 1994). Again, however, although these agents seem particularly effective at reducing the non-neurologic manifestations of Behçet's disease, their effect on the meningoencephalitis remains to be proven.

Acknowledgement

I am very grateful to Dr. Ralf Siedenberg who helped me translate Behçet's original manuscript (1937).

REFERENCES

ADAMANTIADES, B.: Sur un cas d'iritis à hypopyon récidivante (cited by Alemá 1978). Ann. Oculist. 168 (1931) 271–278.

AKMAN-DEMIR, F.G., M. ERAKSOY, I.H. GURVIT, G. SARUHAN-DIRESKENELI and O. ARAH.: Paroxysmal dysarthria and ataxia in a patient with Behçet's disease. J. Neurol. 242 (1995) 344–347.

AL DALAAN, S.R. AL BALAA, K. EL-RAMAHI, Z. AL KAWI, S. BOHLEGA, S. BAHABRI and M.A. AL-JANADI: Behçet's disease in Saudi Arabia. J. Rheumatol. 21 (1994) 658–659.

ALEMÁ, G.: Behçet's disease. In: P.J. Vinken and G.W. Bruyn (Eds.), Handbook of Clinical Neurology, Vol. 34. Amsterdam, North Holland Publishing (1978) 475–512.

ALEMÁ, G. and A. BIGNAMI: Involvement of the nervous system in the Behçet's disease. In: International Symposium on Behçet's Disease. New York, Karger (1966) 52–66.

ALM, L. and L. ÖBERG: Animal experiment in connection with Behçet's syndrome: preliminary report (cited by Pallis and Fudge 1956). Nord. Med. 25 (1945) 603–604.

ALPSOY, E., E. YILMAZ and E. BASARAN: Interferon therapy for Behçet's disease. J. Am. Acad. Dermatol. 31 (1994) 617–619.

ASSAAD-KHALIL, S., M. ABOU-SEIF, S. ABOU-SEIF, F. EL-SEWY and M. EL SAWY: Neurologic involvement in Behçet's disease: clinical genetic and computed tomographic study. In: P. Godeau and B. Wechsler (Eds.), Behçet's Disease. Proc. 6th Int. Conf. Behçet's Dis., Paris. Amsterdam, Elsevier Science Publishers (1993) 409–414.

BALCON, E., S. OZGUN, T. AKOGLU and H. YAZICI: Antibodies to keratin components in patients with Behçet's disease as compared to other rheumatological disorders. In: B. Wechsler and P. Godeau (Eds.), Behçet's Disease. Proc. 6th Int. Conf. Behçet's Dis., Paris. Amsterdam, Elsevier Science Publishers (1993) 87–90.

BANNA, M. and K. EL RAMAHI: Neurologic involvement in Behçet disease: imaging findings in 16 patients. Am. J. Neuroradiol. 12 (1991) 791–796.

BEHÇET'S DISEASE RESEARCH COMMITTEE OF JAPAN: Behçet's disease: guide to diagnosis of Behçet's disease. Jpn. J. Ophthalmol. 18 (1974) 291–294.

BEHÇET, H.: Geschwüre am Mund, am Auge und an den Genitalien. Dermatol. Wochenschr. 36 (1937) 1152–1153.

BEHÇET, H.: Considérations sur les lésions aphteuses de la bouche et des parties génitales ainsi que sur les manifestations oculaires d'origine probablement parasitaire et observations concernant leur foyer d'infection. Bull. Soc. Fr. Dermatol. Syph. 45 (1938) 420–433.

BEHÇET, H.: A propos d'une entité morbide dûe probablement à un virus spécial donnant lieu à une infection généralisée se manifestant par des poussées récidivantes en trois régions principales et ocasionnant en particulier des iritis répétés. Bull. Soc. Fr. Dermatol. Syph. 46 (1939) 476–687.

BEHÇET, H.: Some observations on the clinical picture of the so-called triple symptom complex. Dermatologica 81 (1940) 73–83.

BENLI, S., G. NURLU, T. KANSU, O. SARIBAS and M. DURGUNER: Multimodal evoked potentials, electroneuromyography and clinical correlations in neuro-Behçet disease. In: B. Wechsler and P. Godeau (Eds.), Behçet's Disease. Proc. 6th Int. Conf. Behçet's Dis., Paris. Amsterdam, Elsevier Science Publishers (1993) 419–424.

BESANA, C., G. COMI, A.D. MASCHIO, L. PRADERIO, A. VERGANI, S. MEDAGLINI, V. MARTINELLI, F. TRIULZI and T. LOCATELLI: Electrophysiological and MRI evaluation of neurological involvement in Behçet's disease. J. Neurol. Neurosurg. Psychiatry 52 (1989) 749–754.

BOHLEGA, S., M.S. AL KAWI, S. OMER, D. MCLEAN, B. STIGSBY, A. AL DALAAN, K.M. EL RAMAHI and S. AL-BALAA: Neuro-Behçet's: clinical syndromes and prognosis. In: P. Godeau and B. Wechsler (Eds.), Behçet's Disease. Proc. 6th Int. Conf. Behçet's Dis., Paris. Amsterdam, Elsevier Science Publishers (1993) 429–434.

BOSCH, J.A., M. VALES, R. SOLANS, J. MONTALBAN and M. VILARDELL: Skin hyperreactivity in patients with benign intracranial hypertension as an early manifestation of Behçet's disease. Br. J. Rheumatol. 34 (1995) 184.

BUSKILA, D., D.D. GLADMAN, J. GILMORE and I.E. SDALIT: Behçet's disease in a patient with immunodeficiency virus infection. Ann. Rheum. Dis. 50 (1991) 115–116.

CHAJEK, T. and M. FAINARU: Behçet's disease. Report of 41 cases and review of the literature. Medicine (Baltimore) 54 (1975) 179–196.

CHAMBERLAIN, M.A.: Behçet's syndrome in 32 patients in Yorkshire. Ann. Rheum. Dis. 36 (1977) 491–499.

CRESPO, J., J. RIBEIRO, E. JESUS, A. MOURA, C. REIS and A. PORTO: Behçet's disease – particular features at the central zone of Portugal. In: P. Godeau and B. Wechsler (Eds.), Behçet's Disease. Proc. 6th Int. Conf. Behçet's Dis., Paris. Amsterdam, Elsevier Science Publishers (1993) 207–210.

DAIF, A., A. AWADA, S. AL RAJEH, M. ABDULJABBAR, A. RAHAMAN, A. TAHAN, R. OBEID and T. MALIBARY: Cerebral venous thrombosis in adults. Stroke 26 (1995) 1192–1993.

DENMAN, A.M., E. GRAHAM, L. HOWE, E.J. DENMAN and S. LIGHTMAN: Low dose thalidomide treatment of Behçet's syndrome. In: B. Wechsler and P. Godeau (Eds.), Behçet's Disease. Proc. 6th Int. Conf. Behçet's Dis., Paris. Amsterdam, Elsevier Science Publishers (1993) 649–652.

DILSEN, N., M. KONICE and O. ARAL: Our diagnostic criteria of Behçet's disease – an overview. In: T. Lehner and G.D. Barnes (Eds.), Recent Advances in Behçet's Disease. Int. Congr. Ser. 103. Amsterdam, Excerpta Medica (1986) 171–180.

DILSEN, G., A. YAHMAN, H. GURVIT, P. SERDAROGLU, N. DILSEN, M. KONICE, O. ARAL, L. OCAL and M. INANC: Autonomic dysfunction in Behçet's disease with and without nervous system involvement. In: B. Wechsler and P. Godeau (Eds.), Behçet's Disease. Proc. 6th Int. Conf. Behçet's Dis., Paris. Amsterdam, Elsevier Science Publishers (1993) 443–446.

EL RAMAHI, K. and M.S. AL KAWI: Papilloedema in Behçet's disease: value of MRI in diagnosis of dural sinus thrombosis. J. Neurol. Neurosurg. Psychiatry 54 (1991) 826–827.

EVANS, A.D., D.A. PALLIS and J.D. SPILLANE: Involvement of the nervous system in Behçet's syndrome: report of three cases and isolation of virus. Lancet ii (1957) 349–353.

FALGA-TIRADO, C., J. ORDI-ROS, P. PEREZ-PEMAN, E. CURURULL-CANOSA and J. BOSCH-GIL: L'Hermitte's sign in Behçet's disease. Br. J. Rheum. 34 (1995) 184–185.

FEIGENBAUM, A.: Description of Behçet's syndrome in the Hippocratic third book of endemic diseases. Br. J. Ophthalmol. 40 (1956) 355–357.

FRANCE, R., R.N. BUCHANAN, M.W. WILSON and M.B. SHELDON: Relapsing iritis with recurrent ulcers of the mouth and genitalia (Behçet-syndrome) (Review with report of an additional case). Medicine (Baltimore) 30 (1951) 335–355.

FRIERE, E., J. CORREIA, B. LEAO, C. VASCONCELOS, P. TORRES, H. BRANCO and M. DA SILVA: Behçet's disease report on the experience in an internal medicine department. In: B. Wechsler and P. Godeau (Eds.), Behçet's Disease. Proc. 6th Int. Conf. Behçet's Dis., Paris. Amsterdam, Elsevier Science Publishers (1993) 201–206.

GHARIBDOOST, F., F. DAVATCHI, M. SHAHRAM, C. AKBARIAN, H. CHAMS, P. CHAMS, P. MANSOORI and A. NADJI: Clinical manifestations of Behçet's disease in Iran – analysis of 2176 cases. In: P. Godeau and B. Wechsler (Eds.), Behçet's Disease. Proc. 6th Int. Conf. Behçet's Dis., Paris. Amsterdam, Elsevier Science Publishers (1993) 153–158.

HAMURYNDAN, V., F. MORAL, S. YURDAKUL, C. MAT, Y. TUZUN, Y. OZYAZGAN, H. DIRESKENELI, T. AKOGLU and H. YAZICI: Systemic interferon α2b treatment in Behçet's syndrome. J. Rheumatol. 21 (1994) 1098–1100.

HAMZA, M. and R. BARDI: Familial Behçet's disease. In: P. Godeau and B. Wechsler (Eds.), Behçet's Disease. Proc. 6th Int. Conf. Behçet's Dis., Paris. Amsterdam, Elsevier Science Publishers (1993) 391–794.

HAMZA, M., M. OUESLATI and B. HAMIDA: Peripheral neuropathy in Behçet's disease. In: B. Wechsler and P. Godeau (Eds.), Behçet's Disease. Proc. 6th Int. Conf. Behçet's Dis., Paris. Amsterdam, Elsevier Science Publishers (1993) 653–654.

HENTATI, F., M. FREDJ, N. GHARBI and M.B. HAMIDA: Clinical and biological aspects of Neuro-Behçet's (NB) in Tunisia. In: P. Godeau and B. Wechsler (Eds.), Behçet's Disease. Proc. 6th Int. Conf. Behçet's Dis., Paris. Amsterdam, Elsevier Science Publishers (1993) 415–418.

ISGBD (INTERNATIONAL STUDY GROUP FOR BEHÇET'S DISEASE): Criteria for diagnosis of Behçet's disease. Lancet 335 (1990) 1078–1080.

ISOGAI, E., H. ISOGAI, S. KOTAKE, K. YOSHIKAWA, A. ICHISHI, S.

KJOSAKA, N. SATO, S. HAYASHI, K. OGUMA and S. OHNO: Autoantibodies against Borrelia burgdorferi in patients with uveitis. Am. J. Ophthalmol. 112 (1991) 23–30.

ISOGAI, E., K. YOKOTA, N. FUJII, S. HAYASHI, K. KIMURA, H. ISOGAI, N. ISHII, Y. YAMAKAWA, H. NAKAJIMA, S. OHNO, S. KOTAKE and K. OGUMA: Characterization and functional properties of Streptococcus sanguis isolated from patients with Behçet's disease. In: B. Wechsler and P. Godeau (Eds.), Behçet's Disease. Proc. 6th Int. Conf. Behçet's Dis., Paris. Amsterdam, Elsevier Science Publishers (1993) 73–82.

JENSEN, T.: Sur les ulcérations aphteuses de la muqueuse de la bouche et de la peau génitale combinées avec les symptômes oculaires. Acta Derm. Venereol. 22 (1941) 64–79.

JORIZZO, J.I.: Behçet's disease. Arch. Dermatol. 122 (1986) 556–557.

JORIZZO, J.I., J. ABERNETHY, W.I. WHITE, H.C. MANGELSDORF, C.C. ZOUBOULIS, R. SARICA, K. GAFFNEY, C. MAT, H. YAZICI, A. AL ISLAAN, S.H. ASSAAD-KHALIL, F. KANEKO and E.A.F. JORIZZO: Mucocutaneous criteria for the diagnosis of Behçet's disease: an analysis of clinicopathologic data from multiple international centers. J. Am. Acad. Dermatol. 32 (1995) 968–976.

KANSU, T., P. KIRKALI, E. KANSU and T. ZILELI: Optic neuropathy in Behçet's disease. J. Clin. Neuroophthalmol. 9 (1989) 277–280.

KANSU, T., E. KANSU, T. ZILELI and P. KIRKALI: Neuroophthalmologic manifestations of Behçet's disease. J. Neuroophthalmol. (1991) 7–11.

KNAPP, P.: Beitrag zur Symptomatologie und Therapie der rezidivierenden Hypopyoniritis und der begleitenden aphtösen Schleimhauterscheinungen. Schweiz. Med. Wochenschr. 71 (1941) 1288–1290.

KÖTTER, I., H. DÜRK, G. FLERBECK, U. PLEYER, M. ZIERHUT and J.G. SAAL: Behçet's disease in 39 German and mediterranean patients. In: P. Godeau and B. Wechsler (Eds.), Behçet's Disease. Proc. 6th Int. Conf. Behçet's Dis., Paris. Amsterdam, Elsevier Science Publishers (1993) 197–200.

KOZIN, F., V. HAUGHTON and G.C. BERNHARD: Neuro-Behçet disease: two cases and neuroradiologic findings. Neurology 27 (1977) 1148–1152.

LAKHANPAL, S., K. TANI, J.T. LIE, K. KATOH, Y. ISHIGATSUBO and T. OHOKUBO: Pathologic features of Behçet's syndrome: a review of Japanese autopsy registry data. Hum. Pathol. 16 (1985) 790–795.

LANG, R.A., R.M. LAXER, P. THORNER, M. GREENBERG and E.D. SILVERMAN: Pediatric onset of Behçet's syndrome with myositis: case report and literature review illustrating unusual features. Arthr. Rheum. 33 (1990) 418–419.

LEE, E.-S., S. LEE, D. BANG, Y.H. CHO and S. SOHN: Detection of herpes simplex virus DNA by polymerase chain reaction in saliva of patients with Behçet's disease.

In: B. Wechsler and P. Godeau (Eds.), Behçet's Disease. Proc. 6th Int. Conf. Behçet's Dis., Paris. Amsterdam, Elsevier Science Publishers (1993) 83–86.

MAHRLE, G. and H.-J. SCHULZE: Recombinant interferon-gamma (rIFN-gamma) in dermatology. J. Invest. Dermatol. 95 (1990) 132S–137S.

MASON, R.M. and C.G. BARNES: Behçet's syndrome with arthritis. Ann. Rheum. Dis. 28 (1969) 95–103.

MATHEWS, W.B., E.D. ACHESON, J.R. BATCHELOR and R.O. WELLER: In: W.B. Mathews (Ed.), McAlpine's Multiple Sclerosis. Edinburgh, Churchill Livingston (1984).

MATSUMURA, N. and Y. MIZUSHIMA: Leucocyte movement and colchicine treatment in Behçet's disease. Lancet ii (1975) 813.

MCLEAN, B.N. and D.M. THOMPSON: Oligoclonal banding of IgG in CSF, blood–brain barrier function and MRI findings in patients with sarcoidosis, systemic lupus erythematosus and Behçet's disease involving the nervous system. J. Neurol. Neurosurg. Psychiatry 48 (1995) 548–554.

MCMENEMEY, W.H. and B.L. LAWRENCE: Encephalomyelopathy in Behçet's disease. Lancet ii (1957) 253–358.

MIZUKI, N., H. INOKO, H. ANDO, S. NAKAMURA, K. KASHIWASE, H. AKAZA, Y. FUJINO, K. MASUDA, M. TAKIGUCHI and S. OHNO: Behçet's disease associated with one of the HLA-B51 subantigens, HLA-B*5101. Am. J. Ophthalmol. 116 (1993) 406–409.

MIZUKI, N., N. INOKO, M. ISHIRAHA, H. ANDO, S. NAKAMURA, M. NISHIO and S. OHNO: A complete type patient with Behçet's disease associated with HLA-B*5102. Acta Ophthalmol. 72 (1994) 757–758.

MIZUSHIMA, Y., N. MATSUMURA, M. MORI, T. SHIMIZU, B. FUKUSHIMA, Y. MIMURA, K. SAITO and S. SUIGIURA: Colchicine in Behçet's disease. Lancet 12 (1977) 1037.

MOCHIZUKI, M. and JAPANESE FK506 STUDY GROUP ON REFRACTORY UVEITIS IN BEHÇET'S DISEASE: In: B. Wechsler and P. Godeau (Eds.), Behçet's Disease, Proc. 6th Int. Conf. Behçet's Dis., Paris. Amsterdam, Elsevier Science Publishers (1993) 655–660.

MOORE, S.F. and J.D. O'DUFFY: Lack of association between Behçet's disease and major histocompatibility complex class II antigens in an ethnically diverse North American caucasoid patient group. J. Rheumatol. 13 (1986) 771–773.

MOTOMURA, S., T. TABIRA and Y. KUROIWA: A clinical comparative study of multiple sclerosis and neuro-Behçet's syndrome. J. Neurol. Neurosurg. Psychiatry 43 (1980) 210–213.

MUNKE, H., F. STROCKMANN and G. RAMADORI: Possible association between Behçet's syndrome and chronic hepatitis C. N. Engl. J. Med. 332 (1995) 400–401.

MUTLU, S. and C. SCULLY: The person behind the eponym: Hulûsi Behçet (1889–1948). J. Oral Pathol. Med. 23 (1994) 289–290.

NAKAE, K., F. MASAKI, T. HASHIMOTO, G. INABA, M. MOCHIZUKI and T. SAKANE: Recent epidemiological features of Behçet's disease in Japan. In: P. Godeau and B. Wechsler (Eds.), Behçet's Disease. Proc. 6th Int. Conf. Behçet's Dis., Paris. Amsterdam, Elsevier Science Publishers (1993) 145–1562.

NISHIMURA, M., K. SATOH, M. SUGA and M. ODA: Cerebral angio- and neuro-Behçet's syndrome: neuroradiological and pathological study of one case. J. Neurol. Sci. 106 (1991) 19–24.

NUMAN, F., C. ISLAK, R. BERKMEN, H. TUZUN and O. COKYUK-SET: Behçet disease: pulmonary arterial involvement in 15 cases. Radiology 192 (1994) 465–468.

O'DUFFY, J.D.: Critères proposés pour le diagnostic de la maladie de Behçet et notes thérapeutiques. Rev. Med. 36 (1974a) 2371–2379.

O'DUFFY, J.D.: Suggested criteria for diagnosis of Behçet's disease (BD). J. Rheumatol. 1, Suppl. 1 (1974b) 18.

O'DUFFY, J.D.: Summary of international symposium on Behçet's disease. J. Rheumatol. 5 (1978) 229–233.

O'DUFFY, J.D.: Prognosis in Behçet's syndrome. In: N. Dilsen, M. Konice and C. Ovul (Eds.), Behçet's Disease. Amsterdam, Elsevier Science Publishers (1979) 191–195.

O'DUFFY, J.D.: Vasculitis in Behçet's disease. Rheum. Dis. Clin. N. Am. 16 (1990) 423–431.

O'DUFFY, J.D. and N.P. GOLDSTEIN: Neurologic involvement in seven patients with Behçet's disease. Am. J. Med. 61 (1976) 170–171.

O'DUFFY, J.D., A. CARNEY and S. DEODHAR: Report of 10 cases, 3 with new manifestations. Ann. Intern. Med. 75 (1971) 561–570.

O'DUFFY, J.D., H.F. TASWELL and L.R. ELVEBACK: HL-A antigens in Behçet's disease. J. Rheumatol. 3 (1976) 1–3.

O'NEILL, T.W., A.S. RIGBY, S. MCHUGH, A.J. SILMAN and C. BARNES: Regional differences in clinical manifestations of Behçet's disease. In: P. Godeau and B. Wechsler (Eds.), Behçet's Disease. Proc. 6th Int. Conf. Behçet's Dis., Paris. Amsterdam, Elsevier Science Publishers (1993) 159–164.

OHNO, S., E. NAKAYAMA, S. SUIGIURA, K. ITAKURA, K. AOKI and M. AIZAWA: Specific histocompatibility antigens associated with Behçet's disease. Am. J. Ophthalmol. 80 (1975) 636–637.

PALLIS, C.A. and B.J. FUDGE: The neurological complications of Behçet's syndrome. Arch. Neurol. Psychiatry 74 (1956) 1–14.

PEDERSEN, A., H.O. MADSEN, B.F. VESTERGAARD and L.P. RYDER: Varicella-zoster virus DNA in recurrent aphthous ulcers. Scand. J. Dent. Res. 101 (1993) 311–313.

RESEARCH COMMITTEE ON BEHÇET'S DISEASE (JAPAN):

Behçet's disease: guide to diagnosis of Behçet's disease Jpn. J. Ophthalmol. 18 (1974) 282–290.

RUBINSTEIN, L.J. and H. URICH: Meningo-encephalitis of Behçet's disease: case report with pathological findings. Brain 86 (1963) 151–160.

SAKAMOTO, M., K. AKAZAWA, Y. NISHIOKA, H. SANUI, J. INOMATA and Y. NOSE: Prognostic factors of vision in patients with Behçet disease. Ophthalmology 102 (1995) 317–321.

SCHERRER, M.A., N. VITRAL and F. OREFICE: Clinical study of Behçet's disease in Brazil. In: P. Godeau and B. Wechsler (Eds.), Behçet's Disease. Proc. 6th Int. Conf. Behçet's Dis., Paris. Amsterdam, Elsevier Science Publishers (1993) 181–184.

SCHOTLAND, D.L., S.M. WOLF, H.H. WHITE and H.V. DUBIN: Neurologic aspects of Behçet's disease. Am. J. Med. 34 (1963) 544–553.

SERDAROGLU, P., H. YAZICI, C. OZDEMIR, S. YURDAKUL, S. BAHAR and E. ATKIN: Neurologic involvement in Behçet's syndrome. Arch. Neurol. 46 (1989) 265–269.

SEZER, F.N.: The isolation of a virus as the cause of Behçet's disease. Am. J. Ophthalmol. (1953) 301–315.

SHAFAIE, N., F. SHAHRAM, F. DAVATCHI, M. AKBARIAN, F. GHARIBDOOST and A. NADJI: Behçet's disease in children. In: P. Godeau and B. Wechsler (Eds.), Behçet's Disease. Proc. 6th Int. Conf. Behçet's Dis., Paris. Amsterdam, Elsevier Science Publishers (1993) 381–383.

SHAHRAM, F., F. DAVATCHI, H. CHAMS, M. AKBARIAN and F. GHARIBDOOST: Low dose pulse cyclophosphamide (LDP) in ophthalmologic lesions of Behçet's disease. In: B. Wechsler and P. Godeau (Eds.), Behçet's Disease. Proc. 6th Int. Conf. Behçet's Dis., Paris. Amsterdam, Elsevier Science Publishers (1993) 683–686.

SHAKIR, R.A., K. SULAIMAN, R. KAHN and M. RUDWAN: Neurological presentation of neuro-Behçet's syndrome: clinical categories. Eur. Neurol. 30 (1990) 249–253.

SHARIEF, M.K., M. PHIL, R. HENTGES and E. THOMAS: Significance of CSF immunoglobulins in monitoring neurologic disease activity in Behçet's disease. Neurology 41 (1991) 1398–1401.

SHIGETA, T.: Recurrent iritis with hypopyon and its pathological findings (cited by Shimizu et al. 1979). Acta Soc. Ophthalmol. Jpn. 28 (1924) 516.

SHIMIZU, T., G.E. EHRLICH, G. INABA and K. HAYASHI: Behçet disease (Behçet syndrome). Semin. Arthr. Rheum. 8 (1979) 223–224.

SOYLU, M., T.R. ERSOZ and E. ERKEN: Circulating corneal epithelium antibodies and immune complexes in immunologic eye disorders. Ann. Ophthalmol. 25 (1993) 231.

SPADARO, M., G. SOLDATI, M.E. TERRACCIANO, P. GIUNTI, M. ACCORINTI, M. LA CAVA and P. PIVETTI-PEZZI: A neurological and electrophysiological study on Behçet's disease patients. In: B. Wechsler and P. Godeau (Eds.), Behçet's Disease. Proc. 6th Int. Conf. Behçet's Dis., Paris. Amsterdam, Elsevier Science Publishers (1993) 425–428.

TAKEUCHI, A., M. KODAMA, T. TAKATSU, H. HASHIMOTO and H. MIYASHITA: Mononeuritis multiplex in incomplete Behçet's disease: a case report and the review of the literature. Clin. Rheumatol. 8 (1989) 375–380.

TANG, F., R. HUANG, Y. DONG and N. ZHANG: Anti-PPD antibodies in Chinese Behçet's disease. Chin. Med. Sci. J. 6 (1991) 239–240.

TAYLOR, P.V., M.A. CHAMBERLAIN and J.S. SCOTT: Auto-reactivity in patients with Behçet's disease. Br. J. Rheumatol. 32 (1993) 908–910.

UHLENBROCK, D. and S. SEHLEN: The value of T1-weighted images in the differentiation between MS, white matter lesions and subcortical arteriosclerotic encephalopathy (SAE). Neuroradiology 31 (1989) 203–212.

WADIA, N. and E. WILLIAMS: Behçet's syndrome with neurological complications. Brain 80 (1957) 59–71.

WATANABE, N., H. SETO, S. SATO, M. SIMIZU, Y.-W. WU, M. KAGEYAMA, K. NOMURA and M. KAKISHITA: Brain SPECT with neuro-Behçet disease. Clin. Nucl. Med. 20 (1995) 61–64.

WECHSLER, B., M. VIDAILHET, J.C. PIETTE, M.G. BOUSSER, D. ISOLA, O. BLETRY and P. GODEAU: Cerebral venous thrombosis in Behçet's disease. Neurology 42 (1992) 614–618.

WHITWELL, G.P.B.: Recurrent buccal and vulval ulcers with associated embolic phenomena in skin and eye. Br. J. Dermatol. 46 (1934) 414–419.

WILKINS, M.R., R.I. GOVE, S.D. ROBERS and M.J. KENDALL: Behçet's disease presenting as benign intracranial hypertension. Postgrad. Med. J. 62 (1986) 39–41.

WOLF, S.H., SCHOTLAND, D.L. and L.L. PHILLIPS.: Involvement of nervous system in Behçet's syndrome. Arch. Neurol. 12 (1965) 315–325.

YAMAMOTO, S., H. TOYOKAWA, J. MATSUBARA, H. YANAI, U. INABA, K. NAKAE and M. ONO: A nation-wide survey of Behçet's disease in Japan. Jpn. J. Ophthalmol. 18 (1974) 282–290.

YAMAMOTO, J.H., M. MINAMI, G. INABA, K. MASUDA and M. MOCHIZUKI: Cellular autoimmunity to retinal specific antigens in patients with Behçet's disease. Br. J. Ophthalmol. 77 (1993) 584–589.

YAMAMOTO, J.H., Y. FUJINO, C. LIN, M. NIEDA, T. JUJI and K. MASUDA: S-antigen specific T-cell clones from a patient with Behçet's disease. Br. J. Ophthalmol. 78 (1994) 927–32.

YAZICI, H., Y. TUZUN, H. PAZARLI, B. YALCIN, S. YURDAKUL and A. MUFTUOGLU: The combined use of HLA-B5 and the pathergy test as diagnostic markers of Behçet's disease in Turkey. J. Rheumatol. 7 (1980) 206–210.

YAZICI, H., Y. TUZUN, H. PAZARLI, S. YURKADUL, Y. OZYAZGAN, H. OZDOGAN, S. SERDAROGLU, M. ERSANLI, B.Y. ULKU and A.U. MUFTUOGLU: Influence of age of onset and patient's sex on the prevalence and severity of manifestations of Behçet's syndrome. Ann. Rheum. Dis. 43 (1984) 783–789.

YOKOTA, K., S. HAYASHI, N. FUJII, K. YOSHIKAWA, S. KOTAKE, E. ISOGAI, S. OHNO, Y. ARAKI and K. OGUMA: Antibody response to oral streptococci in Behçet's disease. Microbiol. Immunol. 36 (1992) 815–822.

ZHANG, X.-Q.: (Cited by ISGBD 1990.) Chin. J. Intern. Med. 19 (1980) 15–20 (in Chinese).

II. Immune System Disorders

Handbook of Clinical Neurology, Vol. 27 (71): Systemic Diseases, Part III
M.J. Aminoff and C.G. Goetz, editors

The AIDS dementia complex and human immunodeficiency virus type 1 infection of the central nervous system

RICHARD W. PRICE

Department of Neurology, San Francisco General Hospital, San Francisco, CA, U.S.A.

Human immunodeficiency virus type one (HIV-1) is a retrovirus that causes a chronic or 'slow' infection. Its natural host is man and its primary target is the immune system. Over a variable number of years infection leads to the acquired immunodeficiency syndrome (AIDS) and, eventually, death in almost all of those infected. The course of this infection may be complicated by an array of neurological diseases affecting virtually every component of the central and peripheral nervous system and involving a variety of pathogenetic processes, including particularly opportunistic infections and neoplasms. The broad range of these complications is reviewed in other contributions to this volume and elsewhere (Snider et al. 1983; McArthur and McArthur 1986; Simpson and Tagliati 1994; Price 1996a). Likewise, excellent reviews of the biology, epidemiology, pathogenesis and systemic complications can be found in a number of textbooks and reviews (Broder et al. 1994; DeVita et al. 1997; Sande and Volberding 1997). This chapter deals with central nervous system (CNS) complications that are more directly linked to HIV-1, itself, rather than to a secondary opportunistic process, and focuses chiefly on the most common and important of these, the AIDS dementia complex (ADC). The latter is a syndrome of 'subcortical' dementia that concomitantly alters cognition, motor control and behavior, and characteristically manifests during the late phase of HIV-1 infection.

Because ADC is a component of systemic HIV-1 infection, this chapter begins with a brief considera-

tion of selected facets of systemic and CNS HIV-1 infections, emphasizing issues that are important to diagnosis, pathogenesis and treatment of ADC. Additionally, following a more lengthy consideration of ADC, the chapter also briefly discusses some other, less common, neurological complications of HIV-1 infection that may be caused or provoked by HIV-1 rather than another opportunistic infection.

HIV-1 AND SYSTEMIC INFECTION

HIV-1 is classified among the *lentiviruses* (Gonda et al. 1985), based upon genetic and biological similarities to several animal retroviruses. These include visna-maedi virus, one of the prototypes included by Sigurdsson in his conceptualization of slow infections (Sigurdsson et al. 1957). As with visna virus and the more closely related simian immunodeficiency virus (SIV), the double-stranded RNA of the complex HIV-1 genome encodes structural, enzyme and regulatory genes that together exploit host cells to assure its persistence and subsequent replication and transmission (Folks and Hart 1997). The initial step in cell infection involves interaction of the virion coat glycoprotein, gp120, with primary (CD4) and secondary (members of a family of chemokine receptors) cell surface receptors. By restricting the spectrum of infected cells, these surface interactions determine the predominating phenotype of the disease, including its defining profile of progressive cellular immunodeficiency and resultant opportunistic infections. Host-to-host virus transmission occurs during homo- and heterosexual contact, through passage from the mother to fetus or newborn, and by direct introduction of infected blood in the course of illicit drug use, needle stick or transfusion of blood and blood products; these restricted modes of transmission, in turn, explain the epidemiology and circumscribed individual patient risk of infection.

While the course of infection and progression to disease is slow, the underlying process of systemic infection is remarkably dynamic. Studies in the last few years have shown that infection, even during the period of 'clinical latency' when individuals feel well and are free of opportunistic infections, is usually chronically active with continuous production of new virus (Embretson et al. 1993; Pantaleo et al. 1993; Ho et al. 1995; Wei et al. 1995; Perelson et al. 1996). Indeed, both HIV-1 and CD4 cell turnover occur rapidly, with half-lives measured in days despite a disease course that is measured in years. Slowly, though, the immune system is altered, leading to deficiencies in critical defenses against certain invading microbes, principally those utilizing intracellular sites of replication. Concomitant with depression of cell-mediated defenses, some immune processes are 'activated' and lead to up-regulation in the production of certain cytokines that are produced by immune cells, principally macrophages (Fauci 1995; Pantaleo and Fauci 1995; Haynes et al. 1996). Both of these aspects of the immune disregulation caused by HIV-1, the deficiency and the immune activation, are probably important in the development of ADC as discussed below.

Another aspect of HIV-1 infection with practical implications is that the virus undergoes considerable genetic change during the course of infection (Coffin 1995). Because of the poor fidelity of the virus-coded RNA-dependent DNA polymerase or reverse transcriptase (RT), the mutation rate of HIV-1 during replication is high. Indeed, because of this ongoing mutation and selection, the host is considered to be infected by a virus 'swarm' or *quasispecies* that exhibits considerable diversity. The quasispecies undergoes constant genetic drift depending both on chance and the combined selective pressures of immune defenses, available cell types susceptible to infection, and antiviral drug therapies. These genetic changes can be traced using phylogenetic maps (McCutchan et al. 1996).

Staging systemic infection

Staging of disease progression and activity is a central aspect of clinical diagnosis and management of HIV-1-infected patients. This initially relied upon clinical manifestations – for example, development of thrush and certain other 'minor' sequelae defined the AIDS-related complex (ARC) while major infections such as cerebral toxoplasmosis or *Pneumocystis carinii* pneumonia (PCP) clinically defined AIDS (Center for Disease Control 1987). However, in recent years such clinical definitions have been supplemented and even supplanted by laboratory measurements. Two of these laboratory parameters are now an essential part of clinical practice, at least in the developed world: (1) the blood CD4$^+$ T-lymphocyte count and (2) quantitative measurement of the HIV-1 RNA content (so-called *viral load*) in plasma using nucleic acid amplification techniques (Saag et al. 1996). These two serve complementary functions. The CD4$^+$ count provides a measure of the accumulated damage to the immune system and tells where the patient currently is in the course of the infection. This, in turn, is critical in determining susceptibility to opportunistic diseases and other complications of HIV-1 which are highly *stage-associated*. The importance of this measure is indicated by the use of a CD4$^+$ count of 200 cells/mm^3 or less as a criterion for the diagnosis of AIDS in the revised Centers for Disease Control case definition (Center for Disease Control and Prevention 1992). As discussed below, this susceptibility applies to ADC as well. In a useful analogy, Coffin (1996) has likened the CD4$^+$ count to the *distance* a train has traveled to its destination – in this case toward a state of severe immunodeficiency and eventual death.

The plasma viral load has more recently become a part of everyday clinical practice and serves as a measure of the *activity of infection* in the individual patient (Saag et al. 1996). It has been shown to provide a powerful marker of prognosis, independent of the CD4$^+$ count (Mellors et al. 1995, 1996). In the same analogy alluded to above, the plasma viral load is compared to the *velocity* of the train and indicates how rapidly the patient is moving toward end-stage immunodeficiency and death. Hence, for example, a patient with a viral load in the millions is predicted to progress more rapidly in his or her immunodeficiency than one with a thousand or less copies per ml.

The major use of the viral load in clinical practice is to monitor the need for and effectiveness of anti-retroviral therapy (Hammer 1996). Clinicians now 'treat the viral load', aiming, if possible, to reduce the number of copies of HIV-1 RNA in plasma to unde-tectable levels (using current techniques with detec-tion sensitivities to 200–500 copies/ml although advancing technologies will likely soon routinely measure levels down to 20–50 copies or perhaps less) after initiation of treatment. Similarly, a rising vi-ral load is used as an indicator of the emergence of antiviral drug resistance and the need to change the treatment regimen. The advent and ready availability of this quantitative technology, in concert with the de-velopment of potent antiviral drugs and drug combi-nations, has revolutionized both the conceptualization of HIV-1 infection and its treatment.

Antiretroviral therapy

Because the details of contemporary antiretroviral therapy are beyond the scope of this presentation and likely to continue to evolve considerably over the life of this volume, I will emphasize only some general concepts. The major classes of drugs now in use tar-get viral genes and include RT inhibitors of two types, nucleosides and non-nucleosides, and HIV-1 pro-tease inhibitors (Hammer 1996). Among the nucleo-side RT inhibitors is the first effective specific HIV-1 antiviral drug, zidovudine (formerly azidothymidine or AZT); at least four other nucleosides (didanosine, zalcitabine, stavudine and lamivudine) are now ap-proved for use in the United States. Two non-nucleo-side RT inhibitors (nevirapine and delaverdine) are also approved; however, because resistance to these two develops rapidly, these drugs are almost never given singly or even with one other drug. The third class of drugs is the protease inhibitors which target the HIV-1 protease involved in processing viral pep-tides, an essential step in the maturation of infectious progeny virions. This class of drugs is particularly po-tent, often rapidly reducing the plasma viral load by 100-fold. Thus far, four of these are available for clinical use in the United States (saquinavir, indinavir, ritonavir and nelfinavir).

As use of these drugs has matured, certain general principles have emerged (Hammer 1996). Antiretro-viral drugs are now used almost exclusively in combi-nations of three or four at one time. A related prin-ciple is that when drugs are introduced or changed, this is done simultaneously rather than by the stag-gered initiation of one drug at a time. There are at least three reasons for this combined use. First, the effects of the drugs are often synergistic or additive, and therefore multiple drugs more rapidly and pro-foundly reduce the viral load. Second, their toxicities are often different and allow combinations to be given for long periods without untoward side effects. Third and most important, resistance patterns for most of the drugs differ. As a result of the afore-mentioned high mutation rate of HIV-1 and the high turnover rate of viral replication during active infection, resistant mutations are readily selected during drug therapy and indeed provide a major reason for drug failure over time (Richman 1996). Simultaneous use of multi-ple drugs with different molecular sites of action and different resistance mutations delays the resurgence of replication related to these mutations. Viruses harboring multiple resistance mutations may be less viable and less virulent. In fact, the major mutations associated with resistance to each of the drugs now in use have been defined and can be detected by geno-typic screening methods (Hammer 1996).

Newer technologies for the automated screening of multiple resistance mutations have been introduced as potential guides to treatment, though their clinical utility and cost-effectiveness have not yet been clearly demonstrated.

Returning to the train analogy, the objective of this type of aggressive treatment is to slow the forward progress of HIV-1 infection for as long as possible, and even allow it to roll backwards away from its final destination. Fortunately for an appreciable number of patients such regression does occur, indicated by the restoration of CD4+ cell levels and reduction in rates of subsequent infections and death. Current anti-HIV-1 regimens now often include a protease inhibi-tor, two nucleosides and perhaps a non-nucleoside RT inhibitor. While very often effective, and at times dramatically so, these combinations often entail diffi-cult treatment schedules for patients because of re-quirements that some drugs be taken twice and others 3 times per day, some with and others with-out food, and some cannot be administered simulta-neously. When these drugs are added to multiple other prophylactic and therapeutic drugs, the disci-pline required for rigid compliance is truly formid-able. There are also important interactions in the metabolic pathways of these and other drugs that can impact on efficacy. Finally, their expense precludes worldwide use. Thus these therapies, for all their wonder, are not the final answer to the AIDS epidemic.

HIV-1 INFECTION OF THE CNS

While clearly not as essential as immune system infection, entry into and infection within the CNS appears to be a fundamental aspect of the biology of HIV-1. Indeed, CNS exposure to HIV-1 likely occurs in most individuals at the time of the initial viremia following primary exposure and may continue throughout the course of infection. Studies of cerebrospinal fluid (CSF) in clinically well subjects early in infection have shown: (1) abnormalities of routine laboratory parameters, including cell count, total protein, and immunoglobulin; (2) local, intrablood–brain barrier synthesis of anti-HIV-1 antibody; and (3) culture isolation of virus or detection of viral nucleic acid after PCR amplification (Ho et al. 1985; Resnick et al. 1985, 1988; Goudsmit et al. 1986; Marshall et al. 1988; McArthur et al. 1988; Elovaara et al. 1993; Conrad et al. 1995; Pratt et al. 1996). These abnormalities have been noted in fully functional, asymptomatic patients who have remained well during follow-up care for years. These incidental abnormal levels in cell count, protein, immunoglobulin, oligoclonal bands, and HIV-1 detection must be taken into account when interpreting CSF results obtained for other diagnostic purposes or in following therapy. Also, although this early and continuous exposure is well documented in the case of leptomeningeal infection as reflected in CSF abnormalities, it is not as clearly established that similar brain infection also occurs, although the propitious observation in one iatrogenically infected patient (Davis et al. 1992) and observations of minor histologic changes in brains of others (Gray et al. 1992) suggest that this does occur though the viral burden is likely very low (Bell et al. 1993).

However, such infection, involving both CSF and brain, is characteristically asymptomatic. While headache, malaise and fatigue may complicate the seroconversion-related illness of primary infection, symptoms are usually transient and similar to manifestations accompanying the viremic phase of other acute viral infections, even infections without CNS invasion such as influenza (Carr and Cooper 1997). Some of these patients may manifest an aseptic meningitis syndrome with meningismus and CSF pleocytosis above 20 cells/mm^3. Only rarely has more severe CNS disease, including postinfectious encephalomyelitis, been reported (Carne et al. 1985). Several studies have shown that individuals with these findings are neurologically normal at the time of and subsequent to such testing, and there are no data to indicate that more marked early CSF infection is a predictor of the symptomatic brain infection accompanied by neurological dysfunction that may later manifest in some as ADC.

With the advent of potently effective antiretroviral therapy, attention has now focussed on the full range of CNS HIV-1 infection, both this early continuous, but asymptomatic, infection and the later productive infection that will be discussed below. This is for two reasons. First, because 'sequestered' infection behind the blood–brain and blood–CSF barriers may be important in efforts to prevent or treat ADC or aseptic meningitis; secondly and more broadly, such infection, even if asymptomatic, may have important implications for efforts to treat aggressively, and potentially to eradicate HIV-1 from the infected host (Richman 1996). If HIV-1 replication can proceed autonomously in these compartments where antiviral drug concentrations may be subtherapeutic, then not only will control of this local infection be inadequate but it may provide a particularly favorable setting for the emergence of resistant viral strains. In cell culture, resistant strains are often produced under similar conditions, in the presence of low concentrations of inhibiting drugs. These therapeutic implications have also highlighted the potential importance of the question of whether early asymptomatic CNS HIV-1 infection is largely the result of replication in cells which are 'trafficking through' these compartment without local amplification or permanent residence, or whether these entering cells can establish local foci of independent infection that cannot only persist in the brain, but evolve and produce resistant mutants. If the former is true, then CNS-penetrating drugs are not needed during the early phases of infection or for viral eradication, while if the latter is the case, then therapeutic drug concentrations must be achieved in the CNS throughout the course of treatment. This issue is taken up again later in this chapter in the discussion of antiviral treatment for ADC.

AIDS DEMENTIA COMPLEX (ADC)

ADC is one of the most common and clinically important CNS complications of late HIV-1 infection (Price et al. 1988). It is a source of great morbidity that is considered with dread by those at risk who have seen others suffer its effects, and in its more severe form is associated with a limited survival (Neaton et al. 1994). While its pathogenesis remains enigmatic in several important aspects, ADC is thought by most to relate to HIV-1 itself, rather than to another opportunistic infection, and most likely to relate to HIV-1 infection within the CNS.

Definition and terminology

ADC was identified early in the AIDS epidemic as a common and novel CNS disorder (Horowitz et al. 1982; Gopinathan et al. 1983; Snider et al. 1983; Britton and Miller 1984), although its salient clinical characteristics were not clearly defined until a few years later (Navia et al. 1986a,b). On the basis of its core clinical features that include impairment of attention and concentration, slowing of mental speed and agility, concomitant slowing of motor speed, and apathetic behavior, ADC is classified among the subcortical dementias (Benson 1987). While a variety of terminologies have been proposed since its initial recognition, including subacute encephalitis, subacute encephalopathy, HIV encephalopathy and HIV dementia, and there is some controversy regarding which designation is most appropriate, I continue to favor the ADC terminology and its associated staging system – hence, its use in this chapter. ADC is a clinically coherent syndrome that exhibits some case-to-case variation in individual symptoms and signs, but always within a circumscribed and recognizable spectrum. The three components of the term embody central features of the condition. *AIDS* emphasizes its morbidity and poor prognosis, particularly when its severity is at the stage 2 level, a severity comparable to other clinical AIDS-defining complications of HIV-1 infection. *Dementia* designates the acquired and persistent cognitive decline with preserved alertness that usually dominates the clinical presentation and determines its principal disability. Finally, *complex* was included to emphasize that this is a disease that not only impairs the intellect but also concomitantly alters motor performance and, at times, behavior. This involvement of the nervous system beyond cognition provides evidence of a wider involvement of the CNS than, for example, the dementia of Alzheimer's disease, and a condition that has more in common with some other neurodegenerative conditions such as multisystem atrophy and progressive supranuclear palsy. Additionally, myelopathy may be an important, indeed predominating, aspect of ADC, and organic psychosis may also be a feature in a subset of patients. These manifestations are therefore also encompassed within this term. By contrast, neither neuropathy nor functional psychiatric disturbance are included in ADC.

An empirically derived, five-step ADC staging system has proved valuable in describing its severity in both clinical and investigative settings (Table 1) (Price et al. 1988; Sidtis and Price 1990). This staging is applied once the diagnosis of ADC is made and is based on the degree of functional incapacity in cognitive and motor activities of work and daily living. Staging can usually be readily derived from the patient's history and corroborated by bedside examination without need for special clinical or laboratory tests, and provides a simple vocabulary for communicating ADC severity. Application of this staging designation assumes that the degree of functional impairment relates to ADC and is not confounded by another CNS disorder, either related (for example, primary CNS lymphoma) or unrelated (for example, effects of ongoing or past drug use or head trauma) to HIV-1 infection. The stage 0.5 designation is useful for classifying both patients in whom there are neurological symptoms or signs without functional impairment (subclinical disease) and those in whom mild impairment may possibly relate to another condition that cannot be clearly distinguished (equivocal ADC); it provides a way to deal with subjects who are neither entirely normal nor functionally impaired. By contrast, stage 1 ADC designates definite dysfunction, though sufficient only to render work or daily living more difficult without major incapacity. Stages 2–4 are applied to increasingly more severe dysfunction and establish a degree of neurological severity sufficient to meet criteria for a clinical AIDS diagnosis (Center for Disease Control 1987).

Committees sponsored by the World Health Organization (WHO) (World Health Organization 1990) and American Academy of Neurology (AAN)

TABLE 1

AIDS dementia complex staging (Price et al. 1988; Sidtis et al. 1990).

ADC stage	Characteristics
Stage 0: normal	Normal mental and motor function
Stage 0.5: equivocal/ subclinical	Either minimal or equivocal *symptoms* of cognitive or motor dysfunction characteristic of ADC, or mild signs (snout response, slowed extremity movements), but *without impairment of work or capacity to perform activities of daily living* (ADL). Gait and strength are normal
Stage 1: mild	Unequivocal evidence (symptoms, signs, neuro-psychological test performance) of functional intellectual or motor impairment characteristic of ADC, but able to perform *all but the more demanding aspects of work or ADL.* Can walk without assistance
Stage 2: moderate	Cannot work or maintain the more demanding aspects of daily life, but able to perform *basic activities of self care.* Ambulatory, but may require a single prop
Stage 3: severe	*Major intellectual incapacity* (cannot follow news or personal events, cannot sustain complex conversation, considerable slowing of all output), *or motor disability* (cannot walk unassisted, requiring walker or personal support, usually with slowing and clumsiness of arms as well)
Stage 4: end stage	*Nearly vegetative.* Intellectual and social comprehension and responses are at a rudimentary level. Nearly or absolutely mute. Paraparetic or paraplegic with double incontinence

(Janssen et al. 1991) have proposed alternative terminologies for ADC. These have both disadvantages and advantages. Among the former are: (1) the awkwardness of the general term chosen for the full spectrum of severity – *HIV-1-associated cognitive/motor complex*; (2) omission of a term equivalent to stage 0.5 ADC; and (3) designation of stage 1 as *HIV-1-associated minor cognitive/motor disorder*. Those suffering stage 1 ADC usually do not consider their condition to be 'minor'. A potential advantage of this other terminology is separation of those patients with predominating cognitive impairment (HIV-1-associated dementia) from those with predominating myelopathy (HIV-1-associated myelopathy) which may have both a clinical and pathological basis. On the other hand, while this distinction can usually be made on clinical grounds, it is not always clear-cut, and there is considerable overlap in the clinical findings in many patients. Additionally, abbreviations may be difficult since *HAM* is already used for HTLV-1-associated myelopathy. For these and other reasons, I have continued to use the simpler ADC vocabulary to describe patients with this syndrome, designating a subset as suffering a myelopathic variant when their signs and symptoms indicate

isolated spinal cord dysfunction. Fortunately, the WHO/AAN terminology can be 'translated' into the ADC staging upon which it is founded (Price 1997).

Clinical presentation of ADC

Although the severity and relative prominence of some symptoms and signs compared to others may vary among individual patients, the general character of ADC adheres to a cohesive pattern that can be considered to involve three functional categories: cognition, motor performance and behavior (Navia et al. 1986a). Table 2 provides an outline of some of the early and late manifestations. Of the three categories, cognitive and motor dysfunction are the most helpful in characterizing patients and in defining diagnosis; it is for this reason that they provide the basis of ADC staging which omits behavioral criteria. While the symptoms and signs form a continuum of cognitive and motor abnormalities, it is useful to separately consider milder and more severe affliction.

Stage 0.5 and 1 ADC

Cognitive impairment usually underlies patients' earliest symptoms and is the principal aspect that causes

TABLE 2

Salient clinical manifestations of ADC.

	Early	Late
Cognition	Inattention, reduced concentration, slowing of processing, difficulty changing mental sets	Global dementia
Motor performance	Slowed movements, clumsiness, ataxia	Paraplegia, quadraparesis, urinary incontinence
Behavior	Apathy, dulled personality, agitation and mania in a subgroup	Mutism

friends and relatives to notice that there is something wrong. Mildly afflicted patients most often have difficulty attending to more complex tasks at work or at home. They need to make lists, sometimes very detailed, of the day's activities. They lose track of actions (for example, they may leave the water boiling, or get up to go to another room and then forget why they did so) or of conversations in mid-sentence ('What was I saying?'). Processing unrelated or complex thoughts becomes slower and less facile. While similar lapses can trouble many normal people especially in the face of fatigue or generalized illness, in ADC patients they intrude upon and disrupt smooth daily function to a greater degree. Multistaged tasks become difficult: the waiter can no longer keep verbal orders straight when he arrives at the kitchen; the computer programmer cannot recall what line of code she intended to write next and must now outline each step in detail before proceeding; the avid reader needs to reread paragraphs or pages to the point that an activity that once gave great pleasure is abandoned. When such dysfunction is very mild, it may be difficult to substantiate the basis for these complaints by bedside examination, and it is important to apply tests that are sensitive to these abnormalities, including particularly tests requiring concentration, change of sets, and timed performance. Because it was constructed for other conditions, the standard Mini-Mental Status (Folstein et al. 1975; Jones et al. 1993) is often not sufficiently sensitive to distinguish these patients as abnormal at this point, though when they do perform abnormally, it is usually on reversals (reversing a five-letter word like *world*, or subtracting from 100 by 7s), complex sequential tasks (placing the right thumb on the left ear and sticking out the tongue) or remembering three objects. Notably, even when these and other tasks are performed accurately, the output may be slow and prompting may be re-

quired. Indeed, the examiner should be attentive to these characteristics and not rely solely upon scoring the accuracy. Folstein and colleagues have suggested a verbal adaptation of the Trail Making B test as a screen for the characteristic difficulty in rapid sequential effort (Jones et al. 1993), and Power and colleagues have proposed a bedside screen that examines some of the salient features (Power et al. 1995).

While motor *symptoms* are far less common during this early phase, individuals relying upon rapid or fine coordination may note a change. For example, the guitarist may no longer be able to keep up with a difficult piece or the athlete may be slowed to below a competitive level. An inquiring history may discover a change in handwriting or, less commonly, clumsiness in tying shoes or buttoning a shirt. Moreover, even in those without overt symptoms, motor *signs* may be detected on examination. These include slowing of attempts at rapid opposition of the thumb and forefinger, rotation of the wrist or tapping of the toe. Slowing of ocular saccades may also be noted along with interruption of smooth ocular pursuits. While the gait may be generally steady, it is often slow, and rapid turns may be interrupted by an extra step or performed hesitantly. Reflexes are often abnormal. The deep tendon stretch reflexes, including importantly the jaw jerk, are frequently hyperactive, though the ankle jerks may be relatively less active when there is concomitant polyneuropathy. Babinski signs may be present and other 'pathological' release signs may also be detected; of these, the snout response is relatively frequent and particularly helpful when present in young patients.

The time course and onset of milder ADC are variable. It may begin insidiously or abruptly and progress more rapidly to a higher stage, or it may continue to evolve slowly or even remain static for some period. Some mildly afflicted patients state that they have 'never been the same' cognitively since they developed early symptoms of HIV-1 infection. Whether the underlying disease process remains active in this setting is unclear, though the clinical response to antiretroviral therapy in some suggests that this chronicity is not simply the residuum of a monophasic event in the past.

From a diagnostic viewpoint, it is usually most important to obtain a careful and reliable history, including a history from the patient's associates who have witnessed his or her capacity at home or at work, and to perform a careful examination at the bedside searching for evidence of early cognitive and motor slowing. The history should pursue other possible contributions to the altered cognitive and motor dysfunction and establish the time course of their development and progression. If all of this is inconclusive, quantitative 'neuropsychological' testing, as discussed below, may be of ancillary help both in documenting abnormality and detailing whether the character of abnormality is indicative of ADC. If this and ancillary laboratory studies are inconclusive, then serial observation may be helpful.

In patients with mild ADC, missed diagnosis usually results from overlooking the important aspects of the history. Functional difficulties may erroneously be considered as a 'normal' part of systemic illness by patients and their caregivers despite the fact that most AIDS patients perform quite normally even in the late stage of HIV-1 infection. As a result, the exercise of differential diagnosis in this milder group often involves confirming that ADC is indeed present and that it accounts for significant symptoms. Particularly important is the distinction of ADC from clinical depression which can produce similar complaints but carries distinct therapeutic implications. Hypochondriasis and anxiety in those understandably worried about body function may also lead to similar complaints. The abnormal findings on neurological examination, particularly motor slowing and pathological reflexes, help to differentiate these, though one must be certain that these signs are not caused by another preexisting or acquired condition, such as prior head trauma or the effects of alcohol or illicit drug use. The laboratory studies discussed below may be of ancillary value as well.

Stage 2–4 ADC

In contrast to the need to establish that 'organic' neurological disease is present in those with mild ADC, in those with more advanced disease the principal exercise is in being certain that abnormal function relates to ADC rather than to another CNS disease. Cognitive function in these subjects is clearly abnormal and obviously impairs functional status, though in patients with stage 2 or even 3 ADC who can maintain the civilities of casual conversation and personal interaction it too can be missed by a cursory history and examination. However, with more careful questioning of both the patient and associates, it is clear that stage 2 patients are too slow or forgetful to work, manage the household or often even to maintain their own medications. They may get lost walking or driving and cannot be relied upon to prepare meals, much less to balance the checkbook. On examination a broader array of cognitive 'domains' is afflicted. The bedside Mini-Mental Status is now often abnormal (Folstein et al. 1975). The components mentioned earlier, assessments of attention and concentration, are frequently impaired, and patients have trouble attending sufficiently to recall three objects after 5 or 10 min. Timed activities are further slowed and complex drawings may be simplified. With increased severity, there is frank disorientation to time and place. Surprisingly, in many though not all patients with stage 2 and 3 ADC, judgment remains preserved and they are able to make or assist with decisions.

At this point motor abnormalities may also become more clearly symptomatic and obvious to the patient's family and associates. Walking may be so slow that exercise is limited by time or sufficiently unsteady to require a cane or someone alongside to prevent falling. Hyper-reflexia and pathological reflexes are now also virtually always present, and gait instability and slowness are clearly evident even when walking straight as well as when turning. With further progression, ambulation constantly requires someone to balance and support the patient or is entirely precluded (stage 3 or 4). Thinking and speaking also become slower and the content more impoverished. Concomitant behavioral changes may become more evident. Patients are often noted to have lost their 'sparkle'. They appear duller and less vivacious. If left alone they may sit still without spontaneously offering conversation, but only answering briefly in response to questions. This poverty of output and apathy may be mistaken for depression, but in most of these patients dysphoria is absent, and disinterest and lack of initiative are the predominating aspects of behavior without sadness. A striking variant that manifests in a small minority of patients includes agitation with features of mania (Navia et al. 1986a; Boccellari et al. 1988; Harris et al. 1991; Fernandez and Levy 1993; Sewell et al. 1994a,b). These patients usually exhibit a background of confusion that remains even as the hyperactivity is controlled by medication. When patients then progress even further, paraplegia or near quadriplegia may develop, at times in flexion and usually with associated incontinence of bladder and bowel. Complete or nearly complete mutism with only rudimentary cognition characterizes the end phase (stage 4) of the disease.

The myelopathic variant of ADC has a distinct pathological substrate, known as vacuolar myelopathy (Petito et al. 1985, 1994). This can often be distinguished clinically when the patient's gait dysfunction is disproportionately affected in comparison to the intellect. Some of these patients may become wheelchair bound with normal or near normal cognition. In others the myelopathy is combined with cognitive difficulty. While the lower extremities are more severely affected than the arms, there is no distinct segmental level of spinal cord dysfunction. Rather there is a gradual caudal increase in abnormality: knee tendon reflexes are more active than those in the arms, and gait or toe tapping is worse than hand coordination or rapid finger movements. Mild sensory loss is common and is usually worse distally in the feet with impaired vibration and position sense most common; sensation is usually normal over the trunk and in the upper extremities. Since myelopathy is frequently combined with neuropathy, the cause of sensory loss in individual patients may not always be clear, and one often relies on the ankle tendon jerks to indicate the relative contribution of myelopathy (increased ankle jerks) or neuropathy (decreased ankle jerks) in the presence of increased patellar stretch reflexes. Clinical presentation usually includes an ataxic and, sometimes, spastic gait. As with some other myelopathies, the ataxia may seem worse than can be accounted for by the loss of position sense in the toes and feet. Babinski signs are usually present as well, accompanying the hyperactive deep tendon reflexes. Urinary dysfunction with precipitous micturition may develop early. When compounded by the inability to ambulate and reach the toilet quickly, incontinence may become a problem. However, simple incontinence without any warning is more unusual in those without more severe cognitive difficulty and is not an invariable feature, even in those who cannot walk without assistance.

While the pathology of vacuolar myelopathy is usually greatest in the upper thoracic spinal cord, the hands are usually not wholly spared. They exhibit a degree of motor slowing and, less often, clumsiness. Indeed, there may be evidence of superspinal involvement with a hyperactive jaw jerk, although this more rostral involvement is otherwise asymptomatic; thus, these patients do not customarily manifest dysarthria or other overt cranial nerve abnormalities. The degree of cognitive impairment is variable – in some the myelopathy seems to exist in near isolation, while in others both brain and spinal cord are concomitantly affected. It is this blurring of clinical presentations that makes diagnosis of vacuolar myelopathy indefinite in an appreciable number of patients with gait difficulty and justifies its inclusion in a more global diagnostic term such as ADC or HIV-1-associated cognitive/motor complex.

Diagnosis of stage 2–4 ADC is both an inclusionary and exclusionary exercise. It is important to understand that the distinct features of the syndrome in the setting of late HIV-1 infection allow a *positive* diagnosis. The combination of the characteristic cognitive dysfunction and symmetrical motor abnormalities usually readily allows tentative diagnosis on the basis of the history and examination alone, and there are few other conditions that mimic the typical and uncomplicated presentation. However, some patients with primary CNS lymphoma located deep in the frontal white matter near the lateral ventricles, particularly when bilateral, may manifest a similar progressive dementia accompanied by motor slowing, apathy, impoverished speech output and minimal lateralization. Hydrocephalus can also produce a similar picture. In contrast, toxic and metabolic encephalopathies usually lead to a reduced level of arousal that is parallel to the cognitive deficiency rather than altering the latter in the face of full wakefulness as in ADC. However, these disorders may coexist with and exacerbate ADC, and the combination can confuse diagnosis. ADC patients appear to be more sensitive to the side effects of neuroleptics, and subclinical ADC may explain the sensitivity of AIDS patients more generally to the extrapyramidal side effects of these drugs (Hriso et al. 1991; Factor et al. 1994; Sewell et al. 1994a,b).

Cytomegalovirus (CMV) encephalitis is one of the diagnostically more difficult conditions complicating late HIV-1 infections and may be difficult to distinguish from ADC (Cinque et al. 1992; Kalayjian et al. 1993; Holland et al. 1994; Cohen 1996). However, while its features may vary, CMV encephalitis is more often accompanied by blunted arousal than ADC and probably more frequently causes seizures, particularly therapeutically intractable seizures, than ADC. Hyponatremia has also been noted more frequently in CMV encephalitis. An uncommon form of cerebral toxoplasmosis, the so-called *encephalitic* form (Navia et al. 1986c; Gray et al. 1989; Arendt et al. 1991), may be similarly confused with ADC as well as with CMV encephalitis. This is caused by the acute development of multiple small toxoplasma abscesses throughout the brain and may present with subacute global encephalopathy with few or no focal abnormalities. However, once again, consciousness is usually depressed. Neurosyphilis is commonly considered in the differential diagnosis, though how often it clinically mimics ADC is uncertain (Berger 1991; Holtom et al. 1992; Katz et al. 1993; Simon 1994; Malone et al. 1995; Marra et al. 1996). This may depend upon the frequency of syphilis in the particular clinical setting, and, although it should always be considered in the differential diagnosis of ADC and, indeed, the neurologist and AIDS clinician should always be alert to the possibility of overt or latent neurosyphilis, neurosyphilis is probably an uncommon cause of isolated dementia in AIDS patients in most developed countries. Wernicke's encephalopathy may also occur in AIDS as a result of nutritional deficiency, though consciousness is usually altered during the acute phase along with ocular and other distinct abnormalities (Schwenk et al. 1990; Butterworth et al. 1991; Boldorini et al. 1992).

The time course of more severe ADC is also variable. While vacuolar myelopathy may be more insidious in onset and gradually progressive over many weeks, both this and the cognitive impairment of ADC may develop more rapidly and evolve over only one or a few weeks. Hence the early designation of 'subacute encephalopathy'. More characteristic is development of overt symptoms and signs and their worsening over a period of months. This protracted time course may be helpful in distinguishing ADC from some of the more rapidly evolving encephalitides like CMV and toxoplasmosis, since the latter often develop and progress over a few days. The symptoms and signs of ADC, and more particularly the myelopathy, may also either first appear or exacerbate in the setting of another systemic illness, such as PCP, and then either remain fixed or subsequently improve to some degree during convalescence.

The differential diagnosis of ADC-related myelopathy is also relatively narrow in those with HIV-1 infection. HTLV-I and, as more recently reported, HTLV-II can cause a clinically similar myelopathy that is symmetrical and without segmental focality (Osame et al. 1987, 1990; Vernant et al. 1987; Berger et al. 1991; Harrington et al. 1993; Jacobson et al. 1993; Murphy et al. 1993). While the epidemiologies of HIV-1 and these other retroviruses overlap, particularly HTLV-I and HIV-1 in the Caribbean, southern United States and parts of Africa, there is no evidence that one of these infections exacerbates or accelerates the other. HTLV-I myelopathy, most notably when it evolves rapidly after transfusion, is more often accompanied by back pain and characterized by more marked spasticity than vacuolar myelopathy. Viral serology, CSF profile, the presence of immunosuppression and other ancillary findings usually allow distinction. In contrast to these conditions, other myelopathies in AIDS patients have a more focal and segmental presentation. This includes the myelopathy of varicella zoster virus (VZV) which frequently, though not always, complicates temporally proximate herpes zoster with its characteristic rash and develops at or near the spinal cord segment corresponding to the involved dermatome (Devinsky et al. 1991). The rare cases of spinal cord toxoplasmosis and lymphoma, including both primary CNS lymphoma (intramedullary focus in the spinal cord) and metastatic systemic lymphoma (located in the epidural or meningeal spaces with compression or invasion of the cord), also cause segmental myelopathies (Harris et al. 1990; Fairley et al. 1992; Henin et al. 1992).

While this chapter deals with adult ADC, a similar syndrome complicates pediatric AIDS where it may be a proportionally more important disorder in the sense that it both occurs more commonly than in adults and results in even greater morbidity (Belman 1994). It can manifest as a progressive encephalopathy with reversal of milestones and subsequent microcephaly or can be static with developmental delay. Clinical presentation is with combined cognitive and motor dysfunction. Its etiology and pathogenesis are likely similar to the adult condition, though altered somewhat by involvement of the developing brain. Treatment also follows a similar approach.

Laboratory studies in ADC

Clinical laboratory studies are often useful in evaluating patients with suspected ADC, both by revealing the characteristic abnormalities noted commonly in ADC, and perhaps even more importantly, by detecting other conditions in the differential diagnosis. Several types of laboratory tests are commonly deployed in this setting, with neuroimaging, CSF examination and formal neuropsychological testing being the most helpful.

Neuroimaging
Neuroimaging is usually an essential component of the evaluation of AIDS patients with CNS dysfunction, including those with suspected ADC. Principally this is used to detect evidence of other diagnoses such as the mass lesions of primary CNS lymphoma or the ependymal signal changes of CMV encephalitis. However, imaging can also detect abnormalities associated with ADC. Anatomical imaging, including both computed tomographic (CT) scanning and magnetic resonance imaging (MRI) nearly always show evidence of cerebral atrophy with widened cortical sulci and enlarged ventricles in ADC patients, particularly in those with stages 2–4 (Navia et al. 1986a; Post et al. 1988, 1991; Dal Pan et al. 1992; Gelman and Guinto 1992; Petty 1994). Basal ganglia are also reduced in volume (Aylward et al. 1993). Additionally, MRI in some patients detects T2-weighted abnormalities in the hemispheric white matter and, less commonly, the basal ganglia or thalamus (Jarvik et al. 1988; Post et al. 1988; Jakobsen et al. 1989; Moeller and Backmund 1990; Power et al. 1993). These signal changes can be patchy or 'fluffy' in appearance, or in the white matter may have a more homogeneous 'ground-glass' appearance. While these signal changes are sometimes referred to in the radiology community as 'HIV encephalopathy', neither these imaging abnormalities nor the cerebral atrophy are diagnostic of ADC. Both may be present in some asymptomatic patients, and their severity does not always correspond to that of clinical dysfunction. Hence, one needs to be cautious in equating neuroimaging findings with the clinical syndrome. In children with AIDS, mineralization of the basal ganglia is often prominent (Belman et al. 1986). Despite its sometimes striking pathology, vacuolar myelopathy is usually not detected by spinal MRI, although advances in imaging technology may change this in the future.

Abnormalities in metabolic brain imaging in ADC have also been reported, although these currently are more applicable to research studies than to clinical diagnosis or care. Magnetic resonance spectroscopy (MRS) has demonstrated decreased levels of the neuronal marker, *N*-acetyl neuraminic acid (NAA), in the later phases of ADC and increases in choline peaks somewhat earlier (Jarvik et al. 1993; Barker et al. 1995; Chang 1995; Tracey et al. 1996). Since recent studies suggest that these abnormalities can be reversed by antiviral treatment, it is possible that this technology might eventually be used not only for diagnosis but also to monitor therapy (Vion-Dury et al. 1995). Positron emission tomography (PET) and single photon emission computed tomography (SPECT) scanning have also been used to study ADC, but at present their practical diagnostic utility is also limited (Rottenberg et al. 1987; Ajmani et al. 1991; Kuni et al. 1991; Catafau et al. 1994). PET is costly and has not yet been shown to have diagnostic sensitivity or even specificity, and SPECT is not rigorously quantitative; neither technology has been carefully assessed for clinical use. As with MRS, PET studies showing improvement in cerebral metabolism have been used to corroborate therapeutic effect (Brunetti et al. 1989).

CSF analysis
Because CSF frequently reveals 'background' abnormalities in neurologically asymptomatic HIV-1 infection, routine studies are generally not very useful in establishing an ADC diagnosis. These abnormalities relate to the afore-mentioned early and perhaps continuous exposure of the CNS to HIV-1 and the resultant host responses. They include elevations in protein content, increased immunoglobulins, locally produced antibodies to HIV-1 and mononuclear cell response. Thus, while ADC patients may exhibit a mild pleocytosis (5–15 lymphocytes and mononuclear cells/mm^3) and elevated protein (usually

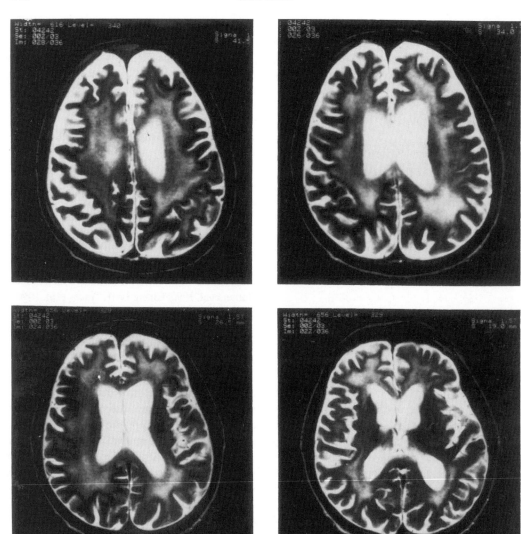

Fig. 1. The montage of T2-weighted images shows abnormalities commonly found in ADC, including increased signal attenuation in the white matter in an asymmetrical pattern and the brain atrophy evidenced by the enlarged lateral ventricles and cerebral sulci.

up to about 60 mg/dl), this does not distinguish them from non-ADC HIV-1-infected patients. Indeed, as HIV-1 infection progresses to its late stage, the CSF pleocytosis tends to decrease in AIDS patients with and without ADC (Marshall et al. 1988).

However, certain assessments are important in differential diagnosis. The latter include the VDRL and treponema-specific serologies in the diagnosis of neurosyphilis, cryptococcal antigen and culture in meningitis, and polymorphonuclear-predominant pleo-

cytosis that is common in CMV-related poly-radiculopathy and occasionally noted in CMV encephalitis as well. PCR and other amplification techniques have more recently been applied to CSF in the diagnosis of CMV, herpes simplex virus types 1 and 2 (HSV-1 and HSV-2) and VZV, all of which may enter into the ADC differential diagnosis (Gilden et al. 1994; Lakeman et al. 1995; Cohen 1996).

Until recently, detection of HIV-1 in CSF has also provided little diagnostic help. Because of background, ongoing asymptomatic infection of the lepto-meninges, HIV-1 can be cultured from CSF of asymptomatic patients. Similarly, non-quantitative PCR can detect viral DNA or RNA in the absence of neurological abnormalities (Syndulko et al. 1994). Quantitative detection of the core protein, p24, by immunoassay is perhaps somewhat better, but is insensitive (Brew et al. 1994; Royal et al. 1994). This antigen is most consistently detected only in patients with severe ADC (stages 3 and 4), but even in this group who characteristically presents little diagnostic problem, it is detected in only about half and is usually not detected in those with milder ADC for whom a laboratory test would be more helpful. More recently, quantitative viral RNA amplification techniques (so-called *viral load* assays) have begun to be applied to CSF, and initial studies suggest a correlation with the presence and severity of ADC (Brew et al. 1997), although this finding is unspecific since individuals with other CNS conditions and even some who are neurologically normal may show relatively high RNA levels in CSF. The utility of this technology in diagnosis and therapeutic monitoring should become clearer in the near future.

Another type of specialized assay involves the measurement of markers of immune activation in the CSF. This includes assessments of β_2-microglobulin (β_2M) and neopterin levels. These are both markers of immune activation, principally macrophage activation, that are elevated in the blood of AIDS patients in late infection and can be used as prognostic predictors, although they are weaker predictors than viral load (Fahey et al. 1990). Their CSF concentrations are elevated in ADC and generally parallel clinical se-verity, though they are also non-specific and elevated in other CNS infections and in primary CNS lymphoma (Brew et al. 1989, 1990, 1992; Griffin et al. 1991). However, in the absence of these other conditions, assessment of these markers may be useful in confirming the presence and perhaps activity of ADC. They may also be useful in assessing therapy, though this issue has not been rigorously and prospectively studied.

Neuropsychological testing

Formal neuropsychological testing may serve as a useful extension of the clinical neurological examination in evaluating ADC in selected patients. Indeed, these tests can be viewed as a quantitative neurological examination and for this reason have provided the principal endpoint measure for clinical trials of ADC treatment (Price and Sidtis 1990; Sidtis et al. 1993; Sidtis 1994). While performances on such tests are not diagnostically specific and individual test performance is variable and can be affected by a variety of factors (age, education, prior drug or alcohol use, head injury, etc.), if compared to appropriate norms and carefully interpreted in the particular clinical context, formal examination can help to determine whether symptoms truly reflect abnormal neurological function, and whether the character of such dysfunction conforms to that of ADC. In general, this is most useful in patients with milder disease (stages 0.5–1 and, less frequently stage 2); it is in this group that defining the degree and type of abnormality is a particularly important aspect of diagnosis. Table 3 lists the tests that have been used by the AIDS Clinical Trials Group

TABLE 3

'Neuropsychological' quantitative tests used to assess ADC progression in ACTG clinical trials (Sidtis 1994).

Digit Symbol Substitution (subtest of the WAIS-R (Wechsler 1981))

Trail Making Test, parts A and B (subtest of Halstead–Reitan Neuropsychological Battery (Reitan and Wolfson 1985)

Grooved Pegboard, right and left hand (Lezak 1995)

Timed Gait

Finger Tapping, right and left hand (subtest of Halstead–Reitan Neuropsychological Battery (Reitan and Wolfson 1985)

Auditory Verbal Learning Test (Lezak 1995)

The tests are heavily weighted to assess motor speed and fine movement, along with capacity for attention/concentration and rapid processing of changing sets. All but the Auditory Verbal Learning Test are components of the NPZ-8, a normalized summary score that is used as a principal outcome variable in ADC studies.

(ACTG) in dementia protocols; they indicate the types of tests that are useful in measuring performance abnormalities in these patients and in following them over time for either worsening (natural history) or therapeutic improvement. Other types of tests may be included in diagnostic evaluations in which relative affliction or preservation of function in domains not commonly altered by mild ADC (for example, language capabilities) may be helpful in distinguishing this from other neurodegenerative processes. However, in both the clinic and clinical trial settings it is usually unnecessary for patients to undergo lengthy and exhaustive (and exhausting) 'batteries' that add little diagnostic or staging information.

Other assessments

Neurophysiological measures are not commonly used in assessing ADC patients except in the face of particular symptoms or signs. Thus, while EEG abnormalities have been documented in ADC, their lack of diagnostic specificity decreases their utility other than in patients with suspected or known seizures (Holtzman et al. 1989; Wong et al. 1990; Van Paesschen et al. 1995). Likewise, evoked potentials may also be abnormal in an appreciable number of ADC patients, but sensory evoked potentials aid principally in documenting the presence of vacuolar myelopathy and perhaps monitoring its therapy (Arendt et al. 1993; Frank and Pahwa 1993; Birdsall et al. 1994). Brain biopsy is not usually justified in ADC patients but may be needed for diagnosis where another process, such as another type of viral encephalitis, is suspected on the basis of atypical clinical presentation or imaging findings.

Epidemiology and natural history of ADC

While early estimates of the incidence and prevalence of ADC have differed widely, related in part to variable criteria for diagnosis (including thresholds of severity for inclusion) and different methods of case ascertainment, it is now recognized that ADC develops principally in the context of late HIV-1 infection and associated severe immunosuppression. Its prevalence in a group accordingly varies depending upon the characteristics of the population sampled. For example, the prevalence is far lower when the population is composed of a broad spectrum of HIV-1-infected subjects and includes asymptomatic sero-

positives with preserved immunity, than when a group of cases is culled from a referral neurology clinic or selected as part of an autopsy series (Janssen et al. 1989, 1992; Goodwin et al. 1990; McArthur et al. 1993; Bacellar et al. 1994; Neaton et al. 1994).

More recently, prospective studies have provided some clarification of the epidemiology and natural history of ADC, at least with respect to its more severe forms. Thus, data from the Multicentered AIDS Cohort Study (MACS) that followed a selected group of gay men, including a subset that seroconverted during the course of the study, have clarified the strong relationship between ADC incidence and low CD4$^+$ T-lymphocyte counts (Bacellar et al. 1994). In this study, the incidence rate over a 5-year period was 7.34 cases per 100 person years for subjects with CD4$^+$ counts $\leqslant$100, 3.04 cases in those with counts of 101–200, 1.31 for counts of 201–350, 1.75 for counts of 351–500 and 0.46 for counts >500. Diagnostically, this means that more severe ADC is principally a disease found in advanced HIV-1 infection, although uncommonly it may develop in those with relatively preserved helper lymphocyte counts. Pathogenetically, these data suggest that severe immunosuppression has a strong 'permissive' effect on the development of ADC, but is neither alone sufficient (since many severely immunosuppressed patients do not develop the disorder) nor absolutely necessary for ADC to manifest. Data from the Community Programs for Clinical Research on AIDS (CPCRA) following AIDS patients in a series of treatment protocols show this same association and additionally show that the development of stage 2 or greater ADC is associated with limited survival (Neaton et al. 1994). The 6-month cumulative mortality of 97 ADC patients among an overall group of 3382 HIV-1-infected subjects followed in this program was 67%. This was nearly 3 times the mortality rate for PCP and closer to the 6-month mortality rates of other neurological diseases which were also high (nearly 85% for PML, 70% for primary CNS lymphoma, and 51% for cerebral toxoplasmosis). This poor prognosis for survival likely relates in part to the late stage of HIV-1 infection in which ADC develops and hence the concomitant susceptibility to other lethal complications of immunosuppression, and also to the limited effectiveness of available treatments and to the vulnerability of neurological debility. The tendency for relatives and care givers to 'give up' on such patients in the face of severe neurological impairment may also influence this short survival.

Pathology and pathogenesis

While the fundamental light microscopic neuropathological findings in the brains of those with ADC were described more than a decade ago (Navia et al. 1986b), the importance of individual features, their relationship to clinical manifestations and their underlying mechanisms remain uncertain in a number of fundamental aspects (Budka et al. 1991). Indeed, despite steady expansion in detailing and refining these neuropathological and, subsequently, the virological findings in ADC, a number of fundamental questions remain incompletely answered. Most notable is the central question of how HIV-1 infection leads to brain injury (Price et al. 1988; Price 1994). Since the mechanisms are likely to be elucidated in coming years, this section emphasizes general issues and mechanisms revealed by clinical studies rather than details of experimental findings and speculations which are reviewed elsewhere (Epstein and Gendelman 1993; Lipton and Gendelman 1995; Price 1996b).

Neuropathology and evidence of brain infection
Brain atrophy is common at the time of post-mortem examination, though perhaps not as readily appreciated as by neuroimaging during life. A number of histologic abnormalities are present in ADC patients, although with inexact clinical correlation. In general, routine light microscopic abnormalities most prominently affect subcortical structures and include: (1) diffuse white matter pallor and associated gliosis (astrocytosis and microgliosis); (2) multinucleated cell encephalitis; and (3) vacuolar myelopathy. Less common are diffuse or focal spongiform changes in the cerebral white matter and small areas of necrosis (Navia et al. 1986b; Petito et al. 1986; Rosenblum 1990; Budka 1991; Gray et al. 1991a). The most common of these abnormalities is the central astrocytosis and diffuse white matter pallor that, in isolation, seems to associate with milder ADC. Inflammation is characteristically scant and consists of a few perivascular lymphocytes and brown-pigmented macrophages accompanying the astrocytosis. This is also accompanied by evidence of blood–brain barrier disruption (Power et al. 1993).

Multinucleated cell encephalitis is a characteristic but not invariant finding in patients with more severe ADC (Navia et al. 1986b; Brew et al. 1995). It is so named because of the presence of HIV-1-infected multinucleated cells derived from fused macrophages and microglia clustered around small blood vessels or appearing more isolated in the parenchyma. They may also be associated with neighboring macrophage and microglial reaction, and with local edema and white matter rarefaction, and are concentrated most often in the white matter and deep gray structures with a distribution that may parallel the selective pathology of multisystem atrophy (Kure et al. 1990). On histological grounds alone this finding justifies the term HIV-1 encephalitis as amply demonstrated by immunohistochemical and in situ hybridization studies showing that these multinucleated cells are infected by the AIDS retrovirus (Shaw et al. 1985; Gabuzda et al. 1986; Koenig et al. 1986; Stoler et al. 1986; Wiley et al. 1986; Pumarole-Sune et al. 1987; Vazeux et al. 1987; Michaels et al. 1988). However, infection of macrophages and microglia is not always accompanied by cell fusion and multinucleation, and the term HIV-1 encephalitis can therefore be applied more broadly when special techniques are applied to pathological study. Because these cells produce core proteins (including p24) and glycoproteins (including gp41), and because they occur in clusters that suggest cell-to-cell spread, it is reasonably presumed that they are *productively* infected cells, i.e., they are making progeny virus and propagating infection in the brain.

More recent studies using immunohistochemical staining to detect regulatory gene products, notably *nef*, and sensitive nucleic acid amplification combined with histology (PCR combined with in situ hybridization) have shown a more promiscuous infection involving astrocytes and perhaps, though more controversially, even neurons and other cells (Saito et al. 1994; Tornatore et al. 1994a; Bagasra et al. 1996; Nuovo and Alfieri 1996; Takahashi et al. 1996). It is likely that infection in these neuroectodermal cells is non-productive, and contributes little or not at all to the amplification of brain infection, yet it may be pathogenically important. This is consistent with cell culture studies that have demonstrated low-level infections of astrocytic and other neuroectodermal cells and cell lines, involving a non-CD4 virus–cell interaction; under these conditions only very low levels of progeny virus and various amplification or rescue techniques are needed to demonstrate its continued presence (Tornatore et al. 1994b). If neurons are indeed ever infected, such infection is likely scant. Nonetheless, it is clear that neurons may be altered in ADC brains, and both selected loss of cortical neurons and alterations in their dendritic structures have been documented (Masliah et al. 1994, 1996). Other studies also suggest that neurons and astrocytes are undergoing apoptosis (Petito and Roberts 1995).

Although inflammation with multinucleated cells also may occur in the spinal cord, clinical evidence of spinal cord dysfunction is most commonly associated with vacuolar myelopathy (Petito et al. 1985). This myelopathy pathologically resembles subacute combined degeneration resulting from vitamin B_{12} deficiency, but levels of this vitamin are generally normal in serum. Although the incidence of vacuolar myelopathy overlaps with that of other pathologic abnormalities found in the brain, it does not correlate with local productive HIV-1 infection (Rosenblum et al. 1989; Dal Pan et al. 1994; Petito et al. 1994; Tan et al. 1995).

Pathogenesis

Evidence from a variety of observations supports a central role for HIV-1 in the ADC, particularly in the subset of patients with more severe brain dysfunction (Price et al. 1988; Price 1994, 1995a,b; Brew et al. 1995). As noted above, numerous studies have now documented HIV-1 productive infection of macrophages and related microglia and perhaps non-productive infection of a wider variety of cells, including particularly astrocytes (Shaw et al. 1985; Gabuzda et al. 1986; Koenig et al. 1986; Stoler et al. 1986; Wiley et al. 1986; Pumarole-Sune et al. 1987; Vazeux et al. 1987; Michaels et al. 1988; Saito et al. 1994; Tornatore et al. 1994b; Nuovo and Alfieri 1996). Despite these findings, there is still some controversy regarding the correlation of this viral load with the presence and severity of ADC (Brew et al. 1995; Glass et al. 1995; Wiley and Achim 1995). Characteristically, it is in the more severely affected patients (stages 2 and greater) that multinucleated cell encephalitis or productive HIV-1 infection is commonly found, but there are some cases of severe clinical disease without a high viral burden and other examples of seemingly abundant infection without severe ante-mortem clinical dysfunction. An additional important aspect of brain infection is the observation that the viral 'strains' cultured or even directly cloned from brains of these patients exhibit a macrophage-tropic phenotype and may even have genetic characteristics that favor replication in *brain* macrophages and microglia (Sharpless et al. 1992; Westervelt et al. 1992; O'Brien 1994; Power et al. 1994).

Because of the seeming discrepancy between the severity of clinical deficits on the one hand, and these pathologic and virologic features on the other, indirect pathogenetic processes relating infection and brain injury have been proposed (for reviews, see Price and Brew 1988; Epstein and Gendelman 1993; Price 1994, 1995a,b; Lipton and Gendelman 1995; Tyor et al. 1995). To explain how uninfected neuroectodermally derived elements of the brain, including neurons and oligodendrocytes, might be injured, investigators have hypothesized that macrophage and microglial infection *drives* a chain of pathological processes that eventuate in neuronal dysfunction. Models have been presented outlining pathogenic sequences in which infected cells either elaborate neurotoxic molecules or provide signals that secondarily activate more complex cell circuits that eventuate in neurotoxicity. Viral gene products, most notably the external glycoprotein, gp120, have been shown to exert neurotoxic effects in cell culture or in animal models (Lackner et al. 1991; Toggas et al. 1994; Lipton and Gendelman 1995); other viral gene products, including gp41 and the products of the *tat* and *nef* regulatory genes, may also be toxic (Sabatier et al. 1991; Werner et al. 1991). Some of these effects are exerted directly on neurons while others involve intermediary cells that serve to transduce and amplify these signals to eventuate in neurotoxicity. Particularly important in these sequences are activation of cytokine circuits, largely in macrophages and perhaps also astrocytes. This type of pathogenic cytokine activation, particularly when chronic as in ADC, may require a systemic 'setting' of cytokine profiles as a result of systemic HIV-1 infection as discussed earlier. In the broad sense these toxic cytokine pathways and their neuropathological sequelae can therefore be considered as immunopathological reactions that link infection to neurotoxicity. More proximate to the neuron, these various virus-coded and cytokine-related reactions may converge to involve pathways implicated in other types of neurodegeneration such as activation of *N*-methyl-D-aspartate receptors and production of nitric oxide (Dawson et al. 1993; Dawson and Dawson 1994). The more recent demonstration of in vivo non-productive infection of astrocytes has also provoked theories of virus-induced dysfunction of these cells as possibly contributory. These theories and the accumulating observations that support and provide mechanistic details have been reviewed recently (Genis et al. 1992; Price 1994; Wesselingh et al. 1994; Lipton and Gendelman 1995). The importance of understanding the individual mechanisms involved in these reactions relates to the possibility that they may provide therapeutic targets.

If the pathogenesis of cerebral injury in ADC remains somewhat enigmatic, the mechanisms responsible for vacuolar myelopathy are even more so. Since productive HIV-1 infection is not found in association with the spinal cord pathology and because the latter phenotypically mimics the pathologies of cobalamin deficiency and nitric oxide toxicity, a shared biochemical pathway has been hypothesized though not yet identified (Petito et al. 1994).

Treatment

Strategies to treat ADC have followed the guidelines suggested by concepts of pathogenesis (Price 1995a). Thus, if brain injury results from the linking of brain infection to endogenous cytokine-linked neuropathic processes, then treatment efforts might attempt to interrupt these processes at various vulnerable points. Foremost among these approaches has been the use of antiretroviral drug therapy, following the major avenue of treating systemic disease. If systemic HIV-1 infection, and more particularly brain infection, is the prime mover in ADC pathogenesis, then suppressing this infection should result in amelioration of 'downstream' neuropathic steps. Additionally, if CNS HIV-1 infection is also important in efforts either to eradicate HIV-1 or to suppress emergence of resistant strains within the CSF, then attention to this aspect of therapy might have broader implications. Unfortunately, direct studies of CNS antiviral therapy are relatively few and have certainly lagged behind studies of systemic treatment. Nonetheless, a number of studies have now indicated that zidovudine has both therapeutic and prophylactic benefit in ADC and HIV-1 brain infection. These include controlled clinical trials in adults and children as well as anecdotal

case reports and population 'experiences' using a variety of measures ranging from neuropsychological performance to pathological observations (Pizzo et al. 1988; Schmitt et al. 1988; Portegies et al. 1989; Sidtis et al. 1993; Vago et al. 1993; Gray et al. 1994).

These observations provide important precedent in showing that ADC can be prevented and treated, and that treatment cannot only halt progression but actually reverse some symptoms and signs. However, they fail to provide further specific guidelines for ADC treatment that conform to the more advanced contemporary approaches to systemic HIV-1 treatment using multi-drug combinations. Indeed, efficacy of most of the other available antiviral drugs has not been examined in this context. As noted earlier, a potentially crucial issue in this setting is the CNS penetration of antiretroviral drugs. If indeed ADC is caused by active and self-sustaining HIV-1 replication within the CNS, then it is important that drug combinations achieve therapeutic concentrations across the blood–brain barrier. Other than zidovudine (Burger et al. 1993) the CNS penetration of the antiretroviral drugs has not been rigorously studied. Hopefully, because of rekindled interest in CNS HIV-1 infection, this ignorance will be remedied in the near future. From existing information, it appears that zidovudine crosses the blood–brain barrier best among the nucleosides, but a new nucleoside now in clinical trial and designated 1592U89 by the manufacturer (Glaxo Wellcome) appears to have similar penetration in addition to high antiviral potency (Harrigan et al. 1997). Also, d4T appears to penetrate relatively well (Foudraine et al. 1997), and the non-nucleoside nevirapine may achieve similar unbound drug levels in blood and CSF (M. Meyers, personal communication). While it is rumored that the protease inhibitors penetrate poorly, minimal data are available at this time, particularly data that take into account protein binding of the drug. In the setting of early infection when penetration of the CNS by infected cells may result in only transient infection, blood–brain barrier penetration may be less important and thus not necessary to prevent or eradicate CNS infection.

What is the clinician to do with this limited information? Since treatment will continue to evolve rapidly in the future, it is likely that current recommendations will soon become obsolete. However, for the present, reasonable guidelines for treatment of ADC should include use of potent drug combinations that suppress patients' systemic infection as described in the first section of this chapter. In ADC patients efforts should be made to include at least one and probably two or more drugs that penetrate the blood–brain barrier. Whether higher doses of some of the drugs might be useful to assure therapeutic concentrations within the CNS remains to be evaluated. Upcoming clinical trials aimed at evaluating prevention and treatment of ADC and at assessing virological efficacy reflected in CSF, along with pharmacokinetic studies assessing CSF drug penetration, should provide more rationale guidance in the future.

Additional approaches to treatment may be classified as *adjuvant therapies* since they target processes beyond the infection. These can be divided into three classes: anti-inflammatory, neuroprotective and compensatory (Price 1995b). The first targets the cytokine activations, the second aims to stabilize the neuron or other target cells and the third to overcome functional deficits less directly. Some of these avenues are being explored by clinical trial. For example, pentoxifylline and thalidomide have been proposed as anticytokine measures, and two neuroprotective strategies are being assessed: the *N*-methyl-D-aspartate receptor inhibitor, memantine, in a controlled cooperative trial, and the calcium channel blocker, nimodipine, in an initial phase 1 study. These follow the rationale suggested by cell culture and experimental animal studies (Lipton 1994; Lipton and Stamler 1994; Lipton and Gendelman 1995). Compensatory measures now include the use of amphetamines and the like to overcome the psychomotor slowing, though without a clear study establishing their clinical efficacy (Fernandez and Levy 1994). Also included under this general category are important efforts to structure the environment and to provide emotional support and comfort care as needed to soften some of the suffering of these patients (Boccellari and Zeifert 1994).

OTHER CLINICAL MANIFESTATIONS OF CNS HIV-1 INFECTION

Early phases of CNS infection

As noted earlier, headache may commonly accompany the malaise of the seroconversion illness of initial infection, sometimes with a more frank syndrome of aseptic meningitis, and postinfectious encephalomyelitis may be a rare sequela. During the period of 'asymptomatic seropositivity', CNS disease is likewise very uncommon (Carne et al. 1985; Tindall and Cooper 1991). This is in some contrast to peripheral nervous system diseases that may be more common in these phases as discussed elsewhere in this volume. An unusual exception to this is a presumed autoimmune demyelinating disorder affecting the CNS. This multiple sclerosis-like illness has been reported in HIV-1-infected patients in the clinically latent phase of infection in the setting of preserved CD4[+] T-lymphocyte counts (Berger et al. 1989; Gray et al. 1991b). Its presentation may include remissions and exacerbations, along with corticosteroid responsiveness. Although these cases may represent the concurrence of two diseases, more likely, as with the demyelinating neuropathies, they relate to an autoimmune process triggered by HIV-1 infection.

Late HIV-1 infection

In addition to ADC, late HIV-1 infection can also be complicated by aseptic meningitis and headache. Aseptic meningitis presumably relates to direct HIV-1 infection of the meninges although it cannot be certain that other unidentified infections do not play a role (Hollander and Stringari 1987; Hollander et al. 1994). Clinically this may segregate into two types: an acute and a chronic form. Both usually present in the transitional phase (CD4[+] T-lymphocyte count <200–500/mm^3) or, less frequently, in the phase of overt AIDS (CD4[+] cell count <200/mm^3). Both are accompanied by meningeal symptoms (e.g., headache and photophobia), although meningeal signs (e.g., nuchal rigidity) are more characteristic of the acute type. Cranial nerve palsies can complicate the course, affecting cranial nerves V, VII, and VIII, with Bell's palsy sometimes recurring. The CSF shows mild mononuclear pleocytosis, usually with normal or mildly depressed glucose and slightly elevated protein levels. While the syndrome itself is benign, it may indicate a poor prognosis in relation to impending progression to AIDS in some patients. The effects of antiretroviral therapies on this disorder have not been tested.

Headache also may occur in AIDS patients without accompanying meningeal reaction. In some patients, this occurs in the setting of a complication of systemic illness (for example, PCP) and resolves when that illness is treated. In others, however, a precipitating infection is not identified, and the term *HIV headache* has been used to describe this condition, which may be severe and protracted (Brew and Miller 1993). Although some of these patients seem to be helped by tricyclic antidepressants, others are not and may even require narcotic analgesics.

REFERENCES

AJMANI, A., E. HABTE-GABR, M. ZARR, V. JAYABALAN and S. DANDALA: Cerebral blood flow SPECT with Tc-99m exametazine correlates in AIDS dementia complex stages. A preliminary report. Clin. Nucl. Med. 16 (1991) 656–659.

ARENDT, G., H. HEFTER, C. FIGGE, E. NEUEN-JAKOB, H.W. NELLES, C. ELSING and H.J. FREUND: Two cases of cerebral toxoplasmosis in AIDS patients mimicking HIV-related dementia. J. Neurol. 238 (1991) 439–442.

ARENDT, G., H. HEFTER and H. JABLONOWSKI: Acoustically evoked event-related potentials in HIV-associated dementia. Electroencephalogr. Clin. Neurophysiol. 86 (1993) 152–160.

AYLWARD, E.H., J.D. HENDERER, J.C. MCARTHUR, P.D. BRETTSCHNEIDER, G.J. HARRIS, P.E. BARTA and G.D. PEARLSON: Reduced basal ganglia volume in HIV-1-associated dementia: results from quantitative neuroimaging. Neurology 43 (1993) 2099–2104.

BACELLAR, H., A. MUNOZ, E.N. MILLER, B.A. COHEN, D. BESLEY, O.A. SELNES, J.T. BECKER and J.C. MCARTHUR: Temporal trends in the incidence of HIV-1-related neurologic diseases: multicenter AIDS cohort study, 1985–1992. Neurology 44 (1994) 1892–1900.

BAGASRA, O., E. LAVI, L. BOBROSKI, K. KHALILI, J. PESTANER, R. TAWADROS and R. POMERANTZ: Cellular reservoirs of HIV-1 in the central nervous system of infected individuals: identification by the combination of *in situ* polymerase chain reaction and immunohistochemistry. AIDS 10 (1996) 573–585.

BARKER, P.B., R.R. LEE and J.C. MCARTHUR: AIDS dementia complex: evaluation with proton MR spectroscopic imaging. Radiology 195 (1995) 58–64.

BELL, J., A. BUSUTTIL, J. IRONSIDE, S. REBUS, Y. DONALDSON, P. SIMMONDS and PEUTHERER: Human immunodeficiency virus and the brain: investigation of viral load in pre-AIDS individuals. J. Infect. Dis. 168 (1993) 919–924.

BELMAN, A.: HIV-1-associated CNS disease in infants and children (Rev.). Res. Publ. – Assoc. Res. Nerv. Ment. Dis. 72 (1994) 289–310.

BELMAN, A.L., G. LANTOS, D. HOROUPIAN, B.E. NOVICK, M.H. ULTMANN, D.W. DICKSON and A. RUBINSTEIN: AIDS: calcification of the basal ganglia in infants and children. Neurology 36 (1986) 1192–1199.

BENSON, D.: The spectrum of dementia: a comparison of the clinical features of AIDS dementia and dementia of the Alzheimer's type. Alzheimer Dis. Assoc. Disord. 1. 4 (1987) 217–220.

BERGER, J.: Neurosyphilis in human immunodeficiency virus type 1-seropositive individuals. A prospective study. Arch. Neurol. 48 (1991) 700–702.

BERGER, J., W. SHEREMATA, L. RESNICK ET AL.: Multiple sclerosis-like leukoencephalopathy revealing human immunodeficiency virus infection. Neurology 39 (1989) 324–329.

BERGER, J.R., S. RAFFANTI, A. SVENNINGSSON, M. MCCARTHY, S. SNODGRASS and L. RESNICK: The role of HTLV in HIV-1 neurologic disease. Neurology 41 (1991) 197–202.

BIRDSALL, H.H., L.N. OZLUOGLU, H.L. LEW, J. TRIAL, D.P. BROWN, M.J. WOFFORD, J.F. JERGER and R.D. ROSSEN: Auditory P300 abnormalities and leukocyte activation in HIV infection. Otolaryngol. Head Neck Surg. 110 (1994) 53–59.

BOCCELLARI, A. and P. ZEIFERT: Management of neurobehavioral impairment in HIV-1 infection (Rev.). Psychiatr. Clin. North Am. 17 (1994) 183–203.

BOCCELLARI, A., J.W. DILLEY and M.D. SHORE: Neuropsychiatric aspects of AIDS dementia complex: a report on a clinical series. Neurotoxicology 9 (1988) 381–389.

BOLDORINI, R., L. VAGO, A. LECHI, F. TEDESCHI and G.R. TRABATTONI: Wernicke's encephalopathy: occurrence and pathological aspects in a series of 400 AIDS patients. Acta Biomed. Ateneo Parmense 63 (1992) 43–49.

BREW, B.J. and J. MILLER: Human immunodeficiency virus-related headache. Neurology 43 (1993) 1098–1100.

BREW, B., R. BHALLA, M. FLEISHER, M. PAUL, A. KHAN, M. SCHWARTZ and R. PRICE: Cerebrospinal fluid β2-microglobulin in patients infected with human immunodeficiency virus. Neurology 39 (1989) 830–834.

BREW, B., R. BHALLA, M. PAUL, H. GALLARDO, J. MCARTHUR, M. SCHWARTZ and R. PRICE: Cerebrospinal fluid neopterin in human immunodeficiency virus type 1 infection. Ann. Neurol. 28 (1990) 556–560.

BREW, B.J., R.B. BHALLA, M. PAUL, J.J. SIDTIS, J.J. KEILP, A.E. SADLER, H. GALLARDO, J.C. MCARTHUR, M.K. SCHWARTZ and R.W. PRICE: Cerebrospinal fluid beta2-microglobulin in patients with AIDS dementia complex: an expanded series including response to zidovudine treatment. AIDS 6 (1992) 461–465.

BREW, B., M. PAUL, G. NAKAJIMA, A. KAHN, H. GALLARDO and R. PRICE: Cerebrospinal fluid HIV-1 p24 antigen and culture: sensitivity and specificity for AIDS-dementia complex. J. Neurol. Neurosurg. Psychiatry 57 (1994) 784–789.

BREW, B., M. ROSENBLUM, K. CRONIN and R. PRICE: The AIDS dementia complex and HIV-1 brain infection: clinical–virological correlations. Ann. Neurol. 38 (1995) 563–570.

BREW, B.J., L. PEMBERTON, P. CUNNINGHAM and M.G. LAW: Levels of HIV-1 RNA in cerebrospinal fluid correlate with AIDS dementia. J. Infect. Dis. 175 (1997) 963–966.

BRITTON, C. and J. MILLER: Neurologic complications in acquired immunodeficiency syndrome (AIDS). Neurol. Clin. 2 (1984) 315–339.

BRODER, S., T.C.J. MERIGAN and D. BOLOGNESI (Eds.): Textbook of AIDS Medicine. Baltimore, MD, Williams and Wilkins (1994).

BRUNETTI, A., G. BERG, G. DICHIRO, R. COHEN, R. YARCHOAN, P. PIZZO, S. BRODER, J. EDDY, M. FULHAM, R. FINN ET AL.: Reversal of brain metabolic abnormalities following treatment of AIDS dementia complex with 3'-azido-2',3'-dideoxythymidine (AZT, zidovudine): a PET-FDG study. J. Nucl. Med. 30 (1989) 581–590.

BUDKA, H.: Neuropathology of human immunodeficiency virus infection (Rev.). Brain Pathol. 1 (1991) 163–175.

BUDKA, H., C.A. WILEY, P. KLEIHUES, J. ARTIGAS, A.K. ASBURY, E.S. CHO, D.R. CORNBLATH, M.C. DAL CANTO, U. DEGIRO-LAMI, D. DICKSON ET AL.: HIV-associated disease of the nervous system: review of nomenclature and proposal for neuropathology-based terminology (Rev.). Brain Pathol. 1 (1991) 143–152.

BURGER, D., C. KRAAIJEVELD, P. MEENHORST, J. MULDER, C. KOKS, A. BULT and J. BEIJNED: Penetration of zidovudine into the cerebrospinal fluid of patients infected with HIV. AIDS 7 (1993) 1581–1587.

BUTTERWORTH, R.F., C. GAUDREAU, J. VINCELETTE, A.M. BOURGAULT, F. LAMOTHE and A.M. NUTINI: Thiamine deficiency and Wernicke's encephalopathy in AIDS. Metab. Brain Dis. 6 (1991) 207–212.

CARNE, C., A. SMITH, S. ELKINGTON ET AL.: Acute encephalopathy coincident with seroconversion for anti-HTLV-III. Lancet ii (1985) 1206.

CARR, A. and D.A. COOPER: Primary HIV infection. In: M.A. Sande and P.A. Volberding (Eds.), The Medical Management of AIDS. Philadelphia, PA, W.B. Saunders (1997) 89–106.

CATAFAU, A.M., M. SOLA, F.J. LOMENA, A. GUELAR, J.M. MIRO and J. SETOAIN: Hyperperfusion and early technetium-99m-HMPAO SPECT appearance of central nervous system toxoplasmosis. J. Nucl. Med. 35 (1994) 1041–1043.

CENTER FOR DISEASE CONTROL: Revision of the CDC surveillance case definition for acquired immunodeficiency syndrome. M.M.W.R. 36 (1987) 1S–14S.

CENTER FOR DISEASE CONTROL AND PREVENTION: 1993 revised classification system for HIV infection and expanded surveillance case definition for AIDS among adolescents and adults. M.M.W.R. 41 (1992) 1–19.

CHANG, L.: In vivo magnetic resonance spectroscopy in HIV and HIV-related brain diseases. Rev. Neurosci. 6 (1995) 365–378.

CINQUE, P., L. VAGO, M. BRYTTING, A. CASTAGNA, A. ACCORDINI, V.A. SUNDQVIST, N. ZANCHETTA, A.D. MONFORTE, B. WAHREN, A. LAZZARIN ET AL.: Cytomegalovirus infection of the central nervous system in patients with AIDS: diagnosis by DNA amplification from cerebrospinal fluid. J. Infect. Dis. 166 (1992) 1408–1411.

COFFIN, J.: HIV population dynamics in vivo: implications for genetic variation, pathogenesis, and therapy. Science 267 (1995) 483–489.

COFFIN, J.M.: HIV viral dynamics. AIDS 10, Suppl. 3 (1996) S75–S84.

COHEN, B.A.: Prognosis and response to therapy of cytomegalovirus encephalitis and meningomyelitis in AIDS. Neurology 46 (1996) 444–450.

CONRAD, D.I., P. SCHMID, K. SYNDULKO, E.J. SINGER, R.M. NAGRA, J.J. RUSSELL and W.W. TOURTELLOTTE: Quantifying HIV-1 RNA using the polymerase chain reaction on cerebrospinal fluid and serum of seropositive individuals with and without neurologic abnormalities. J. Acquir. Immune Defic. Syndr. 10 (1995) 425–435.

DAL PAN, G.J., J.H. MCARTHUR, E. AYLWARD, O.A. SELNES, T.E. NANCE-SPROSON, A.J. KUMAR, E.D. MELLITS and J.C. MCARTHUR. Patterns of cerebral atrophy in HIV-1-infected individuals: results of a quantitative MRI analysis. Neurology 42 (1992) 2125–2130.

DAL PAN, G.J., J.D. GLASS and J.C. MCARTHUR: Clinicopathologic correlations of HIV-1-associated vacuolar myelopathy: an autopsy-based case-control study. Neurology 44 (1994) 2159–2164.

DAVIS, L.E., B.L. HJELLE, V.E. MILLER, D.L. PALMER, A.L. LLEWELLYN, T.L. MERLIN, S.A. YOUNG, R.G. MILLS, W. WACHSMAN and C.A. WILEY: Early viral brain invasion in iatrogenic human immunodeficiency virus infection. Neurology 42 (1992) 1736–1739.

DAWSON, T.M. and V.L. DAWSON: gp120 neurotoxicity in primary cortical cultures (Rev.). Adv. Neuroimmunol. 4 (1994) 167–173.

DAWSON, V., T. DAWSON, G. UHL and S. SNYDER: Human immunodeficiency virus type 1 coat protein neurotoxicity mediated by nitric oxide in primary cortical cultures. Proc. Natl. Acad. Sci. USA 90 (1993) 3256–3259.

DEVINSKY, O., E. CHO, C. PETITO and R. PRICE: Herpes zoster myelitis. Brain 114 (1991) 1181.

DEVITA, V.T.J., S. HELLMAN and S.A. ROSENBERG: AIDS: Etiology, Diagnosis, Treatment and Prevention, 4th edit. Philadelphia, PA, Lippincott-Raven (1997).

ELOVAARA, I., E. NYKYRI, E. POUTIAINEN, L. HOKKANEN, R. RAININKO and J. SUNI: CSF follow-up in HIV-1 infection: intrathecal production of HIV-specific and unspecific IGG, and beta-2-microglobulin increase with duration of HIV-1 infection. Acta Neurol. Scand. 87 (1993) 388–396.

EMBRETSON, J., M. ZUPANIC, J. RIBAS ET AL.: Massive covert infection of helper T lymphocytes during the incubation period of AIDS. Nature 362 (1993) 359–362.

EPSTEIN, L.G. and H.E. GENDELMAN: Human immunodeficiency virus type 1 infection of the nervous system: pathogenetic mechanisms (see comments) (Rev.). Ann. Neurol. 33 (1993) 429–436.

FACTOR, S.A., G.D. PODSKALNY and K.D. BARRON: Persistent neuroleptic-induced rigidity and dystonia in AIDS dementia complex: a clinico-pathological case report. J. Neurol. Sci. 127 (1994) 114–120.

FAHEY, J., J. TAYLOR, R. DETELS, B. HOFMANN, R. MELMED, P. NISHANIAN and J. GIORGI: The prognostic value of cellular and serologic markers in infection with human immunodeficiency virus type 1. N. Engl. J. Med. 322 (1990) 166–172.

FAIRLEY, C.K., J. WODAK and E. BENSON: Spinal cord toxoplasmosis in a patient with human immunodeficiency virus infection (Rev.). Int. J. STD AIDS 3 (1992) 366–368.

FAUCI, A.S.: Host factors in the immunopathogenesis of human immunodeficiency virus (HIV) disease. Paper presented at the 24th Annual Keystone Symposium: HIV Pathogenesis, Keystone, CO (1995).

FERNANDEZ, F. and J.K. LEVY: The use of molindone in the treatment of psychotic and delirious patients infected with the human immunodeficiency virus. Case reports. Gen. Hosp. Psychiatry 15 (1993) 31–35.

FERNANDEZ, F. and J.K. LEVY: Psychopharmacology in HIV spectrum disorders (Rev.). Psychiatr. Clin. North Am. 17 (1994) 135–148.

FOLKS, T.M. and C.E. HART: The life cycle of human immunodeficiency virus type 1. In: V.T. DeVita Jr., S. Hellman and S.A. Rosenberg (Eds.), AIDS: Biology, Diagnosis, Treatment and Prevention. Philadelphia, PA, Lippincott-Raven (1997) 29–43.

FOLSTEIN, M.F., S.E. FOLSTEIN and P.R. MCHUGH: Mini-mental state. J. Psychiatr. Res. 86 (1975) 189–198.

FOUDRAINE, N., F. DE WOLF, R. HOETELMANS, P. PORTEGIES, J. MAAS and J. LANGE: CSF and serum HIV-RNA levels during AZT/3TC and d4T/3TC treatment. Paper presented at the 4th Conference on Retroviruses and Opportunistic Infections, Washington, DC (1997).

FRANK, Y. and S. PAHWA: Serial brainstem auditory evoked responses in infants and children with AIDS. Clin. Electroencephalogr. 24 (1993) 160–165.

GABUZDA, D., D. HO, S. DE LA MONTE ET AL.: Immunohistochemical identification of HTLV-III antigen in brains of patients with AIDS. Ann. Neurol. 20 (1986) 289.

GELMAN, B. and F.J. GUINTO: Morphometry, histopathology, and tomography of cerebral atrophy in the acquired immunodeficiency syndrome. Ann. Neurol. 31 (1992) 32–40.

GENIS, P., M. JETT, E.W. BERNTON, T. BOYLE, H.A. GELBARD, K. DZENKO, R.W. KEANE, L. RESNICK, Y. MIZRACHI, D.J. VOLSKY, L.G. EPSTEIN and H.E. GENDELMAN: Cytokines and arachidonic metabolites produced during human immunodeficiency virus (HIV)-infected macrophage-astroglia interactions: implications for the neuropathogenesis of HIV disease. J. Exp. Med. 176 (1992) 1703–1718.

GILDEN, D.H., B.R. BEINLICH, E.M. RUBINSTIEN, E. STOMMEL, R. SWENSON, D. RUBINSTEIN and R. MAHALINGAM: Varicella-zoster virus myelitis: an expanding spectrum. Neurology 44 (1994) 1818–1823.

GLASS, J.D., H. FEDOR, S.L. WESSELINGH and J.C. MCARTHUR: Immunocytochemical quantitation of human immunodeficiency virus in the brain: correlations with dementia. Ann. Neurol. 38 (1995) 755–762.

GONDA, M., F. WONG-STAAL, R. GALLO, J. CLEMENTS, O. NARAYAN and R. GILDEN: Sequence homology and morphologic similarity of HTLV III and visna virus, a pathogenic lentivirus. Science 227 (1985) 173–177.

GOODWIN, G., A. CHISWICK, V. EGAN, D. ST CLAIR and R. BRETTLE: The Edinburgh cohort of HIV-positive drug abusers: auditory event-related potentials show progressive slowing in patients with Centers for Disease Control stage IV disease. AIDS 4 (1990) 1243–1250.

GOPINATHAN, G., L. LAUBENSTEIN, B. MONDALE and R. KRIGEL: Central nervous system manifestations of the acquired immunodeficiency (AID) syndrome in homosexual men. Neurology 33, Suppl. 2 (1983) 105.

GOUDSMIT, J., E. WOLTERS, M. BAKKER ET AL.: Intrathecal synthesis of antibodies to HTLV-III in patients without AIDS or AIDS related complex. Br. Med. J. 292 (1986) 1231.

GRAY, F., R. GHERARDI, E. WINGATE ET AL.: Diffuse 'encephalitic' cerebral toxoplasmosis in AIDS: report of four cases. J. Neurol. 236 (1989) 273.

GRAY, F., H. HAUG, L. CHIMELLI, C. GENY, A. GASTON, F. SCARAVILLI and H. BUDKA: Prominent cortical atrophy with neuronal loss as correlate of human immunodeficiency virus encephalopathy. Acta Neuropathol. (Berl.) 82 (1991a) 229–233.

GRAY, F., L. CHIMELLI, M. MOHR ET AL.: Fulminating multiple sclerosis-like leukoencephalopathy revealing human immunodeficiency virus infection. Neurology 41 (1991b) 105–109.

GRAY, F., M.C. LESCS, C. KEOHANE, F. PARAIRE, B. MARC, M. DURIGON and R. GHERARDI: Early brain changes in HIV infection: neuropathological study of 11 HIV seropositive, non-AIDS cases. J. Neuropathol. Exp. Neurol. 51 (1992) 177–185.

GRAY, F., L. BELEC, C. KEOHANE, P. DE TRUCHIS, B. CLAIR, M. DURIGON, A. SOBEL and R. GHERARDI: Zidovudine therapy and HIV encephalitis: a 10-year neuropathological survey. AIDS 8 (1994) 489–493.

GRIFFIN, D.E., J.C. MCARTHUR and D.R. CORNBLATH: Neopterin and interferon-gamma in serum and cerebrospinal fluid of patients with HIV-associated neurologic disease. Neurology 41 (1991) 69–74.

HAMMER, S.: Advances in antiretroviral therapy and viral load monitoring. AIDS 10, Suppl. 3 (1996) S1–S11.

HARRIGAN, R., C. STONE, P. GRIFFIN, S. BLOOR, M. TISDALE, B. LARDER and C.T. TEAM: Antiretroviral activity and resistance profile of the carbocyclic nucleoside HIV reverse transcriptase inhibitor 1592U89. Paper presented at the program and abstracts of the IV Conference on Retroviruses and Opportunistic Infections, Washington, DC (1997).

HARRINGTON, W.J.J., W. SHEREMATA, B. HJELLE, D.K. DUBE, P. BRADSHAW, S.K.H. FOUNG, S. SNODGRASS, G. TOEDTER, L. CABRAL and B. POIESZ: Spastic ataxia associated with human T-cell lymphotropic virus type II infection. Ann. Neurol. 33 (1993) 411–414.

HARRIS, M.J., D.V. JESTE, A. GLEGHORN and D.D. SEWELL: New-onset psychosis in HIV-infected patients (see comments) (Rev.). J. Clin. Psychiatry 52 (1991) 369–376.

HARRIS, T.M., R.R. SMITH, J.R. BOGNANNO and M.K. EDWARDS: Toxoplasmic myelitis in AIDS: gadolinium-enhanced MR. J. Comput. Assist. Tomogr. 15 (1990) 809–811.

HAYNES, B.F., G. PANTALEO and A.S. FAUCI: Toward an understanding of the correlates of protective immunity to HIV infection. Science 271 (1996) 324–327.

HENIN, D., T.W. SMITH, U. DE GIROLAMI, M. SUGHAYER and J.J. HAUW: Neuropathology of the spinal cord in the acquired immunodeficiency syndrome. Hum. Pathol. 23 (1992) 1106–1114.

HO, D., M. SARNGADHARA, L. RESNICK ET AL.: Primary human T lymphotropic virus type III infection. Ann. Intern. Med. 103 (1985) 880.

HO, D.D., A.U. NEUMANN, A.S. PERELSON, W. CHEN, J.M. LEONARD and M. MARKOWITZ: Rapid turnover of plasma virions and CD4 lymphocytes in HIV-1 infection. Nature 373 (1995) 123–126.

HOLLAND, N.R., C. POWER, V.P. MATHEWS, J.D. GLASS, M. FORMAN and J.C. MCARTHUR: Cytomegalovirus encephalitis in acquired immunodeficiency syndrome (AIDS). Neurology 44 (1994) 507–514.

HOLLANDER, H. and S. STRINGARI: Human immunodeficiency virus-associated meningitis: clinical course and correlations. Am. J. Med. 83 (1987) 813–816.

HOLLANDER, H., D. MCGUIRE and J.H. BURACK: Diagnostic lumbar puncture in HIV-infected patients: analysis of 138 cases. Am. J. Med. 96 (1994) 223–228.

HOLTOM, P.D., R.A. LARSEN, M.E. LEAL and J.M. LEEDOM: Prevalence of neurosyphilis in human immunodeficiency virus-infected patients with latent syphilis. Am. J. Med. 93 (1992) 9–12.

HOLTZMAN, D., D. KAKU and Y. SO: New onset seizures associated with human immunodeficiency virus infection: causation and clinical features in 100 cases. Am. J. Med. 87 (1989) 173.

HOROWITZ, S., D. BENSON, M. GOTTLIEB, I. DAVOS and J. BENTSON: Neurological complications of gay-related immunodeficiency disorder. Ann. Neurol. 12 (1982) 80.

HRISO, E., T. KUHN, J.C. MASDEU and M. GRUNDMAN: Extrapyramidal symptoms due to dopamine-blocking agents in patients with AIDS encephalopathy. Am. J. Psychiatry 148 (1991) 1558–1561.

JACOBSON, S., T. LEHKY, M. NISHIMURA, S. ROBINSON, D.E. MCFARLIN and S. DHIB-JALBUT: Isolation of HTLV-II from a patient with chronic, progressive neurological disease clinically indistinguishable from HTLV-I-associated myelopathy/tropical spastic paraparesis. Ann. Neurol. 33 (1993) 392–396.

JAKOBSEN, J., C. GYLDENSTED, B. BRUN ET AL.: Cerebral ventricular enlargement relates to neuropsychological measures in unselected AIDS patients. Acta Neurol. Scand. 79 (1989) 59.

JANSSEN, R., A. SAYKIN, L. CANNON, J. CAMPBELL, P. PINSKY, N. HESSOL, P. O'MALLEY, A. LIFSON, L. DOLL, G. RUTHERFORD and J. KAPLAN: Neurological and neuropsychological manifestations of HIV-1 infection: association with AIDS-related complex but not asymptomatic HIV-1 infection. Ann. Neurol. 26 (1989) 592–600.

JANSSEN, R., D. CORNBLATH, L. EPSTEIN, R. FOA, J. MCARTHUR, R. PRICE ET AL.: Nomenclature and research case definitions for neurologic manifestations of human immunodeficiency virus-type 1 (HIV-1) infection. Report of a Working Group of the American Academy of Neurology AIDS Task Force (Rev.). Neurology 41 (1991) 778–785.

JANSSEN, R.S., O.C. NWANYANWU, R.M. SELIK and J.K. STEHR-GREEN: Epidemiology of human immunodeficiency virus encephalopathy in the United States. Neurology 42 (1992) 1472–1476.

JARVIK, J., J. HESSELINK, C. KENNEDY ET AL.: Acquired immunodeficiency syndrome: magnetic resonance patterns of brain involvement with pathologic correlation. Neurology 45 (1988) 731.

JARVIK, J.G., R.E. LENKINSKI, R.I. GROSSMAN, J.M. GOMORI, M.D. SCHNALL and I. FRANK: Proton MR spectroscopy of HIV-infected patients: characterization of abnormalities with imaging and clinical correlation. Radiology 186 (1993) 739–744.

JONES, B.N., E.L. TENG and M.F. FOLSTEIN: A new bedside test of cognition for patients with HIV infection. Ann. Intern. Med. 119 (1993) 1001–1004.

KALAYJIAN, R.C., M.L. COHEN, R.A. BONOMO and T.P. FLANIGAN: Cytomegalovirus ventriculoencephalitis in AIDS. A syndrome with distinct clinical and pathologic features (Rev.). Medicine 72 (1993) 67–77.

KATZ, D.A., J.R. BERGER and R.C. DUNCAN: Neurosyphilis. A comparative study of the effects of infection with human immunodeficiency virus. (Publ. erratum in Arch. Neurol. 50 (1993) 243–249.)

KOENIG, S., H. GENDELMAN, J. ORENSTEIN, M. DAL CANTO, G.

PEZESHKPOUR, M. YUNGBLUTH, F. JANOTTA, A. AKSAMIT, M. MATIN and A. FAUCI: Detection of AIDS virus in macrophages in brain tissue from AIDS patients with encephalopathy. Science 233 (1986) 1089–1093.

KUNI, C.C., F.S. RHAME, M.J. MEIER, M.C. FOEHSE, R.B. LOEWENSON, B.C. LEE, R.J. BOUDREAU and R.P. DUCRET: Quantitative I-123-IMP brain SPECT and neuropsychological testing in AIDS dementia. Clin. Nucl. Med. 16 (1991) 174–177.

KURE, K., W. LYMAN, K. WEIDENHEIM and D. DICKSON: Cellular localization of an HIV-1 antigen in subacute AIDS encephalitis using an improved double-labeling immunohistochemical method. Am. J. Pathol. 136 (1990) 1085–1092.

LACKNER, A.A., S. DANDEKAR and M.B. GARDNER: Neurobiology of simian and feline immunodeficiency virus infections (Rev.). Brain Pathol. 1 (1991) 201–212.

LAKEMAN, F.D., R.J. WHITLEY and a.t.N.I.O.A.a.I.D.C.A.S. GROUP: Diagnosis of herpes simplex encephalitis: application of polymerase chain reaction to cerebrospinal fluid from brain-biopsied patients and correlation with disease. J. Infect. Dis. 171 (1995) 857–863.

LEZAK, M.: Neuropsychological Assessment, 3rd Edit. New York, Oxford University Press (1995).

LIPTON, S.A.: Neuronal injury associated with HIV-1 and potential treatment with calcium-channel and NMDA antagonists. Dev. Neurosci. 16 (1994) 145–151.

LIPTON, S.A. and H.E. GENDELMAN: Seminars in medicine of the Beth Israel Hospital, Boston. Dementia associated with the acquired immunodeficiency syndrome (Rev.). N. Engl. J. Med. 332 (1995) 934–940.

LIPTON, S.A. and J.S. STAMLER: Actions of redox-related congeners of nitric oxide at the NMDA receptor. Neuropharmacology 33 (1994) 1229–1233.

MALONE, J.L., M.R. WALLACE, B.B. HENDRICK, A. LAROCCO, E. TONON, S.K. BRODINE, W.A. BOWLER, B.S. LAVIN, R.E. HAWKINS and E.C.I. OLDFIELD: Syphilis and neurosyphilis in a human immunodeficiency virus type-1 seropositive population: evidence for frequent serologic relapse after therapy. Am. J. Med. 99 (1995) 55–63.

MARRA, C.M., D.W. GARY, J. KUYPERS and M.A. JACOBSON: Diagnosis of neurosyphilis in patients with human immunodeficiency virus type-1. J. Infect. Dis. 174 (1996) 184–189.

MARSHALL, D., R. BREY, W. CAHILL, R. HOUK, R. ZAJAC and R. BOSWELL: Spectrum of cerebrospinal fluid findings in various stages of human immunodeficiency virus infection. Arch. Neurol. 45 (1988) 954–958.

MASLIAH, E., C.L. ACHIM, N. GE, R. DE TERESA and C.A. WILEY: Cellular neuropathology in HIV encephalitis (Rev.). Res. Publ. Assoc. Res. Nerv. Ment. Dis. 72 (1994) 119–131.

MASLIAH, E., N. GE, C. ACHIM, R. DE TERESA and C. WILEY:

Patterns of neurodegeneration in HIV encephalitis. J. Neuro-AIDS 1 (1996) 161–173.

MCARTHUR, J., B. COHEN, H. FARZADEGAN, D. CORNBLATH, O. SELNES, D. OSTROW, R. JOHNSON, J. PHAIR and B. POLK: Cerebrospinal fluid abnormalities in homosexual men with and without neuropsychiatric findings. Ann. Neurol. 23 (Suppl.) (1988) S34–S37.

MCARTHUR, J.C., D.R. HOOVER, H. BACELLAR, E.N. MILLER, B.A. COHEN, J.T. BECKER, N.M. GRAHAM, J.H. MCARTHUR, O.A. SELNES, L.P. JACOBSON ET AL.: Dementia in AIDS patients: incidence and risk factors. Multicenter AIDS Cohort Study. Neurology 43 (1993) 2245–2252.

MCARTHUR, J.H. and J.C. MCARTHUR: Neurological manifestations of acquired immunodeficiency syndrome. J. Neurosci. Nurs. 18 (1986) 242–249.

MCCUTCHAN, F.E., M.O. SALMINEN, J.K. CARR and D.S. BURKE: HIV-1 genetic diversity. AIDS 10 (Suppl.) (1996) S13–S20.

MELLORS, J.W., L.A. KINGSLEY, C.R. RINALDO, J. TODD, B. HOO, R. KOKKA and P. GUPTA: Quantitation of HIV-1 RNA in plasma predicts outcome after seroconversion. Ann. Intern. Med. 122 (1995) 573–579.

MELLORS, J., P. GUPTA, R. WHITE, J. TODD and L. KINGSLEY: Prognosis in HIV-1 infection predicted by the quantity of virus in plasma. Science 272 (1996) 1167–1170.

MICHAELS, J., R. PRICE and M. ROSENBLUM: Microglia in the human immunodeficiency virus encephalitis of acquired immune deficiency syndrome: proliferation, infection and fusion. Acta Neuropathol. (Berl.) 76 (1988) 373–379.

MOELLER, A.A. and H.C. BACKMUND: Ventricle brain ratio in the clinical course of HIV infection. Acta Neurol. Scand. 81 (1990) 512–515.

MURPHY, E.L., J. FRIDEY, J.W. SMITH, J. ENGSTROM, R.A. SACHER, K. MILLER, J. GIBBLE, J. STEVENS, R. THOMSON, D. HANSMA, J. KAPLAN, R. KHABBAZ AND G. NEMO: HTLV-associated myelopathy in a cohort of HTLV-I and HTLV-II-infected blood donors. The REDS Investigators. Neurology 48 (1997) 315–320.

NAVIA, B., B. JORDAN and R. PRICE: The AIDS dementia complex: I. Clinical features. Ann. Neurol. 19 (1986a) 517–524.

NAVIA, B., E.-W. CHO, C. PETITO and R. PRICE: The AIDS dementia complex: II. Neuropathology. Ann. Neurol. 19 (1986b) 525–535.

NAVIA, B., C. PETITO, J. GOLD ET AL.: Cerebral toxoplasmosis complicating the acquired immune deficiency syndrome: clinical and neuropathological findings in 27 patients. Ann. Neurol. 19 (1986c) 224–238.

NEATON, J., D. WENTWORTH, F. RHAME, C. HOGAN, D. ABRAMS and L. DEYTON: Methods of studying interventions. Considerations in choice of a clinical endpoint for AIDS clinical trials. Stat. Med. 13 (1994) 2107–2125.

NUOVO, G. and M. ALFIERI: AIDS dementia is associated with massive, activated HIV-1 infection and concomitant expression of several cytokines. Mol. Med. 2 (1996) 358–366.

O'BRIEN, W.: Genetic and biologic basis of HIV-1 neurotropism. In: R.W. Price and S.W. Perry (Eds.), HIV, AIDS and the Brain. New York, Raven Press (1994) 47–70.

OSAME, M., M. MATSUMOTO, K. USUKU, S. IZUMO, N. IJICHI, H. AMITANI, T. MITSUTOSHI and A. IGATA: Chronic progressive myelopathy associated with elevated antibodies to human T-lymphotropic virus type I and adult T-cell leukemia-like cells. Ann. Neurol. 21 (1987) 117–122.

OSAME, M., A. IGATA, M. MATSUMOTO, M. KOHKA, K. USUKU and S. IZUMO: HTLV-I-associated myelopathy (HAM). Treatment trials, retrospective survey and clinical and laboratory findings. Hematol. Rev. 3 (1990) 271–284.

PANTALEO, G. and A. FAUCI: New concepts in the immunopathogenesis of HIV infection. Annu. Rev. Immunol. 13 (1995) 487–512.

PANTALEO, G., C. GRAZIOSI, J. DEMAREST ET AL.: HIV-1 infection is active and progressive in lymphoid tissue during the clinically latent stage of disease. Nature 362 (1993) 355–358.

PERELSON, A., A. NEUMANN, M. MARKOWITZ, J. LEONARD and D. HO: HIV-1 dynamics in vivo: virion clearance rate, infected cell life-span, and viral generation time. Science 271 (1996) 1582–1586.

PETITO, C.K. and B. ROBERTS: Evidence of apoptotic cell death in HIV encephalitis. Am. J. Pathol. 146 (1995) 1121–1130.

PETITO, C., B. NAVIA, E. CHO ET AL.: Vacuolar myelopathy pathologically resembling subacute combined degeneration in patients with acquired immunodeficiency syndrome (AIDS). N. Engl. J. Med. 312 (1985) 874–879.

PETITO, C., E.-S. CHO, W. LEMANN, B. NAVIA and R. PRICE: Neuropathology of acquired immunodeficiency syndrome (AIDS): an autopsy review. J. Neuropathol. Exp. Neurol. 45 (1986) 635–646.

PETITO, C.K., D. VECCHIO and Y.T. CHEN: HIV antigen and DNA in AIDS spinal cords correlate with macrophage infiltration but not with vacuolar myelopathy. J. Neuropathol. Exp. Neurol. 53 (1994) 86–94.

PETTY, R.K.: Recent advances in the neurology of HIV infection (Rev.). Postgrad. Med. J. 70 (1994) 393–403.

PIZZO, P., J. EDDY, J. FALLON ET AL.: Effect of continuous intravenous infusion of zidovudine (AZT) in children with symptomatic HIV infection. N. Engl. J. Med. 319 (1988) 889–896.

PORTEGIES, P., J.Î.A.L. DE GANS, J.M. DERIX, H. SPEELMAN, M. BAKKER, S. DANNER and J. GOUDSMIT: Declining incidence of AIDS dementia complex after introduction of zidovudine treatment. Br. Med. J. 299 (1989) 819–821.

POST, M., L. TATE, R. QUENCER ET AL.: CT, MR, and patho-

logy in HIV encephalitis and meningitis. Am. J. Roentgenol. 151 (1988) 373.

POST, M.J., J.R. BERGER and R.M. QUENCER: Asymptomatic and neurologically symptomatic HIV-seropositive individuals: prospective evaluation with cranial MR imaging (Comments). Radiology 178 (1991) 131–139.

POWER, C., P.A. KONG, T.O. CRAWFORD, S. WESSELINGH, J.D. GLASS, J.C. MCARTHUR and B.D. TRAPP: Cerebral white matter changes in acquired immunodeficiency syndrome dementia: alterations of the blood–brain barrier. Ann. Neurol. 34 (1993) 339–350.

POWER C., J.C. MCARTHUR, R.T. JOHNSON, D.E. GRIFFIN, J.D. GLASS, S. PERRYMAN and B. CHESEBRO: Demented and nondemented patients with AIDS differ in brain-derived human immunodeficiency virus type 1 envelope sequences. J. Virol. 68 (1994) 4643–4649.

POWER, C., O.A. SELNES, J.A. GRIM and J.C. MCARTHUR: HIV Dementia Scale: a rapid screening test. J. Acquir. Immune Defic. Syndr. 8 (1995) 273–278.

PRATT, R.D., S. NICHOLS, N. MCKINNEY, S. KWOK, W.M. DANKNER and S.A. SPECTOR: Virological markers of human immunodeficiency virus type 1 in cerebrospinal fluid of infected children. J. Infect. Dis. 174 (1996) 288–293.

PRICE, R.W.: Understanding the AIDS dementia complex (ADC). The challenge of HIV and its effects on the central nervous system (Rev.). Res. Publ. Assoc. Res. Nerv. Ment. Dis. 72 (1994) 1–45.

PRICE, R.W.: AIDS dementia complex and HIV-1 brain infection: a pathogenetic framework for treatment and evaluation. Curr. Top. Microbiol. Immunol. 202 (1995a) 33–54.

PRICE, R.W.: The cellular basis of central nervous system HIV-1 infection and the AIDS dementia complex: introduction. J. Neuro-AIDS 1 (1995b) 1–28.

PRICE, R.W.: AIDS dementia complex: a complex, slow virus 'model' of acquired genetic neurodegenerative disease. Cold Spring Harb. Symp. Quant. Biol. (1996a) 759–770.

PRICE, R.W.: Neurological complications of HIV infection. Lancet 348 (1996b) 445–452.

PRICE, R.W.: Management of neurologic complications of HIV-1 infection and AIDS. In: M.A. Sande and P.A. Volberding (Eds.), The Medical Management of AIDS. Philadelphia, PA, WB Saunders (1997) 197–216.

PRICE, R.W. and B. BREW: The AIDS dementia complex. J. Infect. Dis. 158 (1988) 1079–1083.

PRICE, R.W. and J.J. SIDTIS: Evaluation of the AIDS dementia complex in clinical trials. J. AIDS 3, Suppl. (1990) S51–S60.

PRICE, R.W., B. BREW, J. SIDTIS, M. ROSENBLUM, A. SCHECK and P. CLEARY: The brain in AIDS: central nervous system HIV-1 infection and AIDS dementia complex. Science 239 (1988) 586–592.

PUMAROLE-SUNE, T., B. NAVIA, C. CORDON-CARDO, E.-S. CHO and R. PRICE: HIV antigen in the brains of patients with the AIDS dementia complex. Ann. Neurol. 21 (1987) 490–496.

REITAN, R. and D. WOLFSON: The Halstead–Reitan Neuropsychological Test Battery: Theory and Clinical Interpretation. Phoenix, AZ, Neuropsychology Press (1985).

RESNICK, L., F. DIMARZO-VERONESE, J. SCHUPBACH, W.W. TOURTELLOTTE, D.D. HO, F. MULLER, P. SHAPSHAK, M. VOGT, J.E. GROOPMAN, P.D. MARKHAM ET AL.: Intrablood–brain-barrier synthesis of HTLV-III-specific IgG in patients with neurologic symptoms associated with AIDS or AIDS-related complex. N. Engl. J. Med. 313 (1985) 1498–1504.

RESNICK, L., J. BERGER, P. SHAPSHAK and W. TOURTELLOTTE: Early penetration of the blood–brain-barrier by HIV. Neurology 38 (1988) 9–14.

RICHMAN, D.: HIV therapeutics (viewpoints). Science 272 (1996) 1886–1887.

ROSENBLUM, M.: Infection of the central nervous system by the human immunodeficiency virus type 1: morphology and relation to syndromes of progressive encephalopathy and myelopathy in patients with AIDS. Pathol. Ann. 25 (1990) 117–169.

ROSENBLUM, M., A. SCHECK, K. CRONIN, B. BREW, A. KHAN, M. PAUL and R. PRICE: Dissociation of AIDS-related vacuolar myelopathy and productive human immunodeficiency virus type 1 (HIV-1) infection of the spinal cord. Neurology 39 (1989) 892–896.

ROTTENBERG, D.A., J.R. MOELLER, S.C. STROTHER, J.J. SIDTIS, B.A. NAVIA, V. DHAWAN, J.Z. GINOS and R.W. PRICE: The metabolic pathology of the AIDS dementia complex. Ann. Neurol. 22 (1987) 700–706.

ROYAL, W.R., O.A. SELNES, M. CONCHA, T.E. NANCE-SPROSON and J.C. MCARTHUR: Cerebrospinal fluid human immunodeficiency virus type 1 (HIV-1) p24 antigen levels in HIV-1-related dementia. Ann. Neurol. 36 (1994) 32–39.

SAAG, M., M. HOLODNIY, D. KURITZKES, W. O'BRIEN, R. COOMBS, M. POSCHER, D. JACOBSEN, G. SHAW, D. RICHMAN and P. VOLBERDING: HIV viral load testing in clinical practice (Commentary and Rev.). Nature Med. 2 (1996) 625–629.

SABATIER, J.-M., E. VIVES, K. MARBROUK, A. BENJOUAD, H. ROCHAT, A. DUVAL, B. HUE and E. BAHRAOUI: Evidence for neurotoxic activity of tat from human immunodificiency virus type 1. J. Virol. 65 (1991) 961–967.

SAITO, Y., L. SHARER, L. EPSTEIN, J. MICHAELS, M. MINTZ, M. LOUDER, K. GOLDING, T. CVETKOVICH and B. BLUMBERG: Overexpression of nef as a marker for restricted HIV-1 infection of astrocytes in postmortem pediatric central nervous system tissues. Neurology 44 (1994) 474–481.

SANDE, M.A. and P.A. VOLBERDING: The Medical Management of AIDS, 5th Edit. Philadelphia, PA, W.B. Saunders (1997).

SCHMITT, F., J. BIGLEG, R. MCKINNIS ET AL.: Neuropsychological outcome of azidothymidine (AZT) in the treatment of AIDS and AIDS-related complex: a double blind, placebo-controlled trial. N. Engl. J. Med. 319 (1988) 1573–1578.

SCHWENK, J., G. GOSZTONYI, P. THIERAUF, J. IGLESIAS and E. LANGER: Wernicke's encephalopathy in 2 patients with acquired immunodeficiency syndrome. J. Neurol. 237 (1990) 445–447.

SEWELL, D.D., D.V. JESTE, J.H. ATKINSON, R.K. HEATON, J.R. HESSELINK, C. WILEY, L. THAL, J.L. CHANDLER and I. GRANT: HIV-associated psychosis: a study of 20 cases. San Diego HIV Neurobehavioral Research Center Group. Am. J. Psychiatry 151 (1994a) 237–242.

SEWELL, D.D., D.V. JESTE, L.A. MCADAMS, A. BAILEY, M.J. HARRIS, J.H. ATKINSON, J.L. CHANDLER, J.A. MCCUTCHAN and I. GRANT: Neuroleptic treatment of HIV-associated psychosis. H.N.R.C. group. Neuropsychopharmacology 10 (1994b) 223–229.

SHARPLESS, N., W. O'BRIEN, E. VERDIN, C. KUFTA, I. CHEN and M. DUBOIS-DALCQ: Human immunodeficiency virus type 1 tropism for brain microglial cells is determined by a region of the env glycoprotein that also controls macrophage tropism. J. Virol. 66 (1992) 2588–2593.

SHAW, G., M. HARPER, B. HAHN, L. EPSTEIN, D. GAJDUSEK, R. PRICE, B. NAVIA, C. PETITO, C. O'HARA, J. GROOPMAN, E.-S. CHO, J. OLESKE, F. WONG-STAAL and R. GALLO: HTLV-III infection in brains of children and adults with AIDS encephalopathy. Science 227 (1985) 177–182.

SIDTIS, J.J.: Evaluation of the AIDS dementia complex in adults (Rev.). Res. Publ. Assoc. Res. Nerv. Ment. Dis. 72 (1994) 273–287.

SIDTIS, J.J. and R.W. PRICE: Early HIV-1 infection and the AIDS dementia complex (comment). Neurology 40 (1990) 323–326.

SIDTIS, J.J., C. GATSONIS, R.W. PRICE, E.J. SINGER, A.C. COLLIER, D.D. RICHMAN, M.S. HIRSCH, F.W. SCHAERF, M.A. FISCHL, K. KIEBURTZ ET AL.: Zidovudine treatment of the AIDS dementia complex: results of a placebo-controlled trial. AIDS Clinical Trials Group. Ann. Neurol. 33 (1993) 343–349.

SIGURDSSON, B., P. PALSSON and H. GRIMSON: Visna, a demyelinating transmissible disease of sheep. J. Neuropathol. Exp. Neurol. 16 (1957) 389–403.

SIMON, R.: Neurosyphilis. Neurology 44 (1994) 2228–2230.

SIMPSON, D. and M. TAGLIATI: Neurologic manifestations of HIV infection. Ann. Intern. Med. 121 (1994) 7769–7785.

SNIDER, W., D. SIMPSON, S. NIELSON, J. GOLD, C. METROKA and J. POSNER: Neurological complications of acquired immune deficiency syndrome: analysis of 50 patients. Ann. Neurol. 14 (1983) 403–418.

STOLER, M., T. ESKIN, S. BENN, R. ANGERER and L. ANGERER: Human T-cell lymphotropic virus type III infection of the central nervous system – preliminary in situ analysis. J. Am. Med. Assoc. 256 (1986) 2360–2364.

SYNDULKO, K., E.J. SINGER, J. NOGALES-GAETE, A. CONRAD, P. SCHMID and W.W. TOURTELLOTTE: Laboratory evaluations in HIV-1-associated cognitive/motor complex (Rev.). Psychiatr. Clin. North Am. 17 (1994) 91–123.

TAKAHASHI, K., S. WESSELINGH, D. GRIFFIN, J. MCARTHUR, R. JOHNSON and J. GLASS: Localization of HIV-1 in human brain using polymerase chain reaction/in situ hybridization and immunocytochemistry. Ann. Neurol. 39 (1996) 705–711.

TAN, S.V., R.J. GUILOFF and F. SCARAVILLI: AIDS-associated vacuolar myelopathy. A morphometric study. Brain 118 (1995) 1247–1261.

TINDALL, B. and D. COOPER: Primary HIV infection: host responses and intervention strategies. AIDS 5 (1991) 1.

TOGGAS, S.M., E. MASLIAH, E.M. ROCKENSTEIN, G.F. RALL, C.R. ABRAHAM and L. MUCKE: Central nervous system damage produced by expression of the HIV-1 coat protein gp120 in transgenic mice (see comments). Nature 367 (1994) 188–193.

TORNATORE, C., R. CHANDRA, J.R. BERGER and E.O. MAJOR: HIV-1 infection of subcortical astrocytes in the pediatric central nervous system. Neurology 44 (1994a) 481–487.

TORNATORE, C., K. MEYERS, W. ATWOOD, C. CONANT and E. MAJOR: Temporal patterns of human immunodeficiency virus type 1 transcripts in human fetal astrocytes. J. Virol. 68 (1994b) 93–102.

TRACEY, I., C. CARR, A. GUIMARAES, J. WORTH, B. NAVIA and R. GONZALEZ: Brain choline-containing compounds are elevated in HIV-positive patients before the onset of AIDS dementia complex: a proton magnetic resonance spectroscopic study. Neurology 46 (1996) 783–788.

TYOR, W., S. WESSELINGH, J. GRIFFIN, J. MCARTHUR and D. GRIFFIN: Unifying hypothesis for the pathogenesis of HIV-associated dementia complex, vacuolar myelopathy, and sensory neuropathy. J. Acquir. Immune Defic. Syndr. 9 (1995) 379–388.

VAGO, L., A. CASTAGNA, A. LAZZARIN, G. TRABATTONI, P. CINQUE and G. COSTANZI: Reduced frequency of HIV-induced brain lesions in AIDS patients treated with zidovudine. J. Acquir. Immune Defic. Syndr. 6 (1993) 42–45.

VAN PAESSCHEN, W., C. BODIAN and H. MAKER: Metabolic abnormalities and new-onset seizures in human immunodeficiency virus-seropositive patients. Epilepsia 36 (1995) 146–150.

VAZEUX, R., N. BROUSSE, A. JARRY, D. HENIN, C. MARCHE, C. VEDRENNE, J. MIKOL, M. WOLFF, C. MICHON and W. ROZENBAUM: AIDS subacute encephalitis: identification of HIV-infected cells. Am. J. Pathol. 126 (1987) 403–410.

VERNANT, J.C., L. MAURS, A. GESSAIN, F. BARIN, O. GOUT, J.M. DELAPORTE, K. SANHADJI, G. BUISSON and G. DE-THE: Endemic tropical spastic paraparesis associated with human T-lymphotropic virus type I: a clinical and seroepidemiological study of 25 cases. Ann. Neurol. 21 (1987) 123–130.

VION-DURY, J., F. NICOLI, A.M. SALVAN, S. CONFORT-GOUNY, C. DHIVER and P.J. COZZONE: Reversal of brain metabolic alterations with zidovudine detected by proton localised magnetic resonance spectroscopy (Letter). Lancet 345 (1995) 60–61.

WECHSLER, D: Wechsler Adult Intelligence Scale Revised. New York, Psychological Corporation, 1981.

WEI, X., S.K. GHOSH, M.E. TAYLOR, V.A. JOHNSON, E.A. EMINI, P. DEUTSCH, J.D. LIFSON, S. BONHOEFER, M.A. NOWAK, B.H. HAHN, M.S. SAAG and G.M. SHAW: Viral dynamics in human immunodeficiency virus type 1 infection. Nature 373 (1995) 117–122.

WERNER, T., S. FERRONI, T. SAERMARK, R. BRACK-WERNER, R. BANATI, R. MAGER, L. STEINAA, G. KREUTZBERG and V. ERFLE: HIV-1 nef protein exhibits structural and functional similarity to scorpion peptides interacting with K$^+$ channels. AIDS 5 (1991) 1301–1308.

WESSELINGH, S.L., J. GLASS, J.C. MCARTHUR, J.W. GRIFFIN and D.E. GRIFFIN: Cytokine dysregulation in HIV-associated neurological disease (Rev.). Adv. Neuroimmunol. 4 (1994) 199–206.

WESTERVELT, P., D. TROWBRIDGE, L. EPSTEIN, Y. LI, B. HAHN, G. SHAW, R. PRICE and L. RATNER: Macrophage tropism determinants of HIV-1 in vivo. J. Virol. 66 (1992) 2577–2582.

WILEY, C.A. and C.L. ACHIM: Human immunodeficiency virus encephalitis and dementia. Ann. Neurol. 38 (1995) 559–560.

WILEY, C., R. SCHRIER, J. NELSON, P. LAMPERT and M. OLDSTONE: Cellular localization of human immunodeficiency virus infection within the brains of acquired immune deficiency patients. Proc. Natl. Acad. Sci. USA 83 (1986) 7089–7093.

WONG, M., N. SUITE and D. LABAR: Seizures in human immunodeficiency virus infection. Arch. Neurol. 47 (1990) 640.

WORLD HEALTH ORGANIZATION: 1990 World Health Organization consultation on the neuropsychiatric aspects of HIV-1 infection. AIDS 4 (1990) 935–936.

Handbook of Clinical Neurology, Vol. 27 (71): Systemic Diseases, Part III
M.J. Aminoff and C.G. Goetz, editors

CHAPTER 13

Opportunistic infections of the nervous system in AIDS

JOSEPH R. BERGER[1] and BRUCE A. COHEN[2]

[1]*Departments of Neurology and Internal Medicine, University of Kentucky Medical Center, Lexington, KY, and*
[2]*Department of Neurology, Northwestern University School of Medicine, Chicago, IL, U.S.A.*

'As it takes two to make a quarrel,
so it takes two to make a disease,
the microbe and its host.'

Charles V. Chapin (1856–1941)
Papers, 'The Principles of Epidemiology'

Not unexpectedly, the number of opportunistic infections of the nervous system that are observed in association with human immunodeficiency virus (HIV-1) infection is quite extensive. The spectrum of these infections is broad, including representatives of virus, bacteria, fungi and unicellular and multicellular parasites. Table 1 lists the central nervous system (CNS) infections commonly detected in a composite autopsy series of 926 patients with the acquired immunodeficiency syndrome (AIDS) from seven separate studies (Kure et al. 1991).

TABLE 1

Frequency of CNS opportunistic infection at the time of autopsy (Kure et al. 1991).

Cytomegalovirus	15.8%
Toxoplasmosis	13.6%
Cryptococcus	7.6%
Progressive multifocal leukoencephalopathy	4.0%
HSV encephalitis	1.6%
Candidiasis	1.1%
HZV encephalitis	0.6%
Histoplasmosis	0.4%
Tuberculosis	0.3%
Aspergillosis	0.3%

This chapter is devoted to the common and not-so-common infections of the central and peripheral nervous system seen in patients with AIDS. It has been organized by the nature of the opportunistic infection, namely, virus, bacteria, fungi, and parasites.

VIRUS

HERPES VIRUS

Cytomegalovirus (CMV)

Human cytomegalovirus (CMV) is a ubiquitous herpes virus acquired throughout life. In the U.S.A. 60–80% of adults have serologic evidence of infection and 90% or more of HIV-infected patients acquire CMV (Bale 1984; Drew et al. 1981; Drew 1988). Primary infection is usually asymptomatic in young healthy adults but may be associated with a transient mononucleosis-like syndrome (Klemola et al. 1969).

Neuropathologic series have described a range of pathology from isolated cytomegalic cells to severe necrotizing hemorrhagic encephalitis, myelitis or neuritis. In the brain, cytomegalocytes may be found in isolation or in association with microglial nodules. The most common neuropathologic pattern is a diffuse microglial nodular encephalitis disseminated in deep gray and to a lesser degree in white matter (Morgello et al. 1987; Vinters et al. 1989). Molecular diagnostic techniques increase the sensitivity of CMV detection and may identify CMV antigens or DNA in one-third or more of AIDS brains, localized in microglial nodules, macrophages, astrocytes, oligodendrocytes, neurons, ependymal cells, endothelia, and meninges (Wiley and Nelson 1988; Schmidbauer et al. 1989a; Belec et al. 1990).

A necrotizing ventriculitis is seen in about 10% of cases with acute and chronic inflammatory infiltrates, cytomegalic cells, associated vasculitis and meningitis with involvement of choroid plexus and cranial nerve roots and centrifugal extension of inflammation. Dystrophic calcification occurs in some instances (Morgello et al. 1987; Vinters et al. 1988; Kalayjian et al. 1993).

Focal necrotizing encephalitis and myelitis due to CMV result in discrete areas of parenchymal injury with variable degrees of associated inflammatory cell infiltration, cytomegalic cells and hemorrhages. Multiple foci in both deep gray and white matter regions are present, often beyond what is apparent by imaging studies obtained in life. Macrophage infiltration may be prominent and vascular involvement leads to infarction and hemorrhage (Morgello et al. 1987; Vinters et al. 1988; Gungor et al. 1993).

CMV polyradiculomyelitis is usually characterized by a necrotizing radiculitis with associated vasculitis and segmental thrombosis, and a polymorphonuclear or mixed meningitis which may extend into the spinal cord (Eidelberg et al. 1986a; Mahieux et al. 1989; Miller et al. 1990). Necrotizing myelitis may be associated (Jacobsen et al. 1988; Miller et al. 1990). Demyelination of nerve or spinal cord may be seen in conjunction with necrotizing radiculitis (Moskowitz et al. 1984; Bishopric et al. 1985).

Peripheral nerves may be affected by a multifocal necrotizing neuritis and arteritis with polymorphonuclear infiltration resulting in endoneurial necrosis. Cytomegalic cells may or may not be present. Involvement is characteristically asymmetric (Said et al. 1991; Roullet et al. 1994). CMV multifocal neuritis may affect cranial nerves (Small et al. 1989). Necrotizing optic neuritis may result from extension of CMV retinitis (Grossnik-Laus et al. 1987). Dorsal root ganglionitis has been found in association with necrotizing myelitis due to CMV (Tucker et al. 1985). Cytomegalocytes containing CMV-specific antigens have been found in Schwann cells (Bishopric et al. 1985; Grafe and Wiley 1989), and demyelinating

neuropathy has been attributed to CMV (Robert et al. 1989; Cornford et al. 1992a; Morgello and Simpson 1994).

The frequency with which CMV may be found in neuromuscular tissue appears to increase with survival following diagnosis of AIDS. In a series of 115 adult autopsies, 27% had light microscopic evidence of CMV in perineurial, epineural, perimysial or epimysial sites, usually in endothelia. The frequency increased from 19% in those having AIDS for 3 months or less to 46% in those having AIDS for 2 years or longer (Cornford et al. 1992a).

The pathologic features are most suggestive of hematogenous dissemination to the CNS and peripheral nerve. Subsequent extension of infection may involve seeding by virus entering CSF as well as by contiguous extension (Morgello et al. 1987; Wiley and Nelson 1988).

Clinical features

CMV encephalitis. Small case series and case reports have identified a number of clinical presentations of CMV encephalitis in AIDS. A subacute diffuse encephalopathy evolving over weeks is characterized by impairment of cognition and sensorium, apathy and withdrawal from normal activities. Confusion and disorientation contrast with HIV encephalopathy in which sensorium tends to be preserved, and the course evolves more rapidly. Neurologic examination reveals slowing of cognition, impairment of memory and attention and variable motor features, including hyper-reflexia, ataxia and weakness (Holland et al. 1994). Progressive CMV encephalitis has been reported in an HIV-infected infant acquiring CMV infection in utero (Curless et al. 1987).

Patients with necrotizing encephalitis may complain of headache and localized symptoms. Fever and focal neurologic findings may be present on examination (Edwards et al. 1985; Laskin et al. 1987; Masdeu et al. 1988). A patient with hypopituitarism due to focal CMV infection of the hypothalamus has been described (Sullivan et al. 1992). Rarely, CMV may present as a mass lesion (Dyer et al. 1995). When ventriculitis is present, cranial neuropathies, nystagmus and progressive ventricular enlargement may be seen (Kalayjian et al. 1993). Presentation as a brainstem encephalitis, with internuclear ophthalmoplegia, ataxia and vertical gaze paresis in one patient, and tetraparesis and sixth nerve palsy in another, has been described. Both patients progressed to confusion and obtundation and were found to have multifocal necrotizing encephalitis with ventriculitis in one case, and microglial nodular encephalitis with ventriculitis in the other (Fuller et al. 1989).

Acute onset of neurologic deficit may occur in patients with cerebral infarctions, resulting from CMV vasculitis, and may progress in a step-wise fashion (Kieburtz et al. 1993). Acute subarachnoid hemorrhage (Hawley et al. 1983) and intracerebral hemorrhage (Hawley et al. 1983; Dyer et al. 1995) have also been described.

Virtually all patients with CMV encephalitis will have systemic infection. Encephalitis may occur despite maintenance antiviral therapy for CMV retinitis (Schwarz et al. 1990; Berman and Kim 1994). Autopsy studies of patients with CMV encephalitis commonly disclose involvement of other organs (Kalayjian et al. 1993; Holland et al. 1994). Patients with CMV encephalitis may have mixed clinical patterns and may have associated CMV myelitis, polyradiculitis or multifocal neuritis.

Cerebral imaging studies are of limited sensitivity and even less specificity in patients with CMV encephalitis. Often they are negative or show non-specific atrophy in patients with diffuse nodular encephalitis. Ependymal or meningeal enhancement may be suggestive of the diagnosis when present (Fig. 1) (Grafe et al. 1990; Kalayjian et al. 1993; Holland et al. 1994), and areas of focal infarction or necrosis may be visualized (Masdeu et al. 1988; Grafe et al. 1990). Diffuse white matter hyperintensity on MRI has been regarded by some investigators to be a prominent feature (Miller et al. 1997) of CMV encephalitis. Patients with enhancing mass lesions due to CMV (Dyer et al. 1995; Huang et al. 1997) and with hemorrhagic mass lesions (Dyer et al. 1995) have been reported. Progressive ventricular enlargement may suggest ventriculitis (Kalayjian et al. 1993).

CSF findings also vary. Most patients have non-specific protein elevation. Glucose may be normal or decreased. Leukocytes may be absent, or a modest lymphocytesis may be present (Holland et al. 1994). A prominent pleocytosis with polymorphonuclear leukocytes may occur in patients with ventriculitis (Kalayjian et al. 1993). CMV can rarely be cultured from CSF of patients with subacute encephalitis (Dix et al. 1985) or meningoencephalitis (Edwards et al. 1985).

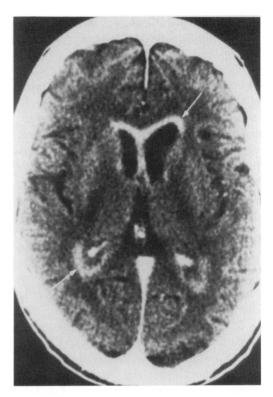

Fig. 1. CMV ventriculoencephalitis. CT of the brain using double-dose delayed contrast technique reveals marked periventricular enhancement in CMV ventriculoencephalitis.

Necrotizing myelitis. Necrotizing myelitis due to CMV in HIV-infected patients is most commonly seen in association with polyradiculitis (Moskowitz et al. 1984; Tucker et al. 1985; Jacobsen et al. 1988). Occasional cases of necrotizing myelitis in the absence of a typical polyradiculitis syndrome have been described presenting with acute or progressive paraplegia and disturbances of urinary and rectal sphincter functions. Reflexes are preserved or enhanced in the legs unless concurrent neuropathy is present, and a sensory level may be demonstrable (Said et al. 1991, case 5; Gungor et al. 1993). Pathologic studies may reveal necrosis of both gray and white matter structures (Vinters et al. 1989).

A patient with a demyelinating myelopathy involving the posterior columns associated with astrocytic gliosis, spongy changes with foamy macrophages and numerous cytomegalic cells has been described. Necrosis or perivascular inflammation was not seen. Clinical features were compatible with a myelopathy (Moskowitz et al. 1984, case 1).

Polyradiculomyelitis. CMV polyradiculomyelitis in HIV-infected patients presents subacutely over days to a few weeks. Initial symptoms of paresthesias or dysesthetic pain localized to perineal and lower extremity regions are followed by a rapidly progressive paraparesis with hypotonia and diminished or absent lower extremity reflexes. Urinary retention is characteristic and rectal sphincter incontinence is common. Variable sensory findings are overshadowed by the motor features. Babinski reflexes and a sensory level may indicate an associated myelitis. With time, symptoms progress by ascending to involve the upper limbs and sometimes the cranial nerves (Tucker et al. 1985; Eidelberg et al. 1986; Behar et al. 1987; Miller et al. 1990; Cohen et al. 1993; So and Olney 1994).

CSF is usually characterized by a polymorphonuclear pleocytosis and a prominent elevation of protein. Hypoglycorrhachia is often present. CMV was cultured from CSF in about half of reported cases (Cohen et al. 1993; So and Olney 1994), though culture may take weeks (Tucker et al. 1985).

MRI may be normal or reveal enhancement of the conus medullaris, cauda equina, meninges and nerve roots (Bazan et al. 1991; Talpos et al. 1991; Whiteman et al. 1994). Electrophysiologic studies reveal features of an axonal neuropathy with acute denervation changes. Variable slowing of nerve conduction may also be present (Behar et al. 1987; Miller et al. 1990).

The appearance of acute cauda equina syndrome in a patient with AIDS is suggestive of CMV when polymorphonuclear pleocytosis is present in CSF; however, the syndrome is not pathognomonic. Other conditions which may produce a cauda equina syndrome in AIDS patients include lymphomatous meningitis, syphilis, toxoplasmosis, other herpes viruses, cryptococcus or bacterial meningitis.

Multifocal neuritis. A progressive multifocal motor and sensory neuropathy which evolves over weeks to months has been demonstrated to result from CMV infection in AIDS. Initial paresthesias and dysesthesias are quickly followed by prominent motor weakness which involves both upper and lower limbs asymmetrically. Neurogenic atrophy may

be prominent. Electrophysiologic studies reveal features of an axonal neuropathy with diminished sensory and motor nerve amplitudes and denervation changes. CMV viremia is often present. CSF may be normal or show non-specific protein elevation. Nerve biopsy reveals a necrotizing neuritis with mononuclear and polymorphonuclear infiltrates and cytomegalocytes localized around endoneurial capillaries in nerve trunks and roots. Necrotizing arteritis may be present (Robert et al. 1989; Said et al. 1991; Roullet et al. 1994). Cranial nerves may be similarly affected (Small et al. 1989). A patient with a multifocal neuropathy revealing both necrotizing and demyelinating pathology has been described (Morgello and Simpson 1994). Neuromuscular pathology due to CMV was found in 27% of a series of 115 AIDS autopsies, predominantly localized to perineurial and epineurial regions (Cornford et al. 1992a).

Diagnosis

Diagnosis of neurologic CMV infections during life in AIDS patients has been hindered until very recently by the rarity of positive cultures from CSF and the limited sensitivity and specificity of other diagnostic studies. Treatment has often been empiric, when histopathologic confirmation of the diagnosis could not be obtained. Recent development of molecular diagnostic techniques and their application to CSF analysis presently offer a new opportunity to diagnose and treat patients with CMV infection of the nervous system.

Detection of CMV DNA in CSF following polymerase chain reaction (PCR) amplification has been found to be highly sensitive and specific for CMV infection of the nervous system in retrospective studies (Cinque et al. 1992; Gozlau et al. 1992c; Wolf and Spector 1992; Clifford et al. 1993). A prospective study found sensitivity and specificity of CSF PCR to be over 90% with positive and negative predictive values of 86 and 97% respectively (Gozlan et al. 1995). Quantification of CMV DNA levels has been correlated with histopathologic evidence of CNS infection in brain or spinal cord with the highest levels found in patients with ventriculoencephalitis (Arribas et al. 1995). In contrast, a retrospective autopsy series found low specificity of CSF detection of CMV DNA for CMV encephalitis, though limited spinal cord and radicular tissue could be assessed (Achim et al. 1994).

In patients with polymorphonuclear CSF pleocytosis, detection of the CMV lower matrix phosphoprotein pp65 within neutrophil nuclei provided a rapid diagnostic test for CMV (Revello et al. 1994). Large atypical cells believed to be macrophages revealed positive immunoperoxidase staining with antisera directed at CMV in a patient determined to have CMV encephalitis and polyradiculitis (Marmaduke et al. 1991). In situ hybridization with digoxigenin-labeled CMV probes detected CMV DNA in CSF cells of six patients with suspected CMV encephalitis, one of whom was confirmed at autopsy. Similar detection of CMV DNA in peripheral blood neutrophils was present in all six cases (Musiani et al. 1994).

Therapy

At the time of writing, two antiviral agents with efficacy against CMV in AIDS patients are available for use. Both are virustatic, requiring continued suppressive maintenance therapy during which viral resistance may emerge. Optimal regimens for treatment of CMV neurologic disease in AIDS are yet to be devised.

Ganciclovir (9-(1,3,-dihydroxy-2-propoxymethyl)-guanine) is an acyclic nucleoside which requires intracellular phosphorylation for efficacy. The first phosphorylation step is dependent on a viral phosphotransferase and the accumulation within infected cells imparts some selectivity to the agent (Matthews and Boehme 1988). Clinical studies have established efficacy against CMV DNA polymerase in patients with CMV retinitis and gastroenteritis. Hematologic toxicity, particularly neutropenia and thrombocytopenia, is a common problem in the use of ganciclovir (Collaborative DHPG Treatment Study Group 1986; Laskin et al. 1987; Dieterich et al. 1988). Resistance related to deficient production of the viral phosphotranferase mediating the first intracellular phosphorylation has been shown to emerge during continued suppressive therapy (Erice et al. 1989).

Foscarnet (trisodium phosphonoformate) is also virustatic, exerting its effect against CMV DNA polymerase without requiring phosphorylation. It has some efficacy against the reverse transcriptase of HIV as well (Palestine et al. 1991) and showed a benefit on survival compared to ganciclovir in patients treated for CMV retinitis (Studies on Ocular Complications of AIDS Research Group 1992). Renal toxicity is the most significant adverse effect, though

seizures and paresthesias may also be seen. Anemia and electrolyte disturbances and mild myelosuppression which is aggravated by concomitant antiretroviral therapy may also occur (Palestine et al. 1991). While foscarnet is active against ganciclovir-resistant strains of CMV, resistance to this agent may also emerge. Concurrent use of both agents may provide additional efficacy in patients with CMV retinitis or gastrointestinal disease despite failure of monotherapy with each (Dieterich et al. 1993).

Little is presently known about the neuropharmacology of ganciclovir and foscarnet. Ganciclovir appears to reach a mean concentration in CSF of about 40% of that in plasma and may reach concentrations necessary to reduce plaque-forming units of most CMV isolates by 50% (ED_{50}) at common current doses of 2.5 mg/kg every 12 h. Concentration in brain compared to serum may be similar. Penetration of foscarnet into CSF has varied from 27% of plasma concentration to 66% after prolonged therapy using samples obtained 1 h after infusion. Levels attained were above the ED_{50} for CMV in tissue culture (McCutchan 1995).

No prospective studies on treatment of neurologic CMV infections have been carried out to date. Case reports have been mixed, though some indications of potential efficacy can be found. In CMV encephalitis, response to ganciclovir in a patient with biopsy-proven focal CMV encephalitis has been documented (Sullivan et al. 1992). Encephalitis has occurred, however, despite maintenance ganciclovir therapy for CMV retinitis, suggesting the emergence of viral resistance (Schwarz et al. 1990; Berman and Kim 1994). The addition of foscarnet reversed obtundation and improved CSF inflammation in a similar patient transiently, but recrudescence followed reduction in dosage several weeks later (Enting et al. 1992). In patients with ventriculoencephalitis, response has been poor. Many of these patients were already exposed to ganciclovir because of prior CMV retinal or gastrointestinal disease (Kalayjian et al. 1993), although one patient failed to respond to primary initiation of therapy (Price et al. 1992). A patient who developed encephalitis 1 week following initiation of ganciclovir therapy for CMV retinitis responded to the addition of foscarnet at split doses of 180 mg/kg/day and was maintained on an alternating regimen of ganciclovir and foscarnet until cardiopulmonary death 7 months later (Peters et al. 1992).

Another patient who presented with CMV meningo-encephalitis as his first opportunistic infection responded to ganciclovir with survival exceeding 1 year (Cohen 1996).

In patients with polyradiculomyelitis, a number of reports have noted response to prompt initiation of ganciclovir therapy. Response could be slow, requiring several months before neurologic improvement was seen (Graveleau et al. 1989; Miller et al. 1990; Cohen et al. 1993; Kim and Hollander 1993b). Other patients have failed to respond despite initiation of therapy (Jacobsen et al. 1988; De Gans et al. 1990). Progression of polyradiculomyelitis in spite of ganciclovir has been shown to be associated with CMV resistance (Cohen et al. 1993). Addition of foscarnet resulted in gradual improvement in a recently reported case (Decker et al. 1994). Improvement in two patients followed combined therapy with both agents after ganciclovir in one and foscarnet in the other failed as monotherapy (Karmochkine et al. 1994).

Multifocal neuropathy responds initially to therapy with ganciclovir; however, relapse of neuropathy or development of CMV polyradiculitis or encephalitis appears common. As many as 75% of patients with multifocal neuropathy may die of disseminated CMV infection (Said et al. 1991; Roullet et al. 1994).

Limited sensitivity of imaging studies and slow recovery in responsive patients has hindered decision-making on modification of therapy. The best guidance for therapeutic efficacy may come from serial CSF studies. In patients with polyradiculomyelitis, resolution of polymorphonuclear leukocytosis and hypoglycorrhachia occurs quickly. Persistence appears indicative of viral resistance (Cohen et al. 1993). The availability of PCR analysis for CMV DNA in CSF offers a means of serial monitoring, and small recent series suggest that this may become a useful marker of therapeutic response or failure (Cinque et al. 1995; Cohen 1996).

HERPES SIMPLEX VIRUS (HSV)

HSV-1 produces sporadic necrotizing encephalitis in apparently normal adults with a predilection to involve the temporal and inferior frontal regions of the brain. HSV-2 produces aseptic meningitis which may be recurrent (Craig and Nahmias 1973). Despite the widespread prevalence and neurotropism of HSV, clinical reports of neurologic infections in HIV-infected patients are sparse and may be confounded by concurrent opportunistic processes, particularly CMV. It has been suggested that a vigorous immune response may be an important element in the pathophysiology of neurologic HSV infection and that the absence of normal immune responses may modify neuropathology. An anergic patient with Hodgkin's disease and HSV encephalitis was noted to have limited neuronal loss and myelin damage (Price et al. 1973). Another individual developed chronic encephalopathy while on steroid therapy. Biopsied brain tissue grew HSV but revealed no necrotizing features and only modest inflammation (Sage et al. 1985).

Necrotizing temporal encephalitis has been reported occasionally in HIV-infected patients. Some atypical features may be present, such as the lack of a diffuse meningeal reaction and abundant viral inclusions several weeks after onset (Tan et al. 1993). Concurrent CMV ventriculoencephalitis and HSV-1 necrotizing temporal encephalitis have been described. Immunohistochemical and in situ hybridization studies revealed two discrete processes (Vital et al. 1995). Two patients with lymphadenopathy and HIV infection had temporal encephalitis with necrosis and intense inflammatory responses. HSV-2 grew from cultures of biopsied brain (Dix et al. 1985).

The clinical presentations in these patients were acute, with fever, headache, confusion and lethargy, seizures, and in one case hemiparesis. EEG revealed focal slowing or periodic discharges over the involved temporal and frontal regions. Computed tomography (CT) scans were normal in two and abnormal in one, revealing a localized hypodensity in the involved temporal lobe. CSF was acellular in two and showed 11 mononuclear cells/mm^3 and 19 erythrocytes/mm^3 in another. Protein was elevated in two cases, normal in one, and glucose was normal in two. Cultures were unrevealing and PCR analysis failed to reveal HSV DNA in CSF of one patient obtained after 11 days of therapy. In each case diagnosis was established histopathologically, from biopsy tissue in three instances and at autopsy in one (Dix et al. 1985; Tan et al. 1993; Vital et al. 1995). PCR amplification techniques may allow rapid detection of HSV-1 DNA in CSF of patients with herpetic encephalitis (Rowley et al. 1990).

Several patients with ventriculoencephalitis have been found to have both CMV and HSV in areas of ependymal and subependymal necrosis. In one patient, necrotizing retinitis due to both CMV and HSV was present in association with a necrotizing ventriculitis. Immunohistochemistry identified extensive CMV staining in necrotic regions while HSV staining in the brain occurred mainly in the endothelia of small arterioles and large venules (Pepose et al. 1984). In three other cases, immunostaining also revealed evidence of both viruses in necrotic regions but evidence of CMV appears to have been more extensive (Laskin et al. 1987).

Clinical presentations in these patients were acute delirium and seizures in one, subacute progressive ataxia, weakness, seizures and lethargy over 6 weeks in another, and gradual deterioration over several months in a third, who also had treated CNS toxoplasmosis. The fourth patient was not noted to have a neurologic illness in life (Pepose et al. 1984; Laskin et al. 1987).

A patient with HIV infection and HSV encephalitis localized to the brainstem has been described, presenting with acute ataxia, fever and disorientation. MRI revealed several foci of increased signal on T2-weighted images without mass effect. Fatal progression ensued despite therapy with acyclovir. Autopsy revealed diffuse HSV infection of oligodendroglia, most intensely in the brainstem and cerebellum, though other regions also showed staining. Demyelination with relative preservation of axons and minimal inflammation was observed. The HSV isolate recovered was found to be 5-fold more virulent than a laboratory strain when inoculated into mice (Hamilton et al. 1995). HSV DNA has also been identified in neuronal nuclei of an HIV-infected patient with microglial nodular brainstem encephalitis by in situ hybridization (Schmidbauer et al. 1989b).

A diffuse meningoencephalitis in HIV-infected patients has been associated with HSV-2. The clinical presentation is non-specific with fever, headache, lethargy or delirium, tremors, or seizures in variable combinations. Imaging studies may be normal or, as in one case, reveal subdural fluid collections. A patient with central diabetes insipidus and HSV-2 meningoencephalitis has been described. CSF may yield a lymphocytic pleocytosis with reported cell counts as high as 720/mm^3. Erythrocytosis may be present. Protein is usually elevated and may be markedly increased. In one patient IgM antibodies to HSV were found in CSF and the CSF/serum index was ele-

vated, indicating local production. HSV can be cultured from CSF though cultures are often unrevealing. Brain biopsy in one patient yielded HSV-2 on culture. Concurrent herpetic lesions may be present in patients with meningitis (Dahan et al. 1988; Gateley et al. 1990; Madhoun et al. 1991). HSV-1 (Yamamoto et al. 1991) and HSV-2 (Picard et al. 1993; Cohen et al. 1994; Tedder et al. 1994) DNA have been identified through PCR amplification in CSF of patients with recurrent meningitis.

HSV myelitis and radiculitis

Herpes genitalis has been shown to cause a lumbosacral radiculitis producing paresthesias and neuralgia in the perineum and lower extremities and urinary retention (Caplan et al. 1977). Meningitis and ascending myelitis may occur in association with (Craig and Nahmias 1973; Wiley et al. 1987a) or in the absence of (Klastersky et al. 1972; Bergstrom et al. 1990) vesicular eruptions. HSV-2 has been identified by immunohistochemical staining in spinal cord, chronically inflamed peripheral nerves and dorsal root ganglia (Wiley et al. 1987a). HIV-infected patients with apparent HSV-2 lumbosacral radiculitis (Madhoun et al. 1991) and ascending myelitis have been described (Britton et al. 1985). Concurrent CMV infection was present in a patient presenting with a subacute polyradiculomyelitis. CSF revealed a polymorphonuclear pleocytosis with depressed glucose and elevated protein. At autopsy, CMV infection involved the brain, spinal cord and some cranial nerves. HSV-2 staining was present by immunohistochemistry to a lesser degree. HSV-2 was cultured from cervical spinal cord, suggesting dual infection. The strain was identical to that cultured from a perirectal ulcer (Tucker et al. 1985).

A patient with necrotizing anterior spinal arteritis and associated myelomalacia had Cowdry type A intranuclear inclusions which stained positively for HSV-2 by immunohistochemistry. A small focus of CMV-positive cells was seen and thought to be a superinfection. The clinical presentation evolved over 3 months and consisted of low back pain, paresthesias and neuralgia in the perineum and lower extremities followed by asymmetric leg weakness and urinary retention. Reflexes were lost in the involved limbs and a thoracic sensory level was present. CSF was acellular with elevated protein and normal glucose. CSF cultures were unrevealing (Britton et al. 1985).

Treatment

Acyclovir (9-(2-hydroxyethoxymethyl)guanine) is the initial antiviral agent of choice for patients with HSV neurologic infection. Acyclovir is phosphorylated intracellularly to its active triphosphate form with the first phosphorylation dependent on a virus-specified thymidine kinase. The triphosphate form binds HSV DNA polymerase and acts as a chain terminator (Whitley and Gnann 1993).

The efficacy of acyclovir therapy of HSV neurologic infection has been variable in HIV-infected patients. Progression of HSV encephalitis has occurred in some cases in spite of therapy (Tan et al. 1993; Hamilton et al. 1995). Acyclovir treatment of cutaneous lesions failed to prevent progressive myelitis in a patient with HSV-2 infection (Britton et al. 1985). Improvement of encephalopathy but not an associated radiculomyelitis occurred in a patient with perianal HSV-2 (Madhoun et al. 1991). In contrast, stabilization of HSV-2 encephalitis occurred in two patients given acyclovir following progression on vidarabine therapy (Dix et al. 1985).

Treatment of HSV infections is complicated by the emergence of acyclovir-resistant strains which have altered or absent thymidine-kinase activity (Erlich et al. 1989; Englund et al. 1990). Acyclovir-resistant isolates have been thought to lack neurovirulence; however, a recent case of relapsing and ultimately fatal meningoencephalitis due to HSV-2 has been reported in which serial viral isolates demonstrated the emergence of acyclovir resistance during therapy. The addition of vidarabine was associated with transient improvement; however, rapid deterioration ensued 1 week later and repeat CSF yielded HSV lacking thymidine kinase activity (Gateley et al. 1990). An alternative therapy is foscarnet, which has been successfully used to treat acyclovir-resistant mucocutaneous lesions in HIV-infected individuals (Chatis et al. 1989; Safrin et al. 1991; Safrin 1992).

VARICELLA ZOSTER VIRUS (VZV)

Decline in immune competence is commonly associated with recurrent VZV, which initially presents as segmental radiculitis in most instances. The age-adjusted risk of VZV radiculitis in HIV-infected individuals was 17 times that of a non-infected homosexual control population, although occurrence did not predict faster progression to AIDS (Buchbinder

et al. 1992). A community-based study in New York City reported that VZV radiculitis in a group of homosexual men was associated with the development of AIDS in almost 73% (Melbye et al. 1987). The frequency of CNS zoster in autopsy series of AIDS patients has ranged from 2 to 4.4% (Petito et al. 1986; Gray et al. 1994).

The manifestations of segmental VZV radiculitis are similar in HIV and non-infected individuals, with a cutaneous eruption characterized by pain and vesicles on an erythematous base in a dermatomal distribution. The trunk is most commonly affected, followed by the face and extremities. Pain may occur without cutaneous lesions, or cutaneous dissemination may occur, suggesting viremia (Gelb 1993). VZV DNA has been detected in circulating mononuclear cells during clinical VZV eruptions (Gilden et al. 1989).

VZV may cause Ramsay–Hunt syndrome or herpes zoster oticus with a vesicular eruption in the auricle and external auditory canal and facial weakness (Mishell and Applebaum 1989). Segmental myoclonus may precede or occur in association with VZV radiculitis (Koppel and Daras 1992).

VZV infection of the CNS in AIDS patients may produce a variety of neuropathologies. A recent classification identified five clinico-pathologic patterns: multifocal encephalitis, ventriculitis, acute meningomyeloradiculitis with necrotizing vasculitis, focal necrotizing myelitis, and vasculopathy resulting in cerebral infarction. Multiple patterns occurred in individual patients, all of whom were immunosuppressed with CD4 counts below 300/mm^3 (Gray et al. 1994). Of clinical significance, a history of cutaneous VZV was absent in up to one-third of cases (Morgello et al. 1988; Chretien et al. 1993; Gray et al. 1994; Moulignier et al. 1995).

VZV encephalitis

A leukoencephalitis with pathologic, radiologic and clinical features resembling progressive multifocal leukoencephalitis has been demonstrated in several HIV-infected patients. Discrete ovoid or round lesions are found in white matter with a predilection for gray–white junctions and periventricular regions. Lesions may be confluent and may occur in brainstem and cerebellar sites in addition to the cerebral hemispheres. Central cavitation necrosis with peripheral edema, reactive astrocytosis, and microglial proliferation comprise the architecture of the lesions and relative sparing of axons may or may not be seen in demyelinated regions. A necrotizing ventriculitis with associated vasculitis, similar to that caused by CMV, may be present. Cowdry type A inclusion bodies are found in astrocytes, oligodendroglia, macrophages, endothelial cells, ependyma and neurons. When cutaneous zoster is associated, it may be recent or remote in occurrence and is often recurrent (Ryder et al. 1986; Gilden et al. 1988; Morgello et al. 1988; Gray et al. 1992, 1994). The location of lesions in the periventricular regions and gray–white junctions suggests hematogenous dissemination to the CNS (Morgello et al. 1988; Gray et al. 1994). Other cases where VZV is present in restricted distribution along discrete pathways indicate that transsynaptic spread of VZV also occurs (Rosenblum 1989; Rostad et al. 1989).

The most frequent clinical presentation in HIV-infected patients with VZV encephalitis is a diffuse subacute encephalopathy with headache, fever, cognitive deficits, lethargy or delirium, seizures and variable focal deficits, evolving over weeks (Ryder et al. 1986; Morgello et al. 1988). Acute encephalitis may be seen (Gray et al. 1992) and a case of chronic progressive encephalitis evolving over 18 months has been reported (Gilden et al. 1988). Isolated brainstem encephalitis has been described (Rosenblum 1989; Moulignier et al. 1995) and a case of ophthalmic zoster followed by chronic progressive dissemination through the visual system (with some involvement of adjacent structures) over an 11-month period has been documented (Rostad et al. 1989).

Cerebral imaging studies are of limited value. Both CT and MRI scans may be normal (Ryder et al. 1986; Morgello et al. 1988; Poscher 1994), or show non-specific lesions with or without peripheral contrast enhancement (Gilden et al. 1988; Rostad et al. 1989; Gray et al. 1992; Moulignier et al. 1995). Serial imaging studies may be normal at first or show non-specific lesions with subsequent appearance or progressive enlargement indicative of an evolving process (Gilden et al. 1988; DeAngelis et al. 1994).

CSF may be normal (Moulignier et al. 1995) but more often reveals features of a viral meningitis, such as moderate lymphocytic pleocytosis, marked pleocytosis or acellular CSF. Protein elevation may be moderate or substantial. Glucose concentrations are normal (Dix et al. 1985; Ryder et al. 1986; Gilden et al. 1988; Morgello et al. 1988; Gray et al. 1992;

Posher 1994). On occasion, VZV can be cultured from CSF, though it is infrequent (Dix et al. 1985). VZV-specific IgG antibodies may be detected by immunofluorescent assay in CSF of patients with meningoencephalitis (Poscher 1994). PCR amplification of VZV DNA in CSF offers a sensitive and specific means of diagnosis (Rozemberg and Lebon 1991; Shoji et al. 1992).

VZV myelitis

In HIV-infected patients, the interval between cutaneous eruptions and the onset of myelopathy may be measured in months. Recurrent cutaneous lesions at multiple dermatomal levels may precede CNS disease (Devinsky et al. 1991; Gilden et al. 1994; Snoeck et al. 1994). Myelopathy may appear prior to the cutaneous eruption (Gomez-Tortosa et al. 1994) or as an acute or subacute meningomyeloradiculitis in the absence of a cutaneous eruption (Vinters et al. 1988; Chretien et al. 1993; Gray et al. 1994).

A neuropathologic series including both HIV-infected and non-HIV-infected cases with VZV myelitis revealed extensive hemorrhagic necrotizing myelitis with vasculitis and thrombosis in dorsal root ganglia, and abnormalities of posterior nerve roots varying from lymphocytic infiltration to hemorrhagic necrosis. The most extensive pathologic changes were in the dorsal horns, ranging from focal necrosis with mild capillary proliferation and wedge-shaped demyelination to extensive necrotizing vasculitis and myelitis (Devinsky et al. 1991). Similar necrotizing myeloradiculitis and vasculitis have been found in HIV-infected patients presenting with rapidly progressive radiculomyelopathy. Cowdry type A inclusion bodies may not be seen even in the fulminant cases, or may occur so rarely as to hinder neuropathologic diagnosis. Immunohistochemistry and in situ hybridization techniques are more sensitive. Myelitis may be diffuse or restricted to segments associated with cutaneous eruptions (Vinters et al. 1988; Chretien et al. 1993; Gray et al. 1994).

Clinical presentations of VZV myelitis in HIV-infected patients may be acute or subacute with progressive weakness, sensory impairment and urinary sphincter dysfunction, often preceded by localized pain. Evolution of symptoms may occur over days to weeks and may then accelerate rapidly to paraplegia (Vinters et al. 1988; Gilden et al. 1994; Gomez-

Tortosa et al. 1994; Snoeck et al. 1994). An acute, rapidly progressive course may occur in some cases culminating in paraplegia and urinary retention. Polyradicular involvement may produce loss of lower extremity reflexes, resulting in a picture similar to CMV polyradiculomyelitis (Vinters et al. 1988; Chretien et al. 1993; Gray et al. 1994).

MRI may reveal non-enhancing intramedullary abnormalities (Gilden et al. 1994), or prominent enlargement of the spinal cord with associated signal abnormalities (Chretien et al. 1993), but may be normal (Gomez-Tortosa et al. 1994) or reveal features of meningitis (Snoeck et al. 1994).

CSF may be normal or reveal non-specific protein elevation and lymphocytic pleocytosis in more gradually evolving cases (Devinsky et al. 1991; Gilden et al. 1994; Gomez-Tortosa et al. 1994). In patients with more rapidly progressive necrotizing myelitis, CSF may contain a mixed lymphocytic and polymorphonuclear pleocytosis with predominance of either cell type (20–200 cells/mm^3), erythrocytosis (400–7300 cells/mm^3), and marked protein elevations (to 780 mg/dl) (Vinters et al. 1988; Chretien et al. 1993; Gray et al. 1994; Snoeck et al. 1994). VZV may be recovered from CSF (Snoeck et al. 1994), but cultures are generally unrevealing.

VZV vasculopathy

Vascular lesions associated with VZV may involve large or small vessels and may be inflammatory or bland. They often occur in conjunction with other manifestations of CNS VZV in HIV-infected patients (Gray et al. 1994). Herpes zoster ophthalmicus may be complicated by contralateral hemiparesis resulting from infarction of arteries of the circle of Willis ipsilateral to the involved trigeminal nerve in both HIV-infected and uninfected patients. The interval between acute ophthalmic zoster and the infarction may be as long as 1 year in HIV-infected individuals (Zaraspe-Yoo et al. 1984; O. Eidelberg et al. 1986; Pillai et al. 1989; Carneiro et al. 1991; Rousseau et al. 1993) and may be associated with VZV encephalitis (Gray et al. 1994; Amlie-Lefond et al. 1995) or meningomyeloradiculitis (Gray et al. 1994). Recurrent cerebral infarctions may be the presenting manifestation of VZV encephalitis in HIV-infected individuals in the absence of cutaneous zoster (Amlie-Lefond et al. 1995).

Neuropathologic studies reveal non-inflammatory

vascular lesions of large leptomeningeal or small deep penetrating arteries, with intimal or fibromuscular proliferation and focal thrombosis (O. Eidelberg et al. 1986; Morgello et al. 1988; Gray et al. 1994; Amlie-Lefond et al. 1995). In other individuals, necrotizing arteritis with transmural inflammation may involve large and small vessels (Gilden et al. 1988; Gray et al. 1994; Amlie-Lefond et al. 1995). Immunohistochemical staining may reveal VZV in sites of vascular necrosis (Gray et al. 1994). Kleinschmidt-De-Masters et al. (1996) have classified the patterns of VZV encephalitis into: (1) large/medium vessel vasculopathy with bland or hemorrhagic infarction; (2) small vessel vasculopathy with mixed ischemic/demyelinative lesions; and (3) ventriculitis/periventriculitis. These are not mutually exclusive with one form typically predominating.

Imaging studies reveal infarction patterns in small or large vessel territories (O. Eidelberg et al. 1986; Rouseau et al. 1993). The combination of multifocal ischemic and hemorrhagic infarctions with deep white matter lesions may be suggestive of VZV in the context of AIDS (Amlie-Lefond et al. 1995). Cerebral arteriography may show features of vasculitis (Carneiro et al. 1991; Rousseau et al. 1993). CSF may be normal in VZV-related cerebral infarction, though more commonly non-specific protein elevation is present. Modest pleocytosis may be present and glucose levels are normal or mildly decreased (O. Eidelberg et al. 1986; Carneiro et al. 1991). PCR amplification techniques allow identification of zoster DNA in CSF samples from patients with VZV vasculopathy (Amlie-Lefond et al. 1995).

Therapy

Treatment of CNS zoster in HIV-infected patients has been disappointing in cases reported to date. Available treatments are virustatic, and diagnosis in life has been extremely difficult. Progression of disease has occurred despite acyclovir or ganciclovir in patients with myelopathy (Chretien et al. 1993; Gilden et al. 1994), though one patient survived without improvement of his paraplegia (Gomez-Tortosa et al. 1994). Patients with acute encephalitis have also failed to respond to ganciclovir or acyclovir (Gray et al. 1992; Amlie-Lefond et al. 1995). One patient presenting with multiple cerebral infarctions was diagnosed by identification of VZV DNA in CSF after PCR amplification and appears to have stabilized on acyclovir therapy continued in maintenance dosages of 2400 mg/day, though reported follow-up is limited to 5 months (Amlie-Lefond et al. 1995). Two patients with zoster retinitis and vasculopathy appear to have stabilized on high doses of oral acyclovir (3000 and 3200 mg/day) for as long as 2 years; however, one appeared to develop VZV encephalitis when the dose was reduced to 2000 mg/day (Rousseau et al. 1993). Four patients with meningitis due to VZV responded to intravenous acyclovir or ganciclovir. In each case MRI scans of the brain revealed no parenchymal lesions (Poscher 1994).

A patient with VZV meningoradiculitis who was found to develop resistance to acyclovir during therapy was treated with foscarnet, resulting in stabilization of his neurologic symptoms, though no improvement was seen. Subsequent dosage reduction of foscarnet from 180 to 90 mg/kg/day was followed by the occurrence of necrotizing retinitis and isolation of VZV from CSF. Despite increase in foscarnet dosage, gradual clinical deterioration ensued (Snoeck et al. 1994). Vidarabine has activity against VZV, but has failed to prevent encephalitis in HIV-infected patients with ophthalmic zoster (Cole et al. 1984). At present, foscarnet is being recommended for patients suspected to have acyclovir-resistant cutaneous zoster (Balfour et al. 1994) and should be considered for patients with CNS VZV. Future evaluation of combination therapy and development of new agents may provide more effective treatment. Earlier diagnosis through PCR analysis of CSF specimens may allow initiation of therapy prior to the development of irreversible necrotizing encephalitis or myeloradiculitis.

JC VIRUS (JCV)

Until the AIDS epidemic, experience with progressive multifocal leukoencephalopathy (PML) was limited. A comprehensive review of PML published in 1984 found only 230 reported cases (Brooks and Walker 1984). Within 1 year of the initial description of AIDS in 1981, PML was recognized as an associated disorder (Gottlieb et al. 1981; Masur et al. 1981; Siegal et al. 1981; Miller et al. 1982; Bedri et al. 1983). Current estimates suggest that approximately 4–5% of all HIV-infected individuals will develop PML (Berger et al. 1987). This formerly rare disease, once regarded as a clinical curiosity by most neurologists, has lately become remarkably common.

By middle adulthood, 80–90% of the population has IgG antibodies against JCV and seroconversion rates have exceeded 90% in some urban areas (Walker and Padgett 1983). No disease has been convincingly associated with acute infection, although Blake et al. (1992) reported meningoencephalitis identified by a rise of IgM titers to JCV in a 13-year-old girl. Although the overwhelming majority of people have antibodies to JCV by adulthood, indicating prior exposure to the virus, the occurrence of PML in the absence of cellular immunodeficiency is quite extraordinary. Indeed, it is but a small minority of persons with underlying impairment of cellular immunodeficiency who ultimately develop disease, suggesting that the presence of JCV and immunodeficiency are not by themselves sufficient conditions for the development of the disorder.

Prior to the AIDS epidemic, the male : female ratio of PML approximated 1:1. The AIDS epidemic transformed this ratio from 5:1 (Holman et al. 1991) to 8:1 (Berger et al. 1998). However, it is likely that the changing pattern of infection with increasing numbers of women affected by AIDS as opposed to homosexual men will return this ratio towards parity. Furthermore, instead of affecting chiefly elderly individuals (Brooks and Walker 1984), as was observed in studies prior to AIDS, PML has become a disease of the young and middle-aged populations affected by AIDS. The greatest incidence is in individuals between the ages of 20 and 50 years (Brooks and Walker 1984; Berger 1998). PML is rarely observed in immunosuppressed children, perhaps chiefly the result of the lower percentages of children who have been exposed to JCV. However, despite its rarity in this age group, it has been described in both HIV-infected children (Henson et al. 1991a, b; Berger et al. 1992c; Berger 1998) and those with other underlying causes of immunodeficiency (Katz et al. 1994).

The spread of JCV is postulated to be by respiratory means (Shah 1990). The high prevalence of antibodies in the adult population and the rarity of PML in children support the contention that PML is the consequence of reactivation of JCV in individuals who have become immunosuppressed. Additionally, high titers of IgM antibody specific for JCV, anticipated if PML were the result of acute infection, are not observed (Padgett and Walker 1983a).

The first three patients described by Aström et al. (1958) in their seminal description of PML had either chronic lymphocytic leukemia or lymphoma and, until the early part of the last decade, the vast majority of patients with PML had lymphoproliferative disorders as the underlying cause of their immunosuppression (McArthur et al. 1993). Lymphoproliferative diseases remained the most common underlying illness for the development of PML until the AIDS epidemic and, in some communities where the incidence of AIDS is small, is still the likeliest underlying disorder. In a review of 69 pathologically confirmed cases and 40 virologically and pathologically confirmed cases of PML performed in 1984, Brooks and Walker (1984) found lymphoproliferative diseases accounted for 62.2% of the cases. In that series immune deficiency states were seen in 16.1%, but AIDS represented only 3% of the total number of cases.

AIDS has been estimated to be the underlying disease for PML in 55% to more than 85% of all current cases (Major et al. 1992). Based on reporting of AIDS to the Centers for Disease Control (CDC) between 1981 and June 1990, 971 of 135,644 (0.72%) individuals with AIDS were reported to have PML (Holman et al. 1991). This is likely to be an underestimate, since to be included in the CDC AIDS reporting system PML cases must have pathologic confirmation. A study of PML among patients with AIDS in the San Francisco Bay area estimated a prevalence of PML of 0.3% (Gillespie et al. 1991). The findings of these investigators suggested that PML in HIV-infected patients was underestimated by as much as 50% (Gillespie et al. 1991). The intrinsic nature of mortality data, inaccurate diagnosis, and incomplete reporting may affect these estimates. Other types of studies suggest that the incidence of PML in AIDS cases is substantially higher than that reported by the CDC, with estimates of 1–5% in clinical studies and as high as 10% in pathologic series (Krupp et al. 1985; Stone et al. 1986; Berger et al. 1987a; Kure et al. 1991; Kuchelmeister et al. 1993; Whiteman et al. 1993). In one large, retrospective, hospital-based, clinical study (Berger et al. 1987a), PML occurred in approximately 4% of patients hospitalized with AIDS. In a combined series of seven neuropathologic studies comprising a total of 926 patients with AIDS (Kure et al. 1991), 4% had PML. Similarly, a neuropathology series from Switzerland detected PML in more than 7% of their pa-

tients dying with AIDS (Lang et al. 1989). A pathologic study based on 548 consecutive, unselected autopsies between 1983 and 1991 performed on patients with AIDS by the Broward County (Florida) Medical Examiner revealed that 29 (5.3%) had PML confirmed at autopsy (Whiteman et al. 1993). Yet another recent neuropathologic review, based on autopsies between 1985 and 1992, found 21 (9.8%) cases of PML in 215 individuals dying with AIDS (Kuchelmeister et al. 1993), but may have been influenced by the numerous referral cases from outside that study center (Kuchelmeister et al. 1993). The increasing frequency with which PML has been observed since the inception of the AIDS epidemic is attested to by a 12-fold increase in the disorder when the 4-year intervals 1981–1984 and 1991–1994 were compared in a series from south Florida (Berger 1998). Of 156 cases of PML in this series, only two were associated with immunodeficient states other than AIDS (Berger 1998). Bacellar et al. (1994), studying the temporal trends in incidence of HIV-related neurologic disease in the Multicenter AIDS Cohort, noted that the average annual incidence rate of PML was 0.15/100 person-years with a yearly rate of increase of 24% between 1985 and 1992. After adjusting for the CD4 lymphocyte count, there appeared to be no residual calendar trends in the incidence rate of PML (Bacellar et al. 1994).

Macroscopically, the cardinal feature of PML is demyelination (Fig. 2). Demyelination may, on rare occasion, be unifocal, but typically occurs as a multifocal process. These lesions may occur in any location in the white matter; however, they have a predilection for the parieto-occipital regions. Not infrequently, lesions involve gray matter (Von Einsiedel et al. 1993) and are also found involving cerebellum, brainstem and, exceptionally, the spinal cord (Bauer et al. 1969; Kuchelmeister et al. 1993; Von Einsiedel et al. 1993). In an autopsy series of 21 cases, 17 cases showed PML foci in infratentorial structures, 13 cases in cerebellum, 13 cases in brainstem, and 10 cases in both regions (Kuchelmeister et al. 1993). The lesions range in size from 1 mm to several centimeters (Aström et al. 1958; Richardson 1970); larger lesions are not infrequently the result of coalescence of multiple smaller lesions.

The histopathologic hallmarks of PML are a triad (Aström et al. 1958; Richardson 1970) of multifocal demyelination, hyperchromatic, enlarged oligoden-

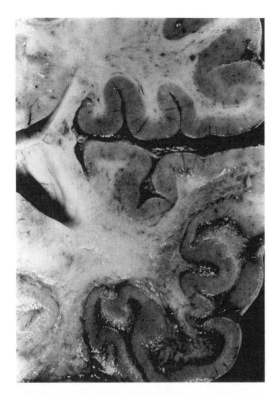

Fig. 2. Progressive multifocal leukoencephalopathy. Large demyelinated lesion of the subcortical frontal white matter in progressive multifocal leukoencephalopathy.

droglial nuclei, and enlarged bizarre astrocytes with lobulated hyperchromatic nuclei. The latter may be seen to undergo mitosis and appear to be quite malignant. In situ hybridization for JCV antigen allows for detection of the virion in the infected cells. Electron microscopic examination will reveal the JCV in the nucleus of the oligodendroglial cells. These virions measure 28–45 nm in diameter and appear singly or in dense crystalline arrays (Zu Rhein and Chou 1965; Zu Rhein 1967). Less frequently, the virions are detected in reactive astrocytes and they are uncommonly observed in macrophages that are engaged in removing the affected oligodendrocytes (Mazlo and Herndon 1970; Mazlo and Tariska 1982).

Even though neuropathologic findings of PML do not reveal fundamental differences between cases with AIDS and non-AIDS PML, the former group, according to some investigators, tends more frequently to present with extensive lesions having particularly destructive, necrotizing character

(Schmidbauer 1990; Kuchelmeister et al. 1993). Some investigators (Kuchelmeister et al. 1993; Berger 1998) have suggested that AIDS-associated PML may present infratentorial lesions more frequently than non-AIDS PML cases, although others have not found a substantial difference (Von Einsiedel et al. 1993).

Clinical features

The clinical hallmark of PML is the presence of focal neurologic disease associated with radiographic evidence of white matter disease generally occurring in the absence of mass effect. Emphasis needs to be placed on the focal features of this disease, particularly those that are apparent on clinical examination. The presence of focal findings on neurologic examination is quite helpful in distinguishing this disorder from HIV dementia, which may be radiographically very similar. The most common presentations include weakness, gait abnormalities, speech disorders, visual deficits and cognitive abnormalities. Table 1 summarizes the initial neurologic manifestations in four separate clinical series of patients with PML unrelated to AIDS (Brooks and Walker 1984) and AIDS-associated PML (Berger et al. 1987a; Von Einsiedel et al. 1993; Berger 1998).

Weakness is typically a hemiparesis, but monoparesis, hemiplegia, and quadriparesis may be observed with progression of disease. On occasion, the patient may present with a rapidly progressive flaccid hemiparesis that over time evolves into a spastic weakness. In older series, weakness is present in more than 80% of patients at the time of diagnosis (Arthur et al. 1989), but the improvement in the diagnostic modalities (in particular, MRI) and better familiarity with the disorder promotes earlier diagnosis. Other motor disturbances, including impaired dexterity, may also be observed. Limb and trunk ataxia resulting most often from cerebellar involvement is detected in 10%. A high proportion of patients with AIDS-associated PML will ultimately develop cerebellar lesions. Nearly one-third of patients have cerebellar signs at the time of diagnosis (Brooks and Walker 1984; Von Einsiedel et al. 1993). Extrapyramidal disease, at least at onset, is rare, but bradykinesia and rigidity may be detected in a substantial minority of patients with advanced disease (Richardson 1970, 1974). Dystonia and severe dysarthria have also been observed as a consequence of lesions in the basal ganglia (Singer et al. 1993, 1994). Not unexpectedly, lesions due to PML in the basal ganglia are chiefly a reflection of involvement of myelinated fibers coursing through this region rather than involvement of the deep gray matter (Whiteman et al. 1993). For the most part, the presentation of the AIDS patient with PML does not appear to be substantially different from that of patients with PML complicating other immunosuppressive conditions, except perhaps for a higher frequency of focal motor deficits, dysarthria and limb incoordination (Krupp et al. 1985; Berger et al. 1987a; Berger 1998). The latter two may be a reflection of a greater frequency of infratentorial lesions in AIDS-related PML.

Neuro-ophthalmic symptoms occur in up to 50% of patients with PML and are the presenting manifestation in 30–45% (Brooks and Walker 1984; Bachman 1993, personal communication). The most common visual deficits are homonymous hemianopsia or quadrantanopsia due to lesions of the optic radiations. Prior to AIDS and the introduction of MRI, cortical blindness was present at the time of diagnosis in up to 8% and occurred more commonly as the disease progressed (Brooks and Walker 1984). Cortical blindness does not appear to be as frequent in the AIDS population as was reported in previous series (Berger 1998). Other ophthalmic manifestations include optic aphasia, alexia without agraphia, and ocular motor abnormalities (Bachman 1993, personal communication; Berger 1998).

The spectrum of cognitive changes observed is quite broad. Unlike the slowly evolving, global dementia of HIV-associated dementia complex (HIV dementia), the mental impairments of PML are often more rapidly advancing and typically occur in conjunction with focal neurologic deficits. Among the abnormalities seen are personality and behavioral changes, poor attention, motor impersistence, memory impairment, dyslexia, dyscalculia, and the alien hand syndrome. A global dementia occurring in the absence of focal neurologic disease is rarely the presenting manifestation of PML (Brooks and Walker 1984; Von Einsiedel et al. 1993). Disturbances of speech and language, including both dysarthria and aphasia, are fairly common in AIDS-associated PML. Aphasia has been noted in as many as 10% of patients with PML (Brooks and Walker 1984).

Other clinical manifestations that are observed less frequently are sensory disturbances, headache, vertigo and seizures (Brooks and Walker 1984; Berger et al. 1987; Von Einsiedel et al. 1993; Berger 1998). Seizures may be seen and have been attributed to lesions that affect cortical gray matter (Von Einsiedel et al. 1993), although Von Einsiedel and colleagues suggest that AIDS-related PML cases present more frequently with seizures, hemiparesis and dysarthria than non-AIDS-associated PML. The results of a large clinical study tend to confirm these observations (Berger 1998).

In AIDS patients, as in those with other underlying diseases, PML usually progresses inexorably to death within a mean of 4 months (Brooks and Walker 1984; Berger et al. 1987). Small clinical studies suggested that approximately 80% of patients with PML succumb within 1 year from the time of diagnosis (Kuchelmeister et al. 1983; Brooks and Walker 1984; Berger et al. 1987; Von Einsiedel et al. 1993; Moore and Chaisson 1996); however, the largest series to date (Berger 1998) indicates that 80% die within 6 months of the diagnosis and the median survival is 3.5 months.

On occasion, individuals with PML experience clinical, radiographic and pathologic recovery, which in some instances may be complete. Often this includes full neurologic recovery in the absence of specific therapeutic intervention (Berger and Mucke 1988; Berger 1998). In our own experience, five pathologically confirmed AIDS patients with PML have exhibited both neurologic recovery and survival in excess of 1 year (Berger 1998). In one patient, survival from time of diagnosis was 92 months. Another patient had total remyelination in the previously affected areas at the time of autopsy more than 2 years after the diagnosis was first established (Berger 1996). Survival exceeding 1 year was observed in 9% of pathologically proven cases in our experience (Berger 1998). There was no correlation with treatment regimens, including the use of cytosine arabinoside. However, there appeared to be an association with PML as the heralding manifestation of AIDS, higher CD4 counts (often >200 cells/mm^3), inflammatory infiltration on biopsy, and contrast enhancement on CT scan or MRI (Berger 1998). Others have described an improvement in AIDS-associated PML in association with the use of aggressive antiretroviral therapy (Fiala 1988; Conway et al. 1990; Singer et al. 1994).

Diagnosis

The diagnosis of PML is strongly supported by radiographic imaging, but currently confirmation requires brain biopsy. CT of the brain reveals hypodense lesions of the affected white matter that generally do not enhance with contrast and almost always exhibit no mass effect. These lesions may have a 'scalloped' appearance as a result of the subcortical arcuate fibers lying directly beneath the cortex (Whiteman et al. 1993). With a higher sensitivity, MRI shows patchy or confluent hyperintense lesions on T2-weighted images in the affected region (Fig. 3). As with CT scan, contrast enhancement is an exception; however, contrast enhancement has been observed with both brain imaging techniques in approximately 5–10% of pathologically confirmed cases of PML (Whiteman 1993; Berger 1998). The enhancement observed is typically faint and peripheral. Mass effect in association with PML may be seen on rare occasions (Berger 1998). When present, it is usually quite subtle (Berger 1998).

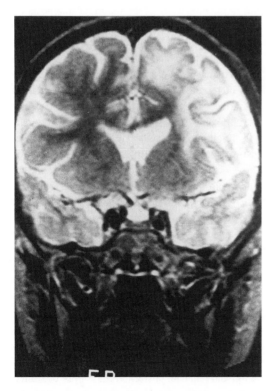

Fig. 3. Progressive multifocal leukoencephalopathy. T2-weighted MRI reveals extensive hyperintense signal abnormalities in the subcortical white matter.

The lesions of PML have a predilection for the frontal and parieto-occipital lobes (Whiteman et al. 1993), but may occur virtually anywhere. Lesions may be unifocal, but, in general, are multiple and bihemispheric. Involvement of the basal ganglia, external capsule and posterior fossa structures (cerebellum and brainstem) is not uncommon (Whiteman et al. 1993). One-third to one-half of all patients will eventually have involvement of the brainstem or cerebellum and in 5–10% of patients the disease activity was isolated to these structures (Berger and Mucke 1988; Yoshimura and Oka 1990; Berger et al. 1998).

Other diseases may cause white matter disease, especially in association with HIV infection. The demyelination observed with HIV dementia may be radiographically indistinguishable from that of PML. Clinically, however, PML is associated with focal neurologic disease and is more rapidly progressive. Radiographic distinctions include a greater propensity of PML lesions to involve the subcortical white matter, its hypointensity on T1-weighted images, its rare enhancement and more frequent occurrence of infratentorial lesions (Whiteman et al. 1993). CMV may also cause demyelinating lesions. Typically these lesions are located in the periventricular white matter and centrum semiovale, and subependymal enhancement is observed (Sze and Zimmermann 1988; Bowen and Post 1991). MRI images similar to those seen in PML have recently been described in a patient with dementia and extrapyramidal secondary to systemic lupus erythematosus (Kaye et al. 1992).

Cerebrospinal fluid and other studies

With the exception of PCR performed on CSF for the presence of JCV, other studies applied to CSF are non-diagnostic. The routine studies performed on CSF are usually normal (Brooks and Walker 1984; Berger et al. 1987a; Von Einsiedel et al. 1993). In patients with PML complicating HIV infection, the CSF abnormalities typically reflect those observed as a consequence of the HIV infection. These abnormalities may include a mononuclear pleocytosis ($\leqslant$20 cells/mm^3), elevated protein ($\leqslant$65 mg/dl) and borderline low glucose (Navia et al. 1986b; Marshall et al. 1988). PML in the absence of AIDS may be associated with a slight elevation in the CSF protein, a mild lymphocytic pleocytosis and the presence of myelin basic protein (Brooks and Walker 1984; Berger et al. 1987a).

Until recently confirmation of PML relied exclusively on typical histopathologic changes and detection of JCV in brain samples from biopsies or at autopsy. JCV can be detected by electron microscopy or isolated in cell cultures, viral antigens can be detected by immunocytochemistry, and viral DNA can be detected by in situ hybridization or PCR (Zu Rhein and Chou 1965; Padgett et al. 1971; Tornatore et al. 1992; Moret et al. 1993). By electron microscopy using negative staining technique, papova-like particles were observed in two of three CSF samples from patients with clinical and neuroradiologic evidence of PML but not in 12 controls with AIDS (Orefice et al. 1993).

The recent application of PCR to CSF samples is promising in establishing the diagnosis pre mortem with less invasive procedures than brain biopsy (Weber et al. 1990). Two earlier encouraging reports were able to detect JCV DNA with 100% specificity in more than three-fourths of CSF samples from patients with PML (Gibson et al. 1993; Moret et al. 1993). Gibson et al. (1993) detected JCV DNA in 10 of 13 CSF samples from patients with previously confirmed PML, while no amplification was obtained in 42 CSF samples from patients without PML. In a second study, CSF samples from 12 AIDS patients were examined by PCR. All nine samples from patients with PML diagnosis amplified JCV DNA products, but five controls and three AIDS patients without PML did not show amplification (Moret et al. 1993). Based on a series of 110 CSF samples, 28 PML cases and 82 controls, Weber et al. (1994) reported 82% sensitivity and 100% specificity for diagnosis of PML with PCR. Yet another large cohort, 156 individual CSF samples, revealed a 92% sensitivity and specificity (McGuire et al. 1994). With an overall sensitivity of 61%, Aksamit and Kost (1994) reported a specificity above 99% in a large sample size, 470 CSF samples. False positive samples might depend on the stage of the disease or on technical variation, such as set of primers and amount of CSF analyzed (Gibson et al. 1981; Moret et al. 1993; Weber et al. 1994). Despite these variations and the limited clinical experience with this technique, PCR is likely to prove a sensitive and highly specific diagnostic tool for confirming PML. If its promise is upheld with larger, more intensive studies, it will doubtlessly reduce the need for brain biopsy to establish the diagnosis.

Serum antibodies are not helpful in establishing the diagnosis, since 80% or more of the population show seropositivity to antibodies against JCV by adulthood (Taguchi et al. 1982). JCV DNA was detected in the peripheral blood lymphocytes by PCR in nearly 30% of HIV-infected patients without PML (Dubois 1996). Therefore, this test is not particularly useful in diagnosing the disorder. The electroencephalogram may show focal slowing but, like other studies, is also non-diagnostic.

Differential diagnosis

The large increase in the incidence of PML in the last decade has been due to the AIDS epidemic, and therefore the majority of PML cases will present in AIDS patients. With increasing frequency, clinicians find themselves confronted by HIV-infected patients with cognitive impairment and a cranial MRI showing 'hyperintense signal abnormalities on T2-weighted image (T2WI) characteristic of PML' due to the HIV dementia (AIDS dementia complex). It is the presence of these white matter lesions detected on MRI that frequently leads to the incorrect diagnosis of PML. HIV dementia may be the initial manifestation of AIDS in up to 3% of adult AIDS patients (Janssen et al. 1992; McArthur et al. 1993), has an estimated annual incidence of close to 7%, and will affect one-third or more of AIDS patients before their death (McArthur et al. 1993).

Cardinal features include an insidiously progressive psychomotor slowing, impaired memory and apathy (Price and Brew 1988; McArthur et al. 1993). Early complaints of forgetfulness, difficulty concentrating and manipulating complex tasks, problems with reading, general slowness, headache, and fatigue are classic. Because of the advanced degree of immunosuppression, AIDS patients with HIV dementia or PML generally exhibit similar constitutional features, including wasting, global alopecia, oral thrush and hairy leukoplakia, seborrheic dermatitis and generalized lymphadenopathy. The patient with HIV dementia commonly has slow mental processing (bradyphrenia), abnormalities of saccadic and pursuit eye movements, diminished facial expression, low-volume, poorly articulated speech, impaired coordination and balance, postural tremor, poor dexterity and a slow clumsy gait. Unlike PML, focal neurologic findings are uncharacteristic and suggest an alternative diagnosis. CSF examination is most valuable in eliminating the possibility of other disorders. Pathologic examination reveals brain atrophy and meningeal fibrosis. The most common histopathologic feature of this illness is white matter pallor, associated with an astrocytic reaction chiefly distributed perivascularly in periventricular and central white matter (Navia et al. 1986). There is no evidence of myelin breakdown or loss of myelin basic protein. Multinucleate giant cells secondary to virus-induced macrophage fusion are the pathologic hallmark of the disease (Sharer 1992). Other pathologic features include microglial nodules, diffuse astrocytosis, and perivascular mononuclear inflammation (Navia et al. 1986).

In HIV dementia, the most commonly reported abnormality on CT of the brain is cerebral atrophy; however, low density white matter abnormalities are also frequently observed. CT scan is quite helpful in ruling out focal mass lesions as a cause of a patient's altered mental status (Berger et al. 1994). On MRI, large areas of white matter lesions are observed diffused over a large area, typically in the centrum semiovale and periventricular white matter (Olsen et al. 1988; Post et al. 1988). Less commonly, localized involvement with ill-defined margins (patchy) or small foci less than 1 cm in diameter (punctate) are observed (Olsen et al. 1988). These white matter abnormalities are frequently mistaken for PML and the history, clinical findings and, to a lesser extent, CSF parameters are quite helpful in distinguishing between the two disorders (Griffin et al. 1991; Royal et al. 1994). The clinician needs to be mindful that these conditions are not mutually exclusive and that both conditions may coexist in the same patient.

HIV dementia and PML are not the only disorders of white matter occurring in AIDS in the absence of mass producing lesions. Incidental white matter abnormalities are not uncommonly observed in HIV-infected individuals and do not appear to have any clinical significance (McArthur et al. 1990). Among the disorders in the radiographic differential diagnosis an acute, diffuse, rapidly fatal leukoencephalopathy has been reported. Others include: (i) an HIV-associated granulomatous angiitis; (ii) a multifocal necrotizing leukoencephalopathy with a predilection for the pons; (iii) a relapsing and remitting illness clinically indistinguishable from multiple sclerosis; and (iv) CMV ventriculoencephalitis and other viral opportunistic infections (Olsen et al. 1988; Post et al. 1988).

Treatment

Unequivocally effective therapy of PML has remained elusive, whether specific antiviral therapy directed at the JCV or attempts to enhance cellular immunity. A variety of treatment regimens has been proposed on the basis of anecdotal reports and small series. The observation that PML may remain stable for long periods of time or even remit in the rare patient (Stam 1966; Kepes et al. 1975; Padgett and Walker 1983; Price et al. 1983; Sima et al. 1983; Berger and Mucke 1988; Embry et al. 1988) highlights the inadequacies of anecdotal reports suggesting the value of a specific therapy. The rarity of PML prior to the AIDS epidemic precluded practical therapeutic trials.

Nucleoside analogues, by interfering with the synthesis of viral DNA, have proven effective in the treatment of some viral diseases. Several nucleoside analogues have been tried in the treatment of PML with varying degrees of anecdotal success. Early experience with cytosine arabinoside (ARA-C, cytarabine), a drug chiefly used in the treatment of myeloproliferative disorders, has been mixed (Castleman et al. 1972; Bauer et al. 1973; Conomy 1974). Rapid and sustained improvement in neurologic symptoms was reported by Bauer et al. (1973) with ARA-C administered intravenously as 60 mg/m^2/day and intrathecally as 10 mg/m^2. The patient described by Marriott et al. (1975) showed more delayed, but similarly sustained improvement following ARA-C 2 mg/kg/day on 5 consecutive days every 3 weeks. Similar anecdotal reports of various degrees of improvement have been reported by others (Conomy et al. 1974; Buckman and Wiltshaw 1976; Rockwell et al. 1976; Peters et al. 1980; O'Riordan et al. 1990; Portegies et al. 1991). These regimens used either intrathecal and/or intravenous administration of ARA-C. The clinical observations regarding the potential efficacy of ARA-C in PML are supported by the recently acquired in vitro data on human fetal brain tissue infected with JCV. Major et al. (1992) have determined that cytosine b-D-arabinofuranoside at a concentration of 25 mg/ml of culture effectively suppresses JCV replication.

Enthusiasm for the use of ARA-C in PML should be tempered by its toxicity profile and the lack of a consistent salutory effect. A study of intrathecal ARA-C administered as 10 mg/m^2/day for 3 days with repeat dosing at variable intervals in 26 AIDS patients with PML revealed a salutory effect of 60% that was sustained in 50% for up to 2 years and was transient (less than 6 months) in the remainder (Britton et al. 1992). However, some case reports suggest a total lack of efficacy of cytosine arabinoside administered either solely intravenously (Castleman et al. 1972; Smith et al. 1982) or in combination with intrathecal therapy (Van Horn et al. 1978). A large collaborative effort was orchestrated through the National Institutes of Health AIDS Clinical Trials Group comparing high-dose antiretroviral therapy alone or in combination with either intravenous or intrathecal ARA-C in treating PML. The results of this study failed to confirm any benefit of either intrathecal or intravenous ARA-C in comparison to placebo (Hall 1998).

Other nucleoside analogues do not appear to treat PML successfully. Wolinsky et al. (1976) noted the failure of a 14-day course of adenosine arabinoside (ARA-A; vidarabine), 20 mg/kg/day in two patients with PML. Similar failures of ARA-A therapy in the treatment of PML have also been described (Rand et al. 1977; Walker 1978). Tarsy et al. (1973) had no success with a combination of prednisone and intrathecal idoxuridine (5-iodo-2'-deoxyuridine) 2 mg/kg/ 12 h. The studies of Kerr et al. (1993) have demonstrated the efficacy of the antineoplastic drug camptothecin, a DNA topoisomerase I inhibitor in blocking JCV replication in vitro by means of pulsed doses employed in amounts that were non-toxic to cells.

Because of their antiviral activity, presumably the result of their ability to stimulate natural killer (NK) cells (Von Einsiedel et al. 1993), interferons have been proposed as potential therapeutic agents in the treatment of PML. α-Interferon has established efficacy in the treatment of other papova virus-related diseases (Weck et al. 1988). In an open label trial of the safety and efficacy of recombinant α-interferon 2A administered as 3 million units subcutaneously daily with a gradual increment (typically by 3 million units every third day) in the treatment of HIV-associated PML, 2 of 17 patients had survival extending greater than 1 year (Berger et al. 1992a). No patient had a dramatic reversal in neurologic function. In one patient, combined therapy of intravenous adenosine arabinoside and β-interferon (Tashiro et al. 1987) showed no efficacy. However, intrathecal β-interferon 1 million units weekly for a total of 19 weeks

and thereafter monthly was associated with modest improvement in her clinical picture and MRI (Tashiro et al. 1987).

The rationale for the use of low-dose heparin sulfate as an adjunct in the treatment of PML is based on the model of the pathogenesis of PML proposed by Houff and Major (1988), which postulates that PML is the result of activated JCV-infected B lymphocytes crossing the blood–brain barrier (BBB) and initiating new areas of neuroglial infection throughout the course of the disease. Heparin sulfate has been shown to prevent activated lymphocytes from crossing the BBB in animal models by stripping the lymphocyte glycoprotein cell surface receptors for cerebrovascular endothelial cells. By the time of diagnosis, the virus is well established in the brain; therefore, this therapy is unlikely to be of any value in established disease, but may be useful as a means of prophylaxis in high-risk patients.

Other agents, either alone or in combination with nucleoside analogues, have been tried in treatment of PML. No value has been demonstrated with corticosteroid therapy alone or in conjunction with other agents (Van Horn et al. 1978; Tarsy et al. 1973). In theory, recovery of the underlying immunologic disorder should be associated with recovery from PML and that has been observed on rare occasions in individuals who have recovered from the illness or condition that had resulted in immunosuppression. A stabilization of the neurologic deficits in a patient treated with tilorone, an immune enhancer (Selhorst et al. 1978), is tempered by the converse observation (Dawson 1982), which noted no improvement following the cessation of immunosuppressive therapy in a patient with PML and myasthenia gravis. A recent report (Conway et al. 1990) suggests that PML occurring in association with HIV infection may respond to zidovudine. A dramatic improvement followed the administration of zidovudine 200 mg every 4 h and worsening followed a reduction in dose to 200 mg every 8 h. A return to prior higher zidovudine doses resulted in neurologic stability (Conway et al. 1990). Zidovudine may affect levels of the *tat* protein that have been demonstrated to transactivate JCV (Tada et al. 1990). In our experience, zidovudine use in AIDS-associated PML, even in high doses (≥1000 mg/day) has been devoid of significant benefit. There is limited experience with the other antiretroviral agents.

Another form of potential therapy is the use of antisense oligonucleotides. These molecules can be designed to bind selectively to a target region of messenger RNA and prevent its translation into protein. The development and trials of such agents for PML awaits further investigation.

BACTERIA

Mycobacterium tuberculosis

HIV infection and AIDS predispose to infection with tuberculosis. In some populations, one-third or more of the cases of tuberculosis occur in association with HIV infection (Pitchenik et al. 1987). These patients are more often *Mycobacterium tuberculosis* culture negative and the majority are not known to be seropositive at the time of diagnosis of tuberculosis (Pitchenik et al. 1987). The incidence of tuberculosis occurring in association with AIDS is related to risk group. Pitchenik et al. (1987) observed that AIDS patients with tuberculosis in south Florida were more often younger (ages 25–44 years) and were more often of African-American or of Haitian ancestry. Intravenous drug abusers also appear to be at greater risk (Sunderam et al. 1986). The percentage of AIDS patients with tuberculosis varies with locale (Nunn and McAdam 1988), with the highest incidence in the U.S.A. among HIV-infected Haitians in south Florida, 60% of whom have tuberculosis (Pitchenik et al. 1984). High rates of tuberculosis among HIV-infected persons have been found worldwide (Nunn and McAdam 1988). Some investigators have estimated rates of 10–18% for mycobacteriosis occurring in all patients with AIDS (Pitchenik et al. 1984).

Tuberculous meningitis

The clinical spectrum of CNS tuberculosis with HIV infection includes meningitis, cerebral abscesses, and tuberculomas (Bishburg 1986), though meningitis appears to be the most common of these neurologic complications. The commonest clinical manifestations include seizures, altered mental status and fever with meningismus (Bishburg 1986). Although HIV infection appears to increase the risk for meningitis with *M. tuberculosis*, it does not appear to alter the clinical manifestations, response to therapy or prognosis of

TABLE 2

The neurologic complications of tuberculosis (after Wood 1988).

Meningeal
    Purulent tuberculous meningitis
        1. Disseminated miliary tuberculosis
        2. Focal caseating plaques
        3. Inflammatory caseous meningitis
        4. Proliferative meningitis
    Serous tuberculous meningitis

Cerebral
    Tuberculous encephalopathy
    Tuberculoma
    Tuberculous brain abscess
    Cerebral infarction
    Tuberculosis of the skull

Spinal
    Spinal tuberculoma
    Tuberculous spinal abscess
    Tuberculous spinal osteomyelitis
    Necrotizing myelopathy
    Tuberculous radiculomyelitis

the disease (Berenguer et al. 1992). Typically, tuberculous meningitis presents as a subacute meningitis. It is preceded by a period of 2–8 weeks of non-specific symptoms, which include malaise, anorexia, fatigue, fever, myalgias, and headache. The prodromal symptoms in infants include irritability, drowsiness, poor feeding and abdominal pain. Eventually, the headache worsens, becoming continuous. In one large series, headache was reported by 86% of patients (Kent 1993). Neck stiffness is reported by about one quarter of patients, but meningismus is detected in higher numbers at the time of examination (Kent 1993). Bulging fontanelles develop in infants. The patient becomes increasingly irritable. Nausea, vomiting, confusion and seizures ensue. Psychosis may also be observed (Daif 1992). A history of tuberculosis is obtained in approximately one-half of cases of childhood tuberculous meningitis (Lincoln 1947b) and in up to 12% of adult cases (Barrett-Connor 1967; Traub et al. 1984; Ogawa et al. 1987).

Although some series report that fever is absent in up to 20% of patients with tuberculous meningitis, a low-grade fever is typically present. The frequent exception is the elderly patient who may present simply with headache, confusion or other neurologic disturbance in the absence of fever. Teoh and Humphries (1991) state that, in their experience, the fever is rarely in excess of 39°C when the tuberculous involvement is limited to the meninges alone.

Meningismus is noted in the adult patients, but is not invariable (Barrett-Connor 1967; Traub et al. 1984; Ogawa et al. 1987). Infants will often exhibit neck retraction and bulging fontanelles. Cranial nerve palsies are common. They have been reported on admission in 15–30% of children (Steiner and Portugaleza 1973; Idriss et al. 1976) and 15–40% of adults (Haas et al. 1977; Klein et al. 1985). Papilledema is frequently observed and, on occasion, fundoscopic examination may reveal choroid tubercles, which are yellow lesions with indistinct borders present either singly or in clusters. Their presence is convincing evidence of the disease, but they appear in only about 10% of cases of tuberculous meningitis not associated with miliary tuberculosis (Illingworth et al. 1956; Lincoln et al. 1960). Visual impairment should not immediately be ascribed solely to tuberculous meningitis. Optochiasmatic arachnoiditis, tuberculomas compressing the optic nerves and ethambutol toxicity in treated patients need to be considered (Teoh and Humphries 1991). Ophthalmoplegia may result from involvement of the third, fourth, or sixth cranial nerves. Alternatively, ophthalmoplegia due to parenchymal lesions resulting in gaze palsies and internuclear ophthalmoplegia may be observed, the consequence of vasculitic lesions or tu-

berculoses. Facial palsy and hearing loss attend the less frequent involvement of the seventh and eighth cranial nerves, respectively. In descending order of frequency, the sixth cranial nerve is most commonly affected, followed by the third, fourth, seventh, second, eighth, tenth, eleventh, and twelfth (Lincoln et al. 1960; Traub et al. 1984).

Among the neurologic signs observed are hemiparesis and hemiplegia and a wide variety of movement disorders, including chorea, hemiballismus, athetosis, cerebellar ataxia and myoclonus. These abnormalities are usually the consequence of infarction due to tuberculous vasculitis, but may occur as the result of mass lesions from tuberculoma or tuberculous abscesses. Involuntary movements may be observed in as many as 13%, occur more commonly in children and, in rare instances, may persist after treatment (Udani et al. 1971). Seizures, either focal or generalized, may occur during the acute illness or months after treatment. A higher incidence of intracranial tuberculomas (Bishberg et al. 1986; Dubé et al. 1992) and tuberculous brain abscesses (Velasco-Martinez et al. 1995) has been reported in this population, typically characterized by ring-enhancing lesions. These lesions may occur as the heralding manifestation of HIV infection and may also be seen in the absence of positive tuberculin skin tests and abnormal chest radiographs (Velasco-Martinez et al. 1995). Tubercu-

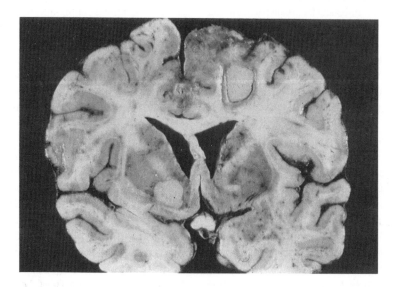

Fig. 4. Tuberculous meningitis and tuberculomas. Meningeal exudate and two well-circumscribed lesions with gaseous necrosis of the basal ganglia and left parasylvian region in tuberculous meningitis.

lomas and tuberculous abscesses (Fig. 4) may result in seizures as well, but most often present as focal neurologic disturbances resulting from mass lesions that may lead to brain herniation. In rare instances, paraparesis may attend an associated myelopathy due to proliferative granulomatous meningitis secondary to *M. tuberculosis* (Vlcek et al. 1984) or may occur as a consequence of spinal abscess (Doll et al. 1987).

Diagnosis

Radiographic studies. CT or MRI of the head may reveal thickening and enhancement of the meninges, particularly the basilar meninges, with hydrocephalus, infarction (Bullock and Welchman 1982; Witrack and Ellis 1985), edema (often located periventricularly) (Bullock et al. 1982), and mass lesion due to associated tuberculoma or tuberculous abscess. In all patients, whether HIV-infected (Villoria et al. 1992) or not, hydrocephalus appears to be the single most common abnormality seen in tuberculous meningitis on CT or MRI. It has been reported in 80% of patients (Offenbacher et al. 1991) and was detected in 100% of 30 children with CNS tuberculosis in another (Waecker and Connor 1990). A recent study found ventricular dilatation in 52% of patients with culture-positive or presumptive tuberculous meningitis and tuberculomas in 16% (Davis et al. 1993). In a study of 35 HIV-infected persons with proven intracranial tuberculosis, 51% had hydrocephalus (Villoria et al.1992). The degree of hydrocephalus correlates with the duration of the disease (Bhargava et al. 1982). Enhancement of the meninges is observed in approximately 50–60%, and infarctions, the third most common CT finding, are seen in about one-fourth of patients (Offenbacher et al. 1991; Villoria et al. 1992). The latter is a consequence of reactive endarteritis obliterans occurring as a consequence of the bathing of arteries that course through the thick, gelatinous exudate located at the base of the brain. Arteritis is seen in approximately 30–40% of cases with basilar meningitis (Leiguarda et al. 1988) and the middle cerebral artery and small perforating branches to the basal ganglia are the vessels most often affected (Sheller and Des Prez 1986). Radiographic imaging of the diffuse infiltration of the brain parenchyma by small granulomas (<5 mm) in the course of miliary tuberculosis may reveal multiple small, contrast-enhancing intraparenchymal lesions (Gee et al. 1992;

Kchouk et al. 1992; Eide et al. 1993). In a series of AIDS patients studied by Villoria et al. (1992), 37% exhibited parenchymal enhancement.

Distinguishing infarction from inflammation due to tuberculosis may be difficult on cranial MR (Offenbacher et al. 1991). Angiography is characterized by a hydrocephalic pattern of the vessels, narrowing of the vessels at the base of the brain, and narrowed or occluded small- and medium-sized vessels with few collaterals (Lehrer 1966). Radiographic clues to the presence of intracranial tuberculosis include multiloculated abscesses, cisternal enhancement, basal ganglia infarction and communicating hydrocephalus which are not findings associated with primary central nervous system lymphoma (PCNSL) or toxoplasma encephalitis (Whiteman et al. 1995). SPECT scans are negative, helping to distinguish these lesions from PCNSL.

Laboratory diagnosis

Routine laboratory studies provide few clues to the diagnosis of tuberculosis. Although the erythrocyte sedimentation rate and the peripheral white cell count are often elevated, they need not be. In fact, leukopenia may also be observed and the differential white blood cell count is not particularly helpful. Hyponatremia may indicate the syndrome of inappropriate antidiuretic hormone (Smith and Godwin-Austin 1980). In the presence of miliary dissemination, cultures of extraneural sites, such as the bone marrow and lung, may be positive.

The chest X-ray in adult patients reveals abnormalities consistent with pulmonary tuberculosis in 25–50% of patients (Stockstill and Kauffman 1983; Traub et al. 1984; Clark et al. 1986), but is often more frequently revealing in children where radiographic changes have been noted in 50–90% (Lincoln et al. 1960; Smith 1975; Sumaya et al. 1975; Idriss et al. 1976). Radiographic findings include miliary disease, apical scarring, hilar adenopathy and a Ghon complex. Miliary disease is observed in 25–50% of adults (Haas et al. 1977; Stockstill and Kauffman 1983; Ogawa et al. 1987) and 15–25% of children (Zarabi et al. 1971; Smith 1975; Delage and Dusseault 1979) with tuberculosis meningitis, but was more common in the era before the availability of effective antituberculous therapy. Skin testing for delayed hypersensitivity to tuberculosis with purified protein derivative (PPD) is not invariably positive either. PPD positivity has been reported in 40–65%

of adults with tuberculosis meningitis (Haas et al. 1977; Clark et al. 1986; Ogawa et al. 1987) and in 85–90% of children (Steiner and Portugaleza 1973; Sumaya et al. 1975; Idriss et al. 1976). Therefore, both the chest X-ray and the PPD may be negative in the face of tuberculous meningitis, so a high index of suspicion for the diagnosis must be held for the prompt initiation of therapy.

CSF analysis is pivotal in diagnosing tuberculous meningitis. As with other forms of bacterial meningitis, the opening pressure should be elevated. However, approximately half of both adult and pediatric patients have normal opening pressures at the time of study (Singhal et al. 1975; Ogawa et al. 1987; Leiguarda et al. 1988). The CSF may appear xanthochromic due to elevated protein concentrations, and spinal block may result in the Froin's syndrome, clot formation of CSF after removal of any red blood cells due to the presence of high concentration of serum proteins, including fibrinogen. Reported median CSF white blood cell counts range from 63 (Traub et al. 1984) to 283 cells/mm^3 (Barrett-Connor 1967). On occasion, the CSF white cell count may be normal (Klein et al. 1985; Ogawa et al. 1987) or, at the other extreme, exceed 4000 cells/mm^3 (Karadanis and Shulman 1976). In the early stages of infection, a significant number of polymorphonuclear cells may be observed, but over the course of several days to weeks they are typically replaced by lymphocytes. The persistent predominance of polymorphonuclear cells may result in mistaken diagnosis (Pardiwalla et al. 1992; Mizutani et al. 1993). On occasion in patients with AIDS, the CSF white cell count may be normal (Bishburg et al. 1986) or acellular (Laguna 1992), emphasizing the value of brain biopsy in establishing the diagnosis. Rarely, large percentages of other cells, such as plasma cells and eosinophils, may be seen with tuberculous meningitis. However, their presence should suggest some other underlying process. In general, an elevated CSF protein is the rule, with values usually in the 100–200 mg/dl range (Kent et al. 1993); values may occasionally exceed 1–2 g/dl. However, in one large series of AIDS patients, the CSF protein was reported to be normal in 43% (Berenguer et al. 1992). The CSF glucose is low, with values usually less than 50% of serum glucose and median values of CSF glucose reported between 18 mg/dl (Barrett-Connor 1967) and 45 mg/dl (Haas et al. 1977). In culture-proven cases, the CSF glucose

tends to be lower than in presumptive tuberculous meningitis (Traub et al. 1984; Ogawa et al. 1987). Low CSF chloride and elevated CSF C-reactive protein (Vaishnavi et al. 1992) are non-specific markers of tuberculous meningitis with limited application.

The hallmark of diagnosis of tuberculous meningitis is the demonstration of M. tuberculosis in the CSF. Typically, no more than 25% of cases have identifiable M. tuberculosis when acid-fast stains are performed on spun specimens of the CSF (Hinman 1967; Sumaya et al. 1975; Klein et al. 1985) and in one study (Davis et al. 1993), acid-fast stains of CSF sediment were positive in only 4%. The percentages of positive smears may improve with increased numbers of specimens (Kennedy and Fallon 1979). Cultures of the CSF for M. tuberculosis are not invariably positive either. Rates of positivity for clinically diagnosed cases range from 25% (Traub et al. 1984) to 70% (Alvarez and McCabe 1984) in patients without AIDS, but are considerably lower in the presence of HIV infection. Rates of positive CSF cultures have varied between 3% and 10% in this population (Small et al. 1991; Berenguer et al. 1992). Mycobacterial cultures require several weeks before they are positive.

The frequency of false negative CSF stains and cultures for M. tuberculosis and the long duration for positive cultures to become available have served as an impetus to the development of alternative diagnostic tests on the CSF that are based on the detection of the organism immunologically or by PCR, detection of antibody to M. tuberculosis, or the detection of substances in the CSF that are believed to be unique to M. tuberculosis (Daniel et al. 1987). With respect to the latter, measurement of the radiolabeled bromide partition ratio following the administration of oral or intravenous (^{84}Br) ammonium bromide was one of the earliest tests employed (Smith et al. 1955). It has been reported to have both a sensitivity and specificity on the order of 90% (Mann et al. 1982; Coovadia et al. 1986). Gas-liquid chromatography has been used to identify a basic indole in the CSF believed to be relatively specific for tuberculosis meningitis (Brooks et al. 1977). Tests of the CSF for the presence of tuberculostearic acid, a component of the mycobacteria cell wall, have been reported to have sensitivity and specificity in excess of 90% (French et al. 1987), and those of CSF adenosine deaminase (ADA) have been reported between 73–100% and

71–99%, respectively (Mann et al. 1982; Coovadia et al. 1986; Ribera et al. 1987). In a study of 180 adults with meningitis of various etiologies, CSF ADA levels ⩾10 IU/l had a sensitivity of 48% and a specificity of 100% in patients with tuberculous meningitis (Lopez-Cortes et al. 1995). Antibodies directed to *M. tuberculosis* can be detected immuno-logically with enzyme linked immunosorbent assay (ELISA) with varying success (Kalish et al. 1983; Watt et al. 1988). Similarly, a variety of immunologic techniques have been employed to detect *M. tuberculosis* antigen in the CSF with high levels of sensitivity and specificity reported (Sada et al. 1983; Krambovitis et al. 1984, Kadival et al. 1987; Radhakrishnan and Mathai 1990). More recently, PCR has been applied to CSF specimens to detect *M. tuberculosis* DNA (Kaneko et al. 1990; Narita et al. 1992; Donald et al. 1993). It has proven to be an effective diagnostic tool for tuberculous meningitis in AIDS patients (Folgueira et al. 1994). This test is becoming increasingly available and may remain positive 4 or more weeks after the initiation of treatment (Donald et al. 1993). With the exception of the radiolabeled bromide partition ratio, the results of all these tests are available within hours, a distinct advantage over the long wait for mycobacterial cultures.

Therapy

The antimicrobials used in the treatment of tuberculous meningitis can be divided into first- and second-line agents. The former include isoniazid, rifampin, pyrazinamide, ethambutol, and streptomycin. Peak levels of *isoniazid* (INH) following oral administration exceed levels needed to inhibit most strains of *M. tuberculosis* in vitro (0.025–0.05 µg/ml) by 100-fold (Mandell and Sande 1985). Penetration through inflamed meninges is excellent, with CSF concentrations 90% that of serum; however, in the absence of inflammation, the penetration is about 20% of serum levels (Forgan-Smith et al. 1973). Drug toxicity reactions with INH include hepatotoxicity, peripheral neuropathy when pyridoxine is not co-administered, and alterations in mental state. The usual daily dose is 5–10 mg/kg/day. *Rifampin* achieves high serum levels after oral administration, and CSF levels are approximately 20% of serum levels in the presence of meningeal inflammation (D'Oliveira 1972). Adverse reactions include renal, hematologic and hepatic disorders. The latter seem to be particularly frequent in

individuals receiving INH concomitantly. The usual dose is 10–20 mg/kg/day in children and 10 mg/kg/day in adults, up to 600 mg/day. *Pyrazinamide* has its effect on intracellular rather than extracellular organisms and penetrates the CSF extremely well. Hepatotoxicity is the chief toxicity, and the usual dose is 20–35 mg/kg/day. *Ethambutol* achieves CSF concentrations of 10–50% of serum levels in the presence of meningeal inflammation (Bobrowitz 1972). The chief toxicity of this tuberculostatic drug is optic neuritis, which develops in as many as 1% of persons on the currently recommended dose of 15–25 mg/kg/day. Careful attention to visual acuity and color perception is required of patients on this drug. *Streptomycin* was the initial drug demonstrated to be efficacious in the treatment of tuberculous meningitis. It must be given parenterally and requires meningeal inflammation for penetration, with CSF levels approximately one quarter of that of the serum (Zintel et al. 1945). Vestibular disorders due to ototoxicity are its chief adverse effect. Currently recommended doses are 20–40 mg/kg/day in children and 1 g/day in adults. The second-line agents for the treatment of tuberculous include *para-aminosalicylic acid* (PAS), *cycloserine* and the *aminoglycosides*: amikacin and kanamycin. There are a number of recommended regimens for the treatment of tuberculous meningitis (Table 3) that depend on whether the *M. tuberculosis* is likely to be drug-resistant or not.

Prognosis

A method for staging the severity of the tuberculous meningitis was developed by the British Medical Research Council in 1948 (Medical Research Council 1948) for purposes of classification in the initial trials of streptomycin in tuberculous meningitis (Table 4). This schema has proven useful for grading the initial severity of the illness and for purposes of prognosis.

In the absence of treatment, tuberculous meningitis is invariably fatal in AIDS, although rare spontaneous recoveries had been reported prior to the AIDS epidemic. Following the introduction of isoniazid in 1952, the survival rates of tuberculous meningitis approached their present rates of 70–80% (Lepper and Spies 1963). Poor prognosis has been correlated with advanced stages by the classification of the British Medical Research Council: extremes of age (Smith and Vollum 1956; Lincoln et al. 1960; Lepper and Spies 1963; Weiss and Flippin 1965; Hinman

TABLE 3

Chemotherapeutic regimens for tuberculous meningitis (Zuger and Lowy 1991).

Drug	Dose	Frequency	Duration
Low probability of drug resistance			
A. Isoniazid	300 mg	Daily	6 months
Rifampin	600 mg	Daily	6 months
Pyrazinamide	15–30 mg/kg	Daily	2 months
B. Isoniazid	300 mg	Daily	9 months
Rifampin	600 mg	Daily	9 months
Ethambutol	25 mg/kg	Daily	2 months
or			
Streptomycin	1 g	Daily	2 months
C. Isoniazid	300 mg	Daily	2 months
	900 mg	Twice weekly	8 months
Rifampin	600 mg	Daily	1 month
	600 mg	Twice weekly	8 months
High probability of drug resistance			
A. Isoniazid	300 mg	Daily	1 year
Rifampin	600 mg	Daily	1 year
Pyrazinamide	15–30 mg/kg	Daily	2 months
Ethambutol	25 mg/kg	Daily	2 months
or			
Streptomycin	1 g	Daily	2 months

B. In cases of documented drug resistance, chemotherapy must be tailored to the demonstrated sensitivities.

TABLE 4

Medical Research Council classification of tuberculous meningitis.

Stage I (early)
   Non-specific symptoms and signs
   Consciousness undisturbed
   No focal neurological signs

Stage II (intermediate)
   Consciousness disturbed but not comatose or
     delirious
   Signs of meningeal irritation
   Minor focal neurological signs, e.g., cranial nerve
     palsies

Stage III (advanced)
   Seizures
   Abnormal movements
   Stupor or coma
   Severe neurological deficits, e.g., paresis

1967), coexistent miliary disease (Lorber 1954; Wasz-Hockert and Donner 1962; Lepper and Spies 1963), extraordinarily high CSF protein levels with spinal block (Weiss and Flippin 1965), and markedly reduced CSF glucose (Wasz-Hockert and Donner 1962).

In AIDS patients, a CD4 lymphocyte count of less than 22 cells/mm^3 and illness lasting more than 14 days before hospital admission were poor prognostic signs (Berenguer et al. 1992). The overall mortality in one large series was 33% in the HIV-infected population with tuberculous meningitis, but mortality directly attributable to the meningitis was 21%, equal to that of a non-HIV-infected control group (Berenguer et al. 1992). In another study (Dube et al. 1992) comparing tuberculous meningitis in patients with and without HIV infection, the only significant difference between the two groups was the increased incidence of intracerebral mass lesions in the HIV-infected group.

Within 2 weeks of the initiation of effective treatment, clinical improvement is noted. However, it is not unusual to observe transient worsening in clinical and CSF parameters following the initiation of therapy, probably occurring as an immunologic phenomenon akin to the Jarisch–Herxheimer reaction in syphilis. Resolution of the fever may require weeks. CSF glucose levels return to normal within 2 months in 50% of patients and within 6 months in almost all, whereas the CSF pleocytosis requires more than 6 months to resolve in 25% and the CSF protein remains elevated in 40% at this time (Lepper and Spies 1963; Barrett-Connor 1967).

Continued alteration in level of consciousness in the early stages of the disease may be the result of concomitant hydrocephalus or hyponatremia. Later, 2–18 months after the initiation of adequate antituberculous therapy, enlarging intracranial tuberculomata may result in clinical deterioration. Teoh et al. (1986) found that the organisms isolated from these enlarging tuberculomata were sensitive to the antibiotic employed and suggested that their appearance was an immunologic phenomenon. Surgery can usually be avoided with the administration of corticosteroids and continued antituberculous therapy.

The rate of sequelae among survivors varies in different series. In children, neurologic sequelae are seen in approximately 25% of survivors (Lorber 1954, 1961); however, higher rates have been observed (Sumaya et al. 1975). In adults, rates of neurologic sequelae approximating that in children have been noted (Barrett-Connor 1967; Haas et al. 1977; Bateman et al. 1983; Ogawa et al. 1987). These sequelae include cognitive disturbances, seizures, hemiparesis, ataxia, and persistent cranial nerve palsies (Smith and Vollum 1956; Donner and Wasz-Hockert and Donner 1962; Ogawa et al. 1987). Visual impairment often accompanies optic atrophy (Mooney 1956).

Atypical mycobacteria

Atypical (non-tuberculous) mycobacteria were recognized as pathogens in the 1950s and, until the AIDS epidemic, chiefly resulted in localized pulmonary disease in middle-aged persons with pre-existing lung disease (Wolinsky 1979). *Mycobacterium avium intracellulare* (MAI) and other atypical mycobacterium infections occur frequently in AIDS and are often extrapulmonary in nature (Pumarola-Sune et al. 1987). As these organisms are frequently isolated from soil, the route of infection is believed to be environmental. In the AIDS patient, symptomatic disease may be the result of reactivation of latent infection. The most common atypical mycobacterium observed in the AIDS patient is MAI; however, other atypical mycobacteria are also seen (Nunn and McAdam 1988), including brain abscess and meningitis in association with *My-cobacterium kansasii* (Haas et al. 1977; Gordon et al. 1992; Bergenet al. 1993) and at least one case of leprosy in association with AIDS (Lamfers et al. 1987). In reality, these are two distinct organisms, but they are bacteriologically very similar and difficult to distinguish. MAI is diagnosed pre mortem in 15–20% of AIDS patients and in as many as 50% post mortem (Hawkins et al. 1986). In the latter, recovery of the organism is chiefly from the spleen and lymph node, but dissemination elsewhere is not uncommon. Symptoms of MAI typically include fever, malaise, night sweats, generalized weakness, and weight loss (Hawkins et al. 1986; Young et al. 1986). In one study, MAI was the most common bacteremia observed in patients with AIDS (Whimbey et al. 1986). Despite the frequency of mycobacterial infection, involvement of the CNS is not as common as may be expected and is most frequently due to *M. tuberculosis*.

In contrast to *M. tuberculosis*, patients with MAI have single or multiple mass lesions twice as commonly as meningitis. In a study from New York, MAI was cultured from the CSF in 15 of 16 cases with atypical mycobacterial meningitis complicating AIDS (Jacob et al. 1993). The other case was due to *M. fortuitum* (Jacob et al. 1993). CT characteristics are diverse and, in general, mirror those seen in tuberculous meningitis in the absence of HIV infection. Hydrocephalus and meningeal enhancement were observed on CT in approximately one-half of all patients in one study in which intravenous drug abusers constituted more than 90% of the population (Villoria et al. 1992). Radiographic studies are often helpful in suggesting the presence of CNS tuberculosis.

Optimal regimens in the treatment of CNS disease due to atypical mycobacteria such as MAI have not been precisely defined. A four-drug regimen is required for treating MAI (Kemper 1992). Current recommendations include using azithromycin (500–1000 mg/day) and clarithromycin (500–1000 mg/day) in combination with ethambutol (15 mg/kg/day) or clofazimine (100 mg/day). Alternative regimens include the use of ciprofloxacin and rifampicin. Further study will be required to determine the best regimen. There are insufficient data to determine the duration of therapy in patients with CNS atypical tuberculosis, and in general, treatment of this disorder is disappointing.

Syphilis

The incidence of neurosyphilis is difficult to evaluate because of underreporting, difficulty in establishing the diagnosis, diverse clinical presentations, and arrest of the illness by treatment with antibiotics for other infections (Koffman 1956; Hooshmand et al. 1972; Joyce-Clark and Molteno 1978; Luxon et al. 1979; Norbeno and Sorenson 1981; Burke and Schaberg 1985). There has been a 3-fold increase in the nationwide incidence of primary and secondary syphilis in the USA from its nadir in 1957–1983 (CDC 1984), and the incidence continues to increase (CDC 1988a). Between 1955 and 1958, the annual rate of infectious (primary and secondary) syphilis fell to approximately 4/100,000 (Fleming 1964), the lowest recorded in the USA, and it changed very little between 1960 and 1985 (Aral and Holmes 1990). However, rates among white men showed sharp increases during this time period and were largely attributed to homosexual practices (Aral and Holmes 1990). Since 1982, the rate of syphilis among homosexual white men has decreased, presumably in response to modified sexual behavior because of AIDS (Aral and Holmes 1990). However, the incidence of infectious syphilis in other segments of the population in the USA and elsewhere has begun to climb dramatically.

The recent increase in syphilis has been particularly noticeable among prostitutes, parenteral drug abusers, and their sexual partners (CDC 1988b), but the increases are not limited to these groups (CDC 1988a). The risk groups for AIDS encompass these same groups (Guinan et al. 1984; Friedland and Klein 1987); thus, it is not surprising to find syphilis and infection with the HIV occurring together. Furthermore, epidemiologic studies have suggested that syphilis, like other genital ulcerative diseases, serves as a cofactor for the acquisition of HIV infection (Guinan et al. 1984; Greenblatt et al. 1988; Holmberg et al. 1988; Simonsen et al. 1988; Stamm et al. 1988; Pepin et al. 1989).

Despite the fact that general reviews of the neurologic complications of HIV infection have reported low rates of neurosyphilis (Snider et al. 1983; Levy et al. 1985; Berger 1987; McArthur 1987), it is surprisingly common in HIV infection. In a retrospective chart review study from south Florida, Katz and Berger (1989) estimated that neurosyphilis, strictly

defined by the presence of reactive CSF VDRL, was diagnosed in approximately 1.5% of HIV-infected patients hospitalized at a large public health hospital in south Florida. An expansion of this study confirmed this estimate (Katz et al. 1993), which concorded remarkably well with the estimates from a prospective longitudinal study from the same institution (Berger 1991). In that study, a history of syphilis and/or a reactive serum FTA-ABS was observed in 43.3% of 180 asymptomatic HIV-seropositive subjects and 50.6% of 77 neurologically symptomatic HIV-seropositive subjects in comparison to 26.1% of HIV-seronegative controls with similar risk factors for HIV infection (Berger 1991). Three (1.8%) neurologically asymptomatic subjects and 3.3% of the neurologically symptomatic subjects had neurosyphilis (Berger 1991). The former is virtually identical to the 2% incidence of a reactive CSF VDRL found by Appleman et al. (1988) in asymptomatic, HIV-infected US servicemen undergoing lumbar puncture. Similarly, in a Brazilian study, Livramento et al. (1989) found 10 (5.9%) cases of neurosyphilis among 170 patients with AIDS and neurologic syndromes. In another study, 9.1% of HIV-infected patients undergoing lumbar puncture because of a reactive serology with no history of recent treatment for syphilis had a reactive CSF VDRL (Holtom et al. 1992).

Clinical features

Anecdotal reports suggest that the clinical manifestations of syphilis may be altered when occurring in association with HIV infection. Among the unusual manifestations reported are peculiar ocular manifestations and rashes, lues maligna, gummas, osteitis and pneumonitis (Marra 1996). False negative serologic responses (Hicks et al. 1987; Strobel et al. 1989), an unusually aggressive neurosyphilis (Johns et al. 1987) and poor therapeutic responses of neurosyphilis to penicillin (Musher et al. 1990) have also been described.

Concomitant HIV infection may significantly alter the natural history of neurosyphilis (Johns et al. 1987; Katz et al. 1989, 1993). Syphilis appears to be not only more aggressive, but also more difficult to treat when it occurs in association with HIV infection (Berry et al. 1987; Musher et al. 1990; Katz et al. 1993). These observations suggest that the host's immune response is critical in controlling this infection. The inability of the HIV-infected patient to establish delayed hypersensitivity to *T. pallidum* may prevent secondary syphilis from evolving to latency or may cause a spontaneous relapse from a latent state. This impairment of delayed hypersensitivity may account for a more rapid progression of neurosyphilis in HIV-infected individuals than would otherwise be expected. *T. pallidum* can be isolated from the CSF of HIV-seropositive patients with primary, secondary, and latent syphilis following currently recommended CDC penicillin therapy (Lukehart 1988). Neurosyphilis has been reported following CDC-recommended therapy for early syphilis HIV-infected individuals (Berry et al. 1987), and it has also been reported that erythromycin has failed to cure secondary syphilis in a patient infected with HIV (Duncan 1989).

In HIV infection, an acute, symptomatic, syphilitic meningitis during the course of secondary syphilis is not uncommon. A decrease in the latent period prior to the development of some neurosyphilitic manifestations, such as meningovascular syphilis and general paresis, has been suggested. The development of meningovascular syphilis within 4 months of primary infection, despite the administration of accepted penicillin regimens (Johns et al. 1987) and the neurologic relapse of syphilis in HIV-infected individuals after appropriate doses of benzathine penicillin for secondary syphilis (Berry et al. 1987), have been reported. In one large study (Katz et al. 1993), the most common forms of neurosyphilis observed with HIV infection were syphilitic meningitis (64% of cases) and meningovascular syphilis (27%). Syphilitic meningitis is characterized by headaches, meningismus, photophobia, impaired vision, cranial nerve palsies (chiefly – in descending order of frequency – VIII, VII, VI, and II), while hearing loss, tinnitus and vertigo may be observed in isolation or in combination. Encephalopathic features resulting from vascular compromise or increased intracranial pressure may be observed. These include confusion, lethargy, seizures, aphasia and hemiplegia. Acute sensorineural hearing loss and acute optic neuritis may occur in association with syphilitic meningitis or independently. Meningovascular syphilis may affect the brain or spinal cord and, although it typically occurs 6–7 years after the initial infection, the latency for this and other forms of neurosyphilis may be considerably shorter in the presence of concomitant HIV infection. The nature of the neurologic features is dependent on the area of the brain or spinal cord affected. Many of the stroke

eponyms described at the turn of the last century were the consequence of meningovascular syphilis producing discrete lesions of the brainstem. The neurologic manifestations include aphasia, hemiparesis, hemianesthesia, diplopia, vertigo, dysarthria and a variety of brainstem syndromes. CT and MRI are invaluable diagnostic aids.

Other unusual manifestations of syphilis that have been reported in association with HIV infection include unexplained fever (Chung et al. 1983), bilateral optic neuritis with blindness (Zambrano et al. 1987), Bell's palsy and severe bilateral sensorineural hearing loss (Fernandeaz-Guerrero et al. 1988), syphilitic polyradiculopathy (Lanska et al. 1988), and syphilitic cerebral gumma presenting as a mass lesion (Berger et al. 1992b) (Fig. 5). Syphilitic meningomyelitis (Berger et al. 1992a) has also been reported and is characterized by slowly progressive weakness and paresthesia of the lower extremities. Eventually, bowel and bladder incontinence and paraplegia supervene. Examination reveals a spastic paraparesis or paraplegia with brisk lower extremity reflexes, loss of the superficial abdominal reflexes and impaired sensory perception with vibratory and position sense being disproportionately affected.

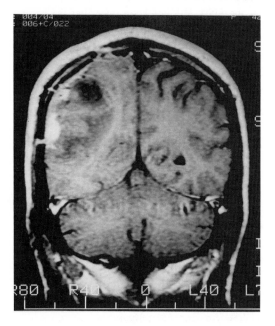

Fig. 5. Neurosyphilis. MRI with gadolinium contrast reveals enhancing dural based lesion with considerable surrounding cerebral edema due to a syphilitic gumma.

Ocular syphilis was seen in 36% of one large cohort with neurosyphilis (Katz et al. 1993). Syphilitic eye disease in AIDS patients has only recently been documented (Zaidman et al. 1983; Zaidman 1986; Carter et al. 1987; Stoumbos and Klein 1987; Zambrano et al. 1987; Passo and Rosenbaum 1988; Becerra et al. 1989; Joyce et al. 1989; Levy et al. 1989). Several comprehensive reviews (Freeman et al. 1984; Gal et al. 1984; Palestine et al. 1984) fail to mention syphilis as an etiology of ocular complaints in AIDS patients. Eye disease may occur in all stages of syphilis infection. Uveitis and perineuritis are more common in early (primary and secondary) syphilis, whereas chorioretinitis, optic atrophy and pupillary abnormalities are found more often in late infection (Spoor et al. 1983). These later manifestations may be signs of progressive disease and require treatment equally as aggressive as that used in the management of neurosyphilis. Conjunctivitis may result from chancre or gummas of the conjunctiva or occur in association with papular syphilides (Spoor et al. 1983). Papilledema may result from the sustained increased intracranial pressure of syphilitic meningitis. Optic perineuritis or optic neuritis may mimic papilledema. Syphilitic retinitis may be difficult to differentiate from that of CMV (Stoumbos and Klein 1987).

As noted, HIV infection may alter the natural history of syphilis as well as the response to treatment. Norris states that 'immunosuppression of T-cell responses can lead to poor clearance of *T. pallidum* and more severe infection' (Norris 1988). Animal experimentation performed over several decades has revealed that immunosuppression from corticosteroid therapy modifies the expression of *T. pallidum* infection. For instance, Turner and Hollander (1957) demonstrated that the administration of cortisone to rabbits infected intradermally with *T. pallidum* (Nichols strain) resulted in quickly enlarging dermal lesions that did not ulcerate and with organisms 'literally swarming over the entire preparation' quite unlike that observed in untreated animals. Corticosteroid therapy has been used to enhance the speed with which *T. pallidum* can be recovered from rabbits who have undergone intratesticular inoculation of potentially infectious material for diagnostic purposes.

The role of immunosuppression on the course of syphilis in humans is uncertain. Two individuals with an unusual presentation of syphilis in the face of im-

munosuppression for organ transplantation have been described (Petersen et al. 1983; Johnson et al. 1988). Syphilitic meningitis, presumably resulting from a recrudescence of previously treated syphilis, has also been described in a 30-year-old homosexual man who was immunosuppressed following a kidney transplant, but also was HIV-infected (Clark and Carlisle 1988). Among the evidence for a significant role of cell-mediated immunity in syphilis is: (1) passive transfer of syphilis-immune serum is only partially protective; (2) untreated, syphilis may progress despite the presence of immobilizing antibodies; (3) delayed hypersensitivity to treponemal antigens develops in late secondary, latent and tertiary syphilis; (4) granulomatous lesions characterize tertiary syphilis; and (5) immunization with killed organisms is usually unsuccessful, whereas immunization with live, attenuated organisms has produced immunity (Musher and Schell 1975). Conceivably, the inability of HIV patients to establish delayed hypersensitivity may prevent secondary syphilis from evolving to latency or may cause a spontaneous relapse from the latent state. The impairment of delayed hypersensitivity probably accounts for the acute development and rapid progression of neurologic disease in AIDS patients, mirroring the nature of other CNS infections in AIDS, such as toxoplasmosis and PML. The pathogen in each of these instances exists in a latent state, becoming evident only after the onset of immunosuppression.

Diagnosing neurosyphilis

Diagnosing neurosyphilis can be quite problematic (see Table 5). Several different diagnostic schemata have been used to establish the diagnosis; however, all have relied on indirect evidence of the presence of *T. pallidum,* as the organism is fastidious and requires the rather cumbersome use of animal inoculation for culture verification. Examples of the diagnostic schemata for neurosyphilis proffered in the literature include:

1. Reactive serum FTA or microhemagglutination *T. pallidum* (MHA-TP) in association with one of the following:
   (a) reactive CSF VDRL;
   (b) CSF pleocytosis (≥ 5 cells/mm^3); or
   (c) increased CSF protein (greater than 46 mg%) (Burke 1985).

2. Reactive serum FTA in association with one of the following:
   (a) neurologic or ophthalmologic disease unexplained by other illnesses;
   (b) reactive CSF FTA and increased CSF cell count (>5 cells/mm³); or
   (c) reactive CSF FTA with progressive neurologic symptoms without other explanation (Hooshmand 1972)
3. Reactive CSF VDRL in the absence of gross blood contamination of the CSF (Katz 1989).

Though it is the most specific test for neurosyphilis, the CSF VDRL, as well as other non-treponemal tests, may be insensitive to the diagnosis of neurosyphilis. In their classic monograph on neurosyphilis, Merritt et al. (1946) stated that the Wasserman and the Davis–Hinton (earlier non-treponemal tests) were negative in a high percentage of patients with neurosyphilis. For instance, the CSF Wasserman or Davis–Hinton were negative in 28% of patients with tabes dorsalis. Other investigators have demonstrated the presence of pathologically confirmed neurosyphilis in the face of a negative CSF VDRL (Burke 1972). Recently, HIV-infected patients have been described

TABLE 5

Diagnosing neurosyphilis in the face of HIV infection.

Definite neurosyphilis
1. + blood treponemal serology, e.g. FTA-ABS, MHA-TP, etc.
2. + CSF VDRL

Probable neurosyphilis
1. + blood treponemal serology
2. − CSF VDRL
3. CSF mononuclear pleocytosis (>20 cells/mm³) or
   < CSF protein (>60 mg/dl)
neurological complications compatible with neurosyphilis, such as cranial nerve palsies, stroke, etc., or evidence of ophthalmological syphilis

Possible neurosyphilis
1. + blood treponemal serology
2. − CSF VDRL
3. CSF mononuclear pleocytosis (>20 cells/mm³) or
   < CSF protein (>60 mg/dl)
no neurological or ophthalmological complications compatible with syphilis

with an initially non-reactive CSF VDRL that reverted to reactive following treatment for neurosyphilis (Feraru and Aronow 1990). Similarly, the absence of a reactive serum non-treponemal study, like the VDRL, is not a reliable means of excluding the diagnosis of neurosyphilis (Jaffe et al. 1978; Wolters 1987) as these studies are often negative in patients with latent or tertiary syphilis. In untreated neurosyphilis, cardiovascular and congenital syphilis, the serum VDRL was non-reactive in 30% (Deacon et al. 1966). Likewise, in a study of neurosyphilis, the serum VDRL was reactive in only 61% (Harner et al. 1968). Specific serum treponemal tests, e.g., FTA-ABS and MHA-TP, are expected to be positive in the presence of neurosyphilis (Jaffe et al. 1978), although Mapelli et al. (1981) reported a rate of only 70% for reactive serum FTA and 65% for *T. pallidum* immobilization tests (TPI) in the association with neurosyphilis.

Conversely, the CSF VDRL is believed to be extremely specific with few falsely reactive studies (Thomas 1964), though patients with falsely reactive CSF VDRL have been reported in association with a spinal ependymoma (Delaney 1976) and meningeal carcinomatosis (Madiedo 1980). A falsely positive CSF VDRL can also be produced by blood contamination of the CSF; however, even at serum VDRL titers of 1:256, gross blood contamination is required to result in a false positive study (Izzat et al. 1971). Parenthetically, as many as 4% of patients with neurosyphilis have a reactive CSF non-treponemal test, but a non-reactive serum test (Dewhurst 1968).

The value of specific treponemal antibody studies on the CSF for diagnosing neurosyphilis remains controversial. Some investigators (Escobar et al. 1970; Davis 1989) have proposed using the CSF FTA-ABS as a sensitive screening modality. The correlation of serum FTA titers to CSF titers suggests the presence of blood diffusion into the CSF (Jaffe et al. 1978; Traviesa et al. 1978), a finding which had previously been demonstrated (Ribault and Colombani 1964). This diffusion may seriously detract from the specificity of the test for diagnosing neurosyphilis. Small amounts of blood contamination of the CSF may also invalidate a reactive CSF FTA-ABS (Davis and Sperry 1979). Other studies have been touted for diagnosing neurosyphilis, such as the MHA-TP (Kinnunen and Hillbom 1986), a 19S IgM to a treponemal antigen (Luger et al. 1972), determination of intrathecal synthesis of anti-treponemal antibody (Van Eijk et al. 1987) and PCR using synthetic DNA probes for *T. pallidum* (Noordhoek et al. 1990).

Relying on non-specific CSF abnormalities in order to diagnose neurosyphilis in the face of HIV infection is fraught with difficulty, as the CSF in HIV-infected patients is frequently abnormal. Marshall et al. (1988) found that 63% of over 400 CSF analyses performed in HIV-infected US servicemen were abnormal. These abnormalities included a CSF pleocytosis, increased CSF protein, and increased CSF IgG levels and IgG synthesis (Marshall et al. 1988), abnormalities that are also observed with neurosyphilis. Muller et al. (1988) proposed measuring intrathecal synthesis of specific IgG in these patients to distinguish between the two disorders; however, a satisfactory means of distinguishing between the two disorders remains to be established.

An approach to diagnosing neurosyphilis in the HIV-infected individual is outlined in Table 5. Marra (1996) has incorporated the CSF FTA-ABS or MHA-TP into an algorithm for the diagnoses of neurosyphilis. If reactive, treatment for neurosyphilis is recommended. The development of primers for *T. pallidum* DNA may ultimately lead to the adoption of CSF PCR for the diagnosis of neurosyphilis.

Treatment

Perhaps no other disease has been as dramatically affected by the discovery of penicillin as syphilis. However, the adequacy of currently recommended treatment regimens remains a question. In fact, there have been no controlled, randomized, prospective studies as to the optimal dose or duration of therapy in neurosyphilis. The treponemicidal level of penicillin is 0.03 µg/ml. Though the organism has been demonstrated to be capable of acquiring plasmids that produce penicillinase, there is no evidence that penicillin has lost its efficacy in the treatment of *T. pallidum*. If penicillin levels become subtherapeutic, the spirochetes begin regenerating within 18–24 h. The Center for Disease Control (CDC) recommends using 2.4 million U of benzathine penicillin intramuscularly at weekly intervals for 3 weeks in the treatment of neurosyphilis, but the recordable penicillin levels in the CSF during treatment fail to reach treponemicidal levels. The concentration of penicillin in the CSF is typically unmeasurable, probably not exceeding 0.0005 µg/ml, which is 1–2% of the serum levels.

Furthermore, viable treponemes have been recovered from the CSF of individuals at the completion of therapy. Another 'recommended' regimen is procaine penicillin 600,000 U intramuscularly daily for 15 days. This regimen, too, may fail to achieve treponemicidal levels of penicillin in the CSF. Ideally, treatment of neurosyphilis should be 12–24 million U of crystalline aqueous penicillin administered intravenously daily (2–4 million U every 4 h) for a period of 10–14 days. This regimen generally requires hospitalization. Because of the expense of treatment, I have occasionally resorted to the placement of an indwelling catheter and self-administered infusions at home in reliable, well-motivated patients. The penicillin should be administered at no less than 4-h intervals to maintain the penicillin levels consistently at or above treponemicidal values, avoiding subtherapeutic troughs that occur when administered at less frequent intervals. An alternative approach to the use of parenteral penicillin is the daily oral administration of amoxicillin 3.0 g and probenicid 1.0 g for 14 days. This regimen achieves treponemicidal levels of amoxicillin in the CSF.

In patients who are penicillin-allergic, erythromycin 500 mg 4 times daily or tetracycline 500 mg 4 times daily for a period of 30 days has been recommended. Erythromycin does not diffuse readily into the brain and CSF, nor has its efficacy been demonstrated in the treatment of neurosyphilis. Similarly, oral therapy with tetracycline yields very low CSF tetracycline concentrations and it, too, has unproven efficacy in the treatment of neurosyphilis. Ideally, in the penicillin-allergic patient with unequivocally established, clinically manifest neurosyphilis, the prudent course is hospitalization, desensitization to penicillin, and subsequent high-dose aqueous penicillin treatment. Recommendations altering the standard follow-up regimens and treatment of HIV-infected patients with syphilis have been suggested (CDC 1988; Kinloch de Loes et al. 1988; Berger 1990). At the present time, it is unknown whether secondary prophylaxis, i.e., continuous administration of penicillin following the initial treatment, is necessary for syphilis. However, careful follow-up is certainly indicated. In general, serum RPR or VDRL titers should decline 4-fold at 6 months and 8-fold by 12 months after treatment of primary and secondary syphilis, and 4-fold at 12 months following treatment for latent syphilis (Romanowski et al. 1991). The decline in serum titers in HIV-infected persons with infectious syphilis (Telzak et al. 1991) or HIV-uninfected persons with neurosyphilis (Nitrini and Spina-França 1987) may not be as brisk. Furthermore, serum titers may not reflect the activity of the CNS disease in HIV-infected persons (Bayne et al. 1986; Berry et al. 1987). With respect to CSF parameters, the cell counts typically normalize within 3–6 months, although CSF protein may remain elevated in approximately 50% at this time (Nitrini and Spina-França 1987). CSF Wasserman tests remained abnormal in 90% of patients at 19–24 months (Nitrini and Spina-França 1987). Other investigators have noted a normalization of CSF cell count and protein within 8 months and reversion to non-reactivity of the CSF VDRL by 10 months in 95% of patients with HIV (Marra et al. 1992). Marra (1996) recommends that after treatment for neurosyphilis in HIV-infected persons, serum nontreponemal tests should be obtained monthly for the first 3 months and then every 3–6 months until nonreactive or repeatedly reactive at a titer of <1:8. The CSF should be re-examined at 3 months after therapy and every 6 months thereafter until normal.

Bartonella

Two members of the genus *Bartonella*, *Bartonella quintana* (formerly *Rochalimaea quintana*) and *Bartonella henselae* (formerly *Rochalimaea henselae*) are agents of severe and potentially fatal disease in HIV-infected persons (Regnery et al. 1995). Infection with *B. henselae* has been associated with traumatic contact with cats (scratches or bites) and domestic cats appear to be a reservoir for this infection (Regnery et al. 1995). *B. henselae* causes several disorders, including bacillary angiomatosis, peliosis hepatis, lymphadenitis, aseptic meningitis with bacteremia, and cat-scratch disease (Wong et al. 1995). Bacillary angiomatosis resulting from *B. henselae* in HIV-infected persons often presents with vascular skin lesions resembling Kaposi's sarcoma, frequently accompanied by fever and anemia (Moore et al. 1995). Other manifestations may include lung nodules with mediastinal adenopathy, peripheral adenopathy, pleural effusions, ascites, and lesions of the liver and spleen (Moore et al. 1995). In some communities, such as the San Francisco Bay area, infection with *B. quintana* appears to be more common than that due to *B. henselae*.

Neurologic disorders associated with *B. henselae* include cerebral and retinal bacillary angiomatosis, cat-scratch-related encephalitis, myelitis, cerebral arteritis, and retinitis (Schwartzman et al. 1994). Contrast enhancing intracerebral mass lesions due to bacillary angiomatosis in which *B. henselae* can be demonstrated pathologically by silver staining have been described in an HIV-infected individual (Sprach et al. 1992). When paired sera and CSF from 50 HIV-infected patients with neurologic disease were screened for the presence of reactive antibodies, Schwartzman et al. (1994) detected *B. henselae*-specific IgG antibody in 32% of serum samples and 26% of CSF specimens. In comparison, only 4–5.5% of HIV-infected persons without neurologic disease had serologic evidence of *B. henselae* infection (Schwartzman et al. 1994). These investigators describe three patients in detail. One patient presented with Bell's palsy and later developed uveitis, other cranial nerve palsies, ataxia, and limb incoordination. Cranial MRI showed extensive white matter disease of both hemispheres evidenced by hyperintense signal abnormalities on T2-weighted image. CSF PCR was also positive for *B. henselae*. In the absence of specific therapy, slow clearing of clinical and radiographic findings was noted (Schwartzman et al. 1994). Two other patients presented with sudden onset of disorientation and hallucinations followed by progressive dementia and death. In both CSF PCR was also positive for *B. henselae* (Schwartzman et al. 1994). The frequency with which *B. henselae* results in neurologic disease in the HIV-infected population requires further study, but the disorder may be more common than its current level of clinical recognition.

Diagnosing the neurologic manifestations of *Bartonella* are important, as it is a treatable disorder. A variety of antimicrobials, including erythromycin, doxycycline and rifampin, have resulted in complete or partial clearing (Knobler et al. 1988; Berger et al. 1989; Schlossberg et al. 1989; Kemper et al. 1990; Schwartzman et al. 1990).

Listeria

Despite the frequent association of *Listeria monocytogenes* infection in individuals with other causes of impaired cell-mediated immunity, such as malignancy and post-organ transplantation, this infection is surprisingly rare in patients with AIDS. This relative infrequency has led to speculation (Jacobs and Murray 1986; Mullin and Sheppell 1987), but no explanation has been forthcoming. Occasionally, *L. monocytogenes* has been reported as a cause of sepsis or meningitis in this population (Kales and Holzman 1990; Decker et al. 1991). In a study encompassing the years 1981–1988, an increased incidence of listeriosis in three medical centers in New York City was attributed to an increased incidence of concomitant, predisposing HIV infection (Kales and Holzman 1990). The clinical manifestations of *L. monocytogenes* in the HIV-infected population did not appear to be different from those in individuals without risk factors for HIV (Kales and Holzman 1990). In addition to acute meningitis (Gould et al. 1986; Koziol 1986; Levy et al. 1988; Kales and Holzman 1990; Decker et al. 1991), *L. monocytogenes* may also result in a chronic meningitis (Gould et al. 1986) and in brain abscesses (Harris 1989; Patey et al. 1989) in patients with AIDS. Penicillin or ampicillin with or without gentamycin is the preferred regimen for *L. monocytogenes* infection (Decker et al. 1991). Penicillin and ampicillin appear to be equally efficacious. Trimethoprim sulfamethoxazole also has been successfully used to treat listerosis in the immunocompromised host (Spitzer et al. 1986).

Nocardia

Like *L. monocytogenes*, *Nocardia asteroides* is an intracellular bacterium that is observed with increased frequency in immunocompromised individuals; however, it has been reported relatively infrequently in AIDS (Kim et al. 1991). Underreporting may be the consequence of the growth properties of *Nocardia*, the presence of other organisms, the common use of sulfonamides for treatment of patients with AIDS, and a low index of suspicion among physicians (Kim et al. 1991). The lung appears to be the most common site of infection with *N. asteroides*. In a study of 30 AIDS patients with nocardiosis, pulmonary disease was observed in 21, extra-pulmonary disease in 8 and pulmonary and extra-pulmonary disease in 1 (Uttamchandani et al. 1994). Brain abscess is the neurologic manifestation most often observed (Holtz et al. 1985; Sharer and Kapila 1985; Adair et al. 1987; Bishburg et al. 1989; Idemyor and Cherubin 1992; Uttamchandani et al. 1994). *Nocardia* may also result in meningitis (Perez-Perez et al. 1990), and

the organism has been isolated from the CSF in the absence of brain abscess (Alamo and Garcia Herruzo 1991). *Nocardia* responds well to sulfonamides, but limited penetration into abscess cavities may necessitate surgical evacuation. Norden et al. (1983) recommend employing non-surgical measures only when the number or distribution of lesions makes surgery technically unfeasible or when the patient's initial response to antimicrobial therapy is excellent (Norden et al. 1983). Delayed diagnosis, extensive disease, and early discontinuation of treatment were associated with poor outcome (Uttamchandani et al. 1994).

Other bacterial infections of the CNS

The spectrum of bacteria that can lead to meningitis and brain abscess in AIDS is extensive. Isolated case reports and small series indicate that meningitis may occur with non-typhoidal *Salmonella* (Fraimow et al. 1990), *Klebsiella pneumonia* (Holder and Halkias 1988), and *Haemophilus influenza* (Steinhart et al. 1992). Brain abscess in the absence of meningitis may also occur with *Salmonella* (Glaser et al. 1985; Holtz et al. 1985).

FUNGI

Cryptococcus neoformans

This is recognized as the second most common opportunistic infection affecting the CNS, and the most common cause of opportunistic meningitis in AIDS patients. The frequency of cryptococcal infections in HIV-infected patients has been estimated at 5–10% prior to widespread use of triazole antibiotics (Dismukes 1988; Powderly 1993). Cryptococcus is the first opportunistic infection in 45–75% of patients in whom it occurs (Kovacs et al. 1985; Chuck and Sande 1989; Clark et al. 1990). The majority of AIDS patients with cryptococcal disease have meningitis; however, pneumonitis, dermatitis, osteomyelitis, myocarditis, pericarditis, gastroenteritis and arthritis also occur. Cryptococcal prostatitis is of special concern because it may serve as a reservoir for recurrence following therapy (Larsen et al. 1989). Most cryptococcal infections in HIV-infected patients occur when CD4 lymphocyte counts fall below 100/ mm^3 (Nightengale et al. 1992; Powderly 1993).

Neuropathology

Cryptococcus neoformans proliferates in the subarachnoid space resulting in leptomeningeal opacification with a gelatinous appearance. In HIV-infected individuals, the inflammatory response may be sparse and discrete, resulting in focal collections of macrophages and formation of giant cells. Focal granulomas composed of macrophages, lymphocytes, giant cells and fungi may be seen. The organisms extend along Virchow–Robin spaces where clusters of fungi may distend the perivascular space into adjacent parenchyma. Intracerebral abscesses and cryptococcomas may be found in some cases. Mucicarmine staining demonstrates the capsule of the organism (Lang et al. 1990; Vinters and Anders 1990).

Clinical features

HIV-infected patients with cryptococcal meningitis most commonly complain of headache and fever, though these symptoms occur in only 67–82% of some large series. Nausea or emesis occurs in about 45% of cases, while photophobia and meningismus are remarkably uncommon, occurring in less than a third of cases. Seizures occur in 4–18%, and cognitive impairment or alterations in sensorium occur in 17–24%. Cranial neuropathies are seen in up to 15%. Focal neurologic deficits are uncommon, occurring in 5–15% of cases. Dizziness, cerebellar ataxia, and syncope have also been noted as presenting features (Kovacs et al. 1985; Zuger et al. 1986; Chuck and Sande 1989; Clark et al. 1990; Rozenbaum and Goncalves 1994).

Visual symptoms may occur in up to 21% of cases and neuro-ophthalmic findings in up to 33% (Jabs et al. 1989; Rozenbaum and Goncalves 1994). Visual symptoms may include transient, abrupt or progressive loss of visual acuity in one or both eyes due to increased intracranial pressure (Denning et al. 1991), necrotizing optic neuropathy from cryptococcal infiltration (Ofner and Baker 1987; Cohen and Glasgow 1993) or compression of the optic nerve with vascular compromise (Lipson et al. 1989). Diplopia may occur intermittently or persistently due to cranial neuropathy, and skew deviation or nystagmus may also be seen (Friedman 1991; Keane 1991, 1993). Papilledema may occur in 1.5–12% of cases (Jabs et al. 1989; Rozenbaum and Goncalves 1994). Visual field deficits may be homonymous, bitemporal or altitudinal (Golnik et al. 1991; Friedman 1991; Garrity et al.

1993). Visual symptoms may be the first indication of recurrent cryptococcal meningitis in patients on maintenance antibiotic therapy (Golnik et al. 1991).

Psychiatric symptoms may also be presenting features of cryptococcal meningitis. Psychosis (Clark et al. 1990) and behavioral changes (Saag et al. 1992) have been noted. Mania as an isolated presenting symptom has been described in two patients, one of whom responded to successful therapy of his meningitis with complete resolution of the manic syndrome (Johannessen and Wilson 1988).

In some cases, no signs of neurologic disease are present and diagnosis results from evaluation of systemic symptoms (Zuger et al. 1986; Rozenbaum and Goncalves 1994).

Radiology

CT imaging is insensitive to cryptococcal meningitis, revealing non-specific atrophy or no abnormalities in most instances (Post et al. 1985; Popovich et al. 1990; Tien et al. 1991). MRI is more sensitive than CT, but still substantially underestimates the lesion burden found on pathologic examination (Matthews et al. 1992).

Contrast enhanced imaging with MRI may reveal meningeal enhancement, but appears to do so less frequently than in non-HIV-infected patients with cryptococcal meningitis. Parenchymal cryptococcomas may appear as enhancing mass lesions within the parenchyma (Zuger et al. 1986; Tien et al. 1991; Andreula et al. 1993). Miliary nodular enhancing leptomeningeal lesions have been noted uncommonly (Tien et al. 1991). In the absence of pathologic confirmation, it is difficult to exclude the possibility of concurrent opportunistic pathology accounting for enhancing mass lesions (Matthews et al. 1992).

Non-enhancing foci (which may be numerous in the basal ganglia and midbrain, displaying signal intensities similar to CSF) represent dilated Virchow–Robin spaces. On pathologic examination these may be filled with clusters of cryptococci and mucinous secretions. These collections have been referred to as gelatinous pseudocysts by some authors (Popovich et al. 1990; Tien et al. 1991; Matthews et al. 1992; Andreula et al. 1993).

Cerebrospinal fluid

In contrast to non-HIV-infected individuals with cryptococcal meningitis, CSF abnormalities may be subtle in AIDS patients. Opening pressure is elevated in about two-thirds and may exceed 500 mm of water. Pleocytosis is lacking or modest in most instances, though occasionally vigorous mixed inflammatory responses are seen. Hypoglycorrhachia and elevated protein levels are found in up to three-fourths of cases.

India ink stains will disclose cryptococci in 70–94% of patients with positive CSF cultures. However, in over 90% of patients with meningitis, cryptococcal antigen titers in CSF are positive (Kovacs et al. 1985; Zuger et al. 1986; Chuck and Sande 1989; Clark et al. 1990; Saag et al. 1992; Rozenbaum and Gonclaves 1994). This means that cryptococci are cultured from CSF in virtually all patients with meningitis; however, occasional individuals with parenchymal cryptococcomas and no meningitis have had sterile CSF (Zuger et al. 1986).

Serologic studies
Cryptococcal antigen titers in serum are usually elevated in patients with cryptococcal meningitis, in some cases even when antigen titers in CSF are negative (Chuck and Sande 1989). The level of serum cryptococcal antigen and changes in response to therapy do not appear reliably to reflect CSF responses (Eng et al. 1986; Clark et al. 1990; Powderly 1993).

Serum cryptococcal antigen titers may reflect disease disseminated disease to other organs, which occurs in conjunction with meningitis in up to one-half of cases. Most common sites for extrameningeal infection are the lungs, blood, urine and bone marrow, but virtually any organ may be involved (Kovacs et al. 1985; Larsen et al. 1989; Clark et al. 1990; Rozenbaum and Goncalves 1994).

Therapy
The optimal antibiotic regimen for cryptococcal meningitis in HIV-infected patients has yet to be devised and authorities vary in their recommendations. There is consensus that the expectation from treatment is control rather than eradication of infection. Early series reported acute mortalities within 6 weeks of the completion of therapy of 18–37 %. Much of this early mortality occurred in the first 2 weeks (Kovacs et al. 1985; Chuck and Sande 1989; Clark et al. 1990; Saag et al. 1992). A variety of treatment regimens was used both within and between studies.

Prior to the AIDS pandemic, the drug of choice for cryptococcal meningitis was amphotericin B, a polyene antibiotic which binds to ergosterol in the cryptococcal membrane, altering its permeability. Renal toxicity, which is related to the cumulative dose administered, is the major limiting adverse effect. During infusion, acute reactions commonly occur with fevers, chills, rigors, headache and nausea, which may be mitigated by pretreatment with hydrocortisone, non-steroidal anti-inflammatory agents, antiemetics and antihistamines. Other potential adverse reactions include seizures, phlebitis, anemia and edema (Sugar et al. 1990). Dosages used have ranged from 0.3 to 1.0 mg/kg/day, with higher doses given for shorter periods before reduction to a maintenance regimen (Zuger et al. 1986; Chuck and Sande 1989; Clark et al. 1990; Kovacs et al. 1990; Larsen et al. 1990; Saag et al. 1992; White et al. 1992; De Lalla et al. 1995).

In studies prior to AIDS, amphotericin B was found to be more effective in treating cryptococcal meningitis when administered with flucytosine, an oral antimycotic metabolized to 5-fluorouracil within fungal cells, which inhibits fungal DNA and RNA synthesis. Flucytosine may also inhibit purine and pyramidine uptake by fungi. Myelosuppression from the usual doses of 100–150 mg/kg/day has been problematic in AIDS patients however, limiting its use, as have gastrointestinal reactions and hepatotoxicity which may also occur (Sugar et al. 1990; Powderly 1993). Large retrospective studies have failed to show survival benefit from combination therapy in AIDS patients, in part due to the impact of toxicity (Chuck and Sande 1989; Clark et al. 1990).

Smaller prospective and retrospective studies have suggested benefits from combinations of flucytosine and amphotericin B at higher doses (Larsen et al. 1990; White et al. 1992; De Lalla et al. 1995). Recently, an open label trial has used flucytosine in combination with fluconazole to obtain comparable response rates to amphotericin (Larsen et al. 1994).

The development of triazole agents, which inhibit cytochrome P450 enzyme activity by limiting fungal ergosterol synthesis (Sugar et al. 1990), has had a major impact on control of cryptococcal infections in AIDS. Fluconazole has a higher CSF penetration than itraconazole, though both have shown efficacy in AIDS patients with cryptococcal meningitis (Stern et al. 1988; Sugar and Saunders 1988; Denning et al. 1989; Nightengale 1995).

Comparisons of triazoles with amphotericin B for

acute therapy of cryptococcal meningitis in AIDS have yielded mixed results to date. The largest trial found no significant difference in overall mortality between fluconazole and amphotericin B, although mortality with fluconazole was concentrated in the first 2 weeks of illness. Doses of the agents and combination therapy with flucytosine were not controlled (Saag et al. 1992). Amphotericin B 0.3 mg/kg/day plus flucytosine 150 mg/kg/day was superior to itraconazole 200 mg/day over 6 weeks in one series (De Gans et al. 1992). Amphotericin B 0.7 mg/kg/day for 1 week followed by 3 × weekly administration for 9 weeks in combination with flucytosine 150 mg/day was superior to fluconazole 400 mg/day in another series (Larsen et al. 1990). Variations in dosages, combinations of medications, severity of illness on entry and definitions of successful outcome among studies limit direct comparisons (Powderly 1993).

Attempts to mitigate toxicity from amphotericin B therapy have included incorporation into liposomes (Coker et al. 1991, 1993) and infusion of amphotericin in a fat emulsion (Leake et al. 1994). Successful responses were obtained in a single patient in the latter instance and a dozen patients in the former. Both regimens were well tolerated.

Studies have not consistently identified patient factors predicting outcome. Neurologic impairment on initiation of therapy (Clark et al. 1990; Sang et al. 1992), cryptococcal antigen titer in CSF (Zuger et al. 1986; Saag et al. 1992), hyponatremia, positive cultures of extrameningeal specimens (Chuck and Sande 1989), abnormal CT scan (Clark et al. 1990), less than 20 cells/mm^3 (Saag et al. 1992) and relapse (Zuger et al. 1986; Chuck and Sande 1989) have been associated with mortality in various reports.

Prevention of relapse

Early series of AIDS patients with cryptococcal meningitis treated only with acute therapy were associated with relapse rates of 50–60% (Kovacs et al. 1985; Zuger et al. 1986; Clark et al. 1990). Open label trials of fluconazole (Sugar and Saunders 1988) and itraconazole (De Gans et al. 1988; Denning et al. 1989) and a retrospective series in which some patients were given amphotericin B or ketoconazole (Chuck and Sande 1989) suggested survival benefit for maintenance therapy. A large placebo-controlled trial documented a 19% persistence of positive CSF

cultures following 6 weeks of primary therapy with amphotericin alone or in combination with flucytosine, and a significant benefit for fluconazole 100–200 mg/day in preventing relapse in patients with sterile CSF (Bozette et al. 1991). Subsequently, a study comparing fluconazole 200 mg/day to amphotericin B 1.0 mg/kg/week found the former agent to be more effective with less toxicity (Powderly et al. 1992b).

Despite maintenance therapy, recurrent cryptococcal meningitis occurs in 2–16% of AIDS patients (Chuck and Sande 1989; Clark et al. 1990; Bozzette et al. 1991; Powderly et al. 1992a). Serum cryptococcal antigen is not a reliable marker for CNS infection (Eng et al. 1986; Bozzette et al. 1991). CSF cryptococcal antigen titers decline with successful therapy, but subsequent rises with recurrence are apparently acute and not predicted by routine monitoring of CSF (Bozzette et al. 1991). Persistence of cryptococcus in the prostate despite sterilization of CSF has been shown (Larsen et al. 1989). Molecular genetic studies have revealed clonal identity of serial isolates obtained from three AIDS patients with recurrent meningitis (Spitzer et al. 1993). Though prospective studies are yet to be done on the natural history of persistently positive cultures in patients with suppression of clinical disease, it seems likely that recurrence results from sequestered organisms in CSF or elsewhere.

Failures of primary or secondary prophylactic therapy might potentially be due to resistant fungi. A case in which resistance to fluconazole appeared to evolve during therapy has been reported (Paugam et al. 1994) and another in which resistance of a *Cryptococcus neoformans var. gattii* species to fluconazole was noted (Peetermans et al. 1993). The latter species is uncommonly reported as a cause of cryptococcal meningitis in HIV-infected patients.

Another small series noted no change in sensitivity of serial isolates to either fluconazole or amphotericin, but noted wide variation in the dose sensitivity to fluconazole among different isolates (Casadevall et al. 1993). Higher doses of fluconazole to 800 mg/day have been reported to be effective salvage therapy in patients failing conventional therapy with either fluconazole or amphotericin (Berry et al. 1992). At least one well characterized isolate with resistance to amphotericin B has been described in a patient who responded to high-dose fluconazole (Powderly et al. 1992a).

Primary prophylaxis

The safety and efficacy of triazole antibiotics has raised the possibility of primary prevention in susceptible immunosuppressed HIV-infected individuals. An open label trial showed a reduced rate of cryptococcal meningitis in patients with CD4 levels ≤ 68/mm^3 given 100 mg/day fluconazole compared to controls from the previous 2 years seen in the same site (Nightengale et al. 1992). A recent review suggested that primary prophylaxis be considered for patients with CD4 counts less than 50/mm^3, but noted that the impact of such therapy on survival or on the susceptibility of various fungi to current agents is uncertain (Pinner et al. 1995).

Therapy-supportive measures

Elevated intracranial pressure is common in patients with cryptococcal meningitis and may result in hydrocephalus. It has been postulated that sustained elevations in intracranial pressure may contribute to early mortality by impairing cerebral circulation (Denning et al. 1991). Therapeutic modalities include drainage by serial lumbar punctures, ventricular drainage, and acetazolamide.

Acute visual loss in patients with HIV-associated cryptococcal meningitis may be transient, abrupt or progressive and may result from increased intracranial pressure (Dennning et al. 1991), necrotizing optic neuropathy due to cryptococcal invasion (Cohen and Glasgow 1993) or constrictive arachnoiditis (Lipson et al. 1989). Patients with papilledema and severely impaired vision may respond to antibiotic therapy alone (Golnik et al. 1991), ventricular shunting of CSF (Tan 1988), or lysis of arachnoid adhesions (Maruk et al. 1988). Optic nerve sheath fenestration resulted in visual recovery of two patients with cryptococcal meningitis who had persistent associated papilledema and visual loss despite antibiotic therapy (Garrity et al. 1993).

Coccidioidomycosis

With intact cellular immunity, granulomatous tissue reactions limit, but do not eradicate, infection, and reactivation may occur with immunosuppression. It is uncertain whether disseminated infection in AIDS patients results from acute infection or reactivation (Bronnimann et al. 1987; Fish et al. 1990; Ampel et al. 1993; Wheat 1995).

HIV-infected patients with CNS coccidioidomycosis present with meningitis (Fish et al. 1990; Galgiani et al. 1993), meningoencephalitis, myelitis or radiculitis (Mischel and Vinters 1995), or cerebral abscesses which may be single or multiple (Jarvik et al. 1988; Levy et al. 1988). The occurrence of CNS disease appears to be uncommon despite immunosuppression, with only 9 cases in a large retrospective review of 77 HIV-infected patients (Fish et al. 1990), and none in a prospective series of 170 patients over a median follow-up of almost 1 year (Ampel et al. 1993).

Neuropathologic reports of CNS coccidioidomycosis in HIV-infected patients describe a necrotizing granulomatous meningitis with abundant fungi, extending along Virchow–Robin spaces to involve underlying brain. Adjacent vessels are affected by extension of inflammation and an endarteritis obliterans may result. Organisms may be found in the adventitia of involved vessels. Cavitary necrosis results in abscess formation. Both granulomatous and suppurative meningitis are described (Jarvik et al. 1988; Mischel and Vinters 1995).

HIV-infected patients with coccidioidal meningitis may have a prominent CSF pleocytosis with elevated protein and hypoglycorrhachia. In five of nine cases in a large review, organisms were cultured from CSF and complement fixation antibodies to Coccidioides were found in the other four. Immunosuppression as measured by CD4 lymphocytes varied, with four of eight patients tested having levels above 200/mm^3 (Fish et al. 1990). Another case was proven at autopsy after CSF studies proved unrevealing (Mischel and Vinters 1995).

Patients with cerebral abscesses will present with headaches and fever and may lack focal neurologic findings. Imaging studies are non-specific and may show small enhancing lesions without mass effect. CSF in patients with abscesses may be normal (Jarvik et al. 1988; Levy et al. 1988).

Limited reports of treatment for CNS coccidioidomycosis in HIV-infected patients reveal some success. Of nine patients reviewed in a large retrospective series, three survived 5–21 months following therapy with amphotericin B, ketoconazole or fluconazole. Mortality for the entire series of 77 patients with HIV infection and coccidioidomycosis was significantly related to CD4 levels (Fish et al. 1990). In a series of 50 patients with coccidioidal meningitis, nine of whom also had HIV infection, fluconazole in doses of 400 mg/day produced responses in six of the nine with survivals of 9–26 months. In two, recrudescent coccidioidomycosis appeared to contribute to their mortality (Galgianni et al. 1993).

Histoplasmosis

Histoplasmosis may occur in 2–5% of AIDS patients from endemic areas compared to less than 1% of those from other regions. In cities with particularly high prevalence, such as Indianapolis, IN, Kansas City, KS, and Nashville and Memphis, TN, up to 25% of AIDS patients may be affected. Histoplasmosis may represent the first AIDS-related illness in up to 75% of such individuals, occurring alone or as a coinfection with another pathogen (Johnson et al. 1988; Wheat et al. 1990b).

Following inhalation, histoplasma transforms into a yeast at body temperature. With intact immune function, the host contains infection by mononuclear phagocytosis and granuloma formation.

The organism parasitizes macrophages, which provide a means of dissemination in immunocompromised individuals. In HIV-infected individuals with histoplasmosis, dissemination appears to occur in 95% of cases, usually when CD4 counts fall below 200/mm^3. HIV-infected patients may acquire primary infection with subsequent dissemination or experience reactivation of previously contained infection as immune compromise occurs (Johnson et al. 1988; Wheat et al. 1990b).

The CNS is affected in up to 20% of cases with a mononuclear meningitis extending to adjacent vessels, resulting in endothelial proliferation, granulomatous vasculitis and fibrinoid necrosis. A granulomatous necrotizing encephalitis may be seen with single or multiple abscesses (Anaissie et al. 1988; Wheat et al. 1990a, b; Weidenheim et al. 1992). In addition to affecting CNS, dissemination from the lungs may produce septicemia with high mortality or involve the skin or almost any organ system (Wheat 1995).

CNS histoplasmosis typically presents with headache and fever, confusion or mental status changes, lethargy or obtundation and cranial neuropathies. Focal neurologic findings may be seen in about 10% and seizures occur in about 10–30% of cases. Meningismus is uncommon, occurring in about 10%.

Stroke syndromes may be the presenting features resulting from meningovasculitis, with thrombosis of basal or meningeal vessels, or septic embolization from infected heart valves (Wheat et al. 1990a; Wheat 1995).

CSF may reveal a lymphocytic pleocytosis with protein elevation and hypoglycorrachia in patients with meningitis; however, some patients have normal CSF or non-specific protein elevation. Isolation of histoplasma takes weeks, when it is recovered. Histoplasma antigen levels are detected in about 40% and antibodies to histoplasma in 60% or more of patients though cross-reactions limit antibody specificity for acute infection (Wheat et al. 1990a, b). Cerebral imaging studies may be normal, or show contrast-enhancing mass lesions, infarction patterns without enhancement or meningeal enhancement (Anaisse et al. 1988; Wheat et al. 1990a; Weidenheim et al. 1992).

Histoplasma capsulatum may be recovered from blood, bone marrow, respiratory secretions or bronchoalveolar lavage fluid in about 85% of HIV-infected patients with disseminated disease (Wheat et al. 1990b; Wheat 1995). In a mixed series of HIV- and non-HIV-infected individuals with histoplasmosis, histoplasma antigen was detected by radioimmuno- or enzyme immunoassays in CSF or serum in about 40%, or in urine in 60% of cases. Cross-reactions may occur with Coccidioides. Antigen elevations may herald recurrent disease (Wheat et al. 1989, 1991). Serologic tests are of limited value in patients from endemic regions, though high levels occur in about 60% of patients with CNS histoplasmosis with or without associated HIV infection (Wheat et al. 1990a, b). Neuropathologic specimens obtained at biopsy or autopsy yield organisms on culture or by demonstration with methenamine silver stains in about 80% of cases (Wheat et al. 1990a, b; Weidenhein et al. 1992).

Prognosis for AIDS patients with CNS histoplasmosis is currently poor, with mortality exceeding 60% in spite of therapy (Wheat et al. 1990b). Amphotericin B in doses up to 1.0 mg/kg/day for initial therapy may be used, but CNS relapse has occurred despite cumulative dosages of 2 g (Weidenheim et al. 1992). The optimal regimen is currently unestablished.

Triazole agents may be of value. Itraconazole in doses of 200–400 mg/day has appeared to prevent relapse in HIV-infected patients with histoplasmosis, though data on CNS involvement were limited. Fluconazole offers better CNS penetration and may be preferable for maintenance therapy in patients with CNS histoplasmosis, though resistance to fluconazole has been seen (Wheat 1995).

For patients with mass lesions, resection and antifungal therapy may be superior to antibiotic therapy alone, though little data exist for AIDS patients with histoplasmomas (Wheat et al. 1990a).

No evidence exists at present to support primary prophylaxis for histoplasmosis in HIV-infected patients. One small study failed to show benefit from fluconazole (Nightengale et al. 1992).

Blastomycosis

Blastomycosis has not been commonly reported in AIDS patients to date, with only 24 documented cases as of 1994. CNS disease occurred in 46%, which is 5–10 times the rate expected from non-HIV-associated case series. The mortality rate of 54% is 5 times the expected rate. Most cases in HIV-infected patients occurred with CD4 counts less than 200/mm^3 (Witzig et al. 1994).

Neuropathologic findings in autopsied cases reveal basilar meningitis or necrotizing arteritis and encephalitis with abscess formation. Organisms may be seen clustered around vessel walls and in meninges. Granuloma formation is often lacking (Fraser et al. 1991; Harding 1991; Pappas et al. 1992; Tan et al. 1993).

Clinical neurologic features are non-specific and include fever and headache, lethargy, progressive obtundation, seizures and variable focal signs. Single or multiple enhancing mass lesions may be present, or a lymphocytic meningitis which may occasionally yield the organism on culture (Harding 1991; Pappas et al. 1992; Witzig et al. 1994). Isolated CNS disease without apparent pulmonary involvement may occur as a presenting manifestation (Pappas et al. 1992) or as a site of recurrence following treatment (Witzig et al. 1994). Most patients are diagnosed from pulmonary specimens obtained at bronchoscopy; however, culture of cutaneous ulcerating lesions, blood, CSF, or cerebral abscess fluid may also yield blastomyces. Histopathologic diagnosis on biopsy or autopsy material may yield organisms before cultures which may require weeks to grow. Serologic studies have not been positive in most reported patients (Pappas et al. 1992; Witzig et al. 1994).

Diagnosed cases of blastomycosis in AIDS patients have been treated initially with amphotericin B followed by maintenance therapy with ketoconazole or fluconazole. One patient was treated initially with ketoconazole. Most patients failed to respond (Fraser et al. 1991; Pappas et al. 1992; Tan et al. 1993; Witzig et al. 1994); however, one individual initially given 800 mg of amphotericin followed by 400 mg/day of ketoconazole maintenance therapy appeared to respond to treatment (Pappas et al. 1992). Itraconazole in doses of 200–400 mg/day has been effective in non-HIV-infected patients with blastomycosis and may prove useful in future AIDS patients (Wheat 1995).

Aspergillus

CNS involvement occurs in one-third to one-half of cases reported in HIV-infected individuals, compared to 10–25% in series of other immunosuppressed patients. *Aspergillus fumigatus* and *Aspergillus flavus* account for most cases of CNS aspergillosis in HIV-infected patients (Singh et al. 1991; Minamoto et al. 1992; Pursell et al. 1992). Spread to the CNS may also occur by septic embolization from colonized heart valves resulting in infarction and formation of mycotic aneurysms or abscesses (Henochowicz et al. 1985; Cox et al. 1990).

Neuropathologic examination reveals a necrotizing vasculitis with hemorrhagic infarctions, and microscopic or macroscopic abscess formation. Meningitis may be present with neutrophilic inflammation. Fungi with regular acute or right-angle branching septate hyphae may be demonstrated (Asnis et al. 1988; Vinters and Anders 1990; Woods and Goldsmith 1990; Carrazana et al. 1991). Dural thrombosis may be the result of extension from infected cranial sinuses (Strauss and Fine 1991; Hall and Farrior 1993). Extension from the lungs may result in vertebral osteomyelitis and meningomyelitis (Woods and Goldsmith 1990) or myelopathy may result from epidural abscess formation (Go et al. 1993).

CNS aspergillosis is difficult to diagnose in life. Patients may present with fever and headache, and changes in sensorium or mentation. Patients with abscesses or infarctions may have focal findings on examination (Woods and Goldsmith 1990; Carrazana et al. 1991; Singh et al. 1991; Pursell et al. 1992). Acute embolic infarction in an AIDS patient with fever and headache may be the presenting feature (Henochowicz et al. 1985). Fever, back pain and features of myelopathy developing over months have been reported in patients with necrotizing meningomyelitis (Woods and Goldsmith 1990) and compressive myelopathy from epidural abscess (Go et al. 1993). Facial neuropathy occurs with otomastoiditis and is often preceded by otalgia and otorhea. Subsequent headache and lethargy are associated with CNS extension (Strauss and Fine 1991; Hall and Farrior 1993a; Lyos et al. 1993).

CSF may show a lymphocytic or neutrophilic pleocytosis (Woods and Goldsmith 1990; Carrazana et al. 1991) or be unremarkable despite basal meningitis (Asnis et al. 1988; Woods and Goldsmith 1990). Cultures often fail to yield the organism.

Cerebral imaging studies may demonstrate enhancing mass lesions typical of abscesses or non-enhancing abnormalities characteristic for infarctions (Hofflin and Remington 1985; Singh et al. 1991; Pursell et al. 1992). White matter lesions may resemble PML (Woods and Goldsmith 1990). CT scans may be normal or show non-specific atrophy in patients with meningitis (Asnis et al. 1988). MRI is more sensitive than CT, but failed to reveal a necrotizing myelitis in one reported case (Woods and Goldsmith 1990).

Diagnosis of invasive aspergillosis is difficult in life. Sputum cultures are insufficient because of the frequency of colonization. Histopathologic demonstration or culture from a sterile space is required (Minamoto et al. 1992). Bronchoalveolar lavage fluid may have a high yield (Lortholary et al. 1993).

Therapy of CNS aspergillosis in diagnosed patients with HIV infection has been discouraging in most instances. Amphotericin B in doses of 0.5–1.0 mg/kg/day has been used without convincing success. One patient with mastoiditis extending to the cerebellum and lateral sinus survived hospitalization and abscess resection but was immediately lost to follow-up (Hall and Farrior 1993). Itraconazole has in vitro activity against aspergillus and has been used successfully in patients with bronchopulmonary disease (Minamoto et al. 1992; Jennings and Hardin 1993). Amphotericin B and itraconazole in combination have resulted in sterilization of cultures from HIV-infected patients with invasive aspergillosis but not documented CNS disease (Lortholary et al. 1993).

Mucormycosis

Infection by Mucorales species – rhizopus, mucor and absidia – are uncommonly reported in HIV-infected patients. Rhinocerebral mucormycosis is most often associated with uncontrolled diabetes. A case has been described in an AIDS patient in association with maxillary and ethmoid sinusitis, who progressed to develop cerebral infarction. Diagnosis was obtained at biopsy of the sinus lesion and therapy with amphotericin B produced clinical stabilization until death 6 months later from CMV infection (Blatt et al. 1991).

Isolated CNS mucormycosis is most often reported in conjunction with intravenous drug use where concurrent HIV infection may be present. Presentation with fever and lethargy with progressive development of focal deficits occurred in reported cases. Imaging studies revealed infarctions or were normal, and CSF, where available, revealed a lymphocytic pleocytosis and elevated protein with no organisms on fungal cultures. Relatively high CD4 lymphocyte counts in some cases suggest that intravenous drug use might have been a more important factor than immunosuppression (Cuadrado et al. 1988; Skolnik and De la Monte 1990). The course of other cases associated with lymphopenia may have been influenced by immunosuppression, though HIV infection can only be presumed (Wetli et al. 1984; Micozzi and Wetli 1985). Therapy with amphotericin has been successful when biopsy established the diagnosis in life (Skolnik and De la Monte 1990).

Candida albicans

Candida albicans is a common mucocutaneous infection in the oropharynx and esophagus of HIV-infected individuals. Candidal fungemia may disseminate infection to other organs, including the CNS, where it may result in meningitis or cerebral abscess formation. Intravenous drug use is an independent risk factor.

Despite the frequency of candidiasis in HIV-infected patients, reports of CNS involvement are sparse; some represent only neuropathologic findings at autopsy, apparently unrecognized in life (Snider et al. 1983; Levy et al. 1985; Petito et al. 1986; Kure et al. 1991). Symptoms may be subtle and unappreciated in life (Koppel et al. 1985).

In cases providing clinical information, features may be non-specific with ring-enhancing mass lesions and focal findings (Pitlik et al. 1983) or meningitis with either neurotrophilic (Ehni and Ellison 1987) or mononuclear (Bruinsma-Adams 1991) predominance. In the latter case, candidal meningitis was the presenting manifestation of AIDS. In candidal meningitis the fungus may be recovered from CSF.

A case of disseminated candidemia resulting in colonization of cerebral arteries resulting in acute cerebral infarction has been reported. CSF was benign, but blood cultures revealed the organism. Treatment with amphotericin B was unsuccessful (Kieburtz et al. 1993).

Treatment of patients with candidal meningitis or abscess with amphotericin has had mixed success. Two patients appeared to respond to surgical drainage and amphotericin (Pitlik et al. 1983) alone or with five flucytosine (Levy et al. 1984). One patient with meningitis responded to amphotericin and five flucytosine with resolution of symptoms and sterilization of CSF, which was maintained for several months by weekly maintenance amphotericin (Ehni and Ellison 1987).

The role of azole therapy for CNS candidiasis is presently unestablished. Reports of responses can be found (Sanchez-Portocarrero et al. 1993), although resistance may emerge in patients on continuing therapy (Heinic et al. 1993).

Sporotrichosis

Sporothrix schenkii is a ubiquitous soil fungus which is found in association with both living and decaying vegetation. It may infect dogs and cats as well as humans, and percutaneous introduction may occur in florists, nursery or forestry workers, landscapers, gardeners, and others with similar exposures.

Clinical infection usually results in a nodular or ulcerating cutaneous lesion which may spread locally through lymphatic channels. Pulmonary disease is less common. Hematogenous dissemination may occur in immune compromised hosts, most commonly producing arthritis but, occasionally, a lymphocytic meningitis (Dismukes 1992c).

Two cases of *Sporothrix schenkii* meningitis in HIV-infected patients with cutaneous sporothrichosis have been reported. One patient developed lymphocytic meningitis with hypoglycorrhachia 2 months

following apparently successful therapy of cutaneous disease with amphotericin B and while on maintenance therapy with fluconazole. Serum and CSF antibodies with an elevated IgG index were present, and the organism was subsequently isolated from CSF. Reinstitution of amphotericin B at 1 mg/kg/day failed to prevent progression, and meningitis due to *Sporothrix schenkii* was confirmed at autopsy (Penn et al. 1992).

A second case occurred in an agricultural worker with a CD4 count of $17/mm^3$. Progressive cutaneous lesions developed despite fluconazole therapy. An MRI done because of hyponatremia revealed non-enhancing lesions in the brainstem, basal ganglia and centrum semiovale. After he developed seizures, despite therapy with amphotericin B, a lumbar puncture revealed a neutrophilic pleocytosis with hypoglycorrhachia and elevated protein. Yeasts were identified on a spun CSF specimen and culture of CSF yielded sporothrix. Addition of high-dose fluconazole, itraconazole and potassium iodide failed to prevent progression. At autopsy, a diffuse meningitis and ventriculitis with thrombotic endarteritis producing infarctions corresponding to the MRI lesions were demonstrated. Yeasts were present in the meninges and infiltrating adjacent vessels. Post-mortem sensitivity testing revealed resistance to the agents employed, but sensitivity to ketoconazole, miconazole and flucytosine (Donabedian et al. 1994).

Cladosporiosis

An intravenous drug user presenting with fever, headache and focal neurologic deficits was found to have multiple contrast-enhancing lesions on CT scan. Empiric therapy for toxoplasmosis failed to prevent progressive deterioration over 7 weeks. Autopsy disclosed cladosporiosis (Colon et al. 1988).

PARASITES

CNS toxoplasmosis

Toxoplasma gondii is the most common cause of cerebral mass lesions in AIDS patients. Estimates of its prevalence vary geographically and among populations in accordance with serologic evidence of exposure, which ranges from 3–45% in the USA to as high as 80% in parts of Europe (Clumeck 1991; Hunter and Remington 1994). Estimates of the frequency of symptomatic toxoplasmosis in the USA range from 6% (Wong et al. 1984) to a third of seropositive individuals (Luft and Remington 1992). In a series from New York City, 2.5% of AIDS patients had CNS toxoplasmosis (Wong 1984). In a treatment study of dideoxyinosine, toxoplasmic encephalitis occurred in 11% of all patients and 25% of those with positive serology and immunosuppression as reflected by CD4 lymphocyte counts less than 100/mm^3 (Oksenhendler et al. 1994).

T. gondii is usually acquired by ingestion of raw or poorly cooked red meat which contains tissue cysts. Cats are the definitive hosts of this obligate intracellular protozoan and excrete oocysts in their feces. Contact with contaminated material provides another source of transmission of infection. The infection has been transmitted by transfusion of infected blood and transplantation of infected organs. Transplacental transmission occurs. Excreted oocysts may survive more than 1 year.

T. gondii organisms released from ingested oocysts enter intestinal mucosal cells, where replication results in trophozoites which disseminate via the bloodstream or lymphatics, infect nucleated host cells and multiply within vacuoles. With continued division, rupture of the cell leads to contiguous spread of infection with progressive necrosis. Both humoral and cellular immune responses are important in containment of the infection, but even with effective responses, eradication of the parasite does not occur. Consequently, *T. gondii* cysts may persist in any organ, though they are particularly common in brain, myocardium, skeletal muscle and lymph nodes. Dormant organisms remain viable (Masur 1992).

Immune competent individuals who acquire toxoplasmosis may have a transient flu-like illness with fever, malaise and lymphadenopathy or, more commonly, suffer no clinical symptoms. With loss of immune competence, reactivation of latent infection leads to the emergence of clinical disease. Because not all AIDS patients with serologic evidence of toxoplasma infection develop clinical disease, it has been suggested that host factors, including genetic susceptibility or variation in strain virulence, may be important pathogenetic elements (Luft and Remington 1992). In murine toxoplasmosis, evidence suggests a role for CD8 lymphocytes in controlling numbers of cysts,

and γ-interferon and interleukin-2 in macrophage activation limiting encephalitis. Microglial activation and perhaps astrocyte induction of CD8 lymphocyte proliferation or participation in antigen presentation may also be important in suppression of toxoplasmic encephalitis (Hunter and Remington 1994).

In most AIDS patients, low levels of IgG antibodies and lack of IgM antibodies to *T. gondii* suggest reactivation of latent infection; however, occasional reported cases are more typical of acute infection (Leport et al. 1988; Renold et al. 1992). Pathologically proven or clinically responsive cases of presumed toxoplasmic encephalitis may occur in 15–20% of AIDS patients (Porter and Sande 1992). Studies on stored sera indicate that at least some cases of seronegative toxoplasmosis may result from loss of previous antibody responses (Renold et al. 1992). Immune competence as indicated by CD4 lymphocyte levels is significantly compromised in the majority of AIDS patients with toxoplasmic encephalitis, with CD4 counts less than 200/mm^3 in about 90% and less than 100/mm^3 in about two-thirds of cases (Porter and Sande 1992).

Neuropathology
Toxoplasmic encephalitis is characterized pathologically by hemorrhagic or coagulative necrosis with a mixed inflammatory reaction. Granulomas may be present and microglial nodules containing encysted organisms or free tachyzoites may be found. Organisms are best demonstrated with immunoperoxidase stains. Hematoxylin–eosin staining may fail to reveal organisms in up to half of cases with positive immunoperoxidase stains (Luft et al. 1984). Vascular involvement is prominent and results in infarction and vasculitis. Less often, small hemorrhages may be found (Israelski and Remington 1989; Cornford et al. 1992). A granular ependymitis may be seen in the ventricular walls (Gray et al. 1988).

Less active lesions may consist of small microglial aggregates with encysted or free organisms. Necrotizing lesions may be circumscribed or poorly demarcated with organisms evident peripherally. As organized abscesses develop, a rim of macrophages surrounds the area of necrosis and rare encysted bradyzoites are found around chronic abscesses (Luft et al. 1984; Farkesh et al. 1986; Navia et al. 1986; Petito et al. 1986; Gray et al. 1988; Lang et al. 1990; Burns et al. 1991; Cornford et al. 1992).

Clinical features

The most common clinical presentation is a subacute illness evolving over days to weeks, characterized by fever, headache, often confusion or cognitive disturbances, and focal neurologic findings on examination, including hemiparesis, ataxia, cranial neuropathies, aphasias, visual field defects and sensory impairments (Navia et al. 1986; Israelski and Remington 1988; Luft and Remington 1992; Porter and Sande 1992; Renold et al. 1992). A patient presenting with panhypopituitarism has been reported (Milligan et al. 1984). Seizures are presenting features in up to 24–29% of patients (Porter and Sande 1992; Renold et al. 1992) and a patient presenting with a diffuse subacute encephalopathy and continuous focal discharges on electroencephalography resembling herpetic encephalitis has been described (Carrazana et al. 1989a).

Meningoencephalitis presenting with diffuse encephalopathy and CSF inflammation in the absence of parenchymal lesions detectable by computerized tomography has been reported (Carramello et al. 1993; Artigas et al. 1994). Subacutely progressive fatal encephalopathy in patients lacking focal findings or radiographic evidence of parenchymal lesions has been found to be due to diffuse toxoplasmic encephalitis with or without positive toxoplasmic serology (Gray et al. 1989). Two cases of diffuse toxoplasmic encephalitis following a more prolonged course more typical of HIV encephalopathy have been reported (Arendt et al. 1991).

The tendency of toxoplasmic encephalitis to occur in the basal ganglia has resulted in a number of descriptions of patients with movement disorders, including chorea, ballistic movements (Nath et al. 1987; Dewey and Jankovic 1989), focal dystonia (Tolge and Factor 1991) and akathisia (Carrazona et al. 1989b). The movements are often unilateral reflecting localization of the underlying lesions and may persist despite resolution of the abscesses following antibiotic therapy. The movements may respond to symptomatic treatment (Nath et al. 1993). Two patients with thalamic pain syndrome due to toxoplasmic abscesses which persisted despite effective antibiotic therapy responded to amitriptyline (Gonzales et al. 1992).

Occasional patients with a brainstem encephalitis due to *T. gondii* have been reported with oculomotor weakness and contralateral ataxia (Kure et al. 1989) or hemiplegia, ipsilateral rubral tremor, or complete external ophthalmoplegia (Darras et al. 1994). Parinaud's syndrome and pineal region abscess have also been reported (Darras et al. 1994; Poon et al. 1994).

Microscopic hemorrhage in necrotic regions is a neuropathologic finding in toxoplasmic encephalitis. Clinically apparent intracerebral hemorrhage has occasionally been noted, perhaps due to toxoplasmic arteritis or endothelial parasitism (Levy et al. 1985; Chaudhari et al. 1989; Trenkwalder et al. 1992).

A patient with progressive obstructive hydrocephalus in the absence of other abnormalities on CT scan was found to have multifocal necrotic lesions in the periaqueductal region and in the basal ganglia and cerebellum (Nola Salas et al. 1987). Another patient with progressive obstructive hydrocephalus was found to have a necrotizing ependymitis due to *T. gondii* (Eggers et al. 1995).

Psychiatric symptoms as an isolated presenting feature of toxoplasmic encephalitis were described in 12 of 53 hospitalized French AIDS patients. Depression, psychomotor slowing, bipolar disease, dementia and schizophrenic syndromes were seen (Linard et al. 1992).

A few patients with toxoplasmic myelitis have been reported. Necrotizing myelitis presenting with rapidly progressive upper limb weakness and paresthesias (Mehren et al. 1988), thoracic pain followed after 2 months by progressive leg weakness and urinary and fecal incontinence (Herskovitz et al. 1989), and an acute conus medullaris syndrome (Harris et al. 1990; Overhage et al. 1990) have been described.

T. gondii may also produce a myositis concurrent with or in the absence of CNS infection. Fever and myalgias with weakness, wasting and elevated creatine kinase levels comprise the clinical features (Gherardi et al.1992).

Radiology

Most AIDS patients with toxoplasmic encephalitis will have multiple discrete mass lesions revealed by imaging procedures. Lesions are most commonly located at the cortical gray-white matter junctions and in the basal ganglia, but also may be found in the brainstem and cerebellum. Following contrast infusion, homogenous or ring enhancement is typical (Fig. 6), though some lesions exhibit marginal enhancement or fail to enhance at all (Levy et al. 1985; Post et al. 1985; Farkash et al. 1986; Navia et al. 1986; Rovira

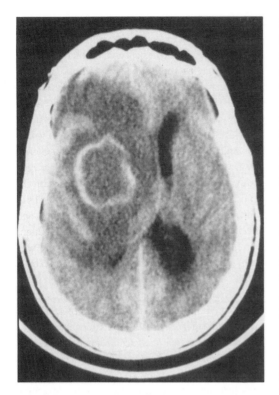

Fig. 6. Toxoplasma encephalitis. Contrast enhanced CT shows a large ring enhancing lesion of the right basal ganglia with surrounding edema and mass effect.

et al. 1991; Porter and Sande 1992; Renold et al. 1992). Double doses of contrast with delayed imaging sequences increase the sensitivity of cranial CT (Post et al. 1985). An algorithm for distinguishing CNS toxoplasmosis from lymphoma has recently been developed by a subcommittee of the American Academy of Neurology (AAN Working Group 1988).

MRI is more sensitive than CT and may reveal additional lesions (Ciricillo and Rosenblum 1990; Porter and Sande 1992). Isolated toxoplasmic abscesses may occur in up to 14% of cases (Porter and Sande 1992; Renold et al. 1992) and false negative MRI scans in patients with toxoplasmic encephalitis have been noted; however, solitary enhancing lesions on MRI were 4 times more likely to be lymphomas than toxoplasmic abscesses in a large series from San Francisco (Ciricillo and Rosenblum 1990).

Effective antibiotic therapy results in parallel clinical and radiologic responses in most patients, with up to 95% showing improvement within 2 weeks (Porter

and Sande 1992). Some radiologic lesions may persist for 6 weeks and resolution may take as long as 6 months (Levy et al. 1986). Radiologic worsening despite clinical improvement may occur in the first 3 weeks of therapy (Luft and Remington 1992).

Cerebrospinal fluid

CSF in patients with cerebral toxoplasmosis is usually unremarkable or non-specific, with modest protein elevation being the most common abnormality. Occasionally, a lymphocytic pleocytosis is seen (Navia et al. 1986; Porter and Sande 1992). Patients with meningoencephalitis or meningomyelitis may show a mixed pleocytosis with prominent protein elevations and normal or depressed glucose levels (Mehren et al. 1988; Herskovitz et al. 1989; Overhage et al. 1990; Caramello et al. 1993). *T. gondii* has occasionally been cultured from CSF of patients with meningoencephalitis (Caramello et al. 1993; Eggers et al. 1995).

IgG antibodies to *T. gondii* may be detected in CSF and may occasionally show 4-fold elevations on serial specimens in the absence of similar changes in serum (Navia et al. 1986; Porter and Sande 1992). Toxoplasma-specific IgG antibody indices may be elevated in about two-thirds of patients with toxoplasmic encephalitis. CSF antibodies are not usually found in patients with serum antibodies but no CNS disease (Potasman et al. 1988; Orefice et al. 1992). IgA antibodies to the P30 surface protein of *T. gondii* were found in two-thirds of CSF samples from patients with acute toxoplasmic encephalitis but only 15% of paired serum samples, and none of a small control group with other opportunistic infections in a recent study (Mastroianni et al. 1994).

Recently, PCR amplification techniques have been used to detect *T. gondii* DNA in CSF and serum. Probes directed at the B1 gene in CSF demonstrated *T. gondii*-specific DNA in two-thirds of patients with toxoplasmic encephalitis but no control patients without CNS disease whether seropositive or not (Schoondermark-Van de Ven et al. 1993). Serial PCR amplification studies to detect the P30 surface protein of *T. gondii* in serum correlated with parasitemia in another recent study but were negative in some patients with toxoplasmic encephalitis (Dupouy-Camet et al. 1993). PCR techniques can also be used to detect *T. gondii* DNA in brain biopsy specimens (Johnson et al. 1993).

Treatment

The current treatment of choice for AIDS patients with CNS toxoplasmosis is a combined regimen of sulfadiazine and pyrimethamine given initially in induction doses of 4–8 g/day and 50–75 mg/day, respectively. A 100–200 mg initial dose of pyrimethamine is often given the first day. To counteract myelosuppression from the pyrimethamine, leukovorin in doses of 10–15 mg/day is given concurrently (Kovacs 1995). Response in most patients with CNS toxoplasmosis is rapid and clinical and radiologic improvements are evident within 2–3 weeks, though occasional patients respond over 6 weeks or longer (Navia et al. 1986a; Porter and Sande 1992).

Patients intolerant to sulfa can be given clindamycin in split doses of 2400–4800 mg/day initially in combination with the pyrimethamine and leukovorin. Efficacy in acute treatment appears to be similar (Danneman et al. 1992; Luft et al. 1993); however, relapse rates in patients on maintenance regimens may be higher in those taking clindamycin compared to those on sulfadiazine (Porter and Sande 1992). Some patients will be intolerant to both sulfa and clindamycin. Alternative options at present include atovaquone (Kovacs 1992; Irribaren et al. 1994), pyrimethamine and azithromycin (Saba et al. 1993), pyrimethamine and dapsone (Ward 1992), pyrimethamine and doxycycline (Hagberg et al. 1993) or pyrimethamine and clarithromycin (Fernandez-Martin et al. 1991). These alternative options have not been shown to have equivalent efficacy to date.

The speed with which clinical and radiographic response occurs permits presumptive diagnosis and empiric initiation of therapy in AIDS patients with typical clinical and radiographic features and positive serology for *T. gondii* (Cohn et al. 1989). Opinions vary for patients lacking toxoplasmic serology or with single lesions by MRI, and we tend to favor early biopsy to establish a firm diagnosis in single lesions in patients with negative serology and in those who fail to respond rapidly to antibiotic therapy directed at *T. gondii*. Induction antibiotic regimens are continued for 6 weeks with repeat imaging studies at 2–3 weeks. Progressive disease at any site should prompt consideration of biopsy to exclude a concurrent or alternate pathology.

Following successful acute therapy, relapse rates of 30–60% occur in patients who discontinue antibiotics (Porter and Sande 1992; Richards et al.

1995) and may occur despite maintenance antibiotics in up to 20% (Porter and Sande 1992; Renold et al. 1992). Patients with persistent enhancement of lesions on neuroimaging studies following initial antibiotic therapy may be at increased risk of recurrence (Laissy et al. 1994). Maintenance regimens continue acute therapy at reduced dosages of 2 g/day sulfadiazine or 1200–1800 mg/day clindamycin with 50–75 mg/day pyrimethamine and leukovorin. Daily maintenance therapy with sulfadiazine and pyrimethamine has been shown to be superior to intermittent regimens in preventing recurrence of toxoplasmic encephalitis (Podzamczer et al. 1995). Pyrimethamine alone is less effective than in combination maintenance regimens with clindamycin or sulfadiazine (De Gans et al. 1992). Data on alternative regimens are presently limited.

The effectiveness of therapy for CNS toxoplasmosis in AIDS prompted interest in primary prophylaxis. Trimethoprim-sulfamethoxazole 160 mg/800 mg taken 2 times daily 2 times weekly was found to prevent toxoplasmic encephalitis in a small uncontrolled series (Carr et al. 1992). Dapsone 50 mg/day plus pyrimethamine 50 mg/week has also been shown to have efficacy in primary prevention (Girard et al. 1993). In contrast, a trial of primary prophylaxis for pneumocystis pneumonia found neither trimethoprim-sulfamethoxazole 3 times weekly with or without pyrimethamine 25 mg/week, nor dapsone 100 mg and pyrimethamine 25 mg/week effective in preventing initial episodes of toxoplasmosis in HIV-infected patients seropositive for *T. gondii* (Mallolas et al. 1993). At present, primary prophylaxis with trimethaprim-sulfamethoxazole or dapsone and pyrimethamine has been recommended for HIV-infected patients with CD4 counts less than 100/mm³ and seropositivity to *T. gondii* (Richards et al. 1995).

Acanthamoeba

Meningoencephalitis resulting from Acanthamoeba or leptomyxid amebic infestation has been reported in HIV-infected patients. Clinical features are nonspecific with fever and headache, with or without focal neurologic findings on examination. Meningismus or signs of increased intracranial pressure may be present. Seizures may be a presenting feature (Wiley et al. 1987b; Gardner et al. 1991; Di Gregorio et al. 1992; Gordon at al. 1992; Tan et al. 1993). Nodular papular or pustular cutaneous lesions precede amebic encephalitis in half the cases (Tan et al. 1993; Sison et al. 1995).

Cerebral imaging studies may show hypodensities with (Wiley et al. 1987; Gardner et al. 1991) or without enhancement (Gordon et al. 1992) following contrast infusion. Lesions vary in size. Mass effect may (Gardner et al. 1991; Gordon et al. 1992, case 2) or may not be present (Gardner et al. 1991, case 3).

CSF may show a neutrophilic pleocytosis with protein elevation (Wiley et al. 1987) or be acellular (Gardner et al. 1991; Gordon et al. 1992). Organisms have not been recovered from CSF in HIV-infected patients to date, and diagnosis has been made histopathologically by brain biopsy (Gordon et al. 1992, case 2) or at autopsy. Cutaneous lesions when present may yield a diagnosis on histopathologic examination (Tan et al. 1993; Sison et al. 1995).

Neuropathologic examination reveals necrotizing arteritis and fibrinoid necrosis (Wiley et al. 1987; Di Gregorio et al. 1992). A thrombo-occlusive angiitis involving thin-walled vessels may be seen (Gardner et al. 1991). A suppurative meningitis may be extensive (Wiley et al. 1987). Amebic trophozoites can be demonstrated with periodic acid Schiff and methenamine silver stains. Encysted organisms may also be seen (Wiley et al. 1987; Gardner et al. 1991; Di Gregorio et al. 1992; Gordon et al. 1992).

A case of meningoencephalitis due to a leptomyxid ameba has also been reported. Clinical presentation and neuropathologic findings were similar to cases described above (Anzil et al. 1991).

It has been suggested that the rapidly progressive course seen in most AIDS patients with acanthamebic encephalitis may be due to diminished ability to mount an effective granulomatous response. While no specific therapy has been identified for acanthamebic encephalitis, a non-HIV-infected patient has been reported to respond to sulfamethazine (Clelund et al. 1982).

Trypanosoma cruzi (Chagas disease)

CNS disease in chronic *T. cruzi* infection is uncommon and is thought to occur predominantly in the setting of immune compromise (Rozemberg and Lebon 1991; Rocha et al. 1994). Despite the widespread prevalence of *T. cruzi* infestation, a limited number of patients with HIV infection and Chagas disease have been reported to date. However, of the 23 cases reviewed in 1994, 20 were found to have multifocal CNS disease. Where noted, most had diminished CD4 lymphocyte counts (Rocha et al. 1994).

The most common clinical presentations of CNS disease were headache and fever, often with focal findings on neurologic examination (Del Castillo et al. 1990; Ferreira et al. 1991; Gluckstein et al. 1992; Metze and Maciel 1993; Solari et al. 1993). One patient presented with progressive ataxia and emesis due to a meningoencephalitis-producing progressive hydrocephalus (Rosemberg et al. 1992).

Cerebral imaging studies in 16 patients revealed focal or multifocal lesions in 15. Following contrast infusion, ring enhancement patterns were common though some non-enhancing or irregular enhancing lesions were also seen. Mass effect was common but not always seen (Rocha et al. 1994).

CSF studies were reported in nine cases and revealed lymphocytic pleocytosis in seven. Moderate-to-marked protein elevation occurred in all eight cases noted, and hypoglycorrhachia was reported in five of eight (Rocha et al. 1994). Antibodies to *T. cruzi* were not found in CSF (Gluckstein et al. 1992; Metze and Maciel 1993).

Empiric therapy for toxoplasmosis was usually initiated on presentation. Correct diagnosis was made histopathologically by biopsy or at autopsy where a multifocal hemorrhagic necrosis with amastigotes of *T. cruzi* were demonstrated intracellularly in glia, macrophages, endothelial cells and less commonly neurons by methenamine silver stains. Free amastigotes were also seen in perivascular and intercellular spaces (Del Castillo et al. 1990; Gluckstein et al. 1992; Solari et al. 1993; Rocha et al. 1994). Purulent meningitis and vasculitis were noted in some instances (Rozemberg and Lebon 1991; Solari et al. 1993).

Therapy in two cases following biopsy diagnosis appeared to be beneficial. One patient stabilized neurologically with residual deficits and radiologic lesions following 4 weeks of nifurtimox before dying of pneumocystis pneumonia several months later (Del Castillo et al. 1990). Another was treated initially with benznidazole 400 mg/day followed by itraconazole and fluconazole 400 mg/day resulting in resolution of fever and stabilization of neurologic symptoms. Maintenance therapy with benznidazole 200 mg/day was associated with clinical stabilization, though with persistent lesions radiographically over a 9-month period (Solari et al. 1993).

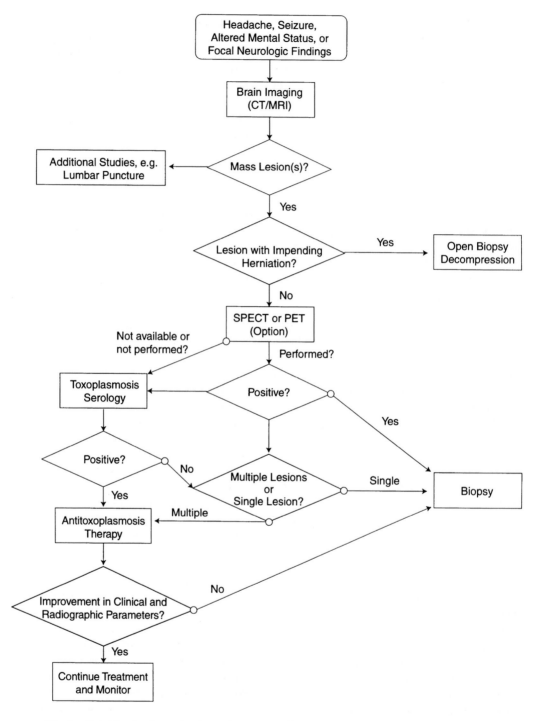

Fig. 7. Algorithm for the evaluation and treatment of intracranial mass lesions in AIDS patients

Despite the frequency of infestation in endemic areas where HIV infection is increasingly being found, CNS opportunistic infection with *T. cruzi* appears uncommon to date. Only one instance was found in a neuropathologic series of 252 HIV autopsy cases from Brazil (Chimelli et al. 1992). Nonetheless, awareness of this parasite, which mimics toxoplasmosis in its clinical and imaging features, should prompt aggressive diagnostic evaluation in patients from endemic areas.

Strongyloides stercoralis

Strongyloides stercoralis is an intestinal nematode found in moist soil in tropical and semi-tropical areas including the southeastern USA. Ingestion results in colonization of the intestine which is usually asymptomatic. Sexual transmission may occur. Both humoral and cellular immunity appear important in control of the infestation and disseminated disease occurs in individuals with immune compromise due to steroids, cytotoxic drugs or systemic disease. A hyperinfection syndrome in which the infective filariform larvae may carry intestinal bacteria to distant sites including the meninges may result (Maayan et al. 1987; Dutcher et al. 1990).

Only a few cases of HIV-infected patients with meningitis associated with the *Strongyloides* hyperinfection syndrome have been reported. In some instances, other risk factors such as steroid and cytotoxic therapy for an AIDS patient with systemic lymphoma were present. The organism was recovered from CSF and stool samples (Dutcher et al. 1990).

Gastrointestinal symptoms may precede neurologic features with anorexia, weight loss, and diarrhea (Armignacco et al. 1989; Harcourt-Webster et al. 1990; Morgello et al. 1993). Subsequent purulent meningitis has been reported due to *E. coli* (Maayan et al. 1987; Armignacco et al. 1989) or *Streptococcus bovis* (Jain et al. 1994).

Thiabendazole therapy has been tried but proved unsuccessful in two instances (Dutcher et al. 1990; Jain et al. 1994) and failed to prevent hyperinfection syndrome in another patient with AIDS (Harcourt-Webster et al. 1991).

Cerebral infestation by *Strongyloides* in two patients produced a granulomatous ependymitis and subependymitis with viable larvae in cortex and white matter in one and white matter and cerebral vessels resulting in infarctions in another. Neither had meningitis nor any inflammatory response to the larvae (Morgello et al. 1993).

Cysticercosis (Taenia solium)

Four cases from a series of 107 HIV-infected patients from Zimbabwe, 13 of whom had intracranial mass lesions, were reported recently. All four presented with headaches and seizures or focal deficits on examination. Multiple cystic lesions were present on CT scans. The frequency of cysticercosis, producing 30% of cerebral mass lesions in an HIV-infected population, was compared to a non-HIV-infected series of 51 intracranial mass lesions in which only 6% were due to neurocysticercosis. The increased frequency might be due to reactivation in a setting of immune suppression (Thornton et al. 1992).

In contrast, a case of asymptomatic neurocysticercosis and cryptococcal meningitis was reported in an AIDS patient with the speculation that immune suppression might prevent the emergence of neurocysticercosis symptoms, resulting from the host inflammatory response to the parasite (White et al. 1995). An autopsy series from Mexico noted only one case of neurocysticercosis in 97 AIDS patients compared to 3 cases in 197 controls, suggesting that the association between HIV infection and neurocysticercosis might be coincidental (Barron-Rodriguez et al. 1990).

ALGAE

A case of meningitis due to *Prototheca wickerhamii* in an HIV-infected intravenous drug user coinfected with cryptococcal meningitis has been reported. Treatment with amphotericin B and 5-fluorocytosine appeared to suppress the fungal infection; however, *P. wickerhamii* was isolated from a subsequent CSF sample. Autopsy disclosed both cryptococcal and protothecal meningitis (Kaminski et al. 1992).

REFERENCES

ABOS, J., F. GRAUS, J.M. MIRO ET AL.: Intercranial tuberculomas in patients with AIDS. AIDS 5 (1991) 461–462.

ACHIM, C.L., R.M. NAGRA, R. WANG, J.A. NELSON and C.A. WILEY: Detection of cytomegalovirus in cerebrospinal fluid autopsy specimens from AIDS patients. J. Infect. Dis. 169 (1994) 623–627.

ADAIR, J.C., A.C. BECK, R.I. APFELBAUM and J.R. BARINGER: Nocardial cerebral abscess in the acquired immunodeficiency syndrome. Arch. Neurol. 449 (1987) 548–550.

AKSAMIT, A.J. and S. KOST: PCR detection of JC virus in PML and control CSF (Abstr.). Neurosci. HIV Infect. Basic Clin. Front. (1994).

ALAMO, C. and J. GARCIA HERRUZO: Nocardiosis in a patient with AIDS. Rev. Clin. Esp. 188 (1991) 83–84.

ALLISON, M.J. and H.P. DALTON: Etiology of meningitis at the Medical College of Virginia, 1961–1965. Va. Med. Mon. 94 (1967) 317–319.

ALVAREZ, S. and W.R. MCCABE: Extrapulmonary tuberculosis revisited: a review of experience at Boston City and other hospitals. Medicine (Baltimore) 63 (1984) 25–54.

AMLIE-LEFOND, C., D.E. KLEINSCHMIDT, B.K. MASTERS, R. MAHALINGAM, L.E. DAVIS and D.H. GILDEN: The vasculopathy of varicella-zoster virus encephalitis. Ann. Neurol. 37 (1995) 784–790.

AMPEL, N.M., C.L. DOLS and J.N. GALGIANI: Coccidioidomycosis during human immunodeficiency virus infection: results of a prospective study in a coccidiodal endemic area. Am. J. Med. 94 (1993) 235–239.

ANAISSIE, E., V. FAINSTEIN, T. SAMO, G.P. BODEY and G.A. SAROSI: Central nervous system histoplasmosis. An unappreciated complication of the acquired immunodeficiency syndrome. Am. J. Med. 84 (1988) 215–217.

ANDREULA, C.F., N. BURDI and A. CARELLA: CNS cryptococcosis in AIDS: spectrum of MR findings. J. Comput. Assist. Tomogr. 17 (1993) 438–441.

ANZIL, A.P., C. RAO, M.A. WRZOLE, G.S. VISVESVARA, J.H. SHER and P.B. KOZLOWSKY: Amebic meningoencephalitis in a patient with AIDS caused by a newly recognized opportunistic pathogen: Leptomyxid ameba. Arch. Pathol. Lab. Med. 115 (1991) 21–25.

APPLEMAN, M.E., D.W. MARSHALL, R.L. BREY ET AL.: Cerebrospinal fluid abnormalities in patients without AIDS who are seropositive for the human immunodeficiency virus. J. Infect. Dis. 158 (1988) 193–199.

ARAL, S.O. and K.K. HOLMES: Epidemiology of sexual behaviour and sexually transmitted diseases. In: Sexually Transmitted Diseases, 2nd Edit. New York, McGraw Hill (1990) 19–36.

ARENDT, G., H. HEFTER, C. FIGGE, E. NEUEN-JACOB, H.W. NELLES, C. ELSING and H.J. FREUND: Two cases of cerebral toxoplasmosis in AIDS patients mimicking HIV-related dementia. J. Neurol. 238 (1991) 439–442.

ARMIGNACCO, O., A. CAPECCHI and P. DE MORI: Strongyloides stercoralis hyperinfection and the acquired immunodeficiency syndrome (Letter). Am. J. Med. 86 (1989) 258.

ARRIBAS, J.R., D.B. CLIFFORD, C.J. FICHTENBAUN, D.L. COMMINS, W.G. POWDERLY and G.A. STORCH: Level of cytomegalovirus (CMV) DNA in cerebrospinal fluid of subjects with AIDS and CMV infection of the central nervous system. J. Infect. Dis. 172 (1995) 527–531.

ARTHUR, R.R., S. DAGOSTIN and K. SHAH: Detection of BK virus and JC virus in urine and brain tissue by the polymerase chain reaction. J. Clin. Microbiol. 27 (1989) 1174–1179.

ARTIGAS, J., G. GROSSE, F. NIEDOBITEK, M. KASSNER, W. RISCH and W. HEISE: Severe toxoplasmic ventriculomeningoencephalomyelitis in two AIDS patients following treatment of cerebral toxoplasmic granuloma. Clin. Neuropathol. 13 (1994) 120–126.

ASNIS, D.S., R.K. CHITKARA, M. JACOBSEN and J.A. GOLDSTEIN: Invasive aspergillosis: an unusual manifestation of AIDS. N.Y. State J. Med. 88 (1988) 653–655.

ASTRÖM, K.E., E.L. MANCALL and E.P. RICHARDSON, JR.: Progressive multifocal leukoencephalopathy: a hitherto unrecognized complication of chronic lymphocytic leukemia and lymphoma. Brain 81 (1958) 93–111.

BACELLAR, H., A. MUÑOZ, E.N. MILLER, B.A. COHEN, D. BESLEY, O.A. SELNES, J.T. BECKER and J.C. MCARTHUR: Temporal trends in the incidence of HIV-1-related neurologic diseases: multicenter AIDS cohort study, 1985–1992. Neurology 44 (1994) 1892–1900.

BACHMAN, D.: Personal communication, Washington, DC (1993).

BALE, J.F., JR.: Human cytomegalovirus infection and disorders of the nervous system. Arch. Neurol. 41 (1984) 310–320.

BALFOUR, H.H., C. BENSON, J. BRAUN, B. CASSENS, A. ERICE, A. FRIEDMAN-KIEN, T. KLEIN, B. POLSKY and S. SAFRIN: Management of acyclovir-resistant herpes simplex and varicella-zoster virus infections. J. Acquir. Immune Defic. Syndr. 7 (1994) 254–260.

BARRETT-CONNOR, E.: Tuberculous meningitis in adults. South. Med. J. 60 (1967) 1061–1067.

BARRON-RODRIGUEZ, L.P., J. JESSURUN and M. HERNANDEZ-AVILA: The prevalence of invasive amebiasis and cysticercosis is not increased in Mexican patients dying of AIDS (Abstr. Th B 524). In: International Conference on AIDS, 6. San Francisco (1990) 253.

BATEMAN, D.E., P.K. NEWMAN and J.B. FOSTER: A retrospective survey of proven cases of tuberculous meningitis in the Northern region, 1970–1980. J. Roy.

Coll. Physicians (Lond.) 17 (1983) 106–110.

BAUER, W., W. CHAMBERLIN and S. HORENSTEIN: Spinal demyelination in progressive multifocal leukoencephalopathy. Neurology 19 (1969) 287.

BAUER, W.R., A.P. TURCI, JR. and K.P. JOHNSON: Progressive multifocal leukoencephalopathy and cytarabine. J. Am. Med. Assoc. 226 (1973) 174–176.

BAYNE, L.L., J.W. SCHMIDLEY and D.S. GOODIN: Acute syphilitic meningitis. Its occurrence after clinical and serological cure of secondary syphilis with penicillin G. Arch. Neurol. 43 (1986) 137–138.

BAZAN, C., III, C. JACKSON, J.R. JINKINS and R.J. BAROHN: Gadolinium-enhanced MRI in a case of cytomegalovirus polyradiculopathy. Neurology 41 (1991) 1522–1523.

BECERRA, L.I., S.M. KSIAZEK, P.J. SAVINO ET AL.: Syphilitic uveitis in human immunodeficiency virus-infected and noninfected patients. Ophthalmology 96 (1989) 1727–1730.

BEDRI, J., W. WEINSTEIN, P. DEGREGORIO ET AL.: Progressive multifocal leukoencephalopathy in acquired immunodeficiency syndrome. N. Engl. J. Med. 309 (1983) 492–493.

BEHAR, R., C. WILEY and J.A. MCCUTCHAN: Cytomegalovirus polyradiculoneuropathy in acquired immune deficiency syndrome. Neurology 37 (1987) 557–561.

BELEC, L., F. GRAY, J. MIKOLL, F. SCARAVILLI, C. MIHIRI, A. SOBEL and J. POIRIER: Cytomegalovirus (CMV) encephalomyeloradiculitis and human immune deficiency virus (HIV) encephalitis: presence of HIV and CMV co-infected multinucleated giant cells. Acta Neuropathol. (Berlin) 81 (1990) 99–104.

BERENGUER, J., J. SOLERA, M.D. DIAZ, S. MORENO, J.A. LOPEZ-HERCE and E. BOUZA: Listeriosis in patients infected with human immunodeficiency virus. Rev. Infect. Dis. 13 (1991) 115–119.

BERENGUER, J., S. MORENO and F. LAGUNA ET AL.: Tuberculous meningitis in patients infected with the human immunodeficiency virus. N. Engl. J. Med. 326 (1992) 668–672.

BERGEN, G.A., B.G. YANGCO and H.M. ADELMAN: Central nervous system infection with Mycobacterium kansasii. Ann. Intern. Med. 118 (1993) 396.

BERGER, J.R: Syphilis of the spinal cord. In: R.A. Davidoff (Ed.), Handbook of the Spinal Cord. New York, Marcel Dekker 5 (1987) 491–538.

BERGER, J.R: Neurosyphilis. In: R.T. Johnson (Ed.), Current Therapy in Neurologic Disease – 3. Philadelphia, PA, Dekker (1990) 143–147.

BERGER, J.R.: Neurosyphilis in HIV-1 seropositive individuals: a prospective study. Arch. Neurol. 48 (1991) 700–702.

BERGER, J.R: Spinal cord syphilis associated with human immunodeficiency virus infection: a treatable myelopathy. Am. J. Med. 92 (1992) 101–103.

BERGER, J.R.: Prolonged survival in pathologically-proven AIDS-associated PML (Abstr.). J. Neurovirol. 2 (1996) 31.

BERGER, J.R. and M. CONCH: Progressive multifocal leukoencephalopathy: the evolution of a disease once considered rare. J. Neurovirol. 1 (1995) 5–18.

BERGER, J.R. and L. MUCKE: Prolonged survival and partial recovery in AIDS-associated progressive multifocal leukoencephalopathy. Neurology 38 (1988) 1060–1065.

BERGER, J.R., M.J.D. POST and R.M. LEVY: AIDS. In: J.O. Greenberg (Ed.), Neuroimaging: A Companion to Adams' and Victor's Principles of Neurology. New York, McGraw-Hill (1984).

BERGER, J.R., B. KASZOVITZ, M.J. POST and G. DICKINSON: Progressive multifocal leukoencephalopathy associated with human immunodeficiency virus infection. A review of the literature with a report of sixteen cases. Ann. Intern. Med. 107 (1987a) 78–87.

BERGER, J.R., L. MOSKOWITZ, M. FISCHL and R.E. KELLEY: Neurologic disease as the presenting manifestation of acquired immunodeficiency syndrome. South. Med. J. 80 (1987b) 683–686.

BERGER, J.R., M. MCCARTHY, L. RESNICK ET AL.: History of syphilis as a cofactor for the expression of HIV infection (Abstr. MAP90). In: Fifth International Conference on AIDS, Montreal (1989a) 93.

BERGER, T.G., J. TAPPERO, A. KAYMEN and P.E. LEBOIT: Bacillary (epithelioid) angiomatosis and concurrent Kaposi's sarcoma in acquired immunodeficiency syndrome. Arch. Dermatol. 125 (1989b) 1543–1547.

BERGER, J.R., M. MCCARTHY, M.A. FLETCHER, N. KLIMAS, E. LO and L. PALL: Serological and cerebrospinal fluid parameters related to syphilis in HIV asymptomatic seropositives (Abstr. WBP 56). In: Fifth International Conference on AIDS, Montreal (1989c) 361.

BERGER, J.R., L. PALL, J.C. MCARTHUR ET AL.: A pilot study of recombinant alpha 2A interferon in the treatment of AIDS-related progressive multifocal leukoencephalopathy (Abstr.). Neurology 42 (1992a) 257.

BERGER, J.R., H. WASKIN, L. PALL, G. HENSLEY, I. IHMEDIAN and M.J.D. POST: Syphilitic cerebral gumma with HIV infection. Neurology 42 (1992b) 1282–1287.

BERGER, J.R., S. SCOTT, J. ALBRECHT, A.L. BELMAN, C. TORNATORE and E. MAJOR: Progressive multifocal leukoencephalopathy in HIV-infected children. AIDS 2 (1992c) 837–841.

BERGER, J.R., L. PALL, M. WHITEMAN and D. LANSKA: Progressive multifocal leukoencephalopathy in HIV infection. J. Neurovirol. 4 (1998) 59–68.

BERGSTROM, J., A. VAHLNE and K. ALESTIG: Primary and recurrent herpes simplex virus type 2 induced meningitis. J. Infect. Dis. 162 (1990) 322–330.

BERMAN, S.M. and R.C. KIM: The development of cytomegalovirus encephalitis in AIDS patients receiving ganciclovir. Am. J. Med. 96 (1994) 415–419.

BERRY, A.J., M.G. RINALDI and J.R. GRAYBILL: Use of high

dose fluconazole as salvage therapy for cryptococcal meningitis in patients with AIDS. Antimicrob. Agents Chemother. 36 (1992) 690–692.

BERRY, C.D., T.M. HOOTEN, A.C. COLLIER and S.A. LUKEHART: Neurologic relapse after benzathine penicillin therapy for secondary syphilis in a patient with HIV infection. N. Engl. J. Med. 316 (1987) 1587–1589.

BHARGAVA, B.S., A.K. GUPTA and P.N. TANDON: Tuberculous meningitis, a CT study. Br. J. Radiol. 55 (1982) 189–196.

BISHBURG, E., G. SUNDERAM, L.B. REICHMAN ET AL.: Central nervous system tuberculosis with the acquired immunodeficiency syndrome and its related complex. Ann. Intern. Med. 105 (1986) 210–213.

BISHBURG, E., R.H. ENG, J. SLIM, G. PEREZ and E. JOHNSON: Brain lesions in patients with acquired immunodeficiency syndrome. Arch. Intern. Med. 149 (1989) 941–943.

BISHOPRIC, G., J. BRUNER and J. BUTLER: Guillain-Barré syndrome with cytomegalovirus infection of peripheral nerves. Arch. Pathol. Lab. Med. 109 (1985) 1106–1108.

BLAKE, K., D. PILLAY, W. KNOWLES, D.W.G. BROWN, P.D. GRIFFITHS and B. TAYLOR: JC virus associated meningoencephalitis in an immunocompetent girl. Arch. Dis. Child. 67 (1992) 956–957.

BLATT, S.P., D.R. LUCEY, D. DEHOFF and R.B. ZELLMER: Rhinocerebral Zygomycosis in a patient with AIDS (Letter). J. Infect. Dis. 164 (1991) 215–216.

BOBROWITZ, E.D.: Ethambutol in tuberculous meningitis. Chest 61 (1972) 629–632.

BOLDORINI, R., S. CRISTINA, L. VAGO, A. TOSONI, S. GUZZETTI and G. COSTANZI: Ultrastructural studies in the lytic phase of progressive multifocal leukoencephalopathy in AIDS patients. Ultrastruct. Pathol. 17 (1993) 599–609.

BOWEN B.C. and M.J.D. POST: Intracranial infections. In: S.W. Atlas (Ed.), Magnetic Resonance Imaging of the Brain and Spine. New York, Raven Press (1991) 501–538.

BOZZETTE, S.A., R.A. LARSEN, J. CHIU, M.A.E. LEAL, J. JACOBSEN, P. ROTHMAN, P. ROBINSON, G. GILBERT, J.A. MCCUTCHAN, J. TILLES, J.M LEEDOM, D.D. RICHMAN and THE CALIFORNIA COLLABORATIVE TREATMENT GROUP: A placebo controlled trial of maintenance therapy with fluconazole after treatment of cryptococcal meningitis in the acquired immunodeficiency syndrome. N. Engl. J. Med. 324 (1991) 580–584.

BRANDON, W.R., L.M. BOULOS and A. MORSE: Determining the prevalence of neurosyphilis in a cohort co-infected with HIV. Int. J. STD AIDS 4 (1993) 99–101.

BRITTON, C.B., R. MESA-TEJADA, C.M. FENOGLIO, A.P. HAYS, G.G. GARVEY and J.R. MILLER: A new complication of AIDS: thoracic myelitis caused by herpes simplex virus. Neurology 35 (1985) 1071–1074.

BRITTON, C.B., M. ROMAGNOLI, M. SISTI and J.M. POWERS: Progressive multifocal leukoencephalopathy and response to intrathecal ARA-C in 26 patients (Abstr.). In: The Proceedings of the 4th Neuroscience of HIV Infection Conference, Amsterdam (1992) 40.

BRONNIMANN, D.A., R.D. ADAM, J.N. GALGIANI, M.P. HABIB, PETERSEN, B. PORTER and J.W. BLOOM: Coccidioidomycosis in the acquired immunodeficiency syndrome. Ann. Intern. Med. 106 (1987) 372–379.

BROOKS, B.R. and D.L. WALKER: Progressive multifocal leukoencephalopathy. Neurol. Clin. 2 (1984) 299–313.

BROOKS, J.B., G. CHOUDHARY, R.B. CRAVEN ET AL.: Electron capture gas chromatography detection and mass spectrum identification of 3-(2′-ketohexyl)indoline in spinal fluids of patients with tuberculous meningitis. J. Clin. Microbiol. 5 (1977) 625–628.

BROWN, S.T., A. ZAIDI, S.A. LARSEN and G.H. REYNOLDS: Serological response to syphilis treatment: a new analysis of old data. J. Am. Med. Assoc. 253 (1985) 1296–1299.

BRUINSMA-ADAMS, I.K.: AIDS presenting as *candida albicans* meningitis: a case report (Letter). AIDS 5 (1991) 1268.

BUCHBINDER, S., M.H. KATZ, N. HESSOL, J.Y. LIU, P.M. O'MALLEY, D. UNDERWOOD and S.D. HOLMBERG: Herpes zoster and human immunodeficiency virus infection. J. Infect. Dis. 166 (1992) 1153–1156.

BUCKMAN, R. and E. WILTSHAW: Progressive multifocal leukoencephalopathy successfully treated with cytosine arabinoside. Br. J. Haematol. 34 (1976) 153–154.

BULLOCK, M.R.R. and J.M. WELCHMAN: Diagnostic and prognostic features of tuberculous meningitis on CT scanning. J. Neurol. Neurosurg. Psychiatry 45 (1982) 1098–1101.

BURKE, A.W.: Syphilis in a Jamaican psychiatric hospital. A review of 52 cases including 17 of neurosyphilis. Br. J. Vener. Dis. 48 (1972) 249–252.

BURKE, J.M. and D.R. SCHABERG: Neurosyphilis in the antibiotic era. Neurology 35 (1985) 1368–1371.

BURNS, D.K., R.C. RISSNER and C.L. WHITE, III: The neuropathology of human immunodeficiency virus infection: the Dallas Texas experience. Arch. Pathol. Lab. Med. 115 (1991) 1112–1124.

CAPLAN, L.R., F.J. KLEEMAN and S. BERG: Urinary retention probably secondary to herpes genitalis. N. Engl. J. Med. 297 (1977) 920–921.

CARAMELLO, P., F. BRUNELLA, A. LUCCHINI, A.M. POLLONO, A. SINICCO and P. GIOANNINI: Meningoencephalitis caused by *toxoplasma gondii* diagnosed by isolation from cerebrospinal fluid in an HIV-positive patient. Scand. J. Infect. Dis. 25 (1993) 663–666.

CARNEIRO, A.V., J. FERRO, C. FIGUEIREDO, L. COSTA, J. CAMPOS, E. FERREIRA and F. DE PADUA: Herpes zoster and contralateral hemiplegia in an African patient infected with HIV-1. Acta Med. Port. 4 (1991) 91–92.

CARR, A., B. TINDALL, B.J. BREW, D.J. MARRIOTT, J.L. HURKNESS, R. PENNY and D.A. COOPER: Low dose trimethoprim-sulfamethoxazole prophylaxis for toxoplasmic encephalitis in patients with AIDS. Ann. Intern. Med. 117 (1992) 106–111.

CARRAZANA, E.J., E. ROSSITCH, JR. and S. SCHACHTER: Cerebral toxoplasmosis masquerading as herpes encephalitis in a patient with the acquired immuno-deficiency syndrome. Am. J. Med. 86 (1989a) 730–732.

CARRAZANA, E.J., E. ROSSITCH, JR. and J. MARTINEZ: Unilateral akathisia in a patient with AIDS and a toxoplasmosis subthalamic abscess. Neurology 39 (1989b) 449–450.

CARRAZANA, E.J., E. ROSSITCH, JR. and J. MORRIS: Isolated central nervous system aspergillosis in the acquired immunodeficiency syndrome. Clin. Neurol. Neurosurg. 93 (1991) 227–230.

CARTER, J.B., R.J. HAMILL and A.Y. MATOBA: Bilateral syphilitic optic neuritis in a patient with a positive test for HIV. Arch. Ophthalmol. 105 (1987) 1485–1486.

CASADEVALL, A., E.D. SPITZER, D. WEBB and M.G. RINALDI: Susceptibilities of serial *Cryptococcus neoformans* isolates from patients with recurrent cryptococcal meningitis to amphotericin B and fluconazole. Antimicrob. Agents Chemother. 37 (1993) 1383–1386.

CASTLEMAN, B., R.E. SCULLY and B.J. MCNEELY: Weekly clinicopathological exercises, case 19-1972. N. Engl. J. Med. 286 (1972) 1047–1054.

CDC: Syphilis – United States, 1983. Morb. Mortal. Wkly. Rep. 33 (1984) 433–441.

CDC: 1984 Tuberculosis Statistics. States and Cities. (HHS Publ. No. [CDC] 85-8429). Atlanta, GA, Centers for Disease Control (1985).

CDC: Early syphilis – Broward County, Florida. Morb. Mortal. Wkly. Rep. 36 (1987) 221–223.

CDC: Tuberculosis control among homeless populations. Morb. Mortal. Wkly. Rep. 36 (1987a) 257–260.

CDC: Tuberculosis in minorities – United States. Morb. Mortal. Wkly. Rep. 36 (1987b) 77–80.

CDC: Tuberculosis in blacks – United States. Morb. Mortal. Wkly. Rep. 36 (1987c) 212–220.

CDC: Tuberculosis and acquired immunodeficiency syndrome – New York City. Morb. Mortal. Wkly. Rep. 36 (1987d) 785–790.

CDC: Diagnosis and management of mycobacterial infection and disease in persons with human immuno-deficiency virus infection. Ann. Intern. Med. 106 (1987e) 204–206.

CDC: Continuing increase in infectious syphilis – United States. Morb. Mortal. Wkly. Rep. (1988a) 3735–3738.

CDC: Relationship of syphilis to drug use and prostitution – Connecticut and Philadelphia, Pennsylvania.

Morb. Mortal. Wkly. Rep. 37 (1988b) 755–764.

CDC: Recommendations for diagnosing and treating syphilis in HIV-infected patients. Morb. Mortal. Wkly. Rep. 37 (1988c) 600–608.

CDC: Recommendations for diagnosing and treating syphilis in HIV-infected patients. Morb. Mortal. Wkly. Rep. 1988, cited in J. Am. Med. Assoc. 259 (1989) 15.

CHAISSON, R.E., G.F. SCHECTER, C.P. THEUER, G.W. RUTHER-FORD, D.F. ECHENBERG and P.C. HOPEWELL: Tuberculosis in patients with the acquired immunodeficiency syndrome. Clinical features, response to therapy, and survival. Am. Rev. Resp. Dis. 136 (1987) 570–574.

CHATIS, P.A., C.H. MILLER, L.E. SCHRAGER and C.S. CRUMPACKER: Successful treatment with foscarnet of an acyclovir resistant mucocutaneous infection with herpes simplex in a patient with acquired immuno-deficiency syndrome. N. Engl. J. Med. 320 (1989) 297–300.

CHAUDHARI, A.B., A. SINGH, S. JIONDAL and T.P. POON: Haemorrhage in cerebral toxoplasmosis. S. Afr. Med. J. 76 (1989) 272–274.

CHIMELLI, L., S. ROSEMBERG, M.D. HAHN, M.B.S. LOPEZ and M.B. NETTO: Pathology of the central nervous system in patients infected with the human immuno-deficiency virus (HIV): a report of 252 autopsy cases from Brazil. Neuropathol. Appl. Neurobiol. 18 (1992) 478–488.

CHRETIEN, F., F. GRAY, M.C. LESCS, C. GENY, M.L. DUBREUIL-LEMAIRE, F. RICOLFI, M. BAUDRIMONT, Y. LEVY, A. SOBEL and H.V. VINTERS: Acute varicella-zoster virus ven-triculitis and meningo-myelo-radiculitis in acquired immunodeficiency syndrome. Acta Neuropathol. (Berlin) 86 (1993) 659–665.

CHUCK, S.L. and M.A. SANDE: Infections with cryptococcus neoformans in the acquired immunodeficiency syndrome. N. Engl. J. Med. 321 (1989) 794–799.

CHUNG, W.M., F.D. PIEN and J.L. GREKIN: Syphilis: a cause of fever of unknown origin. Cutis 31 (1983) 537–540.

CINQUE, P., L. VAGO, M. BRYTTING, A. CASTAGNA, A. ACCORDINI, V.A. SUNDQVIST, N. ZANCHETTA, A. D'A. MONFORTE, B. WAHREN, A. LAZZARIN and A. LINDE: Cytomegalovirus infection of the central nervous system in patients with AIDS: diagnosis by DNA amplification from cerebrospinal fluid. J. Infect. Dis. 166 (1992) 1408–1411.

CINQUE, P., F. BALDANTI, L. VAGO, M.R. TERRENI, F. LILLO, M. FURIONE, A. CASTAGNA, A. D'A. MONFORTE, A. LAZZARIN and A. LINDE: Ganciclovir therapy for cytomegalovirus (CMV) infection of the central nervous system in AIDS patients: monitoring by CMV DNA detection in cerebrospinal fluid. J. Infect. Dis. 171 (1995) 1603–1606.

CIRICILLO, S.F. and M.L. ROSENBLUM: Use of CT and MR imaging to distinguish intracranial lesions and to define the need for biopsy in AIDS patients. J.

Neurosurg. 73 (1990) 720–724.

CLARK, R. and J.T. CARLISLE: Neurosyphilis and HIV infection. South. Med. J. 81 (1988) 1204–1205.

CLARK, R.A., D. GREER, W. ATKINSON, G.T. VALAINIS and N. HYSLOP: Spectrum of cryptococcus neoformans infection in 68 patients infected with human immunodeficiency virus. Rev. Infect. Dis. 12 (1990) 768–777.

CLARK, W.C., J.C. METCALF, M.S. MUHLBAUER., F.C. DOHAN and J.H. ROBERTSON: *Mycobacterium tuberculosis* meningitis: a report of twelve cases and a literature review. Neurosurgery 18 (1986) 604–610.

CLELAND, P.G., R.V. LAWANDE, G. ONYEMELUKWE and H.C. WHITTLE: Chronic amoebic meningoencephalitis. Arch. Neurol. 39 (1982) 56–57.

CLIFFORD, D.B., R.S. BULLER, S. MOHAMMED, L. ROBISON and G.A. STORCH: Use of polymerase chain reaction to demonstrate cytomegalovirus DNA in CSF of patients with human immunodeficiency virus infection. Neurology 43 (1993) 75–79.

CLUMECK, N.: Some aspects of the epidemiology of toxoplasmosis and pneumocystosis in AIDS in Europe. Eur. J. Clin. Microbiol. Infect. Dis. 10 (1991) 177–178.

COHEN, B.A: Prognosis and response to therapy of CMV encephalitis and meningomyelitis in AIDS. Neurology 46 (1996) 444–450.

COHEN, B.A., J.C. MCARTHUR, S. GROHMAN, B. PATTERSON and J.D. GLASS: Neurologic prognosis of cytomegalovirus polyradiculomyelopathy in AIDS. Neurology 43 (1993) 493–499.

COHEN, B.A., A.H. ROWLEY and C.M. LONG: Herpes simplex type 2 in a patient with Mollaret's meningitis: demonstration by polymerase chain reaction. Ann. Neurol. 35 (1994) 112–116.

COHEN, D.B. and B.J. GLASGOW: Bilateral optic nerve cryptococcosis in sudden blindness in patients with acquired immune deficiency syndrome. Ophthalmology 100 (1993) 1689–1694.

COHN, J.A., A. MCMEEKING, W. COHEN, J. JACOBS and R.S. HOLZMAN: Evaluation of the policy of empiric treatment of suspected toxoplasma encephalitis in patients with the acquired immunodeficiency syndrome. Am. J. Med. 86 (1989) 521–527.

COKER, R.J., S.M. MURPHY and J.R.W. HARRIS: Experience with liposomal amphotericin B (AmBisome) in cryptococcal meningitis in AIDS. J. Antimicrob. Chemother. 28 (1991) 319–332.

COKER, R.J., M. VIVIANI, B.G. GAZZARD, B. DU PONT, H.D. POHLE, S.M. MURPHY, J. ATOUGUIA, J.L. CHAMPALIMAUD and J.R.W. HARRIS: Treatment of cryptococcosis with liposomal amphotericin B (AmBisome) in 23 patients with AIDS. AIDS 7 (1993) 829–835.

COLE, E.L., D.M. MEISLER, L.H. CALABIESE, G.N. HOLLAND, B.J. MONDINO and M.A. CONANT: Herpes zoster ophthalmicus and acquired immune deficiency syndrome. Arch. Ophthalmol. 102 (1984) 1027–1029.

COLLABORATIVE DHPG TREATMENT STUDY GROUP: Treatment of serious cytomegalovirus infections with 9-(1,3, dihydroxy-2-propoxymethyl) guanine in patients with AIDS and other immunodeficiencies. N. Engl. J. Med. 314 (1986) 801–805.

COLON, L., G. LASALA, M.D. KANZER, J.E. PARISI, M.L. DE VINATEA and A.M. MACHER: Cerebral cladosporiosis in AIDS (Abstr.). J. Neuropathol. Exp. Neurol. 47 (1988) 387.

COMSTOCK, G.W.: Epidemiology of tuberculosis. Am. Rev. Resp. Dis. 125 (1982) 8–15.

CONOMY, J.P., N.S. BEARD, H. MATSUMOTO and U. ROESSMANN: Cytarabine treatment of progressive multifocal leukoencephalopathy. J. Am. Med. Assoc. 229 (1974) 1313–1316.

CONWAY, B., W.C. HALLIDAY and R.C. BRUNHAM: Human immunodeficiency virus-associated progressive multifocal leukoencephalopathy: apparent response to 3′-azido-3′-deoxythymidine. Rev. Infect. Dis. 12 (1990) 479–482.

COOVADIA, Y.M., A. DAWOOD, M.E. ELLIS, H.M. COOVADIA and T.M. DANIEL: Evaluation of adenosine deaminase activity and antibody to *Mycobacterium tuberculosis* antigen 5 in cerebrospinal fluid and the radioactive bromide partition test for the early diagnosis of tuberculous meningitis. Arch. Dis. Child. 61 (1986) 428–435.

CORNFORD, M.E., H.W. HO and H.V. VINTERS: Correlation of neuromuscular pathology in acquired immunodeficiency syndrome patients with cytomegalovirus infection and zidovudine treatment. Acta Neuropathol. (Berlin) 84 (1992a) 516–529.

CORNFORD, M.E., J.K. HOLDEN, M.C. BOYD, K. BERRY and H. VINTERS: Neuropathology of the acquired immunodeficiency syndrome (AIDS): report of 39 autopsies from Vancouver, British Columbia. Can. J. Neurol. Sci. 19 (1992b) 442–452.

COX, J.N., F. DI DIÓ, G.P. PIZZOLATO, R. LERCH and N. POCHAN: Aspergillus endocarditis and myocarditis in a patient with the acquired immunodeficiency syndrome (AIDS). Virchows Arch. [A] 417 (1990) 255–259.

CRAIG C.P. and A.J. NAHMIAS: Different patterns of neurologic involvement with herpes simplex virus types 1 and 2: isolation of herpes simplex virus type 2 from the buffy coat of two adults with meningitis. J. Infect. Dis. 127 (1973) 365–372.

CUADRADO, L.M., A. GUERRERO, A.L.G. ASENJO, F. MARTIN, E. PALAU and D.G. URRA: Cerebral mucormycosis in two cases of acquired immunodeficiency syndrome. Arch. Neurol. 45 (1988) 109–111.

CURLESS, R.G. and C.D. MITCHELL: Central nervous system tuberculosis in children. Pediatr. Neurol. 7 (1991) 270–274.

CURLESS, R.G., G.B. SCOTT, M. J. POST and J. B. GREGORIOS: Progressive cytomegalovirus encephalopathy following congenital infection in an infant with acquired immunodeficiency syndrome. Child. Nerv. Sys. 3 (1987) 255–257.

D'OLIVEIRA, J.J.G.: Cerebrospinal fluid concentrations of rifampin in meningeal tuberculosis. Am. Rev. Resp. Dis. 106 (1972) 432–437.

DAHAN, P., B. HAETTICH, J.M. LE PARC and J.B. PAOLOGGI: Meningoradiculitis due to herpes simplex virus disclosing HIV infection (Letter). Ann. Rheumat. Dis. 47 (1988) 440.

DAIF, A., T. OBEID, B. YAQUB and M. ABDULJABBAR: Unusual presentation of tuberculous meningitis. Clin. Neurol. Neurosurg. 94 (1992) 1–5.

DANIEL, T.D.: New approaches to the rapid diagnosis of tuberculous meningitis. J. Infect. Dis. (1987) 599–607.

DANNEMANN, B., A. MCCUTCHAN, I. ISRAELSKI, D. ANTENISKIS, C. LEPORT, B. LUFT, J. NUSSBAUM, N. CLUMECK, P. MORLAT, J. CHIU, J.-L. VILDE, M. ORELLANA, D. FEIGAL, A. BARTOK, P. HESELTINE, J. LEEDOM, J. REMINGTON and THE CALIFORNIA COLLABORATIVE TREATMENT GROUP: Treatment of toxoplasmic encephalitis in patients with AIDS. A randomized trial comparing pyrimethamine plus clindamycin to pyrimethamine plus sulfadiazine. Ann. Intern. Med. 116 (1992) 33–43.

DARAS, M., B.S. KOPPEL, L. SAMKOFF and J. MARC: Brainstem toxoplasmosis in patients with acquired immunodeficiency syndrome. J. Neuroimag. 4 (1994) 85–90.

DARMSTADT, G.L. and J.P. HARRIS: Luetic hearing loss: clinical presentation, diagnosis, and treatment. Am. J. Otolaryngol. 10 (1989) 410–423

DASTUR, D.K. and P.M. UDANI: The pathology and pathogenesis of tuberculous encephalopathy. Acta Neuropathol. (Berlin) 6 (1966) 311–326.

DASTUR, D.K., D.K. MANGHANI and P.M. UDANI: Pathology and pathogenetic mechanisms in neurotuberculosis. Radiolog. Clin. N. Am. 33 (1995) 733–752.

DAVIS, L.E. and J.W. SCHMITT: Clinical significance of cerebrospinal fluid tests for neurosyphilis. Ann. Neurol. 25 (1989) 50–55.

DAVIS, L.E. and S. SPERRY: The CSF-FTA test and the significance of blood contamination. Ann. Neurol. 6 (1979) 68–69.

DAVIS, L.E., K.R. RASTOGI, L.C. LAMBERT and B.J. SKIPPER: Tuberculous meningitis in the southwest United States: a community-based study. Neurology 43 (1993) 1775–1778.

DAWSON, D.M.: Progressive multifocal leukoencephalopathy in myasthenia gravis. Ann. Neurol. 11 (1982) 218–219.

DEACON, W.E., J.B. LUCAS and E.V. PRICE: Fluorescent treponemal pallidum antibody absorption (FTA-ABS) test for syphilis. J. Am. Med. Assoc. 198 (1966) 624–628.

DEANGELIS, L., S. WEAVER and M. ROSENBLUM: Herpes zoster (HVZ) encephalitis in immunocompromised patients (Abstr.). Neurology 44 (1994) A332.

DECKER, C.F., G.L. SIMON, R.A. DIGIOIA and C.U. TUAZON: Listeria monocytogenes infections in patients with AIDS: report of five cases and review. Rev. Infect. Dis. 13 (1991) 413–417.

DECKER, C.F., J.H. TARVER III, D.F. MURRAY and G.J. MARTIN: Prolonged concurrent use of ganciclovir and foscarnet in the treatment of polyradiculopathy due to cytomegalovirus in a patient with AIDS (Letter). Clin. Infect. Dis. 19 (1994) 548–549.

DE GANS, J., J.K.M. EEFTINCK SCHATTENKERK and R.J. VAN KETAL: Itraconazole as maintenance treatment for cryptococcal meningitis in the acquired immune deficiency syndrome. Br. Med. J. 296 (1988) 339.

DE GANS, J., P. PORTEGIES, G. TIESSENS, D. TROOST, S.A. DANNER and J.M.A. LANGE: Therapy for cytomegalovirus polyradiculomyelitis in patients with AIDS: treatment with ganciclovir. AIDS 4 (1990) 421–425.

DE GANS, J., P. PORTEGIES, G. TIESSENS, J.K. EEFTINCK SCHATTENKERK, C.J. VAN BOXTEL, R.J. VAN KETEL and J. STAM: Itraconazole compared with amphotericin B plus flucytosine in AIDS patients with cryptococcal meningitis. AIDS 6 (1992a) 185–190.

DE GANS, J., P. PORTEGIES, P. REISS, D. TROOST, T. VAN GOOL and J.M. LANGE: Pyrimethamine alone as maintenance therapy for central nervous system toxoplasmosis in 38 patients with AIDS. J. Acquir. Immune Defic. Syndr. 5 (1992b) 137–142.

DEL CASTILLO, M., G. MENDOZA, J. OVIEDO, R.P.P. BIANCO, A.E. ANSELMO and M. SILVA: AIDS and Chagas' disease with central nervous system tumor-like lesion. Am. J. Med. 88 (1990) 693–694.

DELAGE, G. and M. DUSSEAULT: Tuberculous meningitis in children: a retrospective study of 79 patients, with an analysis of prognostic factors. J. Can. Med. Assoc. 120 (1979) 205–209.

DE LALLA, F., G. PELLIZZER, A. VAGLIA, V. MANFRIN, M. FRANZETTI, P. FABRIS and C. STECCA: Amphotericin B as primary therapy for cryptococcosis in patients with AIDS: reliability of relatively high doses administered over a relatively short period. Clin. Infect. Dis. 20 (1995) 263–266.

DELANEY, P.: False positive serology in cerebrospinal fluid associated with spinal cord tumor. Neurology 26 (1976) 591–593.

DENNING, D.W., R.M. TUCKER, L.H. HANSON, J.R. HAMILTON and D.A. STEVENS: Itraconazole therapy for cryptococcal meningitis and cryptococcosis. Arch. Intern. Med. 149 (1989) 2301–2308.

DENNING, D.W., R.W. ARMSTRONG, B.H. LEWIS and D.A. STEVENS: Elevated cerebrospinal fluid pressures in patients with cryptococcal meningitis and acquired immunodeficiency syndrome. Am. J. Med. 91 (1991) 267–272.

DEVINSKY, O., E.-S. CHO, C.K. PETITO and R.W. PRICE: Herpes zoster myelitis. Brain 114 (1991) 1181–1196.

DEWAY, R.B., JR. and J. JANKOVIC: Hemiballism – hemichorea: clinical and pharmacologic findings in 21 patients. Arch. Neurol. 46 (1989) 862–867.

DEWHURST, K.: The composition of cerebrospinal fluid in the neurosyphilitic psychoses. Acta Neurol. Scand. 45 (1969) 119–123.

DI GREGORIO, C., F. RIVASI, N. MONGIARDO, B. DE RIENZO, S. WALLACE and G.S. VISVESVARA: Acanthamoeba meningoencephalitis in a patient with acquired immunodeficiency syndrome. Arch. Pathol. Lab. Med. 116 (1992) 1363–1365.

DIETERICH, D.T., A. CHACHOUA, F. LAFLEUR and C.WORREL: Ganciclovir treatment of gastrointestinal infections caused by cytomegalovirus in patients with AIDS. Rev. Infect. Dis. 10 (1988) S532–S537.

DIETERICH, D.T., M.A. POLES, E.A. LEW, P.E. MENDEZ, R. MURPHY, A. ADDESSI, J.T. HOLBROOK, K. NAUGHTON and D.N. FRIEDBERG: Concurrent use of ganciclovir and foscarnet to treat cytomegalovirus infection in AIDS patients. J. Infect. Dis. 167 (1993) 1184–1188.

DISMUKES, W.E.: Cryptococcal meningitis in patients with AIDS. J. Infect. Dis. 157 (1988) 624–628.

DISMUKES, W.E.: Cryptococcosis. In: J.B. Wyngaarden, L.H. Smith, Jr. and J.C. Bennett (Eds.), Cecil Text Book of Medicine, 19th Edit. Philadelphia, PA, W.B. Saunders (1992a) 1894–1897.

DISMUKES, W.E.: Histoplasmosis. In: J.B. Wyngaarden, L.H. Smith, Jr. and J.C. Bennett (Eds.), Cecil Text Book of Medicine, 19th Edit. Philadelphia, PA, W.B. Saunders (1992b) 1887–1890.

DISMUKES, W.E.: Sporotrichosis. In: J.B. Wyngaarden, L.H. Smith, Jr. and J.C. Bennett (Eds.), Cecil Text Book of Medicine, 19th Edit. Philadelphia, PA, W.B. Saunders (1992c) 1897–1898.

DIX, R.D., D.E. BREDSEN, K.S. ERLICH and J. MILLS: Recovery of herpesviruses from cerebrospinal fluid of immunodeficient homosexual men. Ann. Neurol. 18 (1985a) 611–614.

DIX, R.D., D.M. WAITZMAN, S. FOLLANSBEE, B.S. PEARSON, T. MENDELSON, P. SMITH, R.L. DAVIS and J. MILLS: Herpes simplex virus type 2 encephalitis in two homosexual men with persistant lymphadenopathy. Ann. Neurol. 17 (1985b) 203–206.

DOLL, D.C., J.W. YARBRO, S.K. PHILLIP and C. KLOTT: Mycobacterial spinal cord abscess with an ascending polyneuropathy (Letter). Ann. Intern. Med. 106 (1987) 333–334.

DONABEDIAN, H., E. O'DONNELL, C. OLSZEWSKI, R.D. MACARTHUR and N. BUDD: Disseminated cutaneous and meningial sporotrichosis in an AIDS patient. Diagn. Microbiol. Infect. Dis. 18 (1994) 111–115.

DONALD, P.R., T.C. VICTOR, A.M. JORDAAN, J.F. SCHOEMAN and P.D. VAN HELDEN: Polymerase chain reaction in the diagnosis of tuberculous meningitis. Scand. J. Infect. Dis. 25 (1993) 613–617.

DONNER, M. and O. WASZ-HOCKERT: Late neurologic sequelae of tuberculous meningitis. Acta Paediatr. 51 (1962) 34–422.

DREW, W.L.: Cytomegalovirus infection in patients with AIDS. J. Infect. Dis. 158 (1988) 449–456.

DREW, W.L., L. MINTZ, R.C. MINER, M. SANDS and B. KETTERER: Prevalence of cytomegalovirus infection in homosexual men. J. Infect. Dis. 143 (1981) 188–192.

DUBOIS, V., M.E. LAFON, J.M. RAGNAUD, J.L. PELLEGRIN, F. DAMASIAO, C. BAUDOUIN, H. MICHAUD and J.A. FLEURY: Detection of JC virus DNA in the peripheral blood leukocytes of HIV-infected patients. AIDS 10 (1996) 353–358.

DUNCAN, W.C.: Failure of erythromycin to cure secondary syphilis in a patient with the human immunodeficiency virus. Arch. Dermatol. 125 (1989) 82–84.

DUPOUY-CAMET, J., S. LAVAREDA DE SOUZA, C. MASLO, A. PAUGAM, A.G. SAIMOT, R. BENAROUS R, C. TOURTE-SCHAEFFER and F. DEROUIN: Detection of *toxoplasma gondii* in venous blood from AIDS patients by polymerase chain reaction. J. Clin. Microbiol. 31 (1993) 1866–1869.

DUTCHER, J.P., S.L. MARCUS, H.B. TANOWITZ, M. WITTNER, J.Z. FUKS and P.H. WIERNIK: Disseminated strongyloidiasis with central nervous system involvement diagnosed antemortem in a patient with acquired immunodeficiency syndrome and Burkitts lymphoma. Cancer 66 (1990) 2417–2420.

DYER, J.R., M.A.H. FRENCH and S.A. MALLAL: Cerebral mass lesions due to cytomegalovirus in patients with AIDS: report of two cases. J. Infect. 30 (1995) 147–151.

EDMOND, R.T.D. and G.S.W. MCKENDRICK: Tuberculosis as a cause of transient aseptic meningitis. Lancet ii (1973) 234–236.

EDWARDS, R.H., R. MESSING and R.R. MCKENDALL: Cytomegalovirus meningoencephalitis in a homosexual man with Kaposi's sarcoma: isolation of CMV from CSF cells. Neurology 35 (1985) 560–562.

EGGERS, C., A. VORTMEYER and T. EMSKOTTER: Cerebral toxoplasmosis in a patient with the acquired immunodeficiency syndrome presenting as obstructive hydrocephalus. Clin. Neuropathol. 14 (1995) 51–54.

EHNI, W.F. and R.T. ELLISON: Spontaneous *Candida albicans* meningitis in a patient with the acquired immune deficiency syndrome (Letter). Am. J. Med. 83 (1987) 806–807.

EIDE, F.F., A.D. GEAN and Y.T. SO: Clinical and radiographic findings in disseminated tuberculosis of the brain. Neurology 43 (1993) 1427–1429.

EIDELBERG, D., A. SOTREL, H. VOGEL, P. WALKER, J. KLEEFIELD and C.S. CRUMPACKER, III: Progressive polyradiculopathy in acquired immune deficiency syndrome. Neurology 36 (1986) 912–916.

EIDELBERG, O., A. SOTREL, D.S. HOROUPIAN, P.E. NEUMANN, T. PUMAROLA-SUNE and R.W. PRICE: Thrombotic cerebral vasculopathy associated with herpes zoster. Ann. Neurol. 19 (1986) 7–14.

EMBRY, J.R., F.G. SILVA, J.H. HELDERMAN, P.C. PETERS and A.I. SAGALOWSKY: Long term survival and late development of bladder cancer in renal transplant patient with progressive multifocal leukoencephalopathy. J. Urol. 139 (1988) 580–581.

ENG, R.H.K., E. BISHBURG, S.M. SMITH and R. KAPILA: Cryptococcal infections in patients with acquired immune deficiency syndrome. Am. J. Med. 81 (1986) 19–23.

ENGLUND, J.A., M.E. ZIMMERMAN, E.M. SWIERKOSZ, J.L. GOODMAN, D.R. SCHOLL and H.H. BALFOUR, JR.: Herpes simplex virus resistance to acyclovir. Ann. Intern. Med. 112 (1990) 416–422.

ENTING, R., J. DE GANS, P. REISS, C. JANSEN and P. PORTEGIES: Ganciclovir/foscarnet for cytomegalovirus meningoencephalitis in AIDS (Letter). Lancet 340 (1992) 559–560.

ERICE, A., S. CHOU, K.K. BIRON, S.C. STANAT, H.H. BALFOUR and M.C. JORDAN: Progressive disease due to ganciclovir resistant cytomegalovirus in immunocompromised patients. N. Engl. J. Med. 320 (1989) 289–293.

ERLICH, K.S., J. MILLS, P. CHATIS, G.J.MERTZ, D.F. BUSCH, S.E. FOLLANSBEE, R.M. GRANT and C.S.CRUMPACKER: Acyclovir-resistant herpes simplex virus infections in patients with the acquired immunodeficiency syndrome. N. Engl. J. Med. 320 (1989) 293–296.

ESCOBAR, M.R., H.P. DALTON and M.J. ALLISON: Fluorescent antibody tests for syphilis using cerebrospinal fluid; clinical correlation in 150 cases. Am. J. Clin. Pathol. 53 (1970) 886–890.

FARKESH, A.E., P.J. MACCABEE, J.H. SHER, S.H. LANDESMAN and G. HOTSON: CNS toxoplasmosis in acquired immune deficiency syndrome. J. Neurol. Neurosurg. Psychiatry 49 (1986) 744–748.

FEHLINGS, M.G. and M. BERNSTEIN: Syringomyelia as a communication of tuberculous meningitis. Can. J. Neurol. Sci. 19 (1982) 84–87

FERARU, E. and H. ARONOW: Neurosyphilis in AIDS patients: initial CSF VDRL may be negative. Neurology 40 (1990) 541–543.

FERNANDEAZ-GUERRERO, M.L., C. MIRANDA, C. CENJOR and F. SANABRIA: The treatment of neurosyphilis in patients with HIV infection (Letter). J. Am. Med. Assoc. 259 (1988) 1495–1496.

FERNANDEZ-MARTIN, J., C. LEPORT, P. MORLAT, M.C. MEYOHAS, J.P. CHAUVIN and J.L. VILDE: Pyrimethamine-clarithromycin combination for therapy of acute toxoplasmic encephalitis in patients with AIDS. Antimicrob. Agents Chemother. 35 (1991) 2049–2052.

FERREIRA, M.S., S.D.A. NISHIOKA, A. ROCHA, A.M. SILVA, R.G. FERREIRA, W. OLIVIER and S. TOSTES, JR.: Acute fatal *Trypanosoma cruzi* meningoencephalitis in a human immunodeficiency virus-positive hemophiliac patient. Am. J. Trop. Med. Hyg. 45 (1991) 723–727.

FIALA, M., L.A.CONE, N. COHEN, D. PATEL, K. WILLIAMS, D. CASAREALE, P. SHAPSHAK and W.W.TOURTELLOTTE:

Responses of neurologic complications of AIDS to 3′-azido-3′deoxythymidine and 9-(1,3-dihydroxy-2-propoxymethyl) guanine. I: clinical features. Rev. Infect. Dis. 10 (1988) 250–256.

FISH, D.G., N.M. AMPEL, J.N. GALGIANI, C.L. DOLS, P.C. KELLY, C.H. JOHNSON, D. PAPPAGIANIS, J.E. EDWARDS, R.B. WASSERMAN, R.J. CLARK, D. ANTONISKIS, R.A. LARSEN, S.J. ENGLENDER and E.A. PETERSEN: Coccidioidomycosis during human immunodeficiency virus infection, a review of 77 patients. Medicine (Baltimore) 69 (1990) 384–391.

FLEMING, W.L.: Syphilis through the ages. Med. Clin. N. Am. 48 (1964) 587–612.

FOLGUEIRA, L., R. DELGADO, E. PALENQUE and A.R. NORIEGA: Polymerase chain reaction for rapid diagnosis of tuberculous meningitis in AIDS patients. Neurology 44 (1994) 1336–1338.

FORGAN-SMITH, R., G.A. ELLARD, D. NEWTON and D.A. MITCHISON: Pyrazinamide and other drugs in tuberculous meningitis. Lancet ii (1973) 374.

FRAIMOW, H.S., G.P. WORMSER, K.D. COBURN and C.B. SMALL: Salmonella meningitis and infection with HIV. AIDS 4 (1990) 1271–1273.

FRASER, V.J., E.J. KEATH and W.G. POWDERLY: Two cases of blastomycosis from a common source: use of DNA restriction analysis to identify strains. J. Infect. Dis. 163 (1991) 1378–1381.

FREEMAN, W.R., C.W. LERNER, J.A. MINES ET AL: A prospective study of the ophthalmologic findings in the acquired immune deficiency syndrome. Am. J. Ophthalmol. 97 (1984) 133–142.

FRENCH, G.L., R. TEOH, C.Y. CHAN, M.J. HUMPHRIES, S.W. CHEUNG and G. O'MAHONY: Diagnosis of tuberculous meningitis by detection of tuberculostearic acid in cerebrospinal fluid. Lancet ii (1987) 117–119.

FRIEDLAND, G.H. and R.S. KLEIN: Transmission of the human immunodeficiency virus. Morb. Mortal. Wkly. Rep. 317 (1987) 1125–1134.

FRIEDMAN, D.I.: Neuro-ophthalmic manifestions of human immunodeficiency virus infection. Neurol. Clin. N. Am. 9 (1991) 55–72.

FULLER, G.N., R.J. GUILOFF, F. SCARAVILLI and J.N. HARCOURT-WEBSTER: Combined HIV-CMV encephalitis presenting with brainstem signs. J. Neurol. Neurosurg. Psychiatry 52 (1989) 975–979.

GAL, A., A. POLLACK and M. OLIVER: Ocular findings in the acquired immunodeficiency syndrome. Br. J. Ophthalmol. 68 (1984) 238–241.

GALGIANI, J.N., A. CATANZARO, G.A. CLOUD, J. HIGGS, B.A. FRIEDMAN, R.A. LARSEN, J.R. GRAYBILL and THE NIAID-MYCOSES STUDY GROUP: Fluconazole therapy for coccidioidal meningitis. Ann. Intern. Med. 119 (1993) 28–35.

GARDNER, H.A.R., A.J. MARTINEZ, G.S. VISVESVARA and A. SOTREL: Granulomatous amebic encephalitis in an AIDS patient. Neurology 41 (1991) 1993–1995.

GARLAND, H.G. and G. ARMITAGE: Intracranial tuber-culoma. J. Pathol. Bacteriol. 37 (1933) 461–471.

GARRITY, J.A., D.C. HERMAN, R. IMES, P. FRIES, C.F. HUGHES and R.J. CAMPBELL: Optic nerve sheath decompression for visual loss in patients with acquired immuno-deficiency syndrome and cryptococcal meningitis with papilledema. Am. J. Ophthalmol. 116 (1993) 472–478.

GATELEY, A., R.M. GANDER, P.C. JOHNSON, S. KIT, H. OTSUKA and S. KOHL: Herpes simplex virus type 2 meningo-encephalitis resistant to acyclovir in a patient with AIDS. J. Infect. Dis. 161 (1990) 711–715.

GEE, G.T., C. BAZAN, III and J.R. JINKINS: Miliary tuberculosis involving the brain: MR findings. Am. J. Roentgenol. 159 (1992) 1075–1076.

GELB, L.D.: Varicella zoster virus: clinical aspects. In: B. Roizman B, R.J. Whitley and C. Lopez (Eds.), The Human Herpesviruses. New York, Raven Press (1993) 281–308.

GHERARDI, R., M. BAUDRIMONT, F. LIONNET, J-M. SALORD, C. DUVIVIER, C. MICHON, M. WOLFF and C. MARCHE: Skeletal muscle toxoplasmosis in patients with acquired immunodeficiency syndrome: A clinical and patho-logical study. Ann. Neurol. 32 (1992) 535–542.

GIBSON, P.E., A.M. FIELD, S.D. GARDNER ET AL.: Occurrence of IgM antibodies against BK and JC polyomaviruses during pregnancy. J. Clin. Pathol. 34 (1981) 674–679.

GIBSON, P.E., W.A. KNOWLES, J.F. HAND and D.W.G. BROWN: Detection of JC Virus DNA in the cerebrospinal fluid of patients with progressive multifocal leuko-encephalopathy. J. Med. Virol. 39 (1993) 278–281.

GILDEN, D.H., R.S. MURRAY, M. WELLISH, B.K. KLEINSCHMIDT-DEMASTERS and A. VAFAI: Chronic progressive varicella-zoster virus encephalitis in an AIDS patient. Neurology 38 (1988) 1150–1153.

GILDEN, D.H., H. DEVLIN, M. WELLISH, R. MAHALINGBHAM, C. HUFF, A. HAYWARD and A. VAFAI: Persistence of varicella-zoster virus DNA in blood mononuclear cells of patients with varicella or zoster. Virus Genes 2 (1989) 299–305.

GILDEN, D.H., B.R. BEFINLICH, E.M. RUBINSTIEN, E. STOMMEL, R. SWENSON, D. RUBINSTEIN and R. MAHALINGHAM: Varicella-zoster virus myelitis: an expanding spec-trum. Neurology 44 (1994) 1818–1823.

GILLESPIE, S.M., Y. CHANG, G. LEMP ET AL.: Progressive multifocal leukoencephalopathy in persons infected with human immunodeficiency virus, San Francisco, 1981–1989. Ann. Neurol. 30 (1991) 597–604.

GIRARD, P-M., R. LANDMAN, C. GAUDEBOUT, R. OLIVARES, A.G. SAIMOT, P. JELAZKO, C. GAUDEBOUT, A. CERTAIN, F. BOVE, E. BOUVET, T. LECOMPTE, J-P. COULAUD and THE PRIO STUDY GROUP: Dapsone-pyrimethamine compared with aerolized pentamidine as primary prophylaxis against pneumocystis carinii preumonia and toxo-plasmosis in HIV infection. N. Engl. J. Med. 328 (1993) 1514–1520.

GLASER, J.B., L. MORTON-KUTE, S.R. BERGER ET AL.: Recurrent *Salmonella typhimurium* bacteremia associated with the acquired immunodeficiency syndrome. Ann. Intern. Med. 102 (1985) 189–193.

GLUCKSTEIN, D., F. CIFERRI and J. RUSKIN: Chagas's disease: another cause of cerebral mass in the acquired immunodeficiency syndrome. Am. J. Med. 92 (1992) 429–432.

GO, B.M., D.J. ZIRING and D.S. KOUNTZ: Spinal epidural abscess due to *aspergillus sp.* in a patient with acquired immunodeficiency syndrome. South. Med. J. 86 (1993) 957–960.

GOLNIK, K.C., S.A. NEWMAN and B. WISPELWAY: Crypto-coccal optic neuropathy in the acquired immune deficiency syndrome. J. Clin. Neuro-Ophthalmol. 11 (1991) 96–103

GOMEZ-TORTOSA, E., I. GADEA, M.I. GEGUNDEZ, A. ESTEBAN, J. RABANO, L. FERNANDEZ-GUERRERO and F. SORIANO: Development of myelopathy before herpes zoster rash in a patient with AIDS. Clin. Infect. Dis. 18 (1994) 810–812.

GONZALES, G.R., S. HERSKOVITZ S, M. ROSENBLUM, K.M. FOLEY, R. KANNER, A. BROWN and R.K. PORTENOY: Central pain from cerebral abscess: thalamic syndrome in AIDS patients with toxoplasmosis. Neurology 42 (1992) 1107–1109.

GORDON, S.M. and H.M. BLUMBERG: *Mycobacterium kansasii* brain abscess in a patient with AIDS. Clin. Infect. Dis. 14 (1992) 789–790.

GORDON, S.M., J.P. STEINBERG, M.H. DU PUIS, P.E. KOZARSKY, J.F. NICKERSON and G.S. VISVESVARA: Culture isolation of acanthamoeba species and leptomyxid amebas from patients with amebic meningoencephalitis, including two patients with AIDS. Clin. Infect. Dis. 15 (1992) 1024–1030.

GOTTLIEB, M.S., R. SCHROFF, H.M. SCHRANKER ET AL.: *Pneumocystis carinii* pneumonia and mucosal candidiasis in previously healthy homosexual men. N. Engl. J. Med. 305 (1981) 1425–1431.

GOULD, I.A., L.C. BELOK and S. HANDWERGER: Listeria mono-cytogenes: a rare cause of opportunisitic infection in the acquired immunodeficiency sydnrome (AIDS) and a new cause of meningitis in AIDS. A case report. AIDS Res. 2 (1986) 231–234.

GOZLAN, J., J-M. SALORD, E. ROULLET, M. BAUDRIMONT, F. CABURET, O. PICARD, M-C. MEYOHAS, C. DUVIVIER, C. JACOMET and J-C. PETIT: Rapid detection of cyto-megalovirus DNA in cerebrospinal fluid of AIDS patients with neurologic disorders. J. Infect. Dis. 166 (1992) 1416–1421.

GOZLAN, J., M.E. AMRANI, M. BAUDRIMONT, D. COSTAGLIOLA, J.M. SALORD, C. DUVIVIER, O. PICARD, M-C. MEYOHAS, C. JACOMET, V. SCHNEIDER-FAUVEAU, J-C. PETIT and E. POULLET: A prospective evaluation of clinical criteria and polymerase chain reaction assay of cerebro-spinal fluid for the diagnosis of cytomegalovirus related neurologic diseases during AIDS. AIDS 9 (1995) 253–260.

GRAFE, M.R. and C.A. WILEY: Spinal cord and peripheral nerve pathology in AIDS: the roles of cytomegalovirus and human immunodeficiency virus. Ann. Neurol. 25 (1989) 561–566.

GRAFE, M.R., G.A. PRESS, D.P. BERTHOTY, J.R. HESSELINK and C.A. WILEY: Abnormalities of the brain in AIDS patients: correlation of post mortem MR findings with neuropathology. Am. J. Roentgenol. 11 (1990) 905–911.

GRAVELEAU, P., R. PEROL and A. CHAPMAN: Regression of cauda equina syndrome in AIDS patient being treated with ganciclovir (Letter). Lancet ii (1989) 511–512.

GRAY, F., R. GHERARDI, C. KEOHANE, M. FAVOLINI, A. SOBEL and J. POIRIER: Pathology of the central nervous system in 40 cases of acquired immune deficiency syndrome (AIDS). Neuropathol. Appl. Microbiol. 14 (1988) 365–380.

GRAY, F., R. GHERARDI, E. WINGATE, J. WINGATE, G. FENELO, A. GASTON, A. SOBEL and J. POIRIER: Diffuse 'encephalitic' cerebral toxoplasmosis in AIDS. J. Neurol. 236 (1989) 273–277.

GRAY, F., M. MOHR, F. ROZENBERG, L. BELEC, M.C LESCS, E. DOURNON, E. SINCLAIR and F. SCARAVILLI: Varicella-zoster virus encephalitis in acquired immunodeficiency syndrome: report of four cases. Neuropathol. Appl. Neurobiol. 18 (1992) 502–514.

GRAY, F., L. BELEC, M.C. LESCS, F. CHRETIEN, A. CIARDI, D. HASSINE, M. FLAMENT-SAILLOUR M, P. DE TRUCHIS, B. CLAIR and F. SCARAVILLI: Varicella-zoster virus infection of the central nervous system in the acquired immune deficiency syndrome. Brain 117 (1994) 987–999.

GREENBLATT, R..M., S.A. LUKEHART, F.A. PLUMMER ET AL.: Genital ulceration as a risk factor for human immunodeficiency virus infection. AIDS 2 (1988) 47–50.

GRIFFIN, D.E., J.C. MCARTHUR and D.R. CORNBLATH: Neopterin and interferon-gamma in serum and cerebrospinal fluid of patients with HIV associated neurologic disease. Neurology 41 (1991) 69–74.

GROSSNIK-LAUS, H.E., E. FRANK and R.L. TOMSAK: Cytomegalovirus retinitis and optic neuritis in acquired immune deficiency syndrome. Opthalmology 94 (1987) 1601–1604.

GUINAN, M.E., P.A. THOMAS, P.F. PINSKY ET AL.: Heterosexual and homosexual patients with the acquired immunodeficiency syndrome: a comparison of surveillance, interview, and laboratory data. Ann. Intern. Med. 100 (1984) 213–218.

GUNGOR, T., M. FUNK, R. LINDE, G. JACOBI, M. HORN and W. KREUZ: Cytomegalovirus myelitis in perinatally acquired HIV. Arch. Dis. Child. 68 (1993) 399–401.

HAAS, E.J., T. MADHAVAN, E.L. QUINN, F. COX, E. FISHER and K. BURCH: Tuberculous meningitis in an urban general hospital. Arch. Intern. Med. 137 (1977) 1518–1521.

HAGBERG, L., B. PALMERTZ and J. LINDBERG: Doxycycline and pyrimethamine for toxoplasmic encephalitis. Scand. J. Infect. Dis. 25 (1993) 157–160.

HALL, C., D. DAFNI, D. SIMPSON ET AL.: Failure of cytarabine in progressive multifocal leukoencephalopathy associated with human immunodeficiency virus infection. N. Engl. J. Med. 338 (1998) 1345–1351.

HALL, P.J. and J.B. FARMER: Aspergillus mastoiditis. Otolaryngol. Head Neck Surg. 108 (1993) 167–170.

HALLERVORDEN, J.: Eigenartige und nicht rubrizierbare Prozesse. In: O. Bumke (Ed.), Handbuch der Geisteskrankheiten, Vol. 2. Die Anatomie der Psychosen. Berlin, Springer (1930) 1063–1107.

HAMILTON, R.L., C. ACHIM, M.R. GRAFE, J.C. FREMONT, D. MINERS and C.A. WILEY: Herpes simplex virus brainstem encephalitis in an AIDS patient. Clin. Neuropathol. 14 (1995) 45–50.

HARCOURT-WEBSTER, J.N., F. SCARAVILLI and A.H. DARWISH: Strongyloides stercoralis hyperinfection in an HIV positive patient. J. Clin. Pathol. 44 (1991) 346–348.

HARDING, C.V.: Blastomycosis and opportunistic infections in patients with acquired immunodeficiency syndrome: an autopsy study. Arch. Pathol. Lab. Med. 115 (1991) 1133–1136.

HARMON, W.E.: Opportunistic infections in children following renal transplantation. Pediatr. Nephrol. 5 (1991) 118–125.

HARNER, R.E., J.L. SMITH and C.W. ISRAEL: The FTA-ABS test in late syphilis. A serological study in 1985 cases. J. Am. Med. Assoc. 203 (1968) 545–548.

HARRIMAN, D.G.F.: Bacterial infections of the central nervous system. In: W. Blackwood and J.A.N. Corsellis (Eds.) Greenfield's Neuropathology. Chicago, IL, Yearbook Medical (1976) 238–268.

HARRIS, J.O., J. MARQUEZ, M.A. SWERDLOFF and I.A. MAGANA: Listeria brain abscess in the acquired immunodeficiency syndrome (Letter). Arch. Neurol. 46 (1989) 250.

HARRIS, T.M., R.R. SMITH, J.R. BOGNANNO and M.K. EDWARDS: Toxoplasmic myelitis in AIDS: gadolinium-enhanced MR. J. Comp. Assist. Tomogr. 14 (1990) 809–811.

HARVEY, R.L. and P.H. CHANDRASEKAR: Chronic meningitis caused by Listeria in a patient infected with human immunodeficiency virus (Letter). J. Infect. Dis. 157 (1988) 1091–1092.

HAWKINS, C.H., J.W.M. GOLD, E. WHIMBEY ET AL.: *Mycobacterium avium* complex infections in patients with the acquired immunodeficiency syndrome. Ann. Intern. Med. 105 (1986) 184.

HAWLEY, D.A., J.F. SCHAEFER, D.M. SCHULZ and J. MULLER: Cytomegalovirus encephalitis in acquired immunodeficiency syndrome. Am. J. Clin. Pathol. 80 (1983) 874–877.

HEINIC, G. S., D.A. STEVENS, D. GREENSPAN, L.A. MACPHAIL, C.L. DODD, S. STRINGARI, S.W. STRULL and H. HOLLANDER:

Fluconazole-resistant Candida in AIDS patients. Oral Surg. Oral Med. Oral Pathol. 76 (1993) 711–715.

HENOCHOWICZ, S., M. MUSTAFA, W.E. LAWRINSON, M. PISTOLE, J. LINDSAY, JR.: Cardiac aspergillosis in acquired immune deficiency syndrome. Am. J. Cardiol. 55 (1985) 1239–1240.

HENSON, J., M. ROSENBLUM and H. FURNEAUX: A potential diagnostic test for PML: PCR analysis of JC Virus DNA. Neurology. 41 (1991a) 338.

HENSON, J., M. ROSENBLUM, R. ARMSTRONG and H. FURNEAUX: Amplication of JC virus DNA from brain and cerebrospinal fluid of patients with progressive multifocal leukoencephalopathy. Neurology 41 (1991b) 11967–11971.

HERSKOVITZ, S., S.E. SIEGEL, A.T. SCHNEIDER, S.J. NELSON, J.T. GOODRICH and G. LANTOS: Spinal cord toxoplasmosis in AIDS. Neurology 39 (1989) 1552–1553.

HICKS, C.N., P.M. BENSON, G.P. LUPTON and E.C. TRAMONT: Seronegative secondary syphilis in a patient infected with human immunodeficiency virus with Kaposi sarcoma. A diagnostic dilemma. Ann. Intern. Med. 107 (1987) 4992–495.

HINMAN, A.R.: Tuberculous meningitis at Cleveland Metropolitan General Hospital. Am. Rev. Resp. Dis. 95 (1967) 670–673.

HOFFLIN, M. and J.S. REMINGTON: Tissue culture isolation of toxoplasma from blood of a patient with AIDS. Arch. Intern. Med. 145 (1985) 925–926.

HOLDER, C.K. and D. HALKIAS: Case report: relapsing, bacteremic Klebsiella pneumoniae meningitis in an AIDS patient. Am. J. Med. Sci. 295 (1988) 55–59.

HOLLAND, N.R., C. POWER, V.P. MATTHEWS, J.D. GLASS, M. FORMAN and J.C. MCARTHUR: Cytomegalovirus encephalitis in acquired immunodeficiency syndrome (AIDS). Neurology 44 (1994) 507–514.

HOLLANDER, H.: Cerebrospinal fluid normalities and abnormalities in individuals infected with human immunodeficiency virus. J. Infect. Dis. 158 (1988) 855–858.

HOLMAN, R.C., R.S. JANSSEN, J.W. BUEHLER, M.T. ZELASKY and W.C. HOOPER: Epidemiology of progressive multifocal leukoencephalopathy in the United States: analysis of national mortality and AIDS surveillance data. Neurology 41 (1991) 1733–1736.

HOLMBERG, S.D., J.A. STEWART, A.R. GERBER ET AL.: Prior herpes simplex virus type 2 infection as a risk factor for HIV infection. J. Am. Med. Assoc. 259 (1988) 1048–1050.

HOLTOM, P.D., R.A. LARSEN, M.E. LEAL and J.M. LEEDOM: Prevalence of neurosyphilis in human immunodeficiency virus-infected patients with latent syphilis. Am. J. Med. 93 (1992) 9–12.

HOLTZ, H.A., D.P. LAVERY and R. KAPILA: Actinomycetales infection in patients with acquired immunodeficiency syndrome. Ann. Intern. Med. 102 (1985) 203–205.

HOOSHMAND, H., M.R. ESCOBAR and S.W. KOPF: Neuro-

syphilis: a study of 241 patients. J. Am. Med. Assoc. 219 (1972) 726–729.

HOUFF, S.A., E.O. MAJOR, D. KATZ ET AL.: Involvement of JC virus-infected mononuclear cells from the bone marrow and spleen in the pathogenesis of progressive multifocal leukoencephalopathy. N. Engl. J. Med. 318 (1988) 301–305.

HOUFF, S.A., D. KATZ, C. KUFTA and E.O. MAJOR: A rapid method for in situ hybridization for viral DNA in brain biopsies from patients with acquired immunodeficiency syndrome (AIDS). AIDS 3 (1989) 843–845.

HSIEH, F.Y., L.G. CHIA and W.C. SHEN: Locations of cerebral infarctions in tuberculous meningitis. Neuroradiology 34 (1992) 197–199.

HUANG, P.P., A.A. MCMEEKING, M.J. STEMPIEN and D. ZAGZAG: Cytomegalovirus disease preventing as a focal brain mass: report of two cases. Neurosurgery 40 (1997) 1074–1078; Discussion 1078–1079.

HUNTER, C.A. and J.S. REMINGTON: Immunopathogenesis of toxoplasmic encephalitis. J. Infect. Dis. 170 (1994) 1057–1067.

IDEMYOR, V. and C.E. CHERUBIN: Pleurocerebral nocardia in a patient with human immunodeficiency virus. Ann. Pharmacother. 26 (1992) 188–189.

IDRISS, Z.H., A.A. SINNO and N.M. KRONFOL: Tuberculous meningitis in childhood. Am. J. Dis. Child. 130 (1976) 364–367.

ILLINGWORTH, R.S. and J. LORBER: Tubercles of the choroid. Arch. Dis. Child. 31 (1956) 467–469.

ILYID, J.P., J.E. OAKES, R.W. HYMAN and F. RAPP: Comparison of the DNAs of varicella-zoster viruses isolated from clinical cases of varicella and herpes zoster. Virology 82 (1977) 345–352.

IRIBARREN, J.A., M. GOENAGA, J. ARRIZABALAGA, F. RODRIGUEZ, M.A. VON WICHMANN, I. HUARTE and C. GARDE: Atovaquone in cerebral toxoplasmosis: preliminary experience Int. Conf. AIDS 10 (1994) 153 (Abstr. PB0626).

ISRAELSKI, D.M. and J.S. REMINGTON: Toxoplasmic encephalitis in patients with AIDS. Infect. Dis. Clin. N. Am. 2 (1988) 429–445.

IZZAT, N.N., J.K. BARTRUFF, J.M GLICKSMAN, W.R. HOLDER and J.M. KNOX: Validity of the VDRL test on cerebrospinal fluid contaminated by blood. Br. J. Vener. Dis. 47 (1971) 162–164.

JABS, D.A., W.R. GREEN, R. FOX, B.F. POLK and J.G. BARTLETT: Ocular manifestations of acquired immune deficiency syndrome. Ophthalmology 96 (1989) 1092–1099.

JACOB, C.N., S.S. HENEIN, A.I. HEURICH and S. KAMHOLZ: Nontuberculous mycobacterial infection of the central nervous system in patients with AIDS. South. Med. J. 86 (1993) 638–640.

JACOBS, J.L. and H.W. MURRAY: Why is *L. monocytogenes* not a pathogen in the acquired immunodeficiency sydrome? Arch. Intern. Med. 146 (1986) 1299–1300.

JACOBSEN, M.A, J. MILLS, J. RUSH, J.J. O'DONNELL, R.G. MILLER,

C. GRECO and M.F. GONZALES: Failure of antiviral therapy for acquired immunodeficiency syndrome related cytomegalovirus myelitis. Arch. Neurol. 45 (1988) 1090–1092.

JAFFE, H.W., S.A. LARSEN, M. PETERS ET AL.: Tests for treponemal antibody in CSF. Arch. Intern. Med. 138 (1978) 252–255.

JAIN, A.K., S. K. AGARWAL and W. EL-SADR: Streptococcus bovis bacteremia and meningitis associated with Strongyloides stercoralis colitis in a patient infected with human immunodeficiency virus. Clin. Infect. Dis. 18 (1994) 253–254.

JANSSEN, R.S., O.C. NWANYANWU, R.M. SELIK and J.K. STEHR-GREEN: Epidemiology of human immunodeficiency virus encephalopathy in the United States. Neurology 42 (1992) 1472–1476.

JARVIK, J.G., J.R. HESSELINK, C. WILEY, S. MERCER, B. ROBBINS and P. HIGGINBOTTOM: Coccidioidomycotic brain abscess in an HIV infected man. West. J. Med. 149 (1988) 83–86.

JENNINGS, T.S. and T.C. HARDIN: Treatment of aspergillosis with itraconazole. Ann. Pharmacol. 27 (1993) 1206–1211.

JOHANNESSEN, D.J. and L.G. WILSON: Mania with cryptococcal meningitis in two AIDS patients. J. Clin. Psychiatry 49 (1988) 200–201.

JOHNS, D.R., M. TIERNEY and D. FELSENSTEIN: Alteration in the natural history of neurosyphilis by concurrent infection with the human immunodeficiency virus. Morb. Mortal. Wkly. Rep. 316 (1987) 1569–1572.

JOHNSON, J.D., P.D. BUTCHER, D. SAVVA and R. E. HOLLIMAN: Application of the polymerase chain reaction to the diagnosis of human toxoplasmosis. J. Infect. 26 (1993) 147–158.

JOHNSON, P.C., N. KHARDORI, A.F. NAJJAR, F. BUTT, P.W.A. MANSELL and G.A. SAROSI: Progressive histoplasmosis in patients with acquired immunodeficiency syndrome. Am. J. Med. 85 (1988) 152–158.

JOHNSON, P.C., S.J. NORRIS, G.P. MILLER ET AL.: Early syphilitic hepatitis after renal transplantation. J. Infect. Dis. (1988).

JOYCE, P.W., K.R. HAYE and M.E. ELLIS: Syphilitic reinitis in a homosexual man with concurrent HIV infection: case report. Genitourin. Med. 65 (1989) 244–253.

KADIVAL, G.V., A.M. SAMUEL, T.B.M.S. MAZARELO and S.D. CHAPARAS: Radioimmunoassay for detecting Mycobacteria tuberculosis antigen in cerebrospinal fluids of patients with tuberculous meningitis. J. Infect. Dis. 155 (1987) 608–611.

KALAYJIAN, R.C., M.L. COHEN, R. BONOMO and T.P. FLANIGAN: Cytomegalovirus ventriculoencephalitis in AIDS. Medicine (Baltimore) 72 (1993) 67–77.

KALES, C.P. and R.S. HOLZMAN: Listeriosis in patients with HIV infection: clincal manifestations and response to therapy. J. Acquir. Immune Defic. Syndr. 3 (1990) 139–143.

KALISH, S.B., R.C. RADIN, D. LEVITZ, R. ZEISS and J.P. PHAIR: The enzyme-linked immunosorbent assay method for IgG antibody to purified protein derivative in cerebrospinal fluid of patients with tuberculous meningitis. Ann. Intern. Med. 99 (1983) 630–633.

KAMINSKI, Z.C., R. KAPILA, L.R. SHARER, P. KLOSER and L. KAUFMAN: Meningitis due to Prototheca wickerhamii in a patient with AIDS. Clin. Infect. Dis. 15 (1992) 704–706.

KANEKO, K., O. ONODERA, T. MIYATAKE and S. TSUHJI: Rapid diagnosis of tuberculous meningitis by polymerase chain reaction (PCR). Neurology 40 (1990) 1617–1618.

KARADANIS, D. and J.A. SHULMAN: Recent survey of infectious meningitis in adults: review of laboratory findings in bacterial, tuberculous and aseptic meningitis. South. Med. J. 69 (1976) 449–457.

KARMOCHKINE, M., J-M. MOLINA, C. SCIEUX, Y. WELKER, F. MORINET, J-M. DECAZES, P. LAGRANGE, L. SCHNELL and J. MODAI: Combined therapy with ganciclovir and foscarnet for cytomegalovirus polyradiculitis in patients with AIDS. Am. J. Med. 97 (1994) 196–197.

KATZ, D.A. and J.R. BERGER: Neurosyphilis in acquired immunodeficiency syndrome. Arch. Neurol. 46 (1989) 895–898.

KATZ, D.A., J.R. BERGER and R.C. DUNCAN: Neurosyphilis: a comparative study of the effect of infection with human immunodeficiency virus. Arch. Neurol. 50 (1993) 243–249.

KATZ, D.A., J.R. BERGER, B. HAMILTON, E.O. MAJOR and M.J. DONOVAN: Progressive multifocal leukoencephalopathy complicating Wiskott-Aldrich Syndrome. Report of a case and review of the literature of progressive multifocal leukoencephalopathy with other inherited immunodeficiency states. Arch. Neurol. 51 (1994) 422–426.

KAYE, B.R., C.M. NEUWELT, S.S. LONDON and J. DEARMOND: Central nervous system systemic lupus erythematosus mimicking progressive multifocal leukoencephalopathy. Ann. Rheum. Dis. 51 (1992) 1152–1156.

KCHOUK, M., F. ZOUITEN, M.H. BEN ROMDHANE, S. TOUIBI and A. ZRIBI: Cerebral miliary tuberculosis. Apropos of 5 cases and review of the literature. J. Radiol. 73 (1992) 589–593.

KEANE, J.R.: Neuroophthalmologic signs of AIDS: 50 patients. Neurology 41 (1991) 841–845.

KEANE, J.R.: Intermittent third nerve palsy with cryptococcal meningitis. J. Clin. Ophthalmol. 13 (1993) 124–126.

KEMPER, C.A., C.M. LOMBARD, S.C. DERESINSKI and L.S. TOMPKINS: Visceral bacillary epithelioid angiomatosis: possible manifestations of disseminated cat scratch disease in immunocompromised host: a report of two cases. Am. J. Med. 89 (1990) 216–222.

KEMPER, C.A., T.C. MENG, J. NUSSBAUM ET AL.: Treatment of

Mycobacterium avium complex bacteremia in AIDS with a four-drug oral regimen. Rifampin, ethambutol, clofazimine, and ciprofloxacin. The California Collaborative Treatment Group. Ann. Intern. Med. 116 (1992) 466–472.

KENNEDY, D.H. and R.J. FALLON: Tuberculous meningitis. J. Am. Med. Assoc. 241 (1979) 264–268.

KENT, S.J., S.M. CROWE, A. YUNG, C.R. LUCAS and A.M. MIJCH: Tuberculous meningitis: a 30 year review. Lancet vii (1993) 987–994.

KEPES, J.J., S.M. CHOU and D L.W. PRICE, JR.: Progressive multifocal leukoencephalopathy with 10 year survival in a patient with nontropical sprue: report of a case with unusual light and electron microscopic features. Neurology 25 (1975) 1006–1012.

KERR, D.A., C.F. CHANG, J. GORDON, M. BJORNSIT and K. KHALILI: Inhibition of human neurotropic virus (JCV) DNA replication in glial cells by camptothecin. Virology 196 (1993) 612–618.

KIEBURTZ, K.D., T.A. ESKIN, L. KETONEN and M.J. TUITE: Opportunistic cerebral vasculopathy and stroke in patients with the acquired immunodeficiency syndrome. Arch. Neurol. 50 (1993) 430–432.

KIM, J., G.Y. MINAMOTO and M.H. GRIECO: Nocardial infection as a complication of AIDS: report of six cases and review. Rev. Infect. Dis. 13 (1991) 624–629.

KIM, Y.S. and H. HOLLANDER: Polyradiculopathy due to cytomegalovirus: report of two cases in which improvement occurred after prolonged therapy and review of the literature. Clin. Infect. Dis. 17 (1993) 32–37.

KINLOCH-DE LOES, S., B. RADEFF and J-H. SUARAT: AIDS meets syphilis: changing patterns of syphiltic infection and its treatment. Dermatologica 177 (1988) 261–264.

KINNUNEN, E. and M. HILLBOM: The significance of cerebrospinal fluid routine screening for neurosyphilis. J. Neurol. Sci. 75 (1986) 205–211.

KLASTERSKY, J., R. CARPEL, J.M. SNOECK, J. FLAMENT and L. THIRY: Ascending nyelitis in association with herpes simplex virus. N. Engl. J. Med. 287 (1972) 182–184.

KLEIN, N.C., B. DAMSKER and S.Z. HIRSCHMAN: Mycobacterial meningitis: retrospective analysis from 1970–1983. Am. J. Med. 79 (1985) 29–34.

KLEINSCHMIDT-DEMASTERS, B.K., C. AMLIE-LEFOND and D.H. GILDEN: The patterns of varicella zoster virus encephalitis. Hum. Pathol. 27 (1996) 927–938.

KLEMOLA, E., R. VON ESSEN and O. WAGER: Cytomegalovirus mononucleosis in previously healthy adults. Ann. Intern. Med. 79 (1969) 267–268.

KNOBLER, E.H., D.N. SILVERS, K.C. FINE, J.H. LEFKOWITCH and M.E. GROSSMAN: Unique vascular lesions associated with human immunodeficiency syndrome. J. Am. Med. Assoc. 260 (1988) 524–527.

KOFFMAN, O.: The changing pattern of neurosyphilis. Can. Med. Assoc. J. 74 (1956) 807–812.

KOPPEL, B.S. and M. DARAS: Segmental myoclonus preceding herpes zoster radiculitis. Eur. Neurol. 32 (1992) 264–266.

KOPPEL, B.S., G.P. WORMSER, A.J. TUCHMAN, S. MAAYAN, D. HEWLETT, JR. and M. DARAS: Central nervous system involvement in patients with acquired immune deficiency syndrome (AIDS). Acta Neurol. Scand. 71 (1985) 337–353.

KOVACS, J.A.: Efficacy of atovaquone in treatment of toxoplasmosis in patients with AIDS. The NIAID clinical center intramural AIDS program. Lancet 340 (1992) 637–638.

KOVACS, J.A.: Toxoplasmosis in AIDS: keeping the lid on. Ann. Intern. Med. 123 (1995) 230–231.

KOVACS, J.A., A.A. KOVACS, M. POLIS, W.C. WRIGHT, V.J. GILL, C.U. TUAZON, E.P. GELMANN, H.C. LANE, R. LONGFIELD, G. OVERTURF, A.M. MACHER, A.S. FAUCI, J.E. PARRILLO, J.E. BENNETT and H. MASUR: Cryptococcosis in the acquired immunodeficiency syndrome. Ann. Intern. Med. 103 (1985) 533–538.

KOZIOL, K., K.S. RIELLY, R.A. BONIN and J.R. SALCEDO: Listeria monocytogenes meningitis in AIDS. Can. Med. Assoc. J. 135 (1986) 43–44.

KRAMBOVITIS E, M.B. MCILLMURRAY, P.E. LOCK, W. HENDRICKSE and H. HOLZEL: Rapid diagnosis of tuberculous meningitis by latex particle agglutination. Lancet ii (1984) 1229–1231

KRUPP, L.B., R.B. LIPTON, M.L. SWERDLOW, N.E. LEEDS and J. LLENA: Progressive multifocal leukoencephalopathy: clinical and radiographic features. Ann. Neurol. 17 (1985) 344–349.

KUCHELMEISTER, K., F. GULLOTTA and M. BERGMANN: Progressive multifocal leukoencephalopathy (PML) in the acquired immunodeficiency syndrome (AIDS). A neuropathological autopsy study of 21 cases. Pathol. Res. Pract. 189 (1993) 163–173.

KURE, K., C. HARRIS, L.S. MORIN and D.W. DICKSON: Solitary midbrain toxoplasmosis and olivary hypertrophy in a patient with acquired inmunodeficiency syndrome. Clin. Neuropathol. 8 (1989) 35–40.

KURE, K., J.F. LLENA, W.D. LYMAN, R. SOEIRO, K.M. WEIDENHEIM, A. HIRANO and D.W. DICKSON: Human immunodeficiency virus-1 infection of the nervous system: an autopsy study of 268 adult pediatric and fetal brains. Hum. Pathol. 22 (1991) 700–710.

LAGUNA, F., M. ADRADOS, A. ORTEGA AND J.M. GONZALEZ-LAHOZ: Tuberculous meningitis with acellular cerebrospinal fluid in AIDS patients. 6 (1992) 1165–1167.

LAISSY, J.P., P. SOYER, C. PARLIER, S. LARIVEN, Z. BENMELHA, V. SERVOIS, E. CASALINO, E. BOUVET, A. SIBERT, F. VACHON and Y. MENU: Persistent enhancement after treatment for cerebral toxoplasmosis in patients with AIDS: predictive value for subsequent recurrence. Am. J. Neuroradiol. 15 (1994) 1773–1778.

LAMFERS, E.J., A.H. BASTIAANS, M. MRAVUNAC and F.H. RAMPEN: Leprosy in the acquired immunodeficiency syndrome. J. Clin. Microbiol. 25 (1987) 154–157.

LANG W., J. MIKLOSSY, J.P. DERUAZ ET AL.: Neuropathology of the acquired immune deficiency syndrome (AIDS): a report of 135 consecutive autopsy cases from Switzerland. Acta Neuropathol. (Berlin) 77 (1989) 379–390.

LANG, W., J. MIKLOSSY, J-P. DERVAZ, G. PIZZOLATO, A. PROBST, T. SCHAFFNER, E. GESSAGA and P. KLEIHUES: Definition and incidence of AIDS associated CNS lesions. Prog. AIDS Pathol. 2 (1990) 89–101.

LANSKA, M.J., D.J. LANSKA and J.W. SCHMIDLEY: Syphilitic polyradiculopathy in an HIV-positive man. Neurology 38 (1988) 1297–1301.

LARSEN, R.A., S. BOZZETTE, A. MCCUTCHAN, J. CHIU, M.A. LEAL, D.D. RICHMAN and THE CALIFORNIA COLLABORATIVE TREATMENT GROUP: Persistant cryptococcus neoformans infection of the prostate after successful treatment of meningitis. Ann. Intern. Med. 111 (1989) 125–128.

LARSEN, R.A., M.A.E. LEAL and L.S. CHAN: Fluconazole compared with amphotericin B plus flucytosine for cryptococcal meningitis in AIDS. A randomized trial. Ann. Intern. Med. 113 (1990) 183–187.

LARSEN, R.A., S.A. BOZZETTE, B.E. JONES, D. HAGHIGHAT, M.A. LEAL, D. FORTHAL, M. BAUER, J.G. TILLES, J.A. MCCUTCHAN and J.M. LEEDOM: Fluconazole combined with flucytosine for treatment of cryptococcal meningitis in patients with AIDS. Clin. Infect. Dis. 19 (1994) 741–745.

LASKIN, O.L., C.M. STAHL-BAYLISS and S. MORGELLO: Concomitant herpes simplex virus type 1 and cytomegalovirus ventriculoencephalitis in acquired immunodeficiency syndrome. Arch. Neurol. 44 (1987a) 843–847.

LASKIN, O.L., D.M. CEDERBERG and J. MILLS for the GANCICLOVIR STUDY GROUP: Ganciclovir for the treatment and suppression of serious infections caused by cytomegalovirus. Am. J. Med. 83 (1987b) 201–207.

LEAKE, H.A., M.N. APPLEYARD and J.P.R. HARTLEY: Successful treatment of resistant cryptococcal meningitis with amphotericin B lipid emulsion after nephrotoxicity with conventional intravenous amphotericin B. J. Infect. 28 (1994) 319–322.

LEIGUARDA, R., BERTHIER, S. STARKSTEIN, M. NOGUES and P. LYLYK: Ischemic infarction in 25 children with tuberculous meningitis. Stroke 19 (1988) 200–204.

LEPORT, C., F. RAFFI, S. MATHERON, C. KATLAMA, B. REGNIER, A.G. SAIMOT, C. MARCHE, C. VEDRENNE and J.L. VILDE: Treatment of central nervous system toxoplasmosis with pyrimethamine/sulfadiazine combination in 35 patients with the acquired immunodeficiency syndrome. Am. J. Med. 84 (1988) 94–100.

LEPPER, M.H. and H.W. SPIES: The present status of the treatment of tuberculosis of the central nervous system. Ann. N.Y. Acad. Sci. 106 (1963) 106–123.

LEVISON, A., J. LUHAN, W.P. MAURELIS and H. HERZON: The effect of streptomycin on tuberculous meningitis. A pathologic study. J. Neuropathol. Exp. Neurol. 9 (1950) 406–419.

LEVY, R.M., V.G. PONS and M.L. ROSENBLUM: Central nervous system mass lesions in the acquired immunodeficiency syndrome (AIDS). J. Neurosurg. 61 (1984) 9–16.

LEVY, R.M., D.E. BREDESEN and M.L. ROSENBLUM: Neurological manifestations of the acquired immunodeficiency syndrome (AIDS): experience at UCSF and review of the literature. J. Neurosurg. 62 (1985) 475–495.

LEVY, R.M., S. ROSENBLOOM and L.V. PERRET: Neuroradiologic findings in the acquired immunodeficiency syndrome (AIDS): a review of 200 cases. Am. J. Neuroradiol. 7 (1986) 833–839.

LEVY, R.M., D.E. BREDESEN and M.L. ROSENBLUM: Opportunistic central nervous system pathology in patients with AIDS. Ann. Neurol. 23 (1988) S7–S12.

LEVY, J.H., R.A. LISS and A.M. MACUITE: Neurosyphilis and ocular syphilis in patients with concurrent human immunodeficiency virus infection. Retina 9 (1989) 175–180.

LINARD, F., P. BEAU, D. SILVESTRI, M. DESI, J. KOREZLIOGLU, N. SEIBEL and J.P. COULAUD: Toxoplasmosis with an onset of isolated psychiatric disturbance (12 cases). Int. Conf. AIDS 8 (1992) 101 (Abstr. PUB 7316).

LINCOLN, E.M.: Tuberculous meningitis in children: serous meningitis. Am. Rev. Tuberc. 56 (1947a) 95–109.

LINCOLN, E.M.: Tuberculous meningitis in children with special reference to serous meningitis. 1. Tuberculous meningitis. Am. Rev. Tuberc. 56 (1947b) 75–94.

LINCOLN, E.M. and E.M. SEWELL: Tuberculosis in Children. New York, McGraw-Hill (1963).

LINCOLN, E.M., S.V.R. SORDILLO and P.A. DAVIES: Tuberculous meningitis in children. A review of 167 untreated and 74 treated patients with special reference to early diagnosis. J. Pediatr. 57 (1960) 807–823.

LIPSON, B.K., W.R. FREEMAN, J. BENIZ, M.H. GOLDBAUM, J.R. HESSELINK, R.W. WEINREB and A.A. SADUN: Optic neuropathy associated with cryptococcal arachnoiditis in AIDS patients. Am. J. Ophthalmol. 107 (1989) 523–527.

LIVRAMENTO, J.A., L.P. MACHADO and A. SPINA-FRANCA: Cerebrospinal fluid abnormalities in 170 cases of AIDS. Arq. Neuropsiquatr. 47 (1989) 326–331.

LOPEZ-CORTES, L.F., M. CRUZ-RUIZ, J. GOMEZ-MATEOS ET AL.: Adenosine deaminase activity in the CSF of patients with aseptic meningitis: utility in the diagnosis of tuberculous meningitis or neurobrucellosis. Clin. Infect. Dis. 20 (1995) 525–530.

LORBER, J.: The results of treatment of 549 cases of tuberculous meningitis. Am. Rev. Tuberc. 69 (1954) 13–25.

LORBER, J.: Long-term follow-up of 100 children who recovered from tuberculous meningitis. Pediatrics 28 (1961) 778–791.

LORTHOLARY, O., M-C. MEYOHAS, B. DUPONT, J. CADRANEL, D. SALMON-CERON, D. PEYRAMOND and D. SIMONIN: Invasive aspergillosis in patients with the acquired immunodeficiency syndrome: report of 33 cases. Am. J. Med. 95 (1993) 177–187.

LUFT, B.J. and J.S. REMINGTON: Toxoplasmic encephalitis in AIDS. Clin. Infect. Dis. 15 (1992) 211–222.

LUFT, B.J., R.G. BROOKS, F.K. CONLEY, R.E. MCCABE and J.S. REMINGTON: Toxoplasmic encephalitis in patients with acquired immune deficiency syndrome. J. Am. Med. Assoc. 252 (1984) 913–917.

LUFT, B.J., R. HAFNER, A.H. KORZUN, C. LEPORT, D. ANTRONISKIS, E.M. BOSLER, D.D. BOURLAND, III, R. UTTAMCHANDANI, J. FUHRER, J. JACOBSEN, P. MORLAT, J-L. VILDE, J.S. REMINGTON and MEMBERS OF THE ACTG 007P / ANRS 009 STUDY TEAM: Toxoplasmic encephalitis in patients with the acquired immunodeficiency syndrome. N. Engl. J. Med. 329 (1993) 995–1000.

LUGER, A., B.L. SCHMIDT, K. STEYNER ET AL.: Diagnosis of neurosyphilis by examination of cerebrospinal fluid. Br. J. Vener. Dis. 48 (1972) 1–10.

LUKEHART, S.A., E.W. HOOK, III, S.A. BAKER-ZANDER, A.C. COOKIER, C.W. CRITHLOW and H.H. HANDSFIELD: Invasion of the central nervous system by Treponema pallidum: implications for diagnosis and treatment. Ann. Intern. Med. 109 (1988) 855–862.

LUXON, L., A.J. LEES and R.J. GREENWOOD: Neurosyphilis today. Lancet i (1979) 90–93.

LYNCH, K.J. and R.J. FRISQUE: Factors contributing to the restricted DNA replicating activity of JC virus. Virology 180 (1991) 306–317.

LYOS, A.T., A. MALPICA, R. ESTRADA, C.D. KATZ and H.A. JENKINS: Invasive aspergillosis of the temporal bone: an unusual manifestation of acquired immunodeficiency syndrome. Am. J. Otolaryngol. 14 (1993) 444–448.

MAAYAN, S., G.P. WORMSER, J. WIDERHORN, E.R. SY, Y.H. KIM and J.A. ERNST: Strongyloides stercoralis hyperinfection in a patient with the acquired immune deficiency syndrome. Am. J. Med. 83 (1987) 945–948.

MACGREGOR, A.R. and M. GREEN: Tuberculosis of the central nervous system with special reference to tuberculous meningitis. J. Pathol. Bacteriol. 45 (1937) 613.

MADHOUN, Z.T., D.B. DU BOIS, J. ROSENTHAL, J.C. FINDLAY and D.C. ARON: Central diabetes insipidus: a complication of herpes simplex type 2 encephalitis in a patient with AIDS. Am. J. Med. 90 (1991) 658–659.

MADIEDO, G., K-C. HO and P. WALSH: False positive VDRL and FTA in cerebrospinal fluid. J. Am. Med. Assoc. 244 (1980) 688–689.

MAHIEUX, F., F. GRAY, G. FENELON, R. GHERARDI, D. ADAMS, A. GUILLARD and J. POIRIER: Acute myeloradiculitis due to cytomegalovirus as the initial manifestition of AIDS. J. Neurol. Neurosurg. Psychiatry 52 (1989) 270–274.

MAJOR E.O., K. AMEMIYA, C. TORNATORE, S. HOUFF and J. BERGER: Pathogenesis and molecular biology of progressive multifocal leukoencephalopathy, the JC virus-induced demyelinating disease of the human brain. Clin. Microbiol. Rev. 5 (1992) 49–73.

MALLOLAS, J., L. ZAMORA, J.M. GATELL, J.M. MIRO, E. VERNET, M.E. VALLS, E. SORIANO and J.C. SAN MIGUEL: Primary prophylaxis for pneumocystis carinii pneumonia: a randomized trial comparing cotrimoxazole, aerosolized pentamidine and dapsone plus pyrimethamine. AIDS 7 (1993) 59–64.

MANDELL, G.L. and M.A. SANDE: Drugs used in the chemotherapy of tuberculosis and leprosy. In: A.G. Gilman, L. Goodman, T.W. Rall and F. Murad (Eds.), The Pharmacological Basis of Therapeutics, 7th Edit. New York, Macmillan (1985) 1199–1218.

MANN, M.D., C.M. MACFARLANE, C.J. VVERBURG and J. WIGGELINKHUIZEN: The bromide partition test and CSF adenosine deaminase activity in the diagnosis of tuberculous meningitis in children. S. Afr. Med. J. 62 (1982) 431–433.

MAPELLI, G., M. PAVONI, T. BELLELLI, W. BARONCINI, A. MANENTE and M. DIBARI: Neurosyphilis today. Eur. Neurol. 20 (1981) 334–343.

MARMADUKE, D.P., J.T. BRANDT and K.S. THEIL: Rapid diagnosis of cytomegalovirus in the cerebrospinal fluid of a patient with AIDS-associated polyradiculopathy. Arch. Pathol. Lab. Med. 115 (1991) 1154–1157.

MARRA, C.M.: Syphilis, human immunodeficiency virus, and the nervous system. In: J.R. Berger and R.M. Levy (Eds.), AIDS and the Nervous System, 2nd Edit. New York, Lippincott-Raven (1996) 677–691.

MARRA, C.M., W.T. LONGSTRETH, JR. and S.A. LUKEHART: Resolution of CSF abnormalities in neurosyphilis: influence of stage and HIV infection (Abstr.). Neurology 42 (1992) 212.

MARRIOTT, P.J., M.D. O'BRIEN, I.C. MACKENZIE ET AL.: Progressive multifocal leukoencephalopathy: remission with cytarabine. J. Neurol. Neurosurg. Psychiatry 38 (1975) 205–209.

MARSHALL, D.W., R.L. BREY, W.T. CAHILL, R.W. HOUK, R.A. ZAJAC and R.N. BOSWELL: Spectrum of cerebrospinal fluid findings in various stages of human immunodeficiency virus infection. Arch. Neurol. 45 (1988) 954–958.

MARUK, C., H. NAKANO, T. SHIMOJI, M. MAEDA and S. ISHII: Loss of vision due to cryptococcal optochiasmatic arachnoiditis and optocurative surgical exploration. Neurol. Med. Chir. (Tokyo) 28 (1988) 695–697.

MASDEU, J.C., C.B. SMALL, L. WEISS, C.M. ELKIN, J. LLENA and R. MESA-TEJADA: Multifocal cytomegalovirus encephalitis in AIDS. Ann. Neurol. 23 (1988) 97–99.

MASTROIANNI, C.M., F. MENGONI, C. VALENTI, R. MARCHESE, V. VULLO and S. DELIA: IgA antibodies to P30 of *toxoplasma gondii* in AIDS patients with cerebral toxoplasmosis. Int. Conf. AIDS 10 (1994) 155 (Abstr. PBO634).

MASUR, H.: Toxoplasmosis. In: J. B. Wyngaarden, L.H. Smith and J.C. Bennett (Eds.), Cecil Textbook of Medicine. Philadelphia, PA, WB Saunders (1992) 1987–1991.

MASUR, H., M.A. MICHELIS, J..B. GREENE ET AL.: An outbreak of community-acquired *Pneumocystis carinii* pneumonia: initial manifestation of cellular immune dysfunction. N. Engl. J. Med. 305 (1981) 1439–1444.

MATTHEWS, T. and R. BOEHME: Antiviral activity and mechanism of action of ganciclovir. Rev. Infect. Dis. 10 (1988) S490–S494.

MATTHEWS, V.P., P.L. ALO, J.D. GLASS, A.J. KUMAR and J.C. MCARTHUR: AIDS-related CNS cryptococcosis: radiologic-pathologic correlation. Am. J. Neuro-Radiol. 13 (1992) 1477–1486.

MAZLO, M. and R.M. HERNDON: Progressive multifocal leukoencephalopathy: ultrastructural findings in two brain biopsies. Neuropathol. Appl. Neurobiol. 3 (1977) 323–339.

MAZLO, M. and I. TARISKA: Are astrocytes infected in progressive multifocal leukoencephalopathy? Acta Neuropathol. (Berlin) 56 (1982) 45–51.

MCARTHUR, J.C.: Neurologic manifestations of AIDS. Medicine (Baltimore) 66 (1987) 407–437.

MCARTHUR J.C., A.J. KUMAR, D.W. JOHNSON ET AL.: Incidental white matter hyperintensities on magnetic resonance imaging in HIV-1 infection. J. AIDS 3 (1990) 252–259.

MCARTHUR, J.C., D.R. HOOVER, H. BACELLAR, E.N. MILLER, B.A. COHEN, J.T. BECKER, N.M.H. GRAHAM, J.H. MCARTHUR, O.A. SELNES, L.P. JACOBSON, B.R. VISSCHER, M. CONCHA and A. SAAH: Dementia in AIDS patients: incidence and risk factors. Neurology 43 (1993) 2245–2252.

MCCUTCHAN, J.A.: Cytomegalovirus infections of the nervous system in patients with AIDS. Clin. Infect. Dis. 20 (1995) 747–754.

MCGUIRE, D., S. BARHITE, H. HOLLANDER and M. MILES: JC Virus DNA in cerebrospinal fluid of human immunodeficiency virus-infected patients: predictive value for progressive multifocal leukoencephalopathy. Ann. Neurol. 37 (1995) 395–399.

MEDICAL RESEARCH COUNCIL: Streptomycin treatment of tuberculous meningitis. Report of the committee on streptomycin in tuberculosis trials. Lancet i (1948) 582–597.

MEHREN, M., P.J. BURNS, F. MAMANI, C.S. LEVY and R. LAURENO: Toxoplasmic myelitis mimicking intramedullary spinal cord tumor. Neurology 38 (1988) 1648–1650.

MELBYE, M., R.J. GROSSMAN, J.J. GOEDERT, M.E. EYSTER and

R.J. BIGGAR: Risk of AIDS after herpes zoster. Lancet i (1987) 728–731.

MERRITT, H.H., R.D. ADAMS and H.C. SOLOMON: Neurosyphilis. New York, Oxford University Press (1946).

METZE, K. and J.A. MACIEL, JR.: AIDS and Chagas's disease (Letter). Neurology 43 (1993) 447–448.

MICOZZI, M.S. and C.V. WETLI: Intravenous amphetamine abuse, primary cerebral mucormycosis and acquired immunodeficiency. J. Forensic Sci. 30 (1985) 504–510.

MILLER, J.R., R.E. BARRETT, C.B. BRITTON ET AL.: Progressive multifocal leukoencphalopathy in a male homosexual with T-cell immune deficiency. N. Engl. J. Med. 307 (1982) 1436–1438.

MILLER, R.F., S.B. LUCAS, M.A. HALL-CRAGS, N.S. BRINK, F. SCARAVILLI, R.J. CHINN, B.E. KENDALL, I.G. WILLIAMS and M.J. HARRISON: Comparison of magnetic resonance imaging with neuropathological findings in the diagnosis of HIV and CMV associated CNS disease in AIDS. J. Neurol. Neurosurg. Psychiatry 62 (1997) 346–351.

MILLER, R.G., J.R. STOREY and C.M. GRECO: Ganciclovir in the treatment of progressive AIDS related polyradiculopathy. Neurology 40 (1990) 569–574.

MILLIGAN, S.A., M.S. KATZ and P.C. CRAVEN: Toxoplasmosis presenting as panhypopituitarism in a patient with the acquired immunodeficiency syndrome. Am. J. Med. 77 (1984) 60–64.

MINAMOTO, G.Y., T.F. BARLAM and N.J. VANDER ELS: Invasive aspergillosis in patients with AIDS. Clin. Infect. Dis. 14 (1992) 66–74.

MISCHEL, P.S. and H.V. VINTERS: Coccidioidomycosis of the central nervous system: neuropathological and vasculopathic manifestations and clinical correlates. Clin. Infect. Dis. 20 (1995) 400–405.

MISHELL, J.H. and E.L. APPLEBAUM: Ramsay-Hunt syndrome in a patient with HIV infection. Otolaryngol. Head Neck Surg. 102 (1990) 177–179.

MIZUTANI, T., N. KUROSWA, Y. MATSUNO, M. MIYAGAWA, S. MATSUYA and T. AIBA: Atypical manifestations of tuberculous meningitis. Eur. Neurol. 33 (1993) 159–162.

MOONEY, A.J.: Some ocular sequelae of tuberculous meningitis. Am. J. Ophthalmol. 41 (1956) 753–768.

MOORE, E.H., L.A. RUSSEL, J.S. KLEIN, C.S. WHITE ET AL.: Bacillary angiomatosis in patients with AIDS: multiorgan imaging findings. Radiology 197 (1995) 67–72.

MOORE, R.D. and R.E. CHAISSON: Natural history of opportunistic disease in an HIV-infected urban clinical cohort. Ann. Intern. Med. 124 (1996) 633–542.

MORET, H., M. GUICHARD, S. MATHERON, C. KATLAMA, V. SAZDOVITCH, J.M. HURAUX and D. INGRAND: Virological diagnosis of progressive multifocal leukoencephalopathy: detection of JC Virus DNA in cerebrospinal fluid and brain tissue of AIDS patients. J. Clin. Microbiol. 31 (1993) 3310–3313.

MORGELLO, S. and D.M. SIMPSON: Multifocal cytomegalovirus demyelinative polyneuropathy associated with AIDS. Muscle Nerve 17 (1994) 176–182.

MORGELLO, S., E.S. CHO and S. NIELSEN: Cytomegalovirus encephalitis in patients with acquired immunodeficiency syndrome. Hum. Pathol. 18 (1987) 289–297.

MORGELLO, S., G.A. BLOCK, R.W. PRICE and C.K. PETITO: Varicella-zoster virus leukoencephalitis and cerebral vasculopathy. Arch. Pathol. Lab. Med. 112 (1988) 173–177.

MORGELLO, S., F.M. SOIFER, C-S. LIN and D.E. WOLFE: Central nervous system *Strongyloides stercoralis* in acquired immunodeficiency syndrome: a report of two cases and review of the literature. Acta Neuropathol. (Berlin) 86 (1993) 285–288.

MORRISON, R.E., S. HARRISON and E.C. TRAMONT: Oral amoxycillin, an alternative treatment for neurosyphilis. Genitourin. Med. 61 (1985) 359–362.

MOSKOWITZ, L.B., J.B. GREGORIOS, G.T. HENSLEY and J.R. BERGER: Cytomegalovirus induced demyelination associated with acquired immunodeficiency syndrome. Arch. Pathology Lab. Med. 108 (1984) 873–877.

MOULIGNIER, A., G. PIALOUX, H. DEGA, B. DUPONT, M. HUERRE and M. BAUDRIMONT: Brain stem encephalitis due to varicella-zoster virus in a patient with AIDS. Clin. Infect. Dis. 20 (1995) 1378–1380.

MULLER, F., M. MOSKOPHIDIS and H. SCHMITZ: Intrathecal synthesis of specific IgG in syphlitic patients with human immunodeficiency virus type 1 infection. J. Neurol. 235 (1988) 252–253.

MULLIN, G.E. and A.L. SHEPPELL: *Listeria monocytogenes* and the acquired immunodeficiency syndrome. Arch. Intern. Med. 147 (1987) 176.

MUSHER, D.M. and R.F. SCHELL: The immunology of syphilis. Hosp. Pract. 10 (1975) 45–50.

MUSHER, D.M., R.F. SCHELL and J.M. KNOX: In vitro lymphocyte response to *Treponema refringens* in human syphilis. Infect. Immun. 9 (1974) 654–657.

MUSHER, D.M., R.J. HAMILL and R.E. BAUGH: Effect of human immunodeficiency virus (HIV) infection on the course of syphilis and on the response to treatment. Ann. Intern. Med. 113 (1990) 872–881.

MUSIANI, M., M. ZERBINI, S. VENTUROLI, G. GENTILOMI, V. BORGHI, P. PIETROSEMOLI, M. PECORARI and M. LA PLACA: Rapid diagnosis of cytomegalovirus encephalitis in patients with AIDS using in situ hybridization. J. Clin. Pathol. 47 (1994) 886–891.

NAHMIAS, A.J., F.K. LEE and S BECKMAN-NAHMIAS: Seroepidemiologic and sociological patterns of herpes simplex virus infection in the world. Scand. J. Infect. Dis. 69 (1990) 19–36.

NARITA, M., Y. MATSUZONO, M. SHIBATA and T. TOGASHI: Nested amplification protocol for the detection of *Mycobacterium tuberculosis*. Acta Paediatr. 81 (1992) 997–1001.

NATH, A., J. JANKOVIC and L.C. PETTIGREW: Movement disorders and AIDS. Neurology 37 (1987) 37–41.

NATH, A., D.E. HOBSON and A. RUSSELL: Movement disorders with cerebral toxoplasmosis and AIDS. Mov. Disord. 8 (1993) 107–112.

NAVIA, B.A., C.K. PETITO, J.W.M. GOLD, E-S. CHO, B.D. JORDAN and R.W. PRICE: Cerebral toxoplasmosis complicating the acquired immune deficiency syndrome: clinical and neuropathological findings in 27 patients. Ann. Neurol. 19 (1986a) 224–238.

NAVIA, B.A., B.D. JORDAN and R.W. PRICE: The AIDS dementia complex: I. Clinical features. Ann. Neurol. 19 (1986b) 517–524.

NAVIA, B.A., E.S. CHO, C.K. PETITO and R.W. PRICE: The AIDS dementia complex: II. Neuropathology. Ann. Neurol. 19 (1986c) 525–535.

NIGHTENGALE, S.D.: Initital therapy for acquired immunodeficiency syndrome-associated cryptococosis with fluconazole. Arch. Intern. Med. 155 (1995) 538–540.

NIGHTENGALE, S.D., S.X. CAL, D.M. PETERSON, S.D. LOSS, B.A. GAMBLE, D.A. WATSON, C.P. MANZONE, J.E. BAKER and J.D. JOCKUSCH: Primary prophylaxis with fluconazole against systemic fungal infections in HIV-positive patients. AIDS 6 (1992) 191–194.

NITRINI, R. and A. SPINA-FRANÇA: Penicilinoterapia intravenosa em altas doses na neurossífilis. Estudo de 62 casos. II. Avaliç o do liquido cefalorraqueano. Arq. Neuropsiquiatr. 45 (1987) 231–241.

NOLLA-SALAS, J., C. RICART, L. D'OLHABERRIAGUE, F. GALI and J. LAMARCA: Hydrocephalus: an unusual CT presentation of cerebral toxoplasmosis in a patient with acquired immunodeficiency syndrome. Eur. Neurol. 27 (1987) 130–132.

NOORDHOEK, G.T., B. WIELES, VAN DER SLUIS and J.D.A. VAN EMBDEN: Polymerase chain reaction and synthetic DNA probes: a means of distinguishing the causative agents of syphilis and yaws? Infect. Immun. 58 (1990) 2011–2013.

NORBENO, O. and P. SORENSON: The incidence and clinical presentation of neurosyphilis in Greater Copenhagen 1974 through 1978. Acta Neurol. Scand. 63 (1981) 237–246.

NORDEN, C.W., F.L. RUBEN and R.SELKER: Nonsurgical treatment of cerebral nocardiosis. Arch. Neurol. 40 (1983) 594–595.

NORRIS, S.J.: Syphilis. In: D.J.M. Wright (Ed.), The Immunology Of Sexually Transmitted Diseases. Boston, MA, Kluwer Academic (1988) 1–32.

NUNN, P.P. and K.P.W.J. MCADAM: Mycobacterial infections and AIDS. Br. Med. Bull. 44 (1988) 801–813.

O'RIORDAN, T., P.A. DALY, M. HUTCHINSON, A.G. SHATTOCK and S.D. GARDNER: Progressive multifocal leukoencephalopathy – remission with cytarabine. J. Infect. 20 (1990) 51–54.

OFFENBACHER, H., F. FAZEKAS, R. SCHMIDT ET AL.: MRI in

tuberculous meningoencephalitis: report of four cases and review of the neuroimaging literature. J. Neurol. 238 (1991) 340–344.

OFNER, S. and R.S. BAKER: Visual loss in cryptococcal meningitis. J. Clin. Neuroophthalmol. 7 (1987) 45–48.

OGAWA, S.K., M.A. SMITH, D.J. BRENNESSEL and F.D. LOWY: Tuberculous meningitis in an urban medical center. Medicine (Baltimore) 66 (1987) 317–326.

OKSENHENDLER, E., I. CHARREAU, C. TOURNERIE, M. AZIHARY, C. CARBON and J.P. ABOULKER: *Toxoplasma gondii* infection in advanced HIV infection. AIDS 8 (1994) 483–487.

OLSEN, W.L., F.M. LONGO, C.M. MILLS and D. NORMAN: White matter disease in AIDS. Findings at MR imaging. Radiology 169 (1988) 445–448.

OREFICE, G., P.B. CARRIERI, A. CHIRIANNI, S. RUBINO, G. LIUZZI, G. NAPOLITANO and A. ROCCO: Cerebral toxoplasmosis and AIDS: clinical, neuroradiological and immunological findings in 15 patients. Acta Neurol. 14 (1992) 493–502.

OREFICE, G., G. CAMPANELLA, S. CICCIARELLO, A. CHIRIANNI, G. BORGIA, S. RUBINO, M. MAINOLFI, M. COPPOLA and M. PIAZZA: Presence of papova-like viral particles in cerebrospinal fluid of AIDS patients with progressive multifocal leukoencephalopathy. An additional test for in vivo diagnosis. Acta Neurol. 15 (1993) 328–332.

OVERHAGE, J.M., A. GREIST and D.R. BROWN: Conus medullaris syndrome resulting from *toxoplasma gondii* infection in a patient with the acquired immunodeficiency syndrome. Am. J. Med. 89 (1990) 814–815.

PADGETT, B.L. and D.L. WALKER: Virologic and serologic studies of progressive multifocal leukoencephalopathy. In: J. Sever and D.L. Madden (Eds.), Polyomaviruses and Human Neurological Disease. New York, Alan R. Liss (1983a).

PADGETT, B.L. and D.L. WALKER: Virologic and serologic studies of progressive multifocal leukoencephalopathy. Prog. Clin. Biol. Res. 105 (1983b) 107–117.

PADGETT, B.L., G.M. ZURHEIN, D.L. WALKER, R.J. ECHROADE and B.H. DESSEL: Cultivation of papova-like virus from human brain with progressive multifocal leukoencephalopathy. Lancet i (1971) 1257–1260.

PALESTINE, A.G., M.M. RODRIGUES, A.M. MACHER ET AL.: Ophthalmic involvement in acquired immunodeficiency syndrome. Ophthalmology 91 (1984) 1092–1099.

PALESTINE, A.G., M.A. POLIS, M.D. DE SMET, B.F. BAIRD, J. FALLOON, J.A. KOVACS, R.T. DAVEY, J.J. ZURLO, K.M. ZUNICH, M. DAVIS, L. HUBBARD, R. BROTHERS, F.L. FERRIS, E. CHEW, J.L. DAVIS, B.J. RUBIN, S.D. MELLOW, J.A. METCALF, J. MANISCHEWITZ, J.R. MINOR, R.B. NUSSENBLATT, H. MASUR and H.C. LANE: A randomized, controlled trial of foscarnet in the treatment of cytomegalovirus reti-

nitis in patients with AIDS. Ann. Intern. Med. 115 (1991) 665–673.

PAPPAS, P.G., J.C. POTTAGE, W.G. POWDERLY, V.J. FRASER, STRATTON, MCKENZIE S, M.L. TAPPER, H. CHMEL, F.C. BONEBRAKE, R. BLUM, R.W. SHAFER, C. KING and W.E. DISMUKES: Blastomycosis in patients with acquired immunodeficiency syndrome. Ann. Intern. Med. 116 (1992) 847–853.

PARDIWALLA, F.K., M.E. YEOLEKAR, S. WAGLE and S.K. BAKSHI: Persistent neutrophilic meningitis. An unusual presentation of tuberculous meningitis. J. Assoc. Phys. India 40 (1992) 632–633.

PASSO, M.S. and J.T. ROSENBAUM: Ocular syphilis in patients with human immunodeficiency virus infection. Am. J. Ophthalmol. 106 (1988) 1–6.

PATEY, O., C. NEDELEC, J.P. EMOND, R. MAYORGA ET AL.: Listeria monocytogenes septicemia in an AIDS patient with a brain abscess. Eur. J. Clin. Microbiol. Infect. Dis. 8 (1989) 746–748.

PAUGAM, A., J. DUPOUY-CAMET, P. BLANCHE, J.P. GANGNEUX, C. TOURTE-SCHAEFER and D. SICARD: Increased fluconazole resistance of cryptococcus neoformans isolated from a patient with AIDS and recurrent meningitis (Letter). Clin. Infect. Dis. 19 (1994) 975–976.

PEETERMANS, W., H. BOBBAERS, J. VERHAEGEN and J. VANDEPITTE: Fluconazole-resistant *cryptococcus neoformans* var gattii in an AIDS patient. Acta Clin. Belg. 48 (1993) 405–409.

PENN, C.C., E. GOLDSTEIN and W.R. BARTHOLOMEW: *Sporothrix schenkii* meningitis in a patient with AIDS (Letter). Clin. Infect. Dis. 15 (1992) 741–743.

PEPIN, J., F.A. PLUMMER, R.C. BRUNHAM ET AL.: The interaction of HIV infection and other sexually transmitted diseases: an opportunity for intervention. AIDS 3 (1989) 3–9.

PEPOSE, J.S., L.H. HILBORNE, P.A. CANCILLA and R.Y. FOOS: Concurrent herpes simplex and cytomegalovirus retinitis and encephalitis in the acquired immune deficiency syndrome (AIDS). Ophthalmology 91 (1984) 1669–1677.

PEREZ-PEREZ, M., P. GARCIA-MARTOS, J.C. ESCRIBANO-MORIANO, and P. MARIN-CASANOVA: *Nocardia caviae* meningitis in a patient with HIV infection (Letter). Rev. Clin. Esp. 187 (1990) 374–375.

PETERS, A.C.B., J. VERSTEEG, G.T.A. BOTS ET AL.: Progressive multifocal leukoencephalopathy: immunofluorescent demonstration of SV40 antigen in CSF cells and response to cytarabine therapy. Arch. Neurol. 37 (1980) 497–501.

PETERS, M., U. TIMM, D. SCHURMANN, H.D. POHLE and D.B. RUF: Combined and alternating ganciclovir and foscarnet in acute and maintenance therapy of human immunodeficiency virus-related cytomegalovirus encephalitis refractory to ganciclovir alone. Clin. Inv. 70 (1992) 456–458.

PETERSEN, L.R., R.H. MEAD and M.G. PERLROTH: Unusual

manifestations of secondary syphilis occurring after orthotoptic liver transplantation. Am. J. Med. 75 (1983) 166–170.

PETITO, C.K., E-S. CHO, W. LEMANN, B.A. NAVIA and R.W. PRICE: Neuropathology of acquired immunodeficiency syndrome (AIDS): an autopsy review. J. Neuropathol. Exp. Neurol. 45 (1986) 635–646.

PICARD, F.J., G.A. DEKABAN, J. SILVA and G.P.A. RICE: Mollaret's meningitis associated with herpes simplex type 2 infection. Neurology 43 (1993) 1722–1727.

PILLAI, S., M.A. MAHMOOD and S.R. LIMAYE: Herpes-zoster ophthalmicus, contralateral hemiplegia and recurrent ocular toxoplasmosis in a patient with acquired immune deficiency syndrome-related complex. J. Clin. Neuroophthalmol. 9 (1989) 229–233.

PINNER, R.W., R.A. HAJJEH and W.G. POWDERLY: Prospects for preventing cryptococcosis in persons infected with human immunodeficiency virus. Clin. Infect. Dis. 21 (1995) S103–S107.

PITCHENIK, A.E., C. COLE, B.W. RUSSELL ET AL.: Tuberculosis, atypical mycobacteriosis, and the acquired immunodeficiency syndrome among Haitian and non-Haitian patients in South Florida. Ann. Intern. Med. 101 (1984) 641–645.

PITCHENIK, A.E., J. BURR, D. FERTEL ET AL.: Human T-cell lymphotropic virus III (HTLV III) seropositivity and related disease among 71 consecutive patients in whom tuberculosis was diagnosed. A prospective study. Am. Rev. Resp. Dis. 135 (1987) 875–879.

PITLIK, S.D., V. FAINSTEIN, R. BOLIVAR, L. GUARDA, A. RIOS, P.A. MANSELL and F. GYORKEY: Spectrum of central nervous system complications in homosexual men with acquired immune deficiency syndrome (Letter). J. Infect. Dis. 148 (1983) 771–772.

PODZAMCZER, D., J.M. MIRO, F. BOLAO, J.M. GATELL, J. COSIN, G. SIRERA, P. DOMINGO, F. LAGUNA, J. SANTAMARIA, J. VERDEJO and the SPANISH TOXOPLASMOSIS STUDY GROUP: Twice weekly maintenance therapy with sulfadiazine–pyrimethamine to prevent recurrent toxoplasmic encephalitis in patients with AIDS. Ann. Intern. Med. 123 (1995) 175–180.

POON, T.P., M. BAHBAHANI, I. MATOSO and B. KIM: Pineal toxoplasmosis mimicking pineal tumor in an AIDS patient. J. Natl. Med. Assoc. 86 (1994) 550–552.

POPOVICH, M.J., R.H. ARTHUR and E. HELMER: CT of intracranial cryptococcosis. Am. J. Neuroradiol. 11 (1990) 139–142.

PORTEGIES, P., P.R. ALGRA, C.E.M. HOLLAR, J.M. PRINS, P. REISS, J. VALK and J.M.A. LANGE: Response to cytarabine in progressive multifocal leukoencephalopathy in AIDS. Lancet 337 (1991) 680–681.

PORTER, S.B. and M.A. SANDE: Toxoplasmosis of the central nervous system in the acquired immunodeficiency syndrome. N. Engl. J. Med. 327 (1992) 1643–1648.

POSCHER, M.E: Successful treatment of varicella zoster virus meningoencephalitis in patients with AIDS: report of four cases and review. AIDS 8 (1994) 1115–1117.

POST, M.J.D., S.J. KURSUNOGLU, G.T. HENSLEY, J.C. CHAN, L.B. MOSKOWITZ and T.A. HOFFMAN: Cranial CT in acquired immunodeficiency syndrome: spectrum of diseases and optimal contrast enhancement technique. Am. J. Roentgenol. 145 (1985) 929–940.

POST, M.J.D., L.G. TATE, R.M. QUENCER ET AL.: CT, MR and pathology in HIV encephalitis and meningitis. Am. J. Roentgenol. 151 (1988) 449–454

POTASMAN, I., L. RESNICK, B. LUFT and J.S. REMINGTON: Intrathecal production of antibodies against toxoplasma gondii in patients with toxoplasma encephalitis and acquired immunodeficiency syndrome (AIDS). Ann. Intern. Med. 108 (1988) 49–51.

POWDERLY, W.G.: Cryptococcal meningitis and AIDS. Clin. Infect. Dis. 17 (1993) 837–842.

POWDERLY, W.G., E.J. KEATH, M. SOKOL-ANDERSON, K. ROBINSON, D. KITZ, J.R. LITTLE and G. KOBAYASHI: Amphotericin B-resistant cryptococcus neoformans in a patient with AIDS. Infect. Dis. Clin. Pract. 1 (1992a) 314–316.

POWDERLY, W.G., M.S. SAAG, G.A. CLOUD, P. ROBINSON, R.D. MEYER, J.M. JACOBSEN, J.R. GRAYBILL, A.M. SUGAR, V.J. MCAULIFFE, S.E. FOLLANSBEE, C.U. TUAZON, J.J. STERN, J. FEINBERG, R. HAFNER, W.E. DISMUKES, NIAID AIDS CLINICAL TRIALS GROUP and NIAID MYCOSES STUDY GROUP: A controlled trial of fluconazole or amphotericin B to prevent relapse of cryptococcal meningitis in patients with the acquired immunodeficiency syndrome. N. Engl. J. Med. 326 (1992b) 793–798.

PRICE, R.W. and B.J. BREW: The AIDS dementia complex. J. Infect. Dis. 158 (1988) 1079–1083.

PRICE, R., N.L. CHERNIK, L. HORTA-BARBOSA and J.B. POSNER: Herpes simplex encephalitis in an anergic patient. Am. J. Med. 54 (1973) 222–227.

PRICE, R.W., S. NIELSEN, B. HORTEN, M. RUBINO, B. PADGETT and D. WALKER: Progressive multifocal leukoencephalopathy: a burnt-out case. Ann. Neurol. 13 (1983) 485–490.

PRICE, T.A., R.A. DIGIOIA and G.L. SIMON: Ganciclovir treatment of cytomegalovirus ventriculitis in a patient infected with human immunodeficiency virus. Clin. Infect. Dis. 15 (1992) 606–608.

PUMAROLA-SUNE, T., B.A. NAVIA, C. CARDON-CARD ET AL.: HIV antigen in the brains of patients with AIDS dementia complex. Ann. Neurol. 21 (1987) 490–496.

PURSELL, K.J., E.E. TELZAK and D. ARMSTRONG: Aspergillus species colonization and invasive disease in patients with AIDS. Clin. Infect. Dis. 14 (1992) 141–148.

RADHAKRISHNAN, V.V. and A. MATHAI: Detection of mycobacterial antigen in cerebrospinal fluid: diagnostic and prognostic significance. J. Neurol. Sci. 99 (1990) 93–99.

RAND, C.W.: Tuberculous abscesses of the brain secondary to tuberculosis of the caecum. Surg. Gynecol. Obstetr. 60 (1935) 229–235.

RAND, K.H., K.P. JOHNSON, L.J. RUBENSTEIN ET AL.: Adenine arabinoside in the treatment of progressive multifocal leukoencephalopathy: use of virus containing cells in the urine to assess response to therapy. Ann. Neurol. 1 (1977) 458–462.

REGNERY, R.L., J.E. CHILDS and J.E. KOEHLER: Infections associated with Bartonella species in persons infected with human immunodeficiency virus. Clin. Infect. Dis. 21 (1995) S94–S98.

RENOLD, C., A. SUGAR, J-P. CHAVE, L. PERRIN, J. DELAVELLE, G. PIZZOLATO, P. BURKHARD, V. GABRIEL and B. HIRSCHEL: Toxoplasma encephalitis in patients with the acquired immunodeficiency syndrome. Medicine (Baltimore) 71 (1992) 224–239.

REVELLO, M.G., E. PERCIVALLE, A. SARASINI, F. BALDANTI, M. FURIONE and G. GERNA: Diagnosis of human cytomegalovirus infection of the nervous system by pp65 detection in polymorphonuclear leukocytes of cerebrospinal fluid from AIDS patients. J. Infect. Dis. 170 (1994) 1275–1279.

RIBAULT, J. and J. COLOMBANI: Le test d'immunofluorescence appliqué au diagnostic de la syphilis. Comparison avec le test de Nelson et la serologic classique. II. Etude de 411 liquides cephalorachidiens. Pathol. Biol. 12 (1964) 276–285.

RIBERA, E., J.M. MARTINEZ-VAZQUEZ, I. OCANA, R.M. SEGURA and C. PASCUAL: Activity of adenosine deaminase in cerebrospinal fluid for the diagnosis and follow-up of tuberculous meningitis in adults. J. Infect. Dis. 155 (1987) 603–607.

RICHARDS, F.O., JR., J.A. KOVACS and B.J. LUFT: Preventing toxoplasmic encephalitis in persons infected with human immunodeficiency virus. Clin. Infect. Dis. 21 (1995) S49– S56.

RICHARDSON, E.P., JR.: Progressive multifocal leukoencephalopathy. In: P.J. Vinken and G.W. Bruyn (Eds.), Handbook of Clinical Neurology, Vol 9. Multiple Sclerosis and Other Demyelinating Diseases. North Holland, NY, Elsevier (1970) 486–499.

RICHARDSON, E.P., JR.: Our evolving understanding of progressive multifocal leukoencephalopathy. Ann. N.Y. Acad. Sci. 230 (1974) 358–364.

RICHARDSON, E.P., JR.: Progressive multifocal leukoencephalopathy 30 years later. N. Engl. J. Med. 318 (1988) 315–316.

ROBERT, M.E., J.J. GERAGHTY, III, S.A. MILES, M.E. CORNFORD and H.V. VINTERS: Severe neuropathy in a patient with acquired immune deficiency syndrome (AIDS). Evidence for widespread cytomegalovirus infection of peripheral nerve and human immunodeficiency virus like immunoreactivity of anterior horn cells. Acta Neuropathol. (Berlin) 79 (1989) 255–261.

ROCHA, A., A.C.O. DE MENSES, A.M. DA SILVA, M.S. FERREIRA,

S.A. NISHIOCA, M.K.N. BURGARELLI, E. ALMEIDA, G. TURCATO, JR., K. METZE and E.R. LOPEZ: Pathology of patients with Chagas' disease and acquired immunodeficiency syndrome. Am. J. Trop. Med. Hyg. 50 (1994) 261–268.

ROCKWELL, D., F.L. RUBEN, A. WINKELSTEIN ET AL.: Absence of immune deficiencies in a case of progressive multifocal leukoencephalopathy. Am. J. Med. 61 (1976) 433–436.

ROMANOWSKI, B., R. SUTHERLAND, G.H. FICK, D. MOONEY and E.J. LOVE: Serological response to treatment of infectious syphilis. Ann. Intern. Med. 114 (1991) 1005–1009.

ROSEMBERG, S., C.J. CHAVES, M.L. HIGUCHI, M.B.S. LOPEZ, L.H.M. CASTRO and L.R. MACHADO: Fatal meningoencephalitis caused by reactivation of trypanosoma cruzi infection in a patient with AIDS. Neurology 42 (1992) 640–642.

ROSENBLUM, M.K: Bulbar encephalitis complicating trigeminal zoster in the acquired immune deficiency syndrome. Hum. Pathol. 20 (1989) 292–295.

ROSTAD, S.W., K. OLSON, J. MCDOUGALL, C-M. SHAW and E.C. ALVORD, JR.: Transsynaptic spread of varicella-zoster virus through the visual system: a mechanism of viral dissemination in the central nervous system. Hum. Pathol. 20 (1989) 174–179.

ROULLET, E., V. ASSUERUS, J. GOZLAN, A. ROPERT, G. SAID, M. BAUDRIMONT, M. EL AMRANI, C. JACOMET, C. DUVIVIER, G. GONZALES-CANALI, M. KIRSTETTER, C. MEYOHAS, O. PICARD and W. ROZENBAUM: Cytomegalovirus multifocal neuropathy in AIDS: analysis of 15 consecutive cases. Neurology 44 (1994) 2174–2182.

ROUSSEAU, F., C. PERRONNE, G. RAGUIN, D. THOUVENOT, A. VIDAL, C. LEPORT and J.L. VILDE: Necrotizing retinitis and cerebral vasculitis due to varicella-zoster virus in patients infected with the human immunodeficiency virus (Letter). Clin. Infect. Dis. 17 (1993) 943–944.

ROVIRA, M.J., M.J.D. POST and B.C. BOWEN: Central nervous system infections in HIV-positive persons. Neuroimaging Clin. N. Am. 1 (1991) 179–200.

ROWLEY, A.H., R.J. WHITLEY, F.D. LAKEMAN and S.M. WOLINSKY: Rapid detection of herpes simplex virus DNA in cerebrospinal fluid of patients with herpes simplex encephalitis. Lancet 335 (1990) 440–441.

ROYAL, W. III, JR., O.A. SELNES, M. CONCHA, T.E. NANCE-SPRONSON and J.C. MCARTHUR: Cerebrospinal fluid human immunodeficiency virus type 1 (HIV-1) p24 antigen levels in HIV-1-related dementia. Ann. Neurol. 36 (1994) 32–39.

ROZEMBERG, F. and P. LEBON: Amplification and characterization of herpes virus DNA in cerebrospinal fluid from patients with acute encephalitis. J. Clin. Microbiol. 29 (1991) 2412–2417.

ROZENBAUM, R. and A.J.R. GONCALVES: Clinical epidemiological study of 171 cases of cryptococcosis. Clin. Infect. Dis. 18 (1994) 369–380.

RYDER, J.W., K. CROEN, B.K. KLEINSCHMIDT-DEMASTERS, J.M. OSTROVE, S.E. STRAUS and D.L. COHN: Progressive encephalitis three months after resolution of cutaneous zoster in a patient with AIDS. Ann. Neurol. 19 (1986) 182–188.

SAAG, M.S., W.G. POWDERLY, G.A. CLOUD, P. ROBINSON, M.H. GRIECO, P.A. SHARKEY, S.E. THOMPSON, A.M. SUGAR, C.U. TUAZON, J.F. FISHER, N. HYSLOP, J.M. JACOBSEN, R. HAFNER, W.E. DISMUKES and the NIAID MYCOSES STUDY GROUP and the AIDS CLINICAL TRIALS GROUP: Comparison of amphotericin B with fluconazole in the treatment of acute AIDS-related cryptococcal meningitis. N. Engl. J. Med. 326 (1992) 83–89.

SABA, J., P. MORLAT, F. RAFFI, V. HAZEBROUCQ, V. JOLY, C. LEPORT and J.L. VILDE: Pyrimethamine plus azithromycin for treatment of acute toxoplasmic encephalitis in patients with AIDS. Eur. J. Clin. Microbiol. Infect. Dis. 12 (1993) 853–856.

SADA, E., G.M. RUIZ-PALACIOS, Y. LOPEZ-VIDAL and S. PONCE DE LEON: Detection of mycobacterial antigens in cerebrospinal fluid of patients with tuberculous meningitis by enzyme-linked immunosorbent assay. Lancet ii (1983) 651–652.

SAFRIN, S.: Treatment of acyclovir-resistant herpes simplex virus infections in patients with AIDS. J. Acq. Immune Defic. Syndr. 5 (1992) S29–S32.

SAFRIN, S., C. CRUMPACKER and P. CHATIS: A controlled trial comparing foscarnet with vidarabine for acyclovir-resistant muco-cutaneous herpes simplex in the acquired immunodeficiency syndrome. N. Engl. J. Med. 325 (1991) 551–555.

SAGE, J.I., M.P. WEINSTEIN and D.C. MILLER: Chronic encephalitis possibly due to herpes simplex virus: two cases. Neurology 35 (1985) 1470–1472.

SAID, G., C. LACROIX and P. CHEMOVILLI: Cytomegalovirus neuropathy in acquired immunodeficiency syndrome: a clinical and pathological study. Ann. Neurol. 29 (1991) 139–146.

SANCHEZ-PORTOCARRERO, J., E. PEREZ-CECILIA, V. ROCA, J. RABANO, P. MARTIN- RABADAN, C. RAMIREZ and E. VARELA DE SEIJAS: Meningitis por candida albicans en 2 adictos a drogas por via parenteral, revision de la literatura. Enferm. Infecc. Microbiol. Clin. 11 (1993) 244–249.

SCHLOSSBERG, D., Y. MORAD, T.B. KROUSE, D.J. WEAR and C. ENGLISH: Culture-proved disseminated cat-scratch disease in acquired immunodeficiency sydnrome. Arch. Intern. Med. 149 (1989) 1437–1439.

SCHMIDBAUER, M., H. BUDKA, W. ULRICH and P. AMBROS: Cytomegalovirus (CMV) disease of the brain in AIDS and connatal infection: a comparative study by histology, immunocytochmistry and in situ hybridization. Acta Neuropathol. (Berlin) 79 (1989a) 286–293.

SCHMIDBAUER, M., H.BUDKA and P. AMBROS: Herpes simplex virus (HSV) DNA in microglial nodular

brainstem encephalitis. J. Neuropathol. Exp. Neurol. 48 (1989b) 645–652.

SCHMIDBAUER, M., H. BUDKA and K.V. SHAH: Progressive multifocal leukoencephalopathy (PML) in AIDS and in the pre-AIDS era: A neuropathological comparison using immunocytochemistry and in situ DNA hybridization for virus detection. Acta Neuropathol. (Berlin) 80 (1990) 375–380.

SCHOONDERMARK-VAN DE VEN, E., J. GALAMA, C. KRAAIJE-VELD, J. VAN DRUTEN, J. MEUWISSEN and W. MELCHERS: Value of the polymerase chain reaction for the detection of toxoplasma gondii in cerebrospinal fluid from patients with AIDS. Clin. Infect. Dis. 16 (1993) 661–666.

SCHWARTZMAN, W.A., A. MARCHEVSKY and R.D. MEYER: Epithelioid angiomatosis or cat scratch disease with splenic and hepatic abnormalities in AIDS: case report and review of the literature. Scand. J. Infect. Dis. 22 (1990) 121–133.

SCHWARTZMAN, W.A., M. PATNAIK, N.E. BARKA and J.B. PETER: Rochalimaea antibodies in HIV-associated neurologic disease. Neurology 44 (1994) 1312–1316.

SCHWARZ, T.F., K. LOESCHKE, I. HANUS, G.JAGER, W. FEIDEN and F.H. STEFANI: CMV encephalitis during ganciclovir therapy of CMV retinitis. Infection 18 (1990) 289–290.

SELHORST, J.B., K.F. DUCY, J.M. THOMAS ET AL.: Remission and immunologic reversals (Abstr.). Neurology. 28 (1978) 337.

SHAH, K.V.: Polyomaviruses. In: B.N. Fields, D.M. Knipe et al. (Eds.), Virology, Vol. 2. New York, Raven Press (1990) 1609–1623.

SHAO, P.P., S.M. WANG, S.G. TUNG ET AL.: Clinical analysis of 445 adult cases of tuberculous meningitis. Chinese J. Tuberc. Resp. Dis. 3 (1980) 131–132.

SHARER, L.R.: Pathology of HIV-1 infection of the central nervous system. A review. J. Neuropathol. Exp. Neurol. 51 (1992) 3–11.

SHARER, L.R. and R. KAPILA: Neuropatholgic observations in acquired immunodeficiency syndrome (AIDS). Acta Neuropathol. (Berlin) 66 (1985) 188–198.

SHELLER, J.R. and R.M. DES PREZ: CNS tuberculosis. Neurol. Clin. 4 (1986) 143–158.

SHOJI, H., Y. HONDA, I. MURAI, Y. SATO, K. OIZUMI and R. HONDO: Detection of varicella-zoster DNA by polymerase chain reaction in cerebrospinal fluid of patients with herpes zoster meningitis. J. Neurol. 239 (1992) 69–70.

SIEGAL, F.P., C. LOPEZ, G.S. HAMMER ET AL.: Severe acquired immunodeficiency in male homosexuals manifested by chronic perianal ulcerative Herpes simplex lesions. N. Engl. J. Med. 305 (1981) 1439–1444.

SIMA, A.A.F., S.D. FINKELSTEIN and D.R. MCLACHLAN: Multiple malignant astrocytomas in a patient with spontaneous progressive multifocal leukoencephalopathy. Ann. Neurol. 14 (1983) 183–188.

SIMONSEN, J.N.,W. CAMERON, M.N. GAKINYA ET AL.: Human immunodeficiency virus infection among men with sexually transmitted diseases: experience from a center in Africa. Morb. Mortal. Wkly. Rep. 319 (1988) 274–278.

SINGER, C., J.R. BERGER, B.C. BOWEN, J.H. BRUCE and W.J. WEINER: Akinetic-rigid sydnrome in a 13-year-old female with HIV related progressive multifocal leukoencephalopathy. Mov. Disord. 8 (1993) 113–116.

SINGER, E.J., G.L. STONER, P. SINGER, U. YOMIYASU, E. LICHT, B. FAHY-CHANDON and W.W. TOURTELLOTTE: AIDS presenting as progressive multifocal leukoencephalopathy with clinical response to zidovudine. Acta Neurolog. Scand. 90 (1994) 443–447.

SINGH, N., V.L. YU and J.D. RIHS: Invasive aspergillosis in AIDS. South. Med. J. 84 (1991) 822–827.

SINGHAL, B.S., S.N. BHAGWATI, A.H. SYED and G.W. LAUD: Raised intracranial pressure in tuberculous meningitis. Neurology (India) 23 (1975) 32–39.

SISON, J.P., K.A. KEMPER, M. LOVELESS ET AL.: Disseminated acanthamoeba infection in patients with AIDS: case reports and review. Clin. Inf. Dis. 20 (1995) 1207–1216.

SKOLNIK, P.R. and S.M. DE LA MONTE: Case records of the Massachusetts General Hospital, case 52-1990. N. Engl. J. Med. 26 (1990) 1823–1833.

SMALL, P., L.W. MCPHAUL, C.D. SOOY, C.B. WOFSY and M.A. JACOBSEN: Cytomegalovirus infection of the laryngeal nerve presenting as hoarseness in patients with acquired immunodeficiency syndrome. Am. J. Med. 86 (1989) 108–110.

SMALL, P.M., G.F. SCHECTER, P.C. GOODMAN, M.A. SAND, R.E. CHAISSON and P.C. HOPEWELL: Treatment of tuberculosis in patients with advanced human immunodeficiency virus infection. N. Engl. J. Med. 324 (1991) 289–294.

SMITH, A.L.: Tuberculous meningitis in childhood. Med. J. Aust. 1 (1975) 57–60.

SMITH, C.R., A.A.F. SIMA, I.E. SALIT and F. GENTILI: Progressive multifocal leukoencephalopathy: failure of cytarabine therapy. Neurology 32 (1982) 200–203.

SMITH, H.V. and R.L. VOLLUM: The treatment of tuberculous meningitis. Tubercle 37 (1956) 301–320.

SMITH, H.V., L.M. TAYLOR and G. HUNTER: The blood–cerebrospinal fluid barrier in tuberculous meningitis and allied conditions. J. Neurol. Neurosurg. Psychiatry 18 (1955) 237–249.

SMITH, J. and R. GODWIN-AUSTEN: Hypersecretion of antidiuretic hormone due to tuberculous meningitis. Postgrad. Med. J. 56 (1980) 41–44.

SNIDER, W.D., D.M. SIMPSON, S. NIELSEN, J.W.M. GOLD, C.E. METROKA and J.B. POSNER: Neurological complications of acquired immune deficiency syndrome: analysis of 50 patients. Ann. Neurol. 14 (1983) 403–418.

SNOECK, R., M. GERARD, C. SADZOT-DELVAUX, G. ANDREI, J.

BALZARINI, D. REYMEN, N. AHADI, J.M. DE BRUYN, J. PIETTE, B. RENTIER, N. CLUMECK and E. DE CLERCQ: Meningo-radiculoneuritis due to acyclovir-resistant varicella zoster virus in an acquired immune deficiency syndrome patient. J. Med. Virol. 42 (1994) 338–347.

SO, Y.T. and OLNEY: Acute lumbosacral polyradiculopathy in acquired immuno deficiency syndrome: experience in 23 patients. Ann. Neurol. 35 (1994) 53–58.

SOLARI, A., H. SAAVEDRA, C. SEPULVEDA, D. ODDO, G. ACUNA, J. LABARCA, S. MUNOZ, G. CUNY, C. BRENGUES, F. VEAS and R.T. BRYAN: Successful treatment of Trypanosoma cruzi encephalitis in a patient with hemophilia and AIDS. Clin. Infect. Dis. 16 (1993) 255–259.

SPITZER, E.D., S.G. SPITZER, L.F. FREUNDLICH and A. CASADEVALL: Persistance of initial infection in recurrent Cryptococcus neoformans meningitis. Lancet 341 (1993) 595–596.

SPITZER, P.G., S.M. HAMMER and A.W. KARCHMER: Treatment of Listeria monocytogenes infection with trimethoprim-sulfamethoxazole: case report and review of the literature. Rev. Infect. Dis. 8 (1986) 427–430.

SPOOR, T.C., P. WYNN, W.C. HARTEL and C.S. BRYAN: Ocular syphilis: acute and chronic. J. Clin. Neuro-ophthalmol. 3 (1983) 197–203.

SPRACH, D.H., L.A. PANTHER, D.R. THORNING, J.E. DUNN ET AL.: Intracerebral bacillary angiomatosis in a patient with human immunodeficiency virus. Ann. Intern. Med. 116 (1992) 740–742.

STAM, F.C.: Multifocal leukoencephalopathy with slow progression and very long survival. Psychiatr. Neurol. Neurochir. 69 (1966) 453–459.

STAMM, W.E., H.H. HANDSFIELD, A.M. ROMPALO ET AL.: The association between genital ulcer disease and acquisition of HIV infection in homosexual men. J. Am. Med. Assoc. 260 (1988) 1429–1433.

STEAD, W.W., J.P. LOFGREN, E. WARREN and C. THOMAS: Tuberculosis as an endemic and nosocomial infection among the elderly in nursing homes. N. Engl. J. Med. 312 (1985) 1483–1487.

STEINER, P. and C. PORTUGALEZA: Tuberculous meningitis in children. Am. Rev. Resp. Dis. 107 (1973) 22–29.

STEINHART, R., A.L. REINGOLD, F. TAYLOR, G. ANDERSON and J.D. WENGER: Invasive Haemophilus influenzae infections in men with HIV infection. J. Am. Med. Assoc. 268 (1992) 3350–3352.

STERN, J.J., B.J. HARTMAN, P. SHARKEY, V. ROWLAND, K.E. SQUIRES, H.W. MURRAY and J.R. GRAYBILL: Oral fluconazole therapy for patients with acquired immunodeficiency syndrome and cryptococcosis: experience with 22 patients. Am. J. Med. 85 (1988) 477–480.

STOCKSTILL, M.T. and C.A. KAUFFMAN: Comparison of cryptococcal and tuberculous meningitis. Arch. Neurol. 40 (1983) 81–85.

STOUMBOS, V.D. and M.L. KLEIN: Syphilitic retinitis in a

patient with acquired immunodeficiency syndrome-related complex. Am. J. Ophthalmol. 1 (1987) 103–104.

STRAUSS, M. and E. FINE: Aspergillus otomastoiditis in acquired immunodeficiency syndrome. Am. J. Otolaryngol. 12 (1991) 49–53.

STROBEL, M., P.H. BEAUCLAIR and J. LACAVE: Seronegative syphilis in AIDS (Letter). Presse Méd. 18 (1989) 1440.

STUDIES OF OCULAR COMPLICATIONS OF AIDS RESEARCH GROUP IN COLLABORATION WITH THE AIDS CLINICAL TRIALS GROUP: Mortality in patients with acquired immunodeficiency syndrome treated with either foscarnet or ganciclovir for cytomegalovirus retinitis. N. Engl. J. Med. 326 (1992) 213–220.

SUGAR, A.M., J.J. STERN and B. DUPONT: Overview: treatment of cryptococcal meningitis. Rev. Infect. Dis. 12 (1990) S338–S348.

SUGAR, M. and C. SAUNDERS: Oral fluconazole as suppressive therapy of disseminated cryptococcosis in patients with acquired immunodeficiency syndrome. Am. J. Med. 85 (1988) 481–489.

SULLIVAN, W.M., G.G. .KELLEY, P. O'CONNOR, P.S. DICKEY, J.H. KIM, R. ROBBINS and G.I. SHULMAN: Hypopituitarism associated with a hypothalamic CMV infection in a patient with AIDS. Am. J. Med. 92 (1992) 221–223.

SUMAYA, C.V., M. SIMEK, M.H.D. SMITH, M.F. SEIDEMANN, G.S. FERRISS and W. RUBIN: Tuberculous meningitis in children during the isoniazid era. J. Pediatr. 87 (1975) 43–49.

SUNDERAM, G., R.J. MCDONALD, T. MANIATIS ET AL.: Tuberculosis as a manifestation of the acquired immunodeficiency syndrome (AIDS). J. Am. Med. Assoc. 256 (1986) 362–366.

SZE, G. and R.D. ZIMMERMANN: The magnetic resonance imaging of infectious and inflammatory disease. Radiol. Clin. N. Am. 26 (1988) 839–859.

TADA, H., J. RAPPAPORT, M. LASHGARI, S. AMINI, F. WONG-STAAL and K. KHALILI: Trans-activation of the JC virus late promoter by the *tat* protein of type 1 human immunodeficiency virus in glial cells. Proc. Natl. Acad. Sci. 87 (1990) 3479–3483.

TAGUCHI, F., J. KAJIOKA and T. MIYAMURA: Prevalence rate and age of acquisition of antibodies aganist JC virus and BK virus in human sera. Microbiol. Immunol. 26 (1982) 1057–1064.

TALPOS, D., R.D. TIEN and J.R. HESSELINK: Magnetic resonance imaging of AIDS related polyradiculopathy. Neurology 41 (1991) 1996–1997.

TAN, B., M. WELDON-LINNE, D.P. RHONE, C.L. PENNING and G.S. VISVESVARA: Acanthamoeba infection presenting as skin lesions in patients with the acquired immunodeficiency syndrome. Arch. Pathol. Lab. Med. 117 (1993) 1043–1046.

TAN, C.T.: Intracranial hypertension causing visual failure in cryptococcus meningitis. J. Neurol. Neurosurg. Psychiatry 51 (1988) 944–946.

TAN, G., L. KAUFMAN, E.M. PETERSON and L.M. DE LA MAZE: Disseminated atypical blastomycosis in two patients with AIDS. Clin. Infect. Dis. 16 (1992) 107–111.

TAN, S.V., R.J. GUILOFF, F. SCARAVILLI, P.E. KLAPPER, G.M. CLEATOR and B.G. GAZZARD: Herpes simplex type 1 encephalitis in acquired immunodeficiency syndrome. Ann. Neurol. 34 (1993) 619–622.

TARSY, D., E.M. HOLDEN, J.M. SEGARRA ET AL.: 5-iodo-2′-deoxyuridine (IUDR): NSC-39661 given intraventricularly in the treatment of progressive multifocal leukoencephalopathy. Cancer Chemother. Rep., Part 1, 57 (1973) 73–78.

TASHIRO, K., S. DOI, F. MORIWAKA and M. NOMURA: Progressive multifocal leucoencephalopathy with magnetic resonance imaging verification and therapeutic trials with interferon. J. Neurol. 234 (1987) 427–429.

TEDDER, D.G., R. ASHLEY R, K.L. TYLER and M.J. LEVIN: Herpes simplex virus infection as a cause of benign recurrent lymphocytic meningitis. Ann. Intern. Med. 121 (1994) 334–338.

TELZAK, E.E., D. ZWEIG, M.S. GREENBERG, J. HARRISON, R.L. STONEBURNER and S. SCHULTZ: Syphilis treatment response in HIV-infected individuals. AIDS 5 (1991) 591–595.

TEOH, R. and M. HUMPHRIES: Tuberculous meningitis. In: H.P. Lambert (Ed.), Infections of the Central Nervous System. Philadelphia, PA, B.C. Dekker (1991) 189–206.

TEOH, R., M.J. HUMPHRIES and G. O'MAHONY: Symptomatic intracranial tuberculoma developing during treatment of tuberculosis: a report of 10 patients and review of the literature. Q. J. Med. 241 (1987) 449–460.

THOMAS, E.W.: Some aspects of neurosyphilis. Med. Clin. N. Am. 48 (1964) 699–705.

THORNTON, C.A., S. HOUSTON and S. LATIF: Neurocysticercosis and human immunodeficiency virus infection, a possible association. Arch. Neurol. 49 (1992) 963–965.

TIEN, R.D., P.K. CHU, J.R. HESSELINK, A. DUBERG and C. WILEY: Intracranial cryptococcosis in immunocompromised patients: CT and MR findings in 29 cases. Am. J. Neuroradiol. 12 (1991) 283–289.

TOLGE, C.F. and S.A. FACTOR: Focal dystonia secondary to cerebral toxoplasmosis in a patient with acquired immune deficiency syndrome. Mov. Disord. 6 (1991) 69–72.

TORNATORE, C., J.R. BERGER, S. HOUFF ET AL.: Detection of JC virus DNA in peripheral lymphocytes from patients with and without progressive multifocal leukoencephalopathy. Ann. Neurol. 31 (1992) 454–462.

TRAUB, M., A.C.F. COLCHESTER, D.P.E. KINGSLEY and M. SWASH: Tuberculosis of the central nervous system. Q. J. Med. 53 (1984) 81–100.

TRAVIESA, D., S.D. PRYSTOWSKY, J.D. NELSON ET AL.: Cerebrospinal fluid findings in asymptomatic patients with reactive serum fluorescent treponemal antibody absorption tests. Ann. Neurol. 4 (1978) 524–530.

TRENKWALDER, P., C. TRENKWALDER, W. FEIDEN, T.J. VOGL, K.M. EINHAUPL and H. LYDTIN: Toxoplasmosis with early intracerebral hemorrhage in a patient with the acquired immunodeficiency syndrome. Neurology 42 (1992) 436–438.

TUCKER, T., R.D. DIX, C. KATZEN, R.L. DAVIS and J.W. SCHIMIDLEY: Cytomegalovirus and herpes simplex ascending myelitis in a patient with acquired immune deficiency syndrome. Ann. Neurol. 18 (1985) 74–79.

TURNER, T.B. and D.H. HOLLANDER: Biology of the Treponematoses. Geneva, WHO (1957).

UDANI, P.M., U.C. PAREKH and D.K. DASTUR: Neurological and related syndromes in CNS tuberculosis. Clinical features and pathogenesis. J. Neurol. Sci. 14 (1971) 314–357.

UTTAMCHANDANI, R.B., G.L. DAIKOS, R.R. REYES ET AL.: Nocardiosis in 30 patients with advanced human immunodeficiency virus infection: clinical features and outcome. Clin. Infect. Dis. 18 (1994) 348–353.

VAISHNAVI, C., U.K. DHAND, R. DHAND, N. AGNIHOTRI and N.K. GANGULY: C-reactive protein, immunoglobulin profile and mycobacterial antigens in cerebrospinal fluid of patients with pyogenic and tuberculous meningitis. J. Hyg. Epidemiol. Microbiol. Immunol. 36 (1992) 317–325.

VAN EIJK, F.V.W., E.C.H. WOLTERS, J.A. TUTUARIMA ET AL.: Effect of early and late syphilis on central nervous system: cerebrospinal fluid changes and neurological deficit. Genitourin. Med. 63 (1987) 77–82.

VAN HORN, G., F.O. BASTIEN and J.L. MOAKE: Progressive multifocal leukoencephalopathy: failure of response to transfer factor and cytarabine. Neurology 28 (1978) 794–797.

VELASCO-MARTINEZ, J.J., A. GUERRERO-ESPEJO, E. GOMEZ-MAMPASO, NAVAS-ELORZA and G. GARCIA-RIBAS: Tuberculous brain abscess should be considered in HIV/AIDS patients. AIDS 9 (1995) 1197–1199.

VILLORIA, M.F., J. DE LA TORRE, F. FORTEA, L. MUNOZ, T. HERNANDEZ and J.J. ALARCON: Intracranial tuberculosis in AIDS: CT and MRI findings. Neuroradiology 34 (1992) 11–14.

VINTERS, H.V. and K.H. ANDERS: Neuropathology of AIDS. Boca Raton, FL, CRC Press (1990) 61–65, 69.

VINTERS, H.V., W.F. GUERRA, L. EPPOLITO, P.E. KEITH III: Necrotizing vasculitis of the nervous system in a patient with AIDS-related complex. Neuropathol. Appl. Neurobiol. 14 (1988) 417–424.

VINTERS, H.V., M.K. KWOK, H.W. HO, K.H. ANDERS, U. TOMIYASU, W.L. WOLFSON and F. ROBERT: Cytomegalovirus in the nervous system of patients with the acquired immune deficiency syndrome. Brain 112 (1989) 245–268.

VITAL, C., E. MONLUM, A. VITAL, M.L. MARTIN-NEGRIER, V. CALES, F. LEGER, M. LONGY-BOURSIER, M. LEBRAS and B. BLOCH: Concurrent herpes simplex type 1 necrotizing encephalitis, cytomegalovirus ventriculoencephalitis and cerebral lymphoma in an AIDS patient. Acta Neuropathol. (Berlin) 89 (1995) 105–108.

VLCEK, B., K.J. BURCHIEL and T. GORDON: Tuberculous meningitis presenting as an obstructive myelopathy: case report. J. Neurosurg. 60 (1984) 196–199.

VON EINSIEDEL, R.W., T.D. FIFE, A.J. AKSAMIT, M.E. CORNFORD, D.L. SECOR, U. TOMIYASU, H.H. ITABASHI and H.V. VINTERS: Progressive multifocal leukoencephalopathy in AIDS: a clinicopathologic study and review of the literature. J. Neurol. 240 (1993) 391–406.

WAECKER, N.J., JR. and J.D. CONNOR: Central nervous system tuberculosis in children: a review of 30 cases. Pediatr. Infect. Dis. J. 999 (1990) 539–543.

WALKER, D.L.: Progressive multifocal leukoencephalopathy: an opportunistic viral infection of the central nervous system. In: P.J. Vinken and G.W. Bruyn (Eds.), Handbook of Clinical Neurology, Vol. 34. Infections of the Nervous System, Part II. Amsterdam, Elsevier/North Holland (1978) 307–329.

WALKER, D.L. and B.L. PADGETT: Progressive multifocal leukoencephalopathy. In: H. Fraenkel-Conrat and R.R. Wagner (Eds.), Comprehensive Virology. New York, Plenum (1983).

WALKER, D.L. and B.L. PADGETT: The epidemiology of human polyomaviruses. In: J.L. Sever and D. Madden (Eds.), Polyomaviruses and Human Neurological Disease. New York, Alan R. Liss (1983) 99–106.

WARD, D.: Dapsone/pyrimethamine for the treatment of toxoplasmic encephalitis. Int. Conf. AIDS 8 (1992) 133 (Abstr. POP3277).

WASZ-HOCKERT, O. and M. DONNER: Results of the treatment of 191 children with tuberculous meningitis in the years 1949–1954. Acta Paediatr. 51 (1962) 7–25.

WASZ-HOCKERT, O. and M. DONNER: Late prognosis in tuberculous meningitis. Acta Paediatr. 141 (1963) 93–102.

WATT, G., G. ZARASPE, S. BAUTISTIA and L.W. LAUGHLIN: Rapid diagnosis of tuberculous meningitis by using an enzyme-linked immunosorbent assay to detect mycobacterial antigen and antibody in cerebrospinal fluid. J. Infect. Dis. 158 (1988) 681–686.

WEBER, T., R.W. TURNER, B. RUF ET AL.: JC virus detected by polymerase chain reaction in cerebrospinal fluid of AIDS patients with progressive multifocal leukoencephalopathy. In: J.R. Berger and R.L. Levy (Eds.), Neurological and Neuropsychological Complications of HIV Infection. Proc. Satellite Meet. Int. Conf. AIDS (1990) 100.

WEBER, T., M. BODEMER, W. ENZENSBERGER, C. EGGERS, S. FRYE, W. HEIDE, W. LUKE and G. HUNSMANN: Enhanced sensitivity of JC Virus DNA detection in CSF by nested primer PCR. 1994. Neuroscience of HIV infection (Abstr.). Basic Clin. Front. (1994).

WECK, P.K., D.A BUDDIN and J.K. WHISNANT: Interferons in the treatment of genital human papillomavirus infections. Am. J. Med. 85 (1988) 159.

WEIDENHEIM, K.M., S.J. NELSON, K. KURE, C. HARRIS, L. BIEMPICA and D.W. DICKSON: Unusual patterns of *histoplasma capsulatum* in a patient with the acquired immunodeficiency virus. Hum. Pathol. 23 (1992) 581–586.

WEISS, W. and H.F. FLIPPIN: The changing incidence and prognosis of tuberculous meningitis. Am. J. Med. Sci. 250 (1965) 46–59.

WETLI, C.V., S.D. WEISS, T.J. CLEARY and E. GYORI: Fungal cerebritis from intravenous drug abuse. J. Forensic Sci. 29 (1984) 260–268.

WHEAT, J.: Endemic mycoses in AIDS: a clinical review. Clin. Microbiol. Rev. 8 (1995) 146–159.

WHEAT, J., R. HAFNER, M. WULFSON, P. SPENCER, K. SQUIRES, W. POWDERLY, B. WONG, M. RINALDI, M. SAAG, R. HAMILL, R. MURPHY, P. CONNOLLY-STRINGFIELD, N. BRIGGS, S. OWENS, NATIONAL INSTITUTE OF ALLERGY AND INFECTIOUS DISEASES CLINICAL TRIALS and MYCOSES STUDY GROUP COLLABORATORS: Prevention of relapse of histoplasmosis with itraconazole in patients with the acquired immune deficiency syndrome. Ann. Intern. Med. 118 (1993) 610–616.

WHEAT, L.J., R.B. KOHLER, R.P. TEWARI, M. GARTEN and M.L.V. FRENCH: Significance of histoplasma antigen in the cerebrospinal fluid of patients with meningitis. Arch. Intern. Med. 149 (1989) 302–304.

WHEAT, L.J., B.E. BATTEIGER and B. SATHAPATAYAVONGS: *Histoplasma capsulatum* infections of the nervous system: a clinical review. Medicine (Baltimore) 69 (1990a) 244–260.

WHEAT, L.J., P.A. CONNOLLY-STRINGFIELD, R.L. BAKER, M.F. CURFMAN, M.E. EADS, K.S. ISRAEL, S.A. NORRIS, D.H. WEBB and M.L. ZECKEL: Disseminated histoplasmosis in the acquired immune deficiency syndrome: clinical findings, diagnosis and treatment, and review of the literature. Medicine (Baltimore) 69 (1990b) 361–374.

WHEAT, L.J., P. CONNOLLY-STRINGFIELD, R. BLAIR, K. CONNOLLY, T. GARRINGER and B.P. KATZ: Histoplasma relapse in patients with AIDS: detection using *histoplasma capsulatum* variety capsulatum antigen levels. Ann. Intern. Med. 115 (1991) 936–941.

WHIMBEY E, J.W.M. GOLD, B. POLSKY ET AL.: Bacteremia and fungemia in patients with the acquired immunodeficiency syndrome. Ann. Intern. Med. 10 (1986) 511–514.

WHITE, A.C., H. DAKIK and P. DIAZ: Asymptomatic neurocysticercosis in a patient with AIDS and cryptococcal meningitis. Am. J. Med. 99 (1995) 101–102.

WHITE, M., C. CIRRINCIONE, A. BLEVINS and D. ARMSTRONG: Cryptococcal meningitis: outcome in patients with AIDS with neoplastic disease. J. Infect. Dis. 165 (1992) 960–963.

WHITEMAN, M., M.J.D. POST, J.R. BERGER, L. LIMONTE, L.G. TATE and M. BELL: PML in 47 HIV+ patients. Radiology 187 (1993) 233–240.

WHITEMAN, M.L.H., B.K. DANDAPANI, R.T. SHEBERT and M.J.D. POST: MRI of AIDS-related polyradiculomyelitis. J. Comp. Assist. Tomogr. 18 (1994) 7–11.

WHITEMAN, M., L. ESPINOZA, M.J. POST, M.D. BELL and S. FALCONE: Central nervous system tuberculosis in HIV-infected patients: clinical and radiographic findings. Am. J. Neuroradiol. 16 (1995) 1319–1327.

WHITLEY, R.J. and J.W. GNANN, JR.: Antiviral therapy. In: B. Roziman, R.J. Whitley and C. Lopez (Eds.), The Human Herpes Viruses. New York, Raven Press (1993) 329–348.

WILEY, C.A. and J.A. NELSON: Role of human immunodeficiency virus and cytomegalovirus in AIDS encephalitis. Am. J. Pathol. 133 (1988) 73–81.

WILEY, C.A., P.D. VAN PATTEN, P.M. CARPENTER, H.C. POWELL and L.J. THAL: Acute ascending necrotizing myelopathy caused by herpes simplex virus type 2. Neurology 37 (1987a) 1791–1794.

WILEY, C.A., R.E. SOFRIN, C.E. DAVIS, P.W. LAMPERT, A.I. BRAUDE, A.J. MARTINEZ and G.S. VISVESVARA: Acanthomoeba meningoencephalitis in an AIDS patient. J. Infect. Dis. 155 (1987b) 130–133.

WITRACK, B.J. and G.T. ELLIS: Intracranial tuberculosis: manifestations on computerized tomography. South. Med. J. 78 (1985) 386–392.

WITZIG, R.S., D.J. HOADLEY, D.L. GREER, K.P. ABRIOLA and R.L. HERNANDEZ: Blastomycosis and human immunodeficiency virus: three new cases and review. South. Med. J. 87 (1994) 715–719.

WOLF, D.G. and S.A. SPECTOR: Diagnosis of human cytomegalovirus central nervous system disease in AIDS patients by DNA amplification from cerebrospinal fluid. J. Infect. Dis. 166 (1992) 1412–1415.

WOLINSKY, E.: Nontuberculous mycobacteria and associated diseases. Am. Rev. Resp. Dis. 119 (1979) 107–159.

WOLINSKY, J.S., K.P. JOHNSON, K. RAND and T.C. MERIGAN: Progressive multifocal leukoencephalopathy: clinical pathological correlates and failure of a drug trial in two patients. Trans. Am. Neurol. Assoc. 101 (1976) 81–82.

WOLTERS, E.C.H.: Neurosyphilis: a changing diagnostic problem. Eur. Neurol. 26 (1987) 23–28.

WONG, B., J.W.M. GOLD, A.E. BROWN, M. LANGE, R. FRIED, M. GRIECO, D. MILDVAN, J. GIRON, M.L. TAPPER, C.W. LERNER and D. ARMSTRONG: Central nervous system toxoplasmosis in homosexual men and parenteral drug abusers. Ann. Intern. Med. 100 (1984) 36–42.

WONG, M.T., M.J. DOLAN, C.P. LATTUADA, JR., R.L. REGNERY ET AL.: Neuroretinitis, aseptic meningitis, and lymphadenitis associated with *Bartonella (Rochalimaea) henselae* infection in immunocompetent patients and patients infected with human immunodeficiency virus type 1. Clin. Infect. Dis. 21 (1995) 352–360.

WOOD, M. and M. ANDERSON: Chronic meningitis. In: Neurological Infections. Philadelphia, PA, W.B. Saunders (1988) 169–248.

WOODS, G.L. and J.C. GOLDSMITH: Aspergillus infection of the central nervous system in patients with acquired immunodeficiency syndrome. Arch. Neurol. 47 (1990) 181–184.

YAMAMOTO, L.J., D.G. TEDDER, R. ASHLEY and M.J. LEVIN: Herpes simplex virus type 1 DNA in cerebrospinal fluid of a patient with Mollaret's meningitis. N. Engl. J. Med. 325 (1991) 1082–1085.

YOSHIMURA, N. and T. OKA: Medical and surgical complications of renal transplantation: diagnosis and management. Med. Clin. N. Am. 74 (1990) 1025–1037.

YOUNG, L.S., C.V. INDERLIED, O.G. BERLIN and M.S. GOTTLIEB: Mycobacterial infection in AIDS patients, with an emphasis on the *Mycobacterium avium* complex. Rev. Infect. Dis. 8 (1986) 1024–1033.

ZAIDMAN, G.W.: Neurosyphilis and retrobulbar neuritis in a patient with AIDS. Ann. Ophthalmol. 18 (1986) 260–261.

ZAIDMAN, G.W., R.S. WEINBERG and W.T. HUMPHREY: Acquired syphilitic uveitis in homosexuals. Ophthalmol-ogy 90 (1983) 106–107.

ZAMBRANO, W., G.M. PEREZ and J. L. SMITH: Acute syphilitic blindness in AIDS. J. Clin. Neuroophthalmol. 7 (1987) 1–5.

ZARABI, M., S. SANE and B.R. GIRDANY: The chest roentgenogram in the early diagnosis of tuberculous meningitis in children. Am. J. Dis. Child. 121 (1971) 389–392.

ZARASPE-YOO, E., R. MILETICH and W.W. TOURTELLOTTE: Herpes zoster ophthalmicus with contralateral hemiplegia in a patient with autoimmune deficiency syndrome (AIDS) (Abstr.). Neurology 34 (1984) 229.

ZINTEL, H.A., H.F. FLIPPIN, A.C. NICHOLS, M.M. WILEY and J.E. RHOADS: Studies on streptomycin in man. I. Absorption, distribution, excretion and toxicity. Am. J. Med. Sci. 210 (1945) 421–430.

ZUGER, A., E. LOUIE, R.S. HOLZMAN, M.S. SIMBERKOFF and J.J. RAHAL: Cryptococcal disease in patients with the acquired immunodeficiency syndrome. Ann. Intern. Med. 104 (1986) 234–240.

ZU RHEIN, G.M. and S.M. CHOU: Particles resembling papova virions in human cerebral demyelinating disease. Science 148 (1965) 1477–1479.

ZU RHEIN, G.M.: Polyoma-like virions in a human demyelinating disease. Acta Neuropathol. (Berlin) 8 (1967) 57–68.

Handbook of Clinical Neurology, Vol. 27 (71): Systemic Diseases, Part III
M.J. Aminoff and C.G. Goetz, editors

CHAPTER 14

Non-infectious complications of HIV infection in the central nervous system

KENDRA PETERSON

Department of Neurology and Neurological Sciences, Stanford University Medical Center, Stanford, CA, U.S.A.

The most common central nervous system (CNS) manifestations of human immunodeficiency virus (HIV) infection occur as a result of direct involvement of the nervous system by HIV itself, or infection of the nervous system by a number of opportunistic organisms. However, a variety of non-infectious disorders also results in CNS dysfunction in many patients with HIV and in fact, taken together, these disorders occur with great frequency. The non-infectious CNS complications of HIV infection are important. First, they often mimic the infectious complications in their presenting clinical features, and they may be difficult to distinguish from the infectious complications on radiologic imaging studies as well. It is therefore essential that the non-infectious complications be considered in the differential diagnosis of patients with HIV who present with CNS dysfunction. Additional diagnostic testing including brain biopsy may be required to establish a specific diagnosis with certainty. Second, the non-infectious CNS complications of HIV may cause significant neurologic disability to affected patients and may significantly affect the survival of patients with HIV. Third, the clinical course and expected outcome of the non-infectious CNS complications of HIV are frequently quite different from those of the usual infectious complications. Fourth, effective treatment strategies may be available for the non-infectious CNS complications that differ dramatically from the strategies used to treat HIV-associated CNS infections. Finally, patients with HIV are prone to develop multi-factorial CNS dysfunction. Even in patients with established CNS infections, the non-infectious complications may coexist. If coexistent non-infectious complications go unrecognized and untreated they may significantly alter the clinical response to otherwise appropriate treatment of CNS infections.

In this chapter, the non-infectious CNS complications of HIV are characterized as either 'cerebrovascular' or 'neoplastic', as these two categories represent the majority of the important non-infectious CNS complications (Table 1). 'Metabolic' derangements, such as hepatic, uremic, or hypoxic encephalopathy that occur in the setting of systemic illness and organ failure are also important considerations in HIV patients with CNS dysfunction. The myriad of pharmaceutical agents used in the treatment of patients infected by HIV may result in CNS dysfunction by a number of mechanisms as well. However, a detailed discussion of the metabolic derangements resulting from either systemic organ failure or drug effects is not included in this context.

CEREBROVASCULAR COMPLICATIONS OF HIV INFECTION

Epidemiology

Cerebrovascular disease is a common CNS complication of HIV infection which must be distinguished

TABLE 1

Non-infectious CNS complications of HIV.

Metabolic encephalopathy
   Organ failure
   Drug toxicity

Cerebrovascular disorders
   Ischemic
      Thrombotic
      Embolic
   Hemorrhagic

Neoplastic
   Metastatic Kaposi's sarcoma
   Lymphoma
      Primary CNS lymphoma
      Metastatic/invasive systemic lymphoma

from infectious manifestations. Cerebrovascular disease is clinically apparent in approximately 5% of patients with HIV, though the range of reported incidence is broad and depends to some extent on the rigor with which the diagnosis is sought. Engström et al. (1989) reviewed 1600 adult patients with HIV and found 10 patients with clinical evidence of stroke, 13 with transient neurologic deficits presumed to be ischemic in nature, and two who had both stroke and transient neurologic events. Snider et al. (1983) found cerebrovascular disease to be the cause of neurologic disability in about 10% of patients with HIV and nervous system disease. The incidence of cerebrovascular disease diagnosed pathologically in autopsy series far exceeds that found in clinical series and may approach 30%. Kure et al. (1991) found about 12% of patients to be affected by cerebrovascular disease in a large series of consecutive autopsies of patients dying with AIDS. Kieburtz et al. (1993) identified cerebral infarcts in the brains of 20% of patients dying with AIDS, Anders et al. (1986) identified cerebral infarcts in 19%, and Mizusawa et al. (1988) identified cerebral infarcts in 29%. Whether the incidence of cerebrovascular disease in patients with AIDS is derived from clinical or autopsy series, the number is unexpectedly high for this group of patients who for the most part fall into a relatively young age range.

Cerebrovascular disease in patients with HIV infection typically presents at a time when patients are severely immunocompromised and have previously experienced AIDS-defining illness. However, it is occasionally the presenting feature of a patient not previously known to have AIDS (Atalia et al. 1992; O'Dell 1992). In the series of Engström et al. (1989) 9/1600 patients developed stroke or transient ischemic attacks as the initial manifestation of AIDS. HIV should be considered in the differential diagnosis of any young patient who develops stroke, regardless of the known risk factors for HIV infection.

In children, the infectious and non-infectious CNS complications of HIV may be proportionately different from those in adults (Park et al. 1990; Kugler et al. 1991; Burns 1992; Kaufman et al. 1992; Philippet et al. 1994). Children are prone to develop the sequelae of direct HIV infection of the brain with resultant diffuse encephalopathy and dementia; however, opportunistic infections of the brain in children with HIV are relatively uncommon. As such, children with AIDS and *focal* nervous system dysfunction are more likely than adults to be suffering from non-infectious complications of HIV including cerebrovascular disorders.

Clinical features

The cerebrovascular complications of HIV infection include the entire spectrum of cerebrovascular syndromes that occur in other settings, although the pathogenesis may be specific to this clinical setting. Cerebral ischemic syndromes are more common than hemorrhagic complications, although both may occur (Table 2). Ischemic events may result from thrombosis of either large or small arteries, or as a result of venous occlusion. Ischemic events may also result from cardioembolic events to large or small vessels. Intracerebral hemorrhage may be the result of underlying coagulopathy, or the development of large or small vessel aneurysms. As such, the clinical presentation of cerebrovascular disease associated with HIV runs the gamut of symptoms associated with cerebrovascular disease in general. Most patients present acutely with 'stroke-like' onset of focal neurologic dysfunction including hemiparesis, sensory loss, ataxia, language dysfunction, other focal cognitive deficits, or abrupt impairment in consciousness. However, because patients with HIV are likely to have other infectious and non-infectious diseases coincident with cerebrovascular disorders, the symptoms of cerebrovascular disease may be masked by other disorders; patients may present instead with more insidious onset of symptoms, fluctuating symp-

TABLE 2

Cerebrovascular disorders associated with HIV infection.

Thrombotic
  Vasculitis
    Infectious (e.g., syphilis, zoster, *Aspergillus*)
    Non-infectious
  Coagulation defects
    DIC
    Anti-phospholipid antibodies
    Abnormalities of clotting proteins
  Venous thrombosis
    Dehydration, sepsis

Embolic
  Non-bacterial thrombotic endocarditis
  Bacterial endocarditis

Hemorrhagic
  Vasculitis
  Hemorrhagic encephalitis
  Thrombocytopenia
    Autoimmune
    Drug-induced
    Marrow replacement
  Coagulopathy
  Mycotic or other acquired aneurysms
  Intratumoral hemorrhage

toms, or more global deficits than might be expected with focal cerebrovascular disease alone. Cerebrovascular disease therefore must be considered as a contributing etiology of CNS dysfunction in patients with HIV infection, whether the presentation is classically 'stroke-like' or more closely fits another clinical pattern. This diagnostic challenge probably explains the discrepancy in the reported incidence of cerebrovascular disease defined by clinical and autopsy studies noted above.

Cerebrovascular thrombosis

The thrombotic complications that occur in patients with HIV may either be indirectly related to infectious etiologies, or occur independently as a result of a variety of clotting disorders and immunological mechanisms. As noted, they may result in either large or small arterial thrombosis or venous occlusion.

Perhaps the most commonly occurring infection to result in cerebrovascular disease is syphilis. Syphilis commonly coexists with HIV infection, and up to 6% of patients with HIV will be found to have a positive CSF VDRL (Berger 1995). Patients with secondary syphilis often develop a mild meningitis (McArthur

and Johnson 1988) which may be asymptomatic. Partially treated or untreated syphilis may progress to meningovascular syphilis, with resultant cerebrovascular complications as well as other tertiary complications of general paresis or tabes dorsalis. The diagnosis of meningovascular syphilis may be difficult to make, unless a strong clinical suspicion exists. Patients with HIV infection frequently have pleocytosis of the cerebrospinal fluid unrelated to syphilis, and serological testing for syphilis may not be positive (Johns et al. 1987; Berger 1989). Nonetheless, meningovascular syphilis may result in serious neurologic compromise (Berger 1991; Holtom et al. 1992). Kase et al. (1988) described a patient who developed a pontine stroke as a result of meningovascular syphilis associated with HIV. Tyler et al. (1994) reported a patient in whom the presenting feature of HIV was a medial medullary infarct which resulted from meningovascular syphilis. Meningovascular syphilis may cause large- or small-vessel arterial thrombosis, typically with acutely developing symptoms referable to the affected arterial distribution. If the diagnosis is made, effective anti-microbial therapy can limit subsequent thrombotic events.

Another important diagnostic consideration in patients with HIV presenting with cerebrovascular disease is *Aspergillus* infection (Woods and Goldsmith 1990). While other fungal infections such as *Cryptococcus* are more common than *Aspergillus* in AIDS patients, *Aspergillus* is especially prone to produce vascular complications. Bland or septic infarcts result from the characteristic vasocentric necrotizing vasculitis that occurs with aspergillosis (Carranza et al. 1991), and hemorrhagic complications may also occur. CNS involvement of *Aspergillus* generally results from hematogenous dissemination from a primary pulmonary infection, although alternatively paranasal sinuses may be infected with direct extension into the brain. Symptoms may be abrupt and focal in onset, or a more diffuse encephalopathy may occur as a result of associated meningitis or brain abscesses. The diagnosis may be difficult to make because, even if there is associated meningitis, the organism may be difficult to obtain from the CSF, and biopsy of focal brain lesions may be required. Unfortunately, despite prompt and aggressive anti-fungal therapy the prognosis for those with immunosuppression and *Aspergillus* infection is exceedingly poor.

Aside from syphilis and *Aspergillus*, other infectious etiologies of CNS vasculitis and resultant vascular thrombosis are unusual but need to be considered. These include herpes zoster and cytomegalovirus infections (Berger et al. 1990). In children with HIV infection, a common pathologic finding of calcific arteriopathy may result in stroke and is thought to be directly related to HIV infection of the brain (Joshi et al. 1987; Kugler et al. 1991).

In addition to the thrombotic cerebrovascular complications of HIV that are secondarily related to CNS infection, there are those that occur as a result of a number of coagulation defects and immunological dysfunction. Primary vasculitis of the cerebrovascular system, or so-called granulomatous angiitis, has been described in patients with AIDS (Yanker et al. 1986; Calabrese et al. 1992). CNS vasculitis of this sort typically affects small parenchymal and leptomeningeal blood vessels. Patients present with a number of clinical syndromes ranging from headache, to fluctuating encephalopathy, to focal or multi-focal strokes with focal neurologic dysfunction. Systemic symptoms of vasculitis such as fevers and weight loss are typically lacking, although these symptoms may be present for other reasons in patients with AIDS. This diagnosis may be difficult to establish clinically and radiologically; confirmation of the diagnosis by angiography is possible in only up to one-third of cases (Calabrese et al. 1992). Biopsy of the meninges and brain parenchyma may be necessary to establish the diagnosis. Unfortunately, pharmacologic immunosuppression is generally contraindicated in patients with HIV infection, so the rare disorder of primary CNS vasculitis is very difficult to treat in this population.

Cerebral thrombotic events in patients with HIV infection also occur as a result of disseminated intravascular coagulation (DIC) associated with systemic illness or sepsis (Berger et al. 1990). DIC may cause a fluctuating encephalopathy associated with multiple focal regions of brain ischemia and typically small vessel infarcts. DIC may also result in hemorrhagic complications (see below). When the syndrome is recognized, usually by abnormalities in coagulation, cerebral DIC is best treated with heparin anticoagulation if intracerebral hemorrhage has not occurred (Bell et al. 1985).

Abnormalities in immune function may predispose patients with HIV to develop thrombotic events as well. Serum anticardiolipin antibodies, which belong to a family of antibodies directed against phospholipids, are identified in elevated titers in about 50% of patients (Rubbert et al. 1994) and may correlate with the severity of AIDS-related illnesses (Sorice et al. 1994). The presence of these antibodies is associated with both arterial and venous thrombotic events, although the mechanism by which this occurs is not entirely known (Canoso et al. 1987; Gallo et al. 1994; Thirumalai and Kirshner 1994). The antibodies may interfere with protein C, protein S, anti-thrombin III, or platelet function (Rubbert et al. 1994; Sorice et al. 1994). Rubbert et al. (1994) found a strong correlation between the presence of anticardiolipin antibodies and cerebral perfusion defects in patients with HIV as assessed by single photon emission computed tomography (SPECT).

Other causes of stroke in young patients without HIV may also be present in patients with HIV and contribute to the development of cerebrovascular disease. Abnormalities in the clotting of proteins have been identified in some patients. Protein S deficiency, which may be associated with stroke in young adults, occurs in many patients with HIV infection (Lafeuillade et al. 1991; Sorice et al. 1994). Sorice et al. also identified antibodies that are directed against protein S. Serum lipoprotein (a) is involved in the development of atherosclerosis and arterial thrombosis and has been shown to be a risk factor for the development of ischemic stroke (Nagayama et al. 1994); it has also been found to be stimulated by infection and may contribute to the development of cerebral infarction (Constans et al. 1993). In patients with cerebral thrombotic events the presence of these specific serum abnormalities may give a clue to the underlying pathogenesis and occasionally suggest specific strategies for treatment that may help avoid repeated vascular insults.

Cerebral venous thrombosis may also occur in patients with HIV infection and result in venous infarcts or hemorrhage. The sagittal sinus is most often involved although any of the venous sinuses may be affected. Rhodes (1987) identified sinus thrombosis at autopsy in 1 of 100 brains of patients dying with AIDS. As in non-HIV-infected individuals, sinus thrombosis associated with HIV infection is most likely to occur in severe cachexia and dehydration associated with sepsis or end-stage illness (Doberson and Kleinschmidt-DeMasters 1994).

Meyohas et al. (1989) described a patient with cerebral vein thrombosis associated with HIV and cytomegalovirus infection. Patients with cerebral venous thrombosis often present with headache and symptoms of increased intracranial pressure and may develop secondary cerebral infarctions or hemorrhages. In those without cerebral hemorrhage, anti-coagulant or anti-thrombotic therapy is probably appropriate.

Cardioembolic complications

A cardiac source of embolus should be considered as a cause of ischemic cerebral infarction in patients with HIV infection (Levy and Bredesen 1988; Berger 1995). Patients present with an abrupt onset of focal neurologic dysfunction in single or multiple vascular territories. Focal or generalized seizures may also result from cardioembolic events. Bacterial endocarditis predisposes patients to septic emboli which may present with ischemic focal neurologic dysfunction. Septic emboli may also result in mycotic aneurysms and hemorrhage (see below). Other sites of emboli may be evident in the fundoscopic examination or systemically. Bacterial endocarditis should especially be considered in the intravenous drug-using HIV-infected patient (who may be more prone to developing subacute bacterial endocarditis).

The possibility of non-infectious cardioembolic etiologies for stroke should be considered in patients with HIV as well. So-called 'marantic' or non-bacterial thrombotic endocarditis occurs in many settings and may be associated with the serious systemic illnesses and neoplasms associated with HIV. This may be a difficult diagnosis to make, presenting with multi-focal or fluctuating disturbances in neurologic function. Frequently, it is difficult to confirm the diagnosis on transthoracic echocardiography, although transesophageal echocardiography may be more sensitive. Some patients have low-grade DIC associated with non-bacterial thrombotic endocarditis. Appropriate treatment, anti-coagulation, may prevent additional embolic events (Rogers et al. 1987a). 'Paradoxical' emboli from a venous source of thrombus should also be considered in patients with embolic strokes who develop ischemic strokes when an intracardiac shunt, such as patent foramen ovale, is present.

Hemorrhagic complications

Intracerebral hemorrhages occur less commonly in patients with HIV than do the ischemic complications. In the series of Kure et al. (1991) about 3% of brains of patients dying with AIDS were found at autopsy to have intracerebral hemorrhage. As with the ischemic complications, the causes can be divided into those that occur secondary to infectious etiologies, and those not directly related to infection.

Any of the infectious vasculitides mentioned that cause ischemic strokes may occasionally cause intracerebral hemorrhage or hemorrhagic infarction as well. *Aspergillus* infection is particularly prone to cause vasocentric necrosis and may directly erode into vessel walls leading to single or multi-focal intracerebral hemorrhage. Toxoplasmosis usually causes focal brain abscesses. However, Trenkwalder et al. (1992) described a patient with HIV and multiple cerebellar hemorrhages who was found at autopsy to have toxoplasmosis encephalitis. Wijdicks et al. (1991) reported a patient in whom the presenting feature of AIDS was multiple cerebral hemorrhages secondary to *Toxoplasma* infection, which had a fulminant course and proved to be fatal.

Large- or small-vessel aneurysms may develop in patients with HIV infection and result in intracerebral or subarachnoid hemorrhages. Mycotic aneurysms develop as a result of septic emboli usually from bacterial endocarditis (Berger 1995). These are usually small, often multiple, peripherally placed aneurysms that result from direct destruction of the vessel wall by invading organisms. Rupture of these small aneurysms may have catastrophic consequences. Treatment usually consists of anti-microbial therapy; surgical intervention may be required (Lerner 1995). Acquired large vessel aneurysms may be associated with HIV infection itself or concurrent opportunistic infections (Sinzobahamvya et al. 1989). Lang et al. (1992) described an 8-year-old child with HIV infection who developed a large aneurysm involving the circle of Willis, but no evidence of intravascular inflammation or infection. Destian et al. (1994) reported a patient with a large inflammatory intracavernous carotid aneurysm that resulted from infection with *Mycobacterium avium* intracellulare. Surgical intervention is usually required for definitive treatment of these large aneurysms whether or not their pathogenesis is clearly infectious.

Intracerebral hemorrhage in patients with HIV infection may also be caused by thrombocytopenia or coagulopathy. Immune thrombocytopenia may occur in patients with HIV infection as a result of antibodies directed against platelet antigens (Stricker et al. 1985). Thrombocytopenia may also occur as a result of marrow infiltration by AIDS-associated lymphoma, or be secondary to drug-induced myelosuppression (Berger 1995). Disseminated intravascular coagulation may result in thrombocytopenia and coagulopathy, and liver failure due to sepsis or other systemic disease may also result in coagulopathy.

Occasionally, intracerebral hemorrhage in patients with AIDS results from intratumoral hemorrhage. Although not especially prone to be hemorrhagic, primary CNS lymphoma or metastatic Kaposi's sarcoma can become hemorrhagic. The clinical presentation of intracerebral hemorrhage depends on the size and location of the hemorrhage, but typically has an abrupt onset of headache and focal neurologic dysfunction or alteration in consciousness.

Treatment

The appropriate treatment of cerebrovascular disorders in patients with HIV depends on identifying a specific underlying etiology. Those complications that are secondary to underlying infection may respond best to antimicrobial therapy, while those related to disorders of coagulation may require anticoagulation. In some instances surgical evacuation of intracerebral hemorrhage may be appropriate and, when possible, correction of thrombocytopenia or coagulopathy can prevent further hemorrhages. If recognized, treatment of cerebrovascular disease associated with HIV often improves the neurologic outcome and prevents recurrence of cerebrovascular complications. Unfortunately in many instances the cerebrovascular complications of HIV coexist with other HIV-associated disorders of the nervous system that are less effectively treated.

NEOPLASTIC COMPLICATIONS OF HIV INFECTION

The disorders in immune function that allow opportunistic infections to develop in patients with HIV infection also permit the development of 'opportunistic' neoplasms. Although the infectious complications of HIV are more common than the neoplastic

complications, neoplasms still cause significant morbidity and mortality in this population. Neoplasms typically occur in patients with AIDS as a late manifestation of the illness. As a result, while the ability to treat opportunistic infections and prolong the lives of HIV-infected individuals has improved over the last several years, the incidence of AIDS-associated neoplasms has also increased (Levine 1992a, 1993). There are two principal neoplasms that are strongly associated with HIV infection: Kaposi's sarcoma and non-Hodgkin's lymphoma (So et al. 1988; Bernstein and Hamilton 1993; Aboulafia 1994). Although currently these are the only neoplasms of significance associated with AIDS, a variety of other neoplasms have been found in occasional patients with AIDS and HIV infection (Spina and Tirelli 1992). With improved treatment of infections and prolongation of survival, these other types of cancer may prove to be of importance in this population as well. In particular, Hodgkin's disease, cervical carcinoma and anorectal carcinomas may be increasingly important (Reynolds et al. 1993; Tirelli et al. 1994). Kaposi's sarcoma and non-Hodgkin's lymphoma are described below in association with their neurologic complications. The neurologic complications of other malignancies that occur rarely in patients with AIDS are not discussed.

Kaposi's sarcoma

Epidemiology
Kaposi's sarcoma is a rare cutaneous neoplasm of mesenchymal origin. In its classic form, it occurs predominantly in elderly men of Eastern European and Mediterranean descent. It involves multicentric cutaneous sites but runs a fairly indolent course without significant visceral involvement. Kaposi's sarcoma has also been found endemically in African populations where it affects young people of both sexes. In the United States, Kaposi's sarcoma was first noted in significant numbers in iatrogenically immunosuppressed patients such as those undergoing organ transplantation. In these immunosuppressed patients the disease is more likely to have an aggressive course with a tendency toward widespread dissemination (Barton et al. 1983; Wang et al. 1993).

Since the beginning of the AIDS epidemic, Kaposi's sarcoma has been noted to occur in epidemic proportions in HIV-infected individuals. Kaposi's sarcoma may be the presenting feature of AIDS and occur prior to the development of opportunistic infections. It is 6 times more common in homosexual men than in other risk groups for AIDS (Levine 1992a; Buonaguro et al. 1994), such as children with HIV infection (Arico et al. 1991). Overall it has been reported in up to half of patients with AIDS (Barton et al. 1983; Buonaguro et al. 1994). It is now also being described in homosexual men who are not infected with HIV and is thought perhaps to be associated with another sexually transmitted agent such as human papillomavirus (Buchbinder and Friedman-Kien 1992).

Clinical features
In patients with HIV infection Kaposi's sarcoma takes a particularly aggressive form. Widespread visceral involvement, particularly involving the gastrointestinal and pulmonary systems, is common (Buchbinder and Friedman-Kien 1992). Lymph node involvement may mimic the lymphadenopathy commonly found in patients with HIV infection and must be distinguished from AIDS-related lymphoma (Lever 1992; Wang et al. 1993).

Neurologic complications
Despite the frequency of Kaposi's sarcoma in patients with HIV infection and its tendency for widespread visceral involvement, the neurologic complications are exceedingly rare, and the medical literature mostly contains isolated case reports. When neurologic complications occur, they are usually related to brain metastases in patients with widely disseminated disease, almost always including pulmonary metastases (Rwomushana et al. 1975; Barton et al. 1983; Gorin et al. 1985). Patients present with focal neurological dysfunction referable to the cerebral involvement. Brain biopsy or surgical excision is essential to establish the diagnosis and exclude other much more common causes of intracerebral lesions in patients with AIDS. Systemic Kaposi's sarcoma may respond to chemotherapy (Levy et al. 1985), although brain metastases are treated with radiotherapy with reasonably good local control. Kaposi's sarcoma also rarely infiltrates peripheral nerves or nerve roots causing localized pain or nerve dysfunction, or may infiltrate the dura and result in focal cerebral or spinal cord compression (Barton et al. 1983). Although these neurologic complications are uncommon, their incidence may increase in the future as patients with HIV infection are treated more effectively for opportunistic infections and survive with Kaposi's sarcoma for longer periods of time.

LYMPHOMA

Lymphoma is the second most common malignancy associated with AIDS. In particular risk groups for HIV infection, especially hemophiliacs, lymphomas are the most common AIDS-associated neoplasms (Gaidano et al. 1994). The great majority of AIDS-related lymphomas are non-Hodgkin's B-cell lymphomas, although T-cell lymphomas and Hodgkin's disease also occur. The AIDS-related lymphomas are clinically and histologically aggressive tumors (Northfelt and Kaplan 1991). They may occur systemically and have neurologic complications, or develop within the nervous system as the primary site.

Systemic lymphoma

Epidemiology
Lymphoma occurs with greatly increased frequency in patients with HIV infection, as compared to the general population. In a study from San Francisco, lymphoma was found to be increased 71 times in patients with AIDS (Reynolds et al. 1993). Lymphoma is often a late complication of HIV, occurring on average 50 months after infection (Gaidano et al. 1994). As such, the Center for Disease Control which records only the initial AIDS-defining illnesses in their incidence rates of AIDS-associated illnesses may grossly underestimate the incidence of lymphoma in AIDS. Nonetheless, it has been estimated that as many as 40% of patients with AIDS who survive at least 36 months will develop lymphoma (Milliken and Boyle 1993; Ling et al. 1994).

Pathobiology
The usual characteristics of systemic lymphomas in patients with AIDS differ from the characteristics of lymphomas in immunocompetent patients. They are most commonly high-grade B-cell lymphomas of the large cell immunoblastic, or less commonly, small non-cleaved (Burkitt's-like) cell types (Milliken and Boyle 1993). T-cell tumors or those with both B- and T-cell lineage may also occur (Hollingsworth et al. 1994). They are clinically aggressive tumors and in 80–90% of patients involve extranodal organs

from the time of presentation. Most commonly they involve the bone marrow and gastrointestinal system, as well as arising initially in the CNS as discussed below (Levine 1992a, 1994). Other less usual sites of involvement may also be present including the heart, oropharynx, or rectum and anus (Gaidano et al. 1994).

The pathogenesis of systemic lymphomas in AIDS is unknown, but likely involves infection with Epstein–Barr virus (EBV). EBV infection may be important in the transformation of lymphocytes (Henderson et al. 1977; Cleary et al. 1985), and its regulation may be impaired in patients with AIDS (Birx et al. 1986). The EBV genome or its protein expression can be identified in about 50% of systemic lymphomas of patients with AIDS (and a much higher proportion of AIDS-related primary CNS lymphomas) (Joske and Knecht 1993; Brousset et al. 1994; Carbone et al. 1994; Levine 1994). Rearrangement of the c-*myc* oncogene is also identified in many AIDS-related lymphomas and may be important in the development of these neoplasms (Levine 1992b, 1994; Joske and Knecht 1993; Knowles 1993).

Clinical features

Clinical involvement of the nervous system by systemic lymphomas occurs in a minority (between 5 and 30%) of patients reported (Ioachim et al. 1991; Ruiz et al. 1994). Subclinical involvement may be more common, and asymptomatic leptomeningeal lymphoma has been diagnosed in as many as 20% of patients with AIDS-related systemic lymphoma (Levine 1992b), and at autopsy in 15–40% (Milliken and Boyle 1993). Nervous system complications of systemic lymphomas are often late events. The incidence of clinically evident CNS metastases may increase as the treatment of systemic lymphomas and opportunistic infections improves and patients' survival is prolonged.

Systemic lymphomas usually involve the nervous system by epidural or meningeal extension. Intraparenchymal spread of tumor is unusual. The clinical symptoms depend on the site of involvement. Epidural spinal involvement may extend from bony metastases, but commonly arises in the epidural space by direct extension from paraspinal nodes or by hematogenous spread. Patients present with back pain and signs of radiculopathy or myelopathy resulting from compression of spinal nerve roots or spinal cord. Cranial epidural involvement may also produce focal cerebral signs or cranial neuropathy. Diagnosis is made with MRI or CT scans or myelography.

As noted, leptomeningeal lymphoma is sometimes asymptomatic. Classically, it produces symptoms of nervous system involvement at multiple levels of the neuraxis (Enting et al. 1994). Multiple cranial neuropathies are common, and polyradicular symptoms or bladder dysfunction frequently occur. Hydrocephalus may result from obstruction of cerebrospinal flow and produce gait disorder, cerebral symptoms or lethargy. CSF examination yields elevated protein and pleocytosis, and malignant cells are frequently identified. When elevated, markers for meningeal lymphoma such as β-glucuronidase or lactate dehydrogenase are non-specific but can support a clinical suspicion of meningeal lymphoma (Shuttleworth and Allen 1980; Rogers et al. 1987b). Myelography may demonstrate thickening and nodularity of involved nerve roots, and MRI may demonstrate diffuse or patchy meningeal enhancement.

Systemic lymphomas may metastasize to the orbit, where, by compression of cranial nerves, extraocular muscles or both, they produce ophthalmoplegia. The ophthalmoplegia is frequently painful and associated with proptosis. Orbital lymphomas may be locally invasive, infiltrating the skull base and causing multiple cranial neuropathies or cerebral involvement (Rahhal et al. 1994). As discussed below, primary CNS lymphoma is prone to develop intraocular, rather than orbital, involvement.

Treatment

Patients with AIDS-related lymphomas may respond favorably to treatment with chemotherapy. A number of chemotherapy regimens have been used with some success in this population (Milliken and Boyle 1993; Sparano et al. 1994). Unfortunately, because of their underlying immunosuppression, most patients do not tolerate the full aggressive chemotherapy regimens that would be used to treat similar lymphomas in non-AIDS patients, and the median survival of patients with AIDS-related systemic lymphomas is only about 5 months. As might be expected, patients who are young, who have relatively high CD4 counts and who are without prior AIDS-defining illness fare better in terms of both their ability to tolerate treatment and their survival (Taillan et al. 1993). Occasional patients appear to have spontaneous regression of their lymphoma (Karnad et al. 1992), although this is probably an exceedingly rare event. Nervous system involvement of systemic lymphoma is treated with irradiation, which may provide prompt and effective palliation of nervous system symptoms and dysfunction. Select patients may benefit from intrathecal chemotherapy as well (Chamberlain and Dirr 1993), although this aggressive treatment is usually not appropriate because of associated illnesses. If treated with radiotherapy, patients usually die as a result of progression of the systemic lymphoma or opportunistic infections rather than as a direct result of the nervous system involvement.

Primary CNS lymphoma

Epidemiology

Lymphoma that originates within the CNS, or primary CNS lymphoma (PCNSL), is a rare neoplasm that accounts for only 1–2% of all brain tumors (Fine and Mayer 1993). In immunocompetent patients PCNSL accounts for only about 2% of all lymphomas. PCNSL is especially prone to develop in patients with a variety of disorders of T-cell immunity including AIDS, congenital immunodeficiency syndromes, Wiskott–Aldrich syndrome, and those with acquired immunodeficiency as occurs iatrogenically after organ transplantation (DeAngelis 1991a). PCNSL also occasionally occurs as a second malignancy in long-term survivors of other cancers (DeAngelis 1991b).

In patients with AIDS, PCNSL is a relatively common illness (Galetto and Levine 1993). About 2–6% of patients with AIDS and clinically evident neurologic dysfunction will be found to have PCNSL (Ruiz et al. 1994), and at autopsy 10–40% of patients dying with AIDS will be found to have PCNSL (Ling et al. 1994). PCNSL lymphoma accounts for 20–25% of all the lymphomas seen in the AIDS population (Milliken and Boyle 1993). After toxoplasmosis, PCNSL is the second most common cause of intracranial mass lesions diagnosed in patients with AIDS, and is the fourth most common cause of death (Ling et al. 1994).

The incidence of PCNSL is increasing in the AIDS population, in patients with other underlying immunosuppression, and in apparently immunocompetent patients as well (DeAngelis 1991a; Miller et al. 1994). The reason for this increase is uncertain. PCNSL is frequently a late manifestation of HIV infection. The disease may become an even more prominent cause of morbidity and mortality as treatment of opportunistic infections becomes more effective and patients survive longer.

Pathobiology

In previous years, the cell of origin that developed into PCNSL was disputed. The tumor was referred to variably in the literature as reticulum cell sarcoma, microglioma, perithelial sarcoma or histiocytic lymphoma. It was gradually recognized that there was great histologic similarity between these tumors and systemic lymphomas. It is now clear that the cell of origin is the lymphocyte and that the tumor is correctly identified as a lymphoma. It is not known, however, why this neoplasm originates in the brain since this organ is devoid of lymphatic tissue (Russell and Rubinstein 1989).

The gross appearance of PCNSL is highly variable. Many are homogenous gray masses which are relatively well circumscribed, while others are more diffuse and infiltrating. There may be areas of necrosis and hemorrhage that appear similar to malignant glial neoplasms. The neoplasm is multifocal in at least a quarter of instances, and this may be even more likely for PCNSL that develops in patients with AIDS. Even in cases in which the lymphoma appears to be unifocal by gross examination, microscopically it is common to find clusters of tumor cells distant from the site of the predominant tumor mass (Russell and Rubinstein 1989).

PCNSL may be located anywhere throughout the CNS including the cerebrum, brainstem, cerebellum and spinal cord. Most often it develops in the cortical parenchyma or the deep gray structures of the brain. PCNSL is often periventricular and may have ependymal extension. Rarely it develops primarily in the meninges, although secondary meningeal involvement from ependymal spread is not uncommon. It is exceptional for this tumor to infiltrate beyond the meninges into the overlying skull. However, the eye is an extension of the CNS, and intraocular lymphoma represents another site of multifocal CNS involvement of lymphoma which develops in about a quarter of affected patients (Peterson et al. 1993). Metastasis to extraneural structures is exceedingly uncommon, except at very late stages of the disease.

The microscopic appearance of PCNSL consists

of a densely cellular perivascular tumor. Despite the characteristic perivascular location, vascular invasion is rarely associated. At the margins of the tumor, more dispersed cells infiltrate the surrounding brain. There may be associated glial reaction and astrocytosis or fibrosis, particularly at the periphery of the tumor (Russell and Rubinstein 1989). Most often the tumor is made up of a homogeneous population of B-cells which are usually of the immunoblastic type (Levine 1992b; Milliken and Boyle 1993). In contrast to other AIDS-related lymphomas, PCNSL is only rarely comprised of small noncleaved cells. Occasionally T-cell lymphomas have also been described (Grant and Von Diemling 1990; Morgello et al. 1990; Mineura et al. 1993; Miller et al. 1994).

Although the reason for the development of PCNSL in patients with AIDS is unknown, EBV is implicated as is true for other AIDS-associated lymphomas. Portions of the EBV genome, its transcripts, or proteins have been identified in PCNSL of patients with AIDS (Hochberg et al. 1983; Bashir et al. 1989, 1993; Hamilton-Dutoit et al. 1993; Auperin et al. 1994). In some studies up to 100% of PCNSL associated with AIDS have been shown to be associated with EBV infection, as opposed to only about 50% of AIDS-associated systemic lymphomas (MacMahon et al. 1991; Levine 1992b; Samoszuk et al. 1993). EBV is much more likely to be pathogenic in AIDS-associated PCNSL than in non-AIDS PCNSL. Although the mechanism by which EBV infection leads to tumorigenesis in PCNSL is unknown; it may involve activation of c-*myc*, *ras*, or other oncogenes (Seremetis et al. 1989; Levine 1994).

Clinical features

AIDS-associated PCNSL is usually a late manifestation of AIDS, developing in patients who have had previous AIDS-defining illnesses, and in those with CD4 lymphocyte counts less than 50 cells/mm³ (Ling et al. 1994). Less commonly it presents in an HIV-infected patient without a previous diagnosis of AIDS. Patients come to medical attention when they develop neurologic symptoms over a period of a few weeks; in patients with non-AIDS-related PCNSL the symptoms are more likely to develop in a more indolent fashion (Fine and Mayer 1993).

Patients present with a variety of symptoms referable to the site or sites of nervous system involvement. They may develop focal weakness, sensory changes or seizures as a result of cortical involvement. Hemiparesis or loss of motor control often results from involvement of deep gray structures. Quite circumscribed lesions may produce precise neurologic dysfunction as in a patient of MacLean et al. (1994) who presented with bilateral fourth cranial nerve dysfunction as the only feature of PCNSL. Meningeal dissemination, which occurs in 15–30% of patients at the time of presentation (Levine 1992b; Miller et al. 1994), is sometimes asymptomatic but more often causes encephalopathy, multiple cranial neuropathies, or spinal radiculopathies. Intraocular involvement, which occurs in about 7% of patients at the time of diagnosis, may produce visual disturbance due to vitreous or retinal involvement (Fine and Mayer 1993; Peterson et al. 1993). However, intraocular lymphoma is rarely the isolated site of PCNSL associated with AIDS. Probably because of the multicentric nature of the disease which occurs more commonly in AIDS patients than in patients without AIDS, at least half of the patients present with a picture of more diffuse encephalopathy with a paucity of focal findings (Levine 1992b).

Diagnosis

Because of the relatively high incidence of PCNSL in patients with AIDS and the variety of clinical presentations, the diagnosis needs to be considered in all AIDS patients who present with CNS complaints, without regard to the presence or absence of focal signs on examination. The appearance of PCNSL on imaging studies such as CT or MRI scans is variable, and indistinguishable from that of toxoplasmosis (Dina 1991; Moschini et al. 1994; Ruiz et al. 1994). Often PCNSL appears on CT as one or more isodense or hyperdense cortical or basal ganglia mass. It usually has indiscreet borders and enhances diffusely and homogeneously with intravenous contrast media. MRI scans often demonstrate a periventricular or cortical mass which in at least 90% of instances enhances with contrast agents, usually in a dense and homogeneous pattern (Fine and Mayer 1993). The pattern of enhancement is variable, however, and up to half of PCNSL associated with AIDS enhance in a ring pattern (Fig. 1). Occasional tumors do not enhance with contrast agents. As noted, AIDS-related PCNSL is especially likely to present

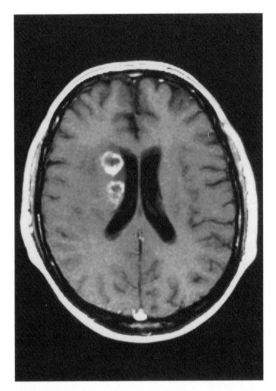

Fig. 1. This enhanced T1-weighted MRI obtained in a 36-year-old man with AIDS demonstrates multi-focal periventricular lesions typical of PCNSL, enhancing in a ring pattern. The primary differential diagnosis for lesions with this appearance would be toxoplasmosis; this patient had biopsy-proven PCNSL.

with multicentric disease. Involvement of the corpus callosum and a periventricular or subependymal location may favor a diagnosis of PCNSL over toxoplasmosis, although this differentiation is not absolute (Ruiz et al. 1994).

CSF examination sometimes establishes a diagnosis of PCNSL in cases in which meningeal dissemination has occurred. This occurs at the time of presentation with a frequency of between 7 and 69% (Fine and Mayer 1993; Miller et al. 1994), the wide range of reported incidence dependent in part on the rigor with which the diagnosis is sought. In many instances a lumbar puncture can be safely performed, and if the procedure yields malignant cells it may prevent the need for a more invasive procedure such as brain biopsy. In addition, a thorough ophthalmological examination including slit lamp examination may show evidence of ocular involve-

ment lending at least circumstantial evidence to the diagnosis of lymphoma. Systemic staging evaluation is of low yield and probably unwarranted in systemically asymptomatic patients, as PCNSL is almost always limited to the CNS at presentation (Miller et al. 1994).

The diagnosis of PCNSL must be considered in order to avoid unnecessary treatment with corticosteroids which may obscure the diagnosis (Geppert et al. 1990). Complete radiographic and pathologic remission may be seen temporarily following corticosteroid treatment, and remission is occasionally seen spontaneously as well. It is rare that patients present with impending cerebral herniation requiring prompt corticosteroids, so in most instances it is possible to avoid corticosteroid therapy. Very rarely, PCNSL may regress spontaneously, obscuring the diagnosis as well (Rubin et al. 1987; Terriff et al. 1992).

The differential diagnosis for a cerebral mass in a patient with AIDS includes toxoplasmosis and, much less commonly, other infectious brain abscesses (e.g. bacterial abscess, cryptococcoma, tuberculoma). It is exceedingly rare for other AIDS-associated neoplasms (Kaposi's sarcoma, systemic lymphoma) to metastasize to the parenchyma of the CNS, and while other primary and metastatic tumors occasionally coexist in patients with AIDS, there is a low incidence of such tumors. Progressive multifocal leukoencephalopathy (PML) and vascular lesions, such as strokes or hemorrhages, are usually easily distinguished by their clinical and radiographic features. The co-existence of PCNSL with other CNS mass lesions, particularly toxoplasmosis, must also be considered due to the propensity of patients with AIDS to develop multiple concurrent illnesses.

Because of the difficulty in distinguishing PCNSL from toxoplasmosis clinically and radiologically, and because toxoplasmosis is a more common cause of intracerebral mass lesions in patients with AIDS, in many institutions patients with AIDS-related CNS mass lesions are treated with antimicrobial agents empirically for an initial period of 10 days to 2 weeks. Clinical and radiologic response to therapy supports the presumptive diagnosis of toxoplasmosis, while no response to therapy or progression of disease prompts further diagnostic procedures, most often stereotactic brain biopsy (Feiden et al. 1993). This approach is reasonable in most patients, although those with rapidly progressive symptoms during the initial treatment period or those with no serological evidence of toxoplasmosis may benefit more from earlier diagnostic intervention.

Both 18-fluorodeoxyglucose positron emission tomography (PET) and thallium SPECT scans have been used in an attempt to distinguish PCNSL from other causes of intracranial mass lesions in AIDS patients (Rosenfeld et al. 1992; Hoffman et al. 1993; O'Malley et al. 1994; Ruiz et al. 1994). Both techniques have demonstrated increased metabolic activity in PCNSL, as is found in other brain tumors such as glial neoplasms. This feature fairly reliably distinguishes these lesions from toxoplasmosis or other infectious CNS lesions. Although these tools are not widely available clinically, their use in the future may encourage earlier brain biopsy in some patients, or prevent unnecessary brain biopsy in others.

Ultimately many AIDS patients who do not respond to empiric anti-toxoplasmosis therapy undergo diagnostic brain biopsy (Feiden et al. 1993; Iacoangeli et al. 1994). This is a relatively well tolerated procedure, and accurately determines a diagnosis in most patients. Surgical decompression or removal of the tumor is not recommended since it does not usually influence the overall outcome of the disease and may worsen the neurologic function of the patient, particularly for those with deep, infiltrating lesions that involve essential brain structures. Since biopsy is generally reserved for those who have progressed despite anti-toxoplasmosis therapy, the most common diagnoses of patients with AIDS undergoing brain biopsy are PCNSL and PML.

Treatment

Regardless of therapy, the overall prognosis for patients with AIDS-associated PCNSL is very poor, and is even less favorable than for patients with AIDS-related systemic lymphomas. This is largely due to the fact that PCNSL is typically a late manifestation of AIDS; three-fourths of patients have had prior opportunistic infections at the time of presentation, as compared to about one-third of those who develop systemic lymphoma (Levine 1992b). Untreated, the survival with AIDS-related PCNSL averages only 1–2 months (Levine 1992b; Fine and Mayer 1993; Ling et al. 1994).

PCNSL is in general a radiosensitive and chemo-sensitive tumor. Whole brain radiotherapy (RT) rather than neuraxis RT is routinely employed in order to prevent the profound bone marrow suppression associated with neuraxis RT. The full prescribed dose of whole brain RT may not be tolerated by some patients with AIDS, which in some instances influences the response to therapy and clinical outcome. RT alone can result in a clinical and radiographic remission in many patients, and up to half may have a complete radiographic remission. Unfortunately the overall survival following RT is only about 4–6 months (Formenti et al. 1989; Baumgartner et al. 1990; Goldstein et al. 1991). This compares to an average survival of about 1 year for immunocompetent patients treated with RT alone for PCNSL. The discrepancy relates to the fact that the great majority of patients with AIDS succumb to opportunistic infections rather than progressive or recurrent PCNSL. More effective treatments for opportunistic infections may make the outlook for AIDS-associated PCNSL a bit more promising in the future.

Most patients with AIDS-related PCNSL would not tolerate the additional immunosuppressive effects of the aggressive multi-agent chemotherapy regimens that are used to treat PCNSL in immunocompetent patients, and chemotherapy is therefore usually not an appropriate option. Occasional patients, however, develop AIDS-related PCNSL earlier in the course of their disease, and for very select patients chemotherapy may be warranted (Forsyth et al. 1994). Chamberlain (1994) treated four patients with whole brain RT and hydroxyurea, as well as combination procarbazine, CCNU and vincristine. In this very select group who had good performance status, limited CNS disease without meningeal dissemination at presentation, and CD4 counts greater than $200/mm^3$, a median survival of 13.5 months was reported. Two of the patients died of progressive PCNSL with meningeal spread, and two died of pneumocystis pneumonia. Although these results are somewhat encouraging, patients eligible for such therapy represent a small minority of patients with AIDS-related PCNSL, in this series 4 of 40 patients. The overall prognosis for AIDS-related PCNSL is not likely to be significantly improved until more effective treatments for the underlying illness exist.

REFERENCES

ABOULAFIA, D.M.: Human immunodeficiency virus-associated neoplasms: epidemiology, pathogenesis, and review of current therapy. Cancer Pract. 2 (1994) 297–306.

ANDERS, K., K.D. STEINSAPIR, D.J. IVERSON, B.J. GLASGOW, L.J. LAYFIELD, W.J. BROWN, P.A. CANCILLA, M.A. VERITY and H.V. VINTERS: Neuropathological findings in the acquired immunodeficiency syndrome (AIDS). Clin. Neuropathol. 5 (1986) 1–20.

ARICO, M., D. CASELLI, P. D'ARGENIO, A.R. DEL MISTRO, M. DEMARTINO, S. LIVADIOTTI, N. SANTORO and A. TERRANGA: Malignancies in children with human immunodeficiency virus type 1 infection. The Italian multicenter study on human immunodeficiency virus infection in children. Cancer 68 (1991) 2473–2477.

ATALIA, A., J. FERRO and F. ATUNES: Stroke in an HIV-infected patient. J. Neurol. 239 (1992) 356–357.

AUPERIN, I., J. MIKOLT, E. OKSENHENDLER, J.B. THIEBAUT, M. BRUNET, B. DUPONT and F. MORINET: Primary central nervous system malignant non-Hodgkin's lymphoma from HIV-infected and non-infected patients: expression of cellular surface proteins and Epstein–Barr viral markers. Neuropathol. Appl. Neurobiol. 20 (1994) 243–252.

BARTON, N.W., B. SAFAI, S.L. NIELSON and J.B. POSNER: Neurologic complications of Kaposi's sarcoma: an analysis of 5 cases and review of the literature. J. Neuro-Oncol. 1 (1983) 333–346.

BASHIR, R.M., N.L. HARRIS, F.H. HOCHBERG and R.M. SINGER: Detection of Epstein–Barr virus in CNS lymphomas by in-situ hybridization. Neurology 39 (1989) 813–817.

BASHIR, R., J. LUKA, K. CHELOHA, M. CHAMBERLAIN and F. HOCHBERG: Expression of Epstein–Barr virus proteins in primary CNS lymphoma in AIDS patients. Neurology 43 (1993) 2358–2362.

BAUMGARTNER, J.E., J.R. RACHLIN, J.H. BECKSTEAD, T.C. MEEKER, R.M. LEVY, W.M. WARA and M.L. ROSENBLUM: Primary central nervous system lymphomas: natural history and response to radiation therapy in 55 patients with acquired immunodeficiency syndrome. J. Neurosurg. 73 (1990) 206–211.

BELL, W.R., N.F. STARKSEN, S. TONG and J.K. PORTERFIELD: Trousseau's syndrome. Am. J. Med. 70 (1985) 423–430.

BERGER, J.R.: Diagnosing neurosyphilis: the value of the cerebrospinal fluid VDRL or lack thereof. J. Clin. Neuroophthalmol. 9 (1989) 234–235.

BERGER, J.R.: Neurosyphilis in human immunodeficiency virus type 1-seropositive individuals: a prospective study. Arch. Neurol. 48 (1991) 700–702.

BERGER, J.R.: AIDS and the nervous system. In: M.J. Aminoff (Ed.), Neurology and General Medicine, 2nd Edit. New York, Churchill Livingstone (1995) 757–778.

BERGER, J.R., J.O. HARRIS, J. GREGORIOS and M. NORENBERG: Cerebrovascular disease in AIDS: a case control study. AIDS 4 (1990) 239–244.

BERNSTEIN, L. and A.S. HAMILTON: The epidemiology of AIDS-related malignancies. Curr. Opin. Oncol. 5 (1993) 822–830.

BIRX D.L., R.R. REDFIELD and G. TOSATO: Defective regulation of Epstein–Barr virus infection in patients with acquired immunodeficiency syndrome (AIDS) or AIDS-related disorders. N. Engl. J. Med. 314 (1986) 874–879.

BROUSSET, P., E. DROUET, D. SCHLAIFER, J. ICART, C. PAYEN, F. MEGGETTO, B. MARCHOU, P. MASSIP and G. DELSO: Epstein–Barr virus (EBV) replicative gene expression in tumour cells of AIDS-related non-Hodgkin's lymphoma in relation to CD4 cell number and antibody titres to EBV. AIDS 8 (1994) 583–590.

BUCHBINDER, A. and A.E. FRIEDMAN-KIEN: Clinical aspects of Kaposi's sarcoma. Curr. Opin. Oncol. 4 (1992) 867–874.

BUONAGURO, L., F.M. BUONAGURO, M.L. TORNESELLO, E. BETH-GIRALDO, E. DEL GUADIO, B. ENSOLI and G. GIRALDO: Role of HIV-1 tat in the pathogenesis of AIDS-associated Kaposi's sarcoma. Antibiot. Chemother. 46 (1994) 62–72.

BURNS, D.K.: The neuropathology of pediatric acquired immunodeficiency syndrome. J. Child Neurol. 7 (1992) 332–346.

CALABRESE, L.H., A.J. FURLAN, L.A. GRAGG and T.J. ROPOS: Primary angiitis of the central nervous system: diagnostic criteria and clinical approach. Cleve. Clin. J. Med. 59 (1992) 293–306.

CANOSO, R.T., L.I. ZON and J.E. GROOPMAN: Anticardiolipin antibodies associated with HTLV-III infection. Br. J. Haematol. 65 (1987) 495–498.

CARBONE, A., A. GLOGHINI, R. VOLPE, M. BOIOCCHI and U. TIRELLI: High frequency of Epstein–Barr virus latent membrane protein-1 expression in acquired immunodeficiency syndrome related Ki-1 (CD30)-positive anaplastic large-cell lymphomas. Italian Cooperative Group on AIDS and Tumors. Am. J. Clin. Pathol. 101 (1994) 768–772.

CARRANZA, E.J., E. ROSSITCH and J. MORRIS: Isolated central nervous system aspergillosis in the acquired immunodeficiency syndrome. Clin. Neurol. Neurosurg. 93 (1991) 227–230.

CHAMBERLAIN, M.C.: Long survival in patients with acquired immune deficiency syndrome-related primary central nervous system lymphoma. Cancer 73 (1994) 1728–1730.

CHAMBERLAIN, M.C. and L. DIRR: Involved-field radiotherapy and intra-Ommaya methotrexate/cytarabine in patients with AIDS-related lymphomatous meningitis. J. Clin. Oncol. 11 (1993) 1978–1984.

CLEARY, M.L., M.A. EPSTEIN, S. FINERTY, R.F. DORFMAN, G.W. BORNKAMM, J.K. KIRKWOOD, A.J. MORGAN and J. SKLAR: Individual tumors of multifocal EB virus-induced malignant lymphomas in tamarins arise from different B-cell clones. Science 228 (1985) 722–724.

CONSTANS, J., J.L. PELLEGRIN, E. PEUCHANT, M.F. DUMON, M. SIMONOFF, M. CLERC, B. LENG and C. CONRI: High plasma lipoprotein (a) in HIV-positive patients. Lancet 341 (1993) 1099–1100.

DEANGELIS, L.M.: Primary central nervous system lymphoma: a new clinical challenge. Neurology 41 (1991a) 619–621.

DEANGELIS, L.M.: Primary central nervous system lymphoma as a second malignancy. Cancer 67 (1991b) 619–621.

DESTIAN, S., H. TUNG, R. GRAY, D.R. HINTON, J. DAY and T. FUKUSHIMA: Giant infectious intracavernous carotid artery aneurysm presenting as intractable epistaxis. Surg. Neurol. 41 (1994) 472–476.

DINA, T.S.: Primary central nervous system lymphoma versus toxoplasmosis in AIDS. Radiology 179 (1991) 823–828.

DOBERSON, M.J. and B.K. KLEINSCHMIDT-DEMASTERS: Superior sagittal sinus thrombosis in a patient with acquired immunodeficiency syndrome. Arch. Pathol. Lab. Med. 118 (1994) 844–846.

ENGSTRÖM, J., D.H. LOWENSTEIN and D. BREDESON: Cerebral infarctions and transient neurologic deficits associated with AIDS. Am. J. Med. 86 (1989) 528–532.

ENTING, R.H., R.A. ESSELINK and P. PORTEGIES: Lymphomatous meningitis in AIDS-related systemic non-Hodgkin's lymphoma: a report of eight cases. J. Neurol. Neurosurg. Psychiatry 57 (1994) 150–153.

FEIDEN, W., K. BISE, U. STEUDE, H.W. PFISTER and A.A. MOLLER: The stereotactic biopsy diagnosis of focal intracerebral lesions in AIDS patients. Acta Neurol. Scand. 87 (1993) 228–233.

FINE, H.A. and R.J. MAYER: Primary central nervous system lymphoma. Ann. Intern. Med. 119 (1993) 1093–1104.

FORMENTI, S.C., P.S. GILL, E. LEAN, M. RARICK, R.R. MEYER, W. BOSWELL, Z. PETROVICH, L. CHAK AND A.M. LEVINE: Primary central nervous system lymphoma in AIDS. Results of radiation therapy. Cancer 63 (1989) 1101–1107.

FORSYTH, P.A., J. YAHOLOM and L.M. DEANGELIS: Combined modality therapy in the treatment of primary central nervous system lymphoma in AIDS. Neurology 44 (1994) 1473–1479.

GAIDANO, G., C. PASTORE, C. LANZA, U. MASSA and G. SAGLIO: Molecular pathology of AIDS-related lymphomas. Biologic aspects and clinicopathological heterogeneity. Ann. Hematol. 69 (1994) 281–290.

GALETTO, G. and A. LEVINE: AIDS-associated primary central nervous system lymphoma. Oncology Core Committee, AIDS Clinical Trials Group. J. Am. Med. Assoc. 269 (1993) 92–93.

GALLO, P., S. SIVIERI, A.M. FERRARINI, B. GIOMETTO, A. RUFATTI, E. RITTER, C. CHIZZOLINI and B. TAVOLATO: Cerebrovascular and neurologic disorders associated with antiphospholipid antibodies in CSF and serum. J. Neurol. Sci. 122 (1994) 97–101.

GEPPERT, M., C.B. OSTERTAG, G. SEITZ and M. KIESSLING: Glucocorticoid therapy obscures the diagnosis of cerebral lymphoma. Acta Neuropathol. (Berl.) 80 (1990) 629–634.

GOLDSTEIN, J.D., D.W. DICKSON, F.G. MOSER, A.D. HIRSCHFELD, K. FREEMAN, J.F. LLENA, B. KAPLAN and L. DAVIS: Primary central nervous system lymphoma in acquired immune deficiency syndrome. A clinical and pathological study with results of treatment with radiation. Cancer 67 (1991) 2756–2765.

GORIN, F.A., J.F. BALE, M. HALKS-MILLER and R.A. SCHWARTZ: Kaposi's sarcoma metastatic to the CNS. Arch. Neurol. 42 (1985) 162–165.

GRANT, J.W. and A. VON DIEMLING: Primary T-cell lymphoma of the central nervous system. Arch. Pathol. Lab. Med. 114 (1990) 24–27.

HAMILTON-DUTOIT, S.J., M. RAPHAEL, J. AUDOUIN, J. DIEBOLD, I. LISSE, C. PEDERSEN, E. OKSENHENDLER, L. MARELLE and G. PALLESEN: In situ demonstration of Epstein–Barr virus small RNAs (EBER 1) in acquired immunodeficiency syndrome-related lymphomas: correlation with tumor morphology and primary site. Blood 82 (1993) 619–624.

HENDERSON, E., G. MILLER, J. ROBINSON and L. HESTON: Efficiency of transformation of lymphocytes by Epstein–Barr virus. Virology 76 (1977) 152–163.

HOCHBERG, F.H., G. MILLER, R.T. SCHOOLEY, M.S. HIRSCH, P. FEORINO and W. HENLE: Central nervous system lymphoma related to Epstein–Barr virus. N. Engl. J. Med. 309 (1983) 745–748.

HOFFMAN, J.M., H.A. WASKIN, T. SCHIFTER, M.W. HANSON, L. GRAY, S. ROSENFELD and R.E. COLEMAN: FDG-PET in differentiating lymphoma from nonmalignant central nervous system lesions in patients with AIDS. J. Nucl. Med. 34 (1993) 567–575.

HOLLINGSWORTH, H.C., M. STETLER-STEVENSON, D. GAGNETEN, D.W. KINGMA, M. RAFFELD and E.S. JAFFE: Immunodeficiency-associated malignant lymphoma. Three cases showing genotypic evidence of both T- and B-cell lineages. Am. J. Surg. Pathol. 18 (1994) 1092–1101.

HOLTOM, P.D., R.A. LARSEN, M.E. LEAL and J.M. LEEDOM: Prevalence of neurosyphilis in human immunodeficiency virus-infected patients with latent syphilis. Am. J. Med. 93 (1992) 9–12.

IACOANGELI, M., R. ROSELLI, A. ANTINORI, A. AMMASSARI, R. MURRI, A. POMPUCCI and M. SCERRATI: Experience with brain biopsy in acquired immune deficiency syndrome-related focal lesions of the central nervous system. Br. J. Surg. 81 (1994) 1508–1511.

IOACHIM, H.L., B. DORSETT, W. CRONIN, M. MAYA and S. WAHL: Acquired immunodeficiency syndrome-associated lymphomas: clinical, pathological, immunological, and viral characteristics of 111 cases. Hum. Pathol. 22 (1991) 659–673.

JOHNS, D.R., M. TIERNEY and D. FELSENSTEIN: Alteration in the natural history of neurosyphilis by concurrent infection with human immunodeficiency virus. N. Engl. J. Med. 316 (1987) 1569–1572.

JOSHI, V.V., B. PAWEL, E. CONNOR, L. SHARER, J.M. OLESKE, S. MORRISON and J. MARIN-GARCIA: Arteriopathy in children with AIDS. Pediatr. Pathol. 7 (1987) 261–268.

JOSKE, D. and H. KNECHT: Epstein–Barr virus in lymphomas: a review. Blood Rev. 7 (1993) 215–222.

KARNAD, A.B., A. JAFFAR and R.H. LANDS: Spontaneous regression of acquired immune deficiency syndrome-related, high-grade, extranodal non-Hodgkin's lymphoma. Cancer 69 (1992) 1856–1857.

KASE, C.S., S.M. LEVITZ, J.M. WOLINSKY and C.A. SULIS: Pontine pure motor hemiparesis due to meningovascular syphilis in human immunodeficiency virus-positive patients. Arch. Neurol. 45 (1988) 832.

KAUFMAN, W.M., C.J. SIVIT, C.R. FITZ, T.A. RAKUSAN, K. HERZOG and R.S. CHANDRA: CT and MR evaluation of intracranial involvement in pediatric HIV infection: a clinical-imaging correlation. Am. J. Neuroradiol. 13 (1992) 949–957.

KIEBURTZ, K.D., T.A. ESKIN, L. KETONEN and M.J. TUITE: Opportunistic cerebral vasculopathy and stroke in patients with the acquired immunodeficiency syndrome. Arch. Neurol. 50 (1993) 430–432.

KNOWLES, D.M.: Biologic aspects of AIDS-associated non-Hodgkin's lymphoma. Curr. Opin. Oncol. 5 (1993) 822–830.

KUGLER, S.L., A. BARZILAI, D.S. HODES, A. STOLLMAN, C.K. KIM, A.C. HYATT and A.M. ARON: Acute hemiplegia associated with HIV infection. Pediatr. Neurol. 7 (1991) 207–210.

KURE, K., J.F. LLENA, W.D. LYMAN, R. SOEIRO, K.M. WEIDENHEIM, A. HIRANO and D.W. DICKSON: Human immunodeficiency virus-1 infection of the nervous system: an autopsy study of 268 adult, pediatric, and fetal brains. Hum. Pathol. 22 (1991) 700–710.

LAFEUILLADE, A., M.C. ALESSI, I. POIZOT-MARTIN, C. DHIVER, R. QUILICHINI, L. AUBERT, J.A. GASTAUT and I. JUHAN-VAGUE: Protein S deficiency and HIV infection. N. Engl. J. Med. 324 (1991) 1220.

LANG, C., G. JACOBI, W. KREUZ, H. HACKER, G. HERMANN, H.G. KEUL and E. THOMAS: Rapid development of giant aneurysm at the base of the brain in an 8-year-old boy with perinatal HIV infection. Acta Histochem. 42 (1992) 83–90.

LERNER, P.I.: Neurologic manifestations of infective endocarditis. In: M.J. Aminoff (Ed.), Neurology and General Medicine, 2nd Edit. New York, Churchill Livingstone (1995) 97–117.

LEVER, A.M.: HIV-1 and HIV-2: overview of disease spectrum. Baillières Clin. Neurol. 1 (1992) 83–101.

LEVINE, A.: Cancer in AIDS (editorial overview). Curr. Opin. Oncol. 4 (1992a) 863–866.

LEVINE, A.M.: Acquired immunodeficiency syndrome-related lymphoma. Blood 80 (1992b) 8–20.

LEVINE, A.M.: AIDS-related malignancies: the emerging epidemic. J. Natl. Cancer Inst. 85 (1993) 1382–1397.

LEVINE, A.M.: Lymphoma complication immuunodeficiency disorders. Ann. Oncol. 5, Suppl. 2 (1994) 29–35.

LEVY, R.M., D.E. BREDESEN and M.L. ROSENBLUM: Neurologic manifestations of the acquired immunodeficiency syndrome (AIDS): experience at UCSF and review of the literature. J. Neurosurg. 62 (1985) 475–495.

LEVY, R.M. and D.E. BREDESEN: Central nervous system dysfuntion in acquired immunodeficiency syndrome. In: M.L. Rosenblum, R.M. Levy and D.E. Bredesen (Eds.), AIDS and the Nervous System. New York, Raven Press (1988) 29–63.

LING, S.M., M. ROACH, D.A. LARSON and W.M. WARA: Radiotherapy of primary central nervous system lymphoma in patients with and without immunodeficiency virus. Ten years of treatment experience at the University of California, San Francisco. Cancer 73 (1994) 2570–2582.

MACLEAN, H., J. IRONSIDE and B. DHILLON: Acquired immunodeficiency syndrome-related primary central nervous system lymphoma. Arch. Ophthalmol. 112 (1994) 269–271.

MACMAHON, E.M.E., J.D. GLASS, S.D. HAYWARD, R.B. MANN, P.S. BECKER, P. CHARACHE, J.C. MCARTHUR and R.F. AMBINDER: Epstein–Barr virus in AIDS-related primary central nervous system lymphoma. Lancet 338 (1991) 967–973.

MCARTHUR, J.C. and R.T. JOHNSON: Primary infection with human immunodeficiency virus. In: M.L. Rosenblum, R.M. Levy and D.E. Bredesen (Eds.), AIDS and the Nervous System. New York, Raven Press (1988) 183–201.

MEYOHAS, M.C., E. ROULLET, C. ROUZIOUX, A. AYMARD, B. PELOSSE, M. ELIASCEIWICZ and J. FROTTIER: Cerebral venous thrombosis and dual primary infection with human immunodeficiency virus and cytomegalovirus. J. Neurol. Neurosurg. Psychiatry 52 (1989) 1010–1011.

MILLER, D.C., F.H. HOCHBERG, N.L. HARRIS, M.L. GRUBER, D.N. LOUIS and H. COHEN: Pathology with clinical correlations of primary central nervous system non-Hodgkin's lymphoma. The Massachusetts General Hospital experience 1958–1989. Cancer 74 (1994) 1383–1397.

MILLIKEN, S. and M.J. BOYLE: Update on HIV and neoplastic disease. AIDS 7 (1993) S203–209.

MINEURA, K., J. SAWATAISHI, T. SASAJIMA, M. KOWADA, A. SUGAWARA and K. EBINA: Primary central nervous system involvement of the so-called 'peripheral T-cell lymphoma'. Report of a case and review of the literature. J. Neuro-Oncol. 16 (1993) 235–242.

MIZUSAWA, H., A. HIRANO, J.F. LLENA and A.SHINTAKU: Cerebrovascular lesions in acquired immune defiency syndrome (AIDS). Acta Neuropathol. (Berl.) 76 (1988) 451–457.

MORGELLO, S., K. MAIESE and C.K. PETITO: T-cell lymphoma of the central nervous system. Arch. Pathol. Lab. Med. 114 (1990) 24–27.

MOSCHINI, M., T. TARTAGLIONE, M. ROLLO, A. FILENI and C. COLOSIMO: Diagnostic imaging of HIV-related intracranial expansive lesions. Rays 19 (1994) 51–67.

NAGAYAMA, M., Y. SHINOHARA and T. NAGAYAMA: Lipoprotein (a) and ischemic cerebrovascular disease in young adults. Stroke 25 (1994) 74–78.

NORTHFELT, D.W. and L.D. KAPLAN: Clinical aspects of AIDS-related non-Hodgkin's lymphoma. Curr. Opin. Oncol. 3 (1991) 872–880.

O'DELL, M.W.: Hemiparesis in HIV infection. Rehabilitation approach. Am. J. Phys. Med. Rehab. 72 (1992) 291–296.

O'MALLEY, J.P., H.A. ZIESSMAN, P.N. KUMAR, B.A. HARKNESS, J.G. TALL and P.F. PIERCE: Diagnosis of intracranial lymphoma in patients with AIDS: value of 201Tl single-photon emission computed tomography. Am. J. Roentgenol. 163 (1994) 417–421.

PARK, Y.D., A.L. BELMAN, T.S. KIM, K. KURE, J. F. LLENA, G. LANTOS, L. BERNSTEIN and D.W. DICKSON: Stroke in pediatric acquired immune deficiency syndrome. Ann. Neurol. 28 (1990) 303–311.

PETERSON, K., K.B. GORDON, M. HEINEMANN and L.M. DEANGELIS: The clinical spectrum of ocular lymphoma. Cancer 72 (1993) 843–849.

PHILIPPET, P., S. BLANCHE, G. SEBAG, G. RODESCH, C. GRISCELLI and M. TARDIEU: Stroke and cerebral infarcts in children infected with human immunodeficiency virus. Arch. Pediatr. Adolesc. Med. 148 (1994) 965–970.

RAHHAL, F.M., D.R. ROSBERGER and M.H. HEINEMANN: Aggressive orbital lymphoma in AIDS. Br. J. Ophthalmol. 78 (1994) 319–321.

REYNOLDS, P., L.D. SAUNDERS, M.E. LAYEFSKY and G.F. LEMP: The spectrum of acquired immunodeficiency syndrome (AIDS)-associated malignancies in San Francisco, 1980–1987. Am. J. Epidemiol. 137 (1993) 19–30.

RHODES, R.H.: Histopathology of the central nervous system in the acquired immunodeficiency syndrome. Hum. Pathol. 18 (1987) 636–643.

ROGERS, L.R., E.S. CHO, S. KEMPIN and J.B. POSNER: Cerebral infarction from non-bacterial thrombotic endocarditis. A clinical and pathological study including the effects of anti-coagulation. Am. J. Med. 83 (1987a) 746–756.

ROGERS, L., S. SCHOLD and K. EASELY: Cerebrospinal fluid tumor markers as an aid to the diagnosis of meningeal metastasis. Proc. Am. Soc. Clin. Oncol. 6 (1987b) 8.

ROSENFELD, S.S., J.M. HOFFMAN, R.E. COLEMAN, M.J. GLANTZ, M.W. HANSON and S.C. SCHOLD: Studies of primary central nervous system lymphoma with fluorine-18-fluorodeoxyglucose positron emission tomography. J. Nucl. Med. 33 (1992) 532–536.

RUBBERT, A., E. BOCK, J. SCHWAB, J. MARIENHAGEN, H. NUSSLEIN, F. WOLF and J.R. KALDEN: Anticardiolipin antibodies in HIV infection: association with cerebral perfusion defects as detected by 99mTc-HMPAO SPECT. Clin. Exp. Immunol. 98 (1994) 361–368.

RUBIN, M., I. LIBMAN, M. BRISSON, M. GOLDENBERG and S. BREM: Spontaneous temporary remission in primary CNS lymphoma. Can. J. Neurol. Sci. 14 (1987) 175–177.

RUIZ, A., W.I. GANZ, M.J. POST, A. CAMP, H. LANDY, W. MALLIN and G.N. SFAKIANAKIS: Use of thallium-201 brain SPECT to differentiate cerebral lymphoma from toxoplasma encephalitis in AIDS patients. Am. J. Neuroradiol. 15 (1994) 1885–1894.

RUSSELL, D.S. and L.J. RUBINSTEIN: Primary cerebral lymphomas. In: Pathology of Tumours of the Nervous System, 5th Edit. Baltimore, MD, Williams and Wilkins (1989) 592–608.

RWOMUSHANA, R.J.W., I.C. BAILEY and S.K. KYALWAZI: Kaposi's sarcoma of the brain. A case report with necropsy findings. Cancer 36 (1975) 1127–1131.

SAMOSZUK, M., V. NGUYEN, F.F. SHADAN and E. RAMZI: Incidence of Epstein–Barr virus in AIDS-related lymphoma specimens. J. AIDS 6 (1993) 913–918.

SEREMETIS, S., G. INGHIRAMI, D. FERRERRO, E.W. NEWCOMB, D.M. KNOWLES, G.P. DOTTO and R. DALLA-FAVERA: Transformation and plasmacytoid differentiation of EBV-infected human B lymphocytes by ras oncogenes. Science 243 (1989) 660–663.

SHUTTLEWORTH, E. and N. ALLEN: CSF β-glucuronidase assay in the diagnosis of neoplastic meningitis. Arch. Neurol. 37 (1980) 684–687.

SINZOBAHAMVYA, N., K. KALANG and W. HAMEL-KALINOWSKI: Arterial aneurysms associated with human immunodeficiency virus (HIV) infection. Acta Chir. Belg. 89 (1989) 185–188.

SNIDER, W.D., D.M. SIMPSON, S. NIELSEN, J.W.M. GOLD, C.E. METROKA and J.B. POSNER: Neurologic complications of acquired immune deficiency syndrome: analysis of 50 patients. Ann. Neurol. 14 (1983) 403–418.

SO, Y.T., A. CHOUCAIR, R.L. DAVIS, W.M. WARA, J.L. ZIEGLER, G.E. SHELINE and J.H. BECKSTEAD: Neoplasms of the central nervous system in acquired immunodeficiency syndrome. In: M.L. Rosenblum, R.M. Levy and D.E. Bredesen (Eds.), AIDS and the Nervous

System. New York, Raven Press (1988) 285–300.

SORICE, M., A. CIRCELLA, F. D'AGOSTINO, M. RANIERI, S. MODRZEWSKA, L. LENTI and G. MARIANI: Protein S and HIV infection. Thromb. Res. 73 (1994) 165–175.

SPARANO, J.A., P.H. WIERNIK, M. STRACK, A. LEAF, N.H. BECKER, C. SARTA, D. CARNEY, R. ELKIND, M. SHAH and E.S. VALENTINE: Infusional cyclophosphamide, doxorubicin and etoposide in HIV-related non-Hodgkin's lymphoma: a follow-up report of a highly active regimen. Leuk. Lymph. 14 (1994) 263–271.

SPINA, M. and U. TIRELLI: Human immunodeficiency viruses as a risk factor in miscellaneous cancers. Curr. Opin. Oncol. 4 (1992) 907–910.

STRICKER, R.B., D.I. ABRAMS, L. CORASH and M.A. SHUMAN: Target platelet antigen in homosexual men with immune thrombocytopenia. N. Engl. J. Med. 313 (1985) 1375–1380.

TAILLAN, B., G. GARNIER, E. FARRARI, A. PESCE, H. VINTI, J.G. FUZIBET and P. DUJARDIN: MACP-B chemotherapy for the treatment of high-grade lymphomas in patients with HIV-1 infection. Acta Haematol. (Basel) 89 (1993) 10–12.

TERRIFF, B.A., P. HARRISON and J.K. HOLDEN: Apparent spontaneous regression of AIDS-related primary CNS lymphoma mimicking resolving toxoplasmosis. J. AIDS 5 (1992) 953–954.

THIRUMALAI, S. and H.S. KIRSHNER: Anticardiolipin antibody and stroke in an HIV-positive patient. AIDS 8 (1994) 1019–1020.

TIRELLI, U., S. FRANCESCHI and A. CARBONE: Malignant tumors in patients with HIV infection. Br. Med. J. 308 (1994) 1148–1153.

TRENKWALDER, P., C. TRENKWALDER, W. FEIDEN, T.J. VOGL, L.M. EINHAUPL and H. LYDTIN: Toxoplasmosis with early intracerebral hemorrhage in a patient with the acquired immunodeficiency sydnrome. Neurology 42 (1992) 436–438.

TYLER, K.L., E. SANDBERG and K.F. BAUM: Medial medullary syndrome and meningovascular syphillis: a case-report in an HIV-infected man and a review of the literature. Neurology 44 (1994) 2228–2230.

WANG, J.C., Y. ROSEN, P.C. GOEL, H. TEPLITZ, M. GOLDBERG and A.E. FRIEDMAN-KIEN: Kaposi's sarcoma presenting as lymphadenopathy in two HIV-negative elderly patients. Am. J. Med. 94 (1993) 342–344.

WIJDICKS, E.F., J.C. BORLEFFS, A.I. HOEPELMAN and G.H. JANSEN: Fatal disseminated hemorrhagic toxoplasmic encephalitis as the initial manifestation of AIDS. Ann. Neurol. 29 (1991) 683–686.

WOODS, G.L. and J.C. GOLDSMITH: Aspergillus infection of the nervous system in patients with AIDS. Arch. Neurol. 47 (1990) 181–184.

YANKER, B.A., P.R. SKOLICK and G.M. SHOUKIMAS: Cerebral granulomatous angiitis associated with isolation of human T-lymphoctrophic virus type III from the nervous system. Ann. Neurol. 20 (1986) 362–364.

Handbook of Clinical Neurology, Vol. 27 (71): Systemic Diseases, Part III
M.J. Aminoff and C.G. Goetz, editors
© 1998 Elsevier Science B.V. All rights reserved

CHAPTER 15

Neuropathies related to HIV infection and its treatment

GARETH J. PARRY

Department of Neurology, University of Minnesota Medical School, Minneapolis, MN, U.S.A.

Early reports stressed the infrequency of involvement of the neuromuscular system in patients harboring the human immunodeficiency virus (HIV) but subsequent experience has clearly established that it is almost ubiquitous. Neuropathy, in one form or another, is the most frequent complication, occurring in the majority of patients. Myopathy is less frequent and is most often seen in patients receiving treatment with zidovudine (AZT). Disorders of the neuromuscular junction and anterior horn cells have been only anecdotally described. In this chapter, the neuropathies seen in association with HIV infection and its treatment will be discussed. Discussion will include the neuropathies associated with cytomegalovirus (CMV) infection, even though they almost certainly result from direct infection with the virus. HIV-related myopathies, including the AIDS-cachexia syndrome and AZT-related myopathy, will be covered in Chapter 16, dealing with the complications of treatment for HIV infection.

EPIDEMIOLOGY OF HIV-ASSOCIATED NEUROPATHIES

The association between neuropathy and HIV infection in seven of nine patients with AIDS was first suggested in 1982, about 1 year after the initial descriptions of the acquired immunodeficiency syndrome (AIDS) (Horowitz et al. 1982). A year later Snider et al. (1983) described eight patients with neuropathy among 50 with neurological complica-

tions seen in 160 patients with AIDS. The condition was a painful sensory neuropathy which was distally located and symmetrical in its distribution. The next report on neuropathy came from Lipkin et al. (1985) with the description of a multifocal neuropathy appearing at an earlier stage of HIV infection, before progression to overt AIDS. Thereafter, there was a rapid proliferation of reports as numerous investigators recognized the frequent association of neuropathy with all stages of HIV infection.

Cornblath et al. (1993) have suggested that neuropathy is the most common neurological disorder in individuals with HIV infection. Prevalence of neuropathy depends on the stage of HIV infection and the method of ascertainment. In Snider et al. (1983) only 16% of patients were affected but subsequent studies have shown this to significantly underestimate the problem. Neuropathy is a rare complication early in the course of infection. In a study comparing 270 HIV-positive but otherwise healthy homosexual men with 193 HIV-negative homosexual controls, there was no difference in the incidence of neuropathic symptoms and signs and only one patient was thought to have neuropathy related to his HIV status (McArthur et al. 1989). Nonetheless, there are many anecdotes of various types of neuropathy occurring immediately following HIV seroconversion (Elder et al. 1986; Piette et al. 1986; Vendrell et al. 1987) and in otherwise healthy HIV-positive patients (Cornblath et al. 1987). Neuropathy becomes much more common in the later stages of the disease. Clinically sig-nificant neuropathy is seen in 30–35% of patients with AIDS, based on evaluation of patients at large HIV clinics (Cornblath and McArthur 1988; So et al. 1988). When patients at all stages of HIV infection are studied, widely divergent estimates of prevalence have been reported; Hall et al. (1991) reported a prevalence of only 30–35% while Gastaut et al. (1989) found neuropathy in 89% of their patients. However, in pa-tients dying from AIDS, close to 100% are found to have abnormalities in peripheral nerves at autopsy (De la Monte et al. 1988; Mah et al. 1988).

There is a striking specificity of the type of neu-ropathy for the different stages of HIV infection, probably reflecting different pathogenetic mecha-nisms (Parry 1988; McArthur et al. 1989). The neuropathies which occur immediately following seroconversion or in the early stages of infection (symptomatic or asymptomatic) are probably auto-immune, occurring at a time when the immune system remains competent. The acute and chronic inflamma-tory demyelinating neuropathies, universally accepted as being autoimmune disorders, occur most often during the asymptomatic stages of infection (CDC stages II and III) as did the one reported case of acute ganglioneuropathy (Elder et al. 1986), another

TABLE 1

HIV-associated neuropathies.

1. Neuropathies following HIV seroconversion
    Guillain–Barré syndrome (GBS)
    Acute sensory ganglioneuropathy
    Bell's palsy
    Brachial neuritis

2. Neuropathies with early HIV infection
    Guillain–Barré syndrome (GBS)
    Chronic inflammatory demyelinating poly-
        radiculoneuropathy (CIDP)
    Mononeuropathy multiplex (secondary to
        vasculitis)

3. Neuropathies with advanced HIV infection
    Distal symmetric polyneuropathy
    Autonomic neuropathy
    CMV-associated neuropathies
        CMV polyradiculoneuropathy
        CMV mononeuropathy multiplex

4. Neuropathies associated with HIV treatment
    2',3'-Dideoxycytidine (ddC)
    2',3'-Dideoxyinosine (ddI)

presumed autoimmune disorder. Vasculitic neurop-athy, which is perhaps due to circulating immune complexes, also occurs during the earlier stages of in-fection (CDC stages III and IVA). In contrast, the neuropathies that occur later (CDC class IVC and D) are clinically and probably pathogenetically distinct. The commonest HIV-related neuropathy, a distal, symmetrical, predominantly sensory neuropathy which is often painful, occurs when the patient is se-verely immunocompromised and increases in preva-lence as the disease advances. The pathogenesis of this neuropathy is unknown but proposed mecha-nisms include infectious, toxic and metabolic dis-orders. Similarly, the neuropathies associated with CMV infection, namely polyradiculoneuropathy and mononeuropathy multiplex, occur only when immune function is sufficiently depressed to allow direct neural invasion of the virus.

CLINICAL SYNDROMES OF HIV-ASSOCIATED
NEUROPATHY

*Neuropathies associated with HIV
seroconversion*

As with any viral infection, the initial infection with HIV-1 may be associated with a non-specific sys-temic illness with fever, lymphadenopathy, rash, diarrhea and other constitutional symptoms and signs. The central nervous system (CNS) may be involved in this initial illness with aseptic meningitis and en-cephalitis or encephalopathy. In addition, a number of peripheral neuropathic syndromes may be seen. Acute unilateral or bilateral facial paralysis, indistin-guishable from Bell's palsy, has been often reported (Levy et al. 1985; Helweg-Larsen et al. 1986; Piette et al. 1986; Belec et al. 1989; Wechsler and Ho 1989). Other cranial neuropathies are occasionally seen but usually in the context of inflammatory demy-elinating polyneuropathies. Acute brachial neuritis (Parsonage–Turner syndrome) (Calabrese et al. 1987) and acute sensory ganglioneuropathy have also been described (Elder et al. 1986). Finally, Guillain–Barré syndrome (GBS) may be seen follow-ing HIV seroconversion as discussed below. Each of these disorders is essentially clinically identical to that occurring following other viral infections and, al-though the exact pathogenesis remains uncertain, most feel that they are autoimmune conditions. If so, they fit the scenario of autoimmune disorders occurring early in the course of infection with HIV. Based on this occasional association with HIV infection, Cornblath et al. (1993) have recommended that all patients with these conditions be checked for HIV infection. However, unless there are risk factors for HIV infection or the neurological disorder is atypical in some fashion (bilateral Bell's palsy, chronic or relapsing course, CSF pleocytosis), the yield of such testing is extremely low.

These neuropathic syndromes accompanying HIV seroconversion tend to be benign and self-limited and usually recover without specific treatment. However, patients with inflammatory neuropathies may benefit from treatment as outlined in the next section.

Inflammatory demyelinating polyneuropathies

The acute and chronic inflammatory demyelinating polyneuropathies are the commonest of the neuropathies which occur during the relatively immunocompetent phase of HIV infection. These neuropathies share many clinical, electrophysiological and pathological features which have been well reviewed in a number of publications and will only be summarized here (see Dyck et al. 1975; McCombe et al. 1987; Barohn et al. 1989). Both are predominantly motor neuropathies with areflexia and elevation of cerebrospinal fluid (CSF) protein concentration. Electrophysiological studies show the changes of multifocal demyelination with severe conduction slowing, temporal dispersion of compound muscle action potentials, conduction block, and a variable degree of associated axonal degeneration. Pathological studies show inflammatory, macrophage-mediated demyelination and axon loss. The acute and chronic variants differ chiefly by virtue of the clinical course: acute inflammatory demyelinating polyneuropathy (Guillain–Barré syndrome (GBS)), is an acute monophasic illness, while chronic inflammatory demyelinating polyneuropathy (CIDP) runs a chronic progressive or relapsing course. Furthermore, GBS recovers spontaneously, even without treatment, whereas CIDP usually requires long-term immunosuppressive treatment. Both are thought to be autoimmune diseases although the evidence, while persuasive, is largely circumstantial.

Another difference between GBS and CIDP is that the former follows an identifiable antecedent ill-

ness in about two-thirds of cases; the most common antecedent event is an infection that is often viral (Parry 1993). One of the viruses that appears to trigger GBS is HIV. The neuropathy may occur 1–2 weeks after the presumptive initial infection (Hagberg et al. 1986; Piette et al. 1986; Vendrell et al. 1987). Since seroconversion may be delayed for several weeks after primary infection, repeat serology should be obtained in patients with HIV risk factors or an atypical presentation. Often an episode of GBS leads to the diagnosis of HIV infection or GBS occurs during the later but still asymptomatic stages of the illness. Similarly, CIDP occurs with a greater than expected frequency in patients infected with HIV, almost always in patients with oligosymptomatic infection. One important clue to the possible association of GBS or CIDP with HIV infection is a persistent CSF pleocytosis (Cornblath et al. 1993). The pleocytosis is usually modest, up to 43 cells/mm^3 in one series (Cornblath et al. 1993), and similar cell counts may be seen in GBS patients without HIV infection (Asbury et al. 1978) so the presence of a CSF pleocytosis per se does not indicate HIV infection. However, in HIV-infected patients with inflammatory neuropathy CSF pleocytosis is the rule, whereas only 10% of HIV-negative patients have cell counts above 10/mm^3 (Parry 1993). Otherwise, GBS and CIDP related to HIV infection closely resemble these neuropathies in HIV-seronegative individuals, with regard to pattern of evolution, clinical features and severity of the neurological disease, electrodiagnostic features, pathology, response to plasmapheresis and high-dose intravenous immunoglobulin (i.v. Ig) and ultimate recovery from the neurological deficits (Cornblath et al. 1987, 1993; Leger et al. 1989).

There is no consensus concerning the best treatment for GBS and CIDP, whether or not it is associated with HIV infection. Plasmapheresis is effective for both, its efficacy having been established in several large studies (Osterman et al. 1984; Guillain–Barré Syndrome Study Group 1985; Dyck et al. 1986; French Cooperative Group 1987). It is effective in the treatment of cases associated with HIV infection (Cornblath et al. 1987; Miller et al. 1985, 1987) and can be used safely since the tubing is disposable. High-dose intravenous immunoglobulin (i.v. Ig) is also effective in the treatment of GBS (Van der Meche et al. 1992) and CIDP (Dyck et al. 1994), including patients with HIV infection (Chimowitz et al.

1989; Malamut et al. 1992; Cornblath et al. 1993). Corticosteroids are effective in treating CIDP (Dyck et al. 1982) but have no role in GBS (Parry 1993). They have been used with great effect in HIV-associated CIDP (Cornblath et al. 1987; Miller et al. 1988a) but there is a widespread reluctance to use long-term corticosteroids in patients already prone to life-threatening opportunistic infections. Because of safety, efficacy, ease of administration and patient acceptance, I recommend high-dose i.v. Ig as the primary treatment of both GBS and CIDP. It is important to stress that many GBS cases do not need immunotherapy; treatment should be reserved for those patients who have lost the ability to walk or in whom there is significant bulbar or respiratory compromise. Even CIDP may occasionally remit spontaneously although subsequent relapse is usual and treatment is ultimately needed in most cases.

Mononeuropathy multiplex

There are several syndromes of mononeuropathy multiplex in patients with HIV infection. These occur mainly in the early stages of infection.

Chronic inflammatory demyelinating polyneuropathy may have a focal or multifocal onset and later progress to the more characteristic picture. In 1985, Lipkin et al. described a heterogeneous group of patients with neuropathy associated with AIDS-related complex. Most of the patients (10/12) had a distinctly multifocal neuropathy which was predominantly sensory and could involve the limbs or trunk. Six patients had evidence of CNS involvement. Prognosis was good, with six patients recovering spontaneously and another improving with plasmapheresis; two were lost to follow-up. Only one patient eventually progressed to AIDS during the 2-year period of followup. Most of these patients probably had inflammatory demyelinating polyneuropathy with a restricted distribution. Three of the patients described by Lipkin had multifocal axonal degeneration, a finding typical of nerve infarction, suggesting that they had an underlying vasculitis although none was found pathologically.

Subsequently, there were several reports of axonal mononeuropathy multiplex occurring in patients with HIV infection and in whom the suspected vasculitis was confirmed (Said et al. 1987, 1988; Lange et al. 1988; Estes et al. 1989; Gherardi et al. 1989; Cornblath et al. 1993). All reports have em-

phasized the pathological features. From the sparse clinical information provided, most patients appear to be in the early stages of infection. There are usually signs of bilateral, distal sensory loss, suggesting either confluence of multiple small focal deficits or, less likely given the early stage of infection, the coexistence of a distal symmetric polyneuropathy. Electrodiagnostic studies show that the essential feature is multifocal axonal degeneration, as expected in an ischemic neuropathy. The neuropathy is usually indolently progressive or stabilizes, either with or without treatment, but improvement has not been reported. As with the inflammatory demyelinating neuropathies, the neuropathy in HIV-associated necrotizing vasculitis does not differ from that associated with other necrotizing vasculitides such as polyarteritis nodosa (Lovelace 1982) or the syndrome of vasculitis confined to peripheral nerve (Kissel et al. 1985; Dyck et al. 1987). The etiology of this syndrome is not known but it may be related to circulating immune complexes comprised of HIV antigen and antibody which are found in the serum of HIV-infected patients. HIV can be found in the blood vessel wall (Said et al. 1987) and in perivascular mononuclear cells (Gherardi et al. 1989) in HIV-associated vasculitic neuropathy.

The third mononeuropathy multiplex syndrome which occurs in HIV-infected patients is very different in that it occurs in advanced AIDS, usually when the CD4[+] count is less than 100/mm^3 (Roullet et al. 1994). It is associated with severe systemic CMV infection and will be discussed in more detail in a later section.

Distal symmetric polyneuropathy

This is, by far, the commonest neuropathy seen in HIV-infected patients. It is not usually seen during the early stages of infection, at least not to a clinically significant degree. However, almost all patients dying of AIDS have this neuropathy by pathological criteria (De la Monte et al. 1988) and up to 60% have clinically significant disease (Miller et al. 1988b; So et al. 1988). The clinical features are highly characteristic (Bailey et al. 1988; Cornblath and McArthur 1988; Miller et al. 1988b). The onset is insidious and the symptoms are always predominantly sensory and in a symmetric and distal distribution. The earliest symptoms are often dysesthesias confined to the soles; these progress slowly but inexorably proximally with time to involve the entire foot and eventually the legs, but seldom do symptoms progress above the knees. The hands are occasionally involved, but only in the very late stages of the illness. The pain may be sufficiently severe to interfere with walking. There may be associated hyperpathia so that the touch of clothing or bed covers is distressing. Usually, pain remains the major feature although it may subside over time as numbness supervenes. Symptomatic weakness rarely occurs; patients may complain of weakness but careful questioning makes it clear that they are limited by pain, not loss of strength. Nonetheless, there is usually mild motor involvement on examination, with atrophy of intrinsic foot muscles and mild weakness of toe flexion and extension but seldom more proximal weakness. The findings on examination are generally very unimpressive. Sensory thresholds are increased for all modalities in the feet and to a lesser extent in the hands. Ankle reflexes are reduced or absent but other reflexes are normal. In fact, there may be increased reflexes from concurrent myelopathy which may coexist, as also may neuropsychological impairment.

Laboratory testing is unhelpful in revealing the etiology of the neuropathy. CSF is usually normal; high protein or pleocytosis strongly suggests coexisting disease of the CNS. The electrodiagnostic studies are characteristic but non-specific. They show that the neuropathy is invariably more widespread than clinically suspected and involves both sensory and motor functions. Sensory nerve action potentials amplitudes are absent or reduced in amplitude, particularly in the legs, but conduction velocities are only mildly slowed. Motor amplitudes are also reduced with mild slowing of conduction velocity. Needle electromyography (EMG) shows distal denervation and reinnervation. Occasionally, the slowing is more prominent than would be expected from axon loss alone, suggesting an element of associated demyelination, but axon loss clearly predominates (Cornblath et al. 1988; Miller et al. 1988b). Pathologically, there is variable axonal degeneration involving both myelinated and unmyelinated axons, with little regeneration. There is mild epineurial and endoneurial perivascular mononuclear cell infiltration (Lipkin et al. 1985; De la Monte et al. 1988; Cornblath et al. 1993). As with the electrodiagnostic findings, the degree of demyelination is occasionally more severe than would be predicted from primary axonal disease alone and suggests that primary demyelination occa-

sionally develops as well and may have a different etiology (De la Monte et al. 1988; Cornblath et al. 1993). The axonal degeneration is accentuated distally, suggesting that this is a classic distal axonopathy. Further support for this concept comes from the findings of axonal degeneration in the rostral portion of the posterior columns (Rance et al. 1988). Mild neuronal loss and inflammation is also seen in the dorsal root ganglia of patients dying of AIDS.

The etiology of the neuropathy has remained elusive, and multiple toxic, nutritional and infectious factors may play a role. The finding of sensory neuronal degeneration has led some to postulate a role for direct viral infection with CMV (Fuller et al. 1989) but the association is insufficiently frequent to support a primary role (Grafe and Wiley 1989; Winer et al. 1989). Cornblath et al. (1993) have postulated a direct role of HIV itself but again the evidence is lacking.

There is no effective treatment. There is one report of improvement during treatment with zidovudine (Yarchoan et al. 1987) but this experience has not been repeated. Results of studies of nerve growth factor (NGF) are eagerly awaited. Symptomatic treatment of pain with standard agents such as amitriptyline, anticonvulsants, mexiletine or topical capsaicin is important.

Autonomic neuropathy

A single case of subacute pandysautonomia has been reported briefly in a patient with advanced HIV infection, in the purported absence of somatic neuropathy (Lin-Greenberg and Taneeja-Uppal 1987). Certainly there was no clinical evidence of weakness or sensory loss but electrodiagnostic testing was not reported and there may have been subclinical somatic involvement. Nonetheless, there was severe autonomic failure without overt somatic involvement, suggesting kinship with the idiopathic form of acute pandysautonomia (Hart and Kanter 1990).

There have been a number of reports of autonomic dysfunction in patients with somatic neuropathy, usually in those with the distal symmetric polyneuropathy and advanced AIDS. Craddock et al. (1987) were the first to suggest a relationship. Four patients had syncope during needle aspiration of the lung. One of these patients and four others were then tested for autonomic dysfunction and all were found to be abnormal, although asymptomatic. Simi-

larly, Cohen and Laudenslager (1989) tested autonomic function in HIV-infected patients who were
asymptomatic for autonomic failure and found that 5
of 10 were abnormal. All but one of the abnormals
had advanced AIDS but none had clinical evidence of
somatic neuropathy. A larger study of autonomic
function in HIV-infected patients at different stages of
disease, comparing results with appropriate controls,
established the presence of autonomic dysfunction
(Freeman et al. 1990). Abnormalities of autonomic
function increased as patients progressed from ARC to
AIDS. Only 6 of 26 patients had concurrent clinical
somatic neuropathy while five were demented and
three had myelopathy. This study confirms the earlier
uncontrolled observations that autonomic dysfunction
is a feature of HIV infection, that it is largely confined to
patients with advanced HIV infection, and that it can
occur in the absence of clinically significant somatic
neuropathy. While none of the patients in any of these
studies had symptoms of autonomic failure, the experience of Craddock et al. (1987) suggests that they may
be at increased risk for severe reactions to procedures
that sometimes cause autonomic instability even in normal individuals. A note of caution was sounded in a
study in which abnormal results of autonomic function
tests in AIDS patients were carefully controlled for
age and coexistent illnesses (Lohmoller et al. 1987).
When these corrections were made, no differences
were found between AIDS patients and controls.

It should come as no surprise that a neuropathy
known to be associated with degeneration of axons
of all classes, including the small myelinated and unmyelinated axons that subserve autonomic function,
should cause abnormalities of autonomic function. It
is perhaps somewhat surprising that an accompanying
somatic neuropathy is so uncommon, but subclinical
neuropathy was not sought in any of the studies. It is
probable that, since the autonomic dysfunction was
largely asymptomatic, a subclinical somatic neuropathy would have been found by appropriate electrodiagnostic testing.

Cytomegalovirus (CMV) neuropathies

Opportunistic infection with CMV is very common in
HIV-infected individuals and is manifested as encephalitis, retinitis, pneumonia and many other organ
infections. Opportunistic infection of peripheral nerve
must now be added to this list. As mentioned above,

CMV has been implicated in the pathogenesis of the
most common HIV-associated neuropathy, the distal,
predominantly sensory neuropathy seen in patients
with advanced HIV infection, but no incontrovertible
evidence exists to support such a role. However,
there are at least two clinical syndromes in which
CMV almost certainly has a role. The most common
and best established is the CMV-associated polyradiculoneuropathy. Less common is CMV-associated mononeuropathy multiplex.

CMV-associated polyradiculopathy

This devastating disorder occurs almost exclusively in
patients in the late stages of HIV infection (Bishopric
et al. 1985; Eidelberg et al. 1986; Behar et al. 1987;
Miller et al. 1990) although occasional cases have
occurred earlier in the course of infection (Crawfurd
et al. 1987) and in at least one patient it was the initial
manifestation (Mahieux et al. 1989). Its highly distinctive clinical features should alert one to early diagnosis
which is critical since early treatment may prevent
progression and perhaps produce improvement
(Graveleau et al. 1989; Fuller et al. 1990; Miller et al.
1990; Novak et al. 1990). The syndrome is predominantly motor, with severe, rapidly evolving, asymmetric leg weakness occurring in all cases. Occasionally
the weakness is heralded by back or buttock pain,
which may be sufficiently severe as to require narcotic
analgesia. The pain may be exacerbated by
straight-leg raising. Patients then develop urinary retention, severe constipation and impotence. Sensory
symptoms and signs are present in almost all cases
but are overshadowed by weakness and sphincter
disturbances. Paresthesias in the legs and in the perineum are common and there may be mild objective
sensory loss over the saddle area, including the genitals. Severe sensory loss is rare (So and Olney
1994). In some cases, presence of a Babinski reflex
suggests spinal cord involvement (Miller et al. 1990),
and occasionally myelopathic features predominate
(Jacobsen et al. 1988). The neurologic deficit begins
in the most caudal segments and spreads rostrally and
may involve the arms and even cranial nerves on rare
occasions. The syndrome evolves rapidly, with patients seeking medical attention usually within 2
weeks.

The CSF picture in these patients is also characteristic. There is invariably a pleocytosis, often of
severe proportions, with occasionally over 1000

cells/mm³. Unlike most viral infections the pleocytosis is usually predominantly polymorphonuclear. The CSF protein concentration is markedly elevated and the glucose is mildly reduced. CMV can be cultured from the CSF or identified by polymerase chain reaction (PCR) amplification of viral antigen in almost all cases. Other investigations are generally unhelpful. The EMG confirms widespread acute axonal degeneration with low amplitude evoked motor responses on nerve conduction studies. Sensory nerve conduction studies are less abnormal in keeping with the radicular localization of the pathology. Amplitudes are normal or mildly reduced, probably due to coexistent HIV-related sensory neuropathy although involvement of the dorsal root ganglion with the acute CMV infection cannot be excluded. Magnetic resonance imaging (MRI) of the spine may be abnormal, showing enlargement of the conus medullaris or clumping of lumbosacral roots and, rarely, enhancement following gadolinium administration (Fig. 1) (Whiteman et al. 1994).

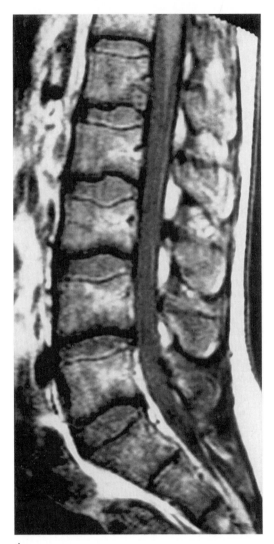 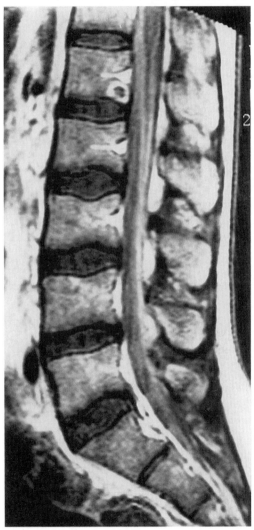

A   B

Fig. 1. MRI scans taken from a patient with CMV polyradiculopathy. (A) The sagittal T1-weighted MR (without contrast) reveals the normal conus. (B) The post-gadolinium MR shows diffuse enhancement along the surface of the conus and throughout the cauda equina. (The MR images are provided with courtesy of Dr. M. Whiteman, Miami University School of Medicine.)

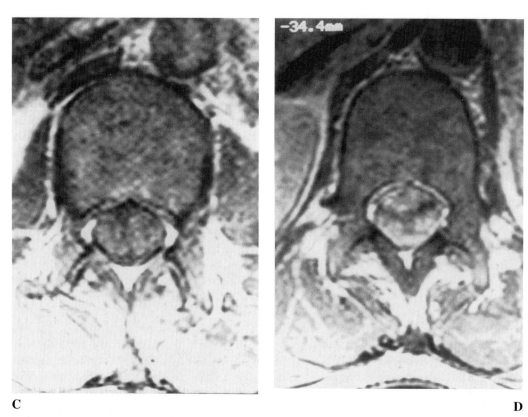

C   D

Fig. 1. (C) The T1-weighted axial MR (without contrast) shows clumping and thickening of the nerve roots. (The MR images are provided with courtesy of Dr. M. Whiteman, Miami University School of Medicine.) (D) The post-gadolinium axial MR shows intense enhancement of the clumped nerve roots. The MR images are provided with courtesy of Dr. M. Whiteman, Miami University School of Medicine.)

Early treatment of this syndrome is imperative if meaningful neurologic function is to be preserved. Ganciclovir, if administered early, may stop progression of the weakness, usually after a delay of 1–2 weeks from the start of treatment (So and Olney 1994). If treatment is started with a few days of the onset of weakness, some improvement may be seen (Miller et al. 1990; So and Olney 1994). It is important to recognize that CMV polyradiculopathy is a complication of advanced AIDS and that prolonged survival is not expected even if improvement of the neurological disorder occurs. In one study, the median survival was 2.7 months and only one patient survived more than 12 months (So and Olney 1994).

Occasional patients develop an acute lumbosacral polyradiculopathy in the absence of CMV infection. Potentially treatable causes include syphilis and mycobacterium tuberculosis, which should be sus-

pected if there is a lymphocytic pleocytosis and low glucose concentration in the CSF (Lanska et al. 1988; Woolsey et al. 1988; Lanska and Lanska 1989). Toxoplasmosis (Mehren et al. 1988), lymphoma (So and Olney 1994) and perirectal abscess (Holtzmann et al. 1989) are other potentially treatable disorders that may present with rapidly progressive leg weakness. So and Olney (1994) also identified another group of patients with lumbosacral polyradiculopathy who could be readily distinguished from those with CMV infection. They had a slower course, less severe weakness and a mononuclear cell CSF pleocytosis. Perhaps most importantly, these patients showed a tendency to improve without specific treatment.

CMV mononeuropathy multiplex
Mononeuropathy multiplex associated with CMV infection was first reported by Fuller et al. (1990b). The

relationship to CMV was suggested by improvement with ganciclovir. A subsequent report (Roullet et al. 1994) identified a group of patients with neuropathy that was either overtly multifocal or was asymmetric in its distribution, suggesting a confluent mononeuropathy multiplex. All patients had advanced AIDS and 10 of 15 had evidence of disseminated CMV infection. The neuropathy evolved acutely or subacutely and was usually initially sensory. Progression to motor involvement occurred in most cases within 3–4 months. Electrophysiological studies showed patchy axonal degeneration. CSF protein was elevated in most cases but usually to a modest degree (< 250 mg/dl in all but one case) and CMV was detected in the CSF by PCR in 90% of cases.

The pathologic changes in these patients were highly characteristic. Perivascular inflammatory cells were seen in nerve and muscle in most cases and, unlike most inflammatory neuropathies, they consisted predominantly of polymorphonuclear leukocytes (Fig. 2). Overt necrotizing vasculitis was seen in one case. Occasionally, specific CMV inclusion bodies were seen in Schwann cells or endoneurial endothelial cells (Fig. 3). Multifocal endoneurial micronecrosis also was occasionally seen.

All patients were treated with specific anti-CMV therapy with either ganciclovir or foscarnet. One patient died within a few days despite treatment, but the rest began to improve within a few weeks and improvement continued for several months before the patients stabilized. Improvement was dramatic; five patients made a near-complete recovery while seven others showed marked improvement, including four who regained the ability to walk. However, three patients later relapsed with evidence of more widespread CMV dissemination, a not uncommon fate in these patients.

This neuropathy progresses more slowly and seems to respond better to treatment than the CMV-related lumbosacral polyradiculopathy. Nonetheless, it is associated with advanced AIDS and the ultimate prognosis is related more to the underlying condition than to the neuropathy.

CMV in painful sensory neuropathy

As already discussed, it has been suggested that CMV infection of peripheral nerves or dorsal root ganglia is the basis for the very common distal symmetric neuropathy seen in patients with AIDS. However, both systemic CMV infection and neuropathy

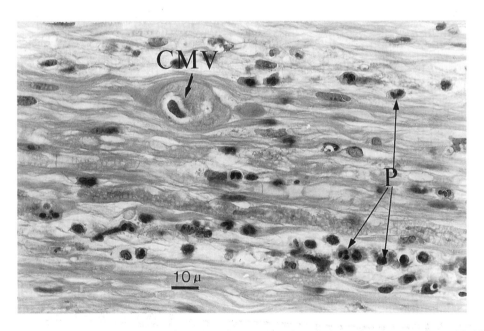

Fig. 2. This light photomicrograph is of a longitudinal section of sural nerve taken from a patient with AIDS and a multifocal neuropathy. The section is taken from paraffin embedded nerve and stained with H and E. A modest endoneurial inflammatory infiltrate is seen (long arrows), consisting mainly of polymorphonuclear leukocytes. A typical intranuclear inclusion of CMV is also present (arrow), probably in a Schwann cell.

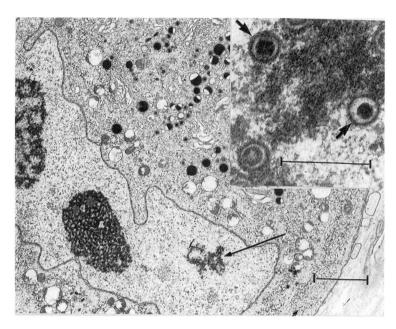

Fig. 3. This electron photomicrograph shows an endothelial cell from an endoneurial capillary. The arrow points to dense chromatin within the nucleus which contains many viral particles which can be seen at higher power in the inset. The osmophilic inclusions within the cytoplasm contain proviral DNA. Bar = 1 μm. (Photomicrographs were provided by Dr. Gerard Said, Universitaire de Bicêtre, Paris.)

are so common in patients with advanced AIDS that the epidemiologic association may be coincidental.

Neuropathies associated with HIV treatment

Neuropathy is an important dose-limiting side effect of treatment of HIV infection with the antiviral agents ddC and ddI. A dose-dependent neuropathy was noted in phase I studies of ddC which was very similar to HIV-associated distal symmetrical polyneuropathy (Yarchoan et al. 1988). The neuropathy was universal in patients receiving 0.03 mg/kg or more but was less common at lower doses and took longer to develop. The earliest symptom was painful dysesthesias in the feet, which inexorably progressed in extent and severity if the drug was not stopped to eventually involve the hands in about half the patients. The sensory thresholds to all modalities were elevated and ankle jerks were reduced or absent, but more proximal reflexes were preserved. There was no evidence of motor involvement. Symptoms continued to worsen for several weeks after the drug was stopped (Dubinsky et al. 1988) but then steadily improved and most patients returned to their premorbid level although the patients with the most severe neu-

ropathy still had symptoms as much as 1 year later. Electrodiagnostic studies show low-amplitude sensory responses with normal motor studies. Abnormalities were accentuated distally, suggesting a sensory neuronopathy with dying back of the most distal portions of the sensory neuron. However, no pathological studies have been reported. An identical neuropathy has been reported in about half the patients receiving ddI (Lambert et al. 1990). Neuropathy does not occur during treatment with zidovudine and a regimen of alternating ddC or ddI with zidovudine has been suggested to treat AIDS patients without incurring the almost inevitable neuropathic complications of ddC and ddI (Dubinsky et al. 1988).

REFERENCES

ASBURY, A.K., B.G. ARNASON, H.R. KARP and D.E. MCFARLIN: Criteria for diagnosis of Guillain–Barré syndrome. Ann. Neurol. 3 (1978) 565–566.

BAILEY, R., A. BALTCH, R. VENKATESH, J. SINGH and M. BISHOP: Sensory motor neuropathy associated with AIDS. Neurology 38 (1988) 886–891.

BAROHN, R.J., J.T. KISSEL, J.R. WARMHOLTS and J.R. MENDELL: Chronic inflammatory demyelinating polyradiculoneuropathy. Clinical characteristics, course, and recommendations for diagnostic criteria. Arch. Neurol. 46 (1989) 878–884.

BEHAR, R., C. WILEY and J.A. MCCUTCHAN: Cytomegalovirus polyradiculoneuropathy in acquired immunodeficiency syndrome. Neurology 37 (1987) 557–561.

BELEC, L., R. GHERARDI, A.J. GEORGES, E. SCHULLER, E. VUILLECARD, B. DI COSTANZO and P.M.V. MARTIN: Peripheral facial paralysis and HIV infection: report of four African cases and review of the literature. J. Neurol. 236 (1989) 411–414.

BISHOPRIC, G., J. BRUNER and J. BUTLER: Guillain–Barré syndrome with cytomegalovirus infection of peripheral nerves. Arch. Pathol. Lab. Med. 109 (1985) 1106–1108.

CALABRESE, L.H., M.R. PROFFITT, K.H. LEVIN, B. YEN-LIEBERMAN and C. STARKEY: Acute infection with human immunodeficiency virus (HIV) associated with brachial neuritis and exanthematous rash. Ann. Intern. Med. 107 (1987) 849–851.

CHIMOWITZ, M.I., A.-M.J. AUDET, A. HALLET and J.J. KELLY, JR.: HIV-associated CIDP. Muscle Nerve 12 (1989) 695–696.

COHEN, J.A. and M. LAUDENSLAGER: Autonomic nervous system involvement in patients with human immunodeficiency virus infection. Neurology 39 (1989) 1111–1112.

CORNBLATH, D.R. and J.C. MCARTHUR: Predominantly sensory neuropathy in patients with AIDS and AIDS-related complex. Neurology 38 (1988) 794–796.

CORNBLATH, D.R., J.C. MCARTHUR, P.G.E. KENNEDY, A.S. WITTE and J.W. GRIFFIN: Inflammatory demyelinating peripheral neuropathies associated with human T-cell lymphotropic virus type III infection. Ann. Neurol. 21 (1987) 32–40.

CORNBLATH, D.R., J.C. MCARTHUR, G.J.G. PARRY and J.W. GRIFFIN: Peripheral neuropathies in human immunodeficiency virus infection. In: P.J. Dyck, P.K. Thomas, J.W. Griffin, P.A. Low and J.F. Poduslo (Eds.), Peripheral Neuropathy. Philadelphia, PA, W.B. Saunders (1993) 1343–1353.

CRADDOCK, C., G. PASVOL, R. BULL, A. PROTHEROE and J. HOPKIN: Cardiorespiratory arrest and autonomic neuropathy in AIDS. Lancet ii (1987) 16–18.

CRAWFURD, E.J.P., P.R.E. BIRD and A.L. CLARK: Cauda equina and lumbar root compression in patients with AIDS. J. Bone Joint Surg. 69B (1987) 36–37.

DE LA MONTE, S.M., D.H. GABUZDA, D.D. HO, R.H. BROWN, E.T. HEDLEYWHITE, R.T. SCHOOLEY, M.S. HIRSCH and A.K. BHAN: Peripheral neuropathy in the acquired immunodeficiency syndrome. Ann. Neurol. 23 (1988) 485–492.

DUBINSKY, R.M., M. DALAKAS, R. YARCHOAN and S. BRODER: Follow-up of neuropathy from 2',3'-dideoxycytidine (Letter). Lancet i (1988) 832.

DYCK, P.J., A.C. LAIS, M. OHTA, J.A. BASTRON, H. OKAZAKI and R.V. GROOVER: Chronic inflammatory polyradiculoneuropathy. Mayo Clin. Proc. 50 (1975) 621–637.

DYCK, P.J., P.C. O'BRIEN, K.F. OVIATT, R.P. DINAPOLI, J.R. DAUBE, J.D. BARTLESON, B. MOKRI, T. SWIFT, P.A. LOW and A.J. WINDEBANK: Prednisone improves chronic inflammatory polyradiculoneuropathy more than no treatment. Ann. Neurol. 11 (1982) 136–141.

DYCK, P.J., J. DAUBE, P. O'BRIEN, A. PINEDA, P.A. LOW, A.J. WINDEBANK and C. SWANSON: Plasma exchange in chronic inflammatory demyelinating polyradiculoneuropathy. N. Engl. J. Med. 314 (1986) 461–465.

DYCK, P.J., T.J. BENSTEAD, D.L. CONN, J.C. STEVENS, A.J. WINDEBANK and P.A. LOW: Nonsystemic vasculitic neuropathy. Brain 110 (1987) 843–854.

DYCK, P.J., W.J. LITCHY, K.M. KRATZ, G.A. SUAREZ, P.A. LOW, A.A. PINEDA, A.J. WINDEBANK, J.L. KARNES and P.C. O'BRIEN: A plasma exchange versus immune globulin infusion trial in chronic inflammatory demyelinating polyradiculoneuropathy. Ann. Neurol. 36 (1994) 838–845.

EIDELBERG, D., A. SOTREL, H. VOGEL, P. WALKER, J. KLEEFIELD and C.S. CRUMPACKER III: Progressive polyradiculopathy in acquired immunodeficiency syndrome. Neurology 36 (1986) 912–916.

ELDER, G., M. DALAKAS, G. PEZESHKPOUR and J. SERVER: Ataxic neuropathy due to ganglioneuritis after probable acute human immunodeficiency virus infection. Lancet ii (1986) 1275–1276.

ESTES, M.L., L.H. CALABRESE, B. YEN-LIEBERMAN, K.H. LEVIN and M.R. PROFFITT: Human immunodeficiency virus (HIV) and necrotizing peripheral nerve vasculitis: an autoimmune phenomenon? (Abstr.) Neurology 39, Suppl. 1 (1989) 293.

FREEMAN, R., M.S. ROBERTS, L.S. FRIEDMAN and C. BROAD-BRIDGE: Autonomic function and human immunodeficiency virus infection. Neurology 40 (1990) 575–580.

FRENCH COOPERATIVE GROUP ON PLASMA EXCHANGE AND GUILLAIN–BARRÉ SYNDROME: Efficacy of plasma exchange in Guillain–Barré syndrome: role of replacement fluids. Ann. Neurol. 22 (1987) 753–761.

FULLER, G.N., J.M. JACOBS and R.J. GUILOFF: Association of painful peripheral neuropathy in AIDS with cytomegalovirus infection. Lancet ii (1989) 937–941.

FULLER, G.N., S.K. GILL, R.J. GUILOFF, R. KAPOOR, S.B. LUCAS, E. SINCLAIR, F. SCARAVILLI and R.F. MILLER: Ganciclovir for lumbosacral polyradiculopathy in AIDS. Lancet i (1990a) 48–49.

FULLER, G.N., C. GRECO and R.G. MILLER: Cytomegalovirus and mononeuropathy multiplex in AIDS (Abstr.). Neurology 40 (1990b) 301.

GASTAUT, J.L., J.A. GASTAUT, J.F. PELLISIER, J.B. TAPKO and O. WEILL: Neuropathies with HIV infection. Prospective study of 56 cases. Rev. Neurol. 145 (1989) 451–459.

GHERARDI, R., F. LEBARGY, P. GAULARD, C. MHIRI, J.F.

BERNAUDIN and F. GRAY: Necrotizing vasculitis and HIV replication in peripheral nerves (Letter). N. Engl. J. Med. 321 (1989) 685–686.

GRAFE, M.R. and C.A. WILEY: Spinal cord and peripheral nerve pathology in AIDS: the roles of cytomegalovirus and human immunodeficiency virus. Ann. Neurol. 25 (1989) 561–566.

GRAVELEAU, P., R. PEROL and A. CHAPMAN: Regression of cauda equina syndrome in AIDS patient being treated with ganciclovir. Lancet ii (1989) 511–512.

GUILLAIN-BARRÉ SYNDROME STUDY GROUP: Plasmapheresis and acute Guillain–Barré syndrome. Neurology 35 (1985) 1096–1104.

HAGBERG, L., B.-E. MALMVALL, L. SVENNERHOLM, K. ALESTIG and G. NORKRANS: Guillain–Barré syndrome as an early manifestation of HIV central nervous system infection. Scand. J. Infect. Dis. 18 (1986) 591–592.

HALL, C.D., C.R. SNYDER, J A. MESSENHEIMER, J.W. WILKINS, W.T. ROBERTSON, R.A. WHALEY and K.R. ROBERTSON: Peripheral neuropathy in a cohort of human immunodeficiency virus-infected patients. Incidence and relationship to other nervous system dysfunction. Arch. Neurol. 48 (1991) 1273–1274.

HART, R.G. and M.C. KANTER: Acute autonomic neuropathy. Two cases and a clinical review. Arch. Intern. Med. 150 (1990) 2373–2376.

HELWEG-LARSEN, S., J. JAKOBSEN, F. BOENSEN and P. ARLIEN-SOBORG: Neurological complications and concomitants of AIDS. Acta Neurol. Scand. 74 (1986) 467–474.

HOLTZMANN, D.M., E. DAVIS and C.M. GRECO: Lumbosacral plexopathy secondary to perirectal abscess in a patient with HIV infection. Neurology 39 (1989) 1400–1401.

HOROWITZ, S.L., D.F. BENSON, M.S. GORDIEB, I. DAVOS and J.R. BENTSON: Neurological complications of gay-related immunodeficiency disorder (Abstr.). Ann. Neurol. 12 (1982) 80.

JACOBSEN, M.A., J. MILLS, J. RUSH, J.J. O'DONNELL, R.G. MILLER, C. GRECO and M.F. GONZALES: Failure of antiviral therapy for acquired immunodeficiency syndrome-related cytomegalovirus myelitis. Arch. Neurol. 45 (1988) 1090–1092.

KISSEL, J.T., A.P. SLIVKA, J.R. WARMOLTS and J.R. MENDELL: The clinical spectrum of necrotizing angiopathy of the peripheral nervous system. Ann. Neurol. 18 (1985) 251–257.

LAMBERT, J.S., M. SEIDLIN, R.C. REICHMAN, C.S. PLANK, M. LAVERTY, G.D. MORSE, C. KNUPP, C. MCLAREN, C. PETTINELLI, F.T. VALENTINE and R. DOLIN: 2',3'-Deoxyinosine (ddI) in patients with the acquired immunodeficiency syndrome or AIDS-related complex. A phase I trial. N. Engl. J. Med. 322 (1990) 1333–1340.

LANGE, D.J., C.B. BRITTON, D.S. YOUNGER and A.P. HAYS: The neuromuscular manifestations of human immunodeficiency virus infections. Arch. Neurol. 45 (1988) 1084–1088.

LANSKA, D.J. and M.J. LANSKA: Treatable causes of meningomyeloradiculitis in individuals infected with human immunodeficiency virus. Arch. Neurol. 46 (1989) 722–723.

LANSKA, M.J., D.J. LANSKA and J.W SCHMIDLEY: Syphilitic polyradiculopathy in an HIV-positive man. Neurology 38 (1988) 1297–1301.

LEGER, J.M., P. BOUCHE, F. BOLGERT, M.P. CHAUNU, M. ROSENHEIM, H.P. CATHALA, M. GENTILINI, J.J. HAUW and P. BRUNET: The spectrum of polyneuropathies in patients infected with HIV. J. Neurol. Neurosurg. Psychiatry 52 (1989) 1369–1374.

LEVY, R.M., D.E. BREDESEN and M.L. ROSENBLUM: Neurological manifestations of the acquired immunodeficiency syndrome (AIDS): experience at UCSF and review of the literature. J. Neurosurg. 62 (1985) 475–495.

LIN-GREENBERG, A. and N. TANEEJA-UPPAL: Dysautonomia and infection with the human immunodeficiency virus (Letter). Ann. Intern. Med. 106 (1987) 167.

LIPKIN, W.I., G. PARRY, D. KIPROV and D. ADAMS: Inflammatory neuropathy in homosexual men with lymphadenopathy. Neurology 35 (1985) 1479–1483.

LOHMOLLER, G., A. MATUSCHKE and F.D. GOEBEL: Testing for neurological involvement in HIV infection. Lancet ii (1987) 1532–1533.

LOVELACE, R.E.: Mononeuritis multiplex in polyarteritis nodosa. Neurology 14 (1982) 434–442.

MAH, V., L.M. VARTAVARIAN, M.-A. AKERS and H.V. VINTERS: Abnormalities of peripheral nerve in patients with human immunodeficiency virus infection. Ann. Neurol. 24 (1988) 713–717.

MAHIEUX, F., F. GRAY, G. FENELON, R. GHERARDI, D. ADAMS, A GUILLARD and J. POIRIER: Acute myeloradiculitis due to cytomegalovirus as the initial manifestation of AIDS. J. Neurol. Neurosurg. Psychiatry 52 (1989) 270–274.

MALAMUT, R.I., N. LEOPOLD and G.J. PARRY: The treatment of HIV-associated chronic inflammatory demyelinating polyneuropathy (HIV-CIDP) with intravenous immunoglobulin (IVIg). Neurology 42, Suppl. 3 (1992) 335.

MEHREN, M., P.J. BURNS, F. MAMANI, C.S. LEVY and R. LAURENO: Toxoplasmic myelitis mimicking intramedullary spinal cord tumor. Neurology 38 (1988) 1648–1650.

MCARTHUR, J.C., B.A. COHEN, O.A. SELNES, A.J. KUMAR, K. COOPER, J.H. MCARTHUR, G. SOUCY, D.R. CORNBLATH, J.S. CHMIEL, M.-C. WANG, D.L. STARKEY, H. GINZBURG, D.G. OSTROW, R.T. JOHNSON, J.P. PHAIR and B.F. POLK: Low prevalence of neurological and neuropsychological abnormalities in otherwise healthy HIV-1-infected individuals: results from the multicenter AIDS cohort study. Ann. Neurol. 26 (1989) 601–611.

MCCOMBE, P.A., J.D. POLLARD and J.G. MCLEOD: Chronic

inflammatory polyradiculoneuropathy. A clinical and electrophysiological study of 92 cases. Brain 110 (1987) 1617–1630.

MILLER, R.G., G. PARRY, W. LANG, R. LIPPERT and D. KIPROV: AIDS-related inflammatory polyradiculoneuropathy: prediction of response to plasma exchange with electrophysiologic testing (Abstr.). Muscle Nerve 8 (1985) 626.

MILLER, R.G., G.J. PARRY, W. PFAEFFL, W. LANG, R. LIPPERT and D. KIPROV: The spectrum of peripheral neuropathy associated with ARC and AIDS. Muscle Nerve 11 (1988a) 857–863.

MILLER, R.G., D.D. KIPROV, G. PARRY and D.E. BREDESEN: Peripheral nervous system dysfunction in acquired immunodeficiency syndrome. In: M.L. Rosenblum, R.M. Levy and D.E. Bredesen (Eds.), AIDS and the Nervous System. New York, Raven Press (1988b) 65–78.

MILLER, R.G., J.R. STOREY and C.M. GRECO: Ganciclovir in the treatment of progressive AIDS-related polyradiculopathy. Neurology 40 (1990) 569–574.

NOVAK, I.S., J.R. TRUJILLO, V.M. RIVERA, R. CONKLIN, F. USSERY III, E.W. STOOL and D.F. PIOT: Ganciclovir in the treatment of CMV infection in AIDS patients with neurologic complications. Neurology 39, Suppl. 1 (1990) 379–380.

OSTERMAN, P.O., J. FAGIUS, G. LUNDEMO, P. PIHLSTEDT, R. PIRS-KANEN, A. SIDEN and J. SAFWENBERG: Beneficial effects of plasma exchange in acute inflammatory polyradiculoneuropathy. Lancet ii (1984) 1296–1299.

PARRY, G.J.: Peripheral neuropathies associated with human immunodeficiency infection. Ann. Neurol. 23, Suppl. (1988) S49–S53.

PARRY, G.J.: Guillain–Barré Syndrome. New York, Thieme (1993).

PIETTE, A.M., F. TUSSEAU, D. VIGNON, A. CHAPMAN, G. PARROT, J. LEIBOWITCH and L. MONTAGNIER: Acute neuropathy coincident with seroconversion for anti-LAV/HTLV-III. Lancet i (1986) 852.

RANCE, N.E., J.C. MCARTHUR, D.R. CORNBLATH, D.L. LAND-STROM, J.W. GRIFFIN and D.L. PRICE: Gracile tract degeneration in patients with sensory neuropathy in AIDS. Neurology 38 (1988) 265–271.

ROULLET, E., V. ASSUERUS, J. GOZLAN, A. ROPERT, G. SAID, M. BAUDRIMONT, M. EL AMRAMI, C. JACOMET, C. DUVIVIER, G. GONZALES-CANALI, M. KIRSTETTER, M.-C. MEYOHAS, O. PICARD and W. ROZENBAUM: Cytomegalovirus multifocal neuropathy in AIDS: analysis of 15 consecutive cases. Neurology 44 (1994) 2174–2182.

SAID, G., C. LACROIX, J.N. ANDRIEU, C. GAUDOVEN and J. LIEBOWITCH: Necrotizing arteritis in patients with inflammatory neuropathy and human immunodeficiency virus (HIV-III) infection (Abstr.).

Neurology 37, Suppl. 1 (1987) 176.

SAID, G., C. LACROIX-CIAUDO, H. FUJIMURA, C. BLAS and N. FAUX. The peripheral neuropathy of necrotizing vasculitis. Ann. Neurol. 23 (1988) 461–465.

SNIDER, W.D., D.M. SIMPSON, S. NIELSEN, J.W.M. GOLD, C.E. METROKA and J.B. POSNER: Neurological complications of acquired immune deficiency syndrome: analysis of 50 patients. Ann. Neurol. 14 (1983) 403–418.

SO, Y.T. and R.K. OLNEY: Acute lumbosacral polyradiculopathy in acquired immunodeficiency syndrome: experience in 23 patients. Ann. Neurol. 35 (1994) 53–58.

SO, Y.T., D.M. HOLTZMAN, D.I. ADAMS and R.K. OLNEY: Peripheral neuropathy associated with acquired immunodeficiency syndrome: prevalence and clinical features from a population-based survey. Arch. Neurol. 45 (1988) 945–948.

VAN DER MECHE, F.G., P.I. SCHMITZ and THE DUTCH GUILLAIN-BARRÉ STUDY GROUP): A randomized trial comparing intravenous immune globulin with plasma exchange in Guillain–Barré syndrome. N. Engl J. Med. 326 (1992) 1123–1129.

VENDRELL, J., C. HEREDIA, M. PUJOL, J. VIDAL, R. BLESA and F. GRAUS: Guillain–Barré syndrome associated with seroconversion for anti-HTLV-III. Neurology 37 (1987) 544.

WECHSLER, A.F and D.D. HO: Bilateral Bell's palsy at the time of HIV seroconversion. Neurology 39 (1989) 747–748.

WHITEMAN, M.L.H., B.K. DANDAPANI, R.T. SHEBERT and M.J. POST: MRI of AIDS-related polyradiculomyelitis. J. Comput. Assist. Tomogr. 18 (1994) 7–11.

WINER, J.B., D. JEFFERIES and S. LIEBOWITZ: Painful peripheral neuropathy and cytomegalovirus pneumonia in AIDS (Letter). Lancet ii (1989) 1330.

WOOLSEY, R.M., T.J. CHAMBERS, H.D. CHUNG and J.D. MCGARRY: Mycobacterial meningomyelitis associated with human immunodeficiency virus infection. Arch. Neurol. 45 (1988) 691–693.

YARCHOAN, R., G. BERG, P. BROUWERS, M.A. FISCHL, A.R. SPITZER, A. WICHMANN, J. GRAFMAN, R.V. THOMAS, B. SAFAI, A. BRUNETTI and C.F. PERNO: Response of human immunodeficiency-virus associated neurological disease to 3'-azido-3'-deoxythymidine. Lancet i (1987) 132–135.

YARCHOAN, R., R.V. THOMAS, J.-P. ALLAIN, N. MCATEE, R. DUBINSKY, H. MITSUYA, T.J. LAWLEY, B. SAFAI, C.E. MYERS, C.F. PERNO, R.W. KLECKER, R.J. WILLS, M.A. FISCHL, M.C. MCNEELY, J.M. PLUDA, M. LEUTHER, J.M. COLLINS and S. BRODER: Phase I studies of 2',3'-dideoxycytidine in severe human immunodeficiency virus infection as a single agent and alternating with zidovudine (AZT). Lancet i (1988) 76–81.

Handbook of Clinical Neurology, Vol. 27 (71): Systemic Diseases, Part III
M.J. Aminoff and C.G. Goetz, editors
© 1998 Elsevier Science B.V. All rights reserved

CHAPTER 16

Neurological complications of treatment for HIV infection

DAVID M. SIMPSON[1] and MICHELE TAGLIATI[2]

[1]*Clinical Neurophysiology Laboratories and Neuro-AIDS Research Program, New York, NY, U.S.A., and*
[2]*Departments of Neurology and Pathology, The Mount Sinai Medical Center, New York, NY, U.S.A.*

The central and peripheral nervous systems are common targets of human immunodeficiency virus type 1 (HIV)-related complications (Simpson and Tagliati 1994). These include infections, neoplasms, vascular complications, neuropathies and myopathies (Berger and Simpson 1997). As more effective treatments of HIV and opportunistic infection improve the survival of patients with the acquired immunodeficiency syndrome (AIDS), the prevalence of neurological disorders will be likely to increase. Multiple pathologies may coexist in the same patient, further complicating the clinical diagnosis. Many drugs used in the treatment of HIV disease and its related complications affect the nervous system (Table 1). Patients with AIDS are particularly vulnerable to drug-related toxicity and allergic reactions (Lee and Safrin 1992). These toxic effects must be properly recognized and evaluated in order to decide whether or not treatment should be discontinued. This chapter will review the neurological complications associated with the treatment of HIV infection.

ANTIRETROVIRAL THERAPY

Antiretroviral agents may cause or contribute to disorders that are similar to some forms of HIV-associated neuropathy and myopathy. It is essential to distinguish between the neuromuscular disorders which are attributable to the toxic effects of nucleoside analogs, and those which are primarily related to underlying HIV disease, in order to provide proper treatment. This section reviews the identification and management of peripheral neuropathy associated with use of the antiretroviral drugs zalcitabine (ddC), didanosine (ddI), and stavudine (d4T). In addition, we discuss the diagnosis and management of HIV myopathy and the role of the antiretroviral zidovudine (ZDV, AZT) in its pathogenesis.

Nucleoside analog-associated neuropathy

Nucleoside analogs are a class of drugs with chemical structures that are similar to the endogenous DNA and RNA nucleosides (Fig. 1). Dideoxynucleosides such as ZDV, ddC, ddI and d4T are synthesized by replacing the hydroxyl group at the 3' position of the deoxyribose ring with various chemical groups. Although the precise mechanism of action is unknown, nucleoside analogs seem to prevent viral replication by competing with endogenous nucleotides for binding sites on reverse transcriptase or by blocking DNA chain elongation (Yarchoan et al. 1990).

Phase I studies of ddC (Yarchoan et al. 1988; Merigan et al. 1989), ddI (Cooley et al. 1990; Lambert et al. 1990) and d4T (Browne et al. 1993) in patients with HIV disease indicated that peripheral neuropathy was a dose-limiting toxic effect.

TABLE 1

Indication and major neurotoxicity of agents used in the treatment of AIDS.

Drug	Indication	Neurotoxicity
Zalcitabine (ddC)	HIV	Peripheral neuropathy
Didanosine (ddI)	HIV	Peripheral neuropathy
Stavudine (d4T)	HIV	Peripheral neuropathy
Zidovudine (ZDV, AZT)	HIV	Myopathy
Acyclovir	HSV	Encephalopathy
Ganciclovir	CMV	Encephalopathy Seizures
Foscarnet	CMV	Seizures
Isoniazid (INH)	TB	Peripheral neuropathy Psychosis Seizures
Ethambutol	TB	Optic neuropathy
Trimethoprim/sulfamethoxazole (TMP/SMX)	PCP	Psychosis Seizures
Dapsone	PCP	Peripheral neuropathy
Amphotericin B (AmB)	Cryptococcosis	Leukoencephalopathy
Ciprofloxacin	MAI/MAC	Anxiety Tremor Seizures
Vinca alkaloids	Lymphoma	Peripheral neuropathy
Methotrexate (MTX)	Lymphoma	Arachnoiditis (IT) Myelopathy Encephalopathy
Cytarabine (Ara-C)	PML/lymphoma	Cerebellar degeneration Myelopathy Encephalopathy Peripheral neuropathy
Radiotherapy (RT)	Lymphoma	Myelopathy Encephalopathy
Thalidomide	Aphthous ulcers	Peripheral neuropathy

HIV = human immunodeficiency virus; HSV = herpes simplex virus; CMV = cytomegalovirus; TB = tuberculosis; PCP = *Pneumocystis carinii* pneumonia; MAI/MAC = *Mycobacterium avium* intracellulare/*Mycobacterium avium* complex; PML = progressive multifocal leukoencephalopathy.

The neuropathy associated with ddC therapy was originally described as a painful stocking-glove axonal sensory-motor neuropathy (Yarchoan et al. 1988). It is characterized by painful dysesthesias of the feet, numbness and subjective distal weakness. Neurological signs include decrease in light touch, temperature, vibratory and proprioceptive sensation, and absent ankle reflexes (Yarchoan et al. 1988; Dubinsky et al. 1989; Merigan et al. 1989). Electrophysiological studies are consistent with axonal degeneration (Yarchoan et al. 1988; Dubinsky et al. 1989; Merigan et al. 1989; Berger et al. 1993). A similar clinical presentation was later described in association with ddI (Cooley et

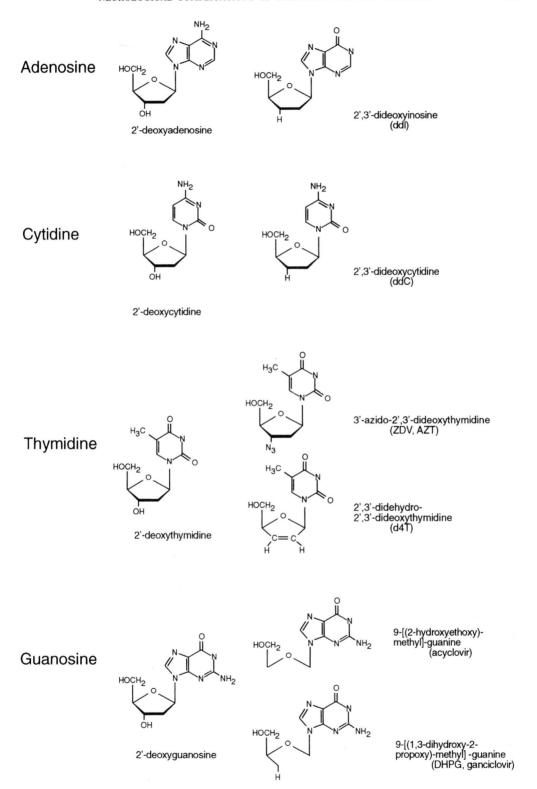

Fig. 1. Chemical structures of endogenous (left) and synthetic (right) nucleoside analogs.

al. 1990; Lambert et al. 1990) and d4T administration (Browne et al. 1993).

Several problems are inherent in determining the incidence of nucleoside analog-related neuropathy (Simpson and Tagliati 1995). The presence of underlying HIV neuropathy must be considered when assessing the presence of a presumed toxic neuropathy. However, there have been relatively few prospective studies with initially devised neurologic endpoints to determine the incidence of distal symmetrical polyneuropathy (DSP) in AIDS. Additionally, diagnostic criteria for neuropathy differ among studies and there has been variability in the training and skill of examiners gathering these data in large, multi-centered clinical trials.

Approximately one-third of patients with AIDS who have not received neurotoxic agents have clinical signs of DSP (So et al. 1988). Electrophysiologic abnormalities indicating DSP are even more frequent, particularly as CD4 counts decline (Rogers et al. 1992; Simpson et al. 1995). The combination of a low CD4 count with neurotoxic drug exposure may place the patient at particularly high risk of DSP (Blum et al. 1993; Fischl et al. 1993). Data from the Multicenter AIDS Cohort Study (MACS) indicate that the incidence of both primary HIV neuropathy and toxic neuropathy have increased between 1988 and 1992 (Bacellar et al. 1994).

Experimental and clinical trial data

Zalcitabine (ddC). Several animal models have demonstrated that ddC administration is associated with clinical, electrophysiologic, and pathologic evidence of neuropathy (Tsai et al. 1989; Arezzo et al. 1990; Anderson et al. 1991). In pigtail macaques infected with the simian AIDS retrovirus serotype 2 (SRV-2), intravenous administration of high-dose ddC (15 mg/kg/day) induced severe toxic effects including peripheral neuropathy (Tsai et al. 1989). Cynomolgus monkeys treated with ddC developed electrophysiologic and histopathological abnormalities consistent with fiber length-dependent, distal axonopathy (Arezzo et al. 1990). In rabbits treated with very high doses of ddC (10–250 mg/kg/day), severe peripheral neuropathy was described after 13–18 weeks of therapy (Anderson et al. 1991).

Decreased nerve conduction velocities and reduced sensory and motor potential amplitudes correlated with primary myelin pathology and less significant axonal lesions (Anderson et al. 1991). These features differ from those described in humans receiving ddC, that have shown clinical and electrophysiologic evidence of distal symmetric axonal neuropathy (Yarchoan et al. 1988; Dubinsky et al. 1989; Merigan et al. 1989; Berger et al. 1993).

The incidence of peripheral neuropathy among different studies of nucleoside analogues is shown in Table 2. The percentage of patients who develop DSP during ddC therapy has ranged from 25 to 66% (Yarchoan et al. 1988; Merigan et al. 1989; Berger et al. 1993; Fischl et al. 1993; Abrams et al. 1994). Neuropathy has only rarely been reported in pediatric populations treated with ddC (Pizzo et al. 1990).

The occurrence of ddC-associated neuropathy is clearly dose-dependent. In a prospective substudy of 52 patients participating in the AIDS Clinical Trial Group (ACTG) protocol 012, all patients who received the highest doses of ddC (0.06 mg/kg and 0.03 mg/kg every 4 h) developed clinical evidence of peripheral neuropathy, with a mean onset of 8 weeks. The time of onset was progressively delayed with lower doses of ddC. Only 2 of 15 patients receiving 0.005 mg/kg every 4 h developed peripheral neuropathy after over 17 weeks of treatment (Berger et al. 1993). The use of intermittent regimens of ddC does not seem to prevent the occurrence of neuropathy. Weekly intermittent doses of ddC (0.03 mg/kg every 4 h) were complicated in 50% of cases by peripheral neuropathy (Skowron et al. 1993).

The severity and progression of symptoms of DSP are also related to the dose of ddC, with a milder clinical presentation in patients who received lower doses of ddC. Upon drug withdrawal, 83% of patients in the ACTG 012 substudy ultimately recovered from neuropathy (Berger et al. 1993). However, the time to recovery varied between 11 and 19 weeks for patients on lower and higher doses of ddC, respectively. Furthermore, patients in the higher-dose groups experienced a prolonged intensification of symptoms lasting 3–6 weeks after ddC withdrawal, termed the 'coasting period'. Similar results were

TABLE 2

Nucleoside analogue-associated neuropathy: clinical trial data.

Study	N3300/ACTG-114	N3492/ACTG-119	N3544	N3447/ACTG-106	Phase I	Phase I	BMS006	BMS A455-019	BMS
Drug	ddC	ddC	ddC	d2dC	ddI	ddI	d4T	d4T	d4T
Daily dosage	2.25 mg	2.25 mg	1.125 mg 2.25 mg	2.25 mg	<750 mg	>750 mg	A . 0.5 mg/kg B. 1.0 mg/kg C. 2.0 mg/kg	80 mg	A. 80 mg B. 40 mg
Median CD4 (cell/mm^3)	90	84	81	70	NA	NA	245	250	40
Subjects	321	59	4085	47	91	79	152	412	12.551
Peripheral neuropathy	35.9%	33.3%	24%	27.6%	34%	51%	6% (dose A) 15% (dose B) 31% (dose C)	14%	23% (dose A) 17% (dose B)

NA = not available.
Adapted from Simpson and Tagliati (1995), with permission.

observed in a substudy of ACTG 155 which evaluated 47 patients treated with ddC (1.0–2.25 mg/day). Moderate to severe peripheral neuropathy developed in 40% of patients, with a median onset of 23.5 weeks. After drug withdrawal, 64% of patients had subjective improvement in symptoms (Blum et al. 1993).

The development of ddC neuropathy is also related to the patient's stage of immunosuppression. A rapid decline in the CD4 cell count increases the risk of ddC-related DSP (Blum et al. 1993). In a study evaluating ddC in 59 patients with advanced HIV infection, 15 (25%) developed peripheral neuropathy. However, 5 of 19 (26%) patients with a CD4 count of less than 50 cells/mm^3 before treatment developed peripheral neuropathy, as compared to 6 of 40 (15%) patients with 50 cells/mm^3 or more (Fischl et al. 1993).

Preliminary data suggested that combination treatment with ddC and ZDV may result in a lower incidence of peripheral neuropathy as compared with ddC monotherapy (Yarchoan et al. 1988; Meng et al. 1992). However, data from the ACTG 155 study (Blum et al. 1993) failed to show significant differences in the incidence of neuropathy between patients receiving ddC alone (23%) and ddC/ZDV in combination therapy (22%).

A reduction of peripheral nerve toxicity was observed in a large study (ACTG 047) in which ZDV and ddC were given as alternating therapy. Alternating monthly high-dose ddC (0.03 mg/kg) and ZDV (1200 mg/day) were complicated by toxic neuropathy in only 14% of the studied individuals. However, the weekly alternation of this same regimen was associated with neuropathy in 41% of cases, suggesting that a sufficient washout period may be required to decrease the incidence of toxic neuropathy (Skowron et al. 1993).

Didanosine (ddI). Most information concerning peripheral neuropathy as a toxic effect of ddI has been derived from phase I, II and open label studies (Cooley et al. 1990; Lambert et al. 1990; Rozencweig et al. 1990; Yarchoan et al. 1990; Kieburtz et al. 1992). The incidence of DSP in phase I ddI studies has ranged from 12 to 34% (Shelton et al. 1992), particularly on high-dose ddI (>12.5 mg/kg/day). In the CPCRA 002 study (Abrams et al. 1994), 14% of patients receiving ddI (500 mg/day) developed neuropathy. In AIDS Clinical Trial Group protocol 175, patients in the ddI monotherapy arm (400 mg/day) had a 3% incidence of DSP (Simpson et al. 1997b).

The neuropathy caused by ddI is clinically similar to ddC neuropathy and HIV-associated DSP. The onset and progression of neurologic toxicity are related both to the daily and cumulative doses of ddI (Rozencweig et al. 1990). Peripheral neuropathy generally appears at daily dosages higher than 12.5 mg/kg/day (Cooley et al. 1990; Lambert et al. 1990; Kieburtz et al. 1992) or cumulative doses higher than 1.5 g/kg (Rozencweig et al. 1990; Yarchoan et al. 1990). However, peripheral neuropathy may occur at lower cumulative doses (1.16 g/kg) in patients with low CD4 cell counts (Rathburn and Martin 1992), as has been observed with ddC therapy. The mean time to onset of ddI neuropathy is 20 weeks (Steiberg et al. 1991). Symptoms progress more rapidly at higher dosages (Yarchoan et al. 1990), but usually resolve within 3–5 weeks after the drug is discontinued. A period of 'coasting' may be experienced before improvement occurs (Kieburtz et al. 1992), similar to that observed with ddC (Berger et al. 1993). Some patients may tolerate reintroduction of ddI at lower doses (Yarchoan et al. 1988; Kieburtz et al. 1992).

Subjects with subclinical neuropathy prior to the initiation of ddI or ddC therapy may have symptoms of neuropathy unmasked following drug exposure (Kieburtz et al. 1992; Berger et al. 1993; Pike and Nicaise 1993). Data from the expanded access trial indicate that among patients who had a history of neuropathy before starting ddI treatment, 24% developed symptomatic neuropathy with ddI treatment, while in patients without a neuropathic history, 15% developed neuropathy (Pike and Nicaise 1993).

Stavudine (d4T). Since 1989, the safety and efficacy of another antiretroviral compound, the pyrimidine analog stavudine (d4T), have been evaluated in several clinical trials. Browne et al. (1993) reported that the principal dose-limiting toxicity of d4T is a sensory peripheral neuropathy, occurring in 55% of patients, with symptoms and signs similar to ddI- and ddC-induced neuropathy. The maximum tolerated dose of d4T in this phase I study was 2.0 mg/kg/day. Data from recent studies clarify the dose–response relationship of d4T-induced neuropathy. In a randomized open trial of 152 patients (median CD4 count = 246 cells/mm³), the incidence of peripheral neuropathy was 6, 15 and 31% in patients receiving 0.5, 1.0 and 2.0 mg/kg/day respectively (Bristol-Myers-Squibb, data on file). Patients with advanced HIV disease were at highest risk for the development of neuropathy, which generally became apparent within 8–16 weeks of therapy. In patients that resumed d4T at one-half dose, 60% were able to tolerate therapy for an average of 9 months.

A controlled trial (A455-019) has compared the efficacy of d4T (40 mg twice per day) to ZDV (600 mg/day) in 822 HIV-infected individuals with a baseline CD4 count between 50 and 500 cells/mm³. The incidence of neuropathy was 14% (58/412) in the d4T group, compared to 4% (18/402) in the ZDV group (A. Pavia, personal communication). In a large, blinded randomized comparison study of patients that failed or were intolerant to other antiretroviral treatments (*n* = 12,551; median CD4 count = 44 cells/mm³), the incidence of peripheral neuropathy was 23% with d4T doses of 40 mg twice per day (approximately 1.0 mg/kg/day) and 17% with 20 mg twice per day (0.5 mg/kg/day). Factors that increased the risk of peripheral neuropathy in this trial include a prior history of neuropathy, a low baseline CD4 count, and a hemoglobin level <11 g/l (R. Murphy, personal communication).

Zidovudine (ZDV). In contrast to ddC, ddI and d4T, there is no evidence that ZDV causes toxic neuropathy (Richman et al. 1987). In the ddC comparative trials N3300 (data on file, Hoffman-LaRoche Laboratories) and N3492 (Fischl et al. 1993), ZDV-associated neuropathy occurred in 18.6 and 0% of patients, respectively. The ACTG 116B/117 study, comparing ZDV (600 mg/day) with ddI (500 or 750 mg/day), surprisingly reported a similar incidence of peripheral neuropathy (14%) in the two treatment regimens (Kahn et al. 1992). The ACTG 047 study reported an incidence of peripheral neuropathy of 17% in a subgroup of patients receiving high-dose ZDV (1200 mg/day). Importantly, the mean CD4 cell count in these patients was 59/mm³ (Skowron et al. 1993).

Patients in advanced stages of immunosuppression may have a similarly high incidence of HIV-related DSP without antiretroviral therapy exposure. Furthermore, the problems in accurate reporting of peripheral neuropathy in large multicenter studies limit the interpretation of these data. Results from the ACTG 155 study indicate that there is not a statistically significant difference in the incidence of neuropathy between ddC- and ZDV-treated groups when only patients with pretreatment CD4 count <50 cells/mm³ are considered (U. Dafni, personal communication). To avoid this confounding effect, Bozzette et al. (1991) evaluated the neuropathic effects of ZDV in a placebo-controlled study of 76 HIV-infected subjects with mean CD4 cell counts close to 400/mm³. There was no evidence of ZDV neurotoxicity as compared with controls. In the ACTG 175 study enrolling subjects with CD4 count of 200–500 cells/mm³, the incidence of neuropathy in the ZDV monotherapy group was 4% (Simpson et al. 1997a).

Pathogenesis

Although the temporal relationship of onset and resolution of peripheral neuropathy with drug intake and discontinuation clearly indicates a neurotoxic effect of nucleoside analogs, the mechanism of this neurotoxicity is unknown. The only available study on human cells demonstrated a delayed cytotoxicity of ddC against T lymphoblasts, associated with a reduced cellular content of mitochondrial DNA (Chen and Chen 1989). This mitochondrial toxicity, attributed to the inhibition of DNA γ-polymerase, is reversible upon removal of the drug. However, studies of ddC effects on murine DNA γ-polymerase reported relatively low levels of enzyme inhibition (Keilbaugh et al. 1990). Species-specific toxicity must be taken into account when interpreting animal models of nucleoside neuropathy (Lipman et al. 1993).

Diagnostic criteria

The clinical similarity between the neuropathy caused by nucleoside analogs and that due to HIV infection may present a diagnostic challenge. Numbness, tingling and pain are present both in AIDS neuropathy and in nucleoside neuropathy. Similarly, both neuropathies affect predominantly distal segments of the limbs, most severely in the lower extremities, while the upper extremities may be relatively spared even in late stages of the disease (Berger et al. 1993). The pattern of onset of symptoms may be useful, as nucleoside neuropathy tends to evolve rapidly while AIDS-related DSP may take weeks to months to develop.

There are several other forms of peripheral neuropathy associated with HIV infection including inflammatory demyelinating polyneuropathy, progressive polyradiculopathy and mononeuropathy multiplex. These must be distinguished from DSP, since their pathogenesis and treatment differ (Simpson and Olney 1992). A targeted neurologic exam revealing diminished ankle reflexes, elevated vibratory threshold in the feet, and graded distal reduction in sensation of pin and temperature is consistent with the diagnosis of DSP. In atypical or challenging cases of HIV-infected patients with neuropathy, neurological consultation should be obtained, with guidance for further testing, including electrophysiological studies, quantitative sensory tests, lumbar puncture, and nerve biopsy.

Management

A search for other underlying causes of neuropathy should be the first step in the management of patients with signs and symptoms of peripheral neuropathy. An appropriate laboratory screen in DSP should include serum assays of glucose, vitamin B_{12}, sedimentation rate and other tests for collagen vascular diseases, thyroid function studies, Lyme titers, serum protein and immunoelectrophoresis. In addition, hereditary neuropathy should be considered and excluded with family history and, if necessary, clinical and electrophysiological examination of family members. Identification and correction of metabolic or nutritional causes should be attempted, while the presence of possible neurotoxins is assessed (Gill et al. 1990; Figg 1991). Because of the difficulty in separating HIV neuropathy from drug-related neuropathy, it is prudent to reduce the potentially offending drug dosage or discontinue treatment for at least 1–2 months in most patients with symptomatic DSP.

The development of nucleoside analog-related neuropathy is dose-related, and patients may tolerate rechallenge with lower doses of the drug. In some patients, subclinical HIV neuropathy may be unmasked by the administration of a neurotoxic antiretroviral agent. The 'gold standard' in the diagnosis of toxic neuropathy is the resolution of symptoms following drug discontinuation (Berger et al. 1993). However, when ddC, ddI or d4T are withdrawn, patients may experience a period of 'coasting', in which the symptoms of neuropathy intensify for several weeks before improving. Patients that have a combination of causes of DSP may have incomplete resolution of symptoms following drug withdrawal.

Initial symptomatic treatment for painful neuropathy consists of analgesics, including non-steroidal anti-inflammatory agents and acetaminophen. Topical capsaicin may also be helpful in a limited percentage of subjects. In patients with persistent and more disabling pain, tricyclic antidepressants may afford added benefit, as has been demonstrated in controlled studies of diabetic neuropathy (Max et al. 1987, 1991). Anticonvulsants, such as carbamazepine and phenytoin, may provide symptomatic relief in some patients. When disabling pain, refractory to the above-mentioned agents, is present, potent narcotics may be required to control symptoms. ACTG protocol 242 revealed no significant difference in pain reduction between amitriptyline, mexiletine and placebo arms (Kieburtz et al. 1997). Controlled clinical trials of topical lidocaine, lamotrigine, and nerve growth factor are under way in the treatment of painful neuropathy associated with AIDS and nucleoside analogs.

HIV myopathy

Following our initial description of a patient with AIDS and polymyositis (Snider et al. 1983), numerous cases of HIV-associated myopathy were reported by several groups (Dalakas et al. 1986; Stern et al. 1987; Gonzales et al. 1988; Simpson and Bender 1988; Manji et al. 1993). The incidence of myopathy in HIV infection has not been established in prospective studies. Furthermore, different authors have used variable criteria for the diagnosis of myopathy. In our experience with a referral-based population, over 20% of the HIV-infected patients diagnosed with a neuromuscular disorder have myopathy as a primary or secondary diagnosis (Tagliati et al. 1994). HIV-associated myopathy may develop in patients at all stages of HIV infection (Lange et al. 1988; Simpson and Bender 1988). When systemic or nutritional factors are excluded, myopathy may occasionally be diagnosed as the cause of a wasting syndrome in HIV-infected patients, characterized by involuntary weight loss and chronic muscle weakness. The importance of determining a specific diagnosis of myopathy in patients with wasting syndrome is evidenced by some patients' improvement following prednisone therapy (Simpson et al. 1990).

Clinical features
The predominant presenting symptom of myopathy is slowly progressive muscle weakness, typically characterized by difficulty in rising from a chair or climbing stairs. Myalgia is present in 25–50% of affected patients (Lange et al. 1988; Simpson and Bender 1988; Simpson et al. 1993a), although it is a non-specific symptom in HIV-infected individuals (Miller et al. 1991). Neurological examination reveals symmetrical weakness of proximal muscle groups, with prominent involvement of neck and hip flexors.

HIV-infected individuals may present with acute rhabdomyolysis characterized by a myopathic syndrome (myalgia, muscle weakness) associated with serum creatine kinase (CK) levels greater than 1500 IU/l (Chariot et al. 1994). The cause of acute rhabdomyolysis in HIV disease is unknown, although investigators have suggested several etiologies including HIV itself, drug toxicity (e.g., ddI, sulfadiazine, pentamidine) and opportunistic infections (Chariot et al. 1994).

Electrodiagnostic and laboratory studies
Electromyography (EMG) is a sensitive diagnostic test in HIV myopathy. In our series of 50 patients with HIV myopathy, 94% had myopathic EMG results characterized by small, brief, and polyphasic motor unit potentials that recruit with full interference patterns. Abnormal irritative activity, such as fibrillation potentials and complex repetitive discharges, was also present in 79% of these cases (Simpson et al. 1993a), as is common in other forms of irritative myopathy. In approximately 50% of our patients with HIV myopathy, nerve conduction abnormalities indicated concurrent

DSP (Simpson and Bender 1988; Tagliati et al. 1994).

The most sensitive serological test for HIV myopathy, as in other primary muscle disorders, is serum creatine kinase (CK) level. CK elevation is one of the classical criteria (clinical, laboratory, EMG, and biopsy) required to establish the diagnosis of HIV-negative polymyositis (Mastaglia and Ojeda 1985). CK levels are usually elevated to a moderate degree in HIV myopathy, with a median level of approximately 500 IU/l (Simpson et al. 1993a). The CK elevation parallels the degree of myonecrosis observed in coincident muscle biopsies, but does not correlate with the degree of clinical weakness (Simpson and Wolfe 1991; Simpson et al. 1995).

Pathology
A spectrum of histopathological findings has been described in HIV-associated myopathy, including inflammatory infiltrates (Fig. 2A) (Dalakas et al. 1986; Lange et al. 1988; Simpson and Bender 1988; Wrzolek et al. 1990), non-inflammatory degeneration (Fig. 2B) (Stern et al. 1987; Simpson and Bender 1988; Wrzolek et al. 1990), nemaline rod bodies (Dalakas and Pezeshkpour 1988; Gonzales et al. 1988; Simpson and Bender 1988), cytoplasmic bodies, and mitochondrial abnormalities (Fig. 2C) (Simpson et al. 1993a; Morgello et al. 1995). The clinical significance of these variable pathological findings is unknown. In our experience, the most common finding on muscle biopsy in HIV-associated myopathy is scattered myofiber degeneration, with occasional associated inflammatory infiltrates. The extent of inflammation is generally less than that observed on HIV-negative polymyositis (Simpson and Wolfe 1991).

Etiology
The pathogenesis of HIV-associated myopathy is unknown. A number of mechanisms have been suggested as contributing to this myopathy. HIV may infect infiltrating cells of the monocyte/macrophage lineage (Chad et al. 1990); however, myofibers remain uninfected (Dalakas and Pezeshkpour 1988; Simpson and Bender 1988; Illa et al. 1991). While more sophisticated assays to detect HIV in tissue, perhaps at a subgenomic level, may yield other information, there is no compel-

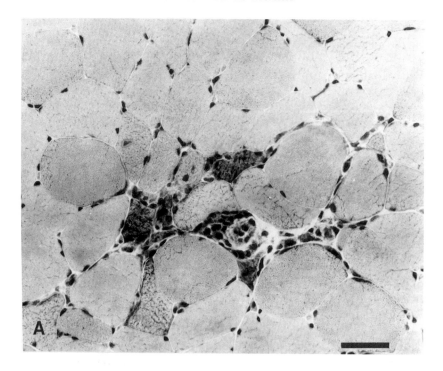

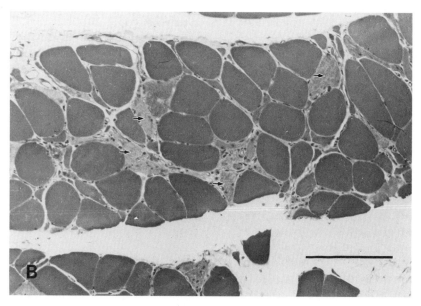

Fig. 2. HIV-associated myopathy. (A) Focal lymphohistiocytic infiltrate in mild HIV-associated myopathy. Modified trichrome-stained cryostat section, quadriceps, initial magnification × 33. Bar = 80 μm. (B) Several atrophic, non-necrotic, degenerating myofibers (arrows) not associated with inflammatory infiltrates. Modified trichrome-stained, cryostat section, quadriceps, initial magnification × 33. Bar = 80 μm.

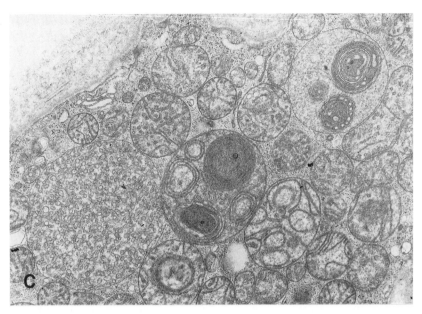

Fig. 2. (C) Mitochondrial abnormalities in degenerating myofibers in HIV-associated myopathy, including double membrane-bounded profiles containing tubular, circinate, and otherwise convoluted cristae. Quadriceps biopsy. Initial magnification × 16,000. (From Simpson and Taliati (1997), with permission.)

ling evidence at this time that HIV infection has a direct role in myopathy.

Immune mechanisms have been proposed in HIV myopathy (Illa et al. 1991; Simpson et al. 1993a), as in other forms of polymyositis (Bohan and Peters 1975; Devere and Bradley 1975). HIV antigens have been localized in macrophages invading muscle (Chad et al. 1990; Illa et al. 1991), suggesting a role for virus-infected inflammatory cells in muscle degeneration. It is possible that toxic factors secreted by macrophages mediate myofiber damage. Tumor necrosis factor (cachectin), a cytokine produced largely by monocytes, causes anorexia and cachexia when injected into laboratory animals (Bentler and Cerami 1987). Lahdevirta et al. (1988) reported that patients with AIDS have elevated cachectin levels while they were normal in asymptomatic HIV-seropositive individuals. The cytokine hypothesis of HIV myopathy has been supported by evidence showing accumulation of interleukin-1α in muscle fibers of HIV-infected patients, most of whom were treated with ZDV (Gherardi et al. 1994).

Several opportunistic agents have been detected in muscles of patients with AIDS, including *Toxoplasma gondii* (Gherardi et al. 1992), CMV (Ho et al. 1989), Microsporidia (Chupp et al. 1993), *Cryptococcus neoformans* (Wrzolek et al. 1990), *Mycobacterium avium intracellulare* (Wrzolek et al. 1990), and *Staphylococcus aureus* (Belec et al. 1991). However, these infections do not play a role in the majority of patients with HIV myopathy.

The role of ZDV

ZDV is a dideoxynucleoside analog of thymidine (Fig. 1) and acts by interfering with viral reverse transcriptase. In 1987, ZDV began receiving widespread use in HIV infection following the results of several large clinical trials demonstrating clinical efficacy (Yarchoan et al. 1987; Volberding et al. 1990). In 1988, four cases of polymyositis in ZDV-treated patients were first reported (Bessen et al. 1988). Three of these patients improved following ZDV withdrawal (one with concomitant corticosteroid therapy). In 1990, Dalakas et al. reported that long-term ZDV therapy was associated with a mitochondrial myopathy characterized by ragged-red myofibers with electron microscopic evidence of mitochondrial abnormalities (Dalakas et al. 1990).

Whether there are distinguishing clinical and

pathological features between HIV- and ZDV-associated myopathies has been the subject of debate (Manji et al. 1993; Simpson et al. 1993a; Dalakas et al. 1994b). ZDV myopathy has been reported to be histologically characterized by atrophic ragged-red fibers (RRFs) with marked myofibrillar alterations including thick filament loss and cytoplasmic bodies (Gherardi et al. 1994). Subsequent to Dalakas' report, findings consistent with mitochondrial degeneration were described by several authors (Mhiri et al. 1991; Chariot et al. 1993; Grau et al. 1993; Peters et al. 1993), leading to the widely accepted hypothesis that mitochondrial toxicity is the specific mechanism of ZDV myopathy (Kuncl and George 1993). However, mitochondrial abnormalities described in ZDV-exposed biopsies (Arnaudo et al. 1991) may also be found in ZDV-naive HIV-infected patients with myopathy (Manji et al. 1993; Simpson et al. 1993a; Morgello et al. 1995). Both ZDV- and HIV-associated myopathy are characterized by severely degenerated and necrotic myofibers (Dalakas et al. 1990; Chariot et al. 1993; Peters et al. 1993; Simpson et al. 1993a; Morgello et al. 1995). That abnormal mitochondria may be present in a wide variety of myopathic disorders is well known (Morgan-Hughes 1986), and the presence or absence of morphologically abnormal mitochondria may not be relevant to the pathogenesis of myopathy.

ZDV has been shown to be a potent inhibitor of γ-polymerase of the mitochondrial matrix in vitro (Simpson, M.V. et al. 1989). Following the description of a mitochondrial myopathy in patients treated with ZDV, evidence of both in vitro mitochondrial abnormalities of human and animal muscle cells exposed to ZDV (Lamperth et al. 1991) and in vivo impairment of muscle energy metabolism in ZDV-treated patients (Soueidan et al. 1992) was reported. Dalakas et al. (1994a) also noted depletion of carnitine and muscle mitochondrial DNA in muscle fibers of ZDV-treated patients with molecular and immunocytochemical techniques (Arnaudo et al. 1991; Pezeshkpour et al. 1991). Other groups have described defects of mitochondrial enzymes, such as cytochrome-c oxidase and reductase in ZDV-treated patients (Mhiri et al. 1991; Chariot et al. 1993), accumulation of cytokines in muscle fiber

mitochondria (Gherardi et al. 1994), and mitochondrial oxidative dysfunction based on data obtained from ^{31}P magnetic resonance (MR) spectroscopy (Weissman et al. 1992).

However, some of these results could not be replicated in studies using similar techniques. For example, in an MR spectroscopy study of AIDS patients with fatigue and myalgia, Miller et al. (1991) were not able to support the hypothesis that ZDV causes a mitochondrial myopathy. Similarly, Herzberg and colleagues could not reproduce the results demonstrating abnormalities in cytochrome-c oxidase (Mhiri et al. 1991) and mitochondrial DNA (Arnaudo et al. 1991) from ZDV exposure (Herzberg et al. 1992). Finally, Reyes et al. (1992) did not observe RRFs in hamsters treated with intraperitoneal ZDV, none of whom developed weakness.

The clinical significance of RRF, or other laboratory evidence of mitochondrial dysfunction in ZDV-treated patients, is unclear. There are no solid epidemiologic data indicating the incidence of myopathy associated with ZDV therapy. None of the large antiretroviral therapy studies (Yarchoan et al. 1987; Volberding et al. 1990) were designed to prospectively establish the diagnosis of myopathy, limiting the utility of retrospective data. In a prospective series of 118 patients (Peters et al. 1993), the incidence of myopathy in the ZDV-treated group was 8% (7 of 88), although the small size of the control group and brief time of follow-up limited the significance of these data. We performed a retrospective analysis of a large primary antiretroviral protocol (ACTG 016), comparing the efficacy of ZDV ($n = 360$) to placebo ($n = 351$). The incidence of myopathy was 3% in the ZDV-treated group, and 0.4% in the placebo group, although this difference did not reach statistical significance (Simpson et al. 1997a). In a prospective analysis in ACTG protocol 175, there was no significant difference in the rate of myalgia or proximal muscle weakness in the ZDV monotherapy versus ddI or combination therapy arms (Simpson et al. 1997b).

The majority of ZDV-treated patients with RRF in Grau's series (Grau et al. 1993) were asymptomatic. Furthermore, the percentage of RRF did not differ in symptomatic and asymptomatic patients. In Dalakas' series (Dalakas et al.

1990), the percentage of RRFs was nearly the same in subjects that improved with ZDV withdrawal (mean = 17.3%) and in those that did not (mean = 15.1%). In our and other authors' experience, there are no clinical features that differentiate HIV from ZDV myopathy (Espinoza et al. 1991; Manji et al. 1993; Simpson et al. 1993a; Gherardi et al. 1994). While some patients with HIV myopathy may improve with ZDV withdrawal (Bessen et al. 1988; Gorard et al. 1988; Panegyres et al. 1988; Dalakas et al. 1990), others do not (Till and MacDonnell 1990; Espinoza et al. 1991; Manji et al. 1993; Simpson et al. 1993a). This is not surprising since even those authors reporting a specific ZDV myopathy also note the coexistence of inflammatory changes identical to HIV polymyositis (Dalakas et al. 1990; Pezeshkpour et al. 1991). Furthermore, ZDV rechallenge in some patients has not reproduced their myopathic symptoms (Panegyres et al. 1988; Gertner et al. 1989; Cupler et al. 1994).

Therapy

Although corticosteroids should be used cautiously in HIV-infected patients because of the risk of further immunosuppression, there is evidence that they provide benefit with tolerable adverse effects in other HIV-associated diseases such as pneumocystis pneumonia (Consensus Statement 1990). Numerous investigators have found corticosteroids to be effective in uncontrolled series of patients with HIV-associated myopathy (Dalakas et al. 1990; Chalmers et al. 1991; Mhiri et al. 1991; Manji et al. 1993). Our randomized, placebo-controlled study of prednisone in HIV-associated myopathy supports these observations (Simpson et al. 1993b). Anecdotal observations suggest that plasmapheresis or intravenous immunoglobulin may have efficacy in the treatment of HIV myopathy, without the immunosuppressive effects of corticosteroids. However, these agents have not been examined in controlled trials.

Since it may be difficult to prospectively identify patients with ZDV myotoxicity, the initial management of patients with significant limb weakness and objective evidence of myopathy includes ZDV withdrawal. The percentage of ZDV-treated patients that show objective improvement in muscle strength following ZDV withdrawal has varied from 18 to 100% (Bessen et al. 1988; Dalakas et al. 1990; Chalmers et al. 1991; Grau et al. 1993; Manji et al. 1993). In our retrospective series, 4 of 15 (26%) patients with myopathy improved in strength after ZDV was discontinued (Simpson et al. 1993a). A management algorithm for HIV-associated myopathies is shown in Fig. 3.

OTHER ANTIVIRAL AGENTS

Acyclovir

Acyclovir is one of the most effective antiviral agents currently available. As a purine nucleoside analog (Fig. 1), it selectively inhibits viral DNA polymerase and prevents replication of herpes virus with little effect on human cells (Bridgen and Whiteman 1983). In HIV-infected and iatrogenically immunocompromised patients, acyclovir is an important agent in the therapy of herpes virus infections including herpes simplex virus (HSV) cutaneous and neurologic infections, varicella zoster (VZV) radiculopathies and encephalopathies, and oral hairy leukoplakia (Straus et al. 1982; Peterslund 1988; Walling et al. 1994). Acyclovir may increase survival of AIDS patients, suggesting a beneficial effect in HIV infection (Stein et al. 1994; Youle et al. 1994).

The safety and efficacy of acyclovir have contributed to the widespread use of this drug (Dorsky and Crumpacker 1987). Adverse effects are rare and include nephrotoxicity, neurotoxicity, rashes, nausea and vomiting (Lee and Safrin 1992; Morris 1994). When neurologic symptoms occur, it may be difficult to determine whether these reflect progression of herpetic infection to the nervous system or an adverse reaction to antiviral therapy.

Acyclovir neurotoxicity is a dose-dependent (more likely for doses higher than 500 mg/m^2 8 hourly, i.v.), self-limited and uncommon event, occurring in approximately 1% of patients (Morris 1994). Mental status abnormalities, including confusion, lethargy and agitation, are the usual presenting symptoms and occur after 1–18 days of therapy (Wade and Meyers 1983; Cohen et al. 1984; Eck et al. 1991; Davenport et al. 1992; Krieble et al. 1993; Adair et al. 1994). Involuntary movements, including myoclonus and action

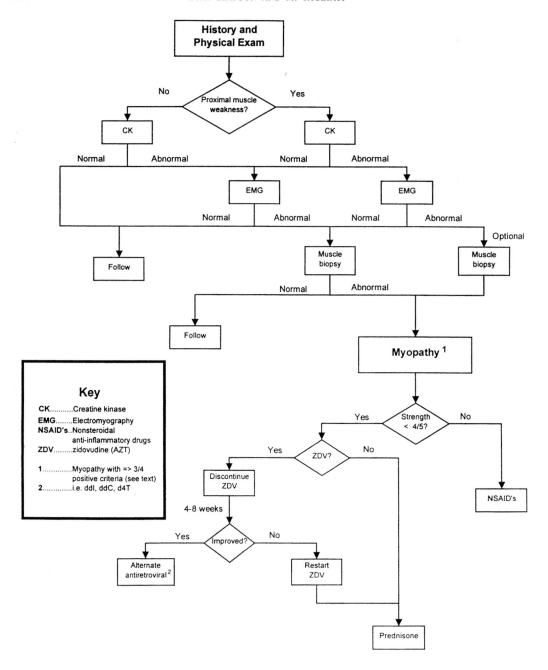

Fig. 3. Diagnosis and management algorithm for patients with HIV-associated myopathy. (Adapted from Simpson and Tagliati (1997), with permission.)

tremor, have been described in 50% of cases. Hallucinations, delirium, ataxia and slurred speech have also been reported. Most of these cases involved elderly patients with malignancies or concurrent renal failure (Adair et al. 1994). The kidney is a major site of acyclovir toxicity which may contribute to acyclovir blood levels above 4.5 μg/ml, which is in the range of neurotoxicity (MacDiarmaid-Gordon et al. 1992; Adair et al. 1994). Prior chemotherapy or radiation therapy may predispose to acyclovir toxicity (Wade and Meyers 1983; Feldman et al. 1988).

Since acyclovir is indicated for viral infections that can involve the CNS, the distinction of acyclovir neurotoxicity from HSV or VZV encephalitis may be difficult. Encephalitic symptoms such as fever, severe headache, seizures, cranial nerve palsies and hemiparesis are rare in acyclovir toxicity (Rashiq et al. 1993). Neuroimaging studies and CSF studies are normal in acyclovir toxicity. EEG may show diffuse slowing abnormalities in acyclovir toxicity (Wade and Meyers 1983), while lateralization is characteristic of HSV encephalitis (Rashiq et al. 1993).

The management of acyclovir neurotoxicity requires assessment of the hydration status and discontinuation of the drug (Rashiq et al. 1993). In most cases symptoms resolve within 2 weeks (Wade and Meyers 1983), but recovery can be accelerated with hemodialysis (Krieble et al. 1993; Adair et al. 1994). When viral encephalitis cannot be excluded, it may be prudent to continue acyclovir given the poor prognosis of untreated encephalitis and the relatively benign course of acyclovir neurotoxicity (Rashiq et al. 1993).

Ganciclovir and foscarnet

Cytomegalovirus (CMV) infection may cause a variety of systemic and neurologic abnormalities in AIDS, including focal encephalitis, ventriculoencephalitis, vasculitis, and radiculomyelitis (Simpson and Tagliati 1994). The management of CMV disease is problematic, since the median survival of patients with CMV encephalitis or radiculomyelitis is only about 5 weeks, and the role of antiviral therapy in altering this rapid course is not yet established. In several case series, patients with CMV neurologic involvement have responded to ganciclovir, foscarnet or a combination of these antiviral medications (Manji et al. 1992; Peters, M. et al. 1992; Kim and Hollander 1993). These two drugs have different toxicity profiles (Markham and Faulds 1994; Wagstaff and Bryson 1994).

Ganciclovir is a 2'-deoxyguanosine analog (Fig. 1) with in vitro and in vivo activity against CMV (Markham and Faulds 1994). Ganciclovir competitively inhibits deoxyguanosine triphosphate (dGTP) incorporation into viral DNA and prematurely blocks DNA chain elongation, similar to other nucleoside analogs (Mar et al. 1983). Intravenous ganciclovir is an effective treatment for AIDS-related CMV retinitis (Collaborative DHPG Treatment Study Group 1986; Holland, G.N. et al. 1989; Jabs et al. 1989), gastrointestinal infection (Dietrich et al. 1993) and pneumonia (Drew 1992).

While neutropenia and thrombocytopenia are the most common adverse events in patients receiving ganciclovir (Faulds and Heel 1990), CNS symptoms have been reported in about 5% of cases (Markham and Faulds 1994). Neurologic symptoms include confusion, somnolence, coma, ataxia, psychosis and hallucinations (DeArmond 1991; Chen et al. 1992). Most of these complications were reported anecdotally and could be related to other opportunistic infections, metabolic disorders or concurrent medications (DeArmond 1991). Generalized tonic-clonic seizures have also been reported in association with ganciclovir (Keay et al. 1987; Barton et al. 1992; Studies of Ocular Complications of AIDS Research Group 1992). The concomitant administration of the antibiotic imipenem-cilastatin may increase the risk of seizures during ganciclovir treatment (DeArmond 1991).

Foscarnet is a pyrophosphate analog with antiviral activity against CMV and other herpes viruses, hepatitis B virus and HIV. The mechanism of action differs from other antivirals in that foscarnet inhibits viral DNA polymerase and HIV reverse transcriptase, preventing cleavage of pyrophosphate from the deoxynucleotide triphosphates needed to elongate the DNA chain (Crumpacker 1992). In immunosuppressed patients, foscarnet therapy has been effective in CMV retinitis (Palestine et al. 1991), gastrointestinal infections, encephalitis, polyradiculopathy, acyclovir-resistant HSV and VZV infections, and hepatitis B (Wagstaff and Bryson 1994).

Dose-limiting adverse effects associated with foscarnet administration include renal impairment (Deray et al. 1989), infusion-related nausea and alterations in serum electrolytes, particularly phosphate, magnesium and calcium (Gearhart and Sorg 1993). Neurologic symptoms are uncommon and include confusion, agitation and muscle tremors (Gearhart and Sorg 1993). Seizures have been reported in 10% of AIDS patients receiving foscarnet (Wagstaff and Bryson 1994), with a rate higher than that observed with ganciclovir therapy (Studies of Ocular Complications of AIDS Research Group 1992). It is not clear whether foscarnet itself or the associated electrolyte abnormalities such as hypocalcemia or hypomagnesemia are responsible for seizures (Palestine et al. 1991; Gearhart and Sorg 1993; Wagstaff and Bryson 1994)

ANTIBIOTICS

Antituberculosis drugs

There is a high incidence of tuberculosis (TB) in HIV infection (Cantwell et al. 1994; Polkey and Rees 1994), particularly in intravenous drug users or individuals from a low socioeconomic class (Chaisson et al. 1987; Cantwell et al. 1994). TB in HIV-seropositive individuals is generally considered the reactivation of infection triggered by HIV-related immunosuppression, as suggested by the increased risk of developing TB with a decreasing CD4 cell count (Selwyn et al. 1989). The clinical course of TB in AIDS is often atypical, with frequent involvement of extra-pulmonary sites and non-specific or unusual pulmonary radiographic features (Chaisson et al. 1987; Polkey and Rees 1994). Although patients with AIDS and TB generally have a favorable response to antituberculosis agents (Chaisson et al. 1987), a high rate of recurrence has been described (Hawken et al. 1993). Furthermore, the frequent occurrence of multidrug-resistant strains of *Mycobacterium tuberculosis* (Weltman and Rose 1994) impairs treatment success and increases the chance of drug toxicity. Adverse drug reactions occur more frequently in AIDS patients than in immunocompetent hosts (Chaisson et al. 1987).

Isoniazid (INH) is the drug of choice for treatment and prophylaxis of TB. Long-term prophylaxis with INH is recommended for all HIV-infected patients who have or are at high risk for *Mycobacterium tuberculosis* infection (Gallant et al. 1994). INH is the hydrazide of isonicotinic

acid and is effective against actively growing tubercle bacilli. INH is usually well tolerated and rarely causes serious adverse effects, the most common of which is hepatitis (Girling 1982). Chronic use of INH may cause peripheral neuropathy, optic neuritis and behavioral symptoms, while seizures have been reported with acute toxicity (Girling 1982).

A predominantly sensory, distal symmetrical peripheral neuropathy was described soon after INH was initially used in the treatment of TB (Biehl and Skavlem 1953; Gammon et al. 1953; Jones and Jones 1953). It is characterized by distal paresthesias in a stocking-glove distribution, usually more severe in the lower limbs. If untreated, deficits may slowly progress to involve position and vibration sense with ataxia, muscle weakness and loss of tendon reflexes (Girling 1982). INH-associated peripheral neuropathy is dose-dependent (Biehl and Nimitz 1954) and is more likely to occur in malnourished patients, pregnant women, patients with chronic liver disease or alcoholism, elderly patients and slow acetylators of the drug (Snider, D.E. 1980; Girling 1982).

Neurological toxicity of INH is due to competitive inhibition of pyridoxal phosphate and pyridoxamine phosphate, the coenzymes derived from pyridoxine, also known as vitamin B_6 (Biehl and Vilter 1954; Wiegand 1956; Price et al. 1957). The inhibition of vitamin B_6 active metabolites causes reduced synthesis of several neurotransmitters including catecholamines, serotonin and γ-aminobutyric acid (GABA) (Girling 1982). Thus, INH neuropathy can be prevented by simultaneous administration of small doses (approximately 6 mg/day) of pyridoxine (Tuberculosis Chemotherapy Center Madras 1963). The supplementary dose of pyridoxine should be low, because pyridoxine may interfere with the antibacterial activity of INH (McCune et al. 1957; Snider, D.E. 1980) and may itself cause a toxic peripheral neuropathy (Schaumburg et al. 1983; Nisar et al. 1990; Berger et al. 1992). Some authors recommend that pyridoxine supplementation should be reserved for patients with nutritional deficiency, pregnant women, and subjects predisposed to neuropathy, such as those with uremia or diabetes (Snider, D.E. 1980). Although HIV-infected patients are predisposed to develop peripheral neuropathy, they do not usually complain of neuropathic symptoms as a side effect of INH therapy (Chaisson et al. 1987; Soriano et al. 1988). Of 166 HIV-infected patients with DSP in our series, none were associated with INH exposure (M. Tagliati, unpublished observations).

Ethambutol. Optic neuropathy has been reported in patients treated with INH, usually in cases of combined therapy with ethambutol (Jimenez-Lucho et al. 1987; Karmon et al. 1989; Russo and Chaglasian 1994). Ethambutol, a bacteriostatic drug used in cases of INH-resistant infections and cases with extrapulmonary or atypical infections, is associated with toxic optic neuropathy (Carr and Henkind 1962). This optic neuropathy is usually dose-dependent and reversible. Dosage levels between 15 and 25 mg/kg/day are considered safe, although several reports of visual dysfunction have been described with lower doses (Russo and Chaglasian 1994). The reported incidence of optic neuropathy ranges between 0.8 and 33% (Russo and Chaglesian 1994). Clinical features include decreased visual acuity, impaired color vision and central scotoma with normal fundus and optic nerve findings. The risk of developing toxic neuropathy with ethambutol increases with renal failure, advanced age and excessive alcohol consumption (Russo and Chaglesian 1994). The management of visual dysfunction during antituberculosis therapy includes immediate suspension of ethambutol and switching to INH (Karmon et al. 1989). Careful monitoring of antituberculosis drug dosage and visual function is recommended to prevent ocular toxicity (Citron and Thomas 1986).

Chronic INH therapy has been reported to cause several psychiatric disturbances, including psychosis (Jackson 1957; Pallone et al. 1993), mania (Chaturvedi and Upadhyaya 1988) and obsessive-compulsive disorder (Bhatia 1990). The most frequently reported symptoms are delusions and visual and auditory hallucinations (Pallone et al. 1993). Risk factors for developing a psychotic reaction to INH include overdosage (higher than 5 mg/kg), previous psychiatric history, age over 50 years, and concurrent disease such as diabetes, alcoholism, hepatic insufficiency and hyper-

thyroidism (Chaturvedi and Upadhyaya 1988; Pallone et al. 1993). The mechanism of INH-associated psychosis is unclear. INH may interfere with the homeostasis of critical neurotransmitters in the CNS including dopamine, serotonin and norepinephrine (Marcus and Coulston 1990). INH-associated psychosis usually responds to discontinuation of the drug and administration of vitamin B_6 and neuroleptics over a period of days (Pallone et al. 1993).

Seizures are a frequent manifestation of acute INH toxicity (Sievers and Herrier 1975; Bredemann et al. 1990; Olson et al. 1994). In a large retrospective study (Olson et al. 1994), 5% of seizures associated with poisoning and drug overdose were attributed to INH. The mechanism of INH-induced seizures seems to be related to the impaired synthesis of the inhibitory neurotransmitter GABA, due to the inhibition of pyridoxal phosphate (Wood and Peesker 1972; Bredemann et al. 1990). GABA deficiency in the brain increases CNS excitability and predisposes to convulsive episodes (Bredemann et al. 1990). Patients with INH-induced seizures should have pyridoxine replacement (Wason et al. 1981; Bredemann et al. 1990). The efficacy of pyridoxine replacement can be enhanced by the removal of INH with hemodialysis or hemoperfusion (Leibenwitz et al. 1989; Bredemann et al. 1990). Anticonvulsant drugs with GABA agonist action, such as barbiturates and benzodiazepines, may be effective in cases of refractory seizures (Bredemann et al. 1990).

Sulfa drugs

TMP-SMZ. Pneumocystis carinii pneumonia (PCP) is one of the most frequent opportunistic infections in patients with AIDS. The incidence of PCP has been declining since 1987, when prophylaxis with pentamidine and trimethoprim-sulfamethoxazole (TMP-SMZ) became common practice (Fischl et al. 1988). TMP-SMZ is the drug of choice for prevention of PCP and a promising preventive tool for *Toxoplasma* encephalitis and common bacterial infections in patients with AIDS (Hardy et al. 1992). A high incidence of adverse reactions to TMP-SMZ occurs in HIV-infected patients (Jung and Paauw 1994).

The spectrum of adverse reactions to TMP-SMZ includes generalized cutaneous eruptions, often with associated fever, gastrointestinal symptoms, neutropenia and thrombocytopenia, liver enzyme abnormalities, and anemia (Jung and Paauw 1994). Adverse neurological reactions are uncommon (Salter 1982; Borucki et al. 1988) and include headache, depression, hallucinations, vertigo, ataxia, tremor and seizures (Frisch 1973; Salter 1982; Borucki et al. 1988). The mechanism of TMP-SMZ-associated neurologic toxicity is unknown. In AIDS patients, neurologic adverse effects may be explained in part by exacerbation of folate deficiency (Smith et al. 1987; Borucki et al. 1988). Neurologic symptoms usually resolve within 3 days after discontinuation of TMP-DMZ treatment (Borucki et al. 1988).

Dapsone (4,4'-diaminodiphenylsulfone) is a sulfone antimicrobial agent which has been used in the treatment of leprosy, dermatitis herpetiformis and other dermatologic disorders. It may also be used in the prophylaxis for PCP pneumonia in patients with HIV infection that are intolerant to sulfonamides (Medina et al. 1990). Despite a lower incidence of adverse effects as compared to TMP-SMZ (Pertel and Hirschtick 1994), severe side effects can occur during dapsone therapy, including methemoglobinemia and hemolytic anemia (Anonymous 1981). Less frequently, cutaneous reactions, nephrotic syndrome and peripheral neuropathy have been described (Millikan and Harrell 1970; Anonymous 1981).

A predominantly motor neuropathy has been attributed to dapsone therapy in several case reports (Saqueton et al. 1969; Rapoport and Guss 1972; Wyatt and Stevens 1972; Du Vivier and Fowler 1974; Epstein and Bohm 1976; Fredericks et al. 1976; Guttman et al. 1976; Koller et al. 1977; Navarro et al. 1989). Progressive weakness is the most common symptom, with an onset of 6 weeks to several years after initiation of dapsone in total doses ranging from 4 to 600 g. Proximal and distal muscles in both upper and lower extremities are variably affected, with sensory symptoms being less prominent (Wyatt and Stevens 1972; Du Vivier and Fowler 1974; Guttman et al. 1976; Koller et al. 1977). Electrophysiological studies

reveal a motor axonopathy, with reduced or absent response following stimulation of involved nerves (Saqueton et al. 1969; Rapoport and Guss 1972; Wyatt and Stevens 1972; Du Vivier and Fowler 1974; Epstein and Bohm 1976; Fredericks et al. 1976; Guttman et al. 1976; Koller et al. 1977). The mechanism of dapsone neurotoxicity is unknown. Slow acetylators may be predisposed to dapsone-associated neurotoxicity (Koller et al. 1977). Discontinuation of the drug usually results in a complete clinical recovery.

Amphotericin B

Fungal infections are common complications in conditions of cellular immunosuppression such as AIDS. Although the most prevalent type of mycotic infection may vary in different geographical sites, cryptococcal meningitis is the most common CNS fungal infection in patients with HIV infection in the United States (Simpson and Tagliati 1994). Amphotericin B (AmB), a polyene macrolide antibiotic, is the most effective antifungal agent currently available, although its use may be limited by frequent adverse reactions. Both infusion-related adverse events (nausea, vomiting, fever, headache, thrombophlebitis) and chronic toxicity (anemia, renal dysfunction) are common and may require discontinuation of the drug. Although neurotoxicity is uncommon, several reports indicate that a fatal leukoencephalopathy can complicate the course of i.v. administration of AmB and its methyl ester derivative (Ellis et al. 1982; Devinsky et al. 1987; Walker and Rosenblum 1992).

Ellis et al. (1982) described a syndrome of progressive lethargy, akinetic mutism, hyperreflexia and sphincter dysfunction complicating the treatment with i.v. amphotericin B methyl ester (AME) in 14 patients. Neuropathologic findings revealed a diffuse leukoencephalopathy with marked axonal loss. Three similar cases of leukoencephalopathy were described in association with i.v. AmB therapy (Devinsky et al. 1987; Walker and Rosenblum 1992). The pathogenesis of AmB- and AME-related leukoencephalopathy is unknown. Polyene macrolide antibiotics, such as AmB and AME, bind to cell membrane sterols and may potentially damage cholesterol-rich myelin (Walker and Rosenblum 1992). However, there is little evidence that these compounds penetrate the blood–brain barrier (BBB) in primates (Lawrence et al. 1980). Alteration of the BBB by cranial irradiation has been proposed as a precondition for the development of AmB neurotoxicity (Devinsky et al. 1987; Walker and Rosenblum 1992). Cranial irradiation, coadministration of neurotoxic drugs (e.g. methotrexate) or underlying CNS mycotic infection may also contribute to the development of AmB-associated leukoencephalopathy (Hoeprich 1982).

Ciprofloxacin

Ciprofloxacin is a fluorinated quinolone structurally related to nalidixic acid and is active against a wide spectrum of bacteria. In patients with AIDS, it is used in the treatment of *Mycobacterium avium* complex infections (Benson 1994). The primary mechanism of action of ciprofloxacin is inhibition of bacterial DNA gyrase (Campoli-Richards et al. 1988). Significant adverse effects are uncommon and include gastrointestinal symptoms, rashes and transient alterations of laboratory values (eosinophilia, transaminases, creatinine and blood urea nitrogen). Symptoms related to CNS stimulatory effects including anxiety, insomnia, tremor, seizures and hallucinations have been reported in less than 5% of patients (Arcieri et al. 1986). The mechanism of CNS effects of ciprofloxacin is unclear. Concomitant administration of theophylline may increase the toxicity of ciprofloxacin (Arcieri et al. 1986).

ANTINEOPLASTIC AGENTS

Several malignancies are associated with HIV infection including Kaposi's sarcoma, non-Hodgkin's lymphoma and cervical cancer (Beral et al. 1991; Rabkin and Blattner 1991). As advances in the treatment of HIV infection improves survival of patients with AIDS, an increase in incidence of other types of cancer is expected (Rabkin and Blattner 1991). Several antineoplastic regimens are currently used in the treatment of HIV-infected patients (Monfardini et al. 1994). Some of these agents have well defined neurotoxic effects.

Vinka alkaloids

Vincristine, vinblastine and vindesine are a group of cytotoxic drugs extracted from the plant *Catharantus roseus* and are effective against leukemias, lymphomas and other neoplasias (Bender and Chabner 1982). The antineoplastic activity of vinka alkaloids is based on their binding to tubulin and consequent interference with the microtubules of the mitotic spindle apparatus (Sahenk et al. 1987). Neurotoxicity is the major dose-limiting adverse effect, with a cumulative dose threshold of about 15–20 mg (Tuxen and Hansen 1994). Neurotoxic effects are similar for the different vinka alkaloids, with vincristine having the highest neurotoxic potential.

The most common target of vincristine neurotoxicity is the peripheral nervous system. Peripheral neuropathies and cranial nerve palsies have been frequently described (Sandler et al. 1969; Casey et al. 1973; Holland, J.F. et al. 1973) and are related to the total dose and duration of vincristine administration. The peripheral neuropathy associated with vincristine therapy is similar to other toxic neuropathies. Distal paresthesias with reduced or absent ankle reflexes are the earliest findings, while objective sensory loss and distal muscle weakness may develop at later stages (MacDonald 1991; Tuxen and Hansen 1994). Electrophysiological studies reveal signs of axonal neuropathy (Casey et al. 1973). Reduction or discontinuation of the drug usually results in partial recovery, although tendon reflexes may remain absent (Legha 1986).

Numerous cranial neuropathies have been reported in association with vincristine therapy (Tuxen and Hansen 1994). Autonomic symptoms include abdominal pain, constipation, urinary retention and orthostatic hypotension (Legha 1986; MacDonald 1991; Tuxen and Hansen 1994). Less commonly, vincristine treatment may be associated with encephalopathy, seizures, and inappropriate secretion of antidiuretic hormone (Suskind et al. 1972; Hurwitz et al. 1988).

The pathogenesis of vinca alkaloid neurotoxicity is likely related to effects on the microtubules, with abnormalities in axoplasmic transport (Sahenk et al. 1987). Patients with pre-existing neuropathy, liver abnormalities, poor nutritional status and age over 60 years may be at increased risk of vincristine neurotoxicity (Tuxen and Hansen 1994). Ganglioside therapy and the ACTH(4–9) analog Org 2766 may prevent the neurotoxicity of vinka alkaloids (Houi et al. 1993; Kiburg et al. 1994).

Methotrexate and cytosine arabinoside

With the use of more aggressive cancer treatment protocols, including methotrexate (MTX), cytosine arabinoside (cytarabine, Ara-C) and radiotherapy (RT), neurotoxicity has become a frequent complication. Since these agents are often used in combination or in rapid succession, it may be difficult to attribute a specific neurological complication to a particular agent. However, characteristic patterns of nervous system dysfunction have been associated with several of these therapies.

MTX, a folate antagonist, is the primary treatment for acute lymphoblastic leukemia (Jonsson and Kamen 1991) and has recently been implemented in the therapy of psoriasis, rheumatoid arthritis and other autoimmune diseases (Cash and Klippel 1994). In patients with AIDS, MTX is used in the chemotherapy of primary CNS lymphoma (Forsyth et al. 1994) and lymphomatous meningitis (Chamberlain and Dirr 1993).

MTX therapy has adverse effects similar to those seen of other antineoplastic drugs, including bone marrow suppression, mucositis and gastrointestinal distress. The intrathecal (i.t.) route of MTX administration may result in acute arachnoiditis with headache, meningismus and signs of increased intracranial pressure in 5–40% of cases (Bleyer 1981). Reversible myelopathy has also been described as a toxic effect of i.t. MTX (Gagliano and Constanzi 1976). An acute lethargic syndrome may follow high-dose i.v. administration of MTX (Yap et al. 1979). Several months following either i.t. or i.v. MTX treatment, severe leukoencephalopathy may occur (Allen et al. 1980). Concurrent cranial irradiation aggravates MTX neurotoxicity (Bleyer 1981).

The mechanisms of MTX neurotoxicity are unclear, since neurons are postmitotic cells. Damage to replicating cells such as astrocytes and endothelial cells may occur, particularly when cranial irradiation is simultaneously used, with damage to the BBB and increased concentrations of MTX in the CNS (Griffin et al. 1977). An alternative hypothesis is based on the evidence that MTX promotes the release of adenosine from fibroblasts and endothelial cells (Cronstein et al. 1993). Since adenosine dilates cerebral blood vessels, slows release of neurotransmitters and the discharge rate of neurons, it may account for some of the neurotoxicity associated with MTX. Children treated with MTX have increased CSF concentrations of adenosine. Furthermore, neurotoxic symptoms may resolve after treatment with aminophylline, an adenosine antagonist (Bernini et al. 1995).

Ara-C is a structural analog of deoxycytidine (Fig. 1) and is currently used as systemic (i.v.) therapy for acute lymphoblastic leukemia and non-Hodgkin's lymphoma and as i.t. therapy for leukemic and lymphomatous meningitis. Ara-C inhibits DNA polymerase and prevents DNA elongation and repair in actively replicating cells (Fram et al. 1985; Kyle et al. 1989). In patients with AIDS, Ara-C may be administered i.t. in the treatment of lymphomatous meningitis (Chamberlain and Dirr 1993). A multicenter clinical trial (ACTG 243) has demonstrated lack of efficacy of i.v. and i.t. Ara-C in the treatment of AIDS-related progressive multifocal leukoencephalopathy (Hall et al. 1998).

Pancytopenia is the most common adverse effect of high-dose Ara-C treatment, requiring transfusion support and i.v. antibiotics. Non-hematologic toxicities involve the liver, gastrointestinal tract, eye and skin (Bolwell et al. 1988). Neurotoxicity is a common and serious complication of Ara-C administration, whether i.t. or i.v. A variety of neurological syndromes have been described including cerebellar dysfunction, myelopathy, encephalopathy and peripheral neuropathy (Baker et al. 1991). An acute cerebellar syndrome including ataxia, dysarthria and nystagmus may appear 3–8 days after initiation of high-dose Ara-C therapy in 8–23% of cases (Winkelman and Hines 1983; Baker et al. 1991; Vogel and Horoupian 1993). This cerebellar syndrome is accompanied by diffuse cerebral white matter hyperintensities on T2-weighted MRI (Vaughan et al. 1993), and pathological findings of Purkinje cell loss in the cerebellar hemispheres and vermis (Winkelman and Hines 1983). In the majority of cases cerebellar symptoms resolve within 1 week, although in 30% of patients these are irreversible. Risk factors for Ara-C neurotoxicity include high-dose regimens (Lazarus et al. 1981), age older than 40 years (Herzig et al. 1987; Rubin et al. 1992), renal dysfunction with serum creatinine >1.2 mg/dl (Rubin et al. 1992), abnormal liver function studies (Nand et al. 1986; Rubin et al. 1992) and a history of previous CNS disorders (Baker et al. 1991). The pathogenesis of cerebellar injury by Ara-C is unknown. Potentially cytotoxic levels of Ara-C can accumulate in the CSF because of a lack of the metabolic enzyme cytidine deaminase (Ho 1973). Although Purkinje cells are postmitotic neurons, Ara-C may be cytotoxic, interfering in the processes of DNA repair and amplification (Baker et al. 1991) or inhibiting a 2'-deoxycytidine-dependent cellular process that is unrelated to DNA synthesis (Wallace and Johnson 1989).

Irreversible myelopathy may occur as a complication of i.t. Ara-C administration, usually in combination with CNS irradiation or i.t. MTX, between 2 days and 6 months following Ara-C administration (Baker et al. 1991). Elevated CSF myelin basic protein levels may be a marker of imminent neurologic dysfunction (Bates et al. 1985). Segmental demyelination of the outer white matter of the spinal cord is the most prominent pathologic finding, suggesting a direct cytotoxicity of Ara-C on tissues adjacent to the CSF (Burch et al. 1988; Baker et al. 1991). Less frequent complications of Ara-C therapy include encephalopathy, seizures, cranial and peripheral neuropathies, variably described as sensory axonal, ascending demyelinating sensorimotor polyneuropathy or brachial plexopathy (Baker et al. 1991).

Radiotherapy

RT is frequently used to treat or prevent CNS neoplasms and can incidentally affect the CNS when administered for extraneural tumors. In AIDS patients, brain RT is used mainly in the treatment of primary CNS lymphoma (Forsyth et al. 1994). Dose-dependent lesions of the brain and the spinal cord are well-recognized adverse events of RT.

Radiation-induced spinal cord injury can present as a transient benign myelopathy or with delayed, more severe damage. The former is common, occurring in 10–15% of cases 4–6 months after incidental RT to the spinal cord during treatment of extraneural tumors (Dropcho 1991). Transient paresthesia radiating down the spine (Lhermitte sign) is the sole symptom, usually without objective neurologic signs (Jones 1964; Dropcho 1991). A severe, progressive myelopathy can occur in 1–12% of patients, 9–18 months after the completion of RT (Jellinger and Sturm 1971; Goldwein 1987). Lower limb numbness, weakness and sphincter dysfunction are the presenting symptoms, progressing to paraplegia or quadriplegia in 50% of patients (Dynes and Smedal 1960). Pathologic findings include demyelination, axonal degeneration and vascular changes in the spinal cord (Dropcho 1991). Although individual differences in sensitivity exist, total spinal doses of less than 4500 cGy given in daily fractions of 180 cGy are relatively safe (Dropcho 1991). Steroids have been reported to stabilize or improve RT myelopathy (Godwin-Austen et al. 1975).

RT-associated encephalopathy can occur during the first days of therapy or several months after completion of whole brain irradiation. The acute form presents with somnolence, headache, nausea and fever. It is frequent among patients with large intracranial masses who received large RT doses (Dropcho 1991). Acute brain toxicity may respond to steroid therapy (Sheline 1980). Delayed forms of brain neurotoxicity associated with RT include focal cerebral necrosis and diffuse cerebral injury (Dropcho 1991). Focal cerebral radiation necrosis occurs several months after completion of RT in 3–9% of patients who survive for 12 months (Marks and Woong 1985; Soffietti et al. 1985). The clinical presentation is non-specific and is difficult to distinguish from recurrence or progression of the primary tumor without pathologic confirmation. The cerebral white matter is predominantly affected, with demyelination and loss of oligodendrocytes, usually accompanied by extensive vascular damage (Rottenberg et al. 1977). Glucocorticoids are effective in some cases (Shaw and Bates 1984). Dif-

fuse cerebral injury is a more common and disturbing effect in those patients who survive more
than 18–24 months after whole brain RT (Dropcho
1991). In 30–50% of cases, progressive dementia
may occur, with neuroimaging evidence of diffuse
cortical atrophy, ventricular dilation and white
matter lesion (Stylopoulos et al. 1988). The use of
focal RT, when applicable, may reduce the risk of
developing diffuse cerebral injury (Dropcho
1991).

OTHER AGENTS

Thalidomide

The sedative thalidomide (γ-phthalimido-
glutarimide) was withdrawn from the United
States market in the early 1960s because of its
teratogenic effects. The drug was later found to
be effective in a number of dermatologic disorders and is currently used as an experimental
drug for Behçet's disease. A placebo-controlled
study indicated that thalidomide is effective in the
treatment of recurrent non-specific aphthous ulcers in HIV-infected patients (Jacobson et al.
1997). The mechanism of these non-sedative effects is still unclear, but may be related to thalidomide's inhibitory effect on cytokines production by monocytes, particularly of tumor necrosis
factor (TNF)-α (Sampaio et al. 1991).

Since the teratogenic effects may be controlled
with adequate contraceptive measures, the most
important adverse events of thalidomide are those
related to its neurotoxicity. Aside from the sedative effect for which it was initially marketed,
thalidomide causes a well-defined peripheral
neuropathy, with an incidence ranging from 1 to
75% of patients treated for 2 or more months
(Florence 1960; Fullerton and O'Sullivan 1968;
Knop et al. 1983; Aronson et al. 1984; Wulff et al.
1985; Hess et al. 1986; Lagueny et al. 1986). The
predominant symptoms of thalidomide neuropathy are sensory and include numbness, hyper-
esthesia and painful paresthesias. Neurological
exam reveals stocking and glove sensory loss and
depressed or absent ankle reflexes (Fullerton and
O'Sullivan 1968; Hess et al. 1986; Lagueny et al.
1986). It may be difficult to distinguish between
thalidomide neuropathy and AIDS-related DSP

(Gorin et al. 1990; Radeff et al. 1990; Youle et al.
1990; Fuller et al. 1991). Motor deficits are rare
(Fullerton and O'Sullivan 1968; Lagueny et al.
1986) and signs of mild pyramidal tract involvement have been occasionally reported (Fullerton
and Kremer 1961). Nerve conduction abnormalities in thalidomide neuropathy indicate sensorimotor axonopathy (Hess et al. 1986; Lagueny et
al. 1986). Electrophysiological abnormalities may
antedate the onset of clinical symptoms, allowing
early detection of patients developing neuropathy
(Clemmensen et al. 1984; Hess et al. 1986;
Lagueny et al. 1986). The pathologic features of
thalidomide-induced neuropathy are those of a
'dying back' process, with axonal degeneration of
the peripheral segments of the longest fibers
(Fullerton and O'Sullivan 1968; Chapon et al.
1985). In addition, degeneration of dorsal root
ganglion cells and dorsal columns has been described (Fullerton and O'Sullivan 1968; Aronson
et al. 1984).

The pathogenetic mechanism of thalidomide
neuropathy is unknown. The structural similarity
of thalidomide with glutamic acid suggests that its
toxicity might be related to altered glutamate function (Fullerton and O'Sullivan 1968). Slow
acetylators have a higher risk of thalidomide neurotoxicity, similar to INH neuropathy (Hess et al.
1986). Thalidomide therapy is contraindicated in
patients with a history of neuropathy, which is a
particular concern in AIDS, where the prevalence
of DSP is over 30% (So et al. 1988). The initial
management of thalidomide neuropathy is drug
discontinuation or dose reduction, although clinical recovery is often delayed or incomplete
(Fullerton and O'Sullivan 1968; Wulff et al. 1985;
Lagueny et al. 1986). The phenomenon of 'coasting', common to many other toxic neuropathies
(Berger et al. 1992, 1993; Kuncl and George
1993), may partially explain delayed improvement after discontinuation of the drug. Other factors that may account for the incomplete recovery
are irreversible nerve damage, and pre-existing
neuropathy.

In summary, the quest for more effective agents
in the treatment of HIV infection and its associated complications has led to an unprecedented
acceleration of development and approval of
novel drugs. It is clear that neurotoxicity is a major

complication of many of these agents. Thus, neurologists must remain involved in AIDS clinical trials and should assist their colleagues in infectious diseases and oncology in the management of these complex patients.

REFERENCES

ABRAMS, D.I., A.I. GOLDMAN, M.S. LAUNER, J.A. KORVIK, J.D. NEATON and L.R. CRANE: A comparative trial of didanosine or zalcitabine after treatment with zidovudine in patients with human immunodeficiency virus infection. N. Engl. J. Med. 330 (1994) 657–662.

ADAIR, J.C., M. GOLD and R.E. BOND: Acyclovir neurotoxicity: clinical experience and review of the literature. South. Med. J. 87 (1994) 1227–1231.

ALLEN, J.C., G. ROSEN, B.M. METHA and B. HORTEN: Leukoencephalopathy following high-dose IV methotrexate chemotherapy with leucovorin rescue. Cancer Treatm. Rep. 64 (1980) 1261–1273.

ANDERSON, T.D., A. DAVIDOVICH, R. ARCEO, C. BROSNAN, J. AREZZO and H. SCHAUMBURG: Peripheral neuropathy induced by 2',3'-dideoxycytidine. A rabbit model of 2',3'-dideoxycytidine neurotoxicity. Lab. Invest. 66 (1991) 63–74.

ANONYMOUS: Adverse reactions of dapsone. Lancet ii (1981) 184–185.

ARCIERI, G., R. AUGUST, N. BECKER, C. BOYLE and E. GRIFFITH: Clinical experience with ciprofloxacin in the USA. Eur. J. Clin. Microbiol. 5 (1986) 220–225.

AREZZO, J., H.H. SCHAUMBURG, C.E. SCHROEDER, M.S. LITWAK and A. DAVIDOVICH: Dideoxycytidine (ddC) neuropathy: an animal model in the cynomolgous monkeys. Neurology 40, Suppl. 1 (1990) 428–429.

ARNAUDO, E., M. DALAKAS, S. SHANSKE, C.T. MORAES, S. DI MAURO and E.A. SCHON: Depletion of muscle mitochondrial DNA in AIDS patients with zidovudine-induced myopathy. Lancet 337 (1991) 508–510.

ARONSON, I.K., R. YU, D.P. WEST, H. VAN DEN BROEK and J. ANTEL: Thalidomide-induced peripheral neuropathy: effect of serum factor on nerve cultures. Arch. Dermatol. 120 (1984) 1466–1470.

BACELLAR, H., A. MUNOZ, E.N. MILLER, B.A. COHEN, D. BESLEY, O.A. SELNES, J.T. BECKER and J.C. MCARTHUR: Temporal trends in the incidence of HIV-1-related neurologic diseases: Multicenter AIDS Cohort Study 1985–1992. Neurology 44 (1994) 1892–1900.

BAKER, W.J., G.L. ROYER and R.B. WEISS: Cytarabine and neurologic toxicity. J. Clin. Oncol. 9 (1991) 679–693.

BARTON, T.L., M.K. ROUSH and L.L. DEVER: Seizures associated with ganciclovir therapy. Pharmacotherapy 12 (1992) 413–415.

BATES, S.E., M.I. RAPHAELSON and R.A. PRICE: Ascending myelopathy after chemotherapy for central nervous system acute lymphoblastic leukemia: correlation with cerebrospinal fluid myelin basic protein. Med. Pediatr. Oncol. 13 (1985) 4–8.

BELEC, L., B. DI COSTANZO, A.J. GEORGES and R. GHERARDI: HIV infection in African patients with tropical pyomyositis. AIDS 5 (1991) 234.

BENDER, R.A. and B.A. CHABNER: Tubulin binding agents. In: B.A. Chabner (Ed.), Pharmacologic Principles of Cancer Treatment. Philadelphia, W.B. Saunders (1982) 256–268.

BENSON, C.: Disseminated *Mycobacterium avium* complex disease in patients with AIDS. AIDS Res. Hum. Retrovirus. 10 (1994) 913–916.

BENTLER, B. and A. CERAMI: Cachectin: more than a tumor necrosis factor. N. Engl. J. Med. 316 (1987) 479–485.

BERAL, V., H. JAFFE and R. WEISS: Cancer surveys: cancer HIV and AIDS. Eur. J. Cancer 27 (1991) 1057–1058.

BERGER, A.R. and D.M. SIMPSON: Neurological complications of AIDS. In: W.M. Scheld, R.J. Whitley and D.T. Durack (Eds.), Infections of the Nervous System, 2nd Edit. Philadelphia, Lippincott-Raven (1997) 255–271.

BERGER, A.R., H.H. SCHAUMBURG, C. SCHROEDER, S. APPEL and R. REYNOLDS: Dose response, coasting, and differential fiber vulnerability in human toxic neuropathy: a prospective study of pyridoxine neurotoxicity. Neurology 42 (1992) 1367–1370.

BERGER, A.R., J.C. AREZZO, H.H. SCHAUMBURG, G. SKOWRON, T. MERIGAN, S. BOZZETTE, D. RICHMAN and W. SOO: 2',3'-Dideoxycytidine (ddC) toxic neuropathy: a study of 52 patients. Neurology 43 (1993) 358–362.

BERNINI, J.C., D.W. FORT, J.C. GRIENER, B.J. KANE, W.B. CHAPPELL and B.A KAMEN: Aminophylline for methotrexate-induced neurotoxicity. Lancet 345 (1995) 544–547.

BESSEN, L.J., J.B. GREENE, E. LOUIE, P. SEITZMAN and H. WEIMBERG: Severe polymyositis-like syndrome associated with zidovudine therapy of AIDS and ARC. N. Engl. J. Med. 318 (1988) 708.

BHATIA, M.S.: Isoniazid-induced obsessive compulsive neurosis. J. Clin. Psychiatry 51 (1990) 387.

BIEHL, J.P. and H.J. NIMITZ: Studies on the use of a high dose of isoniazid. Am. Rev. Tuberc. 70 (1954) 430–444.

BIEHL, J.P. and J.H. SKAVLEM: Toxicity of isoniazid. Am. Rev. Tuberc. 70 (1953) 64.

BIEHL, J.P. and R.W. VILTER: Effect of isoniazid on vitamin B12 metabolism, its possible significance in producing isoniazid neuritis. Proc. Soc. Exp. Biol. Med. 85 (1954) 389–395.

BLEYER, W.A.: Neurologic sequelae of methotrexate and ionizing radiation: a new classification. Cancer Treatm. Rep. 65, Suppl. 1 (1981) 89–98.

BLUM, A., G. DAL PAN, C. RAINES, K. MAYJO and J. MCARTHUR: ddC-related toxic neuropathy: risk factors and natural history (Abstract). Neurology 43, Suppl. 2 (1993) A190–A191.

BOHAN, A. and J.B. PETERS: Polymyositis and dermatomyositis, Part I. N. Engl. J. Med. 292 (1975) 344–347.

BOLWELL, B.J., P.A. CASSILETH and R.P. GALE: High dose cytarabine: a review. Leukemia 2 (1988) 253–260.

BORUCKI, M.J., D.S. MATZKE and R.B. POLLARD: Tremor induced by trimethoprim-sulfamethoxazole in patients with the acquired immunodeficiency syndrome (AIDS). Ann. Intern. Med. 109 (1988) 77–78.

BOZZETTE, S.A., J. SANTANGELO, D. VILLASANA, A. FRASER, B. WRIGHT, C. JACOBSEN, E. HAYDEN, J. SCHNACK, S. SPECTOR and D. RICHMAN: Peripheral nerve function in persons with asymptomatic or minimally symptomatic HIV disease: absence of zidovudine neurotoxicity. J. Acquir. Immune Defic. Syndr. 4 (1991) 851–855.

BREDEMANN, J.A., S.W. KRECHEL and G.W. EGGERS: Treatment of refractory seizures in massive isoniazid overdose. Anesth. Analg. 71 (1990) 554–557.

BRIDGEN, D. and P. WHITEMAN: The mechanism of action, pharmacokinetics, and toxicity of acyclovir – a review. J. Infect. 6, Suppl. (1983) 3–9.

BROWNE, M.J., K.H. MAYER and S.B. CHAFEE: 2′,3′-Didehydro-3′-deoxythymidine (d4T) in patients with AIDS or AIDS-related complex: a phase I trial. J. Infect. Dis. 167 (1993) 21–29.

BURCH, P.A., S.A. GROSSMAN and C.S. REINHARD: Spinal cord penetration of intrathecally administered cytarabine and methotrexate: a quantitative autoradiographic study. J. Natl. Cancer Inst. 80 (1988) 1211–1216.

CAMPOLI-RICHARDS, D.M., J.P. MONK, A. PRICE, P. BENFIELD, P.A. TODD and A. WARD: Ciprofloxacin: a review of its antibacterial activity, pharmacokinetic properties and therapeutic use. Drugs 35 (1988) 373–447.

CANTWELL, M.F., D.E. SNIDER, G.M. CAUTHEN and I.M. ONORATO: Epidemiology of tuberculosis in the United States, 1985 through 1992. J. Am. Med. Assoc. 272 (1994) 535–539.

CARR, R.E. and P. HENKIND: Ocular manifestations of ethambutol. Arch. Ophthalmol. 67 (1962) 566–571.

CASEY, E.B., A.M. JELLIFE, P.M. LEQUESNE and Y.L. MILLET: Vincristine neuropathy: clinical and electrophysiological observations. Brain 96 (1973) 69–86.

CASH, J.M. and J.H. KLIPPEL: Second line drug therapy for rheumatoid arthritis. N. Engl. J. Med. 330 (1994) 1368–1375.

CHAD, D.A., T.W. SMITH, D.A. BLUMENFELD, P.G. FAIRCHILD and U. DEGIROLAMI: HIV-associated myopathy: immunocytochemical identification of an HIV antigen (gp 41) in muscle macrophages. Ann. Neurol. 28 (1990) 579–582.

CHAISSON, R.E., G.F. SCHECTER, C.P. THEUER, G.W. RUTHEFORD, D.F. ECHENBERG and P.C. HOPEWELL: Tuberculosis in patients with the acquired immunodeficiency syndrome. Am. Rev. Resp. Dis. 136 (1987) 570–574.

CHALMERS, A.C., C.M. GRECO and R.G. MILLER: Prognosis in AZT myopathy. Neurology 41 (1991) 1181–1184.

CHAMBERLAIN, M.C. and L. DIRR: Involved field radiotherapy and intraommaya methotrexate/ara-C in patients with AIDS-related lymphomatous meningitis. J. Clin. Oncol. 11 (1993) 1978–1984.

CHAPON, F., B. LECHEVALIER, D.C. DA SILVA, Y. RIVRAIN, B. DUPUY and P. DESCHAMPS: Neuropathies à la thalidomide. Rev. Neurol. 141 (1985) 719–728.

CHARIOT, P., I. MONNET and R. GHERARDI: Cytochrome c reaction improves histopathological assessment of zidovudine myopathy. Ann. Neurol. 34 (1993) 561–565.

CHARIOT, P., E. RUET, F.J. AUTHIER, Y. LEVY and R. GHERARDI: Acute rhabdomyolysis in patients infected by human immunodeficiency virus. Neurology 44 (1994) 1692–1696.

CHATURVEDI, S.K. and M. UPADHYAYA: Secondary mania in a patient receiving isonicotinic acid hydrazide and pyridoxine: case report. Can. J. Psychiatry 33 (1988) 675–676.

CHEN, C.H. and Y.C. CHEN: Delayed cytotoxicity and selective loss of mitochondrial DNA in cells treated with the anti-human immunodeficiency virus compound 2′,3′-dideoxycytidine. J. Biol. Chem. 264 (1989) 11934–11937.

CHEN, J.L., J.M. BROCAVICH and A.Y.F. LIN: Psychiatric disturbances associated with ganciclovir therapy. Ann. Pharmacother. 26 (1992) 193–195.

CHUPP, G.L., J. ALROY, L.S. ADELMAN, J.C. BREEN and P.R. SKOLNIK: Myositis due to Pleistosphora (Microsporidia) in a patient with AIDS. Clin. Infect. Dis. 16 (1993) 15–21.

CITRON, K.M. and G.O. THOMAS: Ocular toxicity from ethambutol. Thorax 41 (1986) 737–739.

CLEMMENSEN, O.J., P. ZANDER OLSEN and K.L. ANDERSEN: Thalidomide neurotoxicity. Arch. Dermatol. 120 (1984) 338–341.

COHEN, S.M.Z., J.A. MINKOVE and J.W. ZEBLEY: Severe but reversible neurotoxicity from acyclovir. Ann. Intern. Med. 100 (1984) 920.

COLLABORATIVE DHPG TREATMENT STUDY GROUP: Treatment of serious cytomegalovirus infections with 9-(1,3-dihydroxy-2-propoxymethyl)guanine in patients with AIDS and other immunodeficiencies. N. Engl. J. Med. 314 (1986) 801–805.

CONSENSUS STATEMENT: On the use of corticosteroids as adjunctive therapy for pneumocystis pneumonia in the acquired immunodeficiency syndrome. N. Engl. J. Med. 323 (1990) 1500–1504.

COOLEY, T.P., L.M. KUNCHES, C.A. SAUNDERS, J.K. RITTER, C.J. PERKINS, C. MCLAREN, R.P. MCCAFFREY and H.A. LIEBMAN: Once-daily administration of 2',3'-dideoxyinosine (ddI) in patients with the acquired immunodeficiency syndrome or AIDS-related complex. N. Engl. J. Med. 322 (1990) 1340–1345.

CRONSTEIN, B.N., D. NAIME and E. OSTAD: The anti-inflammatory mechanism of methotrexate: increased adenosine release at inflamed sites diminishes leucocytes accumulation in an in vivo model of inflammation. J. Clin. Invest. 92 (1993) 2675–2682.

CRUMPACKER, C.S.: Mechanism of action of foscarnet against viral polymerases. Am. J. Med. 92, Suppl. 2A (1992) 3S–7S.

CUPLER, E.J., K. HENCH, C.A. JAY and M. DALAKAS: The natural history of zidovudine (AZT)-induced mitochondrial myopathy (ZIMM) (Abstract). Neurology 44 (1994) A132.

DALAKAS, M. and G.H. PEZESHKPOUR: Neuromuscular diseases associated with human immunodeficiency virus infection. Ann. Neurol. 23, Suppl. (1988) S38–S48.

DALAKAS, M., G.H. PEZESHKPOUR, M. GRAVELL and J.L. SEVER: Polymyositis associated with AIDS retrovirus. J. Am. Med. Assoc. 256 (1986) 2381–2383.

DALAKAS, M., I. ILLA, G.H. PEZESHKPOUR, J.P. LAUKAITIS, B. COHEN and J.L. GRIFFIN: Mitochondrial myopathy caused by long-term zidovudine therapy. N. Engl. J. Med. 322 (1990) 1098–1105.

DALAKAS, M., M.E. LEON-MONZON, I. BERNARDINI, W.A. GAHL and C.A. JAY: Zidovudine-induced mitochondrial myopathy is associated with muscle carnitine deficiency and lipid storage. Ann. Neurol. 35 (1994a) 482–487.

DALAKAS, M., K. KIEBURTZ, R. GHERARDI, P. CHARIOT, J.M. GRAU, J. CASADEMONT and D.M. SIMPSON: HIV or zidovudine myopathy? (Correspondence). Neurology 44 (1994b) 360–364.

DAVENPORT, A., S. GOEL and J.C. MACKENZIE: Neurotoxicity of acyclovir in patients with end-stage renal failure treated with continuous ambulatory peritoneal dialysis. Am. J. Kidney Dis. 20 (1992) 647–649.

DEARMOND, B.: Safety considerations in the use of ganciclovir in immunocompromised patients. Transplant. Proc. 23, Suppl. 1 (1991) 26–29.

DERAY, G., F. MARTINEZ and C. KATLAMA: Foscarnet nephrotoxicity: mechanism, incidence and prevention. Am. J. Nephrol. 9 (1989) 316–321.

DEVERE, R. and W.G. BRADLEY: Polymyositis: presentation, morbidity and mortality. Brain 98 (1975) 637–666.

DEVINSKY, O., W. LEMANN, A.C. EVANS, J.R. MOELLER and D.A. ROTTENBERG: Akinetic mutism in a bone marrow transplant recipient following total-body irradiation and amphotericin B chemoprophylaxis. Arch. Neurol. 44 (1987) 414–417.

DIETRICH, D.T., D.P. KOTLER and D.F. BUSCH: Ganciclovir treatment of cytomegalovirus colitis in AIDS: a randomized, double-blind, placebo-controlled multicenter study. J. Infect. Dis. 167 (1993) 278–282.

DORSKY, D.I. and M.D. CRUMPACKER: Drugs five years later: acyclovir. Ann. Intern. Med. 107 (1987) 859–874.

DREW, W.L.: Cytomegalovirus infection in patients with AIDS. Clin. Infect. Dis. 14 (1992) 608–615.

DROPCHO, E.J.: Central nervous system injury by therapeutic irradiation. Neurol. Clin. 9 (1991) 969–988.

DUBINSKY, R.M., R. YARCHOAN, M. DALAKAS and S. BRODER: Reversible axonal neuropathy from the treatment of AIDS and related disorders with 2',3'-dideoxycytidine (ddC). Muscle Nerve 12 (1989) 856–860.

DUVIVIER, A. and T. FOWLER: Possible dapsone-induced peripheral neuropathy in dermatitis herpetiformis. Proc. Roy. Soc. Med. 67 (1974) 439–440.

DYNES, J.B. and M.I. SMEDAL: Radiation myelitis. Am. J. Roentgenol. 83 (1960) 78.

ECK, P., S.M. SILVER and E.C. CLARK: Acute renal failure and coma after a high dose of oral acyclovir. N. Engl. J. Med. 325 (1991) 1178.

ELLIS, W.G., R.A. SOBEL and S.L. NIELSEN: Leuko-encephalopathy in patients treated with amphotericin B methyl ester. J. Infect. Dis. 146 (1982) 125–137.

EPSTEIN, F.W. and M. BOHM: Dapsone-induced peripheral neuropathy. Arch. Dermatol. 112 (1976) 1761–1762.

ESPINOZA, L.R., J.L. AGUILAR, C.G. ESPINOZA, J. GRESH, J. JARA, L.H. SILVEIRA and M. SELEZNICK: Characteristics and pathogenesis of myositis in human immunodeficiency virus infection. distinction from azidothymidine-induced myopathy. Rheumatol. Dis. Clin. 17 (1991) 117–129.

FAULDS, D. and R.C. HEEL: Ganciclovir: a review of its antiviral activity, pharmacokinetic properties and therapeutic efficacy in cytomegalovirus infections. Drugs 39 (1990) 597–638.

FELDMAN, S., J. RODMAN and B. GREGORY: Excessive serum concentrations of acyclovir and neurotoxicity. J. Infect. Dis. 17 (1988) 385–388.

FIGG, W.D.: Peripheral neuropathy in HIV patient after isoniazid therapy initiated. DICP 25 (1991) 100–101.

FISCHL, M.A., G.M. DICKENSON and L. LA VOIE: Safety and efficacy of sulfamethoxazole and trimethoprim chemoprophylaxis for *Pneumocystis carinii* pneumonia in AIDS. J. Am. Med. Assoc. 259 (1988) 1185–1189.

FISCHL, M.A., R.M. OLSON and S.E. FOLLANSBEE: Zalcitabine compared with zidovudine in patients with advanced HIV-1 infection who received previous zidovudine therapy. Ann. Intern. Med. 118 (1993) 762–769.

FLORENCE, A.L.: Is thalidomide to blame? Br. Med. J. 2 (1960) 1954.

FORSYTH, P.A., J. YAHALOM and L.M. DEANGELIS: Combined-modality therapy in the treatment of primary central nervous system lymphoma in AIDS. Neurology 44 (1994) 1473–1479.

FRAM, R.J. and D.W. KUFE: Effects of 1-β-D-arabinofuranosylcytosine and hydroxyurea on the repair of x-ray induced DNA single-strand breaks in human leukemia blasts. Biochem. Pharmacol. 34 (1985) 2557–2560.

FREDERICKS, E.J., R. KUGELMAN and N. KIRSCH: Dapsone-induced motor polyneuropathy. Arch. Dermatol. 112 (1976) 1158–1160.

FRISCH, J.M.: Clinical experience with adverse reactions to trimethoprim-sulfamethoxazole. J. Infect. Dis. 128, Suppl. (1973) S607–S612.

FULLER, G.N., J.M. JACOBS and R.J. GUILOFF: Thalidomide, peripheral neuropathy and AIDS. Int. J. Sex. Transm. Dis. AIDS 2 (1991) 369–370.

FULLERTON, P.M. and M. KREMER: Neuropathy after intake of thalidomide. Br. Med. J. 2 (1961) 855–858.

FULLERTON, P.M. and D.J. O'SULLIVAN: Thalidomide neuropathy: a clinical, electrophysiological, and histological follow-up study. J. Neurol. Neurosurg. Psychiatry 31 (1968) 543–551.

GAGLIANO, G. and J. CONSTANZI: Paraplegia following intrathecal methotrexate: a report of case and review of the literature. Cancer 37 (1976) 1663–1668.

GALLANT, J.E., R.D. MOORE and R.E. CHAISSON: Prophylaxis for opportunistic infections in patients with HIV infection. Ann. Intern. Med. 120 (1994) 932–944.

GAMMON, G.D., F.W. BURGE and G. KING: Neural toxicity in tuberculosis patients treated with isoniazid (isonicotinic acid hydrazide). Arch. Neurol. Psychiatry 70 (1953) 619.

GEARHART, M.O. and T.B. SORG: Foscarnet-induced severe hypomagnesemia and other electrolyte disorders. Ann. Pharmacother. 27 (1993) 285–289.

GERTNER, E., J.R. THURN, D.N. WILLIAMS, M. SIMPSON, H.H. BALFOUR, F. RHAME and K. HENRY: Zidovudine-associated myopathy. Am. J. Med. 6 (1989) 814–818.

GHERARDI, R., M. BAUDRIMONT and F. LIONNET: Skeletal muscle toxoplasmosis in patients with acquired immunodeficiency syndrome: a clinical and pathological study. Ann. Neurol. 32 (1992) 535–542.

GHERARDI, R., A. FLOREA-STRAT, G. FROMONT, F. PORON, J.-C. SABOURIN and J. AUTHIER: Cytokine expression in the muscle of HIV-infected patients: evidence for interleukin-1-α accumulation in mitochondria of AZT fibers. Ann. Neurol. 36 (1994) 752–758.

GILL, P., M. RARICK and M. BERNSTEIN-SINGER: Treatment of advanced Kaposi's sarcoma using a combination of bleomycin and vincristine. Am. J. Clin. Oncol. 13 (1990) 315–319.

GIRLING, D.J.: Adverse effects of antituberculosis drugs. Drugs 23 (1982) 56–74.

GODWIN-AUSTEN, R.B., D.A. HOWELL and B. WORTHINGTON: Observations on radiation myelopathy. Brain 98 (1975) 557.

GOLDWEIN, J.W.: Radiation myelopathy: a review. Med. Pediatr. Oncol. 15 (1987) 89.

GONZALES, M.F., R.K. OLNEY, Y.T. SO, C.M. GRECO and B. MCQUINN: Subacute structural myopathy associated with human immunodeficiency virus infection. Arch. Neurol. 45 (1988) 585–587.

GORARD, D., K. HENRY and R.J. GUILOFF: Necrotizing myopathy and zidovudine. Lancet i (1988) 1050–1051.

GORIN, I., B. VILETTE, P. GEHANNO and J.P. ESCANDE: Thalidomide in hyperalgic pharyngeal ulceration of AIDS [letter]. Lancet 335 (1990) 1343

GRAU, J., F. MASANES, E. PEDRO and J. CASADEMONT: Human immunodeficiency virus type 1 infection and myopathy: clinical relevance of zidovudine therapy. Ann. Neurol. 34 (1993) 206–211.

GRIFFIN, W., J.S. RASEY and W.A. BLEYER: The effect of photon irradiation on blood–brain barrier permeability to methotrexate in mice. Cancer 40 (1977) 1109–1111.

GUNZLER, V.: Thalidomide in a human immunodeficiency virus (HIV) patients: a review of safety considerations. Drug Saf. 7 (1992) 116–134.

GUTTMAN, L., J.D. MARTIN and W. WELTON: Dapsone motor neuropathy – an axonal disease. Neurology 26 (1976) 514–516.

HALL, C., U. DAFNI, D.M. SIMPSON and AIDS CLINICAL TRIALS GROUP: Failure of cytarabine in progressive multifocal leukoencephalopathy associated with human immunodeficiency virus infection. N. Engl. J. Med. 338 (1998) 1345–1351,

HARDY, W.D., J. FEINBERG and D.M. FINKELSTEIN: A controlled trial of trimethoprim-sulfamethoxazole or aerosolized pentamidine for secondary prophylaxis of *Pneumocystis carinii* pneumonia in patients with acquired immunodeficiency syndrome (AIDS clinical trial group protocol 021). N. Engl. J. Med. 327 (1992) 1842–1848.

HAWKEN, M., P. NUNN and S. GATHUA: Increased recurrence of tuberculosis in HIV-1 infected patients in Kenya. Lancet 342 (1993) 332–337.

HERZBERG, N.H., I. ZORN, R. ZWART, P. PORTEGIES and P. BOLHUIS: Major growth reduction and minor decrease in mitochondrial enzyme activity in culture human muscle cells after exposure to zidovudine. Muscle Nerve 15 (1992) 706–710.

HERZIG, R.H., J.D. HINES and G.P. HERZIG: Cerebellar toxicity with high-dose cytosine arabinoside. J. Clin. Oncol. 5 (1987) 927–932.

HESS, C.W., T. HUNZIKER, A. KÜPFER and H.P. LUDIN: Thalidomide-induced peripheral neuropathy: a prospective clinical, neurophysiological and pharmacogenetic evaluation. J. Neurol. 233 (1986) 83–89.

HO, D.H.W.: Distribution of kinase and deaminase of 1-β-D-arabinofuranosylcytosine in tissues of man and mouse. Cancer Res. 33 (1973) 2816–2820.

HO, H.W., R. BAYLEY, J.M. RHEE and H.V. VINTERS: Neuromuscular pathology in patients with acquired immune deficiency syndrome (AIDS): an autopsy study. J. Neuropathol. Exp. Neurol. 48 (1989) 382.

HOEPRICH, P.D.: Amphotericin B methyl ester and leukoencephalopathy: the other side of the coin. J. Infect. Dis. 146 (1982) 173–176.

HOLDINESS, M.R.: Neurological manifestations and toxicities of the antituberculosis drugs. Med. Toxicol. 2 (1987) 33–51.

HOLLAND, G.N., W.C. BUHLES, B. MASTRE and H.J. KAPLAN: a controlled retrospective study of ganciclovir treatment of cytomegalovirus retinopathy: use of a standardized system for the assessment of disease outcome. Arch. Ophthalmol. 107 (1989) 1759–1766.

HOLLAND, J.F., C. SCHARLAV and S. GAILANI: Vincristine treatment of advanced cancer: a cooperative study of 392 cases. Cancer Res. 33 (1973) 1258–1264.

HOUI, K., S. MOCHIO and T. KOBAYASHI: Gangliosides attenuate vincristine neurotoxicity on dorsal root ganglion cells. Muscle Nerve 16 (1993) 11–14.

HURWITZ, R.L., D.H. MAHONEY, D.L. ARMSTRONG and T.M. BROWDER: Reversible encephalopathy and seizures as a result of conventional vincristine administration. Med. Pediatr. Oncol. 16 (1988) 216–229.

ILLA, I., A. NATH and M. DALAKAS: Immunocytochemical and virological characteristics of HIV-associated inflammatory myopathies: similarities with seronegative patients. Ann. Neurol. 29 (1991) 474–481.

JABS, D.A., C. ENGER and J.G. BARTLETT: Cytomegalovirus retinitis and acquired immunodeficiency syndrome. Arch. Ophthalmol. 107 (1989) 75–80.

JACKSON, S.L.O.: Psychosis due to isoniazid. Br. Med. J. 2 (1957) 743–746.

JACOBSON, J., J. GREENSPAN, J. SPRITZER, D. SIMPSON and

AIDS CLINICAL TRIALS GROUP: Thalidomide for the treatment of oral aphthous ulcers in patients with HIV infection. N. Engl. J. Med. 336 (1997) 1487–1493.

JELLINGER, K. and K. STURM: Delayed radiation myelopathy in man: report of twelve necropsy cases. J. Neurol. Sci. 14 (1971) 389.

JIMENEZ-LUCHO, V.E., R. DEL BUSTO and J. ODEL: Isoniazid and ethambutol as a cause of optic neuropathy. Eur. J. Resp. Dis. 71 (1987) 42–45.

JONES, A.: Transient radiation myelopathy. Br. J. Radiol. 37 (1964) 727.

JONES, W.A. and G.P. JONES: Peripheral neuropathy due to isoniazid. Lancet i (1953) 1073.

JONSSON, O. and B.A. KAMEN: Methotrexate and leukemia. Cancer Invest. 9 (1991) 53–60.

JUNG, A.C. and D.S. PAAUW: Management of adverse reactions to trimethoprin-sulfamexazole in human immunodeficiency virus-infected patients. Arch. Intern. Med. 154 (1994) 2402–2406.

KAHN, J.O., S.W. LAGAKOS and D.D. RICHMAN: A controlled trial comparing continued zidovudine with didanosine in human immunodeficiency virus infection. N. Engl. J. Med. 327 (1992) 581–587.

KARMON, G., H. SAVIR and D. ZEVIN: Bilateral optic neuropathy due to combined ethambutol and isoniazid treatment. Ann. Ophthalmol. 11 (1989) 1013–1017.

KEAY, S., J. BISSETT and T.C. MERIGAN: Ganciclovir treatment of cytomegalovirus infections in iatrogenically immunocompromised patients. J. Infect. Dis. 156 (1987) 1016–1021.

KEILBAUGH, S.A., J.A. MOSCHELLA and C.D. CHIN: Role of mtDNA replication in the toxicity of 3'-azido-3'-deoxythymidine (AZT) in AIDS therapy and studies on other anti-HIV-1 dideoxynucleotides. In: E. Quagliarello, S. Papa, F. Palmieri and C. Saccone (Eds.), Structure, Function and Biogenesis of Energy-Transfer Systems. New York, Elsevier Science (1990) 159–162.

KIBURG, B., A.A. VAN DE LOOSDRECHT and K.M. SCHWEITZER: Effects of ACTH(4–9) analogue, ORG 2766, on vincristine cytotoxicity in two human lymphoma cell lines, U937 and U715. Br. J. Cancer 69 (1994) 497–501.

KIEBURTZ, K.D., M. SEIDLIN and J.S. LAMBERT: Extended follow-up of peripheral neuropathy in patients with AIDS and AIDS-related complex treated with dideoxyinosine. J. AIDS 5 (1992) 60–64.

KIEBURTZ, K.D., C. YIANNOUTSOS, D. SIMPSON and AIDS CLINICAL TRIALS GROUP: A double blind, randomized clinical trial of amitriptyline and mexiletine for painful neuropathy in human immunodeficiency virus infection (Abstract). Ann. Neurol. 42 (1997) 429.

KIM, Y.S. and H. HOLLANDER: Polyradiculopathy due to

cytomegalovirus: report of two cases in which improvement occurred after prolonged therapy and review of the literature. Clin. Infect. Dis. 17 (1993) 32–37.

KNOP, J., G. BONSMANN and A. HAPPLER: Thalidomide in the treatment of sixty cases of chronic discoid lupus erythematosus. Br. J. Dermatol. 108 (1983) 461–466.

KOLLER, W.C., L.K. GEHLMANN, F.D. MALKINSON and F.A. DAVIS: Dapsone-induced peripheral neuropathy. Arch. Neurol. 34 (1977) 644–646.

KRIEBLE, B.F., D.W. RUDY, R.G. GLICK and M.D. CLAYMAN: Case report: acyclovir neurotoxicity and nephrotoxicity – the role for hemodialysis. Am. J. Med. Sci. 305 (1993) 36–39.

KUNCL, R.W. and E.B. GEORGE: Toxic neuropathies and myopathies. Curr. Opin. Neurol. 6 (1993) 695–704.

KYLE, D.W., D. MONROE and D. HENRICK: Effects of 1-β-D-arabinofuranosylcytosine incorporation on eukaryocytic DNA template function. Mol. Pharmacol. 26 (1989) 128.

LAGUENY, A., A. ROMMEL and B. VIGNOLLY: Thalidomide neuropathy: an electrophysiological study. Muscle Nerve 9 (1986) 837–844.

LAHDEVIRTA, J., C. MAURY, A.-M. TEPPO and H. REPO: Elevated levels of circulating cachectin-tumor necrosis factor in patients with AIDS. Am. J. Med. 85 (1988) 289–291.

LAMBERT, J.S., M. SEIDLIN, R.C. REICHMAN, C. PLANK, M. LAVERTY, G. MORSE, C. KNUPP and R. DOLIN: 2',3'-Dideoxyinosine (ddI) in patients with the acquired immunodeficiency syndrome or AIDS-related complex: results of a phase 1 trial. N. Engl. J. Med. 322 (1990) 1333–1340.

LAMPERTH, L., M.C. DALAKAS, F. DAGANI, J. ANDERSON and R. FERRARI: Abnormal skeletal and cardiac muscle mitochondria induced by zidovudine (AZT) in human muscle in vitro and in an animal model. Lab. Invest. 65 (1991) 742–751.

LANGE, D.J., C.B. BRITTON and D.S. YOUNGER: The neuromuscular manifestations of human immunodeficiency virus infections. Arch. Neurol. 45 (1988) 1084–1088.

LAWRENCE, R.M., P.D. HOEPRICH, F.A. JAGDIS, N. MONJI, A.C. HUSTON and C.P. SCHAFFNER: Distribution of doubly radiolabeled amphotericin B methyl ester and amphotericin B in the non-human primate Macaca mulatta. J. Antimicrob. Chemother. 6 (1980) 241–249.

LAZARUS, H.M., R.H. HERZIG and G.P. HERZIG: Central nervous system toxicity of high-dose systemic cytosine arabinoside. Cancer 48 (1981) 2577–2582.

LEE, B.L. and S. SAFRIN: Interactions and toxicities of drugs used in patients wit AIDS. Clin. Infect. Dis. 14 (1992) 773–779.

LEGHA, S.S.: Vincristine neurotoxicity: patho-physiology and management. Med. Toxicol. 1 (1986) 421–427.

LEIBENWITZ, H., L. KREVOLIN and A. SCHWARTZ: Hemoperfusion treatment of massive isoniazid overdose. Del. Med. J. 61 (1989) 71–73.

LIPMAN, J.M., J.A. REICHERT, A. DAVIDOVICH and T.D. ANDERSON: Species differences in nucleotide pool levels of 2',3'-dideoxycytidine: a possible explanation for species-specific toxicity. Toxicol. Appl. Pharmacol. 123 (1993) 137–143.

MACDIARMAID-GORDON, A.R., M. O'CONNOR and M. BEAMAN: Neurotoxicity associated with oral acyclovir in patients undergoing dialysis. Nephron 62 (1992) 280–283.

MACDONALD, D.R.: Neurologic complications of chemotherapy. Neurol. Clin. 9 (1991) 955–967.

MANJI, H., A. MALIN and S. CONNOLLY: CMV polyradiculopathy in AIDS: suggestions for new strategies in treatment [letter]. Genitourin. Med. 68 (1992) 192.

MANJI, H., M.J.G. HARRISON, J.M. ROUND, D.A. JONES, S. CONNOLLY, C.J. FOWLER, I. WILLIAMS and I.V.D. WELLER: Muscle disease, HIV and zidovudine: the spectrum of muscle disease in HIV-infected individuals treated with zidovudine. J. Neurol. 240 (1993) 479–488.

MAR, E.-C., Y.-C. CHENG and E.-S. HUANG: Effect of 9-(1,3-dihydroxy-2-propoxymethyl)guanine on human cytomegalovirus replication in vitro. Antimicrob. Agents Chemother. 24 (1983) 518–521.

MARCUS, R. and M.L. COULSTON: Water-soluble vitamins: the vitamin B complex and ascorbic acid. In: A.G. Goodman, T.W. Rall, A.S. Nies and P. Taylor (Eds.), Goodman and Gilman's Pharmacological Basis of Therapeutics, 8th Edit. New York, Pergamon Press (1990) 1530–1532.

MARKHAM, A. and D. FAULDS: Ganciclovir: an update of its therapeutic use in cytomegalovirus infection. Drugs 48 (1994) 455–484.

MARKS, J.E. and J. WOONG: The risk of cerebral radionecrosis in relation to dose, time and fractionation: a follow-up study. Prog. Exp. Tumor Res. 29 (1985) 210.

MASTAGLIA, F.L. and V.J. OJEDA: Inflammatory myopathies: Part 2. Ann. Neurol. 17 (1985) 317–323.

MAX, M.B., M. CULNANE and S.C. SCHAFER: Amitriptyline relieves diabetic neuropathy pain in patients with normal or depressed mood. Neurology 37 (1987) 589–596.

MAX, M.B., R. KINSHORE-KUMAR and S.C. SCHAFER: Efficacy of desipramine in painful diabetic neuropathy: a placebo-controlled trial. Pain 45 (1991) 3–9.

MCCUNE, R., K. DEUSCHLE and W. MCDERMOTT: Delayed appearance of isoniazid antagonism by pyridoxine in vivo. Am. Rev. Tuberc. 76 (1957) 1100.

MEDINA, I., J. MILLS and G. LEOUNG: Oral therapy for *Pneumocystis carinii* pneumonia in the acquired immunodeficiency syndrome: a controlled trial of trimethoprim-sulfamethoxazole versus trimethoprim-dapsone. N. Engl. J. Med. 323 (1990) 776–782.

MENG, T.C., M.A. FISCHL, A.M. BOOTA, S.A. SPECTOR, D. BENNETT and D.D. RICHMAN: Combination therapy with zidovudine and dideoxycytidine in patients with advanced human immunodeficiency virus infection. A phase I/II study. Ann. Intern. Med. 116 (1992) 13–20.

MERIGAN, T.C., G. SKOWRON, S.A. BOZZETTE, D.D. RICHMAN, R. UTTAMCHANDANI, M. FISCHL and R. SCOOLEY: Circulating p24 antigen levels and responses to dideoxycytidine in human immunodeficiency virus (HIV) infections. Ann. Intern. Med. 110 (1989) 189–194.

MHIRI, C., M. BAUDRIMONT, G. BONNE, C. GENY and R. GHERARDI: Zidovudine myopathy: a distinctive disorder associated with mitochondrial dysfunction. Ann. Neurol. 29 (1991) 606–614.

MILLER, R.G., P.J. CARSON and R.S. MOUSSAVI: Fatigue and myalgia in AIDS patients. Neurology 41 (1991) 1603–1607.

MILLIKAN, L.E. and E.R. HARRELL: Drug reactions to the sulfones. Arch. Dermatol. 102 (1970) 220–224.

MONFARDINI, S., U. TIRELLI and E. VACCHER: Treatment of acquired immunodeficiency syndrome (AIDS)-related cancer. Cancer Treatm. Rev. 20 (1994) 149–172.

MORGAN-HUGHES, J.A.: The mitochondrial myopathies. In: A.G. Engel and B.Q. Banker (Eds.), Myology. New York, McGraw Hill (1986) 1709–1743.

MORGELLO, S., D. WOLFE, E. GODFREY, R. FEINSTEIN, M. TAGLIATI and D.M SIMPSON: Mitochondrial abnormalities in human immunodeficiency virus-associated myopathy. Acta Neuropathol. (Berl.) 90 (1995) 366–374.

MORRIS, D.J.: Adverse effects and drug interactions of clinical importance with antiviral drugs. Drug Saf. 10 (1994) 281–291.

NAND, S., H.L. MESSMORE, JR. and R. PATEL: Neurotoxicity associated with systemic high-dose cytosine arabinoside. J. Clin. Oncol. 4 (1986) 571–575.

NAVARRO, J.C., R.L. ROSALES, A.T. ORDINARIO, S. IZUMO and M. OSAME: Acute dapsone-induced peripheral neuropathy. Muscle Nerve 12 (1989) 604–606.

NISAR, M., S.W. WATKIN, R.C. BUCKNALL and R.A.L. AGNEW: Exacerbation of isoniazid induced peripheral neuropathy by pyridoxine. Thorax 45 (1990) 419–420.

OLSON, K.O., T.E. KEARNEY, J.E. DYER, N.L. BENOWITZ and P.D. BLANC: Seizures associated with poisoning and drug overdose. Am. J. Emerg. Med. 12 (1994) 392–395.

PALESTINE, A.G., M.A. POLIS and M.D. DE SMET: A randomized, controlled trial of foscarnet in the treatment of cytomegalovirus retinitis in patients with AIDS. Ann. Intern. Med. 115 (1991) 665–673.

PALLONE, K.A., M.P. GOLDMAN and M.A. FULLER: Isoniazid-associated psychosis: case report and review of the literature. Ann. Pharmacother. 27 (1993) 167–170.

PANEGYRES, P.K., M. TAN and B.A. KAKULAS: Necrotising myopathy and zidovudine. Lancet i (1988) 1050–1051.

PERTEL, P. and HIRSCHTICK: Adverse reactions to dapsone in persons infected with human immunodeficiency virus. Clin. Infect. Dis. 18 (1994) 630–632.

PETERS, B.S., J. WINER and D.N. LANDON: Mitochondrial myopathy associated with chronic zidovudine therapy in AIDS. Quart. J. Med. 86 (1993) 5–15.

PETERS, M., U. TIMM and D. SCHÜRMANN: Combined and alternating ganciclovir and foscarnet in acute and maintenance therapy of human immunodeficiency virus-related cytomegalovirus encephalitis refractory to ganciclovir alone. A case report and review of the literature. Clin. Invest. 70 (1992) 456–458.

PETERSLUND, N.: Herpes zoster associated encephalitis: clinical findings and acyclovir treatment. Scand. J. Infect. Dis. 20 (1988) 583–592.

PEZESHKPOUR, G.H., I. ILLA and M.C. DALAKAS: Ultrastructural characteristics and DNA immunochemistry in HIV and AZT-associated myopathies. Hum. Pathol. 22 (1991) 1281–1288.

PIKE, I.M. and C. NICAISE: The didanosine expanded access program: safety analysis. Clin. Infect. Dis. 16, Suppl. 1 (1993) S63–S68.

PIZZO, P.A., K. BUTLER and F. BALIS: Dideoxycytidine alone and in an alternating schedule with zidovudine in children with symptomatic human immunodeficiency virus infection. J. Pediatr. 117 (1990) 799–808.

POLKEY, M.I. and P.J. REES: Tuberculosis: current issues in diagnosis and treatment. Br. J. Clin. Pract. 48 (1994) 251–255.

PRICE, J.M., R.R. BROWN and F.C. LARSON: Quantitative studies on human urinary metabolites of tryptophan as affected by isoniazid and deoxypyridoxine. J. Clin. Invest. 36 (1957) 1600.

RABKIN, C.S. and W.A. BLATTNER: HIV infection and cancers other than non-Hodgkin lymphoma and Kaposi's sarcoma. Cancer Surv. 10 (1991) 151–160.

RADEFF, B., R. KUFFER and J. SAMSON: Recurrent aphthous ulcer in patient infected with human immunodeficiency virus: successful treatment with thalidomide. J. Am. Acad. Dermatol. 23 (1990) 523–525.

RAPOPORT, A.M. and S.B. GUSS: Dapsone-induced peripheral neuropathy. Arch. Neurol. 27 (1972) 184–185.

RASHIQ, S., L. BRIEWA and M. MOONEY: Distinguishing acyclovir neurotoxicity from encephalomyelitis. J. Intern. Med. 234 (1993) 507–511.

RATHBURN, R.C. and E.S. MARTIN, III: Didanosine therapy in patients intolerant of or failing zidovudine therapy. Ann. Pharmacother. 2 (1992) 1347–1351.

REYES, M.G., J. CASANOVA and F. VARRICCHIO: Zidovudine myopathy [letter]. Neurology 42 (1992) 1252.

RICHMAN, D.D., M.A. FISCHL, M.H. GRIECO, M.S. GOTTLIEB, P.A. VOLBERDING and O.L. LASKIN: The toxicity of azidothymidine (AZT) in the treatment of patients with AIDS and AIDS-related complex. N. Engl. J. Med. 317 (1987) 192–197.

ROGERS, O.L., M.K. DONOVAN, G.A. PETROFF, K.R. ROBERTSON and C.D. HALL: Electrophysiologic changes over time in the peripheral nerves of subjects infected with the human immunodeficiency virus. Muscle Nerve 15 (1992) 1176.

ROTTENBERG, D.A., N.L. CHERNIK and M.D. DECK: Cerebral necrosis following radiotherapy of extracranial neoplasms. Ann. Neurol. 1 (1977) 339.

ROZENCWEIG, M., C. MCLAREN, M. BELTANGADY, J. RITTER, R. CANETTA, L. SCHACTER, S. KELLEY, C. NICAISE, L. SMALDONE and L. DUNKLE: Overview of phase I trials of 2',3'-dideoxyinosine (ddI) conducted on adult patients. Rev. Infect. Dis. 12, Suppl. 5 (1990) S570–S575.

RUBIN, E.H., J.W. ANDERSEN, D.T. BERG, C.A. SCHIFFER, R.J. MAYER and R.M. STONE: Risk factors for high-dose cytarabine neurotoxicity: an analysis of a cancer and leukemia group B trial in patients with acute myeloid leukemia. J. Clin. Oncol. 10 (1992) 948–953.

RUSSO, P.A. and M.A. CHAGLASIAN: Toxic optic neuropathy associated with ethambutol: implications for current therapy. J. Am. Optom. Assoc. 65 (1994) 332–338.

SAHENK, Z., S.T. BRADY and J.R. MENDELL: Studies on the pathogenesis of vincristine-induced neuropathy. Muscle Nerve 10 (1987) 80–84.

SALTER, A.J.: Trimethoprim-sulfamethoxazole: an assessment of more than 12 years of use. Rev. Infect. Dis. 4 (1982) 196–236.

SAMPAIO, E.P., E.N. SARNO, R. GALILLY, Z.A. COHN and G. KAPLAN: Thalidomide selectively inhibits tumor necrosis factor, a production by simulated human monocytes. J. Exp. Med. 173 (1991) 699–703.

SANDLER, S.G., W. TOBIN and E.S. HENDERSON: Vincristine-induced neuropathy. Neurology 19 (1969) 367–374.

SAQUETON, A.C., A.L. LORINEZ, N.A. VICK and R.D. HAMER: Dapsone and peripheral motor neuropathy. Arch. Dermatol. 100 (1969) 214–217.

SCHAUMBURG, H., J. KAPLAN and A. WINDEBANK: Sensory neuropathy from pyridoxine abuse. N. Engl. J. Med. 309 (1983) 445–448.

SELWYN, P.A., D. HARTEL and V.A. LEWIS: A prospective study of the risk of tuberculosis among intravenous drug-users with human immunodeficiency virus infection. N. Engl. J. Med. 320 (1989) 545–550.

SHAW, P.J. and D. BATES: Conservative treatment of delayed cerebral radiation necrosis. J. Neurol. Neurosurg. Psychiatry 47 (1984) 1338.

SHELINE, G.E.: Therapeutic irradiation and brain injury. Int. J. Radiat. Oncol. Biol. Phys. 6 (1980) 1215.

SHELTON, M.J., A.M. O'DONNEL and G.D. MORSE: Didanosine. Ann. Pharmacother. 26 (1992) 660–669.

SIEVERS, M.L. and R.N. HERRIER: Treatment of active isoniazid toxicity. Am. J. Hosp. Pharm. 32 (1975) 202–206.

SIMPSON, D.M. and A.N. BENDER: Human immunodeficiency virus-associated myopathy: analysis of 11 patients. Ann. Neurol. 24 (1988) 79–84.

SIMPSON, D.M. and R.K. OLNEY: Peripheral neuropathies associated with human immunodeficiency virus infection. In: P.J. Dyck (Ed.), Peripheral Neuropathies: New Concepts and Treatments. Philadelphia, W.B. Saunders (1992) 685–711.

SIMPSON, D.M. and M. TAGLIATI: Neurological manifestations of HIV infection. Ann. Intern. Med. 121 (1994) 769–785.

SIMPSON, D.M. and M. TAGLIATI: Nucleoside analogue-associated peripheral neuropathy in human immunodeficiency virus infection. J. Acquir. Immune Defic. Syndr. 9 (1995) 153–161.

SIMPSON, D.M. and M. TAGLIATI: Neuromuscular syndromes in human immunodeficiency virus disease. In: J.R. Berger and R.M. Levy (Eds.), AIDS and the Nervous System. Philadelphia, Lippincott-Raven (1997) 189–221.

SIMPSON, D.M. and D.E. WOLFE: Neuromuscular complications of HIV infection and its treatment. AIDS 5 (1991) 917–926.

SIMPSON, D.M., A.N. BENDER and J. FARRAYE: Human immunodeficiency virus wasting syndrome may represent a treatable myopathy. Neurology 40 (1990) 535–538.

SIMPSON, D.M., K.A. CITAK, E. GODFREY, J. GODBOLD and D. WOLFE: Myopathies associated with human immunodeficiency virus and zidovudine: can their effects be distinguished? Neurology 43 (1993a) 971–976.

SIMPSON, D.M., J. GODBOLD, J. HASSETT, R. FEINSTEIN and J. GODBOLD: HIV associated myopathy, and the effects of zidovudine and prednisone: preliminary results of placebo-controlled trials (Abstract). Clin. Neuropathol. 12, Suppl. 1 (1993b) S20.

SIMPSON, D.M., P. SLASOR, U. DAFNI, J. BERGER, M. FISCHL and C. HALL: Analysis of myopathy in a placebo-controlled zidovudine trial. Muscle Nerve 20 (1997a) 382–385.

SIMPSON, D.M., D. KATZENSTEIN, S. HAMMER and AIDS

CLINICAL TRIALS GROUP: Neuromuscular function in HIV infection: analysis of a placebo-controlled trial (Abstract). Neurology 48 (1997b) A387.

SIMPSON, M.V., C.D. CHIN, S.A. KEILBOUGH, T.S. LIN and W.H. PRUSOFF: Studies on the inhibition of mitochondrial DNA replication of 3'-azido-3'-deoxythymidine and other deoxynucleoside analogs which inhibit HIV-1 replication. Biochem. Pharmacol. 38 (1989) 1033–1036.

SKOWRON, G., S.A. BOZZETTE, L. LIM, C.B. PETTINELLI, H.H. SCHAUMBURG, C. AREZZO, M.A. FISCHL, W.G. POWDERLY, D.J. GOCKE, D.D. RICHMAN and T.C. MERIGAN: Alternating and intermittent regimens of zidovudine and dideoxycytidine in patients with AIDS or AIDS-related complex. Ann. Intern. Med. 118 (1993) 321–330.

SMITH, I., D.W. HOWELLS, B. KENDALL, R. LEVINSKY and K. HYLAND: Folate deficiency and demyelination in AIDS [letter]. Lancet ii (1987) 215.

SNIDER, D.E.: Pyridoxine supplementation during isoniazid therapy. Tubercle 61 (1980) 191–196.

SNIDER, W., D.M. SIMPSON, S. NIELSEN, C. METROKA and J. POSNER: Neurological complications of acquired immune deficiency syndrome: analysis of 50 patients. Ann. Neurol. 14 (1983) 403–418.

SO, Y.T., D.M. HOLTZMAN and D.J. ABRAMS: Peripheral neuropathy associated with acquired immunodeficiency syndrome: prevalence and clinical features from a population-based survey. Arch. Neurol. 45 (1988) 945–948.

SOFFIETTI, R., R. SCIOLLA and M.T. GIORDANA: Delayed adverse effects after irradiation of gliomas: clinicopathological analysis. J. Neuro-Oncol. 3 (1985) 187.

SORIANO, E., J. MALLOLAS and J.M. GATELL: Characteristics of tuberculosis in HIV-infected patients: a case-control study. AIDS 2 (1988) 429–432.

SOUEIDAN, S., T. SINNWELL, C. JAY, J. FRANK, A. MCLAUGHLIN and M. DALAKAS: Impaired muscle energy metabolism in patients with AZT-myopathy: a blinded comparative study of exercise ^{31}P magnetic resonance spectroscopy (MRS) with muscle biopsy (Abstract). Neurology 42, Suppl. 3 (1992) 146.

STEIBERG, J.P., C.J. GUNTHEL and R.L. WHITE: Outcomes and toxicities on 2',3'-dideoxyinosine (ddI) in the expanded access program (Abstract). In: Proceedings of the 31st Interscience Conference on Antibiotics and Chemotherapy. American Society of Microbiology (1991) 217.

STEIN, D.S., N.M. GRAHAM, L.P. PARK, D.R. HOOVER, J. PHAIR, R. DETELS, M. HO and A.J. SAAH: The effect of the interaction of acyclovir with zidovudine on progression to AIDS and survival. Analysis of data in the Multicenter AIDS Cohort Study. Ann. Intern. Med. 121 (1994) 100–108.

STERN, R., J. GOLD and E.F. DICARLO: Myopathy complicating the acquired immunodeficiency syndrome. Muscle Nerve 10 (1987) 318–322.

STRAUS, S.E., H.A. SMITH and C. BRICKMAN: Acyclovir for chronic mucocutaneous herpes simplex virus infection in immunosuppressed patients. Ann. Intern. Med. 96 (1982) 270–277.

STUDIES OF OCULAR COMPLICATIONS OF AIDS RESEARCH GROUP, in collaboration with THE AIDS CLINICAL TRIALS GROUP: Mortality in patients with the acquired immunodeficiency syndrome treated with either foscarnet or ganciclovir for cytomegalovirus retinitis. N. Engl. J. Med. 326 (1992) 213–220.

STYLOPOULOS, L.A., A.E. GEORGE and M.J. DE LEON: Longitudinal study of parenchymal brain changes in glioma survivors. Am. J. Neuroradiol. 8 (1988) 329.

SUSKIND, R.M., S.W. BRUSILOW and J. ZEHR: Syndrome of inappropriate secretion of antidiuretic hormone produced vincristine toxicity (with bioassay of ADH level). J. Pediatr. 81 (1972) 90–92.

TAGLIATI, M., J. GODBOLD, J. HASSETT, E. GODFREY, J. GRINNELL and D.M. SIMPSON: Neuromuscular disorders in HIV infection: cross-sectional cohort analysis of 250 patients (Abstract). Neurology 44, Suppl. 2 (1994) A367.

TILL, M. and K.B. MACDONNELL: Myopathy with human immunodeficiency virus type I (HIV-1) infection: HIV-1 or zidovudine? Ann. Intern. Med. 113 (1990) 492–494.

TSAI, C., K.E. FOLLIS, M. YARNALL and G.A. BLAKLEY: Toxicity and efficacy of 2',3'-dideoxycytidine in clinical trials of pigtailed macaques infected with simian retrovirus type 2. Antimicrob. Agents Chemother. 33 (1989) 1908–1914.

TUBERCULOSIS CHEMOTHERAPY CENTRE MADRAS: The prevention and treatment of isoniazid toxicity in the therapy of pulmonary tuberculosis. II. An assessment of the prophylactic effect of pyridoxine in low dosage. Bull. World Health Organ. 29 (1963) 457.

TUXEN, M.K. and S.W. HANSEN: Neurotoxicity secondary to antineoplastic drugs. Cancer Treatm. Rev. 20 (1994) 191–214.

VAUGHAN, D.J., J.G. JARVIK, D. HACKNEY, S. PETERS and E.A. STADTMAUER: High-dose cytarabine neurotoxicity: MR findings during the acute phase. Am. J. Neuroradiol. 14 (1993) 1014–1016.

VOGEL, H. and D.S. HOROUPIAN: Filamentous degeneration of neurons. Cancer 71 (1993) 1303–1308.

VOLBERDING, P.A., S.W. LAGAKOS and M.A. KOCH: Zidovudine in asymptomatic human immunodeficiency virus infection: a controlled trial in persons with fewer than 500 CDV-positive cells per cubic mill-imeter. N. Engl. J. Med. 322 (1990) 941–949

WADE, J.C. and J.D. MEYERS: Neurologic symptoms associated with parenteral acyclovir treatment after bone marrow transplantation. Ann. Intern. Med. 98 (1983) 921–925.

WAGSTAFF, A.J. and B.M. BRYSON: Foscarnet: a reappraisal of its antiviral activity, pharmacokinetic properties and therapeutic use in immunocompromised patients with viral infections. Drugs 48 (1994) 199–226.

WALKER, R.W. and M.K. ROSENBLUM: Amphotericin B-associated leukoencephalopathy. Neurology 42 (1992) 2005–2010.

WALLACE, T.L. and E.M. JOHNSON, JR.: Cytosine arabinoside kills postmitotic neurons: evidence that deoxycytidine may have a role in neuronal survival that is independent of DNA synthesis. J. Neurosci. 9 (1989) 115–124.

WALLING, D.M., A.G. PERKINS and J. WEBSTER-CYRIAQUE: The Epstein–Barr virus EBNA-2 gene in oral hairy leukoplakia: strain variation, genetic recombination, and transcriptional expression. J. Virol. 68 (1994) 7918–7926.

WASON, S., P.G. LACOUTURE and F.H. LOVEJOY: Single high-dose pyridoxine treatment for isoniazid overdose. J. Am. Med. Assoc. 246 (1981) 1102–1104.

WEISSMAN, J.D., I. CONSTANTINITIS, P. HUDGINS and D.C. WALLACE: ^{31}P magnetic resonance spectroscopy suggests impaired mitochondrial function in AZT-treated HIV-infected patients. Neurology 42 (1992) 619–623.

WELTMAN, A.C. and D.N. ROSE: Tuberculosis susceptibility patterns, predictors of multidrug resistance, and implications for initial therapeutic regimens at a New York City hospital. Arch Intern. Med. 154 (1994) 2161–2167.

WIEGAND, R.G.: The formation of pyridoxal and pyridoxal 5-phosphate hydrazones. J. Am. Chem. Soc. 78 (1956) 5307.

WINKELMAN, M.D. and J.D. HINES: Cerebellar degeneration caused by high-dose cytosine arabinoside: a clinicopathological study. Ann. Neurol. 14 (1983) 520–527.

WOOD, J.D. and S.J. PEESKER: A correlation between changes in GABA metabolism and isonicotinic acid hydrazide induced seizures. Brain Res. 45 (1972) 489–498.

WRZOLEK, M.A., J.H. SHER, P.B. KOZLOWSKI and C. RAO: Skeletal muscle pathology in AIDS: an autopsy study. Muscle Nerve 13 (1990) 508–515.

WULFF, C.H., H. HAYER, G. ASBOE-HANSEN and H. BRODTHAGEN: Development of polyneuropathy during thalidomide therapy. Br. J. Dermatol. 112 (1985) 475–480.

WYATT, E.H. and J.C. STEVENS: Dapsone induced peripheral neuropathy. Br. J. Dermatol. 86 (1972) 521–523.

YAP, H., G.R. BLUMENSCHEIN and B.S. YAP: High-dose methotrexate for advanced breast cancer. Cancer Treatm. Rep. 63 (1979) 757–761.

YARCHOAN, R., G. BERG and P. BROUWERS: Response of human immunodeficiency virus-associated neurological disease to 2'-azido-3'-deoxythymidine. Lancet i (1987) 132–135.

YARCHOAN, R., C.F. PERNO, R.V. THOMAS, R.W. KLECKER, J.P. ALLAIN, R.J. WILLS, M.A. FISCHL, H. MITSUYA and S. BRODER: Phase I studies of 2',3'-dideoxycytidine in severe human immuno-deficiency virus infection as a single agent and alternating with zidovudine. Lancet i (1988) 76–81.

YARCHOAN, R., J.M. PLUDA and R.V. THOMAS: Long term toxicity/activity profile of 2',3'-dideoxyinosine in AIDS or AIDS-related complex. Lancet 336 (1990) 526–529.

YOULE, M., D. HAWKINS and B. GAZZARD: Thalidomide in hyperalgesic pharyngeal ulceration of AIDS [letter]. Lancet 335 (1990) 1591.

YOULE, M.S., B.G. GAZZARD and M.A. JOHNSON: Effects of high-dose oral acyclovir on herpes virus disease and survival in patients with advanced HIV disease: a double-blind, placebo-controlled study. AIDS 8 (1994) 641–649.

Handbook of Clinical Neurology, Vol. 27 (71): Systemic Diseases, Part III
M.J. Aminoff and C.G. Goetz, editors

Progressive multifocal leukoencephalopathy

JOHN E. GREENLEE

Department of Neurology, University of Utah School of Medicine, Salt Lake City, UT, U.S.A.

Progressive multifocal leukoencephalopathy (PML) is a progressive, almost invariably fatal opportunistic infection of immunocompromised patients, caused by a member of the polyomavirus subgroup of *Papovaviridae,* JC virus (polyomavirus hominis 2). Prior to the advent of the acquired immunodeficiency syndrome (AIDS), PML was an extremely rare condition, even among immunocompromised patients or as a complication of immunosuppression. In this setting, PML tended to be an illness of later adult life, and children were only rarely infected (Zu Rhein et al. 1978; G.L. Stoner et al. 1988; Redfearn et al. 1993). Since 1984, with the AIDS epidemic, the prevalence of PML has risen from 1.5 cases per 10 million individuals in the USA to 6.1 (Holman et al. 1991). Four percent of AIDS patients will develop PML (Berger et al. 1987; G.L. Stoner et al. 1988; Gillespie et al. 1991; Holman et al. 1991). AIDS-PML tends to infect younger adults than was true in the pre-AIDS era and is more likely to affect males (G.L. Stoner et al. 1988). PML is now also known to occur as a complication of childhood HIV infection (Berger et al. 1992; Vandersteenhoven et al. 1992). This chapter will first review current knowledge concerning the clinical and pathologic features of PML and will then discuss the causative agent of PML, JC virus, and its possible role in central nervous system neoplasia.

HISTORICAL PERSPECTIVE

Cases of the disease entity which would later be termed progressive multifocal leukoencephalopathy were described as early as 1930 by Hallervorden, with additional cases being reported by Winkelman and Moore (1941), Bateman et al. (1945) and Christensen and Fog (1955) and reviewed, with retrospective pathologic confirmation, by Astrom et al. (1958). Recognition of PML as a distinct disease entity, however, did not occur until 1958, when Astrom et al. reported three patients with hematological malignancies who developed a fatal disorder characterized pathologically by multifocal areas of demyelination within the cerebrum and brainstem (Astrom et al. 1958). Microscopic examination of demyelinated areas revealed nuclear enlargement and intranuclear inclusions within oligodendrocytes and, unlike the histopathologic findings of any other previously recognized disorder, unusual morphological alteration of astrocytes (Astrom et al. 1958). Cavanagh et al. (1959) provided additional cases of the disorder, all in patients with reticuloendothelial malignancies, and suggested that the disorder might arise as a viral infection in the setting of lymphoma-induced immune deficiency. Richardson (1961) added additional cases of the disorder and, based on the presence of intranuclear inclusions within oligodendrocytes, also suggested that PML might be an infectious process. Evidence for a viral causation of PML was not forthcoming until 1965, however, when Zu Rhein and Chou (1965) and Silverman and Rubinstein (1965) independently described viral particles, identical in size and configuration to murine polyoma virus, in nuclei of oligodendrocytes within PML brains. Padgett et al. (1971) recovered a polyomavirus from PML brain material by use of primary cultures of human fetal brain cells; this isolate was named JC virus, after the initials of the patient from whom it had been cultured. In the next year, Weiner et al. (1972a, b) reported isolation of a second papovavirus, closely resembling the simian agent, SV40 virus, from two cases of PML. Since 1972, JC virus has been consistently recovered from PML material, but the role of an SV40-like virus in the causation of PML has remained at best controversial.

CLINICAL PRESENTATION

PML is almost invariably a complication of impaired immune response (Astrom et al. 1958; Richardson 1961; Padgett and Walker 1983; Walker and Padgett 1983; Holman et al. 1991). Before the beginning of the AIDS epidemic, the disease occurred as an extremely rare complication of three clinical conditions:

(1) leukemias, lymphomas, or other malignant diseases (Astrom et al. 1958; Cavanagh et al. 1959; Richardson 1961; Mancall 1965; Slooff 1966; Woodhouse et al. 1967; Mathews et al. 1976; Allegranza et al. 1981; Tremblay et al. 1982; Padgett and Walker 1983; Choy et al. 1992; Kimel and Schutt 1993);
(2) protracted granulomatous or inflammatory disorders such as tuberculosis, sarcoidosis, or non-tropical sprue (Kepes et al. 1975; Marriott et al. 1975; Smith et al. 1982; Gullotta et al. 1992; Steiger et al. 1993); and
(3) immunosuppression for collagen vascular disease (Slooff 1966; Richardson and Johnson 1975; Sponzilli et al. 1975; Krupp et al. 1985; Choy et al. 1992); or organ transplantation (Manz et al. 1971; Legrain et al. 1974; Selhorst et al. 1978; Egan et al. 1980; Renzik et al. 1981; Flomenbaum et al. 1991; Owen et al. 1995) (Table 1).

TABLE 1

Conditions associated with PML.

Hematological and other malignancies[a]
Leukemias; especially chronic lymphocytic leukemia[b]
Lymphomas
Hodgkin's disease
Reticulum cell sarcoma
Mycosis fungoides
Colon cancer
Infectious and inflammatory conditions
Tuberculosis
Sarcoid
Non-tropical sprue
AIDS
Collagen vascular diseases[a]
Systemic lupus erythematosus
Rheumatoid arthritis
Wegener's granulomatosis
Polyomyositis
Organ transplantation[a]
Kidney
Bone marrow
Liver, heart, combined
Congenital immune deficiencies
Wiskott–Aldrich syndrome
Combined immunodeficiency syndromes

[a]The majority of patients reported with these conditions have been receiving antineoplastic or immunosuppressive agents.
[b]References to specific conditions are provided in the text.

AIDS has since become by far the most frequent condition associated with PML, and the prevalence of PML in AIDS – approaching 2–4% of all AIDS patients – is far greater than in any other predisposing condition (Berger et al. 1987; Gillespie et al. 1991; Holman et al. 1991). Occasionally, PML may be the presenting sign of HIV infection (Jakobsen et al. 1987; Singer et al. 1994).

The onset of PML is usually insidious and is often heralded by focal rather than multifocal neurologic signs (Richardson 1961, 1970; Krupp et al. 1985; Greenlee 1989). Initial symptoms of PML most often indicate cerebral involvement (Astrom et al. 1958; Richardson 1961; Mancall 1965; Cherif et al. 1983; Krupp et al. 1985; Berger et al. 1987; Feiden et al. 1993). Alterations in personality are common: these may initially be subtle but are followed by blunting of intellect and dementia as the disease progresses. Mono- or hemiparesis is common, as is loss of motor praxis or parietal sensory integrative functions. Involvement of the dominant hemisphere may result in expressive or receptive aphasias or, rarely, Gerstmann's syndrome (Iranzo et al. 1992). At times, motor features of the disease may suggest extrapyramidal involvement and may include dystonia (De Toffol et al. 1994; Sweeney et al. 1994) or presentation in children as an akinetic-rigid syndrome (Singer et al. 1993). Neuro-ophthalmologic abnormalities are present in up to 50% of patients and may be the presenting feature of the disease (Greenlee 1989; Lafeuillade et al. 1993; Lewis et al. 1993). Visual difficulties may include progressive visual loss, quadrantic or hemianopic visual field defects, or defects referable to visual association areas or parietal lobes, including visual neglect, Balint's syndrome, and cortical blindness (Lafeuillade et al. 1993; Ayuso-Peralta et al. 1994). Headache is unusual, and the presence of fever suggests the presence of some other or additional condition. PML may occasionally be accompanied by seizures (Moulignier et al. 1995).

In a minority of patients, PML may begin with signs of brainstem involvement as evidenced by difficulties with phonation, swallowing or extraocular movements. PML may also present with cerebellar involvement and loss of cerebellar coordination of motor function (Astrom et al. 1958; Parr et al. 1979; Irie et al. 1992). PML rarely causes symptomatic involvement of the spinal cord – so much so that the presence of spinal cord findings mitigates against the diagnosis of PML or suggests the presence of a second process. Three cases have been reported, however, in which PML involving the spinal cord was felt to result in clinically evident signs (Headington and Umiker 1962; Johnston et al. 1963; Henin et al. 1992; Von Einsiedel et al. 1993): one of these was in the pre-AIDS era, involving a patient with sarcoidosis and tuberculosis (Headington and Umiker 1962; Johnston et al. 1963), and the remaining two involved patients with HIV infection (Henin et al. 1992; Von Einsiedel et al. 1993).

PML, in the great majority of patients, is a remorselessly progressive disease. Most non-AIDS patients with PML die within 1 year. In AIDS-PML the progression of illness is considerably more rapid, and death usually occurs within 4–6 months (Berger et al. 1987; Fong and Toma 1995). There is, however, wide variation in length of survival in both non-AIDS PML and AIDS-PML. Death in non-

AIDS PML has occurred within 2 months, but cases with survival of 8–10 years have also been reported (Hedley-Whyte et al. 1966; Stam 1966; Brun et al. 1973; Kepes et al. 1975; Lortholary et al. 1994), with one case, pathologically diagnosed post mortem, being thought on clinical grounds to have run a course of 20 years and a second, similarly diagnosed, being thought to have progressed over 33 years (Stam 1966; Brun et al. 1973). Clinical remission may occasionally occur, either spontaneously or following reduction of immunosuppressive medications (Brun et al. 1973; Selhorst et al. 1978; Walker 1978; Padgett and Walker 1983; Price et al. 1983). Somewhat surprisingly, rare cases of spontaneous remission have also been described in AIDS-PML (Berger et al. 1987; Berger and Mucke 1988). As will be discussed below, remission of AIDS-PML has also occurred during zidovudine-induced stabilization of the patient's HIV infection (Conway et al. 1990; Von Einsiedel et al. 1993; Lortholary et al. 1994).

Progressive multifocal leukoencephalopathy in children

Prior to the advent of AIDS, PML in children was extremely rare, such that prior to 1980 only two cases of childhood PML had been reported, both in children with severe, combined immunodeficiency syndromes (Zu Rhein et al. 1978; Padgett and Walker 1983; Walker and Padgett 1983). Two additional cases of non-AIDS PML have been reported in adolescents, one in an 18 year old in association with congenital immune deficiency and hemolytic anemia and the other in a 15-year-old boy with Wiskott–Aldrich syndrome (Castaigne et al. 1974; Katz et al. 1994), More recently, however, a number of cases of PML have been reported in children with AIDS (Stoner et al. 1988; Krasinski et al. 1989; Berger et al. 1992; Vandersteenhoven et al. 1992; Redfearn et al. 1993; Singer et al. 1993). In general, the clinical presentation, magnetic resonance imaging (MRI) changes, course, and pathologic features exhibited by these patients were identical to those seen in adults, with early development of focal neurologic findings followed, survival from the underlying HIV infection permitting, by progression to severe neurologic disability, vegetative state, and death. Motor abnormalities have been a prominent feature of the disease in most cases, including development of an akinetic-rigid syndrome in one patient (Singer et al. 1993). Opportunistic infections of the central nervous system are, in general, much less common in pediatric patients with AIDS, and PML would appear to be much less common in childhood AIDS than in adult AIDS (Krasinski et al. 1989; Vandersteenhoven et al. 1992).

NEUROPATHOLOGY

The neuropathologic changes of PML are characterized grossly by myelin loss and microscopically by morphological changes in oligodendrocytes and astrocytes. The clinical features of PML, however, are directly due to destruction of oligodendrocytes, and the resultant demyelination.

PML brains, on cut section, virtually always display involvement of the cerebrum and frequently show evidence of brainstem and cerebellar involvement as well. Within the cerebrum, cortex and deep gray structures usually appear normal, but subcortical and deeper white matter, as well as cerebellum and brainstem, show retraction (Fig. 1), indicating myelin loss, which may be confirmed using histochemical stains of cortical slices (Fig. 2) (Astrom et al. 1958; Richardson 1961, 1970; Zu Rhein 1969; Walker

1978; Richardson and Webster 1983; Schmidbauer et al. 1990).

The histopathologic features of PML have been extensively studied by light microscopy, immunohistology, in situ nucleic acid hybridization methods, and electron microscopy. The pathognomonic histopathologic features of PML are nuclear enlargement or intranuclear inclusions within oligodendrocytes and, in many but not all cases, the presence of histologically atypical, often giant astrocytic cells (Figs. 3 and 4) (Astrom et al. 1958; Richardson 1961; Zu Rhein 1969). The typical PML lesion is an area of demyelination, usually with relative axonal sparing (Astrom et al. 1958; Richardson 1961; Zu Rhein 1969). Oligodendrocytes are greatly reduced in number or absent within the lesion itself, and oligodendrocytes at the edges of the lesion are frequently enlarged, with swollen nuclei or intranuclear inclusions (Fig. 3). Inclusion-bearing oligodendrocytes may occasionally be found in otherwise normal white matter and may also be found within gray matter. A reactive astrocytosis is often present but, in addition, many cases also contain scattered, atypical astrocytes within demyelinated lesions. These astrocytes, which are usually found in or near demyelinated

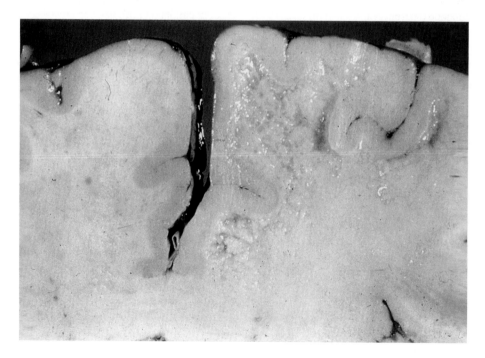

Fig. 1. Cut section of brain from a patient dying with PML, showing retraction of the underlying white matter.

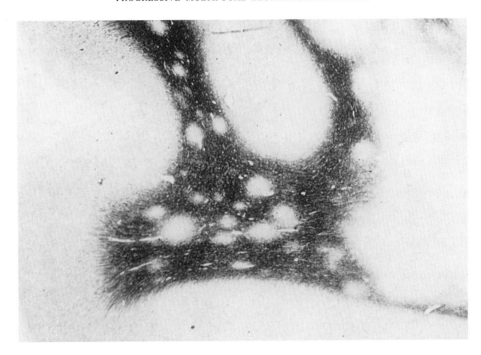

Fig. 2. Myelin stain of brain slice from patient dying of PML, showing multiple foci of demyelination.

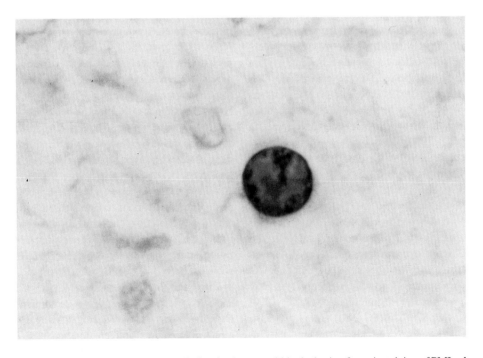

Fig. 3. High-magnification view of infected oligodendrocyte within the brain of a patient dying of PML, showing marked nuclear enlargement.

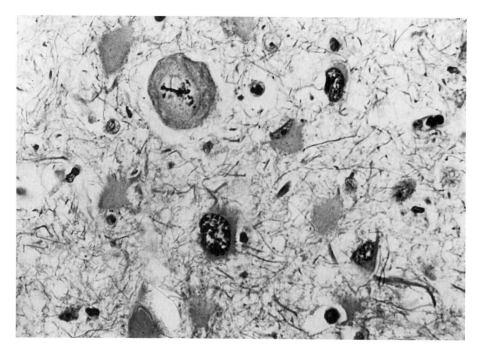

Fig. 4. Section of PML brain within area of demyelination. Several giant astrocytes are present, one of which contains a mitotic figure.

areas, are enlarged and exhibit morphologic features suggestive of cell transformation, with multinucleated forms and mitotic figures (Fig. 4) (Astrom et al. 1958; Richardson 1961; Zu Rhein 1969). Myelin break-down and lipid-laden macrophages become evident as the disease progresses. Extensive inflammation is unusual in non-AIDS PML, although small numbers of lymphocytes may be seen around vessels and in demyelinated areas (Schmidbauer et al. 1990; Vaseux et al. 1990).

The pathology of AIDS-PML is for the most part similar to that seen in non-AIDS cases (Schmidbauer et al. 1990; Vaseux et al. 1990; Neuen Jacob et al. 1993; Von Einsiedel et al. 1993). Demyelination in AIDS-PML is often more extensive than in non-AIDS cases, however (Vaseux et al. 1990), and areas of actual necrosis may be present (Schmid-bauer et al. 1990; Vaseux et al. 1990). Lymphocytic perivascular infiltrates are more frequently evident in AIDS-PML brains (Schmidbauer et al. 1990; Vaseux et al. 1990). AIDS-PML brains may also contain more extensive parenchymal inflammatory in-filtrates, which include lymphocytes, macrophages and multinucleated giant cells (Schmidbauer et al. 1990; Vaseux et al. 1990). Atypical astrocytes may

be infrequent or absent in AIDS-PML brain. Kuchelmeister et al. (1993) have described changes in the cerebellar granule cell layer in six cases of AIDS-PML, consisting of cells with nuclear enlarge-ment to approximately twice the size of normal gran-ule cells. JC virus particles could not be identified in these enlarged granule cells, nor did the cells contain HIV-1 p24 antigen or JC virus nucleic acids. It is un-clear whether or not these enlarged cells represent abortive infection by JC virus or whether the morpho-logic changes observed represent some other patho-logic process.

Studies employing in situ nucleic acid hybridization methods and immunohistochemistry have provided considerable additional information about the distri-bution of JC virus in PML. In situ nucleic acid hybridi-zation studies indicate the presence of JC virus nucleic acids in oligodendrocytes (Fig. 5) and, rarely, in astrocytes (Fig. 6) (Aksamit et al. 1985; Greenlee and Stroop 1988; Aksamit, Jr. 1993). Immunohisto-logic studies have employed antibodies against com-mon polyomavirus antigen (Gerber et al. 1980), monoclonal and polyclonal antibodies to JC virus capsid antigen (Itoyama et al. 1982; Stoner et al. 1988), and human polyomavirus large T antigen, an

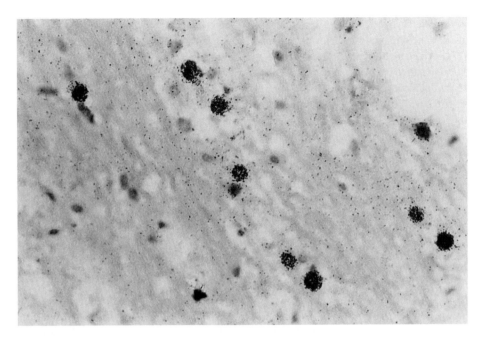

Fig. 5. Section of PML brain probed for JC virus using in situ nucleic acid hybridization methods. Large numbers of exposed emulsion grains, indicative of specific hybridization, are present over the nuclei of oligodendrocytes.

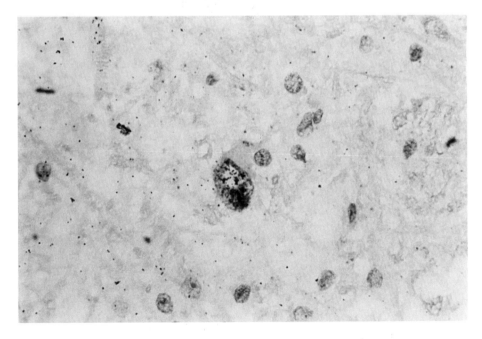

Fig. 6. Section of PML brain probed for JC virus using in situ nucleic acid hybridization methods. Exposed emulsion grains, indicative of specific hybridization, are present over the nuclei of an atypical astrocyte.

early, non-structural protein involved in viral replication and cell transformation (see below) (Itoyama et al. 1982; Greenlee and Keeney 1986; Stoner et al. 1988). JC virus antigens have been routinely detected in nuclei of oligodendrocytes (Fig. 7). Although JC virus structural antigens have been identified within the cytoplasm of macrophages near PML lesions, they have not been described in other cell populations. JC virus T antigen has been found in oligodendrocytes (Fig. 8) and, rarely, within both apparently normal and atypical astrocytes, but not within macrophages (Greenlee and Keeney 1986). Neither JC structural antigens nor JC virus T antigen has been found in neurons, or other cell populations (Gerber et al. 1980; Aksamit et al. 1985, 1986, 1987; Greenlee and Keeney 1986; Stoner et al. 1988; Schmidbauer et al. 1990; Knowles et al. 1991; Aksamit, Jr. 1993). Oligodendrocytes containing JC virus structural or T antigens are much more numerous than are cells with histologically abnormal nuclei and can often be found at some distance from PML lesions, within apparently normal brain.

Astrom and Stoner (1994) have recently described early pathologic changes of PML in brains from individuals with non-AIDS disease. In their study, which employed immunohistochemistry, early lesions were detected which contained atypical astrocytes without apparent infection of oligodendrocytes, raising questions as to whether astrocytes may be the initial cell population affected in PML.

Extensive ultrastructural studies of PML have been carried out by a number of workers, in particular Howatson et al. (1965), Silverman and Rubinstein (1965), Woodhouse et al. (1967), Zu Rhein (1969) and, more recently, Boldorini et al. (1993). Work by Zu Rhein (1969), in the pre-AIDS era, demonstrated polyomavirus particles in oligodendrocytes and, occasionally, within astrocytes. Within oligodendrocytes, polyomavirus virions are present within nuclei (Fig. 9), although cytoplasmic virions may occasionally be seen in cells actually undergoing lysis. Viral particles were not identified in astrocytic nuclei and were never present in the atypical astrocytes pathognomonic for PML; rather, cytoplasmic papovavirus particles were detected in occasional, otherwise normal astrocytes. Viral particles were not seen in microglial cells, neurons, cells of the meninges or pia mater, or endothelial cells. Similar observations were made

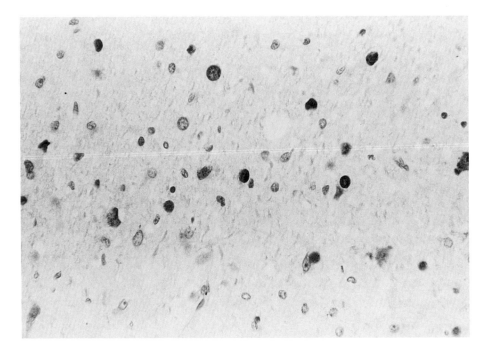

Fig. 7. Section of PML brain stained with antibody directed against polyomavirus common structural antigen. There is intense labeling of multiple infected oligodendrocytes (× 200).

by other investigators (Howatson et al. 1965; Itoyama et al. 1982). Boldorini et al. (1993) used ultrastructural methods to investigate brain material from eight patients with AIDS-PML. These workers, in contrast to earlier investigators, reported polyomavirus virions within neuronal nuclei and cytoplasm in two cases, in neuronal cytoplasm only in three additional cases, and in macrophages in five cases. The presence of cytoplasmic polyomavirus particles with macrophages in PML brains has also been described by Itoyama et al. (1982) and by Mesquita et al. (1992). The presence of papovavirus virions in neurons has not been reported in ultrastructural studies by other investigators or by studies employing immunohistologic or nucleic acid hybridization studies (see below).

The morphologic stages of JC virus replication in oligodendrocytes have been elegantly delineated by Mazlo and Tariska (1980). These investigators demonstrated adherence of polyomavirus virions to the cytoplasmic membrane, entry of viral particles into the cell within vacuoles formed from the cell membrane, movement of virions within vacuoles or within the endoplasmic reticulum to the cell nucleus, and then assembly of viral particles within the cell nucleus. The later stages of viral replication were accompanied by loss of nuclear chromatin, cytoplasmic vacuolation, disappearance of polyribosomes, and finally disintegration of the nuclear and thereafter cellular membranes, with eventual release of virions, usually bound to membrane fragments, into the extracellular space.

DIAGNOSIS

Clinical and radiologic diagnosis

PML is a diagnostic concern in any immunocompromised patient who develops progressive neurologic deficits involving cerebrum, cerebellum or brainstem, especially if the neurologic examination suggests involvement of multiple areas of brain. Suspicion should be particularly high in the setting of known HIV infection or frank AIDS. Hematologic studies and blood chemistries are unhelpful in the diagnosis of PML. Cerebrospinal fluid (CSF) is usually normal but may occasionally contain increased protein or, rarely, a lymphocytic pleocytosis (Astrom

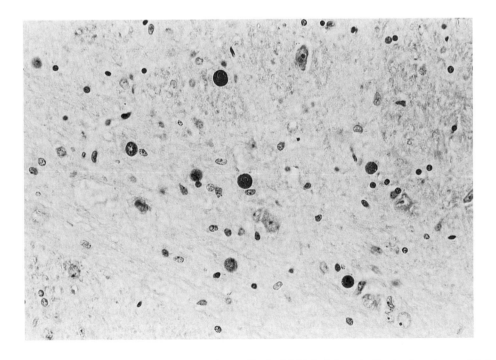

Fig. 8. Section of PML brain stained with anti-T antibody. Multiple infected oligodendrocytes are labeled. (× 200).

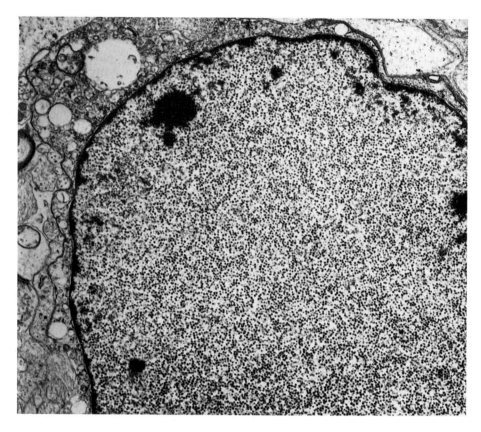

Fig. 9. Electron micrograph of an infected oligodendrocyte undergoing degeneration. The section has been contrasted with uranyl acetate and lead citrate. The nucleus contains large numbers of JC virus particles; in this particular instance, virions are distributed randomly rather than being grouped in crystalline arrays. (Provided by Dr. Jeanette Townsend.)

et al. 1958; Richardson 1961; Greenlee 1989). Serological tests of blood or CSF for anti-JC virus antibodies are not useful in diagnosis, since most individuals have antibody to JC virus, and many immunocompromised individuals with PML may fail to generate a rise in antibody titer (Padgett and Walker 1983). Antibody is frequently absent from the CSF, although detectable CSF titers of anti-JC virus antibody, in the presence of an intact blood–brain barrier, suggest the diagnosis.

MRI is the neuroradiological diagnostic study of choice in PML (Guilleux et al. 1986; Sullivan et al. 1990; Trotot et al. 1990; Irie et al. 1992; Whiteman et al. 1993; Newton et al. 1995; Ng et al. 1995). MRI findings in PML consist of altered signal in subcortical and deep white matter, most clearly evident on T2-weighted scans (Figs. 10 and 11). Some – but not all – PML lesions enhance with intravenous gado-linium (Wheeler et al. 1993; Newton et al. 1995). Rarely, JC virus may produce predominantly subcortical infection without causing visible white matter lesions (Sweeney et al. 1994). In this case, MRI may fail to make the diagnosis. PML lesions may also be detected by computed tomography (CT) (Carroll et al. 1977; Lipton et al. 1988; Wheeler et al. 1993) (Fig. 12). However, changes consistent with demyelination – particularly within brainstem – may be difficult to detect on CT despite clinically evident disease, and lesions as seen on CT are often less extensive than are shown on MRI (Balakrishnan et al. 1990; Ciricillo and Rosenblum 1990; Sullivan et al. 1990). Although both conventional radionuclide scanning and HMPAO-SPECT scanning have been used to diagnose PML, neither procedure can be considered as reliable as either CT or MRI (Mosher et al. 1971; Bowler et al. 1992).

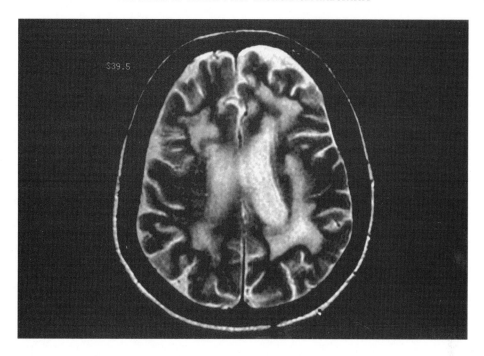

Fig. 10. T2-weighted MRI head scan of a patient with AIDS-PML, showing multiple foci of increased white-matter signal. (Courtesy of Dr. Jay Tsuruda.)

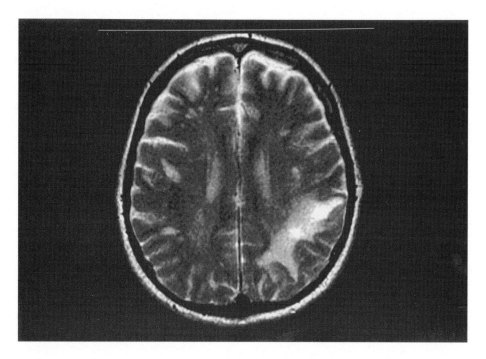

Fig. 11. T2-weighted MRI scan of a patient with PML presenting as confusion, hemianopsia, and neglect in the setting of heart–liver transplant. There is a focus of increased signal in the white matter of the left parietal and occipital lobes. (Courtesy of Dr. Jay Tsuruda.)

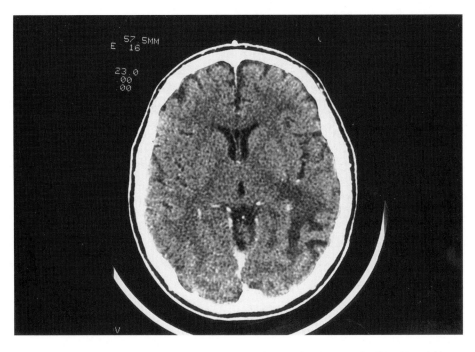

Fig. 12. CT scan of the patient shown in Fig. 11. The area of demyelination seen in Figs. 10 and 11 appears as an area of diminished signal. (Courtesy of Dr. Jay Tsuruda.)

Laboratory diagnosis

For many years, definitive diagnosis of PML during life required brain biopsy with identification of characteristic histologic findings. Accuracy of histologic diagnosis can be increased by identification of JC virus nucleic acids or viral antigens by in situ nucleic acid hybridization methods or immunocytochemistry (Aksamit et al. 1986, 1987; Schmidbauer et al. 1990; Hulette et al. 1991; Prayson and Estes 1993; Iacoangeli et al. 1994) (Figs. 5 and 7). Diagnostic accuracy can be further increased using polymerase chain reaction (PCR) methods to amplify JC virus-specific sequences from paraffin or to detect JC virus DNA using in situ PCR techniques (Schmidbauer et al. 1990; Telenti et al. 1990; Ueki et al. 1994). JC virus has highly specific tissue culture requirements (see below), and viral replication in vitro requires weeks rather than days: these two factors have greatly limited the routine applicability of virus isolation as a clinical diagnostic tool.

A less invasive approach to specific diagnosis of PML has recently been provided by PCR amplification of JC virus nucleic acids directly from CSF. Initial efforts to detect JC virus DNA in CSF were positive in a relatively small percentage of cases (Henson et al. 1991). Subsequent studies, however, have provided accurate diagnosis of PML in 80–90% of cases (Telenti et al. 1992; Gibson et al. 1993; Moret et al. 1993; Vignoli et al. 1993; Weber et al. 1994; McGuire et al. 1995), making PCR of CSF the virological diagnostic study of choice in PML. Serological evidence of JC infection, as mentioned above, is rarely of help in the diagnosis. Similarly, because approximately 33% of immunosuppressed individuals shed JC virus in their urine at some point in time, urinary excretion of JC virus cannot be considered supportive evidence of PML.

TREATMENT

In a small number of cases, PML has stabilized or remitted either spontaneously or following improvement in host immune status (Selhorst et al. 1978; Padgett, Walker, 1983; Ayuso Peralta et al. 1994; Lortholary et al. 1994). In one renal transplant patient, stabilization in PML followed reduction in immunosuppressive therapy and institution of tilorone, a drug believed to inhibit T lymphocytes and stimulate B lymphocytes and interferon production (Selhorst et al. 1978; Padgett and Walker 1983). In this patient, stabilization was accompanied by a rise in titers of serum anti-JC virus antibody and appearance of anti - JC virus antibody in CSF (Padgett and Walker 1983). Stabilization in the course of PML has been reported in one other renal transplant patient in whom immunosuppressive therapy was modified and in one patient without known underlying disease who was treated with tilorone (Padgett and Walker 1983). Conway et al. (1990), von Einsiedel et al. (1993) and Lortholary et al. (1994) have reported patients with AIDS-PML in whom improvement in symptoms followed treatment of the underlying HIV infection with zidovudine. In other instances, however, progression of PML has occurred despite zidovudine therapy (Garrote et al. 1990). Recently, regression of PML lesions has been reported following combined anti-retroviral therapy (Takashima et al. 1996; Elliot et al. 1997; Domingo et al. 1997; Power et al. 1997).

Proven antiviral therapy for PML has not yet been developed, and evaluation of the therapeutic efficacy of any given agent is made difficult by several factors. Firstly, until recently, no detailed controlled trial had been reported in which the efficacy of a given agent was been studied in large numbers of patients. Secondly, rigorous confirmation of the diagnosis of PML has not been made in all cases in which therapeutic agents were reported as being successful or unsuccessful. Thirdly, the therapeutic efficacy of any given agent or combination of agents may well be influenced on a case-to-case basis by the severity of the concomitant immunosuppressed state and by manipulations in immunosuppressive drug regimens or treatment of HIV infection. Finally, the apparent therapeutic effect of any therapeutic agent, in an isolated case report, must be interpreted in the light of the fact that PML occasionally undergoes spontaneous remission, including in AIDS (see above).

Adenine arabinoside (Ara-A, Cytarabine), Acyclovir (Zovirax), and Cytosine arabinoside (Ara-C, Cytarabine) have all been used as therapeutic agents in patients with PML, as has intrathecal interferon-α (Rand et al. 1977; Tashiro et al. 1987; Steiger et al. 1993). Both adenine arabinoside and acyclovir have been used in PML without success (Rand et al. 1977; Bedri et al. 1983; Steiger et al. 1993). Cytosine arabinoside was used for many years as a therapeutic agent in individual cases of both non-AIDS PML and AIDS-PML (Castleman et al. 1972; Bauer et al.

1973; Conomy et al. 1974; Marriott et al. 1975; Rockwell et al. 1976; Van Horn et al. 1978; Peters et al. 1980; Smith et al. 1982; Snider et al. 1983; Kupp et al. 1985; Knight et al. 1988; Portegies et al. 1991; Nicoli et al. 1992; Steiger et al. 1993; de Truchis et al. 1993). Clinical improvement was reported in a few of these cases (Bauer et al. 1973; Conomy et al. 1974; Knight et al. 1988; Portegies et al. 1991; Nicoli et al. 1992). In one patient with PML complicating sarcoidosis, therapy with cytosine arabinoside and acyclovir was without apparent benefit, but improvement in the patient's PML occurred following the combined use of cytosine arabinoside and interferon-α (Steiger et al. 1993). Although these individual case studies were encouraging, cytosine arabinoside has been without benefit in a somewhat larger number of cases (Castleman et al. 1972; Rockwell et al. 1976; Van Horn et al. 1978; Smith et al. 1982; Kupp et al. 1985; de Truchis et al. 1993; Antinori et al. 1994; Fong and Toma, 1995). Three recent studies are of particular importance in assessing the efficacy of cytosine arabinoside in PML. Antinori et al. (1994) followed four patients with AIDS-PML, estimating JC viral burden by PCR analysis of CSF therapy before and after therapy with cytosine arabinoside. In this study, cytosine arabinoside failed to reduce CSF concentrations of JC virus DNA in any patient. Despite the successful use of cytosine arabinoside and interferon-α reported by Steiger et al. (1993), Fong and Toma (1995) did not find this drug regimen of benefit in three AIDS-PML patients. Recently, Hall et al. (1998) have reported the first extensive controlled study of cytosine arabinoside in AIDS-PML. In this multi-institutional study, 57 patients with biopsy-confirmed PML were randomized to receive anti-retroviral therapy alone, anti-retroviral therapy plus intravenous cytosine arabinoside, or anti-retroviral therapy plus intrathecal cytosine arabinoside. There was no difference in survival between any of the three groups, nor was survival in patients treated with anti-retroviral therapy alone different from that previously reported in untreated patients. These data strongly indicate that cytosine arabinoside has no therapeutic role in the treatment of PML. Prolonged survival with regression of lesions has also been reported following the use of combined anti-retroviral therapy along with inhibitors of polyomavirus replication such as camptothecin-related compounds, cidofovir, or the carbocyclic nucleoside analogue, abacavir (O'Reilly, 1997; Vollmer Haase et al. 1997; Sadler et al. 1998). Controlled studies employing these agents, as well as studies of combined anti-retroviral therapy, are currently in progress.

Although the diagnosis of PML can be made with some confidence by MRI and can be confirmed by PCR studies of CSF, it must be kept in mind that other, more easily treatable infectious agents also occur in immunosuppressed patients and are particularly common in patients with AIDS. Such agents, include *Toxoplasma gondii* and, less frequently, *Cryptococcus neoformans, Candida albicans, Mycobacterium tuberculosis* and *Listeria monocytogenes*. These infections may occur together with or in the absence of PML and may at time mimic PML in symptoms and radiologic appearance. Similarly, central nervous system lymphoma may also sometimes mimic PML (Ciricillo and Rosenblum 1990; Morgello et al. 1990; Iacoangeli et al. 1994). Although the usual approach in patients with AIDS is to treat first for *T. gondii* and follow therapeutic effect by serial MRI studies, CT-guided stereotactic or open brain biopsy should be considered to exclude

treatable conditions where therapy for toxoplasmosis has failed, the diagnosis of PML is in question, and there is concern about other, coexisting infections (Feiden et al. 1993; Prayson and Estes 1993; Iacoangeli et al. 1994; Nielsen et al. 1994; Silver et al. 1995).

VIROLOGY

The causative agent of PML, JC virus, is a member of the polyomavirus subgroup of the genus *Papovaviridae*. The virus is a non-enveloped, icosahedral agent with a diameter of 42 nm and contains a genome of supercoiled, double-stranded DNA 5130 base pairs in size. Occasional JC virions may have tubular rather than spherical shape (Fig. 13). The JC virus genome encodes only a single serotype of virus, but extensive variation may occur in the regulatory regions of both BK virus and JC virus. It has been suggested that this variability may account for differences in the species specificity of the virus and also in the cellular tropism of the virus in its interaction with individual patients and in its ability to cause disease (Lynch and Frisque 1990, 1991; Ault and Stoner

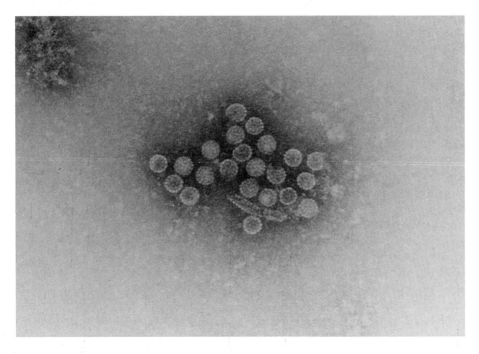

Fig. 13. Electron micrograph of JC virus particles extracted from PML brain, concentrated by ultracentrifugation and stained with 2% phosphotungstic acid. Both spherical and tubular forms of the virus are present.

1992, 1994; Daniel and Frisque 1993; Iida et al. 1993). JC virus shares 75% nucleic acid homology with a second human polyomavirus, BK virus (polyomavirus hominis 1), and 69% homology with the simian agent, SV40 virus (Frisque et al. 1984). Sequences showing greatest divergence between these three agents occur on the late side of the origin of replication (Frisque et al. 1984). Both JC and BK viruses agglutinate human type O erythrocytes, and also erythrocytes of a number of animal species (Mantyjarvi et al. 1972; Padgett et al. 1971, 1977a), a property not shared with SV40 virus.

Synthesis of polyomavirus proteins has been most thoroughly studied in the case of SV40 virus; synthesis of JC virus proteins in infected cells is believed to occur in similar fashion. SV40 virus (or JC virus by analogy) encodes both 'early', or non-structural proteins which control synthesis of viral nucleic acids, and 'late', or structural, proteins which are components of the virion. The two major non-structural proteins, large T antigen ('T' antigen) and small t antigen ('t' antigen), are encoded by overlapping regions of the viral genome. T antigen, an 81 kDa protein, is expressed predominantly in the cell nucleus and is present in more limited fashion within the cell membrane. T antigen is involved in initiation of cellular and viral DNA replication, regulation of synthesis of early viral proteins, activation of transcription of late viral proteins, and maintenance of cell transformation (Butel 1994). T antigen is also capable of forming complexes with a number of cellular proteins including p53, DNA polymerase α, heat shock protein 70, cdc2, cyclin, tubulin, p105 Rb and Rb-related p107 (Butel 1994). The protein also serves as a recognition site for T cell-mediated cytotoxicity. JC virus T antigen is highly cross-reactive with the T antigens of BK virus and SV40 virus. In experimental systems, the presence of large T antigen in cells negative for viral structural proteins has been used as a marker for cell transformation. Small t antigen is associated with intracellular accumulation of viral DNA. Like T antigen, t antigen appears able to bind to tubulin and other cellular proteins and may also be involved in both cell transformation and viral replication (Murphy et al. 1986; Rundell 1987; De Ronde et al. 1989). A third, non-structural protein, agnoprotein, is thought to retard polymerization of VP1 molecules until these have come into contact with viral DNA within the cell nucleus (Jay et al. 1981).

The JC virus particle is composed of three structural proteins, VP1, VP2, and VP3. VP1 is the major structural protein of the virus and contains an antigenic site common to all polyomaviruses (Gerber et al. 1980; Frisque et al. 1984).

Epidemiology of JC virus infection

JC virus occurs in essentially all human populations. By age 5, 10% of children have antibody to JC virus, and by late adult life, 76% of individuals have been infected (Brown et al. 1975; Taguchi et al. 1982; Kitamura et al. 1994). Antibodies to JC virus have been detected in 69% of adults in populations so remote as to have had no contact with either measles or influenza viruses (Brown et al. 1975). Urinary excretion of JC virus, believed to represent reactivated infection, occurs in approximately 13% of patients with leukemia or renal transplantation (Arthur et al. 1983). A 4-fold rise in antibody to JC virus, also consistent with reactivation, occurs in 9–15% of pregnant women but has not been associated with fetal anomalies or serological evidence of fetal infection (Coleman et al. 1980, 1983; Daniel et al. 1981; Andrews et al. 1983).

As mentioned above, two other polyomaviruses have been associated with human infection. By adolescence, over 80% of individuals have antibodies to a second human polyomavirus, BK virus, the prevalence of antibody falling to 53% by late adult life (Gardner 1973; Shah et al. 1973; Taguchi et al. 1982). SV40 virus was a contaminant of the rhesus monkey kidney cells used to prepare the early lots of Salk and Sabin polio vaccines (Shah and Nathanson 1976). Between 10 and 30 million persons received one or more doses of live SV40 virus during immunization against polio in the United States alone (Shah and Nathanson 1976). Subclinical infection was demonstrated in volunteers receiving the virus by both the respiratory and oral routes (Morris et al. 1961; Melnick and Stinebaugh 1962). Reports also exist concerning the presence of antibodies to an SV40-like virus in sera obtained from patients prior to the availability of polio vaccines (Shah et al. 1972; Geissler et al. 1985). Neither of these studies, however, included controls to make certain that the anti-SV40 antibodies detected were not, in fact, cross-reacting antibodies to JC or BK viruses.

Growth of JC virus in vitro

JC was initially isolated using primary cultures of human fetal glial cells (Padgett et al. 1971), and the majority of studies involving isolation of JC virus have been carried out using this tissue culture system. Unfortunately, the use of this tissue culture system presents problems which effectively preclude its routine use in the diagnosis of JC virus infection. These difficulties include the requirement for fetal tissue, the prolonged period of time (sometimes weeks) needed for infection to develop in vitro, and the fact that cytopathic effect is usually subtle at best, so that optimal detection of viral replication may require immunofluorescence staining of infected cultures. Propagation of JC virus has also been reported in a number of other cell culture systems. These have included primary cultures of neonatal human urothelial cells (Beckmann and Shah 1983), fetal astrocytes (Major and Vacante 1989), a continuous cell line of human fetal glial cells transfected with an origin-defective mutant of SV40 virus (Major et al. 1985), Epstein–Barr virus-transformed human B lymphocytes (Atwood et al. 1992), cultured Schwann cells (Assouline and Major 1991), and a continuous line of JC virus-permissive human fetal glial cells by transfecting primary cultures with a replication-defective mutant of JC virus (Mandl et al. 1987). Experience with these culture systems in terms of viral diagnosis is limited. At present, virological diagnosis of PML is usually made by immunohistologic staining of brain material and, more recently, by PCR analysis of CSF (see above).

Pathogenesis of systemic JC virus infection

The route by which JC virus is transmitted in nature is not known. Sundsfjord et al. (1994) were unable to detect BK or JC virus DNA in saliva from 201 children with respiratory infections, in normal adults, or in AIDS patients, causing these authors to postulate that JC virus is most probably acquired orally rather than by respiratory spread. Similarly, very little is known about acute infection by JC virus or about persistence of JC virus in normal persons or immunosuppressed individuals. Acute JC virus in man is believed to be asymptomatic, although Blake et al. (1992) have reported a 13-year-old, immunocompetent girl in whom a chronic meningoencephalitis was accompanied by a rise in titers of antibody to JC virus. Antibody or PCR studies of CSF were not carried out in this case, however, and the relationship between the patient's clinical condition and her apparent JC virus infection remains unclear.

Although JC virus is known to persist in man, the sites of viral persistence have not been fully defined. Chesters et al. (1983), using DNA hybridization methods, identified JC virus DNA in kidneys of 3 of 30 autopsied patients. BK virus DNA was found in kidneys of 10 patients, and both JC and BK virus DNA were present in kidneys of two patients. Neither BK nor JC virus was detected in brains. Dorries and Ter Meulen (1983), employing in situ nucleic acid hybridization methods, identified JC viral nucleic acids within renal tubular cells of a patient dying with PML, and Grinnell et al. (1983) found JC virus nucleic acid sequences in the kidneys from five of seven adult PML patients. JC viral DNA was detected in lung material from one of seven of these patients (Grinnell et al. 1983) and was identified in multiple extraneural organs of two children who developed PML in the setting of severe combined immunodeficiency syndromes (Grinnell et al. 1983). Isolation of JC virus was not reported from any of these extraneural organs, however, and studies employing immunofluorescence staining for JC virus antigens failed to identify infected cell populations. In our own studies, involving PML patients without AIDS, we were unable to detect specific immunoperoxidase labeling of polyomavirus common structural antigen as evidence of productive JC virus infection, in any extraneural organ, although specific staining could be easily detected in brain and spinal cord (Greenlee et al., unpublished observations). More recently, however, Houff et al. (1988), using in situ nucleic acid hybridization methods, have detected JC virus nucleic acids in lymphocytes within spleens, bone marrow, and brains of patients with PML, and Dorries et al. (1994) have identified JC virus genomic sequences in blood leukocytes from immunocompetent individuals.

A number of workers have employed PCR methods to determine whether JC virus DNA could be detected within the central nervous systems of immunologically normal individuals. White et al. (1992), using PCR, amplified JC virus DNA from brains of over 50% of individuals without PML. Similar findings have been reported by Mori et al. (1992). Elsner and Dorries (1992), using PCR methods, identified either JC or BK virus genomic sequences in brains of

12–25% of patients without PML or malignant disease and in approximately 50% of individuals with cancer. These findings are of great interest, since they suggest that JC virus may produce latent infection of brain, and that PML may be a consequence of reactivated infection within the central nervous system.

Actual isolation of JC viruses has not been reported from the urine of immunologically normal individuals. Recent studies employing PCR methodology, however, suggest that asymptomatic excretion of JC virus DNA occurs in 24–27% of immunocompetent individuals, suggesting ongoing urinary shedding of the virus (Markowitz et al. 1993). Surprisingly, the presence of JC virus DNA in urine (as opposed to the presence of intact virions or inclusion-bearing cells) was not increased in HIV-positive individuals (Markowitz et al. 1993). These observations suggest that productive renal infection by JC virus, below limits of resolution of electron microscopic and other previously employed methods of detection, may be a relatively common occurrence in immunologically normal individuals.

Pathogenesis of central nervous system infection by JC virus

The initial events which occur in development of PML are not known. Quinlivan et al. (1992) were able to amplify JC virus DNA from brains of 4 of 13 patients with AIDS but without PML. These findings, together with the detection of JC virus nucleic acid sequences in brains of normal individuals (see above), suggest that PML may begin with reactivation of infection within the central nervous system. Alternatively, however, JC virus could initiate PML by hematogenous spread. Telenti et al. (1992), using PCR, detected JC virus DNA in the blood of one patient with PML, and JC virus has been detected in B lymphocytes and blood leukocytes of PML patients, as well as in B lymphocytes within the central nervous system (Houff et al. 1988; Dorries et al. 1994). It is thus also possible that JC virus could enter the central nervous system within these cell populations, or as an extracellular viremia (Houff et al. 1988; Dorries et al. 1994).

Although reactivation of JC virus infection is common in immunosuppressed patients, the host factors which determine the occurrence of PML in a given patient are not fully understood. Virtually all patients with PML have impaired T lymphocyte function, suggesting that cell-mediated immunity is essential for containment of persistent JC virus infection. Willoughby et al. (1980), in studies prior to the advent of AIDS, found that PML patients exhibited a prominent, specific deficiency in cellular immunity to JC virus in the setting of a more general impairment of lymphocyte-mediated response (Willoughby et al. 1980). In addition, impairment of antibody-mediated immune response is also common in both non-AIDS PML and AIDS-PML, and titers of anti-JC virus antibodies in both groups of PML patients are for the most part essentially identical to antibody titers found in the general population, suggesting that host ability to mount an antibody response against JC virus is significantly impaired in PML (Padgett and Walker 1973, 1983; Walker and Padgett 1983; Bowen et al. 1985). Although impairment of host immune response by HIV is clearly associated with increased susceptibility to PML, other infectious agents, some of which are common in AIDS, may also impair T cell-mediated immune response. These include cytomegalovirus Epstein–Barr virus, and *Toxoplasma gondii.* The roles of these agents in the pathogenesis of PML have not been defined.

Human infection by BK and SV40 viruses

Additional information concerning the pathogenesis of human polyomavirus infections has been obtained from the studies of BK virus and SV40 virus. BK virus has been isolated from the urine of pediatric patients with both hemorrhagic and non-hemorrhagic cystitis (Padgett et al. 1983; Saitoh et al. 1993). BK virus has also been isolated from the urine of a child with acute tonsillitis, and BK virus DNA has been identified in tonsillar tissue from 5 of 12 children with recurrent upper respiratory infections (Goudsmit et al. 1982; Padgett et al. 1983). Goudsmit et al. (1982) found a rise in antibody titers to BK virus in 7 of 17 children admitted to hospital for acute respiratory infections. Mantyjarvi et al. (1973) demonstrated a rise in antibody titers to BK virus in 11 of 66 children with acute respiratory infections. Although sites of BK virus persistence in kidneys or other urogenital tissues have not yet been defined in normal individuals, excretion of BK virus-infected urothelial cells has been repeatedly demonstrated under conditions of immunosuppression (Lecastas et al. 1973; Coleman 1975; Coleman et al. 1980). Urothelial infection by BK virus involving kidneys has been reported in a patient with combined immunodeficiency (Rosen et al. 1983), and BK virus infection involving kidneys, ureters, and bladder has been reported in a patient with malignant lymphoma (Shinohara et al. 1993). The detection of BK virus DNA in tonsillar tissue, as discussed above, suggests that BK virus may persist in human lymphoid tissue as well as renal tissue (Goudsmit et al. 1982), although the identity of the cell populations which might support BK viral persistence in lymphoid tissues has not been established. BK virus has not been associated with PML but has been associated with meningoencephalitis during disseminated infection in a patient with AIDS (Vallbracht et al. 1993).

Interest in human infection by SV40 virus was kindled by the discovery of the virus as a contaminant in early lots of Salk and Sabin polio vaccines. SV40 has not been associated with symptomatic extraneural infection in man. The virus has, however, been shown to produce asymptomatic infection after either oral or respiratory inoculation (Morris et al. 1961; Melnick and Stinebaugh 1962; Horvath and Fornosi 1964). Children receiving SV40-contaminated lots of polio virus vaccine were found in some but not all studies to excrete SV40 virus in their stools for as long as 4 weeks (Sweet and Hilleman 1960; Melnick and Stinebaugh 1962; Horvath and Fornosi 1964; Shah and Nathanson 1976; Goffe et al. 1995). Oral exposure to SV40 virus did not result in antibody response, even in those children who excreted the virus (Melnick and Stinebaugh 1962). SV40 virus was recovered at 3 and 11 days after inoculation from throat swabs in 3 of 35 human volunteers exposed to aerosols of SV40 virus-contaminated stocks of respiratory syncytial virus (Morris et al. 1961). Twenty-two of these 35 patients developed low levels of anti-SV40 antibody. Isolation of SV40 virus from throat swabs and CSF of a child with congenital rubella syndrome was reported by Brandner et al. (1977). However, antibody response to SV40 virus was not present, despite the fact that the child was able to mount an antibody response to rubella virus, suggesting that the SV40 virus isolate may have been a laboratory contaminant.

Animal models of human polyomavirus infection
JC virus does not produce infection in species other than humans, and for this reason, much of the extant information about the pathogenesis of polyomavirus infections is derived from studies of non-human agents. SV40 virus is known to produce renal in-

volvement in its natural host (Ashkenazi and Melnick 1962), but other sites of possible acute or persistent infection have not been studied. Under conditions of impaired host immunity, SV40 virus infection of primates has been associated with fatal interstitial pneumonia and renal tubular necrosis (Sheffield et al. 1980), as well as with a condition which closely resembles PML (Gribble et al. 1975; Holmberg et al. 1977; Horvath et al. 1992).

Two murine agents, polyoma virus and K virus, have been studied to provide information about the pathogenesis of infection with these agents in man. Polyoma virus, the prototype member of the polyomavirus subgroup, produces a protracted, widely disseminated infection in its natural host. This infection is characterized most prominently by renal involvement, but virus can also be recovered from brains (Buffett and Levinthal 1962; Levinthal et al. 1962; Rowe et al. 1962, 1963; Holmberg et al. 1977; Dubensky and Villarreal 1979). Surviving animals develop a latent renal infection which can be reactivated by pregnancy (McCance and Mims 1979). McCance et al. (1983) reported paralysis and wasting in nude (nu/nu) mice inoculated intracranially with the A2 strain of polyoma virus. Brains of symptomatic animals exhibited reduction in numbers of oligodendrocytes, and scattered oligodendrocytes contained intranuclear accumulations of polyoma virus virions. Injury to central nervous system myelin was not evident grossly or histologically but could be detected by electron microscopy (McCance et al. 1983). Harper et al. (1983) attempted to duplicate McCance's work by inoculating nude mice with the LID strain of polyoma virus. These workers observed paralysis in 81 of 95 animals studied. In contrast to the work of McCance et al., however, these workers found that paralysis was invariably associated with spinal cord or nerve root compression by virus-induced bony tumors. It is unclear whether the disparity between these two studies is due to the use of different strains of virus, nor is it known whether infection of oligodendrocytes by the A2 strain of polyoma virus employed by McCance et al. might have resulted in a PML-like illness if the animals had been maintained for longer periods of time.

The interaction of a polyomavirus with its natural host has been most thoroughly studied in the case of a second mouse papovavirus, K virus. In mice less than 6 days of age, K virus produces an overwhelming infection of systemic and pulmonary vascular endothelial cells (Greenlee 1979, 1981). Death results from interstitial pneumonia. Although the virus does not produce demyelination, brains of lethally infected animals demonstrate infection of vascular endothelial cells and scattered oligodendrocytes and, rarely, astrocytes (Greenlee et al., unpublished data). Older animals surviving acute illness also develop a widespread infection of vascular endothelial cells (Greenlee 1981; Greenlee et al. 1991). By 2–3 months after infection, however, viral DNA and viral T and V antigens can be detected in renal tubular epithelial cells (Greenlee et al. 1991). These data are consistent with PCR data concerning urinary excretion of polyomavirus in man (Markowitz et al. 1993) and suggest that polyomaviruses may not cause a truly latent infection but, instead, may persist as productive infections whose activity is sharply limited by host immune response. The immune response to K virus infection involves both B and T cell-mediated immune response: recovery from K virus pneumonia in older mice follows development of antiviral antibody (Mokhtarian and Shah 1980; Greenlee 1981), and resistance to fatal K virus pneumonia can be conferred to newborn mice by passive transfer of antibody or of B cells but not T cells from immune animals (Mokhtarian and Shah 1980). Antibody alone is not sufficient for eradication of infection, however, mice infected with K virus at 8–14 days of age maintain productive infection of scattered cell populations for 2–4 months despite high levels of circulating antibody (Greenlee 1981), and immunosuppression of mice after apparent recovery from K virus infection results in reactivation without significant fall in antibody titers (Greenlee and Dodd 1984; Greenlee et al. 1991). The data derived from studies of murine K virus infection suggest that initial protection against acute polyomavirus infection depends greatly on the ability of the host to mount a prompt antibody response. Cell-mediated immunity, although not essential for containment of acute infection, appears to be required for containment of persistent infection, and failure of T lymphocyte function allows progression of infection by cell-to-cell spread.

The role of SV40 virus

A total of five cases has been reported in which PML was thought to be caused by a virus similar to SV40 virus (Weiner et al. 1972a, b; Scherneck et al. 1981; Brun and Jonsson 1984; Hayashi et al. 1985). One of these cases has since been demonstrated to have been associated with JC virus and not with SV40 virus. Weiner et al. (1972a, b) isolated SV40 virus from two patients with pathologically proven PML. In both instances, SV40 virus was first isolated in primary cultures of African green monkey kidney (AGMK) cells derived from animals without antibody to SV40 virus (Weiner et al. 1972a, b). SV40 virus was subsequently reisolated by inoculation onto primary human fetal brain cultures and by fusion of explant PML brain cultures with BSC-1 cells (Weiner et al. 1972a, b). Confirmation of the presence of SV40 virus in brains of both patients was also achieved by demonstrating that virus extracted from the brain of one patient underwent agglutination by sera monospecific for SV40 virus but not by serum monospecific for JC or BK viruses (Penney et al. 1972). These two cases have remained essentially unique in that exhaustive efforts were made to confirm the presence of SV40 virus, as opposed to JC virus, by then-current techniques; neither case, however, has been re-evaluated using in situ nucleic acid hybridization or PCR methodology.

Three additional cases of apparent SV40-PML have been reported. Brun and Jonsson (1984) reported immunohistologic staining of infected oligodendrocytes in PML brains by anti-SV40 virus antibody but not by antisera specific for JC virus; further studies of this material have not been reported. Hayashi et al. (1985) also diagnosed SV40-PML through the use of immunohistologic techniques. Subsequent molecular analysis of Hayashi's case, however, demonstrated the causative virus to be not SV40 but rather JC virus (Eizuru et al. 1993). Scherneck et al. (1981) reported isolation of SV40 virus in cultures of human fetal brain cells and CV-1 cells inoculated with PML material, but confirmation of these data through molecular or immunohistochemical techniques has not been reported.

POLYOMAVIRUSES AND HUMAN CENTRAL NERVOUS SYSTEM NEOPLASMS

JC, BK, and SV40 viruses are all able to produce central nervous system tumors in experimental animals (Table 2). Some suggestive data also exist which raise questions as to whether JC virus or other

TABLE 2

Neural tumors produced in experimental animals by JC, BK and SV40 viruses.

	Animal species	Route of inoculation	Tumor types	References
JC virus	Hamsters		Papillary ependymoma Choroid plexus papilloma Medulloblastoma Glioblastoma Malignant astrocytoma Primitive neuroectodermal Central neuroblastoma Pineocytoma	Walker et al. 1973; Padgett et al. 1977b; Zu Rhein 1983
		Intraocular	Retinoblastoma Peripheral neuroblastoma	Ohashi et al. 1978
	Squirrel monkeys	Intracerebral	Malignant astrocytoma	London et al. 1983
	Owl monkeys	Intracerebral, intravenous, subcutaneous	Malignant astrocytoma Mixed tumor: malignant, astrocytic and neuroblastic cell types	London et al. 1978; Miller et al. 1984
BK virus	Hamsters	Intracerebral	Papillary ependymoma Choroid plexus papilloma	Costa et al. 1976; Greenlee et al. 1977
SV40 virus	Hamsters	Intracerebral	Papillary ependymoma Choroid plexus papilloma	Sweet and Hilleman 1960; Eddy et al. 1961; Eddy 1962

Adapted from Walker and Frisque (1986).

polyomaviruses might be oncogenic in man. Three epidemiological studies have suggested an increase in the incidence of central nervous system neoplasms possibly attributable to polio vaccine. The first of these studies, by Heinonen et al. (1973), evaluated the offspring of 50,000 pregnancies between 1959 and 1965 and reported a higher incidence of central nervous system tumors in children born of women who had received SV40-contaminated polio virus vaccine during pregnancy than was found in offspring of unvaccinated women. The apparent increase in neoplasms in the vaccinated group, however, was in part due to the inclusion of individuals whose neural tumors consisted of histologic abnormalities of the adrenal glands. A second study, derived from the Connecticut Tumor Registry also raised questions as to an increased incidence in central nervous system tumors, among offspring of children born between 1955 and 1960 (Farwell et al. 1979) and suggested an increase in medulloblastomas and possibly astrocytomas in offspring of women receiving polio vaccine (Farwell et al. 1979). In contrast to these

reports, however, a subsequent study by Mortimer et al. (1981), in which patients exposed to SV40 virus-contaminated polio virus vaccines during infancy and childhood were followed for 17–19 years, failed to detect an increased incidence of brain or other tumors. Similarly, Rosa et al. (1988) failed to detect antibodies to SV40 virus antibody in sera of mothers receiving polio virus vaccine whose offspring developed central nervous system tumors.

Several workers have reported the presence of polyomavirus antigens, infectious virus, or viral nucleic acid sequences in human central nervous system tumors (Greenlee 1989). Weiss et al. (1975) reported the detection of SV40 virus large T antigen in three of eight meningiomas cultured in vitro. SV40 virus was subsequently isolated from one of these tumors, but the virus could not be propagated in vitro, and electron microscopy gave equivocal results (Krieg et al. 1981). Zimmermann et al. (1981) reported the presence of SV40 T antigen in three of five meningiomas. Tabuchi et al. (1978), in investigations of 39 central nervous system tumors, reported immunohistochemical detection of SV40 T antigen in a malignant ependymoma and a choroid plexus papilloma. None of these studies has since been confirmed by identification of SV40 or other polyomavirus DNA within the tumor tissue, however, nor have subsequent investigators been able to duplicate these authors' findings using immunohistochemical methods. Greenlee et al. (1978) used immunofluorescence techniques to study explant cultures of 80 central nervous system tumors, including 14 meningiomas; large T antigen was not identified in any tumor. Kosaka et al. (1980), using anticomplement immunofluorescence methods, were unable to identify polyomavirus large T antigen in any of 69 brain tumors. Guillon et al. (1981) were unable to detect T antigen staining in any of 84 central nervous system tumors. Both the studies by Greenlee et al. and by Guillon et al. noted that occasional human sera produced non-specific nuclear staining which could be confused with weak labeling of T antigen. Takemoto et al. (1974) recovered BK virus from the primary central nervous system reticulum cell sarcoma of a patient with Wiskott–Aldrich syndrome. BK virus T antigen was not detected in frozen sections from the tumor, however, and electron microscopy of the tumor failed to demonstrate viral particles. It remains uncertain as to whether the virus played a causative role in development of the patient's tumor, or whether its presence was due to productive central nervous system infection or simply to the presence of virus within circulating blood. Subsequently, Scherneck et al. (1979) reported isolation of an SV40-like agent from a human glioblastoma, and Meinke et al. (1979) reported isolation of SV40 from a second human meningioma, as well as detection of SV40 genomic sequences in tumor tissue. Detection of SV40 nucleic acid sequences in central nervous system tumors has also been reported by Ibelgaufts and Jones (1982) and by Krieg and Scherer (1984). Bergsagel et al. (1992) reported amplification of DNA sequences homologous with a region of the SV40 virus genome from 10 of 20 choroid plexus papillomas and 10 of 11 ependymomas. In none of these instances has the presence of viral DNA or infectious virus in a given tumor been confirmed by a second laboratory, nor has any of the tumor tissue been studied by methods such as in situ hybridization or in situ PCR which is capable of proving that viral DNA is actually present within neoplastic cells. Furthermore, as discussed above, there is still no definitive proof of the existence of an SV40-like virus of humans. The significance of these data thus remains uncertain.

TABLE 3

Primary central nervous system neoplasms in patients with PML.

Author	Underlying illness	Tumor type	Remarks
Richardson 1961	Not stated	Oligodendroglioma	Asymptomatic tumor found at autopsy
Davies et al. 1973	No known underlying disorder	Lymphoma; possibly Hodgkin's disease	
Castaigne et al. 1974	Congenital immune deficiency and hemolytic anemia	Multifocal gliomas	Gliomas adjacent to PML lesions
Giarusso and Koeppen 1978	Chronic lymphocytic leukemia	Reticulum cell sarcoma	Sarcoma arose in an area of demyelination
Ho et al. 1980	Cancer of the cervix with radiotherapy; renal transplantation	Histiocytic lymphoma	
Egan et al. 1980	Renal transplantation	Reticulum cell sarcoma	
Sima et al. 1983	No known underlying condition	Multifocal gliomas	Gliomas adjacent to PML lesions
Gullotta et al. 1992	Pulmonary tuberculosis	Multifocal gliomas	HIV-negative individual

The most intriguing evidence that human polyoma-viruses might play a role in central nervous system tumors is provided by eight cases in which PML patients were found to have primary central nervous system neoplasms (Richardson 1961; Davies et al. 1973; Castaigne et al. 1974; Giarusso and Koeppen 1978; Egan et al. 1980; Ho et al. 1980; Sima et al. 1983; Gullotta et al. 1992) (Table 3). In four cases, PML was associated with primary central nervous system lymphomas or reticulum cell sarcomas (Davies et al. 1973; Giarusso and Koeppen 1978; Egan et al. 1980; Ho et al. 1980). PML was associated in one patient with an oligodendroglioma (Richardson 1961) and in three other patients with multifocal gliomatous tumors, some of were adjacent to areas of demyelination (Castaigne et al. 1974; Sima et al. 1983; Gullotta et al. 1992). In one of these patients, with PML and multifocal gliomas, in situ nucleic acid hybridization methods demonstrated JC virus nucleic acids within oligodendrocytes but not within tumor cells (Gullotta et al. 1992). In situ hybridization methods do not invariable detect JC virus DNA within PML astrocytes, however, and the failure of this study to detect JC virus in tumor cells does not exclude the possibility of non-productive infection by the virus (Gullotta et al. 1992). Detailed molecular or PCR investigations of the PML-associated lymphomas or of the other two gliomatous tumors have not been reported. The role of JC virus in the causation of these PML-associated tumors thus remains uncertain.

Acknowledgements

This work was supported by the United States Department of Veterans Affairs.

REFERENCES

AKSAMIT, A.J., JR.: Nonradioactive in situ hybridization in progressive multifocal leukoencephalopathy. Mayo Clin. Proc. 68 (1993) 899–910.

AKSAMIT, A.J., P. MOURRAIN, J.L. SEVER and E.O. MAJOR: Progressive multifocal leukoencephalopathy: investigation of 3 cases using in situ hybridization with JC virus biotinylated DNA probe. Ann. Neurol. 18 (1985) 490–496.

AKSAMIT, A.J., J.L. SEVER and E.O. MAJOR: Progressive multifocal leukoencephalopathy: JC virus detection by in situ hybridization compared with immunohisto-chemistry. Neurology 36 (1986) 499–504.

AKSAMIT, A.J. ET AL.: Diagnosis of progressive multifocal leukoencephalopathy by brain biopsy with biotin labelled DNA: DNA in situ hybridization. J. Neuropathol. Exp. Neurol. 46 (1987) 556–566.

ALLEGRANZA, A., R. BOERI, C. MARIANI, A. SGHIRLANZONI and N. PELUCHETTI: A case of progressive multifocal leukoencephalopathy in preleukemic syndrome. Acta Neuropathol. (Berlin) 7 (1981) 192–195.

ANDREWS, C.A., R.W. DANIEL and K.V. SHAH: Serological studies of papovavirus infections in pregnant women and renal transplant recipients. In: J.L. Sever and D.L. Madden (Eds.), Polyomaviruses and Human Neurological Diseases. New York, Alan R. Liss (1983) 133–141.

ANTINORI, A., A. DE LUCA, A. AMMASSARI, A. CINGOLANI, R. MURRI, G. COLOSIMO, R. ROSELLI, M. SCERRATI and E. TAMBURRINI: Failure of cytarabine and increased JC virus-DNA burden in the cerebrospinal fluid of patients with AIDS-related progressive multifocal leukoencephalopathy (Letter). AIDS 8 (1994) 1022–1024.

ARTHUR, R.R., K.V. SHAH, R.H. YOLKEN and P. CHARACHE: Detection of human papovaviruses BKV and JCV in urines by ELISA. In: J.L. Sever and D.L. Madden (Eds.), Polyomaviruses and Human Neurological Diseases. New York, Alan R. Liss (1983) 169–176.

ASHKENAZI, A. and J.L. MELNICK: Induced latent infection of monkeys with vacuolating SV-40 papova virus. Virus in kidneys and urine. Proc. Soc. Exp. Biol. Med. 111 (1962) 367–372.

ASSOULINE, J.G. and E.O. MAJOR: Human fetal Schwann cells support JC virus multiplication. J. Virol. 65 (1991) 1002–1006.

ASTROM, K.E. and G.L. STONER: Early pathological changes in progressive multifocal leukoencephalopathy: a report of two asymptomatic cases occurring prior to the AIDS epidemic. Acta Neuropathol. (Berlin) 88 (1994) 93–105.

ASTROM, K.E., E.L. MANCALL and E.P. RICHARDSON: Progressive multifocal leuko-encephalopathy, a hitherto unrecognized complication of chronic lymphatic leukaemia and Hodgkin's disease. Brain 81 (1958) 93–111.

ATWOOD, W.J., K. AMEMIYA, R. TRAUB, J. HARMS and E.O. MAJOR: Interaction of the human polyomavirus, JCV, with human B-lymphocytes. Virology 190 (1992) 716–723.

AULT, G.S. and G.L. STONER: Two major types of JC virus defined in progressive multifocal leukoencephalopathy brain by early and late coding region DNA sequences. J. Gen. Virol. 73 (1992) 2669–2678.

AULT, G.S. and G.L. STONER: Brain and kidney of progressive multifocal leukoencephalopathy patients contain identical rearrangements of the JC virus promoter/enhancer. J. Med. Virol. 44 (1994) 298–304.

AYUSO-PERALTA, L., F.J. JIMNEZ-JIMENEZ, J. TEJEIRO, A. VAQUERO, F. CABRERA-VALDIVIA, S. MADERO, A. CABELLO and E. GARCIA-ALBEA: Progressive multifocal leukoencephalopathy in HIV infection presenting in Balint's syndrome. Neurology 44 (1994) 1339–1340.

BALAKRISHNAN, J., P.S. BECKER, A.J. KUMAR, S.J. ZINREICH, J.C. MCARTHUR and R.N. BRYAN: Acquired immunodeficiency syndrome: correlation of radiologic and pathologic findings in the brain. Radiographics 10 (1990) 201–215.

BATEMAN, O.J. JR., G. SQUIRES and S.J. THANNHAUSER: Hodgkin's disease associated with Schilder's disease. Ann. Int. Med. 22 (1945) 426–431.

BAUER, W.R., A.P. TUREL, JR. and K.P. JOHNSON: Progressive multifocal leukoencephalopathy and cytarabine: remission with treatment. J. Am. Med. Assoc. 226 (1973) 174–176.

BECKMANN, A.M. and K.V. SHAH: Propagation and primary isolation of JCV and BKV in urinary epithelial cell cultures. In: J.L. Sever and D.L. Madden (Eds.), Polyomaviruses and Human Neurological Diseases. New York, Alan R. Liss (1983) 3–14.

BEDRI, J., W. WEINSTEIN, P. DEGREGORIO and M.A. VERITY: Progressive multifocal leukoencephalopathy in acquired immunodeficiency syndrome. N. Engl. J. Med. 309 (1983) 492–493.

BERGER, J.R. and L. MUCKE: Prolonged survival and partial recovery in AIDS-associated progressive multifocal leukoencephalopathy. Neurology 38 (1988) 1060–1065.

BERGER, J.R., B. KASZOVITZ, J.D. POST and G. DICKINSON: Progressive multifocal leukoencephalopathy associated with human immunodeficiency virus infection. Ann. Intern. Med. 107 (1987) 78–87.

BERGER, J.R., G. SCOTT, J. ALBRECHT, A.L. BELMAN, C. TORNATORE and E.O. MAJOR: Progressive multifocal leukoencephalopathy in HIV-1-infected children. AIDS 6 (1992) 837–841.

BERGSAGEL, D.J., M.J. FINEGOLD, J.S. BUTEL, W.J. KUPSKY and R.L. GARCEA: DNA sequences similar to those of simian virus 40 in ependymomas and choroid plexus tumors of childhood. N. Engl. J. Med. 326 (1992) 988–993.

BLAKE, K., D. PILLAY, W. KNOWLES, D.W. BROWN, P.D. GRIFFITHS and B. TAYLOR: JC virus associated meningoencephalitis in an immunocompetent girl. Arch. Dis. Child. 67 (1992) 956–957.

BOLDORINI, R., S. CRISTINA, L. VAGO, A. TOSONI, S. GUZZETTI and G. COSTANZI: Ultrastructural studies in the lytic phase of progressive multifocal leukoencephalopathy in AIDS patients. Ultrastruct. Pathol. 17 (1993) 599–609.

BOWEN, D.L., H.C. LANE and A.S. FAUCI: Immunopathogenesis of the acquired immunodeficiency syndrome. Ann. Intern. Med. 103 (1985) 703–710.

BOWLER, J.V., P.T. DAVIES and G.D. PERKIN: [99]Tcm HMPAO SPECT in progressive multifocal leucoencephalopathy. Br. J. Radiol. 65 (1992) 447–449.

BRANDNER, G., A. BURGER, D. NEUMANN-HAEFELIN, C. REINKE and H. HELWIG: Isolation of simian virus 40 from a newborn child. J. Clin. Microbiol. 5 (1977) 250–252.

BROWN, P., T. TSAI and D.C. GAJDUSEK: Seroepidemiology of human papovaviruses. Discovery of virgin population and some unusual patterns of antibody prevalence among remote peoples of the world. Am. J. Epidemiol. 102 (1975) 331–340.

BRUN, A. and N. JONSSON: Angiosarcomas in hamsters after inoculation of brain tissue from a case of progressive multifocal leukoencephalopathy. Cancer 53 (1984) 1714–1717.

BRUN, A., E. NORDENFELDT and L. KJELLEN: Aspects on the variability of progressive multifocal leukoencephalopathy. Acta Neuropathol. (Berlin) 24 (1973) 232–243.

BUFFETT, R.F. and J.D. LEVINTHAL: Polyoma virus infection in mice. Pathogenesis. Arch. Pathol. 74 (1962) 512–514.

BUTEL, J.S.: Papovaviruses. In: R.R. McKendall and W.G. Stroop (Eds.), Handbook of Neurovirology. New York, Marcel Dekker (1994) 339–354.

CARROLL, B.A., B. LANE, D. NORMAN and D. ENZMANN: Diagnosis of progressive multifocal leukoencephalopathy by computed tomography. Radiology 122 (1977) 137–141.

CASTAIGNE, P., P. RONDOT, R. ESCOUROLLE, J.L. RIBADEAU-DUMAN, F. CATHALA and J.J. HAUW: Leucoencéphalopathie multifocale progressive et 'gliomes' multiples. Rev. Neurol. 130 (1974) 379–392.

CASTLEMAN, B., R.E. SCULLY and B.U. MCNEELY: Case records of the Massachusetts General Hospital. N. Engl. J. Med. 286 (1972) 1047–1054.

CAVANAGH, J.G., D. GREENBAUM, A.H.E. MARSHALL and L.J. RUBINSTEIN: Cerebral demyelination associated with disorders of the reticuloendothelial system. Lancet ii (1959) 524–529.

CHERIF, A.A., F. DELPUECH, M. HABIB, J.F. PELLISSIER, D. GAMBARELLI, G. SALAMON and R. KHALIL: Leucoencéphalopathie multifocale progressive. Données cliniques scanographiques et neuropathologiques. Rev. Neurol. 139 (1983) 177–186.

CHESTERS, P.M., J. HERITAGE and D.J. MCCANCE: Persistence of DNA sequences of BK virus and JC virus in normal human tissues and in diseased tissues. J. Infect. Dis. 147 (1983) 676–684.

CHOY, D.S., A. WEISS and P.T. LIN: Progressive multifocal leukoencephalopathy following treatment for Wegener's granulomatosis (Letter). J. Am. Med. Assoc. 268 (1992) 600–601.

CHRISTENSEN, E. and M. FOG: A case of Schilder's disease in an adult with remarks to the etiology and pathogenesis. Acta Psychiatr. Scand. 30 (1955) 141–154.

CIRICILLO, S.F. and M.L. ROSENBLUM: Use of CT and MR imaging to distinguish intracranial lesions and to define the need for biopsy in AIDS patients. J. Neurosurg. 73 (1990) 720–724.

COLEMAN, D.V.: The cytodiagnosis of human polyomavirus infection. Acta Cytol. 19 (1975) 93–96.

COLEMAN, D.V., M.R. WOLFENDALE, R.A. DANIEL, N.K. DHANJAL, S.D. GARDNER, P.E. GIBSON and A.M. FIELD: A prospective study of human polyomavirus infection in pregnancy. J. Infect. Dis. 142 (1980) 1–8.

COLEMAN, D.V., S.D. GARDNER, C. MULHOLLAND, V. FRIDIKSDOTTIR, A.A. PORTER, R. LILFORD and H. VALDIMARSSON: Human polyomaviruses in pregnancy. A model for the study of defense mechanisms to virus reactivation. Clin. Exp. Immunol. 53 (1983) 289–296.

CONOMY, J.P., S. BEARD, H. MATSUMOTO and U. ROESSMAN: Cytarabine treatment of progressive multifocal leukoencephalopathy. J. Am. Med. Assoc. 229 (1974) 1313–1316.

CONWAY, B., W.C. HALLIDAY and R.C. BRUNHAM: Human immunodeficiency virus-associated progressive multifocal leukoencephalopathy: apparent response to 3'-azido-3'-deoxythymidine. Rev. Infect. Dis. 12 (1990) 479–482.

COSTA, J., C. YEE, T.S. TRALKA and A.S. RABSON: Hamster ependymomas produced by intracerebral inoculation of a human papovavirus. J. Natl. Cancer Inst. 56 (1976) 863–870.

DANIEL, A.M. and R.J. FRISQUE: Transcription initiation sites of prototype and variant JC virus early and late messenger RNAs. Virology 194 (1993) 97–109.

DANIEL, R., K. SHAH, D. MADDEN and S. STAGNO: Serological investigation of the possibility of congenital transmission of papovavirus JC. Infect. Immun. 33 (1981) 319–321.

DAVIES, J.A., J.T. HOUGHES and D.R. OPPENHEIMER: Richardson's disease (progressive multifocal leukoencephalopathy). Q. J. Med. 42 (1973) 481–501.

DE RONDE, A., C.J. SOL, A. VAN STRIEN, J. TER SCHEGGET and J. VAN DER NOORDAA: The SV40 small t antigen is essential for the morphological transformation of human fibroblasts. Virology 171 (1989) 260–263.

DE TOFFOL, B., M. VIDAILHET, F. GRAY, J.M. BESNIER, P. MENAGE, M.-C. LESCS, P. CHOUTET and A. AUTRET: Isolated motor control dysfunction related to progressive multifocal leukoencephalopathy during AIDS with normal MRI. Neurology 44 (1994) 2352–2355.

DE TRUCHIS, P., M. FLAMENT SAILLOUR, J.A. URTIZBEREA, D. HASSINE and B. CLAIR: Inefficacy of cytarabine in progressive multifocal leucoencephalopathy in AIDS (Letter). Lancet 342 (1993) 622–623.

DOMINGO, P., J.M. GUARDIOLA, A. IRANZO and N. MARGALL: Remission of progressive multifocal leukoencephalopathy after antiretroviral therapy. Lancet 349 (1997) 1554–1555.

DORRIES, K. and V. TER MEULEN: Progressive multifocal leukoencephalopathy: detection of papovavirus JC in renal tissue. J. Med. Virol. 11 (1983) 307–317.

DORRIES, K., E. VOGEL, S. GUNTHER and S. CZUB: Infection of human polyomaviruses JC and BK in peripheral blood leukocytes from immunocompetent individuals. Virology 198 (1994) 59–70.

DUBENSKY, T.W. and L.P. VILLARREAL: The primary site of replication alters the eventual site of persistent infection by polyomavirus in mice. J. Virol. 50 (1984) 541–546.

EDDY, B.E.: Tumors produced in hamsters by SV40. Fed. Proc. 21 (1962) 930–935.

EDDY, B.E., G.S. BORMAN, W.H. BERKELEY and R.D. YOUNG: Tumors induced in hamsters by injection of rhesus monkey kidney cell extracts. Proc. Soc. Exp. Biol. Med. 107 (1961) 191–197.

EGAN, J.D., B.L. RING, M.J. REDING and I.C. WELLS: Reticulum cell sarcoma and progressive multifocal leukoencephalopathy following renal transplantation. Transplantation 29 (1980) 84–86.

EIZURU, Y., K. SAKIHAMA, Y. MINAMISHIMA, T. HAYASHI and A. SUMIYOSHI: Re-evaluation of a case of progressive multifocal leukoencephalopathy previously diagnosed as simian virus 40 (SV40) etiology (published erratum appears in Acta Pathol. Jpn. 43 (1993) 535). Acta Pathol. Jpn. 43 (1993) 327–332.

ELLIOT, B., I. AROMIN, R. GOLD, T. FLANIGAN and M. MILAENO: 2.5 year remission of AIDS-associated progressive multifocal leukoencephalopathy with combined antiretroviral therapy. Lancet 349 (1997) 850.

ELSNER, C. and K. DORRIES: Evidence of human polyomavirus BK and JC infection in normal brain tissue. Virology 191 (1992) 72–80.

FARWELL, J.R., G.J. DOHRMANN, L.D. MARRETT and J.W. MEIGS: Effect of SV40 virus-contaminated polio vaccine on the incidence and type of CNS neoplasms in children: a population-based study. Trans. Am. Neurol. Assoc. 104 (1979) 261–268.

FEIDEN, W., K. BISE, U. STEUDE, H.W. PFISTER and A.A. MOLLER: The stereotactic biopsy diagnosis of focal intracerebral lesions in AIDS patients. Acta Neurol. Scand. 87 (1993) 228–233.

FLOMENBAUM, M.A., J.A. JARCHO and F.J. SCHOEN: Progressive multifocal leukoencephalopathy fifty-seven months after heart transplantation. J. Heart Lung Transplant. 10 (1991) 888–893.

FONG, I.W. and E. TOMA: The natural history of progressive multifocal leukoencephalopathy in patients with AIDS. Canadian PML Study Group. Clin. Infect. Dis. 20 (1995) 1305–1310.

FRISQUE, R.J., G.L. BREAM and M.T. CANNELLA: Human polyomavirus JC genome. J. Virol. 41 (1984) 458–469.

GARDNER, S.D.: Prevalence in England of antibody to human polyomavirus (B.K.). Br. Med. J. 1 (1973) 77–78.

GARROTE, F.J., J.A. MOLINA, C. LACAMBRA, M. MOLLEJO, S. MADERO and T. DEL SER: Ineficacia de la zidovudina (AZT) en la leucoencefalopatia multifocal progresiva (LMP) asociada al sindrome de inmunodeficiencia adquirida (SIDA). [The inefficacy of zidovudine (AZT) in progressive multifocal leukoencephalopathy (PML) associated with the acquired immunodeficiency syndrome (AIDS).] Rev. Clin. Esp. 187 (1990) 404–407.

GEISSLER, E., P. KONZER, S. SCHERNECK and W. ZIMMERMANN: Sera collected before introduction of contaminated polio vaccine contain antibodies against SV40. Acta Virol. 29 (1985) 420–423.

GERBER, M.A., K.V. SHAH, S.N. THUNG and G.M. ZU RHEIN: Immunohistochemical demonstration of common antigen of polyomaviruses in routine histologic tissue sections of animals and men. Am. J. Clin. Pathol. 73 (1980) 794–797.

GIARUSSO, M.H. and A.H. KOEPPEN: Atypical progressive multifocal leukoencephalopathy and primary cerebral malignant lymphoma. J. Neurol. Sci. 35 (1978) 391–398.

GIBSON, P.E., W.A. KNOWLES, J.F. HAND and D.W. BROWN: Detection of JC virus DNA in the cerebrospinal fluid of patients with progressive multifocal leukoencephalopathy. J. Med. Virol. 39 (1993) 278–281.

GILLESPIE, S.M., Y. CHANG, G. LEMP, R. ARTHUR, S. BUCHBINDER, A. STEIMLE, J. BAUMGARTNER, T. RANDO, D. NEAL, G. RUTHERFORD, L. SCHONBERGER and R. JANSSEN: Progressive multifocal leukoencephalopathy in persons infected with human immunodeficiency virus, San Francisco, 1981–1989. Ann. Neurol. 30 (1991) 597–604.

GOFFE, A.P., J. HALE and P.S. GARDNER: Poliomyelitis vaccines. Lancet 1961 (1995) 612–614.

GOUDSMIT, J., P. WERTHEIM-VAN DILLEN, A. VAN STRIEN and J. VAN DER NOORDAA: The role of BK virus in acute respiratory tract disease and the presence of BKV DNA in tonsils. J. Med. Virol. 10 (1982) 91–99.

GREENLEE, J.E.: Pathogenesis of K virus infection in newborn mice. Infect. Immun. 26 (1979) 705–713.

GREENLEE, J.E.: Effect of host age on experimental K virus infection in mice. Infect. Immun. 33 (1981) 297–303.

GREENLEE, J.E.: Progressive multifocal leukoencephalopathy. In: J.S. Remington and M.N. Schwartz (Eds.), Current Clinical Topics in Infectious Diseases, Vol. 10. Boston, MA, Blackwell Scientific (1989) 140–156.

GREENLEE, J.E. and W.K. DODD: Reactivation of persistent papovavirus K infection in immunosuppressed mice. J. Virol. 51 (1984) 425–429.

GREENLEE, J.E. and P.M. KEENEY: Immunoenzymatic labelling of JC papovavirus T antigen in brains of patients with progressive multifocal leukoencephalopathy. Acta Neuropathol. (Berlin) 71 (1986) 150–153.

GREENLEE, J.E. and W.G. STROOP: JC virus nucleic acids in the atypical astrocytes of progressive multifocal leukoencephalopathy (Abstr.). Neurology 38 (Suppl. 1) (1988) 118.

GREENLEE, J.E., O. NARAYAN, R.T. JOHNSON and R.M. HERNDON: Induction of brain tumors in hamsters with BK virus, a human papovavirus. Lab. Invest. 36 (1977) 636–641.

GREENLEE, J.E., L.E. BECKER, O. NARAYAN and R.T. JOHNSON: Failure to demonstrate papovavirus tumor antigen in human cerebral neoplasms. Ann. Neurol. 3 (1978) 479–481.

GREENLEE, J.E., R.C. PHELPS and W.G. STROOP: The major site of murine K-papovavirus persistence and reactivation is the renal tubular epithelium. Microb. Pathog. 11 (1991) 237–247.

GRIBBLE, D.H., C.C. HADEN, L.W. SCHARTZ and R.V. HENRICKSON: Spontaneous progressive multifocal leukoencephalopathy (PML) in macaques. Nature 254 (1975) 602–604.

GRINNELL, B.W., B.L. PADGETT and D.L. WALKER: Distribution of nonintegrated DNA from JC papovavirus in organs of patients with progressive multifocal leukoencephalopathy. J. Infect. Dis. 147 (1983) 669–675.

GUILLEUX, M.H., R.E. STEINER and I.R. YOUNG: MR imaging in progressive multifocal leukoencephlopathy. Am. J. Neuroradiol. 7 (1986) 1033–1035.

GUILLON, F., M. POISSON, G. GIRALDO, E. BETH, A. LORENZO and J.M. HURAUX: Anticorps sériques T SV40 au cours de tumeurs cérébrales de nature diverse: recherche négative chez 84 sujets. Nouv. Presse Méd. 10 (1981) 2584.

GULLOTTA, F., T. MASINI, G. SCARLATO and K. KUCHEL-MEISTER: Progressive multifocal leukoencephalopathy and gliomas in a HIV-negative patient. Pathol. Res. Pract. 188 (1992) 964–972.

HALL, C.B., U. DAFNI, D. SIMPSON, D. CLIFFORD, P.E. WETHERILL, B. COHEN, J. MCARTHUR, H. HOLLANDER, C. YAINNOUTSOS, E. MAJOR, L. MILLAR, J. TIMPONE and ACTC 243 TEAM: Failure of cytosine arabinoside in human immunodeficiency virus-1 associated progressive multifocal leukoencephalopathy. N. Engl. J. Med. (1998).

HALLERVORDEN, J.: Eigenartige und nicht rubrizierbare Prozesse. In: Anonymous, Hand buch des Geisteskrankheiten. Berlin, Springer-Verlag (1930) 1063–1107.

HARPER, J.S., C.J. DAWE, B.D. TRAPP, P.E. MCKEEVER, M. COLLINS, J.L. WOYCIECHOWSKA, D.L. MADDEN and J.L. SEVER: Paralysis in nude mice caused by polyoma-virus-induced vertebral tumors. In: J.L. Sever and D.L. Madden (Eds.), Polyomaviruses and Human Neurological Diseases. New York, Alan R. Liss (1983) 359–367.

HAYASHI, T., A. SUMIYOSHI, M. TANAKA, S. ARAKI and Y.

MINAMISHIMA: Progressive multifocal leukoencephalopathy. A case report with special reference to
SV40 etiology. Acta Pathol. Jpn. 35 (1985) 173–181.

HEADINGTON, J.T. and W.O. UMIKER: Progressive multifocal
leukoencephalopathy: a case report. Neurology 12
(1962) 434–439.

HEDLEY-WHYTE, E.T., B.P. SMITH, H.R. TYLER and W.P.
PETERSON: Multifocal leukoencephalopathy with re-
mission and five-year survival. J. Neuropathol. Exp.
Neurol. 25 (1966) 107–116.

HEINONEN, O.P., S. SHAPIRO, R.R. MONSON, S.C. HARTZ, L.
ROSENBERG and D. SLONE: Immunization during pregnancy against poliomyelitis and influenza in relation to
childhood malignancies. Int. J. Epidemiol. 2 (1973)
229–235.

HENIN, D., T.W. SMITH, U. DE GIROLAMI, M. SUGHAYER and J.J.
HAUW: Neuropathology of the spinal cord in the
acquired immunodeficiency syndrome. Hum. Pathol.
23 (1992) 1106–1114.

HENSON, J., M. ROSENBLUM and H. FURNEAUX: Amplification
of JC virus DNA from brain and cerebrospinal fluid
of patients with progressive multifocal leukoencephalopathy. Neurology 41 (1991) 1967–1971.

HO, K., J.C. GARANCIS, R.D. PAEGLE, M.A. GERBER and W.J.
BORKOWSKI: Progressive multifocal leukoencephalopathy and malignant lymphoma of the brain in a patient
with immunosuppressive therapy. Acta Neuropathol.
(Berlin) 52 (1980) 81–83.

HOLMAN, R.C., R.S. JANSSEN, J.W. BUEHLER, M.T. ZELASKY
and W.C. HOOPER: Epidemiology of progressive multifocal leukoencephalopathy in the United States:
analysis of national mortality and AIDS surveillance
data. Neurology 41 (1991) 1733–1736.

HOLMBERG, C.A., D.H. GRIBBLE, K.K. TAKEMOTO, P.M.
HOWLEY, C. ESPANA and B.E. OSBORN: Isolation of simian
virus 40 from rhesus monkeys (Macaca mulatta)
with spontaneous progressive multifocal leukoencephalopathy. J. Infect. Dis. 136 (1977) 593–596.

HORVATH, B.L. and F. FORNOSI: Excretion of SV40 after oral
administration of contaminated polio vaccine. Acta
Microbiol. Acad. Sci. Hung. 11 (1964) 271–275.

HORVATH, C.J., M.A. SIMON, D.J. BERGSAGEL, D.R. PAULEY,
N.W. KING, R.L. GARCEA and D.J. RINGLER: Simian virus
40-induced disease in rhesus monkeys with simian
acquired immunodeficiency syndrome. Am. J.
Pathol. 140 (1992) 1431–1440.

HOUFF, S.A., E.O. MAJOR, D.A. KATZ, C.V. KUFTA, J.L. SEVER,
S. PITTALUGA, J.R. ROBERTS, J. GITT, N. SAINI and W. LUX:
Involvement of JC virus-infected mononuclear cells
from the bone marrow and spleen in the pathogenesis
of progressive multifocal leukoencephalopathy. N.
Engl. J. Med. 318 (1988) 301–305.

HOWATSON, A.F., M. NAGAI and G.M. ZU RHEIN: Polyoma-
like virions in human demyelinating brain disease.
Can. Med. Assoc. J. 93 (1965) 379–386.

HULETTE, C.M., B.T. DOWNEY and P.C. BURGER: Progressive

multifocal leukoencephalopathy. Diagnosis by in situ
hybridization with a biotinylated JC virus DNA probe
using an automated Histomatic Code-On slide
stainer. Am. J. Surg. Pathol. 15 (1991) 791–797.

IACOANGELI, M., R. ROSELLI, A. ANTINORI, A. AMMASSARI, R.
MURRI, A. POMPUCCI and M. SCERRATI: Experience with
brain biopsy in acquired immune deficiency
syndrome-related focal lesions of the central nervous
system. Br. J. Surg. 81 (1994) 1508–1511.

IBELGAUFTS, H. and K.W. JONES: Papovavirus-related
RNA sequences in human neurogenic tumors. Acta
Neuropathol. (Berlin) 56 (1982) 118–122.

IIDA, T., T. KITAMURA, J. GUO, F. TAGUCHI, Y. ASO, K.
NAGASHIMA and Y. YOGO: Origin of JC polyomavirus
variants associated with progressive multifocal
leukoencephalopathy. Proc. Natl. Acad. Sci. U.S.A.
90 (1993) 5062–5065.

IRANZO, D., J. ROMEU, E. CASTELLOTE and J. COLL: Sindrome
de Gerstmann y leucoencefalopatia multifocal
progresiva en un paciente con SIDA. [Gerstmann's
syndrome and progressive multifocal leukoencephalopathy in a patient with AIDS (Letter).] Rev. Clin.
Esp. 191 (1992) 396–397.

IRIE, T., M. KASAI, N. ABE, K. SETO, T. NAOHARA, K. KAWA-
MURA, T. HIGA, K. SANO, H. TAKAHASHI and K. NAGASHIMA:
Cerebellar form of progressive multifocal leuko-
encephalopathy in a patient with chronic renal failure.
Intern. Med. 31 (1992) 218–223.

ITOYAMA, Y., H.D. WEBSER, N.H. STERNBERGER, E.P.
RICHARDSON, JR., D.L. WALKER, R.H. QUARLES and B.L.
PADGETT: Distribution of papovavirus, myelin-
associated glycoprotein, and myelin basic protein in
progressive multifocal leukoencephalopathy lesions.
Ann. Neurol. 11 (1982) 396–407.

JAKOBSEN, J., N.H. DIEMER, J. GAUB, B. BRUN and S. HELWEG-
LARSEN: Progressive multifocal leukoencephalopathy
in a patient without other clinical manifestations of
AIDS. Acta Neurol. Scand. 75 (1987) 209–213.

JAY, G., S. NOMURA, C.W. ANDERSON and G. KHOURY:
Identification of the SV40 agnogene product: a DNA
binding protein. Nature 291 (1981) 720–722.

JOHNSTON, R.F., R.A. GREEN and J.T. HEADINGTON: Progressive multifocal leukoencephalopathy. Arch.
Intern. Med. 111 (1963) 353–362.

KATZ, D.A., J.R. BERGER, B. HAMILTON, E.O. MAJOR and J.D.
POST: Progressive multifocal leukoencephalopathy
complicating Wiskott–Aldrich syndrome. Arch.
Neurol. 41 (1994) 422–426.

KEPES, J.J., S.M. CHOU and L.W. PRICE: Progressive
multifocal leukoencephalopathy with 10 year survival
in a patient with nontropical sprue. Neurology 25
(1975) 1006–1012.

KIMEL, D.W. and A.J. SCHUTT: Multifocal leukoencephalopathy: occurrence during 5-fluorouracil and leva-
misole therapy and resolution after discontinuation of
chemotherapy. Mayo Clin. Proc. 68 (1993) 363–365.

KITAMURA, T., T. KUNITAKE, J. GUO, T. TOMINAGA, K. KAWABE and Y. YOGO: Transmission of the human polyomavirus JC virus occurs both within the family and outside the family. J. Clin. Microbiol. 32 (1994) 2359–2363.

KNIGHT, R.S., N.M. HYMAN, S.D. GARDNER, P.E. GIBSON, M.M. ESIRI and C.P. WARLOW: Progressive multifocal leukoencephalopathy and viral antibody titers. J. Neurol. 235 (1988) 458–461.

KNOWLES, W.A., I.R. SHARP, L. EFSTRATIOU, J.F. HAND and S.D. GARDNER: Preparation of monoclonal antibodies to JC virus and their use in the diagnosis of progressive multifocal leucoencephalopathy. J. Med. Virol. 34 (1991) 127–131.

KOSAKA, H., Y. SANA, Y. MATSUKADO, T. SAIRENJI and Y. HINUMA: Failure to detect papovavirus-associated T antigens in human brain tumor cells by anti-complement immunofluorescence. J. Neurosurg. 52 (1980) 367–370.

KRASINSKI, K., W. BOROWSKI and R.S. HOLZMAN: Prognosis of human immunodeficiency virus infection in children and adults. Pediatr. Infect. Dis. J. 8 (1989) 216–220.

KRIEG, P. and G. SCHERER: Cloning of SV40 genomes from human brain tumors. Virology 138 (1984) 336–340.

KRIEG, P., E. AMTMANN, D. JONAS, H. FISCHER, K. ZANG and G. SAUER: Episomal simian virus 40 genomes in human brain tumors. Proc. Natl. Acad. Sci. U.S.A. 78 (1981) 6446–6450.

KRUPP, L.B., R.B. LIPTON, M.L. SWERDLOW, N.E. LEEDS and J. LLENA: Progressive multifocal leukoencephalopathy: clinical and radiographic features. Ann. Neurol. 17 (1985) 344–349.

KUCHELMEISTER, K., M. BERGMANN and F. GULLOTTA: Cellular changes in the cerebellar granular layer in AIDS-associated PML. Neuropathol. Appl. Neurobiol. 19 (1993) 398–401.

KUPP, L.B., R.B. LIPTON, M.L. SWERDLOW, N.E. LEEDS and J. LLENA: Progressive multifocal leukoencephalopathy: clinical and radiographic features. Ann. Neurol. 17 (1985) 344–349.

LAFEUILLADE, A., P. PELLEGRINO, T. BENDERITTER and R. QUILICHINI: Cécite corticale révélatrice d'une leuco-encéphalite multifocale progressive au cours du SIDA. [Cortical blindness disclosing progressive multifocal leukoencephalitis in AIDS (Letter).] Nouv. Presse Méd. 22 (1993) 554.

LECASTAS, G., O.W. PROZESKY, J. VAN WYK and H. ELS: Papovavirus in urine after renal transplantation. Nature 241 (1973) 333–334.

LEGRAIN, M., J. GRAVELEAU, S. BRION, J. MIKOL and R. KUSS: Leuco-encéphalopathie multifocale progressive après transplantation rénale. J. Neurosci. 23 (1974) 49–62.

LEVINTHAL, J.D., M. JAKOBOVITS and M.D. EATON: Polyoma disease and tumors in mice: the distribution of viral antigen detected by immunofluorescence. Virology 16 (1962) 314–319.

LEWIS, A.R., L.B. KLINE and N.B. PINKARD: Visual loss due to progressive multifocal leukoencephalopathy in a heart transplant patient. J. Clin. Neuroophthalmol. 13 (1993) 237–241.

LIPTON, R.B., L. KRUPP, D. HOROUPIAN, S. HERSHKOVITZ, J.C. AREZZO and D. KURTZBERG: Progressive multifocal leukoencephalopathy of the posterior fossa in an AIDS patient: clinical, radiographic, and evoked potential findings. Eur. Neurol. 28 (1988) 258–261.

LONDON, W.T., S.A. HOUFF, D.L. MADDEN, D.A. FUCILLO, M. GRAVELL, W.C. WALLEN, A.E. PALMER, J.L. SEVER, B.L. PADGETT, D.L. WALKER and G. ZU RHEIN: Brain tumors in owl monkeys inoculated with a human polyomavirus (JC virus). Science 201 (1978) 1246–1248.

LONDON, W.T., S.A. HOUFF, P.E. MCKEEVER, W.C. WALLEN, J.L. SEVER, B.L. PADGETT and D.L. WALKER: Viral-induced astrocytomas in squirrel monkeys. In: J.L. Sever and D.L. Madden (Eds.), Polyomaviruses and Human Neurological Diseases. New York, Alan R. Liss (1983) 227–237.

LORTHOLARY, O., G. PIALOUX, B. DUPONT, P. TROTOT, R. VAZEUX, J. MIKOL, J.B. THIEBAUT and G. GONZALEZ CANALI: Prolonged survival of a patient with AIDS and progressive multifocal leukoencephalopathy (Letter). Clin. Infect. Dis. 18 (1994) 826–827.

LYNCH, K.J. and R.J. FRISQUE: Identification of critical elements within the JC virus DNA replication origin. J. Virol. 64 (1990) 5812–5822.

LYNCH, K.J. and R.J. FRISQUE: Factors contributing to the restricted DNA replicating activity of JC virus. Virology 180 (1991) 306–317.

MAJOR, E.O. and D.A. VACANTE: Human fetal astrocytes support the growth of the neurotropic human polyomavirus, JCV. J. Neuropathol. Exp. Neurol. 48 (1989) 425–436.

MAJOR, E.O., A.E. MILLER, P. MOURRAIN, R.G. TRAUB, E. DE WIDT and J. SEVER: Establishment of a line of human fetal glial cells that supports JC virus multiplication. Proc. Natl. Acad. Sci. U.S.A. 82 (1985) 1257–1261.

MANCALL, E.L.: Progressive multifocal leukoencephalopathy. Neurology 15 (1965) 693–699.

MANDL, C., D.L. WALKER and R.J. FRISQUE: Derivation and characterization of POJ cells, transformed human fetal glial cells that retain their permissivity for JC virus. J. Virol. 61 (1987) 755–763.

MANTYJARVI, R.A., P.O. ARSTILA and O.H. MEURMAN: Hemagglutination by BK virus, a tentative new member of the papovavirus group. Infect. Immun. 6 (1972) 824–828.

MANTYJARVI, R., O. MEURMAN, L. VIHMA and B. BERLUND: Human papovavirus (B.K.), biological properties and seroepidemiology. Ann. Dis. Res. 5 (1973) 283–287.

MANZ, H.J., H.B. DINSDALE and P.A.F. MORRIN: Progressive multifocal leukoencephalopathy after renal transplantation. Ann. Intern. Med. 75 (1971) 77–81.

MARKOWITZ, R.B., H.C. THOMPSON, J.F. MUELLER, J.A. COHEN and W.S. DYNAN: Incidence of BK virus and JC virus viruria in human immunodeficiency virus-infected and uninfected subjects. J. Infect. Dis. 167 (1993) 13–20.

MARRIOTT, P.J., M.D. O'BRIEN, C.K. MACKENZIE and I. JANOTA: Progressive multifocal leucoencephalopathy remission with cytarabine. J. Neurol. Neurosurg. Psychiatry 38 (1975) 205–209.

MATHEWS, T., H. WISOTZKEY and J. MOOSSY: Multiple central nervous system infections in progressive multifocal leukoencephalopathy. Neurology 26 (1976) 9–14.

MAZLO, M. and I. TARISKA: Morphological demonstraton of the first phase of polyomavirus replication in oligodendroglia cells of human brain in progressive multifocal leukoencephalopathy. Acta Neuropathol. (Berlin) 49 (1980) 133–143.

MCCANCE, D.J. and C.A. MIMS: Reactivation of polyoma virus in kidneys of persistently infected mice during pregnancy. Infect. Immun. 25 (1979) 998–1002.

MCCANCE, D.J., A. SEBESTENY, B.E. GRIFFIN, F. BALKWILL, R. TILLY and N.A. GREGSON: A paralytic disease in nude mice associated with polyoma virus infection. J. Gen. Virol. 64 (1983) 57–67.

MCGUIRE, D., S. BARHITE, H. HOLLANDER and M. MILES: JC virus DNA in cerebrospinal fluid of human immunodeficiency virus-infected patients: predictive value for progressive multifocal leukoencephalopathy. Ann. Neurol. 37 (1995) 395–399.

MEINKE, W., D.A. GOLDSTEIN and R.M. SMITH: Simian virus 40-related DNA sequences in a human brain tumor. Neurology 29 (1979) 1590–1594.

MELNICK, J.L. and S. STINEBAUGH: Excretion of vacuolating SV-40 virus (papova virus group) after ingestion as a contaminant of oral polio vaccine. Proc. Soc. Exp. Biol. Med. 109 (1962) 965–968.

MESQUITA, R., C. PARRAVICINI, M. BJORKHOLM, M. EKMAN and P. BIBERFELD: Macrophage association of polyomavirus in progressive multifocal leukoencephalopathy: an immunohistochemical and ultrastructural study. Case report. APMIS 100 (1992) 993–1000.

MILLER, N.R., P.E. MCKEEVER, W. LONDON, B.L. PADGETT, D.L. WALKER and W.C. WALLEN: Brain tumors of owl monkeys inoculated with JC virus contain the JC virus genome. J. Virol. 49 (1984) 848–852.

MOKHTARIAN, F. and K.V. SHAH: Role of antibody response in recovery from K-papovavirus infection in mice. Infect. Immun. 29 (1980) 1169–1179.

MORET, H., M. GUICHARD, S. MATHERON, C. KATLAMA, V. SAZDOVITCH, J.M. HURAUX and D. INGRAM: Virological diagnosis of progressive multifocal leukoencephalopathy: detection of JC virus DNA in cerebrospinal fluid and brain tissue of AIDS patients. J. Clin. Microbiol. 31 (1993) 3310–3313.

MORGELLO, S., C.K. PETITO and J.A. MOURADIAN: Central nervous system lymphoma in the acquired immunodeficiency syndrome. Clin. Neuropathol. 9 (1990) 205–215.

MORI, M., N. AOKI, H. SHIMADA, M. TAJIMA and K. KATO: Detection of JC virus in the brains of aged patients without progressive multifocal leukoencephalopathy by the polymerase chain reaction and Southern hybridization analysis. Neurosci. Lett. 141 (1992) 151–155.

MORRIS, J.A., K.M. JOHNSON, C.G. AUSILIO, R.M. CHANOCK and V. KNIGHT: Clinical and serologic responses in volunteers given vacuolating virus (SV40) by the respiratory route. Proc. Soc. Exp. Biol. Med. 108 (1961) 56–59.

MORTIMER, E.A., M.L. LEPOW, E. GOLD, F.C. ROBBINS, G.J. BURTON and J.F. FRAUMENI: Long-term follow-up of persons inadvertently inoculated with SV40 as neonates. N. Engl. J. Med. 305 (1981) 517–518.

MOSHER, M.B., G.L. SHALL and J. WILSON: Progressive multifocal leukoencephalopathy: positive brain scan. J. Am. Med. Assoc. 218 (1971) 226–228.

MOULIGNIER, A., J. MIKOL, G. PIALOUX, G. FENELON, F. GRAY and J.B. THIEBAUT: AIDS-associated progressive multifocal leukoencephalopathy revealed by new-onset seizures. Am. J. Med. 99 (1995) 64–68.

MURPHY, C.I., I. BIKEL and D.M. LIVINGSTON: Cellular proteins which can specifically associate with simian virus small t antigen. J. Virol. 59 (1986) 692–702.

NEUEN JACOB, E., C. FIGGE, G. ARENDT, B. WENDTLAND, B. JACOB and W. WECHSLER: Neuropathological studies in the brains of AIDS patients with opportunistic diseases. Int. J. Legal Med. 105 (1993) 339–350.

NEWTON, H.B., M. MAKLEY, A.P. SLIVKA and J. LI: Progressive multifocal leukoencephalopathy presenting as multiple enhancing lesions on MRI: case report and literature review. J. Neuroimag. 5 (1995) 125–128.

NG, S., V.C. TSE, J. RUBINSTEIN, E. BRADFORD, D.R. ENZMANN and F.K. CONLEY: Progressive multifocal leukoencephalopathy: unusual MR findings. J. Comput. Assist. Tomogr. 19 (1995) 302–305.

NICOLI, F., B. CHAVE, J.C. PERAGUT and J.L. GASTAUT: Efficacy of cytarabine in progressive multifocal leukoencephalopathy in AIDS. Lancet 339 (1992) 306.

NIELSEN, C.J., F. GJERRIS, H. PEDERSEN, F.K. JENSEN and P. WAGN: Brain biopsy in AIDS. Diagnostic value and consequence. Acta Neurochir. (Wien) 127 (1994) 99–102.

OHASHI, T., G.M. ZU RHEIN, J.N. VARAKIS, B.L. PADGETT and D.L. WALKER: Experimental (JC virus-induced) intraocular and extraorbital tumors in the Syrian hamster. J. Neuropathol. Exp. Neurol. 37 (1978) 667–676.

O'REILLY, S.: Efficacy of camptothecin in progressive

multifocal leucoencephalopathy (Letter; Comment). Lancet 350 (1997) 291.

OWEN, R.G., R.D. PATMORE, G.M. SMITH and D.L. BARNARD: Cytomegalovirus-induced T-cell proliferation and the development of progressive multifocal leucoencephalopathy following bone marrow transplantation. Br. J. Haematol. 89 (1995) 196–198.

PADGETT, B.L. and D.L. WALKER: Prevalence of antibodies in human sera against JC virus, an isolate from a case of progressive multifocal leukoencephalopathy. J. Infect. Dis. 127 (1973) 467–470.

PADGETT, B.L. and D.L. WALKER: Virologic and serologic studies of progressive multifocal leukoencephalopathy. In: J.L. Sever and D.L. Madden (Eds.), Polyomaviruses and Human Neurological Diseases. New York, Alan R. Liss (1983) 107–117.

PADGETT, B.L., D.L. WALKER, G.M. ZU RHEIN and R.J. ECKROADE: Cultivation of a papova-like virus from human brain with progressive multifocal leukoencephalopathy. Lancet i (1971) 1257–1260.

PADGETT, B.L., C.M. ROGERS and D.L. WALKER: JC virus, a human polyomavirus associated with progressive multifocal leukoencephalopathy: additional biological characteristics and antigenic relationships. Infect. Immun. 15 (1977a) 656–660.

PADGETT, B.L., D.L. WALKER, G.M. ZU RHEIN and J.N. VARAKIS: Differential neurooncogenicity of strains of JC virus, a human papovavirus, in newborn hamsters. Cancer Res. 37 (1977b) 718–720.

PADGETT, B.L., D.L. WALKER, M.M. DESQUITADO and D.U. KIM: BK virus and non-haemorrhagic cystitis in a child (Letter). Lancet i (1983) 770.

PARR, J., S. DIKRAN, D.S. HOROUPIAN and A.C. WINKELMAN: Cerebellar form of progressive multifocal leukoencephalopathy. Can. J. Neurol. Sci. 6 (1979) 123–128.

PENNEY, J.B., L.P. WEINER, R.M. HERNDON, O. NARAYAN and R.T. JOHNSON: Virions from progressive multifocal leukoencephalopathy: rapid serological identification by electron microscopy. Science 178 (1972) 60–62.

PETERS, A.C., J. VERSTEEG, G.T.A.M. BOTS, W. BOOGERD and J. VIELVOYE: Progressive multifocal leukoencephalopathy: immunofluorescent demonstration of simian virus 40 antigen in CSF cells and response to cytarabine therapy. Arch. Neurol. 37 (1980) 497–501.

PORTEGIES, P., P.R. ALGRA, C.E. HOLLAK, J.M. PRINS, P. REISS, J. VALK and J.M. LANGE: Response to cytarabine in progressive multifocal leucoencephalopathy in AIDS. Lancet 337 (1991) 680–681.

POWER, C., A. NATH, F.Y. AOKI and M. DEL BIGIO: Remission of progressive multifocal leukoencephalopathy following splenectomy and retroviral therapy in a patient with HIV infection. N. Engl. J. Med. 336 (1997) 661–662.

PRAYSON, R.A. and M.L. ESTES: Stereotactic brain biopsy for diagnosis of progressive multifocal leukoenceph-

alopathy. South. Med. J. 86 (1993) 1381–4, 1394.

PRICE, R.W., S. NEILSEN, B. HORTEN, M. RUBINO, B. PADGETT and D. WALKER: Progressive multifocal leukoencephalopathy: a burnt-out case. Ann. Neurol. 13 (1983) 485–490.

QUINLIVAN, E.B., M. NORRIS, T.W. BOULDIN, K. SUZUKI, R. MEEKER, M.S. SMITH, C. HALL and S. KENNEY: Subclinical central nervous system infection with JC virus in patients with AIDS. J. Infect. Dis. 166 (1992) 80–85.

RAND, K.H., K.P. JOHNSON, L.J. RUBINSTEIN, J.S. WOLINSKY, J.B. PENNEY, D.L. WALKER, B.L. PADGETT and T.C. MERIGAN: Adenine arabinoside in the treatment of progressive multifocal leukoencephalopathy: use of virus-containing cells in the urine to assess response to therapy. Ann. Neurol. 1 (1977) 458–462.

REDFEARN, A., R.A. PENNIE, J.B. MAHONY and P.B. DENT: Progressive multifocal leukoencephalopathy in a child with immunodeficiency and hyperimmunoglobulinemia M. Pediatr. Infect. Dis. J. 12 (1993) 399–401.

RENZIK, M., J. HALLEUX, E. URBAIN, R. MOUCHETTE, P. CASTERMANS and M. BEUJEAN: Two cases of progressive multifocal leukoencephalopathy after renal transplantation. Acta Neuropathol. (Berlin) 7 (1981) 189–191.

RICHARDSON, E.P.: Progressive multifocal leukoencephalopathy. N. Engl. J. Med. 265 (1961) 815–823.

RICHARDSON, E.P.: Progressive multifocal leukoencephalopathy. In: P.J. Vinken and G.W. Bruyn (Eds.), Multiple Sclerosis and Other Demyelinating Diseases. Handbook of Clinical Neurology, Vol. 9. Amsterdam, North-Holland (1970) 485–499.

RICHARDSON, E.P., JR. and C.P. JOHNSON: Atypical progressive multifocal leukoencephalopathy with plasma cell infiltrates. Acta Neuropathol. (Berlin), Suppl. VI (1975) 345–350.

RICHARDSON, E.P. and H.D. WEBSTER: Progressive multifocal leukoencephalopathy: its pathological features. In: J.L. Sever and D.L. Madden (Eds.), Polyomaviruses and Human Neurological Diseases. New York, Alan R. Liss (1983) 191–203.

ROCKWELL, D., F.L. RUBEN, A. WINKLESTEIN and H. MENDELOW: Absence of immune deficiencies in a case of progressive multifocal leukoencephalopathy. Am. J. Med. 61 (1976) 433–436.

ROSA, F.W., J.L. SEVER and D.L. MADDEN: Absence of antibody response to simian virus 40 after inoculation with killed-poliovirus vaccine of mothers of offspring with neurologic tumors (Letter). N. Engl. J. Med. 318 (1988) 1469.

ROSEN, S., W. HARMON, A.M. KRENSKY, P.J. EDELSON, B.L. PADGETT, B.W. GRINNELL, M.J. RUBINO and D.L. WALKER: Tubulo-intestinal nephritis associated with polyomavirus (BK type) infection. N. Engl. J. Med. 308 (1983) 1192–1196.

ROWE, W.P., J.W. HARTLEY and R.J. HUEBNER: Polyoma and

other indigenous mouse viruses. In: R.T.C. Harris (Ed.), The Problems of Laboratory Animal Diseases. New York, Academic Press (1962) 131–142.

ROWE, W.P., J.W. HARTLEY and R.J. HUEBNER: Polyoma and other indigenous mouse viruses. Lab. Anim. Care 13 (1963) 166–175.

RUNDELL, K.: Complete interaction of cellular 56,000 and 32,000 M proteins with simian virus 40 small-t antigen in productively infected cells. J. Virol. 61 (1987) 1240–1243.

SADLER, M., R. CHINN, J. HEALY, M. FISHER, M.R. NELSON and B.G. GAZZARD: New treatments for progressive multifocal leukoencephalopathy in HIV-1-infected patients. AIDS 12 (1998) 533–535.

SAITOH, K., N. SUGAE, N. KOIKE, Y. AKIYAMA, Y. IWAMURA and H. KIMURA: Diagnosis of childhood BK virus cystitis by electron microscopy and PCR. J. Clin. Pathol. 46 (1993) 773–775.

SCHERNECK, S., M. RUDOLPH, E. GEISSLER, F. VOGEL, L. LUEBBE, H. WAEHLTE, F. WEICKMANN AND W. ZIMMERMAN: Isolation of a SV40-like papovavirus from a human glioblastoma. Int. J. Cancer 24 (1979) 523–531.

SCHERNECK, S., E. GEISSLER, W. JÄNISCH, M. RUDOLPH, F. VOGEL and W. ZIMMERMANN: Isolation of an SV40-like virus from a patient with progressive multifocal leukoencephalopathy. Acta Virol. 25 (1981) 191–198.

SCHMIDBAUER, M., H. BUDKA and K.V. SHAH: Progressive multifocal leukoencephalopathy (PML) in AIDS and in the pre-AIDS era. A neuropathological comparison using immunocytochemistry and in situ DNA hybridization for virus detection. Acta Neuropathol. (Berlin) 80 (1990) 375–380.

SELHORST, J.B., K.F. DUCY, J.M. THOMAS and W. REGELSON: PML: Remission and immunologic reversals (Abstr.). Neurology 28 (1978) 337.

SHAH, K.V. and N. NATHANSON: Human exposure to SV40: review and comment. Am. J. Epidemiol. 101 (1976) 1–11.

SHAH, K.V., F.R. MCCRUMB, R.W. DANIEL and H.L. OZER: Serologic evidence for a simian-virus-40-like infection of man. J. Natl. Cancer Inst. 48 (1972) 557–561.

SHAH, K.V., R.W. DANIEL and R.M. WARZAWSKI: High prevalence of antibodies to BK virus, an SV40-related papovavirus, in residents of Maryland. J. Infect. Dis. 128 (1973) 784–787.

SHEFFIELD, W.D., J.D. STRANDBERG, L. BRAUN, K.V. SHAH and S.S. KALTER: Simian virus 40-associated fatal interstitial pneumonia and renal tubular necrosis in a rhesus monkey. J. Infect. Dis. 142 (1980) 618–622.

SHINOHARA, T., M. MATSUDA, S.H. CHENG, J. MARSHALL, M. FUJITA and K. NAGASHIMA: BK virus infection of the human urinary tract. J. Med. Virol. 41 (1993) 301–305.

SIGURGEIRSSON, B., B. LINDELOF, O. EDHAG and E. ALLANDER: Risk of cancer in patients with dermatomyositis or polymyositis. A population-based study. N. Engl. J. Med. 326 (1992) 363–367.

SILVER, S.A., R.R. ARTHUR, Y.S. EROZAN, M.E. SHERMAN, J.C. MCARTHUR and S. UEMATSU: Diagnosis of progressive multifocal leukoencephalopathy by stereotactic brain biopsy utilizing immunohistochemistry and the polymerase chain reaction. Acta Cytol. 39 (1995) 35–44.

SILVERMAN, L. and L.J. RUBINSTEIN: Electron microscopic observations on a case of progressive multifocal leukoencephalopathy. Acta Neuropathol. (Berlin) 5 (1965) 215–224.

SIMA, A.A.F., S.D. FINDELSTEIN and D.R. MCLACHLAN: Multiple malignant astrocytomas in a patient with spontaneous progressive multifocal leukoencephalopathy. Ann. Neurol. 14 (1983) 183–188.

SINGER, C., J.R. BERGER, B.C. BOWEN, J.H. BRUCE and W.J. WEINER: Akinetic-rigid syndrome in a 13-year-old girl with HIV-related progressive multifocal leukoencephalopathy. Mov. Disord. 8 (1993) 113–116.

SINGER, E.J., G.L. STONER, P. SINGER, U. TOMIYASU, E. LICHT, B. FAHY CHANDON and W.W. TOURTELLOTTE: AIDS presenting as progressive multifocal leukoencephalopathy with clinical response to zidovudine. Acta Neurol. Scand. 90 (1994) 443–447.

SLOOFF, J.L.: Two cases of progressive multifocal leukoencephalopathy with unusual aspects. Psychiatr. Neurol. Neurochir. 69 (1966) 461–474.

SMITH, C.R., A.A.F. SIMA, I.E. SALIT and F. GENTILI: Progressive multifocal leukoencephalopathy: failure of cytarabine therapy. Neurology 32 (1982) 200–203.

SNIDER, W.D., D.M. SIMPSON, S. NIELSEN, J.W.M. GOLD, C.E. METROKA and J.B. POSNER: Neurological complications of acquired immune deficiency syndrome: analysis of 50 patients. Ann. Neurol. 14 (1983) 403–418.

SPONZILLI, E.E., J.K. SMITH, N. MALAMUD and J.R. MCCULLOCH: Progressive multifocal leukoencephalopathy: a complication of immunosuppressive treatment. Neurology 25 (1975) 664–668.

STAM, F.C.: Multifocal leukoencephalopathy with slow progression and very long survival. Psychiatr. Neurol. Neurochir. 69 (1966) 453–459.

STEIGER, M.J., G. TARNESBY, S. GABE, J. MCLAUGHLIN and A.H. SCHAPIRA: Successful outcome of progressive multifocal leukoencephalopathy with cytarabine and interferon. Ann. Neurol. 33 (1993) 407–411.

STONER, G.L., D.K. WALKER and H.D. WEBSTER: Age distribution of progressive multifocal leukoencephalopathy. Acta Neurol. Scand. 78 (1988) 307–312.

STONER, J., D. SOFFER, C.F. RYSCHKEWITSCH, D. WALKER and H.D. WEBSTER: A double-label method detects both early (T antigen) and later (capsid) proteins of JC virus in progressive multifocal leukoencephalopathy brain tissue from AIDS and non-AIDS patients. J. Neuroimmunol. 19 (1988) 223–236.

SULLIVAN, J.M., F.J. HAHN, E. ADICKES, P.Y. HAHN and S. BADAKHSH: Progressive multifocal leukoencephalopathy (PML): CT, MRI, and histopathology correlation. Nebr. Med. J. 75 (1990) 324–328.

SUNDSFJORD, A., A.R. SPEIN, E. LUCHT, T. FLAEGSTAD, O.M. SETERNES and T. TRAAVIK: Detection of BK virus DNA in nasopharyngeal aspirates from children with respiratory infections but not in saliva from immunodeficient and immunocompetent adult patients. J. Clin. Microbiol. 32 (1994) 1390–1394.

SWEENEY, B.J., H. MANJI, R.F. MILLER, M.J.G. HARRISON, F. GRAY and F. SCARAVILLI: Cortical and subcortical JC virus infection: two unusual cases of AIDS associated progressive multifocal leukoencephalopathy. J. Neurol. Neurosurg. Psychiatry 57 (1994) 994–997.

SWEET, B.H. and M.R. HILLEMAN: The vacuolating virus, SV40. Proc. Soc. Exp. Biol. Med. 105 (1960) 420–427.

TABUCHI, K., W.M. KIRSCH, M. LOW, D. GASKIN, J. VAN BUSKIRK, M. SHANE and S. MAA: Screening of human brain tumors for SV40 related T antigen. Int. J. Cancer 21 (1978) 12–17.

TAGUCHI, F., J. KAJIOKA and T. MIYAMURA: Prevalence rate and age of acquisition of antibodies against JC virus and BK virus in human sera. Microbiol. Immunol. 26 (1982) 1057–1064.

TAKASHIMA, K., D. FARRAR, N.D. ELLIOTT, J.D. RICH AND T.P. FLANIGAN: Resolution of AIDS-related opportunistic infections with addition of protease inhibitor treatment (Abstr.). 4th Conference on Retroviruses and Opportunistic Infections 355 (1996) 129.

TAKEMOTO, K.K., A.S. RABSON, M. MULLARKEY, F., R.M. BLAESE, C.F. GARON and D. NELSON: Isolation of papovavirus from brain tumor and urine of a patient with Wiskott–Aldrich syndrome. J. Nat. Cancer Inst. 53 (1974) 1205–1207.

TASHIRO, K., S. DOI, F. MORIWAKA, Y. MARUO and M. NOMURA: Progressive multifocal leukoencephalopathy with magnetic resonance imaging verification and therapeutic trials with interferon. J. Neurol. 234 (1987) 427–429.

TELENTI, A., A.J. AKSAMIT, J. PROPER and T.F. SMITH: Detection of JC virus DNA by polymerase chain reaction in patients with progressive multifocal leukoencephalopathy. J. Infect. Dis. 162 (1990) 858–861.

TELENTI, A., W.F. MARSHALL, A.J. AKSAMIT, J.D. SMILACK and T.F. SMITH: Detection of JC virus by polymerase chain reaction in cerebrospinal fluid from two patients with progressive multifocal leukoencephalopathy. Eur. J. Clin. Microbiol. Infect. Dis. 11 (1992) 253–254.

TREMBLAY, G.F., J.M. ANDERSON and D.L.W. DAVIDSON: Brain biopsy in the diagnosis of cerebral mycosis fungoides. J. Neurol. Neurosurg. Psychiatry 45 (1982) 175–178.

TROTOT, P.M., R. VAZEUX, H.K. YAMASHITA, C. SANDOZ TRONCA, J. MIKOL, C. VEDRENNE, J.B. THIEBAUT, F. GRAY, M. CIKUREL and G. PIALOUX: MRI pattern of progressive

multifocal leukoencephalopathy (PML) in AIDS. Pathological correlations. J. Neuroradiol. 17 (1990) 233–254.

UEKI, K., E.P. RICHARDSON, JR. and D.N. LOUIS: In situ polymerase chain reaction demonstration of JC virus in progressive multifocal leukoencephalopathy, including an index case. Ann. Neurol. 36 (1994) 670–673.

VALLBRACHT, A., J. LOHLER, J. GROSSMANN, T. GLUCK, D. PETERSEN, H.-J. GERTH, M. GENCIC and K. DORRIES: Disseminated BK type polyomavirus infection in an AIDS patient associated with central nervous system disease. Am. J. Pathol. 143 (1993) 29–39.

VANDERSTEENHOVEN, J.J., G. DBAIBO, O.B. BOYKO, C.M. HULETTE, D.C. ANTHONY, J.F. KENNY and C.M. WILFERT: Progressive multifocal leukoencephalopathy in pediatric acquired immunodeficiency syndrome. Pediatr. Infect. Dis. J. 11 (1992) 232–237.

VAN HORN, G., F.O. BASTIAN and J.L. MOAKE: Progressive multifocal leukoencephalopathy: failure of response to transfer factor and cytarabine. Neurology 38 (1978) 794–797.

VASEUX, R., M. CUMONT, P.M. GIRARD ET AL.: Severe encephalitis resulting from coinfections with HIV and JC virus. Neurology 40 (1990) 944–948.

VIGNOLI, C., X. DE LAMBALLERIE, C. ZANDOTTI, C. TAMALET and P. DE MICCO: Detection of JC virus by polymerase chain reaction and colorimetric DNA hybridization assay. Eur. J. Clin. Microbiol. Infect. Dis. 12 (1993) 958–961.

VOLLMER HAASE, J. P. YOUNG and E.B. RINGELSTEIN: Efficacy of camptothecin in progressive multifocal leucoencephalopathy (Letter). Lancet 349 (1997) 1366.

VON EINSIEDEL, R.W., T.D. FIFE, A.J. AKSAMIT, M.E. CORNFORD, D.L. SECOR, U. TOMIYASU, H.H. ITABASHI and H.V. VINTERS: Progressive multifocal leukoencephalopathy in AIDS: a clinicopathologic study and review of the literature. J. Neurol. 240 (1993) 391–406.

WALKER, D.L.: Progressive multifocal leukoencephalopathy: an opportunistic viral infection of the central nervous system. In: P.J. Vinken, G.W. Bruyn and H.L. Klawans (Eds.), Infections of the Nervous System, Part II. Handbook of Clinical Neurology. Amsterdam, North-Holland Publishing Company (1978) 307–329.

WALKER, D.L. and R.J. FRISQUE: The biology and molecular biology of JC virus. In: N.P. Salzman (Ed.), The Papovaviridae. 1. The Polyomaviruses. New York, Plenum Press (1986) 327–377.

WALKER, D.L. and B.L. PADGETT: The epidemiology of human polyomaviruses. In: J.L. Sever and D.L. Madden (Eds.), Polyomaviruses and Human Neurological Diseases. New York, Alan R. Liss (1983) 99–106.

WALKER, D.L., B.L. PADGETT, G.M. ZU RHEIN, A.E. ALBERT and R.F. MARSH: Human papovavirus (JC): induction of brain tumors in hamsters. Science 181 (1973) 674–676.

WEBER, T., R.W. TURNER, S. FRYE, W. LUKE, H.A. KRETZSCHMAR, W. LUER and G. HUNSMANN: Progressive multifocal leukoencephalopathy diagnosed by amplification of JC virus-specific DNA from cerebrospinal fluid. AIDS 8 (1994) 49–57.

WEINER, L.P., R.M. HERNDON, O. NARAYAN, R.T. JOHNSON, K. SHAH, L.J. RUBINSTEIN, T.J. PREZIOSI and F.K. CONLEY: Isolation of virus related to SV40 from patients with progressive multifocal leukoencephalopathy. N. Engl. J. Med. 286 (1972a) 385–390.

WEINER, L.P., R.M. HERNDON, O. NARAYAN and R.T. JOHNSON: Further studies of a simian virus 40-like virus isolated from human brain. J. Virol. 10 (1972b) 147–149.

WEISS, A.F., R. PORTMANN, H. FISCHER, K. ZANG and G. SAUER: Simian virus 40-related antigens in three human meningiomas with defined chromosome loss. Proc. Nat. Acad. Sci. U.S.A. 72 (1975) 609–613.

WHEELER, A.L., C.L. TRUWIT, B.K. KLEINSCHMIDT DEMASTERS, W.R. BYRNE and R.N. HANNON: Progressive multifocal leukoencephalopathy: contrast enhancement on CT scans and MR images. Am. J. Roentgenol. 161 (1993) 1049–1051.

WHITE, F.A., III, M. ISHAQ, G.L. STONER and R.J. FRISQUE: JC virus DNA is present in many human brain samples from patients without progressive multifocal leukoencephalopathy. J. Virol. 66 (1992) 5726–5734.

WHITEMAN, M.L., M.J. POST, J.R. BERGER, L.G. TATE, M.D. BELL and L.P. LIMONTE: Progressive multifocal leukoencephalopathy in 47 HIV-seropositive patients: neuroimaging with clinical and pathologic correlation. Radiology 187 (1993) 233–240.

WILLOUGHBY, E., R.W. PRICE, B.L. PADGETT, D.L. WALKER and B. DUPONT: Progressive multifocal leukoencephalopathy (PML): in vitro cell-mediated immune response to mitogens and JC virus. Neurology 30 (1980) 256–262.

WINKELMAN, N.W. and M.T. MOORE: Lymphogranulomatosis (Hodgkin's disease) of the nervous system. Arch. Neurol. Psychiatry 45 (1941) 304–317.

WOODHOUSE, M.A., A.D. DAYAN, J. BURSTON, I. CALDWELL, J.H. ADAMES, D. MELCHER and H. URICH: Progressive multifocal leukoencephalopathy; electron microscope study of four cases. Brain 90 (1967) 863–870.

ZIMMERMANN, W., S. SCHERNECK, E. GEISSLER and G. NISCH: Demonstration of SV40-related tumor antigen in human meningiomas by different hamster SV40-T-antisera. Acta Virol. 25 (1981) 199–204.

ZU RHEIN, G.M.: Association of papova-virions with a human demyelinating disease (progressive multifocal leukoencephalopathy). Progr. Med. Virol. 11 (1969) 185–247.

ZU RHEIN, G.M.: Studies of JC virus-induced nervous system tumors in the Syrian hamster: a review. In: J.L. Sever and D.L. Madden (Eds.), Polyomaviruses and Human Neurological Diseases. New York, Alan R. Liss (1983) 205–221.

ZU RHEIN, G.M. and S.M. CHOU: Particles resembling papova viruses in human cerebral demyelinating disease. Science 148 (1965) 1477–1479.

ZU RHEIN, G.M., B.L. PADGETT, D.L. WALKER, R.M. CHUN, S.D. HOROWITZ and R. HONG: Progressive multifocal leukoencephalopathy in a child with severe combined immunodeficiency. N. Engl. J. Med. 299 (1978) 256–257.

Handbook of Clinical Neurology, Vol. 27 (71): Systemic Diseases, Part III
M.J. Aminoff and C.G. Goetz, editors

CHAPTER 18

Neurological manifestations of paraproteinemia and cryoglobulinemia

P.H. GORDON, T.H. BRANNAGAN and N. LATOV

Department of Neurology, Neurological Institute, Columbia-Presbyterian Medical Center, New York, NY, U.S.A.

THE MONOCLONAL GAMMOPATHIES

Monoclonal proteins (M-proteins)

Paraproteinemia

Monoclonal paraproteinemia (MP), also called monoclonal gammopathy, is present in many conditions from benign to malignant and occurs non-specifically in chronic inflammatory and infectious diseases (Kyle 1978). Individual clones of antibody-producing cells (B-cells) proliferate and produce excess antibody. These antibodies, called paraproteins or M-proteins, are monoclonal and are identical in heavy and light chain types, idiotype, and antigen specificity (Latov 1984).

MP was first detected by electrophoresis (Longsworth et al. 1939), which became readily available in the 1950s when filter paper was introduced as a supporting medium. Cellulose acetate subsequently supplanted filter paper (Kohn 1957), and today serum is screened with agarose gel electrophoresis. Immunoelectrophoresis is performed when a lymphoproliferative disease (LPD) is suspected, and immunofixation electrophoresis is used to determine immunoglobulin type (Kyle 1995). Bence-Jones protein, or free immunoglobulin light chain in urine, is demonstrated by immunoelectrophoresis or immunofixation of concentrated urine.

MP occurs in 0.7–1.2% of the normal adult population (Axelsson et al. 1966; Saleun et al. 1982;

Malacrida et al. 1987; Vladutiu 1987), is rare under age 50, but increases with each decade: 0.1% in the third decade, 3% in the eighth decade, and 19% after age 95 (Axelsson et al. 1966; Kyle 1993a). The M-proteins, named according to the heavy chain class, are IgG in 61–73%, IgM in 8–10%, and IgA in the remainder (Axelsson et al. 1966; Kyle 1994). The malignant IgG and IgA MP are usually associated with multiple myeloma (MM), whereas the IgM MP are associated with Waldenström's macroglobulinemia (WM), or B-cell leukemia or lymphoma. Central nervous system (CNS) disease in the presence of MP is related to an underlying malignancy or amyloid; disease of the peripheral nervous system is more often associated with non-malignant gammopathies (Latov 1995).

Amyloidosis

Amyloidosis refers to the accumulation of fibrillar proteins or protein–polysaccharide complexes (Barlogie et al. 1992). Because the fibrils are arranged as insoluble β-pleated sheets and accumulate with time, amyloid deposition destroys normal tissue architecture. The capital letter A is used to designate all amyloid proteins and is followed by the letter abbreviation for the protein form. At least 13 different proteins have been identified. Classes include immunoglobulin-derived (AL) amyloidosis, secondary or reactive (AA) amyloidosis, heredo-familial amyloidoses, and senile systemic amyloidosis. β_2-Microglobulin-derived amyloidosis has been recognized more recently in patients undergoing long-term hemodialysis with cuprophane membranes (Barlogie et al. 1992). Secondary amyloidosis (AA), due to deposition of amyloid A, an acute-phase reactant, occurs in chronic infections and inflammatory diseases and does not cause peripheral neuropathy (Kyle and Gertz 1990). The AA fibrils are derived from proteolytic cleavage of a low-molecular-weight protein precursor, serum amyloid A, which is synthesized by hepatocytes and circulates in association with plasma high-density lipoproteins. In the United States, rheumatoid arthritis is the most frequent cause (Cohen 1994). In the hereditary neuropathic amyloidoses, abnormal transthyretin (prealbumin) is deposited in nerves and causes autonomic and peripheral sensory neuropathies. Inheritance is autosomal dominant, except in familial Mediterranean fever (Barlogie et al. 1992).

Primary amyloidosis (AL) is caused by systemic deposition of amyloid and is part of the spectrum of plasma cell dyscrasias. The source of AL amyloid is always a single clone of the B-lymphocyte. Primary amyloidosis is either idiopathic or secondary to MM or WM, and amyloidosis occurs in 15% of patients with MM (Cohen 1985). AL is the most common form of systemic amyloid deposition (Kyle and Gertz 1990). After myeloma, primary systemic amyloidosis is the most common hematologic disease associated with MP. Nine percent of patients with an M-protein at a tertiary referral center had primary amyloidosis (Kyle 1993b). Amyloid deposits in this form are fragments of antibody light chains, which may be produced by macrophage processing (Durie et al. 1982). Median age at onset is 65 years (Kyle and Gertz 1990).

Weakness and weight loss are the most common symptoms of primary amyloidosis (Cohen 1994). Patients also develop macroglossia, purpura, cardiomyopathy, nephrotic syndrome, arthritis, carpal tunnel syndrome, peripheral neuropathy, and autonomic nerve infiltration with postural hypotension, gastrointestinal paresis, and impotence (Kyle and Greipp 1983). Eighty-five percent of patients have MP (Gertz and Kyle 1994), and 15–20% have distal symmetric polyneuropathy (Cohen 1994). More than 95% have a demonstrable clonal excess of plasma cells in the bone marrow (Gertz et al. 1991b). Infiltration of the media and adventitia of small and medium-

sized cerebral arteries is a cause of brain hemorrhage, and CNS deposition may cause dementia (Cosgrove et al. 1985). Rarely, amyloid myopathy causes muscle weakness, stiffness, pain and pseudohypertrophy (Roke et al. 1988; Yamada et al. 1988). Diagnosis should be suspected in patients with these symptoms and verified by Congo Red stain of nerve biopsy or needle aspiration of subcutaneous fat. Rectal biopsy is positive in over 70% (Gertz and Kyle 1994). The proteins show apple-green birefringence under polarized light (Cohen 1994).

Treatment with melphalan and prednisone is used for patients with primary amyloidosis (Gertz et al. 1991a). Colchicine therapy benefits amyloidosis secondary to inflammatory bowel disease (Greenstein et al. 1992), and chlorambucil may benefit amyloidosis associated with juvenile rheumatoid arthritis (Deschenes et al. 1990). Congestive heart failure (Kyle and Greipp 1983) and increased serum β_2-microglobulin (Gertz et al. 1990) carry a poor prognosis, while the presence of peripheral neuropathy as the sole manifestation is associated with a better outcome (Duston et al. 1989). Overall, the median survival is 12 months – from 4 months for those with overt congestive heart failure to 50 months for those with peripheral neuropathy (Kyle et al. 1986).

Cryoglobulinemia

In 1947, Lerner et al. used the term cryoglobulins to denote a group of serum proteins that precipitate when cooled and dissolve when heated, and further characterized the cryoproteins as γ-globulins. Cryoproteins have since been found in chronic infections, lupus, polyarteritis nodosa, viral hepatitis, rheumatoid arthritis, Sjögren's syndrome, and hematologic malignancies including multiple myeloma, Waldenström's macroglobulinemia, CLL, and malignant lymphoma, and cryoglobulins may be found in low levels in normal people (Cream 1972). Autoimmune diseases are the most frequently associated conditions (Brouet et al. 1978). Mixed cryoglobulinemia can be inherited (Nightingale et al. 1981). If the cryoglobulins occur in the absence of an underlying disease, they are called essential.

There are three principal types of cryoglobulinemia: type I is a monoclonal immunoglobulin. In type II, or mixed, the cryoglobulins consist of both polyclonal (usually IgG) and monoclonal (usually IgM) immunoglobulins, with the latter having rheumatoid factor activity against the IgG. Type III, also called mixed, includes polyclonal cryoglobulins which are consistently heterogeneous; they are composed of one or more classes of polyclonal immunoglobulins, and are sometimes non-immunoglobulin molecules such as β_1, C3, or lipoproteins (Brouet et al. 1978). Cryoglobulins are type I in 25%, type II in 26%, and type III in 50% (Brouet et al. 1972). Type III cryoglobulins are associated mainly with infections and collagen vascular diseases, while type I and, to a lesser extent, type II are associated with LPDs, particularly multiple myeloma and Waldenström's macroglobulinemia.

Type I cryoglobulins are typically present in the largest quantities and may even cause a hyperviscosity syndrome. The principal manifestations are circulatory. Patients present with purpura, Raynaud's phenomenon and hemorrhagic infarctions of the digits on exposure to cold. Immune complexes cause vasculitis, arthritis, and nephritis. Liver disease is a constant feature. The clinical features of type II and III cryoglobulinemia, whether essential or secondary, include purpura, arthralgias, hepatosplenomegaly, renal disease, and vasculitis of the skin. Peripheral neuropathy has been reported in 50–70% of patients with mixed cryoglobulinemia (Garcia-Bragado et al. 1988; Ferri et al. 1992).

The cryoglobulin concentration is measured directly by centrifugation (cryocrit) and radial immunodiffusion or indirectly by comparing the serum protein concentration before and after cryoprecipitation. Increasing the concentration of a purified cryoglobulin results in an increase of the temperature at which precipitation occurs (Meltzer and Franklin 1966), and IgM, which predominates in disease, generally undergoes cryoprecipitation at lower concentrations than IgG. Electrostatic (ionic) interactions are a major force in the cold-induced insolubility (Middaugh et al. 1980).

Treatment includes corticosteroids, plasma exchange, and immunosuppressive drugs (Lippa et al. 1986). A vasculitic neuropathy associated with hepatitis C infection may improve with interferon α (Khella et al. 1995).

Neurologic symptoms in cryoglobulinemia include peripheral neuropathy and rare cerebrovascular events (Abramsky and Slavin 1974; Brouet et al. 1978). CNS manifestations are rare, but include meningitis, infarction, intracerebral hemorrhage, my-

elopathy, focal and generalized seizures, deafness, vertigo, visual blurring, and visual loss (Brouet et al. 1974; Gorevic et al. 1980). Encephalopathy, probably due to cryoglobulinemic vasculopathy, may develop and improve with plasma exchange, prednisone, and cyclophosphamide with reduction in serum cryoglobulin (Reik and Korn 1981). Autopsy has shown multiple thrombotic occlusions of small intracerebral blood vessels with adjacent foci of ischemia accompanied by marked demyelination (Abramsky and Slavin 1974). The most common neurologic manifestation is a sensorimotor neuropathy, which occurs in 7–15% of patients with cryoglobulinemia (Khella et al. 1995). Cryoglobulinic neuropathy may be caused by immunologically mediated demyelination, microcirculatory occlusion and vasculitis involving the vasa nervorum. Mononeuritis multiplex associated with HIV infection may improve with plasma exchange (Stricker et al. 1992). Neurologic involvement is more frequent in those with type III cryoglobulinemia (Brouet et al. 1974).

Plasma cell dyscrasia

Plasma cell dyscrasias include a spectrum of diseases characterized by the monoclonal proliferation of lymphoplasmacytic cells in the bone marrow. Typically, these cells produce an M-protein. The disorders include MM, plasmacytoma, WM, primary amyloidosis, and the rare condition of heavy-chain deposition disease. Other LPDs such as lymphoma and chronic lymphocytic leukemia (CLL) rarely produce M-protein (Kyle 1995). Monoclonal gammopathy of undetermined significance (MGUS) is a laboratory abnormality without evidence of malignancy, which sometimes evolves into one of the previously mentioned diseases (Barlogie et al. 1992).

The cause of the monoclonal proliferation of lymphocytes in unknown. Proposed mechanisms include chronic antigenic stimulation, abnormal T-cell regulation, or transformation of immunologically committed B-cells (Potter 1971; Seligman and Brouet 1973; Salmon 1974). Genetic, viral, or chemical factors may be important. *N*- and *K-ras* oncogene mutations are the most frequent molecular lesions in plasma cell dyscrasias, but are not found in MGUS (see below) and solitary plasmacytoma (Corradini et al. 1994). *Ras* mutations represent a late molecular lesion and may be implicated in tumor progression rather than tumor initiation.

Non-malignant monoclonal gammopathy or monoclonal gammopathy of undetermined significance

Non-malignant monoclonal gammopathy or MGUS denotes the presence of MP in a patient who has no evidence of an underlying lymphoproliferative disorder. Three percent of the population older than 70 years and 1% of those older than 50 years have MP (Axelsson et al. 1966; Fine et al. 1972; Saleun et al. 1982). In various series, MGUS accounted for 56–99.5% of patients with MP (Axelsson et al. 1966; Fine et al. 1972; Kahn et al. 1980; Radl 1985; Kyle 1994). The lower frequency reflects the experience in tertiary referral centers to which patients who are ill or with known malignancies are referred, and the higher frequency is from non-hospitalized population studies. The heavy-chain type is IgG in 73%, IgA in 11%, and IgM in 14%; the type of light chain is κ in 62% and λ in 38% (Kyle 1994). Malignant transformation has been reported to occur in 15–17% at 10 years, and 11–33% at 20 years (Fine et al. 1979; Axelsson 1986; Giraldo et al. 1991; Kyle 1994; Van de Poel et al. 1995). However, this might be an overestimate as the relative incidence of non-malignant to malignant monoclonal gammopathy in the general population is 200:1 (Radl 1985). When malignant transformation occurs, it results in MM in 56% of patients, primary amyloidosis in 14%, WM in 12%, and other lymphoproliferative disorders in 8% (Belisle et al. 1990; Kyle 1994). No single factor predicts which patients with monoclonal gammopathy will develop a malignant plasma cell disorder, so periodic clinical evaluation and measurement of the M-protein should be considered.

Myeloma

Multiple myeloma
Multiple myeloma (MM) is the most common plasma cell dyscrasia, affecting approximately 3 in 100,000 Americans each year (Linos et al. 1981), and represents 1% of all malignancies (Ries et al. 1991). The disease was first described in England (MacIntyre 1850) and later became known as Kahler's disease (Kahler 1889). The median age at onset is 60 years, black people are affected twice as often as whites, and there are associations with agricultural occupations, radiation, and benzene exposure (Cuzick 1981; Aksoy et al. 1984; Shimizu et al. 1990; Riedel et al.

1991). MM is caused by neoplastic proliferation of a single line of plasma cells. Initial symptoms and signs are fatigue, weakness, lethargy, hypercalcemia, anemia, bone disease, renal failure and immunodeficiency (Osserman 1959). Most patients with MM show evidence of bone marrow plasmacytosis (usually >15%) and bone disease (90%) (Barlogie et al. 1992), either lytic bone lesions (60%) or generalized osteoporosis (30%) (Bataille et al. 1992). Compression fractures of the spine are common and lead to localized pain, root compression, or spinal cord compression. Frank plasma cell leukemia occurs rarely.

Most patients exhibit a slight nuclear DNA excess of 5–10%, and about 30% of patients with MM have increased c-*myc* RNA expression (Greil et al. 1991). Abnormal karyotypes are seen in 30%, usually at chromosome 11, 1, and 14 (Weh et al. 1993). Interleukin-6 (IL-6) is considered to be a major growth factor for MM (Bataille et al. 1989; Klein et al. 1989), and transgenic C57/BL6 mice carrying the human IL-6 gene show a massive polyclonal plasmacytosis with production of autoantibodies (Hirano et al. 1992). The plasma cell labeling index and β_2-microglobulin levels correlate with cell mass and may be used for prognosis (Bataille et al. 1983; Greipp et al. 1995).

More than 95% of patients have M-proteins in their serum or urine; 55% IgG, 25% IgA, 1% IgD, and 20% only κ or λ light chains (Barlogie et al. 1992). The concentration of the M-component has prognostic significance as it grossly reflects the tumor mass (Durie and Salmon 1975). Uninvolved normal immunoglobulin levels are usually suppressed, accounting for the increased susceptibility to infection (Jacobson and Zolla-Pazner 1986).

For many years, standard therapy for MM has been intermittent cycles of melphalan plus prednisone (Alexanian et al. 1969), which induce a remission in 40% (McLaughlin and Alexanian 1982). The median duration of remission is 2 years. Vincristine, doxorubicin, and dexamethasone (VAD) are effective in cases of melphalan and prednisone resistance (Gregory et al. 1992). In relapsing disease, VAD induces a second remission in 40%. VAD followed by intensive myeloablative therapy and autologous marrow transplantation has a higher frequency of response, but the frequency of early death is also higher and, therefore, survival is not prolonged (Reece et al.

1993). Allogenic bone marrow transplant produces remission in 50%, and a complete response rate of 43%, but the mortality of transplant is 50% (Cavo et al. 1991; Gahrton et al. 1991). As maintenance therapy, interferon has delayed the relapse rate (Oken 1994). In asymptomatic patients, chemotherapy is withheld until there is a risk of a complication (Alexanian and Dimopoulos 1994).

The 5-year relative survival rate is 45.7% for patients with plasmacytoma of bone marrow, 25.9% for MM, and 13% for plasma cell leukemia (Hernandez et al. 1995). The overall median length of survival is 3 years (Kyle 1983; Greipp 1994).

The most common neurologic manifestations are cord and root compression caused by lytic bone lesions (Silverstein and Doniger 1963). Neuropathy may be related to metabolic and toxic insults, chemotherapy, MP, and amyloidosis (Kelly et al. 1981a). CNS invasion may occur in patients with high tumor burden (Durie et al. 1980).

Plasmacytoma

Plasmacytoma arises as part of the presentation of MM, or as a solitary mass. Solitary plasmacytoma, which accounts for 5% of malignant plasma cell diseases (Conklin and Alexanian 1975), is a localized plasma cell tumor that occurs most often in the lung, oral nasopharynx, and nasal sinuses (Robbins and Cotran 1979). MRI of the spine and pelvis is necessary to exclude occult bone or bone marrow disease elsewhere. Generalized myeloma develops in up to one-half of patients with solitary plasmacytoma of bone, usually within 3 years of the diagnosis, presumably as the result of occult disease not detected initially (Dimopolous et al. 1992). Extramedullary plasmacytoma has a lower incidence of conversion to myeloma (Holland et al. 1992).

Eighty-two percent of patients with solitary plasmacytoma of bone have MP (Ellis and Colls 1992). Patients without paraproteinemia at presentation or whose paraprotein decreases after treatment progress to myeloma less often, and failure of the paraprotein to clear after local treatment suggests occult dissemination and predicts later development of overt myeloma (Ellis and Colls 1992). At least 36% of these cases progress to myeloma. Adjuvant chemotherapy does not appear to affect the incidence of conversion.

While 93% of cases respond to radiation therapy, 62% completely (Alexanian 1980; Chak et al. 1987; Dimopoulos et al. 1992), combination therapy may be required to avoid radiation damage to the spinal cord in cases of spine tumor. In instances of local recurrence or dissemination with a progressive course, systemic chemotherapy is used (Delauche-Cavallier 1988). The prognosis is better than for MM, and the median survival is longer than 10 years (Alexanian 1982). Neurologic syndromes arise from nervous system compression and MP. Mass lesion of the brain or calvarium is an unusual presentation.

Osteosclerotic myeloma

Osteosclerotic myeloma is a plasma-cell dyscrasia characterized by sclerotic bone lesions and progressive demyelinating polyneuropathy (Kelly et al. 1983). It accounts for less than 3% of patients with myeloma and produces a focal plasmacytoma in bone. The bone marrow aspirate contains fewer than 5% plasma cells, as opposed to >10% seen in typical myeloma (Kelly et al. 1983). Raised protein levels in CSF, papilledema, and mixed axonal and demyelinating neuropathy by electromyography and nerve biopsy are common features (Soubrier et al. 1994). The M-protein is usually composed of IgA or IgG with λ light chains (Kelly et al. 1983).

Osteosclerotic myeloma may be associated with other systemic manifestations and most patients develop one or more manifestations of the Crow–Fukase syndrome (Crow 1956; Shimpo et al. 1968), also called the POEMS (polyneuropathy, organomegaly, endocrinopathy, M-protein, and skin changes) syndrome (Miralles et al. 1992). Osteosclerosis is present in 3% of patients with MM versus 90% with POEMS syndrome. Skin changes are varied and include hypertrichosis, hyperpigmentation, diffuse skin thickening, hemangiomas, finger clubbing and white nail beds (Waldenström et al. 1978; Miralles et al. 1992). Endocrinopathy produces diabetes, hypothyroidism, and gynecomastia and impotence in men and amenorrhea in women. Organomegaly includes hepatosplenomegaly and generalized lymphadenopathy (Nakanishi et al. 1984).

Diagnosis of osteosclerotic myeloma rests on the demonstration of monoclonal plasma cells in the biopsy of a sclerotic lesion (Takatsuki and Sanada 1983; Nakanishi et al. 1984). Occasionally patients with POEMS syndrome may present without associated osteosclerotic myeloma. Elevated levels of interleukin-6 (Mandler et al. 1992) and tumor necrosis factor-α (Gherardi et al. 1994) are detectable. The link between POEMS and osteosclerotic myeloma remains unexplained (Miralles et al. 1992).

Clinical improvement of the neuropathy with reduced M-protein levels results from treatment of the plasmacytoma with surgical excision, radiation, and prednisolone (Nakanishi et al. 1984; Soubrier et al. 1994). The neuropathy responds to treatment in nearly 50% of patients (Miralles et al. 1992).

Waldenström's macroglobulinemia (WM)

This is a plasmacytoid lymphocytic lymphoma that accounts for 2% of hematologic cancers (Dimopoulos and Alexanian 1994). The median age at diagnosis is 60 years (Barlogie et al. 1992). Symptoms and signs at onset include anemia, bleeding of mucous membranes, MP, and lymphocytosis. Lymphadenopathy, splenomegaly, and hepatomegaly develop eventually and are detected in 40% by CT or MRI of the abdomen (Moulopoulos et al. 1993).

Plasmacytoid lymphocytic proliferation in bone marrow, lymph nodes, or spleen is evident on biopsy, and mast cells are also commonly found in the bone marrow. Bone marrow disease is present in more than 90% of patients (Moulopoulos et al. 1993). Abnormal and complex karyotypes are also common, and some patients have translocations that may activate c-*myc* or *bcl*-2 oncogenes (Han et al. 1983; San Roman et al. 1985).

Production of monoclonal IgM is characteristic, and circulating macroglobulin causes the hyperviscosity syndrome, cryoblobulinemia, amyloidosis, and hemolytic anemia. The hyperviscosity syndrome occurs in 15–50% of patients (Ott 1989; Dimopoulos and Alexanian 1994). While type I cryoglobulins are detected in approximately 15% of patients with overt WM, less than 5% have symptoms (MacKenzie and Fudenberg 1972; Facon et al. 1993). IgM deposition in tissues causes renal disease, amyloidosis, and peripheral neuropathy. Renal disease is less common than in MM, but may complicate amyloidosis or the precipitation of IgM molecules on the glomerular basement membrane (Morel et al. 1970). Five to 10% of patients with macroglobulinemia develop a chronic, predominantly demyelinating, sensorimotor peripheral neuropathy (Kelly et al. 1988). Amyloidosis develops in less than 5% of patients (Gertz et al. 1993). Cardiac, renal,

hepatic, and pulmonary infiltration with amyloid protein are causes of death in such cases.

Treatment of hyperviscosity or cryoglobulinemia requires prompt reduction of monoclonal protein by plasmapheresis (Pimintel 1993). Cytotoxic therapy with alkylating agents such as chlorambucil or cyclophosphamide and glucocorticoids are used for disease control. Fludarabine may be used as a salvage agent for patients with resistant WM (Dimopoulos et al. 1993; Keating et al. 1994). Older age, low hemoglobin, male sex, weight loss, cryoglobulinemia, and cytopenias carry a poor prognosis (Facon et al. 1993; Gobbi et al. 1994). The overall median survival is 5 years. Neurologic symptoms arise from peripheral neuropathy, hyperviscosity, cryoglobulinemia, and rarely CNS infiltration by the malignant cells and amyloidosis.

Chronic lymphocytic leukemia (CLL)

WM and CLL are at different ends of a spectrum of similar lymphoproliferative disorders with marked leukemia uncommon in WM, and MP uncommon in CLL. Since the original report (Minot and Isaacs 1924), CLL, the most common human leukemia in the United States (Tefferi and Phyliky 1992), has been known as an indolent disease that principally affects the elderly. The incidence ranges from rare before the fourth decade to 30.4 cases per 100,000 persons over the age of 80 years (Faguet 1994). The median age at diagnosis is 60 years (Gale and Foon 1985). There is a male:female ratio of 2:1.

The symptoms of CLL are varied. Frequently, the diagnosis is made by detecting lymphocytosis on a routine complete blood cell count. Symptomatic presentation includes fever, night sweats, or weight loss, an increased susceptibility to viral or bacterial infections, and autoimmune hemolytic anemia. Likewise, physical findings range from no abnormalities to lymphadenopathy or organomegaly secondary to lymphocyte infiltration. As the disease progresses, neoplastic B-lymphocytes accumulate, and lymphadenopathy becomes widespread. In advanced cases, non-lymphoid organ infiltration also occurs. Anemia and thrombocytopenia, evidence of bone marrow failure, occur in the most advanced stages. Immunoglobulin abnormalities in CLL include panhypogammaglobulinemia (Faguet et al. 1992) and monoclonal gammopathy, which occurs in 4–31% of patients (Bernstein et al. 1992). Diagnosis is made by detecting an elevated absolute lymphocyte count, greater than 30% replacement of the marrow cellularity by tumor cells, or clonality of blood lymphocytes (Faguet 1994).

Ninety-five percent of patients with CLL exhibit a clonal expansion of B-lymphocytes in the blood, bone marrow, lymph nodes and spleen (Fialkow et al. 1978). Genetic abnormalities are found in 50% of cases of CLL: trisomy 12 and deletions of the long arm of chromosome 13 are most common. Complex karyotypic abnormalities, a high percentage of abnormal metaphases and trisomy 12 are associated with a poor prognosis at all stages (Juliusson and Gahrton 1993; Oscier 1994). The cause of CLL is unknown.

CLL exhibits a wide range of survival, from less than 2 years for patients with bone marrow failure to more than 20 years for patients with non-progressive early-stage disease. Indolent early-stage disease requires no treatment, and patients may survive as long as healthy people (Montserrat et al. 1988). Symptomatic or more advanced disease requires treatment with oral chlorambucil which may be used with prednisone (Han et al. 1973; Catovsky et al. 1991). Fludarabine is used in resistant cases (Montserrat and Rozman 1995). Remission is achieved in 70% (Jaksic and Brugiatelli 1988). Clinical stages (Binet et al. 1981), bone marrow histopathologic findings (Rozman et al. 1984), blood lymphocyte counts (Rozman et al. 1982), lymphocyte doubling time (Montserrat et al. 1986), and cytogenetic abnormalities (Juliusson and Gahrton 1993) are good predictors of survival. Transformation of the disease into large-cell lymphoma (Richter's syndrome) carries a poor prognosis with a median survival of less than 1 year (Robertson et al. 1993). Bone marrow transplants are still experimental (Montserrat and Rozman 1994), and radiation treatment is limited to local palliation.

The major cause of mortality in CLL is recurrent infection (Itala et al. 1992), which results from hypogammaglobulinemia, effects of chemotherapy, and hypersplenism (Molica 1994). The average survival is 6 years from the time of diagnosis (Rai et al. 1975), and the 5-year survival rate is 64% (Hernandez et al. 1995).

Infiltrating CLL may cause peripheral neuropathy, cranial neuropathy, meningitis and epidural compression of the spinal cord (Cramer et al. 1996).

NEUROLOGIC SYNDROMES

Peripheral nervous system disease often results from the paraprotein itself, while disease of the CNS more often is caused by the direct effects of the underlying malignancy. Spinal cord, root, cauda equina, cranial nerve and intracranial compression have been known complications of lymphoproliferative disorders prior to modern methods of detecting MP (Silverstein and Doniger 1963). Leptomeningeal and nerve root infiltration may cause neuropathy, myelopathy, Guillain–Barré syndrome and cranial neuropathy in lymphoma and CLL (Diaz-Arrastia et al. 1992). Disease of the nervous system may also arise from metabolic abnormalities: uremia and hypercalcemia are common causes of encephalopathy in patients with MM. Convulsions are caused by uremia or mass lesion of the brain. Syncope results from severe anemia and autonomic neuropathy. Peripheral neuropathy, entrapment neuropathy, or nervous system compression may all be caused by amyloidosis. Autonomic dysfunction is particularly characteristic and leads to postural hypotension, impotence and incontinence. Ischemic and hemorrhagic infarction arise from coagulopathies (Currie and Henson 1971), infection, vasculitis, and amyloid infiltration of blood vessels. Paraneoplastic polymyositis (Currie et al. 1970) and Guillain–Barré syndrome (Lisak et al. 1977) have long been recognized. Myopathy has been reported in association with WM and antibodies to a muscle fiber surface protein (Al-Lozi et al. 1995). Treatment carries its own set of complications. An acute confusional state associated with use of corticosteroids, delayed encephalo- and myelo-malacia secondary to radiotherapy, and neuropathy caused by vinka alkaloids are examples. Opportunistic infections secondary to immunodeficiency from tumor or treatment include herpes zoster, recrudescence of tuberculosis, bacterial meningitis and progressive multifocal leukoencephalopathy (Currie and Henson 1971). The syndrome of neuromyotonia or Isaacs' syndrome has been reported with plasmacytoma and IgM paraproteinemia (Zifko et al. 1994), and myopathy has been reported with WM and IgM paraproteinemia with activity against a target antigen in muscle (Al-Lozi et al. 1995).

Brain

Disease of the brain and spinal cord more often results from an effect of the malignancy and its treatment than as a consequence of the MP, while MP serves as a marker for the presence of malignancy. Metabolic encephalopathy may result from LPD-induced hypercalcemia, hyponatremia, uremia, infection and hyperviscosity (Cassady 1993). Metabolic disturbances produce seizures more often than does cerebral compression, but impairment of consciousness and seizures may also result from intracranial tumor such as plasmacytoma. Strokes, both ischemic and hemorrhagic, result from hypercoagulable states, non-bacterial endocarditis, septic emboli, thrombocytopenia, and diffuse intravascular coagulation. Sinus venous occlusion may also result from hypercoagulable states. Meningeal infiltration is more common in the leukemias, but may occur in any LPD.

Intracranial tumor

Plasma cell dyscrasias may present as mass lesions in the brain. Solitary plasmacytomas usually occur in the lung, oral nasopharynx, or nasal sinuses (Robbins and Cotran 1979), but the tumors also arise in the skull, brain parenchyma, dura, and in the sella, and MP may be associated (Weiner et al. 1966; Bataille and Sany 1981; Kaneko et al. 1982; Pritchard et al. 1983; Benli and Inci 1995). There are at least 14 reported cases of intracranial solitary dural plasmacytomas, 10 of plasmacytomas of the bones of the base of the skull (Mancardi and Mandybur 1983), and four of solitary intracerebral plasmacytoma (Benli and Inci 1995). A mass may be palpable in tumors arising from the calvarium. Symptoms and signs are from raised intracranial pressure, focal neurologic deficits, and seizures (Stark and Henson 1981). Solitary plasmacytoma of the sphenoid sinus, mimicking pituitary adenoma with diplopia (Losa et al. 1992), and the Foster–Kennedy syndrome, monoclonal gammopathy, and elevated intracranial pressure secondary to olfactory groove tumor (Coppeto et al. 1983) are rare presentations. CT scan shows a homogenously enhancing tumor surrounded by edema which may resemble meningioma. Meningioma and plasmacytoma can both arise at similar sites (Atweh and Jabbour 1982). MRI may show a homogeneous mass with shift, edema and enhancement (Benli and Inci 1995). Treatment consists of surgical resection and radiation therapy (Bindal et

al. 1995), and cure is possible (Kuzeyli et al. 1995), though perioperative complications of hemorrhage and edema have been reported (Conklin and Alexanian 1975). MP may normalize after treatment of solitary plasmacytoma (Pritchard et al. 1983). Bone involvement may suggest non-localized disease, and MM is usually evident at presentation or shortly thereafter in such cases (Bindal et al. 1995). Solitary dural plasmacytoma carries a more benign prognosis, and progression to frank myeloma is rare (Losa et al. 1992).

Symptomatic CNS invasion occurs rarely in all stages of CLL (Cramer et al. 1996). Asymptomatic CNS CLL, however, is common, particularly in later stages, with autopsy studies showing frequencies of 8–71% (Reske-Nielsen et al. 1974; Bojsen-Møller and Nielson 1983; Barcos et al. 1987; Viadana et al. 1987). Presenting features of symptomatic invasion are confusional state, meningitis with cranial nerve palsies, optic neuropathy, and cerebellar signs. MP may be associated. CSF shows a lymphocytic pleocytosis with monoclonal cells. Cytology is frequently falsely negative and immunophenotyping of CSF lymphocytes is necessary (Bennett et al. 1989; Freedman 1990). CSF glucose is normal, and CSF protein may be normal or elevated. Lumbar puncture also shows lymphocytic pleocytosis in asymptomatic cases (Rai et al. 1975). The mechanism by which CLL cells enter the CNS remains unknown. CNS infiltration is associated with a greater number of affected systemic sites, and so CLL infiltration into the CNS may be a random event. Neuropathologic findings include leukemic meningitis, perivascular CLL cells, and lymphocytes extravasated with hemorrhage (Cramer et al. 1996). CLL may extend along perforating vessels from the bone marrow through the dura mater and into the subarachnoid space similar to other leukemias (Azzarelli and Roessmann 1977). CLL rarely manifests itself as a CNS lymphoid mass (Sheaff et al. 1994). In patients with diffuse nervous system disease, progressive multifocal leukoencephalopathy or direct brain infiltration by leukemic cells may be the cause, and a biopsy should be considered because of the different therapeutic approaches (Quitt et al. 1994). Treatment with intrathecal methotrexate increases survival (Cramer et al. 1996). The spinal cord is more commonly involved than the brain.

Neurologic abnormalities in the absence of blood

hyperviscosity are rare in WM, but may be caused by direct brain infiltration (Dutcher and Fahey 1959; Scheithauer et al. 1984; Shimizu et al. 1993). As with invasion by other brain tumors, the presentation includes confusion, poor attention, seizures, and focality. MRI may show an enhancing lesion, and biopsy reveals edema with characteristic lymphoplasmacytic cells in the Virchow–Robin spaces (Shimizu et al. 1993). Symptomatic improvement and reduction of the MP may follow local radiation therapy (Imai et al. 1995).

Isolated amyloidomas rarely involve the CNS (Moreno et al. 1983), the cranial nerves (O'Brien et al. 1994), or the skull base with adjacent brain compression (Unal et al. 1992). MRI shows increased density on CT and hyperintensity of T1-weighted MR, with marked contrast enhancement (J. Lee et al. 1995). Amyloid tumors may be derived from local synthesis by plasma cells (Eriksson et al. 1993) and may recur after surgery. Amyloid of cerebral white matter mimicking multiple sclerosis is a rare presentation (Linke et al. 1992). CNS deposits of β_2-microglobulin in the brain have been documented after 15 years of hemodialysis (Macanec et al. 1992).

Cerebral B-cell lymphoma may produce serum paraprotein (Ellie et al. 1995; S.M. Lee et al. 1995), which causes peripheral polyneuropathy, and myeloma has been reported to involve the temporal bone and cause cranial nerve palsies including facial palsy, hearing loss and tinnitus (Funakubo and Kikuchi 1994).

Hyperviscosity

Hyperviscosity syndromes result from reduced flow in the microcirculation and diminished tissue oxygenation. Etiologies include cryoglobulinemia, hyperfibrinogenemia, polycythemia, leukemia, collagen vascular diseases, sickle cell disease, other hemoglobinopathies, and paraproteinemia (McCallister et al. 1967). Hyperviscosity is present in 30–70% of patents with WM, 5% of whom have neurologic symptoms (Ott 1989), and occurs in 4–25% of patients with MM (Hobbs 1969; Tuddenham et al. 1974; Preston et al. 1980). Some degree of increased blood viscosity is present in 91% of patients with MP (McGrath and Penny 1976).

Impaired circulation affects the heart, retina, hematologic system, as well as the CNS (Ott 1989). Patients present with fatigue, mucous membrane

bleeding, blurred vision, headache, vertigo, hearing loss, ataxia, paresthesias, diplopia, retinopathy, nystagmus, and seizures (Fahey et al. 1965; Bauer 1975; Corral-Alonso and Barbolla 1984). Dementia has been reported as the only symptom (Mueller et al. 1983), and Raynaud's phenomenon occurs in cryosensitive patients. Strokes are ischemic or hemorrhagic (Logothetis et al. 1960), and perivascular extrusion of tumor cells may accompany CNS hemorrhage. Venous sinus occlusion causes venous infarcts (Sigsbee et al. 1979). The fundi may have distended, sausage-shaped retinal veins, hemorrhage, and papilledema (Fahey et al. 1965). Symptoms usually occur when the measured viscosity is at least 4 times that of normal serum (>4.0 mPas), when monoclonal IgM concentrations exceed 3.0 g/dl, or when monoclonal IgA or IgG levels exceed 5.0 g/dl (Alexanian 1977; Ott 1989), but may occur at lower levels in patients with cerebrovascular disease (Pavy et al. 1980; Preston 1981). IgM MP increases blood viscosity more than other immunoglobulin classes because of its higher molecular weight and width. IgM also coats red blood cells (RBC), further compounding impaired microcirculation with RBC aggregation (Dintenfass 1985). Brain pathology shows fibrin thrombi occluding small blood vessels and multiple small infarcts (Kudo et al. 1983; Packer et al. 1985). Symptomatic hyperviscosity requires urgent treatment with plasmapheresis (Buskard et al. 1977; Reinhart et al. 1992).

Leptomeningeal involvement

Meningeal invasion is an uncommon complication of lymphoproliferative disorders (Maldonado 1970), which must be included in the differential diagnosis in a patient with neurologic symptoms and plasma cell dyscrasia. Spread to the meninges results from extension of bony disease, hematogenous metastasis or rarely as an isolated phenomenon without evidence of systemic disease. Symptoms and signs include episodic mental status changes, headache, meningismus, vertigo, cranial nerve palsies, focal weakness, ataxia, and seizures (Leifer et al. 1992). Obstruction of CSF outflow may cause hydrocephalus. Leptomeningeal infiltration occurs in 22% of adults with non-Hodgkin lymphoma, and 20–33% of patients with leukemia (Wiesacker and Koelomel 1979), but is less common in myeloma and macroglobulinemia (Pasmantier and Azar 1969). Lumbar puncture may reveal malignant cells (Leifer et al. 1992). Monoclonal protein in the CSF is not necessarily indicative of leptomeningeal tumor involvement, as small molecular weight paraproteins may cross the blood–brain barrier (Schulman et al. 1980). CSF pleocytosis, elevated protein with or without hypoglycorrhachia, is the rule. When meningeal tumor produces paraprotein which is identical to that in serum, treatment may lower CSF paraprotein levels (Leifer et al. 1992). MRI may show meningeal or cranial nerve enhancement (Chamberlain et al. 1990; Quint et al. 1995). Definitive diagnosis is made by demonstration of malignant cells by CSF cytology with cellular typing or by meningeal biopsy. Infiltration of the meninges may be seen at autopsy with occasional spread to the parenchyma (Spiers et al. 1980). Immunochemical studies of CSF cell surface markers and early biopsy have probably more clinical value than the determination of the humoral CSF parameters (Weller et al. 1992). Only transient benefit occurs with intrathecal chemotherapy and cranial irradiation (Spiers et al. 1980; Leifer et al. 1992). In meningeal myelomatosis, survival from time of diagnosis is 5 months with treatment, and 1 month without treatment (Leifer et al. 1992).

Spinal cord

Spinal cord compression is the most frequently described neurologic complication of myelomatosis (McKissock et al. 1961), and cord compression occurs in 10–15% of patients; early in the course to late in relapse. Back pain is the first symptom, often radicular and worsened with valsalva, and this is followed by pyramidal signs of weakness and stiffness, bowel and bladder involvement and frank sensory level. Cord compression results from ingrowth of tumor through intervertebral foramina or direct extension from vertebrae, vertebral collapse with fragments of bone compressing the cord, and compression by amyloid tumors (Williams et al. 1959). The diagnosis is made by skeletal survey, and MRI or CT myelography. Therapy is wide-margin radiotherapy and high-dose steroids. Surgical decompression is necessary only in radio-resistant cases. Spinal cord compression is a neurologic emergency and treatment is usually futile if begun after progression to paraplegia (Salmon and Cassady 1993). Radiotherapy alone may be used in the palliative management of patients with spinal extradural compression and provides pain

relief in 85% and retained ambulatory ability in 65% (Ampil and Chin 1995).

The spine is also a frequent location for solitary plasmacytoma (Wiltshaw 1976), representing 25–60% of bony origins (Corwin and Lindberg 1979; Bataille et al. 1981). Symptoms are back pain and radicular pain, with signs of spinal cord compression, which complicates 40–71% of cases (Valderrama and Bullough 1968; Delauche-Cavallier et al. 1988). Patients who present with spinal cord compression are on average 10 years younger than those with MM, have involvement of the thoracic cord, and have MP (43%) (Delauche-Cavallier et al. 1988). Treatment with radiation, surgical decompression, and chemotherapy results in a 68.5% survival at 10 years (Bataille et al. 1981), though local recurrence occurs in 5–26% of patients (Wollersheim et al. 1984; Delauche-Cavallier et al. 1988), and dissemination may occur late in the course of the disease 10–20 years after the diagnosis (Alexanian 1980). Local recurrence or dissemination is associated with reappearance or increase in the MP (Delauche-Cavallier et al. 1988). Up to 20% of solitary plasmacytomas of spine evolve into MM (Alexanian 1980).

Other causes of cord compression are amyloid tumors (McAnema et al. 1982), and lymphoma. Radicular and spinal cord compressions are the most common severe neurologic complications of lymphomas, occurring in 4.6% of patients (Monteverde et al. 1991). Compression may occur in both Hodgkin and non-Hodgkin lymphomas, and the lower thoracic and lumbar spine are the areas most frequently involved (Epelbaum et al. 1986). Amyloid masses may present with spinal column masses, with destruction of bone (Arnesen and Manivel 1993). Treatment is as for other causes of malignant compression of the cord. Operative spinal stabilization may be considered in patients with painful spinal instability (Smith et al. 1995).

Motor neuron disease (MND)

A greater than expected association between motor neuron disease (MND), MP, and B-cell LPD, including myeloma, CLL, Hodgkin's disease, WM, and non-Hodgkin lymphoma, may exist (Younger et al. 1991). While MP is present in 1% of the population older than 50, there is a 7.5% frequency of MP in MND (Younger et al. 1990; Sanders et al. 1993; Louis et al. 1996). In addition, the frequency of LPD in MND has been reported to be between 2.5 and 5% (Younger et al. 1990; Rowland et al. 1992), and 46% have MP.

The motor neuron disorders associated with LPD include ALS as well as lower motor neuron syndromes (Gordon et al. 1997). The difference between diseases of the perikaryon (MND) and purely motor peripheral neuropathies is not always clear, however, and some of the patients with lower motor neuron syndromes may have motor neuropathy rather than MND. The distinction is important as motor neuropathy may improve with immunosuppression, whereas MND does not.

Peripheral nerves

The neuropathic syndromes associated with the monoclonal gammopathies are heterogeneous. Approximately 10% of patients with neuropathy of otherwise unknown etiology have a monoclonal gammopathy (Kelly et al. 1981a); 60% have IgM M-proteins, 30% have IgG M-proteins, and 20% IgA M-proteins (Gosselin et al. 1991; Yeung et al. 1991; Suarez and Kelly 1993). The IgM M-proteins frequently have autoreactive specificities against antigens in peripheral nerves and can be divided into several distinct clinical syndromes. The axonal neuropathies which are associated with IgG or IgA λ M-proteins, or with osteosclerotic myeloma, form another distinct syndrome. In other cases, the monoclonal gammopathies are associated with amyloidotic or cryoglobulinemic neuropathy, or by infiltration of nerves by tumor cells. Other patients with monoclonal gammopathies have syndromes resembling chronic inflammatory demyelinating neuropathy (CIDP), whose relationship to the monoclonal gammopathy is unclear.

NEUROPATHY ASSOCIATED WITH IGM
MONOCLONAL GAMMOPATHIES

In patients with neuropathy and IgM M-proteins, the monoclonal gammopathy is usually non-malignant, although it is sometimes associated with WM or CLL. The incidence of peripheral neuropathy in patients with IgM monoclonal gammopathies has been reported to be between 5 and 50% (Logothetis et al. 1960; Harbs et al. 1985; Kyle and Garton 1987; Nobile-Orazio et al. 1987). In most cases, the IgM

M-proteins have autoantibody activity and react with oligosaccharide determinants of glycolipids or glycoproteins (glycoconjugates) concentrated in peripheral nerve. Several distinct syndromes have been recognized and are described below. Occasionally patients present with mononeuritis or mononeuritis multiplex resulting from cryoglobulinemia and vasculitis, or from infiltration of nerve by tumor cells (Logothetis et al. 1960; Aarseth et al. 1961; Ince et al. 1987). In many cases, however, the M-proteins have no identifiable immunologic or biological activity.

Demyelinating neuropathy associated with anti-MAG antibodies

In 50–60% of patients with neuropathy and IgM monoclonal gammopathy, the M-proteins bind to myelin and to an oligosaccharide determinant that is shared by the myelin-associated glycoprotein (MAG), the Po glycoprotein, the peripheral myelin protein PMP22, and the glycolipids sulfoglucuronyl paragloboside (SGPG) and sulfoglucuronyl lactosaminyl paragloboside (SGLPG) (Latov et al. 1988b; Nobile-Orazio et al. 1989; Van den Berg et al. 1996b). The incidence of this type of neuropathy is estimated to be 1–5/10,000 adult population (Latov et al. 1988b). Patients typically present with a slowly progressive distal and symmetrical sensory or sensorimotor neuropathy which affects the arms and legs. An intention tremor may be present, sometimes early in the disease. Cranial nerves and autonomic functions are usually spared. An occasional patient may present with a predominantly motor neuropathy (Antoine et al. 1993; Van den Berg et al. 1996a,b). The cerebrospinal fluid is acellular and protein concentration is usually increased. Visual evoked responses may reveal subclinical involvement of the optic nerves, particularly if the antibodies are present in the cerebrospinal fluid (Barbieri et al. 1987). Monoclonal gammopathies with anti-MAG activity have also been reported to develop following the onset of neuropathy or in Charcot–Marie–Tooth disease, where they may contribute to the neurologic dysfunction (Julien et al. 1987; Gregory et al. 1993; Valldeoriola et al. 1993).

Electrophysiologic studies typically show demyelination or demyelination plus axonal degeneration, and pathologic studies usually show demyelination with deposits of the monoclonal antibodies and complement on affected myelin sheaths. In some nerves, there is widening of the myelin lamellae at the minor dense line, an abnormality which is closely associated with the presence of anti-MAG antibodies (Takatsu et al. 1985; Hays et al. 1988; Vital et al. 1989; Monaco et al. 1990).

Anti-MAG antibodies in patients with neuropathy almost always occur as IgM monoclonal gammopathies. These are usually non-malignant, but may also be associated with WM or CLL. The M-proteins are thought to cause the neuropathy because pathologic studies show demyelination, corresponding to the antigenic specificity of the autoantibodies, and because deposits of anti-MAG M-proteins and complement are found on the affected myelin sheaths. Passive transfer of patients' serum intraneurally into cat nerve induces demyelination (Hays et al. 1987; Willison et al. 1988; Trojaborg et al. 1989), and systemic administration of anti-MAG antibodies in the chicken causes neuropathy and demyelination with the characteristic separation of the myelin lamellae at the minor dense line, similarly to that seen in the human disease (Tatum 1993).

The neuropathy associated with anti-MAG antibodies frequently improves with therapy directed at lowering the autoantibody concentrations. Clinical improvement has been reported to follow therapy with chemotherapy using chlorambucil, cyclophosphamide, or fludarabine (Meier et al. 1984; Latov et al. 1988b; Nobile-Orazio et al. 1988; Leger et al. 1993; Sherman et al. 1994). Plasmapheresis (Haas and Tatum 1988), or intravenous gammaglobulins (i.v. IG) (Cook et al. 1990) may also be of benefit in some patients.

Motor neuropathy and IgM anti-GM1 antibodies

Monoclonal IgM anti-GM1 antibodies were first reported in a patient with IgM monoclonal gammopathy and lower MND syndrome (Freddo et al. 1986a), and in patients with motor neuropathy and multifocal conduction blocks (Pestronk et al. 1988). Other patients with increased titers of monoclonal or polyclonal IgM anti-GM1 antibodies and motor neuropathy or MND were later described (Adams et al. 1991; Pestronk 1991; Apostolski and Latov 1993; Kinsella et al. 1994). Patients with highly elevated anti-GM1 antibody titers typically have progressive weakness with muscle atrophy and fasciculations, and electrophysiologic evidence of denervation. Deep tendon reflexes may be absent or active but without plantar extensor responses. The weakness is frequently asymmetric, involves the arms more than the legs, and bulbar muscles may occasionally be affected. Some of the patients have mild distal paresthesias and sensory loss in the hands or feet, or less commonly, frank sensorimotor neuropathy. Conduction studies frequently show one or more areas of conduction block in motor nerves, although in some patients motor conductions are normal or diffusely slowed. Sensory conductions are typically normal, including the regions of motor conduction block. CSF protein is usually normal but occasionally elevated. Monoclonal gammopathies are less common than polyclonal elevations of anti-GM1 antibodies; 4 of 14 of our patients with highly elevated titers had IgM M-proteins (Kinsella et al. 1994). Serum IgM concentration may be elevated or normal. Estimates of the frequency of increased anti-GM1 antibody titers in patients with motor neuropathy and multifocal motor conduction block range from 18 to 84% (Meier et al. 1984).

Patients with highly elevated IgM anti-GM1 antibodies may constitute a distinct clinical syndrome. The clinical presentation resembles that of MND, but the weakness is reversible, and many of the patients have motor conduction abnormalities. In a recent review of our patients with highly elevated anti-GM1 antibody titers, 5 of 14 patients had a single conduction block, 4 of 14 had multiple conduction blocks, and one had diffusely slowed motor conductions (Kinsella et al. 1994). In 4 of the 14 patients, however, conductions were normal and the diagnosis would have been missed if not for the elevated anti-GM1 antibody titers. The disease is also different from chronic inflammatory demyelinating polyneuropathy (CIDP), as it is purely or predominantly motor, nerve conduction velocities between regions of block are usually normal, the conduction blocks affect motor but not sensory fibers, CSF protein is not commonly elevated, and patients with typical CIDP and sensorimotor neuropathy do not have highly elevated titers of anti-GM1 antibodies.

In most patients, the anti-GM1 antibodies recognize the Gal(B1–3)GalNAc determinant which is shared by asialo GM1 (AGM1) and the ganglioside GD1b. The same determinant is also present on some glycoproteins and is recognized by the lectin peanut agglutinin (PNA). Some of the antibodies, however, are highly specific for GM1 or recognize internal determinants shared by GM2 (Ilyas et al. 1988a; Baba et al. 1989; Kusunoki et al. 1989; Sadiq et al. 1990).

Although GM1 and other Gal(B1–3)GalNAc-bearing glycoconjugates are highly concentrated and widely distributed in the central and peripheral nervous systems, they are mostly cryptic and unavailable to the antibodies. However, anti-GM1 antibodies bind to spinal cord grey matter, and to GM1 on the surface of isolated bovine spinal motor neurons, but not to dorsal root ganglion neurons (Corbo et al. 1993). In peripheral nerve, GM1 ganglioside- and Gal(B1–3)GalNAc-bearing glycoproteins are expressed at the nodes of Ranvier (Corbo et al. 1993). Two of the glycoproteins have been identified as the oligodendroglial-myelin glycoprotein (OMgp) in paranodal myelin, and a versican-like glycoprotein in the nodal gap (Apostolski et al. 1994). The antibodies also bind to the presynaptic terminals at the motor endplate in skeletal muscle, where the antibodies might also exert an effect (Thomas et al. 1989).

Pathologic studies at post mortem in a patient who died with motor neuropathy and elevated titers of anti-GM1 antibodies revealed degeneration of the anterior roots, immunoglobulin deposits on myelin sheaths, and chromatolytic changes in spinal motor neurons (Adams et al. 1993). The predominant involvement of the anterior roots rather than the more distal nerve segments might explain the lack of correlation between the presence of conduction block and the distribution or severity of the weakness in many of the affected patients. The motor neurons may be secondarily involved to the anterior roots, as suggested by the presence of central chromatolysis.

It is not known whether the anti-GM1 antibodies cause or contribute to the disease, or whether they are only an associated abnormality. The binding to motor but not sensory neurons correlates with the clinical syndrome, and GM1 is highly enriched in myelin sheaths of motor nerves and differs in its ceramides in comparison to sensory nerves (Ogawa-Goto et al. 1990, 1992). This might render the anterior roots more susceptible to the autoantibodies' effects. In one study, rabbits immunized with GM1 or Gal(1–3)GalNAc-BSA developed conduction abnormalities with immunoglobulin deposits at the nodes

of Ranvier (Thomas et al. 1991), and in another, serum from a patient with increased titers of anti-GM1 antibodies and IgM deposits at the nodes of Ranvier produced demyelination and conduction block when injected into rat sciatic nerve (Santoro et al. 1992). The human anti-GM1 antibodies have also been shown to bind to and damage or kill mammalian spinal motor neurons in culture (Heiman-Patterson et al. 1993), and to block conduction at the motor endplate (Willison et al. 1994).

In contrast to the IgM anti-GM1 antibodies in patients with chronic motor neuropathies, increased titers of polyclonal IgG or IgA anti-GM1 antibodies are associated with an acute motor axonal neuropathy which is a variant of the Guillain–Barré syndrome. These have been reported to occur following infection with *Campylobacter jejuni* (Yuki et al. 1990; Walsh et al. 1991; McKhann et al. 1993; Van den Berg et al. 1993; Kornberg et al. 1994) which bears GM1-like oligosaccharides (Aspinall et al. 1992; Yuki et al. 1992a; Wirguin et al. 1994), or following parenteral injection of GM1-containing gangliosides (Latov et al. 1991; Nobile-Orazio et al. 1992; Landi et al. 1993). Post-mortem studies in some of the patients who died of the Guillain–Barré syndrome following *Campylobacter jejuni* infection show non-inflammatory degeneration of the anterior roots and chromatolytic changes in spinal motor neurons (McKhann et al. 1993), similar to the chronic disease associated with IgM anti-GM1 antibodies.

In patients with motor neuropathy and highly elevated titers of IgM anti-GM1 antibodies, therapy with chemotherapeutic agents such as chlorambucil, cyclophosphamide or fludarabine (Latov et al. 1988a; Pestronk et al. 1988; Shy et al. 1990; Feldman et al. 1991; Sherman et al. 1994), or with infusion of human i.v. IG (Kaji et al. 1992; Chaudhry et al. 1993; Nobile-Orazio et al. 1993) has been reported to result in clinical improvement.

Predominantly sensory neuropathy with antibodies to GD1b and disialosyl gangliosides

Monoclonal IgM antibodies that react with GD1b and disialosyl gangliosides were first described in a patient with a predominantly sensory demyelinating neuropathy (Ilyas et al. 1985). The monoclonal IgM reacted best with GD3 and GT1b but showed strong cross-reactivity with GD1b and GD2. Several other patients with predominantly sensory neuropathy and monoclonal IgM antibodies to one or more gangliosides of the same series have been reported since (Arai et al. 1992; Daune et al. 1992; Obi et al. 1992; Younes-Chennoufi et al. 1992; Yuki et al. 1992b; Willison et al. 1993). All the monoclonal antibodies reacted strongly with GD1b and, in addition, some showed strong cross-reactivity with GT1b, GD3, GD1a, or GQ1b, or exhibited anti-Pr2 and cold agglutinin activity (Arai et al. 1992; Willison et al. 1993). All had large fiber sensory loss with areflexia, and most had gait ataxia and elevated CSF proteins. The neuropathy was of the demyelinating type, and IgM deposits were not detected in biopsied nerves using direct immunofluorescence. In one of the studies, the monoclonal IgM was found to bind to myelin in cross-sections of normal nerve (Younes-Chennoufi et al. 1992). A mouse monoclonal anti-GD1b antibody was shown to bind to dorsal root ganglia neurons (Kusunoki et al. 1992), possibly explaining the sensory neuropathy. One of the patients developed an acute neuropathy as in the Guillain–Barré syndrome, but following initial improvement, the disease progressed in a stepwise fashion (Yuki et al. 1992a). Response to therapy was generally poor, although some improvement was reported with plasmapheresis and prednisone (Arai et al. 1992; Obi et al. 1992). The antibodies are likely to be responsible for the disease as immunization of rabbits with GD1b induced an experimental sensory ataxic neuropathy (Kusunoki et al. 1966).

Sensory neuropathy and anti-sulfatide antibodies

Monoclonal or polyclonal IgM antibodies to sulfatide have been reported in association with predominantly sensory neuropathy (Pestronk et al. 1991; Quattrini et al. 1992; Nemni et al. 1993; Van den Berg et al. 1993). In several of the patients, the clinical syndrome resembled that of ganglioneuritis or small-fiber sensory neuropathy, and electrophysiologic or nerve biopsy studies were normal and showed no abnormalities. Immunocytochemical studies using the anti-sulfatide antibodies from these patients revealed that the antibodies bound to the surface of rat dorsal root ganglia neurons (Quattrini et al. 1992; Nemni et al. 1993). In several other patients, the anti-sulfatide antibodies cross-reacted with MAG, and patients had a sensorimotor neuropathy, some with demyelination,

widened myelin lamellae, and deposits of IgM on myelin sheaths (Ilyas et al. 1992; Van den Berg et al. 1993; Nobile-Orazio et al. 1994). Anti-sulfatide antibodies may also sometimes cross-react with chondroitin sulfate C (Nemni et al. 1993).

The role of the anti-sulfatide antibodies in the pathogenesis of the associated neuropathies is unknown and has not yet been examined in experimental systems. In immunofluorescence studies, several of the anti-sulfatide antibodies from the patients with demyelinating neuropathy bind to the surface of myelin in peripheral nerve, whereas the antibodies from the patients with ganglioneuritis bind to the surface of dorsal root ganglia neurons but not to unfixed peripheral myelin. It is likely that the anti-sulfatide antibodies differ in their fine specificities and cross-reactivities, and that the antibodies could bind to sensory neurons, myelin, or both, depending on their fine specificities and the orientation of the sulfatide molecule on the surface of the target cells.

Neuropathy and anti-chondroitin sulfate antibodies

Several patients with monoclonal or polyclonal IgM anti-chondroitin sulfate antibodies and predominantly sensory or sensorimotor axonal neuropathy have been described (Sherman et al. 1983; Kabat et al. 1984; Freddo et al. 1985, 1986b; Yee et al. 1989; Quattrini et al. 1991; Nemni et al. 1993). In some, deposit of IgM in the endoneurium was demonstrated by immunofluorescence microscopy. Some of the chondroitin C antibodies cross-reacted with sulfatide (Nemni et al. 1993).

Neuropathy and IgM monoclonal antibodies with other specificities or with no identifiable specificity

Neuropathy associated with monoclonal IgM antibodies to other glycolipids has also been reported. One patient had a monoclonal IgM that bound to sialosyllactosaminyl paragloboside (Baba et al. 1985; Miyatani et al. 1987). Two patients had IgM M-proteins specific for GM2, GM1b-GalNAc, and GD1a-GalNAc (Ilyas et al. 1988b). One patient had a motor neuropathy with antibodies to GD1a (Bollensen et al. 1989). Another monoclonal IgM from a patient with CLL and neuropathy bound to myelin and cross-reacted with denatured DNA and with a conformational epitope of phosphatidic acid and gangliosides (Freddo et al. 1986c; Spatz et al. 1990). Several patients with neuropathy and IgM antibodies to intermediate filaments have also been described (Dellagi et al. 1982).

In other patients with polyneuropathy and IgM monoclonal gammopathy, no demonstrable autoantibody activity can be detected. In these cases, the antibody might be unrelated to the neuropathy or react with some as yet unidentified nerve component. In some of the cases, the monoclonal B-cells may directly infiltrate the peripheral nerves (Ince et al. 1987). Many of these patients respond to immunosuppressive therapy, suggesting that the neuropathies in these patients may also be immune mediated (Kelly et al. 1988).

Neuropathy in patients with IgM M-proteins and cryoglobulinemia

IgM M-proteins may also function as cryoglobulins which precipitate in the cold. In type I cryoglobulinemia, the cryoprecipitate contains the M-protein alone, whereas in type II cryoglobulinemia the M-protein frequently has rheumatoid factor activity and is associated with polyclonal IgG or IgA. Type III cryoglobulinemia is associated with collagen-vascular or chronic inflammatory disease and the cryoprecipitate is composed of polyclonal immunoglobulins (Brouet et al. 1974). The neuropathy in cryoglobulinemia is thought to be due to vasculitis, and the skin is frequently but not always involved. The M-protein may have autoantibody activity in addition to its cryoprecipitability (Chad et al. 1982; Thomas et al. 1992).

NEUROPATHY ASSOCIATED WITH IgG OR IgA MONOCLONAL GAMMOPATHIES

Neuropathy in myeloma

Neuropathy is estimated to occur in 1–13% of all patients with myeloma (Kelly et al. 1981a; Driedger and Pruzanski 1988; Diego Miralles et al. 1992). There is a particular association with osteosclerotic myeloma. Osteosclerosis is found in less than 3% of all myelomas, but approximately 50% of the patients have peripheral neuropathy, which is frequently the pre-senting complaint (Kelly et al. 1983; Driedger and Pruzanski 1988).

In some of the patients, there are associated endocrine abnormalities, with organomegaly, anasarca, hyperpigmentation, hypertrichosis, gynecomastia and hirsutism, epitomized as Crow–Fukase or POEMS syndrome. This syndrome is frequently associated with osteosclerotic myeloma, but also occurs in patients with osteolytic myeloma, extramedullary plasmacytomas, monoclonal or polyclonal gammopathies without evidence for myeloma, or lymphatic hyperplasia, or Castleman's disease.

The neuropathy in osteosclerotic myeloma is typically distal and symmetrical, slowly progressive and involves both sensory and motor fibers. The serum M-proteins are IgG or IgA, and almost always lambda. Occasionally the myeloma is non-secretory without a detectable serum M-protein. CSF protein is usually elevated. Electrophysiologic and pathologic studies typically show axonal degeneration and demyelination (Ohi et al. 1985). Therapy, utilizing radiation for solitary plasmocytomas, or chemotherapy using alkylating agents and corticosteroids for widespread disease may result in improvement of the neuropathy (Donofrio et al. 1984; Alexanian and Dimopoulos 1994). Tamoxifen has been reported to be helpful in some cases (Enevoldson and Harding 1992).

The cause of the neuropathy in osteosclerotic myeloma is unknown; it might be due to the monoclonal immunoglobulins, or to cytokines or other factors with biologic activity secreted by the monoclonal B-cells or plasma cells. The close association with λ light chains suggests a role for the immunoglobulin molecule, possibly through neuronal uptake of light chains (Borges and Busis 1985). Neuropathy has also been described with experimental myeloma (Dayan and Stokes 1972), and following passive transfer of human myeloma antibodies in the mouse (Bessinger et al. 1981). Antibody activity against pituitary tissue was described in one patient with the POEMS syndrome (Reulecke et al. 1988). Secretion of biologically active factors including osteoclastic factors (Mundy et al. 1974), calcitonin (Rousseau et al. 1978), lymphotoxin (Garrett et al. 1987), IL-6 (Mandler et al. 1992), and tumor necrosis factor alpha (Gherardi et al. 1994), and accelerated conversion of androgen to estrogen with elevated serum estrogen levels have also been described in isolated cases (Matsumine 1985).

In other patients with MM, neuropathy may result from nerve compression by bony fractures or plasmacytomas, or from infiltrations of nerves by plasma cells, and present as cranial nerve palsies, mononeuritis, or mononeuritis multiplex (Silverstein and Doniger 1963; Hesselvik 1969). Patients with sensory neuritis or with CIDP have also been described (Kelly et al. 1981a,b). In later stages of MM, complicating factors such as renal failure or cachexia could contribute to or be responsible for the development of neuropathy.

Neuropathy in non-malignant IgG or IgA monoclonal gammopathies

The neuropathy in some cases of non-malignant IgG or IgA monoclonal gammopathies may be caused by some of the same mechanisms that are responsible for the neuropathy in myeloma, because the non-malignant monoclonal gammopathy may progress to overt myeloma (Kyle 1978), and non-malignant gammopathies may be associated with the POEMS syndrome similarly to myeloma (Nakanishi et al. 1984; Miralles et al. 1992). In other cases, patients present with a CIDP-like syndrome, and the association is likely to be coincidental (Read et al. 1978; Noring et al. 1980; Dyck et al. 1991; Bromberg et al. 1992).

Neuropathy in primary systemic amyloidosis

Patients with primary systemic amyloidosis and neuropathy frequently have an associated monoclonal gammopathy, and it is estimated that 25% of patients with monoclonal gammopathy and neuropathy have amyloidosis (Trotter et al. 1977; Kelly et al. 1979; Kyle and Greipp 1983). Patients usually present with symptoms of numbness or painful paresthesias distally in the hands or feet. The neuropathy progresses proximally in a symmetrical fashion and the patients then develop motor weakness, and autonomic symptoms of orthostatic hypotension, bowel and bladder dysfunction, and impotence. In many cases of systemic amyloidosis, patients present with symptoms of neuropathy, but they can also present with systemic diseases such as cardiomyopathy, nephrotic syndrome, or gastrointestinal symptoms. In almost all cases, a monoclonal gammopathy can be discovered in the serum or blood of the affected patients. Amyloid neuropathy has been described in all types of monoclonal gammopathies including in MM, WM, and in non-malignant monoclonal gammopathies, but is more common with IgG or IgA M-proteins. The disease is thought to be caused by the deposition of fragments of immunoglobulin light chains in peripheral nerve and other tissues. The mechanism of light chain deposition in amyloid formation is unknown. The response to therapy is generally poor, but autologous bone marrow transplantation is a promising treatment (Bergethon et al. 1996).

REFERENCES

AARSETH, S., E. OFSTAD and A. TURVIK: Macroglobulinemia Waldenström, a case with hemolytic syndrome and involvement of the nervous system. Acta Med. Scand. 169 (1961) 691–699.

ABRAMSKY, O. and S. SLAVIN: Neurologic manifestations in patients with mixed cryoglobulinemia. Neurology 24 (1974) 245–249.

ADAMS, D., T. KUNTZER, D. BURGER, M. CHOFFLON, M.R. MAGISTRIS, F. REGLI and A.J. STECK: Predictive value of anti-GM1 ganglioside antibodies in neuromuscular diseases; a study of 181 sera. J. Neuroimmunol. 32 (1991) 223–230.

ADAMS, D., T. KUNTZER and A. STECK: Motor conduction block and high titers of anti-GM1 ganglioside antibodies; pathological evidence of a motor neuropathy in a patient with lower motor neuron syndrome. J. Neurol. Neurosurg. Psychiatry 56 (1993) 982–987.

ALEXANIAN, R.: Blood volume in monoclonal gammopathy. Blood 49 (1977) 301.

ALEXANIAN, R.: Localized and indolent myeloma. Blood 56 (1980) 521–525.

ALEXANIAN, R. and M. DIMOPOULOS: The treatment of multiple myeloma. N. Engl. J. Med. 330 (1994) 484–489.

ALEXANIAN, R., A. HAUT, A.U. KHAN, M. LANE, E.M. MCKELVEY, P.J. MIGLIORE, W.J. STUDKEY and H.E. WILSON: Treatment for multiple myeloma. Combination chemotherapy with different melphalan dose regimens. J. Am. Med. Assoc. 208 (1969) 1680–1685.

AL-LOZI, M.T., A. PESTRONK, W.C. YEE and N. FLARIS: Myopathy and paraproteinemia with serum IgM binding to a high-molecular-weight muscle fiber surface protein. Ann. Neurol. 37 (1995) 41–46.

AMPIL, F.L. and H.W. CHIN: Radiotherapy alone for extradural compression by spinal myeloma. Radiat. Med. 13 (1995) 129–131.

ANTOINE, J.C., A. STECK and D. MICHEL: Neuropathie périphérique mortelle a prédominance mortice associée à une IgM monoclonale anti-MAG. Rev. Neurol. 149 (1993) 496–499.

APOLSTOLSKI, S. and N. LATOV: Clinical syndromes associated with anti-GM1 antibodies. Sem. Neurol. 13 (1993) 264–268.

APOSTOLSKI, S., S.A. SADIQ, A. HAYS, M. CORBO, L. SUTURKOVA-MILOSEVIC, P. CHALIFF, K. STEFANSOON, R.G. LEBARON, E. RUOSLAHTI, A.P. HAYS and N. LATOV: Identification of Gal(B1–3)GalNAc bearing glycoproteins at the nodes of Ranvier in peripheral nerve. J. Neurosci. Res. 38 (1994) 134–141.

ARAI, M., H. YOSHINO, Y. KUSANO, Y. YASAKI, Y. OHNISHI and T. MIYATAKE: Ataxic polyneuropathy and anti-Pr2 IgMκ M-proteinaemia. J. Neurol. 239 (1992) 147–151.

ARNESEN, M. and J.C. MANIVEL: Plasmacytoma of the thoracic spine with intracellular amyloid and massive extracellular amyloid deposition. Ultrastruct. Pathol. 17 (1993) 447–453.

ASPINALL, G.O., A.G. MCDONALD, T.S. RAJU, H. PANG, S.D. MILLS, L.A. KURJANCZYK and J.L. PENNER: Serological diversity and chemical structure of *Campylobacter jejuni* low-molecular weight lipopolysaccharides. J. Bacteriol. 174 (1992) 1324–1332.

ATWEH, G.F. and N. JABBOUR: Intracranial solitary extraskeletal plasmacytoma resembling meningioma. Arch. Neurol. 39 (1982) 57–59.

AXELSSON, U.: A 20-year follow-up study of 64 subjects with M-components. Acta Med. Scand. 219 (1986) 519–522.

AXELSSON, U., R. BACKMANN and J. HALLEN: Frequency of pathological proteins (M-components) in 6995 sera from an adult population. Acta Med. Scand. 179 (1966) 235–247.

AZULAY, J.P., O. BLIN, J. POUGET, J. BOUCRAUT, F. BILLE-TURC, G. CARLES and G. SERRATRICE: Intravenous immunoglobulin treatment in patients with motor neuron syndromes associated with anti-GM1 antibodies. A double-blind, placebo-controlled study. Neurology 44 (1994) 429–432.

AZZARELLI, B. and U. ROESSMANN: Pathogenesis of central nervous system infiltration in acute leukemia. Arch. Pathol. Lab. Med. 101 (1977) 203–205.

BABA, H. N. MIYATANI, S. SATO, T. YUASA and T. MIYATAKE: Antibody to glycolipid in a patient with IgM paraproteinemia and polyradioculoneuropathy. Acta Neurol. Scand. 72 (1985) 218–221.

BABA, H., G.C. DAUNE, A.A. ILYAS, A. PESTRONK, D. CORNBLATH, V. CHAUDHRY, J. GRIFFIN and R. QUARLES: Anti-GM1 ganglioside antibodies with differing specificities in patients with multifocal motor neuropathy. J. Neuroimmunol. 25 (1989) 143–150.

BARBIERI, S., E. NOBILE-ORAZIO, L. BALDINI, Z. FAYOUMI, E. MANFREDINI and G. SCARLATO: Visual evoked potentials in patients with neuropathy and macroglobulinemia. Ann. Neurol. 2 (1987) 663–666.

BARCOS, M., W. LANE, G.Z. GOMEZ, T. HAN, A. FREEMAN, H. PREISLER and E. HENDERSON: An autopsy study of 1206 acute and chronic leukemias (1958–1982). Cancer 60 (1987) 827–837.

BARDWICK, P.A., W.J. ZVAIFLER, G.N. GILL, D. NEWMAN, G.D. GREENWAY and D.L. RESNICK: Plasma cell dyscrasia with polyneuropathy, organomegally, endocrinopathy, M-protein and skin changes. The POEMS syndrome. Medicine (Baltimore) 59 (1980) 311–322.

BARLOGIE, B., R. ALEXANIAN and S. JAGANNATH: Plasma cell dyscrasias. J. Am. Med. Assoc. 268 (1992) 2946–2951.

BATAILLE, R. and J. SANY: Solitary myeloma: clinical and prognostic features of a review of 114 cases. Cancer 48 (1981) 845–851.

BATAILLE, R., J. SANY and H. SERRE: Plasmacytomes apparement solitaires des os: aspects cliniques et pronostiques (à propos de 114 cas). Nouv. Presse Méd. 10 (1981) 407–411.

BATAILLE, R., B.G.M. DURIE and J. GRENIER: Serum beta2 microglobulin and survival duration in multiple myeloma: a simple, reliable marker for staging. Br. J. Haematol. 55 (1983) 439–447.

BATAILLE, R., M. JOURDAN, X.G. ZHANG and B. KLEIN: Serum levels of interleukin 6, a potent myeloma cell growth factor, as a reflection of disease severity in plasma cell dyscrasias. J. Clin. Invest. 84 (1989) 2008–2011.

BATAILLE, R., D. CHAPPARD and B. KLEIN: Mechanisms of bone lesions in multiple myeloma. Hematol. Oncol. Clin. N. Am. 6 (1992) 285–295.

BAUER, K.H.: Makroglobulinaemie Waldenström mit Hyperviskositätssyndrom. Med. Welt 26 (1975) 341–348.

BELISLE, L.M., D.C. CASE, JR. and L. NEVEAUX: Quantitation of risk of malignant transformation in patients with benign monoclonal gammopathy (BMG). Blood 76, Suppl. 1 (1990) 342A.

BENLI, K. and S. INCI: Solitary dural plasmacytoma: case report. Neurosurgery 36 (1995) 1206–1209.

BENNETT, J.M., D. CATOVSKY, M.T. DANIEL, G. FLANDRIN, D.A. GALTON, H.R. GRALNICK and C. SULTAN: The French-American-British (FAB) cooperative group. Proposals for the classification of chronic (mature) B and T lymphoid leukaemias. J. Clin. Pathol. 42 (1989) 567–584.

BERGETHON, P.R., M. SKINNER, R.W. SIMMS and R.L. COMENZO: Reversal of the neuropathy in primary (AL) amyloidosis following treatment with high dose melphalan and stem cell rescue. Neurology 46, Suppl. (1996) A449.

BERNSTEIN, Z.P., J.E. FITZPATRICK, A. O'DONNEL, T. HAN, K.A. FOON and A. BHARGAVA: Clinical significance of monoclonal proteins in chronic lymphocytic leukemia. Leukemia 6 (1992) 1243–1245.

BESSINGER, V.A., K.V. TOYKA, A.P. ANZIL, A. FATCH-MOGHADAM, D. NEUMEIER, R. RAUSCHER and K. HEININGER: Myeloma neuropathy: passive transfer from man to mouse. Science 213 (1981) 1027–1030.

BINDAL, A.K., R.K. BINDAL, H. VAN LOVEREN and R. SAWAYA: Management of intracranial plasmacytoma. J. Neurosurg. 83 (1995) 218–221.

BINET, J.L., A. AUGUIER, G. DIGHIERO, C. CHASTANG, H. PIGUET, J. GOASGUEN, G. VAUGIER, G. POTRON, P. COLONA, F. OBERLING, M. THOMAS, G. TCHERNIA, C. JACQUILLAT, P. BOIVIN, C. LESTY, M.T. DUAULT, M. MONCONDUIT, S. BELABBES and F. GREMY: A new prognostic classification of chronic lymphocytic leukemia derived from a multivariate survival analysis. Cancer 48 (1981) 198–206.

BLADE, J., A. LOPEZ-GUILLERMO, C. ROZMAN, F. CERVANTES, C. SALGADO, J.L. AGUILAR, J.L. VIVES-CORRONS and E. MONTSERRAT: Malignant transformation and life expectancy in monoclonal gammopathy of undetermined significance. Br. J. Haematol. 81 (1992) 391–394.

BOJSEN-MØLLER, M. and J.L. NIELSON: CNS involvement in leukaemia. Acta Pathol. Microbiol. Immunol. Scand. Sect. A 91 (1983) 209-216.

BORGES, L.F. and N.A. BUSIS: Intraneuronal accumulation of myeloma proteins. Arch. Neurol. 42 (1985) 690–694.

BOSSENSEN, E., H.I. SCHIPPER and A.J. STECK: Motor neuropathy with activity of monoclonal IgM antibody to GD1a ganglioside. J. Neurol. 236 (1989) 353–355.

BROMBERG, M.B., E.L. FELDMAN and J.W. ALBERS: Chronic inflammatory demyelinating polyradiculoneuropathy: comparison of patients with and without an associated monoclonal gammopathy. Neurology 42 (1992) 1157–1163.

BROUET, J.C., J.P. CLAUVEL, F. DANON, M. DLEIN and M. SELIGMANN: Biologic and clinical significance of cryoglobulins. A report of 86 cases. Am. J. Med. 57 (1974) 775.

BROUET, J.C., F. DANON and M. SELIGMANN: Immuno-chemical classification of human cryoglobulins. In: F. Chenas (Ed.), Cryoproteins. Grenoble, Colloque (1978) 13–19.

BUSKARD, N., D. GALTON and J. GOLDMAN: Plasma exchange in the long-term management of Waldenström's macroglobulinemia. J. Can. Med. Assoc. 117 (1977) 135–137.

CATOVSKY, D., S. RICHARDS and J. FOOKS: CLL trials in the United Kingdom. Leuk. Lymphoma 5, Suppl. (1991) 105–112.

CAVO, M., S. TURA, G. ROSTI, M. GRIMALDI, G. BANDINI, M.Z. BONELLI, E. CALORI, S. RIZZI, M.T. VAN LINT and A. BACIGALUPO: Allogeneic BMT for multiple myeloma (MM). The Italian experience. Bone Marrow Transplant. 7, Suppl. 2 (1991) 31.

CHAD, D., K. PARISER and W.G. BRADLEY: The pathogenesis of cryoglobulinemic neuropathy. Neurology 32 (1982) 725–729.

CHAK, L.Y., R.S. COX, D.G. BOSTWICK and R.T. HOPPE: Solitary plasmacytoma of bone: treatment, progression and survival. J. Clin. Oncol. 5 (1987) 1811–1815.

CHAMBERLAIN, M.C., A.D. SANDY and G.A. PRESS: Leptomeningeal metastasis: a comparison of gadolinium-enhanced MR and contrast-enhanced CT of the brain. Neurology 40 (1990) 435–438.

CHAUDHRY, V., A.M. CORSE and D.R. CORNBLATH: Multifocal motor neuropathy: response to human immune globulin. Ann. Neurol. 33 (1993) 237–242.

COHEN, A.S.: Amyloidosis. In: D.J. McCarty (Ed.), Arthritis and Allied Conditions: a Textbook of Rheumatology. Philadelphia, Lea and Febiger (1985) 1108–1127.

COHEN, A.S.: Amyloidosis. In: K.J. Isselbacher, E. Braunwald, J.D. Wilson, J.B. Martin, A.S. Fauci and D.L. Kasper (Eds.), Harrison's Principles of Internal Medicine, 13th Edit. New York, McGraw-Hill (1994) 1625–1630.

CONKLIN, R. and R. ALEXANIAN: Clinical classification of plasma cell myeloma. Arch. Intern. Med. 135 (1975) 139–143.

COOK, D., M. DALAKAS, A. GALDI, D. BIONDI and H. PORTER: High dose intravenous immunoglobulins in the treatment of demyelinating neuropathy associated with monoclonal gammopathy. Neurology 40 (1990) 212–214.

COPPETO, J.R., M.L.R. MONTEIRO, J. COLLIAS, D. UPHOFF and L. BEAR: Foster–Kennedy syndrome caused by solitary intracranial plasmacytoma. Surg. Neurol. 19 (1983) 267–272.

CORBO, M., A. QUATTRINI and A. LUGARESI: Patterns of reactivity of human anti-GM1 antibodies with spinal cord and motor neurons. Ann. Neurol. 32 (1992) 487–493.

CORBO, M., A. QUATTRINI, N. LATOV and A.P. HAYS: Localization of GM1 and Gal(B1–3)GalNAc antigenic determinants in peripheral nerve. Neurology 43 (1993) 809–816.

CORRADINI, P., M. LADETTO, G. INGHIRAMI, M. BOCCADORO and A. PILERI: N- and K-ras oncogenes in plasma cell dyscrasias. Leuk. Lymphoma 15 (1994) 17–20.

CORRAL-ALONSO, M. and M.L. BARBOLLA: Plasmaferesis y sindrome de hiperviscosidad. Sangre 29 (1984) 940–950.

CORREALE J., D.A. MONTEVERDE, J.A. BUERI and E.G. REICH: Peripheral nervous system and spinal cord involvement in lymphoma. Acta Neurol. Scand. 83 (1991) 45–51.

CORWIN, J. and R.D. LINDBERG: Solitary plasmacytoma of bone vs. extramedullary plasmacytoma and their relationship to multiple myeloma. Cancer 43 (1979) 1007–1013.

CRAMER, S.C., J.A. GLASPY, J.T. EFIRD and D.N. LOUIS: Chronic lymphocytic leukemia and the central nervous system: a clinical and pathological study. Neurology 46 (1996) 19–25.

CREAM, J.J.: Cryoglobulins in vasculitis. Clin. Exp. Immunol. 10 (1972) 117–126.

CROW, R.S.: Peripheral neuritis in myelomatosis. Br. Med. J. 2 (1956) 802–804.

CURRIE S. and R.A. HENSON: Neurological syndromes in the reticuloses. Brain 94 (1971) 307–320.

CURRIE, S., R.A. HENSON, MORGAN and A.J. POOLE: The

incidence of the non-metastatic neurological syndromes of obscure origin in the reticuloses. Brain 93 (1970) 629–640.

DAUNE, G.C., R.G. FARRER, M.C. DALAKAS and R.H. QUARLES: Sensory neuropathy associated with immunoglobulin M to GD1b ganglioside. Ann. Neurol. 31 (1992) 683–685.

DAYAN, A.D. and M.I. STOKES: Peripheral neuropathy and experimental myeloma in the mouse. Nature 236 (1972) 117–118.

DELAUCHE-CAVALLIER, M.C., J.D. LAREDO, M. WYBIER, M. BARD, A. MAZABRAUD, J.L. LE BAIL DARNE, D. KUNTZ and A. RYCKEWAERT: Solitary plasmacytoma of the spine. Long-term clinical course. Cancer 61 (1988) 1707–1714.

DELLAGI, K., J.C. BROUET, J. PERREAU and D. PAULIN: Human monoclonal IgM with autoantibody activity against intermediate filaments. Proc. Natl. Acad. Sci. (USA) 79 (1982) 446–450.

DESCHENES, G., A.M. PRIEUR, F. HAYEM, M. BROYER and M.C. GUBLER: Renal amyloidosis in juvenile chronic arthritis: evolution after chlorambucil treatment. Pediatr. Nephrol. 4 (1990) 463–469.

DIAZ-ARRASTIA, R., D.S. YOUNGER, L. HAIR, G. INGHIRAMI, A.P. HAYS, D.M. KNOWLES, J.G. ODEL, M.R. FETELL, R.E. LOVELACE and L.P. ROWLAND: Neurolymphomatosis: a clinical syndrome re-emerges. Neurology 42 (1992) 1136–1141.

DIEGO MIRALLES, G., J.R. O'FALLON and N.J. TALLEY: Plasma cell dyscrasia with polyneuropathy. N. Engl. J. Med. 327 (1992) 1919–1923.

DIMOPOULOS, M.A. and R. ALEXANIAN: Waldenström's macroglobulinemia. Blood 83 (1994) 1452–1459.

DIMOPOULOS, M.A., J. GOLDSTEIN, J. FULLER, K. DELASALLE and R. ALEXANIAN: Curability of solitary bone plasmacytoma. J. Clin. Oncol. 10 (1992) 587–590.

DIMOPOULOS, M.A., S. O'BRIEN, H. KANTARJIAN, S. PIERCE, K. DELASALLE, B. BARLOGIE, R. ALEXANIAN and M.J. KEATING: Fludarabine therapy in Waldenström's macroglobulinemia. Am. J. Med. 95 (1993) 49–52.

DINTENFASS, L.: Blood Viscosity. Hyperviscosity and Hyperviscosaemia. Lancaster, MTP Press (1985).

DONOFRIO, P.D., J.W. ALBERS, H.S. GREENBERG and B.S. MITCHELL: Peripheral neuropathy in osteosclerotic myeloma: clinical and electrophysiologic improvement with chemotherapy. Muscle Nerve 7 (1984) 137–141.

DRIEDGER, H. and W. PRUZANSKI: Plasma cell neoplasia with peripheral polyneuropathy. Medicine (Baltimore) 59 (1988) 301–310.

DURIE, B.G.M. and S.E. SALMON: A clinical staging system for multiple myeloma: Correlation of measured myeloma cell mass with presenting clinical features and response to treatment and survival. Cancer 36 (1975) 842–854.

DURIE, B.G.M., S.E. SALMON and T.E. MOON: Pretreatment tumor mass, cell kinetics, and prognosis in multiple myeloma. Blood 55 (1980) 364–372.

DURIE B.G.M., B. PERSKY, B.J. SOEHNLEN, T.M. GROGAN and S.E. SALMON: Amyloid production in human myeloma stem-cell culture, with morphologic evidence of amyloid secretion by associated macrophages. N. Engl. J. Med. 307 (1982) 1689–1692.

DUSTON, M.A., M. SKINNER, J. ANDERSON and A.S. COHEN: Peripheral neuropathy as an early marker of AL amyloidosis. Arch. Intern. Med. 149 (1989) 358–360.

DUTCHER, T.F. and J.L. FAHEY: The histopathology of the macroglobulinemia of Waldenström. J. Natl. Cancer Inst. 22 (1959) 887–917.

DYCK, P.J., P.A. LOW and A.J. WINDEBANK: Plasma exchange in polyneuropathy associated with monoclonal gammopathy of undetermined significance. N. Engl. J. Med. 325 (1991) 1482–1486.

ELLIE, E., A. VITAL, A.J. STECK, J. JULIEN, P. HENRY and C. VITAL: High-grade B-cell cerebral lymphoma in a patient with anti-myelin-associated glycoprotein IgM paraproteinemic neuropathy. Neurology 45 (1995) 378–381.

ELLIS, P.A. and B.M. COLLS: Solitary plasmacytoma of bone: clinical features, treatment and survival. Hematol. Oncol. 10 (1992) 201–211.

ENEVOLDSON, T.P. and A.E. HARDING: Improvement in the POEMS syndrome after administration of Tamoxifen. J. Neurol. Neurosurg. Psychiatry 55 (1992) 71–72.

EPELBAUM, R., N. HAIM, M. BEN-SHAHAR, Y. BEN ARIE, M. FEINSOD and Y. COHEN: Non-Hodgkin's lymphoma presenting with spinal epidural involvement. Cancer 58 (1986) 2120–2124.

ERIKSSON, L., K. SLETTEN, L. BENSON and P. WESTERMARK: Tumour-like localized amyloid of the brain is derived from immunoglobulin light chain. Scand. J. Immunol. 37 (1993) 623–626.

FACON, T.M. BROUILLARD, A. DUHAMEL, P. MOREL, M. SIMON, L.P. JOUET, F. BAUTERS and P. FENAUX: Prognostic factors in Waldenström's macroglobulinemia. J. Clin. Oncol. 11 (1993) 1553–1558.

FAGUET, G.B.: Chronic lymphocytic leukemia: an updated review. J. Clin. Oncol. 12 (1994) 1974–1990.

FAGUET, G.B., J.F. AGEE and G.E. MARTI: Clone emergence and evolution in chronic lymphocytic leukemia: characterization of clinical, laboratory and immunophenotypic profiles of 25 patients. Leuk. Lymphoma 6 (1992) 345–356.

FAHEY, J.L., W.F. BARTH and A. SOLOMON: Serum hyperviscosity syndrome. J. Am. Med. Assoc. 192 (1965) 120.

FELDMAN, E.L., M.B. BROMBERG, J.W. ALBERS and A. PESTRONK: Immunosuppressive treatment in multifocal motor neuropathy. Ann. Neurol. 10 (1991) 397–401.

FERRI C., L. LA CIVITA, C. CIRAFISI, G. SICILIANO, G. LONGOMARDO, S. BONBARDIERI and B. ROSSI: Peripheral neuropathy in mixed cryoglobulinemia: clinical and electrophysiologic investigations. J. Rheumatol. 19 (1992) 889–895.

FIALKOW, P.J., V. NAJFELD, A.L. REDDY, J. SINGER and L. STEINMANN: Chronic lymphocytic leukemia: clonal origin in a committed B-lymphocyte progenitor. Lancet 8087 (1978) 444–446.

FINE, J.M., P. LAMBIN and P. LEROUS: Frequency of monoclonal gammapathy (M-components) in 13,400 sera from blood donors. Vox Sang. 23 (1972) 336–343.

FINE, J.M., P. LAMBIN and L.Y. MULLER: The evolution of asymptomatic monoclonal gammapathies: a follow-up of 20 cases over periods of 3–14 years. Acta Med. Scand. 205 (1979) 339–341.

FREDDO, L., A.P. HAYS, W.H. SHERMAN and N. LATOV: Axonal neuropathy in a patient with IgM M-protein reactive with nerve endoneurium. Neurology 35 (1985) 1321–1325.

FREDDO, L., A.P. HAYS, K.G. NICKERSON, L. SPATZ, S. MCGINNIS, G. LIEBERSON, C.Z. VEDELER, M.E. SHY, L. AUTILO-GAMBETTI, F.C. GRAUSS, F. PETITO, L. CHESS and N. LATOV: Monoclonal anti-DNA IgMκ in neuropathy binds to myelin and to a conformational epitope formed by phosphatidic acid and gangliosides. J. Immunol. 137 (1986a) 3821–3825.

FREDDO, L., W.H. SHERMAN and N. LATOV: Glycosaminoglycan antigens in peripheral nerve; studies with antibodies from a patient with neuropathy and monoclonal gammapathy. J. Neuroimmunol. 12 (1986b) 57–64.

FREDDO, L., R.K. YU, N. LATOV, P. DONOFRIO, A.P. HAYS, H.S. GREENBERG, J.W. ALBERS, A.G. ALLESSI, A. LEAVITT, G. DAVAR and D. KEREN: Gangliosides GM1 and GD1b are antigens for IgM M-proteins in a patient with motor neuron disease. Neurology 36 (1986c) 454–458.

FREEDMAN, A.S.: Immunobiology of chronic lymphocytic leukemia. Hematol. Oncol. Clin. N. Am. 4 (1990) 405–429.

FUNAKUBO, T. and A. KIKUCHI: A case of myeloma with facial palsy. Acta Oto-Laryngol. (Stockh.) 511, Suppl. (1994) 2000–2003.

GAHRTON, G., S. TURA, P. LJUNGMANN, B. BELANGER, L. BRANDT, M. CAVO, B. CHAPUIS, A. DE LAURENZI, T. DE WITTE and T. FACON: Allogeneic bone marrow transplantation in multiple myeloma. Bone Marrow Transplant. 325 (1991) 1267.

GALE, R.P. and K.Z. FOON: Chronic lymphocytic leukemia: recent advances in biology and treatment. Ann. Intern. Med. 103 (1985) 101–120.

GARCIA-BRAGADO, F., J.M. FERNANDEZ, C. NAVARRO, M. VILLAR and I. BONAVENTURA: Peripheral neuropathy in essential mixed cryoglobulinemia. Arch. Neurol. 45 (1988) 1210–1214.

GARRETT, I.R., B.G.M. DURIE, G.E. NEDWIN, A. GILLESPIE, T. BRINGMAN, M. SABATINI, D.R. BERTOLINI and G.R. MUNDY: Production of lymphotoxin, a bone resorbing cytokine, by cultured human myeloma cells. N. Engl. J. Med. 317 (1987) 526–532.

GERTZ, M.A. and R.A. KYLE: Amyloidosis: prognosis and treatment. Sem. Arthr. Rheum. 24 (1994) 124–138.

GERTZ, M.A., R.A. KYLE, P.R. GREIPP, J.A. KATZMANN and W.M. O'FALLON: Beta2-microglobulin predicts survival in primary systemic amyloidosis. Am. J. Med. 89 (1990) 609–614.

GERTZ, M.A., R.A. KYLE and P.R. GREIPP: Response rates and survival in primary systemic amyloidosis. Blood 77 (1991a) 257–262.

GERTZ, M.A., P.R. GREIPP and R.A. KYLE: Classification of amyloidosis by the detection of clonal excess of plasma cells in the bone marrow. J. Lab. Clin. Med. 118 (1991b) 33–39.

GERTZ, M.A., R.A. KYLE and P. NOEL: Primary systemic amyloidosis: a rare complication of immunoglobulin M monoclonal gammapathies and Waldenström's macroglobulinemia. J. Clin. Oncol. 11 (1993) 914.

GHERARDI, R.K., CHOUAIB, D. MALAPERT, L. BELEC, L. INTRATOR and J.D. DEGOS: Early weight loss and high serum tumour necrosis factor alpha in polyneuropathy, organomegaly, endocrinopathy, M-protein, skin changes syndrome. Ann. Neurol. 35 (1994) 501–505.

GIRALDO, M.P., D. RUBIO-FELIX, M. PERALLA, J.Z. GRACIA, J.M. BERGUA and M. GIRALT: Gammapatias monoclonales de significado indeterminado: aspectos clinicos, biologicos y evelutivos de 397 casos. Sangre 36 (1991) 377–382.

GOBBI, P.G., R. BETTINI, C. MONTECUCO, L. CAVANNA, S. MORANDI, C. PIERESCA, G. MERLINI, D. BERTOLONI, G. GRIGNANI and U. POZZETTI: Study of prognosis in Waldenström's macroglobulinemia: a proposal for a simple binary classification with clinical and investigation utility. Blood 83 (1994) 2939–2945.

GORDON, P.H., L.P. ROWLAND, D.S. YOUNGER, W.H. SHERMAN, A.P. HAYS, E.D. LOUIS, D.E. LANGE, W. TROJABORG, R.E. LOVELACE, P.L. MURPHY and N. LATOV: Lymphoproliferative disorders and motor neuron disease. Neurology 48 (1997) 1671–1678.

GOREVIC, P.D., H.J. KASSAB, Y. LEVO, R. KOHN, M. MELTZER, P. PROSE and E.C. FRANKLIN: Mixed cryoglobulinemia: clinical aspects and long-term follow-up of 40 patients. Am. J. Med. 69 (1980) 287–308.

GOSSELIN, S., R.A. KYLE and P.J. DYCK: Neuropathy associated with monoclonal gammapathy of undetermined significance. Ann. Neurol. 30 (1991) 54–61.

GREENSTEIN, A.J., D.B. SACHAR, A.K.N. PANDAY, S.H. DIGMAN, S. MEYERS, T. HEIMANN, V. GUMASTE, J.L. WERTHER and H.D. JANOWITZ: Amyloidosis and inflammatory bowel disease: a 50-year experience with 25 patients. Medicine (Baltimore) 71 (1992) 261–270.

GREGORY, R., P.K. THOMAS, R.H.M. KING, P.L.J. HALLAM, S. MALCOLM, R.A.C. HUGHES and A.E. HARDING: Coexistence of hereditary motor and sensory neuropathy type Ia and IgM paraproteinemic neuropathy. Ann. Neurol. 33 (1993) 649–652.

GREGORY, W.M., M.A. RICHARDS and J.S. MALPAS: Combination chemotherapy versus melphalan and prednisolone in the treatment of multiple myeloma: an overview of published trials. J. Clin. Oncol. 10 (1992) 334–342.

GREIL, R., B. FASCHING, P. LOIDL and H. HUBER: Expression of the C-*myc* proto-oncogene in multiple myeloma and chronic lymphocytic leukemia: an in situ analysis. Blood 78 (1991) 180–191.

GREIPP, P.R.: Prognosis in myeloma. Mayo Clin. Proc. 69 (1994) 895–902.

GREIPP, P.R., J.A. LUST, W.M. O'FALLON, J.A. KATZMANN, T.E. WITZIG, R.A. HERNANDEZ, K.J. LAND and R.W. MCKENNA: Leukemias, myeloma, and other lymphoreticular neoplasms. Cancer 75 (1995) 381–394.

HAAS, D.C. and A.H. TATUM: Plasmapheresis alleviates neuropathy accompanying IgM anti-myelin associated glycoprotein paraproteinemia. Ann. Neurol. 23 (1988) 304–396.

HAN, T., E.X. EZDINLI, K. SHIMAOKA and D.V. DESAI: Chlorambucil vs. combined chlorambucil-corticosteroid therapy in chronic lymphocytic leukemia. Cancer 31 (1973) 502–508.

HAN, T., N. SAMADORI, J. TAKEUCHI, H. OZER, E.S. HENDERSON, A. BHARGAVA, J. FITZPATRICK and A.A. SANDBERG: Clonal chromosome abnormalities in patients with Waldenström's and CLL-associated macroglobulinemia. Blood 62 (1983) 525–531.

HARBS, H., M. ARFMANN, E. FRICK, C. HORMANN, U. WURSTER, U. PATZOLD, E. STARK and H. DEICHER: Reactivity of sera and isolated monoclonal IgM from patients with Waldenström's macroglobulinemia with peripheral nerve myelin. J. Neurol. 232 (1985) 43–48.

HAYS, A.P., N. LATOV, M. TAKATSU and W.H. SHERMAN: Experimental demyelination of nerve induced by serum of patients with neuropathy and an anti-MAG IgM M-protein. Neurology 37 (1987) 242–256.

HAYS, A.P., S.L. LEE and N. LATOV: Immunoreactive C3d on the surface of myelin sheaths in neurology. J. Neuroimmunol. 18 (1988) 231–244.

HEIMAN-PATTERSON, T., T. KRUPA, P. THOMPSON, E. NOBILE-ORAZO, A.J. TAHMOUSH and M.E. SHY: Anti-GM1/GD1b M-proteins damage human spinal cord neurons co-cultured with muscle. J. Neurol. Sci. 20 (1993) 38–45.

HERNANDEZ, J.A., K.J. LAND and R.W. MCKENNA: Leukemias, myeloma, and other lymphoreticular neoplasms. Cancer 75 (1995) 381–394.

HESSELVIK, M.: Neuropathological studies on myelomatosis. Acta Neurol. Scand. 45 (1969) 95–108.

HIRANO, T., S. SUEMATSU, T. MATSUSAKA, T. MATSUDA and T. KISHIMOTO: The role of interleukin 6 in plasmacytomagenesis. Ciba Found. Symp. 167 (1992) 188–196.

HOBBS, J.R.: Immunochemical classes of myelomatosis including data from a therapeutical trial conducted by a Medical Research Council working party. Br. J. Haematol. 16 (1969) 599–617.

HOLLAND, J., D.A. TRENKNER, T.H. WASSERMAN and B. FINEBERG: Plasmacytoma. Treatment results and conversion to myeloma. Cancer 69 (1992) 1513–1517.

ILYAS, A.A., R.H. QUARLES, M.C. DALAKAS, P.H. FISHMAN and R.O. BRADY: Monoclonal IgM in a patient with paraproteinemic polyneuropathy binds to gangliosides containing disialosyl groups. Ann. Neurol. 18 (1985) 655–659.

ILYAS, A.A., H.J. WILLISON, M. DALAKAS, J.M. WHITAKER and R.H. QUARLES: Identification and characterization of gangliosides reacting with IgM paraproteins in three patients with neuropathy and biclonal gammopathy. J. Neurochem. 51 (1988a) 851–858.

ILYAS, A.A., S.C. LI, D.K.H. CHOU, Y.T. LI, F.B. JUNGALWALA, M.C. DALAKAS and R.H. QUARLES: Gangliosides GM2, IVGalNAcGm1b, and IVGalNAcGd1a as antigens for monoclonal immunoglobulin M in neuropathy associated with gammopathy. J. Biol. Chem. 263 (1988b) 4369–4373.

ILYAS, A., S.D. COOK, M.C. DALAKAS and F.A. MITHEN: Anti-MAG IgM paraproteins from some patients with polyneuropathy associated with IgM paraproteinemia also react with sulfatide. J. Neuroimmunol. 37 (1992) 85–92.

IMAI, F., K. FUJISAWA, N. KIYA, T. NINOMIYA, Y. OGURA, Y. MIZOGUCHI, H. SANO and T. KANNO: Intracerebral infiltration by monoclonal plasmacytoid cells in Waldenström's macroglobulinemia – case report. Neurol. Med.-Chir. (Tokyo) 35 (1995) 575–579.

INCE, P.G., P.J. SHAW, P.R. FAWCETT and D. BATES: Demyelinating neuropathy due to primary IgM kappa B-cell lymphoma of peripheral nerves. Neurology 37 (1987) 1231–1235.

ITALA, M., H. HELENIUS, J. NIKOSKELAINEN and K. REMES: Infections and serum IgG levels in patients with chronic lymphocytic leukemia. Eur. J. Haematol. 48 (1992) 266–270.

JACOBSON, D.R. and S. ZOLLA-PAZNER: Immunosuppression and infection in multiple myeloma. Sem. Oncol. 13 (1986) 282–290.

JAKSIC, B. and M. BRUGIATELLI: High dose continuous chlorambucil vs intermittent chlorambucil plus prednisone for treatment of B-CLL-IGCI CLL-01 trial. Nouv. Rev. Fr. Hématol. 30 (1988) 437–442.

JULIEN, J., C. VITAL, J.M. VALLAT, A. LAGUENY, X. FERRER and M.J. LEBOUTET: Chronic demyelinating neuropathy with IgM producing lymphocytes in peripheral nerve and delayed appearance of 'benign' monoclonal gammopathy. Neurology 34 (1987) 1387–1389.

JULIUSSON, G. and G. GAHRTON: Cytogenetics in CLL and related disorders. Baillières Clin. Haematol. 6 (1993) 821–848.

KABAT, K.A., J. LIAO, W.H. SHERMAN and E.F. OSSERMAN: Immunological characterization of the specificities of two human monoclonal IgMs reacting with chondroitin sulfates. Carbohydrate Res. 130 (1984) 289–298.

KAHLER, O.: Zur Symptomatologie des multiplen Myeloma: Beobachtung von Albumosurie. Prog. Med. Wschr. 14 (1889) 33.

KAHN, S.N., P.G. RICHES and J. KOHN: Paraproteinaemia in neurological disease: incidence, associations, and classification of monoclonal immunoglobulins. J. Clin. Pathol. 33 (1980) 617–621.

KAJI, R., H. SHIBASAKI and J. KIMURA: Multifocal demyelinating motor neuropathy: cranial nerve involvement and immunoglobulin therapy. Neurology 42 (1992) 506–509.

KANEKO, D., T. IRIKURA, Y. TAGUCHI, H. SEKINO and N. NAKAMURA: Intracranial plasmacytoma arising from the dura mater. Surg. Neurol. 17 (1982) 295–300.

KAY, D.G., C. GRAVEL, F. POTHIER, A. LAPERRIERE, Y. ROBITAILLE and P. JOLICOEUR: Neurological disease induced in transgenic mice expressing the *env* gene of the Cas-Br-e murine retrovirus. Proc. Natl. Acad. Sci. USA 90 (1993) 4358–4352.

KEATING, M.J., S. O'BRIEN, L.E. ROBERTSON, H. KANTARJIAN, M. DIMOPOULOS, P. MCLAUGHLIN, F. CABANILLAS, V. GREGOIRE, Y.Y. LI, V. GANDHI, E. ESTEY and W. PLUNKETT: The expanding role of fludarabine in hematologic malignancies. Leuk. Lymphoma 14, Suppl. 2 (1994) 11–16.

KELLY, J.J., JR., R.A. KYLE, P.C. O'BRIEN and P.J. DYCK: The natural history of peripheral neuropathy in primary systemic amyloidosis. Ann. Neurol. 6 (1979) 1–7.

KELLY, J.J., JR., R.A. KYLE, J.M. MILES, P.C. O'BRIEN and P.J. DYCK: The spectrum of peripheral neuropathy in myeloma. Neurology 31 (1981a) 24–31.

KELLY, J.J., JR., R.A. KYLE, P.C. O'BRIEN, P.J. DYCK: The prevalence of monoclonal gammopathy in peripheral neuropathy. Neurology 31 (1981b) 1480–1483.

KELLY, J.J., JR., R.A. KYLE, J.M. MILES and P.J. DYCK: Osteosclerotic myeloma and peripheral neuropathy. Neurology 33 (1983) 202–210.

KELLY, J.J., JR., L.S. ADELMAN, E. BERKMAN and I. BHAN: Polyneuropathies associated with IgM monoclonal gammopathies. Arch. Neurol. 45 (1988) 1355–1359.

KHELLA, S.L., S. FROST, G.A. HERMANN, L. LEVENTHAL, S. WHYATT, M.A. SAJID and S.S. SCHERER: Hepatitis C infection, cryoglobulinemia and vasculitic neuropathy. Treatment with interferon alpha: case report and literature review. Neurology 45 (1995) 407–411.

KINSELLA, L.J., D.J. LANGE, W. TROJABORG, S.A. SADIQ, D.S. YOUNGER and N. LATOV: Clinical and electrophysiological correlates of elevated anti-GM1 antibody titers. Neurology 44 (1994) 1278–1282.

KLEIN, B., X.G. ZHANG, M. JOURDAN, J. CONTENT, F. HOUSSIAU, L. AARDEN, M. PIECHACZYK and R. BATAILLE: Paracrine but not autocrine regulation of myeloma-cell growth and differentiation by interleukin-6. Blood 73 (1989) 517–526.

KOHN, J.: A cellulose acetate supporting medium for zone electrophoresis. Clin. Chim. Acta 2 (1957) 297–303.

KORNBERG, A.J. and A. PESTRONK: The clinical and diagnostic role of anti-GM1 antibody testing. Muscle Nerve 17 (1994) 100–104.

KORNBERG, A., A. PESTRONK, K. BIESER, T.W. HO, G.M. MCKHANN, H.S. WU and Z. JIANG: The clinical correlates of high-titer IgG anti-GM1 antibodies. Ann. Neurol. 35 (1994) 234–237.

KUDO, H., K. IWATSUJI, M. KANDA, M. KAMEYAMA, T. ADACHI and H. HAIBARA: A case of ovarian non-Hodgkin lymphoma with autoimmune hemolytic anaemia and cerebral infarction. J. Jpn. Soc. Intern. Med. 72 (1983) 908–913.

KUSUNOKI, S., J. SHIMUZU, A. CHIBA, Y. UGAWA, S. HITOSHI and I. KANAZAWA: Experimental sensory neuropathy induced by sensitization with ganglioside GD1b. Ann. Neurol. 39 (1966) 424–431.

KUSUNOKI, S., T. SHIMIZU, K. MATSUMURA, K. MAEMURA and T. MANNEN: Motor dominant neuropathy and IgM paraproteinemia: the IgM M-protein binds to specific gangliosides. J. Neuroimmunol. 21 (1989) 177–181.

KUSUNOKI, S., A. CHIBA, T. TAI and I. KANAZAWA: Localization of GM1 and GD1b antigen in the human peripheral nervous system. Muscle Nerve 16 (1992) 752–756.

KUZEYLI, K., S. DURU, S. CEYLAN and F. AKTURK: Solitary plasmacytoma of the skull. A case report. Neurosurg. Rev. 18 (1995) 139–142.

KYLE R.A.: Monoclonal gammopathy of undetermined significance: natural history in 241 cases. Am. J. Med. 64 (1978) 814–826.

KYLE, R.A.: Long-term survival in multiple myeloma. N. Engl. J. Med. 302 (1983) 314–316.

KYLE, R.A.: Diagnostic criteria of multiple myeloma. Hematol. Oncol. Clin. N. Am. 6 (1992) 347–358.

KYLE, R.A.: Benign monoclonal gammopathy – after 20 to 35 years of follow-up. Mayo Clin. Proc. 68 (1993a) 26–36.

KYLE, R.A: Plasma cell labeling index and beta2-microglobulin predict survival independent of thymidine kinase and C-reactive protein in multiple myeloma. Blood 81 (1993b) 3382–3387.

KYLE, R.A.: Monoclonal gammopathy of undetermined significance. Haematol. Oncol. 8 (1994) 135–141.

KYLE, R.A.: The monoclonal gammopathies. Clin. Chem. 41 (1995) 761–762.

KYLE, R.A. and J.P. GARTON: The spectrum of IgM monoclonal gammopathy in 430 cases. Mayo Clin. Proc. 62 (1987) 719–731.

KYLE R.A. and M.A. GERTZ: Systemic amyloidosis. CRC Crit. Rev. Oncol. Hematol. 10 (1990) 49–87.

KYLE, R.A. and P.R. GREIPP: Amyloidosis (AL). Clinical and laboratory features in 229 cases. Mayo Clin. Proc. 58 (1983) 665–683.

KYLE, R.A., P.R. GREIPP and W.M. O'FALLON: Primary systemic amyloidosis: multivariate analysis for prognostic factors in 168 cases. Blood 68 (1986) 220–229.

LANDI, G., R. D'ALESSANDRO, B.C. DOSSI, S. RICCI, I.L. SIMONE and A. CIOCCONE: Guillain–Barré syndrome after exogenous gangliosides in Italy. Br. Med. J. 307 (1993) 1463–1464.

LANGE, D.J., W. TROJABORG, N. LATOV, A.P. HAYS, D.S. YOUNGER, A. UNCINI, D.M. BLAKE, M. HIRANO, S.M. BURNS, R.E. LOVELACE and L.P. ROWLAND: Multifocal motor neuropathy with conduction block: is it a distinct clinical entity? Neurology 42 (1992) 497–505.

LATOV, N.: Immunological abnormalities associated with chronic peripheral neuropathies: plasma cell dyscrasia and neuropathy. In: P. Behan and F. Spreafico (Eds.), Neuroimmunology. New York, Raven Press (1984) 261–273.

LATOV, N.: Pathogenesis and therapy of neuropathies associated with monoclonal gammopathies. Ann. Neurol. 37, Suppl. 1 (1995) S32–S42.

LATOV, N., W.H. SHERMAN and A.P. HAYS: Peripheral neuropathy and anti-MAG antibodies. CRC Crit. Rev. Neurobiol. 3 (1988a) 301–332.

LATOV, N., A.P. HAYS and P.D. DONOFRIO: Monoclonal IgM with unique specificity to gangliosides GM1 and GD1b, and to lacto-N-tetraose, associated with human motor neuron disease. Neurology 38 (1988b) 763–768.

LATOV, N., C.L. KOSKI and P.A. WALICKE: Guillain–Barré syndrome and parenteral gangliosides (Letter). Lancet 338 (1991) 757.

LEE J., G. KROL and M. ROSENBLUM: Primary amyloidoma of the brain: CT and MR presentation. Am. J. Neuroradiol. 16 (1995) 712–714.

LEE, S.M., P. HARPER, P. LUTHERT and R.A. HUGHES: Primary CNS lymphoma in association with IgM kappa paraproteinaemia and peripheral polyneuropathy: a case report. Eur. Neurol. 35 (1995) 237–239.

LEGER, J.M., E. OKSENHENDLER, A. BUSSEL, C. CHASTANG, A.B. YOUNES-CHENNOUFI, F. DANON, P. BOUCHE and J.C. BROUET: Treatment by chlorambucil with/without plasma exchanges of polyneuropathy associated with monoclonal IgM. Prospective randomized control study in 44 patients. Neurology 43 (1993) A215.

LEIFER, D., T. GRABOWSKI, N. SIMONIAN and Z.N. DEMIRJIAN: Leptomeningeal myelomatosis presenting with mental status changes and other neurologic findings. Cancer 70 (1992) 1899–1904.

LERNER, A.B., C.P. BARNUM and C.J. WATSON: Studies of cryoglobulins; spontaneous precipitation of protein from serum at 5°C in various disease states. Am. J. Med. Sci. 214 (1947) 416–421.

LINKE, R.P., L. GERHARD and F. LOTTSPEICH: Brain-restricted amyloidoma of immunoglobulin lambda-light chain origin clinically resembling multiple sclerosis. Biol. Chem. Hoppe-Seyler 373 (1992) 1201–1209.

LINOS, A., R.A. KYLE, W.M. O'FALLON and L.T. KURLAND: Incidence and secular trend of multiple myeloma in Olmsted County, Minnesota: 1965–77. J. Natl. Cancer Inst. 66 (1981) 17–20.

LIPPA, C.F., D.A. CHAD, T.W. SMITH, M.H. KAPLAN and K. HAMMER: Neuropathy associated with cryoglobulinemia. Muscle Nerve 9 (1986) 626–631.

LISAK, R.P., M. MITCHELL, B. ZWEIMAN, E. ORRECHIO and A.K. ASBURY: Guillain–Barré syndrome and Hodgkin's disease: three cases with immunological studies. Ann. Neurol. 1 (1977) 72–78.

LOGOTHETIS, J., P. SILVERSTEIN and J. COE: Neurological aspects of Waldenström's macroglobulinaemia. Arch. Neurol. 5 (1960) 564–573.

LONGSWORTH, L.G., T. SHEDLOVSKY and D.A. MACINNES: Electrophoretic patterns of normal and pathological human blood serum and plasma. J. Exp. Med. 70 (1939) 399–413.

LOSA, M., M.R. TERRENI, M. TRESOLDI, M. MARCATTI, A. CAMPI, F. TRIULZI, G. SCOTTI and M. GIOVANELLI: Solitary plasmacytoma of the sphenoid sinus involving the pituitary fossa: a case report and review of the literature. Surg. Neurol. 37 (1992) 388–393.

LOUIS, E.D., A.E. HANLEY and T.H. BRANNAGAN: Motor neuron disease, lymphoproliferative disease, and bone marrow biopsy. Muscle Nerve (in press).

MACANEC, K., J. MCCLURE, C.J. BARTLEY, M.J. NEWBOULD and P. ACKRILL: Systemic amyloidosis of beta2 microglobulin type. J. Clin. Pathol. 45 (1992) 832–833.

MACINTYRE, W.: Case of mollities and fragilitas ossium, accompanied with urine strongly charged with animal matter. Med. Chir. Trans. Lond. 33 (1850) 211.

MACKENZIE, M.R. and H.H. FUDENBERG: Macroglobulinemia: an analysis of forty patients. Blood 39 (1972) 874.

MALACRIDA, V., D. DE FRANCESCO, G. BANFI, F.A. PORTA and P.G. RICHES: Laboratory investigation of monoclonal gammopathy during 10 years of screening in a general hospital. J. Clin. Pathol. 40 (1987) 793–797.

MALDONADO, J.E., R.Z. KYLE, J. LUDWIG and H. OKAZAKI: Meningeal myeloma. Arch. Intern. Med. 126 (1970) 660–663.

MANCARDI, G.L. and T.I. MANDYBUR: Solitary intracranial plasmacytoma. Cancer 51 (1983) 2226–2233.

MANDLER, R.N., D.P. KERRIGAN, J. SMART, W. KUIS, P. VILLIGER and M. LOTZ: Castleman's disease in POEMS syndrome with elevated interleukin-6. Cancer (1992) 2697–2703.

MATSUMINE, H.: Accelerated conversion of androgen to estrogen in plasma cell dyscrasia associated with polyneuropathy, anasarca, and skin pigmentation (Letter to Ed.). N. Engl. J. Med. 313 (1985) 1025–1026.

MCANEMA, O.J., M.P. FEELY and W.F. KEALY: Spinal cord compression by amyloid tissue. J. Neurol. Neurosurg. Psychiatry 45 (1982) 1067–1069.

MCCALLISTER, B.D., E.D. BAYRD, E.G. HARRISON and W.F. MCGUCKIN: Primary macroglobulinemia. Am. J. Med. 43 (1967) 394–434.

MCGRATH, M.A. and R. PENNY: Paraproteinemia. Blood

hyperviscosity and clinical manifestation. J. Clin. Invest. 58 (1976) 1155–1162.

MCKHANN, G.M., D.R. CORNBLATH and J.W. GRIFFIN: Acute motor axonal neuropathy: a frequent cause of acute flaccid paralysis in China. Ann. Neurol. 33 (1993) 333–342.

MCKISSOCK, W., W.H. BLOOM and K.Y. CHYNN: Spinal cord compression caused by plasma cell tumors. J. Neurosurg. 18 (1961) 68.

MCLAUGHLIN, P. and R. ALEXANIAN: Myeloma protein kinetics following chemotherapy. Blood 60 (1982) 851–855.

MEIER, C., A. STECK, C. HESS, E. MILONI and L. TSCHOPP: Polyneuropathy in Waldenström's macroglobulinemia. Reduction of endoneurial IgM deposits after treatment with chlorambucil and plasmapheresis. Acta Neuropathol. (Berl.) 64 (1984) 297–301.

MELTZER, M. and E.C. FRANKLIN: Cryoglobulinemia-α study of twenty-one patients. I. IgG and IgM cryoglobulins and factors affecting cryoprecipitability. Am. J. Med. 40 (1966) 828–836.

MIDDAUGH, C.R., E.Q. LAWSON, G.W. LITMAN, W.A. TISEL, D.A. MOOD and A. ROSENBERG: Thermodynamic basis for the abnormal solubility of monoclonal cryoimmunoglobulins. J. Biol. Chem. 255 (1980) 6532–6534.

MINOT, G.P. and R. ISAACS: Lymphatic leukemia. Age, incidence, duration and benefit derived from irradiation. Med. Surg. J. (Boston) 191 (1924) 1–9.

MIRALLES, G.D., J.R. O'FALLON and N. TALLEY: Plasma-cell dyscrasia with polyneuropathy. The spectrum of POEMS syndrome. N. Engl. J. Med. 327 (1992) 1919–1923.

MIYATANI, N., H. BABA, S. SATO, K. NAKAMURA, T. YUASA and T. MIYATAKE: Antibody to sialosyllactosaminylparagloboside in patient with IgM paraproteinemia and polyradiculoneuropathy. J. Neuroimmunol. 14 (1987) 189–196.

MOLICA, S.: Infections in chronic lymphocytic leukemia: risk factors and impact on survival, and treatment. Leuk. Lymphoma 13 (1994) 203–214.

MONACO, S., B. BONETTI and S. FERRARI: Complement dependent demyelination in patients with IgM monoclonal gammopathy and polyneuropathy. N. Engl. J. Med. 322 (1990) 844–852.

MONTSERRAT, E. and C. ROZMAN: Current approaches to the treatment and management of chronic lymphocytic leukaemia. Drugs 47, Suppl. 6 (1994) 1–9.

MONTSERRAT, E. and C. ROZMAN: Chronic lymphocytic leukemia: present status. Ann. Oncol. 6 (1995) 219–235.

MONTSERRAT, E., J. SANCHEZ-BISONO, N. VINOLAS and C. ROZMAN: Lymphoctye doubling time in chronic lymphocytic leukaemia: analysis of its prognostic significance. Br. J. Haematol. 62 (1986) 567–575.

MONTSERRAT, E., N. VINOLAS, J.C. REVERTER and C. ROZMAN: Natural history of chronic lymphocytic leukemia: on

the progression and prognosis of early clinical stages. Nouv. Rev. Fr. Hématol. 30 (1988) 359–361.

MOREL, L. A. BASCH, F. DANON, P. VERROUT and G. RICHET: Pathology of the kidney in Waldenström's macroglobulinemia. N. Engl. J. Med. 283 (1970) 123.

MORENO A.J., J.M. BROWN, T.J. BROWN, G.D. GRAHAM and M.A. YEDINAK: Scintigraphic findings in a primary cerebral amyloidoma. Clin. Nucl. Med. 8 (1983) 528–530.

MOULOPOULOS, L.A., M.A. DIMOPOULOS, D.G. VARMA, J. MANNING, D.A. JOHNSTON, N. LEEDS and H.I. LIBSHITZ: Waldenström macroglobulinemia: MR imaging of the spine and computed tomography of the abdomen and pelvis. Radiology 188 (1993) 669.

MUELLER, J. J.R. HOTSON and J.W. LANGSTON: Hyperviscosity induced dementia. Neurology 33 (1983) 101–103.

MUNDY, G.R., L.G. RAISZ, R.A. COOPER, G.P. SCHECTER and S.E. SALMON: Evidence for the secretion of an osteoclast stimulating factor in myeloma. N. Engl. J. Med. 291 (1974) 1041–1046.

NAKANISHI, T., I. SOBUE, Y. TOYOKURA, H. NISHITANI, Y. KUROIWA, E. SATOYOSHI, T. TSUBAKI, A. IGATA and Y. OZAKI: The Crow–Fukase syndrome: a study of 102 cases in Japan. Neurology 334 (1984) 712–720.

NEMNI, R., R. FAZIO, A. QUATTRINI, I. LORENZETTI, D. MAMOLI and N. CANAL: Antibodies to sulfatide and to chondroitin sulfate C in patients with chronic sensory neuropathy. J. Neuroimmunol. 43 (1993) 79–86.

NIGHTINGALE, S.D., R.P. PELLEY, N.L. DELANEY, W.B. BIAS, M.I. HAMBURGER, L.F. FRIES and A.G. STEINBERG: Inheritance of mixed cryoglobulinemia. Am. J. Hum. Genet. 33 (1981) 735–744.

NOBILE-ORAZIO, E., P. MARMIROLI, L. BALDINI, G. SPAGNOL, S. BARBIERI, M. MAGGIO, N. POLLI, E. POLLI and E. SCARLATO: Peripheral neuropathy in macroglobulinemia: incidence and antigen specificity of M-proteins. Neurology 37 (1987) 1506–1514.

NOBILE-ORAZIO, E., L. BALDINI and S. BARBIERI: Treatment of patients with neuropathy and anti-MAG M-proteins. Ann. Neurol. 34 (1988) 93–97.

NOBILE-ORAZIO, E., E. FRANCOMANO and E. DAVERIO: Anti-myelin associated glycoprotein IgM antibody titers in neuropathy associated with macroglobulinemnia. Ann. Neurol. 26 (1989) 543–550.

NOBILE-ORAZIO, E., M. CARPO, N. MEUCCI, M.P. GRASSI, E. CAPITANI, M. SCIACCO, A. MANGONI and G. SCARLATO: Guillain–Barré syndrome associated with high titers of anti-GM1 antibodies. J. Neurol. Sci. 109 (1992) 200–206.

NOBILE-ORAZIO, E., N. MEUCCI and S. BARBIERI: High dose intravenous immunoglobulin therapy in multifocal motor neuropathy. Neurology 43 (1993) 537–543.

NOBILE-ORAZIO, E., E. MANFREDINI, M. CARPO, N. MEUCCI, S. MONACO, S. FERRARI, B. BONETTI, S. ALLARIA, G. CAVALETTI, F. GEMIGNANI, M. SGARZI, D. PEREGO, L. DURELLI, L. BALDINI and G. SCARLATO: Frequency and clinical correlates of anti-neural IgM antibodies in neuropathy associated with IgM monoclonal gammopathy. Ann. Neurol. (1994).

NORING, L., K.G. KJELLIN and A. SIDEN: Neuropathies associated with disorders of plasmacytes. Eur. Neurol. 19 (1980) 224–230.

OBI, T., S. KUSUNOKI, M. TAKATSU, D. MIZOGUCKI and Y. NISHIMURA: IgM M-protein in a patient with sensory-dominant neuropathy binds preferentially to polysialogangliosides. Acta Neurol. Scand. 86 (1992) 215–218.

O'BRIEN, T.J., P.A. MCKELVIE and N. VRODOS: Bilateral trigeminal amyloidoma: an unusual case of trigeminal neuropathy with a review of the literature. J. Neurosurg. 81 (1994) 780–783.

OGAWA-GOTO, K., N. FUNAMOTO, Y. OHTA, T. ABE and K. NAGASHIMA: Different ceramide compositions of gangliosides between human motor and sensory nerves. J. Neurochem. 55 (1990) 1486–1492.

OGAWA-GOTO, K., N. FUNAMOTO, Y. OHTA, T. ABE and K. NAGASHIMA: Myelin gangliosides of human peripheral nervous system: an enrichment of GM1 in the motor nerve myelin isolated from cauda equina. J. Neurochem. 59 (1992) 1844–1848.

OHI, T., R.A. KYLE and P.J. DYCK: Axonal attenuation and secondary segmental demyelination in myeloma neuropathies. Ann. Neurol. 17 (1985) 255–261.

OKEN, M.M.: Standard treatment of multiple myeloma. Mayo Clin. Proc. 69 (1994) 781–786.

OSCIER, D.G.: Cytogenetic and molecular abnormalities in chronic lymphocytic leukaemia. Blood Rev. 8 (1994) 88–97.

OSSERMAN, E.F.: Plasma-cell myeloma. II. Clinical aspects. N. Engl. J. Med. 261 (1959) 952–960.

OTT, E.: Hyperviscosity syndromes. In: J.F. Toole (Ed.), Handbook of Clinical Neurology: Vascular Diseases, Part III. Amsterdam, Elsevier Science Publishers (1989) 483–492.

PACKER, R.J., L.B. RORKE, B.J. LANGE, K.R. SIEGEL and A.E. EVANS: Cerebrovascular accidents in children with cancer. Pediatrics 76 (1985) 194–201.

PASMANTIER, M.W. and H.Z. AZAR: Extraskeletal spread in multiple plasma cell myeloma: a review of 57 autopsied cases. Cancer 23 (1969) 167–174.

PAVY, M.D., P.L. MURPHY and G. VIRELLA: Paraprotein induced hyperviscosity. A reversible cause of stroke. Postgrad. Med. 68 (1980) 109–112.

PESTRONK, A.: Motor neuropathies, motor neuron disorders, and antiglycolipid antibodies. Muscle Nerve 14 (1991) 927–936.

PESTRONK, A., D.R. CORNBLATH, A.A. ILYAS, H. BABA, R.H. QUARLES, J.W. GRIFFIN, K. ALDERSON and R.N. ADAMS: A treatable multifocal motor neuropathy with antibodies to GM1 ganglioside. Ann. Neurol. 24 (1988) 73–78.

PESTRONK, A., F. LI, J. GRIFFIN, E.L. FELDMAN, D. CORNBLATH, J. TROTER, S. AHU, W.C. YEE, D. PHILLIPS, D.M. PEEPLES and B. WINSLOW: Polyneuropathy syndromes

associated with serum antibodies to sulfatide and myelin associated glycoprotein. Neurology 41 (1991) 357–362.

PIMINTEL, L.: Medical complications of oncologic disease. Emerg. Med. Clin. N. Am. 11 (1993) 407–419.

POTTER, M.: Myeloma proteins (M-components) with antibody-like activity. N. Engl. J. Med. 284 (1971) 831–838.

PRESTON, F.E., K.B. COOKE, M.E. FOSTER, D.A. WINFIELD and D. LEE: Myelomatosis and the hyperviscosity syndrome. Can. Med. Assoc. J. 123 (1980) 731–737.

PRITCHARD, P.B., R.A. MARTINEZ, G.D. HUNGERFORD, J.M. POWERS and P.L. PEROT: Dural plasmacytoma. Neurosurgery 5 (1983) 576–579.

PRUZANSKI, W., R. CHU, N.F. DAMJI, S. GALLER and C.S. NORMAN: Anemia, splenomegaly and hyperviscosity syndrome. Can. Med. Assoc. J. 123 (1980) 731–737.

QUATTRINI, A., R. NEMNI, R. FAZIO, S. IANNACCONE, I. LORENZETTI, F. GRASSI and N. CANAL: Axonal neuropathy in a patient with monoclonal IgM kappa reactive with Schmidt–Lanterman incisures. J. Neuroimmunol. 33 (1991) 73–79.

QUATTRINI, A., M. CORBO and S.K. DHALIWAL: Anti-sulfatide antibodies in neurological disease; binding to rat dorsal root ganglia neurons. J. Neurol. Sci. 112 (1992) 152–159.

QUINT, D.J., R. LEVY and J.C. KRAUSS: MR of myelomatous meningitis. Am. J. Neuroradiol. 16 (1995) 1316–1317.

QUITT, M., I. BAZAC, B. GROSS and E. AGHAI: Fulminant bilateral cerebellar syndrome in a patient with chronic lymphocytic leukemia. Leuk. Lymphoma 15 (1994) 507–510.

RADL, J.: Benign monoclonal gammopathy. In: F. Melchers and M. Potter (Eds.), Mechanisms in B-cell Neoplasia. Berlin, Springer (1985) 221–224.

RAI, K.R., A. SAWITSKY, E.P. CRONKITE, A.D. CHANANA, R.N. LEVY and B.S. PASTERNACK: Clinical staging of chronic lymphocytic leukemia. Blood 16 (1975) 219–234.

READ, D.J., R.I. VANHEGAN and W.B. MATTHEWS: Peripheral neuropathy and benign IgG paraproteinemia. J. Neurol. Neurosurg. Psychiatry 41 (1978) 215–219.

REECE, D.E., M.J. BARNETT, J.M. CONNORS, H.G. KLINGEMANN, S.E. O'REILLY, J.D. SHEPHERD, H.J. SUTHERLAND and G.L. PHILLIPS: Treatment of multiple myeloma with intensive chemotherapy followed by autologous BMT using marrow purged with 4-hydroperoxycyclophosphamide. Bone Marrow Transplant. 11 (1993) 139–146.

REIK, L. and J.H. KORN: Cryoglobulinemia with encephalopathy: successful treatment by plasma exchange. Ann. Neurol. 10 (1981) 488–490.

RESKE-NIELSEN, E., J.H. PETERSEN, J. SOGAARD and K.B. JENSEN: Leukemia of the central nervous system. Lancet i (1974) 211–212.

REULECKE, M., M. DUMAS and C. MEIER: Specific antibody activity against neuroendocrine tissue in a case of

POEMS syndrome with IgG gammopathy. Neurology 38 (1988) 614–616.

REINHART, W.H., O. LUTOLF, U.R. NYDEGGER, F. MAHLER and P.W. STRAUB: Plasmapheresis for hyperviscosity syndrome in macroglobulinemia Waldenström and multiple myeloma: influence on blood rheology and the microcirculation. J. Lab. Clin. Med. 119 (1992) 69–76.

RIEDEL, D.A., L.M. POTTERN and W.A. BLATTNER: Epidemiology of multiple myeloma. In: P.H. Wiernik, G. P. Canellos, R.A. Kyle and C.A. Schiffer (Eds.), Neoplastic Diseases of the Blood, 2nd Edit. New York, Churchill Livingstone (1991) 347–372.

RIES, L.A.G., V.F. HANKEY and B.A. MILLER: Cancer Statistics Review 1973–1988 (DHHS publ. (NIH) no. 91-2789). Washington, DC, US Government Printing Office (1991).

ROBBINS, S.L. and R.S. COTRAN: Pathologic Basis of Disease. Philadelphia, PA, Saunders (1979) 795.

ROBERTSON, L.E., W. PUGH, S. O'BRIEN, H. KANTARJIAN, C. HIRSCH-GINSBERG, A. CORK, P. MCLAUGHLIN, F. CABANILLAS and M.J. KEATING: Richter's syndrome: a report on 39 patients. J. Clin. Oncol. 11 (1993) 1985–1989.

ROKE, M.E., B.D. BOUGHNER, L.C. ANG and G.P.A. RICE: Myopathy in primary systemic amyloidosis. Can. J. Neurol. Sci. 15 (1988) 314–316.

ROUSSEAU, J.J., G. FRANCK, T. GRISAR, M. REZNIK, G. HEYNEN and J. SALMON: Osteosclerotic myeloma with poly-neuropathy and ectopic secretion of calcitonin. Eur. J. Cancer 14 (1978) 133–140.

ROWLAND, L.P.: Ten central themes in a decade of ALS research. Adv. Neurol. 56 (1991) 463–472.

ROWLAND, L.P., W.H. SHERMAN, N. LATOV, D.J. LANGE, T.D. MCDONALD, D.S. YOUNGER, P.L. MURPHY, A.P. HAYS and D. KNOWLES: Amyotrophic lateral sclerosis and lymphoma: bone marrow examination and other diagnostic tests. Neurology 42 (1992) 1101–1102.

ROWLAND, L.P., W.L. SHERMAN, A.P. HAYS, D.J. LANGE, N. LATOV, W. TROJABORG and D.S. YOUNGER: Autopsy-proven amyotrophic lateral sclerosis, Waldenström's macroglobulinemia, and antibodies to sulfated glucuronic acid paragloboside. Neurology 45 (1995) 827–829.

ROZMAN, C., E. MONTSERRAT, E. FELIU, A. GRANENA, P. MARIN, B. NOMDEDEU and J.L. VIVES CORRONS: Prognosis of chronic lymphocytic leukemia: a multivariate survival analysis of 150 cases. Blood 59 (1982) 1001–1005.

ROZMAN, C., E. MONTSERRAT, J.M. RODRIQUEZ-FERNANDES, R. AYATS, T. VALLESPI, R. PARODY, D. PRADOS, M. MOREY and F. GOMIS: Bone marrow histologic pattern – the best single prognostic parameter in chronic lymphocytic leukemia: a multivariate survival analysis of 329 cases. Blood 64 (1984) 642–648.

SADIQ, S.A., F.P. THOMAS, K. KILIDIREAS, S. PROTOPSALTIS, A.P. HAYS, K.W. LEE, S.N. ROMAS, N. KUMAR, L. VAN DEN BERG, M. SANTORO, D.J. LANGE, D.S. YOUNGER, R.E. LOVELACE, W. TROJABORG, W.H. SHERMAN, J.R. MILLER, J.

MINUK, M.A. FEHR, R.I. ROELOFS, D. HOLLANDER, F.T. NICHOLS, H. MITSUMOTO, J.J. KELLEY, T.R. SWIFT, T.L. MUNSAT and N. LATOV: The spectrum of neurologic disease associated with anti-GM1 antibodies. Neurology 40 (1990) 1067–1090.

SALEUN, J.P., M. VICAROIT, P. DEROFF and J.F. MORIN: Monoclonal gammopathies in the adult population of Finistere, France. J. Clin. Pathol. 35 (1982) 63–68.

SALMON, S.E.: Paraneoplastic syndromes associated with monoclonal lymphocytes and plasma cell proliferation. Ann. N.Y. Acad. Sci. 230 (1974) 228–239.

SALMON, S.E. and J.R. CASSADY: Plasma cell neoplasms. In: V.T. DeVita, S. Hellman and S.A. Rosenberg (Eds.), Cancer: Principles and Practice of Oncology, 4th Edit. Philadelphia, PA, Lippincott (1993) 1984–2024.

SAN ROMAN, C.T. FERRO, M. GUZMAN and J. ODRIOZOLA: Clonal abnormalities in patients with Waldenström's macroglobulinemia with special reference to a Burkitt-type t(8;14). Cancer Genet. Cytogenet. 18 (1985) 155.

SANDERS, D.A., L.P. ROWLAND, P.L. MURPHY, D.S. YOUNGER, N. LATOV, W.H. SHERMAN, M. PESCE and D.J. LANGE: Motor neuron diseases and amyotrophic lateral sclerosis: GM1 antibodies and paraproteinemia. Neurology 43 (1993) 418–420.

SANTORO, M., A. UNCINI, M. CORBO, S.M. STAUGAITIS, F.P. THOMAS, A.P. HAYS and N. LATOV: Experimental conduction block induced by serum from a patient with anti-GM1 antibodies. Ann. Neurol. 31 (1992) 385–390.

SCHEITHAUER, B.W., L.J. RUBINSTEIN and M.M. HERMAN: Leukoencephalopathy in Waldenström's macro-globulinemia. Immunohistochemical and electron microscopic observations. J. Neuropathol. Exp. Neurol. 43 (1984) 408–425.

SCHULMAN, P., T. SUN, L. SHARER, P. HYMAN, V. VINCIGUERRA, M. FEINSTEIN, R. BLANCK, M. SUSIN and T.J. DEGNAN: Meningeal involvement in IgG myeloma with cerebrospinal fluid paraprotein analysis. Cancer 46 (1980) 152–155.

SELIGMAN, M.E. and J.C. BROUET: Antibody activity of human myeloma globulins. Sem. Haematol. 10 (1973) 163–177.

SEMBLE, E.L., V.R. CHALLA, D.A. HOLT and E.J. PISKO: Light and electron microscopic findings in POEMS, or Japanese multisystem syndrome. Arthr. Rheum. 29 (1986) 286–291.

SHEAFF, M.T., J.D. VAN DER WALT, J.F. GEDDES and R. MACFARLANE: Prolymphocytic transformation of chronic lymphocytic leukaemia manifesting as a cerebellar lymphoma. Br. J. Neurosurg. 8 (1994) 735–738.

SHERMAN, W.H., N. LATOV, A.P. HAYS, M. TAKATSU, R. NEMNI, G. GALASSI and E.F. OSSERMAN: Monoclonal IgM antibody precipitating with chondroitin sulfate C from patients with axonal polyneuropathy and epidermolysis. Neurology 33 (1983) 192–201.

SHERMAN, W.H., N. LATOV, D.E. LANGE, A.P. HAYS and D. YOUNGER: Fludarabine for IgM antibody mediated neuropathies (abstr.). Ann. Neurol. 36 (1994) 326.

SHIMIZU, D., K. FUJISAWA, H. YAMAMOTO, M. YOSHIKAZU and K. HARA: Importance of central nervous system involvement by neoplastic cells in a patient with Waldenström's macroglobulinemia developing neurologic abnormalities. Acta Haematol. 90 (1993) 206–208.

SHIMIZU, T., H. KATO and W. SCHULL: Studies of the mortality of A-bomb survivors. 9. Mortality, 1950–1985: Part 2. Cancer mortality based on the recently revised doses (DS86). Radiat. Res. 121 (1990) 120–141.

SHIMPO, S., H. NISHITANI and T. TSUNEMATSU: Solitary plasmacytoma with polyneuritis and endocrine disturbances. Nippon Rinsho 26 (1968) 2444–2456.

SHY, M.E., T. HEIMAN-PATTERSON and G.J. PARRY: Lower motor neuron disease in a patient with antibodies against Gal(B1–3)GalNAc in gangliosides GM1 and GD1b: improvement following immunotherapy. Neurology 40 (1990) 842–844.

SIGSBEE, B., M.D.F. DECK and J.B. POSNER: Non-metastatic superior sagittal sinus thrombosis complicating systemic cancer. Neurology 29 (1979) 139–146.

SILVERSTEIN, A. and D.E. DONIGER: Neurologic complications of myelomatosis. Arch. Neurol. 9 (1963) 102–112.

SMITH, S.R., P.W. SAUNDERS and N.V. TODD: Spinal stabilisation in plasma cell disorders. Eur. J. Cancer 31A (1995) 1541–1544.

SOUBRIER, M.J., J.J. DUBOST and J.M. SAUVEZIE: POEMS syndrome: a study of 25 cases and a review of the literature. Am. J. Med. 97 (1994) 543–553.

SPATZ, L.A., K.K. WONG, M. WILLIAMS, R. DESAI, J. GOLIER, J.E. BERMAN, F.W. ALT and N. LATOV: Cloning and sequence analysis of the variable heavy and light chain regions of an anti-myelin/DNA antibody from a patient with peripheral neuropathy and chronic lymphocytic leukemia. J. Immunol. 144 (1990) 2821–2828.

SPIERS, A.S.D., R. HALPERN, S.C. ROSS, R.S. NEIMAN, S. HARAWI and T.E. ZIPOLI: Meningeal myelomatosis. Arch. Intern. Med. 140 (1980) 256–259.

STARK, R.J. and R.A. HENSON: Cerebral compression by myeloma. J. Neurol. Neurosurg. Psychiatry 44 (1981) 833–836.

STRICKER, R.B., K.A. SANDERS, W.F. OWEN, D.D. KIPROV and R.G. MILLER: Mononeuritis multiplex associated with cryoglobulinemia in HIV infection. Neurology 42 (1992) 2103–2105.

SUAREZ, G.A. and J.J. KELLY: Polyneuropathy associated with monoclonal gammopathy of undetermined significance: further evidence that IgM-MGUS neuropathies are different than IgG-MGUS. Neurology 43 (1993) 1304–1308.

TAKATSU, M., A.P. HAYS, N. LATOV, G.M. ABRAMS, W.H. SHERMAN, R. NEMNI, E. NOBILE-ORAZIO, T. SAITO and L. FREDDO: Immunofluorescence study of patients with neuropathy and IgM M-proteins. Ann. Neurol. 18 (1985) 173–181.

TAKATSUKI, K. and I. SANADA: Plasma cell dyscrasia with polyneuropathy and endocrine disorder: clinical and laboratory features of 109 reported cases. Jpn. J. Clin. Oncol. 13 (1983) 543–556.

TATUM, A.H.: Experimental paraprotein neuropathy; demyelination by passive transfer of human IgM anti-MAG. Ann. Neurol. 33 (1993) 502–506.

TEFFERI, A. and R.L PHYLIKY: A clinical update on chronic lymphocytic leukemia. I. Diagnosis and prognosis. Mayo Clin. Proc. 67 (1992) 349–353.

THOMAS, F.P., P H. ADAPON, G.P. GOLDBERG, N. LATOV and A.P. HAYS: Localization of neural epitopes that bind to IgM monoclonal autoantibodies (M-proteins) from two patients with motor neuron disease. J. Neuroimmunol. 21 (1989) 31–39.

THOMAS, F.P., W. TROJABORG, C. NAGY, U. VALLEJOS, M. SANTORO, S.A. SADIQ, N. LATOV and A.P. HAYS: Experimental autoimmune neuropathy with anti-GM1 antibodies and immunoglobulin deposits at the nodes of Ranvier. Acta Neuropathol. (Berl.) 82 (1991) 378–383.

THOMAS, F.P., R.E. LOVELACE and X.S. DING: Vasculitis, axonal degeneration, and demyelinating neuropathy with cryoglobulinemia and anti-MAG monoclonal gammopathy. Muscle Nerve 15 (1992) 891–898.

TROJABORG, W., G. GALASSI, A.P. HAYS, R.E. LOVELACE, M. ALKAITIS and N. LATOV: Electrophysiological study of experimental demyelination induced by serum of patients with IgM M-proteins and neuropathy. Neurology 39 (1989) 1581–1586.

TROTTER, J.L., W.K. ENGEL and T.F. IGNASZAK: Amyloidosis with plasma cell dyscrasia: an overlooked cause of adult onset sensorimotor neuropathy. Arch. Neurol. 34 (1977) 209–214.

TUDDENHAM, E.G.D., J.A. WHITTAKER, J. BRADLEY, J.S. LILLEYMAN and D.R. JAMES: Hyperviscosity syndrome in IgA multiple myeloma. Br. J. Haematol. 27 (1974) 67–76.

UNAL, F., K. HEPGUL, C. BAYINDIR, T. BILGE, M. IMER and I. TURANTAN: Skull base amyloidoma (case report). J. Neurosurg. 76 (1992) 303–306.

VALDERRAMA, J.A.F. and P.G. BULLOUGH: Solitary plasmacytoma of the spine. J. Bone Joint Surg. 50 (1968) 82–90.

VALLDEORIOLA, E., F. GRAUS, A.J. STECK, E. MUNOZ, M. DE LA FUENTE, T. GALLART, T. RIBALATA, J.A. BOMBI and E. TOLOSA: Delayed appearance of anti-myelin associated glycoprotein antibodies in a patient with chronic demyelinating polyneuropathy. Ann. Neurol. 34 (1993) 394–396.

VAN DE POEL, M.H., J.W. COEBERGH and H.F. HILLEN: Malignant transformation of monoclonal gammopathy of undetermined significance among outpatients of a community hospital in southeastern Netherlands. Br. J. Haematol. 91 (1995) 121–125.

VAN DEN BERG, L.H., C. LANKAMP, A.E.J. DE JAGER, N.C. NOTERMANS, P. SODAAR, J. MARRINK, H.J. DE JONG, P.R. BAR and J.H.J. WOKKE: Anti-sulfatide antibodies in peripheral neuropathy. J. Neurol. Neurosurg. Psychiatry 56 (1993) 1164–1168.

VAN DEN BERG, L.H., J. MARRINK, A.E.J. DE JAGER, H.J. DE JONG, G.W. VAN DEN BERG, A.P. HAYS, E. NOBILE-ORAZIO, L.J. KINSELLA, E. MANFRENDINI, M. CORBO, G. ROSOKLIJA, D.S. YOUNGER, R.E. LOVELACE, W. TROJABORG, D.J. LANGE, S. GOLDSTEIN, J.S. DELFINER, S.A. SADIQ, W.H. SHERMAN and N. LATOV: Anti-MAG and anti-SGPG antibodies in neuropathy. Muscle Nerve 19 (1996a) 637–643.

VAN DEN BERG, L.H. A.P. HAYS, E. NOBILE-ORAZIO, E. MANFREDINI, M. CORBO, G. ROSOKLÿA, D.S. YOUNGER, L.J. KINSELLA, R.E. LOVELACE, W. TROJABORG, D.J. LANGE, S. GOLDSTEIN, J.S. DELFINER, S.A. SADIQ, W.H. SHERMAN and N. LATOV: Neuropathy syndromes associated with anti-MAG and anti-SGPG antibodies. Muscle Nerve 19 (1996b) 637–643.

VAN IMHOFF, N. LATOV and S.A. SADIQ: Anti-GM1 antibodies in patients with Guillain–Barré syndrome. J. Neurol. Neurosurg. Psychiatry 55 (1992) 6–11.

VIADANA, E., I.D.J. BROSS and J.W. PICKREN: An autopsy study of 1206 acute and chronic leukemias (1958–1982). Cancer 60 (1987) 827–837.

VITAL, A., C. VITAL, J. JULIEN, A. BAQUEY and A.J. STECK: Polyneuropathy associated with IgM monoclonal gammopathy; immunological and pathological study in 31 patients. Acta Neuropathol. (Berl.) 79 (1989) 160–167.

VLADUTIU, A.O.: Prevalence of M-proteins in serum of hospitalized patients. Physician's response to finding M-proteins in serum protein electrophoresis. Ann. Clin. Lab. Sci. 17 (1987) 157–161.

WALDENSTRÖM, J.G., A. ADNER, K. GYDELL and O. ZETTERVALL: Osteosclerotic 'plasmacytoma' with polyneuropathy, hypertrichosis and diabetes. Acta Med. Scand. 203 (1978) 297–303.

WALSH, F.S., M. CRONIN, S. KOBLAR, P. DOHERTY, J. WINER, A. LEON and R.A.C. HUGHES: Association between glycoconjugate antibodies and Campylobacter infection in patients with Guillain–Barré syndrome. J. Neuroimmunol. 34 (1991) 43–51.

WEH, H.J., K. GUTENSOHN, J. SELBACH, R. KRUSE, G. WACKER-BACKHAUS, D. SEEGER, W. FIEDLER, W. FETT and D.K. HOSSFELD: Karyotype in multiple myeloma and plasma cell leukaemia. Eur. J. Cancer 29 (1993) 1269–1273.

WEINER, L.P., P.N. ANDERSON and J.C. ALLEN: Cerebral plasmacytoma with myeloma protein in the cerebrospinal fluid. Neurology 16 (1966) 615–618.

WELLER, M., A. STEVENS, N. SOMMER, M. SCHABET and H. WIETHOLTER: Humoral CSF parameters in the differential diagnosis of hematologic CNS neoplasia. Acta Neurol. Scand. 86 (1992) 129–133.

WIESACKER, M. and H.W. KOELOMEL: Meningeal involvement in leukemia and malignant lymphomas of adults: incidence, course of disease, and treatment for prevention. Acta Neurol. Scand. 60 (1979) 363–370.

WILLIAMS, H.M., H.D. DIAMOND, L.F. CRAVER and H. PARSONS: Neurological Complications of Lymphomas and Leukemias. Springfield, IL, Charles C. Thomas (1959).

WILLISON, H.J., D.D. TRAPP, J.D. BACHER, M.C. DALAKAS, J.W. GRIFFIN and R.H. QUARLES: Demyelination induced by intraneural injection of human antimyelin associated glycoprotein antibodies. Muscle Nerve 11 (1988) 1169–1176.

WILLISON, H.J., G. PATERSON, J. VEITCH, G. INGLIS and S.C. BARNETT: Peripheral neuropathy associated with monoclonal IgM anti-Pr2 cold agglutinin. J. Neurol. Neurosurg. Psychiatry 56 (1993) 1178–1181.

WILLISON, H.J., M. ROBERTS, G. O'HANLON, G. PATERSON, A. VINCENT and J. NEWSOM-DAVIS: Human monoclonal anti-GM1 ganglioside antibodies interfere with neuromuscular transmission. Ann. Neurol. 36 (1994) 289.

WILTSHAW, E.: The natural history of extramedullary plasmacytoma and its relation to solitary myeloma of bone and myelomatosis. Medicine (Baltimore) 55 (1976) 217–238.

WIRGUIN, I., L.J. SUTURKOVA-MILOSEVIC, P. DELLA-LATTA, T. FISHER, R.H. BROWN and N. LATOV: Monoclonal IgM antibodies to GM1 and asialo-GM1 in chronic neuropathies cross-react with Campylobacter jejuni lipopolysaccharides. Ann. Neurol. 35 (1994) 698–703.

WOLLERSHEIM, H.C.H., R.S.G. HOLDRINET and C.S. HAANEN: Clinical course and survival in 16 patients with localized plasmacytoma. Scand. J. Hematol. 32 (1984) 423–428.

WOODRUFF, R.K., J.M. WHITTLE and L.P. MALPAS: Solitary plasmacytoma. Cancer 43 (1979) 2340–2347.

YAMADA, M., H. TSUKAGOSHI and S. HATAKEYAMA: Skeletal muscle amyloid deposition in AL- (primary or myeloma-associated), AA- (secondary), and pre-albumin-type amyloidosis. J. Neurol. Sci. 85 (1988) 223–232.

YEE, W.C., A.F. HAHN, S.A. HEARN and A.R. RUPAR: Neuropathy in IgM paraproteinemia; immunoreactivity to neural proteins and chondroitin sulfate. Acta Neuropathol. (Berl.) 78 (1989) 57–64.

YEUNG, K.B., P.K. THOMAS, R.H. KING, H. WADDY, R.G. WILL, R.A. HUGHES, N.A. GREGSON and S. LEIBOWITZ: The clinical spectrum of peripheral neuropathies associated with benign monoclonal IgM, IgG and IgA paraproteinemia. Comparative clinical, immunological and nerve biopsy findings. J. Neurol. 238 (1991) 383–391.

YOUNES-CHENNOUFI, A.B., J M. LEGER, J.J. HAUW, J.L. PREUD'HOMME, P. BOUCHE, P. AUCOUTURIER, H.

RATINAHIRANA, C. LUBETZKI, O. LYON-CAEN and N. BAUMANN: Ganglioside GD1b is the target antigen for a biclonal IgM in a case of sensory-motor axonal neuropathy. Ann. Neurol. 32 (1992) 18–23.

YOUNGER, D.S., L.P. ROWLAND, N. LATOV, W. SHERMAN, M. PESCE, D.J. LANGE, W. TROJABORG, J.R. MILLER, R.E. LOVELACE, A.P. HAYS and T.S. KIM: Motor neuron disease and amyotrophic lateral sclerosis: relation of high CSF protein content to paraproteinemia and clinical syndromes. Neurology 40 (1990) 595–599.

YOUNGER, D.S., L.P. ROWLAND, N. LATOV, A.P. HAYS, D.J. LANGE, W. SHERMAN, G. INGHIRAMI, M.A. PESCE, D.M. KNOWLES, J. POWERS, J.R. MILLER, M.R. FETELL and R.E. LOVELACE: Lymphoma, motor neuron diseases, and amyotrophic lateral sclerosis. Ann. Neurol. 29 (1991) 78–86.

YUKI, N., H. YOSHINO, S. SATO and T. MIYATAKE: Acute axonal polyneuropathy associated with anti-GM1 antibodies following Campylobacter enteritis. Neurology 40 (1990) 1900–1902.

YUKI, N., S. HANDA and T. TAKI: Cross-reactive antigen between nervous tissue and bacterium elicits Guillain–Barré syndrome: molecular mimicry between ganglioside GM1 and lipopolysaccharide from Penner's serotype 19 of Campylobacter jejuni. Biomed. Res. 13 (1992a) 451–453.

YUKI, N., N. MIYATANI, S. SATO, Y. HIRABAYASHI, M. YAMAZAKI, N. YOSHIMURA, Y. HAYASHI and T. MIYATAKE: Acute relapsing sensory neuropathy associated with IgM antibody against B-series gangliosides. Neurology 42 (1992b) 686–689.

ZIFKO, U., M. DRICEK, E. MACHACEK, K. JELLINGER and W. GRISOLD: Syndrome of continuous muscle fiber activity and plasmacytoma with IgM paraproteinemia. Neurology 44 (1994) 560–561.

III. Multi-Organ Disease

Handbook of Clinical Neurology, Vol. 27 (71): Systemic Diseases, Part III
M.J. Aminoff and C.G. Goetz, editors

CHAPTER 19

Neurological manifestations of sarcoidosis

ALLAN KRUMHOLZ[1] and BARNEY J. STERN[2]

[1]*Department of Neurology, University of Maryland School of Medicine, Baltimore, MD; and* [2]*Department of Neurology, Emory University School of Medicine, Atlanta, GA, U.S.A.*

Sarcoidosis is a multisystem granulomatous disorder of unknown etiology. Young adults are most commonly affected. Typical presentations include bilateral hilar adenopathy, pulmonary infiltration, skin, and eye lesions. The neurological manifestations of sarcoidosis are termed neurosarcoidosis and occur in about 5% of sarcoidosis patients. Neurosarcoidosis can manifest in a myriad of ways including: cranial neuropathy, aseptic meningitis, mass lesions, encephalopathy/vasculopathy, seizures, hypothalamic-pituitary disorders, hydrocephalus, myelopathy, peripheral neuropathy and myopathy (Wiederholt and Siekert 1965; Delaney 1977; Pentland et al. 1985; Stern et al. 1985; Younger et al. 1988).

Neurosarcoidosis poses major diagnostic problems. This is because its etiology is unknown, its neurological manifestations are diverse, and its diagnosis cannot be readily confirmed by laboratory tests. Absolute confirmation of a presumed diagnosis of neurosarcoidosis requires substantiation with biopsy tissue from an affected region of the nervous system. However, acquiring such tissue from many parts of the nervous system is difficult and sometimes dangerous. Therefore, a diagnosis of neurosarcoidosis is often tentative because it is usually based on the identification of characteristic neurological findings in an individual with proven systemic sarcoidosis as established by clinical, imaging, or histologic findings.

Even after proper diagnosis, treatment of neurosarcoidosis may pose difficulties. Pathologically, sarcoidosis is characterized by an intense immune mediated process that damages tissues at the sites of disease activity, but the precise nature of this inflammatory reaction is not well understood. Moreover, although corticosteroids are regarded as the foundation of treatment, they are not always successful. Some patients with neurosarcoidosis prove refractory to conventional therapy, and approximately 5–10% die (Stern et al. 1985; Luke et al. 1987; Krumholz et al. 1991). Optimal management of patients with neurosarcoidosis, therefore, requires a good understanding of the broad clinical spectrum of neurosarcoidosis, appreciation of the ways to best confirm a diagnosis, and awareness of the full range of treatment options.

HISTORY

Sarcoidosis was first described in 1877 by Sir Jonathan Hutchinson, but his early descriptions reported only the cutaneous manifestations. He named the disease Mortimer's malady, after one of his patients (Fig. 1) (James and Williams 1985; Leiberman et al. 1985). By the turn of the 20th century, Caesar Boeck, a dermatologist, reported that the disorder could involve not only skin but also lymph nodes. Because of its histologic similarity to sarcoma he called the disease 'multiple benign sarkoid', from which the modern term sarcoidosis is derived (Leiberman et al. 1985). Boeck demonstrated that the unifying feature

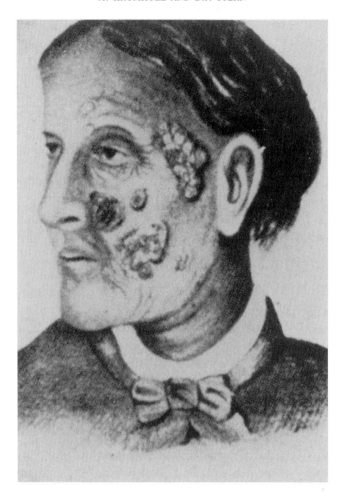

Fig. 1. Hutchinson first described sarcoidosis in 1877 in a patient with cutaneous manifestations. He termed the disease 'Mortimer's malady' after one of his first patients shown here.

of sarcoidosis was epithelioid cell granulomas which could involve different organs, a view that serves as the basis of current thinking about sarcoidosis. In 1936 Jorgen Schaumann advanced the concept that rather than affecting only solitary organ systems such as the skin or lymph nodes, sarcoidosis was instead a systemic disease characterized by the presence of granulomas in multiple organs, and he proposed the term 'lymphogranuloma benignum'. Today, it is well recognized and established that not only can sarcoidosis involve many different organ systems but that it is capable of attacking any body tissue (Leiberman et al. 1985).

Neurological involvement by sarcoidosis was first reported in 1917 by Heerfordt in his original account of what he described as 'uveo-parotid fever'. In that

report, he expressly associated cranial nerve palsies with this sarcoidosis (Heerfordt 1909). Numerous individual case reports and small series subsequently enlarged the known clinical spectrum of neurosarcoidosis. By 1948, Colover was able to collect information on 118 published cases of neurosarcoidosis (Colover 1948). Since then, numerous additional case reports and series attest to the wide range of the neurological manifestations of sarcoidosis. Sarcoidosis may affect any part of the nervous system but typically presents in distinctive clinical patterns.

Our ability to diagnose the presence of systemic sarcoidosis has improved greatly over the years. For example, the introduction of chest X-rays permitted easier diagnosis of sarcoidosis and showed that it was much more common than previously suspected. In

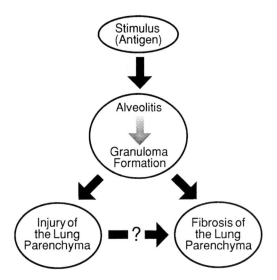

Fig. 2. Postulated mechanisms of pulmonary damage in sarcoidosis.

and magnetic resonance imaging (MRI) have greatly enhanced our ability to diagnose and monitor the progress of both sarcoidosis generally and neurosarcoidosis in particular.

Advances in our understanding of the pathophysiology and immunopathology of systemic sarcoidosis have influenced the treatment of neurosarcoidosis. From its earliest descriptions, sarcoidosis was considered an inflammatory disorder. In the early 1950s Sones and Siltzbach furthered this concept when they were among the first to show that anti-inflammatory corticosteroids had a beneficial effect in sarcoidosis (Deremee 1985a,b). Corticosteroids are still the mainstay of treatment for both sarcoidosis and neurosarcoidosis. Furthermore, recent studies demonstrate that the pathology of sarcoidosis relates directly to the activation of inflammatory cells, particularly CD4 lymphocytes, that aggregate at sites of disease activity (Figs. 2 and 3). This raises the possibility that other forms of immunotherapy could also be useful for treating sarcoidosis. Several such agents are currently being considered and studied as alternatives to corticosteroid therapy for neurosarcoidosis, particularly when the disease proves resistant to conventional treatment (Agbogu et al. 1995).

1976, an association between elevated levels of serum angiotensin converting enzyme (SACE) and active sarcoidosis was discovered (Leiberman 1985). Moreover, dramatic improvements in imaging techniques such as computerized axial tomography (CAT)

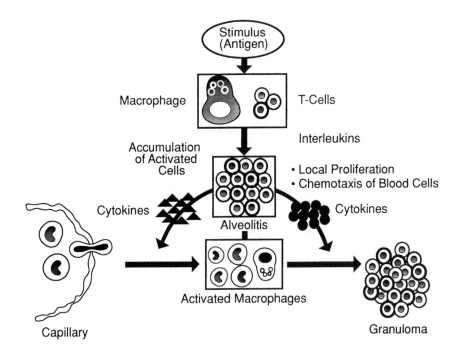

Fig. 3. Immunologic mechanisms judged to active in the pathogenesis of pulmonary sarcoidosis.

We have learned a great deal about neurosarcoidosis and its management. Still, neurosarcoidosis is a significant problem, and despite tremendous advances in diagnosis and better insights into pathophysiology and treatment, it still poses major challenges.

SARCOIDOSIS

Because diseases for which a cause is not known such as sarcoidosis are difficult to define with assurance, their diagnoses are often based on establishing characteristic descriptive profiles. For sarcoidosis even this approach has been difficult. Its varied manifestations and similarity to other granulomatous disorders of both an infectious and a non-infectious nature, combined with the lack of a specific confirmatory diagnostic test, have made reaching consensus on a standard definition for sarcoidosis especially difficult. The first internationally accepted definition of sarcoidosis was proposed in 1975 and is of value. It states that: 'Sarcoidosis is a multisystem granulomatous disorder of unknown etiology, most commonly affecting young adults and presenting most frequently with bilateral hilar adenopathy, pulmonary infiltration, skin or eye lesions. The diagnosis is established most securely when clinical and radiographic findings are supported by histologic evidence of widespread noncaseating epithelioid-cell granulomas in more than one organ . . . .' (James et al. 1976).

Although not all aspects of this proposed full definition are still widely accepted, the view that sarcoidosis is a systemic granulomatous disorder of unknown etiology diagnosed on the basis of clinical and histologic evidence is correct. The original definition, which is actually considerably longer than the excerpted quote above, has been rightly criticized for proposing that sarcoidosis is a consequence of impaired immunologic function. Instead, it is now accepted that the cause of sarcoidosis is not impaired immunity, but rather heightened immunologic activity. Still, apart from that criticism, this early definition is valuable because it offers a simple, operational definition of a very complex and confusing disease.

Sarcoidosis usually presents between the ages of 20 and 40 years. However, it also occurs in children (Weinberg et al. 1983) and older populations. It appears to have similar clinical manifestations in all age groups. Intrathoracic structures are most commonly

TABLE 1

Percentage frequency of organ involvement in sarcoidosis.

Manifestation	Frequency
Intrathoracic	87
Hilar nodes	72
Lung parenchyma	46
Upper respiratory tract	6
Dermatologic	
Skin	18
Erythema nodosum	15
Ocular	15
Lacrimal	3
Parotid	6
Splenomegaly	10
Peripheral lymphadenopathy	28
Bone	3
Cardiac	3
Hepatomegaly	10
Hypercalcemia	13
Neurologic	5
Hematologic, endocrinologic, gastrointestinal and genitourinary	Rare

affected (87% of patients), followed by lymph node, skin, and ocular disease (15–28% of patients) (Table 1). Although it is generally understood that a diagnosis of sarcoidosis is most secure when it is based on histologic confirmation, studies of Nordic manuscripts on sarcoidosis, a major source of research and information about the disease, show that on average 30% of the patients described lack histologic confirmation, and the diagnosis is often based solely on clinical and radiologic findings (Teirstein and Lesser 1983).

Involvement of any organ by sarcoidosis is possible and may occur with or without major symptoms. Sarcoidosis may be asymptomatic, or it can present with constitutional symptoms and pulmonary or extrapulmonary manifestations. Anatomic presence of the disease in an organ often occurs without overt clinical evidence of dysfunction. It is estimated that 20–40% of patients are asymptomatic at presentation, their disease being discovered by routine chest radiography (Teirstein and Lesser 1983).

When present, specific symptoms are often related to the extent of organ involvement. In the case of pulmonary disease, this may relate to the mechanical interference of sarcoidosis with normal lung function. In particular, respiratory symptoms such as cough, dyspnea, chest pain and nasal complaints are noted as the presenting symptom of sarcoidosis in 30–50% of individuals. Constitutional symptoms are described in 20–30% of patients and include fever, fatigue, anorexia or weight loss. Systemic symptoms are said to be more common among blacks (Katz 1983; Israel and Kataria 1985).

In less than 10% of sarcoidosis patients the onset of symptoms is neither of a systemic nor a pulmonary nature (Israel and Kataria 1985). Neurological manifestations of sarcoidosis are in this category. Other forms of extrapulmonary sarcoidosis include skin lesions, lymphadenopathy, parotid gland masses, liver or spleen enlargement, and ocular and cardiac involvement. Erythema nodosum is the most common dermatologic form of sarcoidosis and lupus pernio is the most characteristic skin lesion (Katz 1983; Israel and Kataria 1985). Lupus pernio presents as a persistent, bluish indurated lesion with a predilection for the nose, cheeks, ears and lips (Sharma 1984).

An international staging system has been established for sarcoidosis based on chest X-ray findings. Stage 0 disease is defined as extrathoracic disease and a normal chest X-ray. Approximately 10% of patients with sarcoidosis have stage 0 disease. In stage I disease patients demonstrate bilateral hilar adenopathy but no lung infiltrates on chest radiographs, and approximately 40–60% of patients present in this way. Stage II disease is defined as bilateral hilar adenopathy with associated lung infiltrates; it is reported in 15–30% of patients presenting with sarcoidosis. Finally, in stage III disease, which accounts for the remaining 10–15% of cases, patients demonstrate only pulmonary infiltrates on chest X-ray without hilar adenopathy. Some add a stage IV which is defined as even more extensive lung involvement than stage III (Tanoue et al. 1994).

Many types of blood studies have been reported to be abnormal in sarcoidosis patients, but there is no specific or highly sensitive diagnostic test yet available for sarcoidosis. For example, patients may demonstrate anemia, leukopenia, thrombocytopenia, hypergammaglobulinemia, hypercalcemia, and hepatic or renal dysfunction. One blood test, serum angiotensin converting enzyme (SACE), has been of particular interest. Active sarcoidosis may cause an elevation in SACE, which can then serve as a marker of the disease (Leiberman 1985). SACE is thought to be produced by alveolar macrophages and epithelioid giant cells and, in effect, reflects the 'granulomatous load' of a patient. SACE, however, is neither extremely sensitive, with just 50–60% of active sarcoidosis patients showing abnormalities, nor very specific, because it is also often elevated in patients with other conditions such as liver disease, diabetes mellitus, hyperthyroidism, systemic infection, malignancy and Gaucher's disease (Table 2). Nevertheless, although much controversy exists as to the proper place of the SACE assay in the diagnosis of sarcoidosis, most investigators accept that it is a useful marker for systemic disease activity (Leiberman 1985).

Hypercalcuria is also reported in sarcoidosis. It may be associated with hypercalcemia, nephrocalcinosis or renal calculi. The mechanism of this disorder is thought possibly to be related to excessive gastro-

TABLE 2

Disease states associated with elevated SACE levels.

Disease	n	SACE (mean ± S.D.)	Frequency of elevations (%)
Controls	172	22.6 ± 6	
Sarcoidosis	300	44.7 ± 19	75
Gaucher's disease	9	120.6 ± 73	100
Hyperthyroidism	29	37.3 ± 3.2	81
Berylliosis	4	34.6 ± 6	75
Leprosy	53	34.0 ± 14	53
Cirrhosis of liver	151	30.8 ± 13	28
Diabetes mellitus	265	26.8 ± 8.7	24–32
Silicosis	144	46.6 ± 12.1	21

intestinal calcium absorption in patients with sarcoidosis (Katz 1983; Israel and Kataria 1985).

Most patients with systemic sarcoidosis have a good prognosis. For approximately two-thirds, the disease resolves spontaneously without major difficulties. This benign course is most common in asymptomatic patients with only hilar adenopathy on chest X-ray; they have a 70–80% likelihood of spontaneous remission (Fanburg 1983). However, for one-third of patients, symptoms persist or the disease progressively worsens. Pulmonary dysfunction is the major problem for most patients with a persistent or progressive clinical course, and 15–20% of such sarcoidosis patients have some degree of permanent loss of lung function, as manifested clinically by fibrotic changes, such as honeycombing and retraction, on chest X-ray (Fanburg 1983; Deremee 1985; Leiberman 1985).

Mortality in systemic sarcoidosis is reported as below 5%. Deaths are most often due to respiratory failure and sometimes associated with cardiac problems such as cor pulmonale and heart failure. However, death due to extrathoracic sarcoidosis affecting such organ systems as the kidneys or the nervous system is also well described (Fanburg 1983; Israel 1983; Katz 1983; Deremee 1985).

The cornerstone of therapy for all forms of sarcoidosis is corticosteroids. However, debate as to the precise indications for treatment continues. Indications for treatment are unclear because many patients with sarcoidosis are asymptomatic at the time of presentation and the rate of spontaneous resolution of sarcoidosis is high. In addition, the clinical presentations and course are so varied that treatment studies, particularly well controlled studies, have been difficult to perform. However, based on experience, there is little disagreement that corticosteroids have often proven very effective to rapidly suppress some of the acute manifestations of sarcoidosis. Moreover, chronic therapy is widely considered to be necessary to limit the recurrence or the progression of disease in some patients. However, the value of corticosteroids for the long-term treatment of sarcoidosis remains scientifically unsubstantiated, and corticosteroids are recognized not to be a curative treatment for all patients with sarcoidosis and may not change the ultimate disease course. Corticosteroid treatment seems most clearly indicated for patients with significant functional impairment in any organ system, particularly with significant pulmonary, cardiac, ocular or central nervous system (CNS) damage (Fanburg 1983; Deremee 1985; James 1994).

PATHOPHYSIOLOGY

Although the precise etiology of sarcoidosis remains unknown, major strides have been made in understanding its pathogenesis. For example, there is growing evidence that sarcoidosis is caused by heightened immune processes at sites of disease activity (Rocklin 1983; Kataria 1985; Daniele et al. 1986; Semenzato and Agostini 1994). This contrasts sharply with earlier concepts that related sarcoidosis to impaired immunity and to generalized anergy (Leiberman et al. 1985; Thomas and Hunninghake 1987).

Our current understanding of the immunopathology of sarcoidosis derives largely from studies of pulmonary sarcoidosis. Such investigations have been aided by the technique of bronchoalveolar lavage that allows examination of inflammatory cells from sites of major lung activity (Hunninghake and Crystal 1981; Daniele 1983; Rocklin 1983; Kataria 1985; Semenzato and Agostini 1994). The initial lesion in pulmonary sarcoidosis appears to be an alveolitis, an inflammation of the alveolar structures of the lung (Figs. 2 and 3). This inflammation provides the appropriate environment for granuloma formation. Evidence indicates that these granulomas contain activated mononuclear cells that primarily have a secretory rather than a phagocytic role. There is also activation of other inflammatory cells, particularly CD lymphocytes, that congregate at sites of disease activity and secrete various cytokines, including interleukin-2 (IL-2), interleukin-1 (IL-1), interferon-γ, and tumor necrosis factor (Rocklin 1983; Kataria 1985; Semenzato and Agostini 1994). Indeed, there is immunopathologic concordance between the cells obtained from bronchoalveolar lavage fluid and cells isolated from open lung biopsies in sarcoidosis patients with active pulmonary disease (Kern and Daniele 1985; Thomas and Hunninghake 1987). These immunopathologic mechanisms are certainly not just limited to pulmonary sarcoidosis. Undoubtedly, processes similar to those found in the lung underlie the pathogenesis of other forms of sarcoidosis, including neurosarcoidosis (Fig. 4).

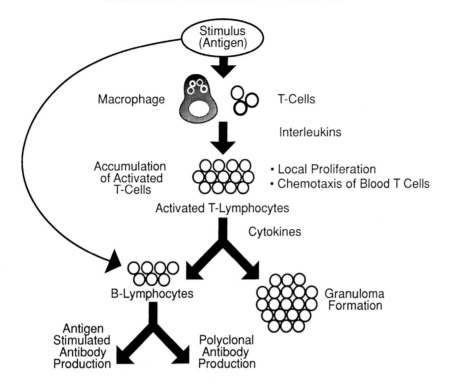

Fig. 4. Immunopathogenic mechanisms postulated as responsible for development of the granulomatous inflammation found in neurosarcoidosis.

Despite the lack of research addressing the specific immunopathogenesis of neurosarcoidosis, there is reason to believe that it is similar to that of systemic sarcoidosis. Therefore the information available from studies of bronchoalveolar lavage fluid and other sites of inflammation represents a reasonable model of processes found in the nervous system. Although there are occasional discrepant results among investigators, a consensus about the immunologic mechanisms behind sarcoidosis is emerging.

Sarcoidosis can be thought of as an inflammatory response to an as yet unidentified antigen (Fig. 2) (Hunninghake and Crystal 1981; Weissler 1994). As described elsewhere in this chapter, there has long been a suspicion that sarcoidosis is an inflammatory disorder; fever, malaise, weight loss, cutaneous anergy, and polyclonal hyperglobulinemia are all consistent with this hypothesis. The central pathologic hallmark of sarcoidosis, the granuloma, consists of macrophages, macrophage-derived epithelioid cells, and multi-nucleated giant cells which secrete cytokines (Figs. 3 and 4) (Myatt et al. 1994). About this central core exist CD4 and CD8 lymphocytes, B lymphocytes, plasma cells, and fibroblasts, in proportions that vary with time.

There is an accumulation of CD4 cells at sites of active inflammation (Thomas and Hunninghake 1987). As reviewed by Weissler (1994), these cells bear markers suggesting that they are acting as activated memory cells. The CD4 cells exhibit class II major histocompatibility antigen and IL-2 and 1,25-dihydroxyvitamin D_3 receptors and spontaneously release IL-2, a cytokine that promotes lymphocyte proliferation, interferon-γ, which activates macrophages, and 1,25-dihydroxyvitamin D_3. The 1,25-dihydroxyvitamin D_3 receptor is expressed more on CD8 than CD4 cells. Since the binding of 1,25-dihydroxyvitamin D_3 to its receptor inhibits cellular proliferation, the increased CD4:CD8 ratio may be maintained in part by the excess vitamin D receptor on the CD8 cells (Biyoudi-Vouenze et al. 1991). The expression of CD69 (very early), HLA-DR (late), and VLA-1 (very late) lymphocyte markers suggest persistent cell activation at inflammation sites (Hol et al. 1993).

The lymphocytes are though to be stimulated by

antigen presentation by activated macrophages present at sites of inflammation (Semenzato and Agostini 1994). The macrophages exhibit class II major histocompatibility antigen and the alpha chain of the IL-2 receptor and release tumor necrosis factor-α, interleukin-6 (IL-6), IL-1β, granulocyte-macrophage colony-stimulating factor, fibronectin, 1,25-vitamin D, PGE_2 and ACE (Steffen et al. 1993; Bost et al. 1994; Fireman and Topilsky 1994). Prostaglandin E_2 decreases granuloma size whereas tumor necrosis factor and IL-1 increase granuloma size (Pueringer et al. 1993).

Fibroblasts release a factor that helps maintain lymphocyte activation, as well as IL-6 which suppresses fibroblast proliferation (Fireman and Topilsky 1994). Interferon-γ diminishes fibrosis by downregulating collagen synthesis and decreasing fibroblast proliferation (Prior and Haslam 1992).

Polyclonal B cells aggregate about granulomas (Fazel et al. 1992). Occasional plasma cells are also found, producing IgM and IgA more than IgG. κ and λ light chains are present. Presumably local cytokine production leads to the proliferation and differentiation of the B cells, which then contribute to the inflammatory process.

Increased levels of intercellular adhesion molecule 1 (ICAM-1) and increased expression of LeuCAM (CD11/CD18 family), which mediate leukocyte adhesion, may be present in some, but not all, sarcoidosis patients (Shakoor and Hamblin 1992; Schaberg et al. 1993; Fireman and Topilsky 1994). Macrophages exhibit an increased expression of ICAM-1 and leukocyte function-associated antigen-1, which may promote the ability of the macrophage to activate T cells (Melis et al. 1991; Smith and Deshazo 1992; Striz et al. 1992).

Genetic analysis of the CD4 T cell receptor, and its potential relationship to class II major histocompatibility molecules, has been pursued to search for clues as to the antigen or antigens inciting the inflammation (Weissler 1994). If sarcoidosis were caused by a specific antigen, whether exogenous or endogenous, there might be an over-representation of specific CD4 markers expressed in patients with sarcoidosis. Indeed, an abundance of V (variable) β3, β5, β8, β14, β15, α2.3, γ9, and δ1 lymphocyte phenotypes is described (Forman et al. 1993; Fireman and Topilsky 1994; Klein et al. 1995).

An association between Vα2.3 and HLA-DR3(17),DQ2 haplotype has been documented (Grunewald et al. 1994). There is also a bias in the expression of C (constant) β1 elements (Tamura et al. 1991) and an increased percentage of γ/δ cells (Raulf et al. 1994) with the Vδ1 marker (Forrester et al. 1993). Unfortunately, the inciting antigen or antigens, however, remains unknown. Of critical importance is whether chronic inflammation requires continuous antigen presentation or whether an intrinsic genetic predisposition allows persistent inflammation if the inciting antigen is no longer locally available. Nonetheless, evidence suggests that CD4 cells are persistently activated and proliferate in a limited clonal response to a specific repertoire of as yet unknown foreign or native antigens (Hol et al. 1993; Semenzato et al. 1993). Some potential or suspect antigens are discussed in the epidemiology section of this chapter.

As the reactive monocytes and macrophages form granulomas, ultimately irreversible, obliterative fibrosis can develop (Fig. 2). Furthermore, small foci of ischemic necrosis can be found, probably as a consequence of vascular compromise due to perivascular inflammation. Importantly, these granulomas are not specific for sarcoidosis and indistinguishable or nearly identical lesions occur in a variety of other conditions that must be excluded before a diagnosis of sarcoidosis can be made with certainty (Thomas and Hunninghake 1987).

EPIDEMIOLOGY

The prevalence of sarcoidosis is estimated to be in the order of 60/100,000 population, while the annual incidence is approximately 11/100,000 population. In certain populations sarcoidosis may even be more common. However, the exact prevalence and incidence of sarcoidosis is difficult to validate because there is no single diagnostic test for sarcoidosis. Differences in case-finding methods also undoubtedly account for some differences in reported frequencies for sarcoidosis and make comparisons of its rate of occurrence in different populations difficult. For instance, studies that include only symptomatic patients actually underestimate the true prevalence of sarcoidosis because 20–40% of people shown to have sarcoidosis are completely symptom free at the time of initial diagnosis and are discovered by routine or screening chest X-rays (Teirstein and Lesser 1983). Reports of disease frequency that rely on the findings of chest radiographic surveys probably give the best indication of the frequency of sarcoidosis because pulmonary involvement is by far the most common manifestation of sarcoidosis. However, some autopsy series suggest that the true prevalence of sarcoidosis may be even higher (Teirstein and Lesser 1983). In contrast, series that depend solely on symptomatic patients suggest a relatively lower rate of sarcoidosis.

Population differences for sarcoidosis have been described and may exist. For instance, there is a reported increased incidence of sarcoidosis in United States blacks compared to whites; the disease also seems to be more severe in blacks (James 1983; Teirstein and Lesser 1983). Certain areas of the world, such as Sweden, also seem to have a higher incidence of sarcoidosis while it appears to be quite rare in other areas such as China or Southeast Asia (Tanoue et al. 1994). These studies also raise the possibility that there may be a genetic predisposition to the development of sarcoidosis. Indeed, sarcoidosis does seem to occur with greater likelihood in some families, but as yet there is no well defined genetic pattern determined and no consistent mode of inheritance has been discovered (James 1983; Teirstein and Lesser 1983). In addition, although there are some indications that sarcoidosis may be slightly more common in women, overall there seems to be no clear sex predilection (James 1983; Teirstein and Lesser 1983).

Sarcoidosis can occur at any age. Although the likelihood of sarcoidosis is greatest in the third and fourth decades, the youngest reported patient is a 28-month-old child (Teirstein and Lesser 1983; Weinberg et al. 1983). The exact frequency in children is difficult to assess because most countries understandably limit the use of routine screening chest radiography in children. Clinically, when sarcoidosis affects children, the distribution of organ involvement appears similar to that of reported adult cases, and the patterns of neurosarcoidosis seem analogous as well (Weinberg et al. 1983).

Overall, our view of the neurological manifestations of sarcoidosis is probably somewhat skewed towards the more serious manifestations of the disease. This is because neurosarcoidosis is a rare disorder and virtually all series and case reports are biased by patient selection issues. Individual case reports characteristically detail only specific, rare or extremely unusual clinical presentations. Even large series are derived from either autopsy data, which are obviously biased towards the most severe types of problems; or referral centers, which again tend to see the more unusual, complex or particularly difficult patients.

There are no definite provoking or exacerbating factors reliably associated with sarcoidosis. The granulomatous lesions in sarcoidosis are similar to those caused by infectious agents such as mycobacteria and fungi, inorganic materials such as zirconium and beryllium, and those seen in association with hypersensitivity reactions to various organic agents (Tanoue et al. 1994). It has been speculated that perhaps infectious agents, or organic agents, such as pine pollen, or inorganic substances may be etiologic factors for sarcoidosis (Desai and Simon 1985; James 1994; Tanoue et al. 1994). In particular, interest has centered on the possibility that sarcoidosis is the consequence of an unusual host response to a common antigen or infectious agent. Of the various possible etiologies, mycobacterial infections have received the most attention.

The potential relationship of sarcoidosis to tuberculosis is an important one that has evolved over the years. Initially, sarcoidosis and tuberculosis were distinguished as separate diseases by various diagnostic tests. Skin testing for tuberculosis exposure, staining tissue for mycobacteria organisms, and culture of mycobacteria from tissue and body fluids were the mainstays of differential diagnosis between these diseases. Based on the absence of evidence of mycobacterial infection in most patients with sarcoidosis, the clear dichotomy between the two diseases had been thought to be well established during the 1970s and 1980s. Still a perplexing challenge has developed with the possibility that mycobacterial infection, manifesting in an atypical fashion, may cause sarcoidosis.

With the advent of DNA techniques to detect mycobacterial infection, the issue of the relationship of sarcoidosis and mycobacterial infection has been reopened. Mitchell et al. (1985) reported evidence for the presence of *M. tuberculosis* in sarcoidosis tissue samples using a DNA/RNA hybridization technique. Hybridization was 4.8 times more likely in sarcoidosis patients than in controls. Saboor et al. (1992), using the polymerase chain reaction (PCR), found *M. tuberculosis* DNA in 10 of 20 patients with sarcoidosis and an additional 20% of patients had non-tuberculosis mycobacterial DNA detected. On the other hand, also using the PCR, Bocart et al. (1992) detected evidence of *M. tuberculosis* in only 2 of 24 patients and Popper et al. (1994) found *M. tuberculosis* in only 2 of 15 sarcoidosis patients. The presence of other mycobacteria species could not be documented (Bocart et al. 1992). Similar findings were reported by Gerdes et al. (1992) and Ghossein et al. (1994). The PCR has been applied to CSF samples in patients with neurosarcoidosis, but only one of six samples from two patients was positive for *M. tuberculosis* (Liedtke et al. 1993).

The debate on the relationship of sarcoidosis and mycobacterial infection continues (Joyce-Brady 1992). Most patients with sarcoidosis are not clearly infected with a mycobacterial species. Yet, in some patients, an immunologic reaction to a mycobacterial antigen might lead to sarcoidosis (Ghossein et al. 1994; Popper et al. 1994). A non-culturable, cell wall-deficient mycobacterium may be responsible, in occasional patients, for granulomatous inflammation (Mitchell et al. 1992; Saboor et al. 1992; Popper et al. 1994). However, if mycobacteria are present in a majority of sarcoidosis patients, they must be present in such exceedingly small numbers that they are unable to be readily detected (Bocart et al. 1992).

PCR analysis may be useful in the occasional patient in whom a distinction between sarcoidosis and tuberculous infection is difficult. The rapidity of PCR compared to conventional culture techniques can be an important consideration. However, the most useful setting in which to apply PCR may be in the evaluation of 'sarcoidosis' patients with progressive disease in spite of 'optimal' immunosuppressive disease. The ability to exclude mycobacterial infection in these patients can be reassuring before proceeding with more intense immunosuppressive therapy.

Despite the fact that a specific etiologic agent has not been established to cause sarcoidosis, it appears that sarcoidosis probably relates to an initiating agent persisting at the site of disease activity for a period of time and triggering a hypersensitivity or otherwise amplified type of immune reaction in a potential, perhaps genetically susceptible, host (Figs. 2, 3 and 4).

NEUROPATHOLOGY

Non-caseating granulomas and the accompanying diffuse mononuclear cell infiltrates that are characteristic of sarcoidosis can be found in any part of the neuraxis including peripheral nerve or muscle (Fig. 5). The most common site of inflammation is the meninges, especially in the basal region of the brain (Figs. 6 and 7) (Delaney 1977; Waxman and Sher 1979). Sarcoid granulomas can be distributed widely or be concentrated in one or more areas to form a mass. Although sarcoidosis is not usually considered to be a primary vasculitis (Moore and Cupps 1983), arteriolar and venous infiltrations do occur and can lead to infarction (Douglas and Maloney 1973; Delaney 1977; Urich 1977; Caplan et al. 1983; Mirfakhraee et al. 1986). The granulomatous inflammation found pathologically can correlate directly with clinical signs and symptoms or may be subclinical and unexpressed (Waxman and Sher 1979; Manz 1983).

CNS involvement can be conceptualized as an inflammatory process primarily affecting the leptomeninges (Fig. 7). Inflammation may spread along the Virchow–Robin perivascular spaces to invade the brain or spinal cord or remain more localized to involve the cranial nerves (Waxman and Sher 1979; Trombley et al. 1981; Mirfakhraee et al. 1986). Presumably at an early stage much of the active inflammation is reversible. Ultimately, as the disease progresses, irreversible fibrosis can develop which leads to permanent neurologic damage and persisting neurologic deficits.

Inflammation can also extend to the CSF pathways, leading to hydrocephalus. Brain or spinal cord disease can take the appearance of discrete granulomatous mass lesions or a diffuse encephalopathy/vasculopathy. The hypothalamic region is the most common site for parenchymal disease.

Granulomas are apparent in the epineurium and perineurium of peripheral nerve in symptomatic patients. The endoneurium may contain a mononuclear cell infiltrate. Perivascular and vascular inflammation can be seen in the epineural and perineurial vessels. All nerve fiber sizes can be affected. There seems to be a predominantly axonal neuropathy with only a minor component of segmental demyelination. The exact mechanism of peripheral nerve damage in sarcoidosis seems to vary, but it includes injury resulting from vascular compromise, direct compression from granulomas, and immunologic factors that may affect the peripheral nerve axons (Galassi et al. 1984).

Muscle pathology is common in sarcoidosis. Muscle biopsy of symptomatic patients reveals typical non-caseating granulomas, and there may be

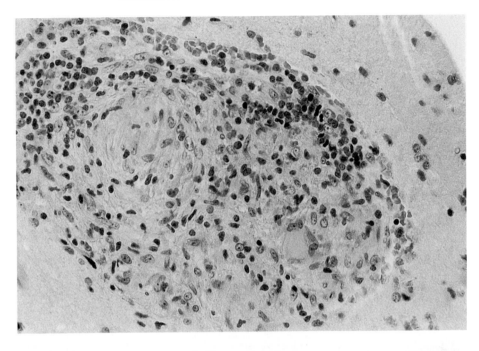

A

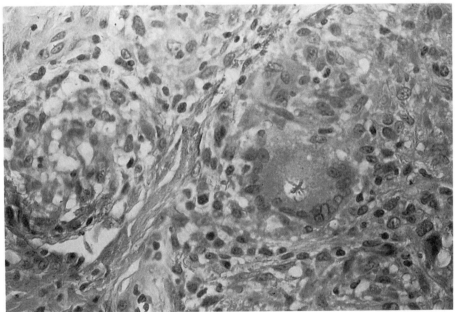

B

Fig. 5. (A) Photomicrograph at 200× magnification of brain showing an intraparenchymal non-caseating or non-necrotizing sarcoid granuloma. (B) Photomicrograph at 400× magnification of a sarcoid granuloma in brain demonstrating a multinucleated giant cell.

more diffuse inflammation with muscle fiber degeneration and regeneration and fibrosis (Hewlett and Brownell 1975; Ando 1989). Moreover, asymptomatic non-caseating granulomas have been found in up to one-half of all sarcoidosis patients having a muscle biopsy (Stjernberg et al. 1981).

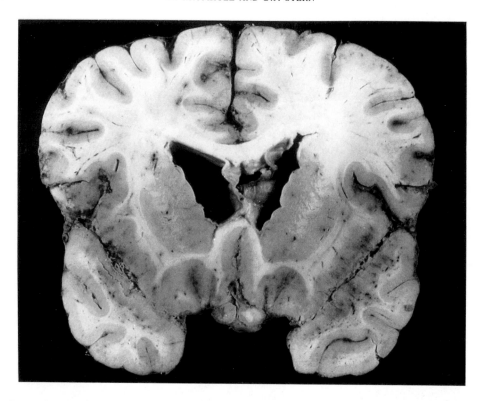

Fig. 6. Coronal mid-frontal section of the brain of a patient with neurosarcoidosis showing thickening and inflammatory changes of the basal meninges and optic region (courtesy of Dr. Steven C. Bauserman).

NEUROLOGIC MANIFESTATIONS

Neurologic symptoms are the presenting feature of sarcoidosis in one-half of individuals with neurosarcoidosis (Stern et al. 1985). Some three-quarters of patients destined to develop neurologic disease do so within 2 years of becoming afflicted with sarcoidosis. The approximate frequency of the various neurologic complications is presented in Table 3. Only rarely do patients with neurosarcoidosis have no evidence of disease in other organ systems such as the lung (Wiederholt and Siekert 1965; Delaney 1977; Pentland et al. 1985; Stern et al. 1985; Oksanen 1986; Luke et al. 1987). However, systemic disease may not always be evident early in a patient's clinical course and, in some instances, it can be difficult to find.

Although the range of clinical manifestations of sarcoidosis is exceptionally varied, most patients present in characteristic ways what can be systematically organized (Table 1). Similarly, the neurologic manifestations of sarcoidosis can be classified, as presented in Table 3. In dealing with neurosarcoidosis

TABLE 3

Neurosarcoidosis.

Clinical manifestation	Approximate frequency (%)
Cranial neuropathy	50–75
Facial palsy	25–50
Aseptic meningitis	10–20
Hydrocephalus	10
Parenchymal disease	
Endocrinopathy	10–15
Mass lesion(s)	5–10
Encephalopathy/vasculopathy	5–10
Seizures	5–10
Neuropathy	5–10
Myopathy	10

patients, it is useful to approach patients using such a classification. However, it should be remembered that one-third to one-half of neurosarcoidosis patients develop more than one neurologic manifestation of their disease.

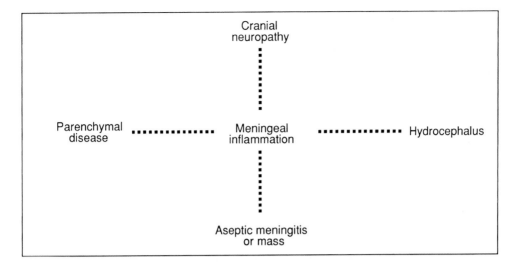

Fig. 7. CNS sarcoidosis: a clinico-pathological construct.

Cranial neuropathy

The most frequent neurologic complication of sarcoidosis is cranial neuropathy (Stern et al. 1985). This occurs in approximately three-quarters of patients with neurosarcoidosis. Any cranial nerve can be affected, and over one-half of patients have multiple cranial nerve lesions (Stern et al. 1985). By far the most commonly affected cranial nerve is the facial or VIIth nerve.

Cranial nerve I. Olfactory nerve dysfunction can occur secondary to meningeal sarcoidosis involving the subfrontal region. However, anosmia or hyposmia may also result from local nasal granulomatous invasion by sarcoidosis, and, therefore, in a patient with olfactory complaints, an otolaryngological evaluation is necessary before attributing impaired olfaction to CNS disease.

Cranial nerve II. Optic nerve involvement in sarcoidosis is much less frequent than other ocular manifestations of sarcoidosis such as uveitis. Optic nerve involvement can present with an acute or subacute, painful, or chronic, painless visual loss (Graham et al. 1986). This visual loss may be due to bulbar or retrobulbar invasion of the optic nerve by granulomas, compression of the optic nerve by a granulomatous mass, or optic atrophy. Optic disk edema may be secondary to papilledema due to sarcoidosis induced increased intracranial pressure or as a direct local invasive effect of sarcoidosis. A chiasmal syndrome also has been reported (Gelwain et al. 1988).

Cranial nerves III, IV, and VI. Disorders of ocular motility may follow involvement of the III, IV, or VI cranial nerves in the granulomatous process (Stern et al. 1985). Typically, these nerves are damaged in their extra-axial course in the subarachnoid space as they traverse the meninges. However, they also may be involved by local orbital disease, and rarely brainstem intra-axial CNS pathways can be affected by sarcoidosis (Kirkham and Kline 1976). Uncommonly, disordered ocular motility may be due to sarcoidosis involving the extraocular muscles themselves (Obenauf et al. 1978). Occasionally, pupillary dysfunction is caused by neurosarcoidosis (Henkind and Gottlieb 1973; Kirkham and Kline 1976; Cohen and Reinhardt 1982; Poole 1984). For instance, Horner's syndrome due to disruption of the cervical sympathetic nerves has been reported (Cohen and Reinhardt 1982).

Cranial nerve V. Trigeminal nerve disease may present as facial numbness or, rarely, trigeminal neuralgia (Stern et al. 1985). Headache may also represent trigeminal nerve dysfunction intracranially. Involvement of the muscles of mastication is unusual.

Cranial nerve VII. Of all the cranial nerve syndromes, peripheral facial cranial nerve (VII) palsy is the most common, and it is also the single most frequent neurological manifestation of sarcoidosis. It develops in 25–50% of all patients with neurosarcoidosis. Although usually unilateral, bilateral facial palsy also can occur presenting with either simultaneous or sequential paralysis. Over half of all patients with facial palsy also have other forms of nervous system involvement. In patients with a solitary facial palsy, spinal fluid examination typically has been reported as normal, but in individuals in whom there are other associated forms of neurosarcoidosis, the spinal fluid examination is reported to be abnormal in 80% of patients. The specific reason for a facial nerve palsy in sarcoidosis may vary. Rarely is the facial palsy caused by parotid inflammation. More likely, the nerve is compromised as it traverses the meninges and subarachnoid space, and as suggested by Oksanen (1986) and Oksanen and Salmi (1986), facial paresis is probably due to intra-axial sarcoidosis induced inflammation involving the facial nerve. In general the prognosis for the facial palsy is good with over 80% of patients having a good outcome in terms of recovery of facial function (Stern et al. 1985; Luke et al. 1987).

Cranial nerve VIII. Eighth cranial nerve involvement is the second most common form of cranial nerve dysfunction in sarcoidosis. Sarcoidosis may involve the auditory or vestibular portions of the nerve. When either loss of hearing or vestibular dysfunction occur, symptoms may be sudden or insidious and often fluctuate over time (Hybels and Rice 1976; Jahrsdoerfer et al. 1981; Juozevicius and Rynes 1986). If hearing loss occurs, it is typically a sensorineural loss. As with facial nerve palsy, bilateral VIII nerve disease may occur. In fact, bilateral VII and VIII nerve dysfunction is highly suggestive of neurosarcoidosis (Wiederholt and Siekert 1965; Stern et al. 1985; Oksanen and Salmi 1986).

Cranial nerves IX and X. Glossopharyngeal and vagus nerve involvement by sarcoidosis cause dysphagia and dysphonia. Hoarseness is more commonly due to laryngeal nerve dysfunction from intrathoracic disease than CNS inflammation involving the vagus nerve (El-Kassimi et al. 1990; Tobias et al. 1990).

Cranial nerves XI and XII. Eleventh and twelfth cranial nerve disease can occur but seems to be quite rare in sarcoidosis.

Meningeal disease

Meningeal disease is seen in approximately 10–20% of patients with neurosarcoidosis and can present as aseptic meningitis or, less commonly, as a meningeal or dural mass lesion. Aseptic meningitis is characterized by headache, meningismus, and a sterile CSF with a predominantly mononuclear pleocytosis and may be a recurrent problem in some patients with neurosarcoidosis (Plotkin and Patel 1986). Hypoglycorrhachia, or low CSF sugar concentration, is occasionally found, and there is often an elevation of CSF protein. Since meningeal involvement in sarcoidosis is a common pathologic finding, it is surprising that aseptic meningitis is not more common. When meningeal sarcoid mass lesions occur, they can mimic intracranial tumors such as meningiomas (Figs. 8 and 9) (Israel 1983; Osenbach et al. 1986; Sethi et al. 1986).

Hydrocephalus

Hydrocephalus is noted in about 10% of neurosarcoidosis patients and can pose great danger. Hydrocephalus is a potentially lethal complication of sarcoidosis. Patients with acute hydrocephalus may die suddenly from increased intracranial pressure, and even patients with chronic hydrocephalus have the potential to decompensate acutely. Patients with acute hydrocephalus characteristically present with headache, altered mentation or consciousness, and impaired gait. On examination papilledema or other signs of raised intracranial pressure can be found. Acute decompensating hydrocephalus is a medical emergency that necessitates prompt diagnosis and treatment. Once clinically suspected, the diagnosis of hydrocephalus is best substantiated with imaging studies such as cranial CT or MRI (Fig. 10).

The hydrocephalus in patients with neurosarcoidosis can be either of the communicating or noncommunicating type. Chronic basilar meningitis with obliteration of subarachnoid cerebrospinal fluid (CSF) flow is a major cause for communicating hydrocephalus. In addition, infiltration of the ventricular system by granulomas, granulomatous compression of the aqueduct, or outlet obstruction of the fourth ventricle by granulomas may all cause non-communicating types of hydrocephalus. Both inflammation or fibrosis due to sarcoidosis can impair CSF flow caus-

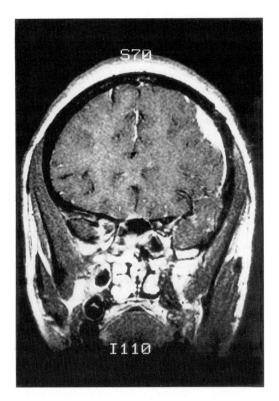

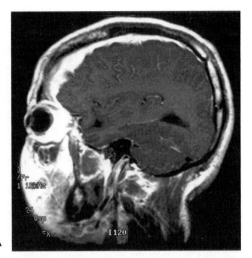

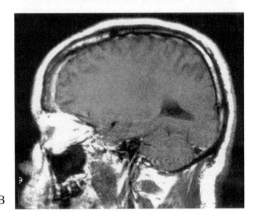

Fig. 8. MRI, coronal, T1-weighted image with gadolinium enhancement, demonstrating a convexity sarcoidosis mass lesion that was initially mistaken for a meningioma.

Fig. 9. (A) MRI, sagittal, T1 with gadolinium, showing a frontal extracerebral mass. (B) MRI, sagittal, T1 with gadolinium, showing same patient as (A) with substantial reduction in the size of the mass following therapy with high-dose corticosteroids.

ing ventricular enlargement (Lukin et al. 1975; Schlitt et al. 1986). Enhancement of the meninges on imaging studies suggests an active inflammatory process which might be expected to respond to corticosteroid treatment (Fig. 11). In contrast, lack of such enhancement may suggest a predominantly fibrotic reaction that would be less likely to respond to anti-inflammatory medications.

CNS parenchymal disease

Parenchymal brain disease is reported in about 50% of patients with neurosarcoidosis and can present in several forms. Hypothalamic dysfunction is the most common manifestation of CNS parenchymatous disease. When hypothalamic dysfunction occurs it usually involves the neuroendocrinological system or the so called 'vegetative functions' such as temperature regulation, appetite, thirst, sleep and libido. However, neuroendocrinological disease in sarcoidosis can also

occur secondary to pituitary disease. Any of the neuroendocrinological systems can be affected by sarcoidosis (Stern et al. 1985).

Endocrine disorders
Characteristically, endocrinologic dysfunction in sarcoidosis is due to either a hypothalamic or pituitary granulomatous mass or a more diffuse 'local' encephalopathy (Stuart et al. 1978; Ismail et al. 1985; Capellan et al. 1990; Chapelon et al. 1990; Missler et al. 1990). This can sometimes be seen with imaging studies (Fig. 12). Given the predilection of the sarcoidosis for the basal meninges (Fig. 7), the relative frequency of such endocrinologic disturbances is not

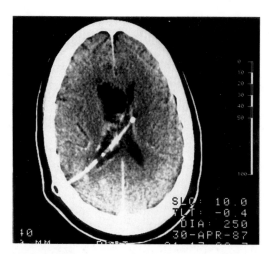

Fig. 10. CT scan in the axial plane in a patient with neurosarcoidosis who presented with symptomatic hydrocephalus requiring placement of a ventricular shunt.

surprising (Brust et al. 1977). Potential endocrinological manifestations of neurosarcoidosis include thyroid disorders, disorders of cortisol metabolism, and sexual dysfunction. An elevated serum prolactin level, found in 3–32% of patients with sarcoidosis, may be an indication of hypothalamic dysfunction. In fact, because neuroendocrinological involvement is relatively common in individuals with CNS neurosarcoidosis, patients with more than just an isolated facial palsy probably merit a thorough endocrinologic evaluation with specific attention to hypothalamic hypothyroidism, hypocortisolism, and hypogonadism.

Hypothalamic disorders that affect vegetative functions vary considerably. A disorder of thirst is the most common hypothalamic disorder related to neurosarcoidosis and is attributed to a change in the hypothalamic 'osmostat' (Stuart et al. 1978). More rarely, the syndrome of inappropriate secretion of antidiuretic hormone or diabetes insipidus can occur (Kirkland et al. 1983). Neurosarcoidosis induced disruptions of hypothalamic function also have been described as causing disorders of appetite, libido, temperature control and weight regulation (Heffernan et al. 1971; Brust et al. 1977; Lipton et al. 1977; Stuart et al. 1978).

Mass lesions

Intraparenchymal mass lesions due to sarcoidosis may present in two ways: an isolated mass (Fig. 13)

or masses occurring in any cerebral area, or multiple cerebral nodules. Such multiple nodules may actually represent inflammatory reaction in the Virchow–Robin spaces. Subdural plaque-like masses may also occur and are discussed in the above section on meningeal manifestations of neurosarcoidosis. Calcifications may also be seen. Although historically, intraparenchymal mass lesions were considered rare, CT and MRI have shown parenchymatous disease to be more frequent than previously thought.

Encephalopathy/vasculopathy

The diffuse encephalopathy and vasculopathy associated with neurosarcoidosis are not well understood manifestations of sarcoidosis. Moreover, it is often difficult both clinically or pathologically to separate clearly between these entities. In fact, these two manifestations of neurosarcoidosis frequently coexist. For these reasons, we find it best to consider them as a single overlapping entity, sarcoidosis-associated encephalopathy/vasculopathy, but recognize that in individual patients one form or the other may predominate (Stern et al. 1985).

The diffuse encephalopathy/vasculopathy found in neurosarcoidosis can involve the cerebral hemispheres or basilar regions. Patients may present with a delirium, personality change, or isolated memory disturbance due to focal or diffuse parenchymal inflammation (Fig. 14) (Ho et al. 1979; Cordingley et al. 1981; Thompson and Checkley 1981). Clinical findings correlate with the extent of enhancement on imaging studies and increased signal intensity on T2-weighted MRI. Sarcoidosis of the CNS has also been associated with a multifocal relapsing encephalopathy.

Encephalopathic patients may have perivascular inflammation or granulomas infiltrating both arteries and veins and extending into brain parenchyma. Several investigators have observed granulomatous small-vessel arteritis in patients with neurosarcoidosis (Meyer et al. 1953; Herring and Urich 1969). Large arteries also have been involved in the granulomatous process (Fig. 15). Although disease is rarely evident on angiography, angiographic changes suggestive of vasculitis as well as an ill-defined occlusive process have been documented (Brown et al. 1989; Corse and Stern 1990).

Transient ischemic attacks and ischemic stroke due to neurosarcoidosis have been reported (Caplan

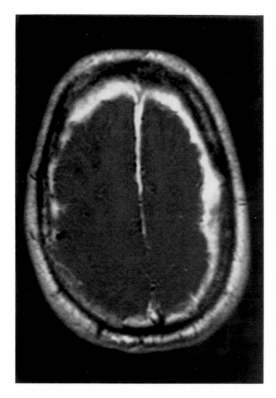 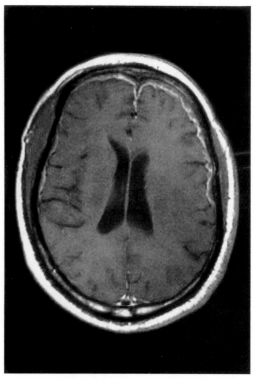

A **B**

Fig. 11. (A) MRI, axial, T1 with gadolinium, demonstrates marked dural enhancement due to sarcoidosis. (B) MRI, axial, T1 with gadolinium, in the same patient as in (A), shows decrease in dural enhancement after corticosteroid treatment.

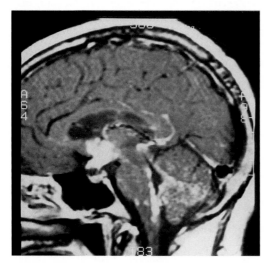

Fig. 12. MRI, sagittal section, T1 with gadolinium, showing hypothalamic and pituitary involvement by sarcoidosis.

et al. 1983; Sethi et al. 1986; Brown et al. 1989; Corse and Stern 1990). Ischemic stroke is usually a consequence of inflammation involving large or small arteries, but other causes include compressive perivascular mass lesions and emboli from sarcoidosis associated cardiomyopathy or cardiac arrhythmias. Caplan et al. (1983) emphasized the arterial and venous involvement of the meninges and parenchyma in the angiitic form of sarcoidosis and related this to observable perivascular lesions in the optic fundus. Also, dural sinus obstruction causing intracranial hypertension has been related to inflammation from sarcoidosis (Byrne and Lawton 1983; Chapelon et al. 1990).

Seizures

Seizures are another important manifestation of CNS parenchymal disease due to neurosarcoidosis. They have been reported in up to 20% of patients with

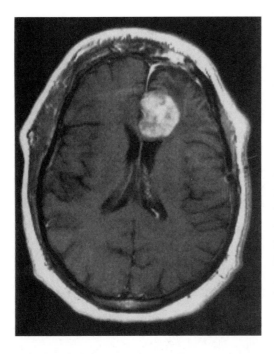

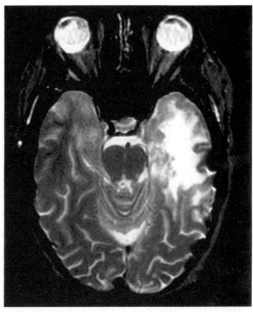

Fig. 13. MRI, axial, T1 with gadolinium, demonstrating a frontal intracerebral mass that was proven by biopsy to be neurosarcoidosis.

Fig. 14. MRI, axial, T2-weighted image, shows a large area of abnormality in the temporal lobe that proved at biopsy to be sarcoidosis manifesting with a focal encephalopathy/ vasculopathy.

neurosarcoidosis and may be focal or generalized. Seizures have been correlated with a poor prognosis in neurosarcoidosis (Delaney 1980). However, the cause of this poor prognosis is not directly due to seizures. Instead, the poor prognosis of neurosarcoidosis patients with seizures is due to the fact that seizures are an indicator for the presence of severe CNS parenchymal disease or hydrocephalus. This more severe pathology actually accounts for the poor outcomes because of a high risk for progressive or recurrent disease or death (Delaney 1980; Krumholz et al. 1991). Importantly, seizures in patients with neurosarcoidosis are relatively easy to control if the underlying CNS inflammatory process can be effectively treated (Krumholz et al. 1991).

Myelopathy

Spinal cord involvement is another form of CNS parenchymal sarcoidosis. Spinal cord sarcoidosis may manifest as intramedullary, intradural extramedullary, or extradural granulomatous disease (Junger et al. 1993; Olek et al. 1995). Intramedullary spinal

cord disease can also present with a myelitis that is analogous to the encephalopathy/vasculopathy of intracranial sarcoidosis. Intraspinal mass lesions due to sarcoidosis present with a non-specific imaging appearance (Fig. 16). Also, spinal arachnoiditis may occur. In addition, sarcoidosis may present as a radiculopathy, polyradiculopathy, or cauda equina syndrome. Finally spinal cord sarcoidosis may appear, typically in the late stages of spinal cord disease, as focal spinal cord atrophy (Junger et al. 1993; Olek et al. 1995).

Peripheral neuropathy

Although sarcoidosis commonly affects cranial nerves, peripheral neuropathy is less frequently described. Still, a variety of peripheral neuropathies are reported in sarcoidosis including chronic sensorimotor, pure motor or sensory neuropathies, mononeuritis multiplex, and acute Guillain–Barré-like syndromes (acute demyelinating polyneuropathy) (Zuniga et al. 1991). The most common form of peripheral neuropathy appears as an axonal sensorimotor neuropathy (Zuniga et al. 1991). Sarcoidosis neuropathy

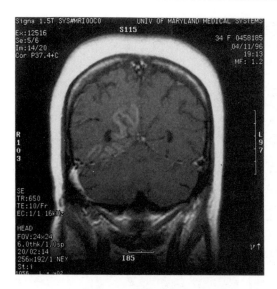

Fig. 15. MRI, coronal, T1-weighted image with gadolinium, showing gyral enhancement in the distribution of a posterior cerebral artery branch stroke.

typically begins months to years following an initial diagnosis of sarcoidosis, but, in some instances, symptoms of neuropathy precede the discovery of systemic sarcoidosis. The neuropathy is usually mild and classically manifests with distal paresthesias, decreased vibration and proprioception sensation, and reduced ankle jerk reflexes (Zuniga et al. 1991). Neuropathy has been attributed to epineural and perineural granulomas with an associated granulomatous vasculitis, producing an axonal degeneration with associated demyelination. Granulomas are also found in the endoneurium and associated with primary segmental demyelination in patients with sensorimotor neuropathy (Nemni et al. 1981). Nerve damage may be due to granulomatous vasculitis, compressive effects of granulomas, local effects of inflammation, direct injury by sarcoidosis, or some form of immunological injury to the nerve.

Myopathy

Sarcoidosis can also directly involve muscles and present with myopathy of various types. Manifestations of sarcoidosis myopathy include acute, subacute or chronic weakness, muscle pain, and palpable muscle nodules. Severe muscle disease can also result in fibrosis and cause contractures. Sarcoidosis

may also manifest with muscular atrophy or occasional pseudohypertrophy. Muscle involvement by non-caseating granulomas can be demonstrated with muscle biopsy. However, incidental non-caseating granulomas have been found in blind muscle biopsy specimens in as many as 50% of sarcoidosis patients without clinical evidence of muscle disease. Differentiating between sarcoidosis myopathy, polymyositis or dermatomyositis, and granulomatous myopathy also may be difficult, and it is important to evaluate other organ systems carefully for evidence of sarcoidosis (Hewlett and Brownell 1975; Stjernberg et al. 1981; Ando et al. 1985).

DIAGNOSIS

Diagnostically, there are principally two clinical scenarios with which patients with neurosarcoidosis present, and they occur with about equal frequency: (1) The patient without established sarcoidosis presents with a clinical picture suggestive of neurosarcoidosis. In this situation the major goal is to establish the presence of systemic sarcoidosis. (2) The patient with already established systemic sarcoidosis develops neurologic symptoms. Here, one should focus on confirming that the neurological problem is due to neurosarcoidosis rather than some other cause. In each case the diagnostic approach may be somewhat different.

Establishing systemic sarcoidosis

When a patient without documented systemic sarcoidosis develops a clinical syndrome suggestive of neurosarcoidosis, confirming evidence for sarcoidosis should be sought in other organ systems. Such systemic disease can best be documented when a thorough, systematic evaluation based on the known natural history of sarcoidosis is undertaken (Table 1). In particular, sarcoidosis most frequently affects intrathoracic structures (87% of patients), followed by lymph node, skin, and ocular disease (15–28% of patients). Consequently, histologic support for a diagnosis of sarcoidosis should be pursued following leads obtained from the patient's clinical evaluation and these statistics. Since corticosteroid therapy can mask signs of systemic disease, treatment should be postponed, if possible, while a search for systemic disease is initiated.

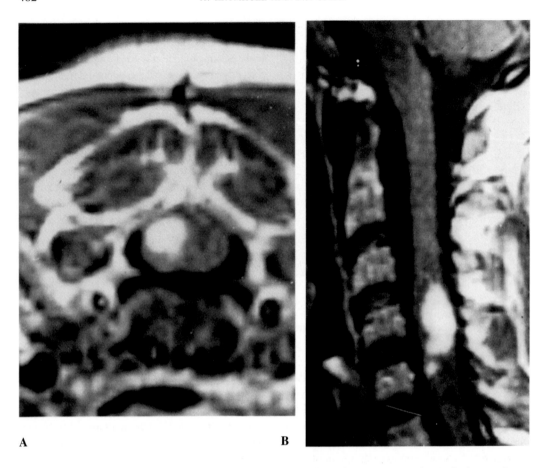

A   **B**

Fig. 16. MRI of the spine, (A) axial, (B) sagittal, with gadolinium, demonstrates an intraspinal mass due to sarcoidosis.

Pulmonary involvement is so common in sarcoidosis that this should be the first organ system to consider when attempting to establish the presence of systemic sarcoidosis. Indeed, nearly 90% of patients with sarcoidosis are reported to show radiographic evidence of pulmonary involvement (Rodan and Putman 1983). Still, although an abnormal chest roentgenogram is often seen in sarcoidosis and can be supportive evidence for that diagnosis, these chest X-ray findings are not necessarily specific or pathognomonic for sarcoidosis. Additional evidence to support pulmonary involvement can be obtained from pulmonary function testing including diffusion capacity. Moreover, when chest X-rays or pulmonary function studies suggest pulmonary involvement, a diagnosis of sarcoidosis is confirmed by obtaining histologic evidence of sarcoidosis with a transbronchial biopsy.

Apart from the lungs, other organ systems also deserve consideration. Because of ease of access, early consideration should be given to a skin or lymph node biopsy of suspicious sites. Clinical information should also guide the evaluation of other organ systems for evidence of systemic sarcoidosis. For instance, nasal mucosal, conjunctival, lacrimal gland, liver, and muscle biopsies can be considered on the basis of individualized clinical assessments. In particular, if a patient has impaired smell or taste, nasal or olfactory nerve disease might be present. In that situation, an endoscopic nasal and sinus examination can show abnormal mucosa amenable to biopsy. Whereas, if dry eyes or mouth are noted, lacrimal, parotid, or salivary gland involvement should be suspected.

In addition, a thorough ocular examination is indicated to search for uveitis, retinal periphlebitis, or superficial lesions for conjunctival biopsy. Ocular findings in sarcoidosis include lacrimal gland inflammation, conjunctival nodules, iritis, uveitis, retinal lesions (vascular sheathing, granulomas, vascular occlusions, and chorioretinitis), and optic disk pathology (edema, nodules, granulomas, and atrophy) (Sharma 1984; James and Williams 1985).

The Kveim–Siltzbach test is not widely available and no longer much used. However, historically it was yet another means for diagnosing systemic sarcoidosis. Kveim–Siltzbach reagent is a suspension derived from the spleen of a patient with sarcoidosis. This suspension is injected intracutaneously and when positive will produce a cutaneous nodule that, on biopsy, will reveal non-caseating granulomas. While this represents a very specific and sensitive test, it does require a period of 4–6 weeks before biopsy can be performed, making this a poor test for patients with acute, severe neurosarcoidosis who may require very prompt corticosteroid treatment (Israel 1983).

Gallium-67 scanning has been promoted as a valuable imaging method for initial detection of systemic sarcoidosis (Chapelon et al. 1990), but it has limited utility for longitudinal clinical follow-up because of its expense and the potential radiation exposure (Turner-Warwick et al. 1986). Detection of bilateral inflammation in the lacrimal, minor salivary, and parotid glands on a gallium scan is especially suggestive of sarcoidosis (Savolaine and Schlembach 1990).

Various laboratory measures have been described as abnormal in sarcoidosis. Although none is highly specific for sarcoidosis, they can be of some value. These abnormalities include increased SACE, various immunologic studies including increased serum γ-globulins and hematologic disorders such as anemia, leukopenia and thrombocytopenia. Metabolic derangements such as hypercalcemia, hypercalciuria, and hepatic and renal dysfunction are all described as part of sarcoidosis. The most specific laboratory test associated with sarcoidosis is the SACE. However, its sensitivity is not that high, with just 50–60% of active sarcoidosis patients showing abnormalities. In addition, SACE is not very specific because it is abnormal in other conditions such as liver disease, diabetes mellitus, hyperthyroidism, systemic infection, malignancy and Gaucher's disease (Table 2) (Leiberman 1985).

Confirming neurosarcoidosis

The second category of neurosarcoidosis patients are those with well documented systemic sarcoidosis who develop neurologic disease suspected to be due to neurosarcoidosis. In general they require less of a systemic assessment, but they still merit careful appraisal because neurosarcoidosis can be confused with many other neurological diseases. It is sometimes not safe to confirm neurosarcoidosis with biopsy tissue from the nervous system, but for some patients this may be necessary.

Indeed, these patients deserve consideration of disease entities that may mimic neurosarcoidosis, particularly infection and neoplasia (Table 4). Once such disorders are excluded, the patient can reasonably be treated for neurosarcoidosis. Still, if a patient does not respond to treatment as expected, the diagnosis of neurosarcoidosis should be revisited and a more extensive evaluation considered to exclude other etiologies (Table 4).

Although not specifically diagnostic, there are tests that support a presumptive diagnosis of neurosarcoidosis. Brain imaging studies can be particularly helpful to confirm the presence, classify the nature, and monitor the treatment of neurosarcoidosis. The preferred imaging technique is MRI with contrast enhancement (Sherman and Stern 1990). Unenhanced T1-weighted images provide less useful information than T2 studies. With T2-weighted imaging, areas of increased signal intensity can better be appreciated, especially in the periventricular distribution. Contrast administration helps by demonstrating leptomeningeal enhancement as well as parenchymal abnormalities (Figs. 8, 9 and 11–16). Enhancement presumably reflects a breakdown of the blood–brain barrier and implies active inflammation. Spine MRI can visualize intramedullary disease, which appears as an enhancing fusiform enlargement, focal or diffuse enhancement, or atrophy (Fig. 16) (Junger et al. 1993; Olek et al. 1995). Enhancing nodules or thickened or matted nerve roots also can be appreciated in MRI images of the cauda equina. Muscle MRI can demonstrate areas of inflammation that are amenable to directed biopsy.

Spinal fluid analysis is another useful method for assessment, diagnosis, and staging of neurosarcoidosis. Over 50% of patients with CNS sarcoidosis will have some CSF abnormality (Stern et al. 1985;

Oksanen 1986). Reported abnormalities include an elevated CSF pressure, a high protein level, hypoglycorrhachia, and a predominantly mononuclear pleocytosis of up to several hundred cells. In addition, some patients have oligoclonal bands in the CSF or an elevated IgG index. However, none of these abnormalities is specific for neurosarcoidosis.

Newer CSF assays may prove more specific for CNS sarcoidosis. CSF angiotensin converting enzyme (ACE) activity tends to be raised in some 50% of untreated patients with CNS sarcoidosis (Oksanen et al. 1985b), although abnormalities are also seen in the presence of infection and malignancy. The CSF ACE level may be abnormal even with steroid therapy, but less consistently than in untreated patients. The degree of elevation of CSF ACE may parallel the clinical course (Chan Seem et al. 1985; Oksanen et al. 1985a, b). CSF ACE is thought to be produced by CNS granulomas, especially those near the meninges. A normal CSF ACE assay does not exclude the diagnosis of neurosarcoidosis. Moreover, CSF ACE's diagnostic value is further limited because assay methodology and normative values have as yet not been well standardized.

In regard to other specialized CSF studies, elevated CD4:CD8 lymphocyte ratios are found in some patients with CNS sarcoidosis (Juozevicius and Rynes 1986; Stern et al. 1987). The utility of this analysis remains to be determined. Lysozyme and β_2-microglobulin are elevated in the serum of some patients with systemic sarcoidosis. These proteins are elevated in the CSF of 75 and 53%, respectively, of patients with CNS sarcoidosis and do not correlate with serum levels (Oksanen et al. 1985a). The CSF levels of these proteins also may fluctuate with neurologic status (Oksanen et al. 1985a).

None of the aforementioned CSF analyses, even the most specialized or immunologically sophisticated, is pathognomonic of CNS sarcoidosis. Infection or neoplasia may cause similar changes and must always be considered. Yet, after exclusion of other causes, such CSF abnormalities may be supportive of a diagnosis of neurosarcoidosis. The utility of these tests and other assays in potentially differentiating CNS sarcoidosis from other diseases, such as multiple sclerosis, merits further study.

Angiography usually has little to contribute in the evaluation of CNS sarcoidosis (Brooks et al. 1982; Brown et al. 1989). Granulomatous masses are typi-

cally avascular (Kendall and Tatler 1978). Rarely, angiographic abnormalities are found such as smooth narrowing of the caudal internal carotid artery (Kendall and Tatler 1978), tapering occlusion of the anterior cerebral artery (Corse and Stern 1990), or multiple small vessel segmental narrowings suggesting an arteritis (Lawrence et al. 1974). Patients with a diffuse encephalopathy or stroke-like episodes are the most likely to demonstrate clinically relevant angiographic findings (Oksanen 1986).

Evoked potentials, visual, auditory and somatosensory, can be useful in evaluating some patients with neurosarcoidosis. Visual evoked potentials (VEPs) can reveal abnormalities of the anterior visual pathways. VEPs are often abnormal in patients with symptomatic optic nerve disease and may be abnormal in some patients with CNS sarcoidosis but no clinical evidence of optic nerve dysfunction (Streletz et al. 1981; Stern et al. 1985; Oksanen and Salmi 1986). Streletz et al. (1981) report VEP abnormalities in a high proportion of patients with neurosarcoidosis. Even some patients with sarcoidosis but without ocular or neurological disease were found to have abnormal VEPs. Their study suggests that subclinical neurosarcoidosis may be more common than previously realized. However, other observations of VEPs do not confirm a high incidence of abnormalities in asymptomatic patients (Stern et al. 1985). Some investigators suggest, instead, that VEP abnormalities in particular and evoked potential disturbances in general are useful in the evaluation of patients with symptomatic neurosarcoidosis because they demonstrate relevant abnormalities that may be used to monitor the clinical course of the disease (Stern et al. 1985).

Similarly, brainstem auditory evoked potentials (BAEPs) are often abnormal in neurosarcoidosis patients with brainstem or eighth nerve symptoms and can be abnormal in neurologically ill patients without overt disease in these areas (Oksanen and Salmi 1986). However, they are rarely, if ever, abnormal in patients without clinically evident CNS sarcoidosis. The degree of abnormality of an evoked potential can fluctuate depending on disease activity, and an abnormal pattern can even normalize (Oksanen and Salmi 1986).

Somatosensory evoked responses have not been comprehensively studied in the assessment of patients with sarcoidosis. Preliminary evidence suggests that their clinical utility is similar to VEP and BAEP in confirming the involvement of a specific sensory system that may be clinically affected (Stern et al. 1992).

Nerve conduction studies in patients with sarcoidosis neuropathy usually reveal changes compatible with an axonal neuropathy, though slowing can be more pronounced and suggestive of demyelinating disease (Challenor et al. 1984; Galassi et al. 1984). Electromyography can demonstrate a denervation pattern in patients with a neuropathy or radiculopathy and myopathic changes in patients with a symptomatic myopathy (Ando et al. 1985). MRI may reveal a characteristic 'star shaped' pattern for muscle nodules (Otake et al. 1990). Muscle or nerve biopsy is informative if the diagnosis of neuromuscular disease is in doubt and can be targeted to an area of enhancement.

Patients with CNS sarcoidosis should be carefully questioned about symptoms relating to neuroendocrinologic or hypothalamic dysfunction, since problems in these areas are the most common brain parenchymal disorders found in sarcoidosis. Inquiry should focus on alterations in menses, libido and potency as well as the presence of galactorrhea. Excessive thirst can be caused by hypothalamic damage that affects the normal osmostatic mechanism, or diabetes insipidus, hypercalcemia, hypercalciuria, and corticosteroid induced diabetes mellitus. Alterations in body temperature, sleep and appetite also can develop as a consequence of hypothalamic sarcoidosis. Patients with CNS symptoms other than transient cranial nerve palsies or aseptic meningitis should undergo a neuroendocrinologic evaluation including thyroid function tests (hypothalamic hypothyroidism needs to be considered), prolactin, testosterone or estradiol, FSH and LH, and cortisol assays.

Neurosarcoidosis, because of its varied manifestations, is in the differential diagnosis of many unexplained neurological syndromes. The diagnosis of sarcoidosis is most secure when based on pathology and when more than one organ system can be documented to be involved. However, since tissue from the nervous system is difficult to secure for pathologic analysis and other tests are not diagnostic of neurosarcoidosis, the diagnosis of neurosarcoidosis, despite the best efforts, must sometimes remain tentative.

TREATMENT

There are no scientifically rigorous studies comparing various treatments for neurosarcoidosis. However, most experts agree that corticosteroid therapy is the mainstay of treatment and is indicated for any patient without a specific contraindication to it. Still, decisions about such issues as the optimal therapeutic dose and duration of therapy should be guided by the patient's clinical course, expected natural history or prognosis, and adverse treatment effects. A treatment paradigm is given in Fig. 17.

Treatment with corticosteroids is widely accepted and recommended for all forms of neurosarcoidosis. In support of this, many individual case reports and series provide evidence that in the short term it can produce impressive responses and alleviate symptoms. Still, it is not absolutely certain that treatment changes the natural history and long-term course of neurosarcoidosis. A major theoretical goal for long-term treatment with corticosteroids is to diminish the irreversible fibrosis that can develop and to minimize tissue ischemia that might result from perivascular inflammation. However, once corticosteroid or other immunotherapy is begun, it need not continue indefinitely, particularly not at extremely high doses. Theoretically, with time, the inflammatory process can recede allowing immunosuppressive therapy to be withdrawn. Recommended treatment regimens for the various manifestations of neurosarcoidosis are detailed below.

Peripheral facial nerve palsy

The most common neurologic manifestation of neurosarcoidosis is a peripheral facial nerve palsy. The facial weakness may improve without any specific treatment. A controlled trial of treatment has not been done, and the long-term prognosis for these patients seems favorable. Still, it seems reasonable, even in this situation, to give a short course of 2 weeks of prednisone treatment for a peripheral facial nerve palsy due to sarcoidosis. The recommended first week's prednisone dose is in the range of 0.5–1.0 mg/kg/day (or 40–60 mg/day), followed by a gradual reduction leading to discontinuation of prednisone over the second week. General supportive care, as used for any patient with a peripheral facial nerve palsy, should be provided, with special attention

TABLE 4

Other diagnostic considerations in patients with suspected neurosarcoidosis.

Neurosarcoidosis feature	Selected diagnostic considerations	
Cranial neuropathy Optic neuropathy	Multiple sclerosis Giant cell arteritis Wegener's granulomatosis Polyarteritis nodosa Meningioma	
Cranial nerves III, IV, VI	Horner's syndrome (due to mediastinal and cervical adenopathy) Graves' ophthalmopathy Wegener's granulomatosis	
Trigeminal neuropathy Peripheral facial palsy	Sjögren's syndrome Bell's palsy *Borrelia burgdorferi* infection (Lyme disease) Parotitis Wegener's granulomatosis	
Eighth nerve dysfunction	Cogan's syndrome Vogt–Koyanagi–Harada syndrome	
Recurrent laryngeal nerve palsy	Mediastinal adenopathy	
Meningeal disease Aseptic meningitis	Cryptococcosis HIV infection Tuberculosis *Borrelia burgdorferi* infection Behçet's syndrome	Vogt–Koyanagi–Harada syndrome Mollaret's meningitis Wegener's granulomatosis Brucellosis Meningeal carcinomatosis
Meningeal mass	Meningioma Fibrosclerosis	
Parenchymal disease Mass lesion	Lymphoma Glioma Metastases Tuberculoma Wegener's granulomatosis	Lymphomatoid granulomatosis Dysgerminoma Pinealoma Reticulum cell sarcoma
Spinal cord dysfunction	Multiple sclerosis Intramedullary tumor	

Encephalopathy and vasculopathy	Fungal, bacterial, tuberculosis infection
	Granulomatous angiitis
	HIV infection
	Sjögren's syndrome
	Systemic lupus erythematosus
	Multiple sclerosis
	Polyarteritis nodosa
	Wegener's granulomatosis
		Lymphomatoid granulomatosis
		Cerebral emboli due to:
		  Left atrial myxoma
		  Sarcoid cardiomyopathy
		Borrelia burgdorferi infection
		Whipple's disease
		Behçet's disease
		Vogt–Koyanagi–Harada syndrome

Peripheral neuropathy	Guillain–Barré syndrome Polyarteritis nodosa Sjögren's syndrome Wegener's granulomatosis Lymphomatoid granulomatosis *Borrelia burgdorferi* infection	
Myopathy	Polymyositis Dermatomyositis Granulomatous myopathy	

Table 4 continued

Neurosarcoidosis feature	Selected diagnostic considerations
CSF analysis	
Hypoglycorrhachia	Carcinomatous and tuberculous meningitis
Elevated IgG index	Multiple sclerosis
	Sjögren's syndrome
	Progressive multifocal leukoencephalopathy
	HIV infection
Oligoclonal bands	Multiple sclerosis
	Sjögren's syndrome
	HIV infection
Elevated angiotensin converting enzyme	CNS infections
	CNS malignancy
	Behçet's disease
Increased CD4/CD8 ratio	? Multiple sclerosis
	? Aseptic meningoencephalitis
Increased β_2-microglobulin	Infection
	Malignancy (especially lymphoma)
	Syphilis
	AIDS dementia complex
Eosinophilia	Lymphoma
	Tuberculosis
	Syphilis
	Parasites
	Foreign body reaction
CT scan	
Basilar enhancement	Tuberculous, fungal and bacterial meningitis
	Meningeal carcinomatosis
Convexity mass(es)	Meningioma
Parenchymal mass(es)	Metastasis
	Glioma
	Toxoplasmosis
	Lymphoma
Orbital disease	Optic neuritis
	Malignancy
	Graves' ophthalmopathy
	Orbital pseudotumor
MRI	
Periventricular white matter lesions	Multiple sclerosis
	Sjögren's syndrome
	Progressive multifocal leukoencephalopathy
	Borrelia burgdorferi infection
Meningeal enhancement	Infectious (TB, fungi); tumor
Angiography	
Arteritis	Granulomatous angiitis
	Drug abuse (sympathomimetics)
	Granulomatous angiitis associated with HIV infection
Evoked potentials	
Abnormal VEPs	Multiple sclerosis
	Sjögren's syndrome
Abnormal BAEPs	Multiple sclerosis
	Sjögren's syndrome

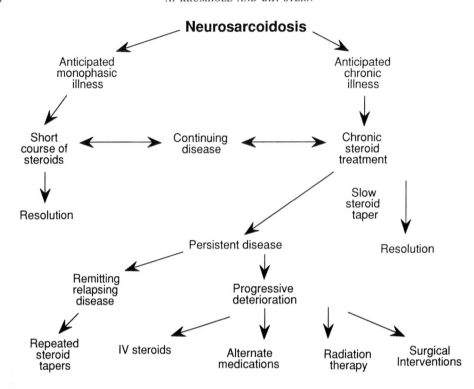

Fig. 17. Treatment paradigm for patients with neurosarcoidosis.

directed at protecting the eye. Occasionally a patient who was improving will relapse as the dose of prednisone is decreased. In that situation, the dose can be increased for a week and a slower taper attempted.

Other cranial nerve palsies and aseptic meningitis

Patients with other cranial neuropathies and aseptic meningitis can be managed with a corticosteroid protocol similar to that described for peripheral facial nerve palsy. However, often more than 2 weeks of therapy is needed. In particular, patients with optic neuropathy or cranial nerve VIII dysfunction may need more prolonged, aggressive therapy. However, even prolonged, aggressive therapy may not prevent irreversible optic and eighth nerve damage.

Aseptic meningitis may also respond to a short 2-week course of prednisone, 0.5–1.0 mg/kg/day. However, the goal of therapy should not be complete clearing of an asymptomatic CSF pleocytosis. Attempts to completely normalize the spinal fluid can needlessly expose patients to the adverse effects of the high doses of corticosteroids. Indeed, there is no evidence that clearing the CSF of an abnormal cellular response necessarily corresponds to clinical well-being in patients with aseptic meningitis who are otherwise asymptomatic.

Hydrocephalus

Asymptomatic ventricular enlargement usually does not require therapy. Mild, symptomatic hydrocephalus can respond to corticosteroid treatment; and prolonged therapy is often appropriate. Life-threatening hydrocephalus or corticosteroid-resistant hydrocephalus requires ventricular shunting or a ventricular drain. Unfortunately, patients can evolve from having mild hydrocephalus to severe, life-threatening disease quite rapidly. Therefore, patients and care-providers should be well educated as to the symptoms of acute progressive hydrocephalus and know how to obtain prompt emergency care. At times, high-dose i.v. corticosteroid therapy (methylprednisolone 20 mg/kg/day for 3 days) can stabilize a patient with life-threatening hydrocephalus, although usually prompt surgical intervention with a ventricular drain or ventriculoperitoneal shunt is necessary (Fig. 10). Shunt placement is not without risk in this patient population, which is why 'prophylactic' shunting in asymptomatic patients with hydrocephalus is not readily advocated. Shunt obstruction from the inflamed CSF and ependyma is common, and placement of a foreign object in the CNS of an immunosuppressed host predisposes to infection.

CNS parenchymal disease

Corticosteroid therapy can improve the status of patients with a diffuse encephalopathy/vasculopathy or a CNS mass lesion (Fig. 8). Only rarely does immunosuppressive treatment improve neuroendocrine dysfunction or vegetative symptoms.

Corticosteroid treatment for CNS parenchymal disease and other severe neurologic manifestations of sarcoidosis usually starts with prednisone 1.0–1.5 mg/kg/day. The higher doses are used in patients with particularly severe disease. Such patients should be observed on high corticosteroid doses for 2–4 weeks to ascertain the clinical response. Also, these types of patients often require more prolonged therapy, and so even when appropriate to lower it, prednisone should still be tapered very slowly. The prednisone dose can be tapered by 5 mg decrements every 2 weeks as the clinical course is monitored. Neurosarcoidosis tends to exacerbate at a low prednisone dose approximating 10 mg or less daily. Some patients exhibit an individual therapeutic lower limit or dosage below which worsening can almost be predicted.

Once a dose of prednisone of 10 mg daily can be achieved, the patient should be evaluated for evidence of clinical as well as subclinical worsening of disease. Clinical disease can be monitored by symptoms, but subclinical disease can also be followed. For patients with CNS disease, an enhanced MRI scan is useful. Intense enhancement in the meninges, for example, suggests that neurosarcoidosis is active and further decreases in corticosteroid dose may lead to a clinical exacerbation. Other manifestations of neurosarcoidosis can be evaluated for subclinical deterioration on an individualized basis, for instance, evaluating nerve conduction studies or a serum creatine kinase level. On the other hand, persistent mild CSF abnormalities are usually not an indication for continuing or escalating high-dose corticosteroid therapy, since patients can remain quite functional in spite of an abnormal CSF. Efforts to 'normalize' the CSF often require powerful immunosuppression, with its attendant adverse effects.

If the sarcoidosis appears quiescent, then even a low daily prednisone dose of about 10 mg can be tapered further by 1 mg every 2–4 weeks. If a patient has a clinical relapse, the dose of prednisone should be doubled (unless the dose is very modest, in which case a prednisone dose of 10–20 mg daily can be prescribed). The patient should then be observed for approximately 4 weeks before another taper is contemplated. Patients may require multiple cycles of higher and lower corticosteroid dosage during attempts to taper medications. Still, this effort is warranted since the disease can become quiescent and without attempts at withdrawing medication, patients may be needlessly exposed to the harmful side effects of long-term, high-dosage corticosteroids.

Daily dosing of prednisone is usually superior to alternate day therapy. However, if a patient has been clinically stable for several months on a modest daily prednisone dose, an attempt can be made to gradually change the patient to alternate day therapy.

Seizures

In sarcoidosis patients with seizures, the seizures are generally not a major limiting problem and can usually be well controlled with antiepileptic medication if the underlying CNS inflammatory reaction can be effectively treated. However, seizures have been shown to be an indication of or marker for the presence of parenchymatous involvement of the brain which is in itself a very serious manifestation of neurosarcoidosis (Luke et al. 1987; Krumholz et al. 1991). Consequently, it is important to recognize this clinical relevance of seizures in patients with neurosarcoidosis. Seizures are a useful, often early warning sign that a patient may have one of the more serious forms of neurosarcoidosis such as an intracranial mass lesion (Krumholz et al. 1991).

Peripheral neuropathy and myopathy

Patients with a peripheral neuropathy or myopathy can respond to a several-week course of corticosteroids, usually beginning with prednisone 1 mg/kg/day (or approximately 60 mg/day). Here, too, prolonged treatment may be indicated. Corticosteroids should be tapered slowly, as discussed above.

Other issues

Critically ill patients may be considered for high-dose intravenous (i.v.) corticosteroid therapy (Soucek et al. 1993). A short course of methylprednisolone 20 mg/kg/day i.v. for 3 days, followed by prednisone 1.0–1.5 mg/kg/day for 2–4 weeks, is occasionally warranted. Aggressive therapy can also clarify whether a patient on a modest corticosteroid dose can incrementally improve. The clinical response should be closely observed and pre-defined measures of improvement should be carefully monitored. One or two target measures such as a specific clinical sign, symptom, functional assessment or neuro-diagnostic test can be used to judge the response to treatment. For instance, the results of psychometric tests or a timed walk can be used for clinical assessment in some patients, while in patients with an intracranial mass, MRI imaging may be a helpful neurodiagnostic measure. This type of treatment regimen can be used to judge a response over a relatively short time period. However, one caveat must be mentioned: if the patient has a CNS mass lesion unresponsive to high-dose i.v. corticosteroids, surgical resection may be appropriate in life-threatening situations.

Surgical considerations

There are several clinical scenarios in which neurosurgical intervention may be necessary for patients with neurosarcoidosis. First, without biopsy confirmation of affected neural tissue, a diagnosis of neurosarcoidosis must remain somewhat in question. Biopsy of peripheral nerve or muscle for such a purpose is relatively easy to obtain. In contrast, CNS tissue for pathologic examination is much more difficult to secure in such patients. Despite this difficulty, if a patient with presumptive CNS sarcoidosis is not responding to treatment as anticipated, consideration should be given to biopsy of a demonstrated CNS lesion so as to exclude other disorders that may be confused with neurosarcoidosis and to confirm the diagnosis of sarcoidosis (Fidler et al. 1993; Cheng et al. 1994). There are two other major situations in which surgical intervention may be appropriate in neurosarcoidosis: patients with intracranial mass lesions unresponsive to corticosteroids, or with hydrocephalus.

Mass lesion

Patients without known systemic sarcoidosis who develop a brain or spinal cord mass are frequently biopsied for determination of the diagnosis. If pathologic examination suggests non-caseating granulomas, appropriate cultures should be obtained. Also, since granulomas can surround some malignancies, adequate biopsy samples are encouraged (Peeples et al. 1991). However, surgical excision of the mass should usually be avoided since surgery is rarely curative and patients have been known to develop new major fixed neurologic deficits or deteriorate from surgical excision of a granulomatous mass. After biopsy, an empiric trial of high-dose corticosteroid treatment is appropriate for most such patients. Furthermore, patients should be evaluated for systemic sarcoidosis, but it is possible that corticosteroid therapy can mask the presence of systemic disease.

If a patient with known sarcoidosis develops a CNS mass, an empiric trial of corticosteroid therapy is appropriate (Fig. 8), especially if infection and malignancy are excluded by CSF examination and other means. If the patient does not respond to corticosteroid therapy, a biopsy should be pursued. In either of these scenarios, if a mass progressively enlarges in spite of corticosteroid therapy, surgical exploration should be strongly considered to evaluate the possibility of an infection or malignancy (Peeples et al. 1991). Some of the important entities that can be mistaken for sarcoidosis are listed in Table 4.

Hydrocephalus

Patients with symptomatic hydrocephalus not responsive to corticosteroid therapy need to be considered for ventricular shunting as previously discussed. The surgical management of such patients can be quite challenging because of the potential for shunt obstruction and the risk of infection. In addition, cerebral ventricles or parts of ventricles can sometimes become 'trapped' or isolated from the rest of the ventricular system, necessitating the placement of multiple shunts.

GENERAL SUPPORTIVE CARE

Patients with neurosarcoidosis, and particularly those receiving treatment with immunosuppressive agents, require close attention to their general medical well-being. Potential adverse effects of treatment should be carefully sought. For example, an exercise and dietary program are often very beneficial to help avoid the weight gain associated with high-dose and long-term corticosteroid treatment. Rehabilitation services should be utilized as appropriate. Depression is not uncommon and treatment can be helpful.

Therapy of endocrinologic disturbances is important. In particular, hypothyroidism and hypogonadism are major problems in neurosarcoidosis and should be treated. In addition, since patients are often on protracted, low-dose corticosteroid regimens, a supplemental dose of corticosteroid is appropriate during periods of intercurrent illness or stress. Hyperglycemia is a potential side effect of long-term, high-dose corticosteroid treatment, but, fortunately, is not usually associated with ketoacidosis. Exercise and dietary programs are useful to manage the hyperglycemia, but occasionally oral hypoglycemic agents or insulin therapy may need to be considered.

Another concern is that patients with sarcoidosis are at risk for osteoporosis. In particular, corticosteroid therapy can cause osteoporosis. Moreover, treatment of osteoporosis is a challenge because sarcoidosis can cause hypercalcemia and hypercalcuria. Handling of osteoporosis in patients with neurosarcoidosis may require reducing corticosteroid dosage when possible, supplementation with calcium and vitamin D, hormonal treatment and use of other agents. Screening should be done for osteoporosis, and when appropriate monitoring with bone density measurements should be considered. Serial measurements of bone density may be particularly useful in judging an individual patient's progress. One potential option for neurosarcoidosis patients with osteoporosis is treatment with Deflazacort, a corticosteroid preparation with a low propensity for causing osteoporosis. In addition, salmon calcitonin nasal spray is an effective treatment for osteoporosis. Since the management of osteoporosis is a rapidly changing field, it is suggested that when osteoporosis is identified an endocrinologist may need to become involved (Schneyer 1995).

It is obvious that patients with refractory neurosarcoidosis are at risk from both the adverse effects of the sarcoidosis inflammatory process and the dangers of treatment side effects and complications. Consequently, if a patient with neurosarcoidosis is doing poorly and is refractory to treatment, not only should the original diagnosis of sarcoidosis be questioned, but the patient should be evaluated for intercurrent complications, as outlined in Table 5.

ALTERNATE TREATMENTS

Treatment alternatives to corticosteroids must sometimes be considered for patients with neurosarcoidosis. Experience in this area is limited and optimal strategies are not well established. Indications for the use of alternate treatments include contraindications to corticosteroids as initial therapy, serious adverse chronic corticosteroid effects, and progressive disease activity in spite of aggressive corticosteroid therapy. Medication alternatives to corticosteroids that have been used to treat sarcoidosis include cyclosporine, azathioprine, methotrexate, cyclophosphamide, and chlorambucil. One non-drug treatment that has also been reported to have some limited success is radiation therapy. These agents or radiation therapy have been not been studied in a scientifically controlled manner against placebos or comparable treatments. Such trials are difficult because neurosarcoidosis is an extremely rare disorder with very varied clinical presentations. Consequently, it has been difficult to conduct controlled treatment studies for either conventional or alternative therapies.

Practically, consideration should be given to introducing alternative therapy whenever a patient shows signs of serious corticosteroid side effects or requires frequent large increases in corticosteroid dosage to control symptoms. Alternative treatment with an immunosuppressive agent or irradiation is a logical adjunctive therapy for refractory neurosarcoidosis given what we understand of the immunopathogenic mechanisms of the disease. Alternative therapy may allow a gradual decrease in corticosteroid dosage to pre-

TABLE 5

Some neurologic complications associated with sarcoidosis and its therapy.

Cryptococcal meningitis
Tuberculous meningitis
Progressive multifocal leukoencephalopathy
Herpes simplex encephalitis
Corticosteroid myopathy
Inclusion body myositis
Spinal epidural lipomatosis

vent or minimize corticosteroid complications, often without deterioration in clinical status. Rarely, however, can corticosteroids be completely eliminated. Still, some patients may continue to deteriorate despite combination therapy and the reported mortality in neurosarcoidosis remains a substantial 5–10% (Stern et al. 1985; Luke et al. 1987; Krumholz et al. 1991).

An algorithm for the use of alternative treatment for refractory neurosarcoidosis is presented in Fig. 18. The choice of alternative treatment should be determined by the potential adverse effects of the therapy and the extent of systemic disease. It is wise to choose an agent whose adverse effects spare an organ or organ system that may already be compromised (Fig. 18). For example, cyclosporine usually should be avoided in patients with significant hypertension or renal disease, while azathioprine may not be the best choice for patients with liver or hematologic problems.

Cyclosporine has been studied more than any other immunosuppressive medication and has been found to be beneficial to some patients with severe neurosarcoidosis refractory to corticosteroid therapy (Stern et al. 1992). Cyclosporine inhibits the amplification of the CD4 cell immune response, which is suspected to be of importance in the pathogenesis of sarcoidosis (Kahan 1989). Although it is ineffective as isolated treatment for pulmonary sarcoidosis, it is an effective adjunct to corticosteroid therapy for refractory neurosarcoidosis (Stern et al. 1992). Other immunosuppressive agents offer less specific antagonism of the immune system.

A trial of cyclosporine plus corticosteroids is reasonable to consider in patients in whom an adequate response to corticosteroid treatment alone has not been achieved. Cyclosporine can be started at 4 mg/kg/day in two divided doses; the trough blood level can be monitored on a monthly basis, and the dose adjusted to maintain a 'therapeutic' range. The blood level does not seem to correlate with the clinical response, but it is reasonable to avoid values outside the target range. Depending on the assay used, the therapeutic range will vary; therefore the physician

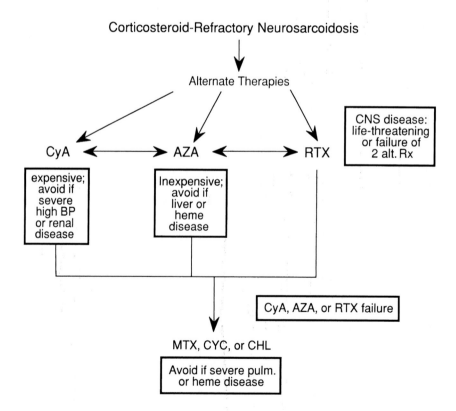

Fig. 18. Treatment paradigm for patients with corticosteroid-refractory neurosarcoidosis.

should follow the laboratory recommendations. Adverse effects should be monitored, especially blood pressure, renal function, and serum magnesium (since hypomagnesemia can predispose to seizures).

Another medication option is azathioprine. Azathioprine is a reasonable second agent, especially because of its moderate cost. It can be started at a dose of 50 mg/day for 1 week and if this is tolerated, the dose can be progressively increased to 2–3 mg/kg/day. Blood and liver profiles should be monitored and the lymphocyte count should be lowered to approximately 1000 mm^3 in patients on corticosteroids. An increase in the red blood cell mean corpuscular volume to above 100 is indicative of an adequate azathioprine effect. Some 10% of patients develop an acute idiosyncratic reaction with fever and abnormal liver function necessitating discontinuation of the drug.

In comparison to cyclosporine and azathioprine, there is not as much experience with the other immunosuppressive medication alternatives to corticosteroid treatment for neurosarcoidosis. Potential options include chlorambucil, cyclophosphamide or methotrexate. In general, side effects to these alternative medications are limited, predictable and respond to withdrawal of the offending agent. It is even possible to restart the medication in some cases without recurrent side effects.

One of the major advantages of alternative immunosuppressant medications is that they may enable a gradual reduction of corticosteroid to about 15–30% of the stabilizing corticosteroid dose thus offering a 'corticosteroid sparing' effect (Stern et al. 1992). Even with the use of alternative medication, there is the possibility of recurrent symptoms when prednisone is decreased below 10 mg/day. Attempts to withdraw corticosteroid totally may result in worsening symptoms, suggesting that alternative immunosuppressant medication is best used as an adjunct to corticosteroids. On the other hand, some patients do quite well on corticosteroids alone after alternative medication has been withdrawn, and some can be maintained on the alternative agent alone.

Another viable option for some patients with refractory CNS neurosarcoidosis is radiation therapy. The small number of studied and reported patients treated with CNS radiation precludes any definite conclusions about efficacy. Some case reports suggest a beneficial response, especially if total nodal and craniospinal irradiation is done. A total dose of 19.5 Gy with daily fractionation of 1.5 Gy has been suggested. Total nodal irradiation has also been administered in refractory patients (Ahmad et al. 1992). It appears that although radiation therapy can sometimes be of benefit, continued immunosuppressive therapy is usually necessary.

Although it is not possible to predict with absolute certainty which patients with neurosarcoidosis will have disease refractory to conventional corticosteroid treatment, there are patients with a particularly high-risk clinical profile. For instance, patients with CNS parenchymatous disease such as mass lesions or extensive encephalopathy/vasculopathy are at especially high risk. Such patients might benefit from the prompt use of adjunctive alternative treatment should they become refractory to corticosteroids or develop intolerable side effects. Response to a particular alternative therapy can only be determined by trial since a good clinical response is highly individualized and patients may show a good response to one agent even after an initial failure with another.

Most patients resistant to multiple therapies have a poor prognosis. If, on the other hand, an alternative medication is discontinued because of adverse drug effects, it is advisable to try other immunosuppressive medications before irradiation, as such patients may respond to other drugs. If a patient is stable for several months on prednisone and an alternate treatment, a slow taper of the alternate treatment can be pursued as the clinical course is monitored.

PROGNOSIS

The clinical course and prognosis for neurosarcoidosis varies but is somewhat predictable. For example, some two-thirds of patients have a monophasic neurologic illness; the remainder have a chronically progressive or remitting-relapsing course. Those with a monophasic illness typically have an isolated cranial neuropathy, most often involving the facial nerve, or an episode of aseptic meningitis. Those with a chronic course usually have CNS parenchymal disease, hydrocephalus, multiple cranial neuropathies (especially involving cranial nerves II and VIII), peripheral neuropathy, and myopathy (Luke et al. 1987). Patients with CNS parenchymal disease or hydrocephalus are at highest risk of death, either from the inflammatory process itself or complications of therapy. Mortality of neurosarcoidosis is approximately 5–10% (Stern et al. 1985; Luke et al. 1987; Krumholz et al. 1991).

Since most patients with neurosarcoidosis are treated with immunosuppressive agents, it is impossible to determine the untreated natural history of the disorder (Stern et al. 1985). Moreover, since treatment with corticosteroids clearly benefits some patients, it would be inappropriate to withhold treatment for patients with neurosarcoidosis. Although therapy can certainly improve patients in the short term, it is not certain exactly how it changes the ultimate outcome of the disease. Even in severely ill or impaired patients the inflammatory process may spontaneously subside over time. Other patients with remitting-relapsing or progressive disease can become severely incapacitated even with aggressive treatment.

Despite limitations in our understanding of the natural history of sarcoidosis, treatment with corticosteroids does seem to benefit many patients with neurosarcoidosis. Even more importantly, patients benefit most from a comprehensive approach to care based on an understanding of the full clinical spectrum of neurosarcoidosis, an appreciation of the whole range of treatment options, and the anticipation of complications, such as those relating to corticosteroid treatment.

NOTE ADDED IN PROOF

One recent series describes promising results with methotrexate to treat patients with sarcoidosis refractory to treatment with corticosteroids or patients who could not tolerate the side effects of corticosteroids (Lower and Baughman 1990, 1995). Also, this same group reports good results treating neurosarcoidosis patients with methotrexate (Lower et al. 1997). In that series many patients who failed treatment with methotrexate were then treated with i.v. cyclophosphamide. Although the findings reported in these studies are promising and important to note, it is also important to recognize that these observations are based on unrandomized and uncontrolled trials (Newman et al. 1997). Further studies will be necessary to confirm these initially promising observations and to establish optimal dosing regimens.

Also, preliminary observations by one of the authors (BJS) have demonstrated in two patients with otherwise 'isolated' CNS sarcoidosis that whole-body 2-fluoro-2-deoxy-glucose (EDG) positron emission tomography (PET) can reveal sites of systemic inflammation by sarcoidosis that are not apparent with gallium scanning and other imaging techniques. This observation led to positive diagnostic biopsies in both patients that have spared these individuals invasive CNS diagnostic procedures. Unfortunately such FDG-PET results are not specific for sarcoidosis. For instance, lymphoma can have a similar appearance, and biopsy is still necessary for diagnosis (Larson 1994).

REFERENCES

AGBOGU, B.N., B.J. STERN, C. SEWELL and G. YANG: Therapeutic considerations in patients with refractory neurosarcoidosis. Arch. Neurol. 52 (1995) 875–879.

AHMAD, K., Y.H. KIM, A.R. SPITZER, A. GUPTA, I.H. HAN, A. HERSKOVIC and W.A. SAKR: Total nodal radiation in progressive sarcoidosis. Am. J. Clin. Oncol. 15 (1992) 311–313.

ANDO, D.G., J.P. LYNCH, III and J.C. FANTONE, III: Sarcoid myopathy with elevated creatine phosphokinase. Am. Rev. Respir. Dis. 131 (1985) 298–300.

ANDO, M.: Role of macrophages in the pathogenesis of sarcoidosis. Sarcoidosis 6 (1989) 76–78.

BIYOUDI-VOUENZE, R., J. CADRANEL, D. VALEYRE, B. MILLERON, A.J. HANCE and P. SOLER: Expression of 1,25(OH)2D3 receptors on alveolar lymphocytes from patients with pulmonary granulomatous diseases. Am. Rev. Respir. Dis. 143 (1991) 1376–1380.

BOCART, D., D. LECOSSIER, A. DE LASSENCE, D. VALEYRE, J.P. BATTESTI and A.J. HANCE: A search for mycobacterial DNA in granulomatous tissues from patients with sarcoidosis using the polymerase chain reaction. Am. Rev. Respir. Dis. 145 (1992) 1142–1148.

BOST, T.W., D.W.H. RICHES, B. SCHUMACHER, P.C. CARRE, T.Z. KHAN, J.A.B. MARTINEZ and L.S. NEWMAN: Alveolar macrophages from patients with beryllium disease and sarcoidosis express increased levels of mRNA for tumor necrosis factor-α and interleukin-6 but not interleukin-1. Am. Rev. Respir. Cell. Mol. Biol. 10 (1994) 506–513.

BROOKS, B.S., T.E. GAMMAL and G.D. HUNGERFORD: Radiologic evaluation of neurosarcoidosis: role of computed tomography. Am. J. Neuroradiol. 3 (1982) 513–521.

BROWN, M.M., A.J. THOMPSON, J.A. WEDZICHA and M. SWASH: Sarcoidosis presenting with stroke. Stroke 20 (1989) 400–405.

BRUST, J.C.M., R.S. RHEE, C.R. PLANK, M. NEWMARK, C.P. FELTON and L.D. LEWIS: Sarcoidosis, galactorrhea, and amenorrhea: 2 autopsy cases, 1 with Chiari–Fromel syndrome. Ann. Neurol. 2 (1977) 130–137.

BYRNE, J.V. and C.A. LAWTON: Meningeal sarcoidosis causing intracranial hypertension secondary to dural sinus thrombosis. Br. J. Radiol. 56 (1983) 755–757.

CAPELLAN, J.I., L. CUELLAR OLMEDO, J. MARTINEZ MARTIN, M. DEL MAR MARIN, M. GARCIA VILANUEVA, F. MARIN ZARZA and H. DE LA CALLE BLASCO: Intrasellar mass with hypopituitarism as a manifestation of sarcoidosis. J. Neurosurg. 73 (1990) 283–286.

CAPLAN, L., J. CORBETT, J. GOODWIN, C. THOMAS, D. SHENKER and N. SCHATZ: Neuro-ophthalmologic signs in the angiitic form of neurosarcoidosis. Neurology 33 (1983) 1130–1135.

CHALLENOR, Y.B., C.P. FELTON and J.C.M. BRUST: Peripheral nerve involvement in sarcoidosis: an electrodiagnostic study. J. Neurol. Neurosurg. Psychiatry 47 (1984) 1219–1222.

CHAN SEEM, C.P., G. NORFOLK and E.G. SPOKES: CSF angiotensin-converting enzyme in neurosarcoidosis. Lancet i (1985) 456–457.

CHAPELON, C., J.M. ZIZA, J.C. PIETTE, Y. LEVY, G. RAGUIN, B. WECHSLER, M.O. BITKER, O. BLETRY, D. LAPLANE and M.G. BOUSSER: Neurosarcoidosis: signs, course and treatment in 35 confirmed cases. Medicine (Baltimore) 69 (1990) 261–276.

CHENG, T.M., B.P. O'NEILL, B.W. SCHEITHAUER and D.G. PIEPGRAS: Chronic meningitis: the role of meningeal or cortical biopsy. Neurosurgery 34 (1994) 590–595.

COHEN, D.M. and R.A. REINHARDT: Systemic sarcoidosis presenting with Horner's syndrome and mandibular paresthesia. Oral Surg. 53 (1982) 577–581.

COLOVER, J.: Sarcoidosis with involvement of the nervous system. Brain 71 (1948) 451–475.

CORDINGLEY, G., C. NAVARRO and J.C.M. BRUST: Sarcoidosis presenting as senile dementia. Neurology 31 (1981) 1148–1151.

CORSE, A.M. and B.J. STERN: Neurosarcoidosis and stroke (letter). Stroke 21 (1990) 152–153.

DANIELE, R.: Abnormalities of the humoral immune system in sarcoidosis. In: B.L. Fanburg (Ed.), Sarcoidosis and Other Granulomatous Diseases of the Lung. New York, Marcel Dekker (1983) 225–242.

DANIELE, R.P., M.D. ROSSMAN, J.A. KERN and J.A. ELIAS: Pathogenesis of sarcoidosis. State of the art. Chest 89, Suppl. (1986) 174S–177S.

DELANEY, P.: Neurologic manifestations in sarcoidosis: review of the literature, with a report of 23 cases. Ann. Intern. Med. 87 (1977) 336–345.

DELANEY, P.: Seizures in sarcoidosis: a poor prognosis. Ann. Neurol. 7 (1980) 494.

DEREMEE, R.: Chest roentgenology of sarcoidosis. In: J. Lieberman (Ed.), Sarcoidosis. Orlando, FL, Grune and Stratton (1985a) 117–135.

DEREMEE, R.A.: The treatment of sarcoidosis. In: J. Lieberman (Ed.), Sarcoidosis. Orlando, FL, Grune and Stratton (1985b) 195–204.

DESAI, S.G. and M.R. SIMON: Epidemiology of sarcoidosis. In: J. Lieberman (Ed.), Sarcoidosis. Orlando, FL, Grune and Stratton (1985) 25–37.

DOUGLAS, A.C. and A.F.J. MALONEY: Sarcoidosis of the central nervous system. J. Neurol. Neurosurg. Psychiatry 36 (1973) 1024–1033.

EL-KASSIMI, F.A., M. ASHOUR and R. VIJAYARAGHAVAN: Sarcoidosis presenting as recurrent left laryngeal nerve palsy. Thorax 45 (1990) 565–566.

FANBURG, B.L.: Treatment of sarcoidosis. In: B.L. Fanburg (Ed.), Sarcoidosis and Other Granulomatous Diseases of the Lung. New York, Marcel Dekker (1983) 381–388.

FAZEL, S.B., S.E.M. HOWIE, A.S. KRAJEWSKI, D. LAMB and P. SOLER: B lymphocyte accumulations in human pulmonary sarcoidosis. Thorax 47 (1992) 964–967.

FIDLER, H.M., G.A. ROOK, N.M. JOHNSON and J. MCFADDEN: *Mycobacterium tuberculosis* DNA in tissue affected by sarcoidosis. Br. Med. J. (1993) 546–549.

FIREMAN, E.M. and M.R. TOPILSKY: Sarcoidosis: an organized pattern of reaction from immunology to therapy. Immunol. Today 15 (1994) 199–201.

FORMAN, J.D., R.F. SILVER, J.T. KLEIN, E.J. BRITT, P.P. SCOTT, S.A. SCHONFELD, C.J. JOHNS and D.R. MOLLER: T cell receptor variable beta-gene expression in the normal lung and in active pulmonary sarcoidosis. Chest 103, Suppl. 2 (1993) 78S.

FORRESTER, J.M., L.S. NEWMAN, Y. WANG, T.E.J. KING and B.L. KOTZIN: Clonal expansion of lung V delta 1+ T cells in pulmonary sarcoidosis. J. Clin. Invest. 91 (1993) 292–300.

GALASSI, G., M. GIBERTONI and A. MANCINI: Sarcoidosis of the peripheral nerve: clinical, electrophysiological and histological study of two cases. Eur. Neurol. 23 (1984) 459–465.

GELWAIN, M.J., R.I. KELLEN and R.M. BURDE: Sarcoidosis of the anterior visual pathway: successes and failures. J. Neurol. Neurosurg. Psychiatry 51 (1988) 1473–1480.

GERDES, J., E. RICHTER, S. RUSCH-GERDES, V. GREINERT, J. GALLE, M. SCHLAAK, H.D. FLAD and H. MAGNUSSEN: Mycobacterial nucleic acids in sarcoid lesion. Lancet 339 (1992) 1536–1537.

GHOSSEIN, R.A., D.G. ROSS, R.N. SALOMON and A.R. RABSON: A search for mycobacterial DNA in sarcoidosis using the polymerase chain reaction. Am. J. Clin. Pathol. 101 (1994) 733–737.

GRAHAM, E.M., C.J.K. ELLIS and M.D. SANDERS: Optic neuropathy in sarcoidosis. J. Neurol. Neurosurg. Psychiatry 49 (1986) 756–763.

GRUNEWALD, J., O. OLERUP, U. PERSSON, M.B. OHERN, H. WIGZELL and A. EKLUND: T-cell receptor variable region gene usage by CD4+ and CD8+ T cells in bronchoalveolar lavage fluid and peripheral blood of sarcoidosis patients. Proc. Natl. Acad. Sci. 91 (1994) 4965–4969.

HEERFORDT, C.: Über eine 'Febrid uveo-parotidea subchronica' an den Glandula parotis und der Uvea des Auges lokalisiert und haufig mit Paresencerebrospinaler Nerven kompliziert. Albrecht v. Graefe's Arch. Ophthalmol. 70 (1909) 254–273.

HEFFERNAN, A., M. CULLEN and R. TOWERS: Sarcoidosis of the hypothalamus. Hormones 2 (1971) 1–12.

HENKIND, P. and M.B. GOTTLIEB: Bilateral internal ophthalmoplegia in a patient with sarcoidosis. Br. J. Ophthalmol. 57 (1973) 792.

HERRING, A. and H. URICH: Sarcoidosis of the central nervous system. J. Neurol. Sci. 9 (1969) 405–422.

HEWLETT, R.H. and B. BROWNELL: Granulomatous myopathy: its relationship to sarcoidosis and polymyositis. J. Neurol. Neurosurg. Psychiatry 38 (1975) 1090–1099.

HO, S.U., R.A. BERENBERG and K.S. KIM: Sarcoid encephalopathy with diffuse inflammation and focal hydrocephalus shown by sequential CT. Neurology 29 (1979) 1161–1165.

HOL, B.E.A., R.Q. HINTZEN, R.A.W. VAN LIER, C. ALBERTS, T.A. OUT and H.M. JANSEN: Soluble and cellular markers of T cell activation in patients with pulmonary sarcoidosis. Am. Rev. Respir. Dis. 148 (1993) 643–649.

HUNNINGHAKE, G.W. and R.G. CRYSTAL: Pulmonary sarcoidosis: a disorder mediated by excess helper T-lymphocyte activity at sites of disease activity. N. Engl. J. Med. 305 (1981) 429–434.

HYBELS, R.L. and D.H. RICE: Neuro-otologic manifestations of sarcoidosis. Laryngoscope 86 (1976) 1873–1878.

ISMAIL, F., J.L. MILLER and S.E. KAHN: Hypothalamic–pituitary sarcoidosis. S. Afr. Med. J. 97 (1985) 139–142.

ISRAEL, H.L.: Diagnostic value of the Kveim reaction. In: B.L. Fanburg (Ed.), Sarcoidosis and Other Granulomatous Diseases of the Lung. New York, Marcel Dekker (1983) 273–286.

ISRAEL, H.L. and Y.P. KATARIA: Clinical aspects of sarcoidosis. In: J. Lieberman (Ed.), Sarcoidosis. Orlando, FL, Grune and Stratton (1985) 7–23.

JAHRSDOERFER, R.A., E.G. THOMPSON, M.M. JOHNS and R.W. CANTRELL: Sarcoidosis and fluctuating hearing loss. Ann. Otol. Rhinol. Laryngol. 90 (1981) 161–163.

JAMES, D.G.: Genetics and familial sarcoidosis. In: B.L. Fanburg (Ed.), Sarcoidosis and Other Granulomatous Diseases of the Lung. New York, Marcel Dekker (1983) 135–146.

JAMES, D.G.: State-of-the art lecture: etiology of sarcoidosis. In: O.P. Sharma (Ed.), Proc. 14th Int. Conference on Sarcoidosis and Other Granulomatous Disorders. Milan, Edizioni Bongraf (1994) 43–58.

JAMES, D.G. and W.J. WILLIAMS: Sarcoidosis and Other Granulomatous Disorders. Philadelphia, PA, W.B. Saunders (1985).

JAMES, D.G., J. TURIAF, Y. HOSODA, W.J. WILLIAMS, H.L.

ISRAEL, A.C. DOUGLAS and L.E. SILTZBACH: Description of sarcoidosis: report of the subcommittee on classification and definition. Ann. N.Y. Acad. Sci. 278 (1976) 742.

JOYCE-BRADY, M.: Tastes great, less filling. The debate about mycobacteria and sarcoidosis. Am. Rev. Respir. Dis. 145 (1992) 986–987.

JUNGER, S.S., B.J. STERN and S.R. LEVINE: Intramedullary spinal sarcoidosis: clinical and magnetic resonance imaging characteristics. Neurology 43 (1993) 333–337.

JUOZEVICIUS, J.L. and R.I. RYNES: Increased helper/suppressor T-lymphocyte ratio in the cerebrospinal fluid of a patient with neurosarcoidosis. Ann. Intern. Med. 104 (1986) 807–808.

KAHAN, B.D.: Cyclosporine. N. Engl. J. Med. 321 (1989) 1781–1782.

KATARIA, Y.P.: Immunology of sarcoidosis. In: J. Lieberman (Ed.), Sarcoidosis. Orlando, FL, Grune and Stratton (1985) 39–63.

KATZ, S.: Clinical presentation and natural history of sarcoidosis. In: B.L. Fanburg (Ed.), Sarcoidosis and Other Granulomatous Diseases of the Lung. New York, Marcel Dekker (1983) 3–36.

KENDALL, B.E. and G.L.V. TATLER: Radiological findings in neurosarcoidosis. Br. J. Radiol. 51 (1978) 81–92.

KERN, J. and R. DANIELE: Bronchoalveolar lavage in sarcoidosis. In: J. Lieberman (Ed.), Sarcoidosis. Orlando, FL, Grune and Stratton (1985) 179–188.

KIRKHAM, T.H. and L.B. KLINE: Monocular elevator paresis, Argyll Robertson pupils and sarcoidosis. Can. J. Ophthalmol. 11 (1976) 330–335.

KIRKLAND, J.L., D.J. PEARSON, C. GODDARD and I. DAVIES: Polyuria and inappropriate secretion of arginine vasopressin in hypothalamic sarcoidosis. J. Clin. Endocrinol. Metab. 56 (1983) 269–272.

KLEIN, J.T., T.D. HORN, J.D. FORMAN, R.F. SILVER, A.S. TEIRSTEIN and D.R. MOLLER: Selection of oligoclonal V-beta-specific T cells in the intradermal response to Kveim–Siltzbach reagent in individuals with sarcoidosis. J. Immunol. 154 (1995) 1450–1460.

KRUMHOLZ, A., B.J. STERN and E.G. STERN: Clinical implications of seizures in neurosarcoidosis. Arch. Neurol. 48 (1991) 842–844.

LARSON, S.M.: Cancer or inflammation? A holy grail for nuclear medicine. J. Nucl. Med. 35 (1994) 1653–1654.

LAWRENCE, W.P., T.E. GAMMAL and W.H. POOL, JR.: Radiological manifestations of neurosarcoidosis: report of three cases and review of literature. Clin. Radiol. 25 (1974) 343–348.

LEIBERMAN, J.: Angiotensin-converting enzyme (ACE) and serum lysozyme in sarcoidosis. In: J. Lieberman (Ed.), Sarcoidosis. Orlando, FL, Grune and Stratton (1985) 145–159.

LEIBERMAN, J., Y.P. KATARIA and YOUNG, JR.: Historical

perspective of sarcoidosis. In: J. Lieberman (Ed.), Sarcoidosis. Orlando, FL, Grune and Stratton (1985) 1–5.

LIEDTKE, W., A. MAY, R. DUX, P.M. FAUSTMANN and C.W. ZIMMERMANN: Mycobacterium tuberculosis polymerase chain reaction findings in neurosarcoidosis. J. Neurol. Sci. 120 (1993) 118–119.

LIPTON, J.M., J. KIRKPATRICK and R.N. ROSENBERG: Hypothermia and persisting capacity to develop fever. Arch. Neurol. 34 (1977) 498–504.

LOWER, E.E. and R.P. BAUGHMAN: The use of low dose methotrexate in refractory sarcoidosis. Am. J. Med. Sci. 299 (1990) 153–157.

LOWER, E.E. and R.P. BAUGHMAN: Prolonged use of methotrexate for sarcoidosis. Arch. Intern. Med. 155 (1995) 846–851.

LOWER, E.E., J.P. BRODERICK, T.G. BROTT and R.P. BAUGHMAN: Diagnosis and management of neurological sarcoidosis. Arch. Intern. Med. 157 (1997) 1864–1868.

LUKE, R.A., B.J. STERN, A. KRUMHOLZ and C.J. JOHNS: Neurosarcoidosis: the long-term clinical course. Neurology 37 (1987) 461–463.

LUKIN, R.R., A.A. CHAMBERS and M. SOLEIMANPOUR: Outlet obstruction of the fourth ventricle in sarcoidosis. Neuroradiology 10 (1975) 65.

MANZ, H.J.: Pathobiology of neurosarcoidosis and clinicopathologic correlation. Can. J. Neurol. Sci. 10 (1983) 50–55.

MELIS, M., M. GIOMARKAI, E. PACE, G. MALIZIA and M. SPATAFORA: Increased expression of leukocyte function associated antigen-1 (LFA-1) and intercellular adhesion molecule-1 (ICAM-1) by alveolar macrophages of patients with pulmonary sarcoidosis. Chest 100 (1991) 910–916.

MEYER, J., J. FOLEY and D. CAMPAGNA-PINTO: Granulomatous angiitis of the meninges in sarcoidosis. Arch. Neurol. 69 (1953) 587–600.

MIRFAKHRAEE, M., M.J. CROFFORD, F.C. GUINTO, JR., H.J. NAUTA and V.W. WEEDN: Virchow–Robin space: a path of spread in neurosarcoidosis. Radiology 158 (1986) 715–720.

MISSLER, U., M. MACK and G. NOWACK: Pituitary sarcoidosis. Klin. Wochenschr. 68 (1990) 342–345.

MITCHELL, J.C., J.L. TURK and D.N. MITCHELL: Detection of mycobacterial rRNA in sarcoidosis with liquid-phase hybridisation. Lancet 339 (1992) 1015–1017.

MITCHELL, J.D., P.L. YAP, L.A. MILNE, P.J. LACHMANN and B. PENTLAND: Immunological studies on the cerebrospinal fluid in neurological sarcoidosis. J. Neuroimmunol. 7 (1985) 249–253.

MOORE, P.M. and T.R. CUPPS: Neurological complications of vasculitis. Ann. Neurol. 14 (1983) 155–167.

MYATT, N., G. COGHILL, K. MORRISON, D. JONES and I.A. CREE: Detection of tumor necrosis factor in sarcoidosis and tuberculosis granulomas using situ hybridization. J. Clin. Pathol. 47 (1994) 423–426.

NEMNI, R., G. GALASSI and M. COHEN: Symmetric sarcoid polyneuropathy: analysis of a sural nerve biopsy. Neurology 31 (1981) 1217–1223.

NEWMAN, L.S., S.R. CECILE and L.A. MAIER: Sarcoidosis. N. Engl. J. Med. 336 (1997) 1224–1234.

OBENAUF, C.D., H.E. SHAW and C.F. SYDNOR: Sarcoidosis and its ophthalmic manifestations. Am. J. Ophthalmol. 86 (1978) 648–655.

OKSANEN, V.: Neurosarcoidosis: clinical presentations and course in 50 patients. Acta Neurol. Scand. 73 (1986) 283–290.

OKSANEN, V. and T. SALMI: Visual and auditory evoked potentials in the early diagnosis and follow-up of neurosarcoidosis. Acta Neurol. Scand. 74 (1986) 38–42.

OKSANEN, V., F. FYHRQUIST, C. GRONHAGEN-RISKA and H. SOMER: CSF angiotensin-converting enzyme in neurosarcoidosis. Lancet i (1985a) 1050–1051.

OKSANEN, V., F. FYHRQUIST, H. SOMER and C. GRONHAGEN-RISKA: Angiotensin converting enzyme in cerebrospinal fluid: a new assay. Neurology 35 (1985b) 1220–1223.

OLEK, M., R. JAITLY and A. KUTA: Sarcoidosis and spinal cord atrophy. Neurologist 1 (1995) 240–243.

OSENBACH, R.K., B. BLUMENKOPF, H. RAMIREZ, JR. and J. GUTIERREZ: Meningeal neurosarcoidosis mimicking convexity en-plaque meningioma. Surg. Neurol. 26 (1986) 387–390.

OTAKE, S., T. BANNO and S. OHABA: Muscular sarcoidosis: findings at MR imaging. Radiology 176 (1990) 145–148.

PEEPLES, D.M., B.J. STERN, V. JIJI and K.S. SAHANI: Germ cell tumors masquerading as central nervous system sarcoidosis. Arch. Neurol. 48 (1991) 554.

PENTLAND, B., J.D. MITCHELL, R.E. CULL and M.J. FORD: Central nervous system sarcoidosis. Quart. J. Med. 56 (1985) 457–465.

PLOTKIN, G.R. and B.R. PATEL: Neurosarcoidosis presenting as chronic lymphocytic meningitis. Pa. Med. 89 (1986) 36–37.

POOLE, C.J.M.: Argylle Robertson pupils due to neurosarcoidosis: evidence for site of lesion. Br. Med. J. 289 (1984) 356.

POPPER, H.H., E. WINTER and G. HOFLER: DNA of Mycobacterium tuberculosis in formalin-fixed, paraffin-embedded tissue in tuberculosis and sarcoidosis detected by polymerase chain reaction. Am. J. Clin. Pathol. 101 (1994) 738–741.

PRIOR, C. and P.L. HASLAM: In vivo levels and in vitro production of interferon-gamma in fibrosing interstitial lung diseases. Clin. Exp. Immunol. 88 (1992) 280–287.

PUERINGER, R.J., D.A. SCHWARTZ, C.S. DAYTON, S.R. GILBERT and G.W. HUNNINGHAKE: The relationship between alveolar macrophage TNF, IL-1, and PGE2 release, alveolitis, and disease severity in sarcoidosis. Chest 103 (1993) 832–838.

RAULF, M., V. LIEBERS, C. STEPPERT and X. BAUR: Increased gamma/delta-positive T-cells in blood and broncho-alveolar lavage of patients with sarcoidosis and hypersensitivity pneumonitis. Eur. Respir. J. 7 (1994) 140–147.

ROCKLIN, R.: Cell-mediated immunity in sarcoidosis. In: B.L. Fanburg (Ed.), Sarcoidosis and Other Granulomatous Diseases of the Lung. New York, Marcel Dekker (1983) 203–224.

RODAN, B.A. and C.E. PUTMAN: Radiological alterations in sarcoidosis. In: B.L. Fanburg (Ed.), Sarcoidosis and Other Granulomatous Diseases of the Lung. New York, Marcel Dekker (1983) 37–75.

SABOOR, S.A., N.M. JOHNSON and J. MCFADDEN: Detection of mycobacterial DNA in sarcoidosis and tuberculosis with polymerase chain reaction. Lancet 339 (1992) 1012–1015.

SAVOLAINE, E.R. and P.J. SCHLEMBACH: Gallium scan diagnosis of sarcoidosis in the presence of equivocal radiographic and CT findings: value of lacrimal gland biopsy. Clin. Nucl. Med. 15 (1990) 198–199.

SCHABERG, T., M. RA, H. STEPHAN and H. LODE: Increased number of alveolar macrophages expressing surface molecules of the CD11/CD18 family in sarcoidosis and idiopathic pulmonary fibrosis is related to the production of superoxide anions by these cells. Am. Rev. Respir. Dis. 147 (1993) 1507–1513.

SCHLITT, M., E.R. DUVALL, J. BONNIN and R.B. MORAWETZ: Neurosarcoidosis causing ventricular loculation, hydrocephalus, and death. Surg. Neurol. 26 (1986) 67–71.

SCHNEYER, C.R. Abnormal bone metabolism in a neurological setting. Neurologist 1 (1995) 259–272.

SEMENZATO, G. and C. AGOSTINI: State-of-the art lecture: immunopathologies of sarcoidosis. In: O.P. Sharma (Ed.), Proc. 14th Int. Conference on Sarcoidosis and Other Granulomatous Disorders. Milan, Edizioni Bongraf (1994) 59–71.

SEMENZATO, G., R. ZAMBELLO, L. TRENTIN and C. AGOSTINI: Cellular immunity in sarcoidosis and hypersensitivity pneumonitis. Chest 103, Suppl. (1993) 139S–143S.

SETHI, K.D., T. EL GAMMAL, B.R. PATEL and T.R. SWIFT: Dural sarcoidosis presenting with transient neurologic symptoms. Arch. Neurol. 43 (1986) 595–597.

SHAKOOR, Z. and A.S. HAMBLIN: Increased CD11/CD18 expression on peripheral blood leucocytes of patients with sarcoidosis. Clin. Exp. Immunol. 90 (1992) 99–105.

SHARMA, O.P.: Sarcoidosis: Clinical Management. London, Butterworth (1984).

SHERMAN, J.L. and B.J. STERN: Sarcoidosis of the CNS: comparison of unenhanced and enhanced MR images. Am. J. Neuroradiol. 11 (1990) 915–923.

SMITH, D.L. and R.D. DESHAZO: Integrins, macrophages, and sarcoidosis. Chest 102 (1992) 659–660.

SOUCEK, D., C. PRIOR and G. LOEF: Successful treatment of spinal sarcoidosis by high-dose intravenous methylprednisolone. Clin. Neuropharmacol. 16 (1993) 464–467.

STEFFEN, M., J. PETERSEN, M. OLDIGS, A. KARMEIER, H. MAGNUSSEN, H.G. THIELE and A. RAEDLER: Increased secretion of tumor necrosis factor-alpha, interleukin-1-beta, and interleukin-6 by alveolar macrophages from patients with sarcoidosis. J. Allergy Clin. Immunol. 91 (1993) 939–948.

STERN, B.J., A. KRUMHOLZ, C. JOHNS, P. SCOTT and J. NISSIM: Sarcoidosis and its neurological manifestations. Arch. Neurol. 42 (1985) 909–917.

STERN, B.J., D.E. GRIFFIN, R.A. LUKE, A. KRUMHOLZ and C.J. JOHNS: Neurosarcoidosis: cerebrospinal fluid lymphocyte subpopulations. Neurology 37 (1987) 878–881.

STERN, B.J., S.A. SCHONFELD, C. SEWELL, A. KRUMHOLZ, P. SCOTT and G. BELENDIUK: The treatment of neurosarcoidosis with cyclosporine. Arch. Neurol. 49 (1992) 1065–1072.

STJERNBERG, N., S. CAJANDER and H. TRUEDSSON: Muscle involvement in sarcoidosis. Acta Med. Scand. 209 (1981) 213–216.

STRELETZ, L.J., R.A. CHAMBERS and S.H. BAE: Visual evoked potentials in sarcoidosis. Neurology 31 (1981) 1545–1549.

STRIZ, I., Y.M. WANG, O. KALAYCIOGLU and U. COSTABEL: Expression of alveolar macrophage adhesion molecules in pulmonary sarcoidosis. Chest 102 (1992) 882–886.

STUART, C.A., F.A. NEELON and H.E. LEBOVITZ: Hypothalamic insufficiency: the cause of hypopituitarism in sarcoidosis. Ann. Intern. Med. 88 (1978) 589–594.

TAMURA, N., D.R. MOLLER, B. BALBI and R.G. CRYSTAL: Differential usage of the T-cell antigen receptor-chain constant region C1 element by lung T-lymphocytes of patients with pulmonary sarcoidosis. Am. Rev. Respir. Dis. 143 (1991) 635–639.

TANOUE, L.T., R. ZITNIK and J.A. ELIAS: Systemic sarcoidosis. In: G.L. Baum and E. Wolinsky (Eds.), Textbook of Pulmonary Diseases. 1. Boston, MA, Little, Brown (1994) 745–774.

TEIRSTEIN, A.S. and M. LESSER: Worldwide distribution and epidemiology of sarcoidosis. In: B.L. Fanburg (Ed.), Sarcoidosis and Other Granulomatous Diseases of the Lung. New York, Marcel Dekker (1983) 101–134.

THOMAS, P.D. and G.W. HUNNINGHAKE: Current concepts of the pathogenesis of sarcoidosis. Am. Rev. Respir. Dis. 135 (1987) 747–760.

THOMPSON, C. and S. CHECKLEY: Short term memory deficit in a patient with cerebral sarcoidosis. Br. J. Psychiatry 139 (1981) 160–161.

TOBIAS, J.K., S.M. SANTIAGO and A.J. WILLIAMS: Sarcoidosis as a cause of left recurrent laryngeal nerve palsy. Arch. Otolaryngol. Head Neck Surg. 116 (1990) 971–972.

TROMBLEY, I.K., S.S. MIRRA and M.L. MILES: An electron microscopic study of central nervous system sarcoidosis. Ultrastruct. Pathol. 2 (1981) 257–267.

TURNER-WARWICK, M., P.L. HASLEM and W. MCALLISTER: Do measurements of bronchoalveolar lymphocytes and neutrophils, serum angiotensin-converting enzyme and gallium uptake help the clinician to treat patients with sarcoidosis? Ann. N.Y. Acad. Sci. 465 (1986) 387–394.

URICH, H.: Neurosarcoidosis or granulomatous angiitis: a problem of definition. Mt. Sinai J. Med. 44 (1977) 718–725.

WAXMAN, J.S. and J.H. SHER: The spectrum of central nervous system sarcoidosis: a clinical and pathologic study. Mt. Sinai J. Med. 46 (1979) 309–317.

WEINBERG, S., H. BENNETT and I. WEINSTOCK: CNS manifestations of sarcoidosis in children. Clin. Pediatr. 22 (1983) 447–481.

WEISSLER, J.C.: Southwestern internal medicine conference. Sarcoidosis: immunology and clinical management. Am. J. Med. Sci. 307 (1994) 233–245.

WIEDERHOLT, W.C. and R.G. SIEKERT: Neurological manifestations of sarcoidosis. Neurology 15 (1965) 1147–1154.

YOUNGER, D.S., A.P. HAYS, J.C. BRUST and L.P. ROWLAND: Granulomatous angiitis of the brain. An inflammatory reaction of diverse etiology. Arch. Neurol. 45 (1988) 514–518.

ZUNIGA, G., A.H. ROPPER and J. FRANK: Sarcoid peripheral neuropathy. Neurology 41 (1991) 1558–1561.

Handbook of Clinical Neurology, Vol. 27 (71): Systemic Diseases, Part III
M.J. Aminoff and C.G. Goetz, editors

CHAPTER 20

Neurological complications of amyloidosis

DIANA M. ESCOLAR and JOHN J. KELLY

Department of Neurology, The George Washington University Medical Center, Washington, DC, U.S.A.

Amyloidosis results from a variety of different disease processes leading to the deposition of the typical β-pleated sheet fibrils. These fibrils are formed by components of various proteins, which are the result of overproduction or abnormal degradation of circulating precursor proteins or of the production of abnormal precursor proteins due to genetic abnormalities.

Much has been learned about this waxy, apparently amorphous, eosinophilic material which was named amyloid by Virchow in 1853. He believed it was composed of polysaccharides (starch-like or cellulose-like). All varieties of amyloid have similar physical properties. Amyloid stains pink with Congo Red and has a distinctive apple-green birefringence under polarized light (Fig. 1). It also stains metachromatically with methyl violet and crystal violet (Elghetany and Saleem 1988). Electron microscopy, the most specific diagnostic method, shows amyloid to be typically composed of felt-like arrays of rigid, linear, non-branching, aggregated fibrils with a hollow core (Cohen and Calkins 1959; Cohen 1967; Glenner 1980a, b) which measure 7.5–10 nm and have an indefinite length stretching beyond the EM field. The fibrils are formed by polypeptide chains deposited perpendicular to the axis of the fibril, forming a meridional or cross-β-pleated sheet (Fig. 2). All amyloid fibrils (ALs) have this specific twisted β-pleated sheet configuration (Glenner 1980a, b)

which is responsible for the optical properties of the ALs and for the characteristic resistance of the fibrils to dissolution by physiologic solutions and to proteolytic digestion (Glenner 1980a, b; Elghetany and Saleem 1988)

Another protein, amyloid P (for pentagonal component), is ubiquitous in all systemic forms of amyloidosis, but distinct from AL proteins. Amyloid P constitutes about 10% of the amyloid material (Glenner 1980a, b) and consists of a parallel pair of pentagonally structured subunits, each approximately 23,000 Da. Its identical counterpart in serum, serum amyloid P (SAP) component, produced by the liver, is part of a family of proteins, called pentraxins, that include C-reactive proteins and hamster female protein. SAP has been found in all vertebrate species examined. Its stable evolutionary conservation suggests that this protein might have an important, as yet unknown, physiologic role (Cohen and Skinner 1990). SAP serum levels are normal in patients with primary and familial amyloidosis, but are often elevated in those with secondary (AA) amyloidosis. SAP binds in a calcium-dependent fashion to the ALs, though this has never been linked pathogenetically to the amyloidogenic properties of the amyloid-precursor proteins. Antibodies against the amyloid P component are very useful to demonstrate the presence of amyloid, as is the use of scintigraphy with ^{123}I-labeled SAP component (Hawkins et al. 1990).

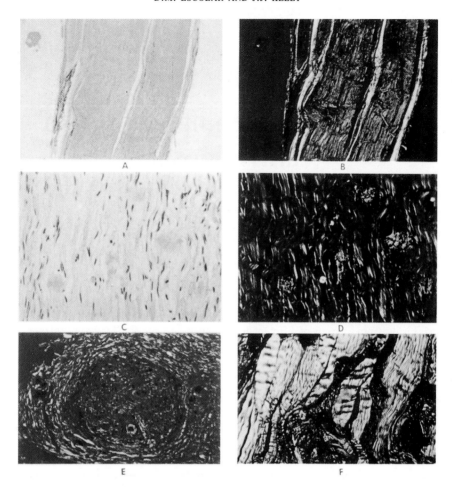

Fig. 1. Four patterns of amyloid deposition in nerve. A: diffuse deposition of amyloid throughout the nerve seen with Congo Red stain. B: typical green birefringence under polarized light. C: globular deposits of amyloid displacing relatively normal nerve fibers. D: section C viewed under polarized light. E: vascular deposition of amyloid. Note that only amyloid stained with Congo Red gives a green birefringence under polarized light, despite collagen and fibrin being birefringent as well. F: amyloid deposition causing compression neuropathy at the carpal tunnel. (From Cohen and Rubinow 1984, reprinted by permission.)

MOLECULAR CLASSIFICATION OF AMYLOIDOSIS

In the past, the classification of the amyloidoses was based on their clinical presentation, organ involvement and the geographical distribution of the familial cases (Glenner 1980a, b). The most recent classification is much simpler and is based on the molecular composition of the amyloid deposits (Haan and Peters 1994), which is determined by immunohistochemical testing. Clinically, however, it is still useful to divide these disorders into localized and systemic forms (Table 1). In localized amyloidosis, amyloid deposition can be seen in the brain (hereditary cerebral hemorrhage with amyloidosis of the Icelandic and Dutch types and Alzheimer's disease), or in the endocrine system (medullary carcinoma of the thyroid). Systemic amyloidosis includes familial amyloid polyneuropathy, primary and myeloma associated amyloidosis, secondary amyloidosis, senile amyloidosis and hereditary renal amyloidosis. The focus of this chapter is those systemic forms of amyloidosis that commonly affect the nervous system, namely primary and myeloma associated amyloidosis and familial amyloid polyneuropathies (Staunton 1991).

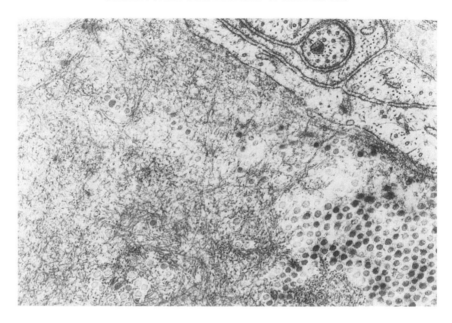

Fig. 2. Ultrastructural aspect of amyloid in neuropathy: entangled ALs, some intermingled with collagen fibers. (From Vital and Vallet 1987, reprinted by permission.)

TABLE 1

Classification of amyloidosis.

Type	Amyloid protein/gene defect	Other nomenclature
Primary amyloidosis	Ig L-chains (κ, λ)	AL
Secondary amyloidosis	protein A	AA

Hereditary amyloidosis
  With peripheral nervous
  system manifestations
	TTR variants
	chromosome 18: point mutations,
	most common: TTR Met-30
		hereditary amyloid polyneuropathy:
		a – upper limbs: FAP II: Indiana (Swiss origin)
		b – lower limbs: FAP I: Portuguese, Japanese,
		  Swedish, USA (Illinois (German),
		  West Virginia), English, Jewish

	gelsolin chromosome 9: point mutation	cranial neuropathy and lattice corneal dystrophy: FAP IV (Finland)
	apolipoprotein 1 chromosome 11: point mutation	polyneuropathy (lower limb), nephropathy, gastric ulcer

With CNS
  manifestations
	cystatin C	Icelandic type of hereditary cerebral amyloid
		  angiopathy

Nephrotic	protein A	familiar Mediterranean fever Muckle–Wells
Cardiopathic	TTR	Danish, Appalachian
Senile amyloidosis	β-amyloid	$A\beta_1$: brain: Alzheimer's disease
	TTR	AS_{c1}: senile cardiac amyloid
	atrial natriuretic peptide	IAA: isolated atrial amyloid
Endocrine	calcitonin	AE_t: medullary thyroid carcinoma
	islet amyloid polypeptide	AE_p: islet of Langerhans
Dialysis arthropathy	β_2-microglobulin	$A\beta_2M$

PRIMARY SYSTEMIC AMYLOIDOSIS

Primary systemic amyloidosis is a rare disease (incidence 8.9/million/year) that affects older people. The median age at diagnosis is about 63 and almost two-thirds of the patients are males (Kyle and Bayrd 1975; Kyle and Greipp 1983; Kyle and Dyck 1993). No underlying condition is found in primary amyloidosis except for a non-malignant monoclonal amyloidogenic light chain immunoglobulin (Ig) which leads to amyloid deposition in various tissues. Weakness, fatigue and weight loss are the most frequent initial symptoms. Congestive heart failure (CHF) and nephrotic syndrome typically dominate the clinical picture and are usually responsible for the patient's death. Peripheral neuropathy (PN) is the most common neurological manifestation, presenting as the initial symptom in about 15% of these patients (Kelly et al. 1979; Kyle and Greipp 1983).

Multiple myeloma (MM) has an incidence of 3/100,000/year. Amyloid deposition occurs in approximately 20% of these patients, depending on the type of L-chain that is produced. In MM, neurological involvement is most often manifested by root pain with compression of the spinal cord and cauda equina, occurring in 5% of the patients (Kyle 1992). PN associated with MM is uncommon, and only about 30–40% of these PNs are related to amyloidosis (Kelly et al. 1981). When PN occurs in the setting of myeloma due to amyloid deposition, it is clinically indistinguishable from that in primary amyloidosis and therefore the clinical features will be discussed together.

Amyloid fibril (AL) composition

In primary and myeloma associated amyloidosis, the ALs are composed of monoclonal light chain Ig, which are glycoproteins produced by plasma cells. The Ig molecule is composed of two heavy (H) and two light (L) polypeptide chains. The L-chains are either lambda (λ) or kappa (κ). Each L-chain has a variable (V_L) and a constant (C_L) sequence. The major protein component in primary amyloidosis is a monoclonal λ or κ type V_L) or a complete L-chain Ig fragment (Glenner and Harada 1971; Glenner 1980a, b; Kyle and Greipp 1983). L-chains of the λ type are more frequently associated with amyloidosis than the κ class (2:1) despite the fact that in myeloma the secreted monoclonal protein is most frequently of the κ

class. λ L-chains have a β-pleated configuration, whereas κ L-chains acquire this configuration only with heat precipitation, which may explain the greater frequency of λ L-chains in amyloidosis (Kyle and Bayrd 1975). There is a greater than expected prevalence of the λ_{IV} subclass in primary amyloidosis (Habermann and Montenegro 1980; Solomon et al. 1982). Only 15–20% of the L-chains (especially λ) are amyloidogenic. The exact property of these proteins that increases their affinity for tissue deposition is not known. It is also unclear if the amyloidogenic Ig fragment is the result of aberrant de novo production of L-chains or a defect in proteolysis or both. It has been postulated that the monoclonal L-chains are absorbed and abnormally processed by macrophages to produce the ALs which are then resistant to dissolution and proteolysis (Durie et al. 1982; Vital and Vital 1984).

Peripheral neuropathy (PN)

Clinical findings

PN is the presenting manifestation of systemic amyloidosis or MM in about 15% of patients (Kelly et al. 1979; Kyle and Greipp 1983; Duston et al. 1989). PN can be a major clinical component, or be obscured by systemic disease and diagnosed only as a result of neurological examination or electromyography (EMG) when the patients present with other complaints. In the first group, time elapsed from the beginning of the symptoms to the established diagnosis of amyloidosis is much longer compared to the second group (Kelly et al. 1979; Duston et al. 1989). Part of the delay in the diagnosis is that patients generally do not seek medical attention as rapidly for paresthesias as for symptoms of cardiac failure, nephrotic syndrome or severe orthostasis.

In either case, the neuropathy is usually distal, symmetrical and progressive, presenting mainly with sensory symptoms. There is often unequivocal sensory dissociation, with temperature and pain sensations affected to a greater degree and earlier than proprioception. The typical patient presents with painful dysesthesias of the distal legs. The arms may become involved soon after. Symptoms of median mononeuropathy due to compression in the carpal tunnel (carpal tunnel syndrome – CTS) can precede or accompany the generalized symptoms. Distal weakness and muscle atrophy, although typically present, are not as prominent and rarely become severe (Kelly et al. 1979). In a recent review (Kyle and Gertz 1990), CTS was found to be the presenting feature of primary amyloidosis in 24% of the patients, almost the same frequency as cardiac failure, and slightly more frequent than PN and autonomic dysfunction. Nephrotic syndrome, however, was the most common presentation (32%).

Autonomic failure is manifested by orthostatic hypotension with fainting, gastrointestinal disturbances, bladder dysfunction and impotence (Gaan et al. 1972; Kelly et al. 1979). These symptoms manifest early in amyloidosis. Postural hypotension with fainting is the most common, and at times most disabling sign of autonomic involvement.

Examination of the patient presenting with symptoms of PN usually discloses symmetrical decrease of temperature and pain sensation in the distal lower extremities, with often paradoxical retention of light touch and proprioception. Some patients, however, have equal involvement of all sensory modalities or even predominant large fiber sensory involvement (Kelly et al. 1979). Mild to moderate distal weakness and atrophy of intrinsic foot muscles are usually present. Deep tendon reflexes are usually decreased or absent distally. Skin trophic changes, with reduced sweating, are often noted, as well as orthostatic hypotension not accompanied by an increase in heart rate (typical of neurogenic orthostasis). The clue to the diagnosis is the sensory dissociation, autonomic findings and frequent involvement of the heart, bowel or kidneys. Despite amyloid deposition, enlarged palpable nerves are uncommon (Kelly et al. 1979).

Electromyographic findings

EMG is useful in the evaluation of these patients. The findings are characteristic and can often help to differentiate the PN of primary amyloidosis from other paraproteinemia associated PNs, such as those associated with monoclonal gammopathy of unknown significance (MGUS) and osteosclerotic myeloma (Kelly 1983). Typical EMG findings are consistent with a symmetrical, primarily axonal neuropathy affecting the longest axons. Conduction velocities are usually minimally slowed despite reduction in motor responses and often absent sensory nerve responses (Melgaard and Nielsen 1977; Kelly 1983). When sensory nerve action potentials are present; there is little prolongation of distal latencies or mild delay proportional to the slowing of conduction velocities, except in the presence of superimposed CTS. Routine nerve conduction studies early in the disease may not show such clear sensory findings since small myelinated and unmyelinated sensory fibers are affected first and more severely, and these are usually not recordable by conventional techniques. In vivo intraneural recordings of small fibers can be helpful, although usually not practical. Sympathetic skin responses, recording of R–R interval variation in the supine, sitting and standing positions, with breathing and during Valsalva maneuver are measurements of autonomic nervous system functioning and can be easily done in the EMG laboratory with the newest EMG equipment. These can be abnormal early in the disease and even in asymptomatic familial amyloidosis carriers (Ando et al. 1992) and should be performed when a patient with suspected amyloidosis is referred to the EMG laboratory for evaluation (McLeod 1992). More sophisticated autonomic evaluation can be done by specialized laboratories and includes, among others, Doppler flowmetry, tilt test and pupillary function test (Ando et al. 1992; McLeod 1992).

Needle examination usually shows distal and symmetric chronic and active denervation and reinnervation changes. Abnormalities are usually less in the upper limb. Proximal muscles may rarely show myopathic findings in those patients with associated amyloid myopathy.

Pathogenesis

The pathogenesis of neuropathy in primary amyloidosis is unclear. Amyloid deposits are typically found surrounding the endoneurial capillaries and in the walls of small vessels in the epineurium (Thomas and King 1974; Kelly et al. 1979). Nodules of amyloid can also be seen indenting myelinated fibers and producing bulbous swellings of the fibers adjacent to the point of compression. Amyloid may also be deposited in linear streaks in the epineurium remote from vessels (Dyck and Lambert 1969). Finally, amyloid may be found deposited in both dorsal root and autonomic ganglia (Dyck and Lambert 1969; Kyle and Dyck 1993).

From these morphological observations, three theories have evolved to explain the nerve damage. One of the oldest is ischemic nerve damage with subsequent axonal degeneration (Kernohan and Woltman 1942). However, although the vasa nervorum are clearly affected by surrounding amyloid deposits, vessels are not occluded. Furthermore, the early selective involvement of the unmyelinated nerve fibers and the diffuse and symmetrical nature of the lesions make ischemia unlikely. Moreover, no nerve infarcts have been reported (Dyck and Lambert 1969). Direct mechanical compression of nerve fibers has also been postulated (Thomas and King 1974). Dyck and Lambert (1969) and Said et al. (1984) demonstrated marked distortion of myelinated fibers. However, this mechanism is unlikely to cause selective unmyelinated fiber loss, since C fiber potentials are the last to disappear during experimental nerve compression (Dyck and Lambert 1969). Amyloid deposition in the proximal dorsal root and autonomic ganglia has also been suggested as the primary site of injury in amyloid PN, aided by an absent blood–brain barrier that permits the passage of the amyloidogenic protein (Verghese 1983). Despite some compelling evidence, it seems unlikely that such a mechanism would explain the entire clinical picture. Direct neurotoxicity of ALs is a possible cause that so far has no experimental support. At the molecular level, endoneurial and perineurial cells, especially macrophages, may influence amyloid formation by enzymatic cleavage of L-chains, while amyloid degradation by phagocytotic uptake is abnormally inefficient (Durie et al. 1982; Sommer and Schröder 1989; Kyle and Gertz 1990).

Pathology

Axonal degeneration of preferentially small myelinated and unmyelinated fibers is the main finding in sural nerve biopsies. Amyloid deposition is found typically forming a cuff around endoneurial vessels, and in the walls of endo- and epineural vessels, which appear thick by H and E staining. Diffuse or linear streaks of amyloid deposits are seen in the epineurium, and in globular shaped deposits in the endoneurium (Dyck and Lambert 1969; Kelly et al. 1979; Kyle and Dyck 1993). As discussed above, these globules have been seen to distort the nerve fibers (Dyck and Lambert 1969). Teased nerve fibers show a predominance of axonal degeneration (Kyle and Dyck 1993). In addition to peripheral nerve involvement, neuron cell counts from the intermediolateral column have been found to be reduced to 50–75% of the control values in patients with primary amyloidosis and orthostatic hypotension (Low et al. 1981).

Electron microscopy shows striking loss of unmy-elinated fibers. By immunohistochemistry, it is possible to identify amyloid as κ or λ L-chain derived (Dalakas 1986; Chalk and Dyck 1992). Immuno-gold-labeled ALs have been found intracellularly in coated and uncoated vesicles in endoneurial macrophages. Although Schwann cells were not found to contain intracellular amyloid, the fibrils were found invading their basal lamina (Sommer and Schröder 1989). Rarely, amyloid polyneuropathy has been associated with a biclonal gammopathy. In one remarkable case, an IgM κ L-chain actively caused a demyelinating neuropathy with widening of the myelin lamellae, while an IgG λ was linked to amyloid forming nodular perivascular and endoneurial deposits (Julien et al. 1984).

Other forms of neurologic involvement

Amyloid myopathy

In 1929, Lubarsh described for the first time macroglossia and firm skeletal muscles in a patient with primary amyloidosis. Although amyloid deposition in the skeletal muscles is often found in AL, and is actually very common in those cases with PN, it is usually asymptomatic and merely a pathological finding. Very rarely, amyloid myopathy can present as a prominent clinical syndrome and as the presenting feature of generalized amyloidosis.

Clinical presentation varies. Most frequently, these patients complain of muscle stiffness, pseudohypertrophy and fatigability. Difficulties in swallowing, hoarseness of the voice or slurred speech and even obstructive sleep apnea can be prominent complaints due to macroglossia (Whitaker et al. 1977; Ringel and Claman 1982). Examination discloses minimal or moderately severe generalized muscle weakness. The muscles are characteristically enlarged, with a very firm, 'wooden-like' feeling. There is resistance to passive movement of the muscles. Palpable nodules can be perceived. Macroglossia can be so prominent as to prevent normal closure of the mouth and mastication. Deep tendon reflexes and sensory modalities are usually preserved. Occasional patients have presented with respiratory failure due to primary involvement of the diaphragm and intercostal muscles (Ashe et al. 1992). Another set of patients can present with a subacute limb-girdle myopathy, with proximal muscle weakness and atrophy, and no evidence of muscle hypertrophy, induration or macroglossia (Jennekens and Wokke 1987; Nadkarni et al. 1995). These patients have proximal muscle weakness and atrophy, including neck extensor muscles, and are indistinguishable clinically from other types of myopathy.

Creatine phosphokinase can be mildly elevated (2–3-fold). Immunological studies have shown the presence of free λ L-chains (Ringel and Claman 1982; Jennekens and Wokke 1987), IgG κ (Ashe et al. 1992) and IgG λ (Nadkarni et al. 1995) and IgA λ (Jennekens and Wokke 1987), but can also be non-revealing (Whitaker et al. 1977).

EMG studies show normal or neuropathic nerve conduction studies. When associated neuropathy is not present, needle examination is usually consistent with a myopathy, with early recruitment of small, polyphasic, short-duration motor unit potentials, and abundant fibrillation potentials and positive waves at rest. The picture thus resembles polymyositis, especially if CK is elevated.

Muscle biopsy characteristically shows very mild muscle fiber changes, with absence of necrosis, phagocytosis or inflammation. Amyloid is found infiltrating the walls of the moderate to large intramuscular vessels (arteries and veins), but with no obstruction of their lumen. In some cases, this and mild deposition in the connective tissue is the only finding. Muscle fibers can be compressed and distorted by the amyloid. Esterase stains may show a few esterase-positive angulated fibers consistent with denervation. Type II more than type I muscle atrophy can be seen (Whitaker et al. 1977). In other reported cases, amyloid was also found surrounding the muscle fibers and deposited in the endomysium. In these cases, ultrastructural studies showed the so-called 'covered muscle fibers'. The sarcolemma had developed elongated folds that occasionally followed the surface of the muscle fiber for some distance (Jennekens and Wokke 1987). In no case was amyloid ever found intracellularly and the myofibrillar structure appeared intact (Ringel and Claman 1982; Jennekens and Wokke 1987). A non-amyloid amorphous substance similar to basal membrane material could be seen in the intramuscular perivascular spaces, more abundant than the ALs (Whitaker et al. 1977). In one study, acetylcholine receptors were seen distributed diffusely along the membrane of those muscle fibers covered by amyloid (Ringel and Claman 1982), suggesting denervation.

Magnetic resonance imaging studies of these patients can be useful, since the appearance of the muscles in amyloid myopathy differs greatly from other myopathies. The T1 and T2 signal characteristics within the muscle are minimally changed, while there is a striking reticulation of the subcutaneous fat (Metzler et al. 1992).

There are several possible explanations for the clinical and laboratory features of amyloid myopathy. First, accumulation of amyloid causing the pseudohypertrophy of the muscles and striking deposit in subcutaneous tissue causes severe muscle induration (Metzler et al. 1992) that could mechanically restrict muscle function. Second, the so-called 'myogenous denervation' (Ringel and Claman 1982) due to interruption of normal propagation of the action potential through the sarcolemma or an intramuscular neuropathy could cause weakness and atrophy and also explain the EMG findings of abundant abnormal spontaneous activity on EMG, in the absence of pathological evidence of muscle necrosis or inflammation. Third, the amyloid deposits could be directly toxic to muscle fibers, although this is not supported by the relative preservation of the myofibrillar structure of the muscle fibers (Ringel and Claman 1982; Jennekens and Wokke 1987).

Miscellaneous neurological involvement

Unusual presentations of peripheral neuropathy
Autonomic nervous system involvement, although more common in familial amyloidosis, can be very severe. Cases with very proximal peripheral nerve and ganglionic involvement can defeat diagnosis until post-mortem examination, as occurred in the case described by Lingenfelser et al. (1992). This patient's pathology showed AL λ deposition in the vagus nerve, spinal cord, sympathetic ganglia and myenteric plexus, in addition to blood vessels of nearly every organ and myocardium.

Rarely, typical amyloid polyneuropathy can be associated with signs of more diffuse upper and lower motor neuron involvement (Abarbanel et al. 1986). An asymmetric presentation of chronically progressive peripheral motor and sensory deficit has been associated with amyloid involvement of lumbosacral plexus and nerve roots, causing distal axonal degeneration. The sacral plexus was markedly enlarged and a fascicular biopsy showed the amyloid deposits. This

could not be characterized further since the deposits
were not found in a peripheral nerve biopsy and there
were no signs of generalized involvement or paraproteinemia (Antoine et al. 1991).

Cranial neuropathy

Although very rare, multiple cranial neuropathies can
be the chief complaint at the time of diagnosis of amyloidosis. These patients present with progressive
asymmetric involvement of cranial nerves, that can
vary from progressive facial weakness, trigeminal
neuropathy, ophthalmoplegia or anosmia to involvement of the lower cranial nerves with subsequent dysphagia and dysarthria (Kelly et al. 1979; Towle and
Maher 1983; Traynor et al. 1991). Most of these patients develop a PN in the course of their illness.
Therapy with prednisone, colchicine and alkylating
agents did not benefit the patients with cranial neuropathies (Traynor et al. 1991).

Systemic involvement

Primary amyloidosis can present with systemic or
generalized symptoms, peripheral nerve symptoms,
cardiac or renal insufficiency or multiple organ involvement. Initial symptoms of fatigue, weakness and
weight loss are prominent. Pain is more frequent in
patients with MM (40%). Eight percent of patients
without MM have diffuse abdominal pain or painful
limb dysesthesias secondary to PN (Kyle and Greipp
1983). Gross bleeding and purpura are also more
common in patients with MM. This is usually seen on
the face, neck and upper eyelid. Skin is very fragile.
The most common initial physical findings are ankle
edema, hepatomegaly (34%), lymphadenopathy,
macroglossia (20%) and orthostatic hypotension
(14%) (Kyle and Greipp 1983).

Cardiac disease

CHF is present as the initial syndrome, or as the disease progresses, in approximately 35% of patients
(Kyle and Greipp 1983). Low output failure is more
commonly due to restrictive cardiomyopathy. Echocardiographic findings of cardiac amyloidosis can
precede initial symptoms and consist of thickening of
the ventricular wall and septum, with a speckled tex-
ture, and abnormal diastolic function by Doppler
studies (Cueto-Garcia et al. 1985). The most common electrocardiographic finding is a low voltage

QIRS complex. The pattern of pseudo-anteroseptal
infarction (no evidence of infarct on autopsy) is also
very frequent (Wright and Calkins 1981; Kyle et al.
1984; Smith et al. 1984). However, amyloid can also
deposit in the coronary arteries and result in angina
pectoris or myocardial infarction (Barth et al. 1970).
Cardiac arrhythmias are common, more frequently
sick sinus rhythm due to amyloid infiltration or fibrosis
of the sinus node. Atrial fibrillation, junctional tachycardia, heart block and premature ventricular complexes are frequent (Kyle and Greipp 1983; Roberts
and Waller 1983). These patients are very sensitive to
digitalis and may develop arrhythmias with low doses.
About 40% of the patients with primary amyloidosis
die of cardiac disease.

Renal involvement

Nephrotic syndrome is present in one-third of the patients with primary amyloidosis, 90% of them at the
time of diagnosis (Kyle and Greipp 1983). ALs are
deposited first in the mesangium of the glomerulus and
extend later along the basal membrane. The degree of
proteinuria does not correlate well with the extent of
amyloid deposition, but does with the extent of tufting
and detachments of epithelial cells (Watanabe and
Saniter 1978; Katafuchi et al. 1984). Severe renal insufficiency is often a late manifestation.

Miscellaneous

Respiratory manifestations are infrequent, despite the
common histological deposit of amyloid in the alveolar septae and blood vessel walls. Patients may present with dyspnea and cough due to interstitial lung
infiltration, but respiratory failure is rarely a cause of
death. Macroglossia, present in about 20% of the patients, may interfere with eating and closing of the
mouth, with constant drooling. If severe enough, obstruction of the airway can occur during sleep or re-
clining positions. The gastrointestinal tract is also
commonly involved in patients with primary amyloidosis. Decreased motility of the small bowel can
lead to pseudo-obstruction. Malabsorption syndrome (present in 5% of AL patients) and diarrhea
are more likely caused by autonomic neuropathy
rather than by direct amyloid infiltration of the intestinal wall (Battle et al. 1979; Kyle and Greipp 1983),
although the pathological changes in the autonomic
nerves are much less severe than in type I familial
amyloidosis (Ikeda et al. 1987b).

There are reports in the literature of direct involvement by amyloid deposition of virtually every segment of the GI tract, causing dysphagia, ulceration of the colon and rectal mucosa with hematemesis and hematochezia, and even perforation of small bowel (Kyle and Greipp 1983).

Bleeding may be a major manifestation in patients with primary amyloidosis. The cause of the bleeding disorder can vary: deficiency of factor X or of vitamin K-dependent clotting factors, increased antithrombin activity, increased fibrinolysis and intravascular coagulation. It has been suggested that amyloid fibrils interact with clotting factors (Kyle and Greipp 1983).

Dermatological manifestations include petechiae, ecchymoses, papules, plaques, nodules, tumors, bullous lesions, alopecia and thickening of the skin (Breathnach and Black 1979). Involvement of the para-articular tissues by amyloidosis can be mistaken clinically as seronegative rheumatoid arthritis (Pascali et al. 1980). Infiltration of the shoulders with amyloid can produce severe pain and swelling, the so-called 'shoulder-pad sign'. Bone lesions are often seen secondary to direct amyloid deposition causing osteolytic lesions or pathological fractures (Khojasteh et al. 1979). Amyloid may involve many other organs (Kyle and Gertz 1990).

Laboratory findings

Primary amyloidosis does not cause any particular hematologic or blood chemistry abnormality, except for monoclonal serum protein abnormalities. Abnormal laboratory findings are the result of complications (i.e., nephrotic syndrome) or underlying MM. Anemia can be seen in about 50% of the patients (Kyle and Bayrd 1975), secondary to gastrointestinal bleeding, MM or renal insufficiency. The factor X level is decreased in less than 5% of the patients and is rarely responsible for bleeding. Prothrombin time can be prolonged, and an inhibitor of thrombin, with subsequent prolongation of the thrombin time, is found in about 60% of the patients (Gastineau et al. 1991). Thrombocytosis can be detected in 5–10% of the patients, probably due to functional hyposplenism. Approximately 80% of the patients have proteinuria (Kyle and Bayrd 1975; Kelly et al. 1979; Kyle and Greipp 1983). The heat test for Bence–Jones protein is usually negative in patients without MM. Serum creatinine levels are above 1.3

mg/dl in one-half of the patients. Liver function tests are usually normal, although alkaline phosphatase may be elevated.

Serum protein electrophoresis is usually abnormal, showing hypogammaglobulinemia in about 15% of patients with MM and 33% of those with systemic amyloidosis (Kyle and Garton 1986). A peak in the γ region, usually of moderate size, can be seen in about 40% of the patients. Those patients with a monoclonal spike of more than 3.0 g/dl are more likely to have overt MM. The γ-globulin peak can be very small, though, and immunoelectrophoresis or immunofixation is necessary to disclose an abnormal monoclonal protein in some patients. Electrophoresis of an adequately concentrated urine specimen usually shows a large albumin peak. As in the serum, a localized globulin band can be seen in almost two-thirds of the patients with MM, but immunoelectrophoresis or immunofixation is usually necessary to reveal a monoclonal L-chain, seen in approximately 75% of patients. In fact, immunofixation of urine in a patient with unexplained nephrotic syndrome can sometimes help establish a diagnosis of amyloidosis (Kyle and Bayrd 1975; Kyle and Greipp 1983).

Diagnosis

The diagnosis of amyloidosis depends on the demonstration of the amyloid deposits in tissues. The most commonly used stain is Congo Red, which produces an apple-green birefringence by polarized light, characteristic of all types of amyloid. This stain, though, is not 100% sensitive, since some neural structures, bone trabeculae and connective tissue can give green birefringence as well (Klatskin 1969). Metachromatic stains can be useful, especially in sural nerve biopsies wherein metachromasia stands out from the background. In occasional cases, all these stains are negative and electron microscopy is needed to demonstrate the diagnostic ALs.

High-resolution scintigraphic studies can reveal even minor deposits of amyloid. This technique uses ^{123}I-labeled purified human SAP, which binds to the amyloid P component associated with all kinds of amyloid (Hawkins et al. 1990). With this method, amyloid deposits can be readily identified but the characterization of the amyloid is only possible by analyzing amyloid from tissue with specific antisera to

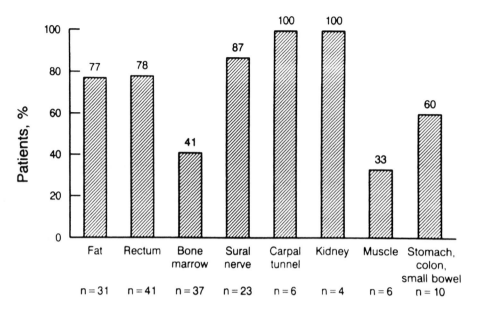

Fig. 3. Results of biopsy from sites of involvement in patients with AL. (From Gertz et al. 1992, reprinted by permission.)

the abnormal proteins (ALκ, ALλ, AA, β_2-microglobulin, TTR, etc.).

The diagnosis of primary amyloidosis should be considered in any patient with a sensorimotor PN who has an abnormal monoclonal protein in serum or urine or a typical clinical picture. Diagnosis can be confirmed by obtaining appropriate tissue for histological investigation (Fig. 3). Abdominal fat aspirate is an uncomplicated procedure that is positive in about 70% of patients (Kyle and Greipp 1983; Gertz et al. 1988; Kyle and Dyck 1993). Overstaining and undercoloration are potential problems with this technique (Gertz et al. 1988). Bone marrow biopsy is positive in only half of patients with amyloidosis but can show an abnormal proliferation of plasma cells, which is important for diagnosis and prognosis of MM. The presence of more than 20% plasma cells, especially if atypical, is associated with overt MM (Kyle and Greipp 1983). Sural nerve biopsy is positive in over 80% of patients with PN and should be considered when other biopsies are negative. Rectal biopsy is also positive in over 80% of patients, comparable to fat aspirate. Care should be taken to include submucosa in the specimen (Kyle and Greipp 1983). If these tissues are negative, then tissue should be obtained from involved organs (renal, liver, carpal tunnel, small intestine, skin, etc.).

Prognosis

As a general principle, patients with a low burden of amyloid in a non-vital organ are destined to survive longer, while those with advanced multiorgan disease do poorly. The median survival of patients with primary amyloidosis is 2 years (Kyle and Dyck 1993). In a recent review of prognosis and treatment in primary amyloidosis, several clinical and laboratory features were found to predict survival (Gertz and Kyle 1994). Patients with PN as the sole manifestation of their disease had the longest survival, ranging from 40 to 56 months (Kelly et al. 1979; Kyle and Greipp 1983; Duston et al. 1989; Gertz and Kyle 1994) (Fig. 4). These patients eventually died from involvement of other organs (Kelly et al. 1979). Conversely, patients with overt CHF or orthostatic hypotension have the shortest survival, usually less than a year (Kyle and Greipp 1983; Gertz and Kyle 1994). Nephrotic syndrome and MM have a detrimental impact on survival as well (Kyle and Greipp 1983).

Other clinical features helpful in the assessment of prognosis are the presence of jaw claudication and hyposplenism. The first is associated with predominant vascular deposits of amyloid and a decreased tendency for large tissue deposits at least early in the course and thus has a better prognosis (Gertz et al.

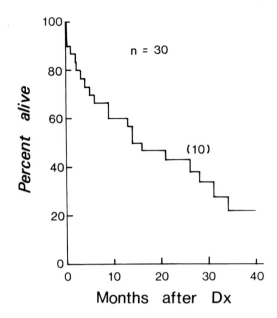

Fig. 4. Survival from time of diagnosis of 30 patients treated for amyloid neuropathy. (From Kelly et al. 1979, reprinted by permission.)

1986). Hyposplenism is associated with hepatomegaly and CHF, hence survival is very poor (Gertz and Kyle 1994).

The median survival of patients with increased serum β_2-microglobulin levels in serum is significantly lower, and it should be measured in all patients with primary amyloidosis (Gertz et al. 1990). In a multivariate analysis of prognostic factors, the presence of CHF, the demonstration of a urinary monoclonal L-chain, hepatomegaly and MM all adversely affected survival during the first year. Next, increased serum creatinine concentration, presence of MM, orthostatic hypotension and a monoclonal serum protein become poor prognostic factors (Kyle et al. 1986).

Treatment

The treatment of primary amyloidosis has had only limited success, in part because of the relative resistance of the amyloidogenic clone to chemotherapy, and in part due to delay in diagnosis, with the subsequent irreversible damage of the compromised organs by the time therapy is initiated. The earlier the treatment is begun, the better the chances of a positive effect. The exact mechanism by which amyloid deposits cause organ damage is undetermined, and it

appears that the relationship is more complex than just the amount of amyloid deposited in the organ. There are several reports in the literature of patients with marked clinical improvement during treatment while biopsy of the affected organ showed increase in amyloid deposits (Kyle et al. 1982; Merlini 1995).

Nevertheless, with the introduction of quantitative scintigraphic and turnover studies with radiolabeled SAP (Hawkins et al. 1990), it has been shown that amyloid deposits can regress rapidly when the supply of the amyloid protein precursor is substantially reduced (Merlini 1995). This constitutes the rationale for alkylating therapy in primary amyloidosis, with the hope of controlling the plasma cell proliferation and production of L-chains. Melphalan and prednisone, with or without colchicine, have been tried in multiple prospective trials (Ravid et al. 1977; Kyle and Greipp 1978; Kyle et al. 1985; Cohen et al. 1987). A nice review of the multiple drug trials can be found in recent publications by Merlini (1995) and Gertz and Kyle (1994).

Overall, melphalan and prednisone treatments, continued for at least 1 year, have resulted in prolonged survival rates. The highest response rate (39%) was obtained in patients with nephrotic syndrome, but with normal serum creatinine levels and no echocardiographic evidence of cardiac amyloidosis. Fifteen percent of patients with cardiac amyloidosis also responded (Kyle and Greipp 1978; Kyle et al. 1985). PN usually does not respond to any treatment (Kelly et al. 1979; Kyle and Greipp 1983; Gertz and Kyle 1994; Merlini 1995) and continues to advance even when other organs show some improvement or arrest of damage. The combination of melphalan and prednisone has been proved superior to colchicine therapy (Kyle et al. 1985), and the addition of colchicine to the above regimen did not add any benefit in one recent randomized trial (Kyle et al. 1994). A trial of alkylating agents and prednisone is warranted in every patient with the diagnosis of primary amyloidosis.

Supportive therapy is important. With the advent of cardiac transplantation, hemodialysis, continual peritoneal dialysis and renal transplantation, survival of patients with primary amyloidosis has improved. A detailed discussion of general therapeutic measures is beyond the scope of this chapter. The interested reader is referred to two very extensive recent reviews of the topic (Gertz and Kyle 1994; Merlini 1995).

The symptoms and findings of PN continue to worsen despite therapy. Dysesthesias tend to disappear as analgesia becomes more prominent. Analgesics, sedatives and tricyclic antidepressants may be helpful. Orthostatic hypotension can be ameliorated with supportive elastic stockings, extending up to the waist. Drugs that cause fluid retention are of little use if the patient has extensive cardiac or renal involvement. These patients, however, frequently have cardiac failure and nephrotic syndrome and require salt restricted diet and diuretics and tolerate fluid expansion poorly.

FAMILIAL AMYLOIDOSIS

Hereditary amyloidosis is relatively uncommon as a clinical problem although its incidence is very high in certain small and circumscribed populations. It is characterized by its great clinical and genetic heterogeneity. The most frequent form is familial amyloid polyneuropathy (FAP), an autosomal dominant disorder. Other varieties (cardiopathic, nephropathic, familial Mediterranean fever) do not have significant neurological manifestations and will not be included in this chapter. An additional form of hereditary generalized amyloidosis, familial oculoleptomeningeal amyloidosis, is a very rare disorder and affects the central nervous system.

In 1952, Andrade described for the first time a hereditary form of polyneuropathy associated with generalized amyloidosis in several families from Porto in Portugal (Andrade 1952). The disease was known in the area as 'mal dos pèsinhos' (foot disease). This endemic disease was characterized by paresis predominantly in the lower extremities, early abnormalities in pain and thermal perception beginning in the lower extremities, gastrointestinal, sexual and sphincter dysfunction. After Andrade's report, more than 500 kindreds have been reported to date in Portugal, the largest focus in the world (Mascarenhas Saraiva et al. 1986). Subsequently a number of other kindreds with similar clinical presentation were reported in Japan (Ikeda et al. 1987a), Sweden (Andersson 1976), USA (Libbey et al. 1984; Kincaid et al. 1989), Italy (Di Iorio et al. 1993), England, Brazil, Greece, Cyprus, France (Holt et al. 1989) and The Netherlands. In the subsequent years, other families with the FAP type I mutation and with a variety of different mutations in the transthyretin gene

were reported. These may have distinct phenotypes, differing mainly from the Portuguese type in the distribution of the neuropathy, and the extent of other organs involved.

The first attempt to classify FAP was rather confusing and based on the geographical distribution of the different kindreds and the name of the person who described them. This prompted a clinical classification of at least five distinctive clinical syndromes (Table 2). FAP type I (Portuguese, Swedish, Japanese) had the clinical features described by Andrade. FAP type II (Indiana/Swiss/Maryland) is manifest first by CTS, followed by neuropathy affecting first the upper extremities and later the lower extremities, and vitreous opacities. Type III (Iowa) is characterized by polyneuropathy affecting mainly the lower extremities, nephropathy and peptic ulcers. Type IV (Finnish) presents with lower cranial neuropathies, lattice corneal dystrophy and skin abnormalities. Appalachian/West Virginia is similar to FAP I, but with significant cardiac involvement. Much progress has been made in recent years on the genetics and chemical characterization of the different amyloid proteins, and a new molecular classification seems nowadays more suitable (Table 3). Nonetheless, the use of this classification may not help in the clinical management of some individual cases. Families with the exact same mutations can have different phenotypes and different disease durations and prognoses, due to different penetrance of the same abnormal gene. The phenotype within members of the same family, though, is usually very homogeneous.

Transthyretin polyneuropathies

Clinical presentation
Familial amyloid polyneuropathy derived from a variant transthyretin (TTR) is the most common group of hereditary amyloidosis. Clinical heterogeneity regarding organ involvement, age of onset and morbidity makes it difficult to precisely define clinical groups. Nonetheless, although simplistic, for practical purposes they can be divided into those with predominant lower limb neuropathy (FAP I) and those with predominant upper limb neuropathy (FAP II). Both are caused by a point mutation in the transthyretin gene on chromosome 18. FAP I in Portugal, Sweden, Japan, England, Brazil and most of the USA is due to a methionine for valine substitution at position

TABLE 2

Clinical characteristics of hereditary amyloid polyneuropathies.

Type	Protein amyloid	Age of onset	Duration (years)	ULN	LLN	AN	CrN	Ophthalmologic signs	Other
I (Portuguese) Andrade 1952	TTR (Met-30)	25–35	7–12	+++	++++	++++	0	vitreous opacities; scalloped pupils	
II (Indiana, Swiss) Rukavina et al. 1956	TTR (Ser-84) Wallace et al. 1988	30–40	14–20	++++	+++	0	0	vitreous opacities	onset with CTS
III (Iowa/van Allen) Van Allen et al. 1969 Nichols et al. 1988	ApoAI	25–45	6 months– 26 years	+++	++++	++++	0	cataracts	peptic ulcer, deafness
IV (Finnish/Meretoja) Kiuru 1992	Gelsolin	25–60	20	0	0	0	++++	lattice corneal dystrophy	blepharochalasis
Appalachian/West Virginia Koeppen et al. 1985	TTR (Ala-60) Koeppen et al. 1990	50–60	2–10	++++	++++	++++	0	none	cardiomyopathy

LLN = lower limb neuropathy; ULN = upper limb neuropathy; AN = autonomic neuropathy; CrN = cranial neuropathy; ++++ = severe; +++ = moderate; 0 = absent.

TABLE 3

Hereditary amyloid polyneuropathy: chemical and genetic features.

Protein	Mutation	Geographic/ethnic origin	Other name
TTR	methionine 30	Portugal, Japan, Sweden, USA, Italy, Brazil, England, The Netherlands	FAP I
	serine 84	Indiana (USA)/Switzerland	FAP II
	histidine 58	Maryland (USA)/Germany	FAP II
	alanine 60	Appalachian/West Virginia (USA)	
	tyrosine 77	Illinois (USA)/Germany	German
	isoleucine 33	Israel	Jewish
Apolipoprotein A1	arginine 26	Iowa (USA)	FAP III
Gelsolin	asparagine 187	Finland, USA, Japan	FAP IV

30 (Met-30). FAP II has a serine instead of isoleucine at position 84 (Ser-84). Indiana type has a valine for isoleucine at position 107 (Uemichi et al. 1994), and histidine for leucine at position 58 occurred in a family from Maryland (Mendell et al. 1990). More than 30 other mutations have been described in recent years, and their phenotypes can resemble type I, II or have predominant cardiac or renal involvement (Table IV).

FAP I is the largest group. It was first described in

TABLE 4

Transthyretin (TTR) hereditary amyloid polyneuropathies.

Aa substitution	Clinical features	Geographical kindred	Other
Methionine 30	LLN, AN, eye	Portugal, Japan, Sweden, USA, Italy, Brazil	FAP I
Isoleucine 33	LLN, eye	Israel	Jewish
Glycine 42	LLN, AN	Japan	FAP I
Arginine 50	LLN	Japan	FAP I
Cystine 114	LLN, eye	Japan	FAP I
Alanine 60	LLN, AN, heart	USA/West Virginia	Appalachian
Serine 84	ULN, CTS, eye	Indiana/Swiss origin	FAP II
Tyrosine 77	PN, kidney	Illinois/German origin	German
Tyrosine 11	LLN, heart, AN	France	
Proline 36	LLN, eye	USA/Greek origin	
Leucine 33	LLN, AN, heart	USA/Polish origin	
Leucine 64	ULN, heart, LLN	USA/Italian origin	not enough FHx
Alanine 71	CTS, PN, AN, eye	France	FAP I, late onset
Valine 107	CTS, PN, AN	USA/German origin	FAP II
Histidine 58	CTS, LLN	USA (Maryland)/German origin	FAP II
Alanine 49	PN, eye, heart	Sicily	
Glutamine 89	CTS, PN, heart	Sicily	
Arginine 10	ULN, AN, eye, heart	USA/Hungarian origin	

LLN = lower limb neuropathy; ULN = upper limb neuropathy; AN = autonomic neuropathy.

1952 in Povoa de Varzin, a fishing town in the Porto region of Portugal (Andrade 1952). This progressive disease is transmitted in an autosomal dominant fashion and has a high mortality rate. The age of onset is variable. In the Portuguese kindreds, it starts in the second or third decade of life, but onset can occur at later ages (Ubbey et al. 1984; Mascarenhas Saraiva et al. 1986; Kincaid et al. 1989; Coelho et al. 1994). Families with later onset have the same mutation as those with earlier onset (Mascarenhas Saraiva et al. 1986) and the question of what causes a delay in the amyloid deposition remains unanswered.. The American kindreds have a median age of onset in the seventh decade, but approximately 20% of the patients are younger than 55, different from primary amyloidosis (Gertz et al. 1992). The Japanese families have a variable age of onset, ranging from the second to the seventh decade (Ikeda et al. 1987a), and the Swedish families have a median age of onset in the sixth decade (Andersson 1976).

The first symptoms are loss of pain and temperature sensation in the lower extremities, which frequently results in self-amputations and painless ulcers. Later on, paresthesias and dysesthesias become prominent, and loss of touch and proprioception is apparent. Over the years, symptoms spread to the upper extremities, but cranial nerves are usually spared, although lower cranial neuropathy can be found in Japanese families in the advanced stages (Ikeda et al. 1987b). Diffuse myalgias can be an early symptom; however, muscle weakness does not develop until much later, accompanied by prominent distal muscle wasting and decreased deep tendon reflexes. In a study of FAP in the USA, autonomic neuropathy ultimately developed in 70% of patients (Gertz et al. 1992). Orthostatic hypotension, fainting spells and gastrointestinal disturbances (diarrhea, incontinence, intestinal pseudo-obstruction, gastric fullness) are frequent and indicate debilitating autonomic dysfunction. Impotence is one of the most constant and earliest symptoms. Ocular manifestations consist of vitreous opacities and occasionally scalloping of the pupils. Cardiac amyloidosis is commonly found at autopsy; however, cardiac failure or disturbances of rhythm are not common in the typical Portuguese and Japanese patients (Andrade 1952; Ikeda et al. 1987a). On the contrary, two-thirds of USA families develop cardiac amyloidosis with CHF and arrhythmia in the course of their disease (Gertz et al. 1992).

Renal involvement is rare in all, except in the Swedish patients (Steen et al. 1982). The prognosis is uniformly fatal, and life expectancy is reduced to 7–10 years after onset of symptoms.

FAP II was first described in a family of Swiss origin residing in Indiana (Rukavina et al. 1956). These patients develop the first symptoms in the fourth to fifth decade. Carpal tunnel symptoms with thenar atrophy due to deposition of amyloid in the flexor retinaculum is the earliest manifestation and can be the sole manifestation of the illness for long periods. A sensorimotor PN eventually develops in the arms and later in the legs. This disease has a more benign course and usually spares the autonomic nervous system. Vitreous opacities are prominent and can lead to blindness. Heart involvement may occur, but is not pronounced. The mutation responsible for this variant is a TTR Ser-84. A similar disease was reported in a family from Maryland (USA) of German ancestry, in which the mutation of TTR consisted of substitution of histidine for leucine at position 58 (Mendell et al. 1990).

Pathophysiology

In 1978, it was shown (Costa et al. 1978) that in FAP, the major component of ALs consisted of TTR fragments. TTR is synthesized in the liver and normally transports 20–25% of serum thyroxine and retinol. It is composed of four identical subunits of 127 amino acids. Its tertiary structure is a β-pleated chain. Eight of these β-pleated monomers form a tetramer (quaternary structure) which is very stable. Most of the TTR variants can be explained by a single base change (point mutation) in one allele on chromosome 18. The most common mutation in Portugal, Japan, America and Sweden is the substitution of methionine for valine at position 30. Many other variants have been described (Table 4). Most of the patients are heterozygous for the mutation (Mita et al. 1986), and hybrid monomers can have amyloidogenic potential as well. Gene dosage does not necessarily appear to influence disease onset, as homozygous individuals for the Met-30 mutation have been described and can be asymptomatic late in life (Holmgren et al. 1988). A mixture of normal and abnormal TTR is found in patients' sera with normal transport function, which implies that both genes are active. There is usually 40% of abnormal and 60% of normal serum TTR, and the opposite ratio is found in the ALs (Benson 1988).

Even though the molecular genetics of the disease has been clarified, the mechanism for amyloidogenesis is still unknown. The amyloidogenic potential of TTR itself is clearly a determinant in amyloid formation but other factors, some likely non-genetic, play a role given the heterogeneity in disease onset observed in families carrying the same mutation. It is interesting that all the mutations found lie on, or near, the surface of the tetramer, which could induce changes in the protein external architecture and permit the tetrameres to join and form fibrils (Benson 1988). Also, the variant TTR has increased resistance to proteolysis. Kinetic analysis in vitro and in vivo of normal and variant prealbumin in serum has shown that there was no difference between the two, even during oxidative damage or inflammation (Ando et al. 1989). This discards the possibility of an increased transfer rate of the variant prealbumin to explain its deposition in tissues as amyloid.

Another possibility is that abnormal processing of the variant TTR might lead to formation of intermediate metabolites. This could account for delayed onset and progression (Varga and Wohlgethan 1988). Tissue elements that bind to the variant TTR and not to the normal one are good candidates for intervening factors in the process and could explain why some tissues are prone to amyloid deposition and others are not (Mascarenhas Saraiva et al. 1986). An immunoassay for TTR variants (using TTR mouse monoclonal antibodies obtained using serum of a patient with FAP) was found to bind to sera from carriers of several amyloidogenic TTR variants associated with PN, but not to normal sera or cardiopathic hereditary amyloidosis (Costa et al. 1993). These antibodies bind to an epitope near the N-terminal side of the TTR monomer. It is proposed that this epitope is exposed by the mutations generating FAP and could be implicated in amyloid deposition in the peripheral nervous system. The proposed theories explaining how amyloid deposition causes organ damage has been discussed above in the primary amyloidosis.

Diagnosis

Diagnosis of amyloidosis depends, as discussed in AL, on the tissue diagnosis. Biopsy sites in FAP with a high yield of finding amyloid are similar to those in AL (Gertz et al. 1992). Carpal tunnel and kidney come close to 100%, while sural nerve is about 90% and subcutaneous fat and rectal biopsy are in the 80–

90% range. Punch skin biopsy can also be an effective method of early diagnosis, positive in 7 of 11 patients with FAP in one study (Rubinow and Cohen 1981). It is interesting that in these patients, amyloid infiltrated blood vessels, sweat glands, dermis and erector pili muscles, but not intracutaneous nerves.

Clinical diagnosis of FAP is straightforward when there is a positive family history and a compatible clinical picture. It becomes more difficult when there is no apparent family involvement. In a review of 1233 cases of FAP from 489 Portuguese families, it was found that in 159 cases neither parent had shown signs of the disease (Coelho et al. 1994). New mutations were not found. In these cases, molecular diagnosis becomes essential since it has been shown that apparently unaffected family members can carry the same Met-30 mutation (Mascarenhas Saraiva et al. 1986; Nakazato et al. 1987a, b).

The molecular diagnosis of TTR variants, including Met-30, Ile-33, Ala-60, Tyr-77, Met-111 and Ile-122, is now possible thanks to the production of recombinant human TTR (Furuya et al. 1991). Methods to study the hereditary amyloidoses consist of isolation of the amyloid subunit proteins from tissue deposits (Susuki et al. 1987), determination of the amino acid structure of the amyloid protein, prediction of a point mutation in the TTR gene to explain a single amino acid substitution and use of restriction fragment length polymorphism analysis to prove the proposed mutation (Benson and Wallace 1989). This last can be done by extracting DNA from the whole blood. It permits one to recognize asymptomatic cases and children of affected individuals and confirms diagnosis in affected individuals. This method is also suitable for prenatal diagnosis (Almeida et al. 1990; Turpin et al. 1992).

Pathology

The primary findings in sural nerve biopsy in FAP are linear deposits of amyloid within the fascicles, and sometimes nodules of amyloid indenting myelinated fibers. Teased-fiber preparations show axons degenerating into linear rows of myelin ovoids with predominant involvement of thin unmyelinated and small myelinated fibers (Dyck and Lambert 1969). Other studies have reported distortion of myelin sheaths, segmental demyelination and Wallerian degeneration adjacent to amyloid deposits (Said et al. 1984).

Studies of the sural nerve pathology in different

stages of the disease have shown that amyloid deposition in early stages is scarce and may not contribute to the abnormalities seen in the axons and myelin at this stage (Coimbra and Andrade 1971a, b). A correlative study (Hanyu et al. 1989) of spinal nerve roots, proximal sciatic nerve, brachial plexus and sural nerve lesions has brought some light to the pathogenesis of amyloidotic nerve lesion. The authors found unaffected spinal nerve roots, but sciatic nerve and brachial plexus had multifocal lesions, with prominent interstitial edema in the endoneurium, mostly adjacent to amyloid deposits. In these regions, the nerve fibers were severely depleted and segmental demyelination was the predominant damage, as shown in teased-fiber preparations. These findings, however, were not seen in the sural nerves, where diffuse fiber loss with axonal degeneration predominated. This correlates with Dyck and Lambert's (1969) findings. It is suggested that multifocal lesions in the proximal portions of the nerves could summate distally to produce a symmetrical polyneuropathy (Hanyu et al. 1989).

Treatment

Treatment of FAP has been limited up to very recently to supportive care. Nowadays, liver transplantation has become the first successful treatment for FAP. The fact that TTR is synthesized in the liver prompted the first experimental liver transplant in Sweden in 1990, with the hope of removing the major source of mutant TTR. The report from 24 Swedish FAP patients who underwent liver transplant has been recently published (Ericzon et al. 1995). Seventeen of 24 patients are alive 3–52 months after liver transplantation. The PN as measured by EMG was unchanged in 13 patients evaluated one or more years after transplant. Improvement was noticed in walking capacity, gastrointestinal symptoms such as diarrhea, constipation and vomiting, urinary problems (some resolved completely) and nutritional status. Adverse factors in the survival and recuperation were poor nutritional status and prolonged disease. From this study, it is concluded that liver transplant should be performed early in the disease, definitely halts neurological decline and improves other symptoms. In the USA, experience with seven patients is similar, although the follow-up period is still relatively short (Skinner et al. 1994). None of the patients has had progression of disease, and some patients have noted decrease in diarrhea.

Gelsolin neuropathy

The hereditary amyloidosis caused by a variant gelsolin protein is clustered mainly in Finland (FAP type IV or Finnish type). A few other families have been reported in the USA (Darras et al. 1986; Gorevic et al. 1991), Japan (Sunada et al. 1992a, b, 1993) and The Netherlands (Haan and Peters 1994). Gelsolin is an actin-binding protein, a normal constituent of the plasma. Most of the familial cases are caused by a single point mutation in chromosome 9, causing asparagine for aspartic acid substitution at position 15 (Haltia et al. 1992; Sunada et al. 1992a, b).

This clinically homogeneous disease has a characteristic triad of neuropathy, ophthalmologic and dermatologic manifestations. The onset is in the third to fourth decade of life, with slow progression, and prolonged, good quality survival until the seventh decade. Patients present with decreased vision, photosensitivity, irritation and dryness of the eyes due to corneal lattice dystrophy. Blepharochalasis and bilateral facial weakness are common, as well as facial myokymia (Kiuru and Seppäläinen 1994). Other cranial neuropathies (V, IX, X and XII) have been reported (Sunada et al. 1992a, b). Eventually, patients develop a mild sensorimotor polyneuropathy, as well as minor autonomic nervous system involvement (Kiuru and Seppäläinen 1992a, b; Kiuru et al. 1994). Amyloid is found in skin, sural nerve and muscle biopsies.

Apolipoprotein A1 neuropathies

Formerly called FAP type III, this entity was first described by Van Allen et al. in 1969 in families in Iowa of Scottish-Irish descent. Clinically, it has similar manifestations to TTR Met-30. These families differ from FAP type I in the presence of significant nephropathy, nerve deafness and peptic ulcer disease. Onset of symptoms is in the third or fourth decade, with an average survival of 12 years after diagnosis. At autopsy, amyloid depositions are very scarce in the kidney and not found in the peptic ulcers. Therefore, several features appear not to be related to amyloid deposits per se. The major constituent of ALs in this type of FAP is apolipoprotein A1, with an arginine for glycine substitution at position 26, secondary to a point mutation in chromosome 11 (Nichols et al. 1988).

REFERENCES

ABARBANEL, J.M., S. FRISHER and A. OSIMANI: Primary amyloidosis with peripheral neuropathy and signs of motor neuron disease. Neurology 36 (1986) 1125–1127.

ALMEIDA, M.R., I. LONGO ALVES, Y. SAKAKI, P.P. COSTA and M.J.M. SARAIVA: Prenatal C diagnosis of familial amyloidotic polyneuropathy: evidence for an early expression of the 1:1 associated transthyretin methionine 30. Hum. Genet. 85 (1990) 623–626.

ALMEIDA, M.R., A. FERLINI, A. FORABOSCO, M. GAWINOWICZ, P.P. COSTA, F. SALVI, R. PLASMATI, C.A. TASSINARI, K. ALTLAND and M.J. SARAIVA: Two transthyretin variants (TTR Ala-49 and TTR Gln-89) in two Sicilian kindreds with hereditary amyloidosis. Hum. Mutat. 1 (1992) 211–215.

ANDERSSON, R.: Familial amyloidosis with polyneuropathy: a clinical study based in patients living in northern Sweden. Acta Med. Scand. 590, Suppl. (1976) 1–64.

ANDO, Y., S. IKEGAWA, A. MIYAZAKI, M. INOUE, Y. MORINO and S. ARAKI: Role of variant prealbumin in the pathogenesis of familial amyloidotic polyneuropathy: fate of normal and variant prealbumin in the circulation. Arch. Biochem. Biophys. 274 (1989) 87–93.

ANDO, Y., S. ARAKI, O. SHIMODA and T. KANO: Role of autonomic nerve functions in patients with familial amyloidotic polyneuropathy as analyzed by laser Doppler flowmetry, capsule hydrograph, and cardiographic R–R interval. Muscle Nerve 15 (1992) 507–512.

ANDRADE, C.: A peculiar form of peripheral neuropathy: familial atypical generalized amyloidosis with special involvement of the peripheral nerves. Brain 75 (1952) 408–427.

ANTOINE, J.C., A. BARIL, C. GUETTIER, F.G. BARRAL, B. BADY, P. CONVERS and D. MICHEL: Unusual amyloid polyneuropathy with predominant lumbosacral nerve roots and plexus involvement. Neurology 41(1991) 206–208.

ASHE, J., C.O. BOREL, G. HART, R.L HUMPHREY, D.A. DERRICK and R.W. KUNCL: Amyloid myopathy presenting with respiratory failure. J. Neurol. Neurosurg. Psychiatry 55 (1992) 162–165.

BARTH, R.F., J.T. WILLERSON, L.M. BUJA, J.L. DECKER and W.C. ROBERTS: Amyloid coronary artery disease, primary systemic amyloidosis and paraproteinemia. Arch. Intern. Med. 126 (1970) 627–630.

BATTLE, W.M., M.R. RUBIN, S. COHEN and W.J. SNAPE, JR.: Gastrointestinal motility dysfunction in amyloidosis. N. Engl. J. Med. 301 (1979) 24–25.

BENSON, M.D.: Hereditary amyloidosis – disease entity and clinical model. Hosp. Pract. 23 (1988) 165–181.

BENSON, M.D. and M.R. WALLACE: Genetic amyloidosis: recent advances. Adv. Nephrol. 18 (1989) 129–138.

BENSON, M.D., A.S. COHEN, K.D. BRANDT and E.S. CATHCART: Neuropathy, M components and amyloid. Lancet i (1975) 10–12.

BENSON, M.D. II, J.C. TURPIN, G. LUCOTTE, S. ZELDENRUST, B. LECHEVALIER and M.D. BENSON: A transthyretin variant (alanine 71) associated with familial amyloidotic polyneuropathy in a French family. J. Med. Genet. 30 (1993) 120–122.

BREATHNACH, S.M. and M.M. BLACK: Systemic amyloidosis and the skin: a review with special emphasis on clinical features and therapy. Clin. Exp. Dermatol. 4 (1979) 517–536.

BRUNI, J., J.M. BILBAO and K.P. PRITZKER: Myopathy associated with amyloid angiopathy. Can. J. Neurol. Sci. 4 (1977) 77–80.

CHALK, C.H. and P.J. DYCK: Application of immunohistochemical techniques to sural nerve biopsies. Neurol. Clin. 10 (1992) 601–613.

CHAMBERS, R.A., W.E. MEDD and H. SPENCER: Primary amyloidosis: with special reference to involvement of the nervous system. Q. J. Med. xxvii (1958) 207–226.

COELHO, T., A. SOUSA, E. LAURENÇO and J. RAMALHEIRA: A study of 159 Portuguese patients with familiar amyloidotic polyneuropathy (FAP) whose parents were both unaffected. J. Med. Genet. 31 (1994) 293–299.

COHEN, A.S.: Amyloidosis. N. Engl. J. Med. 277 (1967) 522–530.

COHEN, A.S. and E. CALKINS: Electron microscopic observations on a fibrous component in amyloid of diverse origins. Nature 183 (1959) 1202–1203.

COHEN, A.S. and A. RUBINOW: Amyloid neuropathy. In: P.J. Dyck et al. (Eds.), Peripheral Neuropathy, 2nd Edit. Philadelphia, PA, Saunders (1984) plate 1, p. 1881.

COHEN, A.S. and M. SKINNER: New frontiers in the study of amyloidosis (Editorial). N. Engl. J. Med. 323 (1990) 542–543.

COHEN, A.S., A. RUBINOW, J.J. ANDERSON, M. SKINNER, J.H. MASON, C. LIBBEY and H. KAYNE: Survival of patients with primary amyloidosis: colchicine-treated cases from 1976 to 1983 compared with cases seen in previous years (1961 to 1973). Am. J. Med. 82 (1987) 1182–1190.

COIMBRA, A. and C. ANDRADE: Familial amyloid polyneuropathy: an electron microscope study of the peripheral nerve in five cases. I. Interstitial changes. Brain 94 (1971a) 199–206.

COIMBRA, A. and C. ANDRADE: Familial amyloid polyneuropathy: an electron microscope study of the peripheral nerve in five cases. II. Nerve fibre changes. Brain 94 (1971b) 207–212.

COSTA, P.M.P., A. TEXEIRA, M.J.M. SARAIVA and P.P. COSTA: Immunoassay for transthyretin variants associated with amyloid neuropathy. Scand. J. Immunol. 38 (1993) 177–182.

COSTA, P.P., A.S. FIGUEIRA and F.R. BRAVO: Amyloid fibril protein related to prealbumin in familiar amyloidotic polyneuropathy. Proc. Natl. Acad. Sci. USA 75 (1978) 4499–4503.

CUETO-GARCIA, L., G.S. REEDER, R.A. KYLE, D. WOOD, G.B. SEWARD, G. NAESSENS, K.P. OSSORD, P.R. GREIPP, W.D. EDWARDS and A.J. TAJIK: Echocardiographic findings in systemic amyloidosis: spectrum of cardiac involvement and relation to survival. J. Am. Coll. Cardiol. 6 (1985) 737–743.

DALAKAS, M.C. and W.K. ENGEL: Amyloid in hereditary amyloid polyneuropathy is related to prealbumin. Arch. Neurol. 38 (1981) 420–422.

DARRAS, B.T., L.S. ADELMAN, J.S. MORA, R.A. BODZINER and T.L. MUNSAT: Familial amyloidosis with cranial neuropathy and corneal lattice dystrophy. Neurology 36 (1986) 432–435.

DI IORIO, G., G. SANGES, A. CERRACCHIO, S. SAMPAOLO, V. SANNINO and V. BONAVITA: Familial amyloidotic polyneuropathy: description of an Italian kindred. Ital. J. Neurol. Sci. 14 (1993) 303–309.

DURIE, B.G., B. PERSKY, B.J. SOEHNLEN, T.M. GROGAN and S.E. SALMON: Amyloid production in human myeloma stem-cell culture, with morphologic evidence of amyloid secretion by associated macrophages. N. Engl. J. Med. 307 (1982) 1689–1692.

DUSTON, M., M. SKINNER, J. ANDERSON and A.S. COHEN: Peripheral neuropathy as an early marker of AL amyloidosis. Arch. Intern. Med. 149 (1989) 358–360.

DYCK, P.J. and E.H. LAMBERT: Dissociated sensation in amyloidosis: compound action potential, quantitative histologic and teased-fiber, and electron microscopic studies of sural nerve biopsies. Arch. Neurol. 20 (1969) 490–507.

ELGHETANY, M.T. and A. SALEEM: Methods for staining amyloid in tissues: a review. Stain Technol. 63 (1988) 201–212.

ERICZON, B.G., O. SUHR, U. BROOME, F. HOLMGREN, F. DURAJ, L. ELEBORG, L WIKSTRÖM, G. NORDEN, S. FRIMAN and C.G. GROTH: Liver transplantation halts the progress of familial amyloidotic polyneuropathy. Transplant. Proc. 27 (1995) 1233.

FITTING, J.W., F. BISCHOFF, F. REGLI and G. CROUSAZ: Neuropathy, amyloidosis and monoclonal gammopathy. J. Neurol. Neurosurg. Psychiatry 42 (1979) 193–202.

FRENCH, J.M., G. HALL, D.J. PARISH and W.T. SMITH: Peripheral and autonomic nerve involvement in primary amyloidosis associated with uncontrollable diarrhoea and steatorrhoea. Am. J. Med. 39 (1965) 277–284.

FURUYA, H., M.J. MASCARENHAS SARAIVA, M.A. GAWINOWICZ, I. LONGO-ALVES, P.P. COSTA, H. SASAKI, I. GOTO and Y. SAKAKI: Production of recombinant human transthyretin with biological activities toward the understanding of the molecular basis of familial amyloidotic polyneuropathy (FAP). Biochemistry 30 (1991) 2415–2421.

GAAN, D., M.P. MAHONEY, D.J. ROWLANDS and A.W. JONES:

Postural hypotension in amyloid disease. Am. Heart J. 84 (1972) 395–400.

GASTINEAU, D.A., M.A. GERTZ, T.M. DANIELS, R.A. KYLE and E.J. BOWIE: Inhibitor of the thrombin time: a common coagulation abnormality. Blood 77 (1991) 2637–2640.

GERTZ, M.A. and R.A. KYLE: Amyloidosis: prognosis and treatment. Semin. Arthr. Rheum. 24 (1994) 124–138.

GERTZ, M.A., R.A. KYLE, W.L. GRIFFING and G.G. HUNDER: Jaw claudication in primary amyloidosis. Medicine (Baltimore) 65 (1986) 173–179.

GERTZ, M.A., C.Y. LI, T. SHIRAHAMA and R.A. KYLE: Utility of subcutaneous fat aspiration for the diagnosis of systemic amyloidosis (immunoglobulin light chain). Arch. Intern. Med. 148 (1988) 929–933.

GERTZ, M.A., R.A. KYLE, P.R. GREIPP, J.A. KATZMANN and W.M. O'FALLON: Beta$_2$-microglobulin predicts survival in primary systemic amyloidosis. Am. J. Med. 89 (1990) 609–614.

GERTZ, M.A., R.A. KYLE and S.N. THIBODEAU: Familial amyloidosis: a study of 52 north American-born patients examined during a 30-year period. Mayo Clin. Proc. 67 (1992) 428–440.

GLENNER, G.G.: Amyloid deposits and amyloidosis: the β-fibrilloses (first of two parts). N. Engl. J. Med. 302 (1980a) 1283–1292.

GLENNER, G.G.: Amyloid deposits and amyloidosis: the β-fibrilloses (second of two parts). N. Engl. J. Med. 302 (1980b) 1333–1343.

GOREVIC, P.D., P.C. MUNOZ, G. GORGONE, J.J. PURCELL, M. RODRIGUEZ, J. GHISO, E. LEVY, M. HALTIA and B. FRANGIONE: Amyloidosis due to a mutation of the Gelsolin gene in an American family with lattice corneal dystrophy type II. N. Engl. J. Med. 325 (1991) 1780–1785.

GRATEAU, G. and M.E. ROUX: Amyloses familiales. Presse Méd. 21 (1992) 1768–1773.

HAAN, J. and W.G. PETERS: Amyloid and peripheral nervous system disease. Clin. Neurol. Neurosurg. 96 (1994) 1–9.

HABERMANN, M.C. and M.R. MONTENEGRO: Primary cutaneous amyloidosis: clinical, laboratory and histopathological study of 25 cases; identification of the gammaglobulins in C$_3$ in the lesions by immunofluorescence. Dermatologica 160 (1980) 240–248.

HALTIA, M., E. LEVY, J. MERETOJA, I. FERNANDEZ-MADRID, O. KOIVUNEN and B. FRANGIONE: Gelsolin gene mutation at codon 187 in familial amyloidosis, Finnish: DNA-diagnostic assay. Am. J. Med. Genet. 42 (1992) 357–359.

HANYU, N., S. IKEDA, A. NAKADAI, N. YANAGISAWA and H.C. POWELL: Peripheral nerve pathological findings in familial amyloid polyneuropathy: a correlative study of proximal sciatic nerve and sural nerve lesions. Ann. Neurol. 25 (1989) 340–350.

HARDING, A.E.: Molecular genetics and clinical aspects of inherited disorders of nerve and muscle. Curr. Opin. Neurol. Neurosurg. 5 (1992) 600–604.

HAWKINS, P.N., J.P. LAVENDER and M.B. PEPYS: Evaluation of systemic amyloidosis by scintigraphy with ^{123}I-labeled serum amyloid-P component. N. Engl. J. Med. 323 (1990) 508–513.

HOLMGREN, G., E. HARTTNER, I. NORDENSON, O. SANDGREN, L. STEEN and E. LUNDGREN: Homozygosity for the trans-thyretin-Met-30 gene in two Swedish sibs with familial amyloidotic polyneuropathy. Clin. Genet. 34 (1988) 333–338.

HOLMGREN, G., B.G. ERICZON, C.G. GROTH, L. STEEN, O. SUHR, O. ANDERSEN, B.G. WALLIN, A. SEYMOUR, S. RICHARDSON, P.N. HAWKINS and M.B. PEPYS: Clinical improvement and amyloid regression after liver transplantation in hereditary transthyretin amyloidosis. Lancet 341 (1993) 1113–1116.

HOLT, I.J., A.E. HARDING, L. MIDDLETON, G. CHRYSOSTOMOU, G. SAID, R.H.M. KING and P.K. THOMAS: Molecular genetics of amyloid neuropathy in Europe. Lancet ii (1989) 524–526.

II, S., S. MINNERATH, K., II, P.J. DYCK and S.S. SOMER: Two-tiered DNA-based diagnosis of transthyretin amyloidosis reveals two novel point mutations. Neurology 41 (1991) 893–898.

IKEDA, S., N. HANYU, M. HONGO, J. YOSHIOKA, H. OGUCHI, N. YANAGISAWA, T. KOBAYASHI, H. TSUKAGOSHI, N. ITO and T. YOKOTA: Hereditary generalized amyloidosis with polyneuropathy. Clinicopathological study of 65 Japanese patients. Brain 110 (1987a) 315–337

IKEDA, S., N. YANAGISAWA, M. HONGO and N. ITO: Vagus nerve and celiac ganglion lesions in generalized amyloidosis: a correlative study of familial polyneuropathy and AL-amyloidosis. J. Neurol. Sci. 79 (1987b) 129–139.

JACOBSON, D.R., F. SANTIAGO-SCHWARZ and J.N. BUXBAUM: Restriction fragment analysis confirms the position 33 mutation in transthyretin from an Israeli patient (SKO) with familial amyloidotic polyneuropathy. Biochem. Biophys. Res. Commun. 153 (1988) 198–202.

JENNEKENS, F.G.I. and J.H.J. WOKKE: Proximal weakness of the extremities as main feature of amyloid myopathy. J. Neurol. Neurosurg. Psychiatry 50 (1987) 1353–1358.

JONES, L.A., J.C. SKARE, J.A HARDING, A.S. COHEN, A. MILUN-SKY and M. SKINNER: Proline at position 36: a new trans-thyretin mutation associated with familial amyloidotic polyneuropathy. Am. J. Hum. Genet. 48 (1991) 979–982.

JULIEN, J.: Les neuropathies amyloïdes familiales. Rev. Neurol. (Paris) 149 (1993) 517–523.

JULIEN, J., C. VITAL, J.M. VALLAT, A. LAGUENY, X. FERRER, C. DEMINIERE, W. LEBOUTER and C. EFFROY: IgM demyelin-ative neuropathy with amyloidosis and biclonal gam-mopathy. Ann. Neurol. 15 (1984) 395–399.

KATAFUCHI, R., T. TAGUCHI, S. TAKEBAYASHI and T. HARADA: Proteinuria in amyloidosis correlates with epithelial detachment and distortion of amyloid fibrils. Clin. Nephrol. 22 (1984) 1–8.

KAUFMAN, B.M.: Primary amyloidosis, paraproteinemia and neuropathy. Proc. Roy. Soc. Med. 69 (1976) 707–708.

KELLY, J.J., JR.: The electrodiagnostic findings in periph-eral neuropathy associated with monoclonal gam-mopathy. Muscle Nerve 6 (1983) 504–509.

KELLY, J.J., R.A. KYLE, P.C. O'BRIEN and P.J. DYCK: The natural history of peripheral neuropathy in primary systemic amyloidosis. Ann. Neurol. 6 (1979) 1–7.

KELLY, J.J., R.A. KYLE, J.M. MILES, P.C. O'BRIEN and P.J. DYCK: The spectrum of peripheral neuropathy on myeloma. Neurology 31 (1981) 24–31.

KERNOHAN, J.W. and H.W. WOLTMAN: Amyloid neuritis. Arch. Neurol. 47 (1942) 132–140.

KHOJASTEH, A., L.T. ARNOLD and M. FARHANGI: Bone lesions in primary amyloidosis. Am. J. Hematol. 7 (1979) 77–86.

KINCAID, J.C., M.R. WALLACE and M.D. BENSON: Late onset familial amyloid polyneuropathy in an American family of English origin. Neurology 39 (1989) 861–863.

KIURU, S.: Familial amyloidosis of the Finnish type (FAF). A clinical study of 30 patients Acta Neurol. Scand. 86 (1992) 346–353.

KIURU, S. and A.M. SEPPÄLÄINEN: Neuropathy in familial amyloidosis, Finnish type (FAF): electrophysiological studies. Muscle Nerve 17 (1994) 299–304.

KIURU, S., E. MATIKAINEN, M. KUPARI, M. HALTIA and J. PALO: Autonomic nervous system and cardiac involvement in familial amyloidosis, Finnish type (FAF). J. Neurol. Sci. 126 (1994) 40–48.

KLATSKIN, G.: Nonspecific green birefringence in Congo Red-stained tissues. Am. J. Pathol. 56 (1969) 1–13.

KOEPPEN, A.H., E.J. MITZEN, M.B. HANS, S.K. PENG and R.O. BAILEY: Familial amyloid polyneuropathy. Muscle Nerve 8 (1985) 733–749.

KOEPPEN, A.H., M.R. WALLACE, M.D. BENSON and K. ALTLAND: Familial amyloid polyneuropathy: alanine-for-threonine substitution in the transthyretin (pre-albumine) molecule. Muscle Nerve 13 (1990) 1065–1075.

KYLE, R.A.: Monoclonal proteins in neuropathy. Neurol. Clin. 10 (1992) 713–734.

KYLE, R.A. and E.D. BAYRD: Amyloidosis: review of 236 cases. Medicine (Baltimore) 54 (1975) 271–299.

KYLE, R.A. and P.A. DYCK: Amyloidosis and neuropathy: In: P.J. Dyck, P.K. Thomas, J.W. Griffin, P.A. Low and J.F. Poduslo (Eds.), Peripheral Neuropathy. Philadelphia, PA, Saunders (1993) 1294–1309.

KYLE, R.A. and J.P. GARTON: Laboratory monitoring of myeloma proteins. Semin. Oncol. 13 (1986) 310–317.

KYLE, R.A. and M.A. GERTZ: Systemic amyloidosis. Crit. Rev. Oncol. Hematol. 10 (1990) 49–87

KYLE, R.A. and P.R. GREIPP: Primary systemic amyloidosis: comparison of melphalan and prednisone versus placebo. Blood 52 (1978) 818–827.

KYLE, R.A. and P.R. GREIPP: Amyloidosis (AL): clinical and laboratory features. Mayo Clin. Proc. 58 (1983) 665–683.

KYLE, R.A., R.D. WAGONER and K.E. HOLLEY: Primary systemic amyloidosis. Resolution of the nephrotic syndrome with melphalan and prednisone. Arch. Intern. Med. 142 (1982) 1445–1447.

KYLE, R.A., P.R. GREIPP, J.P. GARTON and M.A. GERTZ: Primary systemic amyloidosis: comparison of melphalan/prednisone versus colchicine. Am. J. Med. 79 (1985) 708–716.

KYLE, R.A., P.R. GREIPP and W.M. O'FALLON: Primary systemic amyloidosis: multivariate analysis for prognostic factors in 168 cases. Blood 68 (1986) 220–224.

KYLE, R.A., M.A. GERTZ, J.P. GARTON ET AL.: Primary systemic amyloidosis (AL): randomized trial of colchicine vs. melphalan and prednisone vs. melphalan, prednisone and colchicine. In: R. Kisilevsky, M.D. Benson, B. Frangione et al. (Eds.), Amyloid and Amyloidosis. Park Ridge, NJ, Parthenon (1994) 648–650.

LIBBEY, C.A., A. RUBINOW, T. SHIRAHAMA, C. DEAL and A.S. COHEN: Familiar amyloid polyneuropathy. Demonstration of prealbumin in a kinship of German/English ancestry with onset in the seventh decade. Am. J. Med. 76 (1984) 18–24.

LINGENFELSER, T., R.P. LINKE, S. DETTE, W. ROGGENDORF and H. WIETHOLTER: AL amyloidosis mimicking a preferentially autonomic chronic Guillain–Barré syndrome. Clin. Invest. 70 (1992) 159–162.

LOW, P.A., P.A. DYCK, H. OKASAKI, R. KYLE and R.D. FEALET: The splanchnic autonomic outflow in amyloid neuropathy and Tangier disease. Neurology 31 (1981) 461–463.

MAKISHITA, H., M. YAZAKI, M. MATSUDA, S. IKEDA and N. YANAGISAWA: Familial amyloid polyneuropathy type IV (Finnish type): a clinicopathological study. Rinsho Shinkeiga 34 (1994) 431–437.

MASCARENHAS SARAIVA, M.J., P.P. COSTA and D.S. GOODMAN: Genetic expression of a transthyretin mutation in typical and late-onset Portuguese families with familial amyloidotic polyneuropathy. Neurology 36 (1986) 1413–1417.

MCLEOD, J.G.: Invited review: autonomic dysfunction in peripheral nerve disease. Muscle Nerve 15 (1992) 3–13.

MELGAARD, B. and B. NIELSEN: Electromyographic findings in amyloid neuropathy. Electromyogr. Clin. Neurophysiol. 17 (1977) 31–34.

MENDELL, J.R., X.S. JIANG, J.R. WARMOLTS, W.C. NICHOLS and M.D. BENSON: Diagnosis of Maryland/German familial amyloidotic polyneuropathy using allele-specific, enzymatically amplified, genomic DNA. Ann. Neurol. 27 (1990) 553–557.

MERLINI, G.: Treatment of primary amyloidosis. Semin. Hematol. 32 (1995) 60–79.

METZLER, J.P., J.L. FLECKENSTEIN, C.L. WHITE, R.G. HALLER, E.P. FRENKEL and R.G. GRENLEE: MRI evaluation of amyloid myopathy. Skeletal Radiol. 21 (1992) 463–465.

MITA, S., S. MAEDA, M. IDE, T. TSUZUKI, K. SHIMADA and S. ARAKI: Familial amyloidotic polyneuropathy diagnosed by cloned human prealbumin cDNA. Neurology 36 (1986) 298–301.

NADKARNI, N., M. FREIMER and J.R. MENDELL: Amyloid causing a progressive myopathy. Muscle Nerve 18 (1995) 1016–1018.

NAKAZATO, M., M. TANAKA, Y. YAMAMURA, T. KURIHARA, S. MATSUKURA, K. KANGAWA and H. MATSUO: Abnormal transthyretin in asymptomatic relatives in familial amyloidotic polyneuropathy. Arch. Neurol. 44 (1987a) 1275–1278.

NAKAZATO, M., H. SASAKI, H. FURUYA, Y. SAKAKI, T. KURIHARA, S. MATSUKURA, K. KANGAWA and H. MATSUO: Biochemical and genetic characterization of type 1 familial amyloidotic polyneuropathy. Ann. Neurol. 21 (1987b) 596–598.

NEUNDÖRFER, B, J.G. MEYER and B. VOLK: Amyloid neuropathy due to monoclonal gammopathy. A case report. J. Neurol. 21 (1977) 207–215.

NICHOLS, W.C., F.E. DWULET, J. LIEPNIEKS and M.D. BENSON: Variant apolipoprotein A1 as a major constituent of a human hereditary amyloid. Biochem. Biophys. Res. Commun. 156 (1988) 762–768.

OCHIAI, J., S. TOBIMATSU, T. KOBAYASHI, T. KITAMOTO, T. KITAGUCHI, H. FURUYA, I. GOTO and Y. KUROIWA: Nonfamilial prealbumin-type amyloid polyneuropathy. Arch. Neurol. 43 (1986) 1294–1295.

O'CONNOR, C.R., A. RUBINOW, S. BRANDWEIN and A. COHEN: Familiar amyloid polyneuropathy: a new kinship of German ancestry. Neurology 34 (1984) 1096–1099.

PASCALI, E., A. PEZZOLI, M. MELATO and G. ANTONUTTO: Pseudotumoral (para-articular) amyloidosis in non-myelomatous monoclonal gammopathy. Pathol. Res. Pract. 168 (1980) 215–223.

RAVID, M., M. ROBSON and I. KEDAR: Prolonged colchicine treatment in four patients with amyloidosis. Ann. Intern. Med. 87 (1977) 568–570.

RINGEL, S.P. and H.N. CLAMAN: Amyloid-associated muscle pseudohypertrophy. Arch. Neurol. 39 (1982) 413–417.

ROBERTS, W.C. and B.F. WALLER: Cardiac amyloidosis causing cardiac dysfunction: analysis of 54 necropsy patients. Am. J. Cardiol. 52 (1983) 137–146.

RUBINOW, A. and A.S. COHEN: Skin involvement in familial amyloidotic polyneuropathy. Neurology 31 (1981) 1341–1345.

RUKAVINA, J.G., W.D. BLOCK and A.C. CURTIS: Familial primary systemic amyloidosis: an experimental, genetic and clinical study. J. Invest. Dermatol. 27 (1956) 111.

SAID, G., A. ROPERT and N. FAUX: Length-dependent degeneration of fibrils in Portuguese amyloid neuropathy. Neurology 34 (1984) 1025–1032.

SALES-LUIS, M.L., M. GALVÃO, M. CARVALHO, G. SOUSA, M.M. ALVES and R. SERRÃO: Treatment of familiar amyloidotic polyneuropathy (Portuguese type) by plasma exchange (Letter). Muscle Nerve 14 (1991) 377–378.

SARAIVA, M.J.: Recent advances in the molecular pathology of familial amyloid polyneuropathy. Neuromusc. Dis. 1 (1991) 3–6.

SATHER, E., W.C. NICHOLS and M.D. BENSON: Diagnosis of familial amyloidotic polyneuropathy in France. Clin. Genet. 38 (1990) 469–473.

SKARE, J.C., M.J. SARAIVA, I. ALVES, I.B. SKARE, A. MILUNSKY, A.S. COHEN and M. SKINNER: A new mutation causing familial amyloidotic polyneuropathy. Biochem. Biophys. Res. Commun. 164 (1989) 1240–1246.

SKINNER, M., W.D. LEWIS, L.A. JONES, J. KASIRSKY, K. KANE, S. JU, R. JENKINS, R.H. FALK, R.W. SIMMS and A.S. COHEN: Liver transplantation as a treatment for familial amyloidotic polyneuropathy. Ann. Intern. Med. 120 (1994) 133–134.

SMITH, T.A., R.A. KYLE and J.T. LIE: Clinical significance of histopathologic patterns of cardiac amyloidosis. Mayo Clin. Proc. 59 (1984) 547–555.

SOLOMON, A., B. FRANGIONI and E.C. FRANKLIN: Bence Jones proteins and light chains of immunoglobulins: preferential association of the $V_{\lambda IV}$ subgroup of human light chains with amyloidosis AL (λ). J. Clin. Invest. 70 (1982) 453–460.

SOMMER, C. and J.M. SCHRÖDER: Amyloid neuropathy: immunocytochemical localization of intra- and extracellular immunoglobulin light chains. Acta Neuropathol. (Berl.) 79 (1989) 190–199.

STAUNTON, H.: Familial amyloid polyneuropathies. In: P.J. Vinken, G.W. Bruyn, H.L. Klawans and J.M.B.V. De Jong (Eds.), Hereditary Neuropathies and Spinocerebellar Atrophies. Handbook of Clinical Neurology, Vol. 60. Amsterdam, Elsevier Science (1991) 89–115.

STAUNTON, H., P. DERVAN, R. KALE, R.P. LINKE and P. KELLEY: Hereditary amyloid polyneuropathy in north west Ireland. Brain 110 (1987) 1231–1245.

STEEN, L., A. WAHLIN, P. BJERLE and S. HOLM: Renal function in familial amyloidosis with polyneuropathy. Acta Med. Scand. 212 (1982) 233–236.

SUNADA, Y., H. NAKASE, T. SHIMIZU, T. MANNEN and I. KANAZAWA: Gene analysis of Japanese patients with familial amyloidotic polyneuropathy type IV. Rinsho Shinkeiga 32 (1992a) 840–844.

SUNADA, Y., T. SHIMIZU, T. MANNEN and I. KANAZAWA: Familial amyloidotic polyneuropathy type IV (Finnish type): the first description of a large kindred in Japan. Rinsho Shinkeiga 32 (1992b) 826–833.

SUNADA, Y., T. SHIMIZU, H. NAKASE, S. OHTA, T. ASAOKA, S. AMANO, M. SAWA, Y. KAGAWA, I. KANAZAWA and T. MANNEN: Inherited amyloid polyneuropathy type W (gelsolin variant) in a Japanese family. Ann. Neurol. 33 (1993) 57–62.

SUSUKI, T., T. AZUMA, S. TSUJINO, R. MIZUNO, S. KISHIMOTO, Y. WADA, A. HAYASHI, S. IKEDA and N. YANAGISAWA: Diagnosis of familial amyloidotic polyneuropathy: isolation of a variant prealbumin. Neurology 37 (1987) 708–711.

THOMAS, P.K. and R.H.M. KING: Peripheral nerve changes in amyloid neuropathy. Brain 97 (1974) 395–406.

TOWLE, P.A. and E.M. MAHER: Sporadic cranial amyloid neuropathy. Neurology 33, Suppl. 2 (1983) 189.

TRAYNOR, A.E., M.A. GERTZ and R.A. KYLE: Cranial neuropathy associated with primary amyloidosis. Ann. Neurol. 29 (1991) 451–454.

TROTTER, J.L., W.K. ENGEL and T.F. IGNACZAK: Amyloidosis with plasma cell dyscrasia: an overlooked cause of adult onset sensorymotor neuropathy. Arch. Neurol. 34 (1977) 209–214.

TURPIN, J.C., S. BERRICHE, F. DAVID, P. RUME, Y. DUMEZ and G. LUCOTTE: Diagnostic prénatal dans une famille atteinte de neuropathie amyloïde. Presse Méd. 21 (1992) 2152.

UEMICHI, T., J.R. MURREL, S. ZELDENRUST and M.D. BENSON: A new mutant transthyretin (Arg 10) associated with familial amyloid polyneuropathy. J. Med. Genet. 29 (1992) 888–891.

UEMICHI, T., M.A. GERTZ and M.D. BENSON: Amyloid polyneuropathy in two German American families: a new transthyretin variant (Val 107). J. Med. Genet. 31 (1994) 416–417.

UENO, S., T. UEMICHI, N. TAKAHASHI, F. SOGA, S. YORIFUJI and S. TARUI: Two novel variants of transthyretin identified in Japanese cases with familial amyloidotic polyneuropathy; transthyretin (Glu 42 to Gly) and transthyretin (Ser 50 to Arg). Biochem. Biophys. Res. Commun. 169 (1990) 1117–1121.

VAN ALLEN, M.W., J.A. FROHLICH and J.R. DAVIS: Inherited predisposition to generalized amyloidosis: clinical and pathological study of a family with neuropathy, nephropathy and peptic ulcer. Neurology 10 (1969) 10–25.

VARGA, J. and J.R. WOHLGETHAN: The clinical and biochemical spectrum of hereditary amyloidosis. Semin. Arthr. Rheum. 18 (1988) 14–28.

VITAL, A. and C. VITAL: Amyloid neuropathy: relationship between amyloid fibrils and macrophages. Ultrastruct. Pathol. 7 (1984) 21–24.

VITAL, C. and J.M. VALLET: Ultrastructural Study of the Human Diseased Peripheral Nerve, 2nd Edit. Amsterdam, Elsevier Science (1987) Fig. 139, p. 114.

WALLACE, M.R., P.M. CONNEALLY and M.D. BENSON: A DNA

test for Indiana/Swiss hereditary amyloidosis (FAP II). Am. J. Hum. Genet. 43 (1988) 182–187.

WATANABE, T. and T. SANITER: Morphological and clinical features of renal amyloidosis. Virchows Arch. (Pathol. Anat.) 379 (1978) 131–141.

WHITAKER, J.R., K. HASHIMOTO and M. QUINONES: Skeletal muscle pseudohypertrophy in primary amyloidosis. Neurology 27 (1977) 47–54.

WHO-IUIS NOMENCLATURE SUB-COMMITTEE: Nomenclature of amyloid and amyloidosis. Bull. World Health Org. 71 (1993) 105–108.

WRIGHT, J.R. and E. CALKINS: Clinical-pathological differentiation of common amyloid syndromes. Medicine (Baltimore) 60 (1981) 429–448.

ZEMER, I.D., M. PRAS, E. SOHAR, M. MODAN, S. CABILI and J. GAFNI: Colchicine in the prevention and treatment of the amyloidosis of familiar Mediterranean fever. N. Engl. J. Med. 314 (1986) 1001–1005.

Handbook of Clinical Neurology, Vol. 27 (71): Systemic Diseases, Part III
M.J. Aminoff and C.G. Goetz, editors

CHAPTER 21

Neurologic complications of critical illness involving multi-organ failure

J. CLAUDE HEMPHILL III and DARYL R. GRESS

NeuroCritical Care and NeuroVascular Service, Department of Neurology, University of California, San Francisco, CA, U.S.A.

During the past four decades, the medical treatment of critically ill patients has undergone dramatic, rapid, and profound evolution. Individuals who would previously have died as a result of their critical illnesses are being supported and often survive, both with and without long-term sequelae. The diseases which can now be supported and treated involve many and often multiple organ systems and include cardiac disease, hepatic failure, renal failure and sepsis from pulmonary, intra-abdominal, or genitourinary sources. The new successes over the past decades in the treatment of critical illnesses can be attributed to new surgical techniques, the development of new antibiotics, improved hemodynamic monitoring capabilities, and evolving understanding about nutritional support, but the primary reason for the continued evolution of critical care is probably improvement in mechanical ventilation (Petty 1990). By being able to successfully oxygenate and ventilate a critically ill patient for long periods of time using an artificial mechanical ventilator, more time is provided for the underlying disease process to resolve, either with treatment or on its own. However, as patients are supported for long periods of time, ensuing complications of critical illness become more evident. Nosocomial infections, including pneumonia and genitourinary tract infections, have long been recognized as common and significant complications of prolonged treatment of critical illness. Additionally, compromise and failure of organ systems which were not involved at the time of

initial critical illness, such as hepatic or renal dysfunction, have become well recognized and much feared complications of the treatment of critical illness. As the understanding of the complications of long-term treatment of critical illness has also evolved, it has become apparent that the nervous system, both central and peripheral, may be subject to complications and compromise as well. It is only in the past decade that the significant impact that neurologic complications may have on the course of critical medical illness has begun to be recognized (Isensee et al. 1989; Bleck et al. 1993).

Critical intensive care can trace its roots back to the use of mechanical ventilation with negative pressure 'iron lungs' during the polio epidemic of the 1930s–1950s (Ayres 1995). Certainly, neurologic diseases, from stroke to the Guillain–Barré syndrome, may require critical intensive care. However, the most frequent neurologic problems encountered in the critical care setting are those which result from complications of treatment and support of other medical and surgical illnesses (Bleck et al. 1993). *Critical illness* can be defined as an unstable physiologic condition in which small changes may lead to serious deterioration with irreversible organ system damage or death (Ayres 1995). With the evolution of critical intensive care, the term critical illness has often been taken to imply a disease state in which there is dysfunction of more than one organ system. This syndrome of multiple organ system failure may result from a single identifiable unifying cause such as hypoperfusion of the liver and kidneys in the setting of cardiogenic shock. However, more often it represents a syndrome in which multiple organ systems demonstrate dysfunction and the unifying cause is thought to be related to remote effects of sepsis caused by circulating inflammatory mediators and cytokines (Matuschak 1992). Sepsis involves the systemic response to infection by invading microorganisms, be they bacterial, fungal, parasitic, or viral in origin (Ayres 1986). Often, however, the specific invading organism cannot be identified. In fact, blood cultures are negative in about one-half of patients suspected of having sepsis as evidenced by hemodynamic and multi-organ compromise (Ayres 1986). Thus, it may be more appropriate to discuss critical illness and multi-organ system failure in terms of the *systemic inflammatory response syndrome* (SIRS), which is defined as the widespread inflammation that may occur in a variety of disorders even without identified infection (Bone et al. 1992). SIRS may become activated in the setting of sepsis, trauma or burns and may result in multiple systems organ failure (MSOF). Much study has been devoted to determining the causes of parenchymal organ system dysfunction in SIRS including the acute respiratory distress syndrome (ARDS), hepatic failure, renal failure, and hemodynamic compromise (Matuschak 1992); despite this, specific etiologies remain unclear. The central and peripheral nervous systems are frequently involved in SIRS. However, only recently have specific neurologic syndromes in critical illness involving multi-organ failure been categorized and systematically investigated (Bowton 1989; Bolton et al. 1993).

The neurologic complications of critical illness involving multi-organ failure include dysfunction of the central and peripheral nervous systems in the setting of sepsis or SIRS, peripheral nervous system complications related to medications used to treat the underlying critical illness, and complications related to interventions often used in the intensive care unit setting such as placement of invasive hemodynamic monitoring devices like pulmonary artery catheters and arterial catheters. Additionally, neurologic dysfunction related to other organ system failure, for example encephalopathy in hepatic failure or longstanding neuropathy secondary to diabetes, is frequently encountered in the critically ill patient and may make diagnosis of other conditions difficult. Finally, patients in the intensive care unit being treated for non-neurologic critical illnesses may develop neurologic complications related to more conventional neurologic diseases such as ischemic stroke, intracerebral hemorrhage, seizures, brain abscess, meningitis, or Guillain–Barré syndrome. The purpose of this chapter is to introduce the neurologic complications of critical illness involving multi-organ failure which are unique to the intensive care unit setting and do not just represent well known complications of other systemic diseases, such as hepatic encephalopathy. The incidence, clinical presentations, potential mechanisms, and proposed treatments will be reviewed. In this manner, it should be clear that neurologic compromise in the setting of critical illness is a significant cause of morbidity and deserves further study as an important part of SIRS and the treatment of critically ill patients.

SPECIFIC NEUROLOGIC SYNDROMES

Each part of the neuraxis can be affected by a distinct neurologic syndrome which may result during the treatment of critical illness complicated by multi-organ failure. Diffuse central nervous system (CNS) dysfunction in the form of encephalopathy may result from underlying disease such as hepatic or renal dysfunction or may be a result of treatment (Jackson et al. 1985; Bowton 1989). Each portion of the peripheral nervous system may be involved individually or in concert when prolonged weakness occurs in the setting of critical illness. Peripheral neuropathy may result from medications, underlying disease, electrolyte disturbances or as a manifestation of critical illness (Bolton et al. 1993). The neuromuscular junction may be affected by medications such as antibiotics and especially by the use of non-depolarizing neuromuscular blocking agents which may be employed to aid mechanical ventilation (Hansen-Flaschen et al. 1993; O'Connor and Roizen 1993). Muscle may be affected by electrolyte disturbances or by combinations of various drug regimens, such as the combination of corticosteroids with non-depolarizing neuromuscular blockers (Lacomis et al. 1993; Wijdicks 1995).

Because neurologic dysfunction in critically ill patients may involve any or all parts of the neuraxis, the principles of neurologic localization still provide an appropriate approach to the clinical evaluation of these problems. Localization of the site of 'the lesion' with subsequent generation of a differential diagnosis is still the most valid way to approach the neurologic examination of critically ill patients in the intensive care unit. However, there are many barriers to the neurologic examination of these patients which make specific neurologic diagnosis a challenge. Many patients are intubated and are therefore unable to communicate verbally. The significant amount of 'hardware' which accompanies most patients in the intensive care unit, including bladder catheters, endotracheal tubes and ventilators, intravenous and hemodynamic monitoring lines, and ECG and other monitoring devices, can be a significant barrier to the performance of a complete and informative neurologic examination. Additionally, patients are often given sedative drugs or neuromuscular blocking agents with the goal of assisting the patient in tolerating mechanical ventilation and various invasive procedures (Aitkenhead 1989; Hansen-Flaschen et al. 1991). Certainly, these medications may cloud the underlying neurologic examination or result in neurologic complications of their own (Wijdicks 1995). Unfortunately, the principle of Ockham's razor often does not hold in critically ill patients. Frequently, these patients may have multiple diagnoses and multiple ongoing disease processes; in fact, patients do so, by definition, if multi-organ system failure is present. Patients who are immunocompromised from AIDS, cancer with or without chemotherapy, or organ transplantation may be particularly predisposed to these multiple organ system complications; however, all patients with sepsis, burns, or severe trauma are at high risk. Just as multiple organs may be involved, multiple sites along the neuraxis may be affected. All of these factors make the approach to the neurologically compromised critically ill patient with multi-organ system failure a diagnostic and therapeutic challenge. By understanding the basic syndromes that may occur at various sites along the neuraxis in critical illness and by judicious use of neurodiagnostic tests from electrophysiology to radiology, the neurologic complications of critical illness may be more clearly elucidated and understood.

CENTRAL NERVOUS SYSTEM

The first neurologic complications of critical illness involving multi-organ failure to be described involved the CNS (Jeppsson et al. 1981; Jackson et al. 1985).

Because level of consciousness is frequently involved, these are also usually the first recognized by physicians caring for critically ill patients. While focal syndromes from stroke, brain abscess, and trauma occur, global encephalopathy is vastly more common. Spinal cord syndromes are rarer and almost always indicate a focal neurologic process such as spinal cord ischemia or compression from a mass lesion such as tumor or abscess, which is occurring coincident with the critical illness but demands additional attention and possible treatment. Less commonly a concomitant infectious or parainfectious myelitis may occur (Wijdicks 1995).

Brain

Encephalopathy has been recognized as a complication of various systemic illnesses since the origins of medicine. In the 5th century B.C., Hippocrates described 'madness on account of bile' and told of a state like rabies in which a patient said things that could not be comprehended and barked liked a dog (Davidson and Summerskill 1956). Since then, encephalopathy has been known to be a common accompanying feature of hepatic and renal disease and less commonly even of pulmonary and pancreatic disease (Plum and Posner 1980). However, as observation of encephalopathy in critically ill patients has evolved, it has become apparent that many patients with sepsis or SIRS are encephalopathic without an obvious explanation from other specific organ system dysfunction or electrolyte disturbances (Bowton 1989). Thus, the concept of *septic encephalopathy* has evolved.

Septic encephalopathy refers to altered brain function in the setting of sepsis and critical illness without other metabolic or structural explanation (Bolton and Young 1995). Clinically this disorder is apparent as a global encephalopathy without prominent focal findings on neurologic examination. In its milder and early stages it may appear as confusion in an awake patient, with notable alterations in orientation, attention, and concentration. There may be significant fluctuations in level of consciousness and other aspects of cognition, especially during febrile episodes, hypotension, or at night. As the sepsis worsens, especially in the setting of hemodynamically significant alterations, a decreased level of consciousness is apparent and problems with language and cooperation become

TABLE 1

Neurologic syndromes in critically ill patients.

Localization on neuraxis	Syndrome
Central nervous system	
Brain – cerebral hemispheres	global encephalopathy
	septic
	organ system – hepatic
	renal
	medication induced
	sedatives, H_2 blockers, antihypertensives
	drug overdose
	electrolyte disturbances – especially hyponatremia, hypoglycemia
	hypotension/hypoperfusion
	anoxia
	meningitis
	subarachnoid hemorrhage
	Wernicke's
	seizure – post-ictal
	non-convulsive status epilepticus
	hypertensive encephalopathy
	hypothyroidism – myxedema
	focal
	ischemic stroke, intracerebral hemorrhage
	tumor, abscess
	subdural/epidural hematoma, subdural empyema
Brainstem	compression from mass
	ischemic stroke, intraparenchymal hemorrhage
	anoxia
Spinal cord	compression from mass, disk, epidural abscess
	ischemia (hypotensive or embolic)
	polio (anterior horn cell)
Peripheral nervous system	
Peripheral nerves	axonal
	critical illness polyneuropathy
	? neuromuscular blocking agent complication
	metabolic causes – uremia, diabetes
	medication – chemotherapy, antiretroviral agents
	demyelinating – Guillain–Barré syndrome
	chronic inflammatory demyelinating polyneuropathy
Neuromuscular junction	*neuromuscular blocking agents – prolonged effects*
	medications – esp. aminoglycoside antibiotics
	myasthenia gravis, Lambert–Eaton syndrome
Muscle	*septic myopathy*
	cachectic myopathy – with or without disuse atrophy
	acute quadriplegic myopathy
	secondary to corticosteroids +/– non-depolarizing neuromuscular
	blocking agents
	electrolyte disturbances – hypokalemia, hyperkalemia, hypophosphatemia

Complications unique to critical illness are in *italics*.

significant. The neurologic examination, except for mental status evaluation, remains largely intact. Paratonia is frequently found on motor examination and hyperreflexia with Babinski signs may also be present. Release reflexes such as a suck, snout, or grasp can frequently be elicited. Movement disorders such as myoclonus, tremor, and asterixis may be present in these patients (Bolton and Young 1995). While minor focalities may occur in the motor or cognitive examination, certainly the presence of a gross focal deficit such as a hemiparesis, gaze palsy, or paraplegia should alert the examiner to the possibility of stroke, seizure, or other neurologic disorder, as these findings are rare in pure septic encephalopathy. Patients with septic encephalopathy may appear agitated and are frequently given sedative medications such as opiates, benzodiazepines, or antipsychotics (such as haloperidol), which may further exacerbate the encephalopathy and make it difficult to delineate how much of the confusion is related to the illness and how much is related to medications. Eventually, obtundation or coma may ensue. It is important to realize that in patients in deep coma from any cause, meningeal signs may not be present (Plum and Posner 1980); this makes meningitis difficult to exclude on clinical grounds, even when only septic encephalopathy is suspected.

The incidence of encephalopathy in critically ill patients with multi-organ failure from sepsis is quite high. Young et al. (1992) found a 70% incidence of encephalopathy in a prospective study of 69 patients with sepsis. Encephalopathy may be an inevitable component of a septic critical illness if the sepsis is severe enough. While septic encephalopathy may be the most common neurologic complication of critical illness involving multi-organ failure, it is a diagnosis of exclusion. Encephalopathy, of course, may be a component of the neurologic picture of patients in intensive care units with many different disorders. The prognosis of septic encephalopathy itself is related to the prognosis of the underlying critical illness. Several authors have found that patients with septic encephalopathy, especially those with profound encephalopathy or coma, have a much higher mortality than septic patients without encephalopathy (Pine et al. 1983; Sprung et al. 1988). This may be on account of the encephalopathy but more likely serves as a marker for the severity of septic illness in the setting of multiple organ failure. The suggested mortality rate of about 50% in critically ill patients with septic encephalopathy (Sprung et al. 1990; Bolton and Young 1995) is similar to that observed in patients with ARDS (Kollef and Schuster 1995). This demonstrates the relatively poor prognosis of critically ill patients with remote organ failure; prognosis worsens as the number of organ systems affected increases (Beal

and Cerra 1994). Because the treatment of septic encephalopathy is the treatment of the underlying critical illness, it is almost always necessary to undertake a thorough search for any undiagnosed and independently treatable neurologic conditions prior to making the diagnosis of septic encephalopathy.

Other neurologic or metabolic conditions that may arise in critically ill patients, and may therefore need to be excluded prior to making the diagnosis of septic encephalopathy, include meningitis (bacterial, fungal, or viral), encephalitis, brain abscess, subdural empyema, cerebral venous sinus thrombosis (either septic or aseptic), seizure, ischemic stroke (including bacterial endocarditis with septic embolization), malignant hyperthermia, intracerebral hemorrhage, drug overdose, and hypothyroidism. A head CT scan with and without contrast or a head MRI is an appropriate first step to rule out focal cerebral pathology. MRI is better for evaluation of possible cerebral venous sinus thrombosis but may be difficult to perform in critically ill patients, especially those being mechanically ventilated (Ameri and Bousser 1992). Lumbar puncture may be necessary to exclude meningitis or encephalitis. An elevated opening pressure on lumbar puncture, especially in a patient with a non-diagnostic head imaging study, may be an important clue to intracranial hypertension from cerebral venous sinus thrombosis (Karabudak et al. 1990). In patients with coma or profound encephalopathy, we use some type of head imaging, at least a non-contrast head CT scan, to rule out focal intracerebral pathology prior to performing lumbar puncture, because the neurologic examination may be difficult and may be clouded by multiple factors, as previously discussed. Certainly, if meningitis is suspected clinically, antibiotics can be given empirically while tests are being arranged and performed. Head imaging in septic encephalopathy does not reveal any distinct unique abnormalities. Atrophy is sometimes observed, but this may be secondary to the hydration status of the patient, corticosteroid use, or other factors. Lumbar puncture is almost always normal in patients with purely septic encephalopathy. Mild cerebrospinal fluid (CSF) pleocytosis may, however, occur in a variety of clinical conditions, from bacterial endocarditis to a post-seizure state, in the absence of any suspicion of meningitis (Edwards et al. 1983; Lerner 1985). Thus, it is unclear whether the finding of a mild CSF pleocytosis in a patient suspected of having a septic encephalopathy represents a subclinical meningitis or just a reaction to systemic sepsis. In any event, this finding in a critically ill patient would likely necessitate appropriate antimicrobial coverage for meningitis.

Electroencephalography (EEG) is an important tool in the evaluation of the critically ill patient with encephalopathy or other CNS compromise, especially since the neurologic examination may be difficult to interpret and may fluctuate. The typical EEG pattern in septic encephalopathy is one of generalized slowing with a rhythm from 4 to 7 Hz (Bolton and Young 1995). As encephalopathy progresses and worsens, the EEG may demonstrate more slowing, triphasic waves, or a burst-suppression pattern. While it has been suggested that a burst-suppression pattern implies a poor prognosis in other coma states such as hypoxic-ischemic encephalopathy (Simon and Aminoff 1986), in septic encephalopathy the EEG findings do not seem reliable enough to allow accurate prognostication on this basis alone. In fact, Young et al. (1992) have observed patients with septic encephalopathy and a burst-suppression EEG pattern who have made a complete neurologic recovery coincident with recovery of their underlying critical illness.

One of the most important uses for the EEG in the critically ill septic patient with encephalopathy is to exclude seizure, especially non-convulsive status epilepticus, as a cause or contributing factor to the encephalopathy or coma (Tomson et al. 1992). Because patients with non-convulsive status epilepticus may have no motor activity or minor movements unrecognized by medical staff, the EEG may be the only means of detecting this potentially treatable neurologic problem. Additionally, urgent EEG may be the only way to exclude ongoing seizures in a patient who received neuromuscular blocking agents for intubation or other procedures during convulsive status epilepticus. Thus, it seems prudent that almost every patient with significant encephalopathy, and certainly every patient in a coma of unknown etiology, undergo EEG, if for no other reason than to exclude non-convulsive status epilepticus. At present, although only studied sparingly, evoked potential monitoring has not been shown to add any distinct independently useful information beyond that obtained with EEG testing alone (Maekawa et al. 1991).

Standard electrolyte and other laboratory testing, as is usually extensively done in critically ill patients requiring intensive care, may reveal a concomitant cause for ongoing encephalopathy such as renal failure, hepatic failure, hyponatremia, hypernatremia, hypoglycemia, or hypothyroidism. Often it is difficult to determine to what degree each of these factors is contributing to a multifactorial encephalopathy. Only by reversing treatable metabolic conditions or providing temporizing support, as with dialysis, can this usually be clarified. Unfortunately, EEG and other neurologic tests are of limited value in making these distinctions as well. As evidenced by the descriptions of the above neuro-diagnostic tests in septic encephalopathy, there is no present laboratory or neuro-diagnostic test to 'rule in' septic encephalopathy. The diagnosis is made on clinical grounds and these tests are important adjuncts that are useful in excluding other treatable entities.

At present the specific etiology and pathogenesis of septic encephalopathy remain undefined. It is clear, however, that this encephalopathy cannot be explained solely by the failure of other organ systems with resultant metabolic encephalopathies related to those conditions, such as hepatic or uremic encephalopathy. It seems reasonable that some of the same mechanisms and chemical mediators which are responsible for other organ system failures in critical illness, trauma, and sepsis, such as ARDS, hepatic failure, and renal failure, may also play a role in nervous system dysfunction. However, most of the specific characteristics of these putative chemical mediators of SIRS are unknown. Specifically, how they interact with cells to produce the complications of sepsis remains unclear. Substances that have been hypothesized to play a role in the chemical mediation of SIRS in sepsis include tumor growth factor, interleukin-1, -2, and -6, and various other cytokines (Parillo 1993). Interestingly, when interleukin-1 or interleukin-2 are injected intraventricularly in experimental animals, sleep, prominent EEG changes, and probably encephalopathy result (Krueger et al. 1984; Desarro et al. 1990). The manner in which these substances interact with the central or peripheral nervous system, however, remains unclear.

In initial attempts to explain septic encephalopathy, it was proposed that the disorder represented undiagnosed infection of the CNS occurring in the setting of profound systemic sepsis (Jackson et al. 1985). This might take the form of meningitis or a more difficult to diagnose condition such as diffuse cerebral microabscesses. Interestingly, in one retrospective autopsy series (Jackson et al. 1985), 8 of 12 patients with septic encephalopathy demonstrated microabscesses of the cerebral cortex and subcortical white matter, even though there was no suggestion of these findings on prior unenhanced CT scans of the brain and no abnormalities had been found on lumbar puncture. It is, however, unlikely that microabscesses and direct CNS infection are the only explanation for the common finding of encephalopathy in patients with sepsis, especially given that most patients with encephalopathy in the setting of septic and non-septic critical illness who survive do so without neurologic sequelae. Occasionally, unexpected findings occur at autopsy, such as central pontine myelinolysis or ischemic infarcts, but usually the brain is normal except for non-specific findings such as reactive protoplasmic astrocytosis, and any additional findings are insufficient to account for the diffuse encephalopathy (Bolton and Young 1995). Of note, in Bolton and Young's (1995) small prospective autopsy series of critically ill patients with septic encephalopathy, no abnormalities such as microabscesses have been found in the brain.

Several interesting metabolic explanations have emerged and have some merit. It is well known that the immune system, via lymphocytes and macrophages, mediates SIRS of sepsis, probably trauma, and other critical illnesses as well (Bone et al. 1989). These lymphocytes and macrophages may invade tissues, producing capillary leakage and tissue edema, and may have direct effects on cellular metabolism (Bolton and Young 1995). Brain capillary permeability increases early on in sepsis and this may be mediated by these immune cells. Mizock et al. (1990) found elevated phenylalanine concentrations alone in the blood and CSF of patients with septic encephalopathy, whereas patients with hepatic encephalopathy had elevations in all aromatic amino acids measured. Thus, there is evidence that the relative ratio of aromatic to branched-chain amino acids changes as a result of altered blood–brain barrier permeability (Fernstrom and Wurtman 1972; Jeppsson et al. 1981). These substances may then gain access to neuronal receptors and cellular mechanisms in the nervous system from which they had previously been excluded. This is an interesting concept given the recent work demonstrating endogenous benzodiazepine-like substances in the CNS in hepatic encephalopathy (Basile et al. 1991). The presence of these substances and their receptors may be respon-

sible for the response seen in hepatic encephalopathy when the benzodiazepine antagonist, flumazenil, is given (Pomier-Layrargues et al. 1994). Whether the action of flumazenil is specific to hepatic encephalopathy or extends to septic and other metabolic encephalopathies deserves investigation. Additionally, these benzodiazepine-like compounds and other substances may alter the balance of other neurotransmitter systems such as dopamine, norepinephrine, and especially serotonin (Freund et al. 1985).

Whether perfusion of the brain is consistently impaired in sepsis, in the absence of systemic hypotension, is unclear, but some data suggest that this may occur (Bowton et al. 1989; Maekawa et al. 1991). At present there are no convincing data to suggest that primary perfusion abnormalities are the underlying cause of septic encephalopathy, or that increasing CNS perfusion can correct this encephalopathy. Microperfusion defects and functional shunting of blood in the gut and other organ systems are likely important with regard to the systemic effects of sepsis, however (Doglio et al. 1991). Acidosis and impaired oxygen utilization are significant negative factors resulting from systemic sepsis and much work on multi-organ failure has centered around attempts to correct this deficit, systemically or cellularly (Shoemaker et al. 1988; Tuchschmidt et al. 1992). To date, acidosis and oxygen utilization in the brain during sepsis have not been specifically studied, but it seems that these may be important concepts for future evaluation, just as they seem to play an important role in SIRS. Bolton and Young (1995) have suggested that cerebral microperfusion may actually be important in sepsis and septic encephalopathy, especially when the potential role of nitric oxide is considered. Nitric oxide is a gas that acts as a neurotransmitter. It has been implicated as a 'vascular relaxing factor,' and systemic synthesis is increased in sepsis (Palmer et al. 1987; Nava et al. 1991). It also seems to play an important role in the integrity of the blood–brain barrier, probably via its role in regulating vascular tone (Tanaka et al. 1991). While there is presently much enthusiasm regarding the role of nitric oxide in sepsis, and the possible role it plays in alterations of the microcirculation, it is unclear whether the most beneficial efforts would be directed at blocking or at enhancing nitric oxide effects. This is certainly an ongoing area of research and may lead to further understanding of the pathogenesis of septic encephalopathy.

Certainly, septic encephalopathy is not the only cause of delirium in patients in intensive care units. Delirium may be defined as a confusional state marked by a prominent disorder of perception, often with hallucinations, delusions, and sleep disturbances (Adams and Victor 1989). Common causes include medications, metabolic encephalopathies from primary organ dysfunction such as hepatic or uremic encephalopathies, and undiagnosed infection. The term *ICU psychosis* has been used to describe a type of delirium manifest in critically ill patients which may be caused or aggravated by iatrogenic conditions imposed by the environment of the intensive care unit (Briggs 1991). The intensive care unit is often a setting in which frequent noise from voices and alarms, irregularly timed examinations by physicians and nurses, and constant lighting serve to disorient patients from a normal circadian rhythm and disrupt normal sleep patterns. Elderly patients and those with other complicating causes of delirium are at highest risk for exacerbation from these environmental conditions in the intensive care unit. In patients with critical illness involving multi-organ failure, acute delirium and other changes in mental status are common. Because these patients are at such risk for infectious, vascular, and metabolic complications of critical illness, a new delirium should never be attributed solely to an ICU psychosis (Wise and Terrell 1992). As discussed previously, thorough neurologic evaluation, often including a head imaging study, lumbar puncture, and EEG, is essential to identify a new and potentially treatable process. However, it is important to realize that an ICU psychosis may further complicate the diagnosis and management of acute delirium in critically ill patients. On occasion, low doses of an antipsychotic medication such as haloperidol or droperidol may improve this type of delirium. The best approach is to adjust the environment of the intensive care unit to allow adequate uninterrupted sleep, provide environmental cues such as daylight via windows and communication directed at the patient, treat pain adequately, and attempt to limit extraneous noise.

In summary, encephalopathy is the most common neurologic complication in critically ill patients with multi-organ system failure. This encephalopathy may be secondary to individual organ failure (such as hepatic or uremic encephalopathy), may be secondary to drugs, especially those given for sedation or anal-

gesia, or may be a primary manifestation of SIRS of sepsis as in septic encephalopathy. Because of the difficulty of neurologic evaluation in critically ill patients, and because treatment of septic encephalopathy involves treatment of the underlying critical illness and, to date, no other specific interventions, it is essential to rule out other neurologic disease, be it infectious, vascular, or epileptic in origin, before the diagnosis of septic encephalopathy is firmly made. Thus, use of neuro-diagnostic tests such as neuro-imaging, EEG, spinal fluid examination, and serum metabolic evaluation is important. The etiology of septic encephalopathy remains unclear. It is probably multifactorial in nature and may involve microabscesses throughout much of the brain or perfusion defects in the cerebral microcirculation. More likely, however, it involves metabolic derangements that are engendered by the septic state and derive from increased blood–brain barrier permeability to various cytokines, amino acids, and neuromodulatory transmitters. No specific treatment exists and no putative treatments specific for septic encephalopathy have been attempted. The prognosis of a patient with profound septic encephalopathy is related to the prognosis of the underlying critical illness; from 50 to 90% of patients with ARDS or multi-organ failure, with or without septic encephalopathy, die, depending on the number of organ systems involved (Beal and Cerra 1994; Kollef and Schuster 1995). If the patient recovers, full neurologic recovery without sequelae can be expected, even if the patient was previously profoundly impaired or comatose. Further research on the role of various cellular mediators of sepsis, on nitric oxide, and on oxygen utilization by the brain in sepsis will yield more clues as to the specific pathophysiology underlying septic encephalopathy and may suggest treatments based on these cellular mechanisms.

Spinal cord

At present there are no syndromes specific to the spinal cord that have been implicated as unique to critical illness in the setting of multi-organ system failure. This may be because encephalopathy and peripheral nerve weakness make delineation of a concomitant myelopathy difficult. Nevertheless, critically ill patients may suffer a number of different complications involving the spinal cord which may be related or secondary to an underlying critical illness. These can be divided generally into complications related to trauma, ischemia, and infection. Spinal cord trauma may have been present at the time of admission to the intensive care unit but not diagnosed because attention was focused on the other aspects of critical illness. Certainly, any patient with facial abrasions after a fall should be suspected to have a cervical spine injury. Ischemic myelopathies are unfortunately a common complication of aortic aneurysm repair, especially in the setting of hypotension or aneurysmal rupture (Hollier et al. 1992). Infectious or parainfectious myelopathies may be present in patients with viral or even bacterial infection elsewhere, especially pulmonary processes (Dawson and Potts 1991). Spinal cord compression from an epidural abscess is certainly a feared complication that may be difficult to diagnose in a patient with ongoing critical illness. Infectious complications of the spinal cord may be more common in patients who are immunosuppressed (Wijdicks 1995). Spinal cord compression from epidural or subarachnoid spinal hematoma after lumbar puncture should, however, be rare even in a coagulopathic patient since the site of lumbar puncture (usually L_3–L_4) is below the conus medullaris (Owens et al. 1986).

PERIPHERAL NERVOUS SYSTEM

Historically, encephalopathy and CNS dysfunction have been the most commonly identified neurologic complications in critically ill patients. This is probably because confusion, altered mental status, and coma are usually readily apparent even to physicians who are not focused on the neurologic aspects of the illness. However, over the past decade it has become evident that dysfunction of the peripheral nervous system is much more common in critically ill patients than previously thought. In fact, it is actually quite common in patients with prolonged sepsis and critical illnesses lasting several weeks, especially when prolonged mechanical ventilation is required (Spitzer et al. 1992; Bolton and Young 1995). Initially, dysfunction of the peripheral nervous system was suspected in patients who appeared weak and had difficulty being weaned from the ventilator. Often this became apparent after the critical illness had begun to improve and encephalopathy cleared, making neuromuscular examination more reliable. Over the past decade,

there have been many reported cases of neuromuscular complications in patients with critical illness (Zochodne et al. 1987; Hirano et al. 1992; Segredo et al. 1992). These complications may involve each component of the peripheral nervous system, either individually or in combination. Certainly, well-known diseases, such as the Guillain–Barré syndrome, myasthenia gravis, and compression neuropathies, may occur in critically ill patients (Arnason and Asbury 1968; Wijdicks et al. 1994). However, it is evident that most of the neuromuscular complications that occur in critically ill patients with multiple organ failure represent newly described entities that were unknown prior to the evolution of critical care medicine. The precise epidemiology and risk factors for most of these disorders still remain unclear, as do the pathogenetic processes and treatment options.

Diagnosis of a peripheral nervous system disorder in a critically ill patient may be difficult. The initial difficulty is in recognition that something is wrong with the peripheral nervous system. The first suspicion may occur because a patient appears to move too little or is difficult to wean from mechanical ventilation because of apparent respiratory muscle weakness or fatigue. Clinical neurologic evaluation may be difficult. A detailed history is difficult to obtain as patients are often intubated, and family and medical staff have usually been concerned with more acute aspects of the patient's critical illness than with the amount of any spontaneous movement. Encephalopathy, sedative medications, and mechanical devices such as intravenous lines, ECG pads, and catheters may make a detailed neurologic examination difficult to perform and interpret. Finally, in many patients, two or more processes may be occurring concurrently, making differentiation of the relative contributions of encephalopathy and peripheral nerve weakness difficult to discern clinically. For these reasons, electrophysiologic testing of peripheral nerves, neuromuscular transmission and muscles is an exceptionally important adjunct to accurate diagnosis of these conditions.

Each part of the peripheral nervous system appears to be at risk for a unique set of complications in critically ill patients with multiple organ system failure. For this reason, it is appropriate to discuss these disorders based on neurologic localization. However, many of these disorders are newly described and current understanding of them is evolving. Thus, it is likely that there is some overlap in clinical presentation, risk factors, and pathogenetic mechanisms. This should certainly be taken into account when evaluating a patient for peripheral nervous system dysfunction in the setting of critical illness.

Peripheral nerves

Probably the first peripheral disorder unique to critically ill patients with multiple organ system dysfunction was *critical illness polyneuropathy*. This term was used by Zochodne et al. (1987) to describe an axonal polyneuropathy diagnosed in 17 patients with prolonged critical illness involving sepsis and multiple organ failure. The mean length of stay of these patients in the intensive care unit was 68 days. These patients were suspected of having weakness of peripheral nervous system origin because of apparent limb weakness and failure to wean from the ventilator. Neurologic examination demonstrated decreased strength, especially distally, decreased reflexes, and relative sparing of head and neck strength. Sensation, which is often difficult to test accurately in critically ill septic patients, was consistent with distal loss of pain and temperature appreciation. However, it was noted that precise localization based on neurologic examination was difficult because of concomitant encephalopathy and other barriers to complete neurologic evaluation, as discussed earlier. Even so, all the findings were suggestive of a distal symmetric polyneuropathy. Electrophysiologic testing demonstrated decreased amplitudes of sensory nerve action potentials and compound muscle action potentials; conduction velocities and responses to repetitive nerve stimulation were essentially normal. Thus, the clinical and electrophysiologic picture of critical illness polyneuropathy is one of an acquired distal symmetric axonal sensorimotor polyneuropathy. Post-mortem analysis was done on eight of these patients; peripheral nerve examination uniformly demonstrated axonal degeneration and muscle examination showed grouped fiber atrophy consistent with a primary neurogenic cause. CSF examination was unmarkable and non-diagnostic in the 11 patients evaluated. Importantly, the diagnosis of Guillain–Barré syndrome was satisfactorily excluded on the basis of electrophysiologic and spinal fluid evaluations. There were no evident etiologic features and no apparent correlation between serum electrolyte values, cause of sepsis or critical illness or medications received. In fact, the etiology of critical illness polyneuropathy remains unknown.

Since this initial detailed description of critical illness polyneuropathy, there have been many more reports of an axonal polyneuropathy developing in the setting of SIRS (Witt et al. 1991; Wijdicks et al. 1994). It is now apparent that this entity is a much more common complication in critically ill patients than previously thought. Witt et al. (1991) suggest that 70% of patients who develop the septic syndrome also develop critical illness polyneuropathy. Bolton and Young (1995) found neuropathy in 50–75% of patients with moderate to severe septic encephalopathy. The prognosis of the neuropathy is directly related to the prognosis of the underlying critical illness. Since the prognosis of critical illness with multi-system organ failure is death in 50–90% of cases, depending on the number of organ systems involved, a significant number of patients with critical illness polyneuropathy will die from their underlying critical illness. However, the neuropathy nearly always resolves over time if the patient recovers from the underlying illness. Certainly, if the patient's weakness is profound as a result of critical illness polyneuropathy, it may take weeks or even months to resolve completely, and thus may still necessitate prolonged ventilatory support and tracheostomy, with their attendant risks and complications, during the recovery process.

The etiology of critical illness polyneuropathy remains obscure. Because of its association with sepsis and because it is found almost exclusively in patients with multi-organ system failure and encephalopathy, most hypotheses about its etiology have attempted to tie all of these findings together. It does not appear that any individual metabolic factor is sufficient to explain the neuropathy, which is certainly more rapidly progressive than most other metabolically based neuropathies such as those found in diabetes or renal failure. The relationship of this acquired axonal polyneuropathy to prior treatment with non-depolarizing neuromuscular blocking agents (nd-NMBAs) is also unclear. While most patients with critical illness polyneuropathy have not been treated with nd-NMBAs, there has been at least one intriguing report of patients developing polyneuropathy after treatment with vecuronium in the setting of prolonged ventilation, without concomitant sepsis, multi-organ failure, or corticosteroid treatment (Kupfer et al. 1992). Only one of this group of patients, however, showed abnormal electrodiagnostic tests of sensory nerves, suggesting that this is a different entity than critical illness polyneuropathy. No pathology was reported. Certainly, the association of nd-NMBAs with critical illness polyneuropathy is much less strong than their association with the neuromuscular junction and muscle disorders discussed later (Barohn et al. 1994; Giostra et al. 1994). Whether a developing critical illness polyneuropathy may be accelerated in the presence of non-depolarizing neuromuscular blocking agents remains unknown. It seems, however, that there is something significant and unique about the septic state, especially when it has progressed to the level of multi-organ system dysfunction, that is itself neurotoxic. At present, the etiology of multi-organ system failure itself in the presence of profound sepsis remains a mystery. Current postulates implicate systemic release of cytokines as systemic mediators of the inflammatory response in SIRS. Through a cellular mechanism as yet undetermined, different organs are affected. Whether this results from impairment of cellular mechanisms of aerobic metabolism, from increased cellular permeability from lipid bilayer perioxidation by free radicals, or other processes remains speculative (Beal and Cerra 1994). Since peripheral nerve axonal function and axonal transport involve energy-dependent mechanisms, it makes intuitive sense that the same mediators that cause parenchymal organ dysfunction, such as liver and kidney failure or ARDS, might affect peripheral nerves. Also, it may be that selective increased permeability of the blood–nerve barrier to substances such as histamine and serotonin plays a role in the pathogenesis of critical illness polyneuropathy (Lefer 1985). At present, there is active research investigating these various possibilities.

At present, treatment of critical illness polyneuropathy involves treatment of the underlying septic critical illness and supportive care to avoid compression neuropathies and rhabdomyolysis from lack of movement. No specific definitive treatment has yet been shown to alter the course of this neuropathy. Wijdicks (1995) described three patients with critical illness polyneuropathy who failed to respond to intravenous γ-globulin. Although nutritional deficiencies are not as yet definitively linked to development of the neuropathy, proper nutrition (parenteral or enteral)

and physical therapy to maintain range of motion are important aspects of supportive care of the critically ill patient. Because of the difficulty in diagnosing critical illness polyneuropathy solely on clinical grounds, electrophysiology should be performed as soon as weakness is suspected. This is also important to rule out potentially treatable conditions such as the Guillain–Barré syndrome, which may certainly occur in critically ill patients (Arnason and Asbury 1968). While there has been concern that critical illness polyneuropathy may be mistaken for the axonal variant of Guillain–Barré syndrome, the different presentations, especially the lack of autonomic symptoms in critical illness polyneuropathy, should facilitate the distinction on clinical grounds (Wijdicks et al. 1994).

To date, critical illness polyneuropathy is the only specific peripheral nerve disorder that is known to occur as a complication of sepsis and multi-system organ failure in critically ill patients. Whether this disorder is indeed a single diagnostic entitity, or rather a heterogenous group of axonal neuropathies with different specific causes and potential treatments, remains unknown. Additionally it is important to realize that patients with critical illness may have other causes of weakness in which the peripheral nerves are relatively spared and the neuromuscular junction or muscles are the primary site of dysfunction (Op de Coul et al. 1985; Gooch et al. 1991; Segredo et al. 1992).

Even in individuals who do not develop critical illness polyneuropathy, there is notable risk of peripheral nerve impairment during treatment and support of a critical illness. Focal compression neuropathies are more likely to occur in patients with muscle wasting, such as in sepsis, especially when they are less likely to move about in bed because of encephalopathy or physical restriction. Compression stockings designed to prevent deep vein thrombosis may result in peroneal neuropathies, and ulnar neuropathies may occur if elbows are left unpadded. Just as care in positioning patients must be taken during operative procedures in order to avoid compression neuropathies, so must care be exerted in patients requiring prolonged intensive care in order to avoid these complications. Usually, when these neuropathies occur, they are reversible over time, but may restrict later efforts at rehabilitation and recovery (Wijdicks et al. 1994).

Neuromuscular junction

A defect in neuromuscular transmission may be a source of weakness in critically ill patients for a number of reasons. While previously undiagnosed myasthenia gravis may be a consideration, persistent weakness secondary to impaired neuromuscular junction transmission is almost always a result of an acquired disorder secondary to administration of drugs. A number of medications have long been known to impair neuromuscular transmission; these include antibiotics, especially aminoglycosides, and β-blocking agents (Howard 1990). However, the most common medications used in the intensive care unit which impair neuromuscular transmission are the non-depolarizing neuromuscular blocking agents (nd-NMBAs), also known as muscle relaxants. Included in this group of drugs are such agents as pancuronium, vecuronium, rocuronium, and atracurium. As mechanical ventilatory practices have become more complex and sicker patients with more profound lung injury have been treated with prolonged intubation and ventilation, the use of nd-NMBAs has increased. These agents can allow ventilation to proceed with lower peak inspiratory pressures and may decrease the barotrauma injury associated with mechanical ventilation in ARDS. They may aid in improving oxygenation and ventilation by removing any counteractive efforts by patients in respiratory distress or with increased chest wall compliance (Hansen-Flaschen et al. 1991). These medications may be used for prolonged periods of time, often from days to weeks. Because the duration of action of nd-NMBAs is usually no more than 60 min (Miller and Savarese 1990), it was initially thought that prolonged weakness after administration of these drugs would be unlikely. However, there are a number of reports of prolonged weakness due to persistent impairment of neuromuscular blockade even after discontinuation of these agents hours or days earlier (Partridge et al. 1990; Segredo et al. 1992; Prielipp et al. 1995).

Peripheral nerve stimulators, long used in operating rooms, are being increasingly used in the intensive care unit to help physicians assess the degree of paralysis secondary to nd-NMBAs. Peripheral nerve stimulators employ an electrical stimulus usually applied over the ulnar or facial nerve. Strength of contraction of the adductor pollicis or orbicularis oculi muscles, respectively,

usually relative to four consecutive stimuli ('train-of-four'), is used to determine the extent of neuromuscular blockade. In this manner, it is hoped that the minimum effective dose of non-depolarizing neuromuscular blocking agents can be employed (Sharpe 1992).

Segredo et al. (1992) studied 16 patients who had received vecuronium to facilitate mechanical ventilation for at least 2 consecutive days. They used a peripheral nerve stimulator to assess neuromuscular blockade; they also measured plasma concentrations of vecuronium and 3-deacetylvecuronium, the active metabolite of vecuronium. In their group of 16 patients, 7 had prolonged neuromuscular blockade, as measured using the peripheral nerve stimulator, and this lasted from 6 h to 7 days after discontinuation of vecuronium. All patients eventually recovered fully. They attempted to correlate the risk for prolonged neuromuscular blockade with various clinical and laboratory parameters. While some correlation was present between prolonged neuromuscular blockade and female sex, metabolic acidosis, and elevated plasma magnesium concentrations, the most important risk factors for prolonged weakness were the presence of renal failure and high plasma concentrations of 3-deacetylvecuronium. Interestingly, liver failure was not as high a risk factor; this is surprising, given that vecuronium is cleared 80% hepatically and 20% renally. They therefore make the recommendations that only the lowest dose of neuromuscular blocking agent should be used to achieve the desired result and that when these agents are used in the intensive care unit a peripheral nerve stimulator should be used to monitor patients' neuromuscular junction function to maintain at least 1 of 4 twitches on a train-of-four stimulation. It is important to note that in this study, formal nerve conduction studies and electromyography were not performed routinely to exclude other peripheral causes of prolonged weakness. In the one patient who did undergo nerve conduction studies, primary neuromuscular blockade was indicated.

Vecuronium is not the only agent whose use may result in prolonged weakness from persistent neuromuscular blockade. There are reports of prolonged paralysis from pancuronium (Gooch et al. 1991; Giostra et al. 1994), and it is likely that any agent in this class of drug may be associated with this complication. In the report by Gooch et al. (1991) electrophysiologic studies suggested heterogenous causes of the weakness, with some patients showing evidence of a neuropathy while others had predominantly myopathic features with an unclear contribution from persistent neuromuscular junction blockade. It is likely that this group of patients represented a combination of critical illness polyneuropathy, myopathies, and neuromuscular junction blockade, and not a uniform disease entity secondary solely to pancuronium. The important distinction to make is that the syndrome of persistent weakness in the setting of neuromuscular blockade use (or overuse) appears to be secondary solely to continued competitive inhibition of acetylcholine at the postsynaptic nicotinic cholinergic receptor on skeletal muscle. Clinically, patients may appear mildly weak or may be completely paralyzed with absence of any movement, spontaneously initiated respirations, reflexes, or ocular movements (spontaneous or reflex). Pupillary function is spared as these non-depolarizing neuromuscular blockers do not act at muscarinic cholinergic receptors (Miller and Savarese 1990). This syndrome results from prolonged action of a drug which was given for the purpose of inducing weakness or paralysis, and there does not appear to be any structural or functional damage to the peripheral nervous system as a result of this syndrome. Pathologic analysis is lacking, because once this syndrome is recognized and the offending medications are discontinued, full strength is restored, although this may take days. Because neuromuscular junction blockers may be associated with other causes of prolonged weakness, such as myopathy in conjunction with corticosteroid use, monitoring with a peripheral nerve stimulator may be inadequate for prophylactic or diagnostic purposes (Prielipp et al. 1995). Therefore, we recommend full electrophysiologic testing of the peripheral nervous system with nerve conduction studies, repetitive stimulation testing, and electromyography in order to define more precisely the site of the pathology.

Muscle

It has been observed for decades that critically ill patients, especially those with sepsis, frequently undergo muscle wasting, often in the face of seemingly adequate nutritional support. The assumption has been that this represents a catabolic myopathy brought about as a result of multiple factors including elevated cortisol and catecholamine release and other

circulating factors induced by SIRS. In fact there is evidence that activation of lysosomal proteases in muscle by interleukin-1 with resultant prostaglandin E2 production may be in part responsible for generalized muscle breakdown and loss of muscle mass in critically ill patients with multi-organ system failure (Baracos et al. 1983; Clowes et al. 1983). Denervation atrophy has also been postulated to play a role. In this syndrome, known as *cachectic myopathy,* serum creatine kinase levels and electromyography are normal. Muscle biopsy shows type II fiber atrophy (Bolton and Young 1995). Cachectic myopathy was initially thought to be the primary neuromuscular complication of sepsis and responsible for most cases of respiratory muscle fatigue and difficulty in weaning from the ventilator (Roussos and Macklem 1982). However, now that the syndromes of critical illness polyneuropathy and prolonged weakness from nondepolarizing neuromuscular blocking agents have been identified, it is apparent that the causes of respiratory muscle weakness may localize to any part of the peripheral nervous system. In fact, it is now understood that there are primary muscle disorders other than cachectic myopathy which may occur as a complication of critical illness and that these may contribute significantly to prolonged weakness and respiratory fatigue.

Panfascicular muscle fiber necrosis may occur in the setting of profound sepsis (Lannigan et al. 1984). This so-called *septic myopathy* is characterized clinically by weakness progressing to a profound level over just a few days. There may be associated elevations in serum creatine kinase and urine myoglobin. Both EMG and muscle biopsy may be normal initially but eventually show abnormal spontaneous activity and panfascicular necrosis with an accompanying inflammatory reaction, respectively. It is unclear if this septic myopathy represents a distinct clinical entity, is an aggressive variant of cachectic myopathy, or is similar to the next myopathy to be discussed, that which may result from a combination of corticosteroids and neuromuscular blocking agents.

One of the most challenging scenarios regarding mechanical ventilation in the intensive care unit is that of the asthmatic with an acute exacerbation requiring urgent ventilatory support. These patients are treated with high doses of bronchodilators and corticosteroids in an attempt to quell the pulmonary inflammatory response. Additionally, many of these patients have altered lung and chest wall compliance making ventilation difficult, even with a mechanical ventilator. It is in this setting that high doses of non-depolarizing neuromuscular blocking agents, as previously described, may be used for days or even up to a week or more in order to facilitate ventilation (Hansen-Flaschen et al. 1991). There are now several dozen patient reports in the literature regarding prolonged weakness occurring in this setting (Hirano et al. 1992; Barohn et al. 1994; Giostra et al. 1994; Zochodne et al. 1994). All of these patients have been treated with high-dose corticosteroids (usually prednisone 60 mg/ day or its equivalent); over 90% of the patients with this reported complication have been treated with non-depolarizing neuromuscular blocking agents, primarily vecuronium or pancuronium, usually for at least 2 consecutive days. Clinical features include moderate to severe flaccid weakness with onset over 4 days to 2 weeks (Hirano et al. 1992). The clinical spectrum varies from mild weakness to essentially complete quadriplegia. There may be muscular atrophy; reflexes and sensation are preserved. Electromyography classically demonstrates abnormal insertional activity consistent with myopathic features, but may occasionally be normal. Nerve conduction studies are normal except for decreased amplitude of compound motor action potentials; this finding has been interpreted by some authors as demonstrating neuropathic features, but certainly this finding could be indicative of myopathy alone (Daube 1992). Serum creatine kinase values are elevated in up to one-half of cases and muscle biopsy samples are consistent with myopathy, showing widespread muscle fiber atrophy, excess glycogen, myofibril disorganization with selective loss of thick (myosin) filaments, and necrosis with little or no associated inflammation (Danon and Carpenter 1991; Hirano et al. 1992; Hansen-Flaschen et al. 1993); one biopsy sample has shown grouped atrophy most consistent with a neurogenic cause. Given all of these clinical findings, this clinical syndrome has been most commonly interpreted as representing an acute myopathy in the setting of corticosteroid and non-depolarizing neuromuscular blocker use. It has been called *acute quadriplegic myopathy* by Hirano et al. (1992) to emphasize its relatively rapid evolution and the fact that all four limbs may be involved with profound distal and proximal weakness.

The etiology of this disorder remains unknown. It has been known that surgically denervated rat muscle

treated with corticosteroids (dexamethasone) may show loss of myosin with myofilament disorganization (Carpenter et al. 1990). Whether non-depolarizing neuromuscular blockers act as priming agents, effectively exerting a functional denervation on muscle fibers exposed to corticosteroids, or whether these neuromuscular blocking agents have a direct myotoxic effect is unclear. Initial speculation revolved around the steroid structure of vecuronium. It was thought that non-depolarizing neuromuscular blocking agents which did not contain a steroid ring as a primary structure (i.e., atracurium) might be less prone to cause this myopathy. However, as more experience with these other drugs has accrued, it seems that this corticosteroid/neuromuscular myopathy occurs with these agents as well (Miro et al. 1994). Pathologically and clinically this myopathy has some similarities with the classical steroid myopathy which may occur in the setting of prolonged use of corticosteroids. Clinically, however, steroid myopathy is primarily a slowly progressive proximal myopathy. Pathologically, it may show vacuolar changes in both type I and type II fibers with evidence of regeneration (Victor and Sieb 1994). Thus, while steroid myopathy and acute quadriplegic myopathy have some pathologic and clinical similarities, it is unclear whether they represent different ends of the spectrum of a single pathogenetic process or entirely distinct entities.

Acute quadriplegic myopathy has a good prognosis. Mortality is related to the underlying disease. If patients survive their underlying critical illness, the myopathy invariably improves and the patient usually returns to normal. This process may take weeks or months, however, and tracheostomy, prolonged ventilatory support, and transfer to a rehabilitation facility may be necessary. Therefore, it is important to provide reassurance to patients and physicians that the weakness will eventually resolve without residua. At present, it is unclear how to prevent this myopathic complication. By avoiding prolonged use of neuromuscular blocking agents, it is hoped that this complication will occur less frequently. Unfortunately, these agents are sometimes necessary to aid mechanical ventilation. Monitoring with a peripheral nerve stimulator can allow overdosing of non-depolarizing neuromuscular blocking agents to be avoided. However, this is more likely to prevent the complication of prolonged neuromuscular junction blockade than it is to prevent this myopathy. In fact, we experienced a recent case in which neuromuscular blockade was required for mechanical ventilation of a refractory asthmatic patient and monitoring with a peripheral nerve stimulator was provided for the entire week of paralysis. It was only when 4 of 4 twitches were present after a ' train-of-four' stimulation and the patient still appeared profoundly weak that myopathy was considered and later confirmed electrodiagnostically; thus, despite monitoring, the myopathy occurred anyway. As awareness of the complications of use of neuromuscular blocking agents in the intensive care unit is heightened, further experience will help to determine the frequency of this myopathy and hopefully identify measures which can be taken to prevent it.

Thus, peripheral nervous system dysfunction is a common complication in patients with critical illnesses involving multi-organ failure. While these neuropathies, junctional disturbances, and myopathies may be related to the underlying disease or specific organ failure, usually they are a more general complication of SIRS. Exacerbating or predisposing factors include severity of underlying illness, degree of muscle wasting, and medication use, especially non-depolarizing neuromuscular blocking agents. These agents have been implicated to some degree in the pathologic involvement of each part of the peripheral nervous system. Their role in peripheral neuropathy is unclear and remains unconvincing. They are, however, undoubtedly the cause of prolonged neuromuscular junction blockade and certainly a significant predisposing factor, if not cause, of acute quadriplegic myopathy. In essentially all cases, prognosis is related to treatment and prognosis of the underlying critical illness. The fact that only about 50% of patients with critical illness polyneuropathy survive is indicative of the overwhelming mortality associated with sepsis and multi-organ failure, not mortality from the neuropathy. In fact, if patients survive, the critical illness polyneuropathy, prolonged neuromuscular junction blockade, or acquired myopathy usually resolves completely over time. Unfortunately, prolonged weakness predisposes patients to life threatening complications such as pneumonia, urinary tract infection, and pulmonary embolism. Therefore, prevention of these neuromuscular complications seems to be the most prudent approach to their treatment. Early recognition of neuromuscular complications seems to be a first step in their prevention. As under-

standing grows of the cellular derangements which occur in sepsis and the mediators, such as cytokines, which are responsible, the etiologies of these neuromuscular complications will hopefully become more evident and may lead to specific treatments.

COMPLICATIONS OF COMMONLY PERFORMED INTENSIVE CARE UNIT PROCEDURES

In order to facilitate care of the critically ill patient, a number of procedures are commonly performed and these may result in complications, some of which are neurologic. These procedures vary in degree of invasiveness and often involve attempted vascular catheterization based on visual or tactile surface landmarks, not direct vision. The most commonly performed procedures in the intensive care unit (ICU) include endotracheal intubation to allow airway protection or mechanical ventilation, central venous line placement for purposes of vascular access or hemodynamic monitoring, and arterial cannulation for hemodynamic monitoring or ease of phlebotomy. Additionally, patients in the ICU may undergo lumbar puncture, epidural catheterization, placement of an intra-aortic balloon pump or ventricular assist device, cardioversion, or carotid sinus massage. Each of these procedures or maneuvers is associated with its unique set of complications. Fortunately, neurologic complications from these procedures are rare. However, their risk must be weighed when determining the potential benefit to be derived from the procedure.

Serious neurologic complications from endotracheal intubation are exceptionally rare. Theoretically, extension of the neck during intubation could result in cervical spinal cord injury in patients with preexisting spine disease such as cervical vertebral fractures or severe spinal stenosis. Practice has been for trauma patients to have their cervical spines cleared, radiologically or clinically, prior to intubation, to provide axial traction to limit flexion and extension during intubation, or to use an alternative method such as visualization with a fiberoptic bronchoscope to facilitate intubation. In patients with intracranial hypertension, care must be taken during intubation to avoid transient increases in intracranial pressure which might precipitate neurologic deterioration (Messick et al. 1985); however, this is most pertinent to patients with primary neurologic or neurosurgical disease and is

TABLE 2

Neurologic complications unique to critical illness involving multi-organ failure.

Syndrome	Localization	Prevalence	Cause	Time course of recovery*
Septic encephalopathy	cerebral hemispheres	50–75% in sepsis/SIRS	?cytokine mediators of SIRS ?microabscesses ?microperfusion defects	days–a week
Critical illness polyneuropathy	peripheral nerves	up to 70% in sepsis	?cytokine mediators of SIRS ?interaction with nd-NMBAs	weeks–months
Persistent neuromuscular blockade	neuromuscular junction	varies with drug dosage and metabolic state	nd-NMBAs, esp. in renal failure	hours–days
Acute quadriplegic myopathy	muscles	unknown	acute high-dose corticosteroids esp. with nd-NMBAs	weeks–months

*Assuming adequate treatment of the underlying critical illness.

usually not as great a problem in patients with critical illness involving multi-organ failure (except in fulminant hepatic failure with diffuse cerebral edema). Chronic hoarseness after intubation of any patient is a not infrequent complication. This is often more annoying than dangerous and almost always improves over the course of several weeks. A likely etiology is temporary compression of the anterior branch of the recurrent laryngeal nerve against the thyroid cartilage by the endotracheal tube cuff (Cavo 1985). Temporary partial paralysis of the facial nerve has been reported during intubation. This probably results from compression of the nerve against the mandible during a zealous effort to avoid upper airway obstruction by the tongue; complete recovery is the rule (Fuller and Thomas 1956).

Catheterization of a central vein for vascular access or hemodynamic monitoring purposes is associated with a host of potential complications which may result when the procedure is not performed correctly. Pneumothorax and excessive blood loss, sometimes with large hematoma formation, are the most common potential complications, being generally more common with the subclavian approach than with the internal jugular, external jugular, or femoral approaches (Bambauer et al. 1994). The neurologic complications of central venous catheter placement are most notable with the internal jugular approach and usually result from inadvertent puncture of an artery during a line placement attempt. Carotid artery puncture was found to occur in 2% of a recent series of attempted internal jugular vein cannulation (Wijdicks 1995). While occurrence of stroke after carotid puncture is rare, it has been reported on several occasions. Of note, Sloan reported a case of fatal brainstem stroke after inadvertent insertion of a number 8 French cordis-type catheter (a large bore catheter) into a vertebral artery (Sloan et al. 1991). Brown reported a stroke despite anticoagulation after a similar type catheter was placed into the carotid artery (Brown 1982). Urgent surgical exploration is probably indicated should this complication of large-bore catheter arterial insertion occur. Horner's syndrome is also a potential rare complication which may result from local hematoma or local carotid dissection (Garcia et al. 1994). Finally, cerebral air embolism can occur when air is introduced into an open needle or uncapped venous access port (Wijdicks 1995); for this reason, patients should be placed in the Trendelen-

burg position (head down) during internal jugular or subclavian vein cannulation or removal and care should be taken to keep needle lumens covered and venous access ports capped.

Arterial cannulation is a common procedure used to provide access for frequent blood draws and blood pressure monitoring. The radial artery is the most common site for arterial line placement. The primary neurologic complication which may occur in this setting is compression of the median nerve (Marshall et al. 1980). This probably occurs because of compression of the median nerve prior to the transverse carpal ligament by hematoma; multiple puncture attempts and anticoagulation appear to be predisposing risk factors. Brachial artery cannulation or puncture, especially in anticoagulated patients, may result in a profound median nerve compression syndrome secondary to hematoma formation (Luce et al. 1976); urgent evacuation of the hematoma may prevent permanent injury. It is important to realize, however, that these complications are exceptionally rare.

There are many other procedures and maneuvers that critically ill patients may undergo in the intensive care unit setting. It seems that each of these procedures has led to at least one dramatic case report of a devastating neurologic complication associated with it. Introduction of a nasogastric tube into the brain of a patient with severe maxillofacial trauma (Seebacher et al. 1975), massive embolic stroke after application of carotid sinus pressure (Beal et al. 1981), brachial plexopathy after chest tube insertion (Mangar et al. 1991), and epidural hematoma after epidural catheter placement (Horlocker and Wedel 1992) have all been reported. However, prospective series which look at the incidence of these complications have found them to be extremely rare. These reports do serve to demonstrate that critically ill patients are at risk for iatrogenic neurologic complications in addition to complications resulting from their systemic illnesses. These cases also allow some general guidelines to be suggested, such as avoiding carotid massage in patients with carotid bruits and avoiding epidural catheter placement in anticoagulated patients, although the evidence that these precautions safely prevent these complications is often lacking. The experience of the physician performing the procedure and attention to proper technique, even in simple procedures, are the most important factors in avoiding complications (Mansfield et al. 1994).

REFERENCES

ADAMS, R.D. and M. VICTOR: Principles of Neurology, 4th Edit. New York, McGraw-Hill (1989) 324.

AITKENHEAD, A.R.: Analgesia and sedation in intensive care. Br. J. Anaesth. 63 (1989) 196–206.

AMERI, A. and M.-G. BOUSSER: Cerebral venous thrombosis. Neurol. Clin. 10 (1992) 87–111.

ARNASON, B.G. and A.K. ASBURY: Idiopathic polyneuritis after surgery. Arch. Neurol. 18 (1968) 500–507.

AYRES, S.M.: Sepsis and septic shock – synthesis of ideas and proposals for the direction of future research. In: W.J. Sibbald and C.L. Sprung (Eds.), New Horizons: Perspectives on Sepsis and Septic Shock. Fullerton, CA, Society of Critical Care Medicine (1986).

AYRES, S.M.: The promise of critical care: effective and humane care in an era of cost containment. In: W.C. Shoemaker, S.M. Ayres, A. Grenvik and P.R. Holbrook (Eds.), Textbook of Critical Care, 3rd Edit. Philadelphia, PA, W.B. Saunders (1995) 3–7.

BAMBAUER, R., R. INNIGER, K.J. PIRRUNG, R. SCHIEL and R. DAHLEM: Complications and side effects associated with large-bore catheters in the subclavian and internal jugular veins. Artif. Organs 18 (1994) 318–321.

BARACOS, V., H.P. RODEMANN, C.A. DINARELLO and A.L. GOLDBERG: Stimulation of muscle protein degradation and prostaglandin E2 release by leukocytic pyrogen (interleukin-1). N. Engl. J. Med. 308 (1983) 553–558.

BAROHN, R.J., C.E. JACKSON, S.J. ROGER, L.W. RIDINGS and A.L. MCVEY: Prolonged paralysis due to nondepolarizing neuromuscular blocking agents and corticosteroids. Muscle Nerve 17 (1994) 647–654.

BASILE, A.S., R.D. HUGHES, P.M. HARRISON, Y. MURATA, L. PANNELL, E.A. JONES, R. WILLIAMS and P. SKOLNICK: Elevated brain concentrations of 1,4-benzodiazepines in fulminant hepatic failure. N. Engl. J. Med. 325 (1991) 473–478.

BEAL, A.L. and F.B. CERRA: Multiple organ failure syndrome in the 1990s: systemic inflammatory response and organ dysfunction. J. Am. Med. Assoc. 271 (1994) 226–233.

BEAL, M.F., T.S. PARK and C.M. FISHER: Cerebral atheromatous embolism following carotid sinus pressure. Arch. Neurol. 38 (1981) 310–312.

BLECK, T.P., M.C. SMITH, S.J.-C. PIERRE-LOUIS, J.J. JARES, J. MURRAY and C.A. HANSEN: Neurologic complications of critical medical illnesses. Crit. Care Med. 21 (1993) 98–103.

BOLTON, C.F. and G.B. YOUNG: Neurological complications in critically ill patients. In: M.J. Aminoff (Ed.), Neurology and General Medicine, 2nd Edit. New York, Churchill Livingstone (1995) 859–878.

BOLTON, C.F., G.B. YOUNG and D.W. ZOCHODNE: The neurological complications of sepsis. Ann. Neurol. 33 (1993) 94–100.

BONE, R.C., C.J. FISHER, T.P. CLEMMER, G.A. SLOTONAN, C.A. METZ and R.A. BALK: Sepsis syndrome: a valid clinical entity. Methylprednisolone Severe Sepsis Study Group. Crit. Care Med. 17 (1989) 389–393.

BONE, R.C., C.L. SPRUNG and W.J. SIBBALD: Definitions for sepsis and organ failure. Crit. Care Med. 20 (1992) 724–726.

BOWTON, D.L.: CNS effects of sepsis. Crit. Care Clin. 5 (1989) 785–792.

BOWTON, D.L., N.H. BERTELS, D.S. PROUGH and D.A. STUMP: Cerebral blood flow is reduced in patients with sepsis syndrome. Crit. Care Med. 17 (1989) 399–403.

BRIGGS, D.: Preventing ICU psychosis. Nurs. Times 87 (1991) 30–31.

BROWN, C.Q.: Inadvertent prolonged cannulation of the carotid artery. Anesth. Analg. 62 (1982) 150–152.

CARPENTER, S., R. MASSA and G. KARPATI: Depletion and reconstruction of thick myofilaments in steroid treated rat solei after denervation and reinnervation. J. Neurol. Sci. 98 (1990) S377–S378.

CAVO, J.W.: True vocal cord paralysis following intubation. Laryngoscope 95 (1985) 1352–1359.

CLOWES, G.H.A., B.C. GEORGE, C.A. VILLEE and C.A. SARAVIS: Muscle proteolysis induced by a circulating peptide in patients with sepsis or trauma. N. Engl. J. Med. 308 (1983) 545–552.

DANON, M.J. and S. CARPENTER: Myopathy with thick filament (myosin) loss following prolonged paralysis with vecuronium during steroid treatment. Muscle Nerve 14 (1991) 1131–1139.

DAUBE, J.R.: Nerve conduction studies. In: M.J. Aminoff (Ed.), Electrodiagnosis in Clinical Neurology, 3rd Edit. New York, Churchill Livingstone (1992) 283–326.

DAVIDSON, E.A. and W.H.J. SUMMERSKILL: Psychiatric aspects of liver disease. Postgrad. Med. J. 32 (1956) 487–494.

DAWSON, D.M. and F. POTTS: Acute nontraumatic myelopathies. Neurol. Clin. 9 (1991) 585–603.

DESARRO, G.B., Y. MASUDA, C. ASCIOTI, M.G. AUDINO and G. NISTICO: Behavioural and ECoG spectrum changes induced by intracerebral infusion of interferons and interleukin-2 in rats are antagonized by naloxone. Neuropharmacology 29 (1990) 167–179.

DOGLIO, G.R., J.F. PUSAJO, M.A. EGURROLA, G.C. BONFIGLI, C. PARRA, L. VETERE, M.S. HERNANDEZ, S. FERNANDEZ, F. PALIZAS and G. GUTIERREZ: Gastric mucosal pH as a prognostic index of mortality in critically ill patients. Crit. Care Med. 19 (1991) 1037–1040.

EDWARDS, R., L.W. SCHMIDLEY and R.P. SIMON: How often does a CSF pleocytosis follow generalized convulsions? Ann. Neurol. 13 (1983) 460–462.

FERNSTROM, J.D. and R.J. WURTMAN: Brain serotonin content: physiological regulation by plasma neutral amino acids. Science 178 (1972) 414–416.

FREUND, H.R., M. MUGIA-SULLAM, J. PEISER and E. MELAMED: Brain neurotransmitter profile is deranged during sepsis and septic encephalopathy in the rat. J. Surg. Res. 38 (1985) 267–271.

FULLER, J.E. and D.V. THOMAS: Facial nerve paralysis after general anesthesia. J. Am. Med. Assoc. 162 (1956) 645.

GARCIA, E.F., E.F.M. WIJDICKS and B.L. YOUNG: Neurologic complications of internal jugular vein catheterization in critically ill patients. A prospective study in 60 patients. Neurology 44 (1994) 951–952.

GIOSTRA, E., M.R. MAGISTRIS, G. PIZZOLATO, J. COX and J.-C. CHEVROLET: Neuromuscular disorder in intensive care unit patients treated with pancuronium bròmide. Chest 106 (1994) 210–220.

GOOCH, J.L., M.R. SUCHYTA, J.M. BALBIERZ, J.H. PETAJAN and T.P. CLEMMER: Prolonged paralysis after treatment with neuromuscular junction blocking agents. Crit. Care Med. 19 (1991) 1125–1131.

HANSEN-FLASCHEN, J.H., S. BRAZINSKY, C. BASILE and P.N. LANKEN: Use of sedating drugs and neuromuscular blocking agents in patients requiring mechanical ventilation for respiratory failure. J. Am. Med. Assoc. 266 (1991) 2870–2876.

HANSEN-FLASCHEN, J., J. COWEN and E.C. RAPS: Neuromuscular blockade in the intensive care unit: more than we bargained for. Am. Rev. Respir. Dis. 147 (1993) 234–236.

HIRANO, M., B.R. OTT, E.C. RAPS, C. MINETTI, L. LENNIHAN, N.P. LIBBEY, E. BONILLA and A.P. HAYS: Acute quadriplegic myopathy: a complication of treatment with steroids, nondepolarizing blocking agents, or both. Neurology 42 (1992) 2082–2087.

HOLLIER, L.H., S.R. MONEY, T.C. NASLUND, C.D. PROCTOR, W.C. BUHRMAN, R.J. MARINO, D.E. HARMON and F.J. KAZMIER: Risk of spinal cord dysfunction in patients undergoing thoracoabdominal aortic replacement. Am. J. Surg. 164 (1992) 210–213.

HORLOCKER, T.T. and D.J. WEDEL: Anticoagulants, antiplatelet therapy, and neuraxis blockade. Anesthesiol. Clin. North Am. 10 (1992) 1–11.

HOWARD, J.F.: Adverse drug effects on neuromuscular transmission. Sem. Neurol. 10 (1990) 89–102.

ISENSEE, L.M., L.J. WEINER and R.G. HART: Neurologic disorders in a medical intensive care unit: a prospective study. J. Crit. Care 4 (1989) 208–210.

JACKSON, A.C., J.J. GILBERT, G.B. YOUNG and C.F. BOLTON: The encephalopathy of sepsis. Can. J. Neurol. Sci. 12 (1985) 303–307.

JEPPSSON, B., H.R. FREUND, Z. GIMMON, J.H. JAMES, M.F.V. MEYENFELDT and J.E. FISCHER: Blood–brain barrier derangement in sepsis: cause of septic encephalopathy? Am. J. Surg. 141 (1981) 136–142.

KARABUDAK, R., H. CANER, N. OZTEKIN, O.E. OZCAN and T. ZILELI: Thrombosis of intracranial venous sinuses: aetiology, clinical findings and prognosis of 56 patients. J. Neurol. Sci. 34 (1990) 117–121.

KOLLEF, M.H. and D.P. SCHUSTER: The acute respiratory distress syndrome. N. Engl. J. Med. 332 (1995) 27–37.

KRUEGER, J.M., J. WALTER, C.A. DINARELLO, S.M. WOLFF and L. CHEDID: Sleep-promoting effects of endogenous pyrogen (interleukin-10). Am. J. Physiol. 246 (1984) R994–R999.

KUPFER, Y., T. NAMBA, E. KALDAWI and S. TESSLER: Prolonged weakness after long-term infusion of vecuronium bromide. Ann. Intern. Med. 117 (1992) 484–486.

LACOMIS, D., T.W. SMITH and D.A. CHAD: Acute myopathy and neuropathy in status asthmaticus: case report and literature review. Muscle Nerve 16 (1993) 84–90.

LANNIGAN, R., T.W. AUSTIN and J. VESTRUP: Myositis and rhabdomyolysis due to *Staphylococcus* aureus septicemia. J. Infect. Dis. 150 (1984) 784.

LEFER, A.M.: Eicosanoids as mediators of ischemia and shock. Fed. Proc. 44 (1985) 275–280.

LERNER, P.I.: Neurologic complications of infective endocarditis. Med. Clin. North Am. 69 (1985) 385–398.

LUCE, E.A., J.W. FUTRELL, E.F.S. WILGIS and J.E. HOOPES: Compression neuropathy following brachial arterial puncture in anticoagulated patients. J. Trauma 16 (1976) 717–721.

MAEKAWA, T., Y. FUJII, D. SADAMITSU, K. YOKOTA, Y. SOEJIMA, T. ISHIKAWA, Y. MIYAUCHI and H. TAKESHITA: Cerebral circulation and metabolism in patients with septic encephalopathy. Am. J. Emerg. Med. 9 (1991) 139–143.

MANGAR, D., D.L. KELLY, D.O. HOLDER and E.M. CAMPORESI: Brachial plexus compression from a malpositioned chest tube after thoracotomy. Anesthesiology 74 (1991) 780–782.

MANSFIELD, P.F., D.C. HOHN, B.D. FORNAGE, M.A. GREGURICH and D.M. OTA: Complications and failures of subclavian-vein catheterization. N. Engl. J. Med. 331 (1994) 1735–1738.

MARSHALL, G., G. EDELSTEIN and C.A. HIRSHMAN: Median nerve compression following radial arterial puncture. Anesth. Analg. 59 (1980) 953–954.

MATUSCHAK, G.M.: Multiple systems organ failure: clinical experience, pathogenesis, and therapy. In: J.B. Hall, G.A. Schmidt and L.D.H. Wood (Eds.), Principles of Critical Care. New York, McGraw-Hill (1992) 621.

MESSICK, J.M., J.A. NEWBERG, M. NUGENT and R.J. FAUST: Principles of neuroanesthesia for the nonneurosurgical patient with CNS pathophysiology. Anesth. Analg. 64 (1985) 143–174.

MILLER, R.D. and J.J. SAVARESE: Pharmacology of muscle relaxants and their antagonists. In: R.D. Miller (Ed.), Anesthesia, 3rd Edit. New York, Churchill Livingstone (1990) 389–435.

MIRO, O., J.M. GRAU, P. NADAL, C. PICADO, V. PLAZA and A. URBANO-MARQUEZ: Acute myopathy related to the administration of glucocorticoids and neuromuscular blockers. Med. Clin. 103 (1994) 458–460.

MIZOCK, B.A., H.C. SABELLI, A. DUBIN, J.I. JAVAID, A. POULOS and E.C. BACKOW: Septic encephalopathy: evidence for altered phenylalanine metabolism and comparison with hepatic encephalopathy. Arch. Intern. Med. 150 (1990) 443–449.

NAVA, E., R.M.F. PALMER and S. MONCADA: Inhibition of nitric oxide synthesis in septic shock: how much is beneficial? Lancet 338 (1991) 1557–1558.

O'CONNOR, M.F. and M.F. ROIZEN: Use of muscle relaxants in the intensive care unit. J. Intens. Care Med. 8 (1993) 34–36.

OP DE COUL, A.A.W., P.C.L.A. LAMBREGTS, J. KOEMAN, M.J.E. VANPUYENBROEK, H.J. TERLAAK and A.A.W.M. GABREELS-FESTEN: Neuromuscular complications in patients given Pavulon (pancuronium bromide) during artificial ventilation. Clin. Neurol. Neurosurg. 87 (1985) 17–22.

OWENS, E.L., G.W KASTEN and E.A. HESSEL: Spinal subarachnoid hematoma after lumbar puncture and heparinization: a case report, review of the literature, and discussion of anesthetic implications. Anesth. Analg. 65 (1986) 1201–1207.

PALMER, R.M.F., A.G. GERRIGE and S. MONCADA: Nitric oxide release accounts for the biological activity of endothelium-derived relaxing factor. Nature 327 (1987) 524–526.

PARILLO, J.E.: Pathogenetic mechanisms of septic shock. N. Engl. J. Med. 328 (1993) 1471–1477.

PARTRIDGE, B.L., J.H. ABRAMS, C. BAZEMORE and R. RUBIN: Prolonged neuromuscular blockade after long-term infusion of vecuronium bromide in the intensive care unit. Crit. Care Med. 18 (1990) 1177–1179.

PETTY, T.L.: A historical perspective of mechanical ventilation. Crit. Care Clin. 6 (1990) 489–504.

PINE, R.W., M.J. WERTZ, E.S. LENNARD, E.P. DELLINGER, C.J. CARRICO and B.H. MINSHEW: Determinants of organ malfunction or death in patients with intraabdominal sepsis. A discriminant analysis. Arch. Surg. 118 (1983) 242–249.

PLUM, F. and J.B. POSNER: The Diagnosis of Stupor and Coma, 3rd Edit. Philadelphia, PA, F.A. Davis (1980).

POMIER-LAYRARGUES, G., J.F. GIGUERE, J. LAVOIE, P. PERNEY, S. GAGNON, D. D'AMOUR, J. WELLS and R.F. BUTTERWORTH: Flumazenil in cirrhotic patients in hepatic coma: a randomized double-blind placebo-controlled crossover trial. Hepatology 19 (1994) 32–37.

PRIELIPP, R.C., D.B. COURSIN, P.E. SCUDERI, D.L. BOWTON, S.R. FORD, V.J. CARDENAS, J. VENDER, D. HOWARD, E.J. CASALE and M.J. MURRAY: Comparison of the infusion requirements and recovery profiles of vecuronium and cisatracurium 51W89 in intensive care patients. Anesth. Analg. 81 (1995) 3–12.

ROUSSOS, C. and P.T. MACKLEM: The respiratory muscles. N. Engl. J. Med. 307 (1982) 786–797.

SEEBACHER, J., D. NOZIK and A. MATHIEU: Inadvertent intracranial introduction of a nasogastric tube, a complication of severe maxillofacial trauma. Anesthesiology 42 (1975) 100–102.

SEGREDO, V., J.E. CALDWELL, M.A. MATTHAY, M.L. SHARMA, L.D. GRUENKE and R.D. MILLER: Persistent paralysis in critically ill patients after long-term administration of vecuronium. N. Engl. J. Med. 327 (1992) 524–528.

SHARPE, M.D.: The use of muscle relaxants in the intensive care unit. Can. J. Anaesth. 39 (1992) 49–62.

SHOEMAKER, W.C., P. APPEL, H. KRAM, K. WAXMAN and T.S. LEE: Prospective trial of supranormal values of survivors as therapeutic goals in high risk surgical patients. Chest 94 (1988) 1176–1186.

SIMON, R.P. and M.J. AMINOFF: Electrographic status epilepticus in fatal anoxic coma. Ann. Neurol. 20 (1986) 351–355.

SLOAN, M.A., J.D. MUELLER, L.S. ADELMAN and L.R. CAPLAN: Fatal brainstem stroke following internal jugular vein catheterization. Neurology 41 (1991) 1092–1095.

SPITZER, A.K., T. GIANCARLO, L. MAHER, G. AWERBUCH and A. BOWLES: Neuromuscular causes of prolonged ventilator dependency. Muscle Nerve 15 (1992) 682–686.

SPRUNG, C.L., P.N. PEDUZZI, C.H. SHATNEY, M.F. WILSON and L.B. HINSHAW: The impact of encephalopathy on mortality and physiologic derangements in the sepsis syndrome. Crit. Care Med. 16 (1988) 398.

SPRUNG, C.L., P.N. PEDUZZI, C.H. SHATNEY, R.M.H. SCHEIN, M.F. WILSON, J.N. SHEAGREN and L.B. HINSHAW: Impact of encephalopathy on mortality in the sepsis syndrome. The Veterans Administration Systemic Sepsis Cooperative Study Group. Crit. Care Med. 18 (1990) 801–806.

TANAKA, K., K. GOTOH, S. GOMI, S. TAKASHIMA, B. MIHARA, T. SHIRAI, S. NOGAWA and E. NAGATA: Inhibition of nitric oxide synthesis induces a significant reduction in local cerebral blood flow in the rat. Neurosci. Lett. 127 (1991) 129–132.

TOMSON, T., U. LINDBLOM and B.Y. NILSSON: Nonconvulsive status epilepticus in adults: thirty-two consecutive patients from a general hospital population. Epilepsia 33 (1992) 829–835.

TUCHSCHMIDT, J., J. FRIED, M. ASTIZ and E. RACKOW: Elevation of cardiac output and oxygen delivery improves outcome in septic shock. Chest 102 (1992) 216–220.

VICTOR, M. and J. SIEB: Myopathies due to drugs, toxins, and nutritional deficiency. In: A. Engel and C. Franzini-Armstrong (Eds.), Myology, 2nd Edit. New York, McGraw-Hill (1994) 1697–1725.

WIJDICKS, E.F.M.: Neurology of Critical Illness. Philadelphia, PA, F.A. Davis (1995).

WIJDICKS, E.F.M., W.J. LITCHY, B.A. HARRISON and D.R. GRACEY: The clinical spectrum of critical illness polyneuropathy. Mayo Clin. Proc. 69 (1994) 955–959.

WISE, M.G. and C.D. TERRELL: Delirium, psychotic disorders, and anxiety. In: J.B. Hall, G.A. Schmidt and L.D.H. Wood (Eds.), Principles of Critical Care. New York, McGraw-Hill (1992) 1757–1769.

WITT, N.J., D.W. ZOCHODNE, C.F. BOLTON, F. GRAND'MAISON, G. WELLS, G.B. YOUNG and W.J. SIBBALD: Peripheral nerve function in sepsis and multiple organ failure. Chest 99 (1991) 176–184.

YOUNG, G.B., C.F. BOLTON, Y.M. ARCHIBALD, T.W. AUSTIN and G.A. WELLS: The electroencephalogram in sepsis-associated encephalopathy. J. Clin. Neurophysiol. 9 (1992) 145–152.

ZOCHODNE, D.W., C.F. BOLTON, G.A. WELLS, J.J. GILBERT, A.F. HAHN, J.D. BROWN and W.A. SIBBALD: Critical illness polyneuropathy: a complication of sepsis and multiple organ failure. Brain 110 (1987) 819–842.

ZOCHODNE, D.W., D.A. RAMSAY, V. SALY, S. SHELLEY and S. MOFFATT: Acute necrotizing myopathy of intensive care: electrophysiological studies. Muscle Nerve 17 (1994) 285–292.

Handbook of Clinical Neurology, Vol. 27 (71): Systemic Diseases, Part III
M.J. Aminoff and C.G. Goetz, editors

CHAPTER 22

Non-infectious neurologic sequelae of infectious diseases and their treatment

FRED D. LUBLIN and TERRY HEIMAN PATTERSON

Department of Neurology, Allegheny University of the Health Sciences, Philadelphia, PA, U.S.A.

The most important of the non-infectious complications of infection are either immunologically based or due to microbial toxins. Those that are immunologically based likely relate to shared antigenic sequences between microbes and human nervous system tissue, leading to cross-reactivity with the protective immune response to the invading organism. The immunological disorders share certain clinical characteristics: onset most commonly 7–28 days after the infection, absence of evidence for an ongoing infectious process at the time of the neurologic event, absence of evidence of infection in the nervous system structure involved, no relationship between the severity of the infection and the severity of the post-infectious event and rather specific targeting of a portion of the nervous system, either peripheral (PNS) or central (CNS) nervous systems or their subdivisions.

POST-INFECTIOUS SYNDROMES

Acute disseminated encephalomyelitis (ADEM)

Neurologic sequelae of infection have been described for at least 200 years. Acute disseminated encephalomyelitis (ADEM) was first reported in 1790 by Mr. James Lucas, a surgeon in Leeds, UK, who described a case of paraparesis and urinary retention in a young woman recovering from measles. Descriptions of encephalomyelitis following Jennerian vaccination, the first clinical and pathological accounts of post-vaccinal encephalomyelitis, appeared in the 1920s (Spillane and Wells 1964).

The essential diagnostic criterion for these post-infectious illnesses is an acute onset of neurologic dysfunction of primarily CNS origin that is associated with an infectious, usually viral, illness not due to direct infection of the CNS with the agent. As these are sporadic, infrequent illnesses, there has been little organized study of their characteristics, but many anecdotal reports. There are several different terms that have been applied to this group of diseases, depending on the timing of onset of neurologic symptoms in relation to the infection, the nature of the underlying pathology, or the inciting viral pathogen (post-infectious encephalomyelitis, para-infectious encephalomyelitis, post-vaccinal encephalomyelitis, acute disseminated perivenous encephalomyelitis, acute hemorrhagic leukoencephalitis, and others). They nevertheless share similar clinical and pathological features. They are most closely linked by the underlying pathologic abnormality of perivenous inflammatory demyelination. Similarly, these illnesses seem to be indistinguishable clinically and pathologically from the post-vaccinal encephalomyelitis that follows immunizations. Further, these illnesses bear a close resemblance to the animal disease experimental allergic encephalomyelitis (EAE), discussed below, suggesting that all are related to CNS autoimmune mechanisms. For these reasons, in this chapter, we will group all of these terms under the heading acute disseminated encephalomyelitis (ADEM) and discuss any specific distinctions of the various sub-types in the discussion. We will, however, not discuss in detail the post-vaccinal disorders in this chapter, but limit ourselves to the autoimmune consequences of infection only.

Clinical presentation

ADEM may follow any viral infection, even the most common or banal respiratory or gastrointestinal infection. The disorder is most commonly associated with the exanthems of childhood, measles, rubella and varicella. ADEM also has been described following influenza, mumps, herpes simplex, herpes zoster, Coxsackie–Epstein–Barr virus, mycoplasma pneumonia, and smallpox infections and vaccinations (e.g., rabies, rubella, poliomyelitis).

The clinical signs may reflect any combination of involvement of the brain, spinal cord, cranial nerves and spinal roots. Involvement of the brain is most common. Although this is primarily a demyelinating disorder, the clinical picture may reflect signs suggestive of both white and gray matter inflammation, i.e. encephalitis.

The onset usually occurs 3–15 days after an infection or immunization, and in post-exanthematous cases most commonly 2–7 days after the rash (Johnson et al. 1985). The neurologic symptoms occasionally precede the exanthema or may also occur after a prolonged latency. The onset can be either quite abrupt with seizures and progressive deterioration of mental status or there may be a prodromal period of headache, vomiting, meningeal signs and fever (often recurrent fever after initial defervescence of the fever related to the initial infection). The ensuing clinical signs may include hemiparesis, paraparesis, cranial neuropathies, cerebellar ataxia, a variety of movement disorders, and more generalized or multifocal cerebral dysfunction.

The most common parainfectious/postvaccinal complications present as encephalitis (90–96%), myelitis (3–4%) or polyradiculitis (1–7%) (Croft 1969). Optic neuritis may occur alone or in combination with other CNS signs of ADEM. In the case of the encephalitides, the onset is usually abrupt with seizures, sudden obtundation or confusion. The distribution and severity of focal neurologic signs such as motor deficits, movement/coordination disorders, or sensory deficits are variable. Mortality may be as high as 20%. Most patients recover, although many will have residual behavioral abnormalities, dementia or motor deficits (Sriram and Steinman 1984). ADEM is typically monophasic, but recurrent attacks have also been described making the distinction from multiple sclerosis (MS) difficult. As the pathologic findings in ADEM and acute bouts of MS are similar, the clinical picture may also be confusing. However, ADEM more commonly affects cortical functions, such as mental status changes and induction of seizures, than is typically seen in MS.

Several isolated clinical syndromes have been associated with the pathogenic mechanism underlying ADEM. These include acute transverse myelitis, optic neuritis, acute cerebellar ataxia and basal ganglionic syndromes (Sriram and Steinman 1984).

Acute transverse myelitis can occur as a part of the more generalized, multifocal CNS process of ADEM, but may also occur in isolation as the only manifestation of a post-infectious complication. Following measles, rubella or varicella infection, 3–4% of those with neurologic complications had transverse myelitis (Croft 1969). The illness usually starts with spinal pain, followed by ascending motor and sensory symptoms and bladder dysfunction, usually localizing to a high thoracic or cervical level. Deep tendon reflexes are often depressed during the acute period. Recovery is usually slow with variable residual deficits.

A cerebellar syndrome may accompany and persist after more diffuse CNS involvement or occur in isolation. It most commonly occurs after varicella infection (30%), but can be seen with other exanthems. The cerebellar syndrome is most often seen in children. Appendicular cerebellar signs tend to predominate and may be associated with nystagmus, vomiting, and other brainstem signs (Sriram and Steinman 1984). Recovery is usually complete or there may be minimal residua (Peters et al. 1978).

A hemorrhagic, hyperacute form of this disease is referred to as acute hemorrhagic leukoencephalitis (AHLE), known as Weston–Hurst leukoencephalitis. The clinical presentation of AHLE is more fulminant with multifocal cerebral/cortical signs, often rapidly progressing to coma and death.

Epidemiology

The use of inoculations for most of the exanthems and the discontinuation of smallpox vaccination have led to a decrease in the incidence of ADEM. The incidence of ADEM following most viral infections is unknown, as the viral precipitant is usually unidentified. There are statistics for the ADEM associated with the exanthems, but they are becoming dated. Measles is the commonest preceding exanthem in patients with ADEM, occurring at a rate of 1/1000 cases of measles, with a mortality rate of 20%. For varicella, the rate of ADEM is less than 1/10,000 cases, with a 5% mortality rate. For rubella, the rate of ADEM is less than 1/20,000, with a 20% mortality rate (Johnson et al. 1985).

Laboratory assessment and imaging

In ADEM, a mild, predominantly lymphocytic pleocytosis and elevation of protein can be detected in the cerebrospinal fluid (CSF). Granulocytes and erythrocytes are characteristic of the hemorrhagic variant. Elevation of CSF IgG, IgG index and occurrence of oligoclonal bands (OCB) is uncommon and transient in ADEM, in contrast to the high incidence of these alterations and the persistence of OCB in MS. The electroencephalogram (EEG) shows non-specific slowing. Brain imaging in ADEM, most notably magnetic resonance imaging (MRI), generally reveals extensive abnormalities of white matter compatible with demyelination. Simultaneous enhancement with imaging contrast material in all or most of the disseminated lesions may be seen in the acute monophasic ADEM, in contrast to the heterogeneity of the enhancement in the lesions of various ages in MS. Involvement of the basal ganglia and thalami (deep gray matter), patchy or confluent contrast-enhancing cortical lesions, edema of the brainstem or of the spinal cord have also been described in ADEM. These characteristics may also help to differentiate ADEM from MS (Kalman and Lublin 1996).

Pathology

Macroscopic changes seen at autopsy of acute cases of ADEM include congestion and edema. Microscopically, there are inflammatory changes, primarily in a perivenous distribution, involving white matter and, to a lesser extent, gray matter as well. This perivenous inflammation is the hallmark of cell-mediated inflammatory processes. The inflammatory infiltrate is composed of lymphocytes and, in more fulminant cases, neutrophils. The meninges may also show inflammation. Demyelination is usually perivenous, with relative sparing of the underlying axons. These lesions do not usually take the ellipsoid shape of the perivenous plaques of MS (Adams and Duchen 1992). In older lesions there is gliosis.

In AHLE the brain is usually edematous and may demonstrate frank hemorrhages and petechial lesions, mostly in the white matter. Histologically, the lesions center around the venules, resembling ADEM, but with a more acute appearance characterized by neutrophils and red cells. In ADEM, fibrinoid necrosis of small arterioles and veins is also seen accompanied by extravasation of red blood cells and extensive tissue necrosis. Russell (1955) drew attention to the pathologic similarity of AHLE and ADEM, suggesting that they represent differing severities of a pathologic continuum.

Pathogenesis

Several lines of evidence suggest that ADEM does not result from direct viral invasion of the CNS or reactivation of a latent virus, but rather is the consequence of a cell-mediated autoimmune response to a component of CNS tissue.

Studies of CNS tissue in ADEM have neither revealed evidence of viral particles or viral antigen (Gendelman et al. 1984) nor of the cytopathic effect that occurs with viral infection of tissue. There is no evidence of intrathecal viral-specific antibody production. The histologic pattern is not typical of a viral invasion of the brain.

The histopathology is very similar to that seen in EAE, with perivenular lymphocytic infiltration and demyelination. The clinical picture of the acute form of EAE, in which there is development of paralysis about 2 weeks after inoculation with CNS antigen, is quite similar to ADEM as well. A hyperacute form of EAE closely resembles the clinical and pathologic picture seen in AHLE, with a more fulminant onset and prominent hemorrhagic lesions.

This similarity to EAE has led to speculation that ADEM is caused by cross-reactivity between an immunogenic peptide sequence of a microbial pathogen (MBP) and a component of myelin, such that the immune response that develops against the pathogen also produces a cell-mediated autoimmune attack against a component of CNS myelin. Lymphocytes in the CSF and peripheral blood of patients with ADEM may show reactivity to MBP (as opposed to patients with MS, in whom the CSF lymphocytes show reactivity, but not those from peripheral blood) (Lisak and Zweiman 1977; Johnson et al. 1984). Clones of T lymphocytes from brains of some patients dying from ADEM show reactivity to MBP (Hafler et al. 1987). A possible mechanism for auto-immune induction is molecular mimicry between microbes and MBP. Studies have shown that microbial organisms may possess peptide sequences that are recognized by autoreactive MBP T lymphocyte clones (Wucherpfennig and Strominger 1995).

Treatment

There are no well controlled treatment protocols for ADEM and thus systematically validated therapeutic standards are not available. Most of the therapies currently used have been based on empiric treatments of small numbers of patients.

The first line of therapy is directed at stabilizing and supporting potentially seriously ill patients. Seizures will need to be controlled as will increased intracranial pressure. The value of intensive supportive treatment is strongly suggested by several case reports of full recovery, even after prolonged coma.

Immunotherapy. As ADEM is usually a self-limited, monophasic illness, specific immunosuppression should be reserved for those with more severe forms or involvement of critical areas of the CNS. As the disease may progress rapidly, all patients require careful monitoring of disease course. Although the use of corticosteroids has not been proven to be of benefit, corticosteroids or ACTH are commonly used in treating ADEM, based primarily on anecdotal reports. The numerous case reports describing a temporal relationship between administration of corticosteroids and improvement of the disease, or relapses coinciding with the withdrawal of this medication, are not easily ignored. Some have suggested a combination of pulse steroids and plasmapheresis (Stricker et al. 1992; Kanter et al. 1995). In more severely involved patients, especially those with AHLE, consideration can be given to cytostatic therapy, as used in systemic autoimmune disorders. Better understanding of the immunopathogenesis of ADEM should lead to more specific and less toxic therapies directed at the elements of the CNS immune response, i.e., the major histocompatibility complex (MHC) class II molecule, antigen (myelin basic protein, proteolipid protein, myelin-associated glycoprotein, etc.), T cell receptor (TCR), trafficking across the blood–brain barrier (BBB), and cytokine activity in the CNS micro-environment.

If the neurologic sequelae result in disabling deficits, an intensive rehabilitation program may provide further benefit.

The frequency of ADEM has decreased in the last three decades, likely related to the institution of vaccination for many viral diseases (e.g. measles). Rapid advances in biotechnology research may soon solve part of the remaining problems. Nevertheless, neurologists will see occasional ADEM secondary to uneliminated or new infectious agents. Therefore, elaboration of better therapeutic protocols, performance of well designed controlled trials, and introduction of new treatments and better vaccines need to be considered.

Sydenham's chorea

Sydenham's chorea was first described in 1686 by Thomas Sydenham as post-infectious choreic movements occurring in children. It follows rheumatic fever and while the incidence of both Sydenham's chorea and rheumatic fever are declining, it is still important to recognize and treat since treatment is relatively simple, and the risks of untreated rheumatic fever are high (Bruyn and Went 1986; Nausieda 1986; Ayoub 1992).

Clinical manifestations

Sydenham's chorea occurs most commonly in children between the ages of 5 and 15. It begins as involuntary movements weeks to months after group Aβ-hemolytic streptococcal infection, usually of the pharynx. The movement disorder may be either unilateral or bilateral consisting primarily of choreoathetoid movements of the limbs and lasting from 1 to 22 weeks (Nausieda et al. 1980). Rarely, chorea may persist for years (Gibb et al. 1985). While seizures are rare, the choreiform movements are commonly associated with an encephalopathy with irritability, lethargy, confusion, or coma. The incidence of Sydenham's chorea in the US has declined and most recent estimates range from 5 to 32% of rheumatic fever cases (Ayoub 1992). The decline can be attributed to the prompt use of antibiotics for streptococcal infections and prophylactic treatment with penicillin after rheumatic fever to reduce the risk of neurologic or cardiac problems with additional streptococcal infections (Mason et al. 1991).

Pathogenesis

Sydenham's chorea appears to be immune-mediated and antibodies to neuronal cytoplasm antigens in the caudate and subthalamic nuclei that cross-react with group A streptococcal membranes have been demonstrated in patients with rheumatic fever (Husby et al. 1976). There are also streptococcal M proteins which cross-react with basal ganglia (Bronze and Dale 1993).

Differential diagnosis

Sydenham's chorea must be differentiated from other causes of chorea that are infectious or post-infectious. These include: bacterial causes (subacute bacterial endocarditis, neurosyphilis, Lyme disease, tuberculosis); and viral causes (measles, mumps, influenza, cytomegalovirus, subacute sclerosing panencephalitis, HIV, Epstein–Barr virus (mononucleosis), varicella). In addition, other causes of chorea also need to be considered. These include inherited disorders (Wilson's disease, dystonias, Huntington's disease, tics, neuroacanthocytosis), drug-induced syndromes (tardive dyskinesia, cocaine, sympathomimetics, anticonvulsants, lithium, and oral contraceptives), autoimmune disorders (systemic lupus erythematosus, antiphospholipid antibody syndrome, post-vaccinal, MS, and chorea gravidarum), as well as encephalitis, anoxia, thyroid dysfunction, hypocalcemia, and strokes (Padberg and Bruyn 1986).

Diagnostic workup

A complete neurologic examination and diagnostic testing should be performed in any patient with chorea to assess the various causes of chorea. Clinical and family history are essential. It is important to ascertain any recent pharyngeal infections since if there has been recent infection an elevated antistreptolysin O (ASO) remains the best diagnostic test for Sydenham's chorea. However, caution needs to be used since an elevated titer is not specific and can occur in populations with a high prevalence of streptococcal infections and titers may be low if more than 2 months has elapsed since the initial infection. MRI will help rule out structural causes but is usually normal in Sydenham's chorea. Rheumatoid factor and antinuclear antibodies are

usually negative and CSF studies fail to reveal specific abnormalities. EEG may show generalized slowing acutely or after clinical recovery.

Prognosis and complications

Patients generally recover although symptoms may persist for months to years. In children who develop a severe encephalopathy, sequelae such as persistent motor control problems, seizures or cognitive impairment can occur. In patients who have had an episode of Sydenham's chorea, there can be a recurrence of chorea during pregnancy or oral contraceptive use.

Management

While rarely present, if any streptococcal pharyngitis remains, it should be treated with antibiotics. Further, encephalitis is treated with supportive measures when present. Specific treatment of the chorea includes dopamine antagonists, benzodiazepines or valproate (Daoud et al. 1990). If chorea is severe, use of neuroleptics with more specific D2 receptor antagonism (e.g., haloperidol) may be warranted but these agents can produce tardive dyskinesias.

Acute inflammatory demyelinating neuropathy (Guillain–Barré syndrome)

The major clinical features of acute inflammatory demyelinating polyneuropathy (AIDP) were first described by Landry in 1859 as 'acute ascending paralysis'. It was later, in 1916, that the albuminocytologic dissociation was described by Guillain, Barré and Strohl in a report of two soldiers who developed paralysis and areflexia (Asbury 1990). Subsequently, Guillain and Barré as well as many others have enlarged upon the initial reports creating the clinical picture so familiar to all neurologists (Dowling et al. 1987).

Epidemiology

The incidence of AIDP averages from 0.6 to 2.4 cases/100,000 population/year (Alter 1990). It affects all ages, although there may be a minor peak in young adult life with a second larger peak in the 5th–8th decade. Both sexes are affected with only a slight male predominance. There is no seasonal predisposition (Kennedy et al. 1978; Schonberger et al. 1981; Johnson 1982; Kaplan et al. 1983; Beghi et al. 1985). There have been no consistent abnormalities of the histocompatibility loci although there are unconfirmed reports of an over-representation of HLA B8 and A3 as well as DR3 (Gorodezky et al. 1983; Hafez et al. 1985).

Antecedent events are described in two-thirds of patients. Upper respiratory events are most common although 10–20% of patients will describe a gastrointestinal event. These events generally occur 1–3 weeks prior to the onset of neurologic symptoms but can be as long as 6 weeks (Ropper et al. 1991). The preceding event for AIDP is most commonly infectious including viral, bacterial, and mycoplasma infections (see below), but in 5–10% of patients, AIDP may follow a surgical procedure by 1–4 weeks (Stambough et al. 1990; Ropper et al. 1991). Those patients with AIDP following surgery tend to be more severe. AIDP has also been reported following drug therapy with zimelidine (Fagius et al. 1985), gold therapy (Dick and Ramon 1982), D-penicillamine (Knezevic et al. 1984), captopril (Chakraborty and Rudell 1987), streptokinase (Cicale 1987), metrizamide myelography (Gledhill and Verburgh 1986), amitriptyline overdosage (Leys et al. 1987), danazol (Hory et al. 1985), and disulfiram overdose (Rothrock et al. 1984). Malignancies, especially Hodgkin's disease and other lymphomas, have been associated with AIDP (Cameron et al. 1958; Klingon 1965; Lisak et al. 1977). Finally, AIDP has been reported in transplant patients (Drachman et al. 1970), patients receiving fever therapy (Garvey et al. 1955), during pregnancy (Laufenburg and Sirus 1989), and in association with rabies vaccine (McIntyre and Krouse 1949), A/New Jersey influenza vaccine (Schonberger et al. 1981) and tetanus toxoid (Miller and Stanton 1954).

Viral illnesses are most common. CMV is the most common specific viral cause and 15% of AIDP patients demonstrate IgM antibodies to CMV (Dowling and Cook 1981). Other viral illnesses associated with AIDP include Epstein–Barr virus (Rafferty et al. 1954), herpes simplex (Menonna et al. 1977), herpes zoster (Leneman 1966; Dowling and Cook 1981), varicella zoster (Welch 1962), and HIV (Lipkin et al. 1985; Hagberg et al. 1986; Vendrell et al. 1987). In an excellent review of the evidence for viral prodromes, Arnason and Solvien (1993) concluded that the most convincing association of AIDP can be found with CMV, EBV, HIV, and smallpox vaccinia. While less convincing, the authors felt that the association of AIDP with measles virus and varicella zoster virus was still probable.

In addition to viruses, bacterial agents can also precede AIDP. The most common non-viral pathogen is *Campylobacter jejuni*. In fact, 10–20% of patients with AIDP report a gastrointestinal illness and most of these are likely *C. jejuni*. The association between AIDP and *C. jejuni* is very strong and some authors estimate that 20% of cases follow Campylobacter (Arnason and Solvien 1993). In one study, 38% of consecutively studied patients with AIDP had serologic evidence of infection with Campylobacter and almost half of these patients did not report any gastrointestinal symptoms (Kaldor and Speed 1984). Anti-GM1 ganglioside antibodies are detected most frequently in those patients infected with *C. jejuni*. In addition, the structure of the lipopolysaccharide capsule of *C. jejuni* has a GM1 ganglioside-like structure which implicates molecular mimicry as a mechanism in AIDP (Yuki et al. 1993). The next most common non-viral pathogen that can precede AIDP is *Mycoplasma pneumoniae* (Hodges and Perkin 1969; Goldschmidt et al. 1980). Other less common organisms include gram-negative organisms (typhoid, paratyphoid, listeriosis, brucellosis, tularemia and tuberculosis) (Mushinski et al. 1964; Samantray et al. 1977; Vyravanathan and Senanayake 1983; Garcia et al. 1990). Finally, AIDP has also been reported following chlamydial infection, leptospirosis, toxoplasmosis, and malaria (Melnick and Flewett 1964; Morgan and Cawich 1980; Bouchez et al. 1985; Ropper et al. 1991).

Clinical manifestations

AIDP usually presents over several days with an ascending, predominantly motor polyradiculopathy. Progressive and usually symmetric weakness along with areflexia define the clinical picture (Asbury and Cornblath 1990; Ropper et al. 1991; Arnason and Solvien 1993). The severity of weakness is variable and generally it begins in the lower extremities. The distribution can be proximal or distal although proximal is more common. Weakness of trunk, intercostal, and diaphragmatic muscles occurs later. There can be associated cranial nerve involvement and facial diplegia occurs in up to 50% of patients (Ropper et al. 1991). Ten percent of patients can have extra-ocular involvement, with sixth nerve most commonly involved. Bulbar dysfunction including difficulties with swallowing and chewing can occur in up to 40% of patients and be the presenting feature in 5% (Ropper et al. 1991). Bulbar dysfunction is more common in patients who require ventilatory support. Respiratory failure does not occur in the absence of extremity and trunk weakness (Ropper et al. 1991). The presence of respiratory failure correlates with the amount of weakness in shoulder elevation and neck flexion (Ropper et al. 1991). Sensory loss, which occurs in 75% of patients, is variable. While most patients have sensory complaints including distal paresthesias, sensory abnormalities on examination are usually mild. Large fiber modalities (proprioception and vibratory sense) are more affected than small fiber sensation (pain, temperature, and light touch) reflecting the involvement of myelinated fibers. Reflexes are often absent at presentation or may decrease in affected areas over the course of disease progression.

While progressive weakness with areflexia is the hallmark of AIDP, there are several additional common findings and complications that should be noted. Dysautonomia is one of the most common complications, occurring in 65% of patients (Truax 1984; Ropper et al. 1991). The severity of the dysautonomia parallels the severity of weakness and respiratory involvement and increases mortality (Lichtenfeld 1971). Sympathetic and parasympathetic function may be involved and involvement may consist of decreased or increased activity. Tachycardia is the most common abnormality and occurs in 50% of patients (Krone et al. 1983). It is most prominent early in the course of the illness. While the tachycardia is generally benign, bradycardia can lead to sinus arrest and asystole requiring a pacemaker (Krone et al. 1983; Maytal et al. 1989). Other dysautonomias include orthostatic hypotension, hypertension, facial flushing, tightness in the chest, sweating abnormalities, ileus, and both diabetes insipidus and inappropriate ADH secretion (Share 1976; Truax 1984). Thirty to 55% of patients will complain of myalgias and significant pain (Ropper and Shahani 1984). The pain is predominantly in the upper legs, buttocks, and back, may precede the weakness, and tends to occur at night. Urinary retention as well as incontinence can occur in 10–20% of patients due to external sphincter involvement (Kogan et al. 1981). Both papilledema (Shlim and Cohen 1989) and more rarely optic neuritis (Toshniwal 1987) have been reported.

The course of AIDP follows a regular pattern of progression, plateau, and subsequent resolution. In 50% of cases progression of the syndrome occurs over 2 weeks, in 80% of cases over 3 weeks, and in over 90%, the evolution is complete by 4 weeks (Loffel et al. 1977; Asbury and Cornblath 1990; Ropper et al. 1991). There is a subsequent plateau stage that lasts 2–4 weeks (Loffel et al. 1977; Ropper et al. 1991), followed by a recovery phase. While most patients recover over 4–6 months, more severely affected patients can take up to 2 years. Eighty percent of patients will make a satisfactory recovery; 50% will have evidence of neurologic residua although significant neurologic deficit is seen in only 15%; 5% of patients will be permanently disabled. Young age, milder disease, and acute (evolution over 1–3 weeks), but not hyperacute (with progression over days), and improvement within the first week all indicate a better prognosis. Patients that are older, have a hyperacute onset, more severe weakness, and a reduction of the muscle compound action potential of less than 20% have a poorer prognosis. While AIDP is generally a monophasic illness, it can recur in 3–5% of patients and relapse is more likely in patients with a slow evolution (i.e. over 4 weeks).

Clinical variants

Several important clinical variants of AIDP have been described. The Fisher syndrome consists of ophthalmoplegia, ataxia, and areflexia (Fisher 1956). In large series, 5% of AIDP patients can initially present as a Fisher variant (Ropper et al. 1991). These patients demonstrate the classic albumino-cytologic dissociation along with evidence of demyelination of predominantly sensory nerves on electrophysiologic studies. They can develop progressive weakness and sensory disturbances. Preceding events have included EBV and *C. jejuni*.

Ropper (1986) has described a pharyngeal–cervical–brachial variant of AIDP in which involvement is restricted to these areas and does not extend to the lower extremities. The electrophysiologic studies in the limbs can be normal and the reflexes preserved. Spinal fluid may be normal.

In some cases, AIDP can remain a pure motor syndrome (Spaans 1985). These cases are typical except for the lack of sensory involvement. There has also been reported a pure sensory variant (Vallat

1989) although some authors feel it is a separate disorder and part of the spectrum of sensory ganglioneuritis (Arnason and Solvien 1993). A pandysautonomic syndrome has also been described (Young et al. 1975). Finally, there has been an axonal variant described in which the severity is generally worse and recovery poor (Feasby et al. 1986). These variants share CSF findings and electrophysiologic findings.

Etiology and pathogenesis

While the pathogenesis of AIDP has not been established, evidence suggests that it is a cell-mediated autoimmune disease of the peripheral nerves. Experimental allergic neuritis (EAN), an experimental model indistinguishable from Guillain–Barré syndrome (Waksman and Adams 1955), is produced by immunization of animals with peripheral nerve homogenates. The antigen in EAN is P2, a basic protein of peripheral nerve myelin (Brostoff et al. 1977). EAN is a T cell-mediated disorder and the disease can be transferred using lymphoid cells from animals with EAN (Aström and Waksman 1962). Lymphocytes and macrophages act together to destroy myelin within nerves, but the exact sequence of events is unclear. There are synergistic effects of antimyelin antibody and P2-reactive T cells in EAN, suggesting that both cellular and humoral immune mechanisms are involved in the demyelinating process (Spies et al. 1995). While the similarity of clinical presentation, electrophysiology, and pathologic features in AIDP and EAN might suggest that the pathogenesis of both disorders are similar, both the nature of possible inciting antigen(s) and the role that antecedent infections play as a trigger in AIDP remain unknown. It is likely that infectious agents play an indirect role through mechanisms such as molecular mimicry between antigens of infectious agents and components of peripheral nerve myelin (Arnason and Solvien 1993). For instance, antiganglioside antibodies are detected in 20–70% of AIDP patients (Quarles et al. 1990; Svennerholm and Fredman 1990). Anti-GM1 antibodies are more frequently detected in patients on a background of *C. jejuni* infection (Rees et al. 1995) whose lipopolysaccharide has GM1 ganglioside-like structure (Yuki et al. 1993, 1995). Antibody to GQ1b ganglioside has been found to be associated with the Fisher syndrome (Chiba et al. 1993; Willison et al. 1993) and Guillain–Barré syndrome with ophthalmoplegia (Chiba et al. 1993).

Since anti-GQ1b antibodies have limited ability to bind GQ1b at body temperature, it has been suggested that the antigen initiating the immune response is not GQ1b but some cross-reactive glycoprotein (Willison and Veitch 1994). Interestingly, anti-GM1 antibodies have recently been reported to increase the K^+ current (complement-independent) and decrease the Na^+ current (complement-dependent) in isolated rat myelinated nerve fibers (Takigawa et al. 1995); both effects can contribute to conduction block.

What happens after the initial triggering event in AIDP is not well understood. Once AIDP is triggered, it has been suggested that proinflammatory cytokines and adhesion molecules play a role in the pathogenesis of the clinical and pathologic picture (Sharief et al. 1993; Hartung et al. 1994).

Laboratory findings and diagnosis

The mainstay of the diagnosis is the demonstration of an increased spinal fluid protein without mononuclear cells (i.e., albumino-cytologic dissociation) in conjunction with evidence of demyelination on electrophysiologic studies. The spinal fluid protein may remain normal for the first week but then increases over 4–6 weeks, even after the disease is stabilizing (Segurado et al. 1986). Levels often reach greater than 0.1 g/l and sometimes are markedly elevated. The increased protein is likely due to root involvement or breakdown of the BBB. Occasionally, OCB are present. The elevated protein is associated with little to no mononuclear cell infiltration. A pleocytosis should raise the suspicion of HIV infection or other inflammatory etiology.

In addition to the CSF examination, electrophysiologic studies are important in establishing the diagnosis of AIDP. EMG and nerve conduction velocity (NCV) changes are often present even early in the evolution of the disease. The classic findings are those that indicate demyelination and include multifocal conduction block and slowed NCVs with prolonged distal and F-wave latencies. Detailed criteria for demyelination have been outlined by Cornblath (1990). There is early conduction block and reduction in the distal evoked compound muscle action potential (CMAP) amplitude, while changes in nerve conduction occur later. During the first week of the illness, CMAP decreases to about 50% of normal and continues to fall over the next 2–3 weeks. Subsequently, there is a slow recovery of CMAP amplitudes. Conduction velocities are well maintained early, but fall to about 70% during the third week (Albers et al. 1985). Prolongation of F-wave latencies have occurred late in some series (Albers et al. 1985) and as the earliest change in others (Cornblath et al. 1988). Eighty-seven percent of patients will meet the clinical criteria for demyelination at some point during the first 5 weeks (Albers et al. 1985). Variable amounts of axonal damage are reflected by the presence of denervation on needle examination. Changes include decreased recruitment and abnormal spontaneous activity early in the clinical evolution, and polyphasic motor units indicative of sprouting and reinnervation occur later (Albers et al. 1985).

Diagnostic criteria for AIDP have been outlined in detail by Asbury and Cornblath (1990) using key clinical features and laboratory findings. The features that are required for the diagnosis include progressive motor weakness of more than one limb in conjunction with loss of reflexes. Clinical features that are strongly suggestive of the diagnosis include the progression over 2–4 weeks, relative symmetry, mild sensory involvement, cranial nerve involvement (especially the facial nerve), recovery to some degree beginning 2–4 weeks after progression stops, and autonomic dysfunction. Laboratory features supportive of the diagnosis include the presence of albumino-cytologic dissociation in the spinal fluid and electrodiagnostic findings of demyelination (prolonged F-waves, slowed conduction or conduction block, and prolonged distal motor latencies). Features that cast doubt on the diagnosis include marked persistent asymmetry of weakness, persistent bowel or bladder dysfunction, more than 50 mononuclear cells in the CSF or any polymorphonuclear cells, and finally a distinct sensory level.

Other laboratory values, including hematologic and urine studies, are generally normal. However, liver enzymes and CK may be elevated and serum sodium can be decreased if inappropriate ADH secretion occurs.

Pathology
There are variable perivascular edema, mononuclear cell infiltrates, and paranodal as well as segmental demyelination of peripheral nerves. Changes are multifocal with predilection for the nerve roots, sites of entrapment, and distal ends of the nerve (Asbury et al. 1969). In ultrastructural studies, the earliest change is

the paranodal retraction of myelin which results in a widening of the nodal gap (Prineas 1972). As the myelin degenerates, ovoids and phagocytosis of debris become prominent. There have been two types of myelin degeneration identified, both associated with activated macrophages or vesicular disruption in areas not associated with macrophages and possibly mediated by cytokines or other soluble products secreted by lymphocytes and/or macrophages (Brechenmacher et al. 1987). Axonal degeneration can predominate.

Differential diagnosis
AIDP must be differentiated from other acute disorders of the motor unit that cause weakness. The characteristic clinical and laboratory findings in AIDP should generally allow differentiation with minimal difficulty. The main acute myopathic disorder to be considered is periodic paralysis which also presents with acute flaccid weakness. The history of prior episodes, the shorter duration of the weakness and the presence of potassium abnormalities in periodic paralysis help to differentiate it from AIDP.

The acute disorders of the neuromuscular junction that can mimic AIDP include myasthenia gravis, tick paralysis and botulism. Myasthenia gravis characteristically has a history of muscle fatigue with repetitive activity and can be diagnosed using the tensilon test. In botulism there may be a history of tainted food ingestion and a prominence of symptoms of autonomic involvement such as dry mouth, ileus, orthostatic hypotension, and fixed, dilated pupils. Spinal fluid will be normal.

Acute neuropathies that need to be excluded include porphyria, toxins, and polio. Porphyria presents with autonomic involvement including abdominal pain, mental status changes and neuropathy. The toxins that can mimic AIDP include dapsone, organophosphates, hexacarbons, and solvents. Polio is now rare but can occur in non-immunized and partially immunized patients. The lack of sensory symptoms, the presence of meningismus, and a CSF pleocytosis in acute polio help to distinguish it from AIDP.

Treatment
The first issue to address is the safety of the patient. Patients with suspected AIDP should be placed in a monitored setting where their respiratory function can be frequently evaluated and they can be observed for

dysautonomias. Prevention of complications such as respiratory failure and vascular collapse remains the mainstay of management. Ventilatory support and intubation should be carried out if there is a reduced vital capacity of 12–15 ml/kg, pO_2 below 70 mm Hg on inspired room air, or severe oropharyngeal paresis (Ropper and Kehne 1985). In patients with significant dysphagia, alternative feeding programs with either nasogastric feeding or surgically placed feeding tubes should be implemented. The prevention of nosocomial infection is important, because 25% acquire pneumonia and 30% acquire urinary tract infections. Other supportive measures include prophylaxis for pulmonary embolism, adequate nutrition, and prevention of decubitus ulcers and tendon shortening. Once the patient is stable, a systematic rehabilitation program starting with active resistive strengthening exercise is instituted.

Plasmapheresis and i.v. immunoglobulin (i.v. Ig) have been shown to be efficacious in the specific treatment of AIDP. Controlled trials of over 500 patients have demonstrated the efficacy of plasmapheresis in AIDP (Guillain–Barré Syndrome Study Group 1984; Osterman et al. 1984; French Cooperative Group 1987). The time spent in the hospital, on a ventilator, and time to ambulation are shortened in patients treated with plasmapheresis when compared to untreated controls (McKhan et al. 1988). As a rule, five treatments are given over 7–10 days. In cases where plasmapheresis is unavailable or contraindicated, i.v. Ig may well be of benefit. Favorable response to i.v. Ig has been described in case reports and small series (Kleyweg et al. 1988; Shahar et al. 1990). In the Dutch controlled study van der Meche et al. 1992), 52.7% of i.v. Ig treated patients were improved at 4 weeks compared with 34.2% of plasmapheresed patients. The decreased percentage of improved patients in the plasmapheresis group may reflect an increased number of patients with axonal loss as reflected by CMAP amplitudes of less than 3 mV. Similar responses were found to i.v. Ig and pheresis in a recent randomized study of 50 AIDP patients (Bril et al. 1996). Additional studies are needed to confirm the benefit of i.v. Ig in Guillain–Barré syndrome.

Corticosteroid use to treat patients with Guillain–Barré syndrome has shown no benefit in randomized, controlled trials (Hughes et al. 1978; Hughes 1991).

TOXIN-MEDIATED SYNDROMES

Botulism

Clinical descriptions of botulism date back to Hippocrates. In 1897, Van Ermengem isolated the *Bacillus botulinum* (*Clostridium botulinum*) and was the first to recognize that food-borne botulism originates from ingestion of preformed toxin rather than infection with an enteric pathogen. In 1976, enteric colonization with the toxin-forming organism was identified as the cause for infantile botulism, a syndrome in infants of acute afebrile weakness, respiratory collapse, and prolonged, but complete, recovery over weeks to months (Midura and Arnon 1976; Pickett et al. 1976). Enteric colonization leading to clinical botulism has also been reported in adults with severe disturbances of gastrointestinal motility (Bartlett 1986). Adult cases of botulism have been drastically reduced through public health measures in food processing, and infantile botulism is now the most common form of human botulism (Arnon 1986).

Clinical manifestations

Botulism classically presents as a descending, symmetric, flaccid paralysis of all skeletal muscles and many smooth muscles, 2–6 days following the ingestion of botulinum toxin in tainted food, particularly ducks' eggs (Lecour et al. 1988). There is progression of weakness over 1–3 days and quadriplegia with diminished or absent deep tendon reflexes can follow. Bulbar involvement produces external ophthalmoplegia, dysarthria, dysphagia, ptosis, and facial weakness. Accommodative paresis and sixth cranial nerve palsy are frequently early signs (Simcock et al. 1994). Patients often experience a dry mouth and may develop fixed, dilated pupils. Weakness of respiratory muscles develops and can lead to respiratory failure and intubation. Involvement of smooth muscle in the gastrointestinal tract results in constipation or paralytic ileus while bladder involvement causes urinary retention. Paresthesias and asymmetric limb weakness are occasionally observed (Hughes et al. 1981). It is important to recognize that, since the toxin inactivates only cholinergic synapses of the PNS, sensation, mentation, memory, temperature, blood pressure and heart rate should be normal (Haaland and Davis 1980). In general, a more severe clinical picture results from intoxication from type A than from types B or E (Woodruff et al. 1992).

Infantile botulism occurs in infants between 2 weeks and 11 months. The infant may first develop progressively severe constipation that can lead to paralytic ileus (Arnon 1980, 1986; Thompson 1982). The infant may exhibit poor feeding, decreased movement, and lethargy. Flaccid paralysis then develops along with frequent cranial nerve palsies. Infants can demonstrate ptosis, dilated pupils, facial diplegia, and impaired gag reflex along with a weak cry, poor suck, and lethargy (Arnon and Chin 1979; Thompson 1982; Schreiner et al. 1991). Loss of head control may be an early sign. Deep tendon reflexes are decreased or absent. Dry mucous membranes, decreased bowel motility, and urinary retention occur due to involvement of the autonomic nervous system. Weakness of respiratory muscles can progress to respiratory failure. Infants typically recover in the reverse order of the original progression, so that their extremity movement will improve before their respiratory function. Relapse has been reported, even after significant recovery (Glauser et al. 1990).

Patients can also develop botulism following deep wounds that are infected with *C. botulinum*. Signs and symptoms are similar to patients with food-borne botulism. The incubation period ranges from 4 to 51 days (Hikes and Manoli 1981). While the responsible wound is generally deep, botulism has occurred in the context of much less severe wounds including minimal surgical wounds, superficial skin abscesses, tooth abscess, cellulitis, and maxillary sinusitis following intranasal cocaine abuse (MacDonald et al. 1985; Swedberg et al. 1987; Weber et al. 1993).

Epidemiology

C. botulinum is a spore forming, strictly anaerobic, gram-positive bacillus commonly found in soil and water (Dowell 1984). Disease is caused by the toxin produced with infection. Toxin types A, B, E, and F are the main toxins that affect humans. There are distinct geographic distributions of various strains so that strains that produce toxins A, B, and F are usually found in the soil of geographic areas having low rainfall and moderate temperatures. Bacteria producing type A toxin are located primarily west of the Mississippi river and those producing type B toxin east of the Mississippi river (United States Public Health Service 1979), while bacteria which produce type E toxin are found in fresh water marine life and sediment particu-larly in Alaska and the Great Lakes region (United States Public Health Service 1979; Heyward and Bender 1981). In the United States, about 20–30 cases of food-borne botulism and 50–80 cases of infantile botulism occur each year (Morbidity and Mortality Weekly Report 1992). Wound botulism is rare, with only 1–3 cases recognized per year.

Pathogenesis

Botulism is caused by botulinum toxin produced by *Clostridium botulinum* (Dowell 1984).

Botulinum toxin is a family of serologically closely related neurotoxins: A, B, C1, D, E, F and G. Type A toxin is produced as a single-chained polypeptide with a molecular weight of 150,000 (Simpson 1986). The toxin becomes activated through cleavage into a heavy and light chain of about 100,000 and 50,000 molecular weight, respectively (Bandyopadhyay et al. 1987). There are three major domains in the toxin, a receptor-binding site at the carboxy terminus of the heavier chain, a channel-forming domain in the amino terminus of the heavier chain, and an internal toxin contained by the lighter chain. The toxin can be denatured by heating above 80°C (Wright 1955) but is resistant to gastric acid and digestion by enzymes of the gastrointestinal tract. It is the most potent biologic toxin known and an estimated 0.025 ng or 20,000,000 molecules of botulinum toxin are sufficient to kill a mouse (Arnon et al. 1981).

The toxin can be ingested directly (food-borne), *C. botulinum* spores can germinate in the gastrointestinal tract and produce toxin, or wounds can become infected with *C. botulinum* resulting in systemic absorption of the toxin produced. Following systemic absorption, the toxin circulates in the blood and the carboxy terminus of the heavy chain binds to specific receptors of the presynaptic cholinergic synapses in the PNS (Simpson 1986). The CNS is protected since the toxin does not appear to cross the BBB (Sugiyama 1980). Within 30 min of binding, the toxin is internalized through a receptor-mediated endocytosis (Simpson 1980; Middlebrook 1989) and then crosses into the cytoplasm where it cleaves critical exocytotic membrane fusion proteins responsible for the docking and fusion of synaptic vesicles to the presynaptic terminus resulting in altered cholinergic presynaptic stimulus-induced and spontaneous quantal acetylcholine release (Kim et al. 1984; Simpson 1986).

Prevention

Food-borne botulism can be prevented by proper canning techniques. Home-canned foods should be prepared using a pressure cooker or heating contaminated foods above 80°C (Wright 1955). At present, there is no method for preventing colonization of the gastrointestinal tract of infants with *C. botulinum*. Appropriate debridement and cleaning of the wound and administration of antibiotics, such as penicillin, to kill the *C. botulinum* bacteria prevent wound botulism.

Differential diagnosis

The differential diagnosis of food-borne botulism and wound botulism includes Guillain–Barré syndrome, diphtheritic polyneuropathy, tick paralysis, curare poisoning, poliomyelitis, myasthenia gravis, and the Lambert–Eaton syndrome (Davis 1993). Infantile botulism can be mimicked by severe dehydration or electrolyte imbalance, neonatal myasthenia gravis, poliomyelitis, hypothyroidism, tick paralysis, Werdnig–Hoffmann spinal muscular atrophy, Leigh's disease (subacute necrotizing encephalomyelopathy), congenital myopathy, and exposure to toxins such as heavy metals or organophosphates (Brown 1979).

Diagnostic workup

Infants suspected of having infant botulism should first be evaluated for serious systemic or CNS infection and appropriate cultures obtained and prophylactic antibiotics administered. The CSF and blood typically are normal in both infants and adults. EMG usually shows an increment in M-wave amplitude with rapid rates (20–50 Hz) of stimulation (Gutmann and Pratt 1976; Cornblath et al. 1983). At slower rates of repetitive stimulation there are variable responses (Cornblath et al. 1983). Concentric needle electromyography is less specific, with many patients demonstrating short-duration, low-amplitude motor unit potentials and abnormal spontaneous activity (Cornblath et al. 1983). False negatives do occur, however (Graf et al. 1992). Single-fiber electromyography may show increased jitter and blocking. Motor conduction velocities and distal sensory latencies are normal.

The definitive diagnosis of botulism is made by demonstrating the presence of botulinum toxin in serum, stool, or suspected food. The most sensitive diagnostic test for the presence of botulinum toxin is a biologic test administering samples of serum, stool, or food extracts to mice and watching for paralysis and

death (United States Public Health Service 1979). Heating the sample to 100°C or combining it with specific botulinum antitoxin should prevent the animal's death. Isolation of *C. botulinum* organisms from stool, food samples, or wound material supports the diagnosis as well.

Prognosis and complications

Death, usually due to respiratory failure and the resultant complications, occurs in about 20% from food-borne botulism (Horwitz et al. 1977), 15% from wound botulism (Weber et al. 1993), and 5% from infantile botulism (Jagoda and Renner 1990). Complications of all forms of botulism include those that occur in the context of respiratory failure and critical care as well as persistent fatigue that may last for 1–2 years following recovery.

Management

Treatment of botulism begins with supportive care in a monitored setting as soon as the diagnosis is considered or made. Paralysis can occur rapidly and intubation should be undertaken if there is evidence of progressive respiratory failure. Infants as well as adults require close monitoring and supportive care. Many infants experience altered autonomic functions that are generally mild and short-lived including sudden unexplained alterations in heart rate, blood pressure, or skin color. Urinary retention may require intermittent catheterization to prevent bladder infections. A minority of patients develop the syndrome of inappropriate antidiuretic hormone secretion.

Botulinum antitoxin should be given as soon as possible to prevent the progression of weakness in adults. It will not reverse weakness that is already present. Since most cases of human botulism are caused by types A, B, and E, it is possible to give equine anti-A, B, E serum to patients (United States Public Health Service 1979). Infected wounds require debridement and the administration of penicillin along with the botulinum antitoxin (Weber et al. 1993). Further, in all cases, medications such as the aminoglycoside antibiotics, that can cause neuromuscular junction blockade, should be avoided.

Tetanus

Tetanus has been known since the early Egyptians when a patient with trismus and nuchal rigidity after a penetrating skull injury was described in the Edwin Smith Papyrus (Breasted 1930). Tetanospasmin, the toxin responsible for experimental tetanus in animals, was isolated from anaerobic soil bacteria in 1890 and vaccination with the inactivated derivative of this bacterium protected patients from developing tetanus (Behring and Kitasato 1890).

Clinical manifestations

Tetanus is characterized by sustained muscular rigidity and, in severe cases, reflex spasms and dysphagia with respiratory compromise. While in most cases an initial wound can be identified, there are some patients in whom there is no wound found. Once there is entry of bacteria and production of spores, an incubation period varying from a few days to weeks can occur before symptoms are evident. The period of onset between the first symptoms and reflex spasms is variable as well (3–14 days) and the shorter the interval the more severe the syndrome.

Tetanus can occur in either a generalized or local form (Cole and Youngman 1969; Bleck 1991). Disease severity is determined by the temporal progression of symptoms. The earliest signs of generalized tetanus include rigidity of the masseter muscles (trismus, or lockjaw) and facial muscles. There may be straightening of the upper lip with a grimace (risus sardonicus). Localized stiffness near the injury may or may not be present. Subsequently, rigidity progresses to include axial musculatures with involvement of neck, back muscles (opisthotonos), and abdomen. Finally, there is stiffness of limb muscles with relative sparing of distal musculature. In severe cases there are repetitive paroxysmal, violent contractions of involved muscles (reflex spasms) when the patient tries to move or in response to even the slightest external or internal stimuli (fear, hunger, etc.). There can be dysphagia and speech difficulties due to trismus and spasms of the muscles of deglutition. In the most severe cases laryngospasm can lead to respiratory compromise. Autonomic involvement causes predominantly an increase in sympathetic activity with fluctuating blood pressure and heart rate, hyperhidrosis, and hyperthermia. If severe, the dysautonomias can result in respiratory collapse (Udwadia et al. 1992). The disease progresses for up to 14 days following the initial symptoms and improvement usually begins after 4 weeks.

Localized tetanus occurs when the muscular

rigidity is restricted to the wound-bearing extremity and may persist for months. However, it is more common that local tetanus is a forerunner of the generalized form (Bleck 1991).

Etiology

Tetanus is caused by the neurotoxin tetanospasmin, produced by spores of the anaerobic gram-positive rod, *Clostridium tetani*. The spores are introduced into wounds. Under appropriate anaerobic conditions (especially in necrotic wounds), contaminated spores germinate, proliferate, and produce two exotoxins: tetanolysin and tetanospasmin. The clinical relevance of tetanolysin in tetanus is still uncertain. Tetanotoxin can then bind to peripheral nerve terminals (see below) or be spread through the bloodstream to neuromuscular junctions (Price and Griffin 1977). Tetanospasmin is a thermolabile peptide of 151 kDa that is activated by protease cleavage into a heterodimer of one heavy chain (100 kDa) and one light chain (50 kDa) connected by a disulfide bridge (Bergey et al. 1989). This bridge as well as another one on the heavy chain is required for the toxin activity (Kreiglstein et al. 1990). The toxin can be cleaved by papain into three fragments. An A–B fragment is the light chain and the amino terminal end of the heavy chain, joined by a disulfide bridge. The C fragment is a 50-kDa carboxy-terminal polypeptide and contains both a ganglioside binding region at its carboxyl terminus that especially binds to GM3 and GT1 and a region that is responsible for internalization (Colville et al. 1992; Halpern and Loftus 1993). Binding of tetanospasmin to other neuronal glycoproteins, 80 kDa and 116 kDa, has also been demonstrated recently (Schengrund et al. 1992).

Once bound to the nerve terminus, the toxin travels centrally via retrograde axonal transport along the motor nerves and proceeds transsynaptically into presynaptic inhibitory interneurons where it inhibits the release of neurotransmitters (mainly GABA in brainstem and glycine in spinal cord) (Sandberg et al. 1989; Link et al. 1992; Montal et al. 1992; Schiavo et al. 1992; Facchiano et al. 1993). This results in increased muscle activity. Similarly, catecholamine levels may be increased, causing sympathetic overactivity.

Epidemiology

Tetanus is more prevalent in developing countries with approximately 1 million cases (18/100,000) cases occurring annually (Edsall 1975). Almost half of the cases worldwide are neonatal.

Prevention

The mainstay of prevention is proper wound care and vaccination. Boosters are recommended at 10-year intervals throughout adult life. If vaccination or booster injections have occurred within a 5–10-year period preceding an injury no further treatment is required (Gardner and Schaffner 1993). In other cases with uncertain history of vaccination, a complete series of tetanus vaccinations should be administered and if the wound is tetanus prone (severe tissue necrosis, suppuration, and retained foreign bodies) human tetanus immunoglobulin, 250 U intramuscularly, is also needed.

Differential diagnosis

Diseases that can mimic tetanus include strychnine intoxication, malignant neuroleptic syndrome, dystonias, hypocalcemia, and stiff-man syndrome.

Diagnostic workup

Tetanus is usually diagnosed by the characteristic clinical picture. While a history of wound exposure or the presence of a portal of entry supports the diagnosis, only one-third of cultures reveals *C. tetani*. There is no serologic test for toxin in serum or CSF.

Prognosis and complications

Prognosis depends on progression, severity, presence of autonomic instability, portal of entry and age. Common complications include pneumonia, fractures, muscle ruptures, rhabdomyolysis, and renal failure (Luisto and Seppäläinen 1989).

Management

The primary treatment strategies in tetanus include the elimination of the source of toxin, toxin neutralization, control of muscle rigidity and spasms, and ventilatory support (Cole and Youngman 1969; Bleck 1991).

The first steps in treatment include determining the severity of the clinical symptoms and identifying the portal of entry so that the wound can be properly treated. In severe cases it is important to anticipate the need for prophylactic ventilatory support and tube feeding. Once the point of entry is identified, the wound is cleaned and parental infusion of penicillin (10–12 million U daily for 10 days) is begun to eradi-

cate vegetative spores, despite the absence of firm evidence as to its benefit. Metronidazole is an alternative. Prior to wound manipulation, human tetanus immune globulin to neutralize toxin should be administered. The clinical course of the disease is shortened and the mortality reduced by the administration of tetanus immunoglobulin, promptly before manipulating the wound. Equine antitetanus serum (doses up to 10,000–100,000 U) is an alternative in case human tetanus immunoglobulin is not available, but it should be used after allergy testing and desensitization, if needed. A primary immunization series is also required in addition to human tetanus immunoglobulin.

Those with tetanus of moderate severity who still can maintain adequate ventilation should have benzodiazepine treatment, diazepam most preferably, either via nasogastric or i.v. route. These patients can tolerate doses of diazepam as high as 500 mg daily without depression of consciousness. Lorazepam, which has a longer duration of action, can be used as an alternative. In severe cases with ventilatory failure and violent spasms, neuromuscular blockade with atracurium, vecuronium or pancuronium is required. Tracheostomy is usually considered early in these severe cases because once the disease reaches its peak, it takes at least 4 weeks for recovery to occur.

Intrathecal baclofen (infusion or intermittent injections) may be effectively used as a single therapy in moderately severe cases (Saissy et al. 1992a). However, risk of developing central depression with coma and respiratory failure after frequent injections precludes its use as a routine in tetanus treatment. Flumazenil has been used to counteract these adverse effects (Saissy et al. 1992b).

No standard regimen for treatment of autonomic instability has been defined. Combined α- and β-adrenergic blocking agents, e.g., esmolol, labetalol, have been tried with some success. Other alternatives include parenteral administration of morphine, magnesium sulfate, clonidine, atropine, and continuous spinal anesthesia (Bleck 1991).

Diphtheria

Diphtheria (caused by *Corynebacterium diphtheriae*) was likely known to Hippocrates and the first reported epidemics occurred in the 16th century (Medical Research Council 1923). The acute illness begins as a local inflammatory infection of the upper airways but approximately 20% of patients develop cardiomyopathy and neuropathy secondary to the production of an exotoxin (Cross and Sadoff 1988).

Clinical picture

Diphtheria generally begins as a local throat infection. The three major varieties of diphtheria include faucial, laryngeal, and nasal. These syndromes occur after a 2–6-day incubation period with general malaise, irritability, anorexia and aching. These generalized symptoms are accompanied by local symptoms including an exudative membrane adherent to the mucosa of the throat and in severe cases, cervical adenopathy. If the disease is mild, resolution occurs within the first week. However, in the more severe patients, respiratory collapse, cardiac arrhythmias, congestive heart failure and death can occur.

Two types of neuropathic sequela can result from the elaboration of an exotoxin. Firstly there can be a local neuropathy of the cranial nerves that develops between 20 and 30 days from the time of the throat infection with resultant nasal speech, dysphagia, and possible respiratory compromise. The second type of neuropathy is a generalized neuropathy that develops 8–12 weeks following the onset of the infection. The incidence of neuropathy in diphtheria varies between 8 and 66%, usually about 20% (McDonald and Kocen 1993).

Local neuropathy begins with palatal paralysis, nasal speech, and impaired palatal sensation 3–4 weeks after the initial infection. The decrease in palatal sensation and poor cough contribute to respiratory compromise and aspiration. As the syndrome progresses there can be autonomic involvement including poor pupillary accommodation and blurred vision while the pupillary light reflex is spared. Subsequently, between the fifth and seventh weeks after infection, in severe cases, there can be progressive paralysis of pharynx, larynx and diaphragm. This leads to dysphagia, hoarseness, aphonia, and respiratory failure. In the most severe cases, there may be weakness of oculomotor muscles, jaw and facial movements, sternocleidomastoid muscles and tongue.

Generalized neuropathy usually occurs 8–12 weeks after the initial infection although it can occur as early as 3–4 weeks post infection. It tends to be a distal sensory and motor neuropathy with both large and small fiber involvement. In some patients there is a marked sensory ataxia (Walshe 1917–1918). The distal weakness can be accompanied by myalgias and progress to involve proximal and trunk muscles. Reflexes are diminished or absent. Rarely, bowel and bladder involvement occurs. Recovery occurs over days to weeks depending on severity. The neuropathy can be accompanied by severe myalgias. There is usually complete recovery over a period of weeks, with return of reflexes last to occur. CSF has been reported to demonstrate both albumino-cytologic dissociation in some patients as well as pleocytosis in other cases (Solders et al. 1989). Electrophysiology studies are consistent with a demyelinating neuropathy. The changes may not become evident for 2–3 weeks following the onset of weakness.

Treatment

The prompt administration of antitoxin in faucial diphtheria within the first 48 h of onset sharply reduces the incidence of neuropathy. Mortality in both early and late stages appears related most often to cardiac complications, but respiratory failure can occur as well.

REFERENCES

ADAMS, J.H. and L.W. DUCHEN (EDS.): Greenfield's Neuropathology. New York, Oxford University Press (1992) 455–462.

ALBERS, J.W., P.D. DONOFRIO and T.K. MCGONAGLE: Sequential electrodiagnostic abnormalities in acute inflammatory demyelinating polyradiculoneuropathy. Muscle Nerve 8 (1985) 528–539.

ALTER, M.: The epidemiology of Guillain–Barré syndrome. Ann. Neurol. 27 (1990) S7–S12.

ARNASON, B.G.W. and B. SOLVIEN: Acute inflammatory demyelinating polyradiculoneuropathy. In: P.J. Dyck, P.K. Thomas, J.W. Griffin et al. (Eds.), Peripheral Neuropathy, 3rd Edit. Philadelphia, PA, W.B. Saunders (1993) 1437–1497.

ARNON, S.: Infant botulism. Annu. Rev. Med. 31 (1980) 541–560.

ARNON, S.S.: Infant botulism: anticipating the second decade. J. Infect. Dis. 154 (1986) 201–206.

ARNON, S.S. and J. CHIN: The clinical spectrum of infant botulism. Rev. Infect. Dis. 1 (1979) 614–620.

ASBURY, A.K.: Diagnostic considerations in Guillain–Barré syndrome. Ann. Neurol. 9 (1981) S1–S5.

ASBURY, A.K.: Guillain–Barré syndrome: historical aspects. Ann. Neurol. 27 (1990) S2–S6.

ASBURY, A.K. and D.R. CORNBLATH: Assessment of current

diagnostic criteria for Guillain–Barré syndrome. Ann. Neurol. 27 (1990) S21–S24.

ASBURY, A.K., B.G. ARNASON and R.D. ADAMS: The inflammatory lesions in idiopathic polyneuritis. Medicine (Baltimore) 48 (1969) 173–215.

ASTRÖM, K.E. and B.H. WAKSMAN: The passive transfer of experimental allergic encephalomyelitis and neuritis with living lymphoid cells. J. Pathol. Bacteriol. 83 (1962) 89–106.

AYOUB, E.M.: Resurgence of rheumatic fever in the United States. The changing picture of a preventable illness. Postgrad. Med. 92 (1992) 133–142.

BANDYOPADHYAY, S., A.W. CLARK, B.R. DASGUPTA and V. SATHYAMOORTHY: Role of the heavy and light chains of botulinum neurotoxin in neuromuscular paralysis. J. Biol. Chem. 262 (1987) 2660–2663.

BARTLETT, J.C.: Infant botulism in adults (Editorial). N. Engl. J. Med. 315 (1986) 254–256.

BEGHI, E., L.T. KURLAND, D.W. MULDER ET AL.: Guillain–Barré syndrome: clinicoepidemiologic features and effect of influenza vaccine. Arch. Neurol. 42 (1985) 1053–1057.

BEHRING, E. and S. KITASATO: Ueber das Zustandekommen der Diphtherie-Immunität und der Tetanus-Immunität bei Thieren. Dtsch. Med. Wochenschr. 16 (1890) 1113–1114.

BERGEY, G.K., W.H. HABIG, J.I. BENNETT and C.S. LIN: Proteolytic cleavage of tetanus toxin increases activity. J. Neurochem. 53 (1989) 155–161.

BLECK, T.P.: Tetanus. In: W.M. Scheld, R.J. Whitley and D.T. Durack (Eds.), Infections of the Central Nervous System. New York, Raven Press (1991) 603–624.

BOUCHEZ, B., J. POIRRIEZ, G.E. ARNOTT ET AL.: Acute polyradiculoneuritis during toxoplasmosis (Letter). J. Neurol. 231 (1985) 347.

BREASTED, J.H.: The Edwin Smith Surgical Papyrus. Chicago, IL, University of Chicago Press (1930).

BRECHENMACHER, C., C. VITAL, C. DEMINIERE, L. LAURENTJOYE, Y. CASTAING, G. GBIKPI-BENISSAN, J.P. CARDINAUD and J.P. FAVAREL-GARRIGUES: Guillain–Barré syndrome: an ultrastructural study of peripheral nerves in 65 patients. Clin. Neuropathol. 6 (1987) 19–24.

BRIL, V., W.K. ILSE, R. PEARCE, A. DHANANI, D. SUTTON and K. KONG: Pilot trial of immunoglobulin versus plasma exchange in patients with Guillain–Barré syndrome. Neurology 46 (1996) 100–103.

BRONZE, M.S. and J.B. DALE: Epitopes of streptococcal M proteins that evoke antibodies that cross-react with human brain. J. Immunol. 151 (1993) 2820–2828.

BROSTOFF, S.W., S. LEVIT and J.M. POWERS: Induction of experimental allergic neuritis with a peptide from myelin P2 basic protein. Nature 268 (1977) 752–753.

BROWN, L.W.: Differential diagnosis of infant botulism. Rev. Infect. Dis. 1 (1979) 625–628.

BRUYN, G.W. and L.N. WENT: Huntington's chorea. In: P.J. Vinken, G.W. Bruyn and H.L. Klawans (Eds.), Extrapyramidal Disorders. Handbook of Clinical Neurology, Vol. 49. Amsterdam, Elsevier Science (1986) 267–313.

CAMERON, D.G., D.A. HOWELL and J.L. HUTCHINSON: Acute peripheral neuropathy in Hodgkin's disease. Report of a case with histologic features of allergic neuritis. Neurology 8 (1958) 578.

CHAKRABORTY, T.K. and W.J.S. RUDELL: Guillain–Barré neuropathy during treatment with captopril. Postgrad. Med. J. 63 (1987) 221–222.

CHIBA, A., S. KUSUNOKI, H. OBATA, R. MACHINAMI and I. KANAZAWA: Serum anti-GQ1b IgG antibody is associated with ophthalmoplegia in Miller–Fisher syndrome and Guillain–Barré syndrome: clinical and immunohistochemical studies. Neurology 43 (1993) 1911–1917.

CICALE, M.J.: Guillain–Barré syndrome after streptokinase therapy (Letter). South. Med. J. (1987) 1068.

COLE, L. and H. YOUNGMAN: Treatment of tetanus. Lancet i (1969) 1017–1020.

COLVILLE, C.A., M.K. BANSAL, J.H. PHILLIPS and S. VAN HEYNINGEN: The interaction of tetanus toxin with intact bovine adrenal chromaffin cells: binding of toxin and subsequent inhibition of catecholamine release. Biochim. Biophys. Acta 1137 (1992) 264–273.

CORNBLATH, D.R.: Electrophysiology in Guillain–Barré syndrome. Ann. Neurol. 27 (1990) S17–S20.

CORNBLATH, D.R., J.T. SLADKY and A.J. SUMNER: Clinical electrophysiology of infantile botulism. Muscle Nerve 6 (1983) 448–452.

CORNBLATH, D.R., E.D. MELLITS, J.W. GRIFFIN ET AL.: Motor conduction studies in Guillain–Barré syndrome: description and prognostic value. Ann. Neurol. 23 (1988) 354–359.

CROFT, P.B.: Para-infectious and post-vaccinal encephalomyelitis. Postgrad. Med. J. 45 (1969) 392–400.

CROSS, A.S. and J.C. SADOFF: Neurological sequelae to pertussis, diphtheria and tetanus: natural infection and immunization. In: A.A. Harris (Ed.), Microbial Disease. Handbook of Clinical Neurology, Vol. 52. Amsterdam, Elsevier Science (1988) 227–251.

DAOUD, A.S., M. ZAKI, R. SHAKIR and Q. ALSALEH: Effectiveness of sodium valproate in the treatment of Sydenham's chorea. Neurology 50 (1990) 1140–1141.

DAVIS, L.E.: Botulinum toxin: from poison to medicine. West. J. Med. 158 (1993) 25–29.

DICK, D.J. and D. RAMON: The Guillain–Barré syndrome following gold therapy. Scand. J. Rheumatol. 11 (1982) 119–120.

DOWELL, V.R., JR.: Botulism and tetanus: selected epidemiologic and microbiologic aspects. Rev. Infect. Dis. 6 (1984) S202–S207.

DOWLING, P.C. and S.D. COOK: Role of infection in Guillain–Barré syndrome: laboratory confirmation of herpesviruses in 41 cases. Ann. Neurol. 9 (1981) S44.

DOWLING, P.C., B.M. BLUMBERG and S.D. COOK: Guillain–Barré syndrome. In: W.B. Matthews (Ed.), Neuropathies. Handbook of Clinical Neurology, Vol. 51. Amsterdam, Elsevier Science (1987) 239–262.

DRACHMAN, D.A., P.Y. PATTERSON, B.S. BERLIN and J. ROGUSKA: Immunosuppression and the Guillain–Barré syndrome. Arch. Neurol. 23 (1970) 385–393.

EDSALL, G.: Introduction. In: G. Edsall (Ed.), Proc. 4th Intl. Conf. Tetanus, Dakar. Lyon, Foundation Murieux (1975) 19–20.

FACCHIANO, F., F. BENFENATI, F. VALTORTA and A. LUINI: Covalent modification of synapsin I by a tetanus toxin-activated transglutaminase. J. Biol. Chem. 268 (1993) 4591–4688.

FAGIUS, J., P.O. OSTERMAN, A. SIDEN and B.E. WIHOLM: Guillain–Barré syndrome following zimelidine treatment. J. Neurol. Neurosurg. Psychiatry 48 (1985) 65–69.

FEASBY, T.E., J.J. GILBERT, W.F. BROWN, C.F. BOLTON, A.F. HAHN, W.F. KOOPMAN and D.W. ZOCHODNE: An acute axonal form of Guillain–Barré polyneuropathy. Brain 109 (1986) 1115–1126.

FISHER, M.: An unusual variant of acute idiopathic polyneuritis (syndrome of ophthalmoplegia, ataxia and areflexia). N. Engl. J. Med 255 (1956) 57–65.

FRENCH COOPERATIVE GROUP ON PLASMA EXCHANGE IN GUILLAIN–BARRÉ SYNDROME. Efficiency of plasma exchange in Guillain–Barré syndrome: role of replacement fluids. Ann. Neurol. 22 (1987) 753–761.

GARCIA, T., J.C. SANCHEZ, J.F. MAESTRE ET AL.: Brucellosis and acute inflammatory polyradiculopathy. Neurologia 4 (1990) 145.

GARDNER, P. and W. SCHAFFNER: Current concepts: immunization of adults. N. Engl. J. Med. 328 (1993) 1252–1258.

GARVEY, P.H., N. JONES and S.L. WARREN: Polyradiculoneuritis (Guillain–Barré syndrome) following the use of sulfanilamide and fever therapy. J. Am. Med. Assoc. (1955) 115.

GENDELMAN, H.E., J.S. WOLINSKY, R.T. JOHNSON, N.J. PRESSMAN, G.H. PEZESHKPOUR and G.F. BOISSET: Measles encephalomyelitis: lack of evidence of viral invasion of the central nervous system and quantitative study of the nature of demyelination. Ann. Neurol. 15 (1984) 353–360.

GIBB, W.R.G., A.J. LEES and J.W. SCADDING: Persistent rheumatic chorea. Neurology 35 (1985) 101–102.

GLAUSER, T.A., H.C. MAGUIRE and J.T. SLADKY: Relapse of infant botulism. Ann. Neurol. 28 (1990) 187–189.

GLEDHILL, R.F. and A.P. VERBURGH: Exacerbation of Guillain–Barré syndrome following metrizamide myelography (Letter). S. Afr. Med. J. 69 (1986) 663.

GOLDSCHMIDT, B., J. MENONNA, J. FORTUNATO ET AL.: Mycoplasma antibody in Guillain–Barré syndrome

and other neurologic disorders. Ann. Neurol. 7 (1980) 108–112.

GORODEZKY, C., B. VARELA, L.E. CASTRO-ESCOBAR ET AL.: HLA-DR antigens in Mexican patients with Guillain–Barré syndrome. J. Neuroimmunol. 4 (1983) 1.

GRAF, W.D., R.M. HAYS, S.J. ASTLEY and P.M. MENDELMAN: Electrodiagnostic reliability in the diagnosis of infant botulism. J. Pediatr. 120 (1992) 747–749.

GUILLAIN–BARRÉ SYNDROME STUDY GROUP: Plasmapheresis and acute Guillain–Barré syndrome. Neurology 35 (1984) 1096–1104.

GUTMANN, L. and L. PRATT: Pathophysiologic aspects of human botulism. Arch. Neurol. 33 (1976) 175–179.

GUTMANN, L., J.D. MARTIN and W. WALTON: Dapsone motor neuropathy – an axonal disease. Neurology 26 (1976) 514.

HAALAND, K.Y. and L.E. DAVIS: Botulism and memory. Arch. Neurol. 37 (1980) 657–658.

HAFEZ, M., M. NAGATY, Y. AL-TONBARY ET AL.: HLA antigens in Guillain–Barré syndrome. J. Neurogenet. 2 (1985) 285.

HAFLER, D.A., D.S. BENJAMIN, J. BURKS and H.L. WEINER: Myelin basic protein and proteolipid protein reactivity of brain and cerebrospinal fluid-derived T cell clones in multiple sclerosis and postinfectious encephalomyelitis. J. Immunol. 139 (1987) 68–72.

HAGBERG, L., B.E. MALMVALL, L. SVENNERHOLM ET AL.: Guillain–Barré syndrome as an early manifestation of HIV central nervous system infection. Scand. J. Infect. Dis. 18 (1986) 591.

HALPERN, J.L. and A. LOFTUS: Characterization of the receptor-binding domain of tetanus toxin. J. Biol. Chem. 268 (1993) 11188–11192.

HARTUNG, H.-P., K. REINERS, M. MICHELS, R.A. HUGHES, F. HEIDENREICH, J. ZIELASEK, U. ENDERS and K.V. TOYKA: Serum levels of soluble selectin (ELAM-1) in immune-mediated neuropathies. Neurology 44 (1994) 1153–1158.

HEYWARD, L. and T.R. BENDER: Botulism in Alaska, 1947–1980. In: G.E. Lewis, Jr. (Ed.), Biomedical Aspects of Botulism. New York, Academic Press (1981) 285–289.

HIKES, D.C. and A. MANOLI: Wound botulism. J. Trauma 21 (1981) 68–71.

HODGES, G.R. and R.L. PERKIN: Landry Guillain–Barré syndrome associated with *Mycoplasma pneumoniae* infection. J. Am. Med. Assoc. 210 (1969) 2088.

HORWITZ, M.A., J.M. HUGHES and M.H. MERSON: Food-borne botulism in the United States. J. Infect. Dis. 136 (1977) 153–159.

HORY, B., D. BLANC, A. BOILLOT and J. PANOUSE-PERRIN: Guillain–Barré syndrome following danazol and corticosteroid therapy for hereditary angioedema. Am. J. Med. 79 (1985) 111–114.

HUGHES, J.M., J.R. BLUMENTHAL, M.H. MERSON, G.L. LOM-

BARD, V.R. DOWELL and E.J. GANGAROSA, JR.: Clinical features of types A and B food-borne botulism. Ann. Intern. Med. 95 (1981) 442–445.

HUGHES, R.A.C.: Ineffectiveness of high-dose intravenous methylprednisolone in Guillain–Barré syndrome. Lancet 338 (1991) 1142.

HUGHES, R.A.C., J.M. NEWSOM-DAVIS, G.D. PERKIN and J.M. PIERCE: Controlled trial of prednisone in acute polyneuropathy. Lancet ii (1978) 750–753.

HUSBY, G., U. VAN DE RIJN, J.B. ZABRISKIE, Z.H. ABDIN and R.C. WILLIAMS, JR.: Antibodies reacting with cytoplasm of subthalamic and caudate nuclei neurons in chorea and acute rheumatic fever. J. Exp. Med. 144 (1976) 1094–1110.

JAGODA, A. and G. RENNER: Infant botulism: case report and clinical update. Am. J. Emerg. Med. 8 (1990) 318–320.

JAHN, R. and H. NIEMANN: Molecular mechanisms of clostridial neurotoxins. Ann. N.Y. Acad. Sci. 733 (1994) 245–255.

JOHNSON, D.E.: Guillain–Barré syndrome in the US Army. Arch. Neurol. 39 (1982) 21.

JOHNSON, R.T., D.E. GRIFFIN, R.L. HIRSCH, J.S. WOLINSKY, S. ROEDENBECK, I. LINDO DE SORIANO and A. VAISBERG: Measles encephalomyelitis – clinical and immunologic studies. N. Engl. J. Med. 310 (1984) 137–141.

JOHNSON, R.T., D.E. GRIFFIN and H.E. GENDELMAN: Postinfectious encephalomyelitis. Semin. Neurol. 5 (1985) 180–190.

JOY, R.J.T., R. SCALETTAR and D.B. SODEE: Optic nerve and peripheral neuritis. Probable effect of prolonged chloramphenicol therapy. J. Am. Med. Assoc. 173 (1960) 1731.

KALDOR, J. and B.R. SPEED: Guillain–Barré syndrome and *Campylobacter jejuni*: a serologic study. Br. Med. J. 288 (1984) 1867–1870.

KALMAN, B. and F. LUBLIN: Postinfectious encephalomyelitis and transverse myelitis. In: R.T. Johnson and J.W. Griffin (Eds.), Current Therapy in Neurological Disease. St. Louis, MO, Mosby Year Book (1996) 175–178.

KANTER, D.S., D. HORENSKY, R.A. SPERLING, J.D. KAPLAN, M.E. MALACHOWSKI and W.H. CHURCHILL: Plasmapheresis in fulminant acute disseminated encephalomyelitis. Neurology 45 (1995) 824–827.

KAPLAN, J.E., L.B. SCHONBERGER, E.S. HURWITZ ET AL.: Guillain–Barré syndrome in the United States, 1978–1981: additional observations from the national surveillance system. Neurology 33 (1983) 633.

KENNEDY, R.H., M.A. DANIELSON, D.W. MULDER ET AL.: Guillain–Barré syndrome: a 42-year epidemiologic and clinical study. Mayo Clin. Proc. 53 (1978) 93.

KIM, Y.I., T. LOMO, M.T. LUPA and S. THESLEFF: Miniature end-plate potentials in rat skeletal muscle poisoned with botulinum toxin. J. Physiol. (Lond.) 356 (1984) 587–599.

KLEYWEG, R.P., F.G.A. VAN DER MECHE and J. MEULSTEE: Treatment of Guillain–Barré syndrome with high dose gamma globulin. Neurology 38 (1988) 1639–1641.

KLINGON, G.H.: Guillain–Barré syndrome associated with cancer. Cancer 18 (1965) 157–163.

KNEZEVIC, W., J. QUINTNER, F.L. MASTIGLIA and P.J. ZILKO: Guillain–Barré syndrome and pemphigus foliaceus associated with D-penicillamine therapy. Aust. N.Z. J. Med. 14 (1984) 50–52.

KOGAN, B.A., M.H. SOLOMON and A.C. DIOKNO: Urinary retention secondary to Landry–Guillain–Barré syndrome. J. Urol. 126 (1981) 643.

KREIGLSTEIN, K., A. HENSCHEN, U. WELLER and E. HABERMANN: Arrangement of disulfide bridges and positions of sulfhydryl groups in tetanus toxin. Eur. J. Biochem. 188 (1990) 39–45.

KRONE, A., P. REUTHER and U. FUHRMEISTER: Autonomic dysfunction in polyneuropathies: a report of 106 cases. J. Neurol. 230 (1983) 111.

LAUFENBURG, H.F. and S.R. SIRUS: Guillain–Barré syndrome in pregnancy. Am. Fam. Phscn. 39 (1989) 147.

LECOUR, H., M.H. RAMOS, B. ALMEIDA and R. BARBOSA: Foodborne botulism: a review of 13 outbreaks. Arch. Intern. Med. 148 (1988) 578–580.

LENEMAN, F.: The Guillain–Barré syndrome. Arch. Intern. Med. 118 (1966) 139.

LEYS, D., F. PASQUIRE, M.D. LAMBLIN ET AL.: Acute polyradiculopathy after amitriptyline overdosage. Br. Med. J. 294 (1987) 608.

LICHTENFELD, P.: Autonomic dysfunction in the Guillain–Barré syndrome. Am. J. Med. 50 (1971) 772.

LINK, E., L. EDELMANN, J.H. CHOU, T. BINZ, S. YAMASAKI, U. EISEL, M. BAUMERT, T.C. SUDHOF, H. NIEMANN and R. JAHN: Tetanus toxin action: inhibition of neurotransmitter release linked to synaptobrevin proteolysis. Biochem. Biophys. Res. Commun. 189 (1992) 1017–1023.

LIPKIN, I., G. PARRY, D. KIPROV and D. ABRAMS: Inflammatory neuropathy in homosexual men with lymphadenopathy. Neurology 35 (1985) 1479–1483.

LISAK, R.P. and B. ZWEIMAN: In vitro cell-mediated immunity of cerebrospinal-fluid lymphocytes to myelin basic protein in primary demyelinating diseases. N. Engl. J. Med. 297 (1977) 850–853.

LISAK, R.P., M. MITCHELL, B. ZWEIMAN ET AL.: Guillain–Barré syndrome and Hodgkin's disease: three cases with immunologic studies. Ann. Neurol. 1 (1977) 72–78.

LOFFEL, N.B., L.N. ROSSI, M. MUMMENTHALER ET AL.: The Landry–Guillain–Barré syndrome: complications, prognosis, and natural history in 123 cases. J. Neurol. Sci. 33 (1977) 71–79.

LUCAS, J.: An account of uncommon symptoms succeeding the measles; with additional remarks on the infection of measles and smallpox. Lond. Med. J. 11 (1790) 325–331.

LUISTO, M. and A.M. SEPPÄLÄINEN: Electroneuromyographic sequela of tetanus, a controlled study of 40 patients. Electromyogr. Clin. Neurophysiol. 29 (1989) 377–381.

MACDONALD, K.L., G.W. RUTHERFORD, S.M. FRIEDMAN, J.R. DIETZ, B.R. KAYE, G.F. MCKINLEY, J.H. TENNEY and M.L. COHEN: Botulism and botulism-like illness in chronic drug abusers. Ann. Intern. Med. 102 (1985) 616–618.

MASON, T., M. FISHER and G. KUJALA: Acute rheumatic fever in West Virginia: not just a disease of children. Arch. Intern. Med. 151 (1991) 133–136.

MAYTAL, J., L. EVIATAR, S.C. BRUNSON and N. GOOTMAN: Use of demand pacemaker in children with Guillain–Barré syndrome. Pediatr. Neurol. 5 (1989) 303.

MCDONALD, I. and R.S. KOCEN: Diphtheritic neuropathy. In: P.J. Dyck, P.K. Thomas, J.W. Griffin et al. (Eds.), Peripheral Neuropathy, Edit. 3. Philadelphia, PA, W.B. Saunders (1993) 1412–1417.

MCINTYRE, H.D. and H. KROUSE: Guillain–Barré syndrome complicating anti-rabies inoculation. Arch. Neurol. Psychiatry 62 (1949) 802.

MCKHAN, G.M., J.W. GRIFFIN, D.R. CORNBLATH ET AL.: Guillain–Barré syndrome study group: analysis of prognostic factors and the effect of plasmapheresis. Ann. Neurol. 23 (1988) 347–353.

MEDICAL RESEARCH COUNCIL: Diphtheria. London, HMSO (1923).

MELNICK, S.C. and T.H. FLEWETT: Role of infection in the Guillain–Barré syndrome. J. Neurol. Neurosurg. Psychiatry 27 (1964) 395–407.

MENONNA, J., B. GOLDSCHMIDT, N. HAIDRI ET AL.: Herpes simplex virus-IgM specific antibodies in Guillain Barré syndrome and encephalitis. Acta Neurol. Scand. 56 (1977) 223.

MIDDLEBROOK, J.L.: Cell surface receptors for protein toxins. In: L.L. Simpson (Ed.), Botulinum Neurotoxin and Tetanus Toxin. San Diego, CA, Academic Press (1989) 95–119.

MIDURA, T.F. and S.S. ARNON: Identification of *Clostridium botulinum* and its toxin in faeces. Lancet ii (1976) 934–936.

MILLER, H.G. and J.B. STANTON: Neurologic sequelae of prophylactic inoculation. Quart. J. Med. 23 (1954) 1.

MONTAL, M.S., R. BLEWITT, J.M. TOMICH and M. MONTAL: Identification of an ion channel-forming motif in the primary structure of tetanus and botulinum neurotoxins. FEBS Lett. 313 (1992) 12–18.

MORBIDITY AND MORTALITY WEEKLY REPORT: Botulism (foodborne and infant) – by year, United States 1960–1991. Morb. Mortal. Wkly. Rep. 40 (1992) 21.

MORGAN, A.C. and F. CAWICH: Ascending polyneuropathy in leptospirosis – a case study. Ann. Trop. Med. Parasitol. 74 (1980) 567.

MUSHINSKI, J.F., R.M. TANIGUICHI and J.W. STEIFEL: Guillain–Barré syndrome associated with ulceroglandular tularemia. Neurology 14 (1964) 877–879.

NAUSIEDA, P.A.: Sydenham's chorea gravidarum and contraceptive-induced chorea. In: P.J. Vinken, G.W. Bruyn and H.L. Klawans (Eds.), Extrapyramidal Disorders. Handbook of Clinical Neurology, Vol. 49. Amsterdam, Elsevier Science (1986) 359–367.

NAUSIEDA, P.A., B.J. GROSSMAN, W.C. KOLLER, W.J. WEINER and H.L. KLAWANS: Sydenham's chorea: an update. Neurology 30 (1980) 331–334.

OSTERMAN, P.O., J. FAGIUS, G. LUNDEMO ET AL.: Beneficial effects of plasma exchange in acute inflammatory polyradiculopathy. Lancet ii (1984) 1296–1299.

PADBERG, G. and G.W. BRUYN: Chorea: differential diagnosis. In: P.J. Vinken, G.W. Bruyn and H.L. Klawans (Eds.), Extrapyramidal Disorders. Handbook of Clinical Neurology, Vol. 49. Amsterdam, Elsevier Science (1986) 549–564.

PETERS, A.C.B., J. VERSTEEG, J. LINDEMAN and G. BOTA: Varicella and acute cerebellar ataxia. Arch. Neurol. 35 (1978) 769–771.

PICKETT, J., B. BERG, E. CHAPLIN and A. BRUNSTETTER-SHAFER: Syndrome of botulism in infancy: clinical and electrophysiologic study. N. Engl. J. Med. 295 (1976) 770–772.

PRICE, D.L. and J.W. GRIFFIN: Tetanus toxin: retrograde axonal transport of systemically administered toxin. Neurosci. Lett. 4 (1977) 61–65.

PRINEAS, J.W.: Acute idiopathic polyneuritis: an electron microscope study. Lab. Invest. 26 (1972) 133–147.

QUARLES, R.H., A.A. ILYAS and H.J. WILLISON: Antibodies to gangliosides and myelin proteins in Guillain–Barré syndrome. Ann. Neurol. 27 (1990) S48–S52.

RAFFERTY, M., E.E. SCHUMACHER, G.O. GRAIN and E.L. QUINN: Infectious mononucleosis and Guillain–Barré syndrome. Arch. Intern. Med. 93 (1954) 246–253.

REES, J.H., N.A. GREGSON and R.A. HUGHES: Anti-ganglioside GM1 antibodies in Guillain–Barré syndrome and their relationship to *Campylobacter jejuni* infection. Ann. Neurol. 38 (1995) 809–816.

ROPPER, A.H.: Unusual clinical variants and signs in Guillain–Barré syndrome. Arch. Neurol. 43 (1986) 1150.

ROPPER, A.H. and B.T. SHAHANI: Pain in Guillain–Barré syndrome. Arch. Neurol. 45 (1984) 511.

ROPPER, A.H., E.F.M. WIJDICKS and B.T. TRUAX: Guillain–Barré syndrome. Philadelphia, PA, F.A. Davis (1991).

ROTHROCK, J.F., P.C. JOHNSON, S.M. ROTHROCK and R. MERKLEY: Fulminant polyneuritis after overdose of disulfiram and ethanol. Neurology 34 (1984) 357.

RUSSELL, D.S.: The nosological unity of acute haemorrhagic leucoencephalitis and acute disseminated encephalomyelitis. Brain 78 (1955) 369–382.

SAISSY, J.M., J. DEMAZIERE, M. VITRIS, M. SECK, L. MARCOUX,

M. GAYE and M. NDIAYE: Treatment of severe tetanus by intrathecal injections of baclofen without artificial ventilation. Intens. Care Med. 18 (1992a) 241–244.

SAISSY, J.M. M. VITRIS, J. DEMAZAIRE, M. SECK, L. MARCOUX and M. GAYE: Flumazenil counteracts intrathecal baclofen-induced central nervous system depression in tetanus. Anesthesiology 76 (1992b) 1051–1053.

SAMANTRAY, S.K., S.C. JOHNSON, K.V. MATHAI ET AL.: Landry–Guillain–Barré syndrome. A study of 302 cases. Med. J. Aust. 2 (1977) 84–91.

SANDBERG, K., C.J. BERRY, E. EUGSTER and T.B. ROGERS: A role of cGMP during tetanus toxin blockade of acetylcholine release in the rat pheochromocytoma (PC12) cell line. J. Neurosci. 9 (1989) 3946–3954.

SCHENGRUND, C.L., N.J. RINGLER and B.R. DAS GUPTA: Adherence of botulinum and tetanus neurotoxins to synaptosomal proteins. Brain Res. Bull. 29 (1992) 917–924.

SCHIAVO, G., F. BENFENATI, B. POULAIN, O. ROSSETTO, P. POLVERINO DE LAURETO, B.R. DAS GUPTA and C. MONTECUCCO: Tetanus and botulinum toxin-B neurotoxins block neurotransmitter release by proteolytic cleavage of synaptobrevin. Nature 359 (1992) 832–835.

SCHONBERGER, L.B., E.S. HURWITZ, P. KATONA ET AL.: Guillain–Barré syndrome: its epidemiology and associations with influenza vaccination. Ann. Neurol. 9 (1981) S31–S38.

SCHREINER, M.S., E. FIELD and R. RUDDY: Infant botulism: a review of 12 years' experience at the Children's Hospital of Philadelphia. Pediatrics 87 (1991) 159–65.

SEGURADO, O.G., H. KRUGER and H.G. MERTENS: Clinical significance of serum and CSF findings in the Guillain–Barré syndrome and related disorders. J. Neurol. 233 (1986) 202.

SHAHAR, E., E.G. MURPHY and C.M. ROIFMAN: Benefit of intravenously administered immune serum globulin in a patient with Guillain–Barré syndrome. Clin. Lab. Obser. 116 (1990) 141–144.

SHARE, L.: Role of cardiovascular receptors in the control of ADH release. Cardiology 61 (1976) S51.

SHARIEF, M.K., B. MCLEAN and E.J. THOMPSON: Elevated serum levels of tumor necrosis factor in Guillain–Barré syndrome. Ann. Neurol. 33 (1993) 591–596.

SHLIM, D.R. and M.T. COHEN: Guillain–Barré syndrome presenting as high altitude cerebral edema. N. Engl. J. Med 3231 (1989) 545.

SIMCOCK, P.R., S. KELLEHER and J.A. DUNNE: Neuro-ophthalmic findings in botulism type B. Eye 8 (1994) 646–648.

SIMPSON, L.L.: Kinetic studies on the interaction between botulinum toxin type A and the cholinergic neuromuscular junction. J. Pharmacol. Exp. Ther. 212 (1980) 16–21.

SIMPSON, L.L.: Molecular pharmacology of botulinum toxin and tetanus toxin. Annu. Rev. Pharmacol. Toxicol. 26 (1986) 427–453.

SOLDERS, G., I. NENNESMO and A. PERRSON: Diphtheritic neuropathy, an analysis based on muscle and nerve biopsy and repeated electrophysiologic and autonomic function tests. J. Neurol. Neurosurg. Psychiatry 52 (1989) 876.

SPAANS, F.: Guillain–Barré syndrome with exclusively motor involvement. Electroencephalogr. Clin. Neurophysiol. 61 (1985) 15.

SPIES, J.M., J.D. POLLARD, J.G. BONNER ET AL.: Synergy between antibody and P2 reactive T cells in experimental allergic neuritis. J. Neuroimmunol. 57 (1995) 77–84.

SPILLANE, J.D. and C.E.C. WELLS: The neurology of Jennerian vaccination. A clinical account of the neurological complications which occurred during the smallpox epidemic in South Wales in 1962. Brain 87 (1964) 1–44.

SRIRAM, S. and L. STEINMAN: Postinfectious and postvaccinal encephalomyelitis. Neurol. Clin. 2 (1984) 341–353.

STAMBOUGH, J.L., J.G. QUINLAN and J.D. SWANSON: Guillain–Barré syndrome following spinal fusion for adult scoliosis. Spine 15 (1990) 45.

STRICKER, R.B., R.G. MILLER and D.D. KIPROV: Role of plasmapheresis in acute disseminated (postinfectious) encephalomyelitis. J. Clin. Apheresis 7 (1992) 173–179.

SUGIYAMA, H.: Clostridium botulinum neurotoxin. Microbiol. Rev. 44 (1980) 419–448.

SVENNERHOLM, L. and P. FREDMAN: Antibody detection in Guillain–Barré syndrome. Ann. Neurol. 27 (1990) S36–S40.

SWEDBERG, J., T.H. WENDEL and F. DEISS: Wound botulism. West. J. Med. 147 (1987) 335–338.

TAKIGAWA, T., H. YASUDA, R. KIKKAWA ET AL.: Antibodies against GM1 ganglioside affect K^+ and Na^+ currents in isolated rat myelinated nerve fibers. Ann. Neurol. 37 (1995) 436–442.

THOMPSON, J.A.: Infant botulism. Semin. Neurol. 2 (1982) 144–150.

TOSHNIWAL, P.: Demyelinating optic neuropathy with Miller–Fisher syndrome: the case for overlap syndromes with central and peripheral demyelination. J. Neurol. 234 (1987) 353.

TRUAX, B.T.: Autonomic disturbances in the Guillain–Barré syndrome. Semin. Neurol. 4 (1984) 462–468.

UDWADIA, F.E., J.D. SUNAVALA, M.C. JAIN ET AL.: Haemodynamic studies during the management of severe tetanus. Quart. J. Med. 83 (1992) 449–460.

UNITED STATES PUBLIC HEALTH SERVICE: Botulism in the United States, 1899–1977. Handbook for Epidemiologists, Clinicians, and Laboratory Workers. Atlanta, GA, Centers for Disease Control (1979).

VALLAT, J.M.: Sensory Guillain–Barré syndrome (Letter). Neurology 39 (1989) 879.

VAN DER MECHE, F.G.A., P.I.M. SCHMITZ and DUTCH GUILLAIN–BARRÉ STUDY GROUP: A randomized trial comparing intravenous immunoglobulin and plasma exchange in Guillain–Barré syndrome. N. Engl. J. Med. 326 (1992) 1123–1129.

VAN ERMENGEM, E.: A new anaerobic bacillus and its relation to botulism. Rev. Infect. Dis. 1 (1897) 701–719.

VENDRELL, J., C. HEREDIA, M. PUJOL ET AL.: Guillain–Barré syndrome associated with seroconversion for anti-HTLV-III (Letter). Neurology 37 (1987) 544.

VYRAVANATHAN, S. and N. SENANAYAKE: Guillain–Barré syndrome associated with tuberculosis. Postgrad. Med. J. 59 (1983) 516.

WAKSMAN, B.H. and R.D. ADAMS: Allergic neuritis: experimental disease of rabbits induced by the injection of peripheral nervous tissue and adjuvants. J. Exp. Med. 102 (1955) 213–235.

WALSHE, F.M.R.: On the pathogenesis of diphtheritic paralysis. Quart. J. Med. 11 (1917–1918) 191.

WEBER, J.T., H.C. GOODPASTURE, H. ALEXANDER, S.B. WERNER, C.L. HATHEWAY and R.V. TAUXE: Wound botulism in a patient with a tooth abscess: case report and review. Clin. Infect. Dis. 16 (1993) 635–639.

WELCH, R.G.: Chicken-pox and the Guillain–Barré syndrome. Arch. Dis. Child. 37 (1962) 557–559.

WILLISON, H.J., J. VEITCH, G. PATERSON and P.G.E. KENNEDY: Miller Fisher syndrome is associated with serum anti-bodies to GQ1b ganglioside. J. Neurol. Neurosurg. Psychiatry 56 (1993) 204–206.

WILLISON, H.J. and J. VEITCH: Immunoglobulin subclass distribution and binding characteristics of anti-GQ1b antibodies in Miller Fisher syndrome. J. Neuroimmunol. 51 (1994) 159–165.

WOODRUFF, B.A., P.M. GRIFFIN, L.M. MCCROSKEY, J.F. SMART, R.B. WAINWRIGHT, R.G. BRYANT, L.C. HUTWAGNER and C.L. HATHEWAY: Clinical and laboratory comparison of botulism from toxin types A, B, and E in the United States, 1975–1988. J. Infect. Dis. 166 (1992) 1281–1286.

WRIGHT, G.P.: The neurotoxins of Clostridium botulinum and Clostridium tetani. Pharmacol. Rev. 7 (1955) 413–465.

WUCHERPFENNIG, K.W. and J.L. STROMINGER: Molecular mimicry in T cell-mediated autoimmunity: viral peptides activate human T cell clones specific for myelin basic protein. Cell 80 (1995) 695–705.

YOUNG, R.R., A.K. ASBURY, J.L. CORBETT and R.D. ADAMS: Pure pan-dysautonomia with recovery: description and discussion of diagnostic criteria. Brain 98 (1975) 613–636.

YUKI, N., S. SATO, S. TSUJI, T. OHSAWA and T. MIYATAKE: Frequent presence of anti-GQ1b antibody in Fisher's syndrome. Neurology 43 (1993) 414–417.

YUKI, N., Y. ICHIHASHI and T. TAKI: Subclass of IgG antibody to GM1 epitope-bearing lipopolysaccharide of Campylobacter jejuni in patients with Guillain–Barré syndrome. J. Neuroimmunol. 60 (1995) 161–164.

Handbook of Clinical Neurology, Vol. 27 (71): Systemic Diseases, Part III
M.J. Aminoff and C.G. Goetz, editors

CHAPTER 23

Hypothermia: neurologic manifestations and applications

ELAINE J. SKALABRIN[1] and THOMAS P. BLECK[2]

[1]*Department of Neurology and* [2]*Neuroscience Intensive Care Unit, University of Virginia,
Charlottesville, VA, U.S.A.*

Hypothermia, defined as core temperature less than 35°C, occurs when heat dissipation of the body exceeds heat production (Farmer 1992). The purpose of this chapter is first to discuss hypothermia, its relationship to underlying neurologic disease and the clinical manifestations and treatment of accidental hypothermia. Secondly, this chapter will discuss the effect of hypothermia on cerebral function and review the rationale and use of intentional hypothermia as a therapeutic modality against cerebral insult.

The first references to the medical uses of hypothermia are found in association with the military, as cold therapy was initially utilized as a local anesthetic during field surgical procedures. In the 19th century, numerous anecdotal reports of the neurologic manifestations of accidental hypothermia surfaced. In the 1930s, Fay and Smith began experiments in the clinical applications of hypothermia. In normal human subjects, they found that dysarthria occurred at 34°C, retrograde amnesia occurred below 34°C and pupillary reflex was absent at 26°C (Fay and Smith 1941). They then utilized these cooling techniques in the treatment of brain tumors and head injury. Bigelow et al. (1950), in their publication on experimental hypothermia, introduced the idea that hypothermia might be useful in cardiac surgery. Subsequently, Lewis and Taufic (1953) performed the first open heart surgery under hypothermia. In 1957, Rosomoff conducted a series of experiments demonstrating a reduction in canine MCA infarct size following core temperature cooling. Despite these observations supporting the hypothesis that hypothermia protects the brain against the effects of ischemia, the clinical utility of hypothermia has been limited due to the risk of myocardial arrhythmia, blood hypercoagulability and hemodynamic compromise (Goto et al. 1993).

ANATOMY OF THERMOREGULATION

The human core temperature remains nearly constant (within 1°C) between ambient temperatures of 55–140°C (Guyton 1986). Current theories on this exquisite temperature regulation are based on the concept that information arising from disparate thermosensitive regions (e.g., subcutis, spinal cord, preoptic region) is integrated by hypothalamic structures. With this information, the hypothalamus triggers heat production, including shivering, piloerection, vasoconstriction, hunger and thyroxine secretion or heat loss consisting of vasodilation, sweating and the inhibition of heat production mechanisms. Moreover, this hypothalamic regulation is a complex system integrating the autonomic as well as behavioral and hormonal responses to variations in central and environmental temperatures (Janský and Musacchia 1976).

Specific temperature-sensitive neurons located in the anterior hypothalamus and preoptic area alter their firing rates in response to local changes in temperature. This response also varies diurnally and is altered by afferent impulses from peripheral thermoreceptors. Thus, the anterior hypothalamic-preoptic area functions as a thermostat: local warming of the anterior hypothalamic-preoptic area induces sweating and vasodilation; local cooling has the opposite effects (i.e., shivering, vasoconstriction and thyroid activation). Local temperature changes in the posterior hypothalamus and in other brain areas have no effect. However, the physiologic substrates for heat production are controlled by mechanisms in the posterior hypothalamus. The anterior hypothalamus contains mechanisms of regulation of heat loss. Therefore, lesions of the anterior hypothalamus usually interfere with the heat loss mechanism and result in hyperthermia (Martin et al. 1977). Conversely, lesions of the posterior hypothalamus may cause hypothermia by interfering with the shivering and vasoconstriction mechanisms. Large lesions may result in poikilothermia, the fluctuation of more than 2°C in body temperature following ambient temperature change (Fig. 1).

Neuropharmacologic studies indicate that biogenic amines, other neurotransmitters, and neuropeptides are involved in temperature regulation. Injection of opioids in large quantities intraventricularly produces hypothermia in different species, indicating that these peptides may play a role in normal temperature homeostasis (Collu et al. 1988). In addition, central administration of dopamine, acetylcholine, neurotensin, large doses of somatostatin, and small doses of bombesin have all been shown to induce hypothermia (Martin et al. 1977).

CAUSES OF HYPOTHERMIA

General

The most common etiology of hypothermia is cold exposure. This may occur in combination with physical or metabolic exhaustion. Old age, debilitating disease, lack of adequate housing, drug overdose and alcohol ingestion are major contributing factors (Abramowicz 1994). These patients may have an altered perception of ambient temperature and therefore may not seek relief from their environment (Farmer 1992). Alcohol is frequently associated with reports of accidental hypothermia. There are a number of mechanisms that may be responsible for this alcohol-related alteration of thermoregulation. These include peripheral vasodilatory effects, central depressant effects, concomitant peripheral neuropathy, or Wernicke's encephalopathy with hypothalamic involvement (Nakada and Knight 1984). Additional drugs associated with hypothermia include phenothiazines, tricyclic antidepressants, barbiturates, neuromuscular junction blocking agents, lithium, clonidine, anticholinergics, morphine, and acetaminophen overdose (Lieh-Lai et al. 1984). Chronic debilitating diseases such as heart failure, renal failure, hepatic insufficiency, psoriasis, ichthyosis, hypoadrenalism, hypothyroidism, diabetes mellitus, and hypoglycemia have been reported to precipitate hypothermia (Farmer 1992).

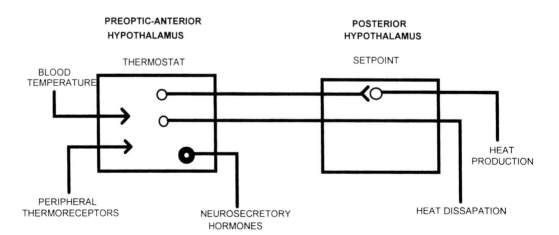

Fig. 1. Hypothalamic temperature regulation.

TABLE 1

Conditions and diseases associated with hypothermia.

Systemic disorders	Neurologic disorders
Hepatic failure	Central failure
Ketoacidosis	Trauma
Heart failure	Tumors
Renal failure	Stroke
Alcoholism	Infection
Cardiopulmonary failure	Subarachnoid hemorrhage
Systemic infection	Sarcoidosis
Pancreatitis	Parkinson's disease
Anorexia nervosa	Multiple system atrophy
Hypoadrenalism	Wernicke's encephalopathy
Diabetes mellitus	Syphilitic arteritis
Hypothyroidism	Hydrocephalus
Hypoglycemia	Encephalocele
Hypopituitarism	Shapiro's syndrome
Environmental exposure	Prader–Willi syndrome
Generalized cutaneous psoriasis, ichthyosis and other exfoliative dermatitides	Menke's kinky hair syndrome
Malnutrition	Peripheral failure
	Spinal cord lesions
Drugs	Autonomic neuropathies
Ethanol	Neuromuscular weakness
Phenothiazines	
Tricyclic antidepressants	
Barbiturates	
Paralytic muscle relaxants	
Lithium	
Clonidine	
Anticholinergics	

Neurologic

Neurologic etiologies for hypothermia can be divided into central and peripheral causes, and the central lesions can be further classified into chronic and paroxysmal (Table 1). Chronic lesions are generally located in the posterior (or entire) hypothalamus and may be due to a variety of pathologic processes. There have, however, been isolated reports of chronic hypothermia secondary to a solitary anterior hypothalamic lesion (Martin et al. 1977). Paroxysmal hypothermia may be accompanied by periodic vasodilation, sweating, lacrimation, nausea, vomiting and, less frequently, bradycardia and altered mentation. Normal body temperature is present between episodes. This syndrome has been historically interpreted as seizure activity; thus the term 'diencephalic epilepsy' has been applied. Although this interpretation is supported by some authors, there are reports of spontaneous paroxysmal or chronic hypothermia which are not clearly attributable to seizure activity (LeWitt et al. 1983).

A rare but recognized syndrome of central paroxysmal hypothermia occurs as Shapiro's syndrome, which comprises of agenesis of the corpus callosum, periodic hypothermia, and hyperhidrosis (Mooradian et al. 1984). Investigation of one such individual suggested a central deficit in temperature regulation, with an abnormally low hypothalamic set point coupled with normal homeothermic reflexes. Therapy with clonidine, a centrally acting α_2 agonist, produced remission of symptoms (Walker et al. 1992).

The primary causes of chronic hypothermia are related to infiltrative or destructive processes of the hypothalamus. Those reported include: tumors, including craniopharyngioma, glioblastoma multiforme, neuroblastoma, lymphoma, angioma, lipoma (Summers et al. 1981) or extension of facial hemangioma; infections (Holm et al. 1988), including tuberculosis, poliomyelitis, syphilis; sarcoidosis (Lipton et al. 1977); multiple sclerosis (Geny et al. 1992); Wernicke's encephalopathy (Tampi and Alexander 1982); gliosis; hydrocephalus; trauma (Ratcliffe et al.

1983); ischemia; systemic lupus erythematosus (SLE) (Kugler et al. 1990); and Parkinson's disease (Martin et al. 1977). In addition, hypothermia has been recognized in association with other syndromes including Menke's kinky hair syndrome, Prader–Willi syndrome (Gunn et al. 1984) and anorexia nervosa (Frankel and Jenkins 1975).

ACCIDENTAL HYPOTHERMIA

Accidental hypothermia is defined as an unintentional decline in core temperature to below 35°C. In the USA, the annual death rate from accidental hypothermia ranges from 2.2 to 4.3/1,000,000 people (US Department of Health and Human Services 1993). Among trauma patients, hypothermia was recorded on the first day of hospitalization in 42% of patients sustaining any moderate trauma (Gentilello and Moujaes 1995). The presentation of accidental hypothermia in the elderly may be subtle (e.g., flat affect or mild dysarthria) and the diagnosis missed unless temperatures are routinely measured. The mortality of hypothermia increases with the presence of concomitant illness (Curley 1995). Lastly, patients may develop hypothermia intraoperatively or postoperatively due to decreased heat production, coupled with heat loss from exposure of deep bodily structures.

The clinical presentation varies with the degree of hypothermia. Mild hypothermia (32–35°C) may be manifested by tachycardia, hypertension, tachypnea, bronchospasm, and shivering. The patient's mental status is usually normal, but may vary considerably. Moderate hypothermia (28–32°C) presents with a progressive decline in the level of consciousness, with paradoxical undressing, confusion, or obtundation. Pupillary dilation is generally present at temperatures below 30°C. Additionally, hypotension, hypoventilation, reduced renal blood flow, absence of shivering and rigid muscle tone may be present. Cardiac arrhythmias (atrial fibrillation, supraventricular tachycardia, or ventricular ectopy) are common below 30°C. Patients with severe hypothermia (26–28°C) may appear clinically dead with unmeasurable vital signs, absent deep tendon reflexes, and unelicitable oculocephalic and corneal reflexes. Spontaneous ventricular fibrillation is present in 50% of patients with severe hypothermia (Farmer 1992). ECG changes include prolonged PR, QRS, and QT intervals.

Osborne's wave is present in approximately 50% of patients at temperatures less than 32°C (Danzl and Pozos 1994). In addition, cardiac contractility and stroke volume decrease. At 28°C cardiac output decreases to less than 50% of baseline (Farmer 1992).

In addition, hypothermia exerts a variety of effects on other major organs. Gastrointestinal motility, hepatic function and pancreatic function all decrease progressively as temperature falls. Renal metabolic activity, glomerular filtration rate and tubular reabsorption of water and solutes also decrease with hypothermia. An initial decline in temperature produces an increased urine volume and decreased urine specific gravity referred to as 'cold diuresis'. The precise mechanism is unknown. This effect may be due either to inhibition of the secretion of anti-diuretic hormone or to a decreased sensitivity of the tubules to the hormone (Maclean and Emslie-Smith 1977).

Hypothermia produces widespread hematologic effects. The hematocrit increases 2% per 1°C decline in temperature. In addition, hypothermia induces thrombocytopenia via sequestration, as well as decreasing platelet activity, as the production of thromboxane B2 is temperature-dependent. Coagulopathy may develop secondary to impaired enzymatic reactions in the coagulation cascade. Conversely, hypercoagulability may occur (Danzl and Pozos 1994). Biggar et al. (1983), utilizing a porcine model, postulated that hypothermia impairs neutrophilic circulation and release from the bone marrow, which may increase the risk of bacterial infection.

Hypothermia shifts the oxyhemoglobin-dissociation curve to the left, resulting in decreased oxygen release from hemoglobin into the tissues. As blood cools, the arterial pH increases and pCO_2 decreases. Thus, a pH of 7.4 and a pCO_2 of 40 torr at a body temperature of 37°C are physiologically equivalent to a pH of 7.5 and a pCO_2 of 30 torr at a body temperature of 30°C. The acid-base equilibrium and a constant $[OH^-]/[H^+]$ ratio are maintained at widely different temperatures. Therefore, the recommended approach to maintaining acid-base balance in hypothermic patients is to utilize uncorrected arterial blood gases, known as the α-stat strategy (Stephan et al. 1992). Although controversial, this approach theoretically allows the arterial pH to increase with decreasing temperatures, and thus optimizes enzymatic function, preserves the normal distribution of metabolic intermediates, and maintains cellular waste disposal. Gradual correction is necessary as the bicarbonate system becomes progressively more efficient as body temperature rises (Danzl and Pozos 1994).

The management of hypothermia first requires accurate monitoring of core temperature, serum acid-base balance, and cardiac and respiratory function. Continuous temperature monitoring with rectal, esophageal, or bladder temperature probes is recommended. The mainstay of treatment for hypothermia is rewarming. Rewarming strategies may be active or passive. Passive rewarming (insulating the patient from further heat loss) is only appropriate in patients with mild hypothermia. Active rewarming is indicated in the presence of core temperatures of less than 32°C, cardiac instability, concomitant risk factors, or failure to respond to passive rewarming. Active external rewarming may be accomplished via a number of techniques, including immersion, radiant heat, forced air, and electric blankets. One major concern during active rewarming by these techniques is the phenomenon of 'core temperature afterdrop'. This is a drop in core temperature as a result of peripheral vasodilation in response to extremity warming. It may result in a decline of mean arterial pressure and peripheral vascular resistance and may be combated by isolating rewarming techniques to the trunk and administering aggressive rehydration. In severe hypothermia, active core rewarming is necessary. Methods of active core rewarming include ventilation with heated humidified air, intravenous infusion, peritoneal or pleural lavage, gastric, colonic or bladder irrigation, and extracorporeal bypass rewarming. These techniques entail a variety of risks and limitations. Extracorporeal bypass rewarming remains the most efficient means of rewarming (Danzl and Pozos 1994). Lastly, excessive pharmacologic manipulation of the vasoconstriction and depressed cardiovascular system should be avoided. In moderate to severe hypothermia, physical exertion, rough handling or invasive procedures can cause ventricular fibrillation. Active rewarming should take place only with adequate monitoring (Abramowicz 1994). In addition, treatment must include correcting underlying causes (i.e., hypothyroidism), preventing complications (i.e., infection, DVT), and detecting occult trauma (Danzl and Pozos 1994) (Table 2).

ELECTROPHYSIOLOGY IN HYPOTHERMIA

Numerous techniques have been employed to monitor cerebral activity during hypothermia. Since cerebral function decreases with temperature, an alteration in electrical activity would be expected. Studying brainstem auditory evoked responses (BAERs), Stockard et al. (1978) reported interwave latency prolongations at 32°C in patients with both spontaneous and induced hypothermia. These abnormalities resolved with rewarming. This relationship was confirmed in humans undergoing hypothermic open heart surgery. Markand et al. (1987) demonstrated an exponential increase in latency of BAER waves I, III, and V with decreasing of temperature. In addition, amplitude rose at 28°C but decreased linearly with further cooling.

Somatosensory evoked potentials (SEPs) have also been utilized for intraoperative monitoring. Stejskal et al. (1980) studied patients undergoing deep hypothermia and found an increase in latency in all waves observed. Russ et al. (1987) found a significant linear correlation between SEP latency and temperature. Hayes et al. (1993) reported that SEP amplitude and latency increased with cooling in hypothermic patients with spinal cord injury. This increase in amplitude suggests that cooling enhances central conduction when conduction deficits are due to focal demyelination. This is probably a consequence of diminished current leakage through partially demyelinated internodal segments increasing the probability that the next node will reach its firing threshold.

During hypothermia, EEG activity slows. Electrocerebral silence may be achieved at nasopharyngeal temperatures of 10–25°C. However, some investigators have reported that the EEG may be characterized by continuous phasic activity at these temperatures. In a quantitative analysis, Levy (1984) utilized power spectrum analysis and demonstrated linear correlations of temperature with the total power and with peak power frequency of the high frequency band. In addition, Reutens et al. (1990) reported the presence of triphasic waves in a patient with accidental hypothermia (rectal temperature 34°C).

TABLE 2

Treatment protocol for hypothermia. (From Larach 1995.)

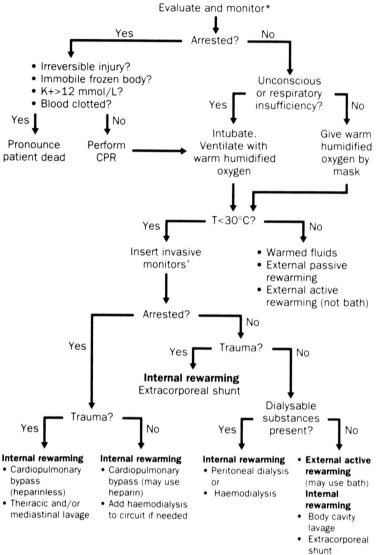

HYPOTHERMIA AND CEREBRAL FUNCTION

To understand the effects of hypothermia on cerebral physiology, animal and human investigations have centered on hypothermia-induced changes of cerebral blood flow (CBF), cerebral metabolism, and other mechanisms that may provide protection against cerebral insult (Fig. 2).

Effect on cerebral blood flow (CBF)

Rosomoff and Holaday (1954) found a decrease of CBF by 6.7%/°C in their canine model at temperatures between 35 and 25°C. In neonates, infants and children undergoing deep hypothermic cardiac surgery, Greeley et al. (1991) demonstrated a linear relationship between reduction in CBF and temperature (Fig. 2).

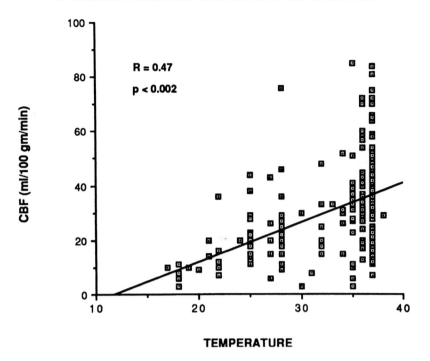

TEMPERATURE

Fig. 2. Plot of the relation of CBF (ml/100 g/min) and temperature (°C) during cardiopulmonary bypass. Regression equation is $y = -17.328 + 1.468x$. (Adapted from Greeley et al. 1989.)

In a study of adults undergoing hypothermic cardiopulmonary bypass, CBF was reduced. However, cerebral metabolism in these patients was reduced to a greater extent, thus resulting in a relatively luxuriant CBF (Croughwell et al. 1992).

Effects on cerebral metabolism

In 1958, experimental work in dogs demonstrated that for every temperature reduction of 10°C, cerebral metabolic rate as measured by oxygen consumption was reduced by one-third to one-half of the initial value (Singer and Bretschneider 1990). This observation is compatible with the known logarithmic relationship between metabolism and temperature during biochemical reactions. This relationship is expressed as a temperature quotient (Q10) for the cerebral metabolic rate of oxygen consumption ($CMRO_2$). The Q10 for $CMRO_2$ is defined as the ratio of two $CMRO_2$ values (higher divided by the lower) over a 10° range. Croughwell et al. (1992) calculated the median temperature coefficient for humans on non-pulsatile cardiopulmonary bypass to be 2.8. Thus, reducing the temperature from 37 to 27°C reduces the cerebral metabolic rate of oxygen consumption by 64%. The Q10 is 2.0–3.0 for most biologic reactions. However, this relationship alone cannot fully explain the observed tolerance to cerebral ischemia seen during deep hypothermia. Specifically, human subjects have undergone complete circulatory arrest for 60 min at 18°C without clinical evidence of cerebral injury. Utilizing a Q10 for $CMRO_2$, one would only predict a 34-min tolerance. In order to address this contradiction, Michenfelder and Milde (1991) designed a series of experiments utilizing canine models. First, they confirmed the findings of Steen et al. (1979) who had reported that between 37 and 28°C, the Q10 = 2.45, but between 27 and 18°C, the mean Q10 doubled to 4.95. Secondly, they proposed that the primary effects of hypothermia on integrated neuronal function (as measured by EEG) occurs between 27 and 17°C. They concluded that the relationship between brain temperature and $CMRO_2$ is complex, involving an interaction between the rates of biochemical reactions and direct effects on cerebral function (Fig. 3). These studies suggest that the basis of cerebral ischemic tolerance during profound hypothermia depends on

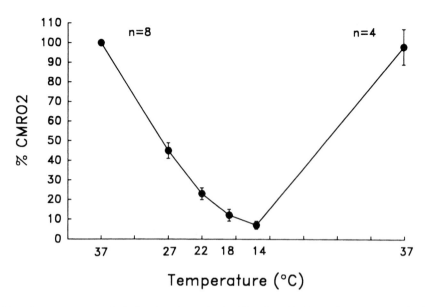

Fig. 3. Effect of temperature on CMRO$_2$. The change in the CMRO$_2$ with change in temperature is plotted as a percent of control (± S.D.). The relationship is neither linear nor exponential. With rewarming see return to control CMRO$_2$. (Adapted from Michenfelder and Mide 1991.)

more than metabolic suppression (Michenfelder and Milde 1991).

Other effects

A number of theories regarding the exact mechanism for the hypothermic cytoprotective effect have been proposed and investigated. Those shown to play a role include the delayed depletion of ATP (Yager et al. 1992), the normalization of protein synthesis (Bergstedt et al. 1993) and alteration in intracellular mediators such as calcium–calmodulin-dependent protein kinase II (Churn et al. 1990), protein kinase C (Cardell et al. 1991) and ubiquitin (Yamashita et al. 1991). Heat shock protein has been shown not to play a significant role (Chopp et al. 1992). Swain et al. (1991), in a study utilizing a sheep model, reported that deep hypothermia can maintain high energy phosphate concentration and intracellular pH in brain, and thus may act to protect against ischemic acidosis. Baker et al. (1991) suggested that the neuroprotective properties of hypothermia may reside in its ability to prevent an increase in the extracellular concentration of amino acids that activate the NMDA receptor complex. Busto et al. (1989c) demonstrated complete inhibition of glutamate release from rats subjected to mild intra-ischemic hypo-

thermia compared to normothermic animals who experienced a 500-fold rise in glutamate after ischemia.

Corbett et al. (1990) demonstrated that both systemic injections of MK-801 (a non-competitive NMDA antagonist) and profound hypothermia protect CA1 neurons against degeneration. However, additional studies indicate that MK-801 conferred only partial protection against ischemic brain injury in normothermic infant rats, whereas combining MK-801 with hypothermia provides total protection against ischemic damage (Ikonomidou et al. 1989). This suggests that the mechanism of ischemic tolerance seen in hypothermia may in part be based on protection against NMDA receptor activity. Hiramatsu et al. (1993), in electrophysiologic investigations, suggested that hypothermia protects against hypoxic damage to excitatory synaptic mechanisms both by prolonging the latency to hypoxic depolarization, and by extending the period of hypoxic depolarization that can be tolerated.

CLINICAL USES OF HYPOTHERMIA

Hypothermia has been in widespread clinical use for over 30 years. The rationale for utilizing hypothermia stems from its ability to alter cerebral function and reduce cerebral metabolism. This ultimately implies an increased tolerance to ischemia.

The study of the pathophysiology of ischemia has lead to the recognition of the phenomenon of 'selective neuronal vulnerability.' Hypoxic-ischemic damage preferentially involves laminae 3, 5, 6 of the cerebral cortex, the hippocampus (especially CA1), the globus pallidus, and the Purkinje cells.

A series of animal studies over the past 15 years has further defined the success and limitations of mild, moderate and profound hypothermia in both diffuse and focal ischemic models at varying times of application (Ginsberg et al. 1992; Maher and Hachinski 1993). Many of these studies yield contradictory conclusions regarding the efficacy of hypothermia. This may in part be related to inaccurate measurement of true brain temperature (Busto et al. 1987).

Hypothermia for protection against global ischemia

The efficacy of profound and moderate hypothermia before and during ischemic injury is well supported by numerous authors (Young et al. 1983; Tisherman et al. 1991; Gillinov et al. 1993). In 1986, O'Connor et al., subjected dogs to 1 h of circulatory arrest with cerebral temperatures of 13°C and found no gross or microscopic evidence of cerebral hypoxia following sacrifice of the animal 7 days after anoxia. Green et al. (1992) in a study of global cerebral ischemia in rats suggested that moderate intra-ischemic hypothermia provides long-lasting protection from behavior deficits as well as neuronal injury. Importantly, this protective effect has also been demonstrated in mild hypothermia. Busto et al. (1989b) showed that lowering the rat brain temperature from 36 to 34°C correlated with a smaller degree of neuronal injury (mild or no damage to the CA1 neurons) after exposure to a global ischemic insult.

However, studies of post-ischemic hypothermia have produced conflicting data. Busto et al. (1989a) showed that moderate hypothermia induced no longer than 5 min after the ischemia completely prevented hippocampal neuronal injury. Chopp et al. (1991) applied whole body hypothermia of 30°C for 2 h immediately following 8 or 12 min of ischemia in rats. Histopathologic examination revealed a one-third reduction of necrotic neurons in CA1/CA2 in the 8-min group, but no protective effect in the 12-

min group. In contrast, Yager et al. (1993) found no difference in cerebral injury between rat pups with post-ischemic recovery under hypothermic compared with normothermic conditions. In 1992, Carroll and Beek suggested that a greater protective effect can be achieved with prolonged post-ischemic hypothermia. When hypothermia was initiated even 1 h after ischemia and maintained for 6 h neuronal injury was attenuated by 49% compared to normothermia. The ischemic time, however, was only 5 min, whereas previous studies had utilized longer ischemic durations.

Hypothermia to protect against focal ischemia

Early studies by Rosomoff (1957) in a canine model elucidated the beneficial effect of hypothermia (24°C) before or 15 min after permanent MCA occlusion. Kader et al. (1992) in the study of focal (MCA) ischemia demonstrated that transient mild hypothermia (30–34°C) significantly decreased infarct volume, even when hypothermia was induced 1 h after the onset of ischemia. Morikawa et al. (1992) found no beneficial effect of hypothermia in a model of permanent focal ischemia, but did demonstrate a 90.5% cortical and 69% striatal infarction size reduction when the brain temperature was decreased from 39 to 30°C in animals undergoing 2 h of reversible ischemia. Similar results were reported in a study utilizing a model of reversible focal ischemia (Ridenour et al. 1992; Goto et al. 1993).

Cerebral effects of hyperthermia

Minamisawa et al. (1990) examined the variable of mild hypothermia and hyperthermia before, during, and after induction of ischemia in rats. Their results confirmed previous findings that a decrease in temperature of only 2°C significantly reduced damage to several selectively vulnerable neuronal populations. An increase in temperature of 2°C significantly enhanced brain damage. In addition, hyperthermia had a tendency to induce infarction in the neocortex and caudoputamen, and cause damage to the substantia nigra pars reticulata. Xue et al. (1992) demonstrated that a rise of 3°C in intra-ischemic brain temperature in rats subjected to focal ischemia significantly increased infarct volume. These results underscore the potentially devastating effects of fever in patients with cerebrovascular disease.

Evidence against the use of hypothermia

A number of studies have demonstrated a worsening of outcome with the use of hypothermia. Factors identified as detrimental include prolonged hypothermia and a rapid rate of cooling. Hypothermia of 24–48 h duration appears to offer less benefit and may be detrimental (Steen et al. 1980). Kirklin postulated that rapid cerebral cooling leads to uneven cerebral cooling and thus, sub-therapeutic protection against ischemia (Kirklin and Barratt-Boyes 1993).

Cardiac surgery

The majority of the prospective and retrospective data on the clinical use of hypothermia arise from the cardiovascular literature. Since its introduction in 1958, the technique of deep hypothermia with circulatory arrest in cardiac surgery has been widely adopted, especially in those conditions in which prolonged arrest time is anticipated (congenital defects and aortic repairs). Recently, these techniques have come under scrutiny, especially with respect to long-term outcome in infants and children.

For most cardiac procedures, mild to moderate hypothermia is used. However, the most dramatic application demonstrating the effects of hypothermia is in deep hypothermia and circulatory arrest, where systemic temperatures of 5°C can be established to allow circulatory arrest for 120 min (Gillinov et al. 1993). Bachet et al. (1991) introduced a technique of 'cold cerebroplegia' utilizing selective brain cooling (carotids cannulated and perfused with cold blood at 6–12°C, while the core temperature is maintained at 25–26°C). In a 5-year series of 54 patients who underwent cardiac surgery utilizing this technique, there was no intraoperative mortality. Transient neurologic deficits were present in 4.3% of their patients. At present, retrospective data indicate that risk factors associated with a less favorable cerebral outcome include prolonged duration of circulatory arrest (usually greater than 60 min), advanced age, rapid cooling (less than 20 min), hyperglycemia, preoperative cyanosis, evidence for increased oxygen extraction before circulatory arrest, and delayed reappearance of EEG (Griepp and Griepp 1992).

Given that infants and children most frequently undergo deep hypothermia with circulatory arrest, long-term intellectual and developmental outcome of patients undergoing this technique is a central issue (Messmer et al. 1976; Clarkson et al. 1980; Blackwood et al. 1986). In follow-up studies of 28 children who had undergone open heart surgery, developmental impairment was associated with cooling periods longer than 20 min (Bellinger et al. 1991).

Choreoathetoid movements may develop in infants within 2 weeks following cardiopulmonary bypass. This has been referred to as post-pump chorea (PPC). Medlock et al. (1993) reviewed 668 cases of children who underwent open cardiac surgery and found 8 (1.2%) developed PPC. They concluded that there was a strong association between PPC, deep hypothermia, and circulatory arrest. The absence of characteristic macroscopic changes suggests a biochemical or microembolic etiology; PPC is frequently associated with developmental delay, and the prognosis for complete resolution is guarded. The cause of choreoathetosis is not clear.

Seizures occur postoperatively in 5–10% of patients in whom the technique of profound hypothermia and circulatory arrest has been used. The seizures are often transient, but may be associated with severe brain injury.

Lastly, techniques to achieve cardioplegia such as topical ice have been associated with phrenic nerve paralysis. Efthimiou et al. (1992) in a prospective, randomized study of 100 patients undergoing open cardiac surgery, found that 34% of patients who received topical ice hypothermia during heart surgery developed diaphragm paralysis due to phrenic nerve cold injury.

New therapy

Numerous cardiovascular surgical techniques have been introduced to decrease the incidence of embolic events, cardiac and neurologic complications. Recent data from a large multicenter randomized trial comparing normothermic and hypothermic cardiopulmonary bypass showed consistent deterioration on psychomotor skills in the early postoperative period which resolved in the 3-month follow-up in both groups. However, there was no difference in incidence or degree of neurologic dysfunction between warm and cold groups (McLean et al. 1994). The technique of normothermic cardiopulmonary bypass, however, may offer less cardiac protection (Buckberg 1994).

Neurosurgery

As in cardiovascular surgery, hypothermia has been utilized to reduce ischemic injury in neurosurgical patients. Its use, however, has not been as widespread or as extensively studied. Lougheed et al. (1955) utilized hypothermia to facilitate a right hemispherectomy. Bottrell et al. (1956) described 22 patients undergoing cerebral aneurysm surgery in whom hypothermia was used to minimize ischemic effects from temporary interruption of cerebral circulation. Nineteen patients survived, 16 with excellent outcomes. Deep hypothermia with cardiopulmonary bypass and circulatory arrest is used selectively at several institutions in surgical management of complex intracranial lesions (Baumgartner et al. 1983). Hypothermia has been successfully utilized during giant aneurysm resection (Silverberg et al. 1981), spinal cord decompression (De la Torre 1981) and carotid endarterectomy. However, there have been very few systematic evaluations of deliberate intraoperative hypothermia beyond preliminary safety studies (Baker et al. 1994).

Neurotrauma

There has been great interest in utilizing hypothermia in patients with severe head trauma to improve both morbidity and mortality. Both animal and human investigations have established the safety and suggest the efficacy of hypothermic therapy.

Clifton et al. (1991) used moderate resuscitative hypothermia to treat rats following a fluid percussion head injury, and demonstrated a significant improvement in behavioral outcome and tissue preservation when cooling was instituted no later than 30 min after the injury. Pomeranz et al. (1993) demonstrated that moderate resuscitative hypothermia produced a significant reduction in the volume of damaged canine brain tissue.

A well-known sequela of closed head injury is delayed traumatic intracerebral hemorrhage. Given that hypothermia is known to affect the coagulation pathway, Resnick et al. (1994) designed a prospective, randomized trial and found no difference in the incidence of delayed traumatic intracerebral hemorrhage between normothermic and moderate hypothermic groups. Marion et al. (1993) randomized 40 consecutively treated patients with severe head trauma to either a normothermia or moderate hypothermia group. They found that hypothermia reduced

ICP (40%) and CBF (26%). $CMRO_2$ was lower during cooling, and higher 5 days after injury. Both groups had similar disabilities and incidences of systemic complications. Phase II trials of moderate hypothermia in severe brain injury provided evidence of improved neurologic outcome with minimal toxicity. In this study, 46 patients with severe brain injury were randomized to either a normothermia or a hypothermia group. Cooling was begun within 6 h of injury with the use of cooling blankets. There were no cardiac or coagulation complications. Seizure incidence was lower in the hypothermia group. Sepsis was more common in the hypothermia group, but this was not statistically significant (Clifton et al. 1993). Efficacy trials of systemic hypothermia in head injury are currently underway.

The clinical and animal data currently available clearly justify continued pursuit and refinement of the clinical induction of hypothermia for the treatment of cerebral insult. Hypothermia may prove to be an important treatment modality in cerebrovascular disease. Controlled randomized trials utilizing mild to moderate hypothermia in the post-ischemic period are needed to determine safety and efficacy. In addition, a better understanding of the physiology of temperature control may lead to novel hypothermic-inducing drugs.

REFERENCES

ABRAMOWICZ, M. (ED.): Treatment of hypothermia. Med. Lett. Drugs Ther. 36 (1994) 116–117.

BACHET, J., D. GUILMET, B. GOUDOT, J.L. TERMIGNON ET AL.: Cold cerebroplegia. A new technique of cerebral protection during operations on the transverse aortic arch. J. Thorac. Cardiovasc. Surg. 102 (1991) 85–93.

BAKER, A.J., M.H. ZORNOW, M.R. GRAFE, M.S. SCHELLER, S.R. SKILLING, D.H. SMULLIN and A.A. LARSON: Hypothermia prevents ischemia-induced increases in hippocampal glycine concentrations in rabbits. Stroke 22 (1991) 666–673.

BAKER, K.Z., W.L. YOUNG, J.G. STONE, A. KADER, C.J. BAKER and R.A. SOLOMON: Deliberate mild intraoperative hypothermia for craniotomy. Anesthesiology 81 (1994) 361–367.

BAUMGARTNER, W.A., G.D. SILVERBERG, A.K. REAM, S.W. JAMIESON, J. TARABEK and B.A. REITZ: Reappraisal of cardiopulmonary bypass with deep hypothermia and circulatory arrest for complex neurosurgical operations. Surgery 94 (1983) 242–249.

BELLINGER, D.C., G. WERNOVSKY, L.A. RAPPAPORT, J.E. MAYER, JR., A.R. CASTANEDA, D.M. FARRELL, D.L. WESSEL, P. LANG, P.R. HICKEY, R.A. JONAS and J.W. NEWBURGER: Cognitive development of children following early repair of transposition of the great arteries using deep hypothermic circulatory arrest. Pediatrics 87 (1991) 701–707.

BERGSTEDT, K., B.R. HU and T. WIELOCH: Postischaemic changes in protein synthesis in the rat brain: effects of hypothermia. Exp. Brain Res. 95 (1993) 91–99.

BIGELOW, W.G., J.C. CALLAGHAN and J.A. HOPPS.: General hypothermia for experimental intracardiac surgery. Ann. Surg. 132 (1950) 131–537.

BIGGAR, W.D., D. BOHN and G. KENT: Neutrophil circulation and release from bone marrow during hypothermia. Infect. Immun. 40 (1983) 708–712.

BLACKWOOD, M.J.A., K. HAKA-IKSE and D.J. STEWARD: Developmental outcome in children undergoing surgery with profound hypothermia. Anesthesiology 65 (1986) 437–440.

BOTTRELL, E.H., W.M. LOUGHEED, J.W. SCOTT and S.E. VANDEWATER: Hypothermia and interruption of carotid, or carotid and vertebral circulation, in the surgical management of intracranial aneurysms. J. Neurosurg. 13 (1956) 1–42.

BUCKBERG, G.D.: Normothermic blood cardioplegia. Alternative or adjunct? J. Thorac. Cardiovasc. Surg. 107 (1994) 860–867.

BUSTO, R., W.D. DIETRICH, M.Y.-T. GLOBUS, I. VALDÉS, P. SCHEINBERG and M.D. GINSBERG: Small differences in intraischemic brain temperature critically determine the extent of ischemic neuronal injury. J. Cereb. Blood Flow Metab. 7 (1987) 729–738.

BUSTO, R., W.D. DIETRICH, M.Y.-T. GLOBUS and M.D. GINSBERG: Postischemic moderate hypothermia inhibits CA1 hippocampal ischemic neuronal injury. Neurosci. Lett. 101 (1989a) 299–304.

BUSTO, R., W.D. DIETRICH, M.Y.-T. GLOBUS and M.D. GINSBERG: The importance of brain temperature in cerebral ischemic injury. Stroke 20 (1989b) 1113–1114.

BUSTO, R., M.Y.-T. GLOBUS, W.D. DIETRICH, E. MARTINEZ, I. VALDÉS and M.D. GINSBERG: Effects of mild hypothermia on ischemia-induced release of neurotransmitters and free fatty acids in rat brain. Stroke 20 (1989c) 904–910.

CARDELL, M., F. BORIS-MÖLLER and T. WIELOCH: Hypothermia prevents the ischemia-induced translocation and inhibition of protein kinase C in the rat striatum. J. Neurochem. 57 (1991) 1814–1817.

CARROLL, M. and O. BEEK: Protection against hippocampal CA1 cell loss by post-ischemic hypothermia is dependent on delay of initiation and duration. Metab. Brain Dis. 7 (1992) 45–50.

CHOPP, M., H. CHEN, M.O. DERESKI and J.H. GARCIA: Mild hypothermia intervention after graded ischemic stress in rats. Stroke 22 (1991) 37–43.

CHOPP, M., Y. LI, M.O. DESEKI, S.R. LEVINE, Y. YOSHIDA and J.H. GARCIA: Hypothermia reduces 72-kDa heat-shock protein induction in rat brain after transient forebrain ischemia. Stroke 23 (1992) 104–107.

CHURN, S.B., W.C. TAFT, M.S. BILLINGSLEY, R.E. BLAIR and R.J. DELORENZO: Temperature modulation of ischemic neuronal death and inhibition of calcium/calmodulin-dependent protein kinase II in gerbils. Stroke 21 (1990) 1715–1721.

CLARKSON, P.M., B.A. MACARTHUR, B.G. BARRATT-BOYES, R.M. WHITLOCK and J.M. NEUTZE: Developmental progress after cardiac surgery in infancy using hypothermia and circulatory arrest. Circulation 62 (1980) 855–861.

CLIFTON, G.L., J.Y. JIANG, B.G. LYETH, L.W. JENKINS, R.J. HAMM and R.L. HAYES: Marked protection by moderate hypothermia after experimental traumatic brain injury. J. Cereb. Blood Flow Metab. 11 (1991) 114–121.

CLIFTON, G.L., S. ALLEN, P. BARRODALE, P. PLENGER, J. BERRY, S. KOCH, J. FLETCHER, R.L. HAYES and S.C. CHOI: A phase II study of moderate hypothermia in severe brain injury. J. Neurotrauma 10 (1993) 263–271.

COLLU, R., G.M. GREGORY and G.R. VAN LOON (EDS.): Clinical Neuroendocrinology. Cambridge, Blackwell Scientific (1988) 526–528.

CORBETT, D., S. EVANS, C. THOMAS, D. WANG and R.A. JONAS: MK-801 reduced cerebral ischemic injury by inducing hypothermia. Brain Res. 514 (1990) 300–304.

CROUGHWELL, N., L.R. SMITH, T. QUILL, M. NEWMAN, W. GREELEY, F. KERN, J. LU and J.G. REVES: The effect of temperature on cerebral metabolism and blood flow in adults during cardiopulmonary bypass. J. Thorac. Cardiovasc. Surg. 103 (1992) 549–554.

CURLEY, F.J.: Hypothermia: a critical problem in the intensive care unit. J. Intens. Care Med. 10 (1995) 1–2.

DANZL, D.F. and R.S. POZOS: Accidental hypothermia. N. Engl. J. Med. 331 (1994) 1756–1760.

DE LA TORRE, J.C.: Spinal cord injury: review of basic and applied research. Spine 6 (1981) 315–335.

EFTHIMIOU, J., J. BUTLER, C. WOODHAM, S. WESTABY and M.K. BENSON: Phrenic nerve and diaphragm function following open heart surgery: a prospective study with and without topical hypothermia. Quart. J. Med. 85 (1992) 845–853.

FARMER, J.C.: Hypothermia and hyperthermia. In: J.M. Civetta and R. Taylor (Eds.), Critical Care, 3rd Edit. Philadelphia, PA, Lippincott (1992) 1419–1423.

FAY, T. and G.W. SMITH: Observations on reflex responses during prolonged periods of human refrigeration. Arch. Neurol. Psychiatry 45 (1941) 215–222.

FRANKEL, R.J. and J.S. JENKINS: Hypothalamic-pituitary function in anorexia nervosa. Acta Endocrinol. (Copenh.) 78 (1975) 209–221.

GENTILELLO, L.M. and S. MOUJAES: Treatment of hypothermia in trauma victims: thermodynamic considerations. J. Intens. Care Med. 10 (1995) 5–14.

GENY, C., P.F. PRADAT, J. YULIS, S. WALTER ET AL.: Hypothermia, Wernicke encephalopathy and multiple sclerosis. Acta Neurol. Scand. 86 (1992) 632–634.

GILLINOV, A.M., J.M. REDMOND, K.J. ZEHR, J.C. TRONCOSO, S. ARROYO, R.P. LESSER, A.W. LEE, R.S. STUART, B.A. REITZ, W.A. BAUMGARTNER and D.E. CAMERON: Superior cerebral protection with profound hypothermia during circulatory arrest. Ann. Thorac. Surg. 55 (1993) 1432–1439.

GINSBERG, M.D., L.L. STERNAU, Y.-T. GLOBUS, W.D. DIETRICH and R. BUSTO: Therapeutic modulation of brain temperature: relevance to ischemic brain injury. Cerebrovasc. Brain Metab. Rev. 4 (1992) 189–225.

GOTO, Y, N.F. KASSELL, K.-I. HIRAMATSU, S.W. SOLEAU and K.S. LEE: Effects of intraischemic hypothermia on cerebral damage in a model of reversible focal ischemia. Neurosurgery 32 (1993) 980–985.

GREELEY, W.J., R.M. UNGERLEIDER, F.H. KERN, O.F.G. BRUSIN, L.R. SMITH AND J.G. REVES: Effects of cardiopulmonary bypass on cerebral blood flow in neonates, infants, and children. Circulation 80, Suppl. (1989) I-212.

GREELEY, W.J., F.H. KERN, R.M. UNGERLEIDER, J.L. BOYD, III ET AL.: The effect of hypothermic cardiopulmonary bypass and total circulatory arrest on cerebral metabolism in neonates, infants, and children. J. Thorac. Cardiovasc. Surg. 101 (1991) 783–794.

GREEN, E.J., W.D. DIETRICH, F. VAN DIJK, R. BUSTO, C.G. MARKGRAF, P.M. MCCABE, M.D. GINSBERG and N. SCHNEIDERMAN: Protective effects of brain hypothermia on behavior and histopathology following global cerebral ischemia in rats. Brain Res. 580 (1992) 197–204.

GRIEPP, E.B. and R.B. GRIEPP: Cerebral consequences of hypothermic circulatory arrest in adults. J. Cardiac Surg. 7 (1992) 134–155.

GUNN, T.R., S. MACFARLANE and L.I. PHILLIPS: Difficulties in the neonatal diagnosis of Menke's kinky hair syndrome – trichopoliodystrophy. Clin. Pediatr. 23 (1984) 514–516.

GUYTON, A.C.: Textbook of Medical Physiology, 7th Edit. Philadelphia, PA, W.B. Saunders (1986) 849–860.

HAYES, K.C., J.T. HSIEH, P.J. POTTER, D.L. WOLFE ET AL.: Effects of induced hypothermia on somatosensory evoked potentials in patients with chronic spinal cord injury. Paraplegia 31 (1993) 730–741.

HIRAMATSU, K.-I., N.F. KASSELL and K.S. LEE: Thermal sensitivity of hypoxic responses in neocortical brain slices. J. Cereb. Blood Flow Metab. 13 (1993) 395–401.

HOLM, I.A., J.F. MCLAUGHLIN, K. FELDMAN and E.F. STONE: Recurrent hypothermia and thrombocytopenia after severe neonatal brain infection. Clin. Pediatr. 27 (1988) 326–329.

IKONOMIDOU, C., J.L. MOSINGER and J.W. OLNEY: Hypothermia enhances protective effect of MK-801 against hypoxic/ischemic brain damage in infant rats. Brain Res. 487 (1989) 184–187.

JANSKÝ, L. and X.J. MUSACCHIA (EDS.): Regulation of Depressed Metabolism and Thermogenesis. Springfield, IL, Charles C. Thomas (1976).

KADER, A., M.H. BRISHMAN, N. MARAIRE, J.-T. HUH and R.A. SOLOMON: The effect of mild hypothermia on permanent focal ischemia in the rat. Neurosurgery 31 (1992) 1056–1061.

KIRKLIN, J.W. and B.G. BARRATT-BOYES (EDS.): Cardiac Surgery: Morphology, Diagnostic Criteria, Natural History, Techniques, Results, and Indications, Vol. 1, 2nd Edit. New York, Churchill Livingstone (1993).

KUGLER, S.L., D.T. COSTAKOS, A.M. ARON and H. SPIERA: Hypothermia and systemic lupus erythematosus. J. Rheumatol. 17 (1990) 680–681.

LARACH, M.G.: Accidental hypothermia. Lancet 345 (1995) 493–498.

LEVY, W.J.: Quantitative analysis of EEG changes during hypothermia. Anesthesiology 60 (1984) 291–297.

LEWIS, F.J. and M. TAUFIC.: Closure of atrial septal defects with aid of hypothermia: experimental accomplishments and the report of one successful case. Surgery 33 (1953) 52–59.

LEWITT, P.A., R.P. NEWMAN, H.S. GREENBERG, L.L. ROCHER ET AL.: Episodic hyperhidrosis, hypothermia, and agenesis of corpus callosum. Neurology 33 (1983) 1122–1129.

LIEH-LAI, M.W., A.P. SARNAIK, J.F. NEWTON, J.N. MICELI ET AL.: Metabolism and pharmacokinetics of acetaminophen in a severely poisoned young child. J. Pediatr. 105 (1984) 125–128.

LIPTON, J.M., J. KIRKPATRICK and R.N. ROSENBERG: Hypothermia and persisting capacity to develop fever. Occurrence in a patient with sarcoidosis of the central nervous system. Arch. Neurol. 34 (1977) 494–504.

LOUGHEED, W.M., W.H. SWEET, J.C. WHITE and W.R. BREWSTER: The use of hypothermia in surgical treatment of cerebral vascular lesions. A preliminary report. J. Neurosurg. 12 (1955) 240–255.

MACLEAN, D. and D. EMSLIE-SMITH (EDS.): Accidental Hypothermia. London, Blackwell Scientific (1977).

MAHER, J. and V. HACHINSKI: Hypothermia as a potential treatment for cerebral ischemia. Cerebrovasc. Brain Metab. Rev. 5 (1993) 277–300.

MARION, D.W., W.D. OBRIST, P.M. CARLIER, L.E. PENROD and J.M. DARBY: The use of moderate therapeutic hypothermia for patients with severe head injuries: a preliminary report. J. Neurosurg. 79 (1993) 354–362.

MARKAND, O.N., B.I. LEE, C. WARREN, R.K. STOELTING ET AL.: Effects of hypothermia on brainstem auditory evoked potentials in humans. Ann. Neurol. 22 (1987) 507–513.

MARTIN, J.B., S. REICHLIN and G.M. BROWN (EDS.): Clinical Neuroendocrinology. Philadelphia, PA, Davis (1977).

MCLEAN, R.F., B.I. WONG, C.D. NAYLOR, W.G. SNOW, E.M. HARRINGTON, M. GAWEL and S.E. FREMES: Cardiopulmonary bypass, temperature, and central nervous system dysfunction. Circulation 90 (1994) II250–II255.

MEDLOCK, M.D., R.S. CRUSE, S.J. WINEK, D.M. GEISS, R.L. HORNDASCH, D.L. SCHULTZ and J.C. ALDAG: A 10-year experience with postpump chorea. Ann. Neurol. 34 (1993) 820–826.

MESSMER, B.J., U. SCHALLBERG, R. GATTIKER and A. SENNING: Psychomotor and intellectual development after deep hypothermia and circulatory arrest in early infancy. J. Thorac. Cardiovasc. Surg. 72 (1976) 495–501.

MICHENFELDER, J.D. and J.H. MILDE: The relationship among canine brain temperature, metabolism, and function during hypothermia. Anesthesiology 75 (1991) 130–136.

MINAMISAWA, H., M.-L. SMITH and B.K. SIESJÖ: The effect of mild hyperthermia and hypothermia on brain damage following 5, 10, and 15 minutes of forebrain ischemia. Ann. Neurol. 28 (1990) 26–33.

MOORADIAN, A.D., G.K. MORLEY, R. MCGEACHIE, S. LUNDGREN and J.E. MORLEY: Spontaneous periodic hypothermia. Neurology 34 (1984) 79–82.

MORIKAWA, E., M.D. GINSBERG, W.D. DIETRICH, R.C. DUNCAN, S. KRAYDIEH, M.Y.-T. G and R. BUSTO: The significance of brain temperature in focal cerebral ischemia: histopathological consequences of middle cerebral artery occlusion in the rat. J. Cereb. Blood Flow Metab. 12 (1992) 380–389.

NAKADA, T. and R.T. KNIGHT: Alcohol and the central nervous system. Med. Clin. North Am. 68 (1984) 121–131.

O'CONNOR, J.V., T. WILDING, P. FARMER, J. SHER, M.A. ERGIN and R.B. GRIEPP: The protective effect of profound hypothermia on the canine central nervous system during one hour of circulatory arrest. Ann. Thorac. Surg. 41 (1986) 255–259.

POMERANZ, S., P. SAFAR, A. RADOVSKY, S.A. TISHERMAN, H. ALEXANDER and W. STEZOSKI: The effect of resuscitative moderate hypothermia following epidural brain compression on cerebral damage in a canine outcome model. J. Neurosurg. 79 (1993) 241–251.

RATCLIFFE, P.J., J.I. BELL, K.J. COLLINS, R.S. FRACKOWIAK and P. RUDGE: Late onset post-traumatic hypothalamic hypothermia. J. Neurol. Neurosurg. Psychiatry 46 (1983) 72–74.

RESNICK, D.K., D.W. MARION and J.M. DARBY: The effect of hypothermia on the incidence of delayed traumatic intracerebral hemorrhage. Neurosurgery 34 (1994) 252–255.

REUTENS, D.C., J.W., DUNNE and S.S. GUBBAY: Triphasic waves in accidental hypothermia. Electroencephalogr. Clin. Neurophysiol. 76 (1990) 370–372.

RIDENOUR, T.R., D.S. WARNER, M.M. TODD and A.C. MCALLISTER: Mild hypothermia reduces infarct size resulting from temporary but not permanent focal ischemia in rats. Stroke 23 (1992) 733–738.

ROSOMOFF, H.L.: Hypothermia and cerebral vascular lesion. II. Experimental middle cerebral artery interruption followed by induction of hypothermia. Arch. Neurol. Psychiatry 78 (1957) 454–464.

ROSOMOFF, H.L. and D.A. HOLADAY: Cerebral blood flow and cerebral oxygen consumption during hypothermia. Am. J. Physiol. 179 (1954) 85–88.

RUSS, W., J. STICHER, H. SCHELD and G. HEMPELMANN: Effects of hypothermia on somatosensory evoked responses in man. Br. J. Anaesth. 59 (1987) 1484–1491.

SILVERBERG, G.D., B.A. REITZ and A.K. REAM: Hypothermia and cardiac arrest in the treatment of giant aneurysms of the cerebral circulation and hemangioblastoma of the medulla. J. Neurosurg. 55 (1981) 337–346.

SINGER, D. and H.J. BRETSCHNEIDER: Metabolic reduction in hypothermia: pathophysiological problems and natural examples – part 1. Thorac. Cardiovasc. Surg. 38 (1990) 205–211.

STEEN, P.A., E.H. SOULE and J.D. MICHENFELDER: Detrimental effect of prolonged hypothermia in cats and monkeys with and without regional cerebral ischemia. Stroke 10 (1979) 522–529.

STEEN, P.A., J.H. MILDE and J.D. MICHENFELDER: The detrimental effects of prolonged hypothermia and rewarming in the dog. Anesthesiology 52 (1980) 224–230.

STEJSKAL, L., V. TRAVNICEK, K. SOUREK and J. KREDBA: Somatosensory evoked potentials in deep hypothermia. Appl. Neurophysiol. 43 (1980) 1–7.

STEPHAN, H., A. WEYLAND, S. KAZMAIER, T. HENZE, S. MENCK and H. SONNTAG: Acid-base management during hypothermic cardiopulmonary bypass does not affect cerebral metabolism but does affect blood flow and neurological outcome. Br. J. Anaesth. 69 (1992) 51–57.

STOCKARD, J.J., F.W. SHARBROUGH and J.A. TINKER: Effects of hypothermia on the human brainstem auditory response. Ann. Neurol. 3 (1978) 368–370.

SUMMERS, G.D., A.C. YOUNG, R.A. LITTLE, H.B. STONER ET AL.: Spontaneous periodic hypothermia with lipoma of the corpus callosum. J. Neurol. Neurosurg. Psychiatry 44 (1981) 1094–1099.

SWAIN, J.A., T.J. MCDONALD, JR., R.C. ROBBINS and R.S. BALABAN: Relationship of cerebral and myocardial intracellular pH to blood pH during hypothermia. Heart Circ. Physiol. 29 (1991) H1640–H1644.

TAMPI, R. and W.S. ALEXANDER: Wernicke's encephalopathy with central pontine myelinolysis presenting with hypothermia. N.Z. Med. J. 95 (1982) 343–344.

TISHERMAN, S.A., P. SAFAR, A. RADOVSKY, A. PEITZMAN, G. MARRONE, K. KUBOYAMA and V. WEINRAUCH: Profound hypothermia (<10°C) compared with deep hypothermia (15°C) improves neurologic outcome in dogs after two hours' circulatory arrest induced to enable resuscitative surgery. J. Trauma 31 (1991) 1051–1061.

U.S. DEPARTMENT OF HEALTH AND HUMAN SERVICES: Hypothermia-related deaths – Cook County, Illinois, November 1992–March 1993. Morb. Mortal. Wkly. Rep. 42 (1993) 917–919.

WALKER, B.R., J.A. ANDERSON and C.R. EDWARDS: Clonidine therapy for Shapiro's syndrome. Quart. J. Med. 82 (1992) 235–245.

YAGER, J.Y., R.M. BRUCKLACHER, D.J. MUJSCE and R.C. VANNUCII: Cerebral oxidative metabolism during hypothermia and circulatory arrest in newborn dogs. Pediatr. Res. 32 (1992) 547–552.

YAGER, J., J. TOWFIGHI and R.C. VANUCCI: Influence of mild hypothermia on hypoxic-ischemic brain damage in the immature rat. Pediatr. Res. 34 (1993) 525–529.

YAMASHITA, K., Y. EGUCHI, K. KAJIWARA and H. ITO: Mild hypothermia ameliorates ubiquitin synthesis and prevents delayed neuronal death in gerbil hippocampus. Stroke 22 (1991) 1574–1581.

YOUNG, R.S.K., T.P. OLENGINSKI, S.K. YAGEL and J. TOWFIGHI: The effect of graded hypothermia on hypoxic-ischemic brain damage: a neuropathologic study in the neonatal rat. Stroke 14 (1983) 929–934.

XUE, D., Z.-G. HUANG, K.E. SMITH and A.M. BUCHAN: Immediate or delayed mild hypothermia prevents focal cerebral infarction. Brain Res. 587 (1992) 66–72.

Handbook of Clinical Neurology, Vol. 27 (71): Systemic Diseases, Part III
M.J. Aminoff and C.G. Goetz, editors

Hyperthermia: heat strokes/ neuroleptic malignant syndrome/ malignant hyperthermia

BASIM YAQUB and SALEH AL DEEB

*Division of Neurology, Department of Clinical Neurosciences, Riyadh Armed Forces Hospital,
Riyadh, Saudi Arabia*

Heat stroke (HS) is a model of thermal injury in man. It is an ancient and modern disease seen in all age groups and in different parts of the world. It occurs in heat waves (Ellis 1972; Jones et al. 1982; Kilbourne et al. 1982; Anonymous 1993, 1994), in certain industries (Wyndham 1966), and is seen in soldiers (Porter 1993; Gardner and Kark 1994), joggers or marathon runners even in temperate countries (O'Donnell 1977; Beard et al. 1979; Hart et al. 1980; Epstein et al. 1995). Neuroleptic malignant syndrome (NMS) was first described by Preston (1959) and later by Delay et al. (1960). Since then, it has been widely reported with an estimated incidence of 0.02–3.2% worldwide (Itoh et al. 1977; Haberman 1978; McAllister 1978; Grunhaus et al. 1979; Caroff 1980; Singh 1983; Bristow and Kohen 1993; Persing 1994). Recent reviews showed lack of awareness of the syndrome and confusion with regard to its diagnostic criteria. Malignant hyperthermia (MH) is a rare disease and occurs in genetically susceptible subjects after exposure to certain anaesthetic drugs or stressful exercise (Britt and Kalow 1970; Gronert 1980; Moochhala et al. 1994). It is due to an abnormal gene regulating the function of the sarcoplasmic reticulum that inhibits calcium binding (Roewer 1991; Urwyler et al. 1991; Byers and Krishna 1992; Ording and Bendixen 1992; Allen 1993; Urwyler and Hartung 1994). The original assumption of autosomal dominant inheritance is now widened to include multifactorial inheritance (Denborough et al. 1962; Ellis et al. 1978; Roewer 1991; Urwyler et al. 1991; Byers and Krishna 1992; Allen 1993).

The three disorders constitute the spectrum of a clinical syndrome in which thermal stress and disturbances of thermal regulation occur. Although they have similarities in their clinical picture, biochemical changes and complications, their aetiopathogenesis is different and, accordingly, their management (Yaqub et al. 1989).

HEAT STROKE

Aetiopathogenesis

The body core temperature is maintained at 37°C by a thermoregulatory mechanism that controls heat production and heat loss. Heat is generated by metabolic activities, physical exercise and from the environment if the environmental temperature is above the body temperature (Stolwijk and Nadel 1973). The body dissipates heat by sensible loss, such as conduction, convection and radiation, and by insensible loss, such as evaporation of sweat from the skin surface and water vapour through the lungs. When the ambient temperature is higher than the core's temperature, evaporation of sweat is the most common way to dissipate heat (Robinson and Robinson 1954). The evaporation of sweat depends on the vapour pressure gradient between sweat and surrounding air, and upon the air speed. This means that when the relative humidity in the air is low (dry air), the pressure gradient will be high and sweat will evaporate quickly. This depends also on the air speed. However, if the relative humidity of the surrounding air is high, the evaporation of sweat will be hindered and sweat will accumulate without any heat loss. The thermoregulatory system has a special regulator in the hypothalamus, which interfaces between the afferent pathways from the temperature sensors and the afferent pathways to the temperature correction affectors, which regulate heat producing and heat dissipating processes. There is probably a set point determining at which this thermoregulatory function works. This is not fixed, but is the product of dynamic equilibrium between heat production and heat loss. Some of the drugs which interfere with temperature regulation may be acting directly on the temperature sensor or indirectly on altering the intensities of excitation or inhibition convergent influences from elsewhere in the central nervous system (CNS) (Bligh 1981). Fever-causing compounds are called pyrogens. Exogenous pyrogens result in the production of endogenous pyrogens, including a heat labile protein with a molecular weight of approximately 15,000 Da, which may be present in the plasma as a trimer. Endogenous pyrogens are carried in the circulation from its various sites of production to the brain, where they act on neurons in the pre-optic area (Vaughn et al. 1980). During heat load, the body defends itself by the upward shift of thermoregulatory 'set point', which decreases the demand for cutaneous circulation and improves blood supply to the vital organs (Attia et al. 1983). The other mechanism to handle the heat load and prevent damage to the brain is the activation of a selective venous cooling system. The blood collected in the capillary bed of the face and the forehead is cooled by sweat and flows directly to the cavernous sinus. During dehydration and when somatic sweat secretion is inhibited, the facial sweat rate remains equal to those of controls, which may be the explanation why tympanic temperature is usually lower than core temperature (Cabanac 1993).

In HS, one or more of the following may occur: (1) sweating becomes ineffective in cooling the body due to high relative humidity, slow or no air movement, or overcrowding; (2) sweat glands are exhausted, sweating stops and no heat is lost; (3) sweat glands are defective due to cholinergic system dysfunction or drugs and cannot produce any large amount of sweat (Sarnquist and Larsson 1973). Excessive heat gain in unacclimatized persons is not proportionally accompanied by heat dissipation, resulting in a steady rise in body temperature exceeding the critical 'set point' leading to HS and, if untreated, to death. The core temperature at which this occurs is not certain, but clinically, HS occurs if the temperature exceeds 40°C. A temperature of 43.5°C reportedly produces brain death, although we have successfully treated patients with a core temperature of 43.9°C (Yaqub et al. 1986).

The effect of rising body heat is determined by the level to which the temperature rises and its duration, as well as the resulting metabolic acidosis and cellular hypoxia (Wyndham 1973). Hyperpyrexia itself causes the initial damage to the cell by denaturation of enzymes, liquefaction of membrane lipids and damage to mitochondria (Shibolet et al. 1967). Thermal cellular injury and circulatory changes associated with acute HS result in widespread tissue injury to the heart, kidney, liver, blood coagulation system but, most dramatically, to the CNS (Malamud et al. 1946). The mediation of metabolic changes and tissue damage is not fully understood. Recent evidence suggests the involvement of endotoxins and cytokines, especially tumour necrosis factor α (TNF-α), interleukin 1α (IL-1α) and interleukin-6 (IL-6) (Bouchama et al. 1993b). Chang (1993) reported significantly high levels of circulating IL-1, TNF-δ and IL-6. Positive correlations were demonstrated between the body temperature and the level of IL-1β, and the cooling time and the level of serum IL-1β. The damage to the nervous system is due to direct thermal effect. Brain dopamine has been implicated as a mediator of brain neuronal damage resulting from ischaemic injury. Augmented IL-1 production and cerebral ischaemia occurred during the onset of HS. An increase in hypothalamic dopamine release and a decrease in local cerebral blood flow occurring during HS has been reported. Pretreatment with IL-1 receptor antagonist (IL-1ra) attenuates the HS formation by reducing hypothalamic dopamine release. The same findings were reported by Lin et al. (1991) during experimental HS in rabbits. The thermal insult seems to be selective and the cerebellum, basal ganglion, hypothalamus and limbic system are more involved than other structures of the cerebral hemisphere or brainstem. The thermal insult to the cerebellum leads to degeneration of Purkinje cells with pyknotic nuclei, chromatolytic changes and swollen dendrites. This underlies the pancerebellar syndrome (Lefkowitz et al. 1983; Yaqub 1987; Yaqub et al. 1987, 1989). Involvement of the sympathetic system and damage to the anterior hypothalamus leads to the absence of sweating and constricted pupils (Brooks and Koizumi 1980; Noback 1981). The peripheral function of the autonomic nervous system is intact and can produce sweating on stimulation (Yaqub et al. 1987). Indirect involvement of the cerebral hemispheres, brainstem or both, with widespread petechial haemorrhages due to coagulation defect and cerebral oedema, leads to altered consciousness and coma (Wilcox 1920; Wilbur and Stevena 1937; Malamud et al. 1946; Grove and Isaacson 1949; Zelman and Guillan 1970; Shibolet et al. 1976). The effect of thermal trauma on the peripheral nervous system and cranial nerves is controversial. Mehta and Baker (1970) suggested direct trauma to the peripheral nerves, basing their hypothesis on slowing of motor conduction velocity. In a fatal case with flaccid paralysis, the post-mortem showed the thermal lesions in the anterior horn cells but not in the peripheral nerves. The motor nerve conduction velocities of various nerves in this case were normal (Delgado et al. 1985). We did brainstem auditory evoked potentials (BAEPs) on victims of HS, which showed loss of waves I and II but preservation of subsequent waves (III and V) indicating the insult to be in the cochlea, the cochlear nerve or the nucleus. The effect of thermal insult on other nerves was lacking in all of our patients (Al-Harthi et al. 1986; Yaqub et al. 1986, 1987; Al-Aska et al. 1987a, b; Yaqub 1987, 1988). It is not clear how muscle is affected by heat. Myoglobinuria and renal failure were not a feature in our patients and the rise in creatinine kinase (CK) may be attributed to muscle rigidity, shivering, or convulsions. This differs from the exertional heatstroke of joggers, in whom myoglobinuria and renal failure are common (Hart et al. 1980).

Other organs are also affected by direct thermal trauma, especially the heart, the liver, the kidney, the lung and the coagulation system (Hassanein et al. 1991; Lelis et al. 1992; Akhtar et al. 1993; Feller and Wilson 1994).

Clinical picture

There are two types of HS, the 'classical HS', which is due to (exogenous) thermal insult, and the 'exertional HS', which is due to internal heat production (L.E. Hart et al. 1980; G.R. Hart et al. 1982; Yaqub et al. 1986). HS is an environmental health hazard for Muslims when the Mecca pilgrimage (Hajj) occurs in the summer months. Hajj is a holy ritual for Muslims and has to be performed during the first 2 weeks of Dul-Hajja of the Arabic calendar (hijra) on a pilgrimage route to Mecca. As the Hijra year is 11 days shorter than the Gregorian year, Hajj (pilgrimage) occurs at different seasons. When it occurs in summer months (May–September), HS is a common occurrence and is unlikely to be preventable as more than 2 million of unacclimatized Muslims of different age groups, nationalities, colour, health records and social background gather to perform their religious duties. Overcrowding, lack of sleep, strenuous exercise and insufficient water services and sanitation are the main contributing factors in HS occurrence at the Mecca pilgrimage where ambient temperatures may reach 48°C (Yaqub et al. 1986).

HS victims usually show a characteristic clinical picture. They are restless, confused or comatose with a hot dry skin and a rectal temperature of 40°C or more. HS usually starts with sudden collapse and disturbed CNS function ranging from drowsy state to deep coma. The collapse occurs early in the disease. The collapse is usually not preceded by heat exhaustion or heat cramps. The pupils are usually constricted and, if the patient is comatose, areflexic; brainstem reflexes (pupillary, corneal, cephalo-ocular) may be lost. The plantar response is usually extensor or equivocal. Impairment of consciousness is usually reversible in the first few hours after cooling. None of our patients showed corporeal or facial sweating, although profuse sweating has been reported (Wilcox 1920; Malamud et al. 1946; Shibolet et al. 1967). Convulsions and shivering occurred only after cooling, and it is not easy to differentiate between them in unconscious patients. The neurological manifestations of 87 patients seen at a Mecca pilgrimage are summarized in Table 1. Patients with temperatures higher than 42°C needed a longer time for cooling and had a worse outcome than those below 42°C (Yaqub et al. 1986).

TABLE 1

Neurological manifestations seen in 87 patients.

Neurological findings	No. of patients	Percentage
Disturbance of consciousness level	87	100
Deep coma, areflexia, absent brainstem reflexes (ocular, corneal, cephalo-ocular) 'brainstem death-like' picture	25	29
Constricted pupils	87	100
Reactive	62	71
Non-reactive	25	29
Automatic complex movement (chewing, swallowing, lip smacking, etc.)	17	20
Body shivering (during cooling)	6	7
Convulsions (after cooling)	5	6
Pancerebellar syndrome	2	2
Pyramidal lateralizing signs (hemiplegia)	2	2

Patients with 'exertional HS' develop all the above symptomatology but with the addition of rhabdomyolysis and renal failure characteristically associated with hypokalemia or normokalemia instead of hyperkalemia (Wang et al. 1995).

Diagnosis

Diagnosis of HS should exclude other causes of coma, e.g., head trauma, particularly when developed in association with sports or military exercise (Savdie et al. 1991). The accompaniment of high fever (> 40°), loss of sweating, and small pupils are the key symptomatology for diagnosis. CK is usually raised in all cases. CK levels are usually higher in 'exertional' than in 'classic' HS. CK isoenzyme estimations are more reliable in reflecting the extent of tissue damage than total enzyme assays. The sensitivity of the isoenzymes may help in posing an early and accurate diagnosis when there is a history of heat exposure and diagnosis is not certain (Van der Linde et al. 1992). Hypophosphatemia is also reported to be associated with HS syndromes; it is independent of parathyroid hormone level or renal tubular dysfunction (Bouchama et al. 1991b).

BAEPs are of limited diagnostic value in non-structural lesions of the brainstem (Starr 1976). However, in patients with coma and absent brainstem reflexes mimicking brainstem death, they are of diagnostic value. We reported seven 'brainstem death-like' patients (Yaqub 1988) in whom BAEPs showed preservation of waves III and V in all and unilateral or bilateral loss of wave I in four patients (Fig. 1). It is the reverse pattern of that seen in brainstem death where there is absence of all waves (I–V) or preservation of wave I (and probably wave II) only and absence of the subsequent waves (III and V) (Starr 1976).

Treatment

Early recognition and prompt treatment reduce the high mortality. Rapid cooling and supporting vital functions are the essential factors in the management of HS (Yaqub et al. 1986; Clowes and O'Donnell 1987). In the past, rapid cooling was done by immersing the patient in a tub of iced water while the skin was massaged vigorously to prevent vasoconstriction (Ferris et al. 1938). This method has many disadvantages. The rapid cooling of the skin from immersion in iced water induces shivering and causes intense vasoconstriction of the cutaneous vessels, which impedes the transfer of heat from the body core to the surface, essential to eliminate heat load (Wyndham et al. 1959). Also, there is no evidence that massaging the skin will prevent vasoconstriction. Immersion in iced water is uncomfortable, unpleasant and unclean because patients may vomit or urinate during cooling (Khogali and Weiner 1980). A body cooling unit (BCU) used for rapid evaporative cool-

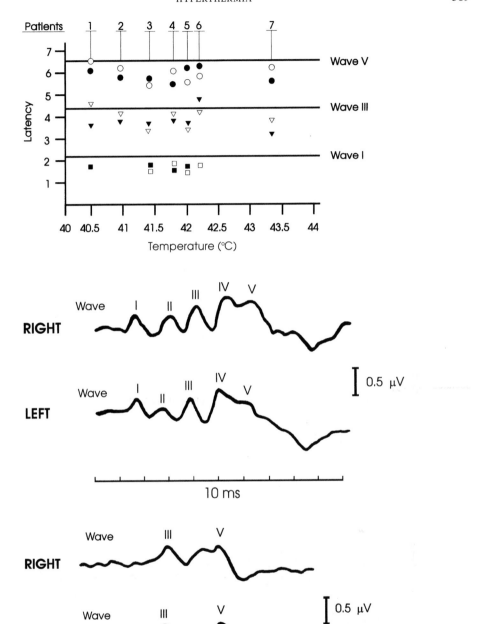

Fig. 1. (A) The latency in ms of all waves in the seven patients and its relationship to the level of core temperature. The upper limit of normal value (mean + 3 S.D.) of waves I, III and V is shown. Patients 3 and 7 died. Key: wave I: ■ = right, □ = left; wave III: ▼ = right, ▽ = left; wave IV: ● = right, ○ = left. (B) BAEPs of patient 3. All waves are seen (I–V) and are of normal latency. (C) BAEPs of patient 7. Wave I is absent bilaterally. Waves III and V are preserved and have normal latency.

ing from a warm skin has been in use since 1979 in the Heat Stroke Treatment Centers established in various places around Mecca for this purpose (Weiner and Khogali 1980). We questioned the practicality and the cost effectiveness of this system and described a simple conventional method for evaporative cooling (Al-Harthi et al. 1986; Al-Aska et al. 1987b). Also, we designed a simple cooling bed (King Saud University Cooling Bed (KSU-CB)). The cooling rate achieved with the KSU-CB was about twice that of the Weiner and Khogali BCU (1980). It is cheap, portable, simple and can be used in any centre for heat wave victims (Al-Aska et al. 1987a; see Fig. 2).

While cooling is the key factor for therapy, many drugs have been tried. Dantrolene was tried without any evidence of influence on the outcome (Bouchama et al. 1991a; Orser 1992; Watson et al. 1993). Naloxone 10 mg/kg, as an endorphin antagonist, was shown to be effective in improving the outcome in rats, but has not been tried in man (Panjwani et al. 1991). Ketanserin (a selective serotonin receptor type 2 antagonist) or indomethacin (an inhibitor of prostaglandin synthesis) markedly reduced cerebral oedema forma-

tion after HS in animals. Their value in man is not yet evaluated (Lee-Chiong and Stitt 1995).

Outcome

The overall mortality of our patients was 11–12.5%, comparable with mortality reports from Heat Stroke Treatment Centers or hospitals in Saudi Arabia (5–18%), but less than those reported from elsewhere (Gauss and Meyer 1917; Malamud et al. 1946; Shibolet et al. 1967; Khogali and Weiner 1980; Khogali et al. 1983; Al-Harthi et al. 1986; Yaqub et al. 1986, 1989; Al-Aska et al. 1987a, b; Yaqub 1987; Torre-Cisneros et al. 1992). Patients with poor prognosis are those with core temperatures of 42°C or above (Yaqub et al. 1986) and a cooling time of 70 min (Al-Aska et al. 1987b) or more (Fig. 3). Torre-Cisneros et al. (1992) reported a worse prognosis with prolonged coma, photomotor, oculomotor and corneal abolition, areflexia and plantar extensor response.

Recovery, if it takes place, is usually without sequelae; however, some patients develop the pancerebellar syndrome, which is irreversible and on CT or

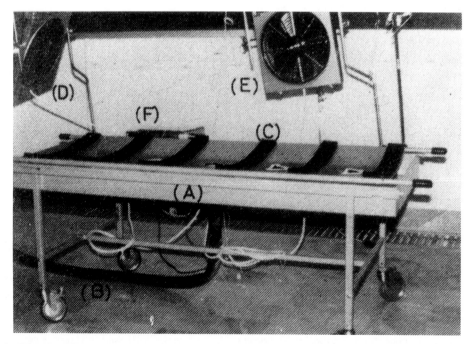

Fig. 2. King Saud University cooling bed: (A) stainless steel pan; (B) drain pipe; (C) rubber strips fixed on steel stretcher; (D) and (E) electric fans adjustable in 3 dimensions; (F) panel for temperature probes, and single electricity outlet for whole unit. (Courtesy of Al-Aska et al. 1987a.)

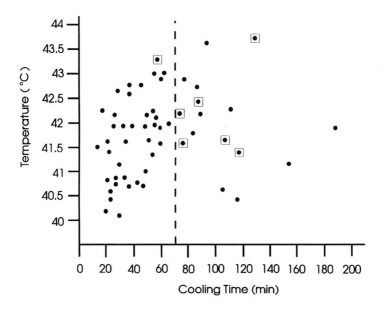

Fig. 3. Relationship of the final outcome to the temperature and cooling time of 56 heat stroke patients. Key:
• = patients who recovered; ▣ = patients who died.

MR imaging is based on generalized cerebellar atrophy (Yaqub et al. 1987; Biary et al. 1995). Despite these changes, the patients usually improve (Fig. 4), most likely due to adaptation and adjustment (Yaqub et al. 1987). Spastic paraparesis due to transverse myelopathy or flaccid paralysis due to anterior horn cell disease is also described (Lin et al. 1991).

Prevention

'Exertional' and 'classic' HS cannot be completely prevented. Clothing, as an interactive barrier, affects thermal balance (Antunano and Nunneley 1992; Pascoe et al. 1994). Limited sun exposure, proper use of sunscreens, adequate fluid intake and electrolyte replacement and acclimatisation are the key factors to prevent HS (Helzer-Julin 1994). Acclimatisations involve a complex of adaptations which include decreased heart rate and rectal temperature, perceived exertion and increased plasma volume and sweat rate. To optimize heat acclimatisation, one should maintain fluid–electrolyte balance and exercise at intensities exceeding 50% of high maximal aerobic power for 10–14 days. Once acclimatisation has been achieved, inactivity results in a decay of favourable adaptations after a few days or weeks (Armstrong and Maresh 1991; Nielsen 1994).

NEUROLEPTIC MALIGNANT SYNDROME

Aetiopathogenesis

The exact pathogenesis of NMS is still controversial, but the current view is that it is due to depletion of dopaminergic neurotransmitters in the basal ganglia (Gibb and Lees 1985). This leads to a reduction of central dopaminergic drive in the striatum and hypothalamus, the thermostat of the body (Henderson and Wooten 1981). The motor symptoms and the autonomic nervous system disturbances of NMS may be related to a relative glutamatergic transmission excess as a consequence of dopaminergic insufficiency. The explanation why some patients have the syndrome may be explained by the D_1 and D_2 dopamine receptor distribution in the neuroleptic non-responsive and neuroleptic responsive, where the neuroleptic non-responsive is associated with significant 20–50% elevation of D_2 autoreceptor across all the midbrain dopamine cell groups (Qian et al. 1992). However, this cannot be the sole pathogenesis as Collins et al. (1987) observed that when a control muscle is incubated with neuroleptic drugs, it develops hypercontractility to halothane and other drugs, which is a characteristic of MH susceptible muscle. This indicates that a raised myoplasmic calcium level

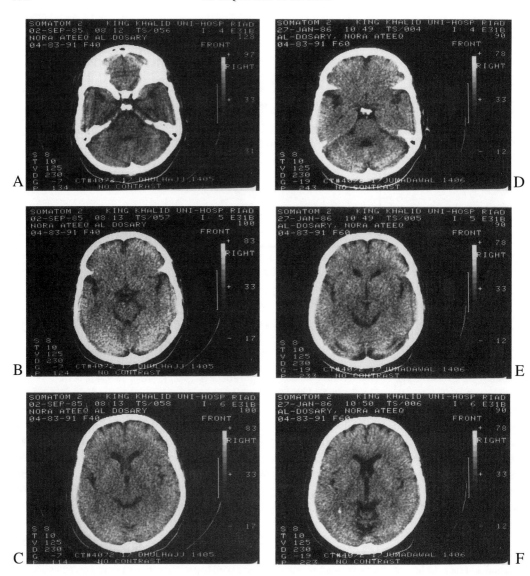

Fig. 4. CT scan appearance: (A–C) 7 days after HS showed normal appearance; (D–F) 5 months later showed atrophy in the region of both cerebellar hemispheres and the vermis, with enlargement of superior and cerebellopontine angle cisterns, visualization of cisterna magna and widening of the fourth ventricle. There was no change in the cerebral hemispheres or the brainstem.

is also important in NMS. The biochemical basis of this has been recently defined. Neuroleptic drugs are calmodulin antagonists and inhibit calcium ATPase in sarcoplasmic reticulum, increase the release of calcium from the sarcoplasmic reticulum into the myoplasm and inhibit the intake of calcium in sarcoplasmic reticulum (Caroff et al. 1983). This is also supported by the reported morphological changes in the muscles of NMS similar to those seen in patients with MH, and the interesting observation that NMS

occurs more commonly in individuals who are susceptible to MH. However, they are pharmacologically distinct implying that cross-reactivity between triggering agents is unlikely to occur (Denborough 1982; Denborough et al. 1984).

Most patients who develop the syndrome are schizophrenics or suffering from affective disorders, illnesses which are believed to be due to deranged neurotransmitter function (Levension 1985). It is also documented in patients with pre-existing brain damage (Lazarus 1992), communicating hydrocephalus and microcephaly (Landry and Latour 1993), mental retardation (Boyd 1992, 1993; Rao et al. 1992), or cortical brain resection (Haynes et al. 1994). These patients are more susceptible because of reduction in marginal stores of dopamine in hypothalamus and basal ganglia resulting from the dopamine blocking activity of neuroleptics. They may develop the syndrome even if exposed to low doses of neuroleptics. It can be caused by most dopamine depleting drugs: butyrophenones (Gibb and Lees 1985), chlorpromazine (Itoh et al. 1977), cyclobenzaprine (Theoharides et al. 1995), domperidone (Spirit et al. 1992), droperidol (Ratan and Smith 1993), fluphenazine (Walker 1959; Henry et al. 1971; Meltzer 1978; Basu 1991; Cape 1994), haloperidol (Weinberger and Kelly 1977; Bernstein 1979; Geller and Greydanus 1979), metoclopramide (Shaw and Matthews 1995), nemonapride (Kubota 1993), perphenazine (Zohar et al. 1992), phenothiazine (Pettigrew et al. 1974), sulpiride (Kiyatake et al. 1991), tiapride (Xerri et al. 1994), trifluoperazine (Preston 1959; Walker 1959), and trimeprazine (Moyes 1973). It is also reported in cocaine abuse (Rodnitzky and Keyser 1992; Yamazaki et al. 1994) and with new atypical neuroleptics, such as remoxipride (Koponen et al. 1993) and clozapine (Anderson and Powers 1991; Vetter et al. 1991; Lieberman and Safferman 1992; Vanelle 1992). Clozapine is a potent atypical neuroleptic drug with a unique lack of extrapyramidal side effects. Its special pharmacological profile, with weak affinity for dopamine type 1 and 2 receptors, was the reason for its introduction in NMS. Nevertheless, several case reports indicated that clozapine can cause NMS (Anderson and Powers 1991; Vetter et al. 1991; Lieberman and Safferman 1992; Vanelle 1992; Buckley and Meltzer 1993; Thornberg and Ereshefsky 1993), but this is controversial and Weller and Kornhuber (1992b, 1993) concluded that the alleged NMS induction by clozapine monotherapy needs further substantiation and that clozapine should be considered as the drug of choice if a psychotic relapse necessitates neuroleptic treatment for patients with a history of NMS.

The reaction is idiosyncratic and may occur when a neuroleptic drug is used for a short period, after the second dose or after taking the drug for months or even years. More than half of the patients can resume the same neuroleptic drugs later without complica-

tions. This suggests that NMS may not predictably develop even in predisposed individuals upon neuroleptic exposure and additional co-factors must be present for the full syndrome to evolve.

NMS is not solely related to neuroleptics and the terminology is probably a misnomer; it is documented in patients with Parkinson's disease after sudden withdrawal of the dopaminergic drugs, levodopa or other dopamine agonists such as amantadine (Toru et al. 1981; Friedman et al. 1985; Rainer et al. 1991; Kornhuber and Weller 1993). It was anecdotally reported in a Parkinson woman during the premenstrual period, where she experienced two episodes without withdrawing or changing the drug regimen (Mizuta et al. 1993). NMS was also reported in Hallervorden–Spatz disease in absence of neuroleptic drugs (Hayashi et al. 1993).

Clinical picture

NMS is a rare but potentially fatal disorder characterized by mental status changes, muscle rigidity, hyperthermia, autonomic dysfunction, and typical laboratory findings. Psychomotor excitement, refusal of food, dehydration, weight loss exceeding 1 kg/week and oral administration of higher doses of neuroleptics are reported to be the combination of risk factors for NMS, especially in patients with bipolar affective disorder or those treated with neuroleptic injections. Systematic examination of early signs and the progression of symptoms in NMS facilitate prompt recognition and interventions to abort the syndrome in its incipient stage. Changes in the mental status or rigidity were the initial manifestations of NMS in 82.3% of the cases. Hyperthermia and autonomic dysfunction are likely to be observed later. Violent psychomotor excitement and aggressiveness may be the initial symptoms of mental changes, to be followed by clouding of consciousness from stupor to coma. Rigidity includes axial and limb muscles. Rigidity is abolished or decreased by muscle relaxants and curare, confirming a presynaptic central origin. Fever is accompanied by profuse sweating. Although fever is a cardinal sign of the syndrome, it can be delayed (Velamoor et al. 1994). The autonomic dysfunctions include tachycardia, diaphoresis and incontinence of urine, stools or both. They are occasionally severe and may be accompanied by episodes of tracheal spasm, apnoea, sinus bradycardia or sinus arrest

(Parry et al. 1994; Schneiderhan and Marken 1994). Rhabdomyolysis may occur. This may lead to potential fatal complications of hyperkalemia, disseminated intravascular coagulation, and acute renal failure (Shiono et al. 1992; Yamazaki and Ogawa 1992). NMS is usually preceded by prodromal signs for 1–3 days, which include progressive bradykinesia and mental withdrawal, sometimes with mutism and cataplexy. Symptoms of NMS then appear abruptly and, in some cases, lead to a dramatic course. If neuroleptic withdrawal is quick, they may last for a few days to 2 weeks, but if diagnosis and neuroleptic discontinuation are not carried out early, the outcome may be grave.

Diagnosis

Although NMS is described as having four classic signs (fever, rigidity, autonomic instability and altered consciousness), no agreed criteria exist for the diagnosis of the syndrome in terms of severity or combination of these signs and, as awareness of the syndrome has increased, milder or incomplete varieties (formes frustes) have been detected without the full-blown picture (Bristow and Kohen 1993; Sanders and Trewby 1993). NMS does not now seem quite as malignant as originally thought. Some authors believe that it is at one end of a scale of effects induced by neuroleptics with parkinsonism or dystonia being the initial effects; others believe that the term should be reserved for the full-blown syndrome, which has the features of an idiosyncratic reaction more akin to malignant hyperpyrexia (Granner and Wooten 1991; Bristow and Kohen 1993). Though the syndrome is now more widely recognised, the diversity of its clinical features may not always be appreciated and may lead to diagnostic confusion with other more common disorders, such as septicaemia (Sanders and Trewby 1993), midbrain infarction (Sanders and Trewby 1993), or lethargic encephalitis (Breggin 1993; Dekleva and Husain 1995), all of which show rigidity and fever. The 'serotonin syndrome' is a relatively new syndrome in which toxic hyperserotonergic states can result from the interaction of serotonergic agents and monoamine oxidase inhibitors. The most frequent clinical features are changes in mental status, restlessness, myoclonus, hyperreflexia, diaphoresis, shivering, and tremor. The presumed pathophysiological mechanism involves brainstem and spinal

cord activation of the serotonin receptor 1A type. Discontinuation of the suspected serotonergic agent and institution of supportive measures are the primary treatment (Sternbach 1991; Keltner and Harris 1994).

The relationship between catatonia, especially acute and subacute lethal catatonia and NMS, is ill understood (Monchablon Espinoza 1991). Many cases of NMS are reported to be preceded by catatonic state. Patients with catatonia may be on neuroleptic medications. Lethal catatonia may be accompanied by fever, which makes differentiation from NMS more difficult (White and Robins 1991; Turcyznski 1993; Miller et al. 1994; Raja et al. 1994). Differentiation from NMS is important as the treatment is different in lethal catatonia, where electroconvulsive therapy is necessary and life saving (Schneiderhan and Marken 1994).

There is no confirmatory diagnostic test for NMS. CK levels are raised in all patients. In at least 60% the level exceeds 1000 IU/l (Maniam and Rahman 1994). However, CK is non-specific and may be raised in patients who become pyrexial while on psychotropics (in some more than 1000 IU/l); none of these has NMS but a concurrent infection responding to appropriate antibiotic therapy (O'Dwyer and Sheppard 1993). A level of CK over 1000 IU/l and rising over repeated assays should alert the physician for diagnosis of NMS (Wilhelm et al. 1994). Exclusive use of CK as a diagnostic criterion may lead to overdiagnosis of NMS. The present consensus holds that if CK levels exceed 1000 IU, NMS should be excluded on neurological grounds. If repeat CKs normalise over 72 h, NMS is unlikely.

Single photon emission computed tomography (SPECT) brain scans using [123]I-*N*-isopropyl-*p*-iodoamphetamine (^{123}I-IMP) showed abnormalities of basal ganglia related to the development of NMS. These abnormalities disappeared after improvement of NMS (Nisjima et al. 1994). CT or MR are not helpful in the diagnosis, but MR, in some cases, showed lesions resembling those of patients with hypertensive encephalopathy (Becker et al. 1992).

Treatment

Treatment consists primarily of early recognition of the syndrome, discontinuation of triggering drugs, management of fluid balance, temperature reduction, and monitoring for complications. Use of dantrolene, amantadine or both may be indicated in more severe, prolonged, or refractory cases (Goedkoop and Carbaat 1982; Weller and Kornhuber 1992a; Caroff and Mann 1993a). Electroconvulsive therapy has been used successfully in some cases, especially when catatonia persists after acute treatment of the syndrome (Verwiel et al. 1994). Amantadine was introduced in the management of NMS because of its beneficial effects in Parkinson's disease. While the dopaminomimetic effects of amantadine are weak under experimental conditions, recent studies have confirmed that amantadine is an antagonist at the N-methyl-D-aspartate (NMDA) type of glutamate receptor. Two lines of evidence suggest that amantadine or other NMDA receptor antagonists could be effective drugs for the reversal of NMS symptoms. First, glutamate antagonists restore the balance between glutamatergic and dopaminergic systems when dopaminergic transmission has been antagonized by neuroleptic drugs. Second, by virtue of their effects against rigor and spasticity, NMDA antagonists may reduce increased muscle tone and prevent rhabdomyolysis (Weller and Kornhuber 1992a). Anticholinergic drugs such as i.v. procyclidine are also used with some success.

Outcome

As a result of these measures, mortality from NMS has declined in recent years from 25% before 1984 to 11.6% thereafter (Persing 1994). The recovery from NMS is usually without sequelae; however, some cases are reported to develop organic amnestic disorder or pancerebellar syndrome due to loss of Purkinje cells with reduction of granule cells in most areas of cerebellar hemispheres and vermis (Van Harten and Kemperman 1991; Naramoto et al. 1993).

Prevention

Neuroleptics may be safely reintroduced within few days of recovery from an NMS episode. Although a significant risk of recurrence does exist, dependent in part on the time elapsed since recovery as well as on dose or potency of neuroleptics used, patients are better treated safely with atypical neuroleptic drugs, such as clozapine (Weller and Kornhuber 1992b, 1993b; Caroff and Mann 1993a).

MALIGNANT HYPERTHERMIA

Aetiopathogenesis

MH is a pharmacogenetic disorder of skeletal muscle. In man, MH is inherited in an autosomal dominant fashion with variable penetrance (Allen 1993); in swine, the principal model for MH, it is inherited in a recessive fashion (Rosenberg and Fletcher 1994). Those with MH susceptibility are usually asymptomatic except in the presence of certain 'triggering' volatile anaesthetic agents such as halothane, isoflurane, enflurane and the muscle relaxant succinylcholine. Upon such exposure, hypermetabolism, increased CO_2 production, acidosis, muscle rigidity, rhabdomyolysis and hyperthermia occur (Lovstad et al. 1995). The overall incidence of MH is low (perhaps 1:50,000–1:71,000 anaesthetics), males more than females, children more than adults (Grassano et al. 1992; Johnson and Edleman 1992; Strazis and Fox 1993; Rosenberg and Fletcher 1994). Children display a paradoxical increase in jaw muscle tone to succinylcholine which can be a normal phenomenon but occasionally presages MH. MH is characterized by an acute hypercatabolic reaction of the muscles in response to the triggering effect of certain anaesthetic drugs or physical stress. There are mainly two categories of MH. The first resembles HS and is stress induced, occurring after strenuous exercise in hot weather or after shivering in winter (Gronert et al. 1980; Hackl et al. 1991). The second and more common is anaesthetic induced. It can vary from a mild reaction to a potentially fatal one. Potent halogenated anaesthetics, such as halothane, isoflurane and enflurane or the skeletal muscle relaxant succinylcholine are the common triggering drugs causing alteration of intracellular Ca^{2+} homeostasis in skeletal muscle (Rosenberg and Fletcher 1994; Treves et al. 1994; Lovstad et al. 1995). The defect responsible for the disease may lie within the mechanism controlling the release of Ca^{2+} from sarcoplasmic reticulum via the ryanodine receptor (RYR1) Ca^{2+} channel.

Central core disease (CCD), closely related to MH, is an autosomal dominant myopathy clinically distinct from MH. In a large kindred in which the gene for CCD was segregating, two-point linkage analysis gave a maximum lod score between the CCD locus and the RYR1 locus of 11.8, with no recombination. Mutation within RYR1 is responsible for MH, and RYR1 is also a candidate locus for CCD. A combination of physical mapping using a radiation-induced human–hamster hybrid panel and multipoint linkage analysis in the Centre d'Etude du Polymorphisme Humain families established the marker order and sex-average map distances (in centimorgans) on the background map as D19S75-(5.2)-D19S9-(3.4)-D19S191-(2.2)-RYR1(1.7)-D19S190-(1.6)-D19S47-(2.0)-CYP2B. Recombination was observed between CCD and the markers flanking RYR1. These linkage data are consistent with the hypothesis that CCD and RYR1 are allelic. The most likely position for CCD is near RYR1, with a multipoint lod score of 11.4 between D19S191 and D19S190, within the same interval as MH RYR1 (Mulley et al. 1993). Healthy members of families with CCD could be at risk of having MH. The results of the linkage study present no evidence for genetic heterogeneity of CCD (Schwemmles et al. 1993).

Abnormalities in the Ca^{2+} release channel of skeletal muscle sarcoplasmic reticulum, the ryanodine receptor, have been implicated in the cause of both the porcine and human MH by physiological and biochemical studies and genetic linkage analysis. In swine, a single founder mutation in the RYR1 gene can account for all cases of MH in all breeds, but a series of different RYR1 mutations is uncovered in human families with MH. Moreover, lack of linkage between MH and RYR1 in some families indicates a heterogeneous genetic basis for the human syndrome (Mickelson et al. 1992). In SR membranes isolated from MH pigs, the Arg615–Cys615 ryanodine receptor mutation is now shown to be directly responsible for an altered tryptic peptide map, due to the elimination of the Arg615 cleavage site. Furthermore, trypsin treatment released 86–99 kDa ryanodine receptor fragments encompassing residue 615 from the SR membranes. The 86–99 kDa domain containing residue 615 is near the cytoplasmic surface of the ryanodine receptor and likely near important Ca^{2+} channel regulatory sites. Defects in the gene encoding the RYR1 localized on human chromosome 19q13.1 have been proposed to be responsible for MH (Iles et al. 1992). Linkage of MH to RYR1 is, however, not observed in all human families with MH. Accordingly, other abnormal genes that may cause the condition are being sought (MacLennan 1992).

Additional loci on chromosomes 17q and 7q have been suggested (Levitt et al. 1992). Linkage of the MHS phenotype, with markers defining a 1 cm interval on chromosome 3q13.1, was found (Sudbrak et al. 1995). The C1840-T mutation in the human ryanodine receptor gene is a rare abnormality in MH families (Steinfath et al. 1995). The Gly 243 Arg and Gly 314 Arg mutations were seen in some families with MH (Keating et al. 1994). Other families with MH were linked to 19q13.1, 19q12–q13.2, 17q11.2–q24. These families showed Arg substitute for Gly 248. Others that have polymorphic substitution failed to segregate with MH implying that they represent polymorphism with little or no defect on the function of RYR1 gene (Ellis et al. 1978; Gillard et al. 1992).

In mild cases, there are series of metabolic processes leading to heat production, CO_2 and lactic acid generation with oxygen consumption (Britt 1982; Flewellen and Nelson 1982; Lopez et al. 1985). In more severe cases, a number of reactions occur with consumption of oxygen and production of CO_2 and lactic acid and activation of the contractile apparatus of the muscle with utilization of ATP. This leads to muscle contraction which is not abolished by muscle relaxants or curare. Once ATP is completely depleted, the various pumping mechanisms of the cell membrane for Na^+ and K^+ fail. Any attempt to arrest the process at this stage is futile (Britt 1982).

MH may occur postoperatively in any MH-susceptible patients. It should be excluded in patients with congenital malformations, especially arthrogryposis multiplex congenita (Hopkins et al. 1991), or 4P syndrome (Wolf–Hirschhorn syndrome) (Chen et al. 1994) or carnitine palmitoyl transferase deficiency (Vladutiu et al. 1993). It can occur after operations in patients with MELAS syndrome (Itaya et al. 1995) or acute lymphatic leukaemia (Sailer et al. 1991). It is also reported after i.v. lidocaine injection for cardiac arrhythmias (Taksukawa et al. 1992).

Patients with MH rarely have physical or laboratory signs of muscle disease. However, scattered cases were reported to have myopathies or muscle-related problems, such as acute rhabdomyolysis (Poels et al. 1991) or idiopathic persistently elevated CK, suggesting an association of MH with a variety of neuromuscular diseases (Wedel 1992). This association is very strong and supported by clinical and laboratory evidence in the case of CCD (Wedel 1992). MH has been reported to occur in some neuromuscular disorders, such as King–Denborough and Becker–Duchenne dystrophy. It is less likely to occur in myotonia congenita, sudden infant death syndrome, or limb girdle dystrophy (Wedel 1992).

The relationship between neuromuscular disease and MH susceptibility to caffeine and halothane in vitro contracture test (i.v. CT), according to the European Malignant Hyperthermia Group Protocol, was performed in 60 patients with neuromuscular disorder confirmed by muscle biopsy. Two test results were classified as MH susceptible (MHS), 10 as MH equivocal (MME) and 48 as MH negative (MHN). The large number of equivocal results is thought to indicate the lack of specificity of the individual components of this test in patients with clinical or histological evidence of neuromuscular disease. The increased in vitro sensitivity to the drugs tested may nevertheless provide some explanation for several in vivo 'MH-like reactions' reported frequently in these patients. These reactions are likely to be based on pathophysiological mechanisms different from those responsible for a true MH crisis. In vitro halothane and caffeine contracture tests are usually negative in patients with neuromuscular diseases. The same has been reported in arthrogryposis multiplex congenita (Gronert et al. 1992).

Clinical picture

MH is a rare clinical syndrome triggered by specific anaesthetic agents. The reaction occurs in the operating theatre usually 15 min after the anaesthetic triggers have been given, but it can be as late as 2 h. Hypercapnia, acidosis hyperpyrexia, muscle rigidity, rhabdomyolysis and renal failure develop quickly and, if left untreated, cardiac arrest occurs. Consciousness is usually preserved and, if the patient is treated promptly, the recovery is usually complete (Rosenberg and Fletcher 1994).

Postoperative pyrexia without the other features should not be always interpreted as MH as it may be due to other cases, especially infections. Atypical or abortive MH is described, especially in patients with unexplained recurrent rhabdomyolysis (Denborough et al. 1984; Poels et al. 1991) or severe exercise induced myolysis (Hackl et al. 1991).

Masseter spasm may be an early sign of MH, but not invariably, as it is common in paediatric otolaryngologic procedures and occurs in 1% of anaesthe-tised children. Although the children may have positive results in vitro to halothane and caffeine tests, they rarely develop MH. However, they should be treated as such. Those patients who develop MH usually have mild myotonia or mild clinical and laboratory abnormalities.

Diagnosis

The definitive diagnosis in suspected susceptible individuals is revealed by exposing an intact muscle fibre (biopsy) to caffeine and halothane in varying concentrations with the aid of i.v. CT. An abnormal contracture response is hypothesized to be the result of an increase in the release of calcium ion from the sarcoplasmic reticulum in response to neuronal stimulation leading to a hypermetabolic state (Johnson and Edleman 1992). The test results are classified as susceptible to MHS, MHN and MHE, with an abnormal caffeine result designated MHE_c or an abnormal halothane result designated MHE_h. While the MHS and MHN groups are diagnostically reliable, the equivocal group is not, with false negatives or false positives (Mortier and Breucking 1993). They represent about 15% of patients tested (Isaacs and Badenhorst 1993) and some are called 'K-type'. The 'K-type' designation is used to describe a patient being investigated for MH when concurrent administration of caffeine and halothane in vitro induces muscle contracture (rigidity, spasm), but when halothane or caffeine given separately produces a normal response. In some centres, K-type individuals are deemed susceptible to MH (Ellis et al. 1992).

In addition to i.v. CT, the 'ryanodine contracture test' has been proposed to improve discrimination between MHS and MHN patients. The ryanodine used is usually a mixture consisting of high-purity ryanodine (HPR) and 9,21-dehydroryanodine (DHR). The effects of both substances were investigated in concentrations of 2, 5 and 10 µmol/l. With all concentrations, contractures appeared earlier in MHS than in MHN muscles, but these differences were significant at all contracture levels with HPR only. Moreover, with the smallest concentration (2 µmol/l), the best discrimination between MHS and MHN was observed (Wappler et al. 1994). The 'ryanodine contracture test' has been proposed to be used in MHE in addition to i.v. CT to reduce the numbers of this group. By using HPR in concentrations of 2 µmol/l, the MHE patients could be assigned to MHS or MHN following their ryanodine-induced contractures (Wappler et al. 1994).

There is speculation that MH might represent a generalized defect in physical membrane properties either at rest or inducible by fluidizing agents. MH might conveniently be detected by examining physical membrane properties of easily accessible cells rather than by the cumbersome method of muscle biopsy. Erythrocytes were isolated and membrane physical properties examined using conventional, widely available, steady-state fluorescence polarization techniques. Erythrocyte membranes were evaluated with multiple probes both in the basal condition and following fluidization with either increasing temperature or two concentrations of a fluidizing alcohol. No discernible differences were detectable between MH positive or negative patients indicating no evidence for a generalized membrane defect in MH with a conclusion that erythrocyte membrane physical properties, by these techniques, are of no use in the preoperative screening for this disorder (Cooper and Meddings 1991). The effect of halothane and caffeine on cytoplasmic free calcium concentration Ca^{2+} in mononuclear cells was assessed and showed no evidence of calcium-induced calcium release (CICR) mechanism. As the main cause of the typical MH is the abnormality in the CICR mechanism, it seems difficult to screen MH susceptibility by using blood mononuclear cells (Kawana et al. 1992).

The effect of halothane on the regulation of blood platelet-free cystosolic calcium was investigated in Quin-2-loaded cells from MHS patients and healthy controls. The resting level of free cytosolic calcium was slightly (but statistically significantly) enhanced in platelets from MHS patients as compared to controls. Halothane induced a dose-dependent, rapid Ca^{2+} release from intracellular stores both in normal and MHS derived cells, but the resulting increase in cytosolic calcium was significantly higher in MHS patients. In platelets from controls, a complete reversibility of the halothane effect could be observed within 30–45 min. The cytosolic Ca^{2+} transients in platelets from patients were different from those of controls either in a higher initial peak or in a diminished decline velocity or in both. The basal Ca^{2+} permeability of the platelet plasma membrane was very low. Generally, halothane caused a dose-dependent increase in Ca^{2+} permeability. However, the influx of external calcium

was significantly higher in platelets from MHS patients than in controls. These results, contrary to those obtained on RBC and mononuclear cells, may indicate a generalized membrane effect (Fink et al. 1992).

Using sonographic examination of the thigh and calf in patients with proven susceptibility to MH and control patients, no consistent and reliable differences were found between control and MH patients (Antognini et al. 1994). Phosphorus magnetic resonance spectroscopy (^{31}P-MRS) in vivo has been recently suggested as a non-invasive diagnostic test in MH susceptibility. However, differences between protocols and also within subjects may have led to inconsistent MRS abnormalities reported during and after exercise. Payen et al. (1993) conducted a study to detect discriminating abnormalities in the leg muscles using in vivo ^{31}P-MRS during the rest period. On the basis of i.v. CT, 14 patients shown to be MHS and 22 patients MHN were compared to 36 control subjects. A score of MRS combined abnormalities was calculated from a stepwise discriminant function analysis. The MHS group had a significantly ($P <$ 0.01) higher inorganic phosphate (P1) to phosphocreatine (PCr) (Pi/PCr) value than MHN and control groups. The MHS group also exhibited a significant phosphodiesters (PDE) to PCr (PDE/PCr) value than both MHN and control groups. Combining both MRS parameters, MHS patients demonstrated abnormal MRS tests. Conversely, MHN patients and controls had normal MRS results (score value > or = 1.65). The sensitivity of the test was 93%, and specificity was 95%. So, ^{31}P-MRs could be useful for distinguishing non-invasively between MHS and MHN patients if several MRS parameters are combined; this approach appears to be more reliable and easier than that used during exercise.

MH susceptibility is caused in some families by inherited variation in a gene located on the short arm of chromosome 19 near to, or identical with, the ryanodine receptor gene (RYR1); this is expressed in skeletal muscle as a calcium release channel of the sarcoplasm reticulum. In other families, a gene in this location is excluded, but the locations of the genes involved have not yet been defined. As genetic heterogeneity could not be excluded, the use of DNA markers to replace i.v. CT in the diagnosis of MH susceptibility cannot be recommended yet (Ball et al. 1993).

Treatment

Identifying individuals at risk, avoiding exposure to potential anaesthetic triggering agents in MH-susceptible patients, and promptly recognizing and treating unexpected MH episodes are the primary means of reducing morbidity and mortality from MH. Interested and informed clinicians and families are the patient's best allies against MH (Kaus and Rockoff 1994). MH can be reversed or pretreated with dantrolene sodium. Myoplasmic-free Ca^{2+} was measured in MH patients before and after the i.v. administration of a cumulative dantrolene dose of 0.5, 1.5, and 2.5 mg/kg. In the MH subjects, dantrolene induced a dose-dependent reduction in myoplasmic-free Ca^{2+} by specifically acting at the ryanodine receptor binding site (Lopez et al. 1992). Dantrolene is quite a safe drug, but some side effects may occur, such as visual symptoms, subjective muscle weakness of the extremities, dizziness, and fatigue. These occurred more commonly in MHS patients than in MHN subjects. In patients recovering from an episode of MH to whom dantrolene has been administered, these side effects should be considered. Although the presence of these side effects does not outweigh the usefulness of the drug in cases of MH, this is not so in deciding whether to administer dantrolene prophylactically before surgical procedures in known or suspected MH patients (Wedel et al. 1995). If it is decided to give the drug prophylactically, dantrolene can be given orally 4 days before the operation at a dose of 75 mg. A further dose of 25 mg may be given as premedication on the day of the operation. I.v. dantrolene may be administered 1.2 mg/kg during the operation and 0.6 mg/kg on the day after surgery (Goto et al. 1993). Azumolene, 3–5 times more potent than dantrolene, has been shown to be effective in the treatment of MH in swine (El-Hayek et al. 1992). The 5-HT_{2A} receptor antagonists ritanserine (0.5–10 mg/kg i.v.) or ketanserine (0.5–10 mg/kg i.v.) have been compared to dantrolene in porcine MHS. Dantrolene exerted a therapeutic effect, whereas neither ritanserin nor ketanserin were effective. This indicates that 5-HT is not critically involved in the mechanisms of halothane-induced MH (Löscher et al. 1994). Nifedipine or verapamil, calcium channel antagonists, have no therapeutic value for the treatment of MH (Foster and Denborough 1993).

Outcome

With dantrolene therapy, treatment of MS should be successful and the mortality rate should be close to zero. Surprisingly enough, reports of deaths due to MH continue to be published up to the present day. Fatality rates have decreased to 10% since 1985 (Strazis and Fox 1993). Analysis of case reports reveals the following reasons for the discrepancy between the expectations and the clinical reality: (1) delay in early diagnosis due to lack of MH-sensitive monitoring (i.e., capnometry, pulse oximetry, blood gas analysis); (2) preoccupation with non-specific facets of therapy, such as cooling, change of the anaesthesia machine, transfer of the patient to the intensive care unit or the administration of drugs which have been shown ineffective in treating MH; (3) administration of insufficient amounts of dantrolene and delayed start of specific therapy due to failure to have immediate access to i.v. dantrolene; (4) failure to increase minute ventilation immediately after making the diagnosis to meet elevated metabolic demands (Schulte-Sasse and Eberlein 1991).

Prevention

The anaesthetic technique chosen for an MH-susceptible patient should include drugs that do not trigger MH, while providing stress-free conditions. The i.v. anaesthetic propofol does not trigger MH in susceptible patients or experimental animals, suggesting that there are important differences between the effects of propofol and of inhalation anaesthetics on Ca^{2+} regulation in MH-susceptible muscle. Total i.v. anaesthesia with propofol-fentanyl was well tolerated by the patients (Klotz 1991; Rivolta et al. 1992; Fruen et al. 1995). The safety of propofol as an induction and maintenance in this category of patients has been demonstrated. General anaesthesia using ketamine, midazolam, Diprivan and nitrous oxide are also quite safe in MH (Allen et al. 1993; Adnet 1994).

In conclusion, the three disorders have different pathogeneses. MH is hereditary, NMS is a dopaminergic neurotransmitter dysregulation, while HS represents environmental heat intoxication. Multisystem organ insult will follow if the process is not arrested early enough. Disturbance of consciousness is an early feature of HS and NMS, but late if ever in MH.

TABLE 2

Pathogenesis and typical clinical features of heat syndromes.

	Malignant hyperthermia	Neuroleptic malignant syndrome	Heat stroke
Pathogenesis	Excessive calcium from sarcoplasmic reticulum	Reduction of dopaminergic drive in the striatum and hypothalamus	Heat gain exceeds dissipation with insult to thermoregulatory centre (the hypothalamus)
Autonomic system a) Sweating b) Pupil	Central failure Profuse Normal	Central failure Profuse Normal	Central failure No sweating Constricted
Disturbance of consciousness	Late if any	Mutism and cataplexy	Early with sudden disturbance of consciousness ranging from drowsiness to coma
Muscle tone	Severe muscle contraction	Severe extrapyramidal rigidity	Muscle contraction
Convulsions	Late if any	Absent	Occur usually after cooling
Shivering	Absent	Absent	Occurs mainly during cooling
Diagnosis	i.v. CT	History, $\uparrow$ CK and isoenzymes	History, $\uparrow$ CK and isoenzymes
Treatment	Dantrolene	Withdrawal of neuro-leptic dantrolene	Evaporative cooling
Mortality rate	10%	10%	5–10%
Cause of death	Early: cardiac arrhythmia Late: acid base dis-turbances, coagula-tion defect, renal and cardiac failure	Acid base imbalance, coagulation defect, renal and cardiac failure	Brain oedema and petechial brain haemorrhages in classical HS; renal failure, disseminated intravascular coagulation and rhabdo-myolysis in exertional HS

The recovery is usually without sequelae, and the mortality has been reduced significantly to ± 10% in the three syndromes.

Although different in pathogenesis, they form a spectrum of one disorder (Table 2) with scattered reports of neuroleptics causing NMS in MHS patients or being an important risk factor in precipitating HS. Anaesthetic agents may precipitate MH in patients who previously suffered from NMS or exertional or classical HS. Exertional or classical HS may occur more often in patients who have MHS or developed NMS (Denborough et al. 1984; Hackl et al. 1991; Poels et al. 1991; Amore and Cerisoli 1992; Caroff and Mann 1993b; Heiman-Patterson 1993; Nimo et al. 1993; Calore et al. 1994).

REFERENCES

ADNET, P.: Use of Diprivan in muscular diseases and malignant hyperthermia. Ann. Fr. Anesth. Reanim. 13 (1994) 490–493.

AKHTAR, M.J., M. AL-NOZHA, S. AL-HARTHI and M.S. NOUH: Electrocardiographic abnormalities in patients with heat stroke. Chest 104 (1993) 411–414.

AL-ASKA, A.K., H. ABU-AISHA, B. YAQUB, S.S. AL-HARTHI and A. SALLAM: Simplified cooled bed for heatstroke. Lancet i (1987a) 381.

AL-ASKA, A.K., B.A. YAQUB, S.S. AL-HARTHI and A. AL-DALAAN: Rapid cooling in management of heat stroke: clinical methods and practical implications. Ann. Saudi Med. 7 (1987b) 135–138.

AL-HARTHI, S.S., B.A. YAQUB, M. AL-NOZHA, A.K. AL-ASKA and M. SERAJ: Management of heat stroke patients by rapid cooling at Mecca pilgrimage (Hajj 1404) comparing a conventional method with a body cooling unit. Saudi Med. J. 7 (1986) 369–376.

ALLEN, G.C.: Malignant hyperthermia and associated disorders. Curr. Opin. Rheumatol. 5 (1993) 719–724.

ALLEN, G.C., L.J. BYFORD and F.M. SHAMJI: Anterior mediastinal mass in a patient susceptible to malignant hyperthermia. Can. J. Anaesth. 40 (1993) 46–49.

AMORE, M. and M. CERISOLI: Heat stroke and hyperthermias. Ital. J. Neurol. Sci. 13 (1992) 337–341.

ANDERSON, E.S. and P.S. POWERS: Neuroleptic malignant syndrome associated with clozapine use. J. Clin. Psychiatry 52 (1991) 102–104.

ANONYMOUS: Heat-related deaths – United States, 1993. Morbid. Mortal. Wkly. Rep. 42 (1993) 558–560.

ANONYMOUS: Heat-related deaths – Philadelphia and United States, 1993–1994. Morbid. Mortal. Wkly. Rep. 43 (1994) 453–455.

ANTOGNINI, J.F., M. ANDERSON, M. CRONAN, J.P. MCGAHAN and G.A. GRONERT: Ultrasonography: not useful in detecting susceptibility to malignant hyperthermia. J. Ultrasound Med. 13 (1994) 371–374.

ANTUNANO, M.J. and S.A. NUNNELEY: Heat stroke in protective clothing: validation of a computer model and the heat-humidity index (HHI). Aviat. Space Environ. Med. 63 (1992) 1087–1092.

ARMSTRONG, L.E. and C.M. MARESH: The introduction and decay of heat acclimatisation in trained athletes. Sports Med. 12 (1991) 302–312.

ATTIA, M., M. KHOGALI, G. EL-KHATIB, M.K. MUSTAFA, N.A. MAHMOUD, A.N. ELDIN and K. GUMAA: Heat stroke: an upward shift of temperature regulation set point at an elevated body temperature. Int. Arch. Occup. Environ. Health 53 (1983) 9–17.

BALL, S.P., H.R. DORKINS, F.R. ELLIS, J.L. HALL, P.J. HALSALL, P.M. HOPKINS, R.F. MUELLER and A.D. STEWART: Genetic linkage analysis of chromosome 19 markers in malignant hyperthermia. Br. J. Anaesth. 70 (1993) 70–75.

BASU, J.: An unusual presentation of neuroleptic malignant syndrome. J. Indian Med. Assoc. 89 (1991) 16.

BEARD, M.E., J.W. HAMER, G. HAMILTON and A.H. MASLOWSKI: Jogger's heat stroke. N.Z. Med. J. 89 (1979) 159–161.

BECKER, T., J. KORNHUBER, E. HOFMANN, M. WELLER, C. RUPPRECHI and H. BECKMANN: MRI white matter hyperintensity in neuroleptic malignant syndrome (NMS) – a clue to pathogenesis? J. Neural Transm. 90 (1992) 151–159.

BERNSTEIN, R.A.: Malignant neuroleptic syndrome: an atypical case. Psychosomatics 20 (1979) 840–846.

BIARY, N., M.M. MADKOUR and H. SHARIF: Post heat stroke parkinsonism and cerebellar dysfunction. Clin. Neurol. Neurosurg. 97 (1995) 55–57.

BLIGH, J.: Iavliaetsia li tsentral'nyi kontrol' postoianstva temperatury primeron gomeostazisa? (Obsuzhdenie teorri i dokazatel'stv). Fiziol. Zh. SSSR Im. I.M. Sechenova 67 (1981) 1068–1078.

BOUCHAMA, A., A. CAFEGE, E.B. DEVOL, O. LABDI, K. EL-ASSIL and M. SERAJ: Ineffectiveness of dantrolene sodium in the treatment of heat stroke. Crit. Care Med. 19 (1991a) 176–180.

BOUCHAMA, A., A. CAFEGE, W. ROBERTSON, S. AL-DOSSARY and A. EL-YAZIGI: Mechanisms of hypophosphatemia in humans with heat stroke. J. Appl. Physiol. 71 (1991b) 328–332.

BOYD, R.D.: Recurrence of neuroleptic malignant syndrome via an inadvertent rechallenge in a woman with mental retardation. Ment. Retard. 30 (1992) 77–79.

BOYD, R.D.: Neuroleptic malignant syndrome and mental retardation: review and analysis of 29 cases. Am. J. Ment. Retard. 98 (1993) 143–155.

BREGGIN, P.R.: Parallels between neuroleptic effects and lethargic encephalitis: the production of dyskinesias and cognitive disorders. Brain Cogn. 23 (1993) 8–27.

BRISTOW, M.F. and D. KOHEN: How 'malignant' is the neuroleptic malignant syndrome? In early mild cases it may not be malignant at all. Br. Med. J. 307 (1993) 1223–1224.

BRITT, B.A.: Malignant hyperthermia – a review. In: A.S. Milton (Ed.), Handbook of Experimental Pharmacology, Pyretics and Antipyretics. Heidelberg, Springer-Verlag (1982) 547–615.

BRITT, B.A. and W. KALOW: Malignant hyperthermia. A statistical review. Can. Anaesth. Soc. J. 18 (1970) 293–315.

BROOKS, C.M. and K. KOIZUMI: The hypothalamus and control of integrative processes. In: V.B. Mountcastle (Ed.), Medical Physiology. St. Louis, MO, Mosby (1980) 941–942.

BUCKLEY, P.F. and H.Y. MELTZER: Clozapine and NMS. Br. J. Psychiatry 162 (1993) 566.

BYERS, D.J. and G. KRISHNA: Malignant hyperthermia. Semin. Pediatr. Surg. 1 (1992) 88–95.

CABANAC, M.: Selective brain cooling in humans: 'fancy' or fact? FASEB J. 7 (1993) 1143–1147.

CALORE, E.E., M.J. CAVALIERE, N.M. PEREZ, D.H. RUSSO, A. WAKAMATSU and D. RAZZOUK: Hyperthermic reaction to haloperidol with rigidity, associated to central core disease. Acta Neurol. (Napoli) 16 (1994) 157–161.

CAPE, G.: Neuroleptic malignant syndrome – a cautionary tale and a surprising outcome. Br. J. Psychiatry 164 (1994) 120–122.

CAROFF, S.M.: The neuroleptic malignant syndrome. J. Clin. Psychol. 41 (1980) 79–83.

CAROFF, S.N. and S.C. MANN: Neuroleptic malignant syndrome. Med. Clin. North Am. 77 (1993a) 185–202.

CAROFF, S.M. and S.C. MANN: Neuroleptic malignant syndrome and malignant hyperthermia. Anaesth. Intens. Care 21 (1993b) 477–478.

CAROFF, S.N., H. ROSENBERG and J.C. GERBER: Neuroleptic malignant syndrome and malignant hyperthermia. Lancet i (1983) 244.

CHEN, J.C., R.K. JEN, Y.W. HSU, Y.B. KE, J.J. HWANG, K.H. WU and T.T. WEI: 4-P syndrome (Wolf–Hirschhorn syndrome) complicated with delay onset of malignant hyperthermia: a case report. Acta Anaesthesiol. Sin. 32 (1994) 275–278.

CLOWES, C.H., JR. and T.F. O'DONNELL, JR.: Heat stroke. N. Engl. J. Med. 291 (1987) 564–567.

COLLINS, S.P., M.D. WHITE and M.A. DENBOROUGH: The effect of calmodulin antagonist drugs on isolated sarcoplasmic reticulum from malignant hyperpyrexia susceptible swine. Int. J. Biochem. 19 (1987) 819–826.

COOPER, P. and J.B. MEDDINGS: Erythrocyte membrane fluidity in malignant hyperthermia. Biochim. Biophys. Acta 1069 (1991) 151–156.

DEKLEVA, K.B. and M.M. HUSAIN: Sporadic encephalitis lethargica: a case treated successfully with ECT. J. Neuropsychiatr. Clin. Neurosci. 7 (1995) 237–239.

DELAY, J., P. PICHOT and M.T. LEMPERIER: Un neuroleptique majeur non phénothiazinique et non réserpinique, l'halopéridol, dans le traitement des psychoses. Ann. Med. Psychol. 118 (1960) 627–630.

DELGADO, G., T. TUNON, J. GALLEGO and J. VILLANUEVA: Spinal cord lesions in heat stroke. J. Neurol. Neurosurg. Psychiatry 48 (1985) 1065–1067.

DENBOROUGH, M.A.: Heat stroke and malignant hyperpyrexia. Med. J. Aust. 1 (1982) 204–205.

DENBOROUGH, M.A., J.F. FOSTER and R.H. LOWELL: Anaesthetic death in family. Br. J. Anaesth. 34 (1962) 395–396.

DENBOROUGH, M.A., S.P. COLLINS and K.C. HOPKINSON: Rhabdomyolysis and malignant hyperpyrexia. Br. Med. J. 288 (1984) 1878.

EL HAYEK, R., J. PARNESS, H.H. VALDIVIA, R. CORONADO and K. HOGAN: Dantrolene and azumolene inhibit [³H]-PH200-110 binding to porcine skeletal muscle dihydropyridine receptors. Biochem. Biophys. Res. Commun. 187 (1992) 894–900.

ELLIS, F.R.: Mortality from heat illness and heat aggravated illness in United States. Environ. Res. 5 (1972) 1–58.

ELLIS, F.R., P.A. CAIN and D.G.F. HARRIMAN: Multifactorial inheritance of malignant hyperthermia susceptibility. In: J.A. Aldrete and B.A. Britt (Eds.), Second International Symposium on Malignant Hyperthermia. New York, Grune and Stratton (1978) 329–338.

ELLIS, F.R., P.J. HALSALL and P.M. HOPKINS: Is the 'K-type' caffaine-halothane responder susceptible to malignant hyperthermia? Br. J. Anaesth. 69 (1992) 468–470.

EPSTEIN, Y., E. SOHAR and Y. SHAPIRO: Exertional heatstroke: a preventable condition. Isr. J. Med. Sci. 31 (1995) 454–462.

FELLER, R.B. and J.S. WILSON: Hepatic failure in fatal exertional heatstroke. Aust. N.Z. J. Med. 24 (1994) 69.

FERRIS, E.B. JR., M.A. BLANKENHORN and H.W. ROBINSON: Heat stroke: clinical and chemical observations on 44 cases. J. Clin. Invest. 17 (1938) 249–262.

FINK, H.S., J.G. HOFMANN, H. HENTSCHEL and U. TILL: Abnormalities in the regulation of blood platelet free cytosolic calcium in malignant hyperthermia. I. Human platelets. Cell Calcium 13 (1992) 149–155.

FLEWELLEN, E.H. and T.E. NELSON: Masseter spasm induced by succinyl choline in children: contracture testing for malignant hyperthermia: report of six cases. Can. Anaesth. Soc. J. 29 (1982) 42–49.

FOSTER, P.S. and M.A. DENBOROUGH: The effect of calcium channel antagonists and BAY K 8644 on calcium fluxes of malignant hyperpyrexia-susceptible muscle. Int. J. Biochem. 25 (1993) 495–504.

FRIEDMAN, J.H., S.S. FEINBERG and R.G. FELDMAN: A neuroleptic malignant-like syndrome due to levodopa therapy withdrawal. J. Am. Med. Assoc. 254 (1985) 2792–2795.

FRUEN, B.R., J.R. MICKELSON, T.J. ROGHAIR, L.A. LITTERER and C.F. LOUIS: Effects of propofol on Ca²⁺ regulation by malignant hyperthermia-susceptible muscle membranes. Anesthesiology 82 (1995) 1274–1282.

GARDNER, J.W. and J.A. KARK: Fatal rhabdomyolysis presenting as mild heat illness in military training. Milit. Med. 159 (1994) 160–163.

GAUSS, H. and K.A. MEYER: Heat stroke: report of 158 cases from Cook County Hospital, Chicago. Am. J. Med. Sci. 154 (1917) 554–564.

GELLER, B. and D.E. GREYDANUS: Haloperidol induced comatose state with hyperthermia and rigidity in adolescence: two case reports with a literature review. J. Clin. Psychiatry 40 (1979) 102–103.

GIBB, W.R.G. and A.J. LEES: The neuroleptic malignant syndrome – a review. Quart. J. Med. 56 (1985) 421–429.

GILLARD, E.F., K. OTSU, J. FUJII, C. DUFF, S. DE LEON, V.K. KHANNA, B.A. BRITT, R.G. WORTON and D.H. MACLENNAN: Polymorphisms and deduced amino acid substitutions in the coding sequence of the ryanodine receptor (RYR1) gene in individuals with malignant hyperthermia. Genomics 13 (1992) 1247–1254.

GOEDKOOP, J.G. and P.A. CARBAAT: Treatment of neuroleptic malignant syndrome with dantrolene. Lancet ii (1982) 49–50.

GOTO, S., K. OGATA, T. FUJIE, T. FUJIGAKI, H. NAKAMURA, T. YUKINARI and O. SHIBATA: Caesarean section in a patient with past history of fulminant malignant hyperthermia. Masui – Jpn. J. Anesthesiol. 42 (1993) 271–275.

GRANNER, M.A. and G.F. WOOTEN: Neuroleptic malignant syndrome or parkinsonism hyperpyrexia syndrome. Semin. Neurol. 11 (1991) 228–235.

GRASSANO, M.T., T. BERSANO, G. RADESCHI, G. GORGONI, M. TORTA and F. GORGERINO: Epidemiologia dell'ipertermia maligna in Piemonta e Valle d'Aosta. Studio condotto con due differenti metodiche. Minerva Anestesiol. 58 (1992) 453–457.

GRONERT, G.A.: Malignant hyperthermia. Anesthesiology 50 (1980) 395–423.

GRONERT, G.A., R.L. THOMPSON and B.M. ONOFRIO: Human malignant hyperthermia. An awake episode and correction by dantrolene. Anaesthesia 59 (1980) 377–378.

GRONERT, G.A., W. FOWLER, G.H. CARDINET, III, A. GRIX, JR., W.G. ELLIS and M.Z. SCHWARTZ: Absence of malignant hyperthermia contractures in Becker–Duchenne dystrophy at age 2. Muscle Nerve 15 (1992) 52–56.

GROVE, I. and N. ISAACSON: The pathology of hyperpyrexia: observation at autopsy in 17 cases of fever therapy. Am. J. Pathol. 25 (1949) 1029–1046.

GRUNHAUS, L., R. SANCOVICI and R. RIMON: Neuroleptic malignant syndrome due to depot fluphenazine. J. Clin. Psychiatry 40 (1979) 99–100.

HABERMAN, M.L.: Malignant hyperthermia: an allergic reaction to thioridazine therapy. Arch. Intern. Med. 138 (1978) 800–801.

HACKL, W., M. WINKLER, W. MAURITZ and P. SPORN: Muscle biopsy for diagnosis of malignant hyperthermia susceptibility in two patients with severe exercise-induced myolysis. Br. J. Anaesth. 66 (1991) 138–140.

HART, G.R., R.J. ANDERSON, C.P. CRUMPLER, A. SHULKIN, G. REED and J.P. KNOCHEL: Epidemic classical heat stroke: clinical characteristic and course of 28 patients. Medicine (Baltimore) 61 (1982) 189–197.

HART, L.E., B.P. EGIER, A.G. SHIMIZU, P.M. TAUDAN and J.R. SUTTON: Exertional heat stroke: the runner's nemesis. Can. Med. Assoc. J. 122 (1980) 1144–1150.

HASSANEIN, T., J.A. PERPER, L. TEPPERMAN, T.E. STARZL and D.H. VAN THIEL: Liver failure occurring as a component of exertional heat stroke. Gastroenterology 100 (1991) 1442–1447.

HAYASHI, K., E. CHIHARA, T. SAWA and Y. TANAKA: Clinical features of neuroleptic malignant syndrome in basal ganglia disease. Spontaneous presentation in a patient with Hallervorden–Spatz disease in the absence of neuroleptic drugs. Anaesthesia 48 (1993) 499–502.

HAYNES, G.R., J. GOTTESMAN, B.H. DORMAN and N.H. BRAHEN: Postoperative hyperthermia in a patient having cortical brain resection. South. Med. J. 87 (1994) 399–401.

HEIMAN-PATTERSON, T.D.: Neuroleptic malignant syndrome and malignant hyperthermia. Important issues for the medical consultant. Med. Clin. North Am. 77 (1993) 477–492.

HELZER-JULIN, N.: Sun, heat, and cold injuries in cyclists. Clin. Sports Med. 13 (1994) 219–234.

HENDERSON, V.W. and G.F. WOOTEN: Neuroleptic malignant syndrome. A pathogenic role for dopamine receptor blockade? Neurology 31 (1981) 132–137.

HENRY, P., M. BARAT and M. BOURGEOIS: Syndrome malin mortel succédant à une injection d'emblée d'oenenthate de fluphenazine. Presse Méd. 79 (1971) 1350.

HOPKINS, P.M., F.R. ELLIS and P.J. HALSALL: Hypermetabolism in arthrogryposis multiplex congenita. Anaesthesia 46 (1991) 374–375.

ILES, D.E., B. SEGERS, L. HEYTENS, R.C. SENGERS and B. WIERINGA: High-resolution physical mapping of four microsatellite repeat markers near the RYR1 locus on chromosome 19q13.1 and apparent exclusion of the MHS locus from this region in two malignant hyperthermia susceptible families. Genomics 14 (1992) 749–754.

ISAACS, H. and M. BADENHORST: False-negative results with muscle caffeine halothane contracture testing for malignant hyperthermia. Anesthesiology 79 (1993) 59.

ITAYA, K., O TAKAHATA, K. MAMIYA, T. SAITO, S. TAMAKAWA, Y. AKAMA, M. KUBOTA and H. OGAWA: Anesthetic management of two patients with mitochondrial encephalopathy, lactic acidosis and stroke-like episodes (MELAS). Masui – Jpn. J. Anesthesiol. 44 (1995) 710–712.

ITOH, H., N. OHTSUKA, K. OGITA, G. YAGI, S. MIURA and Y. KOGA: Malignant neuroleptic syndrome – its present status in Japan and clinical problems. Folia Psychiatr. Neurol. Jpn. 31 (1977) 559–576.

JOHNSON, C. and K.J. EDLEMAN: Malignant hyperthermia: a review. J. Perinatol. 12 (1992) 61–71.

JONES, T.S., A.P. LIANG, E.M. KILBOURNE, M.R. GRIFFIN, P.A. PATRIARCA, S.G. WASSILAK, R.J. MULLAN, R.F. HERRICK, H.D. DONNELL JR., K. CHOI and S.B THACKER: Morbidity and mortality associated with the July 1980 heat wave in St. Louis and Kansas City MO. J. Am. Med. Assoc. 247 (1982) 3327–3331.

KAUS, S.J. and M.A. ROCKOFF: Malignant hyperthermia. Pediatr. Clin. North Am. 41 (1994) 221–237.

KAWANA, Y., M. IINO, H. OYAMADA and M. ENDO: Effects of halothane, caffeine and ryanodine on the intracellular calcium store in blood mononuclear cells. Masui – Jpn. J. Anesthesiol. 41 (1992) 727–732.

KEATING, K.E., K.A. QUANE, B.M. MANNING, M. LEHANE, E. HARTUNG, K. CENSIER, A. URWYLER, M. KLAUSNITZER, C.R. MULLER and J.J. HEFFRON: Detection of a novel RYR1 mutation in four malignant hyperthermia pedigrees. Hum. Mol. Genet. 3 (1994) 1855–1858.

KELTNER, N. and C.P. HARRIS: Serotonin syndrome: a case of fatal SSRI/MAOI interaction. Perspect. Psychiatr. Care 30 (1994) 26–31.

KHOGALI, M. and J.S. WEINER: Heat stroke: report of 18 cases. Lancet ii (1980) 276–278.

KHOGALI, M., H. EL-SAYED and M. AMAR: Management and therapy regimen during cooling and in the recovery room at different heat stroke treatment centres. In: M. Khogali and J.R. Hales (Eds.), Heat Stroke and Temperature Regulation. Sydney, Academic Press (1983) 149–156.

KILBOURNE, E.M., K. CHOI, K.S. JONES and S.B. THACKER: Risk factors for heat stroke. A case-control study. J. Am. Med. Assoc. 247 (1982) 3332–3336.

KIYATAKE, I., K. YAMAJI, I. SHIRATO, M. KUBOTA, S. NAKA-YAMA, Y. TOMINO and H. KOIDE: A case of neuroleptic malignant syndrome with acute renal failure after the discontinuation of sulpiride and maprotiline. Jpn. J. Med. 30 (1991) 387–391.

KLOTZ, R.W.: Propofol anesthesia in the malignant hyperthermia susceptible patient. Nurse Anesth. 2 (1991) 33–35.

KOPONEN, H.J., U.M. LEPOLA and E.V. LEINONEN: Neuroleptic malignant syndrome during remoxipride treatment. A case report. Eur. Neuropsychopharmacol. 3 (1993) 517–519.

KORNHUBER, J. and M. WELLER: Neuroleptic malignant syndrome. Curr. Opin. Neurol. 7 (1994) 353–357.

KUBOTA, T.: Neuroleptic malignant syndrome induced by nemonapride. Acta Neurol. 15 (1993) 142–144.

LANDRY, P. and J. LATOUR: Neuroleptic malignant syndrome in communicating hydrocephaly and microcephaly. J. Clin. Psychopharmacol. 131 (1993) 72–74.

LAZARUS, A.: Neuroleptic malignant syndrome and pre-existing brain damage. J. Neuropsychiatr. Clin. Neurosci. 4 (1992) 185–187.

LEE-CHIONG, T.L., JR. and J.T. STITT: Heat stroke and other heat-related illnesses. Postgrad. Med. 98 (1995) 31–33.

LEFKOWITZ, D., C. FORD, C. RICH, J. BILLER and L. MCHENRY, JR.: Cerebellar syndrome following neuroleptic induced heat stroke. J. Neurol. Neurosurg. Psychiatry 46 (1983) 183–185.

LELIS, M., G. DO CARMO, L. CALDEIRA, J. ABREU, L. FREITAS and M. DOROANA: Febrile coma and disseminated intravascular coagulation following heat stroke. Acta Med. Port. 5 (1992) 215–218.

LEVENSION, J.L.: Neuroleptic malignant syndrome. Am. J. Psychiatry 142 (1985) 1137–1145.

LEVITT, R.C., A. OLCKERS, S. MEYERS, J.E. FLETCHER, H. RO-SENBERG, H. ISAACS and D.A. MEYERS: Evidence for the localization of a malignant hyperthermia susceptibility locus (MHS2) to human chromosome 17q. Genomics 14 (1992) 562–566.

LIEBERMAN, J.A. and A.Z. SAFFERMAN: Clinical profile of clozapine: adverse reactions and agranulocytosis. Psychiatr. Quart. 63 (1992) 51–70.

LIN, J.J., M.K. CHANG, Y.D. SHEU, K.S. TING, S.C. SUNG and T.QI. LIN: Permanent neurologic deficits in heat stroke. Chung Hua i Hsueh Tsa Chih – Chin. Med. J. 47 (1991) 133–138.

LOPEZ, J.R., L. ALAMAO and C. CAPUTO: Intracellular ionized Ca^{++} concentration in muscles from humans with malignant hyperthermia. Muscle Nerve 3 (1985) 355–358.

LOPEZ, J.R., A. GERARDI, M.J. LOPEZ and P.D. ALLEN: Effects of dantrolene on myoplasmic free (Ca^{2+}) measured in vivo in patients susceptible to malignant hyperthermia. Anesthesiology 76 (1992) 711–719.

LÖSCHER, W., C. GERDES and A. RICHTER: Lack of prophylactic or therapeutic efficacy of 5-HT2A receptor antagonists in halothane-induced porcine malignant hyperthermia. Naunyn-Schmiedeberg's Arch. Pharmacol. 350 (1994) 365–374.

LOVSTAD, R.Z, P. HALVORSEN, P.A. STEEN and S. LINDAL: Malignant hyperthermia – still a current and dangerous problem. Tidsskr. Nor. Laegeforen. 115 (1995) 1494–1498.

MACLENNAN, D.H.: The genetic basis of malignant hyperthermia. Trends Pharmacol. Sci. 13 (1992) 330–334.

MALAMUD, N., W. HAYMAKER and R.P. CUSTER: Heat stroke: a clinico-pathological study of 125 fatal cases. Milit. Surg. 99 (1946) 394–449.

MANIAM, T. and M.A. RAHMAN: All elevated creatine kinase is not neuroleptic malignant syndrome. Med. J. Malaysia 49 (1994) 252–254.

MCALLISTER, R.G.: Fever, tachycardia and hypertension with acute catatonic schizophrenia. Ann. Intern. Med. 138 (1978) 1154–1156.

MEHTA, A. and R. BAKER: Resistant neurological deficit in heat stroke. Neurology 20 (1970) 336–340.

MELTZER, H.Y.: Rigidity, hyperpyrexia and coma following fluphenazine enanthate. Psychopharmacology (Berlin) 29 (1978) 337–346.

MICKELSON, J.R., C.M. KNUDSON, C.F. KENNEDY, D.I. YANG, L.A. LITTERER, W.E. REMPEL, K.P. CAMPBELL and C.F. LOUIS: Structural and functional correlates of a mutation in the malignant hyperthermia-susceptible pig ryanodine receptor. FEBS Lett. 301 (1992) 49–52.

MILLER, C.E., H. GILBERT and O. MORALES: Lethal catatonia following temporomandibular joint surgery: a case report. J. Oral Maxillofac. Surg. 52 (1994) 510–512.

MIZUTA, E., S. YAMASAKI, M. NAKATAKE and S. KUNO: Neuroleptic malignant syndrome in a parkinsonian woman during the premenstrual period. Neurology 43 (1993) 1048–1049.

MONCHABLON ESPINOZA, A.J.: Sindromes catatonicos agudos-subagudos con perdida de peso. Su evolucion favorable con TEC. Acta Psiquiatr. Psicol. Am. Lat. 37 (1991) 65–71.

MOOCHHALA, S.M., W.T. TAN and T.L. LEE: The genetic basis of malignant hyperthermia. Ann. Acad. Med., Singapore 23 (1994) 475–478.

MORTIER, W. and E. BREUCKING: Diagnosis of malignant hyperthermia susceptibility. 1. The significance of in vitro susceptibility tests. Anaesthesist 42 (1993) 675–683.

MOYES, O.: Malignant hyperpyrexia caused by trimeprazine. Br. J. Anaesth. 45 (1973) 1163–1164.

MULLEY, J.C., H.M, KOZMAN, H.A. PHILLIPS, A.K. GEDEON, J.A. MCCURE, D.E. ILES, R.G. GREGG, K. HOGAN, F.J. COUCH and D.H. MACLENNAN: Refined genetic localization for central core disease. Am. J. Hum. Genet. 52 (1993) 398–405.

NARAMOTO, A., N. KOIZUMI, N. ITOH and H. SHIGEMATSU: An autopsy case of cerebellar degeneration following lithium intoxication with neuroleptic malignant syndrome. Acta Pathol. Jpn. 43 (1993) 55–58.

NIELSEN, B.: Heat stress and acclimatisation. Ergonomics 37 (1994) 49–58.

NIMO, S.M., B.W. KENNEDY, W.M. TULLET, A.S. BLYTH and J.R. DOUGALL: Drug-induced hyperthermia. Anaesthesia 48 (1993) 892–895.

NISJIMA, K., M. MATOBA and T. ISHIGURO: Single photon emission computed tomography with 123I-IMP in three cases of the neuroleptic malignant syndrome. Neuroradiology 36 (1994) 281–284.

NOBACK, C.R. (Ed.): The Human Nervous System: Basic Principles and Neurobiology, 3rd Edit. New York, McGraw-Hill (1981) 378–379.

O'DONNELL, T.J., JR.: The haemodynamic and metabolic alterations associated with acute stress injury in marathon runners. Ann. N.Y. Acad. Sci. 301 (1977) 262–269.

O'DWYER, A.M. and N.P. SHEPPARD: The role of creatine kinase in the diagnosis of neuroleptic malignant syndrome. Psychol. Med. 23 (1993) 323–326.

ORDING, H. and D. BENDIXEN: Malignant hyperthermia. Nord. Med. 107 (1992) 12–14.

ORSER, B.: Dantrolene sodium and heat stroke. Crit. Care Med. 20 (1992) 1192–1193.

PANJWANI, G.D., M.K. MUSTAFA, A. MUHAILAN, I.S. ANEJA and A. OWUNWANNE: Effect of hyperthermia on somatosensory evoked potentials in the anaesthetized rat. Electroencephalogr. Clin. Neurophysiol. 80 (1991) 384–391.

PARRY, A.K., L.P. ORMEROD, G.W. HAMLIN and P.T. SALEEM: Recurrent sinus arrest in association with neuroleptic malignant syndrome. Br. J. Psychiatry 164 (1994) 689–691.

PASCOE, D.D., L.A. SHANLEY and E.W. SMITH: Clothing and exercise. I: Biophysics of heat transfer between the individual, clothing and environment. Sports Med. 8 (1994) 38–54.

PAYEN, J.F., J.L. BOSSON, L. BOURDON, C. JACQUOT, J.F. LE BAS, P. STIEGLITZ and A.L. BENABID: Improved noninvasive diagnostic testing for malignant hyperthermia susceptibility from a combination of metabolites determined in vivo with ^{31}P-magnetic resonance spectroscopy. Anesthesiology 78 (1993) 848–855.

PERSING, J.S.: Neuroleptic malignant syndrome: an overview. S. Dakota J. Med. 47 (1994) 51–55.

PETTIGREW, R.T., J.M. GALT, C.M. LUDGATE and A.M. SMITH: Clinical effects of whole body hyperthermia in advanced malignancy. Br. Med. J. 4 (1974) 679–682.

POELS, P.J., E.M. JOOSTEN, R.C. SENGERS, A.M. STADHOUDERS, J.H. VEERKAMP and A.M. BENDERS: In vitro contraction test for malignant hyperthermia in patients with unexplained recurrent rhabdomyolysis. J. Neurol. Sci. 105 (1991) 67–72.

PORTER, A.M.: Heat illness and soldiers. Milit. Med. 158 (1993) 606–609.

PRESTON, J.: Central nervous system reactions to small doses of tranquilizers: report of one death. Am. Pract. Dig. Treat. 10 (1959) 627–630.

QIAN, Y., B. HITZEMANN and R. HITZEMANN: D1 and D2 dopamine receptor distribution in the neuroleptic nonresponsive and neuroleptic responsive lines of mice, a quantitative receptor autoradiographic study. J. Pharmacol. Exp. Ther. 261 (1992) 341–348.

RAINER, C., N.A. SCHEINOST and E.J. LEFEBER: Neuroleptic malignant syndrome. When levodopa withdrawal is the cause. Postgrad. Med. 89 (1991) 175–178.

RAJA, M., M.C. ALTAVISTA, S. CAVALLARI and L. LUBICH: Neuroleptic malignant syndrome and catatonia. A report of three cases. Eur. Arch. Psychiatr. Clin. Neurosci. 243 (1994) 299–303.

RAO, K.A., A. ADLAKHA and T. MELOY: Tremor, confusion, and autonomic dysfunction of Down syndrome. Hosp. Pract. (Off.) 27 (1992) 215–216.

RATAN, D.A. and A.H. SMITH: Neuroleptic malignant syndrome secondary to droperidol. Biol. Psychiatry 34 (1993) 421–422.

RIVOLTA, M., R. RIEDO, M. OSTALDO, G. FONTANA and F. MOTTA: The use of propofol in a female patient predisposed to malignant hyperthermia (central core disease). Minerva Anesthesiol. 58 (1992) 219–221.

ROBINSON, S. and A.H. ROBINSON: Chemical composition of sweat. Physiol. Rev. 34 (1954) 202–220.

RODNITZKY, R.L. and D.L. KEYSER: Neurologic complications of drugs. Tardive dyskinesias, neuroleptic malignant syndrome, and cocaine-related syndromes. Psychiatr. Clin. North Am. 15 (1992) 491–510.

ROEWER, N.: Malignant hyperthermia today. Anästh. Intensivther. Notfallmed. 26 (1991) 431–449.

ROSENBERG, H. and J.E. FLETCHER: An update on the malignant hyperthermia syndrome. Ann. Acad. Med. Singapore 23 (1994) 84–97.

SAILER, R., B. HINRICHS and K. MANTEL: Malignant hyperthermia in a child with acute lymphatic leukemia. Anaesthesist 40 (1991) 298–301.

SANDERS, B.P. and P.N. TREWBY: The neuroleptic malignant syndrome: a missed diagnosis? Br. J. Clin. Pract. 47 (1993) 170–171.

SARNQUIST, F. and C.P. LARSSON, JR.: Drug induced heat stroke. Anesthesiology 39 (1973) 348–350.

SAVDIE, E., H. PREVEDOROS, A. IRISH, C. VICKERS, A. CONCANNON, P. DARVENIZA and J.R. SUTTON: Heat stroke following Rugby League football. Med. J. Aust. 155 (1991) 636–639.

SCHNEIDERHAN, M.E. and P.A. MARKEN: An atypical course of neuroleptic malignant syndrome. J. Clin. Pharmacol. 34 (1994) 325–334.

SCHULTE-SASSE, U. and H.J. EBERLEIN: Reasons for the persistent lethality of malignant hyperthermia and recommendations for its reduction. Anaesthesiol. Reanim. 16 (1991) 202–207.

SCHWEMMLES, S., K. WOLFF, L.M. PALMUCCI, T. GRIMM, F. LEHMANN-HORN, C. HUBNER, E. HAUSER, D.E. ILES, D.H. MACLENNAN and C.R. MULLER: Multipoint mapping of the central core disease locus. Genomics 17 (1993) 205–207.

SHAW, A. and E.E. MATTHEWS: Postoperative neuroleptic malignant syndrome. Anaesthesia 50 (1995) 246–247.

SHIBOLET, S.B., R. COLL, T. GILAT and E. SOHAR: Heat stroke: its clinical picture and mechanism in 36 cases. Quart. J. Med. 36 (1967) 525–548.

SHIBOLET, S., M. LANCASTER and Y. DANON: Heat stroke: a review. Aviat. Space Environ. Med. 47 (1976) 280–301.

SHIONO, A., M. HAYASHI, H. YAMANAKA, H. YAJIMA and J. KOYA: A case of neuroleptic malignant syndrome with acute renal failure. Hinyokika Kiyo – Acta Urol. Jpn. 38 (1992) 249–252.

SINGH, A.N.: Neuroleptic malignant syndrome. Br. Med. J. 287 (1983) 129.

SPIRIT, M.J., W. CHAN, M. THIEBERG and D.B. SACHAR: Neuroleptic malignant syndrome induced by domperidone. Dig. Dis. Sci. 37 (1992) 946–948.

STARR, A.: Auditory brainstem responses in brain death. Brain 99 (1976) 543–544.

STEINFATH, M., S. SINGH, J. SCHOLZ, K. BECKER, C. LENZEN, F. WAPPLER, A. KOCHLING, N. ROEWER and J. SCHULTE AM ESCH: C1840-T mutation in the human skeletal muscle ryanodine receptor gene: frequency in northern German families susceptible to malignant hyperthermia and the relationship to in vitro contracture response. J. Mol. Med. 73 (1995) 35–40.

STERNBACH, H.: The serotonin syndrome. Am. J. Psychiatry 148 (1991) 705–713.

STOLWIJK, J.A.J. and E.R. NADEL: Thermoregulation during positive and negative work exercise. Fed. Proc. 32 (1973) 1607–1613.

STRAZIS, K.P. and A.W. FOX: Malignant hyperthermia: a review of published cases. Anesth. Analg. 77 (1993) 297–304.

TATSUKAWA, H., J. OKUDA, M. KONDOH, M. INOUE, S. TERASHIMA, S. KATOH and K. IDA: Malignant hyperthermia caused by intravenous lidocaine for ventricular arrhythmia. Arch. Intern. Med. 31 (1992) 1069–1072.

THEOHARIDES, T.C., R.S. HARRIS and D. WECKSTEIN: Neuroleptic malignant-like syndrome due to cyclobenzaprine? J. Clin. Psychopharmacol. 15 (1995) 79–81.

THORNBERG, S.A. and I. ERESHEFSKY: Neuroleptic malignant syndrome associated with clozapine monotherapy. Pharmacotherapy 13 (1993) 510–514.

TORRE-CISNEROS, J., R.A. FERNANDEZ DE LA PUEBLA GIMENEZ, J.A. JIMENEZ PEREPEREZ, J. LOPEZ MIRANDA, J.L. VILLANUEVA MARCOS, A. BLANCO MOLINA and F. PEREZ-

JIMENEZ: The early prognostic assessment of heat stroke. Rev. Clin. Esp. 190 (1992) 439–442.

TORU, M., O. MATSUDA, K. MAKIGUCH and K. SUGANO: Neuroleptic malignant-like syndrome state following a withdrawal of anti-parkinsonian drugs. J. Nerv. Ment. Dis. 169 (1981) 324–327.

TREVES, S., F. LARINI, P. MENEGAZZI, T.H. STEINBERG, M. KOVAL, B. VILSEN, J.P. ANDERSEN and F. ZORZATO: Alteration of intracellular Ca^{2+} transients in COS-7 cells transfected with the cDNA encoding skeletal-muscle ryanodine receptor carrying a mutation associated with malignant hyperthermia. Biochem. J. 301 (1994) 661–665.

TURCYZNSKI, J.: Trudnosci diagnostyczne oraz skutecznosc elektrowstrzasow w leczeniu ostrej smiertelnej katatonii. Psychiatr. Pol. 27 (1993) 535–543.

URWYLER, A. and E. HARTUNG: Malignant hyperthermia. Anaesthesist 43 (1994) 557–569.

URWYLER, A., K. CENSIER, M.D. SEEBERGER, J. DREWE, J.M. ROTHENBUHLER and F. FREI: Diagnosis of susceptibility for malignant hyperthermia using in-vitro muscle contraction testing in Switzerland. Schweiz. Med. Wochenschr. 121 (1991) 566–571.

VAN DER LINDE, A., A.J. KIELBLOCK, D.A. REX and S.E. TERBLANCHE: Diagnostic and prognostic criteria for heat stroke with special reference to plasma enzyme and isoenzyme release patterns. Int. J. Biochem. 24 (1992) 477–485.

VAN HARTEN, P.N. and C.J. KEMPERMAN: Organic amnestic disorder: a long-term sequel after neuroleptic malignant syndrome. Biol. Psychiatry 29 (1991) 407–410.

VANELLE, J.M.: Clozapine: an exclusive treatment? Encéphale 18 (1992) 441–445.

VAUGHN, L.K., W.L. VEALE and K.E. COOPER: Antipyresis: its effect on mortality rate of bacterially infected rabbits. Brain Res. Bull. 5 (1980) 69–73.

VEDIE, C., F. HEMMI and G. KATZ: Atypical neuroleptic malignant syndrome. Encéphale 20 (1994) 355–359.

VELAMOOR, V.R., R.M. NORMAN, S.N. CAROFF, S.C. MANN, K.A. SULLIVAN and R.E. ANTELO: Progression of symptoms in neuroleptic malignant syndrome. J. Nerv. Ment. Dis. 182 (1994) 168–173.

VERWIEL, J.M., B. VERWEY, C. HEINIS, J.E. THIES and F.H. BOSCH: Succesvolle elektroconvulsietherapie bij een zwangere vrouw met het maligne neuroleptica-syndroom. Ned. Tijdschr. Geneeskd. 138 (1994) 196–199.

VETTER, P., D. PROPPE and S. HOPPE-SEYLER: Neuroleptisches malignes Syndrom (NMS) unter Clozapinmonotherapie und benigne Hyperthermie bei abklingendem NMS unter Clozapin. Nervenarzt 62 (1991) 55–57.

VLADUTIU, G.D., K. HOGAN, I. SAPONARA, L. TASSINI and J. CONROY: Carnitine palmitoyl transferase deficiency in malignant hyperthermia. Muscle Nerve 16 (1993) 485–491.

WALKER, M.F.C.: Simulation of tetanus by trifluoperazine overdosage. Can. Med. Assoc. J. 81 (1959) 109–110.

WANG, A.Y., P.K. LI, S.F. LUI and K.N. LAI: Renal failure and heat stroke. Ren. Fail. 17 (1995) 171–179.

WAPPLER, F., N. ROEWER, C. LENZEN, A. KOCHLING, J. SCHOLZ, M. STEINFATH and J. SCHULTE AM ESCH: High-purity ryanodine and 9,21-dehydroryanodine for in vitro diagnosis of malignant hyperthermia in man. Br. J. Anaesth. 72 (1994) 240–242.

WATSON, J.D., C. FERGUSON, C.J. HINDS, R. SKINNER and J.H. COAKLEY: Exertional heat stroke induced by amphetamine analogues. Does dantrolene have a place? Anaesthesia 48 (1993) 1057–1060.

WEDEL, D.J.: Malignant hyperthermia and neuromuscular disease. Neuromusc. Dis. 2 (1992) 157–164.

WEDEL, D.J., J.G. QUINLAN and P.A. IAIZZO: Clinical effects of intravenously administered dantrolene. Mayo Clin. Proc. 70 (1995) 241–246.

WEINBERGER, D.R. and J.J. KELLY: Catatonia and malignant syndrome: a possible complication of neuroleptic administration. Report of a case involving haloperidol. J. Nerv. Ment. Dis. 165 (1977) 263–268.

WEINER, J.S. and M. KHOGALI: A physiological body-cooling unit for treatment of heat stroke. Lancet i (1980) 507–509.

WELLER, M. and J. KORNHUBER: A rationale for NMDA recepter antagonist therapy of the neuroleptic malignant syndrome. Med. Hypotheses 38 (1992a) 329–333.

WELLER, M. and J. KORNHUBER: Clozapine rechallenge after an episode of neuroleptic malignant syndrome. Br. J. Psychiatry 16 (1992b) 855–856.

WELLER, M. and J. KORNHUBER: Clozapine: a neuroleptic at risk of provoking neuroleptic malignant syndrome (NMS) or an alternative neuroleptic with positive NMS case histories? Fortschr. Neurol. Psychiatr. 61 (1993) 217–222.

WHITE, D.A. and A.H. ROBINS: Catatonia: harbinger of the neuroleptic malignant syndrome. Br. J. Psychiatry 158 (1991) 419–421.

WILBUR, E. and J. STEVENA: Morbid anatomic changes following artificial fever, with report of autopsies. South. Med. J. 30 (1937) 286–289.

WILCOX, W.H.: The nature, prevention and treatment of heat pyrexia. Br. Med. J. 1 (1920) 392–397.

WILHELM, K., J. CURTIS, V. BIRKETT and J. KENNEY-HERBERT: The clinical significance of serial creatine phospho-kinase estimations in acute ward admissions. Aust. N.Z. J. Psychiatry 28 (1994) 453–457.

WYNDHAM, C.H.: A survey of the causal factors in heat stroke and their prevention in gold mining industry. J. S. Afr. Inst. Mining Metallurg. 1 (1966) 245–258.

WYNDHAM, C.H.: The physiology of exercise under heat stroke. Annu. Rev. Physiol. 35 (1973) 193–220.

WYNDHAM, C.H., N.B. SYNDROM and H.M. COOKE: Methods of cooling subjects with hyperpyrexia. J. Appl. Physiol. 14 (1959) 771–776.

XERRI, B., M.M. LEFEVRE and J.B. PAOLAGGI: Neuroleptic malignant syndrome induced by a single injection of tiapride. Rev. Rhum. Mal. Ostéoartic. 61 (1994) 362.

YAMAZAKI, K., S. KATAYAMA, T. IWAI and K. HIRATA: A case of malignant syndrome with leukoencephalopathy due to cocaine abuse. Rinsho Shinkeigaku – Clin. Neurol. Jpn. 34 (1994) 582–586.

YAMAZAKI, Y. and N. OGAWA: Successful treatment of levodopa-induced neuroleptic malignant syndrome (NMS) and disseminated intravascular coagulation (DIC) in a patient with Parkinson's disease. Arch. Intern. Med. 31 (1992) 1298–1302.

YAQUB, B.A.: Neurologic manifestations of heatstrokes at the Mecca pilgrimage. Neurology 37 (1987) 1004–1006.

YAQUB, B.A.: Brainstem auditory evoked potentials in heat stroke: diagnostic and prognostic value. Saudi Med. J. 9 (1988) 49–53.

YAQUB, B.A., S.S. AL-HARTHI, I.O. AL ORAINEY, M.A. LAAJAM and M.T. OBEID: Heat stroke at the Mekkah pilgrimage: clinical characteristics and course of 30 patients. Quart. J. Med. 59 (1986) 523–530.

YAQUB, B.A., A.K. DAIF and C.P. PANAYIOTOPOULOS: Pancerebellar syndrome in heat stroke: clinical course and CT scan findings. Neuroradiology 29 (1987) 294–296.

YAQUB, B.A., T. OBEID and M.A. SERAJ: Neuroleptic malignant syndrome and heat stroke: spectrum of a clinical syndrome. Saudi Med. J. 10 (1989) 254–259.

ZELMAN, S. and R. GUILLAN: Heat stroke in phenothiazine treated patients: a report of three fatalities. Am. J. Psychiatry 126 (1970) 1787–1790.

ZOHAR, Y., Y.P. TALMI, R. SABO, Y. FINKELSTEIN and A. KORZETS: Neuroleptic malignant syndrome during perphenazine treatment in a patient with head and neck cancer: a case report. Otolaryngol. Head Neck Surg. 106 (1992) 206–208.

IV. Oncology

Handbook of Clinical Neurology, Vol. 27 (71): Systemic Diseases, Part III
M.J. Aminoff and C.G. Goetz, editors

CHAPTER 25

Direct effects of systemic tumors or their metastases on the nervous system

CASILDA BALMACEDA and MICHAEL R. FETELL

The Neurological Institute, Columbia Presbyterian Medical Center, New York, NY, U.S.A.

Systemic malignancy may affect the nervous system in several different ways: direct spread of tumor and compression of neural structures; neurological complications of radiotherapy; complications of drug treatment of the systemic tumor; or paraneoplastic disorders. In a patient with systemic cancer, one may also see cerebrovascular episodes, central nervous system (CNS) infections and both metabolic and nutritional disorders that may have neurologic sequelae.

EPIDEMIOLOGY

Systemic tumors frequently spread to the CNS. Up to 35% of all patients with systemic cancer have CNS metastases at some time. Among the 1 million newly diagnosed systemic tumors each year in the US, about 137,000 metastasize to the CNS. Melanoma has the highest propensity to spread to the CNS, with rates of up to 65%. Other tumors that commonly metastasize to the brain are carcinomas of the breast and lung.

Metastasis to the parenchyma of the brain is the most common neurological complication of cancer. On average, up to 50% of patients with cancer show brain metastases at autopsy (Pickren et al. 1983). The relative incidence of brain metastases may be increasing (Aisner 1992; Galicich et al. 1996) as a result of improved treatment of the primary neoplasm. Also, longer overall survival may permit tumors to grow in sanctuary sites such as the brain. Brain metastases are far more common than primary brain tumors, and compared to the 12,000 patients who die with primary brain tumors each year (Boring et al. 1993), more than 10 times as many (about 130,000 patients) with systemic cancer die with brain metastases every year (Posner and Chernik 1978).

OCCURRENCE OF CNS METASTASES IN RELATION TO THE DIAGNOSIS OF THE PRIMARY TUMOR

Usually, a brain metastasis appears after the systemic tumor has been identified and treated. In these circumstances, when there is an antecedent primary tumor, the brain metastasis is termed *metachronous*. Rarely, the brain metastasis appears synchronously with the systemic tumor. Exceptionally, the CNS tumor is discovered first and is found to be metastatic ('precocious CNS metastasis'). Brain metastases may be single or multiple. In most series, up to 50% of brain metastases are single. The term *solitary metastasis* refers to the brain as the only site of spread.

LOCATION/GENERAL DISTRIBUTION

Some systemic tumors rarely metastasize to the brain. Prostate carcinoma frequently affects the vertebrae, compressing the spinal cord; only rarely does a prostate tumor spread to the brain or spinal cord parenchyma. Somewhat more likely is prostate metastasis to the dura with secondary cerebral compression.

Dural metastases occur with prostate carcinoma, breast carcinoma and sarcomas. In the brain, 66% of systemic tumor spread to the CNS occur in the parenchyma (Zimm et al. 1981); metastasis to leptomeninges, calvarium or dura is less common. In the spine, however, the vertebrae are most frequently affected; leptomeningeal infiltration is uncommon and intramedullary metastasis is rare.

Parenchymal

CT studies reveal that about 49% of patients have a single brain metastasis; 20% have two lesions and 13% have three or more (Delattre et al. 1988a). MRI with gadolinium may detect multiple lesions that were otherwise hidden by edema from a large single lesion (Healey et al. 1987; Russell et al. 1987) Therefore, up to 70% of brain metastases can be treated by focal therapy. With parenchymal metastases, the distribution parallels blood flow to different regions of the brain; only around 3% of the tumors affect the brainstem, 15% involve the cerebellum, and the remainder are supratentorial. Some primary tumors show a predilection for particular areas of the brain: pelvic tumors (uterus, colon, or prostate) tend to metastasize to the cerebellum, and breast tumors spread to the pituitary gland (Takakura et al. 1982; Delattre et al. 1988a).

Tumors of the lung, breast, or melanoma, by metastasizing early to the lungs, acquire access to the systemic circulation. They have the highest incidence of parenchymal brain metastases and brain metastases tend to be multiple (Takakura et al. 1982; Byrne et al. 1983; Boogerd et al. 1993) (Fig. 1).

Dural/subdural

Calvarial and dural lesions are most common in breast and prostate cancer, melanoma, leukemia, neuroblastoma or lymphoma (Demierre and Bernay 1983; Taylor et al. 1984). Neurological symptoms may be produced by compression of the superior sagittal sinus or by compression of the cranial nerves (Greenberg et al. 1981). Rarely, the tumor is large enough to compress the adjacent parenchyma. (Fig. 2 shows an exceptional case of a patient with dural metastasis due to colon carcinoma.) The dura usually acts as a barrier to prevent spread from one compartment to another.

Leptomeningeal

Leptomeningeal metastases occur most commonly in patients with leukemia, lymphoma, melanoma and breast or lung carcinoma (Posner and Chernik 1978; Wasserstrom et al. 1982). Fig. 3 shows extensive leptomeningeal metastases in a patient with breast cancer. Leptomeningeal metastases may occur in isolation, or in conjunction with parenchymal or dural metastases. They are discussed in Chapter 26 of this volume.

Calvarial/base of skull

Metastases to the base of the skull arise in a different pattern from intracerebral metastases. The most common primary tumor that metastasizes to the skull base is breast carcinoma, followed by lung and prostate carcinomas (Greenberg et al. 1981). Breast cancer produces base of skull metastases after a long interval (66 months), whereas on average lung carcinoma produces symptoms after a short interval (3 months). At the time of diagnosis of a skull base lesion, 19/43 (44%) patients have had bone metastases somewhere else. Five typical syndromes are recognized, based primarily on nerve entrapments:

- *Orbital syndrome:* consisting of supraorbital pain, blurred vision, diplopia with either proptosis, ophthalmoplegia, or decreased sensation in V1.
- *Parasellar syndrome:* presents with frontal headaches and ophthalmoparesis. Papilledema is present in about one-third of the patients.
- *Middle fossa syndrome:* is seen with involvement of the Gasserian ganglion or branches of the trigeminal nerve. The most common symptom is sensory loss or pain in the V2 and V3 distribution; lancinating trigeminal neuralgia-like pain is uncommon.
- *Jugular foramen syndrome:* unilateral occipital/ postauricular or glossopharyngeal pain is followed by hoarseness and dysphagia. Glossopharyngeal neuralgia may rarely be accompanied by syncope (Vernet 1916, 1918). By the time of diagnosis, most patients have evidence of glossopharyngeal, vagus and accessory nerve dysfunction.
- *Occipital condyle syndrome:* the clinical picture is characteristic. Occipital pain is worse on flexion of the neck. Palpation over the occiput elicits tenderness and there is a XII nerve palsy.

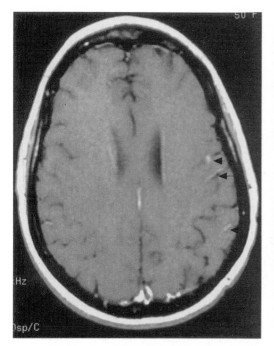

A

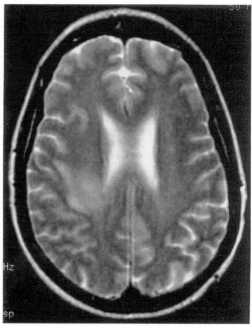

B

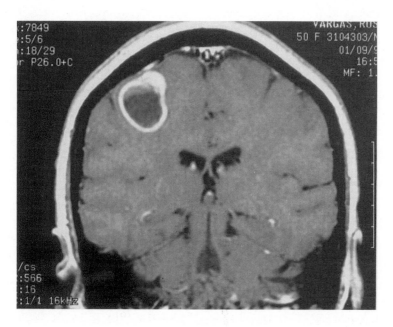

C

Fig.1. Parenchymal brain metastasis of non-small-cell lung carcinoma. (A) T1 gadolinium-enhanced MRI scans show metastases within the sulci in the left fronto-parietal region (arrows). (B) In the same patient, T2 MRI scan shows no abnormality in the region of the sulcal metastases, but increased signal in the right parietal region due to peri-tumoral edema around a cystic brain metastasis seen in (C) on coronal T1 gadolinium-enhanced images. Note that the ring-enhancing lesion is thin walled and simulates an abscess but for the nodular enhancement in the superior medial quadrant. This nodule of tumor differentiates this brain metastasis radiographically from inflammatory disease.

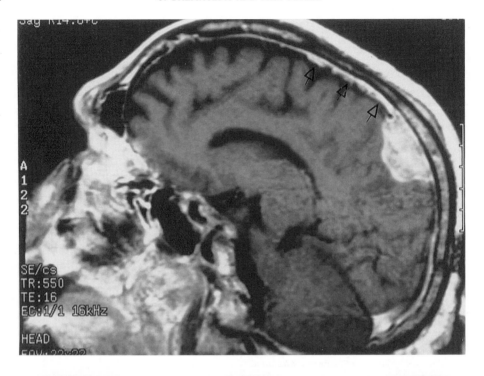

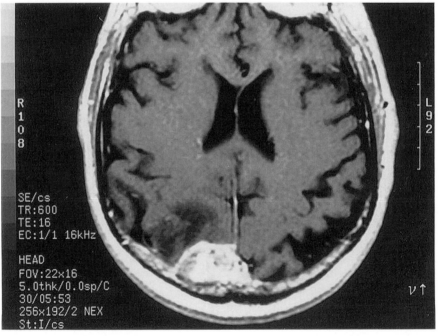

Fig. 2. Dural metastasis of colon carcinoma. (A) T1 gadolinium-enhanced MRI scan shows diffuse meningeal thickening (arrows) and a dural based metastasis compressing the underlying cortex and possibly invading
brain. There is no calvarial invasion. (B) Axial MRI of the same patient shows edema subjacent to the dural
metastasis.

Early localization and treatment are essential. If symptoms are noted for less than a year, up to 87% improve. Only 25% will improve if symptoms were present longer (Vikram and Chu 1979). For the clinician, presentation with any of the above syndromes should raise concern about metastatic disease, even if the patient's primary cancer has been cured. CT is generally more useful than MRI in evaluating the bone destruction caused by tumors at the base of the skull, but occasionally MRI, MRA (magnetic resonance angiogram), MRV or angiography may define associated vascular compromise.

Perineural

The ability of slowly growing tumors to invade the nervous system by spreading along nerve fibers is underappreciated. Perineural spread usually affects the trigeminal nerve, facial nerve, greater auricular nerve, or oculomotor nerve (Morris and Joffe 1983). Squamous carcinoma, basal cell carcinoma and carcinoma of the minor salivary glands are the most common tumors that produce this syndrome. The tumor first invades small peripheral branches of a cranial nerve distally, then infiltrates centripetally along the nerve, and may reach the ganglion in the case of the trigeminal or facial nerves, or the cavernous sinus in the case of the oculomotor nerve. Wide excision of tumor before it spreads is desirable; radiotherapy seems to retard the progression of symptoms in most cases (Trobe et al. 1982; Silbert et al. 1992; Schifter and Barrett 1993; Hayat et al. 1995; Majoie et al. 1995).

Pituitary

Pituitary metastases are uncommon, but not when measured against the overall volume of the gland. Breast carcinoma accounts for 50% of the cases. Pituitary metastases are often found incidentally at autopsy or when a hypophysectomy is performed for pain relief in a patient with breast carcinoma. They are seen at autopsy in up to 1.8% of cancer patients (Abrams et al. 1950) and in up to 9% of patients with breast cancer. Up to 4% of patients with widely metastatic carcinoma have pituitary metastases at autopsy, but this figure rises to 15% of those with metastatic breast carcinoma (Hagerstrand and Shonebeck 1969). The anterior pituitary gland has no direct arterial supply; it is vascularized via the hypothalamo-hy-pophyseal portal system. Almost 80% of metastases affect the posterior pituitary or infundibulum, which derive an abundant blood supply from the inferior hypophyseal arteries (Teears and Silverman 1975; Nelson et al. 1987). About 1% of patients in large surgical series treated with transsphenoidal microsurgery for removal of a symptomatic pituitary tumor had metastatic carcinoma. In contrast to incidental pituitary metastases which are microscopic, symptomatic cases often show a characteristic triad of diabetes insipidus, headache and visual loss (Max et al. 1981). A pituitary mass in a patient with diabetes insipidus and known carcinoma should be considered metastatic until proven otherwise. Metastasis to the pituitary gland can be differentiated from adenoma because symptoms evolve in days or weeks. In exceptional cases a carcinoma metastasizes to a pituitary adenoma (Post et al. 1988). Such 'tumor-to-tumor' metastases are thought to be coincidental.

Pineal

Like pituitary metastases, pineal spread is usually discovered as an incidental finding at autopsy of patients with systemic malignancy and extensive metastatic spread. Rarely, pineal metastases are symptomatic, with symptoms of hydrocephalus (Ortega et al. 1951; Weber et al. 1989; Delahunt et al. 1990; Fetell and Stein 1994). Leptomeningeal seeding frequently accompanies the symptomatic pineal metastasis, because the tumor sits in the suprasellar cistern and sheds cells into the subarachnoid space (Fetell 1988).

Spinal

Spinal metastases can be classified, depending on the site of involvement, as epidural, leptomeningeal, intramedullary, or vertebral. Leptomeningeal metastases are discussed in Chapter 26 of this volume, and this discussion will be limited to vertebral, epidural and intramedullary metastases.

Plexus

In patients with cancer, the most common causes of plexopathies are direct tumor infiltration and injury from radiation therapy. True isolated peripheral nerve metastases are exceedingly rare (Meller et al. 1995). Lung and breast carcinoma are the most commonly

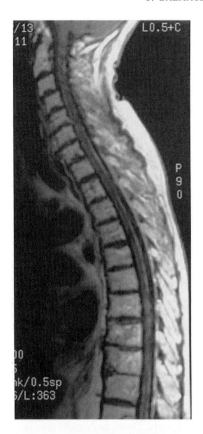

A

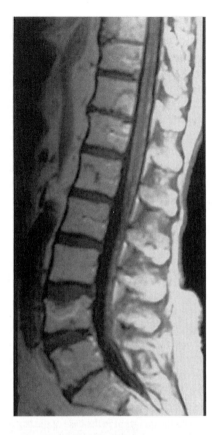

B

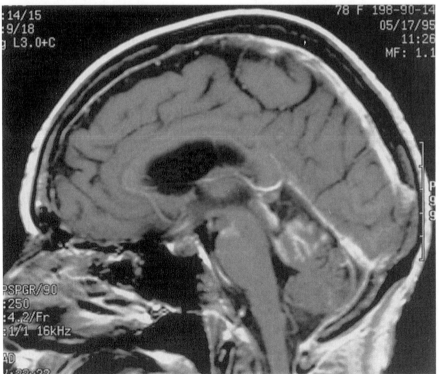

C

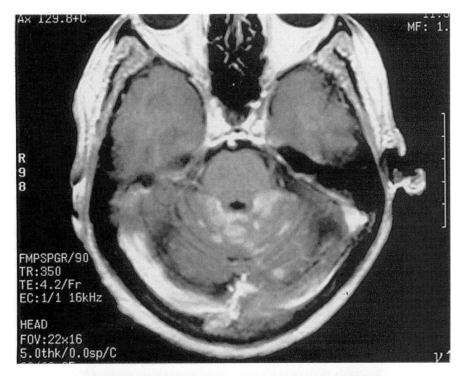

D

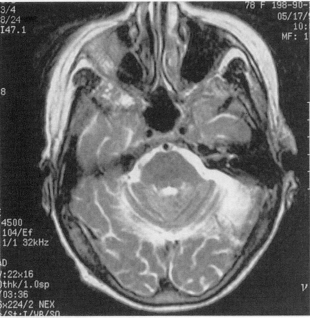

E

Fig. 3. Leptomeningeal metastases of breast carcinoma. (A) T1 gadolinium-enhanced MRI scan of thoracic spine and cervical spine shows diffuse enhancement (of leptomeningeal tumor) on the surface of the cord in a patient with widely metastatic breast carcinoma. (B) T1 gadolinium-enhanced MRI scan of lumbosacral region shows similar findings. (C) T1 gadolinium-enhanced scan of head shows tumor in the interpeduncular region and in the sulci of the cerebellum, and along the subarachnoid space in the posterior fossa. (D) Axial MRI showing contrast-enhancing tumor in the superior cerebellum. (E) This tumor is not seen on this T2 MRI scan, illustrating the utility of contrast enhancement.

associated malignancies (Kori et al. 1981). The presence of pain, involvement of the lower (C7–8, T1) plexus, radiation doses less than 60 Gy and development of symptoms more than 1 year after radiation are all associated with tumor infiltration (Kori et al. 1981). Horner's syndrome is commonly associated with tumor infiltration, whereas lymphedema is usually observed with radiation plexus injury. Widespread myokymia on EMG suggest radiation induced plexopathy. Pain relief in patients with tumor infiltration can be achieved with radiation in at least two-thirds of the patients (Ampil 1985). Surgical exploration of the plexus is sometimes needed to establish the diagnosis (Kori et al. 1981).

PRIMARY TUMOR HISTOLOGIES

Systemic neoplasms vary markedly in their propensity to metastasize to the brain (Table 1, Galicich et al. 1996). In the US, lung carcinoma is the most common source of brain metastases, followed by melanoma, breast carcinoma and leukemia. Some tumors rarely spread to the brain ('neuronophobic tumors') including prostate, GU malignancies, some

sarcomas, and malignancies from the endocrine glands or digestive tract. In about 10% of cases of brain metastases, despite an exhaustive search, the primary site remains unidentified (Tomlinson et al. 1979; Voorhies et al. 1980). In many such cases, a small lung tumor is ultimately found (Debevec 1990).

Melanoma

Up to 46% of patients with malignant melanoma develop brain metastases (Amer et al. 1978; Katz 1981). Brain metastases may be seen most commonly in patients who have responded to systemic therapy, possibly because they survive longer (Legha et al. 1990), usually 27–44 months following diagnosis (Saha et al. 1989). Up to 96% of patients have systemic disease at the time of diagnosis of brain metastases (Byrne et al. 1983). 75% of melanoma metastases to the brain are multiple, and 33–50% show hemorrhage. Deeper primary lesions (Clark's level III–IV or Breslow stage > 3 mm) carry increased risk for metastases. In addition, men and patients with primary lesions in the region of head, neck, or shoulders are at increased risk (Saha et al. 1989).

TABLE 1

Frequency and number of estimated cases of intracranial and brain metastases (USA).

Type of malignancy	No. of expected deaths (1993)	Intracranial metastases		Brain metastases	
		Frequency (%)	n^*	Frequency (%)	n^*
Lung	149,000	41	61,000	35	52,150
Colon, rectum	57,000	8	4,500	6	3,400
Breast	46,300	51	23,500	21	9,600
Liver, pancreas	37,600	6	2,200	1	400
Prostate	35,000	17	6,000	6	2,100
Female genital	24,400	7	1,700	2	500
Esophageal, stomach	24,000	8	1,900	6	1,400
Urinary tract	20,800	21	4,400	17	3,500
Lymphoma	20,500	22	4,500	5	1,000
Leukemia	18,600	48	8,900	8	1,500
Head and neck	11,500	18	2,000	7	800
Melanoma	6,800	65	4,400	49	3,300
Sarcoma	4,150	10	400	5	225
Thyroid	1,050	2.5	25	1.5	15
Others	69,300	5	3,500	2.5	1,750
Total	526,000	25	128,925	15	81,240

*Derived by multiplying frequency by number of expected deaths.
(Reproduced, with permission, from Galicich et al. 1996.)

Lung

Up to 50% of all brain metastases in the US arise from a lung primary (Arbit and Wronski 1995). As the duration of survival from small-cell lung cancer increases, so does the incidence of brain metastases: 10% of patients at diagnosis have brain metastases; an additional 20% develop them during treatment; and up to 50% of patients who survive for 2 years develop brain metastases, approaching the figure of 50% seen at autopsy (Pedersen 1986). Adenocarcinoma of the lung and undifferentiated carcinoma are more likely to metastasize to the brain than squamous cell carcinoma (Burt et al. 1992). About 28% of patients with adenocarcinoma of the lung develop clinical brain metastases, and up to 49% have metastases detected at autopsy (Sorensen et al. 1988). Since 11% of the patients scheduled for surgical resection of lung tumors have asymptomatic brain metastases (Jacobs et al. 1977; Kormas et al. 1992), MRI or CT of the head is strongly recommended in the preoperative evaluation of patients with non-small-cell carcinoma of the lung (NSCCL) (Grant et al. 1988).

Breast

Breast carcinoma is the most common source of brain metastases in women (Zimm et al. 1981; Flowers and Levin 1993), and second only to lung carcinoma as the most common cause of brain metastases overall (Galicich et al. 1996). Up to 20% of women with breast carcinoma develop brain metastases (Di Stefano et al. 1979; Tsukada et al. 1983) and in contrast to brain metastases from other solid tumors such as lung, approximately two-thirds of patients with brain metastases from breast cancer die of the intracranial disease (Zimm et al. 1981; Tsukada et al. 1983). Brain metastases from breast carcinoma may affect several intracranial compartments at the same time, such as the parenchyma, dura, leptomeninges, or a combination of these sites (Kiricuta et al. 1992). Breast carcinoma is the most common cause of leptomeningeal carcinomatosis (Boogerd et al. 1991). A dural-based lesion in a woman with breast carcinoma may be a dural metastasis or a meningioma. Meningiomas are disproportionately frequent in women with breast carcinoma (Bonito et al. 1993). Biopsy or serial imaging studies may be required to clarify the diagnosis.

Gynecologic malignancies

In general, metastasis to the CNS from gynecologic tumors is uncommon, but is thought to be increasing in incidence (Leroux et al. 1991). Despite the propensity to spread to other organs, ovarian carcinoma rarely spreads to the CNS, accounting for only about 1% of all brain metastases (Mayer et al. 1978; Kottke-Marchant et al. 1991). The most common type of gynecologic malignancy metastasizing to the brain is choriocarcinoma, and in some series it accounts for up to 35% of all brain metastases (Adeloye 1982). In 75% of cases there is stage IV disease at the time brain metastases are detected (Rodriguez et al. 1992). Endometrial and cervical carcinoma rarely spread to the CNS (Kottke-Marchant et al. 1991).

Endometrial and cervical carcinoma rarely spread to the CNS (Kottke-Marchant et al. 1991). Cerebral metastases from endometrial carcinoma are rare, seen in less than 2% of patients. In one case pituitary apoplexy was the sole manifestation of the disease; at autopsy, however, there was widespread tumor in many organs. In most cases, however, brain metastases are discovered after the primary diagnosis is known. Brain metastases occur when there is diffuse involvement of many organs, but lung and liver may be spared. This pattern has also been noted in metastases from choriocarcinoma. Survival up to 7 years after diagnosis is much greater than that of cervical or ovarian cancer. Some authors thus justify an aggressive approach to all brain metastases from endometrial carcinoma.

Colon

Carcinoma of the colon is an infrequent cause of brain metastases, seen in up to 4% of patients with brain metastases (Richards and Mikossoch 1963; Weisberg 1970; Potts et al. 1980; Cascino et al. 1983). Most of the patients have extensive systemic disease at the time of presentation, and up to 85% have evidence of pulmonary metastases (Cascino et al. 1983). Lesions in the posterior fossa are over-represented, occurring in up to 35% of the patients (Cascino et al. 1983). In patients with colon carcinoma and liver metastases, the incidence of brain metastases is lower, possibly related to a filtering effect of the liver (Chyun et al. 1980). A common finding on CT is a hyperdense lesion (Cascino et al. 1983). On MRI, it may appear as iso- to hyperintense on T1-weighted images, and hypointense on T2-weighted images. The T2 shortening is thought to be secondary to coagulative necrosis within the tumor (Kovalikova et al. 1987), and not due to hemorrhage, fibrosis, iron, or calcification. Brain metastases from colon carcinoma are relatively radioresistant to conventional radiation (Cascino et al. 1983).

Carcinoma of unknown primary tumor

In about 8% of patients who present with brain metastasis, the initial search fails to identify a primary tumor. The site of origin is eventually identified in only 25% of these cases, and in half the primary tumor is a tiny carcinoma of the lung. In a retrospective study of 43 patients with brain metastases due to an unknown primary, 68% died with the brain as the sole site of disease (Eapen et al. 1988). In contrast to clinically evident systemic tumors, morbidity and mortality are related to the brain metastasis in patients with occult primary tumors, and aggressive treatment of the brain lesion is indicated.

DIAGNOSIS

Clinical diagnosis

Symptoms of brain metastasis are similar to those seen with any space-occupying lesion – headaches, seizures and focal neurological signs. The clinical signs and symptoms of brain metastases are indistinguishable from those of other intracranial expanding masses (Posner 1992) and occur with the following frequency: headaches (53%), motor impairment (66%), mental changes (31%), seizures (15%), ataxia (20%), and papilledema (26%) (Posner and Chernik 1978; Zimm et al. 1981; Cohen et al. 1988). Multiple brain metastases can mimic a progressive encephalopathy or a progressive dementia because headache and seizures are absent (Madow and Alpers 1951).

Headache with brain metastases may be caused by a variety of factors: tumor mass, peritumoral edema, hydrocephalus from posterior fossa metastases, sinus thrombosis due to dural metastases, or leptomeningeal metastasis. Pain-sensitive structures vulnerable to distortion by a metastatic tumor mass include proximal dural arteries, the dura at the base of the skull, the tentorium, and cranial nerves V, VII, IX,

and XI (Ray and Wolff 1940; Rushton and Rooke 1962; Jaeckle 1993). Distortion of distant pain-sensitive structures probably explains why larger tumors are more likely to produce headache, but the degree of peritumoral edema does not correlate with headaches (Forsyth and Posner 1992).

Seizures are the first symptom of 15–20% of brain metastases, but up to 10% additional patients will develop seizures later in their course (Posner 1977; Zimm et al. 1981; Cohen et al. 1988). They occur more frequently in patients with multiple metastases or frontal or temporal lesions. As with primary brain tumors, seizures in patients with brain metastases cause focal symptoms that relate to the area of the brain affected. The type of primary tumor does not seem to correlate with epileptogenicity (of the brain metastasis) (Cohen et al. 1988). For seizures developing early after diagnosis, the best predictor is the presence of a frontal lesion, whereas the occurrence of late seizures correlates with focal hemispheric dysfunction and the presence of multiple hemispheric metastases (Cohen et al. 1988).

A small brain metastasis may be surrounded by marked edema that causes a rapidly deteriorating neurological picture (Fig. 4). Spread of the edema through the white matter tracts, rather than increasing size of the tumor per se, probably accounts for the rapid deterioration. Focal signs and symptoms such as hemiparesis or aphasia are due to direct parenchymal involvement by the tumor, while more generalized symptoms (changes in mental status, headache) result from the edema (Jaeckle 1993). About 10% of the patients with brain metastases have a sudden stroke-like episode (Mandybur 1977), often caused by intratumor hemorrhage or sudden obstruction of CSF flow (Cohen et al. 1988).

Cerebellar metastases, because of markedly increased pressure within the posterior fossa, may present as a neurological emergency requiring immediate surgical intervention (Kitaoka et al. 1990).

Radiographic diagnosis

MRI following the administration of a contrast agent (gadolinium dyethylenetriaminepentaacetic acid – Gd-DTPA) is the most sensitive diagnostic tool for detection of either intraparenchymal or leptomeningeal brain metastases and is more sensitive than double-dose contrast-enhanced CT, particularly for

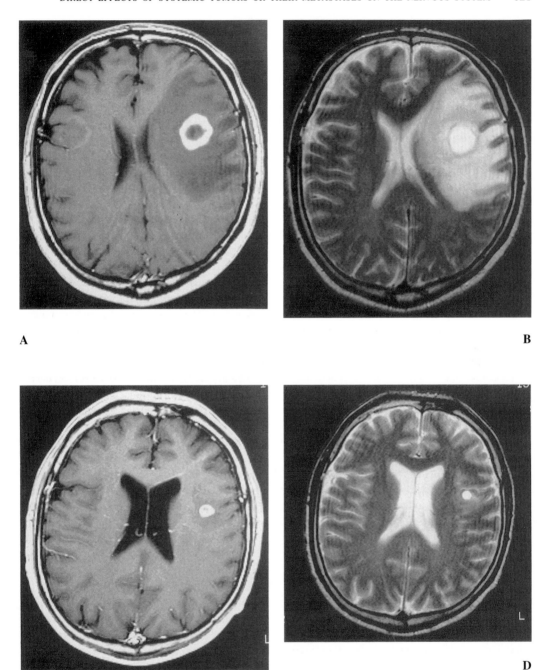

A B

C D

Fig. 4. Effects of stereotactic radiosurgery. (A) T1 gadolinium-enhanced MRI scan of patient with renal carcinoma metastatic to left frontal region. Note the contrast-enhancing ring lesion in the frontal region with dramatic white matter edema best seen on T2 images extending into the subcortical region in a finger-like fashion
(B). Following an initial treatment with stereotactic radiosurgery, the patient received whole brain radiation
therapy. There was marked reduction in the size of the lesion as shown in C and a dramatic reduction in the
surrounding edema (D) 6 months after the therapy was administered.

lesions in the posterior fossa (Fadul et al. 1987). It is also better than CT in differentiating intraparenchymal from dural-based lesions (Davis et al. 1991).

Gadodiamide (Omniscan) is a non-ionic gadolinium chelate, with lower osmolality than Gd-DTPA (Magnevist), and Gd-DOTA (Dotarem). A high-dose gadodiamide technique detected up to 60% more of the small (5 mm diameter) metastases and gave better delineation of the metastases than standard-dose technique (Akeson et al. 1995). Neither the optimal dose of contrast nor the ideal contrast agent for MRI, however, has been determined. The improved sensitivity of high-dose contrast MRI may be cost effective (Mayr et al. 1993), identifying multiple lesions in patients who otherwise would be considered candidates for surgical therapy.

Because of partial volume effects, small lesions require a larger dose of contrast to be detected, and lower doses of contrast agent may underestimate true lesion size (Yuh et al. 1992). Thus high-dose gadolinium injection is advantageous when a single metastasis is detected on standard-dose MRI (particularly if the patient is being considered for resection of the lesion), choosing the ideal lesion for biopsy, or differentiating a posterior fossa brain metastasis from artifact. The utility of high-dose studies in neurologically asymptomatic patients has not been determined. High-dose contrast-enhanced MRI provides additional information that may result in treatment changes in 41% of patients, identifying those who would be less likely to benefit from a craniotomy (Mayr et al. 1993).

Most brain metastases appear as round, well-circumscribed lesions, hypointense on T1-weighted images, and hyperintense on T2. Up to 90% enhance (become hyperintense) after administration of gadolinium contrast. Brain lesions as small as 3 mm can be detected on the contrast-enhanced T1 images, but may be invisible on the pre-contrast examination, emphasizing the need for a contrast study before excluding the presence of brain metastases (Fig. 3E). T2 images are useful for identifying hemorrhagic metastases and visualizing edema surrounding a small metastasis that may otherwise be mistaken for an infarct. Enhancement may be heterogeneous; for example, a dark area in the center of the tumor often represents necrosis within the lesion. A ring-like pattern of enhancement, common in brain metastases, is also seen in brain abscess, but abscesses typically have thinner

rims of contrast enhancement and may show a target-type sign. In general, it is not possible to determine the histology of the brain metastases by MRI characteristics alone. Malignant melanoma is an exception: the lesions are usually isointense to hyperintense on T1-weighted images, and hypointense on T2-weighted images (Atlas et al. 1987; Woodruff et al. 1987). Cerebral metastases from colon carcinoma may display a characteristic hypointensity on T2-weighted images (Suzuki et al. 1993), thought to represent coagulation necrosis (Sillerud et al. 1990)

Some tumors, particularly those of neuroectodermal origin seem to spread to the brain with little or no reaction in surrounding tissue. This syndrome has been called 'miliary brain metastases', referring to a diffuse perivascular distribution of lesions with multiple small tumor nodules. The accompanying clinical syndrome is one of encephalitis without focal neurological signs. The lesions may not even be evident on contrast-enhanced T1 MRI, appearing only as foci of increased T2 signal (Nemzek et al. 1993). Lesions may appear more readily on T2-weighted images, as both edema and tumor have similar signal characteristics, and the lesions surrounded by the edema may appear larger. Areas of edema sometimes show tumor cells on biopsy (Greene et al. 1989).

MRI is limited because it is more expensive than CT, difficult to perform on obtunded patients who are constantly monitored, and cannot be performed on ventilator-dependent patients. Patient with pacemakers or ferromagnetic implants cannot undergo MRI. MRI may not detect early hemorrhage as well as CT (Davis et al. 1991). CT is also better than MRI in detecting lytic bony lesions of the calvarium or base of the skull, while sclerotic or blastic lesions are difficult to detect with either modality (Jaeckle 1991). CT is inferior to MRI in detecting small brain metastases, cortical brain metastases, and posterior fossa brain metastases. When there are multiple brain metastases, MRI usually reveals more lesions than CT (Healey et al. 1987; Russell et al. 1987).

Although brain metastases from lung carcinoma are found in 17–55% of all patients (Richards and Mikossoch 1963), it remains controversial whether asymptomatic patients with lung carcinoma should have cerebral imaging studies (Cole et al. 1994). Brain metastases were detected in none of 30 neurologically asymptomatic patients with lung cancer who had CT and MRI (Cole et al. 1994), arguing strongly against 'prophylactic' brain imaging. Some patients with lung carcinoma are at higher risk and warrant studies: those with (1) mediastinal or disseminated disease (Floyd et al. 1966; Jelenik et al. 1990); (2) adenocarcinoma histology; or (3) non-specific systemic symptoms.

Up to 14% of brain metastases show associated hemorrhage, which may be the first manifestation of the brain metastasis, mimicking a stroke. Hemorrhage occurs with a larger variety of histologic types, although choriocarcinoma, melanoma, lung carcinoma and renal cell carcinoma are the most common. Usually the hemorrhage into the tumor bed results in a focal intraparenchymal hematoma, but it may be massive and may result in subarachnoid, intraventricular, or subdural hemorrhage depending upon the location of the brain metastases. Multiple sites of intracerebral hemorrhage argue strongly in favor of metastatic disease (Gildersleeve et al. 1977), but when a patient without a known malignancy has a single hemorrhagic lesion the differential diagnosis includes primary intracerebral hematoma, primary CNS neoplasm, or a single brain metastasis. Metastasis is more likely than glioma, because hemorrhage into primary gliomas occurs in only 3.7% of cases (Mandybur 1977). It is easy to perform a rapid screen for one of the primary tumors with a chest and abdominal CT, serum β human chorionic gonadotropin, and skin examination. If this is unrevealing, it is reasonable to resect the lesion and to perform neuropathologic diagnosis. Often the hematoma obscures the underlying tumor. Hemorrhagic tumors are aggressive in growth rate and contain numerous abnormal vessels. The survival of these patients is generally poor (Bitoh et al. 1984).

Hemorrhage into a brain metastasis in a patient with a known metastatic tumor is not unusual and may be seen even in the absence of thrombocytopenia or coagulopathies (Graus et al. 1985a, b). When the platelet count is less than 20,000/mm³, hemorrhage may occur randomly in areas of the brain unaffected by tumor spread.

Calcification, rarely seen in brain metastases, may be detected by CT, an appearance that may simulate hemorrhage. This may be seen with colon, lung, and breast carcinomas (Anand and Potts 1982).

Angiograms or MRA (magnetic resonance angiogram) play a small role in the evaluation of brain metastasis except in the preoperative evaluation of vascular tumors such as renal or thyroid carcinoma.

In these tumors, angiography may show the relationship of the tumor to major blood vessels for surgery, or help distinguish hemorrhagic metastases from vascular malformations or other causes.

Positron emission tomography (PET) using the isotope fluorine-18 fluorodeoxyglucose has been used to help distinguish recurrent tumors and radiation necrosis or post-surgical change. Sometimes it can separate low grade tumor and those of higher grade. It cannot, however, distinguish primary and secondary lesions.

EVALUATION OF PATIENTS WITH BRAIN METASTASES

In the patient without a known primary tumor and suspected brain metastases, a rapid and focused screening workup to find the systemic tumor is appropriate. It is preferable to make the diagnosis by biopsy of the primary tumor site, because subsequent therapy is determined by the primary tumor histology. Because most brain metastases originate from either a primary lung tumor or a lesion that has metastasized to the lung, attention is first directed to the chest; chest radiogram is done first and, if negative, chest CT. Physical examination should address potential lesions in the breasts, testicles, prostate, or rectum. Stool should be examined for occult blood (Merchut 1989). If chest CT is normal, the abdomen/pelvis may be examined by CT; if there are no gastrointestinal symptoms, endoscopy is not advised since it is uniformly unrevealing. If the imaging studies suggest a melanoma, it is important to look for acral lentiginous melanoma on the palms and soles, a form of melanoma more common in dark-skinned individuals. Melanoma can also arise from obscure sites such as the eye (uveal tract) or subungual areas. In addition, the primary site may be an amelanotic melanoma or may be a regressed primary lesion. Elevation of the CEA (carcinoembryonic antigen) suggests a bowel malignancy.

If the screening evaluation is unrevealing and there is a single surgically accessible lesion, resection is performed for diagnosis and treatment (see below). When preliminary screening evaluations are unrevealing and multiple lesions are present, however, stereotactic brain biopsy is preferred for making the diagnosis of brain metastases. Stereotactic biopsy is a simple and relatively safe procedure. Making the diagnosis of metastatic tumor is much easier than determining the site of origin. In most cases the biopsy reveals tumor, but the neuropathologist often cannot distinguish the site of origin, e.g. breast vs. lung adenocarcinoma. Immunoperoxidase stains may be of help. It is important to determine the site of origin to plan subsequent chemotherapy.

DIFFERENTIAL DIAGNOSIS (PITFALLS IN DIAGNOSIS)

When a patient with a primary neoplasm develops neurological symptoms, there is a tendency to regard any abnormality on imaging studies of the brain as confirmation of brain metastasis. However, there are many other conditions that can be correctly diagnosed with careful imaging studies. In the differential diagnosis of a single, contrast-enhancing lesion, one should include primary brain tumor, abscess, radionecrosis, and resolving hematoma. When the diagnosis remains ambiguous, a biopsy is often advisable.

In a patient with systemic tumor and immunosuppression due to chemotherapy, pulmonary infections may lead to septic emboli and brain abscesses that can mimic brain metastases (Joss et al. 1981; Kurtzke and Kurland 1983). The diagnosis may be suspected when the lesion occurs in a location atypical for a metastasis, such as the basal ganglia. Malignant gliomas may sometime mimic brain metastases, particularly since they may be multifocal (Sundaresan et al. 1981b). Histologic diagnosis therefore is mandatory for patients with multiple lesions without a known primary tumor. Even in patients with known systemic malignancy, a CNS lesion may be due to a primary tumor or abscess (Patchell et al. 1986).

Demyelinating disease can occasionally appear on CT or MRI as large contrast-enhancing lesions, with mass effect, simulating brain metastases (Peterson et al. 1993). Such patients may present with an acute monophasic illness directly mimicking brain metastases rather than the typical relapsing/remitting course of multiple sclerosis. Although ambiguous lesions may eventually require surgical resection, even the biopsy material may be misinterpreted as neoplastic because of the large reactive astrocytes seen in active demyelinating lesions. Misdiagnosed patients with multiple sclerosis who have been irradiated have a very poor neurological outcome.

Delayed radiation necrosis can present as a discrete mass, which may resemble a brain metastasis radiographically and may even enlarge on subsequent scans (Sundaresan et al. 1981a).

Multiple small contrast-enhancing lesions may be seen in tuberculosis or fungal abscesses. A clue to diagnosis is presence of thin walled rings of contrast enhancement which differ from the thicker and more nodular contrast enhancement seen with metastases.

Strokes may mimic metastatic disease, both clinically and radiographically. A stroke may appear as a discrete lesion, with contrast enhancement. The passage of time may be required to make the distinction between infarct and brain metastasis; with stroke the radiographic findings characteristically will improve over weeks to months. Dural metastases in patients with prostate or breast carcinoma can mimic meningiomas (Senegor 1991). An unusual and potentially confusing phenomenon is 'tumor-in-tumori', a systemic malignancy metastasizing to already existing intracranial tumor such as a glioma or meningioma. The high blood flow to the recipient tumor is thought to favor metastases (Conzen et al. 1986; Zon et al. 1989).

PATHOPHYSIOLOGY OF BRAIN METASTASES

Certain tumors have a propensity to metastasize to particular organs. The organ-specific mode of metastasis holds for the CNS, which is a favored site of metastasis for melanoma, lung and breast carcinoma. Paget (1889) proposed what has been called the 'soil-seed' hypothesis of metastasis, that tumors form metastases preferentially in organs that provide a favorable environment. Although tumor cells may travel in the body to many tissues, they can establish metastatic colonies only in those target tissues that produce certain necessary factors. Ewing (1928) proposed a different explanation for predilection of metastases for particular tissues, holding that distant metastases are directly related to the vascular supply of the target organ. If this were the most important factor, tumors would metastasize in a random fashion, based solely upon vascular access to the target organ.

Clinical studies have shown that the pattern of metastases is not random. For example, in 56 patients with malignant melanoma studied at autopsy, cluster analysis showed that organs similar in developmental origin have a similar propensity for metastasis. In the CNS, metastases to cortex, white, and gray matter correlate with metastases to leptomeninges, another neuroectodermally derived tissue, but not with metastases to dura mater which is derived from mesoderm. CNS metastases correlate negatively with hepatic metastases (De la Monte et al. 1983b). Neuroblastic tumors show similar patterns of metastasis; some tumors predilect meninges, whereas others predilect dura (De la Monte et al. 1983a).

Both paracrine (produced by host tissue cells) and autocrine (produced by the tumor cells themselves) trophic factors have been identified that enhance or inhibit the growth of metastases. In experimental metastasis model systems, one potent trophic factor is a transferrin-like glycoprotein produced from lung tissue-conditioned media (Nicholson 1993).

Metastasis is a multistep process that involves the passage of malignant cells to the target organ, thrombosis in capillaries, adherence to the endothelial cells of the target organ tissue, invasion of the subendothelial matrix, digestion of the basement membrane by proteolytic enzymes, and ultimately passage of tumor cells into the interstitial space (Aznavoorian et al. 1993). In the early stages of malignancy, tumor metastasis is probably dependent upon paracrine growth factors but, as the malignancy advances, it develops autocrine means of growth stimulation (FGF, IGF, TGF) or is no longer dependent upon growth factors. This parallels the clinical observation that in the early metastatic stage tumors may show restricted organ distribution of metastases, but in the latter stages tumors may metastasize freely to multiple organs (Nicholson 1993).

Metastatic spread of systemic tumors to the brain occurs when the primary tumor sheds tumor emboli into the circulation. Since the lung effectively filters larger tumor emboli, in most (but not all) instances a pulmonary lesion is present at the time that the CNS metastasis becomes evident. Support for an embolic mechanism comes from the observed increased incidence of brain metastases in watershed vascular regions compared to other areas of brain (Delattre et al. 1986). Metastases may bypass the lungs when there is a patent foramen ovale, or with pelvic and abdominal tumors they may pass via Batson's venous plexus to the paravertebral and intracranial venous plexi. The other common route of spread of tumors, via lymphatic channels, does not have relevance to the CNS because the brain does not have a lymphatic supply.

Systemic tumors may also affect the CNS by spreading to the adjacent bony structures such as the calvarium or vertebrae, expanding locally, and compressing or invading the CNS secondarily. Head and neck tumors invade the nervous system by growing locally into the skull, either by eroding through bone or infiltrating along cranial nerves at the base of the skull.

TREATMENT

Therapy of brain metastases has to be evaluated in two contexts, local control of brain metastases and overall survival. About one-third to one-half of patients with brain metastases die of neurological causes, hence the survival in most patients will be determined by the extent and responsiveness of systemic disease. This is the most plausible explanation for better prognosis for patients with breast carcinoma; their tumors are more slowly growing or more responsive to therapy.

Glucocorticoids

Since the initial studies in the 1960s (Galicich and French 1961; French and Galicich 1964), glucocorticoids have become one of the most important therapies used to control of signs and symptoms of brain metastases (Ruderman and Hall 1965). Glucocorticoids not only reduce vasogenic edema caused by brain tumors, but that associated with other therapies such as radiation or radiosurgery (Galicich and French 1961). The mechanism by which brain metastases produce edema is not clear, but there is good evidence that the blood–brain barrier is disrupted in the area surrounding the tumor, allowing molecules in the plasma to diffuse into the surrounding brain. Some tumors secrete substances such as vascular permeability factors that induce changes in the adjacent vessels, promoting the formation of edema (Bruce et al. 1987; Delattre and Posner 1990). Although the precise mechanism of action is unknown, glucocorticoids seem to decrease the permeability in the abnormal blood vessels surrounding the metastases, thus decreasing the edema (Shapiro and Posner 1974; Bruce et al. 1987).

Glucocorticoids may have a dramatic effect, often reversing neurologic deficit within 24 h (Ruderman and Hall 1965; Weinstein et al. 1972). Up to 80% of the patients improve neurologically (Galicich and French 1961; Ruderman and Hall 1965; Weinstein et al. 1972). In some instances, however, the effect is more delayed and maximal anti-edema effect is not seen for up to 7 days (Ransohoff 1972; Gutin 1977). Dexamethasone is used most often because of its low mineralocorticoid activity (Weissman 1988). The dosing of glucocorticoids is largely empiric, beginning with a dose of dexamethasone of 16 mg/day when diagnosis is suspected or at the time of surgery (Gutin 1977). If there is no improvement within 2 days, then the dose may be increased. Because of the long half-life of dexamethasone, it can be given once instead of twice a day. Studies on dose–effect relationships are few (Vecht et al. 1994). The dose is maintained during the entire course of radiation, but is titrated to the lowest dose needed to control neurological symptoms and suppress potential brain edema induced by radiation (Weissman et al. 1991). Some studies suggest that patients receiving dexamethasone 4 mg/day during radiotherapy have equivalent relief of neurological symptoms to those receiving 16 mg/day (Vecht et al. 1994), with less side effects. It may be possible to omit corticosteroids entirely in some patients (Wolfson et al. 1994). Only about 20% of patients require chronic administration of steroids to reverse or stabilize their neurological symptoms. Side effects are associated with prolonged therapy (Weissman et al. 1987) and are dose-dependent. They include hypoalbuminemia (Lewis et al. 1986; Weissman et al. 1987), hyperglycemia, proximal myopathy, immunosuppression, peptic ulcer, psychosis, and depression (Eidelberg 1991). Therapy for more than 1 month will suppress the pituitary–adrenal axis and rapid withdrawal of corticosteroids may lead to acute adrenal insufficiency with severe hypotension or generalized weakness and fever that may simulate systemic infection. In patients in whom all therapy has failed, and who are moribund, steroids can provide relief from debilitating headaches or vomiting from increased intracranial pressure.

Anticonvulsants

Anticonvulsant therapy is recommended for patients with seizures, but it remains debatable whether all patients should be treated prophylactically. In a small double-blind prospective randomized trial of patients with brain metastases who had not previously had a seizure, there was no significant difference in seizure frequency of patients who received valproic acid or placebo (Glantz et al. 1994). Anticonvulsant therapy is not without risk. Phenytoin, for example, has 4%

incidence of drug allergy, may cause elevation of hepatic enzymes, alters hepatic metabolism of dexamethasone and chemotherapeutic agents and, when combined with radiotherapy, may provoke a Stevens–Johnson syndrome (Delattre et al. 1988b).

Several chemotherapeutic agents cause seizures either alone or together with other signs of encephalopathy. These include cisplatinum (Berman and Mann 1980), ifosfamide (Pratt et al. 1986), L-asparaginase (Land et al. 1972), and etoposide (Leff et al. 1988). Vincristine may produce hyponatremia with seizures due to SIADH (Legha 1986).

Radiotherapy

One of the most important immediate effects of external radiotherapy is the improvement in neurological symptoms, which can occur in 49–93% of the patients (Borgelt et al. 1980; Coia 1992). Headaches can be relieved in 82% of the cases, motor loss in 74%, and cranial nerve dysfunction in 71% (Coia 1992). The degree of neurologic recovery after external radiotherapy is inversely proportional to the severity of the neurologic dysfunction prior to irradiation (Borgelt et al. 1980).

External beam irradiation delivered to the whole brain is the treatment of choice for multiple brain metastases, single surgically inaccessible lesions, or in patients with uncontrolled systemic disease. The rationale for whole brain radiotherapy (WBRT), even for single metastases, is to eradicate microscopic foci of disease. Radiotherapy is palliative, prolonging survival from an average of 1 month without treatment (2 months with glucocorticoids only) (Posner 1977), to about 4–6 months. Better prognostic groups include patients with breast carcinoma, those with a long interval between diagnosis of systemic tumor and brain metastases, and limited extracranial disease. The precise dose of radiotherapy has not been definitively determined. A dose of 30 Gy of WBRT is usually administered in 10 fractions of 3.0 Gy. Some studies have questioned whether higher doses would be more effective, but in a well-controlled prospective study in which all patients received 32 Gy to the entire brain followed by an additional boost of up to 42.4 Gy, there was no evidence of a dose–response effect with increasing doses of radiation (Sause et al. 1993). An excellent algorithm on the role of radiation has been developed by Coia et al. (1992).

Despite WBRT, less than 10% of patients survive 2 years (Hoskin et al. 1990), and up to half of patients eventually die of progressive brain metastases (Borgelt et al. 1980; Hoskin et al. 1990). Primary tumors vary significantly in their intrinsic responsiveness to radiotherapy. Radiosensitive tumors include germ cell tumors, lymphoma and small-cell lung cancer. Radioresistant tumors include melanoma, renal cell carcinoma, and carcinoma of colon or thyroid (Gay et al. 1987; Maor et al. 1988; Badalament et al. 1990). Whereas overall symptom response rates of brain metastasis to radiotherapy up to 80% (Coia 1992) are seen, brain metastases from melanoma have a response rate of 42% (Katz 1981). NSCCL is considered of intermediate radiosensitivity.

Complications of radiotherapy treatment

Acute complications after radiation are directly related to the dose per fraction and are usually mild or moderate. Hair loss usually starts after 2 weeks of treatment, and skin desquamation can be seen early during treatment. A subacute syndrome of somnolence and fatigue can be seen 1–4 months after radiation (Coia 1992). In some instances, cranial irradiation has been considered to worsen the myelosuppression in patients receiving adjuvant chemotherapy (Lee et al. 1986a, b). A syndrome of erythema multiforme, similar to Stevens–Johnson syndrome, is thought to be associated with the use of phenytoin during cranial irradiation (Delattre et al. 1988b).

Late complications may include dementia, and its incidence varies from 1.9 to 5%, but can be seen in up to 19% of long-term survivors (DeAngelis et al. 1989). MRI or CT changes typically consist of ventricular dilation and central atrophy, and MRI may reveal T2 signal increase in white matter reflecting leukoencephalopathy.

Surgery and radiosurgery

Surgery
The main consideration behind surgery as a potential therapeutic modality for brain metastases is the fact that the lesions are usually well demarcated from the brain parenchyma, thus allowing the surgeon to remove them completely. A second reason to perform surgery, when technically feasible, is the fact that in the majority of the cases radiation by itself is not cura-

tive. Lastly, surgery may be a diagnostic modality when tissue histology has not been obtained.

When both brain and lung lesions are found, removal of both offered an advantage in overall survival over resection of the brain metastases solely (Galicich et al. 1980; Nakagawa et al. 1993). Cushing's group reported some remarkably prolonged survivals, in one instance up to 7 years, in patients who underwent resection of both lung and brain lesions (Grant 1926). In an early series of 22 patients, median survival was 14 months, with stage I lung cancer and a long interval between pulmonary resection and cerebral metastasis conferring a more favorable prognosis (Magilligan et al. 1976). In a larger series of 92 patients, 35 of whom underwent both craniotomy and thoracotomy, median survival was 24.3 months after craniotomy, and two patients were alive and disease-free at 10 and 13 years (Read et al. 1989).

Review of 231 patients with NSCCL who underwent resection for brain metastases showed that incomplete resection or no resection of the brain tumor, male gender, infratentorial location, presence of systemic metastases and age older than 60 years were significantly correlated with shorter survival (Wronski et al. 1993a). Uncontrolled studies showed benefit in terms of longer survival and lower incidence of recurrent CNS disease in patients who underwent surgical resection of their brain metastases followed by radiotherapy (Patchell et al. 1986; Burt et al. 1992). These studies were open to criticism about the selection bias in treating patients with better overall status and smaller, more accessible lesions with surgery.

More recently, a double-blind prospective trial has shown that patients with single brain metastases fare much better if they receive resection of the tumor before starting radiotherapy (median survival = 40 vs. 15 weeks) (Patchell et al. 1990). These patients also remain functionally independent longer (median 38 vs. 8 weeks) and have a longer time to recurrence (>59 vs. 21 weeks). The improvement in function and survival is directly attributable to a reduction in the number of patients who succumbed to a 'neurologic death' – death directly due to recurrent brain metastases. Surgical resection of the brain metastases had no significant effect upon death due to systemic metastasis. Furthermore, 11% of the randomized patients with systemic malignancy operated for a single brain lesion were found to have diagnoses other than brain metastases. A second prospective study by

Vecht et al. (1993) confirmed that combined treatment of single brain metastases (surgery plus radiation) led to a longer survival and a longer functionally independent survival over radiation alone. Patients with progressive extracranial disease had a poor (5-month) survival irrespective of treatment, and radiation alone was recommended by the authors. For patients with a single brain metastasis and stable systemic disease, combined therapy with radiation and chemotherapy was the preferred modality.

Surgical resection of multiple metastases may be helpful and reduces morbidity (Galicich et al. 1996). Other investigators claim that multiple brain metastases and poor neurological performance status imply poor prognosis (Smalley et al. 1992). Few studies address the treatment options for recurrent brain metastases. For patients with recurrent brain metastases from NSSCL, survival was significantly better for those who had a second operation than for those who did not have surgery (Arbit et al. 1995). Although there is no debate that WBRT is the treatment of choice for multiple brain metastases, it is unproved whether or not radiotherapy improves prognosis for single brain metastases after surgical resection. This is the subject of an ongoing trial.

Radiosurgery

Although WBRT of brain metastases produces response rates from 60 to 80% (Sause et al. 1993) and extends survival (Kaplan and Meier 1958; Lang and Slater 1964; Di Stefano et al. 1979; Borgelt et al. 1980; Gelber et al. 1981; Kihlstrom et al. 1992), up to 50% of patients still die of their brain metastases (Patchell et al. 1986). Patients with deep or multiple tumors, affecting an eloquent area of the brain, or those in poor medical condition, are not candidates for surgical intervention. The concept of radiosurgery, developed by Leskell (1951), produces effects similar to surgery with high doses of radiation delivered to a confined region of brain. The spherical nature of most brain metastases makes them ideal candidates for radiosurgery (Phillips et al. 1990). Multiple distant lesions can be treated with minimal dose overlap. Stereotactic radiosurgery (SRS) uses precisely targeted radiation beams to produce very high doses of radiation within a sharply circumscribed volume, minimizing damage to the normal surrounding brain. In the brain, radiosurgery has been administered using a gamma-knife (in which 201 fixed radiation sources

may be activated according to a computerized treatment plan), or using a linear accelerator (in which a single source of radiation is rotated in a complex arc that passes through the target area on each rotation). The two techniques produce equivalent results, but linear accelerator treatment can be used for slightly larger lesions (upper limit = 3.5–4 cm vs. 3.0 cm for the gamma-knife). Since most brain metastases are not infiltrative and have well-defined borders, they are very suitable for radiosurgery treatment.

Radiosurgery has been used to control both single and multiple brain metastases with encouraging results prior to WBRT at diagnosis, or at the time of progression after conventional radiation (Sturm et al. 1987; Loeffler et al. 1990, 1991; Coffey et al. 1991; Fuller et al. 1992; Kihlstrom et al. 1992; Mehta et al. 1992; Alexander and Loeffler 1993). The local control rate in selected studies with radiosurgery was found to be 88% with doses of 16–35 Gy, about twice the 45% local control rates reported with whole brain radiation (Loeffler and Alexander 1993). Local control usually allows a decrease in glucocorticoid requirements (Loeffler and Alexander 1993). In a retrospective review of 116 patients with single brain metastases due to a variety of tumor histologies, 45 of whom had previously failed WBRT, the median survival was 11 months after gamma-knife SRS (or 20 months since diagnosis (Flickinger et al. 1994a, b)). Local control was achieved in 85% of cases, with subsequent local recurrence occurring in only 15%. Of several factors analyzed, only histology was a significant prognosticator of survival, with breast carcinoma faring best. Overall 2-year actuarial tumor control was 67%.

No acute toxicity from radiosurgery was observed, allowing glucocorticoids to be either discontinued completely (61%) or reduced (12%). The patients tolerated the treatment well: in 70% of the cases, their Karnofsky performance status was either better (21%) or stable (49%). The development of perilesional edema after treatment correlated with the size of the treated tumor and the dose of whole brain radiation. Radiographic follow-up of 52 radiosurgically treated metastases in one series (Adler et al. 1992) showed that 29% of the lesions disappeared completely after the treatment, and 50% decreased in size. Two patients had lesions that enlarged radiographically, but biopsy only showed necrosis. In patients autopsied a mean of 10.5 months after treatment, small islands of tumor cells were seen within large areas of necrosis. In some patients, after an initial decrease in the tumor, there was a gradual increase in the lesion, and the perilesional edema. The dominant histological feature was necrosis (Loeffler et al. 1990). The benefits of SRS over conventional surgery are still not established, but it is foreseeable that SRS will be used for small tumors, while surgery will be required for large or hemorrhagic lesions, or those accompanied by significant mass effect.

Melanoma and renal cell carcinoma, considered resistant to WBRT, are paradoxically amongst the histologies responding most favorably to SRS (Fig. 4). In a retrospective review of 23 consecutive patients with melanoma metastatic to the brain treated with SRS, only 2 worsened, 8 were improved neurologically, and 13 were stable. The mean tumor margin dose was 16 Gy. Local tumor control was achieved in 97% of the tumors. In contrast to similar melanoma patients treated with WBRT who have an approximately 2–4-month survival, the median survival was 9 months. Eighteen of 23 patients died because of progression of their systemic disease, and none because of progression of the treated CNS lesions. SRS compared with surgery is less morbid and less expensive (Bremer et al. 1978; Fell et al. 1980; Guazzo et al. 1989; Madajewicz et al. 1991). These results strongly favor a role for SRS in the management of cerebral melanoma.

Current trials are evaluating patients randomized to receive whole brain radiation plus or minus SRS. Another critical question is whether or not the dose of whole brain radiation (to achieve control of 'microscopic disease') may be decreased without compromising local control (Marks 1990). It is unclear if SRS is preferable to surgical resection for a single small brain metastasis (Black 1993), and the role of SRS in the management of multiple brain metastases is not certain. The incidence of long-term side effects of radiosurgery, such as dementia or radiation necrosis, is not known.

Fractionated radiotherapy

Compared to single-fraction SRS techniques shown to be effective for ablation of arteriovenous malformations, a single fraction may not be the most effective treatment for brain metastases. Single large fractions produce greater damage to normal tissue and permit the tumor to repair sublethal radiation damage (Larson et al. 1993). Theoretically, fractionated SRS

is the optimal method of delivering radiotherapy, and with the advent of removable stereotactic head frames, is quite feasible. Existing fractionation schemes vary from one 1.8 Gy/day treatment to multiple daily fractions (Rieke et al. 1993). With stereotactic localization, long overall treatment times designed to spare normal tissue such as skin or mucosa from radiation sequelae may not be required (Brenner and Hall 1994). Ideally, fractionated radiotherapy should consist of large number of fractions (to allow oxygenation and maximize sparing of late responding normal tissues) over a short period of time (to limit tumor repopulation).

The Radiation Therapy Oncology Group (RTOG) conducted a trial of accelerated fractionation (twice/day fractions of 1.6 Gy each) in the treatment of brain metastases (Sause et al. 1993). Overall dose of radiation (total dose of radiation was escalated from 48.0 to 70.4 Gy) did not statistically enhance survival, but an improvement was seen as the dose was increased from 48 to 54.4 Gy. No excessive toxicity with respect to acute or late reflects was seen with dose escalation. This modality of treatment has to be tested against more conventional routines of radiation such as one fraction a day for a total of 10 fractions and 30 Gy.

Chemotherapy

Although there are some data to suggest that the brain metastases may respond to chemotherapy (see below), more often the scenario is one in which the tumor is stable or responsive to chemotherapy at a time when the brain metastases appear. Some investigators believe that adjuvant chemotherapy may prolong survival of patients with breast cancer, and that brain metastases develop within the blood–brain barrier due to the sanctuary status of the CNS (Paterson et al. 1982). In a retrospective study of 417 men with metastatic non-seminomatous germ cell tumors, 6 patients had single brain metastases. Five of the 6 had advanced lung metastases at diagnosis, and all had prolonged survival with surgery and radiotherapy suggesting that in patients with treatable primary tumors the brain may be a sanctuary site for relapse (Gerl et al. 1993).

Brain metastases may develop during systemic chemotherapy. This has been cited as proof of the ineffectiveness of chemotherapy for this disease, which in turn has been blamed upon the presence of the blood–brain barrier (or more specifically a blood–tumor barrier), particularly when chemotherapy has been successful in controlling the systemic tumor. The ineffectiveness of chemotherapy, however, may relate more to tumor insensitivity to chemotherapy rather than poor penetration (Twelves and Souhami 1991). The tumor types that commonly metastasize to the brain (e.g. NSCCL) are not usually chemosensitive. Brain metastases often develop after primary chemotherapy has failed and the tumor has become resistant to otherwise effective agents. This said, there have been several reports of response to systemic chemotherapy of brain metastases of chemosensitive tumors such as breast or small-cell carcinoma of lung or choriocarcinoma (Rustin et al. 1986; Lee et al. 1989). When there are multiple small intracranial brain metastases in patients with a tumor of known chemosensitivity who are not symptomatic neurologically, or in the cases where both systemic and brain metastases are present, it is reasonable to administer chemotherapy first, prior to cranial radiation (Twelves and Souhami 1991).

THE MANAGEMENT OF RECURRENT BRAIN METASTASES

Recurrence of brain metastasis occurs in up to one-third of patients (Sundaresan and Galicich 1985). The majority of recurrences are local (Sundaresan et al. 1988) and occur within a year of initial treatment (Wronski 1992). At recurrence, treatment options are limited by prior radiation and the poor performance status of many of the patients. Only 20% of the patients with recurrences are considered fit for a second surgical resection (Sundaresan et al. 1988). Treatments that have been used at recurrence include a second surgical resection, SRS, brachytherapy (Obbens and Feun 1986) or local chemotherapy infusions. Some authors consider that only those patients whose lesions responded to radiation at diagnosis should be considered for re-irradiation (Kurup et al. 1980; Dritschillo et al. 1981; Cooper et al. 1990). Contrary to the benefits of radiation given at diagnosis, re-irradiation, however, rarely improves the quality of life (Kurup et al. 1980). Response to chemotherapy at the time of recurrence is poor (Feun et al. 1984; Groen et al. 1988). Intra-arterial (carotid) chemotherapy has been used for metastases with up to

50% response rates, but the treatment has lost popularity because of serious toxicity (Feun et al. 1984). Surgical resection at recurrence may lead to neurological improvement in up to 63% of the patients (Sundaresan et al. 1988). Two hundred and forty-four of the patients were alive 2 years after the re-operation; 21 patients had radiation necrosis instead of recurrent tumor. A complete surgical resection, as evidenced by post-operative CT, was possible in 18 of 21 patients. Under these circumstances, surgery at recurrence yields the same risks and side effects as at diagnosis (Sundaresan et al. 1988; Kaye 1992; Bindal et al. 1995; Arbit and Wronski 1995). SRS or brachytherapy has achieved some encouraging results in a select group of patients (Lindquist 1989; Loeffler et al. 1990; Amin et al. 1993).

WHEN TO STOP TREATMENT

There is no consensus on when to stop treatment in patients with brain metastases. For patients with a very poor performance status and whose overall survival is estimated as less than a few weeks, withholding radiation and/or surgery is an alternative. In these patients, corticosteroids and analgesics may offer significant alleviation of symptoms such headaches and lethargy.

SPINAL METASTASES: GENERAL CONSIDERATIONS

Spinal involvement is seen most commonly with breast, lung or prostate malignancies. Spinal involvement at autopsy is seen in up to 60% of the patients with breast cancer (Lenz and Fried 1931), while it occurs in life in about 67% of patients with prostate cancer (Kuban et al. 1986). In patients with lung carcinoma, the most common histology to metastasize to the spine is small-cell carcinoma of lung, seen in about 40% of all cases (Bach et al. 1992). Vertebral metastases may remain localized to the vertebral body or lead to epidural cord compression (ECC). ECC occurs in 5% of patients with cancer (Barron et al. 1959; Lewis et al. 1986; Klein et al. 1991). It is usually seen in the setting of widespread malignancy; only about 3% of the patients present with ECC as the first sign of metastatic disease (Zelefsky et al. 1992). ECC is a serious complication of cancer; if left untreated, it can lead to devastating neurological symptoms such as para- or quadriplegia, or sphincter incontinence. The most common site of compression is at the thoracic level (70% of the cases), while lumbar or cervical ECC occurs in 20 and 10%, respectively (Gilbert et al. 1978; Stark et al. 1982; Bach et al. 1992). In patients with prostate carcinoma, ECC involved one vertebral body in 26%, two in 42% and three or more levels in 32% (Zelefsky et al. 1992).

PATHOGENESIS

The vertebral column is the most common site of skeletal metastases in patients with cancer (Byrne 1992). Several theories have been proposed. Firstly, bone marrow may have a high concentration of factors that stimulate growth of neoplastic cells (Osborn et al. 1995). Secondly, the Batson's vertebral venous plexus is a preferred site of drainage of the pelvic, abdominal, and thoracic organs when there is enough abdominal compression or occlusion of the inferior vena cava (Coman and Delong 1951; Franks 1953; Suzuki et al. 1994). The lymphatic system, which anastomoses with the vertebral venous plexus, may also play a role in the development of spinal metastases. Autopsy studies (Asdourian et al. 1990) show that the location of vertebral destruction corresponds to the location of the vertebral vessels (Algra et al. 1992). The presence of bone marrow in the vertebral body and the concentration of blood vessels in the vertebral body in preference to other parts of the vertebra make the vertebral body vulnerable to spinal metastases (Algra et al. 1992). ECC most commonly is the result of direct extension of tumor originating in the vertebral column. It can also arise from direct extension of metastases in the paravertebral spaces (in up to 15% of the patients) by growing through the neural foramina without affecting the bone. The mechanism of cord injury in ECC is compression of the vertebral venous plexus with resultant cord edema and ultimately ischemia (Ushio et al. 1977; Kato et al. 1985; Manabe et al. 1989).

CLINICAL

Vertebral metastases may present with pain or be completely asymptomatic. Pain is the initial presentation in up to 95% of the patients with ECC (Gilbert et al. 1978). The pain can be axial or radicular and can sometimes be confused with arthritic pain as it may be aggravated by movement. There are several clinical features that help distinguish pain from ECC from that of degenerative joint disease (Byrne 1992). Pain from ECC may occur at any level, while degenerative arthritic pain is usually seen at the lumbar or cervical level (Byrne 1992). The supine position usually alleviates arthritic pain, while that of ECC is usually aggravated (Gilbert et al. 1978). Other symptoms of ECC in decreasing order of frequency are: weakness (76%), autonomic disturbances (57%), sensory dysfunction (51%) (Gilbert et al. 1978). Urinary incontinence is usually a late sign. The first complaints can be as vague as a change in gait. In patients who also have nerve root compression, radicular symptoms referable to the dermatome affected are seen. Occasionally, band-like pain across the chest or abdomen can be misinterpreted as arising from thoracic or abdominal viscera, and not recognized as an early sign of spinal metastases. An approximate location of the ECC can usually be determined by a sensory level on neurological examination.

DIAGNOSTIC EVALUATION

Plain radiographs will reveal vertebral metastases in about 85% of patients with ECC (Gilbert et al. 1978). Loss of the vertebral pedicles is the first sign on plain radiographs (Kricun 1985). This, however, occurs relatively late in the metastatic process, after there has been extensive destruction of the trabecular bone that is difficult to visualize on plain radiographs (Algra et al. 1992). Abnormalities on plain radiographs are common in patients with breast, lung or prostate carcinoma, but rare in patients with lymphoma. Prostate carcinoma typically produces osteoblastic lesions, while breast cancer leads to lytic ones. In patients with symptoms of pain or with neurologic deficits and abnormal radiographs, ECC was found by myelography in 86%, while in those with symptoms and normal radiographs, ECC was found in only 8%. A normal neurological examination in a cancer patient with back pain does not rule out ECC. Bone scans can be sensitive in the detection of vertebral metastases, but are not specific (Portenoy et al. 1989). A spinal CT is more specific than radionuclide scanning in distinguishing vertebral disease from metastases or from degenerative joint disease (Weissman et al. 1985; O'Rourke et al. 1986; Redmond et al. 1988). Spinal CT, however, is not optimal in detecting if the tumor has extended into the epidural space, in which case myelography or MRI are better. While an MRI with contrast is optimal for detection of leptomeningeal disease or intramedullary tumors, a non-contrast MRI usually suffices to detect the vertebral metastases or paravertebral metastases. Myelography can detect sites of ECC, but if there is a complete block by tumor, a spinal puncture may lead to neurological deterioration. Since up to 27% of the patients with localizable ECC show multiple sites of ECC, MRI should be performed of the entire spine.

A sensible operational plan for the management of a patient with possible ECC has been proposed (Portenoy et al. 1989). The most important aspects of management are preservation of neurological function and alleviation of pain. If a patient presents solely with pain, has no neurological deficits and abnormal X-rays, many physicians would irradiate the symptomatic vertebral metastases without any other imaging. Others recommend imaging of the whole spine with MRI or myelogram to identify non-adjacent areas of unrecognized epidural disease which may benefit from early irradiation. If the patient has mild signs or symptoms, or stable symptoms, the MRI/myelogram may be performed the next day; in those with severe or rapidly progressive deficits, MRI or myelogram should be performed the same day as an emergency.

THERAPY

Incidentally found vertebral metastases may not need direct treatment. Those symptomatic with pain but no ECC may be treated with analgesics. Early treatment of those patients with ECC is crucial. The most common treatment for ECC is corticosteroids and radiation. Corticosteroids are known to alleviate pain and improve the neurological deficits in the short term, but it is not clear what the contribution is towards the eventual neurologic recovery (Weissman 1988). The dose of dexamethasone is not agreed upon; many centers start treatment at 4 mg 4 ×/day, other centers have used initial boluses of up to 100 mg, followed by 24 mg, 4 ×/day, and then a tapering schedule of dexamethasone throughout the radiation. A common schema for radiation is 30–40 Gy over 2–4 weeks. Radiation usually includes two vertebral bodies above and two below the level of the lesion. Other therapies include surgery and chemotherapy (Byrne 1992). Some studies have not shown an advantage

for surgery and radiotherapy over radiotherapy alone (Gilbert et al. 1978; Young et al. 1980; Findlay 1984; Posner 1987; Siegal and Siegal 1989), while others have (Bach et al. 1992). The surgical approach is dependent on the location of the lesion: posterior ones are best approached by laminectomy, while anterior ones may require thoracotomy, vertebral body resection and spinal stabilization (Landreaneau et al. 1995). If there is extensive vertebral body involvement a posterior approach may cause further destabilization. Surgery is to be considered in the following circumstances: (a) when the patient presents with ECC, but no primary malignancy is known; (b) when there are worsening symptoms/signs referable to a previously irradiated level; (c) in patients with spinal instability; (d) in patients with neurological deterioration despite radiotherapy and increasing doses of corticosteroids; and (e) in patients with radioresistant tumors. Hormonal manipulation, usually in the form of bilateral orchiectomies to reduce the testosterone level as sole therapy, led to neurologic improvement in a small series of patients with ECC from prostate carcinomas (Flynn and Shipley 1989).

The most important prognostic factor after radiation is the neurologic status of the patient at the time of initiation of therapy. Up to 80% of the patients who were ambulatory at the initiation of treatment remained ambulatory, while only 50% of those who were paraparetic and 10% of those who were plegic became ambulatory (Gilbert et al. 1978; Greenberg et al. 1980; Findlay 1984). Pain relief can be achieved in up to 96% of the patients treated with corticosteroids and radiation (Zelefsky et al. 1992). Patients with compression fractures are less likely to have a favorable response to external radiotherapy (Zelefsky et al. 1992), and in these patients surgery may be a consideration. Survival is most dependent upon control of the underlying malignancy (Bach et al. 1990; Sorensen et al. 1990). Only 9% of a large series of patients with lung carcinoma and ECC survived beyond 12 months (Bach et al. 1992).

Acknowledgments

The authors are grateful to Ms. Annette Claudio for assistance in preparing the manuscript, Dr. Robert De la Paz for help with photography and Dr. Lewis P. Rowland who kindly edited this chapter.

REFERENCES

ABRAMS, H., R. SPIRO and N. GOLDSTEIN: Metastases in carcinoma. Analysis of 1000 autopsied cases. Cancer 3 (1950) 74–85.

ADELOYE, A.: Intracranial metastatic tumors in Ibadan, Nigeria. Surg. Neurol. 17 (1982) 231–232.

ADLER, J., R. COX, I. KAPLAN and D. MARTIN: Stereotactic radiosurgical treatment of brain metastases. J. Neurosurg. 76 (1992) 444–449.

AISNER, J.: Current approaches to small cell lung cancer. Hematol. Oncol. 10 (1992) 7.

AKESON, P., E. LARSSON, D. KRISTOFFERSEN, E. JONSSON and S. HOLTAS: Brain metastases: comparison of gadodiamide injection-enhanced MR imaging at standard and high dose, contrast-enhanced CT and non-contrast-enhanced MR imaging. Acta Radiol. 36 (1995) 300–306.

ALEXANDER, E.I. and J. LOEFFLER: Factors affecting local control in the radiosurgery of brain metastases (Abstr.). J. Neurosurg. 78 (1993) 344A.

ALGRA, R., J. HEIMANS, J. VALK, J. NAUTA, M. LACHNIET and B. VAN KOOTEN: Do metastases in vertebrae begin in the body or the pedicles? Imaging study in 45 patients. Am. J. Roentgenol. 158 (1992) 1275–1279.

AMER, M., M. AL-SARRAF and L. BAKER: Malignant melanoma and central nervous system metastases. Incidence, diagnosis, treatment and survival. Cancer 42 (1978) 660–668.

AMIN, P., M. FIANDACA, W. SEWCHAND, A. WOLF, D. RIGAMONTI and O. SALAZAR: Irradiation with the gamma unit for recurrent, previously irradiated patients (Abstr.). Acta Neurochir. (Wien)122 (1993) 160.

AMPIL, F.L.: Radiotherapy for carcinomatous brachial plexopathy. Cancer 56 (1985) 2185–2188.

ANAND, A. and D. POTTS: Calcified brain metastases: demonstration by computed tomography. Am. J. Neuroradiol. 3 (1982) 527–529.

ARBIT, E. and M. WRONSKI: The treatment of brain metastases. Neurosurg. Quart. 5 (1995) 1–17.

ARBIT, E. and M. WRONSKI: Clinical decision making in brain metastases. Neurosurg. Clin. North Am. 7 (1996).

ARBIT, E., M. WRONSKI, M. BURT and J. GALICICH: The treatment of patients with recurrent brain metastases: a retrospective analysis of 109 patients with non-small cell lung cancer. Cancer 76 (1995) 765–773.

ARMSTRONG, J., M. WRONSKI, J. GALICICH, E. ARBIT, S. LEIBEL and M. BURT: Postoperative radiation for lung cancer metastatic to the brain. J. Clin. Oncol. 12 (1994) 2340–2344.

ASDOURIAN, P., M. WEIDENBAUM and K. DEWALD: The pattern of vertebral involvement in metastatic vertebral breast cancer. Clin. Orthop. 250 (1990) 164–170.

ATLAS, S., R. GROSSMAN, J. GOMORI ET AL.: MR imaging of intracranial metastatic melanoma. J. Comput. Assist. Tomogr. 11 (1987) 577–582.

AZNAVOORIAN, S., A.N. MURPHY, W.G. STETLER-STEVENSON and L.A. LIOTTA: Molecular aspects of tumor cell invasion and metastasis. Cancer 71 (1993) 1368–1383.

BACH, F., B. LARSEN and K. ROHDE: Metastatic spinal cord compression. Occurrence, symptoms, clinical presentations and prognosis in 398 patients with spinal cord compression. Acta Neurochir. (Wien) 107 (1990) 37–43.

BACH, F., N. AGERLIN, J. SORENSEN, T. RASMUSSEN, P. DOMBERNOWSKY and P. SORENSEN: Metastatic spinal cord compression secondary to lung cancer. J. Clin. Oncol. 10 (1992) 1781–1787.

BADALAMENT, R., R. GLUCK and G. WONG: Surgical treatment of brain metastases from renal cell carcinoma. Urology 36 (1990) 112–117.

BARRON, K., A. HIRANO, S. ARAKI and R. TERRY: Experiences with metastatic neoplasms involving the spinal cord. Neurology 9 (1959) 91–106.

BARTH, R. and A. SOLOWAY: Boron neutron capture therapy of primary and metastatic brain tumors. Mol. Chem. Neuropathol. 21 (1994) 139–154.

BERMAN, I. and M. MANN: Seizures and transient cortical blindness associated with cisplatinum diamminedichloride (PDD) therapy in a thirty-year-old man. Cancer 45 (1980) 764–766.

BINDAL, R., R. SAWAYA, M. LEAVENS and J. LEE: Surgical treatment of multiple brain metastases. J. Neurosurg. 79 (1983) 210–216.

BINDAL, R., R. SAWAYA, M. LEAVENS, K. HESS and S. TAYLOR: Reoperation for recurrent metastatic brain tumors. J. Neurosurg. 83 (1995) 600–604.

BITOH, S., H. HASEGAWA, H. OHTSUKI, J. OBASHI, M. FUJIWARA and M. SAKURAI: Cerebral neoplasms initially presenting with massive intracerebral hemorrhage. Surg. Neurol. 22 (1984) 57–62.

BLACK, P.: Solitary brain metastases: radiation, resection or radiosurgery? Chest 103 (1993) 367S–369S.

BONITO, D., L. GIARELLI, G. FALCONIERI, D. BONIFACIO-GORI, G. TOMASIC and P. VIEHL: Association of breast cancer and meningioma. Report of 12 new cases and review of the literature. Pathol. Res. Pract. 189 (1993) 399–404.

BOOGERD, W., A. HART, J. VAN DER SANDE and E. ENGELSMAN: Meningeal carcinomatosis in breast cancer: prognostic factors and influence of treatment. Cancer (1991) 1685–1695.

BOOGERD, W., V. VOS, A. HART and G. BARIS: Brain metastases in breast cancer; natural history, prognostic factors and outcome. J. Neuro-Oncol. 15 (1993) 165–174.

BORGELT, B., R. GELBER, S. KRAMER, L. BRADY, C. CHANG and L. DAVIS: The palliation of brain metastases: the final results of the first two studies by The Radiation Therapy Oncology Group. Int. J. Radiat. Oncol. Biol. Phys. 6 (1980) 1–9.

BORING, C., T. SQUIRES and T. TONG: Cancer statistics. CA Cancer J. Clin. 43 (1993) 7–26.

BREMER, A., C. WEST and M. DIDOLKAR: An evaluation of the surgical management of melanoma of the brain. J. Surg. Oncol. 10 (1978) 211–219.

BRENNER, D. and E. HALL: Stereotactic radiotherapy of intracranial tumors – an ideal candidate for accelerated treatment. Int. J. Radiat. Oncol. Biol. Phys. 28 (1994) 1039–1041.

BRUCE, J., G. CRISCUOLO and M. MERRILL: Vascular permeability induced by protein product of malignant brain tumors: inhibition of dexamethasone. J. Neurosurg. 67 (1987) 880–884.

BURT, M., M. WRONSKI, E. ARBIT and J. GALICICH: Resection of brain metastases from non-small cell lung carcinoma. J. Thorac. Cardiovasc. Surg. 103 (1992) 399–411.

BYRNE, T.: Spinal cord compression from epidural metastases. N. Engl. J. Med. (1992) 614–619.

BYRNE, T., T. CASCINO and J. POSNER: Brain metastases from melanoma. J. Neuro-Oncol. 1 (1983) 313–317.

CASCINO, T., J. LEAVENGOOD, N. KEMENY and J. POSNER: Brain metastases from colon cancer. J. Neuro-Oncol. 1 (1983) 203–209.

CHYUN, Y., E. HAYWARD and J. LOKICH: Metastasis to the central nervous system from colorectal cancer. Med. Pediatr. Oncol. 8 (1980) 305–308.

COHEN, N., G. STRAUSS, R. LEW, D. SILVER and L. RECHT: Should prophylactic anticonvulsants be administered to patients with newly-diagnosed cerebral metastases? A retrospective analysis. J. Clin. Oncol. 6 (1988) 1621–1624.

COIA, L.: The role of radiation therapy in the treatment of brain metastases. Int. J. Radiat. Oncol. Biol. Phys. 23 (1992) 229–238.

COIA, L., N. AARONSON, R. LINGGOOD, J. LOEFFLER and T. PRIESTMAN: A report on the consensus workshop panel on the treatment of brain metastases. Int. J. Radiat. Oncol. Biol. Phys. 23 (1992) 223–227.

COLE, F.J., J. THOMAS, B. WILCOX and H.I. HALFORD: Cerebral imaging in the asymptomatic preoperative bronchogenic carcinoma patient: is it worthwhile? Ann. Thorac. Surg. 57 (1994) 838–840.

COMAN, D. and R. DELONG: Role of vertebral venous system in metastases of cancer to the spinal column. Cancer 4 (1951) 610–618.

CONZEN, M., H. SOLLMAN and R. SCHNABEL: Metastasis of lung carcinoma to intracranial meningioma: case report and review of literature. Neurochirurgia 29 (1986) 206–209.

COOPER, J., A. STEINFELD and J. LERCH: Cerebral metastases: value of reirradiation in selected patients. Radiology 174 (1990) 883–885.

DAVIS, P., P. HUDGINS, S. PETERMAN and J.J. HOFFMAN: Diagnosis of cerebral metastases: double-dose delayed CT vs. contrast-enhanced MR imaging. Am. J. Neuroradiol. 12 (1991) 293–300.

DE LA MONTE, S., G. MOORE and G. HUTCHINS: Nonrandom distribution of metastases in neuroblastic tumors. Cancer 52 (1983a) 915–925.

DE LA MONTE, S.M., G. MOORE and G.M. HUTCHINS: Patterned distribution of metastases from malignant melanoma in humans. Cancer Res. 43 (1983b) 3427–3433.

DEANGELIS, L., J. DELATTRE and J. POSNER: Radiation induced dementia in patients cured of brain metastases. Neurology 39 (1989) 789–796.

DEBEVEC, M.: Management of patients with brain metastases of unknown origin. Neoplasma 37 (1990) 601.

DELAHUNT, B., H. TEOH, V. BALAKRISHMAN, J. NACEY and S. CLARK: Testicular germ cell tumor with pineal metastases. Neurosurgery 26 (1990) 688–691.

DELATTRE, J. and J. POSNER: The blood–brain barrier: morphology, physiology and its changes in cancer patients. In: J. Hildebrand (Ed.), Neurological Adverse Reactions to Anti-Cancer Drugs. Berlin, Springer-Verlag (1990) 3–24.

DELATTRE, J., G. KROL, H. THALER and J. POSNER: Distribution of brain metastases (Abstr.). Ann. Neurol. 20 (1986) 138.

DELATTRE, J., G. KROL, H. THALER and J. POSNER: Distribution of brain metastases. Arch. Neurol. 45 (1988a) 741–744.

DELATTRE, J., B. SAFAI and J. POSNER: Erythema multiforme and Stevens–Johnson syndrome in patients receiving cranial irradiation and phenytoin. Neurology 38 (1988b) 194–198.

DEMIERRE, B. and J. BERNAY: Métastases intracrâniennes du cancer de la prostate. Neurochirurgie 29 (1983) 143–149.

DI STEFANO, A., H. YAP and G. HORTOBAGI: The natural history of breast cancer patients with brain metastases. Cancer 44 (1979) 1913–1918.

DRINGS, P., B. RIZI, G. ABEL and G. VAN KAICK: The frequency of brain metastases in small cell lung cancer (in German). Prax. Klin. Pneumol. 41 (1987) 695–696.

DRITSCHILLO, A., J. BRUCKMAN, J. CASSADY and J. BELLI: Tolerance of brain to multiple courses of radiation therapy. Br. J. Radiol. 54 (1981) 782–786.

EAPEN, L., M. VACHET, G. CATTON, C. DANJOUX, R. MCDERMOT and B. NAIR: Brain metastases with an unknown primary: a clinical perspective. J. Neuro-Oncol. 6 (1988) 31–35.

EIDELBERG, D.: Neurological effects of steroid treatment. In: D. Rottenberg (Ed.), Neurological Complications of Cancer Treatment. Boston, MA, Butterworth-Heinemann (1991) 173–183.

EWING, J.: Neoplastic Diseases, 3rd Edit. Philadelphia, PA, W.B. Saunders (1928).

FADUL, C., K. MISULIS and R. WILEY: Cerebellar metastases: diagnostic and management considerations. J. Clin. Oncol. 5 (1987) 1107–1115.

FAUL, C. and J. FLICKINGER: The use of radiation in the

management of spinal metastases. J. Neuro-Oncol. 23 (1995) 149–161.

FELL, D., M. LEAVENS and C. MCBRIDE: Surgical versus non-surgical management of metastatic melanoma of the brain. Neurosurgery 7 (1980) 238–242.

FETELL, M.R.: Pineal metastases of systemic carcinoma: mechanism of secondary neoplastic meningitis. Neurology 38, Suppl. 1 (1988) 392.

FETELL, M. and B. STEIN: Therapy of pineal region tumors (Abstr.). Neurology 34, Suppl. (1994) 185.

FEUN, L., S. WALLACE and D. STEWART: Intracarotid infusion of cis-diamminedichloroplatinum in the treatment of recurrent malignant brain tumors. Cancer 54 (1984) 794–799.

FINDLAY, G.: Adverse effects of the management of malignant spinal cord compression. J. Neurol. Neurosurg. Psychiatry 47 (1984) 761–768.

FLICKINGER, J., D. KONDZIOLKA, L. LUNSFORD, R. COFFEY, M. GOODMAN and E. SHAW: A multi-institutional experience with stereotactic radiosurgery for solitary brain metastases. Int. J. Radiat. Oncol. Biol. Phys. 28 (1994a) 797–802.

FLICKINGER, J., J. LOEFFLER and D. LARSON: Stereotactic radiosurgery for intracranial malignancies. Oncology 82 (1994b) 81–86.

FLINT, G.: Seizures and epilepsy. Br. J. Neurosurg. 6 (1988) 419–421.

FLOWERS, A. and V. LEVIN: Management of brain metastases from breast carcinoma. Oncology 7 (1993) 21–26.

FLOYD, C., C. STIRLING and I.J. COHN: Cancer of the colon, rectum and anus: review of 1687 cases. Ann. Surg. 163 (1966) 829–837.

FLYNN, D. and W. SHIPLEY: Management of spinal cord compression secondary to metastatic prostate carcinoma. Urol. Clin. North Am. 18 (1989) 145–152.

FORSYTH, P. and J.B. POSNER: Headaches in patients with brain tumors: a study of 111 patients. Ann. Neurol. 32 (1992) 78.

FRANKS, L.: The spread of prostatic carcinoma to the bones. J. Pathol. Bacteriol. 66 (1953) 91–93.

FRENCH, L. and J. GALICICH: The use of steroids for control of cerebral edema. Clin. Neurosurg. 10 (1964) 212–223.

FULLER, B., I. KAPLAN, J. ADLER, R. COX and M. BAGSHAW: Stereotaxic radiosurgery for brain metastases: the importance of adjuvant whole brain irradiation. Int. J. Radiat. Oncol. Biol. Phys. 23 (1992) 413–418.

GALICICH, J. and L. FRENCH: Use of dexamethasone in the treatment of cerebral edema resulting from brain tumors and brain surgery. Am. Pract. Dig. Treat. 12 (1961) 169–174.

GALICICH, J., E. ARBIT and M. WRONSKI: Metastatic brain tumors. In: R. Wilkins and S. Rengachary (Eds.), Neurosurgery. New York, McGraw-Hill (1996) 807–821.

GAY, P., W. LITCHY and T. CASCINO: Brain metastases in hypernephroma. J. Neuro-Oncol. 5 (1987) 51–56.

GELBER, R., M. LARSON, B. BORGET and S. KRAMER: Equivalence of radiation schedules for the palliative treatment of brain metastases in patients with favorable prognosis. Cancer 48 (1981) 1749–1753.

GERL, A., C. CLEMM and W. WILMANNS: Central nervous system as sanctuary site of relapse in patients treated with chemotherapy for metastatic testicular cancer. Clin. Exp. Metast. 12 (1993) 226–230.

GIANNONE, L., D. JOHNSON, K. HANDE and F. GRECO: Favorable prognosis of brain metastases in small cell lung cancer. Ann. Intern. Med. 106 (1987) 386–389.

GILBERT, R., J. KIM and J. POSNER: Epidural spinal cord compression from metastatic tumor: diagnosis and treatment. Ann. Neurol. 3 (1978) 40–51.

GILDERSLEEVE, N., A. KOO and C. MCDONALD: Metastatic tumor presenting as intracerebral hemorrhage. Radiology 124 (1977) 109–124.

GLANTZ, M., M. FRIEDBERG, B. COLE, H. CHOY, L. WAHLBERG, K. FURIE and S. LOUIS: Double-blind, randomized, placebo-controlled trial of anticonvulsant prophylaxis in adults with newly diagnosed brain metastases. Proc. Am. Soc. Clin. Oncol. 13 (1994) 176.

GRANT, D., D. EDWARDS and P. GOLDSTRAW: Computed tomography of brain, chest and abdomen in the preoperative assessment of non-small cell lung cancer. Thorax 43 (1988) 883–886.

GRANT, F.: Concerning intracranial malignant metastases: their frequency and the value of surgery in their treatment. Ann. Surg. 84 (1926) 635–646.

GRAUS, T., G. KROL and K. FOLEY: Early diagnosis of spinal cord epidural metastasis. Correlation of clinical and radiologic findings (Abstr.). Proc. Am. Soc. Clin. Oncol. 4 (1985a) 269.

GRAUS, T., L. ROGERS and J. POSNER: Cerebrovascular complications in patients with cancer. In: Medicine. Baltimore, MD, Williams and Wilkins (1985b) 16–34.

GREENBERG, H., J. KIM and J. POSNER: Epidural spinal cord compression from metastatic tumor: results with a new treatment protocol. Ann. Neurol. 8 (1980) 361–366.

GREENBERG, H., M. DECK, B. VIKRAM, F. CHU and J. POSNER: Metastasis to the base of the skull: clinical findings in 43 patients. Neurology 31 (1981) 530–537.

GREENE, G., P. HITCHON, R. SCHELPER, W. YUH and G. DYSTE: Diagnostic yield in CT guided serial stereotactic biopsy of gliomas. J. Neurosurg. 71 (1989) 494–497.

GROEN, H., E. SMIT, H. HAAXMA-REICHE and P. POSTMUS: Carboplatin as second line treatment for recurrent or progressive brain metastases from small cell lung cancer. Eur. J. Radiat. Oncol. Biol. Phys. 15 (1988) 433–437.

GUAZZO, L., L. ATKINSON, M. WEIDMANN ET AL.: Management of solitary melanoma metastasis of the brain. Aust. N.Z. J. Surg. 59 (1989) 321–324.

GUTIN, P.: Corticosteroid therapy in patients with brain tumors. Natl. Cancer Inst. Monogr. 46 (1977) 151–156.

HAGERSTRAND, I. and J. SHONEBECK: Metastases to the pituitary gland. Acta Pathol. Microbiol. Scand. 75 (1969) 64–70.

HAYAT, G., T. EHSAN, J.B. SELHORST and A. MANEPALI: Magnetic resonance evidence of perineural metastasis. J. Neuroimaging 5 (1995) 122–125.

HEALEY, M., J. HESSELINK, G. PRESS and M. MIDDLETON: Increased detection of intracranial metastases with intravenous Gd-DTPA. Radiology 165 (1987) 619–624.

HOSKIN, P., J. CROW and H. FORD: The influence of extent and local management on the outcome of radiotherapy for brain metastases. Int. J. Radiat. Oncol. Phys. 19 (1990) 111–115.

JACOBS, L., W. KINKEL and R. VINCENT: 'Silent' brain metastases from lung carcinoma determined by computerized tomography. Arch. Neurol. 34 (1977) 690–693.

JAECKLE, K.: Neuroimaging for central nervous system tumors. Semin. Oncol. 18 (1991) 150–157.

JAECKLE, K.A.: Causes and management of headaches in cancer patients. Oncology 7 (1993) 27–31.

JELENIK, J., J.I. REDMOND and J. PERRY: Small cell lung cancer: staging with MR imaging. Radiology 177 (1990) 837–842.

JOSS, R., R. GOLDBERG, J. YATES and I. KRAKOFF: Lung abscesses following corticosteroid therapy for central nervous system metastases. Med. Pediatr. Oncol. 9 (1981) 279–282.

KAPLAN, E. and P. MEIER: Nonparametric estimation from incomplete observation. J. Am. Stat. Assoc. 53 (1958) 457–480.

KATO, A., Y. USHIO, T. HAYAKAWA, K. YAMADA, H. IKEDA and H. MOGAMI: Circulatory disturbance of the spinal cord with epidural neoplasm in rats. J. Neurosurg. 63 (1985) 260–265.

KATZ, H.: The relative effectiveness of radiation therapy, corticosteroids, and surgery in the management of melanoma metastatic to the central nervous system. Int. J. Radiat. Oncol. Biol. Phys. 7 (1981) 852–857.

KAYE, H.: Malignant brain tumors. Late reoperation for metastatic tumors. In: J. Little and I. Awad (Eds.), Reoperative Neurosurgery. Baltimore, MD, Williams and Wilkins (1992) 72–76.

KIHLSTROM, L., B. KARLSSON and C. LINDQUIST: Gamma knife surgery in brain metastases. In: L. Lunsford (Ed.), Stereotactic Radiosurgery Update. New York, Elsevier (1992) 429–434.

KIRICUTA, I., O. KOLBL, J. WILLNER and W. BOHNDORF: Central nervous system metastases in breast cancer. J. Cancer Res. Clin. Oncol. 118 (1992) 542–546.

KITAOKA, K., H. ABE, T. AIDA, M. SATOH, T. ITOH and Y. NAKAGAWA: Follow-up study on metastatic cerebellar tumor surgery: characteristic problems of surgical treatment. Neurol. Med. Chir. 30 (1990) 591–598.

KLEIN, S., R. SANFORD and M. MUHLBAUER: Pediatric spinal epidural metastases. J. Neurosurg. 74 (1991) 70–75.

KORI, S.H., K.M. FOLEY and J.B. POSNER: Brachial plexus lesions in patients with cancer: 100 cases. Neurology 31 (1981) 45–50.

KORMAS, P., J. BRADSHAW and K. JEYASINGHAM: Preoperative computed tomography of the brain in non-small cell bronchogenic carcinoma. Thorax 47 (1992) 106–108.

KOTTKE-MARCHANT, K., M. ESTES and C. NUNE: Early brain metastases in endometrial carcinoma. Gynecol. Oncol. 41 (1991) 67–73.

KOVALIKOVA, Z., M. HOEHN-BERLAGE and K. GERSONDE: Age-dependent variation of T1 and T2 relaxation times of adenocarcinoma in mice. Radiology 164 (1987) 543–548.

KRICUN, M.: Red-yellow marrow conversion: its effect on the location of solitary bone lesions. Skeletal Radiol. 14 (1985) 10–19w.

KUBAN, D., A. EL-MAHDI, S. SIGFRED, P. SCHELLHAMMER and T. BABB: Characteristics of spinal cord compression in adenocarcinoma of the prostate. Urology 28 (1986) 364–369.

KURTZKE, J. and L. KURLAND: The epidemiology of neurologic disease. In: A. Baker and L. Baker (Eds.), Clinical Neurology. Philadelphia, PA, Lippincot-Raven Publishers (1983) 1–143.

KURUP, P., S. REDDY and F. HENDRICKSON: Results of re-irradiation for cerebral metastases. Cancer 46 (1980) 2587–2589.

LAND, V., W. SUTOW and D. FERNBACK: Toxicity of L-asparaginase in children with advanced leukemia. Cancer 30 (1972) 339–347.

LANDREANEAU, F., R. LANDREANEAU, R. KEENAN and P. FERSON: Diagnosis and management of spinal metastases from breast cancer. J. Neuro-Oncol. 23 (1995) 121–134.

LANG, E. and J. SLATER: Metastatic brain tumor: results of surgical and nonsurgical treatment. Surg. Clin. North Am. 44 (1964) 865.

LARSON, D., J. FLICKINGER and J. LOEFFLER: The radiobiology of radiosurgery. Int. J. Radiat. Oncol. Biol. Phys. 25 (1993) 557–561.

LEE, J., T. UMSAWASDI and Y. LEE: Neurotoxicity in long-term survivors of small cell lung cancer. Int. J. Radiat. Oncol. Biol. Phys. 12 (1986a) 313–321.

LEE, J., T. UMSAWADI, H. DHINGRA, H. BARKLEY and W. MURPHY: Effects of brain irradiation and chemotherapy on myelosuppression in small cell lung cancer. J. Clin. Oncol. 4 (1986b) 1615–1619.

LEE, J., W. MURPHY, B. GLISSON, H. DHINGRA, P. HOLOYE and W. HONG: Primary chemotherapy of brain metastasis in small-cell lung cancer. J. Clin. Oncol. 7 (1989) 916–922.

LEE, Y.-T.N.: Malignant melanoma: pattern of metastasis. Cancer J. Clin. 30 (1980) 137–142.

LEFF, R., J. THOMPSON and M. DALY: Acute neurologic dysfunction after high-dose etoposide therapy for malignant glioma. Cancer 62 (1988) 32–35.

LEGHA, S.: Vincristine neurotoxicity: patho-physiology and management. Med. Toxicol. 1 (1986) 421–427.

LEGHA, S., S. GIANAN, N. PAPADOPOULOS and R. BENJAMIN: Brain metastases complicating systemic therapy of patients with metastatic melanoma. Proc. Am. Soc. Clin. Oncol. 9 (1990) 282.

LENZ, M. and J. FRIED: Metastases to the skeleton, brain, and spinal cord from cancer of the breast and the effect of radiotherapy. Ann. Surg. 93 (1931) 278–285.

LEROUX, P., M. BERGER and P. ELLIOT: Cerebral metastases from ovarian carcinoma. Cancer 67 (1991) 2194–2199.

LESKELL, L.: The stereotaxic method and radiosurgery of the brain. Chir. Scand. 102 (1951) 316–319.

LEWIS, D., R. PACKER, B. RANEY, I. RAK, J. BELASCO and B. LANGE: Incidence, presentation, and outcome of spinal cord disease in children with systemic cancer. Pediatrics 78 (1986) 438–443.

LINDQUIST, C.: Gamma-knife surgery for recurrent solitary metastasis of cerebral hypernephroma: case report. Neurosurgery 25 (1989) 576–582.

LOEFFLER, J. and E.I. ALEXANDER: Radiosurgery for the treatment of intracranial metastases. In: E.I. Alexander, J. Loeffler and L. Lunsford (Eds.), Stereotactic Radiosurgery. New York, McGraw-Hill (1993).

LOEFFLER, J., H. KOOY, P. WEN, L. NEDZI and E. ALEXANDER: The treatment of recurrent brain metastases with stereotactic radiosurgery. J. Clin. Oncol. 8 (1990) 576–582.

LOEFFLER, J., E. ALEXANDER, H. KOOY, P. WEN, H. FINE and P. BLACK: Radiosurgery for brain metastases. In: V. DeVita, S. Heliman and S. Rosenberg (Eds.), Principle and Practice of Oncology Updates. Philadelphia, PA, Lippincott (1991) 1–12.

MADAJEWICZ, S., N. CHOWHAN, A. ILIYA, C. ROQUE, R. BEATON and R. DAVIS: Intracarotid chemotherapy with etoposide and cisplatin for malignant brain tumors. Cancer 67 (1991) 2844–2849.

MADOW, L. and B. ALPERS: Encephalitic form of metastatic carcinoma. Arch. Neurol. Psychiatry 65 (1951) 161–173.

MAGILLIGAN, D., J. ROGERS, R. KNIGHTON and J. DAVILA: Pulmonary neoplasm with solitary cerebral metastasis: results of combined excision. J. Thorac. Cardiovasc. Surg. 72 (1976) 690–698.

MAJOIE, C.B., B. VERBEETEN, JR., J.A. DOL and F.L. PEETERS: Trigeminal neuropathy: evaluation with MR imaging. Radiographics 15 (1995) 795–811.

MANABE, S., H. TANAKA, Y. HIGO, P. PARK and T. OHNO: Experimental analysis of the spinal cord compressed by spinal metastasis. Spine 14 (1989) 1308–1315.

MANDELL, L.: The treatment of single brain metastasis

from non oat-cell lung carcinoma. Cancer 58 (1986) 641–649.

MANDYBUR, T.: Intracranial hemorrhage caused by metastatic tumors. Neurology 27 (1977) 650–655.

MAOR, M., A. FRIAS and M. OSWALD: Palliative radiotherapy for brain metastases in renal carcinoma. Cancer 62 (1988) 1912–1917.

MARKS, L.: A standard dose of radiation for 'microscopic disease' is not appropriate. Cancer 66 (1990) 2498–2502.

MAX, M., M. DECK and D. ROTTENBERG: Pituitary metastases: incidence in cancer patients and clinical differentiation from pituitary adenoma. Neurology 31 (1981) 998–1002.

MAYER, R., R. BERKOWITZ and T. GRIFFITHS: Central nervous system involvement by ovarian carcinoma: a complication of prolonged survival with metastatic disease. Cancer 41 (1978) 776–783.

MAYR, N., W. YUH, M. MUHONEN ET AL.: Cost/benefit analysis of high-dose MR contrast studies in the evaluation of brain metastases (Abstr.). In: American Society of Neuroradiology Book of Abstracts (1993) 193–194.

MEHTA, M., J. ROZENTHAL, A. LEVIN, T. MACKIE, S. KUBSAD and M. GEHRING: Defining the role of radiosurgery in the management of brain tumors. Int. J. Radiat. Oncol. Biol. Phys. 24 (1992) 619–625.

MELLER, I., D. ALKALAY, M. MOZES ET AL.: Isolated metastases to peripheral nerves. Cancer 76 (1995) 1829–1832

MERCHUT, M.: Brain metastases from undiagnosed systemic neoplasms. Arch. Intern. Med. 149 (1989) 1076–1080.

MORRIS, J.G.L. and R. JOFFE: Perineural spread of cutaneous basal and squamous cell carcinoma. Arch. Neurol. 40 (1983) 424–429.

MOSER, R., M. JOHNSON and A. YUNG: Metastatic brain cancer in patients with no known primary site. Cancer Bull. 41 (1989) 173–177.

NAKAGAWA, K., T. FUJITA, S. IZUMIMOTO, Y. MIYAWAKI, S. KUBO and Y. NAKAJIMA: Cis-diamminedichloroplatinum (CDDP) therapy for brain metastasis of lung cancer. J. Neuro-Oncol. 16 (1993) 69–76.

NELSON, P., A. ROBINSON and A. MARTINEZ: Metastatic tumors of the pituitary gland. Neurosurgery 21 (1987) 941–944.

NEMZEK, W., V. POIRIER, M. SALAMAT and T. YU: Carcinomatous encephalitis (miliary metastases): lack of contrast enhancement. Am. J. Neuroradiol. 14 (1993) 540–542.

NICHOLSON, G.: Paracrine and autocrine growth mechanisms in tumor metastasis to specific sites with particular emphasis on brain and lung metastasis. Cancer Metast. Rev. 12 (1993) 325–343.

OBBENS, E. and L. FEUN: Treatment options for recurrent brain metastases. Cancer Bull. 38 (1986) 45–48.

O'ROURKE, T., C. GEORGE and J.I. REDMOND: Spinal computed tomography and computed tomographic metrizamide myelography in the early diagnosis of metastatic disease. J. Clin. Oncol. 4 (1986) 576–583.

ORTEGA, P., N. MALAMUD and M. SHIMKIN: Metastases to the pineal body. 15 (1951) 518–528.

OSBORN, J., R. GETZENBERG and D. TRUMP: Spinal cord compression in prostate cancer. J. Neuro-Oncol. 23 (1995) 135–147.

PAGET, S.: The distribution of secondary growths in cancer of the breast. Lancet i (1889) 571–573.

PATCHELL, R., C. CIRRINCIONE, H. THALER, J. GALICICH, J. KIM and J. POSNER: Single brain metastasis: surgery plus radiation or radiation alone. Neurology 36 (1986) 447–453.

PATCHELL, R., P. TIBBS, J. WALSH ET AL.: A randomized trial of surgery in the treatment of single metastases to the brain. N. Engl. J. Med. 322 (1990) 494–500.

PATERSON, A., M. AGARWAL and A. LEES: Brain metastasis in breast cancer patients receiving adjuvant chemotherapy. Cancer 49 (1982) 651–654.

PEDERSEN, A.: Diagnosis of CNS-metastases from SCLS. In: Lung Cancer: Basic and Clinical Aspects. Boston, MA, Martinus Nijhoff (1986) 153–182.

PETERSON, K., M. ROSENBLUM, J. POWERS, E. ALVORD, R. WALKER and J. POSNER: Effect of brain irradiation on demyelinating lesions. Neurology 43 (1993) 2105–2112.

PHILLIPS, M., K. FRANKEL and J. LYMAN: Comparison of different radiation types and irradiation geometrics in stereotactic radiosurgery. Int. J. Radiat. Oncol. Biol. Phys. 18 (1990) 211–220.

PICKREN, J., G. LOPEZ, Y. TZUKADA and W. LANE: Brain metastases: an autopsy study. Cancer Treat. Symp. 2 (1983) 295–313.

PORTENOY, R., B. GALER and O. SALAMON: Identification of epidural neoplasm: radiography and bone scintography in the symptomatic and asymptomatic spine. Cancer 64 (1989) 2207–2213.

POSNER, J.: Management of central nervous system metastases. Semin. Oncol. 4 (1977) 81–91.

POSNER, J.: Back pain and epidural spinal cord compression. Med. Clin. North Am. 71 (1987) 185–205.

POSNER, J.: Management of brain metastases. Rev. Neurol. 148 (1992) 477–487.

POSNER, J. and N. CHERNIK: Intracranial metastases from systemic cancer. Adv. Neurol. 19 (1978) 579–592.

POST, K.D., P.C. MCCORMICK, A.P. HAYS and A.G. KANDJI: Metastatic carcinoma to pituitary adenoma: report of two cases. Surg. Neurol. 30 (1988) 286–292.

POSTMUS, P., E. SMIT and H. HAAXMA-REICHE: Treatment of central nervous system metastases from small cell lung cancer with chemotherapy. Lung Cancer 9 (1993) 281–286.

POTTS, D., G. ABBOTT and J. SNEIDERN: National Cancer Institute study: evaluation of computed tomography

in the diagnosis of intracranial neoplasms. III. Metastatic tumors. Radiology 136 (1980) 657–664.

PRATT, C., A. GREEN and M. HOROWITZ: Central nervous system toxicity following the treatment of pediatric patients with ifosfamide/mesna. J. Clin. Oncol. 4 (1986) 1253–1261.

RANSOHOFF, J.: The effects of steroids on brain edema in man. In: H. Reulen and K. Schurmann (Eds.), Steroids and Brain Edema. Berlin, Springer-Verlag (1972) 211–217.

RAY, B.S. and H.G. WOLFF: Pain sensitive structures of the head and their significance in headache. Arch. Surg. 41 (1940) 813–856.

READ, R., W. BOOP and G. YODER: Management of non-small cell lung carcinoma with solitary brain metastasis. J. Thorac. Cardiovasc. Surg. 98 (1989) 884–891.

REDMOND, J.I., K. FRIEDL, P. CORNETT, M. STONE, T. O'ROURKE and C. GEORGE: Clinical usefulness of an algorithm for the early diagnosis of spinal metastatic disease. J. Clin. Oncol. 6 (1988) 154–157.

RICHARDS, P. and W. MIKOSSOCH: Intracranial metastases. Br. Med. J. 1 (1963) 15–18.

RIEKE, J., D. JONES and M. HAFERMANN: Fractionated stereotactic radiosurgery of brain tumors: results and toxicity in 41 patients. Acta Neurochir. (Wien) 122 (1993) 170.

RODRIGUEZ, G., J. SOPER, A. BERCHUCK, J. OLESON, R. DODGE and G. MONTANA: Improved palliation of cerebral metastases in epithelial ovarian cancer using a combined modality approach including radiation therapy, chemotherapy, and surgery. J. Clin. Oncol. 10 (1992) 1553–1560.

RUDERMAN, N. and T. HALL: Use of glucocorticoids in the palliative treatment of metastatic brain tumors. Cancer 18 (1965) 298–306.

RUSHTON, J.G. and E.A. ROOKE: Brain tumor headache. Headache 2 (1962) 147–152.

RUSSELL, E., G. GEREMIA, C. JOHNSON ET AL.: Multiple cerebral metastases: detectability with Gd-DTPA-enhanced MR imaging. Radiology 165 (1987) 609–617.

RUSTIN, G., E. NEWLANDS and K. BAGSHAWE: Successful management of metastatic and primary germ cell tumors in the brain. Cancer 57 (1986) 2108–2113.

SAHA, S., M. MEYER, E. KREMENTZ, M. RODRIGUEZ, S. HODA and D. CARTER: Central nervous system metastasis in malignant melanoma – a prognostic evaluation. Proc. Am. Soc. Clin. Oncol. 8 (1989) 288.

SAUSE, W., C. SCOTT, R. KRISCH, M. ROTMAN, P. SNEED and N. JANJAN: Phase I/II trial of accelerated fractionation in brain metastases RTOG 85-28. Int. J. Radiat. Oncol. Biol. Phys. 26 (1993) 653–657.

SCHIFTER, M. and A.P. BARRETT: Perineural spread of squamous cell carcinoma involving trigeminal and facial nerves. Oral Surg. Oral Med. Oral Pathol. 75 (1993) 587–590.

SENEGOR, M.: Prominent meningeal 'tail sign' in a patient with metastatic tumor. Neurosurgery 29 (1991) 294–296.

SHAPIRO, W. and J. POSNER: Corticosteroid hormones. Arch. Neurol. 30 (1974) 217–221.

SIEGAL, T. and T. SIEGAL: Current considerations in the management of neoplastic spinal cord compression. Spine 14 (1989) 223–228.

SILBERT, P.L., G.R. KELSALL, J.M. SHEPHERD and S.S. GUBBAY: Enigmatic trigeminal sensory neuropathy diagnosed by facial skin biopsy. Clin. Exp. Neurol. 29 (1992) 234–238.

SILLERUD, L., J. FREYER, M. NEEMAN and M. MATTINGLY: Proton NMR microscopy of multicellular tumor spheroid morphology. Magn. Reson. Med. 16 (1990) 380–389.

SMALLEY, S., E. LAWS, J. O'FALLON, E. SHAW and M. SCHRAY: Resection for solitary brain metastasis. Role of adjuvant radiation and prognostic variables in 229 patients. J. Neurosurg. 77 (1992) 531–540.

SORENSEN, J., H. HENSEN, M. HANSEN and P. DOMBERNOWSKY: Brain metastases in adenocarcinoma of the lung: frequency, risk groups, and prognosis. J. Clin. Oncol. 6 (1988) 1474–1480.

SORENSEN, P., S. BORGENSEN and K. ROHDE: Metastatic epidural spinal cord compression: results of treatment and survival. Cancer 65 (1990) 1502–1508.

STARK, R., R. HENSON and S. EVANS: Spinal metastases: a retrospective survey from a general hospital. Brain 105 (1982) 189–213.

STURM, V., B. KOBER, K. HOVER, W. SCHLEGAL, R. BOESICHER and O. PASTYR: Stereotactic percutaneous single dose irradiation of brain metastases with a linear accelerator. Int. J. Radiat. Oncol. Biol. Phys. 13 (1987) 279–282.

SUNDARESAN, N., J. GALICICH and M. DECK: Radiation necrosis after treatment of solitary intracranial metastases. Neurosurgery 8 (1981a) 329–333.

SUNDARESAN, N., J. GALICICH and T. TOMITA: Computerized tomography findings in multifocal glioma. Acta Neurochir. (Wien) 59 (1981b) 217–226.

SUNDARESAN, N. and J. GALICICH: Surgical treatment of brain metastases. Clinical and computerized tomographic evaluation of the results of treatment. Cancer 55 (1985) 1382–1388.

SUNDARESAN, N., V. SACHDEV, G. DIGIACINTO and J. HUGHES: Reoperation for brain metastases. J. Clin. Oncol. 6 (1988) 1625–1629.

SUZUKI, M., T. TAKASHIMA, M. KADOYA, T. UEDA, F. ARAKAWA and F. UEDA: Signal intensity of brain metastases on T2-weighted images: specificity for metastases from colonic cancers. Neurochirurgia 36 (1993) 151–155.

SUZUKI, T., T. SHIMIZU, K. KUROKAWA, H. JIMBO, J. SATO and H. YAMANAKA: Pattern of prostate cancer metastasis to the vertebral column. Prostate 25 (1994) 141–146.

TAKAKURA, K., H. SANO, S. HOJO and A. HIRANO: Metastatic Tumors of the Central Nervous System. Tokyo, Igaku-Shoin (1982) 112–118.

TAYLOR, H., M. LEFKOWITZ, S. SKOOG, B. MILES, D. MCLEOD and J. COGGIN: Intracranial metastases in prostate cancer. Cancer 53 (1984) 2728–2730.

TEEARS, R. and E. SILVERMAN: Clinicopathologic review of 88 cases of carcinoma metastatic to the pituitary gland. 36 (1975) 216–220.

TOMLINSON, B., R. PERRY and E. STEWART-WYNNE: Influence of site of origin of lung carcinomas on clinical presentation and central nervous system metastases. J. Neurol. Neurosurg. Psychiatry 42 (1979) 82–88.

TROBE, J.D., I. HOOD, J.T. PARSONS and R.G. QUISLING: Intracranial spread of squamous carcinoma along the trigeminal nerve. Arch. Ophthalmol. 100 (1982) 608–611.

TSUKADA, Y., A. FOUAD, J. PICKREN and W. LANE: Central nervous system metastasis from breast carcinoma: autopsy study. Cancer 52 (1983) 2349–2354.

TWELVES, C. and R. SOUHAMI: Should cerebral metastases be treated by chemotherapy alone? Ann. Oncol. 2 (1991) 15–17.

USHIO, Y., N. ARITA, T. HAYAKAWA, H. MOGAMI, H. HASEGAWA and S. BITOH: Chemotherapy of brain metastases from lung carcinoma: a controlled randomized study. Neurosurgery 28 (1991) 201–205.

VECHT, C., H. HAAXMA-REICHE, E. NOORDIJK, G. PADBERG, J. VOORMOLEN and F. HOEKSTRA: Treatment of single brain metastasis: radiotherapy alone or combined with neurosurgery? Ann. Neurol. 33 (1993) 583–590.

VECHT, C., A. HOVESTADT, H. VERBIEST, J. VAN VLIET and W. VAN PUTTEN: Dose-effect relationship of dexamethasone on Kamofsky performance in metastatic brain tumors: a randomized study of doses of 4, 8, and 16 mg per day. Neurology 44 (1994) 675–680.

VERNET, M.: Paralysies laryngées associées. In: Légendre. Lyon, Paris Médical (1916).

VERNET, M.: Syndrome du trou déchiré postérieur (paralysie des nerfs glosso-pharyngien-pneumo-gastrique-spinal). Rev. Neurol. 2 (1918) 117–148.

VIKRAM, B. and F. CHU: Radiation therapy for metastases to the base of the skull. Radiology 130 (1979) 465–468.

VOORHIES, R., N. SUNDARESAN and H. THALER: The single supratentorial lesion: an evaluation of preoperative diagnostic tests. J. Neurosurg. 53 (1980) 364–368.

WASSERSTROM, W., J. GLASS and J. POSNER: Diagnosis and treatment of leptomeningeal metastases from solid tumors. Experience with 90 patients. Cancer 49 (1982) 759–772.

WEBER, P., K. SHEPARD and S. VIJAYAKUMAR: Case report: metastases to the pineal gland. Cancer 63 (1989) 164–165.

WEINSTEIN, J., F. TOY and M. JAFFE: The effect of dexamethasone on brain edema in patients with metastatic brain tumors. Neurology 23 (1972) 121–129.

WEISS, L., B. HAYDOCK, J. PICKREN and W. LANE: Organ vascularity and metastatic frequency. Am. J. Pathol. 101 (1980) 101–113.

WEISBERG, L.: Computerized tomography in intracranial metastases. Arch. Neurol. 36 (1970) 630–634.

WEISSMAN, D.: Glucocorticoid treatment for brain metastases and epidural spinal cord compression: a review. J. Clin. Oncol. 6 (1988) 543–551.

WEISSMAN, D., M. GILBERT, H. WANG and S. GROSSMAN: The use of computed tomography of the spine to identify patients at high risk for epidural metastases. J. Clin. Oncol. 3 (1985) 1541–1544.

WEISSMAN, D., D. DUFER, V. VOGEL and M. ABELOFF: Corticosteroid toxicity in neuro-oncology patients. J. Neuro-Oncol. 5 (1987) 125–128.

WEISSMAN, D., N. JANJAN, B. ERICKSON, F. WILSON, M. GREENBERG and P. RITCH: Twice-daily tapering dexamethasone treatment during cranial radiation for newly diagnosed brain metastases. J. Neuro-Oncol. 11 (1991) 235–239.

WOLFSON, A., S. SNODGRASS, J. SCHWADE, A. MARKOE, H. LANDY and L. FEUN: The role of steroids in the management of metastatic carcinoma to the brain. A pilot prospective trial. Am. J. Clin. Oncol. 17 (1994) 234–238.

WOODRUFF, W.J., W. DJANG, R. MCLENDON ET AL.: Intracerebral malignant melanoma: high-field-strength MR imaging. Radiology 165 (1987) 209–213.

WRONSKI, M.: The outcome of surgical resection of brain metastases from lung cancer in 185 patients. The Memorial Sloan-Kettering Cancer Center experience 1974–1989 (doctoral dissertation). Polish Academy of Science, Medical Research Center, Warsaw (1992).

WRONSKI, M., E. ARBIT, M. BURT and J. GALICICH: Survival after surgical treatment of brain metastases from lung cancer: a follow-up study of 231 patients treated between 1976 and 1991. J. Neurosurg. 83 (1993a) 605–616.

WRONSKI, M., M. ZAKOWSKI, E. ARBIT, W. HOSKINS and J. GALICICH: Endometrial cancer metastasis to brain: report of two cases and a review of literature. Surg. Neurol. 39 (1993b) 355–359.

YOUNG, R., E. POST and G. KING: Treatment of epidural metastases. Randomized prospective comparison of laminectomy and radiotherapy. J. Neurosurg. 53 (1980) 741–748.

YUH, W., J. ENGELKEN, M. MUHONEN, N. MAYR, D. FISHER and J. EHRHARDT: Experience with high-dose gadolinium MR imaging in the evaluation of brain metastases. Am. J. Neuroradiol. (1992) 335–435.

ZELEFSKY, M., H. SCHER, G. KROL, R. PORTENOY, S. LEIBEL and Z. FUKS: Spinal epidural tumor in patients with prostate cancer. Clinical and radiographic predictors of response to radiation therapy. Cancer 70 (1992) 2319–2325.

ZIMM, S., G. WAMPLER, D. STABLEIN, T. HAZRA and H. YOUNG: Intracerebral metastases in solid-tumor patients: natural history and results of treatment. Cancer 48 (1981) 384–394.

ZON, L., W. JOHNS and P. STOMPER: Breast carcinoma metastatic to a meningioma: case report and review of the literature. Arch. Intern. Med. 149 (1989) 959–962.

Handbook of Clinical Neurology, Vol. 27 (71): Systemic Diseases, Part III
M.J. Aminoff and C.G. Goetz, editors

CHAPTER 26

Leptomeningeal metastasis

M. KELLY NICHOLAS and SUSAN M. CHANG

Neuro-Oncology Service, Brain Tumor Research Center, Department of Neurological Surgery, University of California, San Francisco, CA, U.S.A.

Cancer of many types affects the leptomeninges and the spaces they contain. This may occur either as a consequence of metastasis from sites outside the central nervous system (CNS) or via direct extension of intraparenchymal CNS tumors. Rarely, tumors appear to arise within and remain confined to the leptomeninges themselves. Synonymous terms used to describe these conditions include leptomeningeal seeding or metastasis, meningeal carcinomatosis, and carcinomatous or neoplastic meningitis. In the case of hematologic malignancies, reference is often made to meningeal leukemia or lymphoma. No one term adequately reflects the variable findings that may be encountered when cancer affects the leptomeninges. In this chapter, we will refer to all involvement of the leptomeninges by a systemic or primary CNS neoplastic process as leptomeningeal metastasis (LM). We will also briefly describe rare cases of tumors that appear to arise within the leptomeninges themselves.

First described in the 19th century, LM has been recognized with increasing frequency ever since (Eberth 1870). Heightened clinical awareness and improved diagnostic techniques likely account for some of this increase. However, considerable demographic complexities also underlie this finding. For example, there is a marked discrepancy in the relative frequency with which particular tumors metastasize to the leptomeninges. Furthermore, the incidence with which some tumors are encountered is actually diminishing while for others it continues to increase (Yap et al. 1978; Rosen et al. 1982; Littman et al. 1987; Boogerd et al. 1991). These phenomena are explained, in part, by the impact of improved cancer treatments on patient survival and the natural history of the diseases in question. Still, the reasons for the increased incidence of LM remain unexplained.

The morbidity and mortality associated with LM are considerable. Persons of all ages are affected. With the exception of prophylactic therapies, current treatments are palliative. An understanding of the pathophysiology of LM and the difficulties associated with its treatment requires an appreciation of the normal anatomy and physiology of the leptomeninges and cerebrospinal fluid (CSF). Following a review of this anatomy and physiology we will review the clinical presentation, diagnosis, and management of patients with LM. Many reviews of LM focus upon systemic cancers. In this chapter we have also included a discussion of LM arising from primary CNS tumors and those that arise within the LM themselves. This inclusion derives from the observation that all tumors affecting the leptomeninges in a clinically significant way share common features, regardless of their origins.

PATHOPHYSIOLOGY OF LEPTOMENINGEAL METASTASIS

With the rare exception of tumors arising directly from the leptomeninges, metastatic cells must progress through a succession of interdependent stages to reach and then proliferate in the leptomeningeal microenvironment. Both anatomic and molecular biologic factors play important roles in this process. An understanding of the mechanisms underlying metastasis remains a major research goal. We will briefly consider both the anatomic and molecular biologic factors associated with LM. Following this, we will discuss the anatomic and functional consequences of LM as they relate to symptoms and signs of the disease.

Anatomic considerations

Several anatomic specializations serve to restrict functional interactions between the CNS and the rest of the body. Those factors most often cited include the presence of the blood–brain barrier (BBB), the blood–CSF barrier, and the absence of a well developed lymphatic drainage from the CNS (Cserr and Knopf 1992). The contribution of these specializations to the regulation of brain biochemistry and fluid homeostasis is well known. These same factors may influence patterns of metastasis to and from the CNS, as well. However, caution should be exercised in attributing too strong an influence of these factors on patterns of metastasis. For example, although the BBB limits the passage of most plasma proteins from the systemic circulation into the brain, individual cells may traverse the endothelium via receptor-mediated processes, thus gaining direct entry to the brain and surrounding spaces (see below). Similarly, the absence of conventional lymphatic drainage from the CNS is often cited as a reason for the exceptionally low incidence of extra-CNS metastases from primary tumors of the CNS. It should be noted, however, that functional connections do exist between the CNS and systemic lymphatic channels (see Fig. 1) (Cserr and Knopf 1992).

Evidence from experimental models and pathologic review suggests that neoplastic cells may gain access to the leptomeninges and subarachnoid space via several routes. Tumors may extend directly from CNS parenchyma via several pathways. They may also spread from extracranial sites via cranial and peripheral nerves. Rarely, epidural tumors may invade the leptomeninges directly. Finally, tumor cells often metastasize via hematogenous routes, both arterial and venous. These mechanisms of metastasis are outlined in Fig. 2 and summarized in turn below.

Tumors in the CNS parenchyma, both primary and metastatic, may involve the subarachnoid space by direct extension (Mirimanoff and Choi 1987). Alternatively, tumors may spread to the subarachnoid space following extension to the ventricular surface. Among systemic tumors that have metastasized to brain or spinal cord, involvement of the leptomeninges is more often encountered with cerebellar lesions than those metastatic to other sites (Patchell et al. 1986). This may reflect the relatively high degree of contact between cerebellum and subarachnoid space. Another factor may be that metastases elsewhere in the brain often occur at the grey–white matter junction, a considerable distance from both the subarachnoid and ventricular surfaces.

Tumors may gain access to the subarachnoid space via direct extension along cranial and peripheral nerves. Patients with paravertebral and base of skull tumors – either primary or metastatic – represent a sizable proportion of those with LM. Pathologic evaluation confirms the frequent involvement of cranial and peripheral nerves as a mechanism of tumor spread in such cases (Redman et al. 1986).

Hematogenous routes of LM include spread via leptomeningeal veins, extension from choroid plexus of arterially derived metastases, and spread via diploic veins following metastasis to bone marrow of both the skull and vertebral bodies (Shuangshoti et al. 1971; Azzarelli et al. 1984; Smith et al. 1985). Batson (1940) hypothesized that the epidural veins (whose plexus now bears his name) draining parts of the axial skeleton including the skull, vertebrae, and pelvis provided a potential route for CNS metastases. Experiments on laboratory animals have confirmed a role for Batson's plexus in skull and vertebral metastases (Del Regato 1977). Definitive evidence for spread to the leptomeninges or CNS parenchyma is lacking. However, Batson's plexus provides at least an indirect route for LM since metastases to both skull and vertebrae do spread secondarily to involve the leptomeninges.

Molecular biologic considerations

As mentioned above, metastasis of any type requires the successful passage of tumor cells through a series of interdependent events. At the very least, all instances of LM require a loosening of contacts between

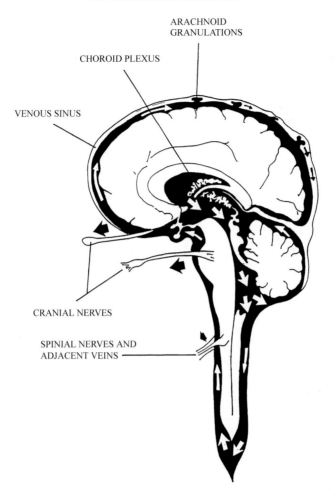

ARACHNOID
GRANULATIONS

CHOROID PLEXUS

VENOUS SINUS

CRANIAL NERVES

SPINIAL NERVES AND
ADJACENT VEINS

Fig. 1. A schematic midsagittal view of the brain and spinal cord surrounded by CSF in the ventricles and sub-arachnoid space (black). The pathways taken by the CSF upon its production in the choroid plexus are indicated by the arrows. Although the majority of the CSF is believed to exit the CNS via the arachnoid granulations over the cortical convexities, there is ample evidence that some CSF reabsorption also occurs at spinal nerve roots via arachnoid granulations in contact with adjacent veins. Still more CSF exits via cranial nerves and drains indirectly into perineural lymphatic channels (see text).

the primary tumor mass and surrounding extracellular matrix (ECM), movement of tumor cells across basement membranes and through the ECM, penetration of the leptomeninges, arrest of cells within the leptomeninges, and additional growth following localization in the new environment. Furthermore, metastases from distant sites must survive in vascular and/or lymphatic spaces en route to the leptomeninges. Once cells have gained access to the leptomeninges and subarachnoid space, they may proliferate in the surrounding CSF and/or adhere to meningeal surfaces where they may grow to variable degrees. From these beginnings, tumors may extend to involve the

perivascular Virchow–Robin spaces and even brain parenchyma itself.

A number of molecules are of potential importance in tumor metastasis. They include, but are not limited to, proteases and cell adhesion molecules (CAMs) including integrins, cadherins, and members of the immunoglobulin gene superfamily. Proteins important in generating cells of the metastatic phenotype are listed in Table 1. Some are discussed further below. It should be stressed that these molecules interact in a complex and tightly regulated manner in both normal and disease states. Thus, it is the relationship between these many factors that determines the

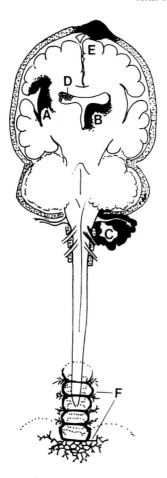

Fig. 2. Anatomic pathways used by tumor cells as they metastasize to the leptomeningeal space. Primary CNS tumors or systemic metastases to brain parenchyma may reach the subarachnoid space via direct extension (A). Tumors in contact with the ventricle may spread subependymally, shed into the CSF and ultimately reach the subarachnoid space (B). Paraspinal and base-of-skull tumors may extend directly into the subarachnoid space via cranial or spinal nerves (C). Cells may gain access to the leptomeninges via leptomeningeal veins (not diagrammed), via the choroid plexus vasculature (D), via diploic veins upon metastasis to surrounding bone (E) and indirectly via Batson's venous plexus of epidural veins (F) that follow the axial skeleton from skull to pelvis.

metastatic phenotype (Stetler-Stevenson et al. 1993; Giese and Westphal 1996).

Proteases
Proteases comprise a large family of enzymes with multiple, highly specialized functions. One family of proteases, the so-called matrix metalloproteinases

(MMPs) (see Table 1) are considered particularly important in mechanisms of metastasis. Important MMPs include collagenases, stromelysins, and gelatinases. MMPs are responsible for ECM degradation, an important early step in metastasis. A clear correlation exists between levels of MMP expression and metastasis in experimental models (Bonfil et al. 1989; Powell et al. 1993). Furthermore, increased levels of these proteins have been found in serum and tumor tissues taken directly from cancer patients (Polette et al. 1991).

It is clear that tumor–host interactions play an important role in mediating MMP activities in both normal and tumor tissues. For example, numerous growth factors and ECM components influence MMP gene expression (Matrisan 1992; Seftor et al. 1992). Once produced, MMPs must be activated in the ECM, a process that occurs to a greater extent in and around tumors than it does in normal tissues (Polette et al. 1991; Gray et al. 1992).

Tissue inhibitors of matrix metalloproteinases (TIMPs) are also important in regulating MMP activity and the metastatic phenotype. Evidence points to overactivity of MMPs and the factors associated with their activation rather than inactivation or underexpression of TIMPs in the generation of cells with metastatic potential. The TIMPs inhibit metastasis in both in vitro and in vivo models (Mignatti et al. 1986; Alvarez et al. 1990). Anti-angiogenic properties have also been demonstrated for the TIMPs (Mignatti et al. 1986; Moses and Langer 1991). Taken together, these data suggest that TIMPs might be used effectively as anti-tumor agents in humans. Indeed, clinical trials with these agents have begun. While none are directed specifically towards the treatment of documented LM, the goal of such therapy is the elimination or reduction of cells with metastatic potential.

Cell adhesion molecules
Numerous molecules that mediate cell–cell and cell–ECM interactions have been identified. Not surprisingly, many of them are implicated in mechanisms of metastasis. One such molecular group, the integrins, comprises a large family of cell surface receptors. For example, the integrin encoding the vitronectin receptor is overexpressed in both melanoma and glioblastoma cell lines (Gladson and Cherech 1991; Gehlsen et al. 1992). Synthetic peptides engineered to inter-

TABLE 1

Molecules important in metastasis*.

Type of molecule	Characteristics
Proteases	
MMPs	Zinc-dependent zymogens,
Interstitial collagenases	secreted as proenzymes,
Stromelysins	and inhibited by TIMPs
Gelatinases	
CAMs	
Integrins	Heterodimers of 14 α
	and 8 β subunits, receptors
	for ECM proteins
Cadherins	Calcium-dependent CAMs
	that may inhibit metastasis
CD44	Hyaluronic acid receptor
Laminin receptor	67 kDa protein over-expressed
	in some metastases

*Stetler-Stevenson et al. 1993; Giese and Westphal 1996.

fere with integrin function inhibit cells in both in vitro and in vivo models (Gehlsen et al. 1988; Saiki et al. 1989). Increased expression of the laminin receptor has also been documented in experimental models of metastasis (Liotta 1986). Blockade of the laminin receptor diminishes the metastatic potential of tumor cells in animal models.

CD44 is another CAM associated with metastasis. Its ligand is hyaluronic acid, a substance found in high concentration in CNS tissues (Kuppner et al. 1992). This may play an important role in both mediating metastases to the CNS and in the dissemination of primary and metastatic tumors within CNS tissue itself (Giese and Westphal 1996).

Functional consequences of leptomeningeal metastases

The symptoms and signs that result from LM are multiple. A handful of pathophysiologic mechanisms, however, underlies many of these findings. First, LM often impairs the normal flow of CSF (Grossman et al. 1982). CSF absorption is impaired by the presence of neoplastic cells with or without an accompanying inflammatory response. Direct infiltration of the arachnoid villi also impedes CSF drainage to the systemic circulation. Alternatively, the outflow of CSF from the ventricular system to the adjoining subarachnoid space may be slowed or blocked. Hydrocephalus – with or without increased intracranial pressure – is a frequent consequence. These abnormalities in CSF dynamics account for many of the most common symptoms and signs encountered in LM such as headache, nausea and vomiting, and changes in vision. They also contribute to laboratory abnormalities and interfere with attempts at treatment. Each of these aspects will be considered in greater detail in the sections that follow.

The presence of neoplastic cells in the leptomeninges – with or without accompanying inflammatory cells – disturbs regional metabolism. This is evidenced by two common abnormal CSF findings: elevated protein and hypoglycorrhachia. Substances contributing to the elevated protein level may be derived from tumor cells, inflammatory cells, and from systemic plasma proteins that enter as a consequence of blood–CSF barrier disruption. The impact of specific intrathecally derived tumor and inflammatory cell products on CNS function remains unknown. However, potential toxicity to underlying CNS structures and functions has been postulated (Wiederkehr et al. 1991; Weller et al. 1992).

Hypoglycorrhachia may result from utilization of glucose by infiltrating cells, diminished glucose utilization by underlying CNS structures, and impaired glucose transport into the CSF (Fishman 1963; Schold et al. 1980). It is important to recognize that this is a consequence of diffuse meningeal involve-

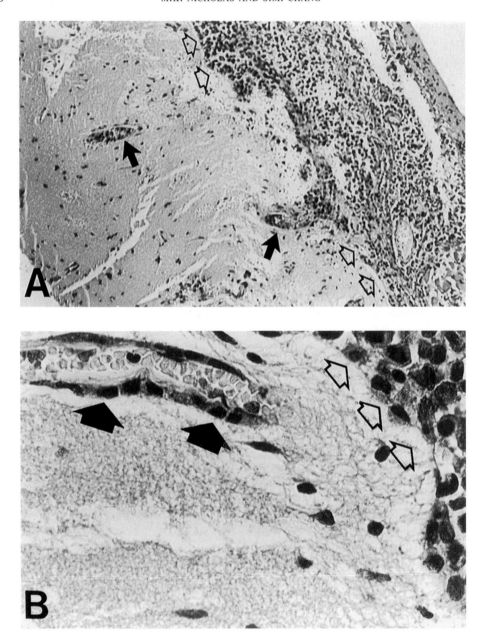

Fig. 3. Photomicrographs of a meningeal biopsy (stained with H and E), from a patient with melanoma affecting the leptomeninges. (A) Malignant and accompanying inflammatory cells can be seen filling the subarachnoid space. Infiltration of the Virchow–Robin spaces surrounding penetrating vessels can be seen (dark arrows). Open arrows indicate the border between cortex and subarachnoid space. (B) A higher powered view demonstrates more clearly the malignant cells in the Virchow–Robin space (dark arrows).

ment that occurs despite normal serum glucose concentrations. The phenomenon of hypoglycorrhachia suggests a potential metabolic competition between tumor and normal cells. There is a rare but well documented neuroendocrine syndrome characterized by isolated weight gain in patients with leukemic LM (Pochedly 1975). The underlying mechanism is hypothesized to be a metabolic competition for glucose between the leukemic cells and hypothalamic neurons critical to weight maintenance.

Focal infiltration of LM into underlying CNS parenchyma is another mechanism that accounts for commonly encountered symptoms and signs. Any CNS structure is vulnerable and symptoms vary accordingly. At times, tumor cells grow into the perivascular spaces, reaching considerable depth but never actually invading CNS parenchyma. An example of this is shown in Fig. 3. Symptoms that may result from either of these processes include seizures, transient ischemic attacks, and stroke (Wasserstrom et al. 1982; Broderick and Cascino 1987; Klein et al. 1989).

INCIDENCE

The incidence of LM is difficult to estimate and varies for different tumor types. Many cases may be asymptomatic or go unrecognized. Thus, incidence rates based upon autopsy series will exceed those determined by clinical criteria alone. For example, although one-half of patients with chronic lymphocytic leukemia will have evidence for LM on post-mortem examination, virtually none will demonstrate symptoms or signs during life (Bojsen-Moeller and Nielsen 1983). Similarly, autopsy series report subarachnoid extension of glioblastoma multiforme in 10–25% of cases although few are clinically apparent (Erlich and Davis 1978; Arita et al. 1994). On the other hand, post-mortem evidence for LM is found in approximately 20% of patients who have systemic tumors and develop neurologic symptoms (Glass et al. 1979). These patients often have coincident intraparenchymal and/or epidural metastases, as well (Bigner and Johnson 1981).

While LM has been reported for virtually every tumor type, several occur with greater frequency than others. Notable among solid systemic tumors are lung and breast cancers and melanoma (Yap et al. 1978; Aroney et al. 1981; Rosen et al. 1982; Amer et al. 1987). Of hematologic malignancies, non-Hodgkin's lymphoma and the acute lymphocytic and myelogenous leukemias predominate (Herman et al. 1979; Meyer et al. 1980). Of primary brain tumors, clinically apparent leptomeningeal involvement is most often seen in primitive neuroectodermal tumors, primary CNS lymphoma, germ cell tumors, pineoblastoma, choroid plexus carcinoma, and ependymoma (Kramer et al. 1994; Balmaceda et al. 1995).

LM may occur at any time following the diagnosis of a systemic or primary CNS neoplasm. Occasionally, leptomeningeal involvement is the first sign of systemic cancer. Estimates of this presentation range from 5 to 38%. In very few instances, however (less than 5%), does the primary tumor go unrecognized after staging procedures (Olsen et al. 1974; Wasserstrom et al. 1982). More often, LM presents as a late complication of cancer, often when systemic disease appears controlled. This may occur in isolation or concurrently with intraparenchymal or dural-based CNS metastases (Gonzalez et al. 1976; Bigner and Johnson 1981). In cases of systemic cancer associated with concurrent intraparenchymal CNS metastases, leptomeningeal involvement is more common if the brain metastasis is in the posterior fossa (Patchell et al. 1986). In the case of primary CNS tumors, leptomeningeal spread is most often associated with local disease progression. As with tumors metastatic to CNS parenchyma, primary CNS tumors of many types originating in the posterior fossa demonstrate a risk for leptomeningeal dissemination. LM is also seen in supratentorial primary CNS tumors in close contact with the CSF such as choroid plexus carcinoma and pineoblastoma.

Several tumor types once encountered with relatively high frequency in the leptomeninges are now seen less often. This is due to the recognition of the potential for LM in particular diseases and the development of successful prophylactic therapies. Acute lymphoblastic leukemia (ALL), for example, once affected the leptomeninges in approximately two-thirds of children with the disease. Prophylactic craniospinal axis irradiation and intrathecal chemotherapy has reduced this incidence to approximately 5% of pediatric patients (Bleyer 1988). Unfortunately, the same measures have had little or no impact on adult patients with ALL (Grossman and Moynihan 1991).

Similarly, patterns of leptomeningeal dissemination in medulloblastoma have been altered by the introduction of craniospinal irradiation. Following focal therapy, medulloblastoma may recur at multiple intra-CNS sites via subarachnoid spread, usually coincident with focal disease progression. Replacement of focal irradiation by treatment of the entire neuraxis in patients at risk for dissemination has had a favorable impact on disease progression and patient survival. In one series, for example, 10-year survival increased from 5 to over 50% in patients with medulloblastoma when focal irradiation was replaced by craniospinal treatment (Landberg et al. 1980). It has been argued that multifocal disease might have been mistaken for leptomeningeal dissemination in these early series since neuraxis staging procedures were not routinely performed prior to 1980 in many institutions (Berger et al. 1995, p. 568).

In contrast to the above examples, LM is increasing for several common systemic solid tumors, most notably lung and breast tumors (Yap et al. 1978; Rosen et al. 1982). Prophylactic treatments do not appear to diminish this risk (Balducci et al. 1984). In cases of small-cell lung cancer, for example, the risk of developing LM increases 50-fold over the first 3 years following diagnosis (Rosen et al. 1982). Concern is often expressed that the incidence of LM in these and other tumor types will continue to rise coincident with improvements in control of primary disease sites.

SYMPTOMS AND SIGNS

The symptoms and signs of LM are highly variable and often multifocal. This is not surprising, given the potential for the disease to affect any level of the neuraxis. Complaints may be of a generalized nature,

reflecting diffuse meningeal irritation or metabolic dysfunction. For example, a patient may present with a mild encephalopathy not otherwise specified. On closer examination, dysfunction of multiple cranial nerves and focal lower motor neuron signs might be discovered. At the other extreme, symptoms may be highly specific, suggesting involvement of a discrete portion of the subarachnoid space. For example, a patient may present with complaints of a numb chin, reflecting isolated involvement of the mental nerve. In such a case, careful examination might or might not reveal additional neurologic signs. As a rule, LM should be suspected in any patient who presents with symptoms and signs referrable to multiple levels of the neuraxis. Table 2 lists common symptoms and signs of LM. Specific abnormalities, symptom complexes, and their causes are discussed further below.

Generalized findings

Patients with LM may present with a number of symptoms for which no single cause can be found. For example, encephalopathy of varying degrees may be encountered. This could be attributed to increased intracranial pressure, cerebral hypoperfusion, meta-

TABLE 2

Symptoms and signs of LM and their associated causes.

Symptoms	Signs	Causes
Headache, nausea, vomiting, visual change, altered mental status or loss of consciousness	Papilledema, abducens palsy, limited upgaze, Cushing's response (hypertension, bradycardia, and respiratory changes) increasing head circumference (children)	Elevated intracranial pressure
Episodic motor or sensory changes, episodic loss of consciousness	Seizures, transient ischemic attacks	Perivascular and intraparenchymal infiltration
Loss of smell or vision, diplopia, facial numbness, facial droop, loss of hearing, dysphagia, dysarthria	Optic neuropathy, paresis of muscles innervated by cranial nerves, sensory loss	Infiltration of cranial nerves and their branches
Neck and back pain, paresthesias, weakness, incontinence	Kernig's and Brudzinski's signs, reflex asymmetry, dermatomal sensory loss, lower motor neuron weakness	Spinal meningeal infiltration with or without inflammation
Diminished memory, dizziness, light-headedness, gait disturbance	Mental status changes, mild ataxia	Metabolic or multifactorial

bolic derangements, or some combination thereof. Similarly, episodic loss of consciousness could represent either seizure activity or plateau waves associated with increased intracranial pressure (Broderick and Cascino 1987). Headaches, too, are usually of a generalized type and often cannot be attributed to a particular cause. The symptoms and signs associated with increased intracranial pressure (see Table 2) may accompany such headaches. Diffuse neck or back pain can usually be attributed to disseminated meningeal infiltration by neoplastic cells with or without accompanying inflammation. Signs and symptoms indistinguishable from those of infectious meningitis are often present. Finally, complaints of gait disturbance are relatively common among patients with LM although particular causes such as distal sensory loss, focal weakness, or frank ataxia are often lacking (Fisher 1982).

Specific (localizing) findings

Many of the symptoms and signs of LM are specific and of localizing value. For example, dysfunction specific to each cranial nerve has been reported (Olsen et al. 1974; Wasserstrom et al. 1982). Some cranial nerves are affected more often than others. For example, paresis of extraocular muscles is common, affecting up to one-third of patients. Facial weakness may affect up to one-quarter of patients at some time in the course of their illness. Facial numbness, on the other hand, is reported in fewer than 5% of patients. Surprisingly, although fewer than 10% of patients report tinnitus or hearing loss, diminished auditory acuity is present in at least 20% of patients. Finally, groups of cranial nerves may be affected simultaneously, often unilaterally, reflecting a metastatic leptomeningeal focus in and around an anatomic site common to them. For example, involvement of the leptomeninges within the cavernous sinus often affects the third, fourth, and sixth cranial nerves together, resulting in multiple unilateral palsies of the extraocular muscles (Ingram et al. 1991).

Neck and back pain, although often diffuse, are at times highly localized. In such cases, a radicular pattern is common, suggesting infiltration of individual nerve roots by neoplastic and/or inflammatory cells. Other signs of focal nerve root involvement include focal weakness, paresthesias, and diminished tendon reflexes. Infiltration of the cauda equina produces a symptom complex that includes bilateral weakness of the lower extremities with associated distal sensory loss, diminished tendon reflexes, and bowel and bladder dysfunction.

LABORATORY EVALUATION

CSF analysis and MR imaging are the principle diagnostic techniques used in the diagnosis of LM. The unequivocal diagnosis of LM requires the visualization of neoplastic cells in the leptomeninges or CSF. This is usually accomplished by CSF analysis, but leptomeningeal biopsies are sometimes performed. MR imaging of the brain and spine have become increasingly sensitive and often aid in the diagnosis of LM. Increasingly, patients with symptoms and signs of LM who lack diagnostic CSF cytology undergo treatment if MR images support the diagnosis (Freilich et al. 1995).

Cerebrospinal fluid analysis

Although the identification of neoplastic cells in CSF samples confirms the diagnosis of LM, false negative results are common. Nevertheless, CSF evaluation remains the most reliable means to a diagnosis of LM. It is rare, in fact, to find an entirely normal CSF profile in patients with LM. Wasserstrom et al. (1982) published results of serial lumbar punctures in 90 patients with LM from a variety of tumor types. In only 3% of patients were all parameters, including opening pressure, protein and glucose levels, cell count and cytology normal on the first specimen. Furthermore, this number decreased to 1% upon repeated evaluations. Table 3 summarizes the CSF

TABLE 3

CSF abnormalities in patients with clinically suspected LM.

Parameter measured	% Positive* (initially/at any time)
Elevated CSF pressure	50/70
Protein elevation	75/90
Hypoglycorrhachia	30/40
Increased cell count	50/70
Positive cytology	50/90
Biochemical markers	varies considerably with tumor type

*Data combined from Wasserstrom et al. (1982), Grossman and Moynihan (1991) and Zachariah et al. (1995).

abnormalities commonly encountered in patients with LM. They are discussed in further detail below.

Elevations of CSF pressure occur in the majority of LM patients, even those without symptoms, signs, or radiographic evidence for hydrocephalus. Because both neoplastic and inflammatory cells in the CSF interfere with its reabsorption, increased pressure can usually be attributed directly to LM. Mass lesions may, however, cause focal obstruction, contributing to the phenomenon. Furthermore, systemic causes for increased intracranial pressure need to be considered in many cancer patients. For example, patients with respiratory failure or systemic venous compression might have elevated CSF pressures as a result of either elevated arterial pCO_2 levels or increased venous pressures.

Protein levels are often elevated and glucose levels depressed in patients with LM. Protein elevation may be the result of blood–CSF barrier incompetence with entry of normally excluded serum proteins into the CSF. Intrathecal protein production by neoplastic and inflammatory cells may contribute. Often the increase is a result of both factors. Protein and immunoelectrophoresis on paired serum and CSF samples can be of value in assessing patients for LM. This is particularly true in the evaluation of immunoglobulin-producing neoplasms (Siegal et al. 1981).

Hypoglycorrhachia occurs often in LM. This may reflect the increased metabolic demands of neoplastic cells, inflammatory cells, or both, on available glucose stores. Alternatively, hypoglycorrhachia may result from impaired of glucose transport proteins active at the BBB and blood–CSF barriers (Fishman 1963). Finally, it has been suggested that cerebral tissues underlying areas of LM might utilize glucose at an increased rate, thus contributing to the overall picture (Fishman 1992, p. 309).

Cell counts vary widely in LM. The CSF may contain innumerable tumor cells with little or no accompanying inflammation. At the other extreme, a marked pleocytosis may be present with few, if any, tumor cells present. Lymphocytes usually comprise the majority of the inflammatory CSF infiltrate. Polymorphonuclear cells and eosinophils, however, may be present, leading one to suspect either bacterial or parasitic infection (Mulligan et al. 1988). Occasionally, reactive lymphocytes in the CSF can be mistaken for malignant cells (Recht et al. 1988; Kappel et al. 1994). The presence of red blood cells (RBCs) in the

CSF usually indicates a traumatic lumbar puncture. However, leptomeningeal tumors can and do bleed. In such instances, CSF xanthochromia is more common than are numerous RBCs themselves. Clearing of RBCs on serially collected tubes of non-xanthochromic CSF generally suffices to rule out anything other than a traumatic procedure.

CSF cytology remains the cornerstone upon which diagnoses of LM are made. CSF should be promptly and properly fixed as false negative results are common. The frequency of positive results increases by collecting serial samples. In the series of Wasserstrom et al. (1982), a positive cytology was obtained on the first lumbar puncture in approximately one-half of the patients. Repeated evaluations increased the frequency to approximately 90%. Pathologists at some institutions divide the CSF into two portions, evaluating unfixed specimens on cytospin preparations and a separate, milipore-filtered fixed specimen. Examples of CSF cytology from several patients are shown in Figs. 4 and 5.

Although emphasis is placed on the incidence of false-negative findings, false-positive cytologies can also be reported. As noted above, reactive lymphocytes may be mistaken for malignant cells, especially in patients with viral infections of the CNS. Positive cytology may also be encountered in patients with in-

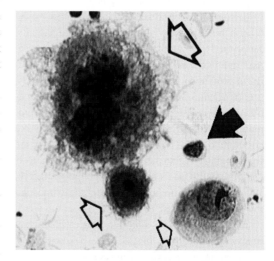

Fig. 4. Malignant cells from the CSF of a patient with a history of adenocarcinoma of the lung. A normal lymphocyte (black arrow) is shown surrounded by three malignant cells (open arrows). Note the marked nuclear and cytoplasmic pleomorphism of the tumor cells.

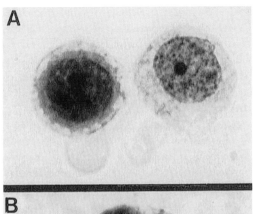

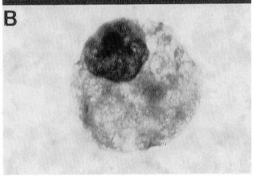

Fig. 5. Malignant cells from the CSF of a patient with adenocarcinoma of the breast. The cells in A were cytocentrifuged from unfixed CSF directly onto slides within minutes of their collection. The cell shown in B was filtered through a Millipore filter prior to staining. In the sample shown, the malignant phenotype of the cells is evident by either technique.

traparenchymal CNS tumors who lack symptoms and signs of LM (Recht et al. 1988). This is especially true if lumbar punctures are performed following CNS surgery. Nevertheless, leptomeningeal involvement is usually found at autopsy in patients with a positive CSF cytology who have not undergone recent surgery (Glass et al. 1979). Some controversy exists as to the treatment such patients should receive.

A number of biochemical markers have been studied in CSF samples from patients with known or suspected LM and are outlined in Table 4. Care must be exercised in their evaluation, however, as the sensitivity and specificity of each vary considerably and CSF levels can be affected by concomitant intraparenchymal metastases and breakdown of the BBB and blood–CSF barriers. Several of the more specific of these markers, such as α-fetoprotein and β-HCG, are useful in monitoring the status of LM after treat-ment of specific tumors. They may rise in the CSF prior to the reappearance of clinically overt or radio-graphically evident disease (Allen et al. 1979). These markers are routinely used at many institutions to follow disease status. Other biochemical markers, the polyamines, for example, may be highly predictive of tumor recurrences in specific diseases but few laboratories are equipped to measure them (Marton et al. 1981).

Lumbar punctures can be performed safely on the majority of patients with LM. Any patient suspected of harboring a mass lesion in either the brain or spinal cord should, however, undergo diagnostic neuro-imaging first. If a mass is identified, a decision regarding lumbar puncture should proceed on a case- by-case basis. Often, steroids can be used to reduce mass effect prior to performance of a lumbar puncture. Patients with suspected abnormalities of coagulation, either as a direct effect of their tumor or as a consequence of treatment, should undergo pre-lumbar puncture laboratory evaluation. Patients who are markedly thrombocytopenic should receive platelet transfusions at the time of lumbar puncture. Any such patient at risk for spinal hemorrhage who develops symptoms and signs following lumbar puncture should also undergo platelet transfusion.

CSF is usually obtained by lumbar puncture. At times, however, lumbar puncture is contraindicated. In these instances, CSF may be accessible from cisternal puncture or from intraventricular drains and reservoirs. Occasionally, CSF obtained from one of these sites will yield a diagnosis of LM when lumbar CSF has not. For example, several series have compared CSF collected from the lumbar and cisternal spaces of the same patients and documented a higher percentage of positive cytologic diagnoses following cisternal puncture (Murray et al. 1983; Rogers et al. 1992).

Caution should be exercised in the comparison of CSF protein and glucose concentrations in samples taken from different levels of the neuraxis. As a rule, protein concentrations are lower and glucose concentrations higher in the ventricle than in the lumbar space (Murray et al. 1983). Finally, all CSF parameters in fluid obtained from Ommaya reservoirs should be interpreted with caution. CSF within these reservoirs does not circulate normally. As a result, elevated protein concentrations and atypical degenerated cells are often encountered.

TABLE 4

Biochemical markers found in the CSF of patients with LM.

Biochemical marker	Tumor(s)	Comments	Reference
α-Fetoprotein	Embryonal carcinoma (EC)	Levels often rise before other signs of recurrence	Allen et al. 1979
β-HCG	EC and choriocarcinoma	As above	As above
Carcinoembryonic antigen (CEA)	Many tumor types	May correlate with disease progression	Yap et al. 1980
β$_2$-Microglobulin	Lymphoma	Positive in infectious meningitis	Koch et al. 1983; Starmans et al. 1977
Melanin	Melanoma	Brown or black CSF	Lups and Haan 1954
Lactate dehydrogenase isoenzyme 5 (LDH-5)	Many tumor types	Positive in some infections	Rogers et al. 1987
Prostate specific antigen	Prostate cancer		Mencel et al. 1994
Polyamines	Medulloblastoma	Few labs equipped	Marton et al. 1981
CA-125	Ovarian cancer		

Meningeal biopsy

Occasionally, patients will undergo meningeal biopsy to arrive at a diagnosis. This may occur in situations where repeated CSF analyses are unrevealing but clinical suspicion remains high. An example of primary leptomeningeal melanoma diagnosed by biopsy is shown in Fig. 3. Increasingly, patients with known cancer and suspected LM are treated without meningeal biopsy if clinical findings and MR imaging support the diagnosis. In rare cases of tumors arising within the leptomeninges themselves, however, a biopsy may be the only means to a diagnosis.

NEUROIMAGING

Neuroimaging plays an increasingly important role in the assessment of LM. Computed tomography (CT) and MR images are both commonly used in the evaluation of patients with known or suspected LM. MR imaging is, without question, more sensitive than CT scanning, but findings consistent with the diagnosis of LM are seen with both techniques. At times, myelography and radionuclide studies are useful as well. These techniques and the findings associated with LM are outlined in Table 5 and discussed further below.

Findings on CT and MR scan that support the diagnosis of LM include communicating hydrocephalus and enhancement of the meninges, cranial and spinal

TABLE 5

Radiologic techniques used in the diagnosis of LM.

Technique	Findings associated with LM
CT	Enhancement at sites of blood–CSF barrier breakdown, hydrocephalus, low-density abnormalities in the periventricular white matter
MRI	Findings similar to those seen on CT scan, but more sensitive to all changes, always superior to CT scan in the evaluation of posterior fossa and spine
Myelography	Largely supplanted by MR imaging, may occasionally be positive when other studies are negative
Nuclear studies	Used in the assessment of CSF flow

nerves, subependymal regions, and CSF-containing cisterns (Lee et al. 1984; Davis et al. 1987; Krol et al. 1988; Rodesch et al. 1990). Once seeded, the leptomeninges may support the growth of sizable tumors causing little diagnostic confusion. Communicating hydrocephalus has several causes in addition to LM, thus the finding is non-specific. Suspicion of the diagnosis increases if the ventricles are symmetrically enlarged and are accompanied by periventricular white matter changes suggesting transependymal CSF absorption. Again, similar abnormalities may be seen as a consequence of radiotherapy or chemotherapy.

Characteristic patterns of enhancement on MR images are shown in Figs. 6–8 and discussed further below. Leptomeningeal enhancement may be seen anywhere along the neuraxis. Fig. 6 demonstrates characteristic patterns of enhancement in brain including diffuse linear enhancement patterns throughout the leptomeninges and punctate nodular enhancement along the surface of the brainstem and pons. Fig. 7 demonstrates the size to which diffuse LM may grow. Cranial nerve enhancement is also seen. Fig. 8 demonstrates the diffuse leptomeningeal enhancement often seen in the leptomeninges surrounding the spinal cord. Patterns of enhancement are often far subtler than those shown. Some advocate the use of increased doses of gadolinium in such cases. Others have found this technique of little additional value (Tam et al. 1996). It should be noted that higher doses of contrast materials may create confu-

sion as normal vessels, particularly in and around the spinal cord, may appear as discrete nodules.

Diffuse meningeal enhancement mimicking LM may be encountered in a number of circumstances. These include infection and hemorrhage (Chang et al. 1988; Phillips et al. 1990). Rare cases of inflammatory arthritis and eosinophilic granuloma have also been reported (Seltzer et al. 1992). Diffuse leptomeningeal enhancement may also be seen following neurosurgical procedures and as a consequence of hydrocephalus (Burke et al. 1990; Schumacher et al. 1994). Finally, a similar pattern is rarely reported following lumbar puncture but should be considered a diagnosis of exclusion (Mittl and Yousem 1994).

Other diagnostic neuroimaging techniques include myelography and radionuclide CSF flow studies. Myelography has been largely replaced by MR imaging. Rarely, myelography will suggest evidence for spinal LM when MR imaging is negative (Krol et al. 1988). Radionuclide flow studies can be useful in defining both the extent of subarachnoid disease and

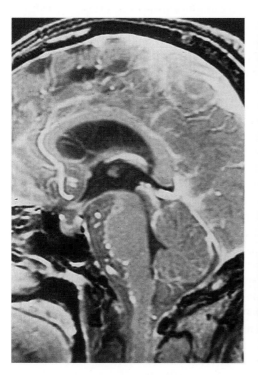
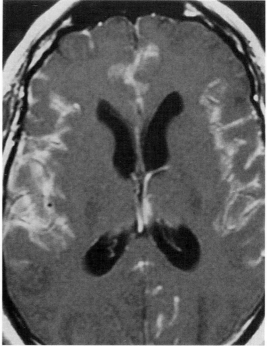

Fig. 6. T1 gadolinium-enhanced sagittal (left panel) and axial (right panel) brain MRI scans from a patient with breast cancer who developed multiple cranial nerve abnormalities on a background of mental status changes. Both panels demonstrate the diffuse leptomeningeal enhancement characteristic of LM. Note the nodular enhancement along the spinal cord and ventral brainstem on the sagittal view. More subtle subependymal enhancement can be seen in the ventricles on the axial view.

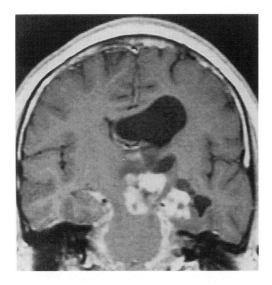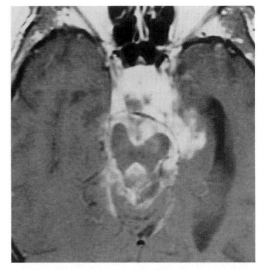

Fig. 7. T1 gadolinium-enhanced coronal (left panel) and axial (right panel) brain MRI scans from a patient with a disseminated juvenile pilocytic astrocytoma. There is considerable distortion of normal brain structure with ventricular trapping.

CSF compartmentalization resulting from blockage of normal CSF pathways (Grossman et al. 1982). Finally, although rarely used for diagnostic purposes, cerebral blood flow studies have demonstrated decreased blood flow in cortical regions underlying leptomeningeal infiltration (Siegal et al. 1985).

GENERAL TREATMENT GUIDELINES

Diagnostic considerations

As stressed above, patients with LM may present with tremendously variable symptoms and signs. Attempts at diagnosis are often frustrating or confusing. Any patient with signs suggesting LM should undergo MR imaging of the potentially affected area(s). Ultimately, the entire neuraxis should be imaged. If MR imaging results are either normal or suspect for LM (see section on Neuroimaging above), a lumbar puncture should follow. If a mass lesion is discovered, lumbar puncture may be delayed until steroid administration has reduced associated edema. At times, lumbar puncture may be completely contraindicated. In some cases, CSF evaluation will necessarily follow craniotomy for such a lesion. When this occurs, delays in CSF evaluation of several weeks are often suggested and even then results are interpreted cautiously.

Patients with a positive CSF cytology should proceed with radio- and chemotherapy, treatment being tailored to the disease in question. In cases of LM from systemic tumors, staging outside the CNS should also be performed as this may impact treatment planning. Occasionally, specific biochemical markers may be positive in the CSF of patients with negative cytology. For instance, a marked increase in CSF α-fetoprotein and β-HCG levels in a patient with embryonal carcinoma might herald disease recurrence. In such a case, treatment might be initiated on these grounds alone. More often, when initial CSF cytology is negative, repeated lumbar punctures for further evaluation are warranted. If at least three CSF evaluations and all MR imagings are normal, patients should be followed clinically. Occasionally, myelography may be of benefit. Finally, in cases where the leptomeninges are clearly involved both clinically and on MR images but no clue to etiology is suggested by history, systemic staging, or CSF analysis, meningeal biopsy should be considered.

Radiotherapy

Most patients with LM receive external beam radiation treatment (EBRT) at some point in the course of their therapy. Other radiation modalities, including stereotactic radiosurgery and intrathecal radiation,

Fig. 8. T1 pre-infusion (left panel) and post-infusion (right panel) MRI scans of the spinal cord from a patient with a remote history of infiltrating ductal carcinoma of the breast who presented with progressive back pain and sensory loss in the lower extremities. Diffuse leptomeningeal enhancement characteristic of LM is seen following the administration of contrast.

may be used in special circumstances (Doge and Hliscs et al. 1984; Moseley et al. 1990). Many patients with systemic leukemias and lymphomas or primary CNS tumors at risk for LM will have received EBRT to some part or all of their neuraxis as part of initial therapy. Options for additional EBRT are limited by potential toxicity in such cases. However, additional palliative treatment to sites of major symptoms may occasionally be considered.

Chemotherapy

Several factors need to be considered in the selection of chemotherapies for LM. First, the agents must have activity against the tumor in question. Second, if administered parenterally, adequate concentrations must be achieved in the leptomeninges. Third, if administered intrathecally, the agents must have a low CNS toxicity profile.

If intrathecal therapy is selected, the route of administration must also be considered. Agents may be injected directly into the subarachnoid space via lumbar or even cisternal puncture. Alternatively, a reservoir can be placed into the lateral ventricle for intraventricular drug administration. While treatments are necessarily individualized, there are advantages to the intraventricular route of administration. Drugs administered to the subarachnoid space by lumbar puncture may or may not be properly placed. Inadvertent epidural or subdural placement of drug has been documented to occur up to 10% of the time following lumbar puncture, even in experienced hands (Larson et al. 1971). Furthermore, because drugs are injected repeatedly, patients are at increased risk for post-lumbar puncture headache and bleeding complications. In addition, lumbar injection almost invariably results in low drug concentrations in the ventricular CSF, even if good distribution occurs throughout the subarachnoid space (Grossman et al. 1982). Finally, MR imaging artifacts that mimic LM may follow lumbar puncture, thus confusing assessment of treatment response.

Drugs administered via an Ommaya reservoir, on the other hand, are guaranteed to reach higher concentrations in the ventricular CSF than those introduced by lumbar puncture. Once a reservoir is placed, patients are relieved of the discomfort associated with repeatedly accessing the CSF. In addition, treatment may be given via an Ommaya reservoir when lumbar puncture is contraindicated, as would be the case in a patient with thrombocytopenia.

The choice of drugs for intrathecal administration is limited by the tolerance of normal CNS tissues to most agents. The most commonly used agents are methotrexate (MTX) and cytarabine (Ara-C). Thiotepa is used less often. Many others have seen limited use or have been tested in animal models. These agents are listed in Table 6. General guidelines for the administration of MTX, Ara-C, and thiotepa follow.

The anatomic and physiologic complexities of the CSF-containing spaces create challenges to treatment. On theoretical grounds, one would predict that intrathecal drug administration would result in better distribution and higher concentrations of drug in the CSF spaces than would parenteral administration.

TABLE 6

Agents used in the intrathecal treatment of LM.

	Reference*
Commonly used agents	
MTX and Ara-C	Yap et al. 1982, Steinherz et al. 1985, Hitchins et al. 1987, Nakagawa et al. 1992
Thiotepa	Gutin et al. 1977, Giannone et al. 1986, Stewart et al. 1987
Infrequently used agents	
Dacarbazine	Champagne and Silver 1992
Diaziquone	Kamen et al. 1982
Colloidal gold	Doge and Hliscs 1984
Lymphokine activated killer cells with IL-2	Shimizu et al. 1987, Samlowski et al. 1993
Ganglioside antibodies	Dippold et al. 1994
Agents being tested in clinical trials	
Intrathecal MTX given with multiagent	DeAngelis et al. 1992
Parenteral chemo-therapy, followed by radiotherapy and post-radiation parenteral chemotherapy for patients with primary CNS lymphoma	Radiation Therapy Oncology Group, protocol RTOG 93-10
DTC 101-DepoFoam™ encapsulated Ara-C administered intra-thecally to patients with LM from any source	Chamberlain et al. 1993
Agents tested in experimental animals	
Melphalan	Friedman et al. 1994
ACNU	Arita et al. 1988
MCNU	Kochi et al. 1994
Topotecan	Sung et al. 1994
Toxin-conjugated antibodies	Zovickian and Youle 1988
Gene therapy	Ram et al. 1994
Trimetrexate	Balis et al. 1986

*References are intended to be representative, not all-inclusive.

While this may be true in principle, it is important to recognize the potential for abnormal CSF flow in cases of LM (Grossman et al. 1982). The following outline of intrathecal chemotherapy administration assumes reasonably normal CSF flow.

MTX is usually administered to persons over 4 years of age as 10–12 mg doses given twice weekly. In general, this will result in sustained therapeutic intra-CSF drug levels (1 μM) for up to 72 h, thus justifying the dosing schedule. Similar levels can be achieved and maintained using smaller, more frequent doses, but this necessitates more frequent manipulation of intraventricular reservoirs and/or lumbar punctures (Strother et al. 1989).

MTX is cleared from the CNS by bulk flow without intrathecal metabolism. Treatment can, in theory, be continued indefinitely and is usually recommended for several months. After 5–8 doses have been administered by this schedule, the interval between doses is increased, first to weekly and finally monthly intervals. Oral leucovorin is routinely administered as protection against systemic toxicity. There is little chance that significant quantities of leucovorin will enter the CSF to interfere with MTX function.

Ara-C, a synthetic pyrimidine nucleoside, can be used in place of or in alternating doses with MTX. 50-mg doses are administered twice weekly, or 30 mg 3 times per week. Therapeutic levels ($>4 \times 10^{-4}$) are achieved at these doses for periods of 72 h or longer (Zimm et al. 1984). Ara-C has been used most extensively in the treatment of leukemia and lymphoma and is less active against most solid tumors.

Thiotepa, an alkylating agent, is used less often and given in 10-mg doses, also twice weekly. Thiotepa is highly lipid soluble and its distribution within the CSF spaces differs from MTX and Ara-C. Nevertheless, in one study median survival did not differ between patients with LM who were randomized to receive either MTX or thiotepa, supporting a role for its activity despite its lipid solubility (Grossman et al. 1993).

Attempts to enhance sustained intrathecal drug delivery to patients with LM have been made. Foam-encapsulated Ara-C has been safely administered to patients with LM and been demonstrated to have activity (Chamberlain et al. 1993). The final results of a phase III study comparing this treatment to the direct intrathecal administration of MTX in patients with LM have not yet been reported.

TABLE 7

Potential toxicities of treatment.

Toxicity (reference)	Features
Leukoencephalopathy (Bleyer 1981) Usually associated with a history of combined radio- and chemotherapies	White matter changes with demyelination, axonal swelling, coagulative necrosis. Highly variable clinical course ranging from subtle cognitive changes to profound morbidity.
Myelopathy (Gagliano and Costanzi 1976) A consequence of intrathecal chemotherapy; considered idiosyncratic.	Spinal cord necrosis without inflammation. Pain followed by motor and sensory deterioration. Recovery variable but generally poor.
Minealizing microangiopathy (Price and Birdwell 1978) A side effect of radiation and possibly chemotherapy.	Degenerative changes of vessel walls, primarily in grey matter. A possible cause of cognitive abnormalities.
Infection Immunity may be impaired by the tumor, its treatment, or both.	Patients often vulnerable to viral, fungal and unusual bacterial infections.
Vascular complications (Graus et al. 1985) Coagulation may be affected by the tumor or its treatment.	Patients susceptible to multiple sequelae including disseminated intravascular coagulation, embolization, infarction, or venous occlusion.

Treatment-associated toxicities

All current therapies with any potential activity against tumors affecting the leptomeninges are also capable of damaging the CNS. Toxicities may be the direct result of treatment. Examples would include radiation- or chemotherapy-induced myelopathy or cognitive change. Injury may also occur more indirectly. For example, infection or bleeding may follow a lumbar puncture or placement of an intraventricular reservoir. Many patients with LM suffer considerable morbidity before subjecting themselves to difficult therapies and tolerate untoward effects poorly. Potential toxicities associated with the treatment of LM are outlined in Table 7.

TUMORS MOST COMMONLY ASSOCIATED WITH LEPTOMENINGEAL METASTASIS

Systemic solid tumors

In theory, any tumor should be capable of LM. Indeed, case reports and limited series are published for a large number of tumor types. It is obvious, however, that LM occurs far more often with some tumor types than others. Breast and lung cancers, for example, metastasize frequently while other common neoplasms, such as those of the prostate, are rarely encountered. Table 8 outlines those tumors reported to metastasize to the leptomeninges. Although exten-

sive, it is not exhaustive and case reports have, undoubtedly, been omitted. The more common tumors associated with LM and their management are discussed below.

Breast cancer

Carcinoma of the breast is among the commonest of solid tumors reported to metastasize to the leptomeninges. Autopsy-based estimates of incidence suggest that 3–40% of breast cancer patients harbor LM (Lee 1983). Coexistent intraparenchymal CNS metastases are often present. LM is often a late complication of breast cancer, but exceptions to this rule are seen (Yap et al. 1978). Recent trends in the initial treatment of breast cancer have affected the patterns of those who present with LM. Freilich et al. (1995) reported the incidence of isolated CNS metastases in 6 of 52 breast cancer patients whose primary tumors responded to treatment with paclitaxel. Four of the six had leptomeningeal involvement. Laboratory data demonstrate low paclitaxel levels in both CSF and brain parenchyma which may account for these isolated CNS relapses.

The treatment of LM in breast cancer usually consists of EBRT to the area(s) of major symptoms with intrathecal and/or parenteral chemotherapy for attempted control of the remainder of the disease in the subarachnoid space and elsewhere (when present). MTX is the drug most commonly used for intrathecal

TABLE 8

Tumors known to metastasize to the leptomeninges.

Tumor type	Reference
Breast cancers**	Yap et al. 1978
Lung cancers**	Aroney et al. 1981
Melanoma**	Amer et al. 1987
Leukemias	
Acute lymphocytic and myelogenous**	Meyer et al. 1980
Chronic lymphocytic	Berry and Jenkin 1981
Lymphomas	
Non-Hodgkin's*	Herman et al. 1979
Hodgkin's disease	Bender and Mayernick 1986
Primary CNS	Balmaceda et al. 1995
Head and neck tumors	Redman et al. 1986
Thyroid cancer	Barnard and Prasons 1969
Prostate cancer	Lyster et al. 1994
Squamous cell carcinoma	Weed and Creasman 1975
Rectal cancer	Bresalier and Karlin 1979
Mycosis fungoides	Lundberg et al. 1976
Gastric carcinoma	Moberg and Reis 1961
Carcinoid	Naggourney et al. 1985
Cervical cancer	Aboulafia et al. 1996
Renal cell cancer	Crino et al. 1995
Transitional cell carcinoma	Hara et al. 1994
PNET	Jennings et al. 1993
Gliomatosis	Dietrich et al. 1993
Pilocytic astrocytoma	Versari et al. 1994
Meningioma	Von Deimling et al. 1995
Germinoma	Ono et al. 1994
Oligodendroglioma	Chen et al. 1995
Retinoblastoma	Bigner and Johnson 1981
Ependymoma	Rezai et al. 1996
Pineocytoma	D'Andrea et al. 1987
Meningioma	Vinchon et al. 1995
Pituitary adenoma	Tonner et al. 1992

*References are intended to be representative, not all-inclusive.
**The most commonly encountered systemic tumors.

therapy. It should be noted, however, that many patients with LM caused by breast cancer also receive parenteral chemotherapy. Some physicians prefer to restrict treatment to the parenteral route, or to forego treatment altogether, citing a lack of evidence for any significant benefit with intrathecal treatment (Clamon and Doebbeling 1987).

Given the state of the literature regarding prognosis in breast cancer-associated LM, it is not surprising to find differences of opinion regarding the approach to treatment. In some series, approximately 25% of patients survive for 1 year following treatment with intrathecal MTX and radiotherapy (Ongerboer de Visser et al. 1983; Siegal et al. 1994). In others, little if any benefit to treatment is demonstrated (Grossman et al. 1993). Furthermore, treatment-associated complication rates can be high (Balm and Hammack 1996). Prognostic factors such as age, systemic disease status and degree of leptomeningeal involvement at diagnosis likely affect the outcomes reported in these relatively small series (Boogerd et al. 1991). Nevertheless, median survival in most series, despite aggressive therapy, seldom exceeds 20 weeks (Grant et al. 1994). In general, patients with few fixed neurologic deficits whose systemic disease is well controlled or slowly growing, tolerate and respond most favorably to treatment. Given the wide range of responses observed, we recommend aggressive therapy in those patients aware of the potential risks and benefits of therapy whose functional status predicts reasonable life quality. Patients with advanced leptomeningeal symptoms sometimes improve transiently with treatment, thus, the option for treatment should at least be considered in these patients.

Lung cancer

The presentation of and treatment approach to patients with LM caused by lung cancer are similar to those observed in those with breast cancer cited above. Although lung tumors of all histologies are capable of seeding the leptomeninges, small-cell lung cancer far exceeds any other type. In fact, a greater percentage of patients with small-cell lung cancer will develop LM than any other group. Improved treatment of systemic disease has resulted in a dramatic increase in the risk for developing LM (Rosen et al. 1982). Unfortunately, attempts at prophylactic treatment of the CNS have not been successful (Balducci et al. 1984). Concomitant brain and epidural metastases are often seen (Aroney et al. 1981).

As is the case with breast cancer patients, focal bulky CNS disease is treated with radiotherapy and the CSF-containing spaces are treated with chemotherapy. Although the CSF seldom clears entirely of malignant cells, even with aggressive intrathecal treatment, the majority of patients with small-cell lung cancer and LM who receive radiation and intrathecal chemotherapy die of progressive systemic disease rather than LM itself or complications of treatment (Rosen et al. 1982). In general, patients with non-small-cell lung cancer respond very poorly to treatment (Grant et al. 1994). In some cases, supportive care may be the most appropriate initial medical intervention.

Melanoma
Melanoma is another disease, increasing in frequency, for which better systemic therapies have been recently devised. Coincident with improved systemic disease control are isolated intra-CNS relapses (Thomas et al. 1982; Mitchell 1989). When the leptomeninges are affected, it is usually in the context of concurrent intracranial mass lesion(s). The pattern of isolated CNS recurrence has been encountered following the administration of both conventional cytotoxic chemotherapies (Thomas et al. 1982) and more novel biomodulators (Mitchell 1989). The potential for bleeding from intraparenchymal lesions often limits therapeutic options in patients with LM from melanoma. As a rule, patients with melanoma-associated LM respond less well to treatment than do those with breast cancer or small-cell lung cancer, leading to greater reservations in recommending aggressive therapy (Grant et al. 1992).

Hematologic malignancies

Taken together, the leukemias and lymphomas comprise a substantial percentage of neoplasms in both children and adults. Leptomeningeal involvement is common among some leukemias and lymphomas and equally uncommon among others. We will consider each in turn below.

Leukemia
Clinically overt leukemic involvement of the leptomeninges was a rare event prior to the development of successful systemic therapies in the 1960s. Soon thereafter, however, the leptomeninges were identified as a common site for recurrence. This was especially true for ALL in children where the risk of developing LM within the first year of systemic treatment was as high as 50%. This fell to approximately 5–10% following the introduction of prophylactic radiation and chemotherapy strategies (Bleyer 1988).

Both radiation and chemotherapy are used in the treatment of children with LM secondary to leukemia. Coexistent bulky disease often affects the CNS and radiation is applied to the areas of greatest involvement. Because children may have received prior radiotherapy, the extent to which radiation can be used may be limited. Fortunately, advances in staging and treatment have substantially reduced the number of children who receive prophylactic radiation (Ganyon et al. 1993).

Options for intrathecal chemotherapy consist of MTX, Ara-C, thiotepa, or some combination thereof. These drugs are administered as described in the sections on solid tumors above. In addition, systemic high-dose chemotherapy that achieves therapeutic CSF concentrations has favorably impacted ALL-associated LM in some instances (Lopez et al. 1985; Ackland and Schilsky 1987). Although prolonged responses occur, of children aggressively treated, approximately one-third will die of progressive CNS disease (Chamberlain 1995). Of the remainder, most will succumb to progressive systemic disease. Thus, much of the treatment remains palliatively directed.

While ALL is commonest in children, adult cases are also reported. More often, adults present with LM following treatment of acute or chronic myeloblastic leukemia. Clinically apparent LM attributable to chronic lymphocytic leukemia is exceedingly rare in patients of any age (Liepman and Votaw 1981). Because the risks for CNS involvement are lower for adults than they are for children, CNS-oriented prophylactic treatment is not usually given. The incidence of LM associated with ALL in adults who do not receive prophylactic therapy is approximately 20%. In acute myeloblastic leukemia incidences of 10% have been reported (Stewart et al. 1981; Kantarjian et al. 1988). In most cases, prophylactic treatment has done little to alter this course (Grossman and Moynihan 1991). However, adult patients with LM associated with acute non-lymphocytic leukemias may respond favorably to treatment (Dekker et al. 1985).

Lymphoma

Lymphomas of all histologies can metastasize to the leptomeninges. However, Hodgkin's disease is rarely involved while LM may complicate the course in 10% of patients with non-Hodgkin's lymphomas (Ersboll et al. 1985). LM may occur at any time in the course of the disease, but most often occurs after initial treatment and in the setting of progressive systemic disease (Mackintosh et al. 1982).

As with most leptomeningeal tumor syndromes, LM associated with lymphoma of any type, if left untreated, carries a poor prognosis. Most patients, however, respond to multimodality therapy with improvement in symptoms and, often, extended life. Following diagnosis and staging, patients usually receive radiation to the most heavily affected areas as well as intrathecal chemotherapy. In addition, steroids, often used to treat symptoms, have a direct cytolytic effect on the tumor cells.

Multimodality therapy results in symptom control or improvement in over 75% of patients and 25% may survive longer than 18 months (Recht et al. 1988). Despite this, median survival ranges from 1 to 8 months (Bunn et al. 1976; Mackintosh et al. 1982; Recht et al. 1988).

Primary CNS tumors

Primary tumors of the CNS have different propensities for spread via the subarachnoid space. Apart from factors such as tumor histology, location, and patient age in some cases, few factors predictive of risk for dissemination have been identified. As mentioned above, the incidence of LM from primary CNS tumors varies considerably for all tumor types. Many of the tumors are uncommon and most of the reports describing the natural history of these tumors span decades when more sensitive diagnostic tests were not available. Not surprisingly, more recent reports tend to describe higher incidences of LM for all tumor types. Finally, as with systemic tumors, autopsy-based series suggest a greater incidence of LM than is clinically apparent for some tumor histologies (Erlich and Davis 1978).

Since most primary CNS tumors present as intracranial mass lesions, there is appropriate reluctance to perform a lumbar puncture for staging purposes at the time of diagnosis. For those tumors with a known propensity for LM, CSF evaluation in conjunction

TABLE 9

Primary tumors with a high propensity to spread in the subarachnoid space*.

Primitive neuroectodermal tumors
Medulloblastoma
Ependymoblastoma
Pineoblastoma
Cerebral neuroblastoma
High-grade glioma in children
Posterior fossa ependymoma
Germ cell tumors
Pineocytoma in children
Choroid plexus carcinoma
Primary CNS lymphoma

*Staging is recommended.

with neuroimaging is recommended. This is important in both prognosis and selection of therapy. Diagnostic confusion may arise if the timing of CSF and neuroimaging evaluation relative to intracranial surgical procedures is ignored. In general, staging investigations, if not performed pre-operatively, should be delayed for 2–3 weeks following surgical procedures to limit this confusion.

Primary CNS tumors are more common in children than adults and recommendations for staging and surveillance of children with these tumors have been proposed (Kramer et al. 1994). The recommendations were based upon the observed risks for LM of particular tumor types. Similar guidelines have not been published for adults with primary CNS tumors. With the exception of pineocytomas and high-grade gliomas, the risks for LM are considered equivalent for these age groups. Those tumors most likely to spread via the subarachnoid space are listed in Table 9 and described in greater detail below.

Primitive neuroectodermal tumor (PNET)

PNETs represent 7–8% of all intracranial tumors and 30% of childhood CNS neoplasms. Leptomeningeal dissemination is common and staging of the neuraxis is recommended at diagnosis and subsequent to initial therapy for surveillance purposes in both children and adults. Evidence for neuraxis dissemination at the time of diagnosis is common, with incidence rates of 20–46% reported in various series (Packer et al. 1985; Allen and Epstein 1988; Deutsch 1988; Prados et al. 1995). In rare instances, LM is reported as the initial manifestation of PNET in the absence of an obvious mass lesion (Jennings et al. 1993). Neuraxis dissemination is an important prognostic factor and affects the choice of therapy for the patient.

As mentioned above, the addition of craniospinal irradiation has had a positive impact on both focal and disseminated disease recurrence in so-called high-risk medulloblastoma patients (Landberg et al. 1980). Risk factors include young age, large tumor size at diagnosis, brainstem involvement, incomplete resection (residual tumor measuring >1.5 cm), tumor outside the posterior fossa, and subarachnoid spread. In children with these risk factors, adjuvant chemotherapy is also of benefit. Significant increases in the event-free 5-year survival rates were seen following the addition of adjuvant chemotherapy to surgery and radiotherapy (Evans et al. 1990). Similar results are suggested for adults (Prados et al. 1995). Obviously, not all patients in these studies had leptomeningeal dissemination at the time of diagnosis. Furthermore, LM was not evident upon treatment failure in all cases. However, these adjuvant therapies do appear to favorably impact overall progression-free survival rates, including those with and at risk for leptomeningeal involvement.

It should be noted that patients who present with overtly disseminated leptomeningeal PNET may respond poorly to treatment. In one series, all patients (7 of 7) whose disease came to diagnosis as a result of diffuse leptomeningeal involvement experienced progressive disease despite craniospinal irradiation and chemotherapy (Jennings et al. 1993). Similarly, survival is poor when recurrent disease is disseminated. In such cases, current treatments are palliative.

Glioma
Gliomas are the most common primary CNS tumors, accounting for more than 50% of the total. Clinically significant leptomeningeal dissemination of infiltrating glial neoplasms is reported to occur in three settings (Cairns and Russell 1931; Moore and Eisinger 1963; Yung et al. 1980; Herman et al. 1995). LM is seen in 20–25% of patients suffering from extensive recurrence of primary tumor (Polmeer and Kernohan 1974; Yung et al. 1980). In fewer than 3% of cases, LM occurs in the absence of progression at the primary site (Delattre 1989). The rarest condition is primary leptomeningeal gliomatosis which occurs in the absence of an intraparenchymal mass (Davila et al.

1993; Dietrich et al. 1993). As is the case with systemic tumors metastatic to brain, gliomas arising in the posterior fossa have a higher incidence of LM than do those located supratentorially (Salazar and Rubin 1976). Again, a marked discrepancy exists between those cases diagnosed clinically and at autopsy (Erlich and Davis 1978). This may derive, in part, from a reluctance to perform lumbar punctures in patients with intracranial mass lesions. In children, the risk of subarachnoid spread of high-grade glioma has been reported at 32% and staging of the neuraxis is recommended (Packer et al. 1985; Grabb et al. 1992). This is not routinely performed in adults. Finally, although most descriptions of LM have referred to tumors of astrocytic origin, cases of oligodendroglioma are also reported (Chen et al. 1995; Rogers et al. 1995).

The vast majority of glial tumors in adults present as focal lesions and recur locally. Thus, initial therapies are directed at the primary site. In those rare cases presenting with leptomeningeal dissemination, there is evidence that the tumors respond to therapies in a manner that parallels their intraparenchymal counterparts (Grant et al. 1992). Thus, those patients with anaplastic astrocytoma and oligodendroglioma respond more favorably to aggressive treatment than do those with glioblastoma multiforme. Duration of response to treatment is, however, poorer in patients with disseminated disease. Because of the relatively higher incidence of leptomeningeal dissemination in children with these tumors, at least one group has recommended craniospinal irradiation in children deemed to be at risk (Grabb et al. 1992).

Ependymoma
Ependymoma is an uncommon tumor, representing 2–8% of all primary CNS malignancies. Half present in the first 2 decades of life. Overall, the risk of subarachnoid spread has been reported at 24% (Kun et al. 1988). Again, those in the posterior fossa are more likely to disseminate than are those originating supratentorially. Salazar et al. (1983) reviewed 50 cases of LM from ependymoma and found that 75% originated in the posterior fossa. Other factors that may predict for LM are younger patients with subtotally resected primary tumors, high-grade and myxopapillary histologies, and a high proliferative index (Rezai et al. 1996). Evidence for leptomeningeal dissemination at diagnosis has been reported at 15% using MRI evaluation (Goldwein et al. 1990). This in-

creases to 22% when imaging and CSF analyses are both employed (Pollack et al. 1995). The finding of dissemination at presentation was not a poor prognostic factor in the latter series (Pollack et al. 1995). LM found after initial diagnosis and treatment of primary disease is almost always accompanied by focal disease progression.

With the exception of completely resected histologically benign tumors, postoperative radiotherapy is now recommended in the initial treatment of ependymoma. It is not clear that craniospinal irradiation prevents the leptomeningeal dissemination of these tumors (Oi and Raimondi 1982). Thus, some controversy surrounds the selection of field size in radiation treatment planning. Nevertheless, craniospinal radiotherapy is often performed on patients over 3 years of age if there is evidence for leptomeningeal disease at presentation (Nazar et al. 1990). Chemotherapy is often used in patients under the age of 3 years in attempts to delay the use of radiotherapy. Chemotherapy has also been administered adjuvantly and at the time of tumor recurrence in patients of all ages (Ganyon et al. 1987; Walker and Allen 1988; Goldwein et al. 1990; Sutton et al. 1990). Most regimens include etoposide and a platinum-based compound. Series have all been small and responses to treatment anecdotal.

Germ cell tumors
Intracranial germ cell tumors are a heterogeneous group of neoplasms of which 50% comprise pure germinomas. They occur for fewer than 5% of all intracranial neoplasms and are most common in the first 2 decades of life. Malignant germ cell tumors are capable of both systemic and leptomeningeal metastasis. Estimates of LM range from 5 to 57%, with large series reporting approximately 10% (Jennings et al. 1985). In a review by Jennings, spinal cord metastases were noted in 11% of germinoma and 23% of endodermal sinus tumors. There has been a higher incidence noted for tumors located in the pineal region than for those in the suprasellar region. As mentioned above, α-fetoprotein and β-HCG levels in the CSF may be useful in the diagnosis and monitoring of these tumors.

Malignant germ cell tumors are, in general, very radiosensitive and radiotherapy is recommended for all patients. Craniospinal irradiation is recommended for any patient with evidence of leptomeningeal seeding (Rao et al. 1981; Edwards et al. 1988). The role of chemotherapy in these tumors varies by histologic subtype and stage of disease. Non-germinomatous malignant germ cell tumors are often treated neoadjuvantly with chemotherapy and with radiotherapy thereafter (Takakura 1985; Sawaya et al. 1990). Similar approaches have been tried for germinomas in attempts to limit the amount of radiation given (Allen et al. 1987). At recurrence, LM is almost always seen in association with local disease progression (Shokry et al. 1985). Treatment options include further surgery, radiotherapy, and chemotherapy. All are palliative.

Pineocytoma
This is a very rare tumor and in most series the number of patients reported is small. The risk of neuraxis dissemination has not been noted for pineocytoma in contrast to the well-known propensity of pineoblastoma (also known as PNET of the pineal region) in most series (Schild et al. 1993). In the pediatric population, however, pineocytomas can behave aggressively and spread throughout CSF pathways (D'Andrea et al. 1987). For this reason, staging of the neuraxis in children with pineocytoma is recommended (Kramer et al. 1994). Subtotally resected pineocytomas are usually treated with radiotherapy. There is some controversy as to how much of the neuraxis should be treated, with some advocating irradiation of the entire ventricular system and others advocating a more limited treatment field (Dattoli and Newell 1990). Systemic chemotherapy has been used in patients with recurrent pineocytoma (D'Andrea et al. 1987). Responses to treatment have been poor.

Primary CNS lymphoma
Primary CNS lymphoma, a non-Hodgkin's lymphoma arising in and confined to the CNS, accounts for approximately 2% of all intracranial tumors. Once rarer than this, the disease has increased in both the immunocompromised patient population and in those with no known associated risk factors (Eby et al. 1988). These tumors present as space occupying intraparenchymal lesions, often in periventricular regions. Intraparenchymal lesions are often multifocal. Because 13% of patients with primary CNS lymphoma have a history of prior cancer, these lesions are often initially attributed to metastases from another source (DeAngelis 1991).

At autopsy, there is contact with the leptomeninges and/or subependymal surface in all cases (Zimmerman 1975). The incidence of leptomeningeal tumor at diagnosis varies, but in a prospective study of 96 patients without the acquired immune deficiency syndrome (AIDS), it was found to be 42% (see this Volume, Chapter 25). In this study, the diagnosis was made by a positive CSF cytology, radiographically defined evidence for disease, meningeal biopsy, or some combination thereof. This may represent an underestimate of the true incidence because corticosteroids, often given prior to CSF analysis and imaging of the spine, result in rapid death of these tumor cells. In the same study, a similar incidence (41%) of leptomeningeal dissemination was documented at the time of tumor recurrence. In other series, the incidence of definite or probable leptomeningeal involvement ranged from 0 to 67% (DeAngelis et al. 1990).

The initial treatment of non-AIDS primary CNS lymphoma has evolved over the past years from a combination of radiotherapy and systemic corticosteroids to include chemotherapy. The intra-arterial administration of neo-adjuvant chemotherapy after BBB disruption has been tested in a small series of patients ($n = 16$) (Neuwelt et al. 1991). Nine of these patients received radiotherapy at some point after chemotherapy. The median survival for the group as a whole was 44.5 months, approximately 4 times that observed for historical control patients treated with radiotherapy and corticosteroids alone. Another treatment strategy recently employed includes the use of chemotherapy both before and after irradiation (DeAngelis et al. 1990). Patients received both intrathecal and parenteral chemotherapy. In a study involving 32 patients, median time to recurrence increased from 10 months (historical controls) to 41 months (treatment group), a 4-fold increase.

In the above-referenced studies, not all patients with primary CNS lymphoma had definite LM. In the study by Balmaceda et al. (1995), however, demonstrable leptomeningeal lymphoma did not affect the probability of achieving a complete remission to therapy, nor the probability of remaining in remission. Treatment that included drugs which penetrate the CSF at therapeutic concentrations was, however, predictive of disease-free survival. Despite the addition of chemotherapy to radiation, relapses continue to be observed in the brain, eyes, and meninges. A cooperative trial utilizing pre-irradiation chemotherapy is under way (Radiation Oncology Group, trial 93-10). The objectives of the trial include: (1) an assessment of tumor response to chemotherapy prior to the use of radiation and (2) a comparison of median survival between those treated according to the protocol and historical controls receiving radiotherapy only.

For patients with AIDS-related primary CNS lymphoma, there are few data on the incidence of leptomeningeal disease at presentation. CSF analysis is rarely helpful in the diagnosis and diagnostic cytologic abnormalities are unusual (So et al. 1986). In general, other disease processes such as opportunistic infections and AIDS dementia preclude the aggressive management of these patients whose median survival remains poor at <3 months (Remick et al. 1990). There is limited evidence for the efficacy of systemic or intrathecal chemotherapy in this patient population (Forsyth et al. 1994).

Choroid plexus tumors
There are two types of choroid plexus tumors, the benign choroid plexus papilloma and the malignant choroid plexus carcinoma. Together, they comprise less than 1% of all primary CNS neoplasms and are most commonly seen in children. CSF seeding by choroid plexus papilloma is rare. In choroid plexus carcinoma, however, diffuse leptomeningeal spread has been reported in 44% of cases (Ausman et al. 1984). Surgical resection is the treatment of choice, but craniospinal radiotherapy and chemotherapy are employed in cases of incomplete resection or when leptomeningeal involvement is evident at the outset. All published series have been small, involving only 2–6 patients. In some patients this has resulted in progression-free survival of longer than 4 years. For others, these therapies have had no impact (Packer et al. 1992).

Other primary CNS tumors
Meningiomas are generally benign tumors carrying a good prognosis, with local recurrence being the most common feature after incomplete removal and in tumors with atypical or anaplastic features. Leptomeningeal spread is uncommon with only a single report of LM in a case of anaplastic meningioma (Vinchon et al. 1995). Rare reports of leptomeningeal dissemination in pituitary adenomas exist as well (Tonner et al. 1992).

Tumors arising within the leptomeninges

Rarely, tumors appear to arise within and remain confined to the leptomeninges. Various histologies are reported, including melanoma, lymphoma, and rhabdomyosarcoma (Aichner and Schuler 1982; Smith et al. 1985; LaChance et al. 1991). These patients present similarly to those whose primary tumors are unknown at diagnosis of leptomeningeal metastasis. They are likely to undergo extensive staging in attempts to locate the primary tumor to no avail. A meningeal biopsy from one such patient with an assumed primary melanoma of the leptomeninges is shown in Fig. 3. This patient presented with multifocal neurologic symptoms and signs, suspicious MR imaging and repeated normal CSF evaluations. Following the diagnosis of leptomeningeal melanoma, he underwent thorough dermatologic and ophthalmologic evaluation but no other evidence for melanoma was found.

REFERENCES

ABOULAFIA, D.M., L.P. TAYLOR, R.D. CRANE, J.L. YON and H. RUDOLPH: Carcinomatous meningitis complicating cervical cancer: a clinicopathologic study and literature review. Gynecol. Oncol. 60 (1996) 313–318.

ACKLAND, S. and R. SCHILSKY: High-dose methotrexate: a critical reappraisal. J. Clin. Oncol. 5 (1987) 2017–2031.

AICHNER, F. and G. SCHULER: Primary leptomeningeal melanoma. Diagnosis by ultrastructural pathology of cerebrospinal fluid and cranial computed CT. Cancer 50 (1982) 1751–1756.

ALLEN, J.C. and F. EPSTEIN: Medulloblastoma and other primary malignant neuroectodermal tumors of the CNS: the effect of patients' age and extent of disease on prognosis. J. Neurosurg. 57 (1988) 1103–1107.

ALLEN, J.C., J. NISSELBAUM, F. EPSTEIN, G. ROSEN and M.K. SCHWARTZ: Alphafetoprotein and human chorionic gonadotropin determination in cerebrospinal fluid. J. Neurosurg. 51 (1979) 368–375.

ALLEN, J.C., J.H. KIM and R.J. PACKER: Neo-adjuvant chemotherapy for newly diagnosed germ cell tumors of the central nervous system. J. Neurosurg. 67 (1987) 65–70.

ALVAREZ, O.A., D.F. CARMICHAEL and Y.A. DECLERCK: Inhibition of collagenolytic activity and metastasis of tumor cells by a recombinant human tissue inhibitor of metalloproteinases. J. Natl. Cancer Inst. 82 (1990) 589–595.

AMER, M.H., M. AL-SARAFF, L.H. BAKER and V.K. VIATKAVICIUS: Malignant melanoma and central nervous system metastases: incidence, diagnosis, treatment, and survival. Cancer 42 (1987) 660–668.

ARITA, N., Y. USHIO, T. HAYKAWA, M. NAGATANI, T.Y. HUANG, S. IZUMOTO and H. MOGAMI: Intrathecal ACNU – a new therapeutic approach against malignant leptomeningeal tumors. J. Neuro-Oncol. 6 (1988) 221–226.

ARITA, N., M. TANEDA and T. HAYAKAWA: Leptomeningeal dissemination of malignant gliomas. Incidence, diagnosis, and outcome. Acta Neurochir. (Wien) 126 (1994) 84–92.

ARONEY, R.S., D.N. DALLEY, W.K. CHUN, D.R. BELL and J.A. LEVI: Meningeal cancer in small cell carcinoma of the lung. Am. J. Med. 71 (1981) 26–32.

AUSMAN, J.I., C. SCHRONTZ, J. CHASON, R.S. KNIGHTON, H. PAK and S. PATEL: Aggressive choroid plexus papilloma. Surg. Neurol. 22 (1984) 472–476.

AZZARELLI, B., L.D. MIRKIN, M. GOHEEN, J. MULLER and C. CROCKETT: The leptomeningeal vein. A site of reentry of leukemic cells into the systemic circulation. Cancer 54 (1984) 1333–1343.

BALDUCCI, L., D.D. LITTLE, T. KHANSUR and M.H. STEINBERG: Carcinomatous meningitis in small cell lung cancer. Am. J. Med. Sci. 287 (1984) 31–33.

BALIS, F.M., C.M. LESTER and D.G. POPLACK: Pharmacokinetics of trimetrexate (NSC352122) in monkeys. Cancer Res. 46 (1986) 169–174.

BALM, M. and J. HAMMACK: Leptomeningeal carcinomatosis. Presenting features and prognostic factors. Arch. Neurol. 53 (1996) 626–632.

BALMACEDA, C., J.J. GAYNOR, M. SUN, J.T. GLUCK and L.M. DEANGELIS: Leptomeningeal tumor in primary central nervous system lymphoma: recognition, significance, and implications. Ann. Neurol. 38 (1995) 202–209.

BARNARD, R.O. and M. PRASONS: Carcinoma of the thyroid with leptomeningeal dissemination following the treatment of a toxic goitre with ^{131}I and methyl thiouracil. Case with a co-existing intracranial dermoid. J. Neurol. Sci. 8 (1969) 299–306.

BATSON, C.V.: Function of vertebral veins and their role in spread of metastases. Ann. Surg. 112 (1940) 138–149.

BENDER, B.L. and D.G. MAYERNICK: Hodgkin's disease presenting with isolated craniospinal involvement. Cancer 58 (1986) 1745–1748.

BERGER, M.S., L. MAGRASSI and F.R. GEYER: Neuronal and neuronal precursor tumors. In: A.H. Kaye and E.R. Laws, Jr. (Eds.), Brain Tumors. Edinburgh, Churchill Livingstone (1995) 561–574.

BERRY, M.P. and R.D. JENKIN: Parameningeal rhabdomyosarcoma in the young. Cancer 48 (1981) 281–288.

BIGNER, S.H. and W.W. JOHNSON: The cytopathology of cerebrospinal fluid. II. Metastatic cancer, meningeal carcinomatosis and primary central nervous system neoplasms. Acta Cytol. 25 (1981) 461–479.

BLEYER, W.A.: Neurological sequelae of methotrexate

and ionizing radiation: a new classification. Cancer Treat. Rep. 65 (1981) 89.

BLEYER, W.A.: Central nervous system leukemia. Pediatr. Clin. North Am. 35 (1988) 789–814.

BOJSEN-MOELLER, M. and J.L. NIELSEN: CNS involvement in leukaemia. An autopsy study in 100 consecutive patients. Acta Pathol. Microbiol. Immunol. Scand. A 91 (1983) 209–216.

BONFIL, R.D., R.R, REDDEL, H. URA, R. REICH, R. FRIDMAN, S.C. HARRIS and J.P. KLEIN-SZANTO: Invasive and metastatic potential of a v-Ha-ras transformed human bronchial epithelial cell line. J. Natl. Cancer Inst. 81 (1989) 587–594.

BOOGERD, W., A.M.A. HART, J.J. VAN DER SANDE and E. ENGELSMAN: Meningeal carcinomatosis in breast cancer, prognostic factors and influence of treatment. Cancer 67 (1991) 1685–1991.

BRESALIER, R.S. and D.A. KARLIN: Meningeal metastasis from rectal carcinoma with elevated cerebrospinal fluid carcinoembryonic antigen. Dis. Colon Rectum 22 (1979) 216–217.

BRODERICK, J.P. and T.L. CASCINO: Nonconvulsive status epilepticus in a patient with leptomeningeal cancer. Proc. Mayo Clin. 62 (1987) 835–837.

BROWN, P.D., A.T. LEVY, I.M. MARGULIES, L.A. LIOTTA and W.G. STETLER-STEVENSON: Independent expression and cellular processing of M_r 72,000 type IV collagenase in human tumorigenic cell lines. Cancer Res. 50 (1990) 6184–6191.

BUNN, P.A., P.S. SCHEIN, P.M. BANKS and V.T. DEVITA: Central nervous system complications in patients with diffuse histiocytic and undifferentiated lymphoma: leukemia revisited. Blood 47 (1976) 310.

BURKE, J.W., A.E. PODRANSKY and W.G. BRADLEY: Meninges: benign postoperative enhancement on MR images. Radiology 174 (1990) 99–102.

CAIRNS, H. and D.S. RUSSELL: Intracranial and spinal metastasis in gliomas of the brain. Brain 54 (1931) 377–420.

CHAMBERLAIN, M.C.: A review of leptomeningeal metastases in pediatrics. J. Child Neurol. 10 (1995) 191–199.

CHAMBERLAIN, M.C., S. KHATIBI, J.C. KIM, S.B. HOWELL, E. CHATELUT and S. KIM: Treatment of leptomeningeal metastasis with intraventricular administration of depot cytarabine (DTC 101). A phase I study. Arch. Neurol. 50 (1993) 261–264.

CHAMPAGNE, M.A. and H.K.B. SILVER: Intrathecal dacarbazine treatment of leptomeningeal malignant melanoma. J. Natl. Cancer Inst. 84 (1992) 1203–1204.

CHANG, K.H., M.H. HAN, J.K. ROH, I.O. KIM, M.C. HAN and C.-W. KIM: Gd-DTPA-enhanced imaging of the brain in patients with meningitis: comparison with CT. Am. J. Neuroradiol. 9 (1988) 1045–1050.

CHEN, R., D.R. MACDONALD and D.A. RAMSAY: Primary diffuse leptomeningeal oligodendroglioma. J. Neurosurg. 83 (1995) 724–728.

CLAMON, G. and B. DOEBBELING: Meningeal carcinomatosis from breast cancer: spinal cord versus brain involvement. Breast Cancer Res. Treat. 9 (1987) 213–217.

CRINO, P.B., R.A. SATER, M. SPERLING and C.D. KATSETOS: Renal cell carcinomatous meningitis: pathologic and immunohistochemical features. Neurology 45 (1995) 189–191.

CSERR, H.F. and P.M. KNOPF: Cervical lymphatics, the blood–brain barrier and the immunoreactivity of the brain: a new view. Immunol. Today 13 (1992) 507–512.

D'ANDREA, D.A., R.J. PACKER, L.B. ROURKE, L.T. BILANIUK, L.N. SUTTON, D.A. BRUCE and L. SCHUT: Pineocytomas of childhood. A reappraisal of natural history and response to therapy. Cancer 59 (1987) 1353–1357.

DATTOLI, M.J. and J. NEWALL: Radiation therapy for intracranial germinoma: the case for limited volume treatment. Int. J. Radiat. Oncol. Biol. Phys. 19 (1990) 429–433.

DAVILA, G., C. DUYCKAERTS, J.P. LAZARETH, M. POISSON and J.Y. DELATTRE: Diffuse primary leptomeningeal gliomatosis. J. Neuro-Oncol. 15 (1993) 45–49.

DAVIS, P., N. FRIEDMANN, S. FRY, J. MALKO, J. HOFFMANN and I. BRAUN: Leptomeningeal metastasis: MR imaging. Radiology 163 (1987) 449–454.

DEANGELIS, L.M.: Primary central nervous system lymphoma as a secondary malignancy. Cancer 67 (1991) 1431–1435.

DEANGELIS, L.M., J. YAHALOM and M.-H. HEINEMANN: Primary CNS lymphoma: combined treatment with chemotherapy and radiotherapy. Neurology 40 (1990) 80–86.

DEANGELIS, L., J. YAHALOM, H.T. THALER and U. KHER: Combined modality therapy for primary CNS lymphoma. J. Clin. Oncol. 10 (1992) 635–643.

DEKKER, A.W., A. ELDERSON, K. PUNT and J.J. SIXMA: Meningeal involvement in patients with acute nonlymphocytic leukemia. Cancer 56 (1985) 2078–2082.

DELLATTRE, J.Y., R.W. WALKER AND M.K. ROSENBLUM: Leptomeningeal gliomatosis with spinal cord or cauda equina compression: a complication of supratentorial glioma in adults. Acta Neurol. Scand. 79 (1989) 133–139.

DEL REGATO, J.A.: Pathways of metastatic spread of malignant tumors. Semin. Oncol. 4 (1977) 33–38.

DEUTSCH, M.: Medulloblastoma: staging and treatment outcome. Int. J. Radiat. Oncol. Biol. Phys. 14 (1988) 1103–1107.

DIETRICH, P.Y., M.S. AAPRO, A. RIEDER and G.P. PIZZOLATO: Primary diffuse leptomeningeal gliomatosis (PDLG): a neoplastic chronic meningitis. J. Neuro-Oncol. 15 (1993) 275–283.

DIPPOLD W., H. BERNHARD and K.H. MEYER ZUM BUSCHENFELDE: Immunological response to intrathecal and systemic treatment with ganglioside antibody R-24 in patients with malignant melanoma. Eur. J. Cancer 30A (1994) 137–144.

DOGE, H. and R. HLISCS: Intrathecal therapy with [198]Au-colloid for meningiosis prophylaxis. Eur. J. Nucl. Med. 9 (1984) 125–128.

EBERTH, C.J.: Zur Entwicklung des Epithelioms (Cholesteatoms) der Pia und der Lunge. Virchows Arch. 49 (1870) 51–63.

EBY, N.L., S. GRUFFERMAN, C.M. FLANNELLY, S.C. SCHOLD, F.S. VOGEL and P.C. BURGER: Increasing incidence of the primary brain lymphoma in the US. Cancer 62 (1988) 2461–2465.

EDWARDS, M.S.B., R.J. HUDGINS, C.B. WILSON, V.A. LEVIN and W.M. WARA: Pineal region tumors in children. J. Neurosurg. 66 (1988) 689–697.

ERLICH, S.S. and R.L. DAVIS: Spinal subarachnoid metastases from primary intracranial glioblastoma. Cancer 42 (1978) 2854–2864.

ERSBOLL, J., H.B. SCHULTZ, B. THOMSEN, N. KEIDING and N.I. NISSEN: Meningeal involvement in non-Hodgkin's lymphoma: symptoms, incidence, risk factors, and treatment. Scand. J. Hematol. 35 (1985) 487–496.

EVANS, A.E., D.T. JENKIN and R. SPOSTO: The treatment of medulloblastoma. Results of a prospective randomized trial of radiation therapy with and without CCNU, vincristine, and prednisone. J. Neurosurg. 72 (1990) 572–582.

FISHER, C.M.: Hydrocephalus as a cause of gait disturbance in the elderly. Neurology 32 (1982) 1358–1363.

FISHMAN, R.A.: Studies on the transport of sugars between blood and cerebrospinal fluid in normal states and in meningeal carcinomatosis. Trans. Am. Neurol. Assoc. 88 (1963) 114–118.

FISHMAN, R.A.: CSF findings in diseases of the nervous system. In: R.A. Fishman (Ed.), Cerebrospinal Fluid in Diseases of the Nervous System. Philadelphia, PA, W.B. Saunders (1992) 309.

FORSYTH, P.A., J. YAHALOM and L.M. DEANGELIS: Combined-modality therapy in the treatment of primary central nervous system lymphoma in AIDS. Neurology 44 (1994) 1473–1479.

FREILICH, R.J., G. KROL and L.M. DEANGELIS: Neuroimaging and cerebrospinal fluid cytology in the diagnosis of leptomeningeal metastasis. Ann. Neurol. 38 (1995) 51–57.

FRIEDMAN, H.S., G.E. ARCHER, R.E. MCLENDON, J.M. SCHUSTER, O.M. COLVIN, A. GUSPARI, R. BLUM, P.A. SAVINA, H.E. FUCHS and D.D. BIGNER: Intrathecal melphalan therapy of human neoplastic meningitis in athymic nude rats. Cancer Res. 54 (1994) 4710–4714.

GAGLIANO, R.G. and J.J. COSTANZI: Paraplegia following intrathecal methotrexate. Cancer 37 (1976) 1663–1668.

GANYON, P.S., L.J. ETTINGER, D. MOEL, S.E. SIEGEL, E.S. BAUM, W. KRIVIT and G.D. HAMMOND: Pediatric phase I trial of carboplatin: a Children's Cancer Study Group report. Cancer Treat. Rep. 71 (1987) 1039–1042.

GANYON, P.S., P.G. STEINHERZ, W.A. BLEYER, A.R. ABLIN, V.C. ALBO, J.Z. FINKELSTEIN, N.J. GROSSMAN, L.J. NOVAK, A.F. PYESMANY and G.H. REAMAN: Improved therapy for children with acute lymphoblastic leukemia and unfavorable presenting features: a follow-up report of the Children's Cancer Group Study CCG-106. J. Clin. Oncol. 11 (1993) 2234–2242.

GEHLSEN, K.R., W.S. ARGRAVES, M.D. PIERSBACHER and E. RUOSLAHTI: Inhibition of in vitro cell invasion by arg-gly-asp-containing peptides. J. Cell Biol. 106 (1988) 925–930.

GEHLSEN, K.R., G.E. DAVIS and P. SRIRAMARAO: Integrin expression in human melanoma cells with differing invasive and metastatic properties. Clin. Exp. Metastasis 10 (1992) 111–120.

GIANNONE, L., F.A. GRECO and J.D. HAINSWORTH: Combination intaventricular chemotherapy for meningeal neoplasia. J. Clin. Oncol. 4 (1986) 68–73.

GIESE, A. and M. WESTPHAL: Glioma invasion in the central nervous system. Neurosurgery 39 (1996) 235–252.

GLADSON, C.L. and D.A. CHERECH: Glioblastoma expression of vitronectin and the alpha-v/B-3 integrin. J. Clin. Invest. 88 (1991) 1924–1932.

GLASS, J.P., M. MELAMED, N.C. CHERNIK and J.B. POSNER: Malignant cells in cerebrospinal fluid (CSF): the meaning of a positive CSF cytology. Neurology 29 (1979) 1369–1375.

GOLDWEIN, J.W., J.M. LEAHEY, R.J. PACKER, L.N. SUTTON, W.J. CURRAN, L.B. RORKE, L. SCHUT, P.S. LITTMAN and G.J. D'ANGIO: Intracranial ependymomas in children. Int. J. Radiat. Oncol. Biol. Phys. 19 (1990) 1497–1502.

GONZALES-VITALE, J.C. and R. GARCIA-BUNUEL: Meningeal carcinomatosis. Cancer 37 (1976) 2906–2911.

GRABB, P.A., L. ALBRIGHT and D. PANG: Dissemination of supratentorial malignant gliomas via the cerebrospinal fluid in children. Neurosurgery 30 (1992) 64–71.

GRANT, R., B. NAYLOR, L. JUNCK and H.S. GREENBERG: Clinical outcome in aggressively treated meningeal gliomatosis. Neurology 42 (1992) 252–254.

GRANT, R., B. NAYLOR, H.S. GREENBERG and L. JUNCK: Clinical outcome in aggressively treated meningeal carcinomatosis. Arch. Neurol. 51 (1994) 457–461.

GRAUS, F., L.R. ROGERS and J.B. POSNER: Cerebrovascular complications in patients with cancer. Cancer 60 (1985) 16–27.

GRAY, S.T., R.J. WILKINS and K. YUN: Interstitial collagenase gene expression in oral squamous cell carcinoma. Am. J. Pathol. 141 (1992) 301–306.

GROSSMAN, S.A. and T.J. MOYNIHAN: Neoplastic meningitis. Neurol. Clin. North Am. 9 (1991) 843–856.

GROSSMAN, S.A., D.L. TRUMP, D.C. CHEN, G. THOMPSON and E.E. CAMARGO: Cerebrospinal fluid flow abnormalities in patients with neoplastic meningitis. An evaluation using ¹¹¹indium-DTPA ventriculography. Am. J. Med. 73 (1982) 641–647.

GROSSMAN, S.A., J.C. RUCKDESCHEL, D.L. TRUMP, T.J. MOYNI-HAN and D.S. ETTINGER: Randomized prospective comparison of intraventricular methotrexate and thiotepa in patients with previously untreated neo-plastic meningitis. Eastern Cooperative Oncology Group. J. Clin. Oncol. 11 (1993) 561–569.

GUTIN, P.H., J.A. LEVI, P.H. WIERNICK and M.D. WALKER: Treatment of malignant meningeal disease with intra-thecal thioTEPA: a phase II study. Cancer Treat. Rep. 61 (1977) 885–887.

HARA, Y., Y. KOBAYASHI, K. GOTO, K. TOZUKA, A. TOKUE and M. MOCHIZUKI: A case of carcinomatous meningitis from transitional cell carcinoma of the urinary bladder. Acta Urol. Jpn. 40 (1994) 1113–1117.

HERMAN, C., W.J. KUPSKY, L. ROGERS, R. DUMAN and P. MOORE: Leptomeningeal dissemination of malignant glioma simulating cerebral vasculitis. Stroke 26 (1995) 2366–2370.

HERMAN, T.S., M. HAMMOND and S.E. JONES: Involvement of the central nervous system by non-Hodgkin's lymphoma: the Southwest Oncology Group experi-ence. Cancer 43 (1979) 390–397.

HITCHINS, R.N., D.R. BELL, R.L. WOODS and J.A. LEVI: A prospective randomized trial of single-agent versus combination chemotherapy in meningeal carcinoma-tosis. J. Clin. Oncol. 5 (1987) 1655–1662.

INGRAM, L.C., D.L. FAIRCLOUGH, W.L. FURMAN, J.T. SAND-LUND, L.E. KUN, G.K. RIVERA and C.H. PUI: Cranial nerve palsy in childhood acute lymphoblastic leukemia and non-Hodgkin's lymphoma. Cancer 67 (1991) 2262–2268.

JENNINGS, M.T., R. GELMAN and F. HOCHBERG: Intracranial germ cell tumors: natural history and pathogenesis. J. Neurosurg. 63 (1985) 155–167.

JENNINGS, M.T., N. SLATKIN, M. D'ANGELO, L. KETONEN, M.D. JOHNSON, M. ROSENFELD, J. CREASY, N. TULIPAN and R. WALKER: Neoplastic meningitis as the presentation of occult primitive neuroectodermal tumors. J. Child Neurol. 8 (1993) 306–312.

KAMEN, B.A., J.S. HOLCENBERG and S.E. SIEGEL: Aziridinyl-benzoquinone (AZQ) treatment of central nervous system leukemia (Letter). Cancer Treat. Rep. 66 (1982) 2105–2106.

KANTARJIAN, H.M., M.J. KEATING, R.S. WALTERS, E.H. ESTEY, K.B. MCCREDIE, T.L. SMITH, W.T. DALTON, A. CORK, J.M. TRUJILLO and E.J. FREIREICH: Acute promyelocytic leukemia. Am. J. Med. 80 (1988) 1784–1789.

KAPPEL, M.H., K.C. MANIVEL and J.J. GOSWIWITZ: Atypical lymphocytes in spinal fluid resembling post-trans-plant lymphoma in a cardiac transplant recipient: a case report. Acta Cytol. 38 (1994) 470–474.

KLEIN, P., E.C. HALEY, G.F. WOOTEN and S.R. VANDENBERG: Focal cerebral infarctions associated with perivas-cular tumor infiltrates in carcinomatous leptomenin-geal metastases. Arch. Neurol. 46 (1989) 1149–1152.

KOCH, T.R., K.M. LICHTENFELD and P.H. WIERNICK: De-tection of central nervous system metastasis with cerebrospinal fluid beta-2-microglobulin. Cancer 52 (1983) 101–109.

KOCHI, M., S. TAKAKI, J. KURATSU, H. SETO, I. KITAMURA and Y. USHIO: Neurotoxicity and pharmacokinetics of ven-triculolumbar perfusion of methyl 6-[3-(2-chloro-ethyl)-3-nitrosoureido]-6-deoxy-alpha-D-glucopyra-noside (MCNU) in dogs. J. Neuro-Oncol. 19 (1994) 239–244.

KRAMER, E.D., L.G. VEZINA, R.J. PACKER, C.R. FITZ, R.A. ZIM-MERMAN and M.D. COHEN: Staging and surveillance of children with central nervous system neoplasms: re-commendations of the neurology and tumor imaging committees of the Children's Cancer Group. Pediatr. Neurosurg. 20 (1994) 254–263.

KROL, G., G. SZE, M. MALKIN and R. WALKER: MR of cranial and spinal meningeal carcinomatosis: comparison with CT and myelography. Am. J. Neuroradiol. 9 (1988) 709–714.

KUN, L.E., E.H. KOVNAR and R.A. SANFORD: Ependymomas in children. Pediatr. Neurosci. 14 (1988) 57–63.

KUPPNER, M.C., E. VANMEIR, T. GAUTHIER, M.T. HAMON and N. DETRIBOLET: Differential expression of the CD44 molecule in human brain tumor. Int. J. Cancer 50 (1992) 572–577.

LANDBERG, T.G., M.L. LINDGREN, E.K. CAVALLIN-STAHL, G.O. SVAHN-TAPPER, G. SUNDBARG, S. GARWICZ. J.A. LAGER-GREN, V.L. GUNNESON, A.E. BRUN and S.E. CRONQVIST: Improvements in the radiotherapy of medulloblas-toma. Cancer 45 (1980) 670–678.

LARSON, S.M., G.L. SCHALL and G. DI CHIRO: The influence of previous lumbar puncture and pneumoencepha-lography on the incidence of unsuccessful radio-isotope cisternography. J. Nucl. Med. 12 (1971) 555–557.

LEE, Y.-T.N.: Breast carcinoma: pattern of metastasis at autopsy. J. Surg. Oncol. 23 (1983) 175–180.

LEE, Y.Y., J.P. GLASS, A. GEOFFRAY and S. WALLACE: Cranial computed tomographic abnormalities in leptomenin-geal metastasis. Am. J. Roentgenol. 1443 (1984) 1035–1039.

LIEPMAN, M.K. and M.L. VOTAW: Meningeal leukemia complicating chronic lymphocytic leukemia. Cancer 47 (1981) 2482–2484.

LIOTTA, L.A.: Tumor invasion and metastases – role of the extracellular matrix. Rhoads Memorial Award Lecture. Cancer Res. 46 (1986) 1–7.

LITTMAN, P., P. COCCIA, W.A. BLEYER, J. LUKENS, S. SIEGEL, D. MILLER, H. SATHER and D. HAMMOND: Central nervous system (CNS) prophylaxis in children with low risk acute lymphoblastic leukemia (ALL). Int. J. Radiat. Oncol. Biol. Phys. 13 (1987) 1443–1449.

LOPEZ, J., E. NASSIF, P. VANNICOLA, J.G. KRIKORIAN and R.P.

AGARWAL: Central nervous system pharmacokinetics of high-dose cytosine arabinoside. J. Neuro-Oncol. 3 (1985) 119–124.

LUNDBERG W.B., E.C. CADMAN and R.T. SKEEL: Leptomeningeal mycosis fungoides. Cancer 38 (1976) 2149–2153.

LUPS, S. and A.M.F.H. HAAN: The Cerebrospinal Fluid. Amsterdam, Elsevier (1954).

LYSTER, M.T., M.S. KIES and T.M. KUZEL: Neurologic complications of patients with small cell prostate cancer. Report of two cases. Cancer 74 (1994) 3159–3163.

MALKIN, M.G., E.S. CIBAS, M. FLEISHER, M.K. SCHWARTZ, M.R. MELAMED and J.B. POSNER: Cerebrospinal fluid flow cytometry, combined with tumor marker analysis, is useful in the diagnosis of leptomeningeal metastasis. Neurology 37, Suppl. (1987) 300.

MARTON, L.J., M.S. EDWARDS, V.A. LEVIN, W.P. LUBICH and C.B. WILSON: CSF polyamines: a new and important means of monitoring patients with medulloblastoma. Cancer 47 (1981) 757–768.

MATRISAN, L.M.: The matrix-degrading metalloproteinases. BioEssays 14 (1992) 455–462.

MCINTOSH, S., E.H. KLATSKIN, R.T. O'BRIEN, G.T. ASPNES, B.L. KAMMERER, C. SNEAD, S.M. KALAVSKY and H.A. PEARSON: Chronic neurologic disturbance in childhood leukemia. Cancer 37 (1976) 853–857.

MENCEL, P.J., L.M. DEANGELIS and R.J. MOTZER: Hormonal ablation as effective therapy for carcinomatous meningitis from prostatic carcinoma. Cancer 73 (1994) 1892–1894.

MEYER, R.J., P.P. FERRERIA and J. CUTTNER: Central nervous system involvement at presentation in acute granulocytic leukemia. Am. J. Med. 68 (1980) 691–694.

MIGNATTI, P., E. ROBBINS and D.B. RIFKIN: Tumor invasion through the human amnion membrane: requirement for a proteinase cascade. Cell 47 (1986) 487–498.

MIRIMANOFF, R.R. and CHOI, N.C.: Intradural spinal metastases in patients with posterior fossa brain metastases from various primary cancers. Oncology 44 (1987) 232–236.

MITCHELL, M.S.: Relapse in the central nervous system in melanoma patients successfully treated with biomodulators. J. Clin. Oncol. 7 (1989) 1701–1709.

MITTL, R.L. and D.M. YOUSEM: Frequency of unexplained meningeal enhancement in the brain after lumbar puncture. Am. J. Neuroradiol. 15 (1994) 633–638.

MOBERG A. and G.V. REIS: Carcinosis meningium. Acta Med. Scand. 170 (1961) 747–755.

MOORE, M.T. and G. EISINGER: Extra primary seeding of glioblastoma multiforme in the subarachnoid space and ependyma. Neurology 13 (1963) 855–865.

MOSELEY, R.P., A.G. DAVIEW, R.B. RICHARDSON, M. ZALUTSKY, S. CARRELL, J. FABRE, N. SLACK, J. BULLIMORE, B. PIZER and V. PAPANASTASSIOU: Intrathecal administration of ^{131}I radiolabelled monoclonal antibody as a treatment for neoplastic meningitis. Br. J. Cancer 62 (1990) 637–642.

MOSES, M.A. and R. LANGER: A metalloproteinase inhibitor as an inhibitor of neovascularization. J. Cell. Biochem. 47 (1991) 230–235.

MULLIGAN, M.J., R. VASU, C.E. GROSSI, E.F. PRASTHOFER, F.M. GRIFFIN, A. KAPILA, J.M TRUPP and J.C. BARTON: Neoplastic meningitis with eosinophilic pleocytosis in Hodgkin's disease: a case with cerebellar dysfunction and a review of the literature. Am. J. Med. Sci. 296 (1988) 322–326.

MURRAY, J.J., F.A. GRECO, S.N. WOLFF and J.D. HAINSWORTH: Neoplastic meningitis: marked variations of cerebrospinal fluid composition in the absence of extradural block. Am. J. Med. 75 (1983) 289–294.

NAGGOURNEY, R.A., R. HEDAYA, M. LINNOILA and P.S. SCHEIN: Carcinoid carcinomatous meningitis. Ann. Intern. Med. 102 (1985) 779–782.

NAKAGAWA, H., A. MURASAWA, S. KUBO, Y. NAKNIMA, S. IZUMOTO and T. HAYAKAWA: Diagnosis and treatment of patients with meningeal carcinomatosis. J. Neurooncol. 13 (1992) 81–89.

NAZAR, G.B., H.J. HOFFMAN, L.E. BECKER, D. JENKIN, R.P. HUMPHREYS and E.B. HENDRICK: Infratentorial ependymomas in childhood: prognostic factors and treatment. J. Neurosurg. 72 (1990) 408–417.

NEUWELT, E.A., D.L. GOLDMAN and S.A. DAHLBORG: Primary CNS lymphoma treated with osmotic blood brain barrier opening: prolonged survival and preservation of cognitive function. J. Clin. Oncol. 9 (1991) 1580–1590.

OI, S. and A.J. RAIMONDI: Ependymoma in children. Pediatric Neurosurgery. In: Surgery of the Developing Nervous System. New York, Grune and Stratton (1982) 419–428.

OLSEN, M.E., N.L. CHERNIK and J.B. POSNER: Infiltration of the leptomeninges by systemic cancer. A clinical and pathologic study. Arch. Neurol. 30 (1974) 122–137.

ONGERBOER DE VISSER, B.W., R. SOMERS, W.H. NOOYEN, P. VAN HEERDE, A.A. HART and J.G. MCVIE: Intraventricular methorexate therapy of leptomenineal metastasis from breast carcinoma. Neurology 33 (1983) 1565–1572.

ONO, M., I. ISOBE, J. UKI, H. KURIHARA, T. SHIMIZU and K. KOHONO: Recurrence of primary intracranial germinomas after complete resection and radiotherapy: recurrence patterns and therapy. Neurosurgery 35 (1994) 615–621.

ORITA, A.T., O. HAYASHIDA, T. NISHIZAKI and H. FUDABA: Malignant meningioma metastasizing through the cerebrospinal pathway. Acta Neurol. Scand. 85 (1992) 368–371.

PACKER, R.J., K.R. SIEGEL, L.N. SUTTON, P. LITTMAN, D.A. BRUCE and L. SCHUT: Leptomeningeal dissemination of primary central nervous system tumors in childhood. Ann. Neurol. 18 (1985) 217–221.

PACKER, R.J., G. PERILONGO, D. JOHNSON, L.N. SUTTON, G. VEZINA, R.A. ZIMMERMAN, J. RYAN, G. REAMAN and L. SCHUT: Choroid plexus carcinoma of childhood. Cancer 69 (1992) 580–585.

PATCHELL, R.A., C. CIRRINCOINE, H.T. THALER, J.H. GALICICH, J.H. KIM and J.B. POSNER: Single brain metastases: surgery plus radiation or radiation alone. Neurology 36 (1986) 447–453.

PHILLIPS, M.E., T.J. RYALS, S.A. KAMBHU and W.T.C. YUH: Neoplastic vs. inflammatory meningeal enhancement with GD-DTPA. J. Comput. Assist. Tomogr. 14 (1990) 536–541.

POCHEDLY, C.: Neurological manifestations in acute leukemia. II. Involvement of cranial nerves and hypothalamus. N.Y. State J. Med. 75 (1975) 715–721.

POLETTE, M., C. CLAVEL, D. MULLER, J. ABECASSIS, I. BINNINGER and P. BIREMBAUT: Detection of mRNAs encoding collagenase 1 and stromelysin 2 in carcinomas of the head and neck by in situ hybridization. Invasion Metastasis 11 (1991) 76–83.

POLLACK, I.F., P.C. GERSZTEN, A.J. MARTINEZ, R.H. LO, B. SHULTZ, A.L. ALBRIGHT, J. JANOSKY and M. DEUTSCH: Intracranial empendymoma of childhood: long term outcome and prognostic factors. Neurosurgery 37 (1995) 655–666.

POLMEER, F.E. and J.W. KERNOHAN: Meningeal gliomatosis. Arch. Neurol. Psychiatry 57 (1974) 593–616.

POWELL, W.C., J.D. KNOX, M. NAVRE, T.M. GROGAN, J. KITTELSON, R.B. NAGLE and G.T. BOWDEN: Expression of the metalloproteinase matrilysin in DU-145 cells increases their invasive potential in severe combined immunodeficient mice. Cancer Res. 53 (1993) 417–422.

PRADOS, M.D., R.E. WARNICK, W.M. WARA, D.A. LARSON, K. LAMBORN and C.B. WILSON: Medulloblastoma in adults. Int. J. Radiat. Oncol. Biol. Phys. 32 (1995) 1145–1152.

PRICE, R.A. and D.A. BIRDWELL: The central nervous system in leukemia. III. Mineralizing microangiopathy and dystrophic calcification. Cancer 42 (1978) 717–728.

PRICE, R.A. and W.W. JOHNSON: The central nervous system in childhood leukemia. I. The arachnoid. Cancer 31 (1973) 520–533.

RAM, Z., S. WALBRIDGE, E.M. OSHIRO, J.J. VIOLA, Y. CHIANG, S.N. MUELLER, R.M. BLAESE and E.H. OLDFIELD: Intrathecal gene therapy for malignant leptomeningeal neoplasia. Cancer Res. 54 (1994) 2141–2145.

RAO, Y., E. MEDINI, R. HASELOW, T. JONES and S. LEVITT: Pineal and ectopic pineal tumors: the role of radiation therapy. Cancer 48 (1981) 708–713.

RECHT, L., D.J. STRAUS, C. CIRRINCIONE, H.T. THALER and J.B. POSNER: Central nervous system metastases from non-Hodgkin's lymphoma: treatment and prophylaxis. Am. J. Med. 84 (1988) 425–435.

REDMAN, B.G., E. TAPAZOGLOU and M. AL-SARAFF: Meningeal carcinomatosis in head and neck cancer. Report of six cases and review of the literature. Cancer 58 (1986) 2656–2661.

REMICK, S.C., C. DIAMOND and J.A. MIGLIOZZI: Primary central nervous system lymphoma in patients with and without the acquired immune deficiency syndrome. A retrospective analysis and review of the literature. Medicine (Baltimore) 69 (1990) 345–360.

REZAI, A.R., H.H. WOO, M. LEE, H. COHEN, D. ZAGZAG and F.J. EPSTEIN: Disseminated ependymoma of the central nervous system. J. Neurosurg. 85 (1996) 616–624.

RODESCH, G., P. VAN BOGAERT, N. MAVROUDAKIS, P.M. PARIZEL, J.J. MARTIN, C. SEGEBARTH, M. VAN VYVE, D. BALERIAUX and J. HILDEBRAND: Neuroradiologic findings in leptomeningeal carcinomatosis: the value interest of gadolinium-enhanced MRI. Neuroradiology 32 (1990) 26–32.

ROGERS, L., S. SCHOLD and K. EASLEY: Cerebrospinal fluid markers as an aid in the diagnosis of meningeal metastasis. Proc. Am. Soc. Clin. Oncol. 6 (1987) 8.

ROGERS, L.R., P.M. DUCHESNEAU, C. NUNEZ, A.J. FISHLEDER, J.K. WIECK, L.J. BAUER and J.M. BOYETT: Comparison of cisternal and lumbar CSF examination in leptomeningeal metastasis. Neurology 42 (1992) 1239–1241.

ROGERS, L.R., M.L. ESTES, S.A. ROSENBLOOM and L. HARROLD: Primary leptomeningeal oligodendroglioma: case report. Neurosurgery 36 (1995) 166–169.

ROSEN, S.T., J. AISNER, R.W. MAKUCH, M.J. MATTHEWS, D.C. IHDE, M. WHITACRE, E.J. GLATSTEIN, P.H. WIERNICK, A.S. LICHTER and P.A. BUNN: Carcinomatous leptomeningitis in small cell lung cancer: a clinicopathologic review of the National Cancer Institute experience. Medicine (Baltimore) 61 (1982) 45–53.

RUTKA, J.T., G. APODACA, R. STERN and M. ROSENBLUM: The extracellular matrix of the central and peripheral nervous systems: structure and function. J. Neurosurg. 69 (1988) 155–170.

SAIKI, I., J. MURATA, J. IIDA, J. NISHI, K. SUGIMURA and I. AZUMA: The inhibition of murine lung metastases by synthetic polypeptides [poly(arg-gly-asp) and poly (tyr-ile-gly-ser-arg)] with a core sequence of cell adhesion molecules. Br. J. Cancer 59 (1989) 194–197.

SALAZAR, O.M. and P. RUBIN: The spread of glioblastoma multiforme as a determining factor in the radiation treated volume. Int. J. Radiat. Oncol. Biol. Phys. 1 (1976) 627–637.

SALAZAR, O.M., H. CASTRO-VITA, P. VAN HOUTTE, P. RUBIN and C. AYGUN: Improved survival in cases of intracranial ependymoma after radiation therapy: late report and recommendations. J. Neurosurg. 59 (1983) 652–659.

SAMLOWSKI, W.E., K.J. PARK, R.E. GALINSKY, J.H. WARD and G.B. SCHUMANN: Intrathecal administration of interleukin-2 for meningeal carcinomatosis due to malignant melanoma: sequential evaluation of intracranial pressure, cerebrospinal fluid cytology, and cytokine induction. J. Immunother. 13 (1993) 49–54.

SAWAYA, R., D.K. HAWLEY, W.D. TOBLER, J.M. TEW and A.A. CHAMBERS: Pineal and third ventricular tumors. In: J. Youmans (Ed.), Neurological Surgery. Philadelphia, PA, W.B. Saunders (1990) 3171–3203.

SCHILD, S.E., B.W. SCHEITHAUER and P.J. SCHAUMBERG: Pineal parenchymal tumors: clinical pathologic and therapeutic aspects. Cancer 72 (1993) 870–880.

SCHOLD, S.C., W.R. WASSERSTROM, M. FLEISHER, M.K. SCHWARTZ and J.B. POSNER: Cerebrospinal fluid biochemical markers of central nervous system metastases. Ann. Neurol. 8 (1980) 597–604.

SCHUMACHER, J.D., R.D. TIEN and H. FRIEDMAN: Gadolinium enhancement of the leptomeninges caused by hydrocephalus: a potential mimic of leptomeningeal metastasis. Am. J. Neuroradiol. 15 (1994) 639–641.

SEFTOR, R.E., E.A. SEFTOR, K.R. GEHLSEN, W.G. STETLER-STEVENSON, P.D. BROWN, E. RUOSLAHTI and M.J. HENDRIX: Role of the alpha v beta 3 integrin in human melanoma cell invasion. Proc. Natl. Acad. Sci. 89 (1992) 1557–1561.

SELTZER, S., A.S. MARK and S.W. ATLAS: CNS sarcoidosis: evaluation with contrast-enhanced MR imaging. Am. J. Neuroradiol. 12 (1992) 1227–1233.

SHIMIZU, K., Y. OKAMOTO, Y. MIYAO, M. YAMADA, Y. USHIO, T. HAYAKAWA, H. IKEDA and H. MOGAMI: Adoptive immunotherapy of human meningeal gliomatosis and carcinomatosis with LAK cells and recombinant interleukin-2. J. Neurosurg. 66 (1987) 519–521.

SHOKRY, A., R.C. JANZER, A.R. VON HOCHSTETTER, M.G. YASARGIL and C. HEDIGER: Primary intracranial germ cell tumors. A clinicopathologic study of 14 cases. J. Neurosurg. 62 (1985) 826–830.

SHUANGSHOTI, S., P. TANGCHAI and M.G. NETSKY: Primary adenocarcinoma of choroid plexus. Arch. Pathol. 91 (1971) 101–106.

SIEGAL, T., J. SHORR, I. LUBETZKI-KORN, D. SOFFER, E. NEPARSTEK, R. TUR-KASPA and O. ABRAMSKY: Myeloma protein synthesis within the CNS by plasma cell tumors. Ann. Neurol. 10 (1981) 271–273.

SIEGAL, T., B. MILDWORF, D. STEIN and E. MELAMED: Leptomeningeal metastases: reduction of regional cerebral blood flow and cognitive impairment. Ann. Neurol. 17 (1985) 100–102.

SIEGAL, T., A. LOSSO and M.R. PFEFFER: Analysis of 31 patients with sustained off-therapy response following combined modality therapy. Neurology 44 (1994) 1463–1469.

SMITH, D.B., A. HOWELL, M. HARRIS, V.H. BRAMWELL and R.A. SELLWOOD: Carcinomatous meningitis associated with infiltrating lobular carcinoma of the breast. Eur. J. Surg. Oncol. 11 (1985) 33–36.

SO, Y.T., L.H.L. BECKSTEAD and R.L. DAVIS: Primary central nervous system lymphoma in acquired im-

mune deficiency syndrome: a clinical and pathological study. Ann. Neurol. 20 (1986) 566–572.

STARMANS, J.J.P., J. VOS and H.J. VAN DER HELM: The β-2-microglobulin content of the cerebrospinal fluid in neurological disease. J. Neurol. Sci. 33 (1977) 45–51.

STEINHERZ, P., B. JEREB and J. GALICHICH: Therapy of CNS leukemia with intraventricular chemotherapy and low dose neuraxis radiotherapy. J. Clin. Oncol. 3 (1985) 1217–1226.

STETLER-STEVENSON, W.G., S. AZNAVOORIAN and L.A. LIOTTA: Tumor cell interactions with the extracellular matrix during invasion and metastasis. Ann. Rev. Cell. Biol. 9 (1993) 541–573.

STEWART, D.J., M.J. KEATING, K.B. MCCREDIE, T.L. SMITH, E. YOUNESS, S.G. MURPHY, G.P. BODEY and E.J. FREIRIECH: Natural history of central nervous system acute leukemia in adults. Cancer 47 (1981) 184–196.

STEWART, D.J., J.A. MAROUN, H. HUGENHOLTZ, B. BENOIT, A. GIRARD, M. RICHARD, N. RUSSELL, L. HUEBSCH and J. DROUIN: Combined intraommaya methotrexate, cytosine arabinoside, hydrocortisone and thio-TEPA for meningeal involvement by malignancies. J. Neurooncol. 5 (1987) 315–322.

STROTHER, D.R., A. GLYNN-BARNHART, E. KOVNAR, R.E. GREGORY and S.B. MURPHY: Variability in the disposition of intraventricular methotrexate: a proposal for rational dosing. J. Clin. Oncol. 7 (1989) 1741–1747.

SUNG, C., S.M. BLANEY, D.E. COLE, F.M. BALLIS and R.L. DEDRICK: A pharmacokinetic model of topotecan clearance from plasma and cerebrospinal fluid. Cancer Res. 54 (1994) 5118–5122.

SUTTON, L.N., G. GOLDWEIN and B. PERILONGO: Prognostic factors in childhood ependymomas. Pediatr. Neurosurg. 16 (1990) 57–65.

TAKAKURA, K.: Intracranial germ cell tumors. Clin. Neurosurg. 32 (1985) 429–444.

TAM, J.K., W.G. BRADLEY, S.K. GOERGEN, D.-Y. CHEN, P.J. PEMA, M.D. DUBIN, L.M. TERESI and J.E. JORDAN: Patterns of contrast enhancement in the pediatric spine at MR imaging with single- and triple-dose gadolinium. Radiology 198 (1996) 273–278.

THOMAS, M.R., W.A. ROBINSON, L.M. GLODE, M.E. DANTAS, H. KOEPPLER, N. MORTON and J. SUTHERLAND: Treatment of advanced malignant melanoma with high dose chemotherapy and autologous bone marrow transplantation. Preliminary results – phase I study. Am. J. Clin. Oncol. 5 (1982) 611–622.

TONNER, D., P. BELDING, S.A. MOORE and J.A. SCHLECHTE: Intracranial dissemination of an ACTH secreting pituitary neoplasm – a case report and review of the literature. J. Endocrinol. Invest. 15 (1992) 387–391.

VERSARI, P., G. TALAMONTI, G. D'ALIBERTI, R. FONTANA, N. COLOMBO and G. CASADEI: Leptomeningeal dissemination of juvenile pilocytic astrocytoma: case report. Surg. Neurol. 41 (1994) 318–321.

VINCHON, M., M.-M. RUCHOUX, J.-P. LEJEUNE, R. ASSAKER and J.-L. CHRISTIAENS: Carcinomatous meningitis in a case of anaplastic meningioma. J. Neurooncol. 23 (1995) 239–243.

VONDEIMLING, A., J.H. KRAUS, A.P. STANGL, R. WELLEN-RUTHER, D. LENARTZ, J. SCHRAMM, D.N. JANIS, V. RAMESH, J. GUSELLA and O.D. WIESTLER: Evidence for subarachnoid spread in the development of multiple meningiomas. Brain Pathol. 5 (1995) 11–14.

WALKER, R. and J.C. ALLEN: Cisplatin in the treatment of recurrent childhood primary brain tumors. J. Clin. Oncol. 6 (1988) 62–66.

WASSERSTROM, W., J.P. GLASS and J.B. POSNER: Diagnosis and treatment of leptomeningeal metastases from solid tumors: experience with 90 patients. Cancer 49 (1982) 759–772.

WEED, J.C. and W.T. CREASMAN: Meningeal carcinomatosis secondary to advanced squamous cell carcinoma of the cervix: a case report. Gynecol. Oncol. 3 (1975) 201–204.

WELLER, M., A. STEVENS, N. SOMMER and H. WIETHOLTER: Tumor necrosis factor-alpha in malignant melanomatous meningitis (Letter). J. Neurol. Neurosurg. Psychiatry 55 (1992) 74.

WIEDERKEHR, F., M.R. BUELER and D.J. VONDERSCHMITT: Analysis of circulating immune complexes isolated from plasma, cerebrospinal fluid, and urine. Electrophoresis 12 (1991) 478–486.

YAP, B.-S., H.-Y. YAP, H.A. FRITSCHE, G. BLUMENSCHEIN and G.P. BODEY: CSF carcinoembryonic antigen in meningeal carcinomatosis from breast cancer. J. Am. Med. Assoc. 244 (1980) 1601–1610.

YAP, H.-Y., B.-S. YAP, C.K. TASHIMA, A. DISTEFANO and G.R. BLUMENSCHEIN: Meningeal carcinomatosis in breast cancer. Cancer 42 (1978) 283–286.

YAP, H.-Y., B.-S. YAP, S. RASMUSSEN, M.E. LEVENS, G.N. HOR-TOBAGYI and G.R. BLUMENSCHEIN: Treatment for meningeal carcinomatosis in breast cancer. Cancer 50 (1982) 219–222.

YUNG, W.-K.A., B.C. HORTEN and W.R. SHAPIRO: Meningeal gliomatosis: a review of 12 cases. Ann. Neurol. 8 (1980) 605–608.

ZACHARIAH, B. S.B. ZACHARIAH, R. VARGHESE and L. BAL-DUCCI: Carcinomatous meningitis: clinical manifestations and management. Int. J. Clin. Pharmacol. Ther. 33 (1995) 7–12.

ZIMM, S., J.M. COLLINS, J. MISER, D. CHATTERJI and D.G. POPLACK: Cytosine arabinoside cerebrospinal fluid kinetics. Clin. Pharmacol. Ther. 35 (1984) 826–830.

ZIMMERMAN, H.M.: Malignant lymphomas of the nervous system. Acta Neuropathol. (Berlin) (Suppl. IV) (1975) 69–74.

ZOVICKIAN, J. and R.J. YOULE: Efficacy of intrathecal immunotoxin therapy in an animal model of leptomeningeal neoplasia. J. Neurosurg. 68 (1988) 767–774.

Handbook of Clinical Neurology, Vol. 27 (71): Systemic Diseases, Part III
M.J. Aminoff and C.G. Goetz, editors

CHAPTER 27

Paraneoplastic syndromes

LAWRENCE M. CHER[1], JOHN W. HENSON[2], ASHA DAS[3] and FRED H. HOCHBERG[2]

[1]*Neurologist/Neuro-Oncologist, Austin and Repatriation Medical Centre, Heidelberg, Victoria, Australia;*
[2]*Neurology Service, Massachusetts General Hospital, Boston, MA, U.S.A.; and*
[3]*Department of Neuro-Oncology, Massachusetts General Hospital, Boston, MA, U.S.A.*

The neurologic paraneoplastic syndromes constitute disorders affecting multiple levels of the nervous system. Certain well recognized syndromes may be confidently diagnosed by informed clinicians aware of the association between malignancy and neurologic difficulties (dermatomyositis and lung cancer, Lambert–Eaton syndrome and small cell cancer, opsoclonus and neuroblastoma). Many of the same syndromes may appear in the absence of cancer or mimic spontaneous brain degeneration (e.g., Purkinje cell cerebellar degeneration) or infections of the neuraxis (herpes simplex encephalitis or subacute spongioform degeneration for example).

The paraneoplastic disorders represent non-metastatic so-called remote complications of cancer. The syndromes often appear prior to the diagnosis of malignancy. Months to years may pass between the onset of depression, memory deficits and hallucinations and the identification of a causative small cell carcinoma of the lung (SCLC). The brain is not the direct target of malignant spread and the symptoms do not reflect the effects of infection, stroke or the complications of chemotherapy. Infections, endocrine and metabolic disorders associated with cancer (e.g., syndrome of inappropriate antidiuretic hormone) are reviewed elsewhere in this series. These syndromes are thought to be immunologically mediated. Since the review in Vol. 38 of the Handbook of Clinical Neurology by Henson and Urich (1979), there has been further delineation of the clinical spectrum, and anti-

body detection has become a useful diagnostic aid. Evidence suggests that the immunologic response limits the growth and spread of the malignancy; in many patients the causative neoplasm is limited in extent and may even spontaneously regress. Patients with these syndromes often harbor antibodies in their serum and cerebrospinal fluid (CSF) that recognize antigens shared by neurons and tumor cells. An appropriate immune response against a tumor antigen cross-reacts with similar antigens expressed by the nervous system. Direct evidence for this has been developed during investigation of the Lambert–Eaton myasthenic syndrome (LEMS). Less proven is the causative role of the immune response in the pathogenesis of paraneoplastic cerebellar degeneration (PCD), paraneoplastic encephalomyelitis (PEM) and sensory neuropathy, paraneoplastic motor system disease and paraneoplastic opsoclonus, myoclonus and ataxia. Clearly, the immune response represents a potential target of therapies (Cher et al. 1995) designed to halt paraneoplastic syndromes. The striking similarities between LEMS and these three paraneoplastic neurological syndromes support a unifying autoimmunity hypothesis:

(1) All four diseases are distinctive clinical syndromes in which patient's CSF and/or serum harbor high titers of antibodies that recognize an antigen shared by the nervous system and the tumor.

(2) Autoantibody synthesis within the central ner-

vous system (CNS) has been demonstrated for each of the three CNS syndromes.

(3) The autoantibodies appear tumor specific. The finding of one of the autoantibodies (e.g., anti-Yo, Hu, Ri) in a patient with a neurological syndrome predicts the identification of an underlying carcinoma of specific histologic type.

(4) The neurological syndrome precedes the diagnosis of cancer in approximately two-thirds of patients. When identified, the tumor is likely indolent (Darnell and DeAngelis 1993).

On occasion a paraneoplastic syndrome develops in the setting of prior cancer which can no longer be identified to be active. In this chapter we provide an exposition of the most common paraneoplastic syndromes. These are organized by neurologic site: peripheral nervous system, the neuromuscular junction, disorders of muscle, the CNS disorders (encephalitis and myelitis), the cerebellar degenerations, opsoclonus and myoclonus, involvements of the spinal cord, eye and miscellaneous locations. In each section care has been taken to provide the reader with a clear depiction of the clinical spectrum of disease in addition to modern concepts of molecular pathophysiology and therapy.

HISTORY

The recognition of paraneoplastic syndromes is recent. Greenfield (1934) described an example under the title 'Subacute cerebellar degeneration' and noted infiltrates of lymphocytes surrounding vessels of brainstem and spinal cord. During the 1960s there appeared reports of 'limbic encephalitis', associated commonly with SCLC. Rarely tumor was not found. The 'cortical cerebellar' degeneration of Brain and Wilkinson (1965) was subdivided (Henson and Urich 1982b) into cancer-related 'degenerative' and 'inflammatory' types. 'The process affects primarily the Purkinje cell layer, in which the loss of neurons ranges in most cases from severe to almost total.' Inflammatory changes were noted elsewhere in the nervous system (see also Bruyn 1987).

Few patients develop these syndromes. Of 1465 cancer patients in the London Hospital, Croft and Wilkinson (1965) found 'carcinomatous neuromyopathy' in 7.2%. Of these, 4.2% were afflicted with neuromyopathy or 'proximal weakness and wasting of muscles associated with diminution or loss of two

or more tendon reflexes'. Three percent of patients had more specific syndromes, 2.3% with either neuropathy, myopathy or myasthenic syndrome and 0.6% with a central paraneoplastic syndrome. Hudson et al. (1993) used a questionnaire to identify one example of cerebellar difficulty among 1000 women with ovarian cancer. Of 150 patients with SCLC (Elrington et al. 1991), two were found to have LEMS and one had an occult subacute sensory neuropathy (SSN). Half of SCLC patients have less specific syndromes, such as neuromuscular or autonomic deficits presenting as mild weakness with weight loss and dry mouth with (16%) distal vibratory loss.

PERIPHERAL NERVOUS SYSTEM

The description by Denny-Brown (1948) of SSN sparked investigation of peripheral neuropathy in cancer patients. The origin of these syndromes is often obscured by the peripheral neurotoxic effects of chemotherapeutic agents, metabolic disorders and inactivity. McLeod (1993) estimates that 5% of cancer patients have clinical neuropathies, 12% have abnormalities on quantitative sensory testing, and 40% have electrophysiologic abnormalities. The neuropathies are often separated into pure disorders of sensation (dorsal root ganglion or nerve) such as SSN, or disorders of sensorimotor dysfunction. The neuropathies of paraproteinemia and myeloma (Pollard 1993) will not be dealt with here.

Subacute sensory neuropathy (SSN)

Denny-Brown's 1948 description provides a lasting example of this process. A boiler cleaner aged 59 years . . . . developed an ache in his right ear radiating into the right side of the neck. After 14 days he noticed numbness of the soles of both feet ascending up the legs. Numbness of his mouth gradually spread to the whole face in several days. The hands were becoming numb . . . . [Three months later] he began to have intermittent tingling sensations in both hands and both feet as well as stabs of pain in the . . . ankles, knees and buttocks. Sensation of pain was absent over the whole face and tongue, and later over the whole body. The corneal reflexes were absent. Sensation to light touch was also absent over most areas of the body, and only impaired deep sensibility and temperature sense remained. Vibration perception remained only on the clavicles, sternum, and upper dorsal spines. Walking was possible only with assistance, as was shuffling and he was grossly ataxic.

SSN occurs alone or in association with PEM. Characterized as a sensory neuropathy, the brunt of the damage is borne by the dorsal root ganglion cell (Fig. 1). Pain and loss of sensation follow. Women are three times more likely to be afflicted (Chalk et al. 1992) and then likely harbor SCLC or breast carcinoma. In 92% SSN presents as asymmetric upper limb pain, paresthesia and sensory loss, and usually precedes the diagnosis of cancer. Lacking a diagnosis, there may be evaluations over several months followed by unwarranted carpal tunnel release or ulnar nerve transposition. Smith (1992) identified the appearance of an ataxic syndrome or a hyperalgesic-ataxic syndrome in the setting of gastrointestinal dysmotility.

Commonly pain, paresthesias and numbness begin distally. Progressive loss of sensation occurs and can involve both the trunk and face. Proprioceptive loss, often severe, is associated with pseudo-athetosis (38% of patients) and a severe sensory ataxia. Pain sensation is lost in many patients. Sensory nerve action potentials are usually absent. The process of deafferentation may produce apparent motor weakness. Intestinal pseudo-obstruction with SSN reflects hypomotility of the small intestine or colon. Anorexia and weight loss precede satiety, vomiting, constipation and autonomic dysfunction. Pseudo-obstruction has been reported as an isolated presentation in association with anti-Hu antibodies (Liang et al. 1994) and SCLC (Condom et al. 1993). Seldom are paraneoplastic syndromes restricted entities. Variants occur, with predominant hyperalgesia (Darnell and DeAngelis 1993) or accompanying features of PEM. Nerve conduction studies show low-amplitude or absent sensory nerve action potential in a patchy distribution. Electrophysiologic compound motor action potentials (CMAPs) are preserved and motor conduction velocities generally are within the normal range.

SSN results from inflammation and degeneration of the dorsal root ganglion cells. Kinesthetic sensory impairment and ataxia reflect large-diameter dorsal root ganglion neuronal damage while small-diameter dorsal root ganglion neuronal loss gives rise to abnormalities of pain and temperature sensation. The initial response to this degeneration is lymphocytic response which gives way to fibrosis (Fig. 1A) and marked secondary demyelinative changes in the dorsal columns and peripheral sensory nerves. In patients with intestinal pseudo-obstruction, neuronal loss occurs in the myenteric plexus where Schwann cell proliferates and lymphocytic infiltrates are noted (Smith 1992). The therapy of this and related conditions is not likely to be successful. Anecdotal responses follow the use of steroids (ST), chemotherapy (CT) of the underlying neoplasm, or plasmapheresis (PE) (Table 1).

Sensorimotor paraneoplastic neuropathy may be acute, subacute or relapsing. The acute neurop-

TABLE 1

The therapy of subacute sensory neuronopathy (SSN).

Disorder	Tumor	Therapy	Response	Author
SSN	SCLC	PE	stable	Anderson et al. 1988b
With encephalo-myelitis		ST, CT	CR	
Paraneoplastic sensory neuropathy	SCLC (19) (5/19 anti-Hu ab)	7 ST	PR (1)	Chalk et al. 1992
SSN	SCLC	treat underlying cancer	stable (2)	Graus et al. 1985
SSN	SCLC	CT, ST, PE	stable (1)	Graus et al. 1992
		ST, PE, cyclophosphamide	stable (1)	
		CT	stable (2)	

CT = chemotherapy; PE = plasma exchange; ST = steroids. Number of patients in parentheses.

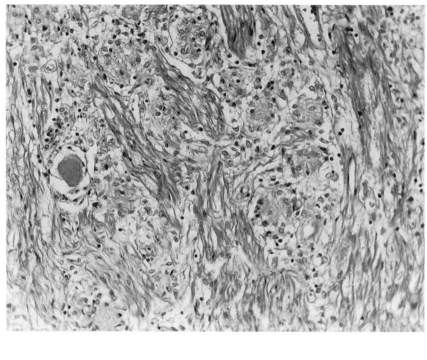

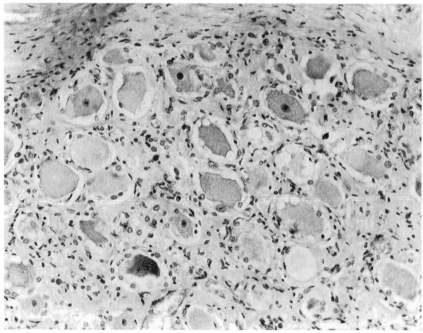

Fig. 1. (A) Dorsal root ganglion of a patient with SSN and SCLC. There is almost complete loss of ganglion cell bodies with replacement by compact groups of satellite cells (arrow = nodules of Nageotte). (B) Normal post-mortem dorsal root ganglion for comparison. (H and E, 160 ×.) (Courtesy of Dr. David Louis, Massachusetts General Hospital.)

athy, similar to acute inflammatory demyelinating polyneuropathy (AIDP) (Klingon 1965; Croft et al. 1967), has been described in a number of malignancies including Hodgkin's disease (Lisak et al. 1977). The nerve pathology has not been well described, but is likely to be demyelinating. On the contrary, axonal involvement characterizes the subacute (progressive) pure sensory (SSN) neuropathy described above. Clinically the neuropathy is distal, with sensory loss and reduced reflexes, and usually involves the lower limbs to a greater extent than the upper limbs. A mild 'terminal neuropathy' has also been described (McLeod 1993). Therapy of the non-sensory neuropathy takes advantage (Table 2) of steroids as well as control of the underlying neoplasm with surgery or chemotherapy. Responses are not common but are identified.

Paraneoplastic vasculitic neuropathy

Oh et al. (1991) reported: 'A 55-year-old pediatrician with a 1-year history of metastatic endometrial carcinoma had been successfully treated with high-dose progesterone at 1000 mg/day. Eight months after the diagnosis of endometrial carcinoma she noticed numbness and lancinating pain below the left knee followed by weakness and left foot drop within 2 weeks. She soon began to have similar though less severe sensory complaints and weakness below the right knee. Initial examination showed asymmetric sensorimotor peripheral neuropathy in the legs .... There was decreased sensation to pin-prick and light touch up to the mid-calf level and a mild diminution of proprioception and vibratory sensation in the feet. Reflexes were absent in the legs.' The causative vas-

TABLE 2

Therapy of paraneoplastic neuropathy.

Neuropathy	Cancer	Therapy	Response	Author
Peripheral neuropathy	breast	ST	brief improvement (1)	Croft 1990a
	stomach	ST	improved (1)	
Pure sensorimotor neuropathy	gastric lymphoma	S	resolution (1)	Enevoldson 1990
Peripheral neuropathy	transitional cell of the renal pelvis	S	CR (1)	Fawcett 1977
Mononeuritis multiplex due to vasculitis	SCLC (2)	ST	stabilized (2)	Johnson et al. 1979
Inflammatory brachial plexopathy	HD	ST	CR (1)	Lachance et al. 1991
Paraneoplastic vasculitic neuropathy	endometrial	cyclophosphamide	improved (1)	Oh et al. 1991
SSN	HD	ST, CT	improved (1)	Sagar 1982
Polyneuropathy	SCLC	CT	marked improvement recurrence coinciding with relapse	Takigawa 1993
Sensorimotor neuropathy	lung	ST, CT	improved then relapse (1)	Vincent et al. 1986
	HD	RT, CT, ST	improved (1)	
	immunoblastic lymphadenopathy	ST	improved (1)	
Sensorimotor neuropathy	SCLC	CT, ST, PE, RT	resolution (1)	Weissman 1989

CT = chemotherapy; PE = plasma exchange; RT = radiotherapy; ST = steroids. Number of patients in parentheses.

culitic neuropathy is usually associated with hematologic malignancy (Kurzrock and Cohen 1993) but there are 36 examples of hypersensitivity vasculitis associated with solid tumors (Kurzrock et al. 1994). The most common is a leukocytoclastic vasculitis, characterized by a cutaneous vasculitis, and pathologically by prominent neutrophilic infiltration of the vessels with a 'leukocytoclastic reaction' due to disruption of leukocyte membranes. More recently a microvasculitis involving both nerve and muscle has been described (Johnson et al. 1979; Vincent et al. 1986) in association with solid tumors (SCLC, renal cell cancer, adenocarcinomas of lung, prostate and endometrium) and lymphomas. Epineurial and perimysial vessels are invaded by T-cells (Matsumuro et al. 1994). These changes produce asymmetric painful, sensorimotor peripheral neuropathy involving either upper or lower limbs. Unlike mononeuritis multiplex, large nerves are not involved. Nerve conduction studies and electromyographic evaluations are consistent with a diffuse axonal peripheral neuropathy. Elevated concentrations of CSF protein are common in this condition – a finding which is unusual for most axonal neuropathies. The diagnosis rests on the demonstration of microscopic inflammation within nerve (or muscle) specimens. In some patients with SCLC there has been an associated encephalomyelitis, and recently the syndrome has been associated with anti-Hu antibody in the absence of encephalomyelitis (Younger et al. 1994). Treatment responses (Table 2) have followed therapy with cyclophosphamide (Oh et al. 1991) and remission of the underlying SCLC (Matsumuro et al. 1994).

Neuromuscular junction disorders

Lambert–Eaton myasthenic syndrome (LEMS)
First described in 1956 (Lambert et al. 1956) in patients with cancer, LEMS occurs either associated with malignancy (most commonly SCLC) or with other autoimmune disorders (thyroid disease, pernicious anemia, vitiligo, and type I diabetes mellitus) (O'Neill et al. 1988). Basic to both is a disorder of the presynaptic voltage-gated calcium channel (VGCC) receptor of the neuromuscular junction.

Clinical features. The striking syndrome is illustrated by the following example. A 65-year-old woman developed dryness of the mouth and slowly evolving weakness of her legs which made walking of stairs or exiting car seats difficult. Muscle cramps were noted in the legs. There were no intellectual or cranial nerve difficulties. Reflexes were initially present but soon disappeared. Electrophysiologic studies revealed decreased compound action potentials. At low rates of repetitive stimulation (10 Hz) there was decrement of the amplitude of the action potential – a change suggestive of myasthenia gravis. At faster rates of stimulation (50 Hz) the potential rose in an incremental fashion. A malignancy was not identified during exhaustive evaluation of the chest and abdomen using both computed scanning and bronchoscopy. Notable improvement followed the institution of mestinon at low doses in addition to 2,4-diaminopyridine. Studies performed (Dr. V. Lennon, Mayo Clinic) revealed antibodies directed against the acetylcholine receptor as well as against the sodium-gated channel receptor. These studies suggested LEMS with features of myasthenia gravis. Eighteen months after presentation she developed right-sided weakness and receptive aphasia. Masses found anterior to the motor cortex and coincidentally in the hilum of the lung were small cell cancer when examined by microscope.

General fatigue commonly precedes weakness. Consultation is heralded by gait dysfunction which follows weakness upon standing. As opposed to myasthenia gravis, ocular symptoms are rare. Occasionally patients may present with respiratory failure (Barr et al. 1993). Autonomic dysfunction is common, but usually confined to cholinergic function, with xerostomia and erectile failure in males. Although sluggish pupillary responses may occur, orthostatic hypotension is not usually a feature. The proximal weakness mostly involving the lower limbs is improved (facilitated) on repetitive or sustained contraction. The weakness perceived on the first examination of the patient improves with sequential contractions. This is best demonstrated in moderately affected muscles. Whereas the muscle contractions in myasthenia gravis saturate the sparse post-synaptic acetylcholine receptors, the contractions in LEMS lead to increased calcium concentration within the nerve terminal and increase in the suboptimal quantal release of acetylcholine. The weak muscles are hyporeflexic and the gait waddling. This combination of proximal weakness and reflex loss is the hallmark of LEMS.

At the root of diagnosis is the neurophysiologic

study. The weakened muscles exhibit reduced amplitude CMAP. Characteristic is facilitation (2-fold increase of CMAP) after rapid stimulation (20–50 Hz) or sustained contraction. Repetitive stimulation at rates of 2 Hz is associated with a decremental response.

Molecular pathophysiology. In LEMS autoantibodies disrupt the function of neuronal VGCC, leading to a reduction in the release of acetylcholine at the neuromuscular junction and at synapses of autonomic nerves. Changes in membrane voltage and depolarization of the neuronal membrane by influx of K^+ ions lead to the opening of calcium channels. Ca^{2+} ions flow into the nerve terminal, and neurotransmitter vesicles fuse with the neuronal membrane, releasing acetylcholine into the neuromuscular junction. The release of each vesicle, or quanta, of acetylcholine into the neuromuscular junction can be measured electrophysiologically as a miniature end-plate potential.

When muscle biopsy specimens from LEMS patients receive a single supramaximal stimulus, there emerge miniature end-plate potentials that are normal in amplitude but reduced in number. This suggested a reduction in the number of quanta of acetylcholine being released into the neuromuscular junction (Elmqvist and Lambert 1968). Identical electrophysiological features were observed in mouse muscle following exposure to immunoglobulin G (IgG) from patients with LEMS. This established (Lang et al. 1981; Prior et al. 1985) that antibodies were the direct cause of the physiological abnormality. The identity of the target of LEMS IgG was suggested by electron micrographs of mouse muscle that was incubated with LEMS IgG. Particles (VGCC) lining each side of the dense bar were clumped and decreased in number (Fukunaga et al. 1983; Fukuoka et al. 1987; Nagel et al. 1988). These changes were dependent upon divalent LEMS IgG (i.e., $F(ab')_2$ fragments). The distance between particle rows and the distance between the aggregated particles were similar to the distance between the two binding sites of an IgG molecule. The identity of the antigens was elucidated by the discovery that LEMS IgG could precipitate [^{125}I]conotoxin associated with VGCC complexes, suggesting that either the VGCC themselves or associated proteins are the targets (Sher et al. 1989). Neuronal VGCC are classified electrophysiologically and pharmacologically as L, N, P, and T types. The calcium channel complex is composed of four subunits (α_1, α_2, β, and

γ). The α_1 subunit forms the transmembrane pore and is the target of conotoxin.

Cultures of SCLC cells, expressing VGCC, can be functionally modulated by LEMS IgG. This demonstration (Roberts et al. 1985) supported the hypothesis that LEMS results from cross-reaction of tumor-directed antibodies with antigens in the nervous system. LEMS IgG could alter VGCC function by lowering the transmembrane flux of Ca^{2+} in a time-dependent and IgG concentration-dependent fashion (Kim and Neher 1988). Patch-clamp experiments demonstrated that individual VGCC were affected in an 'all-or-none' manner. This suggested that the decreased quantal release of acetylcholine resulted from inactivation of some fraction of VGCC. The particle aggregation seen by electron microscopy pointed to VGCC disruption by an immune mechanism in which the rate of turnover of the channels is elevated, as opposed to direct blockade of the channel (Passafaro et al. 1992).

While these results established the autoimmune basis of LEMS, determination of the specific antigenic targets of the LEMS IgG or the mechanism of the resulting VGCC dysfunction has proved more difficult. Early attempts to identify the molecular weights of proteins in crude neuronal membrane extracts by Western blot using LEMS IgG were unsuccessful, possibly due to the low concentration of antigen (Chester et al. 1987). Following purification of VGCC complexes from rat brain synaptosomes (nerve terminals purified from brain homogenates), Western blot analysis revealed a 58 kDa protein that was recognized by LEMS IgG (Leveque et al. 1992). A monoclonal antibody then raised against the 58 kDa protein was used to isolate a cDNA from a rat brain λ gt11 expression library. The identified protein revealed a high degree of homology to those in the family of synaptotagmins, which are Ca^{2+}-sensing proteins in the membranes of synaptic vesicles (Ullrich et al. 1994). While the 58 kDa protein copurified with conotoxin-bound VGCC, it remains unclear how antibodies to synaptotagmin might lead to the aggregation of VGCC seen on electron microscopy.

A more direct approach to the LEMS antigen involved isolation of a cDNA from a human fetal brain expression library by screening with LEMS IgG (Rosenfeld et al. 1993). The predicted protein was the calcium channel β subunit family member. Thus, in agreement with the functional evidence implicating

L.M. CHER ET AL.

TABLE 3

Therapy of Lambert–Eaton myasthenic syndrome (LEMS).

Syndrome	Cancer	Therapy	Response	Author
LEMS	Bronchial	CT	CR (1)	Berglund 1982
LEMS	SCLC	S, RT, ST, PE, 3,4-DAP	CR (1)	Chalk et al. 1990
		RT, 3,4-DAP	CR (1)	
		CT, ST, PE, 3,4-DAP	CR (1)	
		S, RT	improved then relapse (1)	
		CT, 3,4-DAP	CR (1)	
		CT, RT, PE, 3,4-DAP	CR (1)	
		CT, ST, azathioprine, PE, 3,4-DAP, pyridostigmine	improved then relapse (1)	
		CT, ST, PE, 3,4-DAP pyridostigmine	CR (1)	
		CT, PE, 3,4-DAP	CR (1)	
		S, ST, azathioprine, PE	improved then relapse (1)	
		ST, azathioprine, PE	improved then relapse (1)	
		S,ST, azathioprine, PE, 3,4-DAP	improved then relapse (1)	
		ST, 3,4-DAP, pyridostigmine	improved then relapse (1)	
		ST, azathioprine, PE, guanidine	improved then relapse (1)	
LEMS	SCLC	guanidine	PR (1)	Clamon 1984
		CT, RT	CR	
	SCLC	CT	CR (1)	
	SCLC	guanidine	PR (1)	
		RT	CR	
	SCLC	S	PR (1)	
		CT	CR	
	SCLC	S, ST, RT, guanidine, DPH, CT	PR (1)	
LEMS	SCLC	CT	improved (1)	Hawley 1980
LEMS	SCLC	ST	PR (1)	Ingram 1984
		PE	CR	
		CT	PR	
	lung	PE	PR (1)	
LEMS	SCLC	guanidine	improvement (1)	Jablecki 1984
		CT, RT	CR	
LEMS	SCLC	CT, RT, guanidine	CR (1)	Jenkyn 1980
LEMS	SCLC	anticholinesterase	improvement (1)	Kalter 1985
		CT, RT	CR	
Myasthenia gravis	SCLC	anticholinesterase	improved (1)	Leger 1993
LEMS	renal cell, SCLC, endometrial, cervical, basal cell, breast	3,4-DAP	improved (7)	McEvoy 1991
LEMS	SCLC	guanidine, pyridostigmine	PR (1)	Oh et al. 1991
		RT, CT	CR then relapse 8 months prior to tumor recurrence	
	SCLC and breast	pyridostigmine, ST, PE	PR (1)	
		CT, RT	CR	

CT = chemotherapy; PE = plasma exchange; RT = radiotherapy; S = surgery; ST = steroids. Number of patients in parentheses.

multiple antigenic targets, at least two VGCC-related antigens have been identified by molecular cloning techniques (Johnston et al. 1994). These data have yet to clearly elucidate the mechanism by which the antibodies induce VGCC dysfunction.

Therapy (Table 3). This topic has been well reviewed by McEvoy (1994). LEMS is not as responsive to therapy as myasthenia gravis. Therapy involves three different strategies, which must often be combined for the best outcome. SCLC, while rarely curable, is very responsive to chemotherapy or surgery of the cancer, and treatment in patients with paraneoplastic LEMS is associated with improvement. However, many patients still have significant disability. Pyridostigmine has some activity in LEMS. Acetylcholine release can be enhanced by blocking potassium channels involved in terminating the action potential, prolonging the time available for calcium entry into the cell, and enhancing acetylcholine release. The drug 4-aminopyridine improves muscle strength but is associated with an unacceptable lowering of the seizure threshold. An alternative, the experimental agent 3,4-diaminopyridine (3,4-DAP), is well tolerated, has been shown to increase strength and is associated with a much lower incidence of seizures. As the agent has not been formulated for commercial distribution, its use remains restricted to research institutions. LEMS patients, in comparison to myasthenics, respond slowly and incompletely to immunosuppression. Prednisone (ST) given on alternating days, is the usual first-line therapy, taking weeks to offer benefit. Azathioprine has been used with benefit and other immunosuppressive agents (cyclophosphamide and cyclosporin) have efficacy. Plasma exchange (PE), less effective than in myasthenia gravis, provides benefit within 2 weeks and commonly lasts for 6 weeks. Responses are reported following intravenous immunoglobulin (i.v. Ig) therapy. McEvoy (1994) recommends a cascade of therapy commencing with pyridostigmine and treatment of the underlying tumor followed by 3,4-DAP, prednisone/azathioprine, PE or i.v. Ig and lastly cyclosporine, cyclophosphamide or guanidine.

Myopathies

Inflammatory myopathies

The inflammatory myopathies comprise dermatomyositis (DM), polymyositis (PM) and inclusion body myositis (IBM). IBM is not associated with malignancy. DM is an idiopathic inflammatory myopathy associated with characteristic skin changes including a heliotrope violaceous rash on the upper eyelids, Gottron's papules over the knuckles, and poikiloderma in the 'V' of the neck. The muscle changes consist of perivascular inflammation secondary to microangiopathy. Within muscle fibers at the periphery of the fascicle microinfarcts lead to perifascicular atrophy. PM is not associated with the cutaneous abnormalities and the inflammation occurs predominantly within the fascicles (Dalakas 1992).

Both PM and DM may occur alone or in conjunction with other connective tissue diseases. The relationship between PM, DM and malignancy has been extensively reviewed (Callen 1994)

The exact relationship between DM, PM and malignancy has been the subject of controversy. Much of the debate relates to the methods employed for diagnosis and the criteria used to differentiate DM and PM. Bohan and Peter (1975) point out that 'literature reviews' may falsely elevate the true incidence of malignancy with DM and PM as these combinations are both more likely to be reported and result from extensive evaluations likely to disclose an unrelated tumor. Prospective studies identify prior or subsequent cancer in 25% of patients with DM. The rate of cancer and PM is lower and may approach background risk.

Whereas most paraneoplastic syndromes are linked to cancer histologies, this is not the case with inflammatory myopathies. Only ovarian cancer (Barnes 1976) is over-represented in DM and PM. Treatment of the associated malignancy may improve the inflammatory myopathy (Cox et al. 1990). The search for tumor in patients with DM or PM without known malignancy will be discussed later.

Necrotizing myopathy

This syndrome is not common (Smith 1969). Weakness may rapidly progress to respiratory failure and death. There is widespread necrosis of skeletal muscles, with little inflammation (Henson and Urich 1982d, p. 406). In addition, degeneration of intramuscular nerve fibers has been described (Brownell and Hughes 1975).

Emslie-Smith and Engel (1991) described three patients whose necrotizing myopathy was associated with thickened 'pipestem' capillaries, and capillary depletion without inflammatory cells. Notable was microvascular deposition of the complement membrane attack complex. One of these patients had transitional cell carcinoma of the bladder. The myopathy in all patients responded to corticosteroids. It is unclear whether this is the same syndrome described above.

Stiff man syndrome (SMS)

SMS is characterized by chronic rigidity of the body musculature with superimposed painful spasms (McEvoy 1991). Approximately 60% of patients have antibodies against glutamic acid decarboxylase (GAD), often associated with insulin-dependent diabetes mellitus and other organ-specific autoimmune diseases (Solimena et al. 1990). Three women with breast cancer developed SMS associated with autoantibodies directed against a 128 kDa neuronal antigen concentrated at synapses (Folli et al. 1993). SMS has also been described in a patient with Hodgkin's disease (Ferrari et al. 1990) and in a patient with thymoma (Nicholas et al. 1994). Recently the neuronal antigen has been identified as amphiphysin (De Camilli et al. 1993), a protein found in high levels within the brain localized within nerve terminals. SMS in association with cancer has some features not typical of the classic syndrome: stiffness was localized to

the proximal limbs and trunk; improvement with corticosteroids was common (Solimena et al. 1990). Therapy of SMS makes use (Table 4) of muscle relaxant medications, chemotherapy or steroids.

CENTRAL NERVOUS SYSTEM (CNS)

Paraneoplastic encephalomyelitis (PEM)

Encephalomyelitis associated with cancer (Henson et al. 1965) is clearly recognized in association with inflammatory infiltrates occurring throughout the neuraxis. Commonly seen in association with SSN, a number of different clinical syndromes have been described under the rubric of PEM.

Limbic encephalitis

First described by Brierley et al. (1960), the term 'limbic encephalitis' was coined in 1968 by Corsellis et al. Progressive memory deficit, often associated with anxiety and depression, is seen over weeks to months. The memory loss may fluctuate, and there may be episodes of confusion, partial complex seizures and bizarre behavior with hallucinations. Neuropsychological evaluation of patients reveals deficits involving the acquisition and retention of new material

TABLE 4

Therapy of Stiff man syndrome (SMS).

Syndrome	Cancer	Therapy	Response	Author
SMS	SCLC	lioresal	mild improvement (1)	Bateman 1990
SMS	HD	diazepam and baclofen	PR (1)	Ferrari et al. 1990
Cerebellar with antibodies to glutamic acid decarboxylase		CT	CR	
SMS antibody to neuronal protein 128 kDa	breast	tamoxifen, diazepam, baclofen	CR (1)	Folli et al. 1993
SMS and limbic encephalitis	pharyngeal	diazepam, clonazepam, VPA and baclofen	PR (1)	Masson 1987
		ST	CR of SMS	
SMS antibodies to neuronal protein 128 kDa →	breast	ST	PR (1)	Piccolo 1989
		tamoxifen	CR	
	thymoma and MG	no treatment	CR (1)	

CT = chemotherapy; ST = steroids. Number of patients in parentheses.

within the context of intact intelligence, language and praxis (Camara and Chelune 1987). The onset usually occurs over weeks, but may be more acute. The clinical alterations reflect lymphocytic infiltration and neuronal loss within the temporal lobes (Fig. 2). Corsellis et al.'s (1968) report is illustrative: 'The patient . . . had felt weak and ill and had lost 20 lb in weight . . . he had two convulsions followed by a third . . . each fit had started with clonic movements of the left hand and arm before becoming generalized . . . . [One month later] the patient became mildly confused and the next morning awoke with a complete loss of memory for the previous 8 weeks . . . his memory deficit remained the salient feature. He could register information correctly but within two minutes could not recall it. He could do mental arithmetic correctly . . . he interpreted spoken and written requests well, he could name objects and answer written questions and he appreciated the meaning of mime and gesture . . . his personality was well preserved. He could not, however, recall a name after three minutes and his retrograde amnesia now covered several years. He described how on waking each morning he would not know where he was . . . his distant memory was good . . . '. (At autopsy a probable SCLC was found.)

Equally of interest is the description of Burton et al. (1988): 'The patient was a 26-year-old man who, in June 1983, was noted to have increasing memory loss. In August 1983 he suffered a major motor seizure . . . and CT showed an enhancing right temporal lobe lesion. . . over the next 6 months he had progressive decline in short-term memory and long-term memory. . . . In February 1984, a right testicular mass was noted. A radical orchidectomy was done which revealed a non-seminomatous germ cell neoplasm. In June 1984 . . . the neurologic examination revealed impaired mental status with a short attention span, poor short-term memory (none of three objects at 1 minute), and poor long-term memory . . . an open biopsy [of the left temporal lobe] demonstrated neuronal loss, gliosis and perivascular chronic inflammation . . . [Following chemotherapy] [t]here was a gradual improvement in his mental status during therapy with normalization of his short-term and long-term memory. He continued to have difficulty with abstraction and was somewhat emotionally labile. . . .

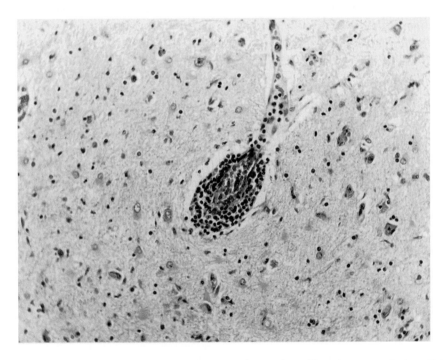

Fig. 2. Perivascular lymphocytic infiltration, with scattered parenchymal lymphocytes and microglial cells in the temporal lobe of a patient with limbic encephalitis and SCLC (H and E, 160 ×). (Courtesy of Dr. Jean Paul Vonsattel, Massachusetts General Hospital.)

In December 1984 his short-term and long-term memory deteriorated to his prechemotherapy condition. He was now easily enraged and was physically combative [and recurrent tumor was found]. . . . During therapy his mental status improved with memory normalization and no further episodes of aggressive or combative behavior.'

Brainstem encephalitis

The described example from the report of Henson and Urich (1982a) provides a carefully wrought statement of the syndrome: 'A 74-year-old man developed a midline cerebellar ataxia, progressive dysarthria and emotional lability over 2 months. No appendicular ataxia was noted. MRI and CSF were normal. No antineuronal antibodies were noted on immunofluorescence. A thymic carcinoma was found and treated with surgery and radiotherapy. Despite therapy, he continued to slowly deteriorate over the next 6 months, developing a pancerebellar syndrome, with nystagmus, cerebellar dysarthria, and dysmetria, together with dysarthria, dysphagia and pathologically brisk reflexes. Memory remained relatively spared.'

Any level of the brainstem may be impaired. Medulla involvement produces vertigo, ataxia, nystagmus, nausea and vomiting. Bulbar palsy may also be a feature. Pontine nuclear difficulties include disorders of ocular motility, internuclear ophthalmoplegia, third, fourth and sixth nerve palsies. Recently described are two patients with prostate cancer whose lost horizontal gaze was in the setting of continuous facial muscle spasms. Pathologic evidence was consistent with brainstem encephalitis (Baloh et al. 1993). A central hypoventilatory syndrome has been described (Kaplan and Itabashi 1974). Features of corticospinal tract involvement are not uncommon nor are extrapyramidal movement disorders including dystonia (Dietl et al. 1982), tremor, dystonia and parkinsonism with loss of neurons within the substantia nigra (Golbe et al. 1989) and chorea and dystonia (Albin et al. 1988).

Myelitis: paraneoplastic anterior horn cell disease

Symptoms related to spinal cord involvement usually occur within the context of extensive neuraxis involvement. Anterior horn cell loss associated with muscle wasting of the upper limbs and weakness may mimic motor neuron disease (MND). The legs and neck may also be involved. The sensory loss that appears usually reflects the coexistence of SSN. Indeed half of SSN cases occur within the setting of PEM.

Dalmau et al. (1992) provide a careful description of an afflicted 70-year-old man who 'with a remote history of renal cancer developed numbness and burning in his hands and feet 1.5 months after diagnosis of anaplastic carcinoma of the prostate. Over the next 2 months there was a relentless progression of symptoms, and after admission to hospital his examination showed wasting and weakness in the legs with areflexia, and severe sensory changes in the arms and legs to T10. Examination of the CSF demonstrated increased protein (101 mg/ml) with no cells . . . he progressed to develop nystagmus on right lateral gaze, facial weakness, urinary retention and hyponatremia, and became bed-bound with increasing sensory and motor deficits. Neurologic examination showed normal mental status, dysarthria, a left afferent pupillary defect, nystagmus on right lateral gaze, right VI and VII cranial nerve palsies, profound lower extremity weakness with fasciculations, and hypotonia and minimal upper extremity weakness. Reflexes were absent, and there was a severe deficit to all sensory modalites in all extremities. The anti-Hu antibody was detected in the serum and CSF. . . . Autopsy showed . . . striking depletion of DRG neurons with secondary degeneration of the dorsal columns. . . . Severe inflammatory changes in the spinal cord gray matter with gliosis and conspicuous neuronal loss. . . .'

Anti-Hu syndrome

Dalmau et al. (1992a) reviewed the features of 71 patients with PEM who harbored the anti-Hu antibody. Three-quarters were associated with SCLC. Surprisingly two patients had prostate cancer. The neurologic disease was the presenting feature in 83% of patients. Seventy-three percent of patients had involvement of multiple levels of the neuraxis including 28% with autonomic dysfunction. The latter was associated with orthostatic hypotension, abnormal pupillary responses, and areas of hyperhidrosis. The sudden death of these patients may reflect autonomic dysregulation. Most common were symptoms asso-

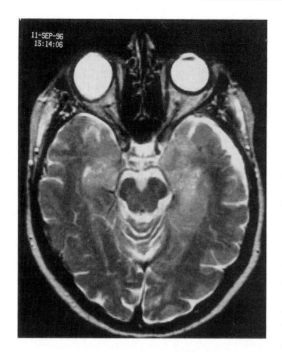

Fig. 3. Non-enhanced MRI scan showing low absorption abnormalities in both mesial temporal lobes in a patient with limbic encephalitis.

ciated with coexistent SSN but one-fifth had motor weakness and wasting. Cerebellar symptoms were associated with inflammatory infiltrates within the deep nuclei of the cerebellum as well as associated neuronal loss.

Progressive neurologic deficits eventually plateau at the stage of severe affliction. Rarely the findings reverse spontaneously (Henson and Urich 1982a). Prior to neuroimaging the diagnosis of many paraneoplastic neurologic syndromes was delayed or made at autopsy. Computed tomography (CT) and magnetic resonance imaging (MRI) do not identify enhancing metastatic tumor deposits in brain and spinal cord. In PEM, the MRI is likely normal but rare abnormalities are noted (Glantz et al. 1994). CT abnormalities, noted in one of seven cases (Fig. 3), consisted of an enhancing mesial temporal mass. MRI can reveal T2 signal abnormalities sometimes with mass effect within the mesial temporal structures. Atrophy may be seen in the brainstem in patients with brainstem encephalitis in the late phase of the illness.

Pathology. Predominant is perivascular lymphocytic inflammation within gray matter and overlying meninges (Fig. 2). There is neuronal loss, gliosis, and

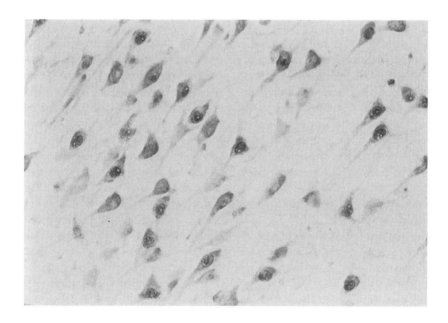

Fig. 4. Normal post-mortem cerebral cortex stained with anti-Hu serum. All neurons are stained, predominantly in the nucleus. Anti-Ri serum produces identical staining (200 ×). (Courtesy of Dr. Josep Dalmau, Memorial Sloan–Kettering Cancer Center.)

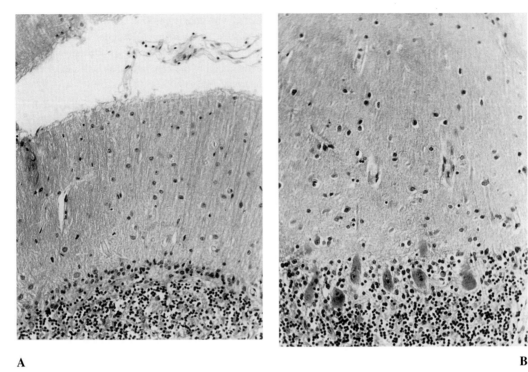

A   **B**

Fig. 5. (A) Severe Purkinje cell loss in a patient with cerebellar degeneration and ovarian cancer. Note the thinning of the molecular layer (which is the top three-quarters of the figure). Bergmann gliosis is seen at the interface between molecular and granular layers. (B) Normal post-mortem cerebellum for comparison shows normal numbers of Purkinje cells as well as full thickness molecular layer. (H and E, 160 ×.) (Courtesy of Dr. David Louis, Massachusetts General Hospital.)

Wallerian degeneration. White matter changes are uncommon. In limbic encephalitis neuronal loss occurs within the pyramidal layer of the hippocampus, with associated gliosis. The amygdala is prominently involved. Operative specimens from patients with mesial temporal lobe abnormalities show gliosis and perivascular inflammation (Burton et al. 1988), which may resolve at the time of post-mortem examination (Ingenito et al. 1990). In the brainstem the medulla bears the brunt of the inflammatory response. Included are the motor and vestibular nuclei, the pontine cranial nerves and the substantia nigra. The myelitis predominantly affects neurons. Anterior horn cells are lost and there occurs prominent denervation atrophy of skeletal muscle.

Molecular pathophysiology. There is strong evidence that the neurological syndrome results from an attack by humoral and cellular arms of the immune system. In PEM/SN with SCLC patients harbor high titers of IgG antibodies against a set of 35–40 kDa

proteins. These proteins are selectively expressed in the nuclei of neurons (Fig. 4) throughout the human nervous system and in all SCLCs, some neuroblastomas, and medulloblastomas (Graus et al. 1985; Dalmau et al. 1990, 1992b). The antibodies are designated 'anti-Hu' (or sometimes as anti-neuronal nuclear antibodies, or ANNA, type 1). As anti-Hu antibodies are present at higher concentrations in CSF than in serum, their synthesis likely occurs within the CNS (Furneaux et al. 1990a, b). The antibodies can be detected within neuronal nuclei of patients with PEM, and there is a correlation between sites of accumulation and those areas of the brain most affected by the paraneoplastic syndrome (Jean et al. 1994). The mechanism for inflammation that occurs is unsettled. The inflammatory infiltrates seen in PEM/SN consist of B-cells, CD4[+] and CD8[+] lymphocytes, and macrophages. Natural killer cells are not abundant, and complement deposits are not detected. Finally, SCLC cells of patients with anti-Hu antibodies express major histocompatibility complexes (MHC I

TABLE 5

Therapy of paraneoplastic encephalomyelitis (PEM).

Syndrome	Cancer	Therapy	Response	Author
Limbic encephalopathy	SCLC	RT, CT	CR (1)	Brennan et al. 1983
Limbic encephalitis	testicular	S, CT	CR	Burton et al. 1988
Memory difficulties and hallucinations	HD	CT, RT	CR (1)	Carr 1982
Paraneoplastic limbic encephalitis	breast	protein A column therapy	CR (1)	Cher et al. 1995
Paraneoplastic limbic encephalitis/sensory neuropathy	lung	ST	CR (1), PR (1)	Dalmau et al. 1992a,b (includes pts from Anderson 1988a)
Limbic encephalitis	SCLC	CT, RT, PE, ST, cyclophosphamide	stable (1)	Graus et al. 1992
Confusion brain biopsy – granulomatous angiitis	HD	CT, RT	CR (1)	Greco 1976
Limbic encephalitis	SCLC	CT, RT	CR (1)	Markham and Abeloff 1982
Paraneoplastic encephalomyeloneuritis anti-Hu antibody	SCLC	ST	improved (1)	Oh et al. 1991
Catatonia/paraneoplastic encephalopathy	SCLC	haldol, clonazepam	reduced confusion, improved catatonia (1)	Tandon 1988
Encephalomyelitis with anti-Hu	SCLC	immunosuppressant, antineoplastic	stabilization (7/14)	Valldeoriola 1992
Anti-Hu syndrome		i.v. Ig	stable (3/19), improved (1/19)	Vega 1994
Cerebellar degeneration/ anti-Yo syndrome		antitumor tx		
Limbic encephalitis	bronchial	CT	MRI resolution (1)	Vollmer 1993

CT = chemotherapy; PE = plasma exchange; RT = radiotherapy; S = surgery; ST = steroids. Number of patients in parentheses.

and II), complexes seldom expressed in the absence of anti-Hu antibodies (Dalmau et al. 1995). Attempts to induce passive transfer of PEM/SN with anti-Hu IgG has not been successful, and the mechanism by which the antibodies might induce neuronal dysfunction is not yet clear.

What is the target of anti-Hu IgG? Screening of a human fetal brain expression cDNA library with anti-Hu serum led to the identification of a 380 amino acid, 48 kDa fusion protein, called HuD (Szabo et al. 1991). Fig. 5 illustrates Western blot analysis of neuronal proteins using anti-Hu sera. HuD contains three regions of homology to the family of RNA-binding proteins, and it specifically binds the AU-rich segments at the 3′ end of cellular mRNAs such as c-Myc and c-Fos that regulate the cell cycle (Liu et al. 1995). Similarly identified was HuC, a gene whose protein is homologous to HuD. Anti-Hu IgG was found to recognize another human RNA-binding protein of very similar amino acid sequence, Hel-N1 (Dropcho and King 1994). These genes undergo alternative mRNA splicing. The distinctive neurological features seen in individual patients may be related to the site-specificity of each gene. All of the proteins have homology to ELAV (embryonic lethal abnormal vision), a Drosophila RNA-binding protein that is required for neuronal development and maintenance (Szabo et al. 1991), and thus constitute a family of neuron-specific RNA binding proteins that may regulate both neuronal and small cell cancer genes. It is hypothesized that inhibition of the function of these proteins leads to neuronal death.

General approaches to the therapy of central paraneoplastic syndromes

In general, the response to therapy of these syndromes has been disappointing. For ease, these responses are identified in Table 5. A number of principles are likely to be important in improving the treatment response. Many patients are treated relatively late at a time where irreversible damage has already occurred. Earlier diagnosis now depends on increased awareness, commercially available molecular probes, and MRI criteria. The outcome is likely to be better at the inflammatory stage before neuronal loss. As spontaneous remissions are few and response to chemotherapy rare early therapy is warranted.

Reports exist of improved neurologic function following treatment of testicular cancer (Burton et al. 1988) ovarian cancer (Mintz and Sirota 1993) and SCLC (Paone and Jeyasingham 1980; Markham and Abeloff 1982; Brennan and Craddock 1983). It is likely that complete remission of the underlying cancer is required to successfully improve the paraneoplastic syndrome. Corticosteroids have benefited children with opsoclonus and neuroblastoma. In adults responses are less convincing but worth trying (Dropcho et al. 1993; Glantz et al. 1994). Scant data exist for drugs such as cyclophosphamide. Glantz et al. (1994) reported responses in PEM and necrotizing myelopathy to PE or i.v. gammaglobulin given with azathioprine or cyclophosphamide. Cocconi et al. (1985) reported rapid incomplete responses to PE. A review of 11 patients with PEM and PCD treated with PE and chemotherapy showed no improvement although two remained stable (Graus et al. 1992). I.v. Ig has been utilized by Moll et al. (1993) to treat a lady with bilateral breast carcinoma, who after surgery and the commencement of adjuvant chemotherapy developed PCD associated with a novel anti-Purkinje cell antibody. PE was begun with little benefit. Following this i.v. Ig was administered (0.4 g/kg for 5 consecutive days), with marked improvement which began on day 4 of the therapy. Glantz et al. (1994) included one patient with limbic encephalitis who responded dramatically to i.v. Ig, with azathioprine maintenance. Most recently we have provided therapy using columns containing protein A immunoadsorbents. The column consists of a silica matrix to which is attached staphylococcal protein A, which avidly binds IgG and immune complexes. We investigated its use in patients with central paraneoplastic syndromes. Following therapy, a patient with POM, who had been bed-bound, was able to walk independently. The neurologic disorder did not recur. We have treated a further six patients (four with POM, two with PEM and one with PCD). Three POM patients had a complete remission, and the fourth obtained significant benefit (Cher et al. 1995). The exact role of protein A immunoadsorption in the other syndromes remains to be ascertained as part of an FDA approved trial that is ongoing.

Paraneoplastic cerebellar degeneration (PCD)

Subacute pancerebellar difficulties develop over weeks to months (Brouwer 1919) in association with malignancies of the ovary and breast, Hodgkin's disease and SCLC. The latter do not usually overlap with the more widespread PEM. PCD remains rare. Hudson et al. (1993) found no cases of PCD in 908 patients with ovarian carcinoma. Nevertheless the characteristic neurologic picture must always raise the suspicion of a paraneoplastic origin. Few exceptions exist (Ropper 1993) and include the ataxic form of Creutzfeldt–Jakob disease.

Initial modest impairment of gait develops over a period of weeks to a few months but is soon followed by incoordination of arms, legs and trunk, dysarthria and often nystagmus. The disorder may rarely begin acutely, with severe disability developing overnight. Downbeating nystagmus has also been noted and may be under-reported as are oscillopsia or vertigo. The initial asymmetric involvement soon leads to limitation to chair or bed. While dementia or anxiety was previously reported as common (Greenfield 1934), a study by Anderson et al. (1988a, b) suggested that motor impairments are the cause of the apparent cognitive changes. As with other syndromes, there may be subtle features of non-cerebellar involvement and two separate syndromes may coexist (Tsukamoto et al. 1993). No abnormalities exist on MRI scans in the acute setting (Anderson et al.

1988a, b), although later cerebellar atrophy can be seen (Wang et al. 1988).

Moll et al.'s (1993) report defines the syndrome and illustrates the utility of both PE and i.v. Ig therapy as well as antineuronal antibody testing: 'A 44-year-old woman with bilateral adenocarcinoma of the breast had surgery [which revealed undifferentiated ductal carcinoma] ... followed by adjuvant chemotherapy consisting of three courses of cyclophosphamide, methotrexate and 5-flourouracil over a period of 2 months. During this treatment, she developed severe disequilibrium. She was unable to walk by herself or to sit up because of vigorous orthostatic tremor. There was also severe intentional tremor of the hands and feet. [An antineuronal antibody, which stained nuclei of Purkinje cells, and granular cells and cells in the molecular layer, did not react with Hu or Yo antigens.] A diagnosis of paraneoplastic cerebellar degeneration was made. . . . [the patient deteriorated following PE] . . . 26 days after onset of neurological symptoms, we started immune globulin . . . 4 days after starting i.v. Ig, signs of cerebellar dysfunction began to improve and the patient was able to sit up independently. One week later she was able to walk by herself. . . . during PE anti-neuronal antibody titer in serum decreased from 1:1600 to below 1:200.'

Pathology. Usually at autopsy examination the cerebellum is atrophic. Most striking is the severe almost total loss of Purkinje cells throughout the cerebellar cortex (Fig. 5). Diminished cells of the granular layer may be seen. Silver staining reveals empty basket fibers whose Purkinje cells have died. Proliferation of Bergmann astrocytes also occurs. Although early descriptions (Henson and Urich 1982b) differentiated the 'purely degenerative' cases and those with inflammation, it is likely that the degree of inflammation depends on the interval between the onset of PCD and death.

Anti-Yo syndrome (Peterson et al. 1992)
Anti-Yo is the designation given to an antibody which binds to a cytoplasmic antigen. On Western blotting the antigen is of 34 and 62 kDa weight. Most notable is the association with ovarian carcinoma (26 of 55 patients), breast cancer (13) and other gynecological malignancies (7). Rare (1 patient) is the association with adenocarcinoma of the lung, or (6 patients) ad-

enocarcinoma of unknown primary. In two-thirds of cases PCD, commonly with downbeat nystagmus, precedes the diagnosis of malignancy. The neurologic syndrome heralds the identification of tumor. Thus Hetzel et al. (1990) diagnosed 7 patients with gynecological cancer without evidence of a palpable pelvic mass or pelvic abnormality on CT or MRI. Similarly Peterson et al. (1992) noted 4 patients whose microscopic gynecological malignancy was discovered only at laparotomy.

PCD and Hodgkin's disease
Hodgkin's disease (HD) rarely involves the CNS. Therefore a cerebellar syndrome in a patient with HD raises the specter of paraneoplasia. The gait disorder tends to be mild and spontaneous remission may occur. In 80% of cases the diagnosis of HD precedes the onset of PCD. The activity of the lymphoma appears not to be a factor. A cytoplasmic antibody to Purkinje cells was demonstrable in only 29% of cases but Western blotting was not able to identify a specific Purkinje cell antigen target.

PCD, SCLC and LEMS
A further subgroup of PCD is associated with SCLC and LEMS and an antibody against the presynaptic calcium channel (Clouston et al. 1992). Nine patients without anti-Purkinje cell or anti-Hu antibodies had either SCLC, LEMS or both. Two had no malignancy and 6 had LEMS. Two of 6 patients with LEMS had signs only of hyporeflexia, and LEMS was diagnosed on electrophysiological data. Antibodies to the VGCC were detected in the serum of 5 of 7 patients tested, and in 1 of 2 samples of CSF. Blumenfeld et al. (1991) described a similar patient whose LEMS responded to PE without benefit of the PCD.

Molecular pathophysiology. Women with subacute cerebellar degeneration often harbor 'anti-Yo' antibody in serum and CSF. Although Yo has been well characterized, the steps required for cerebellar degeneration are unknown. Recently it has been suggested that autoantibodies (in PCD and POM) may be directed against the non-NMDA glutamate receptor and modulate glutaminergic receptor function (Gahring et al. 1995). Anti-Yo IgG has been administered by intraventricular injection to guinea pigs in an attempt to reproduce cerebellar degeneration (Graus

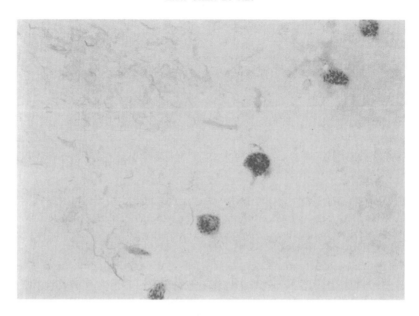

Fig. 6. Normal post-mortem cerebellum stained with anti-Yo serum showing granular cytoplasmic staining (arrow) restricted to Purkinje cells (100 ×). (Courtesy of Dr. Josep Dalmau, Memorial Sloan-Kettering Cancer Center.)

et al. 1991). There was selective uptake by Purkinje cells of both anti-Yo and irrelevant IgG, but the animals failed to develop either clinical difficulties or Purkinje cell injury.

The antibody recognizes an antigen restricted to the cytoplasm of Purkinje cells and to neurons in the deep cerebellar nuclei (see Fig. 6) (Trotter et al. 1976; Greenlee and Brashear 1983; Jaeckle et al. 1983). Immunoelectron microscopy has localized the antigens to ribosomes, granular endoplasmic reticulum, Golgi vesicles and the plasma membrane (Rodriguez et al. 1988). The antibodies also recognize cytoplasmic antigens in tumor tissue from the neurologically afflicted but rarely from the unaffected (Furneaux et al. 1990a, b). The antibodies, polyclonal and usually IgG, are likely synthesized within the CNS (Jaeckle et al. 1985).

Anti-Yo antibodies (also called type I) recognize several antigens by Western blot analysis of purified cerebellar Purkinje cell extracts. Notable is a 62 kDa protein (CDR 62, or 'cerebellar degeneration-related' 62 kDa protein) and a minor 34 kDa species (CDR 34) (Cunningham et al. 1986). All anti-Yo sera recognize CDR 62, whereas only some recognize CDR 34. A commercially available Western blot assay employs recombinant CDR 62. The genes for both proteins have been cloned and characterized. CDR 62 was isolated by screening a HeLa expression cDNA library (HeLa cells, derived from a carcinoma of the uterine cervix, express abundant CDR 62), and the major feature of the deduced amino acid sequence is the presence of a leucine zipper motif (Fathallah-Shaykh et al. 1991). Leucine zippers mediate protein–protein interactions. The CDR 62 gene has been localized to chromosome 16 (Gress et al. 1992). CDR 34 was isolated from a human cerebellar expression library (Dropcho et al. 1987; Furneaux et al. 1989).

Although low levels of the CDR 34 protein are expressed in the heart, its expression is otherwise restricted to the cerebellum and to the tumors of affected patients. The CDR 34 gene, which is located on the X chromosome (Chen et al. 1990), encodes a highly unusual, 233 amino acid protein from a single exon. Ninety percent of the protein is composed of 34 hexapeptide repeats of unknown function. CDR 34 mRNA can be detected in the tumors of patients with anti-Yo associated PCD. The function of the protein is not known.

The therapy of the PCD syndrome (Table 6) has made use of chemotherapy or surgical removal of the underlying cancer, corticosteroids, PE and i.v. Ig.

Paraneoplastic opsoclonus myoclonus (POM) syndromes

This syndrome has existed under numerous synonyms, which has led to some confusion. The term 'opsoclonus' refers to 'dancing eyes', and the myoclonus has been referred to as 'dancing feet'. The clinical features are often characteristic and may wax and wane. The pathologic correlate of the process is uncertain as there is minimal neuronal loss. POM occurs sporadically, after viral infections and medications (Pranzatelli 1992). Although Henson and Urich (1982c) recognized it as a syndrome in association with childhood neuroblastoma (Kinsbourne's disease), its association with malignancy in adults was increasingly recognized (Digre 1986). The latter authors review 58 examples in association with malignancy following which Anderson et al. (1988b) argued that this should be classified as a separate paraneoplastic syndrome. An antibody to 53–61 kDa and 79–84 kDa antigens (anti-Ri) has been identified in a number of patients with POM and breast cancer (Budde-Steffen et al. 1988).

The disorder produces saccadic eye movements in combination with myoclonus and truncal ataxia.

Opsoclonus/ocular flutter

Opsoclonus was first described by Orzechowski (1927) who noted: 'The eyes are in a continual state of agitation, the eyes are shaking and are displaced by very rapid, unequal movements, which generally occur in the horizontal plane. . . . Between the horizontal shaking which always predominates can be seen sudden jerks which occur in other directions. . . . We always find a few symptoms of cerebellar disease . . . but what dominates the clinical picture is . . . the myoclonus, the jerking of the fingers, chin, lips, lids, forehead and even difficulty with the muscles of the trunk and lower extremities' (excerpted from Young et al. 1993).

Characteristically, rapid saccadic eye movements, with no intersaccadic interval ('back-to-back saccades') appear conjugately. In ocular flutter these movements are limited to the horizontal plane. Opsoclonus occurs in multiple planes and is increased by saccadic movements or fixation.

Myoclonus

The myoclonus is usually severe and involves facial muscles, the head and neck, limbs and trunk. Palatal myoclonus gives the voice a wavering quality. Walking or standing and feeding are often impossible in the setting of rapid jerks of the extremities which may seem tremulous.

Other neurologic symptoms

Truncal ataxia and dysmetria may be hidden by the myoclonus, and anxiety and depression can be confusing to the clinician. Rarely the intellect worsens, cognitive impairment ensues and psychiatric referral is necessitated by a fulminant encephalopathy with agitation and hallucinations. Although vertigo and vomiting may occur, cranial nerve and reflex changes are few and sensory abnormalities are not noted. Patients may progress to coma and death (Anderson et al. 1988a, b).

Pathology

The brains of patients with POM rarely reveal loss of Purkinje cells, changes within the dentate nucleus, and demyelination within the cerebellar white matter (Giordana et al. 1989). Often there is a striking lack of neuronal loss and only scattered perivascular lymphocytes despite a devastating neurological picture (Anderson et al. 1988a, b; Young et al. 1993). The immune response may affect neuronal function while sparing the cell and thus is perhaps more likely to be responsive to effective therapy.

Molecular pathophysiology

A subset of patients with POM produces high titers of an antibody ('anti-Ri') that recognizes an antigen in the nuclei of all CNS neurons and in tumor cells of affected patients. Commonly these cancers are in the breast, gynecological sites and of small cell lung origin. The pattern of anti-Ri staining is similar to that of anti-Hu (Fig. 4) (Budde-Steffen et al. 1988; Luque et al. 1991). Tumors from patients without POM do not react with the antibody. Western blot analysis of neuronal extracts demonstrated that anti-Ri antibodies recognize a protein of 55 kDa as well as minor species of 80 kDa. These are different from anti-Hu. Screening of a human cerebellar cDNA expression library yielded a gene encoding a 40 kDa fusion protein of 510 amino acids that is specifically recognized by affinity purified anti-Ri IgG (Buckanovich et al. 1993). The primary sequence of this protein (named Nova) revealed the presence of three copies of a 111 base pairs sequence that has homology of both se-

TABLE 6

Therapy of paraneoplastic cerebellar degeneration (PCD).

Syndrome	Tumor	Therapy	Benefit	Reference
PCD With antibodies	ovary, breast adenoma NSCLC, SCLC	PE cytoxan, ST	PR (2/9) with PE	Anderson et al. 1988a, b
Without antibodies	lung, breast, HD	treat cancer spontaneous (1/24)	PR (4/9) CR (1)	
PCD	colon	spontaneous S	PR (1) CR (1)	Auth 1957
PCD	HD	spontaneous	CR (1)	Cairncross 1980
PCD	ovarian	PE	PR (1)	Cocconi et al. 1985
PCD and LEMS with anti-HU and anti-VGCC	SCLC	i.v. Ig	CR (1)	Counsell 1994
PCD	NSCLC	diagnostic lymph node extirpation	CR (1)	Eekhoff 1985
Cerebellar degeneration	SCLC	CT	PR (2)	Gonzalez 1992
PCD	ovarian	PE, CT, S, RT	stable then pro- gression (1)	Graus et al. 1991
Pancerebellar	HD	mantle RT	improved (1)	Greenberg 1984
PCD anti-Purkinje cell antibodies	SCLC	CT	slight improvement (1)	Greenlee and Brashear 1983
Cerebellar degeneration 6/21 with AB to Purkinje cells	HD	no treatment	CR (1)	Hammack et al. 1992
PCD (n = 16) With Purkinje cell antibody	ovarian	CT (1) with CDDP and cyclo- phosphamide	slight improvement (1)	Hammack et al. 1990
		TAH/BSO and 12 cycles chlorambucil	slight improvement (1)	
Without antibodies (16)	SCLC	XRT and CT	improvement (1)	
	renal	nephrectomy	minimal response (1)	
PCD	chondro- sarcoma	S	CR then relapse (1)	Kearsley 1985
	poorly differ- entiated adenoma of fallopian tube			
Anti-Yo antibody negative	tubal	CT	Improved (1)	Mintz and Sirota 1993
PCD antineuronal antibody	breast	i.v. Ig	CR (1)	Moll et al. 1993
PCD	SCLC	S	CR (1)	Paone and Jeya- singham 1980

Table 6 (*Continued.*)

Syndrome	Tumor	Therapy	Benefit	Reference
PCD	ovarian, breast, endometrial	PE	with transient improvement (4/22) PR (1/22)	Peterson et al. 1992; Anderson et al. 1988a, b
Anti-Yo antibody	no known malignancy	ST	mild transient improvement (1/17)	
		treat underlying tumor	moderate improvement (1)	
PCD	HD	CT	improved (1)	Rewcastle 1963
Spino-cerebellar degeneration	prostate	stilbestrol	improved (1)	Roberts 1967
PCD antibody to Purkinje cell	breast	S, PE	improved (1)	Royal 1987
PCD	HD	PE	CR (1)	Sapra 1981
Cerebellar degeneration	HD	CT	improved (1/3)	Schlake 1989
		CT, RT	improved (1)	
PCD anti-Purkinje cell antibodies	uterine	PE, RT, CT	stable (1)	Tsukamoto et al. 1987
PCD limbic encephalitis antineuronal nuclear antibody of 41 kDa	adenocarcinoma of colon	S	CR of limbic encephalitis and stable cerebellar ataxia (1)	Tsukamoto et al. 1993
PCD with antibody to Purkinje cell	HD	RT	stable disease (1)	Trotter et al. 1976

CT = chemotherapy; PE = plasma exchange; RT = radiotherapy; S = surgery; ST = steroids. Number of patients in parentheses.

quence and spacing with a triplicated sequence in the RNA-binding protein hnRNP K. hnRNP K has been suggested to regulate heterogeneous nuclear RNA (pre-mRNA) splicing. The presence of several splice variants of this protein suggests the presence of a family of Nova proteins. Nova mRNA is expressed specifically in brain and in some types of cancer cells. During development Nova expression is restricted to neurons in the ventral brainstem, ventral spinal cord and cerebellum, a pattern that is consistent with the clinical syndrome seen in patients. Although the distribution of the Nova protein expression in the adult CNS is much more widespread, anti-Ri associated immunological responses appear to target the basis pontis and dorsal mesencephalon (Hormingo et al. 1994).

Therapy of POM
The therapy (Table 7) of POM has depended on control of the underlying tumor, steroids or immunosup-

pression. Children can experience spontaneous remission or respond to ACTH or corticosteroids. Complete resolution of the difficulty is seldom seen. Adults seldom improve spontaneously and experience minimal benefit from steroid treatment. Many adults succumb to the process without improvement.

SPINAL CORD SYNDROMES

Motor neuron disorders: paraneoplastic anterior horn cell disease

Brain et al. (1965) are credited with linking an illness similar to motor neuron disease (MND) with cancer. Three of their 11 patients came to autopsy where changes were limited to the spinal cord, with involvement of anterior horn cells in one, and concomitant loss of myelination of pyramidal tracts in the second. The third had only a motor neuropathy.

TABLE 7

Therapy of paraneoplastic opsoclonus myoclonus (POM).

Syndrome	Cancer	Therapy	Response	Author
Opsoclonus/myoclonus	SCLC	CT	CR (1)	Anderson et al. 1988
		clonazepam	CR (1)	
		S, RT	PR (1)	
Opsoclonus	epidermoid lung	none	PR (1)	Bellur 1975
Opsoclonus/myoclonus	ganglioneuroblastoma	S, RT, CT	CR (1)	Berg 1974
	ganglioneuroblastoma	S, RT, CT, ST,	PR (1)	
	ganglioneuroblastoma	S, CT, ST	PR(1)	
	neuroblastoma	S, CT, ST	PR(1)	
Opsoclonus/myoclonus antibody 53–61 kDa to neurons	breast	ST	improved (1)	Brashear 1990
	ganglioneuroblastoma	S	PR (1)	
Paraneoplastic opsoclonus and ataxia	breast and ?occult lung	prochlorperazine, baclofen, meclizine, ST, CT	PR (1)	Case Records NEJM 1994
Opsoclonus/myoclonus	SCLC, breast	protein A column therapy	CR (2)	Cher et al. 1995
Opsoclonus/myoclonus	neuroblastoma	S, Coleys bacterial toxin	CR (1)	Cushing
Opsoclonus/myoclonus	neuroblastoma	S, RT, CT	CR of opsoclonus and PR of myoclonus and ataxia (1)	Davidson 1968
Opsoclonus/myoclonus/ ataxia/tremor	breast	S	PR (1)	Digre 1986
Opsoclonus/myoclonus	medullary thyroid	clonazepam	PR (1)	Dropcho 1986
Opsoclonus/myoclonus anti-Ri antibody positive	no malignancy	ST and with recurrence cyclophosphamide	CR(1)	Dropcho et al. 1993
Opsoclonus/myoclonus	neuroblastoma	S, RT	CR (1)	Dyken 1968
Nausea/vomiting/ vertigo/paresis of upward gaze, gait ataxia and anti-Ri antibody	breast	lorazepam	PR (1)	Escudero 1993
Opsoclonus/myoclonus	neuroblastoma, ganglioneuroma, ganglioneuroblastoma	ST, RT, CT	CR (7)	Farrelly 1984

Table 7 (*Continued.*)

Syndrome	Cancer	Therapy	Response	Author
Opsoclonus/myoclonus	neuroblastoma	i.v. Ig	resolution (1)	Fisher 1992
Opsoclonus/myoclonus	breast	no treatment	spontaneous resolution (1)	Furman 1988
	SCLC	no treatment	PR (1)	
Oscillopsia, ataxia anti-Ri antibody	breast	S, CT, ST	improved (1)	Glantz et al. 1994
Opsoclonus and cerebellar ataxia with anticerebellar antibodies	breast	S	improved (1)	Greenlee and Brashear 1983
Opsoclonus/myoclonus	sympathicoblastoma	spontaneous	CR (1)	Klingman 1944
Opsoclonus/myoclonus	breast	chlormethiazole	CR (1)	Lago 1992
Ataxia, opsoclonus	breast	CT, ST	improved (1/8)	Luque et al. 1991
Anti-Ri antibody	breast	ST	improved (1/8)	
	adenocarcinoma of unknown	no treatment	stable (1)	
Opsoclonus/myoclonus	neuroblastoma, ganglioneuroblastoma	ACTH, ST	PR (3)	Mitchell 1988
Opsoclonus/ataxia	undifferentiated bronchogenic	thiamine, S	CR of opsoclonus and residual ataxia (1)	Nausieda 1981
Opsoclonus/myoclonus	SCLC	no treatment	spontaneous remission of myoclonus (1)	Ridley 1987
	SCLC (1)	benzodiazepine, clonazepam, diazepam	myoclonus improved (1)	
Opsoclonus/myoclonus	neuroblastoma	S, RT, CT	CR (1)	Solomon 1968
	ganglioblastoma	S, RT	PR (1)	
Opsoclonus/myoclonus /ataxia	ganglioneuroblastoma, neuroblastoma	S, ACTH, RT	improved (6/10); partial (3)	Telander 1989
Opsoclonus/myoclonus	undifferentiated bronchial	CT	CR (1)	
Opsoclonus/myoclonus	breast	baclofen, ST, CT	PR (1)	Wray in Anderson et al. 1988

CT = chemotherapy; PE = plasma exchange; RT = radiotherapy; S = surgery; ST = steroids. Number of patients in parentheses.

Norris and Engel (1965) later identified malignancy in 10% of a series of 130 patients with ALS. Others disputed this association. Brain et al. (1965) reported 11 cases of MND occurring in predominantly solid tumors, in whom only two were autopsied. It appears that a motor neuropathy (with loss of anterior horn cells) can occur, either as part of PEM, or as a less well defined entity. In addition, an associated motor neuropathy with multifocal conduction blocks can be seen. Rarely the MND responds to treatment of the underlying tumor. Rosenfeld and Posner (1991) noted: 'despite the atypia of many of these patients, taken together, they indicate that in extremely rare instances, a syndrome resembling ALS using clinical, electrophysiologic, and in some patients pathologic criteria may occur in patients with an underlying neoplasm in a fashion that suggests that the neoplastic disorder is causal.'

Lymphoma

Recently Younger et al. (1991) has expanded the association between a syndrome similar to MND and non-Hodgkin's lymphoma. Typical features of loss of anterior horn cells were found in 6 of 13 patients undergoing post-mortem examination. An association with paraproteinemia was noted in 3 of 7 patients. In two, this led to discovery of asymptomatic NHL on bone marrow biopsy. Some patients may have upper motor neuron findings as well and the syndrome is often confused with that due to multifocal motor conduction block.

Necrotizing myelopathy

Henson and Urich (1982c) described this as 'the rarest and most obscure neurological complication of malignancy', which may mimic acute transverse myelitis or spinal cord compression. Ojeda's (1984) report is illustrative: 'A 59-year-old white woman was found to have a large cell carcinoma of the right upper lobe of the lung . . . . Irradiation . . . was given. . . . The patient was admitted . . . with a 4-week history of paresthesias of the left leg. A myelogram and CT scan of the thoracolumbar spine were normal [CSF examination was normal.] . . . Steady deterioration followed and there was obvious paraplegia [1 month later] . . . . [At autopsy] the spinal cord was soft and

granular. Horizontal sections . . . at multiple levels revealed necrosis.'

Usually acute and rapidly evolving, the process extends to involve the spinal cord at the thoracic level. The CSF protein is usually elevated. Myelography may show slight widening of the involved cord. MRI in two patients with biopsy proven necrotizing myelopathy showed patchy gadolinium enhancement, and T2 abnormality within the spinal cord which appeared within 2–4 weeks (Glantz et al. 1994). The condition has been described in both solid and hematologic malignancies (Gray et al. 1980). Pathologically widespread spinal cord necrosis is seen with little inflammatory response or vascular occlusion (Ojeda 1984). Some of the reported cases could be explained by prior irradiation or chemotherapy toxicities (intrathecal methotrexate) (Henson and Urich 1982c). Therapy (Table 8) depends on surgical therapy of the underlying tumor (see table that follows) but is not likely to be of benefit.

EYE

Cancer associated retinopathy

Sawyer et al. (1976) described three patients whose SCLC was associated with fluctuating visual symptoms, unobtainable electroretinogram and retinal degeneration. The distinctive clinical triad consists of photosensitivity, ring scotomatous visual field loss and attenuated caliber of retinal arterioles (Jacobson et al. 1990) although this combination is not present in all patients. These features accompany night blindness, and bizarre visual sensations such as 'swarms of bees' and 'spaghetti-like' lines, that may come and go (Thirkill et al. 1993a, b). Loss of color vision may also occur. One patient failed to appreciate her bloody vaginal discharge (Campo et al. 1992) reflecting a small cell endometrial cancer. A similar syndrome has also been described in patients with melanoma (Berson and Lessell 1988) and is also associated with loss of rod function. Pathologically there is widespread degeneration of the outer retinal layers with relative preservation of other retinal layers. Electroretinography shows loss of rod function with sparing of the cones. This is characterized by an absent scotopic response to blue light in the dark adapted state, but a preserved photopic response to

TABLE 8

Therapy of paraneoplastic motor neuron disease (MND).

Syndrome	Cancer	Therapy	Response	Author
Paraneoplastic MND	NSCLC	S	improved (1)	Brain et al. 1965
Paraneoplastic MND	renal cancer	S	improved (1)	Buchanan 1973
Paraneoplastic MND	renal cell carcinoma	S	CR (1)	Evans 1990
	other cancers	treatment of tumor	improved/ stabilized (8)	
Paraneoplastic MND	NSCLC	S	improved (1)	Mitchell 1979
Paraneoplastic MND	lymphoma	CT	stabilized (1)	Norris and Engel 1965
Paraneoplastic MND	NSCLC	RT	improved (1)	Peacock 1979
Paraneoplastic MND	macroglobulinemia	CT	improved (1)	Peters 1968
MND	lymphoma	CT	improved (1)	Younger et al. 1991

CT = chemotherapy; RT = radiotherapy; S = surgery. Number of patients in parentheses.

white light. Dark adaption is also impaired. More recently an antibody directed against a 23 kDa retinal protein has been described in patients with SCLC (Thirkill et al. 1989), which has been identified as recoverin, a photoreceptor cell-specific protein (Thirkill et al. 1992). SCLC expresses this protein when cultured (Thirkill et al. 1993a, b). The syndrome may occur in other malignancies not associated with the recoverin antibody (Scott 1997). Corticosteroid therapy or PE may help some patients, but only preliminary data is available.

MISCELLANEOUS SYNDROMES DESCRIBED AS PARANEOPLASTIC

A number of unusual syndromes have been reported to be paraneoplastic (Table 9).

Screening for malignancy in patients with suspected paraneoplastic disorders

The stimulus for the search for systemic cancer depends on the identification of the paraneoplastic syndrome and the evaluation for circulating antibody. Thus in a patient with PCD in the absence of known cancer the detection of anti-Yo obligates the clinician to perform laparotomy. At the other extreme, history, physical examination and chest X-ray and tests for fecal occult blood is adequate for the patient with MND or inflammatory myopathy. In the latter case repeated examinations may be warranted. In a patient who presents with a clinical syndrome very suggestive of a paraneoplastic syndrome the history should be gleaned for risk factors of malignancy, most importantly cigarette smoking. A family history of breast or

TABLE 9

Miscellaneous paraneoplastic syndromes.

Syndrome	Tumor	References
Neuromyotonia	SCLC, thymoma	Partanen et al. 1980; Walsh 1976; Halbach et al. 1987; Litchy and Auger 1986
Brachial plexopathy	Hodgkin's disease	Lachance et al. 1991
Optic neuropathy	SCLC	Malik et al. 1992
Downbeating nystagmus	breast	Guy and Schatz 1988
Spinocerebellar degeneration 'plus'	Langerhans cell histiocytosis	Goldberg-Stern et al. 1995

colon cancer or other malignancy may be helpful. Clinical examination should concentrate on respiratory, breast, abdominal and testicular examination, as well as a rectal examination and a test of fecal occult blood. A careful examination of lymph nodes is also important. Radiological investigations include chest X-ray, CT of chest, abdomen and pelvis, mammography in women and testicular ultrasound in younger men. Bone marrow biopsy may be considered if the syndrome is associated with lymphoma. A number of serum markers for malignancy are now available and may help to pinpoint the origin of the tumor (Table 10).

Clinical utility of paraneoplastic associated antibodies

The utility of the antibodies in the diagnosis of paraneoplastic neurological syndromes and their associated cancers is now well established. Highly specific assays are necessary to document the presence of these antibodies, and a combination of immunohistochemistry and Western blot analysis is the best approach to their identification (Figs. 4 and 6). These assays employ the patient's serum to test for immunoreactivity with antigens in specific neuronal populations by immunohistochemistry and by Western blot analysis for reactivity to the cloned fusion proteins (i.e., Hu, Yo, and Ri). Western analysis plays an important role in distinguishing antibodies with similar immunohistochemical characteristics. Importantly, the anti-Hu and anti-Ri antibodies both give similar patterns of nuclear neuronal staining and the syndromes associated with these antibodies can have similar clinical presentations, but the associated tumors are very different. Western blot analysis clearly identifies these antibodies. Commercially available Western blot assays which utilize recombinant proteins will fail to identify the presence of novel antineuronal antibodies which are directed against proteins other than Hu, Yo or Ri. Research centers specialize in the characterization of novel antibodies which may in the future prove specific for paraneoplastic syndromes or tumor histologies. In clinical practice it is crucial for the clinician to know whether a patient has one of the three well-characterized antibodies that can be used as highly specific markers of a neurological condition and its associated tumor.

Acknowledgements

We would like to thank Dr. Asa Das who reviewed the therapeutic literature and provided the therapy tables, Dr. Henry M. Furneaux (Memorial Sloan–Kettering Cancer Center, New York, NY) for reading portions of the manuscript, and Dr. Josep Dalmau (Memorial Sloan–Kettering Cancer Center) and Dr. David Louis (Massachusetts General Hospital) for assistance with the figures.

TABLE 10

Tumor markers useful in diagnosis of cancers causing paraneoplastic syndromes.

Cancer	Marker
Prostate	PSA
Ovary	CA 125
	CASA
Germ cell tumors	AFP
	βHCG
	LDH
Choriocarcinoma	βHCG

Source: Pandha and Waxman 1995.

REFERENCES

ALBIN, R.L., M.B. BROMBERG, J.B. PENNEY and R. KNAPP: Chorea and dystonia: a remote effect of carcinoma. Mov. Disord. 3 (1988) 162–169.

ANDERSON, N.E., C. BUDDE-STEFFEN, M.K. ROSENBLUM, F. GRAUS, D. FORD, B.J. SYNEK and J.B. POSNER: Opsoclonus, myoclonus, ataxia, and encephalopathy in adults with cancer: a distinct paraneoplastic syndrome. Medicine (Baltimore) 67 (1988a) 100–109.

ANDERSON, N.E., J.B. POSNER, J.J. SIDTIS, J.R. MOELLER, S.C. STROTHER, V. DHAWAN and D.A. ROTTENBERG: The metabolic anatomy of paraneoplastic cerebellar degeneration. Ann. Neurol. 23 (1988b) 533–540.

BALOH, R.W., S.E. DEROSSETT, T.F. CLOUGHESY, R.W. KUNCL, N.R. MILLER, J. MERRILL and J.B. POSNER: Novel brainstem syndrome associated with prostate carcinoma. Neurology 43 (1993) 2591–2596.

BARNES, B.E.: Dermatomyositis and malignancy: a review of the literature. Ann. Intern. Med. 84 (1976) 68–76.

BARR, C.W., G. CLAUSSEN, D. THOMAS, J.T. FESENMEIER, R.L. PEARLMAN and S.J. OH: Primary respiratory failure as the presenting symptom in Lambert–Eaton myasthenic syndrome. Muscle Nerve 16 (1993) 712–715.

BERSON, E.L. and S. LESSELL: Paraneoplastic night blindness with malignant melanoma. Am. J. Ophthalmol. 106 (1988) 307–311.

BLUMENFELD, A.M., L.D. RECHT, D.A. CHAD, U. DEGIROLAMI, T. GRIFFIN and K.A. JAECKLE: Coexistence of Lambert–Eaton myasthenic syndrome and subacute cerebellar degeneration: differential effects of treatment. Neurology 41 (1991) 1682–1684.

BOHAN, A. and J.B. PETER: Polymyositis and dermatomyositis. N. Engl. J. Med. 292 (1975) 344–348.

BRAIN, W.R. and M. WILKINSON: Subacute cerebellar degeneration associated with neoplasms. Brain 88 (1965) 465–478.

BRAIN, W.R., P.B. CROFT and M. WILKINSON: Motor neurone disease as a manifestation of neoplasm. Brain 88 (1965) 479–500

BRENNAN, L. and P. CRADDOCK: Limbic encephalopathy as a non-metastatic complication of oat cell lung cancer: its reversal after treatment of the primary lung lesions. Am. J. Med. 75 (1983) 518–520.

BRIERLEY, J.B., J.A.N. CORSELLIS, R. HEIRONS and S. NEVIN: Subacute encephalitis of later adult life mainly affecting the limbic areas. Brain 83 (1960) 357–368.

BROUWER, B.: Beitrag zur Kenntnis der chronischen diffusen Kleinhirnerkrankunger. Mendel's Neurol. Zbl. 38 (1919) 674–682.

BROWNELL, B. and J.T. HUGHES: Degeneration of muscle in association with carcinoma of the bronchus. J. Neurol. Neurosurg. Psychiatry 38 (1975) 363–370.

BRUYN, R.P.M.: Paraneoplastic polyneuropathy. In: P.J. Vinken, G.W. Bruyn, H.L. Klawans and W.B. Matthews (Eds.), Handbook of Clinical Neurology, Vol. 51: Neuropathies. Amsterdam, Elsevier Science (1987) 465–473.

BUCKANOVICH, R.J., J.B. POSNER and R.B. DARNELL: Nova, the paraneoplastic Ri antigen, is homologous to an RNA-binding protein and is specifically expressed in the developing motor system. Neuron 11 (1993) 657–672.

BUDDE-STEFFEN, C., N.E. ANDERSON, M.K. ROSENBLUM, F. GRAUS, D. FORD, B.J. SYNEK, S.H. WRAY and J.B. POSNER: An antineuronal autoantibody in paraneoplastic opsoclonus. Ann. Neurol. 23 (1988) 528–531.

BURTON, G.V., D.E. BULLARD, P.J. WALTHER and P.C. BURGER: Paraneoplastic limbic encephalopathy with testicular carcinoma. A reversible neurologic syndrome. Cancer 62 (1988) 2248–2251.

CALLEN, J.P.: Relationship of cancer to inflammatory muscle diseases: dermatomyositis, polymyositis, and inclusion body myositis. Rheum. Dis. Clin. North Am. 20 (1994) 943–953.

CAMARA, E.G. and G.J. CHELUNE: Paraneoplastic limbic encephalopathy. Brain Behav. Immun. 1 (1987) 349–355.

CAMPO, E., M.N. BRUNIER and M.J. MERINO: Small cell carcinoma of the endometrium with associated ocular paraneoplastic syndrome. Cancer 69 (1992) 2283–2288.

CASE RECORDS OF THE MASSACHUSETTS GENERAL HOSPITAL: Case 32-1994. N. Engl. J. Med. 331 (1994) 528–535.

CHALK, C.H., A.J. WINDEBANK, D.W. KIMMEL and P.G. MCMANIS: The distinctive clinical features of paraneoplastic sensory neuronopathy. Can. J. Neurol. Sci. 19 (1992) 346–351.

CHEN, Y.T., W.J. RETTIG, A.K. YENAMANDRA, C.A. KOZAK, R.S.K. CHAGANTI, J.B. POSNER and L.J. OLD: Cerebellar degeneration-related antigen: a highly conserved neuroectodermal marker mapped to chromosomes X in human and mouse. Proc. Natl. Acad. Sci. USA 87 (1990) 3077–3081.

CHER, L.M., F.H. HOCHBERG, J. TERUYA, M. NITSCHKE, R.F. VALENZUELA, J.D. SCHMAHMANN, M. HERBERT, H.D. ROSAS and C. STOWELL: Therapy for paraneoplastic neurologic syndromes in six patients with protein A column immunoadsorption. Cancer 75 (1995) 1678–1683.

CHESTER, K.A., B. LANG, J. GILL, A. VINCENT and J. NEWSOM-DAVIS: Lambert–Eaton syndrome antibodies: reaction with membranes from a small cell lung cancer xenograft. J. Neuroimmunol. 18 (1987) 97–104.

CLOUSTON, P.D., C.B. SAPER, T. ARBIZU, I. JOHNSTON, B. LANG, D.J. NEWSOM-DAVIS and J.B. POSNER: Paraneoplastic cerebellar degeneration. III. Cerebellar degeneration, cancer, and the Lambert–Eaton myasthenic syndrome. Neurology 42 (1992) 1944–1950.

COCCONI, G., G. CECI, G. JUVARRA, M.R. MINOPOLI, T. COCCHI, F. FIACCADORI, A. LECHI and P. BONI: Successful treatment of subacute cerebellar degeneration in ovarian carcinoma with plasmapheresis. A case report. Cancer 56 (1985) 2318–2320.

CONDOM, E., A. VIDAL, R. ROTA, F. GRAUS, J. DALMAU and I. FERRER: Paraneoplastic intestinal pseudo-obstruction associated with high titres of Hu autoantibodies. Virchows Arch. 423 (1993) 507–511.

CORSELLIS, J.A.N., G.J. GOLDBERG and A.R. NORTON: 'Limbic encephalitis' and its association with carcinoma. Brain 91 (1968) 481–496.

COX, N.H., C.M. LAWRENCE, J.A.A. LANGTRY and F.A. IVE: Dermatomyositis: associations and an evaluation of screening investigations for malignancy. Arch. Dermatol. 126 (1990) 61–65.

CROFT, P.B. and M. WILKINSON: The incidence of carcinomatous neuromyopathy in patients with various types of carcinoma. Brain 88 (1965) 427–434.

CROFT, P.B., H. URICH and M. WILKINSON: Peripheral neuropathy of sensorimotor type associated with malignant disease. Brain 90 (1967) 31–65.

CUNNINGHAM, J., F. GRAUS, N. ANDERSON and J.B. POSNER: Partial characterization of the Purkinje cell antigens in paraneoplastic cerebellar degeneration. Neurology 36 (1986) 1163–1168.

DALAKAS, M.C.: Inflammatory myopathies. In: L.P. Rowland and S. DiMauro (Eds.), Handbook of Clinical Neurology, Vol. 62: Myopathies. Amsterdam, Elsevier (1992) 369–390.

DALMAU, J., H.M. FURNEAUX, R.J. GRALLA, M.G. KRIS and J.B. POSNER: Detection of the anti-Hu antibody in the serum of patients with small cell lung cancer – a quantitative Western blot analysis. Ann. Neurol. 27 (1990) 544–552.

DALMAU, J., F. GRAUS, M.K. ROSENBLUM and J.B. POSNER: Anti-Hu associated paraneoplastic encephalomyelitis/sensory neuronopathy. A clinical study of 71 patients. Medicine (Baltimore) 71 (1992a) 59–72.

DALMAU, J., H.M. FURNEAUX, C. CORDON-CARDO and J.B. POSNER: The expression of the Hu (paraneoplastic encephalomyelitis/sensory neuronopathy) antigen in human normal and tumor tissues. Am. J. Pathol. 141 (1992b) 881–886.

DALMAU, J., F. GRAUS, N.K.V. CHEUNG, M.K. ROSENBLUM, A. HO, A. CANETE, J.Y. DELATTRE, S.J. THOMPSON and J.B. POSNER: Major histocompatibility proteins, anti-Hu antibodies, and paraneoplastic encephalomyelitis in neuroblastoma and small cell lung cancer. Cancer 75 (1995) 99–109.

DARNELL, R.B. and L.M. DEANGELIS: Regression of small-cell lung carcinoma in patients with paraneoplastic neuronal antibodies. Lancet 341 (1993) 21–22.

DE CAMILLI, P., A. THOMAS, R. COFIELL, F. FOLLI, B. LICHTE, G. PICCOLO, H.M. MEINCK, M. AUSTONI, G. FASSETTA, G. BOTTAZZO, D. BATES, N. CARTLIDGE, M. SOLIMENA and W.K. KILIMANN: The synaptic vesicle-associated protein amphiphysin is the 128-kD autoantigen of Stiff-Man syndrome with breast cancer. J. Exp. Med. 178 (1993) 2219–2223.

DENNY-BROWN, D.: Primary sensory neuropathy with muscular changes associated with carcinoma. J. Neurol. Neurosurg. Psychiatry 11 (1948) 73–87.

DIETL, H.W., M. PULST, P. ENGELHARDT and P. MEHRACIN: Paraneoplastic brainstem encephalitis with acute dystonia and central hypoventilation. J. Neurol. 227 (1982) 229–238.

DIGRE, K.B.: Opsoclonus in adults. Report of three cases and review of the literature. Arch. Neurol. 43 (1986) 1165–1175.

DROPCHO, E.J. and P.H. KING: Autoantibodies against the Hel-N1 RNA-binding protein among patients with lung carcinoma: an association with type 1 anti-neuronal nuclear antibodies. Ann. Neurol. 36 (1994) 200–205.

DROPCHO, E.J., Y.T. CHEN, J.B. POSNER and L.J. OLD: Cloning of a brain protein identified by autoantibodies from a patient with paraneoplastic cerebellar degeneration. Proc. Natl. Acad. Sci. USA 84 (1987) 4552–4556.

DROPCHO, E.J., L.B. KLINE and J. RISER: Antineuronal (anti-Ri) antibodies in a patient with steroid-responsive opsoclonus-myoclonus. Neurology 43 (1993) 207–211.

ELMQVIST, D. and E.H. LAMBERT: Detailed analysis of neuromuscular transmission in a patient with the myasthenic syndrome sometimes associated with bronchogenic carcinoma. Mayo Clin. Proc. 43 (1968) 689–713.

ELRINGTON, G.M., N.M. MURRAY, S.G. SPIRO, J. NEWSOM-DAVIS: Neurological paraneoplastic syndromes in patients with small cell lung cancer. A prospective survey of 150 patients. J. Neurol. Neurosurg. Psychiatry 54 (1991) 764–767.

EMSLIE-SMITH, A.M. and A.G. ENGEL: Necrotizing myopathy with pipestem capillaries, microvascular deposition of the complement membrane attack complex (MAC), and minimal cellular infiltration. Neurology 41 (1991) 936–939.

FATHALLAH-SHAYKH, H., S. WOLF, E. WONG, J.B. POSNER and H.M. FURNEAUX: Cloning of a leucine-zipper protein recognized by the sera of patients with antibody-associated paraneoplastic cerebellar degeneration. Proc. Natl. Acad. Sci. USA 88 (1991) 3451–3454.

FERRARI, P., M. FEDERICO, L.M. GRIMALDI and V. SILINGARDI: Stiff-man syndrome in a patient with Hodgkin's disease. An unusual paraneoplastic syndrome. Haematologica 75 (1990) 570–572.

FOLLI, F., M. SOLIMENA, R. COFIELL, M. AUSTONI, G. TALLINI, G. FASSETTA, D. BATES, N. CARTLIDGE, G.F. BOTTAZO, G. PICCOLO and P. DE CAMILLI: Autoantibodies to a 128-kd synaptic protein in three women with the stiff-man syndrome and breast cancer. N. Engl. J. Med. 328 (1993) 546–551.

FUKUNAGA, H., A.G. ENGEL, B. LANG, J. NEWSOM-DAVIS and A. VINCENT: Passive transfer of Lambert–Eaton myasthenic syndrome with IgG from man to mouse depletes the presynaptic membrane active zones. Proc. Natl. Acad. Sci. USA 80 (1983) 7636–7640.

FUKUOKA, T., A.G. ENGEL, B. LANG, J. NEWSOM-DAVIS, C. PRIOR and D.W. WRAY: Lambert–Eaton myasthenic syndrome. I. Early morphological effects of IgG on the presynaptic membrane active zones. Ann. Neurol. 22 (1987) 193–199.

FURNEAUX, H.M., E.J. DROPCHO, D. BARBUT, Y.T. CHEN, M.K. ROSENBLUM, L.J. OLD and J.B. POSNER: Characterization of a cDNA encoding a 34-kDa Purkinje neuron protein recognized by sera from patients with paraneoplastic cerebellar degeneration. Proc. Natl. Acad. Sci. USA 86 (1989) 2873–2877.

FURNEAUX, H.F., L. REICH and J.B. POSNER: Autoantibody synthesis in the central nervous system of patients with paraneoplastic syndromes. Neurology 40 (1990a) 1085–1091

FURNEAUX, H.M., M.K. ROSENBLUM, J. DALMAU, E. WONG, P. WOODRUFF, F. GRAUS and J.B. POSNER: Selective expression of Purkinje-cell antigens in tumor tissue from patients with paraneoplastic cerebellar degeneration. N. Engl. J. Med. 322 (1990b) 1844–1851.

GAHRING, L.C., R.E. TWYMAN, J.E. GREENLEE and S.W.

ROGERS: Autoantibodies to neuronal glutamate receptors in patients with paraneoplastic neurodegenerative syndrome enhance receptor activation. Mol. Med. 1 (1995) 245–253.

GIORDANA, M.T., R. SOFFIETTI and D. SCHIFFER: Para-neoplastic opsoclonus: a neuropathologic study of two cases. Clin. Neuropathol. 8 (1989) 295–300.

GLANTZ, M.J., H. BIRAN, M.E. MYERS, J.P. GOCKERMAN and M.H. FRIEDBERG: The radiographic diagnosis and treatment of paraneoplastic central nervous system disease. Cancer 73 (1994) 168–175.

GOLBE, L.I., D.C. MILLER and R.C. DUVOISIN: Paraneoplastic degeneration of the substantia nigra with dystonia and parkinsonism. Mov. Disord. 4 (1989) 147–152.

GOLDBERG-STERN, H., R. WEITZ, R. ZAIZOV, M. GORNISH and N. GADOTH: Progressive spinocerebellar degeneration 'plus' associated with Langerhans cell histiocytosis: a new paraneoplastic syndrome? J. Neurol. Neurosurg. Psychiatry 58 (1995) 180–183.

GRAUS, F., C. CORDON-CARDO and J.B. POSNER: Neuronal antinuclear antibody in sensory neuronopathy from lung cancer. Neurology 35 (1985) 538–543.

GRAUS, F., I. ILLA, M. AGUSTI, T. RIBALTA, F. CRUZ-SANCHEZ and C. JUAREZ: Effect of intraventricular injection of an anti-Purkinje cell antibody (anti-Yo) in a guinea pig model. J. Neurol. Sci. 106 (1991) 82–87

GRAUS, F., F. VEGA, J.Y. DELATTRE, I. BONAVENTURA, R. RENE, D. ARBAIZA and E. TOLOSA: Plasmapheresis and antineoplastic treatment in CNS paraneoplastic syndromes with antineuronal autoantibodies. Neurology 42 (1992) 536–550.

GRAY, F., J.J. HAUW, R. ESCOURELLE and P. CASTAIGNE. Myélopathies nécrosantes et pathologie néoplasique. Trois observations anatomo-cliniques. Rev. Neurol. 136 (1980) 235–246.

GREENFIELD, J.G.: Subacute spinocerebellar degeneration occurring in elderly patients. Brain 57 (1934) 161–176.

GREENLEE, J.E. and H.R. BRASHEAR: Antibodies to cerebellar Purkinje cells in patients with paraneoplastic cerebellar degeneration and ovarian carcinoma. Ann. Neurol. 14 (1983) 609–613.

GRESS, T., A. BALDINI, M. ROCCHI, H. FURNEAUX, J.B. POSNER and M. SINISCALCO: In situ mapping of the gene coding for a leucine zipper DNA binding protein (CDR62) to 16p12–16p13.1. Genomics 13 (1992) 1340–1342.

GUY, J.R. and N.J. SCHATZ: Paraneoplastic downbeating nystagmus: a sign of occult malignancy. J. Clin. Neuro-Ophthalmol. 8 (1988) 269–272.

HALBACH, M., V. HOMBERG and H.J. FREUND: Neuromuscular, autonomic and central cholinergic hyperactivity associated with thymoma and acetylcholine receptor-binding antibody. J. Neurol. 234 (1987) 433–436.

HAMMACK, J., H. KOTANIDES, M.K. ROSENBLUM and J.B. POSNER: Paraneoplastic cerebellar degeneration. II.

Clinical and immunologic findings in 21 patients with Hodgkin's disease. Neurology 42 (1992) 1938–1943.

HENSON, R.A. and H. URICH: Remote effects of malignant disease: certain intracranial disorders. In: P.J. Vinken, G.W. Bruyn and H.L. Klawans (Eds.), Handbook of Clinical Neurology, Vol. 38: Neurological Manifestations of Systemic Diseases, Part I. Amsterdam, North-Holland Publishers (1979) 625–668.

HENSON, R.A. and H. URICH: Encephalomyelitis with carcinoma. In: Cancer and the Nervous System. Oxford, Blackwell Scientific (1982a) 314–345.

HENSON, R.A. and H. URICH: Cortical cerebellar degeneration. In: Cancer and the Nervous System. Oxford, Blackwell Scientific (1982b) 346–367.

HENSON, R.A. and H. URICH: Miscellaneous and possibly fortuitous associations. In: Cancer and the Nervous System. Oxford, Blackwell Scientific (1982c) 432–451.

HENSON, R.A. and H. URICH: Muscular and neuromuscular disorders. In: Cancer and the Nervous System. Oxford, Blackwell Scientific (1982d) 406–431.

HENSON, R.A., H.L. HOFFMAN and H. URICH: Encephalomyelitis with carcinoma. Brain 88 (1965) 449–464.

HETZEL, D.J., C.R. STANHOPE, B.P. ONEILL and V.A. LENNON: Gynecologic cancer in patients with subacute cerebellar degeneration predicted by anti-Purkinje cell antibodies and limited in metastatic volume. Mayo Clin. Proc. 65 (1990) 1558–1563.

HORMINGO, A., J. DALMAU, M.K. ROSENBLUM, M.E. RIVER and J.B. POSNER: Immunological and pathological study of anti-Ri-associated encephalopathy. Ann. Neurol. 36 (1994) 896–902.

HUDSON, C.N., M. CURLING, P. POTSIDES and D.G. LOWE: Paraneoplastic syndromes in patients with ovarian neoplasia. J. Roy. Soc. Med. 86 (1993) 202–204.

INGENITO, G.G., J.R. BERGER, N.J. DAVID and M.D. NORENBERG: Limbic encephalitis associated with thymoma. Neurology 40 (1990) 382.

JACOBSON, M.D., C.E. THIRKILL and S.J. TIPPING: A clinical triad to diagnose paraneoplastic retinopathy. Ann. Neurol. 28 (1990) 162–167

JAECKLE, K.A., A.N. HOUGHTON, S.L. NIELSON and J.B. POSNER: Demonstration of serum anti-Purkinje antibody in paraneoplastic degeneration and preliminary antigenic characterization. Ann. Neurol. 14 (1983) 111.

JAECKLE, K.A., F. GRAUS, A. HOUGHTON, C. CARDON-CARDO, S.L. NIELSON and J.B. POSNER: Autoimmune response of patients with paraneoplastic cerebellar degeneration to a Purkinje cell cytoplasmic protein antigen. Ann. Neurol. 18 (1985) 592–600.

JEAN, W.C., J. DALMAU, A. HO and J.B. POSNER: Analysis of the IgG subclass distribution and inflammatory infiltrates in patients with anti-Hu-associated paraneoplastic encephalomyelitis. Neurology 44 (1994) 140–147.

JOHNSON, P.C., L.A. ROAK, R.H. HAMILTON and J.F. LAGUNA: Paraneoplastic vasculitis of nerve, a remote effect of cancer. Ann. Neurol. 5 (1979) 437–444.

JOHNSTON, I., B. LANG, K. LEYS and J. NEWSOM-DAVIS: Heterogeneity of calcium channel autoantibodies detected using a small-cell lung cancer line derived from a Lambert–Eaton myasthenic syndrome patient. Neurology 44 (1994) 334–338.

KAPLAN, A.M. and H.J. ITABASHI: Encephalitis associated with carcinoma. Central hypoventilation syndrome and cytoplasmic inclusion bodies. J. Neurol. Neurosurg. Psychiatry 37 (1974) 1166–1176.

KIM, Y.I. and E. NEHER: IgG from patients with Lambert–Eaton syndrome blocks voltage-dependent calcium channels. Science 239 (1988) 405–408.

KLINGON, G.H.: The Guillain–Barré syndrome associated with cancer. Cancer 18 (1965) 157–163.

KURZROCK, R. and P.R. COHEN: Vasculitis and cancer. Clin. Dermatol. 11 (1993) 175–187

KURZROCK, R., P.R. COHEN and A. MARKOWITZ: Clinical manifestations of vasculitis in patients with solid tumors. A case report and review of the literature. Arch. Intern. Med. 154 (1994) 334–340

LACHANCE, D.H., B.P. O'NEILL, C.M. HARPER, JR., P.M. BANKS and T.L. CASCINO: Paraneoplastic brachial plexopathy in a patient with Hodgkin's disease. Mayo Clin. Proc. 66 (1991) 97–101.

LAMBERT, E.H., L.M. EATON and E.D. ROOKE: Defect of neuromuscular conduction associated with malignant neoplasms. Am. J. Physiol. 187 (1956) 612–613.

LANG, B., D. WRAY, J. NEWSTOM-DAVIS and A. VINCENT: Autoimmune aetiology for myasthenia (Eaton–Lambert) syndrome. Lancet ii (1981) 224–226.

LEVEQUE, C., T. HOSHINO, P. DAVID, Y. SHOJI-KASAI, K. LEYS, A. OMORI, B. LANG, O. EL FAR, K. SATO, N. MARTIN-MOUTOT, J. NEWSOM-DAVIS, M. TAKAHASHI and M.J. SEAGAR: The synaptic vesicle protein synaptotagmin associates with calcium channels and is a putative Lambert–Eaton myasthenic syndrome antigen. Proc. Natl. Acad. Sci. USA 89 (1992) 3625–3629.

LIANG, B.C., J.W. ALBERS, A.A. SIMA and T.T. NOSTRANT. Paraneoplastic pseudo-obstruction, mononeuropathy multiplex, and sensory neuronopathy. Muscle Nerve 17 (1994) 91–96

LISAK, R.P., M. MITCHELL, B. ZWEIMAN, E. ORRECHIO and A.K. ASBURY: Guillain–Barré syndrome and Hodgkin's disease: three cases with immunological studies. Ann. Neurol. 1 (1977) 72–78

LITCHY, W.J. and R.G. AUGER: Neuromyotonia, normocalcemic tetany, and thymoma: a unique clinical and electrophysiologic presentation. Muscle Nerve 9 (1986) 651.

LIU, J., J. DALMAU, A. SZABO, M. ROSENFELD, J. HUBER and H. FURNEAUX: Paraneoplastic encephalomyelitis antigens bind to the AU-rich elements of mRNA. Neurology 45 (1995) 544–550.

LUQUE, F.A., H.M. FURNEAUX, R. FERZIGER, M.K. ROSENBLUM, S.H. WRAY, S.C. SCHOLD, M.J. GLANTZ, K.A. JAECKLE, H. BIRAN, M. LESSER, W.A. PAULSEN, M.E. RIVER and J.B. POSNER: Anti-Ri: an antibody associated with paraneoplastic opsoclonus and breast cancer. Ann. Neurol. 29 (1991) 241–251.

MALIK, S., A.J. FURLAN, P.J. SWEENEY, G.S. KOSMORSKY and M. WONG: Optic neuropathy: a rare paraneoplastic syndrome. J. Clin. Neuro-Ophthalmol. 12 (1992) 137–141.

MARKHAM, M. and M. ABELOFF: Small-cell lung cancer and limbic encephalitis. Ann. Intern. Med. 96 (1982) 785.

MATSUMURO, K., S. IZUMO, F. UMEHARA, T. ARISATO, I. MARUYAMA, S. YONEZAWA, H. SHIRAHAMA, E. SATO and M. OSAME: Paraneoplastic vasculitic neuropathy: immunohistochemical studies on a biopsied nerve and post-mortem examination. J. Intern. Med. 236 (1994) 225–230.

MCEVOY, K.M.: Stiff-man syndrome. Mayo Clin. Proc. 66 (1991) 300–304.

MCEVOY, K.M.: Diagnosis and treatment of Lambert–Eaton myasthenic syndrome. Neurol. Clin. 12 (1994) 387–399.

MCLEOD, J.G.: Paraneoplastic neuropathies. In: P.J. Dyck, P.K. Thomas, J.W. Griffin, P.A. Low and J.F. Poduslo (Eds.), Peripheral Neuropathy, 3rd Edit. Philadelphia, PA, W.B. Saunders (1993) 1583–1590.

MILLER, N.R.: Paraneoplastic syndromes. In: J.C. Walsh and W.F. Hoyt (Eds.), Clinical Neuro-Ophthalmology, 4th Edit. Baltimore, MD, Williams and Wilkins (1988) 1710–1747.

MINTZ, B.J. and D.K. SIROTA: A case report of neurologic improvement following treatment of paraneoplastic cerebellar degeneration. Mt. Sinai J. Med. 60 (1993) 163–164.

MOLL, J.W., S.C. HENZEN LOGMANS, F.G. VAN DER MECHE and CH.J. VECHT: Early diagnosis and intravenous immune globulin therapy in paraneoplastic cerebellar degeneration (Letter). J. Neurol. Neurosurg. Psychiatry 56 (1993) 112.

NAGEL, A., A.G. ENGEL, B. LANG, J. NEWSOM-DAVIS and T. FUKUOKA: Lambert–Eaton myasthenic syndrome IgG depletes presynaptic membrane active zone particles by antigenic modulation. Ann. Neurol. 24 (1988) 552–558.

NICHOLAS, A.P., A. CHATTERJEE, E.J. DROPCHO, M.M. ARNOLD and S.J. OH: Stiff-man syndrome associated with thymoma. Ann. Neurol. 36 (1994) 294.

NORRIS, JR., F.J. and W.K. ENGEL: Carcinomatous amyotrophic lateral sclerosis. In: W.R. Brain and F.J. Norris, Jr. (Eds.), The Remote Effects of Cancer on the Nervous System. New York, Grune and Stratton, (1965) 24–34.

OH, S.J., R. SLAUGHTER and L. HARRELL: Paraneoplastic vasculitic neuropathy: a treatable neuropathy. Muscle Nerve 14 (1991) 152–156.

OJEDA, V.A.: Necrotizing myelopathy associated with malignancy: a clinicopathologic study of two cases and literature review. Cancer 53 (1984) 1115–1123.

O'NEILL, J.H., N.M.F. MURRAY and J. NEWSOM-DAVIS: The Lambert–Eaton myasthenic syndrome. A review of 50 cases. Brain 111 (1988) 577–596.

ORZECHOWSKI, K.: De l'ataxie dysmétrique des yeux: remarques sur l'ataxie des yeux dite myoclonique (opsoclonie, opsochorie). J. Psychol. Neurol. 35 (1927) 1–18.

PANDHA, H.S. and J. WAXMAN: Tumour markers. Q. J. Med. 88 (1995) 233–241.

PAONE, J. and K. JEYASINGHAM: Remission of cerebellar dysfunction after pneumonectomy for bronchogenic carcinoma. N. Engl. J. Med. 302 (1980) 156.

PARTANEN, V.S.J., H. SOININEN, M. SAKSA, and P. RIEKKINEN: Electromyographic and nerve conduction findings in a patient with neuromyotonia, normocalcemia tetany and small-cell lung cancer. Acta Neurol. Scand. 61 (1980) 216–226.

PASSAFARO, M., F. CLEMENTI and E. SHER: Metabolism of ω-conotoxin-sensitive voltage-operated calcium channels in human neuroblastoma cells: modulation by cell differentiation and anti-channel antibodies. J. Neurosci. 12 (1992) 3372–3379.

PETERSON, K., M.K. ROSENBLUM, H. KOTANIDES and J.B. POSNER: Paraneoplastic cerebellar degeneration. I. A clinical analysis of 55 anti-Yo antibody-positive patients. Neurology 42 (1992) 1931–1937.

POLLARD, J.D.: Neurological complications of plasma cell dyscrasias. In: C.G. Goetz, C.M. Tanner and M.J. Aminoff (Eds.), Handbook of Clinical Neurology, Vol. 63: Systemic Diseases, Part 1. Amsterdam, Elsevier (1993) 391–412.

PRANZATELLI, M.R.: The neurobiology of the opsoclonus-myoclonus syndrome. Clin. Neuropharmacol. 15 (1992) 186–228.

PRIOR, C., B. LANG, D. WRAY and J. NEWSOM-DAVIS: Action of Lambert–Eaton myasthenic syndrome IgG at mouse motor nerve terminals. Ann. Neurol. 17 (1985) 587–592.

ROBERTS, A., S. PERERA, B. LANG, A. VINCENT and J. NEWSOM-DAVIS: Paraneoplastic myasthenic syndrome IgG inhibits 45cA²⁺ flux in a human small cell carcinoma line. Nature 317 (1985) 737–739.

RODRIGUEZ, M., L.I. TRUH, B.P. O'NEILL and V.A. LENNON: Autoimmune paraneoplastic cerebellar degeneration: ultrastructural localization of antibody-binding sites in Purkinje cells. Neurology 38 (1988) 1380–1386.

ROPPER, A.H.: Seronegative, non-neoplastic acute cerebellar degeneration. Neurology 43 (1993) 1602–1605.

ROSENFELD, M.R. and J.B. POSNER: Paraneoplastic motor neuron disease. Adv. Neurol. 56 (1991) 445–459.

ROSENFELD, M.R., E. WONG, J. DALMAU, G. MANLEY, J.B.

POSNER, E. SHER and H.M. FURNEAUX: Cloning and characterization of a Lambert–Eaton myasthenic syndrome antigen. Ann. Neurol. 33 (1993) 113–120.

SAWYER, R.A., J.B. SELHORST, L.E. ZIMMERMAN and W.F. HOYT: Blindness caused by photoreceptor degeneration as a remote effect of cancer. Am. J. Ophthalmol. 81 (1976) 606–613.

SCOTT, C.L., L.M. CHER and J. O'DAY: Cancer associated retinopathy and non small cell lung cancer. J. Clin. Neurosci. 4 (1997) 355.

SHER, E., N. CANAL, G. PICCOLO, C. GOTTI, C. SCOPPETTA, A. EVOLI and F. CLEMENTI: Specificity of calcium channel autoantibodies in Lambert–Eaton myasthenic syndrome. Lancet ii (1989) 640–643.

SMITH, B.: Skeletal muscle necrosis associated with carcinoma. J. Pathol. 97 (1969) 207–210.

SMITH, B.E.: Inflammatory sensory polyganglionopathies. Neurol. Clin. 10 (1992) 735–759.

SOLIMENA, M., F. FOLLI, R. APARISI, G. POZZA and C.P. DE: Autoantibodies to GABA-ergic neurons and pancreatic beta cells in stiff-man syndrome. N. Engl. J. Med. 322 (1990) 1555–1560.

SZABO, A., J. DALMAU, G. MANLEY, M. ROSENFELD, E. WONG, J. HENSON, J.B. POSNER and H.M. FURNEAUX: HuD, a paraneoplastic encephalomyelitis antigen, contains RNA-binding domains and is homologous to ELAV and sex-lethal. Cell 67 (1991) 325–333.

THIRKILL, C.E., P. FITZGERALD, R.C. SERGOTT, A.M. ROTH, N.D. TYLER and J.L. KELTNER. Cancer-associated retinopathy (CAR syndrome) with antibodies reacting with retinal, optic nerve and cancer cells. N. Engl. J. Med. 321 (1989) 1589–1594.

THIRKILL, C.E., R.C. TAIT, N.K. TYLER, A.M. ROTH and J.L. KELTNER: The cancer-associated retinopathy antigen is a recoverin-like protein. Investig. Ophthalmol. Vis. Sci. 33 (1992) 2768–2772

THIRKILL, C.E., J.L. KELTNER, N.K. TYLER and A.M. ROTH: Antibody reactions with retina and cancer-associated antigens in 10 patients with cancer-associated retinopathy. Arch. Ophthalmol. 111 (1993a) 931–937.

THIRKILL, C.E., R.C. TAIT, N.K. TYLER, A.M. ROTH and J.L. KELTNER: Intraperitoneal cultivation of small-cell carcinoma induces expression of the retinal cancer-associated retinopathy antigen. Arch. Ophthalmol. 111 (1993b) 974–978.

TROTTER, J.L., B.A. HENDIN and C.K. OSTERLAND: Cerebellar degeneration with Hodgkin disease. Arch. Neurol. 33 (1976) 660–661.

TSUKAMOTO, T., R. MOCHIZUKI, H. MOCHIZUKI, M. NOGUCHI, H. KAYAMA, M. HIWATASHI and T. YAMAMOTO: Paraneoplastic cerebellar degeneration and limbic encephalitis in a patient with adenocarcinoma of the colon. J. Neurol. Neurosurg. Psychiatry 56 (1993) 713–716.

ULLRICH, B., C. LI, J.Z. ZHANG, H. MCMAHON, R.G.W. ANDER-

SON, M. GEPPERT and T.C. SUDHOF: Functional properties of multiple synaptotagmins in brain. Neuron 13 (1994) 1281–1291.

VINCENT, D., F. DUBAS, J.J. HAUW, P. BODEAU, F. L'HERMITTE, A. BUGE and P. CASTAIGNE: Nerve and muscle microvasculitis. J. Neurol. Neurosurg. Psychiatry 49 (1986) 1007–1010.

WALSH, J.C.: Neuromyotonia: an unusual presentation of intrathoracic malignancy. J. Neurol. Neurosurg. Psychiatry 39 (1976) 1086–1091.

WANG, A., S. LEIBOWICH, P. RIDKER and W. DAVID: Paraneoplastic cerebellar degeneration in a patient with ovarian carcinoma. Am. J. Neuroradiol. 9 (1988) 216–217.

YOUNG, C.A., J.M. MACKENZIE, D.W. CHADWICK and I.R. WILLIAMS: Opsoclonus-myoclonus syndrome: an autopsy study of three cases. Eur. J. Med. 2 (1993) 239–241.

YOUNGER, D.S., L.P. ROWLAND, N. LATOV, A.P. HAYS, D.J. LANGE, W. SHERMAN, G. INGHIRAMI, M.A. PESCE, D.M. KNOWLES, J. POWERS, J.R. MILLER, M.R. FETELL and R.E. LOVELACE: Lymphoma, motor neuron diseases and amyotrophic lateral sclerosis. Ann. Neurol. 29 (1991) 78–86.

YOUNGER, D.S., J. DALMAU, G. INGHIRAMI, W.H. SHERMAN and A.P. HAYS: Anti-Hu-associated peripheral nerve and muscle microvasculitis. Neurology 44 (1994) 181–183.

Index

Prepared by W. van Ockenburg

Entries in this index refer both to the present Volume and to Volumes 1-43 in the preceding series of the Handbook of Clinical Neurology